AF497726

APPLICATION FOR

THE HERD BOOK,

OR ANY PARTICULARS RESPECTING THE WORK,

TO BE ADDRESSED TO

THE SECRETARY,

SHORTHORN SOCIETY OF GREAT BRITAIN AND IRELAND,

· 12 HANOVER SQUARE, LONDON, W.

COATES'S
HERD BOOK:

CONTAINING

THE PEDIGREES

OF

IMPROVED

SHORT-HORNED CATTLE.

VOLUME TWENTY-SIXTH.

[NEW SERIES.]

PUBLISHED BY

**THE SHORTHORN SOCIETY OF THE UNITED KINGDOM
OF GREAT BRITAIN AND IRELAND.**

LONDON:
PRINTED BY TAYLOR AND FRANCIS, RED LION COURT, FLEET STREET.

MDCCCLXXX.

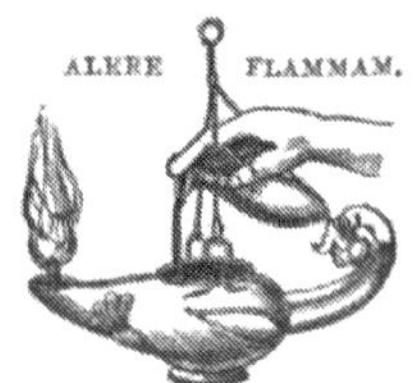

PRINTED BY TAYLOR AND FRANCIS,
RED LION COURT, FLEET STREET.

ERRATA.

VOLUME XXV.

Note.—In publishing this list of Errata the Editing Committee desire to impress upon Breeders the great importance of their writing names and figures *very plainly*, and of carefully checking the accuracy of their entries before sending them for insertion in the Herd Book.

The Committee will be glad if those Breeders who may notice inaccuracies, will immediately forward particulars thereof to the Secretary, with a view to their correction.

Page 42. Cæsar (41175), the property of Mr. G. Miles, Llangattock Park, *not* Mr. G. Miln.

89. Duke of Uist (41436). Sire was Duke of Yardley (36556), *not* Baron Fantail (37790).

132. Harold (41071), calved March 18, 1878, *not* March 8.

152. Laird Gwynne (41782). Colour red and little white, *not* roan.

240. Reversioner (42276), the property of Mr. John Willis, Carperby, *not* Mr. T. Willis.

305. Baroness Hopewell. Name of 1878 C.C. "Baroness Hopeful," *not* Baroness Hopewell.

331. Buttercup. Her 1878 C.C. "Daisy" calved January 12, *not* June 12.

336. Bijou 2nd. Produce calved 1877 *not* 1878, and the B.C. *should be* C.C. (dead).

425. Aline, calved February 24, 1875, *not* February 21.

443. Lady Farewell. Sire of 1878 C.C. "Queen Farewell 3rd" was Duke of Worcester 3rd (39796), *not* Marquis of Worcester 3rd (40315).

462. Scothern Duchess. Name of 1878 C.C. "Scothern Duchess 3rd," *not* "Scothern Duchess 2nd."

479. "Ferula." Breeder Mr. J. Greenwood, Swarcliffe, *not* Mr. F. B. Greenwood.

495. Kathleen. No. of sire, "Baron Killerby," is 27949, *not* 26492.

Page 506. Cassandra. Her 1878 B.C. " King of Carthage " calved January 17, *not* July 17.

506. Charmer Duchess Echter. Her 1878 C.C. " Venusta 2nd " calved June 15, 1878, *not* January 15.

518. Lady Macbeth 3rd, calved July 16, 1868, *not* July 15, 1870.

529. Oxford Belle 2nd. Names of 1877 and 1878 Cow Calves *should be* Oxford Belle 4th and 5th, *not* Belle of Oxford 4th and 5th.

555. Dewdrop. Sire of 1878 C.C. " Dewdrop 3rd " was Wideawake (27805), *not* Visigoth (35907).

619. Princess Janette 3rd. Name of 1878 B.C. " Prince John 3rd," *not* Prince John 2nd.

629. Tulip. Her 1878 C.C. " Tulip 4th " calved February 26, *not* January 26.

643. Christine 6th. Breeder Mr. G. Graham, The Oaklands, *not* Mr. T. G. Curtler.

643. Countess Fawsley. Her 1878 B.C. " Earl of Buckinghamshire 2nd," calved April 1, *not* July 1.

650. Red Rose of Braemar. Name of 1878 C.C. " Red Rose of Maplewell," *not* Red Rose of Maplewell 5th.

654. Picotee. Sire of 1878 C.C. Medlar was *not* Prince Royal (24859), but a bull named " Young Baron " not entered, but eligible for entry in the Herd Book.

663. Duchess of Towneley 5th. Name of 1878 B.C. " Derwent King," *not* Duke of Siddington.

667. Francesca. Colour of 1876 C.C. " Frisky," red and little white, *not* red.

691. Eugenie 6th. Colour of 1878 C.C. " Eugenie 7th," roan, *not* red and white.

REFERENCE TO BREEDERS.

HER MOST GRACIOUS MAJESTY THE QUEEN, WINDSOR PARK ;
275, 276, 277, 313, 384, 502, 548, 649, 671.

His Royal Highness the Prince of Wales, Sandringham, King's Lynn ;
8, 104, 157, 160, 250, 276, 277, 278, 342, 449, 466.

A.

Abell, J., Higham ; 164, 278.
Ackers, B. St. John, Prinknash Park ; 89, 147, 216, 278, 279, 280.
Addison, S. T., Ellenhall ; 66, 145.
Adkins, G. C., The Lightwoods ; 439.
Adkins, H., Edgbaston ; 453.
Adkins, J. C., Milcote ; 68, 121, 126, 170, 280, 443.
Adwick, T., Staythorpe, 357.
Ailesbury, Marquis of, Savernake ; 262, 284, 285.
Aitchison, P., West Garleton ; 280.
Alexander, A. J., Kentucky, U.S.A. ; 432.
Alexander, J., Limavady ; 280.
Alexander, S. M., Roe Park ; 280, 281.
Allan, J., Billie Mains ; 158, 281.
Allan, J., Crieffvechter ; 281, 282.
Allen, B. H., Clifford Priory ; 32, 117, 282, 283.
Allen, G., Knightley Hall ; 78, 140, 322, 332, 334, 346, 468, 615, 643, 683.
Allen, G., Unicarville ; 33, 35, 124, 283, 284, 394, 538.
Allen, S. H., Eastover ; 128, 284, 285.
Allen, T., Thurmaston Lodge ; 588.
Allsopp, H. (now Sir Henry Allsopp, Bart.), Hindlip Hall ; 16, 58, 115, 116, 119, 133, 135, 154, 205, 227, 251, 253, 285, 286, 287, 288.

Andrews, J., Carnesune ; 288, 289.
Andrews, T., Ardara ; 289.
Angas, J. H., Collingrove, Angaston, South Australia ; 289.
Angerstein, W., Weeting Hall ; 44, 452, 495, 559, 589, 611, 622, 697.
Angus, G., and Exors. of, Broomley ; 85, 290, 570, 572, 581, 662.
Angus, Messrs., Broomley ; 467, 496, 553.
Angus, J., Bearl ; 23, 26, 37, 41, 119, 168, 221, 290.
Annandale and Sons, Lintzford ; 44, 192, 194, 248.
Annett, J. W., Houndalee ; 290.
Archdall, E. M., Crock-na-Crieve ; 290, 291.
Archdall, N. M., Crock-na-Crieve; 283, 291, 361, 363, 468, 469, 645.
Arkell, D., Butler's Court ; 16, 144, 291, 292.
Arkell, T., Draycott ; 293.
Arkell, W., Dudgrove ; 36, 244, 291, 292, 293, 314, 549, 672, 682.
Arkell, W., jun., Hatherop ; 292, 293.
Arklay, R., Ethiebeaton ; 162, 293, 533, 534.
Armistead, J. F., Cobb Wall House ; 246, 293.
Armstrong, S., Gally House ; 36, 109, 293, 294, 339, 521, 539.
Armstrong, Sir W. G., Cragside ; 241, 294, 500, 685.
Ashburner, G., Low Hall ; 52, 74, 80, 106, 110, 150, 151, 230, 295, 296, 338, 487.
Ashburner, R. W., Low Hall ; 296.

O.

P.

R.

For full Addresses of the Breeders, refer to the List of Members at the end of the Volume.

BULLS.

(42636)   A 1,

Roan, calved February 9, 1879, bred by Mr. F. Morice, Springfield ; got by Undergraduate (35835), dam (Emma 3rd) by Count Montijo (21495), g. d. (Emma) by Hero of Thorndale (18061), gr. g. d. (Emilia 2nd) by Prince Albert (15084), — (Emilia) by Young Shaftoe (9625), — by Humber (7102), — by Guardian (3947), — by Firby (1040), — by Hector (1104), — by Regent (544). — Old York.

(42637)   AARIFI PASHA,

Roan, calved May 18, 1879, bred by Mr. G. Garne, Churchill Heath ; got by Grand Duke of Geneva 2nd (31288), dam (Charming Nonsuch) by Charming King (30698), g. d. (Nonsuch 6th) by Monitor (24615), gr. g. d. (Nonsuch 4th) by Lord Oxford (20214), — (Nonsuch 3rd) by Ninth Duke of Oxford (17738), — by Earl of Warwick (11412), — by Duke of Cornwall (5947), — by Velocipede (5552), — by Sir Thomas (2636), — by Frederick (1060), — by Harold (291), — by North Star (459), — by Favourite (252), — by Favourite (252), — by Favourite (252), — by Broken Horn (95), — bred by Mr. Best, of Manfield.

(42638)   ABBOT OF RYEDALE,

Roan, calved February 28, 1878, bred by Mr. W. P. Horne, Moulton ; got by Ryedale Duke (35435), dam (Miss Marvel) by Lothair (29231), g. d. (Marvel of Peru) by Frederick First Fruits (23989), gr. g. d. (Marvel) by Bumper (19371 — (Marvelmore) by Bumper (19371), — by Sunbeam (15360).

B

(42639)   **ABRINCHAN,**
Roan, calved April 12, 1879, bred by Mr. Evan Baillie, Dochfour; got by
Abbot of Windsor (32903), dam (Princess Maud) by Knight of Killerby (29000),
g. d. (Princess Royal) by Roan Windsor (24967), gr. g. d. (Victoria) by Duke of
Hamilton (19618), — (Playful) by Inkermann (14730), — by Lord Barrington
(9308), — by Sir Robert (5180), — by Sir Robert (5180), — by Son of Farmer
(2001), — by Son of Mentor (2300), — by Mars.

(42640)   **ACADEMICIAN,**
Red, calved January 29, 1878, bred by Mr. G. V. Hart, Kilderry, the property
of Mr. C. T. McCausland, Drenagh; got by Viscount Studley (35903), dam
(Amethyst) by Herdsman (26374), g. d. (Precious Jewel) by Northern Chief
(22429), gr. g. d. (Jewel) by The Beau (32660), — (Barmaid) by Victor (13949),
— by Xerxes (11064), — by Governor (3916), — by Young Conservative (8981),
— by Emperor (6974), — by Comus (911), — by Sir Charles (592).

(42641)   **ACHILLES,**
Red, calved October 8, 1879, bred by the Rev. E. T. Williams, Caldicot Parsonage, the property of Mr. D. Lawrence, Llangibby Castle; got by Orlando
(40411), dam (Charmian) by Earl of Pamflete (28517), g. d. (Chaplet) by Lygon
(24494), gr. g. d. (Columbine) by Young Duke of Cambridge (17708), — (Cowslip) by Rivers (10716), — by Mozart (11830), — by Stirling (5330), — by
Young Frederick (3836), — by Favourite (3768), — by Tathwell Studley
(5401), — by a Son of Waddingworth (668).

(42642)   **ACROBAT,**
Red and white, calved November 11, 1879, bred by Colonel R. Loyd Lindsay,
Lockinge Park; got by Earl of Horton 11th (36588), dam (Azalea) by Duke of
Cerisia (30937), g. d. (Aurora) by Chanter (19423), gr. g. d. (Alice) by Wise-
tonian (17244), — (Acorn) by Magician 2nd (10486), — by Norfolk (9442), —
by Fourth Duke of Northumberland (3649), — by Mortham (4512), — by Anti-
Radical (1642), — by North Star (459).

(42643)   **ACTON COMET,**
White, calved March 14, 1878, bred by Mr. J. Nichols, Iron Acton; got by
Lord Aberdeen (31610), dam (Countess) by Reflector (29760), g. d. (Clio) by
Coronet (23624), gr. g. d. (Calliope) by Lamp of Lothian (16356), — (Cherish)
by Archduke (17316), — by Koh-i-noor (11642), — by Douglas (12715), —
by Ned Poins (6241), — by Old Strickland (11870).

(42644)   **ACTOR,**
White, calved August 27, 1878, bred by Mr. R. Reynell, Killynon; got by
Albion (36112), dam (Hawthorn) by Royal Prince (27384), g. d. (Sweetbrier)
by Nimrod (13388), gr. g. d. (Charlotte) by Selim (6454), — (Rebecca) by
Rex (6385), — by Sir Thomas Fairfax (5196), — by Ambo (1636), — by
Memnon (2295), — by Pilot (496), — by Agamemnon (9), — by Burrell's Bull
of Burdon.

(42645)   **ADJUTANT GWYNNE,**
Red, calved January 10, 1879, bred by Mr. W. Bolton, The Island; got by
Albion (36122), dam (Augusta Gwynne) by Lieutenant General (31600), g. d.
(Annie Gwynne) by King Richard (26523), gr. g. d. (Ally Gwynne) by Grey
Gauntlet (19908), — (Moll Gwynne) by Equinox (17810), — by Young Usurer

(10985), — by Sir Thomas Fairfax (5196), — by Wallace (5586), — by Wellington (2824), — by Marmion (406), — by Merlin (430), — by Layton (366), — by Phenomenon (491), — by Favourite (252), — by Favourite (252), — by Hubback (319), — by Snowdon's Bull (612), — by Waistell's Bull (669), — by Masterman's Bull (422), — by the Studley Bull (626).

(42646) ADMIRAL,

White, calved June 28, 1878, bred by the Exors. of Mr. F. Jordan, Eastburn House, the property of Mr. F. Dunn, Ebor House ; got by Earl of Waterloo 4th (33820), dam (Alba) by Cynthius (25864), g. d. (Acacia) by Olliver (24685), gr. g. d. (Agate) by Fitz James (19755), — (Agatha) by Saville (16915), — by Apollo (9899), — by Puritan (9523), — by Count Conrad (3510), — by The Miller (5452), — by Reveller (2528), — by Hector (1105), — by Swap (2719), — by Emperor (1013), — by Ryedale (2582), — by Windsor (698).

(42647) ADMIRAL HORNBY,

Red and white, calved January 7, 1878, bred by Mr. J. G. Attwater, Britford, the property of Captain D. H. Mytton, Garth ; got by Duke of Ozleworth (31008), dam (Queen Darlington) by Lord Darlington 2nd (29096), g. d. (Queen of the Empire) by Prince of the Empire (20578), gr. g. d. (Fair Flower) by Master Warlaby (18362), — (Titania) by Homer (14714), — by Fitz Leonard, Jun. (14553), — by Young Rufus (13649), — by Constitution (12634), — by Young Comet (1853), — descended from Jolly's Bull (4115).

(42648) ADMIRAL OF THE FLEET,

Red and white, calved October 20, 1879, bred by Mr. T. Willis, Carperby ; got by Admiral Windsor (32912), dam (Windsor's Hyacinth) by Windsor's Prince (32881), g. d. (Camelia Windsor) by Windsor Fitz-Windsor (25458), gr. g. d. (Camelia) by Royal Alfred (18748), — (Mayflower) by Knight of the Garter (13124), — by Frederick (11489), — by Spendthrift (10867), — by The Irishman (5446), — by Borderer (3191), — by Paganini (2405), — by Forester (1055), — by Frederick (1060), — by Planet (502), — by Comet (155).

(42649) ADMIRAL RODNEY,

Red and white, calved December 30, 1879, bred by Captain D. H. Mytton, Garth ; got by Admiral Hornby (42647), dam (Agnes) by Vespasian (32759), g. d. (Gem) by Berwick (23411), gr. g. d. (Glossary 2nd) by Hardware (19919), — (Glossary) by Hazel (26352), — by Harlequin (26340), — by Emperor (11438), — by Priam (15079).

(42650) ADONIS,

Roan, calved May 10, 1878, bred by Mr. G. D. Beresford, The Palace, Armagh ; got by Pilgrim Father (35037), dam (Hester) by Knight of the Thistle (26555), g. d. (Heroine 3rd) by Viceroy (21020), gr. g. d. (Heroine) by Valiant (10989), — (Lady Harris) by Lord John (11731), — by Sir John Sinclair (5165), — by Collard (3419), — by Matchem (2281), — by Cato (119), — by Jupiter (342), — by George (273), — by Chilton (136), — by Irishman (329), — by B (45).

(42651) AFGHAN,

Roan, calved July 28, 1879, bred by Mr. C. H. Cock, Bridgefoot ; got by Duke of Worcester 5th (39797), dam (Simplicity) by Cherry Duke (25752), g. d. (Selina) by Earl of Darlington (21636), gr. g. d. (Snowdrop) by Gentleman

(12937), — (Virgin) by Charles 1st (8947), — by Belvedere 4th (3130), — by Panton Favourite (4646), — by Plenipotentiary (2436), — by Mameluke (2257), — by Prime Minister (2454), — by Surprise (2716).

(42652) **AIRDRIE'S KIRKLEVINGTON DUKE 3RD,**
Roan, calved June 9, 1879, bred by Mr. G. Fox, Elmhurst Hall; got by Twenty-fourth Duke of Airdrie (36460), dam (Kirklevington Duchess 6th) by Second Duke of Claro (21576), g. d. (Kirklevington 18th) by Third Lord Oxford (22200), gr. g. d. (Kirklevington 10th) by Delhi (15865), — (Kirklevington 8th) by General Canrobert (12926), — by Earl of Derby (10177), — by Earl of Liverpool (9061), — by Duke of Northumberland (1940), — by Belvedere (1706), — by Son of Second Hubback (2683), — a Cow of Mr. Bates's, descended from the stock of Mr. Maynard, of Eryholme.

(42653) **AJAX,**
Roan, calved May 15, 1879, bred by Mr. R. Reynell, Killynon; got by Agamemnon (39357), dam (Violet 4th) by Lord Spencer (26738), g. d. (Violet 3rd) by Dr. McHale (15587), gr. g. d. (Portia) by Paddy Hopewell (16677), — (Violet) by Baron Warlaby (7813), — by Burley Fairfax (6822), — by Monarch (4495), — by Invalid (4076), — by Satellite (1420), — by Cato (119), — by Jupiter (342), — by George (273), — by Chilton (136), — by Irishman (329), — by B. (45).

(42654) **ALBION,**
White, calved June 20, 1877, bred by Mr. G. Blackwell, Kingscote, the property of Mr. J. Kendal, Trotshill; got by Emperor of Gipsies 2nd (33851), dam (Springflower 3rd) by Severn Boy (32473), g. d. (Wallflower 17th) by Bleeding Heart 5th (25639), gr. g. d. (Wallflower 5th) by Chaffcutter (12572), — (Wallflower 3rd) by Oregon (8371), — by Bristol (7852).

(42655) **ALBION,**
Roan, calved May 16, 1879, bred by Mr. G. W. Lambart, Beau Parc; got by Jupiter (38477), dam (Alba) by Rupert (29902), g. d. (Anglia) by British Sailor (23472), gr. g. d. (Glory) by Royal Standard (40644), — (Fame) by Famous (12857), — by Tom Jones (40825), — by Cato (12566), — by Narcissus (4540).

(42656) **ALEC,**
White, calved August 12, 1879, bred by Mr. A. Garfit, Scothern; got by Grand Duke 25th (34065), dam (Alice Hawthorn) by Heydon Duke 3rd (31371), g. d. (Alonzo Premium) by Alonzo (19219), gr. g. d. (Red Light Premium) by Julius Cæsar (18122), — (Premium) by Monarch (18414), — by Statesman (5321), — by Raby (2474), — by Son of Cossack (925).

(42657) **ALEXANDER THE GREAT,**
Roan, calved March 10, 1879, bred by Mr. W. Handley, Green Head; got by Sir Arthur Windsor (35541), dam (Edith 9th) by Pammon (22486), g. d. (Edith 6th) by Carlisle (14244), gr. g. d. (Edith) by U. X. L. (9764), — (Lady Melbourne) by Bachelor (5770), — by Lord Melbourne (4264), — by Sillery (5131), — by Young Western Comet (1575), — by Young Western Comet (1575), — by Western Comet (689), — by Son of Favourite (252).

(42658) **ALEXANDER THE GREAT,**
Roan, calved September 3, 1879, bred by the Duke of Northumberland, Alnwick Castle; got by Sir Raymond (40716), dam (Red Blossom) by Rifleman (27283),

g. d. (Maid of Aln) by Melsonby (18380), gr. g. d. (Young Jessy) by George 3rd (16147), — (Violet) by Shaftoe (5107), — by Richardi (4944), — Cowslip, bought of Mr. Richardson, of Woodhouses.

(42659)   ALFRED,

Roan, calved August 28, 1876, bred by Mr. J. Jardine, Dryfeholm; got by Knight of Derwent (31536), dam (Wild Agnes) by Wild Boy (25447), g. d. (Dahlia) by Pizarro (20497), gr. g. d. (Venus) by Master Annandale (14916), — (Sprightly) by Young Earl (14467), — by Snowy Down (8607), — by Prince Albert (4778), — by Son of Cumberland (5256), — by Exmouth (3747), — by Prince (4765), — by Leopold (2199).

(42660)   ALIMONDI,

Red, calved April 8, 1879, bred by Messrs. J. and E. Tindall, Knapton Hall; got by Astral Prince (39386), dam (Rosette 3rd) by Cecil (25725), g. d. (Rosette) by Prince Louis (20563), gr. g. d. (Tamarind) by Cavendish (15745), — (Bloom) by My Lord (14972), — by Villiers (13959), — by Liberator (7140), — by Prince Albert (4791), — by Sligo (5210), — by Bulmer (1760).

(42661)   ALLIANCE,

White, calved January 17, 1879, bred by Mr. J. Stratton, Alton Priors; got by Royal James (35387), dam (Lady Mary) by Jack Frost (31425), g. d. (Queen Mary) by Young Windsor (17241), gr. g. d. (Duchess of Glo'ster 6th) by King John (14763), — (Duchess of Glo'ster) by The Red Duke (8694), — by Lottery (4280), — by Lottery (4280), — by Phœnix (6290).

(42662)   ALLITHWAITE,

Roan, calved March 2, 1879, bred by Colonel Gunter, Wetherby Grange, the property of Mr. Maude, Blawith; got by Oxford Cherry Duke 2nd (34972), dam (Oxford Maid) by Eighteenth Duke of Oxford (25995), g. d. (Milkmaid) by Marquis (20290), gr. g. d. (Dairymaid) by Le Moor (22090), — (Barmaid) by Oxford (15035), — by Red Richmond (13578), — by Earl Stanhope (5966), — by Lord Stanley (4269), — by Young Matchem (4425), — by Fairfax (1023), — by Son of Marske (418), — by Palmsun (7311), — by a Bull of Mr. Maynard's, — by Son of Favourite.

(42663)   AMEER,

Red and white, calved May 1, 1878, bred by Mr. R. P. Maxwell, Finnebrogue; got by Czarowitz 4th (33497), dam (Daphne) by Menotti (37089), g. d. (Delia 2nd) by Prince Victor (20606), gr. g. d. (Delia) by Prince Andrew (18588), — (Minerva) by Musician (13362), — by Billy the Bean (12471), — by Ballinasloe (12429), — by Superb Fairfax (5357), — by Jupiter (20039), — by Maximus (2284), — by Monarch (4494), — by Western Comet (689), — by Primus (4763), — by Son of Favourite (252), — by Cupid (177).

(42664)   ANCHOR,

Roan, calved December 23, 1879, bred by Mr. W. Handley, Green Head; got by Master Harbinger (40324), dam (Arabella) by Sir Arthur Windsor (35541), g. d. (Ammonia) by Prince Arthur (29597), gr. g. d. (Ammons) by Sir Walter Trevelyan (25179), — by General Garibaldi (21813), — by Tenant Farmer (13828).

(42665)   ANDREW LAMMIE,

Red, calved April 29, 1878, bred by Mr. W. Mackie, Petty Fyvie; got by Baron Lowther (33062), dam (Dairymaid) by Lord Charles (31634), g. d. (Housemaid)

by Blair Athol (25638), gr. g. d. (Milkmaid) by Duke (19593), — (Duchess) by
Sovereign (17020), — by Duke of Leinster (10155), — by Willie (9835), — by
Son of Borderer (809), — by Edrom (1956), — by Reformer (2502), — by Raby
(2473), — by Sir Rowland (1455).

(42666) ANTONY,
Roan, calved May 13, 1878, bred by Colonel Gunter, Wetherby Grange, the property of Mr. Padbury, Perth, West Australia ; got by Second Duke of Glo'ster
(28392), dam (Antonia 2nd) by Baron Oxford 3rd (25579), g. d. (Antonia) by
Duke of Darlington (21586), gr. g. d. (Antoinette) by Fourth Duke of Thorndale (17750), — (America) by Marmaduke (14897), — by Grand Duke 2nd
(12961), — by Fusilier (11499), — by Third Duke of York (10166), — by Second
Cleveland Lad (3408), — by Duke of Cleveland (1937), — by Belvedere (1706),
— a Cow of Mr. Bates's, of Kirklevington.

(42667) APRIL FOOL,
Roan, calved April 1, 1879, bred by Mr. H. Bettridge, East Hanney ; got by
Burgundy (37926), dam (Medusa) by Marquis of Sockburn (34787), g. d. (Medora) by Masterpiece (24561), gr. g. d. (Miss Peel) by Cynric (19542), — (Miss
Ambler) by Royal Oak (16870), — by Bashaw (12449), — by Lord George
(9314), — by Manager (8271), — by Raffler (7391), — by Gazer (7030), — by
a Bull of Mr. Champion's, of Blyth.

(42668) AP WATERLOO,
Red and white, calved January 21, 1879, bred by Major H. Platt, Gorddinog ;
got by Edward Waterloo (38246), dam (Charlotte 5th) by Favourite (33895),
g. d. (Charlotte 4th) by Oxford Wild Eyes 2nd (32038), gr. g. d. (Charlotte 3rd)
by New York (22412), — (Charlotte) by Duke of Wellington (12776), — by
Monk (11824), — by Dan O'Connell (9011), — by Lord Adolphus Fairfax (4249),
— by Young Favourite (3770), — by Waterloo (2816), — by Lawnsleeves (365),
— by Phenomenon (491), — by Favourite (252), — by Favourite (252), — by
Favourite (252), — by Hubback (319), — by Snowdon's Bull (612), — by Waistell's Bull (669), — by Masterman's Bull (422), — by the Studley Bull (626).

(42669) ARCHDUKE 2ND,
Roan, calved March 15, 1879, bred by Mr. E. Lythall, Radford Hall ; got by
Tablet (37558), dam (Archduchess) by Duke Frederick (25929), g. d. (Agate) by
Second Duke of Airdrie (19600), gr. g. d. (Agatha) by The Corsair (15378),
— (Lady Augusta) by Pestalozzi (10603), — by Belshazzar (1703), — by
Camden (1776), — by Abraham (2905), — by Young George (3885), — by
George (276).

(42670) ARCHDUKE OF OXFORD,
Roan, calved January 26, 1879, bred by Lord Penrhyn, Penrhyn Castle ; got by
Duke of Glo'ster 7th (39735), dam (Archduchess of Oxford) by Grand Duke 20th
(31281), g. d. (Grand Duchess of Oxford 7th) by Lord Oxford (20214), gr. g. d.
(Grand Duchess of Oxford) by Grand Duke 3rd (16182), — (Countess of Oxford)
by Earl of Warwick (11412), — by Fourth Duke of York (10167), — by Second
Duke of Northumberland (3646), — by Short Tail (2621), — by Matchem (2281),
— by Young Wynyard (2859).

(42671) ARCHDUKE OF OXFORD,
Red and white, calved March 11, 1879, bred by Sir John Swinburne, Bart., Capheaton ; got by Duke of Oxford 27th (33709), dam (Archduchess Oxford) by

Oxford Beau 3rd (32013), g. d. (Archduchess Claro) by Second Duke of Claro (21576), gr. g. d. (Archduchess Keadby) by Twenty-fourth Duke of Oxford (23781), — (Archduchess 2nd) by Seventh Duke of Oxford (17741), — by Lord Lochinvar (14826), — by Edward (6963), — by Agon (2941), — by Wauldby (2818), — by Darrington, — by Adonis (8), — by Ryedale (2582), — by Lancaster (360).

(42672)   ARCHER,

Roan, calved in March 1878, bred by Mr. W. Harland, Blois Hall, the property of Mr. J. Radcliffe, Stearsby; got by Hutton Conyers (34193), dam (Fern 5th) by Clarion (33393), g. d. (Fern 4th) by Victorious (30226), gr. g. d. (Favourite 3rd) by Knight of the Thistle (20086), — (Favourite 2nd) by Faithful (16024), — by Marton (14914), — by Favour (9114), — by Young Cheshireman (17555), — by Valentine Junior (6630), — by Sir William (5203), — by Noble (4577), — — by Imperial (2151), — by Fairfax (1023), — by Young Warlaby (5598), — by Young Dimple (971), — by Snowball (2648), — by Layton (2190).

(42673)   ARGUS,

Red, calved November 3, 1879, bred by the Rev. E. T. Williams, Caldicot Parsonage; got by Orlando (40411), dam (Chaplet) by Lygon (24494), g. d. (Columbine) by Young Duke of Cambridge (17708), gr. g. d. (Cowslip) by Rivers (10716), — (Raspberry) by Mozart (11830), — by Stirling (5330), — by Young Frederick (3836), — by Favourite (3768), — by Tathwell Studley (5401), — by a Son of Waddingworth (668).

(42674)   ARTAXERXES,

Red and white, calved July 9, 1879, bred by Mr. J. Christy, Boynton Hall; got by Lord Oxford 7th (38645), dam (Azalea 2nd) by Oxford's Baron (32030), g. d. (Azalea) by Duke of Babraham (25934), gr. g. d. (French Aster) by Duke of Grafton (21594), — (German Aster) by Royal Arch (18749), — by Archduke 2nd (15588), — by Young Weathercock (15495), — by General Elliott (10266), — by Young Locksley (22111), — by Sherborne (10805), — by Prince (4772), — by Stanhope (5315), — by Son of Sir Kenneth (1450), — by Pilot (1319), — by Thorpe (1515), — by Waistell's Bull (1567).

(42675)   ARTHUR WELLESLEY,

Roan, calved April 20, 1879, bred by Mr. R. H. Gould, Didmarton; got by Prince Arthur (38892), dam (Welcome 3rd) by Count Bickerstaffe 2nd (25838), g. d. (Welcome 2nd) by Henchman (21918), gr. g. d. (Welcome) by General Havelock (16120), — (Welsh Pride) by Royal Blood (13638), — by Udolpho (13970), — by Cotherstone (6903), — by Augustus (6752), — by Velocipede (5552), — by Priam (2452), — by Jorry (4007), from the stock of Mr. Booth.

(42676)   ASHFIELD,

White, calved April 27, 1876, bred by Mr. R. Stratton, The Duffryn; got by Earl of Killerby (33802), dam (Arrah-na-Pogue) by Royal Prince (27384), g. d. (Miss Molly 2nd) by Red Rover (24923), gr. g. d. (Miss Molly) by Puck (18655), — (Lady Rachel) by Duke of Bedford (11378), — by Coriander (6895), — by a son of Prince George (2464), — by Matchem (2281).

(42677)   ASTON BEAU,

Roan, calved November 19, 1879, bred by Mr. H. Webb, Streetly Hall; got by

Baron Aston (37770), dam (Houri) by Heydon Duke (28850), g. d. (Fair Maid) by Second Duke of Claro (21576), gr. g. d. (Queen of the Harem) by Duke of Kent (19619), — (Medora) by Lord Althorpe (14800), — by James 2nd (8173), — by Mahomed (6170), — by Fanatic (1996), — by Spectator (2688), — by Sir Roger-de-Coverley (5187), — by Albion (1619), — by Son of Favourite (252).

(42678)   ATTRACTIVE KING,
Red and white, calved December 12, 1879, bred by Mr. T. Pears, Hackthorne; got by King of Trumps (31512), dam (Attraction) by Robin (24968), g. d. (Alice Buckingham) by Royal Buckingham (20718), gr. g. d. (Anna Maria) by Sir Roger (16991), — (Adelaide) by The Squire (12217), — by Lambton (9273), — by General Washington (6036), — by Welham (5618), — by Roman (2559), — by Columella (904), — by Albion (14), — by Palmflower (480), — by Blyth Comet (85), — by Neswick (1266), — by Favourite (1033).

(42679)   AUGUSTUS,
Red and white, calved April 4, 1879, bred by Mr. W. F. Budds, Courtstown; got by Grey Friar (36733), dam (Augusta) by Red Duke (27643), g. d. (Keopolina) by Professor Miller (18649), gr. g. d. (Cherry Ripe) by Cœur-de-Lion (15783), — (Cherry Blossom) by The Templar 2nd (13879), — by South Durham (9676), — by Pilot (7333), — by Young Sir Harry Liddell (7503), — by Mickley (7234), — by Adonis (2933), — by Lord Prudhoe (8251), — by Hollon's Bull (313).

(42680)   BABINGLY DUKE,
Red and white, calved November 26, 1879, bred by H.R.H. the Prince of Wales, Sandringham; got by Marquis of Oxford 2nd (37055), dam (Blythesome Eyes) by Third Duke of Hillhurst (30975), g. d. (Wild Eyes Duchess) by Grand Duke 9th (19879), gr. g. d. (Wild Eyes 19th) by Lablache (16353), — (Wild Eyes 18th) by Solon (13766), — by Second Duke of Oxford (9046), — by Fourth Duke of Northumberland (3649), — by Duke of Northumberland (1940), — by Belvedere (1706), — by Emperor (1975), — by Wonderful (700), — by Cleveland (145), — by Butterfly (104), — by Hollon's Bull (313), — by Mowbray's Bull (2342), — by Masterman's Bull (422), — descended from M. Dobison's stock.

(42681)   BABRAHAM BARON,
Roan, calved January 2, 1879, bred by Mr. H. Webb, Streetly Hall, the property of Mr. M. Slater, Junior, Trumpington; got by Havering Baron (38413), dam (Babraham Beauty) by Prolific 2nd (27213), g. d. (Beauty) by Cheltenham (12588), gr. g. d. (Woodbine) by Lord of the North (11743), — (Welcome) by Paris (7314), — by King of Trumps (4156), — by Vanguard (5545), — by Red Rover (4903), — by Anticipation (750), — by Emperor (1014), — by Young Windsor (699), — by Windsor (698).

(42682)   BABRAHAM DUKE,
Red and white, calved August 1, 1873, bred by Mr. C. A. Barnes, Charleywood, the property of Mr. G. W. Daniell, St. Leonards, Blandford; got by Lord Eglintoun (31652), dam (Babraham Duchess 2nd) by Duke of Grafton (21594), g. d. (Babraham Duchess) by Guelder Rose (19910), gr. g. d. (Young Celia 6th) by Young Duke of Cambridge (14433), — (Celia) by Third Duke of Northumberland (3647), — by Bashaw (1692), — by Helmsman (2109), — by Columella (904), — by Regent (544), — by Palatine (478), — by Palmflower (480), — by Patriot (486), — by Driffield (223), — by C. Holmes's Bull (314).

(42683)   BACCHUS 2ND,

Roan, calved May 23, 1879, bred by Mr. C. W. Griffin, Werrington; got by Beaconsfield (42767), dam (Blush 8th) by Telemachus 10th (35728), g. d. (Blush 2nd) by Baron Torr (23380), gr. g. d. (Blush) by Claxton (21433), — (Bride) by Game Boy (14585), — by Joe Miller (10355), — by Warden (15482), — by Kaffer (14754), — by Grant's Bull (14647), — bred by Mr. Allington.

(42684)   BADDOW OXFORD,

Red and white, calved February 13, 1879, bred by Mr. R. H. Crabb, Baddow Place; got by Oxford's Baron (32030), dam (Grand Duchess of Oxford 5th) by Oxford's Baron (32030), g. d. (Grand Duchess of Oxford 4th) by Thorndale Duke (27661), gr. g. d. (Grand Duchess of Oxford 2nd) by Fourth Duke of Geneva (25964), — (Grand Duchess of Oxford) by Grand Duke 7th (19877), — by Rembrandt (13587), — by Jasper (11609), — by Second Duke of Oxford (9046), — by Second Duke of Northumberland (3646), — by Belvedere (1706), — by Son of Second Hubback (2683), — a Cow of Mr. Bates's, of Kirklevington.

(42685)   BADMINTON 18TH,

Roan, calved November 8, 1878, bred by Viscountess Ossington, Ossington; got by Sherwood Chieftain (39098), dam (Coquette 8th) by Twenty-first Duke of Oxford (30999), g. d. (Coquette) by Kelstern Grand Duke (22024), gr. g. d. (Charmer) by Monarch (16575), — (Cherry) by West Australian (17223), — by Victor (13951), — by Berkeley (9830), — by Son of Old Strickland (11870).

(42686)   BAILIFF,

Roan, calved August 14, 1875, bred by Mr. J. Bowman, High House, the property of Mr. R. Cousins, Stubbsgill: got by Borough Member (33186), dam (Rosamond Gwynne) by Sir Windsor (22927), g. d. (Rebecca Gwynne) by Knight of Distington (18158), gr. g. d. (Ruth Gwynne) by Exquisite (14524), — (Young Dowager Gwynne) by St. Thomas (10777), — by Prime Minister (2456), — by Wallace (5586), — by Marmion (406), — by Merlin (430), — by Layton (366), — by Phenomenon (491), — by Favourite (252), — by Favourite (252), — by Hubback (319), — by Snowdon's Bull (612), — by Waistell's Bull (669), — by Masterman's Bull (422), — by the Studley Bull (626).

(42687)   BALLINTRAID,

Red and white, calved April 29, 1878, bred by Mr. J. A. Gordon, Udale, the property of Mr. M. Gunn, Culgower; got by Rosario (35315), dam (Honey Bee) by Royal Eden (35363), g. d. (Beeswing) by Surly (32635), gr. g. d. (Honey Comb 3rd) by Trojan Hero (20991), — (Honey Comb) by Prince of Featherstone (29652), — by Sultan (15355), — by Musician (13361), — by Fame (10221), — by Lord Adolphus Fairfax (4249), — by Barmpton Grazier (3091), — by Rex (4942), — by Wellington (680), — by Favourite (252).

(42688)   BANDINELL,

Roan, calved August 16, 1879, bred by the Earl of Zetland, Upleatham; got by Blucher (39475), dam (Adele) by Scots Fusilier (35483), g. d. (Allonette) by Sir James the Rose (15290), gr. g. d. (Alert) by Surrey (17067), — (Arabella) by Apollo (9899), — by Fourth Duke of York (10167), — by Second Cleveland Lad (3408), — by Velocipede (5552), — by Col. Cradock's Son of Rob Roy (557), — bred by Mr. Kitchen, from the stock of Messrs. C. and R. Colling.

(42689) **BANDY,**
Roan, calved September 6, 1872, bred by Mr. W. S. Marr, Uppermill, the property of Mr. R. Garden, North Ythsie; got by Heir of Englishman (24122), dam (Bessie 5th) by Prince Louis (27158), g. d. (Bessie 3rd) by Young Pacha (20457), gr. g. d. (Bessie 2nd) by Sir Hubert (18844), — (Bessie) by Clarendon (14280), — by Sir Arthur (12072), — by Young Ury (10984), — by The Pacha (7612), — by Second Duke of Northumberland (3646), — by Sillery (5131), — by Carleton (843), — by Diamond (205), — by Diamond (205).

(42690) **BANKER,**
Roan, calved November 3, 1877, bred by Mr. F. W. Park, Grove, the property of Mr. G. Grimes, South Leverton; got by Balmoral (36151), dam (Windsor's Walnut) by Windsor's Seal (37688), g. d. (Girdle) by Patrician (24728), gr. g. d. (Gipsy) by The Yeoman (25305), — (Midnight) by Economist (21669), — by Cromwell (19528), — by Old Buck (15017), — by Earl of Dublin (10178), — by Janizary (8175), — by Snowball (8602), — by Caliph (1774), — by Norman (2379), — by White Boy (1580), — by Wyville's Bull, — by a Bull bred by Mr. Charge.

(42691) **BANKER,**
White, calved March 28, 1879, bred by Captain Gandy, Castle Bank, the property of Mr. J. Strong, Culgaith; got by Grand Duke of Morecambe (36722), dam (Elvira 6th) by Grand Duke 10th (21848), g. d. (Elvira 2nd) by Eighth Duke of Oxford (15939), gr. g. d. (Ruby Rose 2nd) by The Baronet (10918), — (Ruby Rose) by Burgundy (7861), — by Pilot (4707), — by Navarino (2352), — by Favourite (256), — by Phenomenon (491), — by Favourite (252), — by Favourite (252), — by Hubback (319), — by Snowdon's Bull (612), — by Waistell's Bull (669), — by Masterman's Bull (422), — by the Studley Bull (626).

(42692) **BANNER BEARER,**
Red and white, calved October 9, 1878, bred by Mr. C. Cooke, Whiteley Green; got by Paul Gwynne (37181), dam (Lancaster 36th) by Lenton Hero (22092), g. d. (Lancaster 26th) by Third Duke of Norfolk (21601), gr. g. d. (Lancaster 10th) by Wonderful (14022), — (Lancaster 7th) by Priam (15079), — by The Queen's Roan (7389), — by Will Honeycomb (5660), — by Spectator (2688), — by Albion (1619), — by Lancaster (360), — by a Son of Windsor (698), — by Comet (155).

(42693) **BANTER,**
Red, calved July 30, 1879, bred by Mr. R. Stratton, The Duffryn; got by Pearl Diver (37182), dam (Badinage) by Charles 1st (33322), g. d. (Brownie) by Majesty (26789), gr. g. d. (Miss Brunette) by Miracle (24602), — (Brunette) by Lord of the Manor (14836), — by Hero of the West (8150), — by Newtonian, — by Phœnix (6290).

(42694) **BARON,**
Roan, calved January 3, 1870, bred by Mr. J. S. Bult, Dodhill House, the property of Mr. J. Edwards, Holway; got by Earl of Fife (23835), dam (Baby) by Augustus Windsor (19248), g. d. (Anemone 4th) by Upstart (9760), gr. g. d. (Anemone) by Allan-a-Dale (7778), — (Ultima) by Little John (4232), — by Caliph (1774), — by Swing (2721), — by Argus (759), — by Defender (194), — by Petrarch (488), — by Own Brother to R. Colling's White Heifer, — by Butterfly (104), — by Globe (278).

(42695) **BARON,**
Roan, calved March 23, 1876, bred by Lord Chesham, Latimer House, the property of Mr. H. Bury, Mangnolls; got by Duke of Oxford 28th (33710), dam (Baroness) by Baron Oxford 4th (25580), g. d. (Bland) by Sir James (22902), gr. g. d. (Blanc Mange) by Magistrate (13274), — (Blanche 5th) by Antinous (12401), — by Killjoy (14759), — by Diamond (5918), — by Norfolk (2377), — by Belvedere (1706), — by Belvedere (1706), — by Lancaster (360), — by Petrarch (488), — by Major (397), — by Chapman's Son of Punch (122), — by Dickson's Grandson of Punch (213), — by Checks (132), — by R. Grimston's Bull (282), — by J. Coates's Bull (148).

(42696) **BARON,**
Roan, calved February 3, 1879, bred by Mr. W. Innes, Meikle Clinterty, the property of Mr. J. Wilson, Mains of Scotstown; got by Merry Baron (43649), dam (Fragrant 5th) by Duke of Lancaster (30986), g. d. (Fragrant 4th) by Baronet (30448), gr. g. d. (Fragrant 3rd) by Lord Granville (24395), — (Fragrant 2nd) by Royal Standard (22803), — by John Bull (11618), — by Matadore (11800), — by Second Duke of Northumberland (3646), — by Mahomed (6170), — by Sillery (5131), — by Carleton (843), — by Diamond (205), — by Diamond (205).

(42697) **YOUNG BARON,**
Roan, calved May 2, 1879, bred by Mr. H. Bury, Mangnolls, the property of Mr. J. Harrison, Much Hoole; got by Baron (42695), dam (Lady Bury) by Rose Duke 2nd (35319), g. d. (Prosperity) by Great Grand Duke of Craven (26315), gr. g. d. (Dora) by Baron Calder (25558), — (Rosette) by Lord Gainford (18232), — by Prince of Prussia (16751).

(42698) **BARON ACOMB,**
Red and white, calved June 9, 1879, bred by Mr. R. Loder, Whittlebury; got by Grand Duke 22nd (34062), dam (Oxford Ida) by Third Duke of Clarence (23727), g. d. (Oxford's Ada) by Lord Oxford 2nd (20215), gr. g. d. (Ada) by General Canrobert (12927), — (Augusta) by Third Duke of York (10166), — by Second Cleveland Lad (3408), — by Duke of Cleveland (1937), — by Belvedere (1706), — a Cow of Mr. Bates's, of Kirklevington.

(42699) **BARON BARRINGTONIA,**
Red, calved June 15, 1878, bred by Mr. G. Underwood, Little Gaddesden; got by Duke of Barringtonia (30925), dam (Buttercup) by Count Glo'ster (23637), g. d. (Blonde) by The Druid (20948), gr. g. d. (Bloomer) by General Pelissier (14605), — (Bloom) by Royal (13636), — by Magnet (10488), — by Fitz-Hardinge (8073), — by Elevator (6969), — by Consul (1868), — by Gazer (7030).

(42700) **BARON BEVERLEY 3RD,**
Roan, calved April 8, 1879, bred by Mr. W. R. Bromet, Cocksford; got by Duke of Clarence 5th (38133), dam (Baroness Beverley 6th) by Baron Turncroft Oxford 2nd (33087), g. d. (Baroness Beverley) by Duke of Clarence (19611), gr. g. d. (Wharfdale Beverley) by Third Duke of Wharfdale (21619), — (Miss Beverley 25th) by Royal Butterfly 16th (20724), — by Ivanhoe (14735), — by Disraeli (10125), — by Lord George Bentinck (9317), — by Second Earl of Beverley (5963), — by Second Cleveland Lad (3408), — Red Darlington.

(42701)   BARON BEVERLEY 4TH,
Roan, calved July 12, 1879, bred by Mr. W. R. Bromet, Cocksford ; got by Duke
of Clarence 5th (38133), dam (Baroness Beverley) by Duke of Clarence (19611),
g. d. (Wharfdale Beverley) by Third Duke of Wharfdale (21619), gr. g. d. (Miss
Beverley 25th) by Royal Butterfly 16th (20724), — (Miss Beverley 4th) by
Ivanhoe (14735), — by Disraeli (10125), — by Lord George Bentinck (9317), —
by Second Earl of Beverley (5963), — by Second Cleveland Lad (3408), — Red
Darlington.

(42702)   BARON BEVERLEY 5TH,
Roan, calved August 11, 1879, bred by Mr. W. R. Bromet, Cocksford ; got by
Duke of Clarence 5th (38133), dam (Baroness Beverley 2nd) by Third Duke of
Tregunter (31026), g. d. (Clarence Beverley) by Duke of Clarence (19611), gr. g. d.
(Miss Beverley 6th) by Ivanhoe (14735), — (Miss Beverley 3rd) by Voltigeur
(13964), — by Disraeli (10125), — by Lord George Bentinck (9317), — by Second
Earl of Beverley (5963), — by Second Cleveland Lad (3408), — Red Darlington.

(42703)   .   BARON BLANCHE,
Red and white, calved February 21, 1878, bred by Mr. J. Wood, Humbleton
Hall ; got by Oxford-le-Grand (29496), dam (Orange Blossom) by Baron Geneva
(25568), g. d. (Blossom) by Sir James (22902), gr. g. d. (Blanche 5th) by Antinous
(12401), — (Blanche 1st) by Killjoy (14759), — by Diamond (5918), — by Norfolk (2377), — by Belvedere (1706), — by Belvedere (1706), — by Lancaster
(360), — by Petrarch (488), — by Major (397), — by Chapman's Son of Punch
(122), — by Dickson's Grandson of Punch (213), — by Checks (132), — by R.
Grimston's Bull (282), — by J. Coates's Bull (148).

(42704)   BARON BRAILES 2ND,
Red, calved February 21, 1879, bred by Mr. T. Nichols, The Grange ; got by
Lord Clarence Waterloo (36926), dam (Formosa) by Duke of Brailes (23724), g. d.
(Gionetta) by Sarawak (15238), gr. g. d. (Smock Frock) by Earl of Dublin
(10178), — (London Pride) by Janizary (8175), — by Snowball (8602), — by
Caliph (1774), — by Norman (2379), — by White Boy (1580), — by Wyville's
Bull, — by a Bull bred by Mr. Charge.

(42705)   BARON BRILLIANT 4TH,
White, calved May 29, 1879, bred by Mr. A. Garfit, Scothern ; got by Grand
Duke 25th (34065), dam (Brilliant Rose 6th) by Lord of Scothern (34626), g. d.
(Brilliant) by May Duke (13320), gr. g. d. (Blanche 3rd) by Antinous (12401),
— (Blanche) by Diamond (5918), — by Norfolk (2377), — by Belvedere (1706),
— by Belvedere (1706), — by Lancaster (360), — by Petrarch (488), — by Major
(397), — by Chapman's Son of Punch (122), — by Dickson's Grandson of Punch
(213), — by Checks (132), — by R. Grimston's Bull (282), — by J. Coates's
Bull (148).

(42706)   BARON CHILFROME,
White, calved January 20, 1878, bred by Mr. F. E. Pope, Great Toller, the property of Mrs. Symes, Kingston Russel ; got by Baron Cossington (37784), dam
(Lady Gentian) by Lord Harry (31666), g. d. (Gentian) by Eighth Duke of York
(23808), gr. g. d. (Gertrude) by Windsor Castle (21118), — (Gilt) by Nottingham (15014), — by The Red Duke (8694), — by Hero of the West (8150), —
by Kenilworth (7118), — by Phœnix (6290).

(42707)   BARON CLARO 3RD,
Red and white, calved July 7, 1878, bred by Mr. A. P. Clear, Maldon; got by Oxford's Baron (32030), dam (Cranberry) by Second Duke of Claro (21576), g. d. (Yewberry) by Seventh Duke of York (17754), gr. g. d. (Hollyberry) by Douglas (12714), — (Snowberry) by Rivers (10716), — by Diamond (5918), — by Albert (2950), — by Stirling (5330), — by Commodore (1858), — by Tathwell Studley (5401), — by Blyth Comet (85).

(42708)   BARON CRAGGS,
Red and white, calved December 19, 1879, bred by Mr. R. Loder, Whittlebury; got by Grand Duke of Oxford 3rd (39953), dam (Carrie Craggs) by Grand Duke of Clarence (28750), g. d. (Carolina 7th) by Second Duke of Collingham (23730), gr. g. d. (Carolina 6th) by Lord Lally (22161), — (Carolina 2nd) by Douglas (12714), — by Second Cleveland Lad (3408), — by Short Tail (2621), — by Son of Second Hubback (2683), — Craggs, bought of Mr. Bates, and descended from the stock of Mr. Maynard, of Eryholme.

(42709)   BARONET,
Roan, calved January 19, 1877, bred by Mr. A. I. Fortescue, Kingcausie, the property of Mr. F. Simmers, Cullerlie; got by Æolus (37716), dam (Lily of the Valley) by Victorious (35883), g. d. (Lily) by Prince Gwynne (27141), gr. g. d. (Truth) by Dumfries (17760), — (Catherine 2nd) by Fourth Duke of Bolton (15915), — by Lovemore (10476), — by Launcelot (6122), — by Eclipse (3684), — by Boughton (7841), — by Young Merlin (6204), — by Midas (435), — by Denton (198).

(42710)   BARONET,
Red, calved March 9, 1879, bred by the Earl of Dartrey, Dartrey House, the property of Mr. W. Johnson, Prumplestown House; got by Young Druid (39699), dam (Buttercup) by King James 2nd (34321), g. d. (Lady Butterfly) by Czar (23670), gr. g. d. (Butterfly) by Master Butterfly 2nd (14918), — (Campanula) by Lord Wiseton (13256), — by Hudson (9228), — by Fairfax Royal (6987), — by The Toucher (6596), — by George (2057), — by Togston (5487), — bred by Mr. Laing, of Longhoughton.

(42711)   BARON EXETER,
Red and white, calved August 31, 1877, bred by Mr. J. Howard, Clapham Park; got by Cambridge Duke 5th (30644), dam (Gipsy Gwynne) by Royal Oxford (27380), g. d. (Nancy Gwynne) by Fourth Duke of Thorndale (17750), gr. g. d. (Betty Gwynne) by Duke of Cambridge (12747), — (Sukey Gwynne) by St. Thomas (10777), — by Prime Minister (2456), — by Marmion (406), — by Merlin (430), — by Layton (366), — by Phenomenon (491), — by Favourite (252), — by Favourite (252), — by Hubback (319), — by Snowdon's Bull (612), — by Waistell's Bull (669), — by Masterman's Bull (422), — by the Studley Bull (626).

(42712)   BARON FAWSLEY,
Roan, calved March 21, 1879, bred by Mr. J. Robinson, Broughton Pastures; got by King Charming (28962), dam (Countess Fawsley) by Lord Waterloo 2nd (26755), g. d. (Camilla) by Touchstone (20986), gr. g. d. (Christine) by Sir James (16980), — (Chrysalis) by Earl of Dublin (10178), — by Grey Friar (9172), — by Fawsley (6004), — by Marcellus (2260), — by Rufus (2570), — by Wellington (683), — by Windsor (698), — by Windsor (698), — by Own Brother to North Star (459).

(42713) **BARON FAWSLEY,**
Roan, calved March 28, 1879, bred by Mr. R. Loder, Whittlebury ; got by
Grand Duke 22nd (34062), dam (Oxford Fawsley 3rd) by Lord Oxford 2nd
(20215), g. d. (Fawsley 3rd) by Grand Duke 4th (19874), gr. g. d. (Coquelicot)
by Duke of Cambridge (12742), — (Blouzelind) by Earl of Dublin (10178), —
by Janizary (8175), — by Caliph (1774), — by Rob Roy (557), — by Satellite
(1420), — by Sir Dimple (594), — by Styford (629).

(42714) **BARON FAWSLEY 2ND,**
Roan, calved July 24, 1879, bred by Mr. R. Loder, Whittlebury ; got by Grand
Duke 22nd (34062), dam (Lady Gertrude Fawsley) by Grand Duke of Kent
(26289), g. d. (Fawsley 3rd) by Grand Duke 4th (19874), gr. g. d. (Coquelicot)
by Duke of Cambridge (12742), — (Blouzelind) by Earl of Dublin (10178), —
by Janizary (8175), — by Caliph (1774), — by Rob Roy (557), — by Satellite
(1420), — by Sir Dimple (594), — by Styford (629).

(42715) **BARON FAWSLEY 3RD,**
Red and white, calved November 13, 1879, bred by Mr. R. Loder, Whittlebury;
got by Grand Duke of Oxford 3rd (39953), dam (Grand Duchess of Fawsley 4th)
by Grand Duke 22nd (34062), g. d. (Geneva Fawsley) by Eighth Duke of Geneva
(28390), gr. g. d. (Fawsley 3rd) by Grand Duke 4th (19874), — (Coquelicot)
by Duke of Cambridge (12742), — by Earl of Dublin (10178), — by Janizary
(8175), — by Caliph (1774), — by Rob Roy (557), — by Satellite (1420), — by
Sir Dimple (594), — by Styford (629).

(42716) **BARON FITZCLARENCE,**
Roan, calved September 21, 1879, bred by Mr. H. Hussey Vivian, Park Wern ;
got by Lord Fitzclarence 16th (36943), dam (Trip Away) by Duke of Fawsley
(28387), g. d. (Thrifty) by Dalesman (25869), gr. g. d. (Transit) by Young
Windsor (17241), — (Truth) by Hermit (14697), — by Lord of the Manor
(14836), — by Hero of the West (8150), — by Corporal (6899), — by Kenilworth (7118).

(42717) **BARON FLOWER 4TH,**
Roan, calved February 5, 1879, bred by Messrs. Dalton, Cummersdale ; got by
Duke of Siddington (38182), dam (Jelly Flower 7th) by Oxford Isis (32024), g. d.
(Jelly Flower 4th) by Thirteenth Duke of Oxford (21604), gr. g. d. (Jelly Flower
3rd) by Specimen (17026), — (Jelly Flower) by Bloody Douglas (12478), — by
Rubini (9586), — by Gainford 2nd (6030), — by Emperor (1974), — by a Bull
bred by Mr. Headlam.

(42718) **BARON GOLDSCHMIDT 5TH,**
Roan, calved April 17, 1878, bred by Mr. R. Crowe, Speeton, the property of
Mr. W. Crowe, Flamborough ; got by Sir Gregory Gwynne (37465), dam
(Madame Goldschmidt 2nd) by Baron Booth (23354), g. d. (Madame Gold-
schmidt) by Master Goldschmidt (20305), gr. g. d. (Carissima) by Cardigan
(12556), — (Carry) by Prince Arthur (13497), — by Nobleman (13392), — by
Newton (2367), — by Young Magog (2247), — by Margrave (2263), — by Sir
Charles (593), — by Sir Dimple (594), — by St. Albans (1412), — by Layton
(366), — by Charge's Grey Bull (872), — by J. Brown's Red Bull (97).

(42719) **BARON GRAY,**
White, calved July 15, 1879, bred by the Duke of Richmond and Gordon,
Gordon Castle ; got by Royal Hope (32392), dam (Lustre 8th) by Duke of

Bowland (21568), g. d. (Lustre 3rd) by Magnum Bonum (13277), gr. g. d.
(Lustre 1st) by Bloomsbury (9972), — (Lustre) by Second Duke of Northumberland (3646), — by Bachelor (1666), — by Sultan (1485), — by North Star
(458).

(42720)   BARON GWYNNE,
Roan, calved April 22, 1879, bred by Lord Skelmersdale, Lathom House, the
property of Mr. J. H. Angas, Collingrove, Angaston, South Australia; got
by Baron Oxford 4th (25580), dam (Minstrel' 5th) by Ninth Duke of Geneva
(28391), g. d. (Minstrel 4th) by Tenth Duke of Oxford (17739), gr. g. d.
(Minstrel 2nd) by Prince of Glo'ster (13517), — (Minstrel) by Count Conrad
(3510), — by Wallace (5586), — by Wellington (2824), — by Marmion (406),
— by Merlin (430), —, by Layton (366), — by Phenomenon (491), — by
Favourite (252), — by Favourite (252), — by Hubback (319), — by Snowdon's
Bull (612), — by Waistell's Bull (669), — by Masterman's Bull (422), — by the
Studley Bull (626).

(42721)   BARON GWYNNE,
Red, calved September 13, 1879, bred by Mr. D. McIntosh, Havering Park; got
by Duke of Havering (33664), dam (Dorothy Gwynne) by Second Duke of
Wellington (28465), g. d. (Flora Gwynne) by Oxford Gwynne (24711), gr. g. d.
(Fairy Gwynne) by Grand Duke 5th (19875), — (Fortuna Gwynne) by Duke of
Leinster (17724), — by Captain Hardinge (10023), — by St. Thomas (10777),
— by Prime Minister (2456), — by Marmion (406), — by Merlin (430), — by
Layton (366), — by Phenomenon (491), — by Favourite (252), — by Favourite
(252), — by Hubback (319), — by Snowdon's Bull (612), — by Waistell's Bull
(669), — by Masterman's Bull (422), — by the Studley Bull (626).

(42722)   BARON GWYNNE,
Red and white, calved December 23, 1879, bred by Mr. R. Loder, Whittlebury;
got by Grand Duke of Oxford 3rd (39953), dam (Grand Duchess Gwynne 2nd)
by Grand Duke 22nd (34062), g. d. (Duchess Gwynne 4th) by Baron Oxford 5th
(27958), gr. g. d. (Duchess Gwynne 2nd) by Third Duke of Clarence (23727), —
(Duchess Gwynne) by Duke of Wetherby (17753), — by Flying Dutchman
(10235), — by St. Thomas (10777), — by Prime Minister (2456), — by Wallace
(5586), — by Marmion (406), — by Merlin (430), — by Layton (366), — by
Phenomenon (491), — by Favourite (252), — by Favourite (252), — by
Hubback (319), — by Snowdon's Bull (612), — by Waistell's Bull (669), — by
Masterman's Bull (422), — by the Studley Bull (626).

(42723)   BARON HAMILTON,
Red and white, calved June 26, 1879, bred by Mr. W. Sheraton, Broom House,
the property of Mr. H. Povey, Ellesmere; got by The Sultan (40811), dam
(Lady Jane Hamilton) by Darius (30860), g. d. (Lady Hamilton 3rd) by Knight
of the Garter (34400), gr. g. d. (Lady Hamilton) by Second Duke of Cumberland
(23735), — (Mary Hamilton) by Viceroy (21019), — by Bannerman (19258),
— by Hickory (14706), — by King John (14763), — by Hero of the West
(8150), — by Bourton Hero (9983), — by Killerby (7124), — by Beggarman
(3118), — by Samson (5081), — by Remus (6381), — by Premier (2449).

(42724)   BARON HAVERING 21st,
Red, calved May 17, 1879, bred by Mr. D. McIntosh, Havering Park; got by
Duke of Havering (33664), dam (Charmer 19th) by Third Duke of Geneva

(23753), g. d. (Charmer 11th) by Baron Killerby (23364), gr. g. d. (Charmer 8th) by Prince James (20554), — (Charmer 7th) by Highthorn (13028), — by Earl of Dublin (10178), — by Little John (4232), — by Caliph (1774), — by Sir Walter (2637), — by Hotspur (1117), — by Coxcomb (928), — by Midas (435), — by Comet (155), — by R. Colling's Son of Favourite (252), — by same Son of Favourite (252), — by Hubback (319).

(42725)   BARON IRIS,
Red and white, calved February 28, 1879, bred by Mr. I. Clark, Heddington; got by Rouser (39025), dam (Iris 6th) by Florist (23962), g. d. (Iris 2nd) by Saunterer (16914), gr. g. d. (Iris) by Accordion (5708), — (Purity) by Despot (1915), — by Napoleon (4531), — by Rival (2534), — by Darlington (956), — by Denton (198), — by Rockingham (560), — by Jobling's Son of Phenomenon (491), — by Phenomenon (491), — by Colonel (152), — by Styford (629).

(42726)   BARON KENTDALE,
Roan, calved June 17, 1878, bred by Mr. W. H. Wakefield, Sedgwick, the property of Mr. R. Sandham, Lancaster; got by Baron Barrington 4th (33006), dam (Welcome) by Dunrobin (28486), g. d. (Banks 3rd) by Frederick Warlaby (23990), gr. g. d. (Banks 1st) by Sir Colin (16957), — (Lancaster) by Leap Year (11677).

(42727)   BARON KIRKLEVINGTON,
Red and white, calved August 17, 1879, bred by Mr. H. Allsopp, Hindlip Hall; got by Duke of Underley (33745), dam (Kirklevington Princess 4th) by Third Duke of Glo'ster (33653), g. d. (Kirklevington Duchess 9th) by Grand Duke of Clarence (28750), gr. g. d. (Duchess of Kent) by Lord Liverpool (22168), — (Kirklevington 14th) by Fourth Duke of Oxford (11387), — by Earl of Derby (10177), — by Earl of Liverpool (9061), — by Duke of Northumberland (1940), — by Belvedere (1706), — by Son of Second Hubback (2683), — a Cow of Mr. Bates's, descended from the stock of Mr. Maynard, of Eryholme.

(42728)   BARON KNIGHTLEY,
Red, calved January 25, 1879, bred by Colonel R. Loyd Lindsay, Lockinge Park; got by Earl of Horton 11th (36588), dam (Blue Bell 2nd) by Count Blanche (33455), g. d. (Blueberry) by Rob Roy (29806), gr. g. d. (Burlesque) by Fawsley Baronet (23920), — (Britannia) by Master Coleshill (18344), — by Sultan (15358), — by Neptune (11847), — by Fanatic (8054), — by Earl of Durham (5965), — by Pedestrian (7321), — by Edrom (1956), — by Scipio (1421), — by Hector (1104), — by Midas (435), — by Marquis (407), — by Chilton (136), — by Ben (70).

(42729)   BARON LECHLADE,
Red and white, calved March 9, 1879, bred by Mr. D. Arkell, Butlers Court; got by Baron Lee 4th (37799), dam (Fillpail 3rd) by Factory Boy (33870), g. d. (Fairmaid) by Royal George (29866), gr. g. d. (Fillpail 2nd) by Mercury (22342), — (Fillpail) by Oxford Don (20451).

(42730)   BARON MAPLEWELL,
Red, calved May 18, 1879, bred by Sir W. H. Salt, Bart., Maplewell, the property of Mr. J. Barrs, Nailstone; got by Fifth Duke of Glo'ster (36494), dam (Maplewell 2nd) by Fifth Lord Oxford (31738), g. d. (Kirklevington 18th) by Third Lord Oxford (22200), gr. g. d. (Kirklevington 10th) by Delhi (15865), — (Kirklevington 8th) by General Canrobert (12926), — by Earl of Derby (10177),

— by Earl of Liverpool (9061), — by Duke of Northumberland (1940), — by Belvedere (1706), — by Son of Second Hubback (2683), — a Cow of Mr. Bates's, descended from the stock of Mr. Maynard, of Eryholme.

(42731)　　　　　BARON MAPLEWELL 2ND,
Red, calved July 17, 1879, bred by Sir W. H. Salt, Bart., Maplewell, the property of Mr. R. Arnold, Shackerstone; got by Fifth Duke of Glo'ster (36494), dam (Maplewell) by Second Duke of Glo'ster (28392), g. d. (Kirklevington 18th) by Third Lord Oxford (22200), gr. g. d. (Kirklevington 10th) by Delhi (15865), — (Kirklevington 8th) by General Canrobert (12926), — by Earl of Derby (10177), — by Earl of Liverpool (9061), — by Duke of Northumberland (1940), — by Belvedere (1706), — by Son of Second Hubback (2683), — a Cow of Mr. Bates's, descended from the stock of Mr. Maynard, of Eryholme.

(42732)　　　　　BARON MARLBORO' 2ND,
Roan, calved December 4, 1879, bred by Mr. G. Garne, Churchill Heath ; got by Grand Duke of Marlboro' (38380), dam (Blanche Geneva 2nd) by Grand Duke of Geneva 2nd (31288), g. d. (Thorndale Blanche 2nd) by Sir Rainald (27485), gr. g. d. (Thorndale Blanche) by Twelfth Duke of Thorndale (26020), — (Blanche 8th) by Clifford (21437), — by Magistrate (13274), — by Antinous (12401), — by Killjoy (14759), — by Diamond (5918), — by Norfolk (2377), — by Belvedere (1706), — by Belvedere (1706), — by Lancaster (360), — by Petrarch (488), — by Major (397), — by Chapman's Son of Punch (122), — by Dickson's Grandson of Punch (213), — by Checks (132), — by R. Grimston's Bull (282), — by J. Coates's Bull (148).

(42733)　　　　　BARON MELDRETH,
White, calved February 2, 1879, bred by Mr. C. Ellis, Meldreth ; got by Meldreth Duke (34832), dam (Rose of Spring) by Lord Oxford 2nd (20215), g. d. (Garland 5th) by Waterloo Duke (21077), gr. g. d. (Sabrina) by May Duke (13320), — (Garland 3rd) by Monk (11824), — by Pestalozzi (10603), — by Hector (4000), — by Emperor (1974), — by Margrave (2263), — by Leopold (2199), — by Hector (2103), — by Traveller (655), — by Surly (2715), — by Colonel (152).

(42734)　　　　　BARON MELTON,
Roan, calved March 17, 1879, bred by Mr. Jonas Webb, Melton Ross ; got by Baron Aston (37770), dam (Lady of the Lea) by Cambridge (25705), g. d. (Lady of the Glen 3rd) by Costa (21487), gr. g. d. (Lady Bird) by Count de Gourcy (17632), — (Luxury) by Cheltenham (12588), — by Horatio (10335), — by Third Duke of Northumberland (3647), — by Velocipede (5552), — by St. Thomas (2636), — by Marske (418), — by Comet (155), — by Tom (652), — by Favourite (1033), — by Hutton's Bull (323), — by Barningham (56).

(42735)　　　　　BARON OF KENT 10TH,
Roan, calved April 17, 1879, bred by Messrs. F. Leney and Sons, Wateringbury ; got by Sixth Duke of Oneida (30997), dam (Countess 5th) by Sixth Duke of Oneida (30997), g. d. (Chorus) by Fourth Duke of Thorndale (17750), gr. g. d. (Charming) by Mameluke (13289), — (Charmer 6th) by Cardinal (11246), — by White Friar (9827), — by Little John (4232), — by Caliph (1774), — by Sir Walter (2637), — by Hotspur (1117), — by Coxcomb (928), — by Midas (435), — by Comet (155), — by R. Colling's Son of Favourite (252), — by same Son of Favourite (252), — by Hubback (319).

(42736)   BARON OF NAWTON,
Roan, calved July 9, 1879, bred by Mr. T. Stamper, Highfield House; got by
Duke of Nawton 3rd (39764), dam (Raspberry 2nd) by Saxton (35478), g. d.
(Raspberry) by Prince Alfred (27107), gr. g. d. (Madam Hardy) by New Year's
Day (16615), — (Sprightly) by President (16708), — by Essex Lad (8039), —
by Vampire (7665), — by Red Rover (4903), — by Alexander (1624), — by a
Bull of Mr. C. Colling's.

(42737)   BARON OXFORD 3RD,
Red and white, calved November 17, 1879, bred by Mr. D. McIntosh, Havering
Park; got by Duke of Havering (33664), dam (Baroness Oxford 5th) by Fifth
Duke of Wetherby (31033), g. d. (Baroness Oxford) by Second Duke of Claro
(21576), gr. g. d. (Lady Oxford 5th) by Third Duke of Thorndale (17749), —
(Lady Oxford 4th) by Grand Duke 2nd (12961), — by The Lord of Eryholme
(12205), — by Third Duke of York (10166), — by Duke of Northumberland
(1940), — by Short Tail (2621), — by Matchem (2281), — by Young Wynyard
(2859).

(42738)   BARON OXFORD 9TH,
Red and white, calved January 7, 1879, bred by the Duke of Devonshire, Holker
Hall; got by Fifth Duke of Wetherby (31033), dam (Baroness Oxford) by
Second Duke of Claro (21576), g. d. (Lady Oxford 5th) by Third Duke of Thorn-
dale (17749), gr. g. d. (Lady Oxford 4th) by Grand Duke 2nd (12961), — (Maid
of Oxford) by The Lord of Eryholme (12205), — by Third Duke of York
(10166), — by Duke of Northumberland (1940), — by Short Tail (2621), — by
Matchem (2281), — by Young Wynyard (2859).

(42739)   BARON OXFORD 10TH,
Roan, calved June 1, 1879, bred by the Duke of Devonshire, Holker Hall; got
by Duke of Barrington (39712), dam (Baroness Oxford 6th) by Fifth Duke of
Wetherby (31033), g. d. (Baroness Oxford) by Second Duke of Claro (21576),
gr. g. d. (Lady Oxford 5th) by Third Duke of Thorndale (17749), — (Lady Oxford 4th) by Grand Duke 2nd (12961), — by The Lord of Eryholme (12205),
— by Third Duke of York (10166), — by Duke of Northumberland (1940), —
by Short Tail (2621), — by Matchem (2281), — by Young Wynyard (2859).

(42740)   BARON OXFORD BARRINGTON,
White, calved August 20, 1878, bred by Mr. J. Harward, Winterfold, the pro-
perty of Mr. H. D. Vavasour, Hazlewood Castle, Tadcaster; got by Baron
Oxford 4th (25580), dam (Lally 10th) by Fifth Lord Wild Eyes (26762), g. d.
(Lally 5th) by Duke of Wetherby (17753), gr. g. d. (Lally 2nd) by Malachite
(18313), — (Lally) by Earl of Derby (10177), — by Earl of Liverpool (9061),
— by Second Duke of Cambridge (3638), — by Belvedere (1706), — by Son of
Herdsman (304), — by Wonderful (700), — by Alfred (23), — by Young
Favourite (6994).

(42741)   BARON OXFORD BARRINGTON 2ND,
Red, calved September 8, 1878, bred by Mr. J. Harward, Winterfold, the property of the Exors. of Mr. J. Harward, Winterfold; got by Baron Turncroft
Oxford 4th (37822), dam (Duchess Lally) by Fifth Duke of Wetherby (31033),
g. d. (Lally 10th) by Fifth Lord Wild Eyes (26762), gr. g. d. (Lally 5th) by
Duke of Wetherby (17753), — (Lally 2nd) by Malachite (18313), — by Earl
of Derby (10177), — by Earl of Liverpool (9061), — by Second Duke of Cam-

bridge (3638), — by Belvedere (1706), — by Son of Herdsman (304), — by Wonderful (700), — by Alfred (23), — by Young Favourite (6994).

(42742)   BARON PAUL,
Roan, calved November 29, 1879, bred by Mr. E. W. Meade-Waldo, Stonewall Park ; got by Baron Aylesby (39397), dam (Pauline 16th) by King Victor (28986), g. d. (Pauline 8th) by Lord Blithe (22126), gr. g. d. (Pauline 6th) by Heir of Windsor (26364), — (Pauline 3rd) by Ravenspur (20628), — by British Boy (11206), — by Hopewell (10332), — by Hamlet (8126), — a Cow bred by Mr. Booth.

(42743)   BARON RING,
Roan, calved February 25, 1879, bred by Mr. A. Mackenzie Lyle, Donaghmore House, the property of Mr. H. Ferguson, Loughill ; got by Lord of the Manor (38641), dam (April Blossom) by Sailor Prince (35442), g. d. (Bessy) by Lord Francis (24393), gr. g. d. (Grisi) by Duke of Montrose (21599), — (De Beriot) by Lumley (16478), — by Master Goldschmidt (13315), — by Marquis (11786), — by The Lord of Gilling (6587), — by Buckingham (3239), — by Son of Chance (868), — by Chance (868), — by Grazier (1085), — by Plutarch (1327), — by Midas (435), — by Grandson of Simon (590).

(42744)   BARON ROTHSCHILD,
Red, calved January 20, 1879, bred by Mr. G. Graham, The Oaklands ; got by Baron Fantail (37790), dam (Cheerful Rose 2nd) by Lord Thorndale (29210), g. d. (Cheerful Rose) by Cambridge Duke 4th (25706), gr. g. d. (Cheerful) by Mystic (20391), — (Coquette) by Mameluke (13289), — by Garrick (11506), — by Earl of Dublin (10178), — by Little John (4232), — by Caliph (1774), — by Sir Walter (2637), — by Hotspur (1117), — by Coxcomb (928), — by Midas (435), — by Comet (155), — by R. Colling's Son of Favourite (252), — by same Son of Favourite (252), — by Hubback (319).

(42745)   BARON RYEDALE 2ND,
Roan, calved January 6, 1877, bred by the Earl of Feversham, Duncombe Park, the property of Mr. J. Sisson, Newfield ; got by Twentieth Duke of Oxford (28432), dam (Belle of Ryedale 3rd) by Second Duke of Tregunter (26022), g. d. (Princess of Battersea) by Mandarin (18317), gr. g. d. (Ballad Singer) by Fifth Duke of Oxford (12762), — (Bloomer) by Ben Nevis (9960), — by Cleveland Lad (3407), — by Triumph (5518), — by Grazier (1085), — by Parrington (4653), — by Baron (58), — by Windsor (698), — out of a grand-daughter of Washington (674).

(42746)   BARON RYEDALE 4TH,
Roan, calved June 10, 1879, bred by the Earl of Feversham, Duncombe Park got by Connaught Knightley (41267), dam (Beeswing) by Orestes (22443), g. d. (Barefoot) by Chanticleer (17530), gr. g. d. (Ballad Singer) by Fifth Duke of Oxford (12762), — (Bloomer) by Ben Nevis (9960), — by Cleveland Lad (3407), — by Triumph (5518), — by Grazier (1085), — by Parrington (4653), — by Baron (58), — by Windsor (698), — out of a grand-daughter of Washington (674).

(42747)   BARON RYEDALE 5TH,
White, calved October 26, 1879, bred by the Earl of Feversham, Duncombe Park ; got by Fifth Duke of Tregunter (33743), dam (Bloom of Ryedale 4th)

by Twentieth Duke of Oxford (28432), g. d. (Beeswing) by Orestes (22443), gr. g. d. (Barefoot) by Chanticleer (17530), — (Ballad Singer) by Fifth Duke of Oxford (12762), — by Ben Nevis (9960), — by Cleveland Lad (3407), — by Triumph (5518), — by Grazier (1085), — by Parrington (4653), — by Baron (58), — by Windsor (698), — out of a grand-daughter of Washington (674).

(42748)   BARON SEGRAVE,

Roan, calved April 10, 1878, bred by Lord Mowbray and Stourton, Stourton, the property of Mr. J. Wildon, Knapton Grange ; got by Lord Oxford Bright Eyes (34648), dam (Rosamond 8th) by Sprightly Lord (30044), g. d. (Rosamond 6th) by Rose Duke (22760), gr. g. d. (Rosamond 5th) by Hesperus (18062), — (Rosamond 4th) by Abbot (14054), — by Voltigeur (12274), — by Hautboy (10305), — by Westmoreland (5633), — by Chancellor (1809), — by Umpire (5530), — by Northern Light (4586), — by Young Alfred (2993), — by Lemon Horns (2196), — by Witham's White Bull (5670).

(42749)   BARON SINNINGTON,

White, calved October 18, 1878, bred by Mr. R. Lesley, Sinnington Lodge ; got by Cleveland 3rd (39600), dam (Martha) by Saxton (35478), g. d. (Lady Jane) by Chilton (25774), gr. g. d. (Lady Elizabeth) by Sir Richard (25166), — (Mary) by Wathstone's Hero (25417), — by White Willie (35991), — by The Friar (15383).

(42750)   BARON SOCKBURN,

Red and white, calved July 28, 1879, bred by Mr. S. Shaw, Brooklands ; got by Lord Oxford Sockburn 2nd (38648), dam (Baroness Blanche) by Baron Water- loo (27977), g. d. (Bantling) by Britannicus (17452), gr. g. d. (Blanchette) by Magistrate (13274), — (Blanche 5th) by Antinous (12401), — by Killjoy (14759), — by Diamond (5918), — by Norfolk (2377), — by Belvedere (1706), — by Belvedere (1706), — by Lancaster (360), — by Petrarch (488), — by Major (397), — by Chapman's Son of Punch (122), — by Dickson's Grandson of Punch (213), — by Checks (132), — by R. Grimston's Bull (282), — by J. Coates's Bull (148).

(42751)   BARON STRAWBERRY,

Roan, calved April 18, 1879, bred by Mr. J. Owen, Middleton ; got by White Prince (40914), dam (Nellie) by Warwickshire Lad (37649), g. d. (Strawberry) by Grand Monarch 2nd (19887), gr. g. d. (Buttercup 2nd) by Don Pedro (19584), — (Young Buttercup) by Alma (14087), — by Concord (11301), — by Second Duke of Lancaster (5951), — by Crichton (3516), — by Ploughboy (4726).

(42752)   BARON TREGUNTER,

Red and white, calved January 23, 1879, bred by Mr. R. Loder, Whittlebury, the property of Messrs. Maclean and Co., Auckland, New Zealand ; got by Duke of Tregunter 7th (38194), dam (Grand Duchess of Fawsley 3rd) by Grand Duke 22nd (34062), g. d. (Lady Gertrude Fawsley) by Grand Duke of Kent (26289), gr. g. d. (Fawsley 3rd) by Grand Duke 4th (19874), — (Coquelicot) by Duke of Cambridge (12742), — by Earl of Dublin (10178), — by Janizary (8175), — by Caliph (1774), — by Rob Roy (557), — by Satellite (1420), — by Sir Dimple (594), — by Styford (629).

(42753)   BARON TURNCROFT BATES 6TH,

Red and white, calved December 30, 1879, bred by the Rev. P. Graham, Turn- croft ; got by Duke of Tregunter 7th (38194), dam (Lady Thorndale Bates) by

Fourth Duke of Thorndale (17750), g. d. (Lady Bates 3rd) by Fourth Duke of Oxford (11387), gr, g. d. (Lady Bates 2nd) by The Buck (13836), — (Lady Bates) by Duke of Glo'ster (11382), — by Fourth Duke of York (10167), — by Second Duke of Oxford (9046), — by Fourth Duke of Northumberland (3649), — by Cleveland Lad (3407), — by Belvedere (1706), — by Son of Herdsman (304) — by Wonderful (700), — by Alfred (23), — by Young Favourite (6994).

(42754)   BARON TURNCROFT FANTAIL 3RD,
Red, calved September 16, 1879, bred by the Rev. P. Graham, Turncroft; got by Duke of Tregunter 7th (38194), dam (Fantail 4th) by Touchstone (20986), g. d. (Fantail) by Barleycorn (17348), gr. g. d. (Fair Helen) by General Canrobert (12927), — (Florella) by Fifth Duke of York (10168), — by Fourth Duke of Northumberland (3649), — by Short Tail (2621), — by Belvedere (1706), — by Son of Young Wynyard (2859), — descended from J. Brown's Old Red Bull (97).

(42755)   BARON WESTOE,
Roan, calved August 27, 1879, bred by Mr. T. Chalk, Linton; got by Baron Aston (37770), dam (Vocalist) by Young England (31110), g. d. (Warbler) by Sinbad (29981), gr. g. d. (Songstress) by Grand Duke 16th (24063), — (Seraphina 3rd) by Royal Essex (18767), — by Fitz-Clarence (11477), — by Sweet William (7571), — by Earl of Essex (6955), — by Stratton (5336), — by Fanatic (1996), — by Red Rover (4902), — by Rufus (2576), — by Emperor (1014).

(42756)   BARON WINSOME 7TH,
Red, calved December 4, 1879, bred by the Duke of Devonshire, Holker Hall; got by Wetherby Winsome 2nd (40905), dam (Winsomedale 2nd) by Second Duke of Tregunter (26022), g. d. (Winsome 9th) by Grand Duke 17th (24064), gr. g. d. (Winsome 2nd) by Lord Oxford (20214), — (Winsome) by Oxford 2nd (18507), — by Crusade (7938), — by Third Duke of York (10166), — by Second Cleveland Lad (3408), — by Duke of Northumberland (1940), — by Belvedere (1706), — by Emperor (1975), — by Wonderful (700), — by Cleveland (145), — by Butterfly (104), — by Hollon's Bull (313), — by Mowbray's Bull (2342), — by Masterman's Bull (422), — descended from M. Dobison's stock.

(42757)   BARON WINSTON,
Red, calved March 16, 1875, bred by Mr. W. Sheraton, Broom House; got by Lilleshall (34456), dam (Winston Lady) by Knight of the Garter (34400), g. d. (Lady Fitz-Harper) by Lord Fitz-Hugh of Wensleydale (26645), gr. g. d. (Lady Amazon) by War Knight (27748), — (Sailor's Bride) by Sinbad (27460), — by Harold (14670), — by Grandson of Emperor 2nd (11438).

(42758)   BARON YORK,
Roan, calved January 3, 1879, bred by Mrs. Mace, Sherborne; got by Thorndale Geneva (39217), dam (Violet 8th) by Kenelm Butterfly (31462), g. d. (Violet 3rd) by Seventh Duke of York (17754), gr. g. d. (Rose of Oxford) by Fourth Duke of Oxford (11387), — (White Rose) by Siddington Duke (15263), — by Harold (10299), — by Leo (4208), — by Henwood (2114), — by Wharfdale (1578), — by Young Marske (419), — by Meteor (432), — by Western Comet (689), — by Favourite (252), — by Cupid (177), — by Grandson of Bolingbroke (280), — by Foljambe (263), — by R. Alcock's Bull (19), — by J. Smith's Bull (608), — by Jolly's Bull (337).

(42759) **BARRINGTON,**
Red, calved December 26, 1877, bred by Mr. J. J. Sharp, Broughton, the property of Messrs. Smith and Freeman, Sherborne; got by Duke of Barrington 5th (33575), dam (May Flower) by Baron York (30500), g. d. (Melodious) by Royalist (27370), gr. g. d. (Modesty) by White Velvet (19146), — (Mohair) by Old Buck (15017), — by Frantic (12897), — by Boccaccio (7838), — by Duke of Rothsay (6943), — by Belshazzar (1703), — by Noble Henry (2374), — by Abraham (2905), — by Mustachios (4527), — by Simon (5134), — by Young George (3885), — by George (276).

(42760) **YOUNG BARRINGTON,**
Roan, calved April 6, 1879, bred by Mr. W. H. Brown, Belbroughton, the property of the Rev. E. T. Williams, Caldicot Parsonage; got by Baron Barrington 6th (33008), dam (Florentia 18th) by Grand Duke 13th (21850), g. d. (Florentia 9th) by Seventh Duke of York (17754), gr. g. d. (Florentia 2nd) by Archduke (17316), — (Felicia) by Koh-i-noor (11642), — by Usurer (9763), — by Zenith (5702), — by Sweet William (5368), — by Roman (2561), — by Monarch (2324), — by Jupiter (342), — by Sir Oliver (605), — by Trunnell (659), — by Favourite (252), — by Favourite (252), — by Dalton Duke (188), — by R. Alcock's Bull (19), — by J. Smith's Bull (608), — by Jolly's Bull (337).

(42761) **BARRISTER,**
Roan, calved May 18, 1877, bred by Mrs. Fowle, Chute Lodge, the property of Mr. S. H. Allen, Eastover; got by J. P. (41731), dam (Bella) by Hercules (28843), g. d. (Blanche) by Lord Lennox (20189), gr. g. d. (Britannia) by Viscount Killerby (19081),— (Barbara) by Napier (13368), — by Protector (10665), — by Rob Roy (7434), — by Raffler (7391).

(42762) **BARRISTER,**
Roan, calved February 18, 1879, bred by Sir R. C. Musgrave, Bart., Eden Hall; got by Royal Cambridge 4th (40624), dam (Baroness Barbon 2nd) by Duke of Kirklevington (33683), g. d. (Baroness Barbon) by Baron Oxford 3rd (25579), gr. g. d. (Princess of Battersea) by Mandarin (18317), — (Ballad Singer) by Fifth Duke of Oxford (12762), — by Ben Nevis (9960), — by Cleveland Lad (3407), — by Triumph (5518), — by Grazier (1085), — by Parrington (4653), — by Baron (58), — by Windsor (698), — out of a grand-daughter of Washington (674).

(42763) **BARTON,**
White, calved March 3, 1871, bred by Mr. R. Stratton, Burderop, the property of Mr. Lyne, Barton; got by James 1st (24202), dam (Dewdrop) by Ivanhoe (18096), g. d. (Gossamer) by Nottingham (15014), gr. g. d. (Spider) by Duke of Gloster (10153), — (Spider) by Hero of the West (8150), — by Lottery (4280).

(42764) **BASSOON,**
Roan, calved November 3, 1879, bred by Lord Bolton, Bolton Hall; got by Heir-at-Law (34124), dam (Blanditia) by Marmion (26821), g. d. (Amorous) by Strawberry Prince (25240), gr. g. d. (Eba) by May Duke (13321), — (Emma 2nd) by Hartforth (9191), — by Free Trader (10246), — by Gainford (2044), — by Magnum Bonum (2243), — by Rob Roy (557), — by Son of Houghton (318), — by Sir Stephen (1456), — by Sedbury (1424).

(42765) BASTO,
Red, calved November 23, 1878, bred by Mr. W. Scott, Glendronach; got by
Ivanhoe (36796), dam (Sweet Myrtle) by Amateur 4th (27882), g. d. (Brides-
maid) by Nobleman (26967), gr. g. d. (Mary) by Cœur-de-Lion (12611), —
(Alma) by Moss Trooper (11827), — by Donside Fairfax (11365), — by Duke
(3630), — by Reveller (2528), — by Grazier (1085), — by Cato (857), — by
Atlas (42), — by Favourite (257), — by Mr. Robinson's Bull (4974), — by
Badsworth (47).

(42766) BAVENO,
Red, calved November 2, 1879, bred by Mr. R. Pinder, Whitwell; got by Prince
William (37283), dam (Sandpiper 6th) by Fiery Star (33914), g. d. (Sandpiper
3rd) by The Stuart (27650), gr. g. d. (Sandpiper 2nd) by President (27088), —
(Sandpiper) by The Briar (15376), — by Francisco (12893), — by Columbus
(10063), — by Prince of Wales (6345), — by Edward (3696), — by Topper
(2768), — by Favourite (3768), — by Windsor (698), — by Windsor (698).

(42767) BEACONSFIELD,
Roan, calved April 29, 1877, bred by Mr. C. W. Griffin, Werrington; got by
Telemachus 10th (35728), dam (Lady Blanche 2nd) by Nestor (24648), or Sugar
Cane (32625), g. d. (Lady Blanche) by Red Butterfly (22690), gr. g. d. (Bella)
by Louis (14861), — (Lady Blanche) by Rivers (10716), — by Mozart (11830),
— by Stirling (5330), — by Young Frederick (3836), — by Favourite (3768),
— by Tathwell Studley (5401), — by Son of Waddingworth (668).

(42768) BEADSMAN,
White, calved July 16, 1879, bred by Mr. J. Angus, Bearl; got by Ben Brace
(30524), dam (Daffodil) by Royal Frederick (35366), g. d. (Daisy 2nd) by Merry
Monarch (22349), gr. g. d. (Daisy 1st) by Alderman (17291), — (Topsy) by
Duke of Tyne (12773), — by Young Hector (7074), — by Young Hastings
(3988), — by Short Legs (5124), — by Ajax (723), — by Hector (2103), — by
Surly (2715), — by Sir Harry (5155), — by Colonel (152), — by Son of Hub-
back (319).

(42769) BEAU BENEDICT,
Roan, calved February 2, 1879, bred by Mr. W. Linton, Sheriff Hutton, the
property of Mr. R. Thompson, Inglewood Bank; got by Paul Potter (38854),
dam (Home Beauty) by Mountain Chief (20383), g. d. (Hand Maid) by May
Day (20323), gr. g. d. (White Rose) by Magnus Troil (14880), — (Miss Henderson) by Magnus Troil (14880), — by Bates (12451), — by Lord Warden
(20233), — by Snyders (7525), — by Duke of York (9049).

(42770) BEAUDESERT,
Roan, calved October 13, 1879, bred by Mr. Y. R. Graham, Yardley Stud
Farm; got by Baron Fantail (37790), dam (Fantail 3rd) by Lord Thorndale
(29210), g. d. (Fantail 2nd) by Costa (21487), gr. g. d. (Fantail) by Barleycorn
(17348), — (Fair Helen) by General Canrobert (12927), — by Fifth Duke of
York (10168), — by Fourth Duke of Northumberland (3649), — by Short Tail
(2621), — by Belvedere (1706), — by Son of Young Wynyard (2859), —
descended from J. Brown's Old Red Bull (97).

(42771) BEAU DESERT DUKE,
Red and white, calved August 16, 1879, bred by Mr. J. Darling, Beau Desert;
got by Duke of Oxford 27th (33709), dam (Emma 9th) by Oxford Beau 4th

(34964), g. d. (Economy) by Third Lord Lally (24408), gr. g. d. (Emilia) by Duke
(17700), — (Emma 5th) by Lord Raglan (13240), — by Emperor (6973), —
by Sir Walter (2639), — by Wharncliffe (2834), — by Kilnshaw (2175), — by
Guy (2084), — by Frederick (1060), — by Pyramus (533), — by Meteor (432).

(42772)   BEAU OF CHILTON,
Red and white, calved November 22, 1879, bred by Mr. E. Freeman, Chilton;
got by King Alfred (36833), dam (Charming Foggathorpe) by Charming Lad
(30699), g. d. (Lady Superior 4th) by Darlington (23679), gr. g. d. (Lady
Superior 3rd) by Prince Leopold (20558), — (Lady Superior) by British Baronet
(15691), — by Cardinal (11246), — by Marquis of Rockingham (10560), — by
Sheldon (8557), — by Benjamin (1710), — by Malbro' (1189), — by Ebor
(997), — by Regent (546), — by North Star (459), — by R. Colling's White
Bull (151), — bred by Mr. R. Colling.

(42773)   BEAU OF OXFORD 5TH,
Red, calved December 20, 1879, bred by Lord Penrhyn, Penrhyn Castle; got
by Grand Duke of Oxford (31293), dam (Fourth Belle of Oxford) by Grand
Duke 11th (21849), g. d. (Third Belle of Oxford) by Marmaduke (14897),
gr. g. d. (Second Belle of Oxford) by Duke of Glo'ster (11382), — (Oxford
17th) by Lamartine (11662), — by Third Duke of York (10166), — by Duke
of Northumberland (1940), — by Short Tail (2621), — by Matchem (2281), —
— by Young Wynyard (2859).

(42774)   BEESWAX,
Red and white, calved May 19, 1879, bred by Mr. G. H. Lett, Millpark; got
by British Lad (41151), dam (Honeycell) by Prince of Munster (32174), g. d.
(Honeysuckle) by Marine (24529), gr. g. d. (Honeycomb) by Roseberry (22757),
— (Nectar) by Hindoo (14707), — by War Eagle (15483), — by Captain
Shaftoe (6833), — by Christoval (3382), — by Ranunculus (2479), — by Albion
(2970), — by Ratify (2481), — by Logic (4248).

(42775)   BEL ALP,
White, calved May 2, 1879, bred by Mr. F. B. Greenwood, Swarcliffe, the property of Mr. G. Carter, Sandhill Farm; got by Knight of Windermere (38533),
dam (Staubach) by Peer of the Realm (27057), g. d. (Simplon) by Grindelwald
(26323), gr. g. d. (Sylvia 3rd) by Royal Buck (25004), — (Sylvia 2nd) by Frank
(19776), — by Melton (18381), — by Keighley (18125), — by Sharow (18819),
— by Governor (7046), — by Prince Alfred (6329), — by Colossus (1847), —
by a Bull of Mr. J. Woodhouse's.

(42776)   BELISARIUS,
Roan, calved April 5, 1867, bred by Mr. J. G. Grove, Castle Grove, the property of Lord Bloomfield, Loughton, Moneygall; got by British Crown (21322),
dam (Nota Bene) by King Arthur (13110), g. d. (Norma) by Druid (10140),
gr. g. d. (Little Red Rose) by Petrarch (7329), — (Florence) by Second Duke
of York (5959), — by Raspberry (4875), — by Young Matchem (4422), — by
Isaac (1129), — by Young Pilot (4702), — by Pilot (496), — by Julius Cæsar
(1143).

(42777)   BELLADRUM,
Red and white, calved April 4, 1879, bred by Lord Lovat, Beaufort Farm; got
by Duke of Beaufort (38122), dam (Beaufort Rose) by Bachelor of Arts (32982),
g. d. (Broadhooks 10th) by Champion of England (17526), gr. g. d. (Broadhooks

9th) by Baronet (15614), — (Broadhooks 6th) by The Baron (13833), — by Prince Edward Fairfax (9506), — by White Bull (5643), — by Young Lady-kirk (4170), — by Albion (731), — by Sirius (598), — by Wellington (679), — by Sultan (631), — by Signior (588).

(42778)   BELL OXFORD,
Red, calved September 16, 1878, bred by Mr. J. S. Robinson, Mount Falinge, the property of Mr. A. Bell, Barrowford ; got by Earl of Waterloo 3rd (36601), dam (Venetia's Bride) by Bridegroom (30584), g. d. (Venetia) by Ravenswood (22682), gr. g. d. (Venom) by Sir Charles (22891), — (Vixen) by Sweetmeat (18952), — by Royal Favourite (15200), — by Sylvan (10907), — by Lictor (6128), — by Gracchus (3917), — by Ganthorpe (2049), — by Midas (7236), — by Grazier (1085).

(42779)   BELTED KNIGHT,
Roan, calved October 14, 1879, bred by Mr. R. Thompson, Inglewood Bank ; got by Brilliant Butterfly (36270), dam (Love Token) by Grand Duke of Fawsley 3rd (31286), g. d. (Farewell) by Royal Westmoreland (35416), gr. g. d. (General's Daughter) by General Haynau (11520), — (Red Rose 6th) by Gainford 2nd (10255), — by Emperor (9084), — by Sir Thomas Newton (6495), — by General (3869).

(42780)   BEN AVON,
White, calved December 8, 1878, bred by Messrs. J. W. and E. Cruickshank, Lethenty, the property of Miss Milne, Otterburn ; got by Balmoral (36151), dam (Bracelet 3rd) by Lord of Rocklands (22183), g. d. (Bracelet 2nd) by Vanguard (21009), gr. g. d. (Bridal) by Buckingham (11219), — (Lady of the Lake) by Northern Light (13398), — by Hopewell (10332), — by Diamond (5918), — by Priam (2452), — by Volunteer (1553), — by Argus (759), — by Pilot (496), — by Remus (550), — by Blucher (82), — by Albion (14), — by Shakespeare (582), — by Easby (232).

(42781)   BENEDICT,
Roan, calved February 7, 1879, bred by Mr. T. M. Hopkins, Lower Wick Farm ; got by Emiliani (41507), dam (Bonny Fawsley 2nd) by Duke of St. Johns (36538), g. d. (Bonny Fawsley) by Baron Fawsley (27936), gr. g. d. (Bonnie Girl) by Cramer (17638), — (Balmoral) by Diamond (14395), — by Belmont (14153), — by Matadore (10516), — by Wizard (6688), — by Wiseton (2848), — by Mercury (2301), — by Hector (1105), — by Regent (544).

(42782)   BENEDICT,
Red, calved June 26, 1879, bred by Mr. T. Willis, Carperby ; got by Windsor Benedict (40933), dam (Caroline Windsor) by Admiral Windsor (32912), g. d. (Charlotte Windsor) by Windsor Fitz-Windsor (25458), gr. g. d. (Charlotte Bronte) by First Fruits (16048), — (Village Maid) by Lyon Playfair (11758), — by Bernardo (8885), — by Billingham (6786), — by Sol (2655), — by Harlsey (2091), — by Maximus (2284).

(42783)   BENEDICT BOOTH,
Red, calved May 24, 1879, bred by Mr. T. Willis, Carperby ; got by Windsor Benedict (40933), dam (Diamond Necklace) by Booth's Royal Seal (28060), g. d. (Windsor's Necklace) by Windsor Fitz-Windsor (25458), gr. g. d. (Rose of Cashmere) by Lord Frederick (22156), — (Rose of Lucknow) by Sir Colin

(16953), — by King Alfred (16334), — by Gipsy King (11532), — by Wilberforce (9830), — by Swintonian (9702), — by Young Symmetry (7577), — by
Shipton (2620), — by Fitz-Champion (7007), — by Young Snowball (7521).

(42784)   BENEDICTINE,
White, calved August 8, 1877, bred by the Earl of Zetland, Aske Hall; got by
King Harry (36841), dam (Bell Benedict) by Royal Benedict (27348), g. d.
(Blue Bell) by Cynthius (25864), gr. g. d. (Beldame) by Cobham (14287), —
(Bellona) by General (14589), — by Tenth Duke of York (11398), — by Belle-
ville (6778), — by Monarch 3rd (4502), — by Young Hastings (3987), — by
Election (1961), — by Barmpton (54), — by Whitenose (692), — by George
(275), — by Ben (70), — by Favourite (252), — by Chapman's Son of Punch
(122).

(42785)   BENEZET,
Red, calved December 20, 1879, bred by Mr. T. Willis, Carperby; got by
Windsor Benedict (40933), dam (Rose Fitz-Clarence) by Major Windsor (34739),
g. d. (Rose Clarence) by Fitz-Clarence (14552), gr. g. d. (Moss Rose) by Gavazzi
(11508), — (Moss Rosebud) by Wilberforce (9830), — by Swintonian (9702),
— by Young Symmetry (7577), — by Shipton (2620), — by Fitz-Champion
(7007), — by Young Snowball (7521).

(42786)   BENJAMIN,
White, calved March 23, 1879, bred by Mr. J. Nicholson, Murton; got by Ben
Lomond (37854), dam (Bright Gem) by White Duke (32849), g. d. (Jewel) by
Orange Prince (27005), gr. g. d. (Jessy) by Prince Albert (18579), — (Bright
Eyes) by Milo (10532), — by Zadkiel (11069), — by Matchem 3rd (4420),
— by Waterloo (2816), — by Borodino (3192), — by Blucher (1725).

(42787)   BENTICK,
Red, calved April 13, 1875, bred by Mr. W. S. Marr, Uppermill; got by Heir
of Englishman (24122), dam (Bessie 6th) by Young Pacha (20457), g. d. (Bessie
2nd) by Sir Hubert (18844), gr. g. d. (Bessie) by Clarendon (14280), — (Helen)
by Sir Arthur (12072), — by Young Ury (10984), — by The Pacha (7612), —
by Second Duke of Northumberland (3646), — by Sillery (5131), — by Carleton
(843), — by Diamond (205), — by Diamond (205).

(42788)   BEN TUDOR,
White, calved April 1, 1879, bred by Mr. J. Angus, Bearl; got by Ben Brace
(30524), dam (Snowdrop 3rd) by Manfred (26801), g. d. (Snowdrop) by Brigade
Major (21312), gr. g. d. (Lily) by Lord Albert (20143), — (Blossom) by
Frederick (14571), — by Red Prince (13576), — by Brilliant (7851), — by
Sir Robert (5180), — by Bachelor (1665), — by Short Legs (5124), — by
Ajax (723), — by Hector (2103), — by Surly (2715),— by Sir Harry (5155),
— by Colonel (152), — by Son of Hubback (319).

(42789)   BEN WYVIS,
Roan, calved February 20, 1879, bred by Mr. G. Marr, Cairnbrogie, the property
of Mr. J. Cran, Kirkton; got by Bromley (36289), dam (Dainty 8th) by Young
Hero (26385), g. d. (Dainty 7th) by Grand Prince (26308), gr. g. d. (Dainty 4th)
by Baron Sebastopol (21241), — (Dainty 3rd) by Fourth Duke of Bolton
(15915), — by Fairfax Hero (9106), — by Second Duke of Northumberland
(3646), — by Emperor (3716), — by Invalid (4076), — by Magnet (392), —
by Palmflower (480).

(42790) **BERKELEY,**
Red and white, calved May 10, 1879, bred by Mr. D. R. Scratton, Ogwell;
got by Lord Turncroft Oxford 2nd (38668), dam (Lady Wild Eyes 10th) by
Duke of Siddington 2nd (33732), g. d. (Lady Wild Eyes 8th) by Grand Duke of
Waterloo (28766), gr. g. d. (Lady Wild Eyes 3rd) by Cherry Grand Duke
(23554), — (Lady Wild Eyes 2nd) by Touchstone (20986), — by Weathercock
(9815), — by Second Cleveland Lad (3408), — by Second Duke of Northumberland (3646), — by Short Tail (2621), — by Emperor (1975), — by Wonderful
(700), — by Cleveland (145), — by Butterfly (104), — by Hollon's Bull
(313), — by Mowbray's Bull (2342), — by Masterman's Bull (422), —
descended from M. Dobison's stock.

(42791) **BERKELEY DUKE OF OXFORD,**
Roan, calved April 29, 1879, bred by Lord Fitzhardinge, Berkeley Castle, the
property of Mr. W. McCulloch, Glenroy, Melbourne; got by Duke of Connaught
(33604), dam (Oxford Duchess 2nd) by Second Duke of Collingham (23730), g. d.
(Lady of Oxford 11th) by Baron of Oxford (23371), gr. g. d. (Lady of Oxford
7th) by Sixth Duke of Thorndale (23794), — (Lady of Oxford 2nd) by Grand
Duke 2nd (12961), — by Third Duke of York (10166), — by Duke of Northumberland (1940), — by Short Tail (2621), — by Matchem (2281), — by Young
Wynyard (2859).

(42792) **BETA,**
Roan, calved March 22, 1879, bred by Mr. C. Wilson, Shotley Park; got by
Cavalier (36328), dam (Violet Leaf) by Masterpiece (31882), g. d. (Russian Violet
by Scotland's Pride (25100), gr. g. d. (Red Violet) by Allan (21172), — (Violet)
by Lord Bathurst (13173), — by Matadore (11800), — by Hudson (9228), — by
Fairfax Royal (6987), — by Inkhorn (6091), — by Grazier (1085), — by Sampson,
— by Wallace (1560).

(42793) **BEVERLEY,**
Roan, calved October 21, 1879, bred by Colonel Gunter, Wetherby Grange, the
property of Mr. W. Thomlinson, Cowthorpe; got by Oxford's Prince (34998),
dam (Miss Beverley 4th) by Cherubim 2nd (30727), g. d. (Miss Beverley 2nd) by
Lieutenant Oxford (24336), gr. g. d. (Miss Beverley 18th) by Royal Butterfly
16th (20724), — (Miss Beverley 6th) by Ivanhoe (14735), — by Voltigeur
(13964), — by Disraeli (10125), — by Lord George Bentinck (9317), — by
Second Earl of Beverley (5963), — by Second Cleveland Lad (3408), — Red
Darlington.

(42794) **BICKMARSH,**
Roan, calved November 20, 1879, bred by Mr. R. Hemming, Bentley Manor;
got by Gambados (39911), dam (Ada 24th) by Normansells (34923), g. d.
(Ada 12th) by Saturnus (29932), gr. g. d. (Ada 9th) by Cavalier (23524), —
(Ada 4th) by Grand Cazique (26282), — by Prince Carl (20538), — by Orthodox 28th (18493), — by Lemnos (13146), — by Earl Stanhope (5966), — by
True Blue (5522), — by Miracle (2321), — by Sir Henry (1446), — by Count
(170), — by Bracken (91), — by Badsworth (47), — by Driffield (223), — bred
by Sir G. Strickland.

(42795) **BIRTH NIGHT,**
Roan, calved November 26, 1877, bred by Mr. J. S. Bult, Dodhill House; got
by Gallant Gay (33983), dam (Bertha 2nd) by Cardinal (28144), g. d. (Bertha)

by Conqueror (21466), gr. g. d. (Anemone 2nd) by Duke of Cambridge (12742),
(Anemone) by Allan-a-Dale (7778), — by Little John (4232), — by Caliph
(1774), — by Swing (2721), — by Argus (759), — by Defender (194), — by
Petrarch (488), — by Own Brother to R. Colling's White Heifer, — by Butterfly
(104), — by Globe (278).

(42796)   BLANCHE'S OXFORD,

Red, calved April 3, 1876, bred by Sir W. H. Salt, Bart., Maplewell, the
property of Mr. E. Jones, Pool Quay Farm ; got by Fifth Lord Oxford (31738),
dam (Blanche 2nd) by Grand Duke 17th (24064), g. d. (Blanche) by General
Napier (24023), gr. g. d. (Blanche Flower) by May Duke 2nd (18372), —
(Balm) by Britannicus (17452), — by Magistrate (13274),— by Antinous (12401),
— by Diamond (5918), — by Norfolk (2377), — by Belvedere (1706), — by
Belvedere (1706), — by Lancaster (360), — by Petrarch (488), — by Major
(397), — by Chapman's Son of Punch (122), — by Dickson's Grandson of
Punch (213), — by Checks (132), — by R. Grimston's Bull (282), — by J.
Coates's Bull (148).

(42797)   BLANCHE WILD EYES,

White, calved January 13, 1878, bred by Mr. S. Lipscomb Seckham, Bletchley
Park, the property of Messrs. Harris and Biggs, Cublington ; got by Duke of
Worcester 3rd (39796), dam (Princess Adelaide) by Duke of Cambridge (25940),
g. d. (Princess Mary) by Grand Duke 9th (19879), gr. g. d. (Bantling) by
Britannicus (17452), — (Blanchette) by Magistrate (13274), — by Antinous
(12401), — by Killjoy (14759), — by Diamond (5918), — by Norfolk (2377),
— by Belvedere (1706), — by Belvedere (1706), — by Lancaster (360), — by
Petrarch (488), — by Major (397), — by Chapman's Son of Punch (122), — by
Dickson's Grandson of Punch (213), — by Checks (132), — by R. Grimston's
Bull (282), — by J. Coates's Bull (148).

(42798)   BLANCHE WILD EYES 2ND,

White, calved March 19, 1879, bred by Mr. S. Lipscomb Seckham, Bletchley
Park ; got by Duke of Worcester 3rd (39796), dam (Princess Adelaide)
by Duke of Cambridge (25940), g. d. (Princess Mary) by Grand Duke 9th
(19879), gr. g. d. (Bantling) by Britannicus (17452), — (Blanchette) by
Magistrate (13274), — by Antinous (12401), — by Killjoy (14759), — by
Diamond (5918), — by Norfolk (2377), — by Belvedere (1706), — by Belvedere
(1706), — by Lancaster (360), — by Petrarch (488), — by Major (397), — by
Chapman's Son of Punch (122), — by Dickson's Grandson of Punch (213), —
by Checks (132), — by R. Grimston's Bull (282), — by J. Coates's Bull (148).

(42799)   BLANCHE WILD EYES 3RD,

Roan, calved August 29, 1879, bred by Mr. S. Lipscomb Seckham, Bletchley
Park ; got by Duke of Worcester 3rd (39796), dam (Fair Adelaide) by Master
Fragrant (34810), g. d. (Princess Adelaide) by Duke of Cambridge (25940),
gr. g. d. (Princess Mary) by Grand Duke 9th (19879), — (Bantling) by Britannicus (17452), — by Magistrate (13274), — by Antinous (12401), — by
Killjoy (14759), — by Diamond (598), — by Norfolk (2377), — by Belvedere
(1706), — by Belvedere (1706), — by Lancaster (360), — by Petrarch (488),
— by Major (397), — by Chapman's Son of Punch (122), — by Dickson's
Grandson of Punch (213), — by Checks (132), — by R. Grimston's Bull (282),
— by J. Coates's Bull (148).

(42800) **BLOOMING BOY,**
Red and white, calved June 17, 1879, bred by Mr. J. R. Singleton, Great
Givendale; got by Duke of Oxford 37th (38171), dam (Blooming Rose) by
Bobby (25648), g. d. (Bloomer) by Marquis (26825), gr. g. d. (Beauty) by
Thorndale Lad (23066), — (Beauty) by Sir Colin (15276), — by Lord Belle-
ville (14804), — by Albert (5729), — by Earl Stanhope (5966), — by Cordilleras
(3484), — by Midas (2309), — by Remus (550).

(42801) **BLUEBEARD,**
Roan, calved July 4, 1879, bred by Colonel R. Loyd Lindsay, Lockinge Park;
got by Earl of Horton 11th (36588), dam (Bluebell) by Prince Rupert (35178),
g. d. (Blueberry) by Rob Roy (29806), gr. g. d. (Burlesque) by Fawsley Baronet
(23920), — (Britannia) by Master Coleshill (18344), — by Sultan (15358), —
by Neptune (11847), — by Fanatic (8054), — by Earl of Durham (5965), —
by Pedestrian (7321), — by Edrom (1956), — by Scipio (1421), — by Hector
(1104), — by Midas (435), — by Marquis (407), — by Chilton (136), — by
Ben (70).

(42802) **BLUSHING DUKE 2ND,**
Red and white, calved October 19, 1876, bred by the Rev. W. Holt Beever, Pencraig Court, the property of Mr. J. Rankin, Bryngwyn; got by Royal Cumberland (27358), dam (Blush) by Prince Henry (18608), g. d. (Donna) by Patriot
(10594), gr. g. d. (Duchess) by Sir Frederick (8577), — (Damsel) by Marquis
(4386), — by Mynheer (1255), — by Enchanter (244), — by Major (398), —
by Windsor (698), — by Favourite (252), — by Punch (531), — by Hubback
(319).

(42803) **BLUSHING DUKE 4TH,**
Roan, calved March 14, 1879, bred by the Rev. W. Holt Beever, Pencraig
Court; got by Grand Duke of Clarence (28750), dam (Royal Blush) by Royal
Cumberland (27358), g. d. (Court Blush) by The Baron (25277), gr. g. d. (Blush)
by Prince Henry (18608), — (Donna) by Patriot (10594), — by Sir Frederick
(8577), — by Marquis (4386), — by Mynheer (1255), — by Enchanter (244),
— by Major (398), — by Windsor (698), — by Favourite (252), — by Punch
(531), — by Hubback (319).

(42804) **BLUSHING DUKE 5TH,**
Red and white, calved May 4, 1879, bred by the Rev. W. Holt Beever, Pencraig
Court; got by Grand Duke of Clarence (28750), dam (Blush) by Prince Henry
(18608), g. d. (Donna) by Patriot (10594), gr. g. d. (Duchess) by Sir Frederick
(8577), — (Damsel) by Marquis (4386), — by Mynheer (1255), — by Enchanter
(244), — by Major (398), — by Windsor (698), — by Favourite (252), — by
Punch (531), — by Hubback (319).

(42805) **BLUSHING DUKE 6TH,**
Roan, calved September 8, 1879, bred by the Rev. W. Holt Beever, Pencraig
Court; got by Twentieth Duke of Oxford (28432), dam (Early Blush) by
Mountain Dew (34879), g. d. (Blush) by Prince Henry (18608), gr. g. d.
(Donna) by Patriot (10594), — (Duchess) by Sir Frederick (8577), — by
Marquis (4386), — by Mynheer (1255), — by Enchanter (244), — by Major
(398), — by Windsor (698), — by Favourite (252), — by Punch (531), — by
Hubback (319).

(42806) BLUSHING DUKE 7TH,

Roan, calved December 7, 1879, bred by the Rev. W. Holt Beever, Pencraig Court; got by Twentieth Duke of Oxford (28432), dam (Carmine) by Mountain Dew (34879), g. d. (Court Blush) by The Baron (25277), gr. g. d. (Blush) by Prince Henry (18608), — (Donna) by Patriot (10594), — by Sir Frederick (8577), — by Marquis (4386), — by Mynheer (1255), — by Enchanter (244), — by Major (398), — by Windsor (698), — by Favourite (252), — by Punch (531), — by Hubback (319).

(42807) BLUSHING JOE,

Red, calved April 21, 1879, bred by Mr. H. D. de Vitre, Charlton House; got by Blushing Duke (37869), dam (Not for Joe) by Hush (28883), g. d. (Lady Culshaw) by Royal Butterfly 17th (22774), gr. g. d. (Rose of Lancashire) by Master Butterfly 4th (14920), — (Young Barmpton Rose) by Richard Cœur-de-Lion (13590), — by Lord John (11731), — by Baron of Ravensworth (7811), — by Raree Show (4874), — by Thick Hock (6601), — by Expectation (1988), — by Belzoni (1709), — by Comus (1861), — by Denton (198).

(42808) BOANERGES,

Roan, calved April 14, 1879, bred by Mr. J. Garsed, The Moorlands; got by Kentish Charmer (38482), dam (Blonde) by Lord of the Lake (26710), g. d. (Blanche) by Duffryn (19592), gr. g. d. (Blossom) by Frank (17874), — (Nancy) by Scapulary (15274), — by Shardeloes (6462), — bred by Mr. T. T. Drake.

(42809) BOB RUTH,

Red and white, calved March 10, 1879, bred by Mr. J. R. Singleton, Great Givendale, the property of Mr. A. Kirby, Market Weighton; got by Duke of Oxford 37th (38171), dam (Mary Ruth) by Bobby (25648), g. d. (Maude) by Patriot (20475), gr. g. d. (Graceful) by Ferdinand (12871), — (Grace) by Surplice (10901), — by Remus (9545), — by Snowball (5227), — by Red High-flyer (2488), — by Old Snowball (2648), — descended from the stock of Mr. Newby, of Thorpe.

(42810) BOHEMIAN,

Red and white, calved July 21, 1879, bred by Mr. R. Jefferson, Preston Hows; got by Banner Bearer (27907), dam (Good Cressida) by Good Fitz (21844), g. d. (Hecuba) by Florian (12887), gr. g. d. (Young Cressida) by Duke of Athol (10150), — (Cressida) by Cossack (1880), — by Miracle (2320), — by Matchem (2281), — by Fitz-Remus (2025), — by Cato (119), — by Whitworth (695).

(42811) BOLD BUCCLEUCH,

Roan, calved January 23, 1879, bred by Mr. W. Scott, Glendronach; got by Ivanhoe (36796), dam (Flower Girl) by Duke of Lancaster (30986), g. d. (Florence 3rd) by Lord Granville (24395), gr. g. d. (Florence) by Lord Ythan 2nd (22238), — (Mary Ann) by Garibaldi (17916), — by Moss Trooper (11827), — by Robin O'Day (4973), — by Holkar (4014), — by Mr. Jopp's Bull (9256), — Kate of Darlington.

(42812) BONDSMAN,

Roan, calved August 7, 1879, bred by Mr. A. Graham, Yanwath Hall; got by The Colonel (35747), dam (Borderess 3rd) by Dean (30863), g. d. (Borderess 1st) by British Prince (19354), gr. g. d. (Borderess) by Borderer (15676), — (Red Rose 4th) by Precedent (11918), — by Election (9075), — by Sir Thomas Newton (6495), — by General (3867).

(42813)   BONNY LAD,
Roan, calved September 25, 1878, bred by Sir Wilfrid Lawson, Bart., Brayton;
got by Wild Eyes Duke (36007), dam (Bonny Lass) by Baslow (30507), g. d.
(Countess of Roseberry) by Seventeenth Duke of Oxford (25994), gr. g. d. (Belinda) by Sir Roger (16991), — (Berrington Lass) by Sir Walter 2nd (10834),
— by Brawith Boy (7846), — by Sir Walter (2639), — by Roseberry (567), —
by Roseberry (567), — by Constitution (166), — by Hastings (293), — by Constellation (163).

(42814)   BORDER CHIEF,
Roan, calved February 12, 1879, bred by Mr. E. Freeman, Chilton; got by
King Alfred (36833), dam (Lady Lucy) by Charming Lad (30699), g. d. (Queen
of the Vale) by Confessor (41260), gr. g. d. (Lady of the Vale) by Bismarck
(23422), — (Eleanor) by Parade (18518), — by Lord Ruby (14851), — by Preston
(20514), — by Monzani (6222), — by Firby (1040), — by Arbutus (1649).

(42815)   BORDERER,
Roan, calved February 3, 1879, bred by Colonel Graham, Grahams Hill; got by
Sweetheart's Duke 2nd (37552), dam (Avice No. 6) by Prelate (29570), g. d.
(Avice) by Mac Turk (14872), gr. g. d. (Lucy) by Heir-at-Law (13005), —
(Sonsie 4th) by Southwick (5282), — by Adam (2920), — by Camden (1776),
— by Noble Henry (2374), — by Young Sir Alexander (2977).

(42816)   BORDER KNIGHT,
Red and white, calved March 28, 1879, bred by Mr. T. Penny, Bartlehill; got
by Earl of Studley 4th (36598), dam (Dewy Eve) by The Hero (37581), g. d.
(Countess 1st) by Duke of Norfolk (33697), gr. g. d. (Bell) by Valiant (19039),
— by Delia's Red Richard (15868).

(42817)   BORDER KNIGHT,
Roan, calved December 8, 1879, bred by Mr. E. Freeman, Chilton; got by King
Alfred (36833), dam (Lady Lucy) by Charming Lad (30699), g. d. (Queen of
the Vale) by Confessor (41260), gr. g. d. (Lady of the Vale) by Bismarck
(23422), — (Eleanor) by Parade (18518), — by Lord Ruby (14851), — by
Preston (20514), — by Monzani (6222), — by Firby (1040), — by Arbutus
(1649).

(42818)   BOSCOBEL,
Roan, calved May 29, 1879, bred by Mr. S. L. Horton, Park House; got by
Marquis of Blandford 6th (41983), dam (Columbine) by Abacot (32900), g. d.
(Maria) by Second Duke of Wharfdale (19649), gr. g. d. (Mary 2nd) by Marc
Antony (14895), — (My Mary) by British Yeoman (8906), — by Adrian (5714),
— by Noble (4577), — by Young Don Juan (3010), — by Young Isaac (2154),
— by Son of Young Dimple (6518), — by Young Dimple (971), — by Snowball
(2648), — by Layton (2190).

(42819)   BRADDYLL DUKE,
White, calved August 3, 1878, bred by Mr. G. Steele, Must Hill; got by
Hartington 3rd (36745), dam (Perfection 2nd) by Duke of Melbourne (28423),
g. d. (Perfection) by Grenadier (21876), gr. g. d. (Neatness) by Aaron (15540),
— (Nectarine) by Crusader (12668), — by Plato (11907), — by Wonderful
(9841), — by Norfolk (2377), — by Tarquin (2734), — by Red Simon (2499),
— by Governor (1077), — by Son of Wellington (680).

(42820)   BRADLEY DUKE,
Red, calved January 26, 1877, bred by Mr. G. M. Dixon, Bradley Hall, the property of Mr. R. H. Frank, Ashbourne Hall ; got by Jack Frost 2nd (43390),
dam (Duchess of Cambridge 5th) by Duke of Lorn (25985), g. d. (Duchess of
Cambridge) by First Duke of Oxford and Glo'ster (19637), gr. g. d. (Lucretia)
by The Briar (15376), — (Lucetta) by Camphine (14225), — by Francisco
(12893), — by Champion (12576), — by Bellerophon (3119), — by Sillery
(5131), — by Champagne (3317), — by Favourite (256), — by Gibson's
Sampson.

(42821)   BRAEMAR,
Roan, calved April 15, 1878, bred by Mr. F. W. Park, Grove ; got by Balmoral
(36151), dam (M.P.'s Rose of the Hills) by M.P. (29398), g. d. (Windsor's Red
Rose) by Windsor Fitz-Windsor (25458), gr. g. d. (Rose of the Valley) by Lord
of the Valley (14837), — (Moss Rose) by Gavazzi (11508), — by Wilberforce
(9830), — by Swintonian (9702), — by Young Symmetry (7577), — by Shipton
(2620), — by Fitz-Champion (7007), — by Young Snowball (7521).

(42822)   BRAMPTON,
Red and white, calved April 8, 1878, bred by Mr. J. N. Beasley, Brampton
House, the property of Lord Overstone, Overstone Park ; got by Duke of
Jantja (33675), dam (Juanita) by Jay (26457), g. d. (Duchess Janette) by
Second Duke of Airdrie (19600), gr. g. d. (Countess Janette) by Lilyvick
(10421), — (Young Jocund) by Japetus (10359), — by Monarch (7249), — by
MacIvor (2237), — by Northern Light (1280), — by Wellington (684), — by
Phenomenon (491), — by Favourite (252), — by Favourite (252), — by Favourite
(252), — by Hubback (319), — by Snowdon's Bull (612), — by Waistell's Bull
(669), — by Masterman's Bull (422), — by the Studley Bull (626).

(42823)   BREASTPLATE,
Red and white, calved March 1, 1879, bred by Mr. J. Radcliffe, Stearsby ; got
by Sir Hugo Irwin (35563), dam (Bracelet 2nd) by Bismarck (28040), g. d.
(Necklace) by British Crown (21322), gr. g. d. (Polly) by Sir Samuel (15302),
— (Mary) by Majestic (13279), — by Harbinger (10297), — by Buckingham
(3239).

(42824)   BRECON,
Roan, calved August 16, 1878, bred by Mr. B. H. Allen, Clifford Priory ; got
by King Rufus (38502), dam (Beetroot) by Macgregor (29241), g. d. (Blithesome) by Barleycorn the Younger (21209), gr. g. d. (Blithe) by Washington
(17213), — (Blanche) by Earl of Chester (9057), — by Earl of Chester (9057),
— by Constellation (7922), — by Powyke Patriot (4746), — by Powyke
Wharfdale (4748), — Old Strawberry.

(42825)   BREIDDEN,
White, calved January 10, 1879, bred by Captain D. H. Mytton, Garth ; got by
Constantine 2nd (33439), dam (Maud) by Vespasian (32759), g. d. (Gem) by
Berwick (23411), gr. g. d. (Glossary 2nd) by Hardware (19919), — (Glossy) by
Hazel (26352), — by Harlequin (26340), — by Emperor (11438), — by Priam
(15079).

(42826)   BRIAN FITZALAN,
Roan, calved May 28, 1879, bred by Mr. J. B. Booth, Killerby ; got by King

Brian (34308), dam (Gipsies' Dorcas) by Prince of the Gipsies (27179), g. d.
(Hope's Dorcas) by Great Hope (24082), gr. g. d. (Royal Dorcas) by Grand
·Royal (19891), — (Princess Dorcas) by Prince Arthur (13497), — by Champion
(11264), — by Belted Will (9952), — by Zenith (5702), — by Guardian (3947),
— by Firby (1040), — by Ivanhoe (1131), — by Regent (544), — by Blyth
Comet (85).

(42827)   BRIDAL FERN,
Red, calved March 2, 1879, bred by Mr. H. Smith, Mountmellick ; got by Royal
Fern (29865), dam (Bride of Lorne) by British Boy (25675), g. d. (Royal
Bridesmaid) by Ravenspur (20628), gr. g. d. (Bella Donna) by Sir Samuel
(15302), — (The Bride) by Baron Warlaby (7813), — by Hamlet (8126), — by
Prince Ernest (7366), — by Bright (1739), — by Shylock (2622), — by Percy
(1314), — by Blucher (84), — by Jupiter (344), — by Marshal Beresford (415),
— by Crocus (932), — by a Bull of Mr. R. Colling's.

(42828)   BRIDEGROOM,
Red, calved March 2, 1879, bred by Mr. J. P. Tynte, Tynte Park, the property
of Mr. H. Smith, Mountmellick ; got by Royal Fern (29865), dam (Bracelet) by
British Boy (25675), g. d. (Spangle) by Sheet Anchor (18820), gr. g. d. (Clara)
by Comrade (14307), — (Grey Aurelia) by Prince Ernest (7366), — by Booth
(11195), — by The Doctor (2748), — by Linton (4227), — by Streamer
(624).

(42829)   BRIDEGROOM ELECT,
Roan, calved May 28, 1879, bred by Mr. T. H. Miller, Singleton Park ; got by
Star Çhieftain (35672), dam (Bridal Gift) by King Charles (24240), g. d.
(Bride of Windsor) by Roan Windsor (24967), gr. g. d. (Familiar Hamilton) by
Duke of Hamilton (19618), — (Familiar 4th) by Red Duke (13571), — by
Master Belleville (11795), — by Fitz-Leonard (7010), — by Velocipede (5552),
— by Young Matchem (2282), — by Jack Tar (1133), — by Pilot (496), — by
Young Albion (15).

(42830)   BRIGADIER STEPHENSON,
Red and white, calved February 23, 1879, bred by Mr. J. Downing, Ashfield ;
got by Robert Stephenson (32313), dam (Brooch) by Knight of the Empire
(22061), g. d. (Bracelet 2nd) by Vanguard (21099), gr. g. d. (Bridal) by Buckingham (11219), — (Lady of the Lake) by Northern Light (13398), — by
Hopewell (10332), — by Diamond (5918), — by Priam (2452), — by Volunteer
(1553), — by Argus (759), — by Pilot (496), — by Remus (550), — by
Blucher (82), — by Albion (14), — by Shakespeare (582), — by Easby (232).

(42831)   BRIGHT FAVOURITE,
White, calved June 10, 1879, bred by Mr. G. Allen, Unicarville ; got by Bright
Baronet (37891), dam (Elfleda) by Blood Royal (23429), g. d. (Mantalini) by
Master Rembrandt (16545), gr. g. d. (Elfleda) by Gainford 4th (11501), —
(Ladylike) by Stars and Stripes (12148), — by Sir Henry (10824), — by Buck-
ingham (3239), — by Marcus (2262), — by Matchem (2281), — by Alderman
(1622), — by Pilot (496), — by Remus (550), — by Sir Charles (592), — by
R. Colling's Son of Favourite (252), — by same Son of Favourite (252), —
Strawberry.

(42832)   BRIGHT JASON,

Red and white, calved October 31, 1879, bred by Mr. T. Rose, Melton Magna ; got by Bright Damon (36268), dam (Bright Ina Rose) by Monarch (31930), g. d. (Bridesmaid) by Cherry Grand Duke 3rd (28174), gr. g. d. (Bright Ringlet) by Ringleader (15164), — (Bright Queen) by Fitz-Clarence (14552), — by British Prince (14197), — by Vanguard (10994), — by Crown Prince (10087), — by Zadig (8796), — by Auld Robin Gray (6753), — by James Chrisp's Bull, — by Burley (1766), — by Isaac (1129), — by Pilot (496), — by Albion (14), — by Lame Bull (359), — by Shipton (587), — by Son of Suworrow (636), — by Son of Twin Brother to Ben (88), — by Twin Brother to Ben (660).

(42833)   BRIGHT MONARCH,

Roan, calved August 6, 1879, bred by Mr. T. H. Miller, Singleton Park ; got by Frederick the Great (36665), dam (Bright Belle) by Royal Benedict (27348), g. d. (Bright Ringlet) by Ringleader (15164), gr. g. d. (Bright Queen) by Fitz-Clarence (14552), — (Bright Princess) by British Prince (14197), — by Vanguard (10994), — by Crown Prince (10087), — by Zadig (8796), — by Auld Robin Gray (6753), — by James Chrisp's Bull, — by Burley (1766), — by Isaac (1129), — by Pilot (496), — by Albion (14), — by Lame Bull (359), — by Shipton (587), — by Son of Suworrow (636), — by Son of Twin Brother to Ben (88), — by Twin Brother to Ben (660).

(42834)   BRIGHT STOCKINGS,

Red and white, calved June 11, 1878, bred by Messrs. T. and J. McElderry, Ballymoney, the property of General, Viscount Templetown, Castle Upton ; got by British Mantalini (36281), dam (Jollity A) by Abab (27850), g. d. (Jollity 4th) by Sir Roderick (22914), gr. g. d. (Jollity) by Carlysle (14245), — (Juliet) by Comet (11298), — by Lord Clarendon (10434), — by Wiseton (2848), — by Firby (1040), — by Ivanhoe (1131), — by Regent (544), — by Blyth Comet (85).

(42835)   BRILLIANT DUKE,

Roan, calved March 1, 1879, bred by Mr. A. Garfit, Scothern, the property of Mr. S. R. C. Ward, Neasham Hill ; got by Grand Duke 25th (34065), dam (Brilliant Rose) by General Napier (24023), g. d. (Brilliant) by May Duke (13320), gr. g. d. (Blanche 3rd) by Antinous (12401), — (Blanche) by Diamond (5918), — by Norfolk (2377), — by Belvedere (1706), — by Belvedere (1706), — by Lancaster (360), — by Petrarch (488), — by Major (397), — by Chapman's Son of Punch (122), — by Dickson's Grandson of Punch (213), — by Checks (132), — by R. Grimston's Bull (282), — by J. Coates's Bull (148).

(42836)   BRITAIN'S FAME,

Roan, calved March 23, 1879, bred by Mr. G. Low, Birtown ; got by Great Fame (34087), dam (Britannia) by Escape (28556), g. d. (Clarina) by Napoleon 3rd (29417), gr. g. d. (Brilliant) by Chieftain (30732), — (Emily) by Billy the Beau (12471), — by Tariff (8653), — by Prince Ernest (7366), — by Marquis of Chandos (6190), — by Prince Paul (4827), — by Monarch (2324), — by Dr. Syntax (220), — by Charles (127), — by St. John (572), — by Chilton (136), — by White Bull (421), — by Favourite (252), — by Dalton Duke (188), — by R. Alcock's Bull (19), — by J. Smith's Bull (608), — by Jolly's Bull (337).

(42837) **BRITISH AMBO,**
Roan, calved May 31, 1879, bred by Messrs. J. Cock and Sons, Coat Green Farm; got by British Gwynne (39503), dam (Isabella 2nd) by Arch Hero (27893), g. d. (Reggi) by Constitution (30793), gr. g. d. (Regina) by King Otho (18148), — (Princess Royal) by Crown Prince (14353), — by Selim (6454), — by Rex (6385), — by Sir Thomas Fairfax (5196), — by Ambo (1636), — by Memnon (2295), — by Pilot (496), — by Agamemnon (9), — by Burrell's Bull of Burdon.

(42838) **BRITISH BARONET,**
Roan, calved September 27, 1879, bred by Mr. E. Hodgson, Blands Wath; got by British Sovereign (36285), dam (Lady Jane) by Young Oxford Gwynne (29492), g. d. (Miss Bland) by Garibaldi (24008), gr. g. d. (Beautiful Star) by Benedict 2nd (25624), — (Red Rose) by Sultan (18938), — by Hopeful (9222), — by Loyal (6154).

(42839) **BRITISH BOY,**
Red, calved September 2, 1877, bred by Mr. J. Whyte, Aldborough, the property of Mr. R. Jefferson, Preston Hows; got by British King (36279), dam (Chamouni) by England's Fame (31111), g. d. (Charity) by England's Hero (26104), gr. g. d. (Clorinde) by Lord Plymouth (24455), — (Clotilde) by Knight Errant (18154), — by Cardigan (12556), — by Noble (4578), — by Newton (2367), — by Young Magog (2247), — by Margrave (2263), — by Sir Charles (593), — by Sir Dimple (594), — by St. Albans (2584), — by Layton (366), — by Charge's Grey Bull (872), — by J. Brown's Red Bull (97).

(42840) **BRITISH BUCK,**
Roan, calved March 11, 1876, bred by Mr. G. Allen, Unicarville, the property of Mr. J. B. Houston, Orangefield; got by British Hero (30604), dam (Magnolia) by Prince Victor (20606), g. d. (Magnolia) by The Baron (13833), gr. g. d. (Model) by Matadore (11800), — (Brunette) by Prince Edward Fairfax (9506), — by Premier (6308), — by Soldier (5239), — by Commodore (3453), — by A-la-Mode (725).

(42841) **BRITISH EARL,**
Red, calved February 27, 1879, bred by Mr. W. Dent, Kaber Fold, the property of Mr. W. Hutchinson, Rookby Scarth; got by Earl of Sheffield (33812), dam (Welcome) by Warlaby (32792), g. d. (Susan) by Third Duke of Athol (12734), gr. g. d. (Mignonette) by Bates (12451), — (Henrietta) by Ingram (9236), — by Liberator (7140), — by Prince Albert (4791), — by Young Matchem (4422), — by Young Red Rover (4904) or Rockingham (2551), — by Whisker (1579), — by Pilot (496).

(42842) **BRITISH GOLD,**
Roan, calved February 24, 1877, bred by Mr. R. Jefferson, Preston Hows, the property of Mr. H. Caddy, Rougholm; got by British Boy (30597), dam (Golden Drop) by Royal Sovereign (22802), g. d. (Golden Rose) by Blood Royal (14169) gr. g. d. (Golden Lady) by Baron Warlaby (7813), — (Golden Rule) by Remus (4932), — by Prince Comet (1342), — by Count (170), — by Constellation (163), — by Young Favourite (255).

(42843) **BRITISH GOLD,**
Roan, calved October 20, 1879, bred by Mr. W. Cleasby, Eden Place, the property of Mr. T. James, Kirkby Stephen; got by British Sovereign (36285), dam

(Miss Soulby) by Duke of Gainford (33644), g. d. (Emily) by Captain Booth (21362), gr. g. d. (Amelia) by Archdeacon (21184), — (Annie) by Fennel (19737), — by Betony (30533), — by Windsor (21117), — by China (19447), — by Vanguard (21010), — by Oswald (20445).

(42844)   BRITISH LION,
Roan, calved April 25, 1878, bred by Mr. W. Arkell, Dudgrove, the property of Mr. W. Arkell, Jun., Hatherop; got by Pompey (35059), dam (Brilliance) by Research (27270), g. d. (Bounty) by Second Baron Wetherby (21243), gr. g. d. (Britannia) by Master Coleshill (18344), — (Blossom) by Sultan (15358), — by Neptune (11847), — by Fanatic (8054), — by Earl of Durham (5965), — by Pedestrian (7321), — by Edrom (1956), — by Scipio (1421), — by Hector (1104), — by Midas (435), — by Marquis (407), — by Chilton (136), — by Ben (70).

(42845)   .BRITISH PILOT,
Roan, calved March 18, 1879, bred by Messrs. J. Cock and Sons, Coat Green Farm; got by British Gwynne (39503), dam (Isabella 1st) by Arch Hero (27893), g. d. (Reggi) by Constitution (30793), gr. g. d. (Regina) by King Otho (18148), — (Princess Royal) by Crown Prince (14353), — by Selim (6454), — by Rex (6385), — by Sir Thomas Fairfax (5196), — by Ambo (1636), — by Memnon (2295), — by Pilot (496), — by Agamemnon (9), — by Burrell's Bull of Burdon.

(42846)   BRITISH STANDARD,
Roan, calved June 1, 1878, bred by Mr. J. Tweedie, Deuchrie; got by British Grenadier (33217), dam (Rose of Warlaby) by Prince of Warlaby (20593), g. d. (Roseleaf) by Le Moor (22090), gr. g. d. (Rosebush) by Oxford (15035), — (Rosy Morn) by Redmond (11978), — by Cotherstone (6903), — by Young Matchem (4425), — by Fairfax (1023), — by Son of Marske (418), — by Palmsun (7311), — by a Bull of Mr. Maynard's, — by Son of Favourite.

(42847)   BRITISH STATESMAN,
Roan, calved in April 1878, bred by Mr. S. Campbell, Kinellar, the property of Mr. J. Russell, Richmond Hill, Ontario, Canada; got by Golden Prince (38363), dam (Rosebud 3rd) by Aberdeen (21142), g. d. (Rosebud) by Scarlet Velvet (16916), gr. g. d. (Thalia) by Earl of Aberdeen (12800), — (Myrtle) by Balmoral (9920), — by Dannecker (7949), — by Ury (17157), — by Heriot (4017), — by Grey Diomed (2076), — by Juniper (1144).

(42848)   BRITISH STATESMAN,
Red and white, calved August 4, 1878, bred by Mr. S. Armstrong, Gally House; got by British Lad (41151), dam (Fanny 18th) by Lord of the Manor (29181), g. d. (Fanny 16th) by Prince Bertram (27119), gr. g. d. (Fanny 15th) by Duke of York (23804), — (Fanny 10th) by Dr. McHale (15887), — by King Alfred (16334), — by Orphan Boy (11878), — by Hamlet (8126), — by General (3871), — by Prince Ernest (7366), — by Prince Arthur (9502), — by Nolan's Comet.

(42849)   BRITISH VICEROY,
White, calved December 11, 1879, bred by the Rev. T. Staniforth, Storrs; got by Royal Stuart (40646), dam (British Queen) by Sovereign (27538), g. d. (British Maid) by British Prince (14197), gr. g. d. (Village Rose) by Blood Royal (14169), — (Village Maid) by Crown Prince (10087), — by Sir Leonard

(10827), — by Pilgrim (4701), — by Second Hubback (1423), — by Frederick (1060), — by Western Comet (689), — by Western Comet (689), — by Western Comet (689), — by Son of Favourite (252).

(42850) BROOMLEY DUKE,
Roan, calved April 16, 1879, bred by Mr. J. Todd, Mereside ; got by Gwynne Duke (34098), dam (Rosamond) by Seventeenth Duke of Oxford (25994), g. d. (Red Rose) by Merry Monarch (22349), gr. g. d. (Bloomer Girl) by Nobleman (20413), — (Bloom) by Red Prince (13576), — by Brilliant (7851), — by Sir Robert (5180), — by Bachelor (1665), — by Short Legs (5124), — by Ajax (723), — by Hector (2103), — by Surly (2715), — by Sir Harry (5155), — by Colonel (152), — by Son of Hubback (319).

(42851) BROTHER INNOCENCE,
Roan, calved October 1, 1879, bred by Captain Ashby, Naseby Woolleys ; got by Earl of Geneva (33794), dam (Inquiry) by Third Duke of Geneva (21592), g. d. (Invoice) by Pan (18516), gr. g. d. (Inquest) by Field Marshal (14545), — (Sultana) by Modbury Premium (11820), — by Vagabond (9765), — by Locksley (4240), — by Phosphorus (4694), — by Antonio (3019), — by Thorpe (2757), — by St. Leger (1414), — by Comus (1861), — by Own Brother to Remus (550).

(42852) BROUGHTON,
Roan, calved March 5, 1879, bred by Mr. J. Angus, Bearl, the property of Messrs. Fewster, Newlands, Ebchester ; got by Ben Brace (30524), dam (Rosebud 14th) by Whiff (30299), g. d. (Rosebud 10th) by Knight of Broomley (24271), gr. g. d. (Rosebud 6th) by Knight of the Grand Cross (22064), — (Rosebud 3rd) by Alderman (17291), — by Tom Boy (9749), — by Exmouth (6983), — by Captain (6830), — by Sir Harry Liddell (5157), — by Lord Prudhoe (8251), — by Sir Harry (5155), — by Hollon's Bull (313).

(42853) BUCCANEER,
Red, calved January 1, 1875, bred by the Duke of Buccleuch, Dalkeith Park, the property of Mr. J. Blyth, Leckiebank ; got by Lord Cecil (26621), dam (Lady's Maid) by Royal Errant (22780), g. d. (Lady of Dalkeith) by Lord Stanley (18275), gr. g. d. (Passion Flower) by Cardigan (12556), — (Patience) by Lord Fanny (13187), — by Noble (4578), — by Master Charley (7215), — by Baronet (1686), — by Young Magog (2247), — by Reformer (2502), — by Margrave (2263), — by Leopold (2199), — by Hector (2103), — by Surly (2715) — by Traveller (655), — by Colonel (152).

(42854) BUCKDEN WATERLOO,
Roan, calved December 23, 1878, bred by Mr. J. Linton, Buckden Wood ; got by Duke of Underley 3rd (38196), dam (Waterloo 3rd) by Colonel Tregunter 7th (30771), g. d. (Water Hen) by Wellesley (25421), gr. g. d. (War Cloud) by Booth Royal (15673), — (War Nymph) by Vanguard (10994), — by Baron Warlaby (7813), — by Fourth Duke of Northumberland (3649), — by Norfolk (2377), — by Waterloo (2816), — by Waterloo (2816).

(42855) BURGHLEY,
Red and little white, calved October 30, 1877, bred by Mr. J. A. M. Cope, Drummilly, the property of the Earl of Dufferin, Clandeboye ; got by Telemachus 11th (37564), dam (Elvira 11th) by Cambridge Duke 5th (30644), g. d. (Elvira 9th)

by Grand Duke 11th (21849), gr. g. d. (Elvira 7th) by Grand Duke 10th (21848),
— (Elvira 2nd) by Eighth Duke of Oxford (15939), — by The Baronet (10918),
— by Burgundy (7861), — by Pilot (4707), — by Navarino (2352), — by
Favourite (256), — by Phenomenon (491), — by Favourite (252), — by
Favourite (252), — by Hubback (319), — by Snowdon's Bull (612), — by
Waistell's Bull (669), — by Masterman's Bull (422), — by the Studley Bull
(626).

(42856)   **BUTTERBOY,**

Roan, calved December 8, 1878, bred by Mr. T. Mace, Eastleach ; got by Duke
of Wellington 7th (33753), dam (Daisy Butterfly) by Duke Geneva (30911), g. d.
(Butterfly Duchess) by Royal Cambridge 2nd (25010), gr. g. d. (Lady Butterfly)
by Royal Butterfly 14th (20722), — (Beauty) by Berwick (17409), — by
Cornwall (14329), — by Colchicum (8963), — by Chieftain (12595), — by Ich
Dien (9234), — by Prince (9494), — by Rufus (1203).

(42857)   **BUTTERMAN,**

Roan, calved May 30, 1878, bred by the Trustees of the late Sir W. S. Maxwell,
Bart., Keir Mains, the property of Mr. J. McQueen, Divers Wells ; got by Prince
of Schleswig (35158), dam (Butternut) by Scotland's Pride (25100), g. d. (Butterfly
12th) by St. Clair (25078), gr. g. d. (Butterfly 5th) by Lord Raglan (13244), —
(Butterfly) by Matadore (11800), — by Report (10704), — by The Pacha (7612),
— by Second Duke of Northumberland (3646), — by Mahomed (6170), — by
Sillery (5131), — by Carleton (843), — by Diamond (205), — by Diamond
(205).

(42858)   **BUZ,**

Red and white, calved February 11, 1879, bred by the Duke of Northumberland,
Alnwick Castle ; got by Sir Raymond (40716), dam (Modesty) by Brigadier
(28078), g. d. (Rosette) by Briton (25686), gr. g. d. (Maid of Aln) by Melsonby
(18380), — (Young Jessy) by George 3rd (16147), — by Shaftoe (5107), — by
Richardi (4944), — Cowslip, bought of Mr. Patterson, of Wood Houses.

(42859)   **CABIN BOY,**

White, calved December 31, 1879, bred by Sir R. C. Musgrave, Bart., Eden Hall ;
got by Baron Underley (37824), dam (Camilla) by Fourteenth Duke of Oxford
(21605), g. d. (Calista 5th) by Richard (16834), gr. g. d. (Calista 3rd) by Scaleby
Champion (15243), — (Lady Macbeth 2nd) by Disraeli (10125), — by Morton
(14963), — by Young Ronald (5007), — by Son of General (3867), — by Don
Quixote (987), — by Pizarro (1323), — by Favourite (256), — by Western
Comet (689).

(42860)   **CABUL,**

Red and white, calved March 23, 1878, bred by Mr. J. Thomson, Newseat of
Dumbreck, the property of Mr. A. I. Fortescue, Kingcausie ; got by Blucher
(33170), dam (Roseberry) by Albert (30370), g. d. (Rosebud) by Lieutenant
(24335), gr. g. d. (Roslin) by Holbein (21943), — (Roan Duchess) by Prince of
Coburg (15100), — by Prince of Coburg (15100), — by Christmas Rose (14274),
— by Duke of Clarence (9040), — by Angus Hero (6745), — by Darlington
(3561).

(42861)   **CABUL,**

Roan, calved March 4, 1879, bred by Mr. S. P. Foster, Killhow ; got by Duke
of Ormskirk (36526), dam (Concord) by Ravensbourn (29716), g. d. (Charmer

14th) by Third Duke of Geneva (23753), gr. g. d. (Rarity) by Costa (21487), — (Ruby) by Lord of the Harem (16430), — by Mameluke (13289), — by Cardinal (11246), — by Earl of Dublin (10178), — by Little John (4232), — by Caliph (1774), — by Sir Walter (2637), — by Hotspur (1117), — by Coxcomb (928), — by Midas (435), — by Comet (155), — by R. Colling's Son of Favourite (252), — by same Son of Favourite (252), — by Hubback (319).

(42862) CABUL,

Roan, calved November 14, 1879, bred by Colonel R. Loyd Lindsay, Lockinge Park ; got by Earl of Horton 11th (36588), dam (Clotilda Rock) by Lord Rockville (34658), g. d. (Clotilde) by Grand Duke of Kent 2nd (28759), gr. g. d. (Chaff 13th) by Second Earl of Walton (19672), — (Clarissa) by Oxford Duke (18508), — by Viceroy (13945), — by Duke of Cornwall (5947), — by Morpeth (7254), — by Helicon (2107), — by Henwood (2114), — by Nestor (452), — by Harold (291), — by Meteor (432), — by Comet (155), — by Cupid (177).

(42863) CABUL,

Roan, calved December 17, 1879, bred by Mr. R. P. Maxwell, Finnebrogue ; got by Ards (40983), dam (Colleen Bawn 2nd) by Czarowitz 4th (33497), g. d. (Colleen Bawn) by Half Sovereign (34104), gr. g. d. (Irish Girl) by Prince Victor (20606), — (Susan) by Prince Andrew (18588), — by Musician (13362), — by Forrest (10240), — by Albion (7771), — by Prince Ernest (7366), — by Wharncliffe, — by Oliver.

(42864) CAER RHYN,

Red and white, calved March 20, 1879, bred by Mr. R. Hemming, Bentley Manor ; got by Corrie (38031), dam (Fairy 6th) by Quintus (29705), g. d. (Fairy 2nd) by Castlereagh (19409), gr. g. d. (Fairy) by Colonel (14294), — (Fatima) by Contract (10071), — by Moreton (22377).

(42865) CÆSAR,

Roan, calved November 6, 1877, bred by Captain D. H. Mytton, Garth, the property of Mr. Hall, Ireland ; got by Constantine 2nd (33439), dam (Miss Dorothy) by Beacon (36225), g. d. (Gem) by Berwick (23411), gr. g. d. (Glossary 2nd) by Hardware (19919), — (Glossy) by Hazel (26352), — by Harlequin (26340), — by Emperor (11438), — by Priam (15079).

(42866) CÆSAR,

Red and white, calved March 8, 1879, bred by Mr. W. Curtis, Fernham ; got by Count of Hanney (42980), dam (Mrs. Curtis) by Second Lord Waterloo (29219), g. d. by Lancaster (24302), gr. g. d. by Lord Walton (22227), — by Midsummer (8314), — descended from the stock of Mr. Tucker, of Bourton.

(42867) CAMBRIDGE DUKE 10TH,

Roan, calved March 3, 1877, bred by Mr. J. Lynn, Church Farm, the property of Messrs. Morrin, Auckland, New Zealand ; got by Twenty-fourth Duke of Airdrie (36460), dam (Red Rose 6th) by Cambridge Duke 4th (25706), g. d. (Red Rose 5th) by Grand Duke 4th (19874), gr. g. d. (Red Rose 3rd) by Englishman (19701), — (Red Rose) by Marmaduke (14897), — by Puritan (9523), — by Third Duke of York (10166), — by Second Cleveland Lad (3408), — by Belvedere (1706), — by Belvedere (1706), — by Second Hubback (1423), — by His Grace (311), — by Yarborough (705), — by Favourite (252), — by Punch (531), — by Foljambe (263), — by Hubback (319).

(42868)   CAMBRIDGE LAD,

Red and white, calved May 25, 1875, bred by Mr. W. Tippler, Roxwell; got by Lord Cambridge (38588), dam (Cambridge Violet) by Cambridge Duke 3rd (23503), g. d. (Violet 12th) by Young Butley Butterfly (39526), gr. g. d. (Violet 6th) by Conte de Vermandois (19507), — (Violet 3rd) by Duke of Suffolk (14456), — by Lord Barrington 1st (13170), — by Fourth Duke of York (10167), — by Second Duke of York (5959), — by Sir Thomas (7516), — by Sultan Selim (2710), — by Satellite (1420), — from the stock of Mr. Smith of Shedlaw.

(42869)   CAMBRIDGE PRINCE,

White, calved March 13, 1879, bred by Sir Wilfrid Lawson, Bart., Brayton; got by Wild Eyes Duke (36007), dam (Lady Cambridge) by Royal Cambridge (25009), g. d. (Lady Oxford) by Lord Chancellor (20160), gr. g. d. (Lady Richmond) by Oxford 2nd (18507), — (Mrs. Richmond) by Duke of Richmond (7996), — by Cotherstone (6903), — by Yellow Boy (5694), — by Red Robin (2492), — by Burley (1766), — by Son of Young Albion (15), — by Sir Marton (1453), — by Palmsun (7311), — by a Bull of Mr. Parrington's.

(42870)   CANDAHAR,

Red and white, calved December 15, 1879, bred by Mr. R. P. Maxwell, Finnebrogue; got by Ards (40983), dam (Miss Hetty Underley) by Baron Underley (30490), g. d. (Hesbia) by Third Earl of Cleveland (28497), gr. g. d. (Harmony) by Duke of Montrose (21599), — (Hoops) by Silk and Scarlet (20806), — by Comet (11298), — by Essex Hero (10207), — by Wiseton (2848), — by Firby (1040), — by Ivanhoe (1131), — by Regent (544), — by Blyth Comet (85).

(42871)   CANDLESHOE,

Red, calved August 27, 1879, bred by Mr. G. Gunnis, Bratoft; got by Syrian Duke (40782), dam (Cosey) by Syrian Prince (35715), g. d. (Coral) by Seventh Duke of York (17754), gr. g. d. (Currant) by Duke of Bolton (12738), — (Miss Baldface 2nd) by Hermit (11574), — by Cramer (6907), — a Cow bred by Mr. Allenby, of Maidenwell.

(42872)   CANLEY CÆSAREWITCH 4TH,

Red, calved in July, 1876, bred by Mr. T. Hands, Canley, the property of Mr. C. Cambwell, Allesley; got by Cæsarewitch 4th (33263), dam (Chance 7th) by Second Duke of Darlington (28378), g. d. (Chance 5th) by Canley Duke (23506), gr. g. d. (Chance 2nd) by Dreadnought (12718), — (Chance) by Cramer 2nd (10084), — by Duke of Richmond (10163), — by Anthony (1640), — by Sportsman (5304), — by Anthony (1640), — by Edgcott (1953), — by Audley (3055), — by Son of Merlin (6522).

(42873)   CANLEY ELFORD 6TH,

Roan, calved May 29, 1878, bred by Mr. T. Hands, Canley, the property of Mr. E. Hawkins, Cruckfield; got by Duke of Elford 2nd (36490), dam (Lavender 7th) by Second Duke of Darlington (28378), g. d. (Lavender 3rd) by Prince Pearl 3rd (20599), gr. g. d. (Lizzie Lavender) by Sir Sam (16992), — (Lavender 2nd) by Captain (10022), — by Barrister (7815), — by Marcellus (6181), — by Major (4338), — by Eclipse (1949), — by Son of Dishley Ketton (982), — by Cossack (925), — by Patriot (486).

(42874) **CAPTAIN BENN,**

Roan, calved November 12, 1877, bred by Mr. J. Angus, Bearl, the property of Messrs. Reed and Stephenson, East Gate ; got by Captain Brace (37938), dam (Royal Strawberry) by Royal Frederick (35366), g. d. (Strawberry 4th) by Roan Chief (27294), gr. g. d. (Strawberry 2nd) by Freeman (14574), — (Strawberry 1st) by Lahore (11659), — by Young Hector (7074), — by Wallace (5588), — descended from C. Colling's Son of Hubback (319).

(42875) **CAPTAIN CECIL,**

Roan, calved April 10, 1878, bred by Mr. J. Currie, Halkerston, the property of Mr. W. Little, Burnfoot ; got by Cecil (36329), dam (Jealousy 13th) by Baronet (25564), g. d. (Jealousy 3rd) by Lord Buckingham (20151), gr. g. d. (Young Jealousy) by Scarlet Velvet (16916), — (Jealousy) by Baron Warlaby (7813), — by General Washington (6036), — by Virgil (7686), — by Son of Emperor (1014), — by Young Favourite (255), — by Young Major (1186).

(42876) **CAPTAIN DARLINGTON 2ND,**

White, calved May 3, 1879, bred by Mr. F. G. Dalgety, Lockerley Hall ; got by Parnassus (38850), dam (Miss Darlington) by Lord Darlington 2nd (31645), g. d. (Ma Belle) by Henchman (21918), gr. g. d. (Melody) by Strephon (17051), — (Dinah) by Glanville (11535), — by Queen's Roan (7389), — by Will Honeycomb (5660), — by Frank (7020), — by Spectator (2688), — by Albion (1619), — by Lancaster (360), — by Son of Windsor (698), — by Comet (155).

(42877) **CAPTAIN DARLINGTON 3RD,**

Red, calved November 2, 1879, bred by Mr. F. G. Dalgety, Lockerley Hall ; got by Parnassus (38850), dam (Miss Darlington 2nd) by Lord Darlington 7th (34519), g. d. (Miss Darlington) by Lord Darlington 2nd (31645), gr. g. d. (Ma Belle) by Henchman (21918), — (Melody) by Strephon (17051), — by Glanville (11535), — by Queen's Roan (7389), — by Will Honeycomb (5660), — by Frank (7020), — by Spectator (2688), — by Albion (1619), — by Lancaster (360), — by Son of Windsor (698), — by Comet (155).

(42878) **CAPTAIN RADSTOCK,**

Roan, calved February 11, 1878, bred by Mr. W. Parkin, Blaithwaite, the property of Messrs. R. and W. Hornsby, Abbey House ; got by Lord Radstock (34656), dam (Lady Blanche 2nd) by Lieutenant Oxford (24336), g. d. (Lady Blanche Hollyhock) by Fourteenth Duke of Oxford (21605), gr. g. d. (Hollyhock 4th) by Richard (16834), — (Hollyhock 3rd) by Palmerston (15041), — by Calcott (14216), — by Mingo (7239), — by General (3867), — by Pizarro (1323).

(42879) **CARACTACUS,**

Roan, calved October 13, 1878, bred by Captain D. H. Mytton, Garth ; got by Constantine 2nd (33439), dam (Joyce) by Vespasian (32759), g. d. (Gem) by Berwick (23411), gr. g. d. (Glossary 2nd) by Hardware (19919), — (Glossy) by Hazel (26352), — by Harlequin (26340), — by Emperor (11438), — by Priam (15079).

(42880) **CARDINAL,**

Red, calved April 7, 1879, bred by Mr. H. Smith, Mountmellick ; got by Earl of Aylesby (38215), dam (Carnation 2nd) by Lord of the South (31730), g. d. (Carnation) by Lord Claud (29086), gr. g. d. (Dagmar) by British Duke (19350), — (Lady Willy) by Knight of Windsor (16350), — by Vanguard (10994), — by Son of William (2840), — by Alderman (2976), — by Waterloo (2816), — by Son of Wynyard (703), — by Buston's Styford (103).

(42881) CARDINAL WINDSOR 2ND,
Red and white, calved September 28, 1879, bred by Mr. T. Stamper, Highfield
House ; got by Cardinal Windsor (37949), dam (Sall) by Sandringham (32452),
g. d. (Saucy Jane) by Manfred (26801), gr. g. d. (Saucy) by Prince Arthur
(22574), — (Sabina) by Hamlet (19914), — by Lord Flounce (14818), — by Lord
Alexander Croft (11707), — by Lord Buchan (9309), — by Sir Thomas (6492),
— by Benjamin (1710).

(42882) CARLISLE,
Red and white, calved August 14, 1879, bred by Mr. H. Smith, Mountmellick ;
got by Earl of Aylesby (38215), dam (Countess 3rd) by Conqueror (30786), g. d.
(Carnation 2nd) by Lord of the South (31730), gr. g. d. (Carnation) by Lord
Claud (29086), — (Dagmar) by British Duke (19350), — by Knight of Windsor
(16350), — by Vanguard (10994), — by Son of William (2840), — by Alderman (2976), — by Waterloo (2816), — by Son of Wynyard (703), — by Buston's Styford (103).

(42883) CARLOS,
Red and white, calved January 14, 1879, bred by Sir J. Swinburne, Bart., Capheaton ; got by Duke of Oxford 27th (33709), dam (Caroline) by Grand Duke
10th (21848), g. d. (Cleopatra 9th) by Lord Oxford (20214), gr. g. d. (Cleopatra
2nd) by Prince of Glo'ster (13517), — (Cleopatra) by Earl of Dublin (10178),
— by Standish (7550), — by Bellerophon (3119), — by Alderman (2976), — by
Waterloo (2816), — by Young Wynyard (2859), — by Son of Simon (590), —
by Buston's Styford (103).

(42884) CARMINE DUKE,
Red, calved March 15, 1878, bred by Mr. J. Webb, Melton Ross, the property of
Messrs. E. and F. Empson, Bonby ; got by Duke of Sesame (33729), dam (Carmine Rose) by Cambridge Duke 4th (25706), g. d. (Camilla) by Waverley 4th
(21084), gr. g. d. (Clematis) by Sir John (12084), — (Clementina) by Clementi
(3399), — by Young Matchem (4422), — by Isaac (1129), — by Young Pilot
(4702), — by Pilot (496), — by Julius Cæsar (1143).

(42885) CARN BREA,
Red and white, calved December 10, 1878, bred by Mr. W. Trethewy, Tregoose,
the property of Mr. Hearle, Tywarnhale ; got by M. C. (31898), dam (Ruth
104th) by British Lion (30609), g. d. (Ruth 23rd) by Rufus (35423), gr. g. d.
(Ruth 9th) by Vandumper (23114), — (Ruth 3rd) by Lord Fingal (11716), —
by Frantick (8088), — by Harold (8131), — by Cedric (3311), — by Nimrod
(4571), — by Grandson of Blyth Comet (85), — by Crispin (174), — by Meteor
(431), — by Meteor (431).

(42886) CARPET-KNIGHT,
Roan, calved November 13, 1879, bred by Mr. R. Blezard, Pool Park ; got by
Viscount Dursley (37636), dam (York's Queen) by Eighth Duke of York (28480),
g. d. (Queen 8th) by Fourteenth Duke of Oxford (21605), gr. g. d. (Queen 4th)
by Scaleby Champion (15243), — (Queen 3rd) by Palmerston (15041), — by
Bracebridge (14187), — by Son of Selim (5703).

(42887) CARRIGEEN,
White, calved January 19, 1878, bred by Messrs. Christy Brothers, Fort Union,
the property of Mr. J. Mulqueen, Cartown House ; got by Lord Broughton

(31626), dam (Queen of Beauty 4th) by Hero of Thorndale (18061), g. d. (Queen of Beauty 2nd) by Duke of Beauford (11377), gr. g. d. (Queen of Beauty) by Shotley 2nd (7494), — (Daisy) by Stokes's Old Bull (8368), — by Young Spectator (8619), — by Phantassie (8389), — by Young Rockingham (8498).

(42888)　　　　　CARTHUSIAN,

Roan, calved August 18, 1878, bred by Messrs. J. W. and E. Cruickshank, Lethenty, the property of Mr. J. Holmes, Greystock; got by Knight of St. Patrick (38520), dam (Lady of the Valley) by Great Hope (24082), g. d. (Ladylike) by Prince Victor (20606), gr. g. d. (Mantalini) by Master Rembrandt (16545), — (Elfleda) by Gainford 4th (11501), — by Stars and Stripes (12148), — by Sir Henry (10824), — by Buckingham (3239), — by Marcus (2262), — by Matchem (2281), — by Alderman (1622), — by Pilot (496), — by Remus (550), — by Sir Charles (592), — by Son of Favourite (252), — by Son of Favourite (252), — Strawberry.

(42889)　　　　　CASANOVA,

Roan, calved November 3, 1879, bred by Mr. R. Blezard, Pool Park; got by Viscount Dursley (37636), dam (Virago) by Notley (31991), g. d. (Vanity) by Grand Duke 7th (19877), gr. g. d. (Viscountess) by Duke of Argyle (11375), — (Baroness) by Master Goldsmidt (13315), — by Vandyke (7669), — by Carcase (3285), — by Tyro (2781), — by Falstaff (1993), — by Dr. Syntax (220), — by Charles (127), — by Henry (301), — by Favourite (252), — by White Bull (421), — by Bolingbroke (86), — by Foljambe (263), — by Hubback (319), — bred by Mr. Maynard.

(42890)　　　　　CATO,

Roan, calved February 2, 1879, bred by Mr. F. Garth, Crackpot; got by David (36425), dam (Bessie Annetta) by Major (31801), g. d. (Annabelle) by Frederick First Fruits (23989), gr. g. d. (Annette Bumper) by Bumper (19371), — (Lady Annetta) by Consolation (14316), — by Whittington (12299), — by Noble (4578), — by Newton (2367), — by Emperor (3716), — by Satellite (1420), — — by Cato (119), — by Jupiter (342), — by George (273), — by Chilton (136), — by Irishman (329), — by B. (45).

(42891)　　　　　CATO,

Red and white, calved June 22, 1879, bred by Mr. W. Bolton, The Island; got by Albion (36112), dam (Lady Frances) by Lieutenant General (31600), g. d. (Fluent) by Chieftain (21419), gr. g. d. (Julia) by Sultan (17053), — (Juliana) by Bridegroom (11203), — by Zadig (8796), — by Earl of Durham (5965), — by St. Helena (5055), — by Emperor (3716), — by Satellite (1420), — by Cato (119), — by Jupiter (342), — by George (273), — by Chilton (136), — by Irishman (329), — by B. (45).

(42892)　　　　　CATO,

Red and white, calved December 9, 1879, bred by the Duke of Devonshire, Holker Hall; got by Wetherby Winsome 2nd (40905), dam (Lady Chisholm) by Twenty-second Duke of Oxford (31000), g. d. (Lady Cressida) by Seventeenth Duke of Oxford (25994), gr. g. d. (Andromache) by Hogarth (13036), — (Young Cressida) by Duke of Athol (10150), — by Cossack (1880), — by Miracle (2320), — by Matchem (2281), — by Fitz-Remus (2025), — by Cato (119), — by Whitworth (695).

(42893)   CAVENDISH,

White, calved August 6, 1872, bred by Mr. W. Angerstein, Weeting Hall; got by Baron Oxford 4th (25580), dam (Caroline) by Grand Duke 10th (21848), g. d. (Cleopatra 9th) by Lord Oxford (20214), gr. g. d. (Cleopatra 2nd) by Prince of Glo'ster (13517), — (Cleopatra) by Earl of Dublin (10178), — by Standish (7550), — by Bellerophon (3119), — by Alderman (2976), — by Waterloo (2816), — by Young Wynyard (2859), — by Son of Simon (590), — by Buston's Styford (103).

(42894)   CECIL,

Red and white, calved November 11, 1877, bred by Mr. R. Thompson, Inglewood Bank, the property of Mr. J. Goulding, Aik Bank; got by Brilliant Butterfly (36270), dam (Cicely) by Grand Duke of Fawsley 3rd (31286), g. d. (Jess) by Fifth Duke of Oxford (23785), gr. g. d. (Jessamine) by Marquis of Cobham (22299), — (Jessie) by Warrior (21067), — by Tweedside (12246), — by Remus (11987), — by Strathmore (6547), — by Sir William (12102), — by Togston (5487).

(42895)   CEDRIC,

Red, calved November 25, 1879, bred by Mr. R. Pinder, Whitwell; got by Burghley (36296), dam (Dame Prim) by Constellation (28243), g. d. (Dame Prudence) by Prince of Rosedale (24837), gr. g. d. (Dame Patience) by Prince of the Realm (22627), — (Dame Quickly) by Valasco (15443), — by British Prince (14197), — by Baron Warlaby (7813), — by Clementi (3399), — by Young Matchem (4422), — by Isaac (1129), — by Young Pilot (4702), — by Pilot (496), — by Julius Cæsar (1143).

(42896)   CEFN PRINCE,

Red and white, calved February 16, 1879, bred by Captain C. R. Conwy, Bodrhyddan; got by Waterloo Prince (35949), dam (Hortensia) by Winterfold (36020), g. d. (Helena) by Lord Red Eyes 2nd (24460), gr. g. d. (Majestic) by Archduke (21185), — (Stately) by Castlereagh (19409), — by Impatience (14723), — by Balkan (14120), — by Normanby (10573), — by The Hon. Charles Fairfax (7605), — by Vulcan (2808), — by Matilda's Son of Emperor (4432), — by Parrington (4653), — by Baron (58), — by Wellington (680), — by Favourite (252), — by Punch (531), — by Foljambe (263), — by Hubback (319).

(42897)   CETYWAYO,

Roan, calved December 9, 1878, bred by Messrs. Annandale and Sons, Lintzford; got by Crown Prince (38059), dam (Little Emily) by Beverley Butterfly 2nd (28022), g. d. (Lady) by Benjamin (19303), gr. g. d. (Queen Elizabeth) by Lord Nelson (13207), — (Wealthy) by Son of Fitz-Maurice (36084).

(42898)   CHAMPION,

Roan, calved January 16, 1877, bred by Mr. J. Smith, Balmain, the property of Mr. J. Wallace, Balbegne; got by Sir James (35567), dam (Charm) by Master of the Mint (31878), g. d. (Chance) by Premier (32074), gr. g. d. (Sweet Red Rose) by Prince Patrick (16760), — (Sweetest Rose) by Richard Cœur-de-Lion (13590), — by Tyne (7653), — by Wolsington (2852), — by Brandon (3236), — by Acton (15551), — by Twin Brother to Ben (660).

(42899)   CHAMPION OF FERNHAM,

Roan, calved March 18, 1879, bred by Mr. W. Curtis, Fernham; got by Count of Hanney (42980), dam (Red Rose 2nd) by Duke of Flamborough (25960),

g. d. (Red Rose) by Manager (24521), gr. g. d. (Vanity 3rd) by Romeo (22754), — (Vanity 2nd) by Napier (14973), — by The Dandy (13846), — by St. John (27413), — by Sir Walter 2nd (10834), — by Major (4338).

(42900)   CHANTICLEER 2ND,

Roan, calved December 23, 1878, bred by Mr A. H. Longman, Shendish, the property of Mr. E. Byron, Coulsdon Court; got by Duke of Hillhurst 2nd (39748), dam (Jacob's Ruby) by Lord Gwynne (29113), g. d. (Ruby) by Lord of the Harem (16430), gr. g. d. (Cornelian) by Mameluke (13289), — (Coral) by Cardinal (11246), — by Earl of Dublin (10178), — by Little John (4232), — by Caliph (1774), — by Sir Walter (2637), — by Hotspur (1117), — by Coxcomb (928), — by Midas (435), — by Comet (155), — by R. Colling's Son of Favourite (252), — by same Son of Favourite (252), — by Hubback (319).

(42901)   CHARLEY,

White, calved in March 1860, bred by Mr. J. Morton, Skelsmergh Hall; got by Sovereign (15324), dam (Claret 2nd) by Prelate (11919), g. d. (Claret) by Brewster (7847), gr. g. d. (Clara) by Tomboy (5494), — (Nancy) by Red Rover (8467), — by Eclipse (8021), —by Chance (7890).

(42902)   CHARLTON,

Roan, calved March 25, 1879, bred by Mr. H. D. de Vitre, Charlton House, the property of Mr. G. E. Pym, Reigate; got by Grand Duke of Kent 2nd (28759), dam (Not for Joe) by Hush (28883), g. d. (Lady Culshaw) by Royal Butterfly 17th (22774), gr. g. d. (Rose of Lancashire) by Master Butterfly 4th (14920), — (Young Barmpton Rose) by Richard Cœur-de-Lion (13590), — by Lord John (11731), — by Baron of Ravensworth (7811), — by Raree Show (4874), — by Thick Hock (6601), — by Expectation (1988), — by Belzoni (1709), — by Comus (1861), — by Denton (198).

(42903)   CHARMING DUKE 5TH,

Red, calved January 30, 1879, bred by Messrs. F. Leney and Sons, Wateringbury; got by Sixth Duke of Oneida (30997), dam (Duchess 3rd) by Grand Duke 15th (21852), g. d. (Countess) by Knightley Grand Duke (24268), gr. g. d. (Chorus) by Fourth Duke of Thorndale (17750), — (Charming) by Mameluke (13289), — by Cardinal (11246), — by White Friar (9827), — by Little John (4232), — by Caliph (1774), — by Sir Walter (2637), — by Hotspur (1117), — by Coxcomb (928), — by Midas (435), — by Comet (155), — by R. Colling's Son of Favourite (252), — by same Son of Favourite (252), — by Hubback (319).

(42904)   CHARMING DUKE 6TH,

Red and white, calved September 4, 1879, bred by Messrs. F. Lency and Sons, Wateringbury; got by Sixth Duke of Oneida (30997), dam (Duchess 4th) by Grand Duke of Kent (26289), g. d. (Duchess 2nd) by Grand Duke 15th (21852), gr. g. d. (Countess) by Knightley Grand Duke (24268), — (Chorus) by Fourth Duke of Thorndale (17750), — by Mameluke (13289), — by Cardinal (11246), — by White Friar (9827), — by Little John (4232), — by Caliph (1774), — by Sir Walter (2637), — by Hotspur (1117), — by Coxcomb (928), — by Midas (435), — by Comet (155), — by R. Colling's Son of Favourite (252), —by same Son of Favourite (252), — by Hubback (319).

(42905)   ·CHARMING DUKE 7TH,

White, calved December 2, 1879, bred by Messrs. F. Lency and Sons, Wateringbury; got by Duke of Rothesay (36534), dam (Charming Duchess 6th) by Second

Duke of Collingham (23730), g. d. (Twin Duchess 3rd) by Knightley (22051), gr. g. d. (Twin Duchess) by Fourth Duke of Thorndale (17750), — (Charming) by Mameluke (13289), — by Cardinal·(11246), — by White Friar (9827), — by Little John (4232), — by Caliph (1774), — by Sir Walter (2637), — by Hotspur (1117), — by Coxcomb (928), — by Midas (435), — by Comet (155), — by R. Colling's Son of Favourite (252), — by same Son of Favourite (252), — by Hubback (319).

(42906) CHELMSFORD,
Red, calved April 20, 1879, bred by Mr. R. J. Mackay, Burgie Lodge, the property of Mr. J. Fletcher, Rosehaugh ; got by Charm's Windsor (36342), dam (Red Rose) by King Charming (34309), g. d. (Rosebud) by Baron Cecil (27921), gr. g. d. (Primrose) by Rover (29836), — (Rose 4th) by Sovereign 2nd (30032), — (Rose 2nd) by a Shorthorn Bull purchased at the Highland Society's Show in 1856, — (Rose) by Snowflake, — Lady Emma, a Shorthorn Cow purchased in 1852.

(42907) CHERRY BARRINGTON,
Roan, calved May 5, 1879, bred by Mr. C. C. Garner, The Wolds ; got by Duke of Barrington 6th (33576), dam (Chloris) by Lord Tregunter (31758), g. d. (Cherry Ripe) by Cardinal York (25719), gr. g. d. (Cherry) by Monarch (16575), — (Cherry Blossom) by Mosquito (14964), — by Viceroy (13944), — by King Richard (26524), — by Highnam, — a Cow purchased at Castlecomb, 1848.

(42908) CHERRY BAY,
Red, calved January 3, 1879, bred by Mr. T. Kirkham, Biscathorpe House ; got by Cherry Gwynne 4th (36350), dam by Simple Simon (32487), g. d. by Rufus (37409), gr. g. d. by Aurelius (14111), — by Touchstone (23084), — by Cardinal (10026).

(42909) CHERRY BISCATHORPE,
Red, calved March 24, 1879, bred by Mr. T. Kirkham, Biscathorpe House ; got by Cherry Gwynne 4th (36350), dam by Hemlock (28842), g. d. by Ranby (27230), gr. g. d. by Cock of the Walk (15782), — by Aurelius (14111), — by Touchstone (23084), — by Cardinal (10026).

(42910) CHERRY BOUNCE,
Red and white, calved July 7, 1879, bred by Mr. T. Rose, Melton Magna ; got by Bright Damon (36268), dam (Cerise) by Water Gnome (35940), g. d. (Cherry) by Ravenspur (20628), gr. g. d. (Sweetbriar) by Nimrod (13388), — (Charlotte) by Selim (6454), — by Rex (6385), — by Sir Thomas Fairfax (5196), — by Ambo (1636), — by Memnon (2295), — by Pilot (496), — by Agamemnon (9), — by Burrell's Bull of Burdon.

(42911) CHERRY BRANDY,
Red, calved August 3, 1878, bred by Mr. A. Baird, Robeston Hall, the property of Mr. H. D. Fussell, Amworth Castle ; got by Earl of Horton 5th (33797), dam (Cherry) by Duke of Kent (30979), g. d. (Comfit) by Fitz-Roy (16058), gr. g. d. (Comely) by Duncan (15952), — by Swindon (17071), — by Ranunculus (22676), — by Albion (2967), — by Memnon (2294).

(42912) CHERRY CORNER,
Red, calved March 20, 1879, bred by Mr. T. Kirkham, Biscathorpe House ; got by Cherry Gwynne 4th (36350), dam by Simple Simon (32487), g. d. by Ranby (27230), gr. g. d. by Cock of the Walk (15782), — by Aurelius (14111), — by Touchstone (23084), — by Cardinal (10026), — by Neptune (7273).

(42913)   CHERRY COUNT,

Roan, calved June 5, 1879, bred by Mr. J. Singleton, Teresa Cottage; got by Thorndale Cherry Lad (40818), dam (Countess of Oxford 5th) by Baron Jocelyn (33055), g. d. (Countess of Oxford 2nd) by Telemachus (27603), gr. g. d. (Lady Oxford) by Imperial Oxford (18084), — (Sandpiper) by The Briar (15376), — by Francisco (12893), — by Columbus (10063), — by Prince of Wales (6345), — by Edward (3696), — by Topper (2768), — by Favourite (3768), — by Windsor (698), — by Windsor (698).

(42914)   CHERRY DUKE,

Red, calved March 14, 1879, bred by Mr. N. Milne, Dryhope, the property of Mr. W. Smith, Melkington; got by Earl of Kelso (38228), dam (Cherry Duchess) by Thorndale Duke 2nd (32712), g. d. (Lady Cherry) by Oxford Cherry (24709), gr. g. d. (Lady Mary) by Redmond (20646), — (Young Sophy) by MacTurk (14872), — by Heir-at-Law (13005), — by Hudibras (10339), — by Baron of Ravensworth (7811), — by Baronet (1686), — by Spectator (2688), — by Fitz-Remus (2025), — by Whitworth (695).

(42915)   CHERRY DUKE 3RD,

Roan, calved April 25, 1879, bred by Mr. J. H. Casswell, Laughton; got by Duke of Elmhurst (39731), dam (Cherry Lunesdale) by Lord Lunesdale Bates (34592), g. d. (Cherry Ripe) by Cambridge Duke 4th (25706), gr. g. d. (Cherry) by Grand Duke 6th (19876), — (Cheerful) by Mystic (20391), — by Mameluke (13289), — by Garrick (11506), — by Earl of Dublin (10178), — by Little John (4232), — by Caliph (1774), — by Sir Walter (2637), — by Hotspur (1117), — by Coxcomb (928), — by Midas (435), — by Comet (155), — by R. Colling's Son of Favourite (252), — by same Son of Favourite (252), — by Hubback (319).

(42916)   CHERRY DUKE 7TH,

Roan, calved January 9, 1879, bred by Messrs. Dalton, Cummersdale; got by Duke of Cumberland (38137), dam (Cherry Rose 2nd) by Oxford Isis (32024), g. d. (Cherry Rose) by Duke of Cambridge (28369), gr. g. d. (Cheritta) by Fourteenth Duke of Oxford (21605), — (Cherry 2nd) by Stapleton (15336), — by Bracebridge (14187), — by Mingo (7239), — by Lowther (9338), — by Snowball (1465), — by Pizarro (1323).

(42917)   CHERRY DUKE 8TH,

Roan, calved December 8, 1879, bred by Messrs. Dalton, Cummersdale; got by Duke of Siddington (38182), dam (Cherry Rose 2nd) by Oxford Isis (32024), g. d. (Cherry Rose) by Duke of Cambridge (28369), gr. g. d. (Cheritta) by Fourteenth Duke of Oxford (21605), — (Cherry 2nd) by Stapleton (15336), — by Bracebridge (14187), — by Mingo (7239), — by Lowther (9338), — by Snowball (1465), — by Pizarro (1323).

(42918)   CHERRY DUKE 11TH,

Red and white, calved May 15, 1879, bred by Lord Penrhyn, Penrhyn Castle; got by Grand Duke of Oxford (31293), dam (Cherry Duchess 23rd) by Oxford Beau (29485), g. d. (Cherry Duchess 15th) by Second Duke of Grafton (25968), gr. g. d. (Cherry Duchess 11th) by Duke of Geneva (19614), — (Cherry Duchess 6th) by Grand Duke 3rd (16182), — by Grand Duke 2nd (12961), — by Grand Duke (10284), — by Sheldon (8557), — by The Colonel (5428), — by Thorp (2757), — by Pirate (2430), — by Houghton (318), — by Marshal Blucher (416), — from the stock of Messrs. Wright and Charge.

(42919)   CHERRY DUKE OF HILLHURST,

Roan, calved February 27, 1878, bred by Mr. J. W. Larking, Cansiron Farm, late the property of Mr. E. Cazalet, Fairlawn; got by Third Duke of Hillhurst (30975), dam (Cherry Queen) by Baron Oxford 5th (27958), g. d. (Cherry Princess) by General Napier (24023), gr. g. d. (Cherry Duchess 8th) by Grand Duke 3rd (16182), — (Cherry Duchess 6th) by Grand Duke 3rd (16182), — by Grand Duke 2nd (12961), — by Grand Duke (10284), — by Sheldon (8557), — by The Colonel (5428), — by Thorp (2757), — by Pirate (2430), — by Houghton (318), — by Marshal Blucher (416), — from the stock of Messrs. Wright and Charge.

(42920)   CHERRY DUKE OF HILLHURST 2ND,

Red, calved March 6, 1879, bred by Mr. L. H. Wraith, Newfield; got by Third Duke of Hillhurst (30975), dam (Cherry Queen) by Baron Oxford 5th (27958), g. d. (Cherry Princess) by General Napier (24023), gr. g. d. (Cherry Duchess 8th) by Grand Duke 3rd (16182), — (Cherry Duchess 6th) by Grand Duke 3rd (16182), — by Grand Duke 2nd (12961), — by Grand Duke (10284), — by Sheldon (8557), — by The Colonel (5428), — by Thorp (2757), — by Pirate (2430), — by Houghton (318), — by Marshal Blucher (416), — from the stock of Messrs. Wright and Charge.

(42921)   CHERRY DUKE OF HOLKER,

Roan, calved December 9, 1879, bred by the Duke of Devonshire, Holker Hall; got by Duke of Glo'ster 7th (39735), dam (Cherry Duchess of Hillhurst) by Third Duke of Hillhurst (30975), g. d. (Cherry Queen) by Baron Oxford 5th (27958), gr. g. d. (Cherry Princess) by General Napier (24023), — (Cherry Duchess 8th) by Grand Duke 3rd (16182), — by Grand Duke 3rd (16182), — by Grand Duke 2nd (12961), — by Grand Duke (10284), — by Sheldon (8557), — by The Colonel (5428), — by Thorp (2757), — by Pirate (2430), — by Houghton (318), — by Marshal Blucher (416), — from the stock of Messrs. Wright and Charge.

(42922)   CHERRY GRAND DUKE 9TH,

Roan, calved April 10, 1879, bred by Mr. R. E. Oliver, Sholebroke Lodge; got by Grand Duke 30th (38373), dam (Cherry Grand Duchess 10th) by Grand Duke 25th (34065), g. d. (Cherry Grand Duchess 6th) by Grand Duke 19th (28746), gr. g. d. (Cherry Grand Duchess 2nd) by Second Duke of Claro (21576), — (Cherry Grand Duchess) by Grand Duke 7th (19877), — by Grand Duke 3rd (16182), — by Grand Duke 2nd (12961), — by Grand Duke (10284), — by Sheldon (8557), — by The Colonel (5428), — by Thorp (2757), — by Pirate (2430), — by Houghton (318), — by Marshal Blucher (416), — bred from the stock of Messrs. Wright and Charge.

(42923)   CHERRY GWYNNE 3RD,

Red and white, calved August 4, 1878, bred by Mr. D. R. Scratton, Ogwell; got by Sixth Cherry Duke (30705), dam (Double Gwynne) by Rufus (27397), g. d. (Dolly Gwynne) by Duke of York (14461), gr. g. d. (Young Dowager Gwynne) by St. Thomas (10777), — (Dowager Gwynne) by Prime Minister (2456), — by Wallace (5586), — by Marmion (406), — by Merlin (430), — by Layton (366), — by Phenomenon (491), — by Favourite (252), — by Favourite (252), — by Hubback (319), — by Snowdon's Bull (612), — by Waistell's Bull (669), — by Masterman's Bull (422), — by the Studley Bull (626).

(42924) CHERRY PIE,

Red, calved February 22, 1879, bred by Mr. T. Kirkham, Biscathorpe House ; got by Cherry Gwynne 4th (36350), dam by Clarion (30748), g. d. by Ranby (27230), gr. g. d. by Touchstone (23084), — by Cardinal (10026), — by Neptune (7273).

(42925) CHERRY PIE,

Red and white, calved March 25, 1879, bred by Mr. R. Burrow, Wrayton Hall, the property of Mr. J. Crook, Inskip ; got by Cherry Bumper (37973), dam (Playmate 4th) by Lord Darlington (26634), g.d. (Playmate 2nd) by Lord John (18846), gr. g. d. (Playmate) by Duke of Tortworth (14459), — (Playful) by Rothersthorpe (10745), — by Monarch (7249), — by Eclipse (3684), — by Magnum Bonum (2243), — by Spectre (2689), — by MacGregor (2235), — by Goldfinder (2064), — by Star (2699), — by Boughton (90).

(42926) CHERRY PRINCE,

Red and little white, calved September 4, 1879, bred by Mr. A. Graham, Yanwath Hall ; got by The Colonel (35747), dam (Cherry Gwynne 2nd) by Duke Ferdinand (25928), g.d. (Cherry Gwynne) by General Gwynne (19840), gr. g.d. (Light of Day) by Cherry Duke 4th (17552), — (Lovely Lass) by Canrobert (12539), — by Florian (12887), — by Ploughboy (7338), — by Sir Peter Laurie (9650), — by Young Favourite (6003), — by Lord Melbourne (4264), — by Favourite (3772), — by Stephen (5324), — by Young Western Comet (1575), — by Son of Layton (366), — by Layton (366), — by Simon (590).

(42927) CHERRY PRINCE 6TH,

Red and white, calved January 28, 1879, bred by Mr. S. L. Horton, Park House ; got by Marquis of Blandford 6th (41983), dam (Waterloo Cherry 4th) by Idsal (31404), g. d. (Waterloo Cherry) by Kirbythore Waterloo (24263), gr. g. d. (Aurora 2nd) by Brutus (17469), — (Aurora) by Constantine (14318), — by Prince of Wales (4833), — by Hector (16250), — by Prince Ernest (4818), — by Sir William (2640), — by Young Rockingham (2549), — by Wellington (2824), — by Northumberland (464), — by Buston's Styford (103), — by Lame Bull (358), — by Bolingbroke (86).

(42928) CHERRY PRINCE 10TH,

Roan, calved December 1, 1879, bred by Messrs. Dalton, Cummersdale ; got by Duke of Siddington (38182), dam (Cherry 6th) by Royal Cumberland (27358), g. d. (Cressida 2nd) by Fourteenth Duke of Oxford (21605), gr. g. d. (Christobel) by Richard (16834), — (Cherry 3rd) by Scaleby Champion (15243), — by Stapleton (15336), — by Bracebridge (14187), — by Mingo (7239), — by Lowther (9338), — by Snowball (1466), — by Pizarro (1323).

(42929) CHERRY STONE,

Roan, calved March 7, 1879, bred by Mr. C. E. Lyon, Johnson Hall, the property of Mr. R. H. Frank, Ashbourne Hall ; got by Baron Turncroft Siddington (37823), dam (Queen Cherry) by Cherry Duke (25752), g. d. (Queen of Airdrie) by Second Duke of Airdrie (19600), gr. g. d. (Queen of Hearts) by May Duke (13320), — (Queen Bess) by Third Duke of Oxford (9047), — by Freeman (10244), — by Hamlet (8128), — by Short Tail (2621), — by Emperor (1974), — by Young Lancaster (361), — by St. Albans (2584), — by Lawnsleeves (365), — bred by Messrs. James, of Stamford.

(42930) CHERRY TOM,

Red, calved December 19, 1878, bred by Mr. T. Kirkham, Biscathorpe House; got by Cherry Gwynne 4th (36350), dam by Simple Simon (32487), g. d. by Hemlock (28842), gr. g. d. by Ranby (27230), — by Cock of the Walk (15782), — by Aurelius (14111), — by Touchstone (23084).

(42931) CHERRY VALLEY,

Red, calved November 19, 1878, bred by Mr. T. Kirkham, Biscathorpe House; got by Cherry Gwynne 4th (36350), dam by Simple Simon (32487), g. d. by Hemlock (28842), gr. g. d. by Ranby (27230), — by Cock of the Walk (15782), — by Touchstone (23084), — by Cardinal (10026).

(42932) CHERRY WALNUT,

Red, calved January 28, 1876, bred by Sir G. R. Philips, Bart., Weston Park, the property of Mr. Penson, Foxcote; got by Cherry Grand Duke 5th (30712), dam (Duchess of Brailes 3rd) by Grand Duke 15th (21852), g. d. (Miss Knightley) by Bull's Run (19368), gr. g. d. (Gionetta) by Sarawak (15238), — (Smock Frock) by Earl of Dublin (10178), — by Janizary (8175), — by Snowball (8602), — by Caliph (1774), — by Norman (2379), — by White Boy (1580), — by Wyville's Bull, — by a Bull bred by Mr. Charge.

(42933) CHIEF CONSTABLE,

Roan, calved February 26, 1879, bred by Mr. J. Thompson, Anlaby; got by Baron Turncroft Oxford 3rd (36209), dam (Duchess of Cambridge 6th) by Second Squire of Waterloo (25218), g. d. (Duchess of Cambridge 5th) by Actor (17274), gr. g. d. (Duchess of Cambridge 2nd) by Titus Rembrandt (15415), — (Duchess of Cambridge) by Rembrandt (13587), — by Lord Grey (10446), — by Cavaignac (10033), — by Sir Walter (2639), — by Cordilleras (3484), — by Young Remus (2523), — by Remus (550), — by Hollings (2131), — by His Honour (2126), — by Partner (2409), — by Hutton's Bull (2145), — Lady.

(42934) CHIEFTAIN,

Red, calved April 6, 1876, bred by Mr. W. Duthie, Collynie, the property of Mr. G. Shepherd, Tarves; got by Frederick Fitz-Windsor (31196), dam (Clara 2nd) by Marmaduke (20284), g. d. (Clara) by Damian (17660), gr. g. d. (Rose Leaf) by Lord Scarbro' (14852), — (Jessie) by Moss Trooper (11827), — by Duke of Clarence (9040), — by Bowmont (3200), — by Son of Exmouth (3747), — by Mr. Robson's Bull (9562), — by Prince (4765), — by Wellington (2824).

(42935) CHOP,

White, calved March 24, 1865, bred by the Earl of Cawdor, Stackpole Court, late the property of Mrs. T. H. Vivian, Singleton; got by Dalesman (15849), dam (Rissole) by Cupid (14359), g. d. (Sweetbread) by Belshazzar 2nd (14154), gr. g. d. (Sweetbread) by Napoleon (10552), — by Plenipo (4724), — by Sir John (16985), — by Rex (1375), — by Baron (58), — by a Bull of Mr. Mason's, — by Falstaff (250), — by Irishman (329).

(42936) CHRISTMAS DUKE 4TH,

Roan, calved April 3, 1879, bred by Lord Braybrooke, Audley End; got by Duke of Underley 3rd (38196), dam (Christmas Rose 3rd) by Duke of Rosedale 3rd (33723), g. d. (Christmas Rose 2nd) by Grand Duke 17th (24064), gr. g. d. (Christmas Rose) by Fourth Duke of Thorndale (17750), — (The Beauty) by Puritan (9523), — by Third Duke of York (10166), — by Second Cleveland Lad

(3408), — by Belvedere (1706), — by Belvedere (1706), — by Second Hubback (1423), — by His Grace (311), — by Yarborough (705), — by Favourite (252), | by Punch (531), — by Foljambe (263), — by Hubback (319).

(42937)   CHRISTMAS NUMBER,
Roan, calved December 25, 1879, bred by Mr. G. Garne, Churchill Heath ; got by Grand Duke of Marlboro' (38380), dam (Geneva's Nonsuch) by Grand Duke of Geneva 2nd (31288), g. d. (Charming Nonsuch) by Charming King (30698), gr. g. d. (Nonsuch 6th) by Monitor (24615), — (Nonsuch 4th) by Lord Oxford (20214), — by Ninth Duke of Oxford (17738), — by Earl of Warwick (11412), — by Duke of Cornwall (5947), — by Velocipede (5552), — by Sir Thomas (2636), — by Frederick (1060), — by Harold (291), — by North Star (459), — by Favourite (252), — by Favourite (252), — by Favourite (252), — by Broken Horn (95), — bred by Mr. Best, of Manfield.

(42938)   CHRISTY MINSTREL,
White, calved March 7, 1879, bred by Mr. F. Welsh, The Châlet, the property of Mr. W. Spaight, Derry Castle ; got by Robert Stephenson (32313), dam (Victoria 75th) by Lord of the Manor (29181), g. d. (Victoria 50th) by St. Ringan (27417), gr. g. d. (Violet) by Ravenspur (20628), — (Leonora) by Dr. McHale (15887), — by Fugleman (14580), — by Cecil (12571), — by Baron Martin (12444), — by Comus (12625), — by Second Comet (5101), — by Belzoni (783), — by Satellite (1420), — by Cato (119), — by Pope (514), — by. Favourite (252), — by White Bull (421), — by Favourite (252), — by Dalton Duke (188), — by R. Alcock's Bull (19), — by J. Smith's Bull (608), — by Jolly's Bull (337).

(42939)   CLANSMAN,
Red, calved October 20, 1878, bred by Mr. A. Graham, Yanwath Hall ; got by Welcome Prince (39302), dam (Cherry Gwynne 2nd) by Duke Ferdinand (25928), g. d. (Cherry Gwynne) by General Gwynne (19840), gr. g. d. (Light of Day) by Cherry Duke 4th (17552), — (Lovely Lass) by Canrobert (12539), — by Florian (12887), — by Ploughboy (7338), — by Sir Peter Laurie (9650), — by Young Favourite (6003), —: by Lord Melbourne (4264), — by Favourite (3775), — by Stephen (5324), — by Young Western Comet (1575), — by Son of Layton (366), — by Layton (366), — by Simon (590).

(42940)   CLAPHAM DUKE,
Roan, calved October 27, 1879, bred by Mr. J. Howard, Clapham Park ; got by Marquis of Oxford (34786), dam (Broughton Gwynne) by Royal Broughton (27352), g. d. (Young Daffy Gwynne 5th) by Young Duke of Cambridge 2nd (17709), gr. g. d. (Young Daffy Gwynne 3rd) by Lord Royston 2nd (16449), — (Daffy Gwynne) by Sir Harry (10819), — by Conservative (3472), — by Wallace (5586), — by Marmion (406), — by Merlin (430), — by Layton (366), — by Phenomenon (491), — by Favourite (252), — by Favourite (252), — by Hubback (319), — by Snowdon's Bull (612), — by Waistell's Bull (669), — by Masterman's Bull (422), — by the Studley Bull (626).

(42941)   CLARENCE,
Roan, calved February 1, 1879, bred by Earl Beauchamp, Madresfield Court ; got by Marquis of Blandford 2nd (34779), dam (Sunbeam) by Archduke (21185), g. d. (Sunshine) by Ortolan (18496), gr. g. d. (Gay Lass) by Columbus (17591), — (Fillpail) by Duke William (11400), — by Cotgrave (6901), — by Progress (4839).

(42942) CLARENCE 2ND,

Roan, calved June 10, 1879, bred by Mr. D. R. Scratton, Ogwell; got by Lally's Hillhurst Duke 2nd (38539), dam (Duchess of Clarence 3rd) by Grand Duke 6th (19876), g. d. (Duchess of Clarence) by Duke of Clarence (19611), gr. g. d. (Duchess Nan) by Romulus (15185), — (Duchess Nanny) by Jasper (11609), — by Second Duke of Oxford (9046), — by Second Duke of Northumberland (3646), — by Belvedere (1706), — by Son of Second Hubback (2683), — a Cow of Mr. Bates's, of Kirklevington.

(42943) CLARENCE DUKE,

Roan, calved February 16, 1879, bred by Mr. G. Ashburner, Low Hall; got by Duke of Oxford 41st (38174), dam (Grand Duchess of Clarence) by Grand Duke 9th (19879), g. d. (Duchess of Clarence 6th) by King of the Roses (22043), gr. g. d. (Duchess of Clarence 2nd) by Squire of Clarence (25214), — (Duchess of Clarence) by Duke of Clarence (19611), — by Romulus (15185), — by Jasper (11609), — by Second Duke of Oxford (9046), — by Second Duke of Northumberland (3646), — by Belvedere (1706), — by Son of Second Hubback (2683), — a Cow of Mr. Bates's, of Kirklevington.

(42944) CLAVERHOUSE,

Red, calved March 28, 1879, bred by Messrs. Christy Brothers, Fort Union; got by Lord Broughton (31626), dam (Queen of Beauty 8th) by The Earl (27623), g. d. (Queen of Beauty 5th) by Fairy King (21716), gr. g. d. (Queen of Beauty 4th) by Hero of Thorndale (18061), — (Queen of Beauty 2nd) by Duke of Beauford (11377), — by Shotley 2nd (7494) — by Stokes's Old Bull (8368), — by Young Spectator (8619), — by Phantassie (8389), — by Young Rockingham (8498).

(42945) CLODHOPPER,

Roan, calved October 20, 1874, bred by Mr. T. H. Bland, Dingley Grange; got by Earl of Waterloo 2nd (33819), dam (Graceful) by Rupee (32433), g. d. (Lass o'Gowrie) by White Velvet 2nd (23216), gr. g. d. (Gazelle) by Tom Sayers (20985), — (Graceful Fairy) by Paxford (20480), — by Trajan (10970), — by Nero 3rd (22403), — by Romulus (15184), — by Roman (15179), — by Sultan (5346), — by Aid-de-Camp (722), — by Emperor (1014), — by Major (397), — by Major (397).

(42946) CLOSEBURN,

Roan, calved April 9, 1878, bred by Mr. A. Stobo, Porterstown, the property of Mr. J. Kirkpatrick, Whitespotts; got by Duke of Tyne (38195), dam (Twin Princess) by Prince Arthur Patrick (29600), g. d. (Lively) by Sir William (22925), gr. g. d. (Snowfall) by Tweedside (12246), — (Syren) by Petrarch (7329), — by Ethelred (5990), — by Hecatomb (2102), — by Belvedere 2nd (3127), — by Barmpton (5774), — by Kitt (7127), — by Middleton's Red Bull (438).

(42947) COBWEB,

Red, calved December 26, 1879, bred by the Rev. E. T. Williams, Caldicot Parsonage, the property of Mr. E. Knight, Caldicot; got by Cleon (38005), dam (Cordelia) by Earl of March (31073), g. d. (China Aster) by Lygon (24494), gr. g. d. (China Rose) by Young Duke of Cambridge (17708), — (Concord) by Ariel (15591), — by Ottoman (13442), — by Ruber (13644), — by Brownie

(8909), — by Ranunculus (2479), — by William (2840), — by Childers (1824), — by Richard (1376), — by Jupiter (342), — by Charles (127), — by Windsor (698), — by Chilton (136), — by Colonel (152).

(42948) COLONEL,
Roan, calved November 11, 1878, bred by Messrs. A. and A. Mitchell, Alloa got by Brocklesby (36288), dam (Celeste) by Foggathorpe 1st (31179), g. (Chlorale) by Lord Plymouth (24455), gr. g. d. (Clotilda) by Knight Errant (18154), — (Chloe) by Cardigan (12556), — by Noble (4578), — by Newton (2367), — by Young Magog (2247), — by Margrave (2263), — by Sir Charles (593), — by Sir Dimple (594), — by St. Albans (2584), — by Layton (366), — by Charge's Grey Bull (872), — by J. Brown's Red Bull (97).

(42949) COLONEL,
Roan, calved October 25, 1879, bred by Mr. T. Stamper, Highfield House; got by Duke of Nawton 3rd (39764), dam (Gay Lass) by Merry Monarch (22349), g. d. (Garland) by Frederick (14571), gr. g. d. (Young Cherry) by Lahore (11659), — (Cherry) by Traveller (7646), — by Fitz Maurice (3807), — by Reformer (2502), — by Leopold (2199), — by Sir Harry (5155), — by Traveller (655), — by Colonel (152), — by Son of Hubback (319).

(42950) COLONEL CAREY,
Roan, calved February 15, 1879, bred by Mrs. Mace, Sherborne; got by Duke of Wellington 7th (33753), dam (Sally) by Butter Boy (21349), g. d. (Louisa) by Royal Butterfly 14th (20722), gr. g. d. (Lucy Long) by Orangeman (18485), — (Young Lucy) by General Pelissier (14605), — by Chieftain (12595), — by Morpeth (8325), — by The Prince (7615), — by Colling (902), — by Son of Alexander (1624), — by Grandson of Favourite (252).

(42951) COLONEL LIND,
Red and white, calved April 17, 1879, bred by Mr. B. Hannan, Riverstown; got by St. Ronan (35458), dam (Jenny Lind 17th) by Abercorn (25484), g. d. (Jenny Lind 8th) by Napoleon 3rd (29417), gr. g. d. (Jenny Lind 2nd) by Napoleon (18438), — (Jenny Lind) by Druid (10140), — by Belted Will (6780), — by Regent (2517), — by Borderer (3191), — by Eclipse (1949), — by Togston (5487), — by Bolingbroke (3184), — by Son of Midas (435), — by Twin Brother to Ben (660).

(42952) COLONEL MORAY,
White, calved March 15, 1877, bred by Mr. J. H. Dickson, Saughton Mains, the property of Mr. P. R. Latham, The Kames, Doune; got by Bywell (33261), dam (Merino) by Strathallan (37539), g. d. (Gertrude) by Bridegroom (17441), gr. g. d. (Lady Mary) by Windsor (37683), — (Blythsome) by Major (4344), by Ethelred (5990), — by Thorpe, — by Rennie's Romulus.

(42953) COLONEL SLICK,
Roan, calved August 24, 1879, bred by Lord Rathdonnell, Lisnavagh; got by Anchor (32947), or Lieutenant (43466), dam (James's Nancy) by King James (28971), g. d. (Red Nancy) by Sovereign (27538), gr. g. d. (Nancy) by Baron Warlaby (7813), — (Red Rose) by Hamlet (8126), — by Prince Ernest (7366), — by Vanquish (2793), — by Monarch (4495).

(42954) COLUMBUS,
Roan, calved March 16, 1879, bred by Mr. S. Grave, Mirkholme; got by Young

Catesby (41192), dam (Columbine) by Emperor (23883), g. d. (White Columbine) by Helvellyn (34136), gr. g. d. (Columbine) by Gainford 2nd (6030), — (Lively) by Emperor (1974), — by Wellington (2824), — by Major (2255), — by Thorpe (1515), — by Northumberland (464), — by Bolingbroke (86).

(42955) *COMET,

Roan, calved August 23, 1859, bred by Mr. A. M. Winslow, Putney, Vermont, U.S.A.; got by Prince of Orange 2nd (43835), dam (Pocahontas 4th) by Kirklevington (11640), g. d. (Pocahontas 2nd) by Tam O'Shanter (42475), gr. g. d. (Pocahontas) by North American (6253), — (Princess) by Washington (1566), — by Blaize (76), — by Charles (127), — by Blyth Comet (85), — by Prince (521), — by Patriot (486).

(42956) COMET,

Roan, calved January 1, 1879, bred by Mr. R. P. Maxwell, Finnebrogue, the property of Mr. Percy Smyth, Headborough; got by Woodranger (39340), dam (Peri) by Half Sovereign (34104), g. d. (Cynthia 4th) by Prince Victor (20606), gr. g. d. (Cynthia 2nd) by Musician (13362), — (Cynthia) by Ballinasloe (12429), — by Superb Fairfax (5357), — by Jupiter (20039), — by Maximus (2284), — by Monarch (4494), — by Western Comet (689), — by Primus (4763), — by Son of Favourite (252), — by Cupid (177).

(42957) COMMANDANT,

Roan, calved May 6, 1879, bred by Mr. J. Downing, Ashfield; got by Robert Stephenson (32313), dam (Countess) by The Sutler (23061), g. d. (Coquette) by Comet (11298), gr. g. d. (Norma) by Druid (10140), — (Little Red Rose) by Petrarch (7329), — by Second Duke of York (5959), — by Raspberry (4875), — by Young Matchem (4422), — by Isaac (1129), — by Young Pilot (4702), — by Pilot (496), — by Julius Cæsar (1143).

(42958) COMMISSIONER,

Roan, calved March 8, 1879, bred by Mr. W. Parkin, Blaithwaite; got by Wigton (35995), dam (Lady Blanche 3rd) by Duke of Tynedale (28462), g. d. (Lady Blanche 2nd) by Lieutenant Oxford (24336), gr. g. d. (Lady Blanche Hollyhock) by Fourteenth Duke of Oxford (21605), — (Hollyhock 4th) by Richard (16834), — by Palmerston (15041), — by Calcott (14216), — by Mingo (7239), — by General (3867), — by Pizarro (1323).

(42959) COMYN,

White, calved February 28, 1879, bred by Mr. R. H. Gould, Didmarton; got by Prince Arthur (38892), dam (Charmer 3rd) by Lord Darlington 3rd (31646), g. d. (Charmer 2nd) by Grand Duke of York (24071), gr. g. d. (Charmer) by Monarch (16575), — (Cherry) by West Australian (17223), — by Victor (13951), by Berkeley (7830), — by Son of Old Strickland (11870).

(42960) CONINGSBY,

Roan, calved April 25, 1879, bred by Mr. H. Caddy, Rougholm; got by Student (39176), dam (Clochette) by Royal Fame (35364), g. d. (Castanet 4th) by Royal Knight (25032), gr. g. d. (Castanet 2nd) by Ravenspur (20020), — (Castanet) by Prince Arthur (13497), — by Comet (11298), — by Druid (10140), — by Petrarch (7329), — by Second Duke of York (5959), — by Raspberry (4875), — by Young Matchem (4422), — by Isaac (1129), — by Young Pilot (4702), — by Pilot (496), — by Julius Cæsar (1143).

(42961)   CONJUROR,

Roan, calved April 18, 1878, bred by Mr. J. Findlater, Milltack, the property of Mr. W. Henderson, Mossfield; got by Knight of Windsor (31574), dam (Eleanor) by Oxford (34960), g. d. (Lady Ann Hope) by Lord Cobham (22141), gr. g. d. (Roan Butterfly) by Invincible (24191), — (Miss Butterfly) by Symmetrical (23006), — by Master Butterfly 2nd (14918), — by Procurator (10657), — by Fourth Duke of Northumberland (3649), — by Red Highflyer (2488), — by Pyramid (4852), — by Harry Lorrequer (3985), — by Blucher (84), — by Magnum Bonum (4322), — by Buston's Styford (103), — by Son of Wetherell's Bull (690).

(42962)   CONNAUGHT OXFORD,

Roan, calved March 30, 1879, bred by Mr. H. P. Baxter, Southall Green Farm; got by Wild Duke of Connaught (40921), dam (Lovely) by Oxford Prize (29498), g. d. (Cherry) by Victor 4th (30220), gr. g. d. (Laura) by Castor (17513), — (Lady) by Sir Harry (13730), — by Manchester (14891), — by Splendid (5298), — by Adjutant (2924), — by Sir Alexander (591), — by Stephen (1456), — by Western Comet (689), — by Charge's Grey Bull (872), — by Favourite (252), — by Bartle (777), — descended from the Studley White Bull (627).

(42963)   CONNAUGHT RANGER,

Roan, calved March 13, 1879, bred by Mr. T. R. Hulbert, North Cerney; got by Connaught (39618), dam (Golden) by Earl of Verulam (26077), g. d. (Glitter) by Eighth Duke of York (23808), gr. g. d. (Brilliance) by Lamp of Lothian (16356), — (Duchess of Glo'ster 8th) by His Highness (14708), — by Clarendon (12605), — by The Red Duke (8694), — by Lottery (4280), — by Lottery (4280), — by Phœnix (6290).

(42964)   CONNAUGHT WILD EYES,

Roan, calved March 5, 1879, bred by Mr. H. P. Baxter, Southall Green Farm; got by Wild Duke of Connaught (40921), dam (Wild Eyes 34th) by Bramble 2nd (36259), g. d. (Wild Eyes 30th) by Don John (19583), gr. g. d. (Wild Eyes 27th) by Gainford 5th (12913), — (Wild Eyes 26th) by Second Cleveland Lad (3408), — by Short Tail (2621), — by Emperor (1975), — by Wonderful (700), — by Cleveland (145), — by Butterfly (104), — by Hollon's Bull (313), — by Mowbray's Bull (2342), — by Masterman's Bull (422), — descended from M. Dobison's stock.

(42965)   CONQUEROR,

Roan, calved September 25, 1878, bred by Mr. C. Bayes, Kettering; got by Grand Duke of Darlington (39948), dam (Chaff 9th) by Second Duke of Wetherby (21618), g. d. (Coquette of Oxford) by Marquis of Oxford (18339), gr. g. d. (Clara) by Viceroy (13945), — (Chaff) by Duke of Cornwall (5947), — by Morpeth (7254), — by Helicon (2107), — by Henwood (2114), — by Nestor (452), — by Harold (291), — by Meteor (432), — by Comet (155), — by Cupid (177).

(42966)   CONRAD,

Roan, calved December 3, 1878, bred by Mr. J. Todd, Scalthwaiterigg, the property of Mr. J. Clark, Sizergh Castle Farm; got by British Duke (37901), dam (Brenda) by Prince of Perth (32176), g. d. (Minna) by White Oxford (27797), gr. g. d. (Spot) by Grenadier 2nd (28794), — (Red Rose) by Statesman (22973), — by Wellington (21091), — by Grand Duke of Waterloo (14641), — by Grand Duke of Waterloo (14641), — by Prime Minister (13488), — by Lord Malcolm (10456), — by Son of Elis (3703), — by Vaux (5549).

(42967) CONSORT,
Roan, calved July 30, 1877, bred by Messrs. A. and A. Mitchell, Alloa; got by
Foggathorpe 1st (31179), dam (Corisande) by Lord Plymouth (24455), g. d.
(Coral) by Valasco (15443), gr. g. d. (Carmine) by Prince Arthur (13497),—(Car-
nation) by Newton (2367), — by Young Magog (2247), — by Margrave (2263),
— by Sir Charles (593), — by Sir Dimple (594), — by St. Albans (2584), —
by Layton (366), — by Charge's Grey Bull (872), — by J. Brown's Red
Bull (97).

(42968) CONSTANTINE 5TH,
Roan, calved December 20, 1879, bred by Mr. S. L. Horton, Park House; got
by Marquis of Blandford 6th (41983), dam (Lily) by Prince Charming (27130),
g. d. (Ellie) by Prince Albert (18579), gr. g. d. (Rose of Midsummer) by Sir
Colin (15276), — (Red Rose) by Earl of Scarborough (9064), — by The Old
Red Bull, — by Paganini (2405), — by Paul Jones (8383), — by Pirate (2430),
— by Sedbury (1424), — by Charge's Grey Bull (872).

(42969) CORONET,
Roan, calved February 18, 1878, bred by Messrs. J. and E. Tindall, Knapton
Hall; got by Sampiero (35466), dam (Rosette 3rd) by Cecil (25725), g. d.
(Rosette) by Prince Louis (20563), gr. g. d. (Tamarind) by Cavendish (15745),
— (Bloom) by My Lord (14972), — by Villiers (13959), — by Liberator (7140),
— by Prince Albert (4791), — by Sligo (5210), — by Bulmer (1760).

(42970) CORPORAL,
Roan, calved February 27, 1879, bred by Mrs. C. H. Stopford-Sackville, Drayton
House, the property of Mr. W. Chew, Ringstead; got by Freeman (36666), dam
(Curly) by George 1st (19848), g. d. (Carey) by Tenedos (20936), gr. g. d.
(Countess) by The Sheriff (12216), — (Cowslip) by Prince Ernest (4818), —
by Mowthorpe (2343), — by Ambo (1636), — by Burley (1766), — by Isaac
(1129), — by Pilot (496), — by Albion (14), — by Lame Bull (359), — by
Shipton (587), — by Son of Suwarrow (636), — by Son of Twin Brother to
Ben (88), — by Twin Brother to Ben (660).

(42971) CORSAIR,
Red and white, calved September 19, 1878, bred by Mr. J. Norrish, Gays; got
by Conrad (36392), dam (Duchess of Oxford 3rd) by Aberdeen Lad (32904), g. d.
(Duchess of Oxford 2nd) by King of the Forest (24254), gr. g. d. (Duchess of
Oxford) by Twelfth Duke of Oxford (19633), — (Red Hart) by May Duke
(13320), — by Game Boy (14586), — by Walton (6658), — by Marmion (4383),
— by Young Rubens (5026), — by Matchem 3rd (4420), — by Young Eryholme
(1981), — by Belzoni (1709), — by Comus (1861), — by Denton (198), — by
Henry (301).

(42972) COSTA OF NASEBY,
Roan, calved December 19, 1879, bred by Captain Ashby, Naseby Woolleys;
got by Earl of Geneva (33794), dam (Grace Costa 3rd) by Cherry Fawsley
30711), g. d. (Grace Costa) by Costa (21487), gr. g. d. (Ganza) by Old Buck
15017), — (Smock Frock) by Earl of Dublin (10178), — by Janizary (8175),
— by Snowball (8602), — by Caliph (1774), — by Norman (2379), — by White
Boy (1580), — by Wyville's Bull, — by a Bull bred by Mr. Charge.

(42973) COTSWOLD PRINCE,
Roan, calved November 25, 1879, bred by Mrs. Hampson, Ullen Wood; got by

Grand Duke of Waterloo (28766), dam (Ceres 14th) by Baron Butterfly (25557), g. d. (Ceres 4th) by Lord Wild Eyes 3rd (22235), gr. g. d. (Ceres) by Duke of Wharfdale (19648), — (Red Duchess) by Baron Hopewell (14134), — by Frederick (11489), — by Whittington (12299), — by Second Cleveland Lad (3408), — by Duke of Northumberland (1940), — by Norfolk (2377), — by Belvedere (1706), — by Belvedere (1706), — by Lancaster (360), — by Petrarch (488), — by Major (397), — by Chapman's Son of Punch (122), — by Dickson's Grandson of Punch (213), — by Checks (132), — by R. Grimston's Bull (282), — by J. Coates's Bull (148).

(42974) COUNT,

Red, calved February 25, 1879, bred by Mr. F. Morice, Springfield; got by Undergraduate (35835), dam (Jenny Jones 7th) by Victor (32774), g. d. (Jenny Jones 5th) by Hero of Thorndale (18061), gr. g. d. (Jenny Jones) by Young Shaftoe (9625), — (Jenny Lind) by Zero (8799), — by Old Bull (8368), — by Young Spectator (8619), — by Phantassie (8389), — by Young Rockingham (8498).

(42975) COUNT CLAPHAM,

Roan, calved March 24, 1878, bred by Mr. W. Duthie, Collynie, the property of Mr. J. Moir, Mains of Wardhouse; got by Clapham (37999), dam (Countess 4th) by Gladstone (31253), g. d. (Countess 2nd) by Heir of Englishman (24122), gr. g. d. (Countess) by Marmaduke (20284), — (Fragrant) by John Bull (11618), — by Matadore (11800), — by Second Duke of Northumberland (3646), — by Mahomed (6170), — by Sillery (5131), — by Carleton (843), — by Diamond (205), — by Diamond (205).

(42976) COUNT DE ACTON,

Roan, calved February 5, 1876, bred by Mr. J. Nichols, Iron Acton; got by Weldred (35959), dam (Countess) by Reflector (29760), g. d. (Clio) by Coronet (23624), gr. g. d. (Calliope) by Lamp of Lothian (16356), — (Cherish) by Archduke (17316), — by Koh-i-noor (11642), — by Douglas (12715), — by Ned Poins (6241), — by Old Strickland (11870).

(42977) COUNT ERRANT,

Roan, calved July 20, 1879, bred by the Trustees of the late Sir W. S. Maxwell, Bart., Keir Mains; got by Count Broughton (39639), dam (Lady Errant) by Banner Bearer (27907), g. d. (Ferooza) by Knight Errant (18154), gr. g. d. (Barbary) by Cardigan (12556), — (Barmaid) by Lord Fanny (13187), — by Noble (4578), — by Newton (2367), — by Baronet (1686), — by Reformer (2502), — by Margrave (2263), — by Leopold (2199), — by Hector (2103), — by Surly (2715), — by Traveller (655), — by Colonel (152).

(42978) COUNT FITZCLARENCE 5TH,

Roan, calved August 6, 1879, bred by Mr. F. N. Smith, Wingfield Park; got by Lord Fitzclarence 9th (34548), dam (Red Rose 3rd) by Duke of Monmouth (33694), g. d. (Red Rose 2nd) by Falconer (23907), gr. g. d. (Red Rose) by Crœsus (21506), — (Rosebud) by Midas (22355), — by Mandarin (18319), — by Speculator (13775), — by The Hero (10934), — by Paragon (8378), — by Ganthorpe (2049), — by Midas (7236), — by Grazier (1085).

(42979) COUNT MOLTKE,

Roan, calved June 12, 1879, bred by Mr. W. H. Wakefield, Sedgwick; got by Duke of Holker (38153), dam (Roany 3rd) by Lord of Garsdale (34617), g. d.

(Roany 2nd) by Dunrobin (28486), gr. g. d. (Roany) by Frederick Warlaby (23990), — (Rose Bud 1st) by Squire Stuart (20891), — by Precedent (11918), — by Abraham Parker (9856).

(42980)   COUNT OF HANNEY,
Roan, calved January 30, 1876, bred by Mr. H. Bettridge, East Hanney, the property of Mr. W. Curtis, Fernham; got by Count Blanche (33455), dam (Rosa Ringlet) by Young Duke Jamaica (30976), g. d. (Red-heart Rose) by Artemus Ward (23326), gr. g. d. (Rynil Rose) by A 1 (15538), — (Rosette) by Royal (13636), — by Lord George (9314), — by Fitz-Hardinge (8073), — by Augustus (6751), — by Consul (1868), — by Second Fairfax (8050), — from the stock of Mr. Champion, of Blyth.

(42981)   COUNTRY LAD,
Red, calved November 1, 1879, bred by Mr. J. A. Mumford, Brill House; got by Wild Prince (40928), dam (Country Lass) by Notley (31991), g. d. (Camilla) by Earl of Lancaster (21647), gr. g. d. (Criterion) by Earl Ducie (17767), — (Cricket) by Sorcerer (13772), — by Bristol (7852), — by Morpeth (7254), — by Sir Thomas (2636), — by Sir Thomas (2636), — by Marske (418), — by Comet (155), — by Tom (652), — by Favourite (1033), — by Hutton's Bull (323), — by Barningham (56).

(42982)   COUNT TOWNELEY 2ND,
Roan, calved April 13, 1879, bred by Mr. J. Snarry, Marramatt Farm; got by Oxford Ryedale 2nd (34991), dam (Princess 3rd) by Duke of Towneley (21615), g. d. (Red Princess) by Gipsy Prince (17965), gr. g. d. (Brandy) by Codrington (14290), — (Fox) by Starlight (12146), — by Prince Edward (6334), — by Traveller (6617), — by Charles (3343), — by Commodore (5874), — by Cupid (5900), — by Parson (1306), — by St. John (572), — by Pope (514), — by Chilton (136).

(42983)   COUNT WATERLOO,
Roan, calved February 11, 1879, bred by Mr. H. Allsopp, Hindlip Hall; got by Marquis of Oxford (34786), dam (Waterloo 37th) by Oxford Beau (29485), g. d. (Waterloo 30th) by Third Duke of Wharfdale (21619), gr. g. d. (Waterloo 25th) by Duke of Geneva (19614), — (Waterloo 17th) by Red Knight (11976), — by Grand Duke (10284), — by Third Duke of Oxford (9047), — by Second Cleveland Lad (3408), — by Duke of Northumberland (1940), — by Norfolk (2377), — by Waterloo (2816), — by Waterloo (2816).

(42984)   COURTIER,
Roan, calved April 13, 1879, bred by Mr. W. Duthie, Collynie, the property of Mr. J. Law, New Keig; got by Clapham (37999), dam (Countess 4th) by Gladstone (31253), g. d. (Countess 2nd) by Heir of Englishman (24122), gr. g. d. (Countess) by Marmaduke (20284), — (Fragrant) by John Bull (11618), — by Matadore (11800), — by Second Duke of Northumberland (3646), — by Mahomed (6170), — by Sillery (5131), — by Carleton (843), — by Diamond (205), —by Diamond (205).

(42985)   COURTIER,
Roan, calved August 19, 1879, bred by Mr. W. Hawkes, Thenford; got by Harcourt (39977), dam (Countess) by Fair Thane (31127), g. d. (Cestus) by Windsor Fitz-Windsor (25458), gr. g. d. (Venus) by Lord Frederick (22156) —

(Gentle Breeze) by King Alfred (16334), — by Wilberforce (9830), — by Wharfdale Hero (9821), — by Symmetry (5384), — by Fitz-Champion (7007), — by Young Snowball (7521).

(42986) COURTOWN AFGHAN,
Red and white, calved April 30, 1879, bred by the Earl of Courtown, Courtown House ; got by Prince Milan (38916), dam (Dandy 8th) by Champagne Charley (33314), g. d. (Dandy 5th) by Hermit (26376), gr. g. d. (Dandy 2nd) by Clydesdale (15779), — (Darling 2nd) by Sir Charles Napier (13708), — by Captain (11241), — by Municipal Bill (2344), — by Planet (2432).

(42987) COURTOWN BATTENBERG,
Red and white, calved February 9, 1879, bred by the Earl of Courtown, Courtown House ; got by Prince Milan (38916), dam (Pet) by Earl of Shrewsbury (23847), g. d. (Purity) by Earl of Hardwicke 2nd (19667), gr. g. d. (Prudence) by Sir Charles Napier (13708), — (Phœbe) by Soldier (20864), — by Volcano (21044), — by Municipal Bill (2344), — by Bonaparte (19327), — Phœnix, bought of Mr. De Renzi.

(42988) COURTOWN WARRIOR,
Red, calved June 13, 1879, bred by the Earl of Courtown, Courtown House ; got by Prince Milan (38916), dam (Honey) by Champagne Charley (33314), g. d. (Huntress) by Fudge (21782), gr. g. d. (Diana) by Mutineer (24635), — (Young Virgo) by Eclipse (17791), — by Priam (4758), — by Linton (4227), — by Pirate (1322), — by Adam (717).

(42989) COXSWAIN,
Red, calved April 27, 1879, bred by Mr. F. Stratton, Merdon ; got by Abelard (39346), dam (Junket) by Heir Apparent (24119), g. d. (Julia) by Humphrey Clinker (13055), gr. g. d. (Jenny) by Hero of the West (8150), — (Madam) by Lottery (4280), — by Phœnix (6290).

(42990) CRAGGS DUKE,
Roan, calved October 21, 1877, bred by Mr. F. W. Stone, 7 Stone Buildings, London ; got by Duke Polycherry (33763), dam (Sapho) by Duke of Kent (25979), g. d. (Sapphire) by Standard (22963), gr. g. d. (Mary Stuart) by Standard (22963), — (Lady Stuart) by John O'Groat (18115), — by Grand Sultan (16189), — by Duke of Sussex (12772), — by Lord Foppington (10437), — by Second Cleveland Lad (3408), — by Duke of Northumberland (1940),— by Son of Second Hubback (2683), — Craggs, bought of Mr. Bates, and descended from the stock of Mr. Maynard, of Eryholme.

(42991) CRESSIDA GWYNNE,
Roan, calved May 19, 1878, bred by Mr. R. Thompson, Inglewood Bank ; got by British Duke (37901), dam (Rosamond Gwynne) by Sir Windsor (22927), g. d. (Rebecca Gwynne) by Knight of Distington (18158), gr. g. d. (Ruth Gwynne) by Exquisite (14524), — (Young Dowager Gwynne) by St. Thomas (10777), — by Prime Minister (2456), — by Wallace (5586), — by Marmion (406), — by Merlin (430), — by Layton (366), — by Phenomenon (491), — by Favourite (252), — by Favourite (252), — by Hubback (319), — by Snowdon's Bull (612), — by Waistell's Bull (669), — by Masterman's Bull (422), — by the Studley Bull (626).

(42992)   CRICCIN WATERLOO,

Red, calved April 2, 1879, bred by Mr. R. C. Welsby, Criccin ; got by Edward Waterloo (38246), dam (Agnes 4th) by Favourite (33895), g. d. (Agnes 2nd) by New York (22412), gr. g. d. (Agnes) by Nonsuch (18460), — (Carthagena) by Daisy Duke 2nd (14365), — by Hamlet (8126), — by The Seer (9616), — by Sir Frederick (8577), — by Jerry (4100), — by Marquis (4386), — by Henwood (2114), — by Mynheer (2345), — by Duke Humphrey (230), — by Son of George (107b), — by Badsworth (47).

(42993)   CROOME DIAMOND,

White, calved November 28, 1879, bred by Mr. J. Briscoe, Hill Croome ; got by Duke of St. Johns 2nd (36539), dam (Lady Valentine) by Third Duke of Waterloo (23801), g. d. (Lady Love) by Gold Nugget (16176), gr. g. d. (Lady St. Michael) by Whittlebury (9829), — (Lassy) by Planet (9483) — by Don Juan (1923), — by Cawdor (862), — by Marshal Beresford (415), — by Wellington (683), — by Windsor (698), — by Middleton's Bull (438), — by Chapman's Bull, of Dinsdale.

(42994)   CROSSFELL 36TH,

Roan, calved March 10, 1879, bred by Mr. J. Mitchell, Howgill Castle ; got by Crossfell 7th (39653), dam (Ferronia 5th) by Kirkbythore Wild Eyes 1st (34361), g. d. (Ferronia 3rd) by Lord Derby (26637), gr. g. d. (Ferronia 2nd) by Duke of Lancaster (21597), — (Ferronia 1st) by Freebooter (17882).

(42995)   CROSSFELL 38TH,

Roan, calved April 21, 1879, bred by Mr. J. Mitchell, Howgill Castle ; got by Crossfell 1st (39649), dam (Lily White) by Heart of Gold (26355), g. d. (Lily Shaftoe) by Duke of Lancaster (21597), gr. g. d. (Shaftoe) by Pugnator (18656), — (Silly Shaftoe) by Sir Charles Napier (13714), — by Gainford 2nd (10255), — by Captain Shaftoe (6833), — by Thalberg (5416), — by Gainford (2044).

(42996)   CROSSFELL 39TH,

White, calved April 21, 1879, bred by Mr. J. Mitchell, Howgill Castle ; got by Crossfell 7th (39653), dam (Rose) by Prince James (32148), g. d. (Red Rose) by Lawrence (16363), gr. g. d. (Miss Lizzy 2nd) by Benedict 2nd (25624), — (Miss Lizzy) by Sultan (18938), — by Loyal (6154), — by Hopeful(9222), — a Northumberland Heifer.

(42997)   CROSSFELL 43RD,

Red, calved August 1, 1879, bred by Mr. J. Mitchell, Howgill Castle ; got by Cherry King (36352), dam (Rose of Wharton) by True Briton (35823), g. d. (Rose of Stobars) by Knight of Stobars (40085), gr. g. d. (Tuscan Rose 5th) by Brutus (17469), — (Tuscan Rose 4th) by Sir Colin (16957) — by Sir Kenneth (10826), — by Duke of Richmond (7997), — by Provost (4846), — by Melmoth (2291), — by Holywell (2132), — by Emperor (1974), — by Barmpton (1677), — by St. Albans (2584), — by Simon (590), — by Pope (514).

(42998)   CROSSFELL 44TH,

Red and white, calved August 3, 1879, bred by Mr. J. Mitchell, Howgill Castle ; got by Stanley (37522), dam (Lady Elvira) by Royal Edgar (32380), g. d. (Elvira 13th) by Fourteenth Duke of Oxford (21605), gr. g. d. (Elvira 4th) by Disraeli (10125), — (Elvira 3rd) by Morton (14963), — by Titian (5485), — by Smeaton (5212), — by Champion (3319), — by Emperor (1010), — by Allison's Grey Bull (26), — descended from the Studley White Bull (627).

(42999)   CROSSFELL 47TH,

Red and white, calved September 15, 1879, bred by Mr. J. Mitchell, Howgill Castle; got by Stanley (37522), dam (Fortune's Freedom 2nd) by Wild Indian (32867), g. d. (Fortune's Freedom) by Oxford's Freedom (29502), gr. g. d. (Fortune's Frolic) by Silky Belleville (27453), — (Fortune 3rd) by Marplot (22291), — by Misfortune (18405), — by Leopold (9293), — by Preston's Son of Gainford (2044), — by Colling's Roan Bull, — by Lowther (4283).

(43000)   CROSSFELL 48TH,

Roan, calved September 15, 1879, bred by Mr. J. Mitchell, Howgill Castle; got by Crossfell 16th (39654), dam (Margery 4th) by Kirkbythore Wild Eyes 3rd (36864), g. d. (Margery) by Heart of Gold (26355), gr. g. d. (Gamblesby) by Duke of Lancaster (21597), — (Old Harrison) by Pugnator (18656), — bred by Mr. Harrison, Gamblesby.

(43001)   CROSSFELL 49TH,

White, calved September 28, 1879, bred by Mr. J. Mitchell, Howgill Castle; got by Crossfell 16th (39654), dam (Teaty 2nd) by Kirkbythore Wild Eyes 3rd (36864), g. d. (Teaty) by Trevelyan d'Eden (35813), gr. g. d. (Three Teats) by Duke of Lancaster (21597), — (Horn) by Pugnator (18656).

(43002)   CROSSFELL 50TH,

White, calved October 22, 1879, bred by Mr. J. Mitchell, Howgill Castle; got by Crossfell 16th (39654), dam (Roachy 2nd) by Kirkbythore Wild Eyes 3rd (36864), g. d. (Young Roachy) by Trevelyan d'Eden (35813), gr. g. d. (Old Roachy) by Duke of Lancaster (21597), — (Roachy) by Pugnator (18656).

(43003)   CROSSFELL 51ST,

White, calved December 8, 1879, bred by Mr. J. Mitchell, Howgill Castle; got by Crossfell 17th (39655), dam (Royal Rose) by King of Scotland (34332), g. d. (Christmas Herald) by Monarch (31930), gr. g. d. (Family Herald) by Grand Herald (26301), — (Red Rose 8th) by Precedent (11918), — by Gainford 2nd (10255), — by Emperor (9084), — by Sir Thomas Newton (6495), — by General (3867).

(43004)   CROWNED HEAD,

Roan, calved November 28, 1879, bred by the Rev. T. Staniforth, Storrs; got by Lord Prinknash (34655), dam (Foreign Empress) by Fitz-Royal (26167), g. d. (Foreign Princess) by Prince of Warlaby (15107), gr. g. d. (Flower of Denmark) by Fitz-Clarence (14552), — (Flower Maid) by Vanguard (10994), — by Londesboro' (6142), — by Rinaldo (4949), — by Sir Thomas (2636), — by Sir Alexander (591), — by Marske (418), — by North Star (459), — by Wellington (680), — by Favourite (252), — by Favourite (252), — by Ben (70), — by Hubback (319), — by Snowdon's Bull (612), — by Sir J. Pennyman's Bull (601).

(43005)   CROWN OF BRITAIN,

White, calved February 19, 1878, bred by Mr. W. Talbot Crosbie, Ardfert Abbey, the property of Mr. W. R. Meade, Ballymartle; got by England's Glory (23889), dam (Corona) by Regal Booth (27262), g. d. (Crown of Light) by Northern Light (24670), gr. g. d. (Diadem) by Admiral (12340), — (Florence) by Desmond (10112), — by Old Bull (8368), — by Young Spectator (8619), — by Phantassie (8389), — by Young Rockingham (8498).

(43006) **CROWN PRINCE,**
White, calved September 26, 1879, bred by Mr. W. Handley, Green Head; got
by Alfred the Great (36121), dam (Nicety) by Earl of Derwent (28503), g. d.
(Earl's Flora) by Earl of Eglinton (23832), gr. g. d. (Flora Cobham) by Marquis
of Cobham (22299), — (Flower of Fitz-Clarence) by Alfred Fitz-Clarence (19215),
— by Veteran (13941), — by Waterloo Hero (13981), — by Duke of Richmond
(7996), — by Lord Stanley (4269), — by Velocipede (5552), — by Priam (2452),
— by Jerry (4097), — descended from the stock of Mr. Booth.

(43007) **CRUSADER,**
Red and white, calved April 1, 1878, bred by Mr. R. B. Blyth, Woolhampton;
got by Oxford's Butterfly (34993), dam (Duchess Marie) by Duke of Kennet
(30977), g. d. (Valentine) by Tam O'Shanter (20930), gr. g. d. (Rosa Bonheur)
by Prince Imperial (16740), — (Rosa) by Marmaduke (14897), — by The Beau
(12182), — by Puritan (9523), — by Fanatic (8054), — by Auld Robin Grey
(6753), — by Scrip (2604), — by Burley (1766), — by Isaac (1129), — by
Pilot (496), — by Albion (14), — by Lame Bull (359), — by Shipton (587), —
Son of Suworrow (636), — by Son of Twin Brother to Ben (88), — by Twin
Brother to Ben (660).

(43008) **CULPEPPER,**
Roan, calved March 22, 1879, bred by Mr. R. Hemming, Bentley Manor; got
by Corrie (38031), dam (Ada 22nd) by Quintus (29705), g. d. (Ada 8th) by
Cavalier (23524), gr. g. d. (Ada) by Orthodox 28th (18493), — (Angerona) by
Lemnos (13146), — by Earl Stanhope (5966), — by True Blue (5522), — by
Miracle (2321), — by Sir Henry (1446), — by Count (170), — by Bracken (91),
— by Badsworth (47), — by Driffield (223), — bred by Sir G. Strickland.

(43009) **CULVER DUKE,**
Red and white, calved December 12, 1878, bred by Mr. E. Byrom, Culver; got
by Conrad (36392), dam (Cherry Duchess) by Cherry Duke 6th (30705), g. d.
(Dewdrop) by King Pippin (20068), gr. g. d. (Daffodil) by General (16100), —
(Dewdrop) by Roan Duke (15166), — by Gainford 5th (12913), — by Chilton
(10054), — by Marquis (13303).

(43010) **CULVER DUKE 2ND,**
Red, calved November 9, 1879, bred by Mr. E. Byrom, Culver; got by Lord
Rothesay Sweetheart (40238), dam (Cherry Duchess) by Cherry Duke 6th (30705),
g. d. (Dewdrop) by King Pippin (20068), gr. g. d. (Daffodil) by General (16100),
— (Dewdrop) by Roan Duke (15166), — by Gainford 5th (12913), — by Chilton
(10054), — by Marquis (13303).

(43011) **CULVER GENERAL,**
Roan, calved May 30, 1879, bred by Mr. E. Byrom, Culver; got by Lord Rothe-
say Sweetheart (40238), dam (Fawsleyana) by Earl of Fawsley 3rd (28506), g. d.
(Dewdrop) by King Pippin (20068), gr. g. d. (Daffodil) by General (16100), —
(Dewdrop) by Roan Duke (15166), — by Gainford 5th (12913), — by Chilton
(10054), — by Marquis (13303).

(43012) **CUPID,**
Red, calved August 1, 1878, bred by Mr. R. P. Maxwell, Finnebrogue; got by
Czarowitz 4th (33497), dam (Venus) by Half Sovereign (34104), g. d. (Goddess)
by Jupiter (24230), gr. g. d. (First Fruits) by Prince Victor (20606), — (Diana)

by Musician (13362), — by Billy the Beau (12471), — by Ballinasloe (12429), — by Superb Fairfax (5357), — by Jupiter (20039), — by Maximus (2284), — by Monarch (4494), — by Western Comet (689), — by Primus (4763), — by Son of Favourite (252), — by Cupid (177).

(43013) CUPID,
Roan, calved March 4, 1879, bred by Mr. W. Crickmore, Seething; got by Romulus (39006), dam (Venus) by Bonbon (30560), g. d. (Virginia) by Victoria's Brian (23141), gr. g. d. (Snowdrop) by Virginius (23147), — (Roan Cow) by Garibaldi (21792), — by Merman (13330).

(43014) CYCLONE,
Roan, calved May 6, 1879, bred by Mr. J. Stratton, Alton Priors; got by Proteus (40552), dam (Clematis) by Endymion (31109), g. d. (Cassandra) by James 1st (24202), gr. g. d. (Miss Cassy) by Bude Light (21342), — (Casket) by Hermit (14697), — by Belgrave (8873), — by Duke of St. Albans (6944), — by Will Honeycomb (5660), — by Spectator (2688), — by Son of Sir Roger-de-Coverley (5187), — by Son of Albion (1619), — by Alexander (1624).

(43015) CYPRUS,
Red and white, calved November 10, 1877, bred by Mr. G. Hewer, Ley Gore House, the property of Mr. G. Wood, Croxden Abbey; got by Prince of the Heath (35161), dam (Crimea 4th) by Prince George (24807), g. d. (Crimea 3rd) by Royal Judge (20739), gr. g. d. (Crimea 2nd) by Norman (16628), — (Crimea) by Gloster's Grand Duke (12949), — by Sherborne (10805), — by Locksley (4240), — by Stanhope (5315), — by Prince (4772), — bought in the North.

(43016) CZAROWITZ 5TH,
Roan, calved January 9, 1877, bred by Mr. F. W. Low, Kilshane, the property of the Earl of Clancarty, Garbally; got by Visigoth (35907), dam (Dagmar 3rd) by Woodcraft (25470), g. d. (Dagmar) by Cœur-de-Lion (15783), gr. g. d. (Bride of Apperly) by The Templar 2nd (13879), — (Baroness) by Moss Trooper (13357), — by Tom Boy (9749), — by Exmouth (6983), — by Mickley (7234), — by Sir Harry Liddell (5157), — by Adonis (2933), — by Son of Wellington (683), — by Hollon's Bull (313).

(43017) DADLINGTON DUKE 2ND,
Red, calved April 15, 1878, bred by Mr. J. A. Geary, The Poplars, Dadlington, the property of Mr. J. Grewcock, Stapleton; got by May Lad (37085), dam (Belle of Dadlington) by Baron York 2nd (33113), g. d. (Primrose) by Field Marshal (19742), gr. g. d. (Camelia) by Marengo (13294), — (Scotch Rose) by Plato (11907).

(43018) DAINTY DUKE 5TH,
Red, calved February 12, 1879, bred by Mr. S. P. Foster, Killhow; got by Duke of Glo'ster 7th (39735), dam (Dainty 2nd) by Duke of Hillhurst (28401), g. d. (Dutiful) by Third Duke of Wharfdale (21619), gr. g. d. (Duenna) by Grand Duke 11th (21849), — (Dulcinea) by Duke of Geneva (19614), — by Master Rembrandt (16545), — by Jasper (11609), — by Second Duke of Oxford (9046), — by Second Duke of Northumberland (3646), — by Belvedere (1706), — by Son of Second Hubback (2683), — a Cow of Mr. Bates's, of Kirklevington.

(43019) DAIRY BOY,
Roan, calved June 30, 1878, bred by Sir Wilfrid Lawson, Bart., Brayton; got

by Wild Eyes Duke (36007), dam (Dairy Lass) by Cambridge Duke (28120),
g. d. (Dairy Maid) by Grand Vizier (26313), gr. g. d. (Dairy) by Perth (22513),
— (Dairy) by Duke (14419). — by Son of Comus (12626).

(43020) DAIRYMAN,

Roan, calved June 14, 1878, bred by Mr. W. Burton, Eastoft Hall, the property
of Mr. R. S. Brundell, Leicester House; got by Pearl Diver (35025), dam
(Eastoft Matron) by Telemachus (27603), g. d. (Oxford Matron) by Imperial
Oxford (18084), gr. g. d. (Matron) by The Briar (15376), — (Monitress) by
Ozier (10588), — by Miracle 2nd (2322), — by Woodville (2856), — by Christ-
mas (3380), — by Atlas (42), — by Windsor (698), — by Windsor (698).

(43021) DAIRYMAN,

Roan, calved October 19, 1878, bred by Mr. J. A. M. Cope, Drummilly; the
property of Mr. T. Dawson, Portadown; got by Glo'ster's Gwynne (38360),
dam (Doralice) by Baron Eccleswall (33037), g. d. (Deodora) by Third Duke of
Wetherby (26030), gr. g. d. (Diadem) by Touchstone (20986), — (Dewdrop) by
Earl of Derby (21637), — by Amiens (14095), — by Rubens (15209), — by
Meteor (10526), — by Phœnix (6290), — by Shakespeare (5108), — by Rival
(2534), — by Darlington (956), — by Denton (198), — by Rockingham (560),
— by Jobling's Son of Phenomenon (491), — by Phenomenon (491), — by
Colonel (152), — by Styford (629).

(43022) DAIRYMAN,

Roan, calved April 28, 1879, bred by Sir R. C. Musgrave, Bart., Eden Hall;
got by Baron Underley (37824), dam (Diamond 5th) by Ignoramus (28887),
g. d. (Diamond 2nd) by British Cherry (23461), gr. g. d. (Diana) by Baron of
Ravensworth (17380), — (Diamond) by Lofty (13168), — by Captain Shaftoe
(6833), — by Thalberg (5416), — by Gainford (2044).

(43023) DAISY DUKE,

Roan, calved April 25, 1878, bred by the Rev. W. Holt Beever, Pencraig Court,
the property of Mr. J. Garsed, The Moorlands; got by Second Duke of Siddington (33732), dam (Royal Blush) by Royal Cumberland (27358), g. d. (Court
Blush) by The Baron (25277), gr. g. d. (Blush) by Prince Henry (18608), —
(Donna) by Patriot (10594), — by Sir Frederick (8577), — by Marquis (4386),
— by Mynheer (1255), — by Enchanter (244), — by Major (398), — by
Windsor (698), — by Favourite (252), — by Punch (531), — by Hubback (319).

(43024) DALESMAN,

White, calved June 16, 1879, bred by Mr. A. W. Long, Mint Cottage; got by
Napoleon (37119), dam (Mary Bell 5th) by Hartington (26346), g. d. (Mary
Bell 4th) by Red Windsor (22709), gr. g. d. (Mary Bell 2nd) by Optimas
Gwynne (18483), — (Mary Bell 1st) by Magician (14875), — by Rubini (9586),
— by Emperor (1974), — by Wharton (2833), — by Headlam's Bull.

(43025) DALKEITH,

Roan, calved March 12, 1879, bred by Mr. W. Scott, Glendronach; got by Ivan-
hoe (36796), dam (Dew Drop 2nd) by Grand Forth 2nd (26300), g. d. (Snow
Drop) by Valiant 2nd (25353), gr. g. d. (Dew Drop) by Sovereign (17020), —
(Young Lily) by Frederick (12899), — by Orbliston (16655), — by Duplicate
Duke (6952), — by Sir Walter (2639), — by Son of Emperor (2671), — by
Wonderful (700), — by Cardinal (841), — by Baronet (61), — by Cleveland
(145), — by Butterfly (104), — by Globe (278).

(43026) **DALTON HERO,**
Roan, calved July 6, 1879, bred by Major E. G. S. Hornby, Dalton Hall; got
by Royal Ensign (37388), dam (Sylvan 18th) by Brutus 2nd (30622), g. d. (Sylvan 12th) by General Napier (26238), gr. g. d. (Sylvan 6th) by Viscount Oxford
(25386), — (Sylvan 3rd) by Golden Eclipse (14625), — Sylvan.

(43027) **DAMSON WINE,**
Roan, calved February 23, 1879, bred by Mr. H. Bettridge, East Hanney; got
by Burgundy (37926), dam (Damascus Girl) by Baron Booth 1st (27915), g. d.
(Damask 2nd) by Artemus Ward (23326), gr. g. d. (Damask) by A 1 (15538),
— (Damsel) by Enterprise (11443), — by Patriot (10595), — by a Son of Elevator (6969), — by No Mistake (8357), — by Young Consul (6893), — by
Fairfax (1023), — by Speculation (1472).

(43028) **DANDELION,**
White, calved October 12, 1878, bred by the Hon. O. Duncombe, Waresley Park,
the property of Mr. D. Hartley, Westerdale; got by Sea King (35488), dam
(Countess of Windsor) by Earl of Windsor (28527), g. d. (Foam) by General
Wetherby (24026), gr. g. d. (Patty 3rd) by Hypocrite (19996), — (Patty 2nd)
by Robinson Crusoe (13610), — by Bridegroom (11203), — by Captain Edwards
(8929), — by Almacks (7779), — by Dan O'Connell (3557), — by Young
Grazier (3928), — by Milton (2314), — by Policy (1329), — by Barmpton (54).

(43029) **DANGER,**
Red and white, calved October 25, 1878, bred by Sir W. Miles, Bart., Leigh
Court, the property of Mr. J. W. Paull, Ilminster; got by Wild Duke of Geneva
(36004), dam (Dauntless 21st) by Meteor (29364), g.d. (Dauntless 17th) by Success (27581), gr.g.d. (Dauntless 14th) by Cormorant (19511), — (Dauntless 3rd)
by Franklin (14568), — by Latimer (9280), — by Magnum Bonum (2243), —
from the stock of Colonel Cradock.

(43030) **DARLINGTON DUKE 3RD,**
Roan, calved May 19, 1879, bred by Mr. G. Fox, Elmhurst Hall; got by Twenty-
fourth Duke of Airdrie (36460), dam (Deepdale) by Second Duke of Tregunter
(26022), g.d. (Darlington 17th) by Grand Duke 11th (21849), gr. g. d. (Darlington
13th) by Marmaduke 2nd (22287), — (Darling) by Autocrat 2nd (19250), — by
Fourth Duke of Oxford (11387), — by Percy (9472), — by Thomas (5471), —
by Eryholme (3736), — by Reformer (4914), — by Young Favourite (3770), —
by Wellington (2825).

(43031) **DARLINGTON DUKE 4TH,**
Red and white, calved March 18, 1879, bred by Mr. J. J. Stone, Stoneleigh
Park, the property of the Executors of Mr. J. J. Stone; got by Stoneleigh Duke
4th (39172), dam (Darlington Duchess 3rd) by Oxford Duke (32022), g. d.
(Prioress) by Grand Duke 7th (19877), gr. g. d. (Princess) by Royal Butterfly
5th (18756), — (Diadem) by Marmaduke (14897), — by Fourth Duke of
Oxford (11387), — by Percy (9472), — by Thomas (5471), — by Eryholme
(3736), — by Reformer (4914), — by Young Favourite (3770), — by Wellington
(2825).

(43032) **DARLINGTON LAD,**
Roan, calved March 26, 1878, bred by Mr. R. Parker, The Tarn, the property of
Mr. J. R. Fox, High House; got by Lord Darlington 8th (34520), dam (Blanche

3rd) by Sultan (30088), g. d. (Blanche) by General Napier (24023), gr. g. d. (Blancheflower) by May Duke 2nd (18372), — (Balm) by Britannicus (17452), — by Magistrate (13274), — by Antinous (12401), — by Diamond (5918), — by Norfolk (2377), — by Belvedere (1706), — by Belvedere (1706), — by Lancaster (360), — by Petrarch (488), — by Major (397), — by Chapman's Son of Punch (122), — by Dickson's Grandson of Punch (213), — by Checks (132), — by R. Grimston's Bull (282), — by J. Coates's Bull (148).

(43033)   DARLINGTON YORK,

Roan, calved February 13, 1879, bred by Mr. S. T. Addison, Ellenhall ; got by Duke of Darlington 5th (41377), dam (Miss Emily 6th) by Eighth Duke of York (28480), g. d. (Miss Emily 2nd) by Earl of Hardwicke (14476), gr. g. d. (Young Celia 2nd) by Lord of the South (13216), — (Young Celia) by Lord of the North (11743), — by Third Duke of Northumberland (3647), — by Bashaw (1692), — by Helmsman (2109), — by Columella (904), — by Regent (544), — by Palatine (478), — by Palmflower (480), — by Patriot (486), — by Driffield (223), — by C. Holmes's Bull (314).

(43034)   DAVID GWYNNE,

Roan, calved November 15, 1878, bred by Mr. H. Caddy, Rougholm ; got by Student (39176), dam (Dottie Gwynne) by Sultan (30088) or Royal Fame (35364), g. d. (Daisy Gwynne) by Knight of Distington (18158), gr. g. d. (Dolly Gwynne) by Duke of York (14461), — (Young Dowager Gwynne) by St. Thomas (10777), — by Prime Minister (2456), — by Wallace (5586), — by Marmion (406), — by Merlin (430), — by Layton (366), — by Phenomenon (491), — by Favourite (252), — by Favourite (252), — by Hubback (319), — by Snowdon's Bull (612), — by Waistell's Bull (669), — by Masterman's Bull (422), — by the Studley Bull (626).

(43035)   DAYLIGHT,

Roan, calved June 8, 1879, bred by Mr. H. Smith, Mountmellick ; got by Earl of Aylesby (38215), dam (Daybreak 4th) by Conqueror (30786), g. d. (Daybreak 1st) by Lord Claud (29086), gr. g. d. (Daybreak) by British Duke (19350), — (Warbler) by Knight of Windsor (16350), — by Emperor (12835), — by Barley Sugar (11140), — by Prince Ernest (7366), — by Captain (11241), — by The Doctor (2748), — by Brough (3228), — by Pilot (496).

(43036)   DEAN OF JERSEY,

Red, calved February 6, 1879, bred by Mr. J. Singleton, Teresa Cottage ; got by Duke of Oxford 31st (33713), dam (Lady Villiers) by Earl of the Valley (31086), g. d. (Jessica) by Duke of Waterloo (21616), gr. g. d. (Julia) by Ranter (18666), — (Lady Jersey) by Seventh Duke of York (17754), — by Duke of Glo'ster (11382), — by Lord Warden (7167), — by Fawsley (6004), — by Warden (5595), — by Javelin (4093), — by Blyth (797), — by Wellington (684), — by Phenomenon (491), — by Favourite (252), — by Favourite (252), — by Favourite (252), — by Hubback (319), — by Snowdon's Bull (612), — by Waistell's Bull (669), — by Masterman's Bull (422), — by the Studley Bull (626).

(43037)   DEEPDALE,

Roan, calved April 17, 1879, bred by Mr. C. Cradock, Hartforth ; got by Chaser (37970), dam (Drip Drop) by The Don (35752), g. d. (Strawberry Drop) by

Strawberry Prince (25240), gr. g. d. (Raindrop) by Dairy Prince (17655), — (Eardrop) by Apollo (9899), — by Richmond (13592), — by Free Trader (10246), — by Gainford (2044), — by Magnum Bonum (2243), — by Rob Roy (557), — by Son of Houghton (318), — by Sir Stephen (1456), — by Sedbury (1424).

(43038)   DEFENDER,
Red and white, calved September 2, 1879, bred by Mr. R. C. Morton, Lane House; got by Duke of Havering (33664), dam (Dowager Duchess 4th) by Grand Duke 11th (21849), g. d. (Dowager) by Duke of Geneva (19614), gr. g. d. (Duchess 1st) by Master Rembrandt (16545), — (Duchess Nanny) by Jasper (11609), — by Second Duke of Oxford (9046), — by Second Duke of Northumberland (3646), — by Belvedere (1706), — by Son of Second Hubback (2683), — a Cow of Mr. Bates's, of Kirklevington.

(43039)   DENISON,
White, calved February 10, 1876, bred by Lord Chesham, Latimer House, the property of Mr. W. Nicholson, Basing Park; got by Duke of Oxford 28th (33710), dam (Morisco's Blanche) by Morisco (18421), g. d. (May Fly) by Duke of Moscow (14447), gr. g. d. (Constance) by Selim (6454), — (Helena) by Rex (6385), — by Norfolk (2377), — by Belvedere (1706), — by Belvedere (1706), — by Lancaster (360), — by Petrarch (488), — by Major (397), — by Chapman's Son of Punch (122), — by Dickson's Grandson of Punch (213), — by Checks (132), — by R. Grimston's Bull (282), — by J. Coates's Bull (148).

(43040)   DENTSDALE'S BUTTERFLY PRINCE,
Roan, calved December 18, 1878, bred by Mr. G. J. Bell, The Nook, the property of Mr. R. Errington, Scotby Farm; got by Duke of Dentsdale (33616), dam (Butterfly Princess 18th) by Fourteeenth Duke of Oxford (21605), g. d. (Butterfly Princess 9th) by Scaleby Champion (15243), gr. g. d. (Butterfly Princess 8th) by Palmerston (15041), — (Butterfly Princess) by Wellington (7706), — by Bellerophon (3119), — by Blucher (1725), — by Butterwick (3251).

(43041)   DERWENT KING,
Roan, calved November 4, 1878, bred by Mr. J. R. Singleton, Great Givendale, the property of Messrs. J. and E. Tindall, Knapton Hall; got by Duke of Oxford 37th (38171), dam (Duchess of Towneley 5th) by Beau Siddington (30515), g. d. (Duchess of Towneley 4th) by Red Duke (32253), gr. g. d. (Duchess of Towneley 3rd) by Nestor (26957), — (Duchess of Towneley) by Duke of Towneley (21615), — by Le Moor (22090), — by Oxford (15035), — by Puritan (13541), — by Crusade (7938), — by Cotherstone (6903), — by Yellow Boy (5694), — by Red Robin (2492), — by Burley (1766), — by Son of Young Albion (15), — by Sir Marton (1453), — by Palmsun (7311), — by a Bull of Mr. Parrington's.

(43042)   DEVON GWYNNE,
Red and white, calved May 31, 1879, bred by Mr. C. Williams, Pilton House, the property of Mr. H. Humfrey, Kingstone Farm; got by Lord Ashton Wild Eyes (34485), dam (Lady Gwynne) by Duke of Cambridge (25940), g. d. (Jenny Gwynne) by Duke of York (14461), gr. g. d. (Polly Gwynne) by Flying Dutchman (10235), — (Young Dowager Gwynne) by St. Thomas (10777), — by Prime Minister (2456), — by Wallace (5586), — by Marmion (406), — by Merlin

(430), — by Layton (366), — by Phenomenon (491), — by Favourite (252), — by Favourite (252), — by Hubback (319), — by Snowdon's Bull (612), — by Waistell's Bull (669), — by Masterman's Bull (422), — by the Studley Bull (626).

(43043)   DICK MILCOTE,

Red and white, calved January 8, 1879, bred by Mr. J. C. Adkins, Milcote; got by Hotspur (38440), dam (Dorothy) by Cherry Grand Duke 5th (30712), g. d. (Emmeline) by University (27693), gr. g. d. (Eva 2nd) by Liddington (24334), — (Turvey) by Greville (19907), — by Rebec (15132), — by Meteor (10526), — by Phœnix (6290), — by Shakespeare (5108), — by Rival (2534), — by Darlington (956), — by Denton (198), — by Rockingham (560), — by Jobling's Son of Phenomenon (491), — by Phenomenon (491), — by Colonel (152), — by Styford (629).

(43044)   DIDO DUKE,

Roan, calved March 6, 1878, bred by Mr. J. Cross, Chiswell Hall; got by Third Duke of Hillhurst (30975), dam (Empress 6th) by The Earl (23034), g. d. (Empress) by Alfred (6732), gr. g. d. (Iris) by Accordion (5708), — (Purity) by Despot (1915), — by Napoleon (4531), — by Rival (2534), — by Darlington (956), — by Denton (198), — by Rockingham (560), — by Jobling's Son of Phenomenon (491), — by Phenomenon (491), — by Colonel (152), — by Styford (629).

(43045)   DIOGENES,

White, calved April 23, 1878, bred by Captain D. H. Mytton, Garth; got by Geneva Prince (36691), dam (Early Rose) by Richmond (35269), g. d. (Eglantine) by Manager (24521), gr. g. d. (White Rose 2nd) by Victor (19061), — (White Rose) by Cronstadt (14352), — by Autocrat (19249), — by Warrior (12287), — by Sweetmeat (8645), — Old Strawberry (winner of First Prize, County Show, Shrewsbury, 1844) bred by Dr. Rowley.

(43046)   DON FREDERICK,

Roan, calved January 12, 1879, bred by Colonel R. Loyd Lindsay, Lockinge Park; got by Earl of Horton 11th (36588), dam (Duchess of Flamborough) by Duke of Flamborough (25960), g. d. (Duchess of Lancaster 12th) by Lord Stanley (22217), gr. g. d. (Duchess of Lancaster 7th) by Sarsden Clipper (20787), — (Duchess of Lancaster 4th) by Third Duke of Cambridge (15920), — by Vanguard (10994), — by Duke of York (11393), — by Second Duke of Lancaster (5951), — by Crichton (3516), — by Ploughboy (4726).

(43047)   DOROTHY'S GENEVA,

Roan, calved February 11, 1879, bred by Captain Ashby, Naseby Woolleys; got by Earl of Geneva (33794), dam (Dorothy) by Telemachus 3rd (32650), g. d. (Desdemona) by Grand Duke of Wateringbury (26296), gr. g. d. (Daphne) by Comet (19484), — (Daisy) by Mameluke (13289), — by Kirklevington 4th (14775), — by Alfred (6732), — by Accordion (5708), — by Tathwell (5399),— by Belvedere 2nd (3127), — by Bellerophon (3119), — by Barmpton (5774), — by Kitt (7127), — by Kitt (7127), — by Page's Bull (6269), — by Middleton's Bull (438).

(43048)   DORRINGTON,

Red, calved December 27, 1878, bred by Earl Beauchamp, Madresfield Court;

got by Magnum Bonum (41954), dam (Dolly Varden) by Festival (26147), g. d. (Delight) by Headman (24111), gr. g. d. (Lady Cambridge) by Duke of Cambridge (15921), — (Polly Horton) by Columbus (17591), — by Duke William (11400), — by Cotgrave (6901), — by Progress (4839).

(43049) DOWNSHIRE,
Red and white, calved March 22, 1879, bred by Mr. R. P. Maxwell, Finnebrogue; got by Czarowitz 4th (33497), dam (Daphne) by Menotti (37089), g. d. (Delia 2nd) by Prince Victor (20606), gr. g. d. (Delia) by Prince Andrew (18588), — (Minerva) by Musician (13362), — by Billy the Beau (12471), — by Ballinasloe (12429), — by Superb Fairfax (5357), — by Jupiter (20039), — by Maximus (2284), — by Monarch (4494), — by Western Comet (689), — by Primus (4763), — by Son of Favourite (252), — by Cupid (177).

(43050) DRACO,
White, calved September 2, 1879, bred by Earl Fitzwilliam, Coollattin Park; got by King Lud (34327), dam (Dinah) by Robert Burns (29795), g. d. (Dido) by Lord Stanley (24466), gr. g. d. (Diana) by Ben Lawers (23408), — (Duchess) by Clydesdale (15779), — by Sir Charles Napier (13708), — by Baron of Ashby (19282), — by Municipal Bill (2344), — by Planet (2432).

(43051) DRUID,
Red and white, calved January 12, 1879, bred by Mr. H. Smith, Mountmellick; got by Conqueror (30786), dam (Dunlavin 3rd) by Lord Claud (29086), g. d. (Dunlavin 2nd) by Sheet Anchor 2nd (25121), gr. g. d. (Dunlavin) by British Duke (19350), — (Rosette) by Gipsy Boy (14617), — by Emperor (12835), — by Barley Sugar (11140), — by Booth (11195), — by The Doctor (2748), — by Linton (4227), — by Streamer (624).

(43052) DRUMLEANING,
Red and white, calved December 15, 1877, bred by Mr. J. Strong, Culgaith, the property of Mrs. Stanley, Danceing Gate; got by Earl of Doune (36579), dam (Golden Lady Eliza) by Pure Gold (32226), g. d. (Lady Eliza) by Lord of Nunwick (26702), gr. g. d. (Eliza) by Brigadier (33203), — by General Haynau (11520).

(43053) DRYHOPE GEM,
Roan, calved March 6, 1879, bred by Mr. N. Milne, Dryhope, the property of Mr. J. Muir, Dryhope; got by Earl of Kelso (38228), dam (Red Gem) by Thorndale Duke 2nd (32712), g. d. (Gem) by Freebooter (19789), gr. g. d. (Ruby) by Sam Glen (10780), — (Brenda) by White Prince (12298), — by Ethelred (5000), — by One Eye (13423), — by Bachelor (1000), — from the stock of Mr. J. Chrisp, of Doddington.

(43054) DUFFIELD,
Red, calved February 1, 1879, bred by Mr. J. H. Casswell, Laughton; got by Duke of Barrington 5th (33575), dam (Duchess of Glo'ster) by Grand Duke of Thorndale (31297), g. d. (Lady Walton 2nd) by Earl of Glo'ster (21644), gr. g. d. (Lady Walton) by Third Earl of Walton (19673), — (Levity) by Lord Thoresby (14856), — by Horrox (11591), — by Second Duke of Oxford (9046), — by Cleveland Lad (3407), — by Red Rose Bull (2493), — by Rex (1375,) — bred by Mr. Richardson, of Hart.

(43055) YOUNG DUKE,

Red, calved October 27, 1879, bred by Mr. T. Wynne, Lislea ; got by Grand Duke (41641), dam (Harmony) by Prince Victor (20606), g. d. (Music 2nd) by Dillon (17680), gr. g. d. (Music) by Musician (13362), — (Adelaide) by Kossuth (13128), — by Sir John Sinclair (5165), — by Coriander (6895), — by Escape (11446), — by Carter's Falstaff (1993), — by Whisker (2835), — by Hero, — Vesta.

(43056) DUKE ANTHONY,

Roan, calved December 7, 1879, bred by Mr. J. R. Singleton, Great Givendale ; got by Duke of Oxford 37th (38171), dam (Duchess of Towneley 4th) by Red Duke (32253), g. d. (Duchess of Towneley 3rd) by Nestor (26957), gr. g. d. (Duchess of Towneley) by Duke of Towneley (21615), — (Wallflower) by Le Moor (22090), — by Oxford (15035), — by Puritan (13541), — by Crusade (7938), — by Cotherstone (6903), — by Yellow Boy (5694), — by Red Robin (2492), — by Burley (1766), — by Son of Young Albion (15), — by Sir Marton (1453), — by Palmsun (7311), — by a Bull of Mr. Parrington's.

(43057) DUKE CAROLUS,

White, calved June 27, 1877, bred by Mr. J. Rigg, Wrotham Hill Park ; got by Duke of Rothesay (36534), dam (Grand Duchess Carolina 2nd) by Eighth Duke of Geneva (28390), g. d. (Grand Duchess Carolina) by Grand Duke 10th (21848), gr. g. d. (Carolina 5th) by Seventh Duke of York (17754), — (Carolina 2nd) by Douglas (12714), — by Second Cleveland Lad (3408), — by Short Tail (2621), — by Son of Second Hubback (2683), — Craggs, bought of Mr. Bates, and descended from the stock of Mr. Maynard, of Eryholme.

(43058) DUKE CHARMING,

Roan, calved November 22, 1879, bred by Mr. S. Hughes, The Elms, Stoke Bishop ; got by Duke of Rothesay (36534), dam (Charming Duchess 4th) by Seventeenth Duke of Oxford (25994), g. d. (Twin Duchess 3rd) by Knightley (22051), gr. g. d. (Twin Duchess) by Fourth Duke of Thorndale (17750), — (Charming) by Mameluke (13289), — by Cardinal (11246), — by White Friar (9827), — by Little John (4232), — by Caliph (1774), — by Sir Walter (2637), — by Hotspur (1117), — by Coxcomb (928), — by Midas (435), — by Comet (155), — by R. Colling's Son of Favourite (252), — by same Son of Favourite (252), — by Hubback (319).

(43059) DUKE CHERRY,

Red, calved November 8, 1879, bred by Mr. H. Lovatt, Low Hill ; got by Baron Oxford 5th (27958), dam (Cherry Duchess) by Second Duke of Glo'ster (28392), g. d. (Southwick Cherry Flower) by Red Bates (20634), gr. g. d. (Cherry Bloom) by Heir-at-Law (13005), — (Cherry Blossom) by Roland (2556), — by Pirate (2430), — by Houghton (318), — by Marshal Blucher (416), — bred from the stock of Messrs. Wright and Charge.

(43060) DUKE CRAGGS,

Red, calved May 16, 1879, bred by Sir W. H. Salt, Bart., Maplewell, the property of Mr. R. Taylor, Sigglesthorne Manor; got by Fifth Duke of Glo'ster (36494), dam (Grand Duchess Carolina 3rd) by Duke Lally (36457), g. d. (Grand Duchess Carolina 2nd) by Eighth Duke of Geneva (28390), gr. g. d. (Grand Duchess Carolina) by Grand Duke 10th (21848), — (Carolina 5th) by Seventh Duke of

York (17754), — by Douglas (12714), — by Second Cleveland Lad (3408), — by
Short Tail (2621), — by Son of Second Hubback (2683), — Craggs, bought of
Mr. Bates, and descended from the stock of Mr. Maynard, of Eryholme.

(43061) **DUKE GWYNNE 2ND,**
Roan, calved May 13, 1879, bred by Mr. C. Sharpley, Kelstern Hall, the property
of Mr. L. W. Arkwright, Parndon Hall; got by Grand Duke 27th (34067), dam
(Lady Ann Gwynne) by Friponnier (26208), g. d. (Polly Gwynne 3rd) by Duke
of Cumberland (21584), gr. g. d. (Polly Gwynne) by Flying Dutchman (10235),
— (Young Dowager Gwynne) by St. Thomas (10777), — by Prime Minister
(2456), — by Wallace (5586), — by Marmion (406), — by Merlin (430), — by
Layton (366), — by Phenomenon (491), — by Favourite (252), — by Favourite
(252), — by Hubback (319), — by Snowdon's Bull (612), — by Waistell's Bull
(669), — by Masterman's Bull (422), — by the Studley Bull (626).

(43062) **DUKE HILDA,**
White, calved December 2, 1879, bred by Mr. C. Cradock, Hartforth; got by
Chaser (37970), dam (Mary) by Chatsworth (23546), g. d. (Marigold) by
Maximilian (20322), gr. g. d. (Beauty) by Althorpe (15580), — (Bella) by The
Buck (13836), — by The Beau (12182), — by Duke of Cornwall (5947), — by
Prince Ernest (4818), — by Wallace (5586), — by Wellington (2824), — by
Marmion (406), — by Merlin (430), — by Layton (366), — by Phenomenon
(491), — by Favourite (252), — by Favourite (252), — by Hubback (319), —
by Snowdon's Bull (612), — by Waistell's Bull (669), — by Masterman's Bull
(422), — by the Studley Bull (626).

(43063) **DUKE OF ABERCORN 2ND,**
Red and white, calved March 30, 1879, bred by Mr. M. Kearney, The Ford; got
by Duke of Abercorn (39708), dam (Valentine 5th) by Prince Imperial (29632),
g. d. (Valentine 2nd) by New Year's Gift (24657), gr. g. d. (Valentine 1st) by
Blair Athol (21286), — (Valentine) by Seignor (18806), — by Welcome Guest
(15497), — by Payment (11891), — by Old Tommy (10581), — by Zadig (8796),
— by Prince Albert (4793), — by Sultan Selim (2710), — by Prince Edward
(2462), — by Wellington (683), — by North Star (458).

(43064) **DUKE OF AUDLEY,**
Red and white, calved September 21, 1879, bred by Mr. E. Lythall, Radford
Hall; got by Duke of Rosedale 3rd (33723), dam (Fancy) by Thorndale Duke
(27661), g. d. (Memory) by Fourth Duke of Geneva (25964), gr. g. d. (Memento)
by Old Buck (15017), — (Gundreda) by Earl of Dublin (10178), — by Janizary
(8175), — by Snowball (8602), — by Caliph (1774), — by Norman (2379),
— by White Boy (1580), — by Wyville's Bull, — by a Bull bred by Mr.
Charge.

(43065) **DUKE OF BARRINGTON,**
Red, calved November 3, 1879, bred by Sir W. H. Salt, Bart., Maplewell, the
property of Mr. G. Ward, Nuneaton Fields; got by Fifth Duke of Glo'ster
(36494), dam (Marchioness of Barrington) by Grand Duke 22nd (34062), g. d.
(Grand Duchess of Barringtonia) by Eighteenth Duke of Oxford (25995), gr. g. d.
(Grand Duchess of Barrington) by Grand Duke 7th (19877), — (Countess of
Barrington 2nd) by Ninth Duke of Oxford (17738), — by Grand Duke 3rd
(16182), — by Grand Turk (12969), — by Earl of Derby (10177), — by Earl

of Liverpool (9061), — by Second Duke of Cambridge (3638), — by Belvedere
(1706), — by·Son of Herdsman (304), — by Wonderful (700), — by Alfred
(23), — by Young Favourite (6994).

(43066)   DUKE OF BARRINGTON 10TH,

Roan, calved April 6, 1879, bred by the Duke of Devonshire, Holker Hall; got
by Fifth Duke of Wetherby (31033), dam (Countess ¦of Barrington 5th) by
Grand Duke 10th (21848), g. d. (Countess of Barrington 4th) by Lord Oxford
(20214), gr. g. d. (Countess of Barrington) by Grand Duke 3rd (16182), —
(Laurel) by Grand Turk (12969), — by Earl of Derby (10177), — by Earl of
Liverpool (9061), — by Second Duke of Cambridge (3638), — by Belvedere
(1706), — by Son of Herdsman (304), — by Wonderful (700), — by Alfred
(23), — by Young Favourite (6994).

(43067)   DUKE OF BARRINGTON 10TH,

Red and white, calved August 22, 1879, bred by Mr. H. J. Sheldon, Brailes
House; got by Duke of Rothesay (36534), dam (Duchess of Barrington) by
Duke of Connaught (33604), g. d. (Countess of Barrington 4th) by Lord Oxford
(20214), gr. g. d. (Countess of Barrington) by Grand Duke 3rd (16182), —
(Laurel) by Grand Turk (12969), — by Earl of Derby (10177), — by Earl of
Liverpool (9061), — by Second Duke of Cambridge (3638), — by Belvedere
(1706), — by Son of Herdsman (304), — by Wonderful (700), — by Alfred
(23), — by Young Favourite (6994).

(43068)   DUKE OF BEAUFORT,

Roan, calved November 10, 1879, bred by Mr. W. Bliss, Chipping Norton, the
property of Count Renard, Castle Strehlitz, Upper Silesia; got by Duke of Oxford
36th (38170), dam (Gwynne Princess 8th) by Oxford le Grand (29496), g. d.
(Gwynne Princess 2nd) by Grand Monarch (28774), gr. g. d. (Danee) by King of
the Roses (22043), — (Dolly) by Sir James the Rose (15290), — by The Beau
(12182), — by Duke of Cornwall (5947), — by Prince Ernest (4818), — by
Wallace (5586), — by Wellington (2824), — by Marmion (406), — by Merlin
(430), — by Layton (366), — by Phenomenon (491), — by Favourite (252), —
by Favourite (252), — by Hubback (319), — by Snowdon's Bull (612), — by
Waistell's Bull (669), — by Masterman's Bull (422), — by the Studley Bull
(626).

(43069)   DUKE OF BEDFORD,

Roan, calved November 27, 1878, bred by Mr. J. Briscoe, Hill Croome; got by
Duke of St. Johns 2nd (36539), dam (Nemophila) by Superior (25255), g. d.
(Nosegay Girl) by Plum of Windsor (18556), gr. g. d. (Nosegay) by Bulwark
(14209), — (North 11th) by Koh-i-noor (11642), — by Jacobin (16307), — by
Augustus (3057), — by Ronald (5005), — by Sir Thomas (2636), — by
Francisco (2032), — by Wellington (683), — by Phenomenon (491), — by
Favourite (252).

(43070)   DUKE OF BORDEAUX,

Roan, calved January 9, 1879, bred by Colonel R. Loyd Lindsay, Lockinge
Park; got by Earl of Horton 11th (36588), dam (Virginian Creeper) by Lord
Claret (24376), g. d. (Vesper) by His Majesty (19965), gr. g. d. (Vesta) by His
Highness (14708), — (Venus) by Humphrey Clinker (13055), — by The Red
Duke (8694), — by Hero of the West (8150), — by Lottery (4280), — by Lot-
tery (4280), — by Phœnix (6290).

(43071)   DUKE OF BRAILES 10TH,
Red, calved March 25, 1879, bred by Mr. H. J. Sheldon, Brailes House; got by
Duke of Rothesay (36534), dam (Africa) by Duke of Brailes (23724), g. d.
(America) by Marmaduke (14897), gr. g. d. (Asia) by Grand Duke 2nd (12961),
— (Apricot) by Fusilier (11499), — by Third Duke of York (10166), — by
Second Cleveland Lad (3408), — by Duke of Cleveland (1937), — by Belvedere
(1706), — a Cow of Mr. Bates's, of Kirklevington.

(43072)   DUKE OF BURTON,
Roan, calved November 9, 1879, bred by Major E. G. S. Hornby, Dalton Hall;
got by Royal Ensign (37388), dam (Sylvan 22nd) by Oxford Cherry Duke 2nd
(34971), g. d. (Sylvan 13th) by General Napier (26238), gr. g. d. (Sylvan 8th)
by Viscount Oxford (25386), — (Sylvan 3rd) by Golden Eclipse (14625), —
Sylvan.

(43073)   DUKE OF CAMBRIDGE,
Red and white, calved March 15, 1879, bred by Sir R. C. Musgrave, Bart.,
Eden Hall, the property of Mr. Jonathan Musgrave, Lorton; got by Royal
Cambridge 4th (40624), dam (Diamond 6th) by Duke of Kirklevington (33683),
g. d. (Diamond 5th) by Ignoramus (28887), gr. g. d. (Diamond 2nd) by British
Cherry (23461), — (Diana) by Baron of Ravensworth (17380), — by Lofty
(13168), — by Captain Shaftoe (6833), — by Thalberg (5416), — by Gainford
(2044).

(43074)   DUKE OF CAMBRIDGE,
Red, calved May 8, 1879, bred by Mr. W. Bliss, Chipping Norton; got by Elm-
hurst Prince (41503), dam (Jemima Foggathorpe) by Lord Chief Justice (34507),
g. d. (Oxford Foggathorpe) by Eleventh Duke of Oxford (19632), gr. g. d. (Mrs.
Fagan) by Sir Rainald (27485), — (Gross Lind) by Trafalgar (23086), — by
Baron Lind (17374), — by Norman (13394), — by Third Duke of Oxford (9047),
— by Sheldon (8557), — by Benjamin (1710), — by Malbro' (1189), — by
Ebor (997), — by Regent (546), — by North Star (459), — by R. Colling's
White Bull (151), — bred by Mr. R. Colling.

(43075)   DUKE OF CAROLINA 5TH,
Red and white, calved December 4, 1879, bred by Mr. D. McIntosh, Havering
Park; got by Duke of Havering (33664), dam (Duchess Carolina 2nd) by Third
Duke of Geneva (23753), g. d. (Caroline 5th) by Seventh Duke of York (17754),
gr. g. d. (Carolina 2nd) by Douglas (12714), — (Caroline) by Second Cleveland
Lad (3408), — by Short Tail (2621), — by Son of Second Hubback (2683), —
— Craggs, bought of Mr. Bates, and descended brom the stock of Mr. Maynard,
of Eryholme.

(43076)   DUKE OF CERISIA 4TH,
Red and white, calved July 5, 1879, bred by Mr. H. J. Sheldon, Brailes House;
got by Duke of Rothesay (36534), dam (Cherry Duchess 22nd) by Grand Duke
11th (21849), g. d. (Cherry Duchess 11th) by Duke of Geneva (19614), gr. g. d.
(Cherry Duchess 6th) by Grand Duke 3rd (16182), — (Cherry Duchess 3rd) by
Grand Duke 2nd (12961), — by Grand Duke (10284), — by Sheldon (8557),
— by The Colonel (5428), — by Thorp (2757), — by Pirate (2430), — by
Houghton (318), — by Marshal Blucher (416), — from the stock of Messrs.
Wright and Charge.

(43077) DUKE OF CHARMING LAND 17th,

Roan, calved July 9, 1879, bred by Mr. H. J. Sheldon, Brailes House ; got by Sixth Duke of Oneida (30997), dam (Countess 6th) by Sixth Duke of Oneida (30997), g. d. (Countess 4th) by Grand Duke of Kent (26289), gr. g. d. (Chorus) by Fourth Duke of Thorndale (17750), — (Charming) by Mameluke (13289), — by Cardinal (11246), — by White Friar (9827), — by Little John (4232), — — by Caliph (1774), — by Sir Walter (2637), — by Hotspur (1117), — by Coxcomb (928), — by Midas (435), — by Comet (155), — by R. Colling's Son of Favourite (252), — by same Son of Favourite (252), — by Hubback (319).

(43078) DUKE OF CLAPHAM,

Red and white, calved April 28, 1879, bred by Mr. G. Ashburner, Low Hall, the property of Mr. R. Simpson, Lawkland, Clapham, Yorkshire; got by Duke of Oxford 41st (38174), dam (Oxford's Cherry) by Cherry Duke of Lightburne (36349), g. d. (Oxford's Duchess) by Grand Duke 9th (19879), gr. g. d. (Christmas Oxford) by Fifteenth Duke of Oxford (23776), — (Countess of Oxford) by Tenth Duke of Oxford (17739), — by Jasper (14737), — by Zadoc (11070), — by Hector (6063), — by Favourite (3772), — by Punch (4849), — by Anthony, — by Young Western Comet (1575), — by Son of Layton (366), — by Layton (366), — by Simon (590).

(43079) DUKE OF CLEVELAND,

Red, calved October 16, 1878, bred by Mr. J. Robinson, Broughton Pastures ; got by Thorndale Grand Duke (37593), dam (Rose of Raby 3rd) ¦by Second Duke of Glo'ster (28392), g. d. (Maid of Lorn) by Sixth Duke of Airdrie (19602), gr. g. d. (Rose of Raby) by Lumley (16478), — (Peace) by Grand Duke 2nd (12961), — by Second Cleveland Lad (3408), — by Fourth Duke of Northumberland (3649), — by Second Earl of Darlington (1945), — a Cow of Mr. Bates's, of Kirklevington.

(43080) DUKE OF CLEVELAND 3rd,

Roan, calved February 15, 1879, bred by Messrs. F. Leney and Sons, Wateringbury ; got by Sixth Duke of Oneida (30997), dam (Acomb) by Eighth Duke of Geneva (28390), g. d. (Arethusa) by Duke of Brailes (23724), gr. g. d. (Antoinette) by Fourth Duke of Thorndale (17750), — (America) by Marmaduke (14897), — by Grand Duke 2nd (12961), — by Fusileer (11499), — by Third Duke of York (10166), — by Second Cleveland Lad (3408), — by Duke of Cleveland (1937), — by Belvedere (1706), — a Cow of Mr. Bates's, of Kirklevington.

(43081) DUKE OF CLEVELAND 4th,

Red and white, calved April 25, 1879, bred by Messrs. F. Leney and Sons, Wateringbury ; got by Sixth Duke of Oneida (30997), dam (Arethusa) by Duke of Brailes (23724), g. d. (Antoinette) by Fourth Duke of Thorndale (17750), gr. g. d. (America) by Marmaduke (14897), — (Asia) by Grand Duke 2nd (12961), — by Fusileer (11499), — by Third Duke of York (10166), — by Second Cleveland Lad (3408), — by Duke of Cleveland (1937), — by Belvedere (1706), — a Cow of Mr. Bates's, of Kirklevington.

(43082) DUKE OF CORNWALL 2nd,

Red and white, calved March 12, 1879, bred by the Earl of Dunmore, Dunmore, the property of Sir Curtis M. Lampson, Bart., Rowfant ; got by Marquis of Oxford 2nd (37055), dam (Duchess 114th) by Sixth Duke of Geneva (30959), g. d. (Duchess 97th) by Third Duke of Wharfdale (21619), gr. g. d. (Duchess

92nd) by Fourth Duke of Oxford (11387), — (Duchess 84th) by Archduke
(14099), — by Fourth Duke of Oxford (11387), — by Usurer (9763), — by
Second Duke of Oxford (9046), — by Second Duke of Northumberland (3646),
— by Cleveland Lad (3407), — by Belvedere (1706), — by Second Hubback
(1423), — by Second Hubback (1423), — by The Earl (646), — by Ketton 2nd
(710), — by Comet (155), — by Favourite (252), — by Daisy Bull (186), —
by Favourite (252), — by Hubback (319), — by J. Brown's Red Bull (97).

(43083)   DUKE OF CUMBERLAND 3RD,
Red and white, calved February 5, 1879, bred by Messrs. Dalton, Cummersdale; got by Duke of Siddington (38182), dam (Certainty Duchess) by Royal
Cumberland (27358), g. d. (Thorndale's Grand Duchess) by Grand Duke 6th
(19876), gr. g. d. (Balcony) by Second Duke of Thorndale (17748), — (Berrington Blossom) by Sir Roger (16991), — by May Duke (13320), — by Sir Walter
2nd (10834), — by Brawith Boy (7846), — by Sir Walter (2639), — by Roseberry (567), — by Roseberry (567), — by Constitution (166), — by Hastings
(293), — by Constellation (163).

(43084)   DUKE OF CUMBERLAND 4TH,
Red and white, calved December 22, 1879, bred by Messrs. Dalton, Cummersdale; got by Duke of Siddington (38182), dam (Certainty Duchess) by Royal
Cumberland (27358), g. d. (Thorndale's Grand Duchess) by Grand Duke 6th
(19876), gr. g. d. (Balcony) by Second Duke of Thorndale (17748), — (Berrington Blossom) by Sir Roger (16991), — by May Duke (13320), — by Sir Walter
2nd (10834), — by Brawith Boy (7846), — by Sir Walter (2639), — by Roseberry (567), — by Roseberry (567), — by Constitution (166), — by Hastings
(293), — by Constellation (163).

(43085)   DUKE OF DARLINGTON 6TH,
Roan, calved July 12, 1878, bred by Mr. H. J. Sheldon, Brailes House, the property of Mr. C. Hobbs, Maisey Hampton; got by Duke of Rothesay (36534),
dam (Countess of Darlington) by Duke of Brailes (23724), g. d. (Darlington
12th) by Duke of Geneva (19614), gr. g. d. (Darlington 9th) by Marmaduke
(14897), — (Darlington 8th) by Fourth Duke of Oxford (11387), — by Sir
Hugh (12082), — by Percy (9472), — by Thomas (5471), — by Eryholme (3736),
— by Reformer (4914), — by Young Favourite (3770), — by Wellington (2825).

(43086)   DUKE OF DARLINGTON 7TH,
Red and white, calved July 10, 1879, bred by Mr. H. J. Sheldon, Brailes House,
the property of Messrs. Sturgeon and Sons, South Ockendon Hall; got by Duke
of Rothesay (36534), dam (Countess of Darlington) by Duke of Brailes (23724),
g. d. (Darlington 12th) by Duke of Geneva (19614), gr. g. d. (Darlington 9th)
by Marmaduke (14897), — (Darlington 8th) by Fourth Duke of Oxford (11387),
— by Sir Hugh (12082), — by Percy (9472), — by Thomas (5471), — by Eryholme (3736), — by Reformer (4914), — by Young Favourite (3770), — by
Wellington (2825).

(43087)   DUKE OF DEVONSHIRE,
Red, calved October 9, 1879, bred by Mr. G. Garne, Churchill Heath; got by
Grand Duke of Geneva 2nd (31288), dam (Daphne 3rd) by Second Duke of
Rowley (28441), g. d. (Daphne) by Bull's Bay (23490), gr. g. d. (Dorcas) by
Falstaff (28567), — (Damsel) by Oberon (15016), — by Londonderry (13169),
— by Meteor (10526), — by Phœnix (6290), — by Shakespeare (5108), — by

Rival (2534), — by Darlington (956), — by Denton (198), — by Rockingham (560), — by Jobling's Son of Phenomenon (491), — by Phenomenon (491), by Colonel (152), — by Styford (629).

(43088)   DUKE OF EDINBURGH,

Red, calved November 5, 1878, bred by Mr. W. Scott, Glendronach; got by Norseman (34924), dam (Lady Jane 3rd) by Mountain Chief (34878), g. d. (Lady Jane 2nd) by Plunderer (32070), gr. g. d. (Lady Jane) by Valiant 2nd (25353), — (Lady Ann) by Factor (14525), — by Baillie (32985), — by Aberdeen (2903), — by Monitor (2331), — by Reformer (2503), — by Snowdrop (2653), — by Wellington (2824), — by Midas (435), — by Orion (470), — by Traveller (655), — by Bolingbroke (86).

(43089)   DUKE OF ELFORD 4th,

Red and white, calved February 2, 1879, bred by Lieut.-Colonel Webb, Elford House; got by Grand Gwynne 2nd (38384), dam (Duchess of Elford) by Grand Duke of Lightburne (26290), g. d. (Diligent) by Third Duke of Wharfdale (21619), gr. g. d. (Duenna) by Grand Duke 11th (21849), — (Dulcinea) by Duke of Geneva (19614), — by Master Rembrandt (16545), — by Jasper (11609), — by Second Duke of Oxford (9046), — by Second Duke of Northumberland (3646), — by Belvedere (1706), — by Son of Second Hubback (2683), — a Cow of Mr. Bates's, of Kirklevington.

(43090)   DUKE OF ELFORD 5th,

Red and white, calved July 18, 1879, bred by Lieut.-Colonel Webb, Elford House; got by Duke of Barrington 4th (39712), dam (Countess Nancy 2nd) by Grand Duke 23rd (34063), g. d. (Countess Nancy) by Grand Duke 19th (28746), gr. g. d. (Dignity) by Grand Duke 11th (21849), — (Dora) by Duke of Geneva (19614), — by Master Rembrandt (16545), — by Jasper (11609), — by Second Duke of Oxford (9046), — by Second Duke of Northumberland (3646), — by Belvedere (1706), — by Son of Second Hubback (2683), — a Cow of Mr. Bates's, of Kirklevington.

(43091)   DUKE OF ELMHURST 2nd,

Red, calved June 21, 1879, bred by Mr. G. Fox, Elmhurst Hall; got by Duke of Oxford 39th (38173), dam (Duchess of Airdrie 20th) by Tenth Duke of Thorndale (28458), g. d. (Duchess of Airdrie 9th) by Royal Oxford (18774), gr. g. d. (Duchess of Airdrie 4th) by Fordham Duke of Oxford (31181), — (Duchess of Airdrie 2nd) by Second Duke of Athol (11376), — by Second Duke of Oxford (9046), — by Second Cleveland Lad (3408), — by Short Tail (2621), — by Second Hubback (1423), — by Second Earl (1511), — by Marske (418), — by Ketton 1st (709), — by Comet (155), — by Favourite (252), — by Daisy Bull (186), — by Favourite (252), — by Hubback (319), — by J. Brown's Red Bull (97).

(43092)   DUKE OF FARLETON,

Red and white, calved July 14, 1879, bred by Mr. R. C. Morton, Lane House; got by Duke of Havering (33664), dam (Dowager Duchess 5th) by Third Duke of Geneva (23753), g. d. (Dowager Duchess 4th) by Grand Duke 11th (21849), gr. g. d. (Dowager) by Duke of Geneva (19614), — (Duchess 1st) by Master Rembrandt (16545), — by Jasper (11609), — by Second Duke of Oxford (9046), — by Second Duke of Northumberland (3646), — by Belvedere (1706), — by Son of Second Hubback (2683), — a Cow of Mr. Bates's, of Kirklevington.

(43093)   DUKE OF FLANDERS 7TH,

Roan, calved March 30, 1879, bred by Mr. W. R. Bromet, Cocksford; got by Eighteenth Duke of Oxford (25995), dam (Countess of Flanders 2nd) by Third Duke of Wharfdale (21619), g. d. (Fame) by Fourth Duke of Oxford (11387), gr. g. d. (Flora) by The Beau (12182), — (Fair Light) by Liberal (10418), — by Maunby (7223), — by Wellington (5625), — by Superb (5356), — by Fleetham (2028), — by Admiral (5), — by Young Denton (963), — by Young Albion (15).

(43094)   DUKE OF FLANDERS 8TH,

Roan, calved April 5, 1879, bred by Mr. W. R. Bromet, Cocksford; got by Eighteenth Duke of Oxford (25995), dam (Countess of Flanders) by Second Duke of Wetherby (21618), g. d. (Fame) by Fourth Duke of Oxford (11387), gr. g. d. (Flora) by The Beau (12182), — (Fair Light) by Liberal (10418), — by Maunby (7223), — by Wellington (5625), — by Superb (5356), — by Fleetham (2028), — by Admiral (5), — by Young Denton (963), — by Young Albion (15).

(43095)   DUKE OF FLANDERS 9TH,

Red and white, calved May 30, 1879, bred by Mr. W. R. Bromet, Cocksford; got by Duke of Clarence 5th (38133), dam (Countess of Flanders 6th) by Duke of Tregunter 5th (33743), g. d. (Countess of Flanders) by Second Duke of Wetherby (21618), gr. g. d. (Fame) by Fourth Duke of Oxford (11387), — (Flora) by The Beau (12182), — by Liberal (10418), — by Maunby (7223), — by Wellington (5625), — by Superb (5356), — by Fleetham (2028), — by Admiral (5), — by Young Denton (963), — by Young Albion (15).

(43096)   DUKE OF FLANDERS 10TH,

Roan, calved December 22, 1879, bred by Mr. W. R. Bromet, Cocksford; got by Duke of Clarence 5th (38133), dam (Countess of Flanders 8th) by Duke of Tregunter 5th (33743), g. d. (Countess of Flanders 2nd) by Third Duke of Wharfdale (21619), gr. g. d. (Fame) by Fourth Duke of Oxford (11387), — (Flora) by The Beau (12182), — by Liberal (10418), — by Maunby (7223), — by Wellington (5625), — by Superb (5356), — by Fleetham (2028), — by Admiral (5), — by Young Denton (963), — by Young Albion (15).

(43097)   DUKE OF FLORENCE,

Red and white, calved April 2, 1879, bred by Mr. J. Singleton, Teresa Cottage; got by Lightburne's Duke of Oxford 2nd (38564), dam (Lady Florence 4th) by Baron Holker (30459), g. d. (Lady Florence 3rd) by Eighteenth Duke of Oxford (25995), gr. g. d. (Lady Florence) by Duke of Brailes (23724), — (Countess) by Duke of Cambridge (12742), — by Earl of Dublin (10178), — by Grey Friar (9172), — by Fawsley (6004), — by Marcellus (2260), — by Rufus (2576), — by Wellington (683), — by Windsor (698), — by Windsor (698), — by Own Brother to North Star (459).

(43098)   DUKE OF GLO'STER,

Roan, calved March 27, 1879, bred by Mr. W. Bliss, Chipping Norton; got by Duke of Connaught (33604), dam (Kentish Nonsuch) by Grand Duke of Kent (26289), g. d. (Nonsuch 5th) by Charleston (21400), gr. g. d. (Nonsuch 4th) by Lord Oxford (20214), — (Nonsuch 3rd) by Ninth Duke of Oxford (17738), — by Earl of Warwick (11412), — by Duke of Cornwall (5947), — by Velocipede (5552), — by Sir Thomas (2636), — by Frederick (1060), — by Harold (291), — by North Star (459), — by Favourite (252), — by Favourite (252), — by Favourite (252), — by Broken Horn (95), — bred by Mr. Best, of Manfield.

(43099)   DUKE OF GOLDCLIFF,

Red and white, calved May 22, 1879, bred by Mr. W. Price, Goldcliff; got by
Prince of Goldcliff (43827), dam (Rosebud 2nd) by Master Broomstick (34803),
g. d. (Rosebud) by Lord Darlington 2nd (29096), gr. g. d. (Rose Blossom) by
Duke of Liverpool (23766), — (Rose Leaf) by Duke of Cambridge (15921), —
by Macdonald (13269), — by Robin Hood (4971), — by Selim (2609), — by
The Squire (1512), — by Sovereign (10861).

(43100)   DUKE OF GRAFTON 13TH,

Roan, calved February 3, 1879, bred by Lord Penrhyn, Penrhyn Castle; got by
Grand Duke of Oxford (31293), dam (Grand Duchess of Oxford 3rd) by Thorndale Duke (27661), g. d. (Grand Duchess of Oxford) by Grand Duke 7th (19877),
gr. g. d. (Duchess of Oxford) by Rembrandt (13587), — (Duchess Nanny) by
Jasper (11609), — by Second Duke of Oxford (9046), — by Second Duke of
Northumberland (3646), — by Belvedere (1706), — by Son of Second Hubback
(2683), — a Cow of Mr. Bates's, of Kirklevington.

(43101)   DUKE OF GUNTERSTONE,

Red, calved April 4, 1879, bred by Colonel Gunter, Wetherby Grange; got by
Oxford Cherry Duke 2nd (34972), dam (Mild Eyes 8th) by Fifth Duke of
Wetherby (31033), g. d. (Mild Eyes 5th) by Third Duke of Wharfdale (21619),
gr. g. d. (Mild Eyes) by Stout (17048), — (Young Red Eyes) by Oxford Duke
(15036), — by Duke of Richmond (7996), — by Second Cleveland Lad (3408),
— by Duke of Northumberland (1940), — by Belvedere (1706), — by Emperor
(1975), — by Wonderful (700), — by Cleveland (145), — by Butterfly (104),
— by Hollon's Bull (313), — by Mowbray's Bull (2342), — by Masterman's
Bull (422), — descended from M. Dobison's stock.

(43102)   DUKE OF HADLEY,

Red and white, calved October 14, 1879, bred by Mr. G. Allen, Knightley Hall,
the property of Mr. G. Henshall, Castle Mead; got by Second Duke of Rowley
(28441), dam (Princess Rose) by Lord Kingscote Rosy (34576), g. d. (Patricia)
by Patrician (24728), gr. g. d. (Princess of Oxford) by Baron Oxford (23375), —
(Princess Alice) by Sixth Duke of Oxford (12765), — by The Baronet (10918),
— by Burgundy (7861), — by Prince Albert (4801), — by Chancellor (1809),
— by Lunesdale (2233), — by Grandson of Major Rudd's Venus, — by Barmpton
(54), — by Comet (155), — by Washington (674), — by Son of Cupid (177).

(43103)   DUKE OF HAMILTON 3RD,

Red, calved June 1, 1877, bred by Mr. W. Sheraton, Broom House, the property
of Mr. J. D. Owen, Plasyn Grove; got by Lilleshall (34456), dam (Lady
Hamilton 3rd) by Knight of the Garter (34400), g. d. (Lady Hamilton) by
Second Duke of Cumberland (23735), gr. g. d. (Mary Hamilton) by Viceroy
(21019), — (Mary Seaton) by Bannerman (19258), — by Hickory (14706), —
by King John (14763), — by Hero of the West (8150), — by Bourton Hero
(9983), — by Killerby (7124), — by Beggarman (3118), — by Samson (5081),
— by Remus (6381), — by Premier (2449).

(43104)   DUKE OF HAZLECOTE 52ND,

Roan, calved April 25, 1877, bred by Colonel Kingscote, Kingscote, the pro-
perty of Mr. G. Blackwell, Hazlecote; got by Duke of Hillhurst (28401), dam
(Honey 38th) by Grand Duke of Clarence (28750), g. d. (Honey 26 A) by Second
Duke of Wetherby (21618), gr. g. d. (Honey Flower) by Grand Duke of Oxford

(16184), — (Honeydew) by Viceroy (13945), — by Oregon (8371), — by Premier (7344), — by Bellerophon (3119), — by Alderman (2976), — by Waterloo (2816), — by Young Wynyard (2859), — by Son of Simon (590), — by Buston's Styford (103).

(43105) **DUKE OF HINTON,**
Roan, calved in October 1873, bred by Captain Blathwayt, Dyrham Park, the property of Mr. G. Anstee, Hinton ; got by Duke of Albany (25931), dam (Rosalie) by Diophantus (19570), g. d. (Rosebud) by Mosquito (14964), gr. g. d. (White Rose) by Viceroy (13944), — (Rosemary) by Hotspur (18079), — by Ned Poins (6241), — by Old Strickland (11870), — by Bishop Bains.

. (43106) **DUKE OF HOLDERNESS 3RD,**
White, calved September 26, 1877, bred by Mr. J. Wood, Humbleton Hall ; got by Baron Wild Eyes 2nd (36214), dam (Duchess Nan) by Oxford's Baronet (29499), g. d. (Duchess 5th) by May Duke (31892), gr. g. d. (Duchess 3rd) by Royal Butterfly 3rd (18754), — (Duchess 2nd) by Master Rembrandt (16545), — by Jasper (11609), — by Second Duke of Oxford (9046), — by Second Duke of Northumberland (3646), — by Belvedere (1706), — by Son of Second Hubback (2683), — a Cow of Mr. Bates's, of Kirklevington.

(43107) **DUKE OF HUMBLETON,**
White, calved August 16, 1878, bred by Mr. J. Wood, Humbleton Hall ; got by Grand Duke 27th (34067), dam (Duchess Nan) by Oxford's Baronet (29499), g. d. (Duchess 5th) by May Duke (31892), gr. g. d. (Duchess 3rd) by Royal Butterfly 3rd (18754), — (Duchess 2nd) by Master Rembrandt (16545), — by Jasper (11609), — by Second Duke of Oxford (9046), — by Second Duke of Northumberland (3646), — by Belvedere (1706), — by Son of Second Hubback (2683), — a Cow of Mr. Bates's, of Kirklevington.

(43108) **DUKE OF HUNTSLAND,**
Red and white, calved June 17, 1879, bred by Sir Curtis M. Lampson, Bart., Rowfant ; got by Duke of Underley 2nd (36551), dam (Siddington 12th) by Second Duke of Tregunter (26022), g. d. (Siddington 2nd) by Fourth Duke of Oxford (11387), gr. g. d. (Kirklevington 7th) by Earl of Derby (10177), — (Kirklevington 4th) by Earl of Liverpool (9061), — by Duke of Northumberland (1940), — by Belvedere (1706), — by Son of Second Hubback (2683), — a Cow of Mr. Bates's, descended from the stock of Mr. Maynard, of Eryholme.

(43109) **DUKE OF INVERNESS,**
Red and white, calved April 5, 1879, bred by Lord Lovat, Beaufort Farm ; got by Duke of Beaufort (38122), dam (Broadhooks 10th) by Champion of England (17526), g. d. (Broadhooks 9th) by Baronet (15614), gr. g. d. (Broadhooks 6th) by The Baron (13833), — (Lady Elizabeth Fairfax) by Prince Edward Fairfax (9506), — by White Bull (5643), — by Young Ladykirk (4170), — by Albion (731), — by Sirius (598), — by Wellington (679), — by Sultan (631), — by Signior (588).

(43110) **DUKE OF LANCASTER,**
Roan, calved September 23, 1877, bred by Mr. C. A. Barnes, Solesbridge, the property of Mr. E. George, Napsbury ; got by Lord of Lochaber 2nd (36983), dam (Duchess of Lancaster 14th) by Grand Duke 11th (21849), g. d. (Duchess of Lancaster 11th) by Duke of Geneva (19614), gr. g. d. (Duchess of Lancaster

5th) by Marmaduke (14897), — (Duchess of Lancaster 3rd) by Duke of York (11393), — by Second Duke of Lancaster (5951), — by Crichton (3516), — by Ploughboy (4726).

(43111)　　　DUKE OF LANCASTER,

Roan, calved March 8, 1879, bred by Mrs. Mace, Sherborne ; got by Geneva's Duke (38349), dam (Miss Lancaster) by Duke Geneva (30911), g. d. (Duchess of Lancaster) by Lord Radnor (29198), gr. g. d. (Duchess of Lancaster 7th) by Sarsden Clipper (20787), — (Duchess of Lancaster 4th) by Third Duke of Cambridge (15920), — by Vanguard (10994), — by Duke of York (11393) — by Second Duke of Lancaster (5951), — by Crichton (3516), — by Ploughboy (4726).

(43112)　　　DUKE OF LEICESTER,

Red and white, calved January 25, 1879, bred by Mr. T. Holford, Papillon Hall ; got by Viscount Oxford (40876), dam (Airdrie Duchess 7th) by Duke of Hillhurst 2nd (39748), g. d. (Airdrie Duchess 2nd) by Fourteenth Duke of Thorndale (28459), gr. g. d. (Duchess of Airdrie 10th) by Royal Oxford (18774), — (Duchess of Airdrie 7th) by Clifton Duke (23580), — by Second Duke of Athol (11376), — by Second Duke of Oxford (9046), — by Second Cleveland Lad (3408), — by Short Tail (2621), — by Second Hubback (1423), — by Second Earl (1511), — by Marske (418), — by Ketton 1st (709), — by Comet (155), — by Favourite (252), — by Daisy Bull (186), — by Favourite (252), — by Hubback (319), — by J. Brown's Red Bull (97).

(43113)　　　DUKE OF LICHFIELD,

Red and white, calved November 27, 1878, bred by Mr. G. Fox, Elmhurst Hall, the property of Colonel Lane, King's Bromley Manor ; got by Twenty-fourth Duke of Airdrie (36460), dam (May Rose 2nd) by Miss Butterfly's Son (34860), g. d. (May Rose) by Airdrie (30365), gr. g. d. (Easter Day) by Pilot (32066), — (Poppy) by Ashland (11122), — by Prince Charles 2nd (32113), — by Shakespeare (12062), — by Reformer (2505), — by Belvedere (1706), — by Second Hubback (1423), — by His Grace (311), — by Yarborough (705), — by Favourite (252), — by Punch (531), — by Foljambe (263), — by Hubback (319).

(43114)　　　DUKE OF MILLOM,

Red, calved May 21, 1879, bred by Mr. G. Ashburner, Low Hall, the property of Mr. J. Harker, Salthouse, Millom ; got by Duke of Glo'ster 7th (39735), dam (Oxford's Blanche) by Duke of Oxford (31004), g. d. (Blanche 5th) by Grand Duke 10th (21848), gr. g. d. (Sylph) by Glo'ster (14619), — (Sylvia) by Childers (10052), — by Selim (6454), — by Rex (6385), — by Norfolk (2377), — by Belvedere (1706), — by Belvedere (1706), — by Lancaster (360), — by Petrarch (488), — by Major (397), — by Chapman's Son of Punch (122), — by Dickson's Grandson of Punch (213), — by Checks (132), — by R. Grimston's Bull (282), — by J. Coates's Bull (148).

(43115)　　　DUKE OF NANTWICH,

Red and white, calved April 1, 1879, bred by Baron W. H. von Schröder, The Rookery ; got by Friar of Knowlmere (33975), dam (Acacia) by Cæsar Augustus (25704), g. d. (Acorn) by The Czar (20947), gr. g. d. (Oakleaf) by The Baron (13833), — (Aroma) by Matadore (11800), — by Fitz Adolphus Fairfax (9124), — by Fitz Leonard (7010), — by Sovereign (7539), — Queen, bred by Mr. Robertson, of Ladykirk.

(43116)   DUKE OF NENAGH,

Red and white, calved March 7, 1879, bred by Mr. G. Bolton, 16 Merrion Square North, Dublin ; got by Duke of Underley (33745), dam (Honey Bell 3rd) by Baron Underley (30490), g. d. (Honey Bell 2nd) by Third Duke of Clarence (23727), gr. g. d. (Honey Bell) by Grand Duke of Oxford (16184), — (Helen) by Oregon (8371), — by Premier (7344), — by Bellerophon (3119), — by Alderman (2976), — by Waterloo (2816), — by Young Wynyard (2859), — by Son of Simon (590), — by Buston's Styford (103).

(43117)   DUKE OF NUNWICK,

Roan, calved February 17, 1879, bred by Sir W. C. Trevelyan, Bart., Wallington, the property of Mr. J. Hunter, The Palms ; got by Duke of Oxford 27th (33709), dam (Pearl 14th) by Knight of Killerby (29000), g. d. (Pearl 10th), by British Cherry (23461), gr. g. d. (Pearl 7th) by Lablache 2nd (20092), — (Pearl 6th) by Welcome Guest (15497), — by Euclid (9097), — by Homer (2134), — by Emperor (1974), — by Snowdrop (2653), — by Sir Charles (593), — by St. Albans (1412), — by Cupid (177), — by Simon (590), — by Punch (531), — by Bolingbroke (86).

(43118)   DUKE OF ORLEANS,

Roan, calved February 15, 1879, bred by Colonel R. Loyd Lindsay, Lockinge Park ; got by Earl of Horton 11th (36588), dam (Maid of Orleans) by Duke of Jamaica (23758), g. d. (Roan Duchess) by Glo'ster's Grand Duke (12949), gr. g. d. (Charmer) by Fourth Duke of York (10167), — (Chaplet) by Usurer (9763), — by Duke of Cornwall (5947), — by Morpeth (7254), — by Helicon (2107), — by Henwood (2114), — by Nestor (452), — by Harold (291), — by Meteor (432), — by Comet (155), — by Cupid (177).

(43119)   DUKE OF OXFORD 2ND,

Red and white, calved January 9, 1878, bred by Mr. T. Baker, Harpole, the property of Viscount Clifden, Naseby Farm ; got by Ducal Knight (33552), dam (Duchess of Oxford 8th) by Garibaldi 4th (24009), g. d. (Duchess of Oxford 7th) by Royal Master Butterfly (22789), gr. g. d. (Duchess of Oxford 5th) by Hibernian (19958), — (Duchess of Oxford 4th) by Second Duke of Lancaster (14443), — by Duke of Oxford (11386), — by Sweet William (5368), — by William (2840), — by Mercury (2301).

(43120)   DUKE OF OXFORD 2ND,

Roan, calved December 18, 1879, bred by Mr. E. Williamson, Ramsdell Hall ; got by Duke of Oxford (38169), dam (Cherry Bud) by Duke of Albany (25931), g. d. (Cherry) by Monarch (16575), gr. g. d. (Cherry Blossom) by Mosquito (14964), — (Cherry) by Viceroy (13944), — by King Richard (26524), — by Highman, — a Cow purchased at Castlecomb, 1848.

(43121)   DUKE OF OXFORD 50TH,

Red and white, calved February 19, 1879, bred by the Duke of Devonshire, Holker Hall; got by Duke of Glo'ster 7th (39735), dam (Grand Duchess of Oxford 32nd) by Fifth Duke of Wetherby (31033), g. d. (Grand Duchess of Oxford 22nd) by Baron Oxford 4th (25580), gr. g. d. (Grand Duchess of Oxford 6th) by Imperial Oxford (18084), — (Grand Duchess of Oxford 4th) by Grand Duke of Wetherby (17997), — by Fourth Duke of York (10167), — by Second Duke of Northumberland (3646), — by Short Tail (2621), — by Matchem (2281), — by Young Wynyard (2859).

(43122)   DUKE OF OXFORD 51st,

Red and white, calved April 4, 1879, bred by the Duke of Devonshire, Holker
Hall, the property of the Earl of Lathom, Lathom House; got by Fifth Duke of
Wetherby (31033), dam (Grand Duchess of Oxford 14th) by Grand Duke 10th
(21848), g. d. (Grand Duchess of Oxford 7th) by Lord Oxford (20214), gr. g. d.
(Grand Duchess of Oxford) by Grand Duke 3rd (16182), — (Countess of Oxford)
by Earl of Warwick (11412), — by Fourth Duke of York (10167), — by Second
Duke of Northumberland (3646), — by Short Tail (2621), — by Matchem (2281),
— by Young Wynyard (2859).

(43123)   DUKE OF OXFORD 53rd,

Roan, calved November 24, 1879, bred by the Duke of Devonshire, Holker Hall;
got by Duke of Oxford 43rd (39773), dam (Grand Duchess of Oxford 34th) by
Duke of Oxford 27th (33709), g. d. (Grand Duchess of Oxford 26th) by Baron
Oxford 4th (25580), gr. g. d. (Grand Duchess of Oxford 11th) by Grand Duke
10th (21848), — (Grand Duchess of Oxford 5th) by Priam (18567), — by Earl
of Warwick (11412), — by Fourth Duke of York (10167), — by Second Duke of
Northumberland (3646), — by Short Tail (2621), — by Matchem (2281), — by
Young Wynyard (2859).

(43124)   DUKE OF OXFORD 54th,

Roan, calved November 29, 1879, bred by the Duke of Devonshire, Holker Hall;
got by Duke of Oxford 43rd (39773), dam (Grand Duchess of Oxford 39th) by
Baron Oxford 7th (36199), g. d. (Grand Duchess of Oxford 27th) by Baron Oxford 4th (25580), gr. g. d. (Grand Duchess of Oxford 6th) by Imperial Oxford
(18084), — (Grand Duchess of Oxford 4th) by Grand Duke of Wetherby (17997),
— by Fourth Duke of York (10167), — by Second Duke of Northumberland
(3646), — by Short Tail (2621), — by Matchem (2281), — by Young Wynyard
(2859).

(43125)   DUKE OF RAINTON 7th,

Red, calved February 20, 1879, bred by Sir W. C. Trevelyan, Bart., Wallington,
the property of the Hon. R. Baillie Hamilton, Langton; got by Duke of Oxford
27th (33709), dam (Emma 8th) by Oxford Beau 4th (34964), g. d. (Economy)
by Lord Lally 3rd (24408), gr. g. d. (Emilia) by Duke (17700), — (Emma 5th)
by Lord Raglan (13240), — by Emperor (6973), — by Sir Walter (2639), — by
Wharncliffe (2834), — by Kilnshaw (2175), — by Guy (2084), — by Frederick
(1060), — by Pyramus (533), — by Meteor (432).

(43126)   DUKE OF RICHMOND,

Red, calved February 18, 1879, bred by Mr. W. Lambe, Aubourn; got by Duke
of Charming Land 2nd (36477), dam (Rose of Richmond) by Thorndale Lad
(23066), g. d. (Rose of Delhi) by Great Mogul (14651), gr. g. d. (Rose of
Aubourn) by White Knight (14001), — (White Face) by Frederick (10243), —
by Prince Comet (9505), — by Sovereign (18902), — by Young Favourite
(17836), — by Crispin (174).

(43127)   DUKE OF ROTHER,

Roan, calved May 19, 1879, bred by Mr. T. W. Cadman, Ballifield Hall, the
property of Mr. G. Appleyard, Canklow; got by Duke of Sandale 2nd (38180),
dam (Lady Mary) by Cambridge Duke 4th (25706), g. d. (Lady Julia) by Julius
Cæsar (22008), gr. g. d. (Comet's Farewell) by Great Comet (16192), — (Fillpail)
by Sir John (12084), — by The Bonus (10922).

(43128) DUKE OF STRETTON,

Red, calved March 7, 1878, bred by the Rev. H. O. Wilson, Church Stretton, the property of Mr. T. G. Wainwright, Leebotwood; got by Coventry (33463), dam (Lady 5th) by Grand Duke of Lancaster (31292), g. d. (Lady 3rd) by Sir Roger (16991), gr. g. d. (Lady 2nd) by Warlaby (21063), — (Lady) by Magnus Troil (14880), — by Wildman (14008), — by De Grey (11346), — by General Fairfax (11519), — by Liberator (7140), — by Matchem (4422), — by Young Red Rover (4904) or Rockingham (2551), — by Whisker (1579), — by Pilot (496).

(43129) DUKE OF SUSSEX,

White, calved February 18, 1879, bred by Mr. J. P. Clark, North Ferriby; got by Marquis of Oxford (34786), dam (Gazelle 32nd) by Wildfire (32866), g. d. (Gazelle 25th) by Second Duke of Tregunter (26022), gr. g. d. (Gazelle 11th) by Seventh Duke of York (17754), — (Gazelle 2nd) by Earl of Walton (17787), — by Fourth Duke of Oxford (11387), — by Snowstorm (12119), — by Hampden (8129), — by Leo (4208), — by Henwood (2114), — by Sir Stephen (1456), — by Prince of Waterloo (528), — by Mayflower (425), — by a Bull of Mr. Nicholson's, descended from the stock of Mr. J. Brown, of Aldborough.

(43130) DUKE OF SUSSEX 2ND,

Red and white, calved October 13, 1879, bred by Sir Curtis M. Lampson, Bart., Rowfant; got by Duke of Underley 2nd (36551), dam (Duchess of Glo'ster) by Ninth Duke of Geneva (28391), g. d. (Duchess of Airdrie 14th) by Tenth Duke of Thorndale (28458), gr. g. d. (Duchess of Airdrie 6th) by Clifton Duke (23580), — (Duchess of Airdrie 4th) by Fordham Duke of Oxford (31181), — by Second Duke of Athol (11376), — by Second Duke of Oxford (9046), — by Second Cleveland Lad (3408), — by Short Tail (2621), — by Second Hubback (1423), — by Second Earl (1511), — by Marske (418), — by Ketton 1st (709), — by Comet (155), — by Favourite (252), — by Daisy Bull (186), — by Favourite (252), — by Hubback (319), — by J. Brown's Red Bull (97).

(43131) DUKE OF ULSTER 2ND,

Roan, calved February 25, 1878, bred by Mrs. Fawcett, Scaleby Castle, the property of Mr. W. Rhodes, Lundholme Cottage; got by Grand Duke of Kirklevington (34071), dam (Duchess Ursula) by Eighth Duke of York (28480), g. d. (Ursula 27th) by Grand Duke 13th (21850), gr. g. d. (Ursula 16th) by Seventh Duke of York (17754), — (Ursula 12th) by General Canrobert (12927), — by Duke of Glo'ster (11382), — by Usurer (9763), — by Prince Ernest (4818), — by A-la-Mode (725), — by Ratify (2481), — by Wellington (680), — by Favourite (252), — by Punch (531).

(43132) DUKE OF ULVERSTON 2ND,

Roan, calved December 13, 1877, bred by Mr. M. Kennedy, Stone Cross, the property of Mr. F. B. Owen, Deefield; got by Duke of Oxford 34th (36529), dam (Oxford's Blanche) by Duke of Oxford 23rd (31001), g. d. (Blanche Rose 3rd) by General Napier (24023), gr. g. d. (Olga) by Old Buck (15017), — (Czarina) by May Duke (13319), — by Selim (6454), — by Rex (6385), — by Norfolk (2377), — by Belvedere (1706), — by Belvedere (1706), — by Lancaster (360), — by Petrarch (488), — by Major (397), — by Chapman's Son of Punch (122), — by Dickson's Grandson of Punch (213), — by Checks (132), — by R. Grimston's Bull (282), — by J. Coates's Bull (148).

(43133) DUKE OF VELVET EYES,

Red and white, calved October 20, 1879, bred by Mr. R. Loder, Whittlebury; got by Grand Duke of Oxford 3rd (39953), dam (Velvet Eyes) by Duke of Hill-hurst (28401), g. d. (Roguish Eyes) by Second Earl of Walton (19672), gr. g. d. (Red Eyes) by Duke of Richmond (7996), — (Wild Eyes 23rd) by Second Cleveland Lad (3408), — by Duke of Northumberland (1940), — by Belvedere (1706), — by Emperor (1975), — by Wonderful (700), — by Cleveland (145), — by Butterfly (104), — by Hollon's Bull (313), — by Mowbray's Bull (2342), — by Masterman's Bull (422), — descended from M. Dobison's stock.

(43134) DUKE OF VENICE 2ND,

Red, calved March 19, 1877, bred by Earl Spencer, Althorp Park, the property of Mr. G. Smithson, Northfield; got by Duke of Barrington 2nd (30922), dam (Venice) by Grand Duke 4th (19874), g. d. (Ring) by Rex (16833), gr. g. d. (Veil) by Vaulter (15451), — (White Cow) by Dominie (11361), — by Louis Blanc (10474), — by Lord Warden (7167), — by Zenith (5702), — by Orontes (4623), — by William (2840), — by Childers (1824), — by Richard (1376), — by Jupiter (342), — by Charles (127), — by Windsor (698), — by Chilton (136), — by Colonel (152).

(43135) DUKE OF WALLINGTON 7TH,

Roan, calved May 1, 1879, bred by Sir W. C. Trevelyan, Bart., Wallington, the property of Mr. J. Swann, Bedlington; got by Duke of Oxford 27th (33709), dam (Duchess of Wallington 5th) by Oxford Beau 4th (34964), g. d. (Duchess of Wallington 3rd) by Third Lord Wharfdale (26759), gr. g. d. (Duchess of Wallington) by Second Duke of Wharfdale (19649), — (Amity) by Seventh Duke of York (17754), — by Sixth Duke of Oxford (12765), — by Handel (14663), — by Mozart (11830), — by Diamond (5918), — by Albert (2950), — by Stirling (5330), — by Commodore (1858), — by Tathwell Studley (5401), — by Blyth Comet (85).

(43136) DUKE OF WARCOP 2ND,

White, calved December 15, 1879, bred by Captain T. Chamley, Warcop House; got by Grand Duke of Morecambe (36722), dam (Silver Duchess) by Grand Duke of Kent 2nd (28759), g. d. (Golden Duchess 2nd) by Barrington Oxford (25607), gr. g. d. (Golden Duchess) by Golden Duke (19860), — (Czarina) by Grand Duke 2nd (12961), — by Grand Duke (10284), — by Humber (7102), — by Zenith (5702), — by Roman (2561), — by Childers (1824), — by Richard (1376), — by Jupiter (342), — by Charles (127), — by Windsor (698), — by Chilton (136), — by Colonel (152).

(43137) DUKE OF WARWICK,

Roan, calved May 8, 1878, bred by Mr. W. G. Garne, Broadmoor, the property of Mr. M. E. Von Gronow, Prussia; got by Grand Duke of Geneva 2nd (31288), dam (Duchess of Warwick) by Earl of Warwickshire 3rd (28524), g. d. (Butter-fly's Duchess) by Royal Butterfly 20th (25007), gr. g. d. (Delicacy) by The Druid (20948), — (Destiny) by Progression (16770), — by Enterprise (11443), — by Patriot (10595), — by Son of Elevator (6969), — by No Mistake (8357), — by Young Consul (6893), — by Fairfax (1023), — by Speculation (1472).

(43138) DUKE OF WARWICK 2ND,

Red and white, calved November 9, 1878, bred by Mr. W. G. Garne, Broadmoor; got by Emperor (41508) dam (Duchess of Warwick 3rd) by Grand Duke of

Geneva 2nd (31288), g. d. (Duchess of Warwick) by Earl of Warwickshire 3rd (28524), gr. g. d. (Butterfly's Duchess) by Royal Butterfly 20th (25007), — (Delicacy) by The Druid (20948), — by Progression (16770), — by Enterprise (11443), — by Patriot (10595), — by Son of Elevator (6969), — by No Mistake (8357), — by Young Consul (6893), — by Fairfax (1023), — by Speculation (1472).

(43139) DUKE OF WARWICK 3RD,

Red and white, calved April 18, 1879, bred by Mr. W. G. Garne, Broadmoor; got by Sir Robert Frogmore (40719), dam (Duchess of Warwick) by Earl of Warwickshire 3rd (28524), g. d. (Butterfly's Duchess) by Royal Butterfly 20th (25007), gr. g. d. (Delicacy) by The Druid (20948), — (Destiny) by Progression (16770), — by Enterprise (11443), — by Patriot (10595), — by Son of Elevator (6969), — by No Mistake (8357), — by Young Consul (6893), — by Fairfax (1023), — by Speculation (1472).

(43140) DUKE OF WATERLOO 2ND,

Red, calved March 1, 1879, bred by Mr. T. Horsfall, Limefield House, the property of Mr. W. Duerden, Townhouse, Nelson; got by Seventeenth Duke of Oxford (25994), dam (Waterloo 28th) by Prince Bertram (27119), g. d. (Waterloo 22nd) by Speculator (13775), gr. g. d. (Waterloo 18th) by Bosquet (14183), — (Waterloo 15th) by The Hero (10934), — by Third Duke of Oxford (9047), — by Second Cleveland Lad (3408), — by Duke of Northumberland (1940), — by Norfolk (2377), — by Waterloo (2816), — by Waterloo (2816).

(43141) DUKE OF WATERLOO 5TH,

Roan, calved February 9, 1879, bred by Sir W. C. Trevelyan, Bart., Wallington, the property of Sir W. G. Armstrong, Cragside; got by Duke of Oxford 27th (33709), dam (Waterloo 32nd) by Grand Duke of Lightburne 2nd (26291), g. d. (Waterloo 22nd) by Speculator (13775), gr. g. d. (Waterloo 18th) by Bosquet (14183), — (Waterloo 15th) by The Hero (10934), — by Third Duke of Oxford (9047), — by Second Cleveland Lad (3408), — by Duke of Northumberland (1940), — by Norfolk (2377), — by Waterloo (2816), — by Waterloo (2816).

(43142) DUKE OF WELLINGTON,

White, calved April 13, 1875, bred by Mr. G. Angus, Broomley, the property of Mr. R. Thompson, Burnbank; got by Royal Frederick (35366), dam (Strawberry Blossom) by Prince Charlie (29606), g. d. (Countess) by Young Valasco (23103), gr. g. d. (Strawberry) by Benedict (7828), — (Duchess of Wellington) by Brilliant (7851), — by Young Hastings (3988), — by Wallace (5588), — by President (4752), — by Wellington (683), — by Sir Harry (5155), — by Traveller (655), — by Colonel (152), — by Son of Hubback (319).

(43143) DUKE OF WELLINGTON,

Roan, calved January 1, 1879, bred by Captain T. Chamley, Warcop House; got by Duke of Oxford 31st (33713), dam (Duchess of Waterloo) by Earl of Jersey (23838), g. d. (Countess of Waterloo) by Veteran (13941), gr. g. d. (Waterloo 13th) by Third Duke of Oxford (9047), — (Waterloo 9th) by Second Cleveland Lad (3408), — by Duke of Northumberland (1940), — by Norfolk (2377), — by Waterloo (2816), — by Waterloo (2816).

(43144) DUKE OF WELLINGTON 2ND,

Red, calved July 13, 1879, bred by Mr. W. Bliss, Chipping Norton, the property of Count Renard, Castle Strehlitz, Upper Silesia; got by Baron Turncroft Oxford

4th (37822), dam (Duchess of Wortley) by Duke of Wortley (33759), g. d. (Statira 16th) by Thirteenth Duke of Oxford (21604), gr. g. d. (Statira 5th) by May Duke (13320), — (Statira) by Duke of Glo'ster (11328), — by Balco (9918), — by Sir Launcelot (5166), — by Major (4345), — by Ganthorpe (2049), — by Don Juan (1923), — by Shylock (2622).

(43145) **DUKE OF WESTLAND,**
Red and white, calved May 6, 1879, bred by Mr. F. Stratton, Merdon; got by Abelard (39346), dam (Last Rose of Westland) by King James (28971), g. d. (Cushla Machree) by Duke of Leinster (17724), gr. g. d. (Mavourneen) by Red Rover (24923), — (Sheelagh) by Emperor (12835), — by Duke of Bedford (11378), — by Coriander (6895), — by Son of Prince George (2464), — by Matchem (2281).

(43146) **DUKE OF WETHERBY 7TH,**
Red and white, calved December 27, 1879, bred by Colonel Gunter, Wetherby Grange; got by Duke of Underley 3rd (38196), dam (Duchess 111th) by Second Duke of Barrington (30922), g. d. (Duchess 109th) by Second Duke of Claro (21576), gr. g. d. (Duchess 100th) by Third Duke of Wharfdale (21619), — (Duchess 87th) by Seventh Duke of York (17754), — by Grand Duke of Oxford (16184), — by Fourth Duke of Oxford (11387), — by Usurer (9763), — by Second Duke of Oxford (9046), — by Second Duke of Northumberland (3646), — by Cleveland Lad (3407), — by Belvedere (1706), — by Second Hubback (1423), — by Second Hubback (1423), — by The Earl (646), — by Ketton 2nd (710), — by Comet (155), — by Favourite (252), — by Daisy Bull (186), — by Favourite (252), — by Hubback (319), — by J. Brown's Red Bull (97).

(43147) **DUKE OF WINDMORE,**
Red, calved January 12, 1879, bred by Messrs. W. and R. Harker, Kirkby Stephen, the property of Mr. J. Wearmouth, Windmore End; got by Lord of the Manor (38642), dam (Maria 7th) by Sockburn Lad (30024), g. d. (Maria 3rd) by Oxford (20449), gr. g. d. (Young Maria) by Valentine (17161), — (Maria) by Monitor (13350), — by Loyal (6154), — by Eamont (3672), — by Northumberland (1286).

(43148) **DUKE OF WORCESTER 4TH,**
Red, calved December 17, 1875, bred by Mr. T. Holford, Papillon Hall, the property of Mr. T. W. Tatton, Wythenshawe; got by Grand Duke 23rd (34063), dam (Lady Worcester 5th) by Fourth Duke of Geneva (30958), g. d. (Lady Worcester 4th) by Second Duke of Wetherby (21618), gr. g. d. (Lady Worcester) by Charleston (21400), — (Clear Star) by Marton Duke (22307), — by Red Duke (18676), — by Third Duke of York (10166), — by Second Cleveland Lad (3408), — by Duke of Northumberland (1940), — by Belvedere (1706), — by Emperor (1975), — by Wonderful (700), — by Cleveland (145), — by Butterfly (104), — by Hollon's Bull (313), — by Mowbray's Bull (2342), — by Masterman's Bull (422), — descended from M. Dobison's stock.

(43149) **DUKE OF WORCESTER 6TH,**
Roan, calved December 2, 1877, bred by Mr. T. Holford, Papillon Hall, the property of Mrs. Perry Herrick, Beau Manor Park; got by Grand Duke 23rd (34063), dam (Lady Mild Eyes) by Eighth Duke of York (28180), g. d. (Lady Worcester 10th) by Eighth Duke of Geneva (28390), gr. g. d. (Lady Worcester 2nd) by Charleston (21400), — (Clear Star) by Marton Duke (22307), — by Red

Duke (18676), — by Third Duke of York (10166), — by Second Cleveland Lad (3408), — by Duke of Northumberland (1940), — by Belvedere (1706), — by Emperor (1975), — by Wonderful (700), — by Cleveland (145), — by Butterfly (104), — by Hollon's Bull (313), — by Mowbray's Bull (2342), — by Masterman's Bull (422), — descended from M. Dobison's stock.

(43150)   DUKE OF WRAY,
Roan, calved May 7, 1878, bred by Mr. M. Kennedy, Stone Cross, the property of Mr. M. Bowness, Low Wray; got by Wild Boy of the Dale (39315), dam (Miss Blanche Cole) by Second Duke of Glo'ster (28392), g. d. (Blanche 3rd) by Tenth Duke of Oxford (17739), gr. g.` d. (Blanche) by Dundas (17763), — (Sylph) by Glo'ster (14619), — by Childers (10052), — by Selim (6454), — by Rex (6385), — by Norfolk (2377), — by Belvedere (1706), — by Belvedere (1706), — by Lancaster (360), — by Petrarch (488). — by Major (397), — by Chapman's Son of Punch (122), — by Dickson's Grandson of Punch (213), — by Checks (132), — by R. Grimston's Bull (282), — by J. Coates's Bull (148).

(43151)   DUKE ONEIDA,
Red, calved January 19, 1879, bred by Mr. J. Rigg, Wrotham Hill Park, the property of Mr. H. Webb, Streetly Hall; got by Duke of Ormskirk (36526), dam (Waterloo of Oneida) by Sixth Duke of Oneida (30997), g. d. (Waterloo 37th) by Oxford Beau (29485), gr. g. d. (Waterloo 30th) by Third Duke of Wharfdale (21619), — (Waterloo 25th) by Duke of Geneva (19614), — by Red Knight (11976), — by Grand Duke (10284), — by Third Duke of Oxford (9047), — by Second Cleveland Lad (3408), — by Duke of Northumberland (1940), — by Norfolk (2377), — by Waterloo (2816), — by Waterloo (2816).

(43152)   DUKE SOCKBURN,
Red and white, calved December 23, 1878, bred by Mr. S. Shaw, Brooklands, the property of Mr. C. Howgate, Lightcliffe; got by Lord Oxford Sockburn 2nd (38648), dam (Duchess 10th) by Baron Winsome (30498), g. d. (Duchess 9th) by Grand Duke of Oxford (24070), gr. g. d. (Duchess 8th) by Grand Duke of Lancaster (19883), — (Duchess 6th) by Royal Duke (16865), — by Brennus (8902), — by Duke of Norfolk (5952), — by Cleveland Lad (3407), — by Red Highflyer (2488), — by Sir Charles (5146), — by Harry Lorrequer (3985), — by Blucher (84), — by Magnum Bonum (4322), — by Buston's Styford (103), — by Son of T. Wetherell's Bull (690).

(43153)   DUKE WILD EYES,
Roan, calved December 13, 1879, bred by Mr. R. H. Crabb, Baddow Place; got by Duke of Oxford (39770), dam (Wild Eyes A 2) by Oxford's Baron (32030), g. d. (Wild Eyes A 1) by Baron Oxford 3rd (25579), gr. g. d. (Wild Eyes 24th) by Fourth Duke of Oxford (11387), — (Wild Eyes 22nd) by Wild Duke (19148), — by Lord Barrington 1st (13170), — by Second Duke of Oxford (9046), — by Fourth Duke of Northumberland (3649), — by Duke of Northumberland (1940), — by Belvedere (1706), — by Emperor (1975), — by Wonderful (700), — by Cleveland (145), — by Butterfly (104), — by Hollon's Bull (313), — by Mowbray's Bull (2342), — by Masterman's Bull (422), — descended from M. Dobison's stock.

(43154)   DUKE WILD EYES,
Roan, calved May 9, 1879, bred by Sir W. H. Salt, Bart., Maplewell, the property of Mr. W. Murray, Chesterfield, Ontario, Canada; got by Fifth Duke of Glo'ster

(36494), dam (Wild Duchess of Glo'ster) by Third Duke of Glo'ster (33653), g. d. (Wild Oxford) by Lord Oxford 2nd (20215), gr. g. d. (Wild Eyes 24th) by Lord Barrington 3rd (16382), — (Wild Eyes 16th) by Second Duke of Oxford (9046), — by Fourth Duke of Northumberland (3649), — by Duke of Northumberland (1940), — by Belvedere (1706), — by Emperor (1975), — by Wonderful (700), — by Cleveland (145), — by Butterfly (104), — by Hollon's Bull (313), — by Mowbray's Bull (2342), — by Masterman's Bull (422), — descended from M. Dobison's stock.

.(43155)   DUKE WILD EYES 2ND,

Red, calved July 8, 1879, bred by Sir W. H. Salt, Bart., Maplewell, the property of the Earl of Carysfort, Warmington ; got by Fifth Duke of Glo'ster (36494), dam (Lady Worcester 19th) by Sixth Duke of Geneva (30959), g. d. (Lady Worcester 2nd) by Charleston (21400), gr. g. d. (Clear Star) by Marton Duke (22307), — (Bright Star) by Red Duke (18676), — by Third Duke of York (10166), — by Second Cleveland Lad (3408), — by Duke of Northumberland (1940), — by Belvedere (1706), — by Emperor (1975), — by Wonderful (700), — by Cleveland (145), — by Butterfly (104), — by Hollon's Bull (313), — by Mowbray's Bull (2342), — by Masterman's Bull (422), — descended from M. Dobison's stock.

(43156)   DUSTER,

Roan, calved March 3, 1879, bred by Sir Wilfrid Lawson, Bart, Brayton ; got by Waterloo Lad (40894), dam (Daisy) by Cambridge Duke (28120), g. d. (Daisy) by Perth (22513), gr. g. d. (Daisy) by Lord Durham (16410), — (Cherry) by Comus (12626), — by Devereux (11350), — by Sir Charles Napier (10816). ˙

(43157)   DYNAMITE,

Roan, calved November 7, 1878, bred by Mr. J. Whyte, Aldborough ; got by Gunpowder (28801), dam (Phillis 14th) by Royal Broughton (27352), g. d. (Phillis 6th) by General Hopewell (17953), gr. g. d. (Phillis 2nd) by Red Knight (16809), — (Phillis) by Homer (14714), — by Cardigan (12556), — by Young Rufus (13649), — by Constitution (12634), — by Young Comet (1853), — descended from Jolly's Bull (4115).

(43158)   EARL BENEDICT,

Red, calved July 8, 1879, bred by Mr. T. Willis, Carperby ; got by Windsor Benedict (40933), dam (Rose of the Vale) by Windsor Fitz-Windsor (25458), g. d. (Rose of the Valley) by Lord of the Valley (14837), gr. g. d. (Moss Rose) by Gavazzi (11508), — (Moss Rosebud) by Wilberforce (9830), — by Swintonian (9702), — by Young Symmetry (7577), — by Shipton (2620), — by Fitz-Champion (7007), — by Young Snowball (7521).

(43159)   EARL CLARENDON,

Roan, calved September 14, 1879, bred by Mr. D. B. Holborow, Knockdown House ; got by Lord Hillhurst (38614), dam (Miss Clarence) by Prince of Clarence 2nd (29649), g. d. (Clytie) by Duke of Hazlecote (25969), gr. g. d. (Caroline) by Young Favourite (19729), — (Phœbe) by Hero of the West (8150), — by John (4108), — by Liberty (7141), — by Martin (2279), — by Vertumnus (1546), — by Sir Stephen (1456), — by Prince of Waterloo (528), — by Mayflower (425), — by a Bull of Mr. Nicholson's, descended from the stock of Mr. J. Brown, of Aldborough.

(43160)   **EARL CRISPIN,**
Roan, calved March 28, 1879, bred by Mr. W. Dent, Kaber Fold ; got by Earl of
Sheffield (33812), dam (Lady Crispin) by Warlaby (32792), g. d. (Lady Booth
3rd) by Squire Booth (30049), gr. g. d. (Lady Booth) by Lord of the Hills
(18267), — (Mary) by Majestic (13279), — by Harbinger (10297), — by
Buckingham (3239).

(43161)   **EARL GWYNNE,**
White, calved August 31, 1879, bred by Mr. J. A. M. Cope, Drummilly ; got by
Knight of Raby (34393), dam (Nora Gwynne) by Lieutenant General (31600)
g. d. (Nell Gwynne) by Grey Gauntlet (19908), gr. g. d. (Moll Gwynne) by
Equinox (17810), — (Modesty 5th) by Young Usurer (10985), — by Sir
Thomas Fairfax (5196), — by Wallace (5586), — by Wellington (2824), —
by Marmion (406), — by Merlin (430), — by Layton (366), — by Phenomenon
(491), — by Favourite (252), — by Favourite (252), — by Hubback (319), —
by Snowdon's Bull (612), — by Waistell's Bull (669), — by Masterman's Bull
(422), — by the Studley Bull (626).

(43162)   **EARL OF AYLESBY 3RD,**
Red, calved October 8, 1879, bred by Mr. B. St. John Ackers, Prinknash Park ;
got by King Harold (40053), dam (Bright Dowager) by Duke of York (23804),
g. d. (Bright Queen) by Fitz-Clarence (14552), gr. g. d. (Bright Princess) by
British Prince (14197), — (Bright Dawn) by Vanguard (10994), — by Crown
Prince (10087), — by Zadig (8796), — by Auld Robin Grey (6753), — by
James Chrisp's Bull, — by Burley (1766), — by Isaac (1129), — by Pilot (496),
— by Albion (14), — by Lame Bull (359), — by Shipton (587), — by Son of
Suworrow (636), — by Son of Twin Brother to Ben (88), — by Twin Brother
to Ben (660).

(43163)   **EARL OF BEVERLEY 2ND,**
Roan, calved April 29, 1879, bred by Sir J. Swinburne, Bart., Capheaton ; got
by Lord Cockburn (38594), dam (Miss Beverley 22nd) by Duke of Oxford 31st
(33713), g. d. (Miss Beverley 18th) by Beverley Butterfly (28021), gr. g. d.
(Miss Beverley 11th) by Silver Medal (18828), — (Miss Beverley 4th) by
Ivanhoe (14735), — by Disraeli (10125), — by Lord George Bentinck (9317),
— by Second Earl of Beverley (5963), — by Second Cleveland Lad (3408), —
Red Darlington.

(43164)   **EARL OF BOYLE,**
Red and white, calved May 1, 1879, bred by Mr. F. Stratton, Merdon ; got by
Abelard (39346), dam (Cowslip 15th) by Uncle Ned (19020), g. d. (Cowslip 9th)
by Uncle Tom (13913), gr. g. d. (Cowslip 8th) by Consolation (14316), —
(Cowslip 7th) by Gay Lad (12922), — by Earl of Antrim (10174), — by Duke
of Norfolk (5952), — by Waterloo (2816), — by Kitt (7127), — by Kitt (7127),
— by Page's Bull (6269), — by Middleton's Bull (438).

(43165)   **EARL OF CLARENCE,**
Roan, calved January 28, 1878, bred by Mr. W. Leigh Clare, Raby Vale, the
property of Mr. A. Howcroft, Morley Hall ; got by Telemachus 7th (39190),
dam (Red Rose 3rd) by Grand Duke of Lightburne 2nd (26291), g. d. (Red
Rose) by Grand Duke of Lancaster (19883), gr. g. d. (Cambridge Moss Rose) by
Second Duke of Cambridge (12743), — (Moss Rose) by Lord of Brawith

(10465), — by Laudable (9282), — by Grouchy (6051), — by Sir Thomas Fairfax (5196), — by Young Colling (1843), — by Wilkinson's Red Bull (2838), — by Son of Hollings (2131), — by Partner (2409), — by Hollings (2131), — by R. Alcock's Bull (19).

(43166)   EARL OF GLO'STER,
Roan, calved August 19, 1879, bred by Captain Liddon, Acton Lodge; got by Duke of Oxford 20th (28432), dam (Princess Magenta) by Grand Duke of Clarence (28750), g. d. (Princess Maud) by Sir Knight (25157), gr. g. d. (Alexandra) by Prince Henry (18608), — (Beatrice) by Lord Raglan (13222),— by Patriot (10594), — by Sir Frederick (8577), — by Marquis (4386), — by Mynheer (1255), — by Enchanter (244), — by Major (398), — by Windsor (698), — by Favourite (252), — by Punch (531), — by Hubback (319).

(43167)   EARL OF GRAFTON,
Red and white, calved June 18, 1878, bred by Mr. C. Ashton, Delrow, the property of Mr. H. Squires, Bushey; got by Gay Troubadour (38339), dam (Duchess of Grafton 4th) by Young England (31110), g. d. (Duchess of Grafton 2nd) by Fourth Duke of Geneva (25964), gr. g. d. (Duchess of Grafton) by Sir James (16980), — (Duchess of Glo'ster) by Duke of Glo'ster (11382), — by Usurer (9765), — by Dan O'Connell (3557), — by Brutus (1752), — by Frederick (1060), — by Cato (1794), — by Son of Wellington (679), — bred by Mr. Robertson, of Ladykirk.

(43168)   EARL OF KIRKLEVINGTON,
Roan, calved January 25, 1879, bred by Mr. R. P. Davies, Horton, the property of Mr. J. Martin, Hawkshead Hall; got by Earl of Horton 10th (36587), dam (Kirklevington Duchess 23rd) by Oxford's King (34997), g. d. (Kirklevington Duchess 16th) by Second Duke of Glo'ster (28392), gr. g. d. (Duchess of Kent) by Lord Liverpool (22168), — (Kirklevington 14th) by Fourth Duke of Oxford (11387), — by Earl of Derby (10177), — by Earl of Liverpool (9061), — by Duke of Northumberland (1940), — by Belvedere (1706), — by Son of Second Hubback (2683), — a Cow of Mr. Bates's, descended from the stock of Mr. Maynard, of Eryholme.

(43169)   EARL OF LEICESTER 19TH,
Roan, calved February 26, 1879, bred by Mr. E. H. Cheney, Gaddesby Hall, the property of Mr. J. Linton, Buckden Wood; got by Third Duke of Glo'ster (33653), dam (Tube Rose 43rd) by Saladin (35461), g. d. (Tube Rose 40th) by Zanoni (37700), gr. g. d. (Tube Rose 4th) by Wolviston (21125), — (Tube Rose 3rd) by Third Duke of Cambridge (5941), — by Earl of Antrim (10174), — by South Durham (5281), — by Bellerophon (3119), — by Belvedere (1706),— by Waterloo (2816), — by Baron (58), — by Phenomenon (491), — by Favourite (252), — by Favourite (252), — by Favourite (252), — by Hubback (319), — by Snowdon's Bull (612), — by Waistell's Bull (669), — by Masterman's Bull (422), — by the Studley Bull (626).

(43170)   EARL OF MAR,
Red and white, calved March 4, 1879, bred by Messrs. Dalton, Cummersdale; got by Duke of Siddington (38182), dam (York's Matchless 23rd) by Eighth Duke of York (28480), g. d. (Matchless 19th) by Royal Cumberland (27358), gr. g. d. (Matchless 17th) by Fourteenth Duke of Oxford (21605), — (Matchless 10th) by Richard (16834), — by Scaleby Champion (15243), — by Stapleton

(15336), — by Grisdale (14654), — by Mingo (7239), — by Snowball (1465), — by Pizarro (1323).

(43171) **EARL OF MORAY,**
Roan, calved March 4, 1878, bred by Mr. C. MacKessach, Cloves, the property of Mr. D. C. Bruce, Broadland ; got by Fairservice (43208), dam (Lady Alice) by Lord Charles (31634), g. d. (Lavinia Alice) by Duke of Bedford (23722), gr. g. d. (Alice Lisle) by Signet Seal (18824), — (Princess Alice) by Lord Elcho (22150), — by Western Star (25430), — by Hareshaw (26339), — by Jacob (6101), — by Proselyte (4842), — by Othello (4631), — by Juniper (1145), — by Mentor (426), — by Lancaster (360), — by Ketton (346), — by George (274).

(43172) **EARL OF NANTWICH,**
Red and white, calved December 31, 1879, bred by Baron W. H. von Schröder, The Rookery ; got by Sultan's Fame (35699), dam (Miss Goldie 2nd) by Baron of Knowlmere (30474), g. d. (Miss Goldie) by Golden Era (31266), gr. g. d. (Miss Eyre) by British Hope (21324), — (Currer Belle) by Royal Alfred (1748), — by First Fruits (16048), — by Lyon Playfair (11755), — by Bernardo (8885), — by Bellingham (6786), — by Sol (2655), — by Harlsey (2091), — by Maximus (2284).

(43173) **EARL OF NUNWICK,**
Red, calved October 28, 1879, bred by Mr. T. Taylor, Hall Garth ; got by Earl of Sheffield (33812), dam (Lady of Nunwick 2nd) by Hubback Junior (31395), g.d. (Lady of Nunwick) by Lord of Nunwick (26702), gr.g.d. (Cherry Blossom) by Lieutenant (18198), — (Dahlia) by Duke (14421), — by Mango (4359), — from the stock of Mr. Pattinson, of Kettleside.

(43174) **EARL OF PEMBROKE,**
Red and white, calved April 4, 1879, bred by Mr. A. Baird, Robeston Hall ; got by Earl of Horton 5th (33797), dam (Ada) by Duke of Kirklevington (25982), g.d. (Adeliza 2nd) by Second Duke of Wetherby (21618), gr. g. d. (Adeliza) by Grand Duke of Oxford (16184), — (Agnes) by Chaffcutter (12572), — by Jemmie (10352), — by Bristol (7852), — by Brunel (7857), — by Leo (4208), — by Ovis (4635), — by Rival (2534), — by Darlington (956), — by Denton (198), — by Rockingham (560), — by Jobling's Son of Phenomenon (491), — by Phenomenon (491), — by Colonel (152), — by Styford (629).

(43175) **EARL OF ROBESTON 6TH,**
Roan, calved September 20, 1879, bred by Mr. A. Baird, Robeston Hall ; got by Earl of Horton 5th (33797), dam (Rhoda 13th) by Lord Darlington 3rd (31646), g. d. (Rhoda 10th) by Second Duke of Claro (21576), gr. g. d. (Rosalie) by Archbishop (17312),   (Rhoda) by Harry of Glo'ster (14674), — by Oregon (8371), — by Monzani (6222), — by Homer (2134).

(43176) **EARL OF RYEDALE,**
Red, calved December 5, 1879, bred by Mr. J. R. Singleton, Great Givendale ; got by Silent Duke (42382), dam (Rose of Ryedale 3rd) by Coriolanus (30795), g. d. (Carnation) by Vesuvius (21017), gr. g. d. (Canary) by Sir Oliver (18854), — (Charm) by Leicester (14787), — by Second Earl of Beverley (5963), — by Cleveland Lad (3407), — by Symmetry (5385), — by Grazier (1085), — by Parrington (4653), — by Baron (58), — by Windsor (698), — out of a Granddaughter of Washington (674).

(43177) **EARL OF SIDDINGTON,**
Roan, calved September 29, 1878, bred by Lord Moreton, Tortworth Court; got
by Third Duke of Hillhurst (30975), dam (Siddington 15th) by Third Duke of
Clarence (23727), g. d. (Siddington 11th) by Second Duke of Tregunter (26022),
gr. g. d. (Siddington 5th) by Seventh Duke of York (17754), — (Siddington) by
Fourth Duke of Oxford (11387), — by Earl of Derby (10177), — by Earl of
Liverpool (9061), — by Duke of Northumberland (1940), — by Belvedere (1706),
— by Son of Second Hubback (2683), — a Cow of Mr. Bates's, descended from
the stock of Mr. Maynard, of Eryholme.

(43178) **EARL PERCY,**
White, calved December 28, 1879, bred by Messrs. Dalton, Cummersdale; got
by Baron Winsome (39449), dam (Gwynne Princess) by Prince of Perth (32176),
g. d. (Rosamond Gwynne) by Sir Windsor (22927), gr. g. d. (Rebecca Gwynne)
by Knight of Distington (18158), — (Ruth Gwynne) by Exquisite (14524),
— by St. Thomas (10777), — by Prime Minister (2456), — by Wallace (5586),
— by Marmion (406), — by Merlin (430), — by Layton (366), — by Phenomenon (491), — by Favourite (252), — by Favourite (252), — by Hubback
(319), — by Snowdon's Bull (612), — by Waistell's Bull (669), — by Masterman's Bull (422), — by the Studley Bull (626).

(43179) **EARL SANDALE,**
Red and white, calved December 7, 1879, bred by Mr. T. W. Cadman, Ballifield
Hall, the property of Mr. G. J. Young, Claxby House; got by Duke of Sandale 2nd (38180), dam (Fillpail) by Knight of Lothian (31545), g. d. (Lady
Mary) by Cambridge Duke 4th (25706), gr. g. d. (Lady Julia) by Julius Cæsar
(22008), — (Comet's Farewell) by Great Comet (16192), — by Sir John (12084),
— by The Bonus (10922).

(43180) **EARL TUNSTALL 5TH,**
Roan, calved November 5, 1879, bred by the Rev. W. Sneyd, Keele Hall; got
by Duke of Oxford (38169), dam (Thorndale Duchess 2nd) by Eighth Duke of
Geneva (28390), g. d. (Thorndale Duchess) by Twelfth Duke of Thorndale (26020),
gr. g. d. (Waterloo Duchess 2nd) by Waterloo Duke (21077), — (Columbine) by
Lord of the Harem (16430), — by Glo'ster's Grand Duke (12949), — by Fourth
Duke of York (10167), — by Usurer (9763), — by Duke of Cornwall (5947), — by
Morpeth (7254), — by Helicon (2107), — by Henwood (2114), — by Nestor (452),
— by Harold (291), — by Meteor (432), — by Comet (155), — by Cupid (177).

(43181) **EARL VICTOR 2ND,**
White, calved December 5, 1879, bred by Captain Blathwayt, Dyrham Park;
got by Oxford's King (34997), dam (Verbena 10th) by Lord Tregunter (31758),
g. d. (Verbena 8th) by Cherry Prince (28182), gr. g. d. (Verbena 5th) by Wolfran
(25469), — (Verbena 2nd) by Lord Liverpool (22168), — by Mosquito (14964),
— by Fourth Duke of Oxford (11387), — by Usurer (9763), — by Second Duke
of York (5959), — by Sir Thomas (7516), — by Sultan Selim (2710), — by
Satellite (1420), — from the stock of Mr. Smith, of Shedlaw.

(43182) **EDITOR,**
Roan, calved March 17, 1875, bred by Mr. J. C. Bowstead, Hackthorpe Hall,
the property of Mr. O. Wallis, Bradley Hall; got by Monarch (31930), dam
(Family Herald) by Grand Herald (26301), g. d. (Red Rose 8th) by Precedent

(11918), gr. g. d. (Red Rose 6th) by Gainford 2nd (10255), — (Red Rose 5th) by Emperor (9084), — by Sir Thomas Newton (6495), — by General (3867).

(43183)   EDWARD SENIOR,
Red and white, calved December 24, 1878, bred by Mr. H. Lees, Pickhill Hall; got by Edward Waterloo (38246), dam (Hoiden) by Earl of Killerby (33802), g. d. (Romping Girl) by Hotspur (28878), gr. g. d. (Reckless) by Mandarin (26799), — (Rispah) by Duke of Grafton (21594), — by Carolus (14246), — by King Arthur (13110), — by Preston (8408), — by South Star (7538), — by Van Amburgh (5543), — by Plenipo (4724), — by Young Matchem (4425), — by Sir Thomas (2636), — by Marske (418), — by Comet (155), — by Tom (652), — by Favourite (1033), — by Hutton's Bull (323), — by Barningham (56).

(43184)   EFFINGHAM,
Red, calved February 19, 1878, bred by Mr. P. Stevenson, Rainton, the property of Mr. G. Brown, Greenfield House, Raskelfe; got by Duke of Edlingham 3rd (36489), dam (Lobisa) by Oxford's Duke 2nd (34994), g. d. (Loo Duchess) by Duke Royal (21623), gr. g. d. (Loo) by Cromwell (17640), — (Lily Bell) by Horrox (11591), — by Second Duke of Oxford (9046), — by Cleveland Lad (3407), — by Red Rose Bull (2493), — by Rex (1375), — bred by Mr. Richardson, of Hart.

(43185)   EGMONT,
White, calved April 1, 1878, bred by Mr. J. W. Philips, Heybridge; got by His Lordship (38433), dam (Flower of Belgium) by Royal Prince (27384), g. d. (Flower of Denmark) by Fitz-Clarence (14552), gr. g. d. (Flower Maid) by Vanguard (10994), — (Flower Girl) by Londesboro' (6142), — by Rinaldo (4949), — by Sir Thomas (2636), — by Sir Alexander (591), — by Marske (418), — by North Star (459), — by Wellington (680), — by Favourite (252), — by Favourite (252), — by Ben (70), — by Hubback (319), — by Snowdon's Bull (612), — by Sir J. Pennyman's Bull (601).

(43186)   ELCHO CLUNY,
Red and white, calved April 16, 1879, bred by Mrs. Gordon, Cluny Castle; got by Friar of Knowlmere (33975), dam (Flora 2nd) by Masterman (26864), g. d. (Flora) by Squire of Bushey (20889), gr. g. d. (Tulip) by Richard Cœur-de-Lion (13590), — (Opulent) by Abraham Parker (9856), — by Tom Boy (9750), — by Magician (7185), — by Vivian (5575), — by Belvedere 2nd (3126), — by Alive O' (2995), — by Young Alive O' (2996), — by Eclipse (236), — by Charge's Grey Bull (872), — by Paddock Bull (477).

(43187)   ELECTOR,
Roan, calved February 17, 1879, bred by Mr. W. Thompson, Moresdale; got by British Duke (37901), dam (Princess Royal) by Edgar (19080), g. d. (Lady Alberta) by Albert Victor (23293), gr. g. d. (Village Girl) by Hamlet (18017), — by Eclipse (17791), — by Albert (11093).

(43188)   ELLEN'S DUKE,
White, calved January 16, 1879, bred by Mr. G. Scoby, Beadlam Grange; got by Duke of Tregunter 5th (33743), dam (Ellen) by Grand Monarch (28774), g. d. (Eveline) by King of the Roses (22043), gr. g. d. (Emily) by Cobham (14287), — (Viscountess) by Fourth Duke of York (10167), — by Second Cleveland Lad (3408), — by Velocipede (5552), — by Col. Cradock's Son of Rob Roy (557), — bred by Mr. Kitchen, from the stock of Messrs. C. and R. Colling.

(43189)   ELPHINSTONE CLUNY,
Roan, calved May 26, 1879, bred by Mrs. Gordon, Cluny Castle; got by Sir
Windsor Broughton (27507), dam (Flora 12th) by Baron of Knowlmere (30474),
g. d. (Flora 2nd) by Masterman (26864), gr. g. d. (Flora) by Squire of Bushey
(20889), — (Tulip) by Richard Cœur-de-Lion (13590), — by Abraham Parker
(9856), — by Tom Boy (9750), — by Magician (7185), — by Vivian (5575), —
by Belvedere 2nd (3126), — by Alive O' (2995), — by Young Alive O' (2996),
— by Eclipse (236), — by Charge's Grey Bull (872), — by Paddock Bull (477).

(43190)   EMPEROR,
Roan, calved December 9, 1872, bred by Mr. J. Kelly, Lissalea House; got by
Napoleon 3rd (29417), dam (Crocus) by Chieftain (30732), g. d. (Blush 2nd) by
Beau of Killerby (6770), gr. g. d. (Duchess) by King of the Valley (9270), —
(Star) by Liberal (2201), — by Voltaire.

(43191)   EMPEROR,
Roan, calved February 8, 1878, bred by Mr. G. Hewer, Ley Gore House, the
property of Mr. J. Cooke, Malpas; got by Prince of the Heath (35161), dam
(Evenlode 3rd) by Paritor (27040), g. d. (Evenlode 2nd) by Leygore (18191),
gr. g. d. (Young Evenlode) by Havelock (14676), — (Evenlode) by Young Bruin
(12504), — by Sleet (13757), — by Snowball (8602), — by Locksley (4240), —
by Ambô Dexter (1636), — by Burley (1766), — by Sir Alexander (591), — by
Norman (1277), — by Palmsun (1302).

(43192)   EMPEROR,
White, calved January 12, 1879, bred by Mr. T. Nash, Featherstone; got by
Duke of Goldsmith (41386), dam (Empress) by Phosphate (32064), g. d. (Helen)
by Paris (29523), gr. g. d. (Pheasant) by Provost (24878), — (Water Wave 17th)
by Wandering Willie (19110), — by Chieftain (14267), — by Homer (13038),
— by Prince John (9508), — by Vampire (6632), — by Senator (2610), — by
Columella (904), — by Shakespeare (1429), — by Blyth Comet (85), — by
Neswick (1266), — by Favourite (1033).

(43193)   EMPEROR ALEXANDER,
Red and white, calved February 1, 1877, bred by Mr. W. Curtis, Fernham; got
by Aeronaut 2nd (36103), dam (Empress 4th) by Plebeian (29556), g. d. (Empress 2nd) by Duke of Towneley's Aide-de-Camp (23797), gr. g. d. (Empress 1st)
by Prince of Thorndale (24844), — (Gem) by Bacchus (17339), — by Duke
Valorous (15951), — by Humber (7102), — by Guardian (3947), — by Mercury (2301), — by Monarch (2324), — by St. Albans (2584), — by Jupiter
(342), — by Sir Oliver (605), — by Trunnell (659), — by Favourite (252), —
by Favourite (252), — by Dalton Duke (188), — by R. Alcock's Bull (19), —
by J. Smith's Bull (608), — by Jolly's Bull (337).

(43194)   EMPEROR BARRINGTON,
Roan, calved July 8, 1879, bred by the Executors of Mr. J. Harward, Winter-
fold; got by Baron Barrington 6th (33008), dam (Empress of York 4th) by
Eighth Duke of Geneva (28390), g. d. (Empress of York 3rd) by Third Duke of
Claro (23729), gr. g. d. (Empress of York) by Seventh Duke of York (17754),
— (Europa) by Cupid (14359), — by Twelfth Duke of York (12784), — by
Belshazzar 2nd (14154), — by Napoleon (10552), — by Plenipo (4724), — by
Sir John (16985), — by Rex (1375), — by Baron (58), — by a Bull of Mr.
Mason's, — by Falstaff (250), — by Irishman (329).

(43195)   EMPEROR OF KIRKLEVINGTON,

Roan, calved March 18, 1878, bred by Mr. J. Graham, Calthwaite Hall, the property of Mr. Milligan, Johnby; got by Grand Duke of Kirklevington (34071), dam (Empress 3rd) by Royal Cumberland (27358), g. d. (Empress) by Oxford Duke (15036), gr. g. d. (Emma 4th) by Emperor (6973), — (Emma 3rd) by Sir Walter (2639), — by Wharncliffe (2834), — by Kilnshaw (2175), — by Guy (2084), — by Frederick (1060), — by Pyramus (533), — by Meteor (432).

(43196)   ENGINEER,

Roan, calved April 30, 1877, bred by Mr. J. Nelson, Maiden Hill, the property of Mr. T. Preston, Skirwith; got by Golden Duke (36705), dam (Lady Gay) by Golden Duke (31265), g. d. (Lady Wilson 2nd) by Golden Dick (24047), gr. g. d. (Lady Wilson) by General Wilson (16143), — (Pearl) by The Pope (13869), — by Mango (4359), — by Pea Bloom (9469), — by Young Seagull (5100), — by Archibald (1652).

(43197)   ENGLAND'S PRIDE,

Roan, calved April 11, 1878, bred by Mr. J. Nelson, Maiden Hill; got by Golden Duke (36705), dam (Lady Gay) by Golden Duke (31265), g. d. (Lady Wilson 2nd) by Golden Dick (24047), gr. g. d. (Lady Wilson) by General Wilson (16143), — (Pearl) by The Pope (13869), — by Mango (4359), — by Pea Bloom (9469), — by Young Seagull (5100), — by Archibald (1652).

(43198)   ENGLISHMAN,

Roan, calved August 29, 1879, bred by Mr. H. Smith, Mountmellick; got by Conqueror (30786), dam (English Sylvia) by England's Glory (23889), g. d. (Sylvia 6th) by Ravenspur (20628), gr. g. d. (Sylvia 4th) by Sir James (16980), — (Sylvia 3rd) by Royal Duke (18766), — by Orator (15026), — by Baron Norton (11155), — by Lord Anthony Stanley (10432), — by Major, — Old Finne.

(43199)   EPISTLE,

Red and white, calved June 21, 1878, bred by Mr. W. Hewer, Sevenhampton, the property of Mr. G. H. Marshall, Patrington; got by British Prince (37907), dam (Epitaph 26th) by Westrop (35969), g. d. (Epitaph 22nd) by Lord Lyon (24424), gr. g. d. (Epitaph 16th) by General Johnson (21815), — (Epitaph 6th) by Viscount Killerby (19081), — by Lord Durham (9311), — by Humber (7102), — by Guardian (3947), — by William (2840), — by Richard (1376), — by Jupiter (342), — by Charles (127), — by Windsor (698), — by Chilton (136), — by Colonel (152).

(43200)   ERL KING,

Red and white, calved May 18, 1879, bred by Mr. J. A. M. Cope, Drummilly; got by Glo'ster's Gwynne (38360), dam (Elvira 12th) by Cambridge Duke 5th (30644), g. d. (Elvira 9th) by Grand Duke 11th (21849), gr. g. d. (Elvira 7th) by Grand Duke 10th (21848), — (Elvira 2nd) by Eighth Duke of Oxford (15939), — by The Baronet (10918), — by Burgundy (7861), — by Pilot (4707), — by Navarino (2352), — by Favourite (256), — by Phenomenon (491), — by Favourite (252), — by Favourite (252), — by Hubback (319), — by Snowdon's Bull (612), — by Waistell's Bull (669), — by Masterman's Bull (422), — by the Studley Bull (626).

(43201)   ERSKINE CLUNY,

Red, calved March 20, 1879, bred by Mrs. Gordon, Cluny Castle; got by Lollius

Booth (34470), dam (Flora 10th) by Watchman 2nd (27756), g. d. (Flora 3rd) by
Masterman (26864), gr. g. d. (Flora) by Squire of Bushey (20889), — (Tulip)
by Richard Cœur-de-Lion (13590), — by Abraham Parker (9856), — by Tom Boy
(9750), — by Magician (7185), — by Vivian (5575), — by Belvedere 2nd (3126),
— by Alive O' (2995), — by Young Alive O' (2996), — by Eclipse (236), — by
Charge's Grey Bull (872), — by Paddock Bull (477).

(43202)   ESCA,
Red, calved March 11, 1879, bred by Mr. J. A. M. Cope, Drummilly ; got by
Knight of Raby (34393), dam (Elvira 11th) by Cambridge Duke 5th (30644),
g. d. (Elvira 9th) by Grand Duke 11th (21849), gr. g. d. (Elvira 7th) by Grand
Duke 10th (21848), — (Elvira 2nd) by Eighth Duke of Oxford (15939), — by
The Baronet (10918), — by Burgundy (7861), — by Pilot (4707), — by Navarino
(2352), — by Favourite (256), — by Phenomenon (491), — by Favourite (252),
— by Favourite (252), — by Hubback (319), — by Snowdon's Bull (612), — by
Waistell's Bull (669), — by Masterman's Bull (422), — by the Studley Bull (626).

(43203)   ESKDALE,
Roan, calved February 19, 1879, bred by Mr. A. Cruickshank, Sittyton, the
property of Mr. J. Stott, Greenheads ; got by Pride of the Isles (35072), dam
(Evening Star) by Royal Duke of Glo'ster (29864), g. d. (Morning Star) by
Champion of England (17526), gr. g. d. (Grandiflora) by Lord Sackville (13249),
— (Flora) by Fairfax Royal (6987), — by Premier (6308), — by Saturn (5089),
— by Favourite (6997), — by Grindon (3942), — bred by Mr. Rennie, of
Phantassie.

ˈ(43204)   ESMOND,
White, calved March 16, 1879, bred by Mr. J. W. Philips, Heybridge ; got by
Royal Saxon (39057), dam (Flower of Belgium) by Royal Prince (27384), g. d.
(Flower of Denmark) by Fitz-Clarence (14552), gr. g. d. (Flower Maid) by Van-
guard (10994), — (Flower Girl) by Londesboro' (6142), — by Rinaldo (4949),
— by Sir Thomas (2636), — by Sir Alexander (591), — by Marske (418), — by
North Star (459), — by Wellington (680), — by Favourite (252), — by
Favourite (252), — by Ben (70), — by Hubback (319), — by Snowdon's Bull
(612), — by Sir J. Pennyman's Bull (601).

(43205)   ETHELRED,
Roan, calved July 15, 1878, bred by Mr. J. Close, Holmescales ; got by Knight of
the Bath (38521), dam (Eva) by Royal Duke Hamilton (29863), g. d. (Evaline)
by Whiff (30299), gr. g. d. (Lady Fragrant) by Roan Windsor (24967), — (Vivia)
by Bit of Booth (17418), — by Majestic (16492), — by Ambo (12391), — by
Lord George (10439), — by Vanguard (10994), — by Hamlet (8126), — by Sir
Roger (7515), — by Studley Royal (5342).

(43206)   FAIR DAY,
Roan, calved August 7, 1879, bred by Mr. W. Faulkner, Rothersthorpe, the
property of Mr. Hutchinson, Bozeat; got by Fair Thane (31127), dam (Formosa)
by Rosehill (29833), g. d. (Fancy) by Knight of Branches (20076), gr. g. d.
(Fame) by Rufus (15216), — (Wildbine) by Admiral (9861), — by Royalty
(6420), — by Teetotal (5410), — by Rebel (4882), — by Coxcomb (928), — by
Minor (441), — by Son of Phenomenon (491), — by Traveller (655), — by
Colonel (152), — by R. Colling's Son of Broken Horn (95), — by R. Colling's
Son of Hubback (319).

(43207) **FAIRPLAY,**
Red, calved July 3, 1879, bred by Mr. W. Hawkes, Thenford, the property of
Messrs. F. and H. Tanner, Chesterton ; got by Harper (38407), dam (Frolicsome)
by Lord Blythe (22126), g. d. (Frolic) by Prince Rupert (20605), gr. g. d.
(Fuchsia) by Hatcliffe (12997), — (Rosy Gem) by Lincoln (10422), — by Pro-
tection (10662), — by Thornton (5477), — by Barmpton Grazier (3091), — by
Midas (1230), — by Young Favourite (254), — by Expectation (247).

(43208) **FAIRSERVICE,**
Roan, calved January 18, 1876, bred by Mr. R. Bruce, Manor House Farm, Gt.
Smeaton, the property of Mr. J. MacKessach, Balnaferry ; got by Baron Killerby
(27949), dam (Fair Tyne) by Duke of Tyne (17751), g. d. (Julia 3rd) by Hydra
(19995), gr. g. d. (Julia) by Talpa (15365), — (Annie) by Picotee (15063), — by
Garioch Lad (17938), — by Bucephalus (6784), — by Crusader (934), — by
Sultan (1485), — by Mars (411), — by North Star (458).

(43209) **FAIRY KING,**
Roan, calved March 7, 1876, bred by Mr. T. Doran, Thomastown, the property of
Mr. J. Kelly, Lissalea House ; got by Flagstaff (39884), dam (Maid of the Isles)
by Prince Arthur (29598), g. d. (Young Lavender) by Cholera (23567), gr. g. d.
(Lavender) by Solway 2nd (13767), — by Barclay (9925), — by Romeo (10738),
— by William (4159), — by. Adonis (719), — by Charles (127), — by Neswick
(1266).

(43210) **FAIRY KING,**
Red and white, calved May 19, 1878, bred by Mr. F. J. Moss, Stainfield Hall ;
got by Ruby King (37407), dam (Fairy Queen) by Dunbar (31045), g. d. (Daisy
Queen) by Chieftain (21421), gr. g. d. (Roan Mantle) by General Hopewell
(17953), — (Red Mantle) by Highthorn (13028), — by Trustthorpe (15432), —
by Argus (3033), — by Remus (4932), — by Prince Comet (1342), — by Count
(170).

(43211) **FALCON,**
Red and white, calved March 10, 1879, bred by Mr. R. H. Gould, Didmarton ;
got by Prince Arthur (38892), dam (Fancy 5th) by General Clarence (28689), g. d.
(Fancy 4th) by Chaffcutter 2nd (25728), gr. g. d. (Fancy 3rd) by Hector (26357),
— (Fancy 2nd) by Royal Oxford (18773), — by Norval (26991), — by a Bull of
Mr. Rich's, Didmarton.

(43212) **FANTAIL DUKE,**
Roan, calved October 28, 1879, bred by Sir John Swinburne, Bart., Capheaton ;
got by Duke of Oxford 27th (33709), dam (Fantail's Duchess) by Ninth Duke
of Geneva (28391), g. d. (Fantail) by Barleycorn (17348), gr. g. d. (Fair Holon)
by General Canrobert (12927), — (Florella) by Fifth Duke of York (10168), —
by Fourth Duke of Northumberland (3649), — by Short Tail (2621), — by
Belvedere (1706), — by Son of Young Wynyard (2859), — descended from
J. Brown's Old Red Bull (97).

(43213) **FARMER'S BOY,**
Roan, calved June 27, 1879, bred by Sir Wilfrid Lawson, Bart., Brayton ; got
by Wild Eyes Duke (36007), dam (Dairy Lass) by Cambridge Duke (28120), g. d.
(Dairy Maid) by Grand Vizier (26313), gr. g. d. (Dairy) by Perth (22513), —
(Dairy) by Duke (14419), — by Son of Comus (12626).

H

(43214)   FARNLEY PRINCE 2ND,
Roan, calved January 12, 1879, bred by Mr. G. Taylor, Stanton Prior, the
property of Mr. Kempe, Bromsgrove; got by Waterloo Prince (35949), dam
(Miss Fawkes) by Golden Duke 2nd (21837), g. d. (Lady Ada) by Lord Cobham
(20164), gr. g. d. (Ada) by Nabob (11834), — (Lady Anna) by Zadig (8796), —
by Lord Durham (9311), — by St. Helena (5055), — by Emperor (3716), — by
Satellite (1420), — by Cato (119), — by Jupiter (342), — by George (273), —
by Chilton (136), — by Irishman (329), — by B. (45).

(43215)   FARNLEY PRINCE 2ND,
Red and white, calved April 15, 1879, bred by Mr. R. Thompson, Inglewood
Bank ; got by Brilliant Butterfly (36270), dam (Farnley Princess) by Prince of
Perth (32176), g. d. (Cambridge Moss Rose 2nd) by Barrington Oxford (25607),
gr. g. d. (Cambridge Moss Rose) by Second Duke of Cambridge (12743), — (Moss
Rose) by Lord of Brawith (10465), — by Laudable (9282), — by Grouchy (6051),
— by Sir Thomas Fairfax (5196), — by Young Colling (1843), — by Red Bull
(2838), — by Son of Hollings (2131), — by Partner (2409), — by Hollings
(2131), — by R. Alcock's Bull (19).

(43216)   FAUNUS,
White, calved April 1, 1877, bred by Mr. W. Robinson, Edge Hill, the property
of Mr. R. Postlethwaite, Pennington ; got by Prince Alfred (29593), dam (Flower
Bud) by Lord Napier (26688), g. d. (Flower Child) by Cherry Prince (23555),
gr. g. d. (Flower Damsel) by Breast Plate (19337), — (Flower Maid) by Vanguard
(10994), — by Londesboro' (6142), — by Rinaldo (4949), — by Sir Thomas
(2636), — by Sir Alexander (591), — by Marske (418), — by North Star (459),
— by Wellington (680), — by Favourite (252), — by Favourite (252), — by Ben
(70), — by Hubback (319), — by Snowdon's Bull (612), — by Sir J. Pennyman's
Bull (601).

(43217)   FAVOURITE,
White, calved February 26, 1877, bred by the Earl of Lichfield, Shugborough ;
got by Sir Arthur (44013), dam (Parallel) by Buccaneer (25693), g. d. (Penny-
worth) by Prince Consort (22583), gr. g. d. (Penny Royal) by Royal Oak (16870),
— (Peach) by Havelock (14676), — by Valiant (7662), — by Fitz-Hardinge
(8073), — by Lord John (4259), — by Young Consul (6893), — by Newnham
(2365), — by Satellite (1420), — by Jupiter (342), — by Sir Oliver (605), — by
Trunnell (659), — by Favourite (252), — by Favourite (252), — by Dalton Duke
(188), — by R. Alcock's Bull (19), — by J. Smith's Bull (608), — by Jolly's
Bull (337).

(43218)   FAVOURITE,
Roan, calved April 22, 1879, bred by Mr. D. Hartley, Westerdale ; got by
Robinson Crusoe (39001), dam (Beauty) by Sixth Duke of Wetherby (28474),
g. d. (Rose Bud) by Waxwork (23190), gr. g. d. (Rachel) by Prince of I. (35149),
— (Rose of Shamrock) by Sepoy (35499), — by Hero (18055).

(43219)   FAWSLEY DUKE 3RD,
Roan, calved October 3, 1878, bred by Sir Curtis M. Lampson, Bart., Rowfant ;
got by Duke of Underley 2nd (36551), dam (Oxford Fawsley 4th) by Grand Duke
of Kent (26289), g. d. (Oxford Fawsley 2nd) by Lord Oxford 2nd (20215), gr.
g. d. (Fawsley 3rd) by Grand Duke 4th (19874), — (Coquelicot) by Duke of

Cambridge (12742), — by Earl of Dublin (10178), — by Janizary (8175), — by Caliph (1774), — by Rob Roy (557), — by Satellite (1420), — by Sir Dimple (594), — by Styford (629).

(43220) FAWSLEY KING,
Roan, calved July 21, 1879, bred by Mr. A. Graham, Yanwath Hall; got by The Colonel (35747), dam (Flora) by Waterloo Commander (32811), g. d. (Faulder 2nd) by Hackthorpe (22889), gr. g. d. (Fairy Flower) by Baronet (19274), — (Emily) by Duke (14421), — by General Haynau (11520).

(43221) FELIX KILLERBY,
Roan, calved February 21, 1879, bred by the Duke of Richmond and Gordon, Gordon Castle, the property of Mr. R. Turner, Arradoul; got by Baron Killerby (27949), dam (Destiny) by Felix Booth (23925), g. d. (Mystery) by Whipper-In (19139), gr. g. d. (Flirt) by Magnum Bonum (13277), — (Romp) by Bloomsbury (9972), — by The Pacha (7612), — by Second Duke of Northumberland (3646), — by Mahomed (6170), — by Sillery (5131), — by Carleton (843), — by Diamond (205), — by Diamond (205).

(43222) FENNEL DUKE 4TH,
Roan, calved December 16, 1877, bred by Mr. J. Woodhouse, Scale Hall, the property of Mr. J. Bridson, Belle Isle; got by Fennel Duke (36639), dam (Pink 2nd) by Duke of Lancaster (25984), g. d. (Pink) by Lord Beaumont (16384), gr. g. d. (Primrose) by Dragoon (14412), — (Strawberry) by Peter Parker (13468), — by Brandy (9984).

(43223) FERGUS,
Roan, calved September 18, 1879, bred by Mr. W. Hawkes, Thenford; got by Harcourt (39977), dam (Fragrance) by Royal Commander (29857), g. d. (Frolicsome) by Lord Blithe (22126), gr. g. d. (Frolic) by Prince Rupert (20605), — (Fuchsia) by Hatcliffe (12997), — by Lincoln (10422), — by Protection (10662), — by Thornton (5477), — by Barmpton Grazier (3091), — by Midas (1230), — by Young Favourite (254), — by Expectation (247).

(43224) FIELD MARSHAL,
Roan, calved March 29, 1871, bred by Mr. J. Evans, Uffington, the property of Mr. G. Kelsall, Marton; got by Manager (24521), dam (Sweetbriar) by Victor (19061), g. d. (Mistletoe) by Sir Charles (16946), gr. g. d. (Nectarine) by Cronstadt (14352), — (Nancy) by Young Pompey (13480), — by Son of David (1912).

(43225) FIELD MARSHAL,
Red and white, calved October 2, 1878, bred by Messrs. N. Russell and Son, Northallerton, the property of Mr. W. Mitchell, Cleasby; got by Royal Halnaby (39011), dam (Florence) by Manfred (26801), g. d. (Flora) by Lord Blithe (22126), gr. g. d. by Emperor of the Sea (19695), — by Ascanius (17326).

(43226) FIREFLY 2ND,
Roan, calved March 25, 1878, bred by the Earl of Bective, Underley Hall, the property of Mr. Holliday, Silloth; got by Twenty-fourth Duke of Airdrie (36460), dam (Butterfly Princess 22nd) by Royal Cumberland (27358), g. d. (Lady Butterfly Princess 13th) by Richard (16834), gr. g. d. (Butterfly Princess 6th) by D'Israeli (10125), — (Butterfly Princess) by Wellington (7706), — by Bellerophon (3119), — by Blucher (1725), — by Butterwick (3251).

(43227)   FIRST FRUITS,

Red and white, calved August 4, 1879, bred by Mr. J. P. Tynte, Tynte Park, the property of Mr. H. Smith, Mountmellick; got by Martello (40319), dam (First Love) by Royal Fern (29865), g. d. (Sybil) by Bright Spur (25671), gr. g. d. (Hebe) by Hohenlohe (18074), — (Crinoline) by Comrade (14307), — by Barley Sugar (11140), — by Romulus (6408), — by California (17487), — by Sockburn (6509), — by Monarch (2324), — by Planet (2432).

(43228)   FITZ-CLARENCE BENEDICT,

Roan, calved September 2, 1879, bred by Mr. T. Willis, Carperby; got by Windsor Benedict (40933), dam (Bashful Bride) by K. C. B. (26492), g. d. (Blushing Bride) by Fitz-Clarence (14552), gr. g. d. (Maiden's Blush) by Gipsy King (11532), — (Maid of Masham) by Bernardo (8885), — by Billingham (6786), — by Sol (2655), — by Harlsey (2091), — by Maximus (2284).

(43229)   FITZROY,

Red, calved February 17, 1876, bred by Mr. W. S. Marr, Uppermill, the property of Mr. Cowie, Cromlybank; got by Cherub 4th (33359), dam (Fragrance 2nd) by Gold Digger (24044), g. d. (Fragrance) by Lord Granville (24395), gr. g. d. (Fragrant 2nd) by Royal Standard (22803), — (Fragrant) by John Bull (11618), — by Matadore (11800), — by Second Duke of Northumberland (3646), — by Mahomed (6170), — by Sillery (5131), — by Carleton (843), — by Diamond (205), — by Diamond (205).

(43230)   FITZ-ROYAL,

Roan, calved April 20, 1879, bred by Mr. G. Yeats, Studley; got by Royal Manfred (40640), dam (Clarice) by Clarion (33393), g. d. (Isabella) by Christmas Present (28201), gr. g. d. (Bella) by Emperor (23875), — (Cherry) by Perseus (18533), — by Cotherstone (6903), — by Lofty (2217), — by a Son of Lindrick (1170), — by Shylock (2622).

(43231)   FLAGBEARER,

Roan, calved July 5, 1879, bred by Mr. J. Tyacke, Merthen; got by Vain Glory (39248), dam (Flora) by Orlando (32000), g. d. (Fury) by Hector (24117), gr. g. d. (Fuchsia) by Favourite (19730), — (Miss Fairfax) by Red Duke (16803), — by Sir Roger de Coverley (12095), — by Fairfax.

(43232)   FLAG OF ELSWICK,

Roan, calved May 9, 1879, bred by Mr. J. Thompson, Elswick; got by Goldstick (38366), dam (Sunray) by Florio (28618), g. d. (Sultana) by Knight of the Harem (24278), gr. g. d. (Sunray) by Lord Buckingham (20152), — by Troy (19020), — by Miser (13341).

(43233)   FLAG OF WESTMORELAND,

Red and white, calved February 25, 1879, bred by Mr. A. Graham, Yanwath Hall; got by The Colonel (35747), dam (Fairmaid 3rd) by Dean (30863), g. d. (Fairy Queen) by Hackthorpe (21889), gr. g. d. (Fairy Flower) by Baronet (19274), — (Emily) by Duke (14421), — by General Haynau (11520).

(43234)   FLAG OF YANWATH,

Roan, calved August 10, 1879, bred by Mr. A. Graham, Yanwath Hall; got by The Colonel (35747), dam (Fair Lady) by Waterloo Commander (32811), g. d. (Faulder 2nd) by Hackthorpe (21889), gr. g. d. (Fairy Flower) by Baronet (19274), — (Emily) by Duke (14421), — by General Haynau (11520).

(43235) FLAGSTAFF,

Roan, calved May 9, 1878, bred by Mr. G. Cather, Carrichue; got by Flag of Ireland (28613), dam (Fleda 5th) by King Oberon (26510), g. d. (Fleda 4th) by Oberon (22438), gr. g. d. (Fleda 2nd) by British Prince (14197), — (Fleda) by Baron Warlaby (7813), — by Prince Ernest (7366), — by Bright (1739), — by Shylock (2622), — by Percy (1314), — by Blucher (84), — by Jupiter (344), — by Marshal Beresford (415), — by Crocus (932), — by a Bull of Mr. R. Colling's.

(43236) FLAVIUS 2ND,

Red and white, calved December 26, 1879, bred by Mr. F. Dodd, Rush Court; got by Lord Darlington 14th (40149), dam (Florentina 2nd) by Grand Prince of Claro (28781), g. d. (Flower Girl) by Duke of Albany (25931), gr. g. d. (Florentia 9th) by Seventh Duke of York (17754), — (Florentia 2nd) by Archduke (17316), — by Koh-i-noor (11642), — by Usurer (9763), — by Zenith (5702), — by Sweet William (5368), — by Roman (2561), — by Monarch (2324), — by Jupiter (342), — by Sir Oliver (605), — by Trunnell (659), — by Favourite (252), — by Favourite (252), — by Dalton Duke (188), — by R. Alcock's Bull (19), — by J. Smith's Bull (608), — by Jolly's Bull (337).

(43237) FLOWER BOY,

White, calved July 18, 1878, bred by Mr. F. B. Greenwood, Swarcliffe, the property of Sir W. Worsley, Bart., Hovingham; got by Hautboy (36748), dam (Fraxinella) by High Constable (34158), g. d. (Flora) by Grindelwald (26323), gr. g. d. (Florence) by Gamecock (19807), — (Fairy) by Humphrey Clinker (13055), — by Lord Newton (18251), — by Son of Kentucky (9266), — by Barton (8871), — by Young Forester (3822).

(43238) FLOWER OF DAY,

Roan, calved April 28, 1879, bred by Mr. T. Chalk, Linton; got by Viscount Worcester (40881), dam (Flower 2nd) by Hamlet (31335), g. d. (Flower) by Duke Homfrey (19599), gr. g. d. by Havelock (16231), — by Atlas (15596).

(43239) FOREST BRIAR,

Red and white, calved September 7, 1879, bred by the Rev. J. Burdon, Castle Eden; got by Lord of the Forest (38638), dam (Sweetbriar) by Red Errant (24912), g. d. (Burnett Rose) by Young Freedom (21777), gr. g. d. (White Rose) by Statesman (18927), — (Melody) by Richard Cœur-de-Lion (13590), — by Brilliant (7851), — by Young Hector (7074), — by Bachelor (1665), — by Wallace (5588), — by Leopold (2199), — by Sir Harry (5155), — by Traveller (655), — by Colonel (152), — by Son of Hubback (319).

(43240) FORESTER,

Red, calved December 22, 1879, bred by Mrs. Severne, Thenford House; got by Petrarch (42136), dam (Frivolity) by Cherry Butterfly (23550), g. d. (Filigree) by Grand Duke 7th (19877), gr. g. d. (Festoon) by Kirklevington 4th (14775), — (Fringe) by Londonderry (13169), — by Broughton Hero (6811), — by Rockingham (2550), — by Remus (2524), — by Juniper (1144), — by White Comet (1582), — by Wright's Grandson of Favourite (2073), — by Cattley's Grey Bull (1798).

(43241) FOREST KING,

Roan, calved January 10, 1879, bred by Mr. T. Dargue, Whale Farm; got by Knight of the Forest (31556), dam (Warlaby Duchess) by Knight of Warlaby

(31571), g. d. (Windsor Duchess 2nd) by Shirley Fitz-Windsor (27451), gr. g. d. (Duchess Hamilton) by Duke of Hamilton (19618), — by Master Hopewell (14929), — by Benedict (7828).

(43242) FOREST KING,
Roan, calved March 20, 1879, bred by Mr. E. Robinson, Nafferton ; got by Halloo (34105), dam (Red Rose) by Rose Duke 2nd (32341), g. d. (Bloomer 4th) by Young Valentine (27697), gr. g. d. (Bloomer 2nd) by Trustee (23089), — (May Flower) by Prince Tom (15110), — by Galaor (12914), — by Lord Grey (10446), — by Maltster (9362), — by Splendid (5298), — by Adjutant (2924), — by Ambo (1636), — by a Bull of the Earl of Carlisle's.

(43243) FOREST KNIGHTLEY,
Roan, calved March 20, 1879, bred by Mr. N. Milne, Dryhope, the property of Mr. J. Muir, Dryhope ; got by Earl of Kelso (38228), dam (Lady Knightley) by Grand Knight (26302), g. d. (Lady Mary) by Redmond (20646), gr. g. d. (Young Sophy) by MacTurk (14872), — (Sophy) by Heir-at-Law (13005), — by Hudibras (10339), — by Baron of Ravensworth (7811), — by Baronet (1686), — by Spectator (2688), — by Fitz-Remus (2025), — by Whitworth (695).

(43244) FOREST RANGER,
Roan, calved December 20, 1878, bred by Mr. J. Morton, Fences Farm, the property of Mr. E. R. M. Pratt, Ryston Hall ; got by Hogarth 3rd (40003), dam (Favourite Broughton) by Broughton Royal (33236), g. d. (Favourite) by Forester (19767), gr. g. d. (Spot) by Grand Prince (16187), — (Markham Favourite) by Lord of Lindsay (13210), — by Sharp's Bull.

(43245) FORTUNATUS,
Roan, calved December 5, 1879, bred by Mr. R. Stratton, The Duffryn ; got by Pearl Diver (37182), dam (Chloe) by James 3rd (26450), g. d. (Celia) by Coronet (23623), gr. g. d. (Cleopatra 3rd) by King John (14763), — (Virginia) by Hero of the West (8150), — by Lottery (4280), — by Lottery (4280), — by Phœnix (6290).

(43246) FRANK,
Red and white, calved February 26, 1877, bred by Major McCraith, Loughloher, the property of Mr. W. Johnson, Prumplestown House ; got by Great Fame (34087), dam (Red Fanny 2nd) by Richard Gwynne (32229), g. d. (Red Fanny) by Prince of the Realm (22627), gr. g. d. (Fanny 9th) by Fugleman (14580), — (Fanny 3rd) by Chieftain (7898), — by Mayo (16555), — by Peel (4672), — by Duke (3633).

(43247) FRANK,
Red and white, calved March 5, 1879, bred by Mr. F. Worrall, Barrow-in-Furness, the property of Mr. G. Hardy, Timperley ; got by Duke of Oxford 34th (36529), dam (Jessy 35th) by Paris (20469), g. d. (Jessy 32nd) by Cerdic (19415), gr. g. d. (Jessy 31st) by Master Gwynne (16539), — (Jessy 23rd) by General Havelock (14598), — by Despot (11348), — by Pantaloon (9467), — by Tommy Lad (6611), — by Young Rockingham (2549), — by Monitor (2331), — by Young Rockingham (2549), — by Major (2255), — by a Bull of Mr. Newby's, — by Northumberland (464), — by Buston's Styford (103), — by Bolingbroke (86).

(43248)   FRED VOKES,

White, calved November 28, 1877, bred by Mr. E. Randall, East Down Lodge, the property of Mr. E. Martin, East Down Lodge; got by Spectator (37508), dam (Miss Vokes) by Fra Diavolo (26190), g. d. (Water Wave 6th) by Wandering Willie (19110), gr. g. d. (Kathleen) by Prince John (9508), — (Carnation) by Lord John (4257), — by Archduke (3026), — by Darlington (3561), — by Count (170), — by Young Alfred (740), — by St. John (572), — by Windsor (698), — by Cupid (177).

(43249)   FREEBOOTER,

Red and white, calved January 27, 1879, bred by Mr. W. A. Mitchell, Auchnagathle, the property of Mr. R. Arklay, Ethiebeaton; got by Duke of Chamburgh (36052), dam (Foxglove) by Magnet (31793), g. d. (Fairy) by Forth 4th (28636), gr. g. d. (Countess) by Monarch (31929), — (Mysie) by Old England (24681), — by California (12528).

(43250)   FREEMAN,

Red and white, calved February 23, 1879, bred by Mr. J. Moffat, Ballyhyland, sold to the Provincial Government of Brabant, Belgium; got by Royalist (37396), dam (Freedom) by Protector (35188), g. d. (Frantic) by Wideawake (27805), gr. g. d. (Freak) by White Chieftain (21096), — (Frisk) by Sir Colin (16960), — by Barley Sugar (11140), — by Solway (7530), — by Prince Ernest (7366), — by The General (9146), — by Prince Paul (4827), — by Booth (11195), — by Monarch (2324), — by Emperor (3716).

(43251)   FREEMASON,

Red and white, calved April 28, 1878, bred by Mr. C. Hobbs, Maisey Hampton; got by Duke of Hazlecote 36th (33671), dam (Matchless 9th) by Lord Hastings (29115), g. d. (Matchless) by Earl of Walton (17787), gr. g. d. (Matchless) by Sir Richard (15298), — (Miss Bloomer) by Siddington Duke (15263), — by Guizot (9174), — by Guizot (9174), — by Actor (6701).

(43252)   FRESHMAN,

Red and white, calved November 4, 1879, bred by Mr. J. Singleton, Teresa Cottage; got by Lord Cockburn (38594), dam (Lady Cambridge) by Cambridge Duke 5th (30644), g. d. (Lady Jocelyn) by Grand Duke 10th (21848), gr. g. d. (Elvira 10th) by Richard (16834), — (Elvira 4th) by D'Israeli (10125), — by Morton (14963), — by Titian (5485), — by Smeaton (5212), — by Champion (3319), — by Emperor (1010), — by Allison's Bull (26).

(43253)   FRUGAL,

Red, calved January 7, 1879, bred by Mr. J. Parker, Ingleby; got by Duke of Kent (33680), dam (Simple) by Wellington (30288), g. d. (Sylph) by Lord McDonald (22175), gr. g. d. (Symphony) by War Prince (19114), — (Songstress) by Son of Vanguard (10994), — by War Eagle (12283), — by Strauss (10888), — by Thurgarton (7632), — by Strelley (5339), — by Alexander (2978), — by Dishley (3593), — by Young Warrior (9807), — by Son of Chilton (136), — by Young North Star (9449).

(43254)   FRUITFUL,

Red, calved February 27, 1879, bred by Mr. W. S. Marr, Uppermill, the property of Mr. J. Smith, Balmain; got by Cherub 4th (33359), dam (Fragrance 2nd) by Gold Digger (24044), g. d. (Fragrance) by Lord Granville (24395), gr. g. d. (Fragrant 2nd) by Royal Standard (22803), — (Fragrant) by John Bull

(11618), — by Matadore (11800), — by Second Duke of Northumberland (3646),
— by Mahomed (6170), — by Sillery (5131), — by Carleton (843), — by Diamond
(205), — by Diamond (205).

(43255)　　　　　　　　　　FUSILIER,
Roan, calved January 27, 1879, bred by H.R.H. the Prince of Wales, Sandringham; got by Frederic (38317), dam (Festive) by Homer (34170), g. d. (Feather)
by Theodorus (27639), gr. g. d. (Fledge) by Zealot (25480), — (Fleecy) by Mentor
(20333), — by Plato (18552), — by Pioneer (11904), — by The Captain (5422),
— by Hero (4021), — by Percy (1314), — by Lord Grantham's Son of Comet
(155).

(43256)　　　　　　　　　　FUSILIER,
Red and white, calved October 17, 1879, bred by Mr. B. Hannan, Riverstown;
got by Premier Lind (40474), dam (Jenny Lind 19th) by Abercorn (25484), g. d.
(Jenny Lind 5th) by Ancient Briton (19225), gr. g. d. (Jenny Lind 2nd) by
Napoleon (18438), — (Jenny Lind) by Druid (10140), — by Belted Will (6780),
— by Regent (2517), — by Borderer (3191), — by Eclipse (1949), — by Togston
(5487), — by Bolingbroke (3184), — by Son of Midas (435), — by Twin Brother
to Ben (660).

(43257)　　　　　　　　　　GALLANT,
Roan, calved April 3, 1877, bred by Mr. J. M. Graham, Battleby, the property
of Messrs. Robertson and Sons, Bellechin; got by Schoolmaster (39086), dam
(Gazelle) by Lord Lansdowne (29128), g. d. (Graceful) by Baronet (15614),
gr. g. d. (Grandiflora) by Lord Sackville (13249), — (Flora) by Fairfax Royal
(6987), — by Premier (6308), — by Saturn (5089), — by Favourite (6997), —
by Grindon (3942), — bred by Mr. Rennie.

(43258)　　　　　　　　　　GALLANT BUTTERFLY,
Red and white, calved November 17, 1879, bred by Mr. R. Thompson, Inglewood
Bank; got by Brilliant Butterfly (36270), dam (Duchess of Lancaster 7th) by
Baron Ribblesdale (21238), g. d. (Duchess of Lancaster 5th) by Inglewood
(20006), gr. g. d. (Duchess of Lancaster 2nd) by Precedent (11918), — (Lancaster Belle) by Louis Napoleon 2nd (13259), — by Duke of Lancaster (10929),
— by North Star (9447), — by Lord Adolphus Fairfax (4249), — by Thick Hock
(6601), — by Expectation (1988), — by Belzoni (1709), — by Comus (1861), —
by Denton (198).

(43259)　　　　　　　　　　GAMBLER,
Roan, calved March 3, 1879, bred by Mr. J. Nelson, Maiden Hill; got by
Marquis (41979), dam (Lady Rachel) by Golden Dick (24047), g. d. (Lady
Wyndham) by General Wyndham (24029), gr. g. d. (Lady Wilson) by General
Wilson (16143), — (Pearl) by The Pope (13869), — by Mango (4359), — by Pea
Bloom (9469), — by Young Seagull (5100), — by Archibald (1652).

(43260)　　　　　　　　　　GAME BOY,
Roan, calved May 18, 1878, bred by the Earl of Bective, Underley Hall, the property of Mr. T. W. Cadman, Ballifield Hall; got by Third Duke of Hillhurst
(30975), dam (Gazelle 29th) by Second Duke of Tregunter (26022), g. d. (Gazelle
8th) by Seventh Duke of York (17754), gr. g. d. (Gazelle 2nd) by Earl of
Walton (17787), — (Selina) by Fourth Duke of Oxford (11387), — by Snowstorm (12119), — by Hampden (8129), — by Leo (4208), — by Henwood

(2114), — by Sir Stephen (1456), — by Prince of Waterloo (528), —· by Mayflower (425), — by a Bull of Mr. Nicholson's, descended from the stock of Mr.
J. Brown, of Aldborough.

(43261) GANYMEDE,
Roan, calved March 7, 1879, bred by Mr. J. Stratton, Alton Priors; got by
Abelard (39346), dam (Miss Gertrude) by James 1st (24202), g. d. (Miss Glanville 3rd) by Buckingham (15700), gr. g. d. (Miss Glanville 2nd) by Waterloo
(11025), — (Roseanne) by Hero of the West (8150), — by Lottery (4280), —
by Phœnix (6290).

(43262) GARIBALDI,
White, calved May 6, 1878, bred by Mr. W. Hicks, St. Austell, the property of
Mr. J. J. Nicholls, Rialton Barton; got by Monarch Gwynne (37103), dam
(Beatrice 2nd) by Sailor Boy (37419), g. d. (Beatrice) by Orlando (32000),
gr. g. d. (Beatrice) by Sir Edward (25137), — (Bloomer) by Red Duke (16803),
— by Prince of Wales (13520), — by Sir Roger de Coverley (12095), — by
Kenilworth (7118), — by Phœnix (6290).

(43263) GARNETT LAD,
Roan, calved August 28, 1878, bred by Mr. W. H. Wakefield, Sedgwick, the
property of Viscount Templetown, Castle Upton; got by Duke of Holker (38153),
dam (Garnett 9th) by Baron Barrington 4th (33006), g. d. (Garnett 4th) by
Dunrobin (28486), gr. g. d. (Garnett 1st) by General Garibaldi (21813), —
(Lancaster) by Faust (16033).

(43264) GARRYOWEN,
Red and white, calved March 11, 1876, bred by Mr. F. W. Low, Kilshane; got
by Nobleman (31979), dam (Limerick Lass 6th) by Wideawake (27805), g. d.
(Limerick Lass 2nd) by Star of the West (22972), gr. g. d. (Lady Clarina) by
Volunteer (15478), — (Lady Camilla) by Norfolk (9442), — by Rex (6385), —
by Sir Thomas Fairfax (5196), — by Ambo (1636), — by Memnon (2295), —
by Pilot (496), — by Agamemnon (9), — by Burrell's Bull of Burdon.

(43265) GASCONY,
Red and white, calved May 5, 1879, bred by Mr. R. S. Drought, Passage West;
got by Sir Gwyon (44036), dam (Oriental Daisy) by Irish Baron (31417), g. d.
(Oriental) by Castle Grove (19408), gr. g. d. (Queen of the East) by Nobleman
(18457), — (Circassia) by Lamp of Lothian (16356), — by Matadore (11800),
— by Lord of Brawith (10465), — by Amateur (3007), — by Belshazzar (1703),
— by Abraham (2905), — by Simon (5134), — by Young George (3885), — by
George (276).

(43266) GAUTAMA,
Roan, calved February 24, 1878, bred by Mr. W. Talbot Crosbie, Ardfert Abbey,
the property of Mr. M. King, Strangemore; got by Royal Fitz-Rose (37390),
dam (Peri) by Northern Light (24670), g. d. (Circassia) by Lamp of Lothian
(16356), gr. g. d. (Lady Barcroft) by Matadore (11800), — (Malvina) by Lord of
Brawith (10465), — by Amateur (3007), — by Belshazzar (1703), — by Abraham (2905), — by Simon (5134), — by Young George (3885), — by George (276).

(43267) GENERAL 3RD,
Red, calved February 8, 1878, bred by Mr. T. W. Watson, Lubbenham, the property of Mr. W. Newton, High Croft Lodge; got by The General (35768), dam

(Lady Audrey) by First Lord (33923), g. d. (Canzonet) by Brilliant (33210), gr. g. d. (Cathay) by Faithful (33872), — (Cerito) by Forester (19767), — by Monarch (20371), — by Newton (20403), — by Norfolk (7280), — by Glaucus (3898), — by Twin Bull (20996), — by Finedon (19746), — by Rob Roy (556).

(43268)   GENERAL BRASSEY,

Roan, calved August 28, 1877, bred by Mr. T. H. Bland, Dingley Grange, the property of Mr. J. J. Brown, Shangton ; got by Earl of Waterloo 2nd (33819), dam (Blooming Bride) by Lord Collingham (29089), g. d. (Bridesmaid) by Fourth Duke of Thorndale (17750), gr. g. d. (Blushing Bride) by Great Mogul (14651), — (The Bride) by Young Lochinvar (13164), — by Bridegroom (11203), — by Lansdown (9277), — by Sir Robert (7510), — by Charles (1815), — by Stapleton (2698), — by Rob Roy (557).

(43269)   GENERAL GRANT,

Red and white, calved January 14, 1879, bred by Mrs. Mace, Sherborne ; got by Thorndale Geneva (39217), dam (Violet 7th) by Duke Geneva (30911), g. d. (Violet 3rd) by Seventh Duke of York (17754), gr. g. d. (Rose of Oxford) by Fourth Duke of Oxford (11387), — (White Rose) by Siddington Duke (15263), — by Harold (10299), — by Leo (4208), — by Henwood (2114), — by Wharf-dale (1578), — by Young Marske (419), — by Meteor (432), — by Western Comet (689), — by Favourite (252), — by Cupid (177), — by Grandson of Bolingbroke (280), — by Foljambe (263), — by R. Alcock's Bull (19), — by J. Smith's Bull (608), — by Jolly's Bull (337).

(43270)   GENERALISSIMO,

Roan, calved February 9, 1879, bred by Messrs. W. and H. Morley, East Gate Farm, the property of Mr. Castle, Liverpool ; got by Broomley Victor (37917), dam (Lady Flora 2nd) by Master Blithe (29313), g. d. (Lady Flora) by Protector (20609), gr. g. d. (Cleasby Daisy) by King Consort (16335), — (Lilladale) by Emperor (14498), — by a Son of Belleville (6778).

(43271)   GENERAL LORN,

Roan, calved November 9, 1879, bred by Mr. G. Ashburner, Low Hall ; got by Duke of Oxford 41st (38174), dam (Bride of Lorn) by Sockburn Lad (30024), g. d. (Lady Oxford) by Tenth Duke of Oxford (17739), gr. g. d. (Lily) by Hope (13042), — (Daisy) by Duke of Richmond (8000), — by Bachelor (5770), — by Romulus (6405), — by Sillery (5131), — by Young Western Comet (1575), — by Western Comet (689), — by Son of Favourite (252).

(43272)   GENERAL ROBERTS,

Roan, calved March 10, 1878, bred by Mr. S. Stewart, Sandhole ; got by Con-stellation (38027), dam (Myrtle) by Alliance (32937), g. d. (Duchess) by Scarlet Royal (27433), gr. g. d. (Wild Rose) by Banks of Don (37759), — (Wildfire 2nd) by King of Sardinia (16338), — by The Prince (17113), — by Earl of Aberdeen (12800), — by Balmoral (9920), — by Dannecker (7949), — by Ury (5536), — by Satellite (1420), — by Alfred (1412), — by Blucher (1725).

(43273)   GENERAL ROBERTS,

Roan, calved February 28, 1879, bred by Mrs. Mace, Sherborne ; got by Geneva's Duke (38349), dam (Violet 4th) by Seventh Duke of York (17754), g. d. (Rose of Oxford 2nd) by Fourth Duke of Oxford (11387), gr. g. d. (White Rose) by

Siddington Duke (15263), — (Orphan) by Harold (10299), — by Leo (4208), — by Henwood (2114), — by Wharfdale (1578), — by Young Marske (419), — by Meteor (432), — by Western Comet (689), — by Favourite (252), — by Cupid (177), — by Grandson of Bolingbroke (280), — by Foljambe (263), — by R. Alcock's Bull (19), — by J. Smith's Bull (608), — by Jolly's Bull (337).

(43274)　　　　　GENERAL ROBERTS,
Red, calved April 7, 1879, bred by Mr. W. Scott, Glendronach; got by Norseman (34924), dam (Helena) by Jeweller (34254), g. d. (Lady Jane) by Valiant 2nd (25353), gr. g. d. (Lady Ann) by Factor (14525), — (Agnes) by Baillie (32985), — by Aberdeen (2903), — by Monitor (2331), — by Reformer (2503), — by Snowdrop (2653), — by Wellington (2824), — by Midas (435), — by Orion (470), — by Traveller (655), — by Bolingbroke (86).

(43275)　　　　　GENERAL ROBERTS,
Roan, calved October 11, 1879, bred by Colonel R. Loyd Lindsay, Lockinge Park; got by Earl of Horton 11th (36588), dam (Romola) by Lord Napier (26691), g. d. (Rosetta) by Costa (21487), gr. g. d. (Rosette) by Prince of Prussia (16752), — (Red Rose) by Horatio (10335), — by Third Duke of Northumberland (3647), — by Velocipede (5552), — by Sir Thomas (2636), — by Marske (418), — by Comet (155), — by Tom (652), — by Favourite (1033), — by Hutton's Bull (323), — by Barningham (56).

(43276)　　　　　GENERAL ROBERTS,
Red, calved December 21, 1879, bred by Mr. R. P. Maxwell, Finnebrogue; got by Ards (40983), dam (Roan Myrtle) by Harry Underley (36744), g. d. (White Myrtle) by Prince Victor (20606), gr. g. d. (Myrtle) by Dillon (17680), — (Myrtle) by Musician (13362), — by Beau Belle (9940), — by Ballinasloe (12429), — by Rupture.

(43277)　　　　　GENERAL SEYMOUR 2ND,
Red and white, calved April 15, 1879, bred by Mr. T. W. Cadman, Ballifield Hall; got by Duke of Sandale 2nd (38180), dam (Lady Seymour 2nd) by Knight of Derby (31535), g. d. (Lady Seymour) by General Seymour (28697), gr. g. d. (Countess of Wragby)by Sir Roger (16991), — (Columbine) by Lambton (9273), — by General Washington (6036), — by Plenipo (4724), — by Ratify (2481), — by Wellington (680), — by Favourite (252), — by Punch (531).

(43278)　　　　　GENERAL WOLFE,
Roan, calved April 5, 1878, bred by Mr. N. Milne, Dryhope, the property of Mr. M. Frier, Nether Kidston; got by Earl of Kelso (38228), dam (Rosetta 4th) by Brilliant Cherry (28085), g. d. (Rosetta 2nd) by Viscount Stanley (23152), gr. g. d. (Rosetta) by Sam Glon (10780), — (Little Moss Rose) by Skelton (9667), — by Earl of Durham (5965), — by Newton (2367), — by Emperor (1974), — bred from the Chilton stock.

(43279)　　　　　GENEVA PRINCE 5TH,
Red and white, calved January 12, 1879, bred by Mr. W Munton, Banbury, the property of Mr. J. H. Gardner, Upper Boddington; got by Prince of Geneva 4th (37250), dam (Fleur d'Amour 3rd) by Pretender (29577), g. d. (Fleur d'Amour) by Waverley 3rd (21083), gr. g. d. (Feu de Joie) by Dusty Miller (17765), — (Beauty) by Abydos (14055), — by The Fop (27630), — Culworth Beauty, bred by Mr. G. Horne, Trafford Bridge.

(43280) GENEVA PRINCE 7TH,
Roan, calved May 20, 1879, bred by Mr. W. Munton, Banbury; got by Prince of Geneva 4th (37250), dam (Louise 2nd) by Prince of Wales (32189), g. d. (Louise) by Fashion (23913), gr. g. d. (Lilia) by Dusty Miller (17765), — (Blink Bonny) by Abydos (14055), — Culworth Beauty, bred by Mr. G. Horne, Trafford Bridge.

(43281) GENEVA PRINCE 8TH,
Roan, calved September 15, 1879, bred by Mr. W. Munton, Banbury, the property of Mr. W. Chamberlin, Adderbury; got by Prince of Geneva 4th (37250), dam (Berrymoor Rose 3rd) by Crown Prince (30825), g. d. (Berrymoor Rose) by Roan Silk (27307), gr. g. d. (Christmas Rose) by Ploughboy (18555), — (Berrymoor Lass) by Abydos (14055), — Culworth Beauty, bred by Mr. G. Horne, Trafford Bridge.

(43282) GENEVA PRINCE 9TH, ·
Red and white, calved December 30, 1879, bred by Mr. W. Munton, Banbury; got by Prince of Geneva 4th (37250), dam (Counsellor's Memento) by Baron Puck (35168), g. d. (Counsellor's Heiress) by Roan Silk (27307), ˙gr. g. d. (Counsellor's Grand Daughter) by Guy Fawkes (21886), — (Counsellor's Daughter) by Royal Counsellor (20725), — by Abydos (14055), — Culworth Beauty, bred by Mr. G. Horne, Trafford Bridge.

(43283) GENEVA'S HART,
Roan, calved August 30, 1879, bred by Sir Pryse Pryse, Bart., Gogerddan; got by Marquis of Geneva (31838), dam (Fawn) by Count Bickerstaffe (23630), g. d. (Daisy 2nd) by Baron Trent (19285), gr. g. d. (Eleanor) by The General (13856), — (Helen) by Homer 2nd (13039), — by Adolphus (12356), — by Northumberland (4595).

(43284) GERAD GWYNNE,
Roan, calved January 8, 1879, bred by Mr. S. P. Foster, Killhow; got by Duke of Underley (33745), dam (Opal Gwynne) by Baron Oxford 4th (25580), g. d. (Ora Gwynne) by Grand Duke of Lightburne (26290), gr. g. d. (Orange Gwynne) by Grand Duke 5th (19875), — (Ophelia Gwynne) by May Duke (13320), — by Duke of Glo'ster (11382), — by Conservative (3472), — by Wallace (5586), — by Marmion (406), — by Merlin (430), — by Layton (366), — by Phenomenon (491), — by Favourite (252), — by Favourite (252), — by Hubback (319), — by Snowdon's Bull (612), — by Waistell's Bull (669), — by Masterman's Bull (422), — by the Studley Bull (626).

(43285) GLACIER,
White, calved March 15, 1879, bred by the Duke of Northumberland, Alnwick Castle, the property of the Owners of Backworth Colliery, Newcastle; got by Sir Raymond (40716), dam (Julia) by White Prince (35988), g. d. (Constance) by Mayor of Windsor (31897), gr. g. d. (Louisa) by Briton (25686), — (Myrtle) by President (20510), — by George 3rd (16147), — by Cleveland Lad (3407), — Cherry, bred by the Duke of Northumberland.

(43286) GLADSTONE,
Roan, calved December 2, 1878, bred by Mr. S. Campbell, Kinellar; got by Golden Prince (38363), dam (Nonpareil 28th) by Sir Christopher (22895), g. d. (Nonpareil 24th) by Lord Sackville (13249), gr. g. d. (Nonpareil 23rd) by The

Baron (13833), — (Nonpareil 17th) by Matadore (11800), — by Prince Edward Fairfax (9506), — by Diamond (5918), — by Young Frederick (3836), — by Commodore (1858), — by Tathwell Studley (5401), — by Blyth Comet (85).

(43287) GLEN,

Roan, calved July 18, 1879, bred by Mr. A. C. Innes, Dromantine; got by Retriever (43899), dam (Lady Andrew) by Sir Andrew (37451), g. d. (Countess) by Count Robert (30812), gr. g. d. (Artful) by Knight of the Grand Cross (31558), — (Artemis) by Alabama (30366), — (Zenobia) by Syntax (32645).

(43288) GLENFALLOCH,

Roan, calved April 6, 1878, bred by Messrs. J. W. and E. Cruickshank, Lethenty, the property of Mr. J. C. Toppin, Musgrave Hall; got by Knight of St. Patrick (38520), dam (Vain Woman) by Merry Monarch (22349), g. d. (Virtue) by Valasco (15443), gr. g. d. (Lady Georgina) by Knight Errant (18154), — (Georgie) by Prince George (13510), — by Hopewell (10332), — by Warrior (12287).

(43289) GOLD DUST,

Roan, calved March 7, 1878, bred by Colonel Fisher, Castle Grogan, the property of Sir H. W. Gore Booth, Bart., Lissadell; got by Rembrandt (42273), dam (Golden Tint) by Neutral Tint (31971), g. d. (Golden Link) by Goldstick (28732), gr. g. d. (Olga) by Prince Imperial (20553), — (Duchess) by Lord of the Valley (10467), — (Princess) by Oliver (2386).

(43290) GOLDEN BANNER,

Red and white, calved January 20, 1879, bred by Mr. S. Armstrong, Gally House; got by British Lad (41151), dam (Golden Chain) by St. Ringan (27417), g. d. (Golden Drop) by Royal Sovereign (22802), gr. g. d. (Golden Rose) by Blood Royal (14169), — (Golden Lady) by Baron Warlaby (7813), — by Remus (4932), — by Prince Comet (1342), — by Count (170), — by Constellation (163), — by Young Favourite (255).

(43291) GOLDEN CROWN,

Red, calved February 9, 1879, bred by Mrs. Pery, Coolcronan House; got by Royal Crown (40628), dam (Good Manners) by Prince Christian (22581), g. d. (Ladylike 4th) by Ravenspur (20628), gr. g. d. (Ladylike 2nd) by Prince Regent (18637), — (Ladylike) by Stars and Stripes (12148), — by Sir Henry (10824), — by Buckingham (3239), — by Marcus (2262), — by Matchem (2281), — by Alderman (1622), — by Pilot (496), — by Remus (550), — by Sir Charles (592), — by R. Colling's Son of Favourite (252), — by Son of Favourite (252), — Old Strawberry.

(43292) GOLDEN LION,

Roan, calved July 9, 1878, bred by Mr. A. Cruickshank, Sittyton; got by General Windsor (28701), dam (Golden Queen) by Champion of England (17526), g. d. (Gold Mint) by The Baron (13833), gr. g. d. (Pure Gold) by Young Fourth Duke (9037), — (Star Pagoda) by Duplicate Duke (6952), — by Robin O'Day (4973), — by Sir Walter (2639), — by Young Jerry (8177), — by Roseberry (567), — by Roseberry (567), — by Constellation (163), — by Hastings (293), — by Hastings (293), — by Leopold (372).

(43293) GOLDEN TIME,

Roan, calved April 17, 1879, bred by Mr. A. Metcalfe, Park House; got by Royal Gift (39040), dam (Golden Queen) by Asteroid (21193), g. d. (Golden

Arch) by The Druid (18981), gr. g. d. (Golden Halo) by Hopewell (10332), —
(Golden Lady) by Baron Warlaby (7813), — by Remus (4932), — by Prince
Comet (1342), — by Count (170), — by Constellation (163), — by Young
Favourite (255).

(43294)   GOLDFINCH,
Roan, calved October 28, 1877, bred by Mr. A. Cruickshank, Sittyton, the property of Sir W. Forbes, Bart., Fintray House; got by Grand Vizier (34086),
dam (Guineas) by Prince Imperial (22595), g. d. (Golden Chain) by Lord
Raglan (13244), gr. g. d. (Gold Mint) by The Baron (13833), — (Pure Gold) by
Young Fourth Duke (9037), — by Duplicate Duke (6952), — by Robin O'Day
(4973), — by Sir Walter (2639), — by Young Jerry (8177), — by Roseberry
(567), — by Roseberry (567), — by Constellation (163), — by Hastings (293),
— by Hastings (293), — by Leopold (372).

(43295)   GOLDSMITH,
Roan, calved February 13, 1879, bred by Mr. W. Bolton, The Island, the
property of Mr. J. Moffat, Ballyhyland; got by Albion (36112), dam (Golden
Ringlet) by Prince of Munster (32174), g. d. (Golden Ring) by King Richard
(26523), gr. g. d. (Golden Rose) by Blood Royal (14169), — (Golden Lady) by
Baron Warlaby (7813), — by Remus (4932), — by Prince Comet (1342), —
by Count (170), — by Constellation (163), — by Young Favourite (255).

(43296)   GOLDSMITH,
Roan, calved April 22, 1879, bred by Mr. G. Ashburner, Low Hall, the property
of Mr. J. Newby, Loweswater; got by Duke of Oxford 41st (38174), dam
(Oxford's Gem) by Oxford 4th (24706), g. d. (Double Oxford) by Oxford (20449),
gr. g. d. (Lady Oxford) by Tenth Duke of Oxford (17739), — (Lily) by Hope
(13042), — by Duke of Richmond (8000), — by Bachelor (5770), — by Romulus
(6405), — by Sillery (5131), — by Young Western Comet (1575), — by
Western Comet (689), — by Son of Favourite (252).

(43297)   GOLIATH,
Red and white, calved March 16, 1879, bred by Messrs. W. and G. Bird, Volis;
got by Old Daisy Bull (38811), dam (Newton Butterfly) by Red Duke (29730),
g. d. (Lady Butterfly) by Sir Charles Knightley (27466), gr. g. d. (Kent Butterfly) by Second Duke of Kent (19620), — (Miss Thornton) by Welcome Guest
(15947), — by Aaron Smith (12331), — by Priam (10630), — by South Comet
(6528), — by Short Tail (2621), — by Wyville (5692), — by Attratum (3053).

(43298)   GOOD NEWS,
Roan, calved March 26, 1878, bred by Mr. W. Butler, Urlingford; got by Great
Fame (34087), dam (Tidings 2nd) by Sam (32448), g. d. (Tidings) by Master
Harbinger (18352), gr. g. d. (Love) by Master Harbinger (18352), — (Loo) by
Victory (19074), — (Lotus 2nd) by Berrington Boy (11173), — (Leda) by Claret
(10057), — by Son of Cupid, — by Exotic (16012), — by Duke (1934), — by
Isaac (1129), — by Blucher (6793), — by Cecil (120).

(43299)   GOOD REPUTE,
Red, calved February 8, 1878, bred by Mr. J. Jones, Ballyloughlan, the property of Viscount Bangor, Castleward; got by Great Fame (34087), dam (Roan
Welcome 2nd) by Richard Gwynne (32299), g. d. (White Welcome) by Vavasour
(35852), gr. g. d. (Roan Welcome) by Master Harbinger (18352), — (Sibyl 2nd)

by First Fruits (16049), — by Bannerman (12435), — by Claret (10057), — by Sir Charles (7501), — by Son of Cupid, — by Exotic (16012).

(43300) GOPSALL DUKE,
Roan, calved February 19, 1878, bred by Earl Howe, Gopsall Hall; got by Duke of Hazlecote 38th (36498), dam (Monthly Rose 9th) by Burleigh (25697), g. d. (Monthly Rose 6th) by Admiral (21146), gr. g. d. (Monthly Rose 3rd) by Guardsman (14656), — (Monthly Rose) by Duncombe (11402), — by Second Duke of Oxford (9046), — by Robin Hood (8491), — by Simon (5135), — by Crispin (174), — by Fisher's Old Bull (3799), — by Grandson of Favourite (252), — by Grandson of Favourite (252), — descended from the stock of Mr. Cornforth, of Barforth

(43301) GORDON GWYNNE,
Roan, calved June 12, 1879, bred by Mr. H. Caddy, Rougholm; got by Second Duke of Glo'ster (28392), dam (Agatha Gwynne) by Waterloo Cherry (27763), g. d. (Theresa Gwynne) by Knight of Distington (18158), gr. g. d. (Sally Gwynne) by Lablache (11656), — (Young Dowager Gwynne) by St. Thomas (10777), — by Prime Minister (2456), — by Wallace (5586), — by Marmion (406), — by Merlin (430), — by Layton (366), — by Phenomenon (491), — by Favourite (252), — by Favourite (252), — by Hubback (319), — by Snowdon's Bull (612), — by Waistell's Bull (669), — by Masterman's Bull (422), — by the Studley Bull (626).

(43302) GOSHAWK,
Roan, calved June 10, 1879, bred by Mr. R. S. Bruere, Braithwaite Hall; got by St. Swithin's Star Drop (40667), dam (Silver Lily Flower) by Booth's Satellite (30564), g. d. (Pearl Flower) by Booth's Royal Signet (28061), gr. g. d. (Welcome Flower) by The Sutler (23061), — (Windsor's Lilac Flower) by Windsor (14013), — by Sylvan King (13819), — by Leo (13150), — by The Silkey Laddie (10947), — by St. Martin (8526), — by Rouge (5012), — by Cleveland (3404), — by Burton (3250).

(43303) GOUNOD,
Roan, calved February 23, 1879, bred by Mr. T. M. Hopkins, Lower Wick Farm; got by Second Duke of Rowley (28441), dam (Georgiana 18th) by Duke of Wetherby 6th (33756), g. d. (Georgiana 15th) by Thirteenth Duke of Oxford (21604), gr. g. d. (Georgiana 12th) by Earl of Glo'ster (21644), — (Georgiana 7th) by Fourth Duke of Oxford (11387), — by General Canrobert (12926), — by St. Bernard (15227), — by Lord George Bentinck (10444), — by King Pippin (14769), — by Earl Stanhope (5966).

(43304) GRAND CHERRY,
Red, calved December 16, 1879, bred by Mr. G. Murton Tracy, Redlands; got by Hillhurst Cherry Duke (39999), dam (Cherry Branch) by Cherry Prince 4th (25765), g. d. (Eba) by May Duke (13321), gr. g. d. (Emma 2nd) by Hartforth (9191), — (Emma 1st) by Freetrader (10246), — by Gainford (2044), — by Magnum Bonum (2243), — by Rob Roy (557), — by Son of Houghton (318), — by Sir Stephen (1456), — by Sedbury (1424).

(43305) GRAND DUKE,
Roan, calved March 3, 1877, bred by Sir R. Peel, Bart., Drayton Manor, the property of Mr. J. Hopper, Killeythorpe; got by The Druid (35753), dam

(Isabella 8th) by Red Knight (29741), g. d. (Young Isabella) by Count Bickerstaffe (23630), gr. g. d. (Isabella 5th) by Daybreak (11338), — (Isabella 2nd)
by Alfred (6732), — by Accordion (5708), — by Despot (1915), — by Napoleon
(4531), — by Rival (2534), — by Darlington (956), — by Denton (198), — by
Rockingham (560), — by Son of Phenomenon (491), — by Phenomenon (491),
— by Colonel (152), — by Styford (629).

(43306)   GRAND DUKE 36TH,
Roan, calved March 13, 1879, bred by Mr. R. E. Oliver, Sholebroke Lodge ; got
by Grand Duke 30th (38373), dam (Grand Duchess 25th) by Second Duke of
Tregunter (26022), g. d. (Grand Duchess 23rd) by Grand Duke 7th (19877),
gr. g. d. (Grand Duchess 17th) by Imperial Oxford (18084), — (Grand Duchess
10th) by Grand Duke 3rd (16182), — by Prince Imperial (15095), — by Grand
Duke (10284), — by Cleveland Lad (3407), — by Belvedere (1706), — by Second
Hubback (1423), — by Second Hubback (1423), — by The Earl (646), — by
Ketton 2nd (710), — by Comet (155), — by Favourite (252), — by Daisy Bull
(186), — by Favourite (252), — by Hubback (319), — by J. Brown's Red
Bull (97).

(43307)   GRAND DUKE 37TH,
Roan, calved May 23, 1879, bred by the Earl of Bective, Underley Hall ; got
by Duke of Underley (33745), dam (Grand Duchess 23rd) by Grand Duke 7th
(19877), g. d. (Grand Duchess 17th) by Imperial Oxford (18084), gr. g. d. (Grand
Duchess 10th) by Grand Duke 3rd (16182), — (Grand Duchess 5th) by
Prince Imperial (15095), — by Grand Duke (10284), — by Cleveland Lad
(3407), — by Belvedere (1706), — by Second Hubback (1423), — by Second
Hubback (1423) — by The Earl (646), — by Ketton 2nd (710), — by Comet
(155), — by Favourite (252), — by Daisy Bull (186), — by Favourite (252), —
by Hubback (319), — by J. Brown's Red Bull (97).

(43308)   GRAND DUKE 39TH,
Red and white, calved June 25, 1879, bred by Mr. R. E. Oliver, Sholebroke
Lodge ; got by Duke of Underley 3rd (38196), dam (Grand Duchess 28th) by
Third Duke of Clarence (23727), g. d. (Grand Duchess 17th) by Imperial Oxford
(18084), gr. g. d. (Grand Duchess 10th) by Grand Duke 3rd (16182), — (Grand
Duchess 5th) by Prince Imperial (15095), — by Grand Duke (10284), — by
Cleveland Lad (3407), — by Belvedere (1706), — by Second Hubback (1423),
— by Second Hubback (1423), — by The Earl (646), — by Ketton 2nd (710),
— by Comet (155), — by Favourite (252), — by Daisy Bull (186), — by
Favourite (252), — by Hubback (319), — by J. Brown's Red Bull (97).

(43309)   GRAND DUKE 40TH,
Roan, calved September 15, 1879, bred by Mr. R. E. Oliver, Sholebroke Lodge ;
got by Grand Duke 30th (38373), dam (Grand Duchess 31st) by Third Duke of
Clarence (23727), g. d. (Grand Duchess 17th) by Imperial Oxford (18084), gr. g. d.
(Grand Duchess 10th) by Grand Duke 3rd (16182), — (Grand Duchess 5th) by
Prince Imperial (15095), — by Grand Duke (10284), — by Cleveland Lad
(3407), — by Belvedere (1706), — by Second Hubback (1423), — by Second
Hubback (1423), — by The Earl (646), — by Ketton 2nd (710), — by Comet
(155), — by Favourite (252), — by Daisy Bull (186), — by Favourite (252),
— by Hubback (319), — by J. Brown's Red Bull (97).

(43310) *GRAND DUKE OF AIRDRIE,

Roan, calved June 10, 1877, bred by Messrs. Avery and Murphy, Detroit, Michigan, U.S.A., the property of Sir H. Allsopp, Bart., Hindlip Hall; got by Fordham Duke of Oxford 4th (41569), dam (Airdrie Duchess 5th) by Duke of Oneida 2nd (33702), g. d. (Airdrie Duchess 2nd) by Fourteenth Duke of Thorndale (28459), gr. g. d. (Duchess of Airdrie 10th) by Royal Oxford (18774), — (Duchess of Airdrie 7th) by Clifton Duke (23580), — by Second Duke of Athol (11376), — by Second Duke of Oxford (9046), — by Second Cleveland Lad (3408), — by Short Tail (2621), — by Second Hubback (1423), — by Second Earl (1511), — by Marske (418), — by Ketton 1st (709), — by Comet (155), — by Favourite (252), — by Daisy Bull (186), — by Favourite (252), — by Hubback (319), — by J. Brown's Red Bull (97).

(43311) GRAND DUKE OF BARRINGTONIA 5TH,

Red and white, calved August 31, 1879, bred by Mr. R. E. Oliver, Sholebroke Lodge; got by Grand Duke 30th (38373), dam (Grand Duchess of Barringtonia 2nd) by Third Duke of Clarence (23727), g. d. (Grand Duchess of Barrington) by Grand Duke 7th (19877), gr. g. d. (Countess of Barrington 2nd) by Ninth Duke of Oxford (17738), — (Countess of Barrington) by Grand Duke 3rd (16182), — by Grand Turk (12969), — by Earl of Derby (10177), — by Earl of Liverpool (9061), — by Second Duke of Cambridge (3638), — by Belvedere (1706), — by Son of Herdsman (304), — by Wonderful (700), — by Alfred (23), — by Young Favourite (6994).

(43312) GRAND DUKE OF BARRINGTONIA 6TH,

Red, calved November 8, 1879, bred by Mr. R. E. Oliver, Sholebroke Lodge; got by Grand Duke 30th (38373), dam (Grand Duchess of Barringtonia 3rd) by Grand Duke 21st (34061), g. d. (Grand Duchess of Barrington) by Grand Duke 7th (19877), gr. g. d. (Countess of Barrington 2nd) by Ninth Duke of Oxford (17738), — (Countess of Barrington) by Grand Duke 3rd (16182), — by Grand Turk (12969), — by Earl of Derby (10177), — by Earl of Liverpool (9061), — by Second Duke of Cambridge (3638), — by Belvedere (1706), — by Son of Herdsman (304), — by Wonderful (700), — by Alfred (23), — by Young Favourite (6994).

(43313) GRAND DUKE OF DARLINGTON 5TH,

Roan, calved October 3, 1879, bred by Mr. R. Loder, Whittlebury; got by Grand Duke of Oxford 3rd (39953), dam (Duchess of Darlington 2nd) by Second Duke of Tregunter (26022), g. d. (Darlington 19th) by Third Duke of Wharfdale (21619), gr. g. d. (Darlington 15th) by Grand Duke 11th (21849), — (Darlington 12th) by Duke of Geneva (19614), — by Marmaduke (14897), — by Fourth Duke of Oxford (11387), — by Sir Hugh (12082), — by Percy (9472), — by Thomas (5471), — by Eryholme (3736), — by Reformer (4914), — by Young Favourite (3770), — by Wellington (2825).

(43314) GRAND DUKE OF EDEN,

Roan, calved September 18, 1879, bred by Captain Gandy, Castle Bank, the property of Mr. T. Blenkarn, Keisley; got by Grand Duke of Morecambe (36722), dam (Elvira 7th) by Marquis 2nd (31826), g. d. (Elvira 6th) by Grand Duke 10th (21848), gr. g. d. (Elvira 2nd) by Eighth Duke of Oxford (15939), — (Ruby Rose 2nd) by The Baronet (10918), — by Burgundy (7861), — by Pilot (4707), — by Navarino (2352), — by Favourite (256), — by Phenomenon

(491), — by Favourite (252), — by Favourite (252), — by Hubback (319), — by Snowdon's Bull (612), — by Waistell's Bull (669), — by Masterman's Bull (422), — by the Studley Bull (626).

(43315)   GRAND DUKE OF HAZLECOTE,

Roan, calved April 22, 1879, bred by Mr. T. Stamper, Highfield House; got by Duke of Hazlecote 23rd (30970), dam (Queen of Roses) by Snowstorm (35616), g. d. (White Rose) by Wathstone's Hero (25417), gr. g. d. (Charity) by Gold Dust (17973), — (Faith) by Ferdinand (12871), — by Surplice (10901), — by Bachelor (3071), — by Navigator (1260).

(43316)   GRAND DUKE OF HILLHURST,

Red and white, calved June 13, 1879, bred by Mr. H. P. Baxter, Southall Green Farm; got by Duke of Hillhurst (28401), dam (Cowslip 4th) by Grand Duke 11th (21849), g. d. (Cherry Cheeks) by Mac Turk (14872), gr. g. d. (Cherry Lips) by Cherry Duke 2nd (14265), — (Cowslip 5th) by Chieftain (10048), — by Duke of Norfolk (5952), — by Waterloo (2816), — by Kitt (7127), — by Kitt (7127), — by Page's Bull (6269), — by Middleton's Bull (438).

(43317)   GRAND DUKE OF KIRKLEVINGTON 4TH,

Red, calved March 22, 1879, bred by Mrs. Fawcett, Scaleby Castle; got by Baron Turncroft Oxford 4th (37822), dam (Kirklevington Duchess 11th) by Second Duke of Glo'ster (28392), g. d. (Kirklevington Duchess 5th) by Second Duke of Claro (21576), gr. g. d. (Duchess of Kent) by Lord Liverpool (22168), — (Kirklevington 14th) by Fourth Duke of Oxford (11387), — by Earl of Derby (10177), — by Earl of Liverpool (9061), — by Duke of Northumberland (1940), — by Belvedere (1706), — by Son of Second Hubback (2683), — a Cow of Mr. Bates's, descended from the stock of Mr. Maynard, of Eryholme.

(43318)   GRAND DUKE OF OXFORD 5TH,

Roan, calved April 28, 1879, bred by Lord Penrhyn, Penrhyn Castle; got by Grand Duke 20th (31281), dam (Grand Duchess of Oxford 7th) by Lord Oxford (20214), g. d. (Grand Duchess of Oxford) by Grand Duke 3rd (16182), gr. g. d. (Countess of Oxford) by Earl of Warwick (11412), — (Oxford 15th) by Fourth Duke of York (10167), — by Second Duke of Northumberland (3646), — by Short Tail (2621), — by Matchem (2281), — by Young Wynyard (2859).

(43319)   GRAND DUKE OF SURMISE,

White, calved April 28, 1879, bred by Sir G. R. Philips, Bart., Weston Park; got by Grand Duke 29th (38372), dam (Lady Ferrers) by Duke of Kingscote (25981), g. d. (Fancy 3rd) by Worth (23244), gr. g. d. (Fancy) by Fourth Duke of Thorndale (17750), — (Surmise) by Duke of Glo'ster (11382), — by Earl of Derby (10177), — by Duke of Sutherland (6945), — by Locomotive (4242), — by Short Tail (2621), — by Gambier (2046), — by Young Wynyard (2859), — by Bulls of Messrs. C. and R. Colling's.

(43320)   GRAND DUKE OF WATERLOO,

Red and white, calved February 23, 1879, bred by Mr. W. McCulloch, Evington; got by Twenty-fourth Duke of Airdrie (36460), dam (Lady Waterloo 28th) by Ninth Duke of Geneva (28391), g. d. (Lady Waterloo 19th) by Second Lord of Waterloo (22198), gr. g. d. (Lady Waterloo 10th) by Lord of Waterloo (18269), — (Lady Waterloo 5th) by Count Waterloo (14343), — by George (12941), — by Third Duke of Oxford (9047), — by Cleveland Lad (3407), — by Norfolk (2377), — by Waterloo (2816), — by Waterloo (2816).

(43321)   GRAND DUKE OF WATERLOO 3RD,

Roan, calved April 4, 1879, bred by Mr. R. E. Oliver, Sholebroke Lodge; got by
Grand Duke 30th (38373), dam (Waterloo 43rd) by Grand Duke of Oxford (31293),
g. d. (Waterloo 35th) by Grand Duke 11th (21849), gr. g. d. (Waterloo 27th)
by Duke of Geneva (19614), — (Waterloo 17th) by Red Knight (11976), — by
Grand Duke (10284), — by Third Duke of Oxford (9047), — by Second Cleveland
Lad (3408), — by Duke of Northumberland (1940), — by Norfolk (2377), —
by Waterloo (2816), — by Waterloo (2816).

(43322)   GRAND DUKE OF WATERLOO 4TH,

Red and white, calved June 10, 1879, bred by Mr. R. E. Oliver, Sholebroke
Lodge; got by Grand Duke 30th (38373), dam (Waterloo Bienvenue) by Oxford
Beau (29485), g. d. (Waterloo 33rd) by Grand Duke 11th (21849), gr. g. d.
(Waterloo 25th) by Duke of Geneva (19614), — (Waterloo 17th) by Red Knight
(11976), — by Grand Duke (10284), — by Third Duke of Oxford (9047), — by
Second Cleveland Lad (3408), — by Duke of Northumberland (1940), — by
Norfolk (2377), — by Waterloo (2816), — by Waterloo (2816).

(43323)   GRAND DUKE OF WORCESTER 2ND,

Red, calved October 24, 1879, bred by Mr. H. Allsopp, Hindlip Hall; got
by Third Duke of Hillhurst (30975), dam (Grand Duchess 29th) by Grand Duke
22nd (34062), g. d. (Grand Duchess 24th) by Third Duke of Clarence (23727),
gr. g. d. (Grand Duchess 17th) by Imperial Oxford (18084), — (Grand Duchess
10th) by Grand Duke 3rd (16182), — by Prince Imperial (15095), — by Grand
Duke (10284), — by Cleveland Lad (3407), — by Belvedere (1706), — by Second
Hubback (1423), — by Second Hubback (1423), — by The Earl (646), — by
Ketton 2nd (710), — by Comet (155), — by Favourite (252), — by Daisy Bull
(186), — by Favourite (252), — by Hubback (319), — by J. Brown's Red Bull
(97).

(43324)   GRANDEE,

Roan, calved March 9, 1879, bred by Mr. E. W. Meade-Waldo, Stonewall Park;
got by King Harry (36841), dam (Guidage) by Blinkhoolie (23428), g. d. (Guid-
ance) by Breast Plate (19337), gr. g. d. (Guide Post) by Booth Royal (15673), —
— (Guide Light) by Vanguard (10994), — by Vanguard (10994), — by Baron
Warlaby (7813), — by Londesboro' (6142), — by Ranunculus (2479), — by
Remus (4932), — by Prince Comet (1342), — by Count (170), — by Constellation (163), — by Young Favourite (255).

(43325)   GRAND GWYNNE 5TH,

Red, calved April 7, 1879, bred by Lieut.-Colonel Webb, Elford House; got by
Grand Gwynne 2nd (38384), dam (Parrot Gwynne) by Grand Duke of Lightburne
(26290), g. d. (Polly Gwynne 5th) by Barrington Oxford 2nd (27987), gr. g. d.
(Polly Gwynne 3rd) by Wild Duke 5th (27807), — (Polly Gwynne 2nd) by
Wild Duke 4th (21107), — by Flying Dutchman (10235), — by St. Thomas
(10777), — by Prime Minister (2456), — by Wallace (5586), — by Marmion
(406), — by Merlin (430), — by Layton (366), — by Phenomenon (491), — by
Favourite (252), — by Favourite (252), — by Hubback (319), — by Snowdon's
Bull (612), — by Waistell's Bull (669), — by Masterman's Bull (422), — by
the Studley Bull (626).

(43326)   GRAND GWYNNE 6TH,

Red, calved August 9, 1879, bred by Lieut.-Colonel Webb, Elford House; got by Duke of Barrington 4th (39712), dam (Bee Gwynne) by Marquis 6th (34777), g. d. (A. Gwynne) by Baron Oxford (23375), gr. g. d. (Gipsy Gwynne) by Grand Duke of Lightburne (26290), — (Goody Gwynne) by Grand Duke 5th (19875), — by May Duke (13320), — by Duke of Cambridge (12747), — by Captain Hardinge (10023), — by St. Thomas (10777), — by Prime Minister (2456), — by Marmion (406), — by Merlin (430), — by Layton (366), — by Phenomenon (491), — by Favourite (252), — by Favourite (252), — by Hubback (319), — by Snowdon's Bull (612), — by Waistell's Bull (669), — by Masterman's Bull (422), — by the Studley Bull (626).

(43327)   GRAND JUNCTION,
Roan, calved June 10, 1877, bred by Mr. M. Savidge, Sarsden Lodge Farm ; got by Masterman (34813), dam (Faith) by British Hope (21324), g. d. (Charity) by Lord of the Harem (16430), gr. g. d. (Cherry Pie) by Fitz-Milton (14554), — (Red Cherry) by Prince of Wales (8432), — by Bucephalus (6816), — by Locksley (4240), — by Stanhope (5315), — a Cow bought in the North.

(43328)   GREAT HEART,
Roan, calved June 26, 1879, bred by Mr. E. W. Meade-Waldo, Stonewall Park ; got by King Harry (36841), dam (Golden Hope) by King Richard (26523), g. d. (Golden Drop) by Royal Sovereign (22802), gr. g. d. (Golden Rose) by Blood Royal (14169), — (Golden Lady) by Baron Warlaby (7813), — by Remus (4932), — by Prince Comet (1342), — by Count (170), — by Constellation (163), — by Young Favourite (255).

(43329)   GREY SOMERSET,
Roan, calved August 10, 1879, bred by Mr. J. H. Braikenridge, The Rookery ; got by Lord Prinknash 2nd (38653), dam (Golden Fruit) by King Richard (26523), g. d. (Golden Locket) by Grey Gauntlet (19908), gr. g. d. (Golden Rose) by Blood Royal (14169), — (Golden Lady) by Baron Warlaby (7813), — by Remus (4932), — by Prince Comet (1342), — by Count (170), — by Constellation (163), — by Young Favourite (255).

(43330)   GRIMALDI,
Red and white, calved September 22, 1879, bred by Mr. H. Allsopp, Hindlip Hall ; got by Third Duke of Hillhurst (30975), dam (Graceful) by Marnhull Duke (29283), g. d. (Grace) by Lord Wharfdale (26761), gr. g. d. (Gipsy) by Mameluke (24520), — (Grateful) by Sir John (25154), — by Benedict (39457), — by Prince (18575).

(43331)   GUARANTEE,
Roan, calved January 2, 1878, bred by Mr. J. Evans, Uffington, the property of Mr. J. P. Wilkes, Oakley House ; got by Geneva Prince (36691), dam (Bertha) by Duke of Lancaster (28411), g. d. (Bride) by Manager (24521), gr. g.d. (Bridesmaid) by Romeo (22754), — (Cassia 4th) by Grand Duke of York (12966), — by The Dandy (13846), — by St. John (27413), — by Sir Walter 2nd (10834), — by Major (4338).

(43332)   GUARDIAN,
White, calved November 10, 1877, bred by Mr. S. H. Loy, Keldhead; got by Cleveland 3rd (39600), dam (Rose of the Keld) by Duke of Ryedale 2nd (33727), g. d. (White Rose) by Shuttlecock (25126), gr. g. d. (White Hind) by Seventh Duke of Oxford (17741), — (Bonnie Annie) by Seventh Duke of Oxford (17741), —

by Clement Cleveland (15775), — by Udolpho (13907), — by Lord Stanley (4269), — by Velocipede (5552), — by Priam (2452), — by Jerry (4097), — from the stock of Mr. Booth.

(43333)　　　　　GUARDSMAN,

Roan, calved March 7, 1876, bred by Mr. B. H. Allen, Clifford Priory; got by Wetherby Lad (32838), dam (Grenadine 2nd) by Lord Wild Eyes 2nd (22234), g. d. (Grenadine) by Rob Roy (20694), gr. g. d. (Grenade) by Titan (17125), — (Amazon) by Lord Thoresby (14856), — by Horrox (11591), — by Cleveland Lad (3407), — a Cow of Mr. Bell's, of Mosbro' Hall.

(43334)　　　　　GWYNNE FITZ ROSE,

Red, calved March 16, 1879, bred by Mr. W. Talbot Crosbie, Ardfert Abbey, sold to Messrs. Troupen Morren and Penay Plunkett for the Provincial Government of Liege, Belgium; got by Royal Fitz-Rose (37390), dam (Light of Gwynne) by Northern Light (24670), g. d. (Lady Gwynne) by Lamp of Lothian (16356), gr. g. d. (Sweet Poll Gwynne) by Duke of Cambridge (12747), — (Peg Gwynne) by Young Benedict (15641), — by Sir Harry (10819), — by Sir Thomas (10777), — by Conservative (3472), — by Chorister (3378), — by Wellington (2824), — by Marmion (406), — by Merlin (430), — by Layton (366), — by Phenomenon (491), — by Favourite (252), — by Favourite (252), — by Hubback (319) — by Snowdon's Bull (612), — by Waistell's Bull (669), — by Masterman's Bull (422), — by the Studley Bull (626).

(43335)　　　　　GWYNNE PRINCE,

White, calved February 12, 1879, bred by the Earl of Zetland, Aske Hall; got by Scots Fusilier (35483), dam (Cynthia 6th) by Telemachus 4th (35723), g. d. (Cynthia 3rd) by Bobby (25648), gr. g. d. (Cynthia 2nd) by Fitz-James (19755), — (Cynthia) by Apollo (9899), — by Lottery (10472), — by Duke of Cornwall (5947), — by Prince Ernest (4818), — by Wallace (5586), — by Wellington (2824), — by Marmion (406), — by Layton (366), — by Phenomenon (491), — — by Favourite (252), — by Favourite (252), — by Hubback (319), — by Snowdon's Bull (612), — by Waistell's Bull (669), — by Masterman's Bull (422), — by the Studley Bull (626).

(43336)　　　　　GWYNNE PRINCE,

White, calved October 17, 1879, bred by Mr. J. Martin, Hawkshead Hall; got by Prince of Lightburne 2nd (37258), dam (Flossy Gwynne 3rd) by Marquis of York (34791), g. d. (Flossy Gwynne) by Twelfth Duke of Oxford (19633), gr. g. d. (Fanny Gwynne) by Grand Duke 5th (19875), — (Faustina Gwynne) by May Duke (13320), — by Young Benedict (15641), — by St. Thomas (10777), — by Prime Minister (2456), — by Marmion (406), — by Merlin (430), — by Layton (300), — by Phenomenon (491), — by Favourite (252), — by Favourite (252), — by Hubback (319), — by Snowdon's Bull (612), — by Waistell's Bull (669), — by Masterman's Bull (422), — by the Studley Bull (626).

(43337)　　　　　HAMLET,

Red and little white, calved November 16, 1878, bred by Mr. C. Hicks, Felstead-bury; got by British Heir (33218), dam (Virtue) by Prince Alexander (35082), g. d. (Virgin) by Duke of York (23804), gr. g. d. (Vesta) by Highthorn (13028), — (Vineleaf) by Hamlet (8126), — by Duke of Cambridge (7987), — by Zohrab (5705), — by Commodore (6881), — by Ormsby (4621), — by Barmpton Major (3092), — by Cossack (925), — by Captain (3271).

(43338)   HARBY,

Roan, calved February 18, 1878, bred by Mr. C. A. Cantlie, Keithmore, the property of Mr. J. Simpson, Clunymore; got by Lord Irwin (29123), dam (Spot
3rd) by Argus (27894), g. d. (Spot) by Blair Athol (25638), gr. g. d. (Buttercup)
by Commander (25811), — (Daisy 2nd) by Guy Fawkes (12981), — by Sir Arthur
(12071), — by The Pacha (7612), — by Second Duke of Northumberland (3646),
— by Sillery (5131), — by Carleton (843), — by Diamond (205), — by Diamond
(205).

(43339)   HARD TIMES,

Roan, calved November 2, 1878, bred by Mr. R. A. Jackson, Clayfield; got by
Lord Berwick (38583), dam (Clara's Daughter) by Grand Duke of Cambridge
2nd (26285), g. d. (Clara) by Pammon (22486), gr. g. d. (Duchess of York) by
Charming Boy (17548), — (Duchess of Kent) by Duke of Richmond (7996), —
by Belshazzar (1703), — by Noble Henry (2374), — by Simon (5134), — by
Young George (3885), — by George (276).

(43340)   HARDY GWYNNE,

Roan, calved March 26, 1878, bred by Mr. G. Fox, Elmhurst Hall, the property of
Mr. A. F. Hurt, Alderwasley; got by Twenty-fourth Duke of Airdrie (36460), dam
(Fanny Gwynne) by Lord Oxford 2nd (29192), g. d. (Fauna Gwynne) by Grand
Duke 5th (19875), gr. g. d. (Faustina Gwynne) by May Duke (13320), — (Flora
Gwynne) by Young Benedict (15641), — by St. Thomas (10777), — by Prime
Minister (2456), — by Marmion (406), — by Merlin (430), — by Layton (366),
— by Phenomenon (491), — by Favourite (252), — by Favourite (252), — by
Hubback (319), — by Snowdon's Bull (612), — by Waistell's Bull (669), — by
Masterman's Bull (422), — by the Studley Bull (626).

(43341)   HAREWOOD,

Roan, calved March 4, 1879, bred by the Earl of Harewood, Harewood House;
got by Oxford Cherry Duke 2nd (34972), dam (Wasp) by Phosphate (32064),
g. d. (Honeycomb 4th) by White Satin (27800), gr. g. d. (Honeycomb) by Prince
of Featherstone (29652), — (Honey Flower) by Sultan (15355), — by Musician
(13361), — by Fame (10221), — by Lord Adolphus Fairfax (4249), — by
Barmpton Grazier (3091), — by Rex (4942), — by Wellington (680), — by
Favourite (252).

(43342)   HAROLD,

Roan, calved March 8, 1879, bred by Mr. W. Lett, Rushock; got by Waterloo
Prince (35949), dam (Elfrida 7th) by Lord Darlington 2nd (29096), g. d.
(Elfrida 3rd) by Duke of Liverpool (23766), gr. g. d. (Elfrida 2nd) by Cupid
(14359), — (Elfrida) by Duke of Ulster (12774), — by Koh-i-noor (11642), —
by Saturn (12051), — Empress, bred by Messrs. Rich, Didmarton.

(43343)   HAROLD,

Red and little white, calved August 2, 1879, bred by Mr. A. Graham, Yanwath
Hall; got by The Colonel (35747), dam (Hazeltop 33rd) by Earl of Eglinton
(23832), g. d. (Hazeltop 19th) by Edgar (19680), gr. g. d. (Hazeltop 12th) by
Lablache 2nd (20092), — (Hazeltop 8th) by Prince of Glo'ster (13517), — by
Abraham Parker (9856), — by Cleatham (5861), — by Majesty (2250), — by
The Duke (5432).

(43344)   HAWKSEYE,

Red, calved April 5, 1879, bred by Earl Beauchamp, Madresfield Court; got by

Nettleworth (38792), dam (Lavinia 3rd) by Marquis of Blandford (34779), g. d. (Lavinia 2nd) by Festival (26147), gr. g. d. (Lavinia) by Diplomatist (19571), — (Bess) by Impatience (14723), — by Hero (8146), — by Major (4338), — by Ormsby (4621), — by Cossack (925), — by Alpha (3004), — by Captain (3271), — by Captain (108).

(43345) **HAYLE,**
Roan, calved January 20, 1879, bred by Mr. D. R. Scratton, Ogwell; got by Duke of Oxford 33rd (36528), dam (Lally Geneva) by Grand Duke of Geneva (28756), g. d. (Lally 7th) by Third Lord Oxford (22200), gr. g. d. (Lally 2nd) by Malachite (18313), — (Lally) by Earl of Derby (10177), — by Earl of Liverpool (9061), — by Second Duke of Cambridge (3638), — by Belvedere (1706), — by Son of Herdsman (304), — by Wonderful (700), — by Alfred (23), — by Young Favourite (6994).

(43346) **HECTOR,**
Roan, calved April 10, 1879, bred by Mr. F. Morice, Springfield; got by Undergraduate (35835), dam (Red Rose) by Vauban (27708), g. d. (Jenny Jones 5th) by Hero of Thorndale (18061), gr. g. d. (Jenny Jones) by Young Shaftoe (9625), — (Jenny Lind) by Zero (8799), — by Old Bull (8368), — by Young Spectator (8619), — by Phantassie (8389), — by Young Rockingham (8498).

(43347) **HEIR OF THORNDALE,**
Red, calved February 28, 1879, bred by Mr. H. Allsopp, Hindlip Hall; got by Duke of Collingham 3rd (38134), dam (Thorndale Maid 3rd) by Marnhull Duke (29283), g. d. (Thorndale Maid) by Third Duke of Thorndale (17749), gr. g. d. (Passion Flower 3rd) by Fourth Duke of Thorndale (17750), — (Passion Flower) by Homer (16277), — by Bashaw (12449), — by Enterprise (11443), — by Young Consul (6893), — by Newnham (2365), — by Satellite (1420), — by Jupiter (342), — by Sir Oliver (605), — by Trunnell (659), — by Favourite (252), — by Favourite (252), — by Dalton Duke (188), — by R. Alcock's Bull (19), — by J. Smith's Bull (608), — by Jolly's Bull (337).

(43348) **HERCULES,**
Roan, calved May 19, 1879, bred by Mr. J. Angus, Bearl; got by Hawthorn (36751), dam (Milk Maid) by Roan Chief (27294), g. d. (Cheerful) by Warlock (19113), gr. g. d. (Cherry Blossom) by Frederick (14571), — (Young Cherry) by Lahore (11659), — by Traveller (7646), — by Fitz-Maurice (3807), — by Reformer (2502), — by Leopold (2199), — by Sir Harry (5155), — by Traveller (655), — by Colonel (152), — by Son of Hubback (319).

(43349) **HERO,**
Roan, calved February 25, 1879, bred by Mr. F. Morice, Springfield; got by Undergraduate (35835), dam (Duchess of Thorndale 6th) by Victor (32774), g. d. (Duchess of Thorndale 5th) by Hero of Thorndale (18061), gr. g. d. (Duchess of Thorndale 3rd) by Hero of Thorndale (18061), — (Gold Lace) by Soubadar (18901), — by Western Wonder (17225), — by Duke of Beauford (11377), — by Humber (7102), — by Guardian (3947), — by Firby (1040), — by Hector (1104), — by Regent (544), — Old York.

(43350) **HERO,**
White, calved March 4, 1879, bred by Mr. R. W. Gaussen, Brookmans Park; got by Duke of Worcester 5th (39797), dam (White Butterfly) by Grand Duke

of Geneva (28756), g. d. (Pride of Aylesford 2nd) by Grand Duke 16th (24063), gr. g. d. (Pride of Aylesford) by Sixth Duke of Airdrie (19602), — (Maid of Aylesford) by Parade (18518), — by Lord Ruby (14851), — by Preston (20514), — by Monzani (6222), — by Firby (1040), — by Arbutus (1649).

(43351) YOUNG HERO,
Red and white, calved June 11, 1879, bred by Mr. C. P. Noel, Bell Hall, the property of Mr. M. Roe, Stoke; got by King Hal (40052), dam (Marigold) by Earl Mortimer (33780), g. d. (Lupine) by Lord Red Eyes 2nd (24460), gr. g. d. (Lydia) by Viscount Middleham (25385), — (Favourite) by Castlereagh (19409), — by Hero of Oxford (18060), — by Prince of Orange (15101), — by Hero (8146), — by Major (4338), — by Ormsby (4621), — by Cossack (925), — by Alpha (3004), — by Captain (3271).

(43352) HEYDON DUKE 10TH,
Roan, calved May 3, 1879, bred by Lord Braybrooke, Audley End, the property of Mr. J. J. Sharp, Broughton; got by Duke of Rosedale 3rd (33723), dam (Heydon Rose 2nd) by Third Duke of Geneva (23753), g. d. (Heydon Rose) by Englishman (19701), gr. g. d. (The Beauty) by Puritan (9523), — (Cambridge Rose 6th) by Third Duke of York (10166), — by Second Cleveland Lad (3408), — by Belvedere (1706), — by Belvedere (1706), — by Second Hubback (1423), — by His Grace (311), — by Yarborough (705), — by Favourite (252), — by Punch (531), — by Foljambe (263), — by Hubback (319).

(43353) HILLHURST CHERRY,
Red and white, calved April 17, 1879, bred by Mr. G. Murton Tracy, Redlands; got by Hillhurst Cherry Duke (39999), dam (Cherry Empress 2nd) by Horsa (34188), g. d. (Cherry Empress) by Cherry Prince 4th (25765), gr. g. d. (Cherry Blanche) by Dairy Prince (17655), — (Sultana) by Sultan (15354), — by Cardinal Wiseman (12560), — by Son of Clementi (3399), — by Magnum Bonum (2243), — by Thorp (2757), — by Pirate (2430), — by Houghton (318), — by Marshal Blucher (416), — from the stock of Messrs. Wright and Charge.

(43354) HILL TULIP LAD 5TH,
Roan, calved January 2, 1879, bred by Mr. C. Sturge, Bewdley; got by Foster Brother (36661), dam (Tulip 3rd) by Saracen (25091), g. d. (Tulip) by General (17947), gr. g. d. (Twilight) by Hector (9200), — (Thistle) by Eclipse (3684), — by Boughton (7841), — by Young Merlin (6204), — by Midas (435), — by Denton (198).

(43355) HIS ROYAL HIGHNESS,
Roan, calved August 10, 1868, bred by Mr. J. Laxton, Morborne; got by Baron Windsor (23391), dam (Judith 4th) by Rebellion (10684), g. d. (Judith 2nd) by Morborne (18420), gr. g. d. (Judith) by Orontes (4623), — (Rosamond) by Alabaster (1616), — by Monarch (2324), — by Satellite (1420), — by Cato (119), — by Jupiter (342), — by George (273), — by Chilton (136), — by Irishman (328), — by B. (45).

(43356) HOGARTH,
Red and white, calved May 25, 1878, bred by Mr. A. Graham, Yanwath Hall; got by The Colonel (35747), dam (Hazeltop) by Dean (30863), g. d. (Harmless) by Midsummer (22358), gr. g. d. (Hardness) by Baronet (19274), — by Duke (14421), — by General Haynau (11520).

(43357)   HONEYCOMB,

Red, calved July 3, 1877, bred by Mr. J. A. M. Cope, Drummilly, the property of Lord Carbery, Laxton Hall; got by Duke of Bradwardine (28365), dam (Honeybell 2nd) by Third Duke of Clarence (23727), g. d. (Honeybell) by Grand Duke of Oxford (16184), gr. g. d. (Helen) by Oregon (8371), — (Honeysuckle) by Premier (7344), — by Bellerophon (3119), — by Alderman (2976), — by Waterloo (2816), — by Young Wynyard (2859), — by Son of Simon (590), — by Buston's Styford (103).

(43358)   HONEYCOMB,

Roan, calved July 4, 1879, bred by Mr. T. Purkis, West Wratting Grange ; got by Wharfdale Oxford (27786), dam (Sweetheart 36th) by Baron Napier (30471), g. d. (Sweetheart 35th) by Patrician (24728), gr. g. d. (Sweetheart 11th) by The Baron (13833), — (Sweetheart 3rd) by Daybreak (11338), — by Accordion (5708), — by Little John (4232), — by Caliph (1774), — by Sir Walter (2637), — by Hotspur (1117), — by Coxcomb (928), — by Midas (435), — by Comet (155), — by R. Colling's Son of Favourite (252), — by same Son of Favourite (252), — by Hubback (319).

(43359)   HONEYMOON,

Red and white, calved July 12, 1879, bred by Mr. J. A. M. Cope, Drummilly ; got by Glo'ster's Gwynne (38360), dam (Honey Lips) by Third Duke of Clarence (23727), g. d. (Honey Pride) by Grand Duke of Oxford (16184), gr. g. d. (Helen) by Oregon (8371), — (Honeysuckle) by Premier (7344), — by Bellerophon (3119), — by Alderman (2976), — by Waterloo (2816), — by Young Wynyard (2859), — by Son of Simon (590), — by Buston's Styford (103).

(43360)   HOPEFUL EARL,

Red and white, calved June 7, 1879, bred by Mr. H. Smith, Mountmellick ; got by Earl of Aylesby (38215), dam (Rose of Hope) by Conqueror (30786), g. d. (Rose of the Heath) by Prince Bertram (27119), gr. g. d. (Young Moss Rose) by Dr. McHale (15887), — (Young Harriette) by Dorrington (17691), — by Hopewell (10332), — by Baron Warlaby (7813), — by Pilgrim (4701), — by Shakespeare (2614), — by Sheridan (2616), — by Normanby (1278), — by a Bull of Mr. W. Smith's, West Rasen.

(43361)   HORNSHAY DUKE 8TH,

Red and white, calved May 14, 1879, bred by Mr. S. Bailey, Hornshay ; got by Duke of Wellington 10th (36554), dam (Baby 2nd) by Cardinal (28144), g. d. (Baby) by Augustus Windsor (19248), gr. g. d. (Anemone 4th) by Upstart (9760), — (Anemone) by Allan-a-dale (7778), — by Little John (4232), — by Caliph (1774), — by Swing (2721), — by Argus (759), — by Defender (194), — by Petrarch (488), — by Own Brother to R. Colling's White Heifer, — by Butterfly (104), — by Globe (278).

(43362)   HOTSPUR MILCOTE,

Red, calved March 4, 1879, bred by Mr. J. C. Adkins, Milcote ; got by Hotspur (38440), dam (Fawsley Rose) by Prince of Fawsley (32171), g. d. (Galleon) by Morocco (16590), gr. g. d. (Polytint) by Earl of Dublin (10178), — (Cornbind) by Janizary (8175), — by Snowball (8602), — by Little John (4232), — by Caliph (1774), — by Rob Roy (557), — by Satellite (1420), — by Sir Dimple (594), — by Styford (629).

(43363)   HOVINGHAM,
White, calved October 7, 1879, bred by Sir W. C. Worsley, Bart., Hovingham; got
by Sir Arthur Ingram (32490), dam (Irwin's Star) by Lord Irwin (29123), g.d.
(Louise) by White Windsor (27803), gr. g. d. (Mushroom) by Earl of Windsor
(17788), — (Beauty 2nd) by Magnus Troil (14880), — by Bates (12451), — by
General Fairfax (11519), — by Liberator (7140), — by Prince Albert (4791),
— by Young Matchem (4422), — by Young Red Rover (4904), or Rockingham
(2551), — by Whisker (1579), — by Pilot (496).

(43364)   HUGH,
Roan, calved February 24, 1879, bred by Mr. J. Morton, Fences Farm ; got by
Hesperus (39994), dam (Julia) by High Sheriff (26392), g. d. (Jewess) by
Rifleman (18704), gr. g. d. (Jewel) by Meteor (18391), — (Judith 3rd) by
Rebellion (10684), — by Orontes (4623), — by Alabaster (1616), — by Monarch
(2324), — by Satellite (1420), — by Cato (119), — by Jupiter (342), — by
George (273), — by Chilton (136), — by Irishman (329), — by B. (45).

(43365)   HUZ,
Roan, calved February 11, 1879, bred by the Duke of Northumberland, Alnwick
Castle ; got by Sir Raymond (40716), dam (Modesty) by Brigadier (28078), g. d.
(Rosette) by Briton (25686), gr. g. d. (Maid of Aln) by Melsonby (18380), —
(Young Jessy) by George 3rd (16147), — by Shaftoe (5107), — by Richardi
(4944), — Cowslip, bought of Mr. Patterson, of Wood Houses.

(43366)   HYDER ALI,
Red, calved May 23, 1878, bred by Mr. A. Scott, Towie Barclay ; got by Roseberry (39016), dam (Lady Grace) by Lord Forth (26649), g. d. (Polly Perkins) by
Viceroy (19054), gr. g. d. (Heroine) by Earl of Aberdeen (12800), — (Bloomer)
by Seafield (9616), — by Sultan (5349), — by Plenipo (4725), — by Abbot
(2899).

(43367)   ICEBERG,
White, calved May 20, 1879, bred by the Duke of Northumberland, Alnwick
Castle; got by Sir Raymond (40716), dam (Selina 8th) by Duke of Tyne (33744),
g. d. (Selina 7th) by Royal Gwynne (22784), gr. g. d. (Selina 2nd) by Royal
Duke (16865), — (Selina) by Guy Faux (12980), — by Young Pompey (13480),
— by Broomborough (9994), — by The Minstrel (8687), — by Bloomsbury
(3171), — by Red Rover (4906), — by Wharfdale (1578), — by Palemon (479),
— by Meteor (432), — by Western Comet (689), — by Favourite (252), — by
Cupid (177), — by Grandson of Bolingbroke (280), — by Foljambe (263), —
by R. Alcock's Bull (19), — by J. Smith's Bull (608), — by Jolly's Bull (337).

(43368)   IDSAL 2ND,
Roan, calved July 14, 1879, bred by Mr. S. L. Horton, Park House ; got by
Marquis of Blandford 6th (41983), dam (Maria Fawsley) by Marquis (31830),
g. d. (Maria) by Second Duke of Wharfdale (19649), gr. g. d. (Mary 2nd) by
Marc Antony (14895), — (My Mary) by British Yeoman (8906), — by Adrian
(5714), — by Noble (4577), — by Young Don Juan (3610), — by Young Isaac
(2154), — by Son of Young Dimple (6518), — by Young Dimple (971), — by
Snowball (2648), — by Layton (2190).

(43369)   IMPERIAL CHERRY,
Red and white, calved July 8, 1879, bred by Mr. G. Murton Tracy, Redlands ; got

by Hillhurst Cherry Duke (39999), dam (Cherry Empress 3rd) by Cherry Duke
of Glo'ster (36348), g. d. (Cherry Empress 2nd) by Horsa (34188), gr. g. d.
(Cherry Empress) by Cherry Prince 4th (25765), — (Cherry Blanche) by Dairy
Prince (17655), — by Sultan (15354), — by Cardinal Wiseman (12560), — by
Son of Clementi (3399), — by Magnum Bonum (2243), — by Thorp (2757), —
by Pirate (2430), — by Houghton (318), — by Marshal Blucher (416), — from
the stock of Messrs. Wright and Charge.

(43370)   IMPERIALIST,
Red, calved July 3, 1878, bred by Mr. T. Hewer, Inglesham ; got by Lothair
(40270), dam (Lady Alice) by The Prince (32697), g. d. (Honeycomb) by Herne's
Oak (21924), gr. g. d. (Honeydew) by Runnimede (15221), — (Alva) by Waterloo (11025), — by Wroughton (8789), — by Young Phœnix (7331).

(43371)   IMPUDENCE,
Red and white, calved March 26, 1879, bred by Mr. G. W. Lambart, Beau Parc ;
got by Jupiter (38477), dam (Blush) by British Sailor (23472), g. d. (The Rose)
by Sir Robert (22912), gr. g. d. (Agga 2nd) by British Flag (21323), — (Ellen)
by Minos (11816), — by Premier (11920), — by Narcissus (4540), — by Linton
(4227).

(43372)   INDUSTRY,
Red, calved July 11, 1877, bred by Captain Ashby, Naseby Woolleys, the property of Mr. G. Stratton, Husbands Bosworth ; got by Fidelio (33912), dam
(Inquiry) by Third Duke of Geneva (21592), g. d. (Invoice) by Pan (18516),
gr. g. d. (Inquest) by Field Marshal (14545), — (Sultana) by Modbury Premium
(11820), — by Vagabond (9765), — by Locksley (4240), — by Phosphorus
(4694), — by Antonio (3019), — by Thorpe (2757), — by St. Leger (1414), —
by Comus (1861), — by Own Brother to Remus (550).

(43373)   IN-ESSE,
Red, calved March 26, 1879, bred by Earl Fitzwilliam, Coollattin Park, sold to
the Provincial Government of Brabant, Belgium ; got by Robert Burns (29795),
dam (Iris) by Lord Stanley (24466), g. d. (Isabel) by Fusilier (24000), gr. g. d.
(Irish Girl) by Barley Sugar (11140), — (Fleda) by Baron Warlaby (7813), —
by Prince Ernest (7366), — by Bright (1739), — by Shylock (2622), — by
Percy (1314), — by Blucher (84), — by Jupiter (344), — by Marshal Beresford
(415), — by Crocus (932), — by a Bull of Mr. R. Colling's.

(43374)   INGLEBY PRINCE,
Roan, calved February 8, 1868, bred by Mr. W. Godson, Normanby ; got by
Lord McDonald (22175), dam by Plum (24757), g. d. by Lord Mar (16424),
gr. g. d. by Comptroller (14300), — by Mario (11779).

(43375)   INGOLDSBY,
Roan, calved March 24, 1879, bred by Earl Fitzwilliam, Coollattin Park, sold to
the Provincial Government of Liege, Belgium ; got by King Lud (34327), dam
(Ingot's Pet) by Robert Burns (29795), g. d. (Ingot) by Lord of the Isles
(26706), gr. g. d. (Indiana) by Fusilier (24000), — (Ireland's Glory) by Clydesdale (15779), — by Barley Sugar (11140), — by Baron Warlaby (7813), — by
Prince Ernest (7366), — by Bright (1739), — by Shylock (2622), — by Percy
(1314), — by Blucher (84), — by Jupiter (344), — by Marshal Beresford (415),
— by Crocus (932), — by a Bull of Mr. R. Colling's.

(43376)   INGOMAR,

Red, calved September 11, 1879, bred by Messrs. J. G. W. and W. R. Lyle, Donaghmore House; got by Lord of the Manor (38641), dam (Maybloom) by Sir William (32540), g. d. (Bessy) by Lord Francis (24393), gr. g. d. (Grisi) by Duke of Montrose (21599), — (De Beriot) by Lumley (16478), — by Master Goldschmidt (13315), — by Marquis (11786), — by The Lord of Gilling (6587), — by Buckingham (3239), — by Son of Chance (868), — by Chance (868), — by Grazier (1085), — by Plutarch (1327), — by Midas (435), — by Grandson of Simon (590).

(43377)   INKERMANN,

Roan, calved May 15, 1877, bred by Mr. J. Johnstone, Halleaths, the property of Mr. J. Jardine, Dryfeholm; got by Magician (34720), dam (Ella) by Keir Butterfly 8th (28951), g. d. (Miss Eliza) by Rifleman (32304), gr. g. d. (Eliza 2nd) by Mac Turk (14872), — (Edith) by Pride (10631), — by Baron of Ravensworth (7811), — by Sir Thomas (5194), — by Noble Henry (2374), — by Abraham (2905), — by Mustachios (4527), — by Simon (5134), — by Young George (3885), — by George (276).

(43378)   INVERTHERNIE,

Red, calved February 11, 1879, bred by Mr. A. Scott, Towie Barclay; got by Roseberry (39016), dam (Jealousy 14th) by Prince Frederick of Cambridge (29621), g. d. (Jealousy 3rd) by Lord Buckingham (20151), gr. g. d. (Young Jealousy) by Scarlet Velvet (16916), — (Jealousy) by Baron Warlaby (7813), — by General Washington (6036), — by Virgil (7686), — by Son of Emperor (1014), — by Young Favourite (255), — by Young Major (1186).

(43379)   IRISH HOPE,

Roan, calved July 22, 1877, bred by Mr. G. Allen, Unicarville, the property of Messrs. T. and J. Macafee, Carrysiskan; got by Irish Hero (36791), dam (Belle of Hope) by Great Hope (24082), g. d. (The Belle of Scarboro') by Blood Royal (23429), gr. g. d. (Regina) by Northern Prince (31986), — (Regalia) by Regenerator, — by Gainford 4th (11501), — by Stars and Stripes (12148), — by Sir Henry (10824), — by Buckingham (3239), — by Marcus (2262), — by Matchem (2281), — by Alderman (1622), — by Pilot (496), — by Remus (550), — by Sir Charles (592), — by R. Colling's Son of Favourite (252), — by same Son of Favourite (252), — Strawberry.

(43380)   IRISH O'CONNOR,

Roan, calved April 8, 1879, bred by Mr. R. Jamison, Movilla; got by Irish Hero (36791), dam (Molly Connor) by Lord Spencer (26738), g. d. (Molly Asthore) by Emperor (12835), gr. g. d. (Miss Molly) by Puck (18655), — (Lady Rachel) by Duke of Bedford (11378), — by Coriander (6895), — by Son of Prince George (2464), — by Matchem (2281).

(43381)   IRISH SOVEREIGN,

Roan, calved May 10, 1878, bred by Mr. W. G. Garne, Broadmoor, the property of M. A. de Wonck, Belgium; got by Royal Fitz-Rose (37390), dam (Soverina) by Irish Baron (31417), g. d. (Sovereign's Beauty) by Royal Sovereign (20800), gr. g. d. (Beam of Beauty) by Lamp of Lothian (16356), — (Lady Cicely) by Admiral (12340), — by Monarch (13346), — by Favour (9114), — by Albert 3rd (6725), — by Valentine (5540), — by Roderick (2554), — by Governor (1077), — by Son of Wellington (680).

(43382) IRONMASTER,
Roan, calved January 29, 1876, bred by Mr. R. Jefferson, Preston Hows, the property of Mr. T. Burgess, Burleydam; got by Forlorn Hope (33957), dam (Lively) by Prince Alfred (29593), g. d. (Laura Maria) by Grand Duke 17th (24064), gr. g. d. (Rita 1st) by Podacres (24760), — (Rita) by Cerdic (19415), — by Len (14788), — by Voltaire (13963), — by Lord George (10439), — by Vanguard (10994), — by Hamlet (8126), — by Sir Roger (7515), — by Studley Royal (5342).

(43383) IRVING,
Red, calved April 28, 1879, bred by Mr. A. Fawkes, Farnley Hall; got by Baron Winsome 6th (33111), dam (Lady Imogene) by The Baron (32654), g. d. (Lady Isola) by Lord Darlington (26633), gr. g. d. (Isabel) by Royal Oak (16873), — (Beauty) by General Bosquet (14591), — by Triumph (8717), — by Matchless (4428), — by Boughton (2868), — by Roman (2559), — by Columella (904), — by Albion (14), — by Cinnamon (139), — by Neswick (1266).

(43384) IRWIN MYRTLE,
Red, calved October 20, 1879, bred by Mr. C. A. Cantlie, Keithmore; got by Lord Irwin (29123), dam (Myrtle 29th) by Baron's Heir (36205), g. d. (Myrtle 21st) by Boanerges (25647), gr. g. d. (Myrtle 20th) by Prince Louis (20560), — (Myrtle 6th) by Cecil (12571), — by Fairfax Hero (9016), — by Duplicate Duke (6952), — by Prince (8414), — by Sir Walter (2639), — by Son of Emperor (2671), — by Wonderful (700), — by Cardinal (841), — by Baronet (61), — by Cleveland (145), — by Butterfly (104), — by Globe (278).

(43385) ISLAND PRINCE,
Red, calved June 17, 1878, bred by Mr. J. B. Houston, Orangefield, the property of Mr. T. Wynne, Lislea; got by British Buck (42840), dam (Island Queen) by Robert Burns (29795), g. d. (Irish Girl 8th) by Chief Justice (28188), gr. g. d. (Irish Girl) by Barley Sugar (11140), — (Fleda) by Baron Warlaby (7813), — by Prince Ernest (7366), — by Bright (1739), — by Shylock (2622), — by Percy (1314), — by Blucher (84), — by Jupiter (344), — by Marshal Beresford (415), — by Crocus (932), — by a Bull of Mr. R. Colling's.

(43386) ISLE AXHOLME 3RD,
Red and white, calved in July, 1879, bred by Mr. S. C. Brunyee, Sand Hall; got by Isle Axholme 2nd (40024), dam (Barbara) by Barbarossa (30415), g. d. (Bianca) by Sir Hugh Rose (20830), gr. g. d. (Lady Blanche) by Wild Watch (12307), — (Virgin) by Hamlet (8126), — by Roan Duke (8486), — by Muley (4519), — by Young Eryholme (1981), — by Wonderful (700), — by Apollo (36), — by Merlin (429), — by Alfred (23), — by Butterfly (104), — by Suworrow (636).

(43387) ISLESMAN,
Roan, calved November 22, 1878, bred by Mr. T. F. Jamieson, Mains of Waterton; got by Lord of the Isles (40218), dam (Blanche 9th) by Pride of the Isles (35072), g. d. (Blanche 6th) by Banner Bearer (27907), gr. g. d. (Red Duchess) by Keir Butterfly 1st (24235), — (Red Roan Duchess) by Duke of Wharfdale (19648), — by Master Frederick (18348), — by Valiant (10989), — by Frederick (11489), — by Whittington (12299), — by Second Cleveland Lad (3408), — by Duke of Northumberland (1940), — by Norfolk (2377), — by Belvedere (1706), — by Belvedere (1706), — by Lancaster (360), — by Petrarch (488),

— by Major (397), — by Chapman's Son of Punch (122), — by Dickson's Grandson of Punch (213), — by Checks (132), — by R. Grimston's Bull (282), — by J. Coates's Bull (148).

(43388) IVANHOE,

Roan, calved July 18, 1878, bred by Mr. E. Robinson, Nafferton, the property of Mr. Mawson; got by Halloo (34105), dam (Eva 3rd) by Master Edmund (29322), g. d. (Eva) by Lord Valentine (26751), gr. g. d. (Beach) by Trustee (23089), — (Birthday) by Prince Tom (15110), — by Horatio (13046), — by Launcelot (11670), — by Maltster (9362), — by Splendid (5298), — by Adjutant (2924), — by Ambo (1636), — by a Bull of the Earl of Carlisle's.

(43389) JACKADANDY,

White, calved April 7, 1879, bred by Sir G. R. Philips, Bart., Weston Park; got by Grand Duke 29th (38372), dam (Jardiniere) by John of Oxford (34267), g. d. (Junket) by Jacket (24199), gr. g. d. (Judith) by Juvenile (22021), — (Judy) by Second Duke of Airdrie (19600), — by Hayman (16245), — by Euxine (12845), — by Lycurgus (7180), — by Sweet William (5368), — by Javelin (4093), — by Blyth (797), — by Wellington (684), — by Phenomenon (491), — by Favourite (252), — by Favourite (252), — by Favourite (252), — by Hubback (319), — by Snowdon's Bull (612), — by Waistell's Bull (669), — by Masterman's Bull (422), — by the Studley Bull (626).

(43390) JACK FROST 2ND,

Red, calved March 29, 1875, bred by Sir J. Rolt, Ozleworth Park, the property of Mr. Wild, Cold Eaton; got by Duke of Hazlecote 23rd (30970), dam (White Frost) by Royal Butterfly 17th (22774), g. d. (Chill) by Champagne (17522), gr. g. d. (Snowball) by John Ford (9253), — (Snowdrop) by Fourth Duke of York (10167), — by Sir Thomas Fairfax (5196), — by Young Sea Gull (5100), — by Crusader (934), — by Sultan (1485), — by White Comet (1582), — by Son of Chilton (136), — by Bolingbroke (86).

(43391) JACK MILCOTE,

Roan, calved August 17, 1879, bred by Mr. J. C. Adkins, Milcote; got by Hotspur (38440), dam (Jantjana 2nd) by Lord Calthorpe (26619), g. d. (Jantjana) by Streamer (25241), gr. g. d. (Jantja 3rd) by Euxine (12845), — (Jantja) by Lycurgus (7180), — by Sweet William (5368), — by Javelin (4093), — by Blyth (797), — by Wellington (684), — by Phenomenon (491), — by Favourite (252), — by Favourite (252), — by Favourite (252), — by Hubback (319), — by Snowdon's Bull (612), — by Waistell's Bull (669), — by Masterman's Bull (422), — by the Studley Bull (626).

(43392) JANIZARY,

Roan, calved December 22, 1879, bred by Colonel R. Loyd Lindsay, Lockinge Park; got by Earl of Horton 11th (36588), dam (Jessica) by Rob Roy (29806), g. d. (Judy) by Second Duke of Airdrie (19600), gr. g. d. (Jantja 9th) by Hayman (16245), — (Jantja 3rd) by Euxine (12845), — by Lycurgus (7180), — by Sweet William (5368), — by Javelin (4093), — by Blyth (797), — by Wellington (684), — by Phenomenon (491), — by Favourite (252), — by Favourite (252), — by Favourite (252), — by Hubback (319), — by Snowdon's Bull (612), — by Waistell's Bull (669), — by Masterman's Bull (422), — by the Studley Bull (626).

(43393)   JANUARY DUKE,

Roan, calved January 3, 1878, bred by Mr. G. Underwood, Little Gaddesden; got by Duke of Jantja (33675), dam (Princess Geneva 6th) by Vespasian (32761), g. d. (Princess Geneva) by Prince of Geneva (24829), gr. g. d. (Guess 3rd) by The Duke (32671), — (Guess 2nd) by The Duke (32671), — by Baronet (11149), — by The Star (10951), — by Admiral (8806), — by Don Juan (1923), — by Cawdor (862), — by Marshal Beresford (415), — by Wellington (683), — by Windsor (698), — by Middleton's Bull (438), — by Chapman's Bull, of Dinsdale.

(43394)   JERICHO,

Red and white, calved June 21, 1879, bred by Colonel R. Loyd Lindsay, Lockinge Park; got by Don Carlos (39693), dam (Josephine) by Baron Wetherby 2nd (21243), g. d. (Jessamine) by Wolfsbane (15518), gr. g. d. (Julia) by Dukedom (12729), — (Juno) by Monarch (13347), — by Evander (6981), — by Zenith (5702), — by William (2840), — by Wiseton (2848), — by Alabaster (1616), — by Son of Esk (1019), — by Regent (1366), — by Wellesley (1571), — by Viscount (666), — by Mac George (1183).

(43395)   JESTER KING,

Red and white, calved July 31, 1878, bred by Mr. J. H. Rockett, Snaith Hall; got by Jester (41718), dam (Cherry Ripe) by Baron Percy (30477), g.d. (Cherry Luck) by Golden Cherry (31263), gr. g. d. (Lady Miney) by Baron Wetherell (19289), — (Lady Minna) by High Sheriff (16271), - - by Omer Pacha (13421), — by Tom of Lincoln (8714), — by Buchan Hero (3238), — by Lord John (4257), — by Berryman (3143), — by Lenny (2197), — by Cappy (1782), — by Western Comet (689).

(43396)   JESUIT,

Roan, calved January 28, 1879, bred by Mr. G. Graham, The Oaklands; got by Duke of Yardley (36556), dam (Jessica 4th) by Lord Thorndale (29210), g. d. (Jessica) by Lodowick (20136), gr. g. d. (Lady Jocelyn) by Seventh Duke of York (17754), — (Lady Jane) by Duke of Glo'ster (11382), — by Lord Warden (7167), — by Fawsley (6004), — by Warden (5595), — by Javelin (4093), — by Blyth (797), — by Wellington (684), — by Phenomenon (491), — by Favourite (252), — by Favourite (252), — by Favourite (252), — by Hubback (319), — by Snowdon's Bull (612), — by Waistell's Bull (669), — by Masterman's Bull (422), — by the Studley Bull (626).

(43397)   JINGO,

Roan, calved January 22, 1879, bred by Sir G. R. Philips, Bart., Weston Park; got by Grand Duke 29th (38372), dam (Joyous) by Cherry Grand Duke 5th (30712), g. d. (Joyful) by Cherry Duke (25752), gr. g. d. (Junia) by Duke of Glo'ster (11382), — (Joan) by Lycurgus (7180), — by Javelin (4093), — by Mac Ivor (2237), — by Northern Light (1280), — by Wellington (684), — by Phenomenon (491), — by Favourite (252), — by Favourite (252), — by Favourite (252), — by Hubback (319), — by Snowdon's Bull (612), — by Waistell's Bull (669), — by Masterman's Bull (422), — by the Studley Bull (626).

(43398)   JOCK,

Red and white, calved July 7, 1879, bred by Earl Beauchamp, Madresfield Court; got by Magnum Bonum (41954), dam (Lavinia) by Diplomatist (19571), g. d.

(Bess) by Impatience (14723), gr. g. d. (Berangeria) by Hero (8146), — (Primrose) by Major (4338), — by Ormsby (4621), — by Cossack (925), — by Alpha (3004), — by Captain (3271), — by Captain (108).

  (43399)   JORROCKS,
Red and white, calved March 20, 1879, bred by Sir G. R. Philips, Bart., Weston Park; got by Grand Duke 29th (38372), dam (Juanita) by Jay (26457), g. d. (Duchess Janette) by Second Duke of Airdrie (19600), gr. g. d. (Countess Janette) by Lilyvick (10421), — (Young Jocund) by Japetus (10350), — by Monarch (7249), — by Mac Ivor (2237), — by Northern Light (1280), — by Wellington (684), — by Phenomenon (491), — by Favourite (252), — by Favourite (252), — by Favourite (252), — by Hubback (319), — by Snowdon's Bull (612), — by Waistell's Bull (669), — by Masterman's Bull (422), — by the Studley Bull (626).

  (43400)   JOSEPHUS VLADIMER,
Roan, calved October 13, 1879, bred by Mr. J. Waind, Ankness; got by Count Vladimer (33461), dam (Gipsy Maid) by Snowstorm (35616), g. d. (Countess) by Wathstone's Hero (25417), gr. g. d. (Cygnet) by Captain (14229), — (Citron) by Captain (14229), — by Chevy Chase (7897), — by Monarch (9407), — by Gainford (2044), — by Samson (5080), — by Young Sovereign (5286), — by White Walton (5652).

  (43401)   JOVIAL BOY,
Roan, calved February 23, 1879, bred by Captain Beak, Somerford; got by Pompey (35059), dam (Joyful) by Second Duke of Wateringbury (26024), g. d. (Jennie Dore) by Purple Emperor (27222), gr. g. d. (Jennie Dawson) by Viscount Killerby (19081), — (Jennie Deans) by Pluralist (10620), — by Sir Joseph Paxton (12088), — by Augustus (6751), — by Raffler (7391).

  (43402)   J. P. D.,
Red and white, calved December 24, 1876, bred by Mr. S. H. Allen, Eastover; got by J. P. (41731), dam (Datura) by Æolus (27861), g. d. (Dorcas) by Clint (33399), gr. g. d. (Dodo) by Star Prince (18925), — by Savernake (35475), — (Violet) by Mazeppa (34831).

  (43403)   JUDGE,
Roan, calved October 24, 1879, bred by Mr. J. Singleton, Teresa Cottage; got by Lord Cockburn (38594), dam (Jessy 38th) by Sixth Duke of Kirklevington (30982), g. d. (Jessy 35th) by Paris (20469), gr. g. d. (Jessy 32nd) by Cerdic (19415), — (Jessy 31st) by Master Gwynne (16539), — by General Havelock (14598), — by Despot (11348), — by Pantaloon (9467), — by Tommy Lad (6611), — by Young Rockingham (2549), — by Monitor (2331), — by Young Rockingham (2549), — by Major (2255), — by a Bull of Mr. Newby's, — by Northumberland (464), — by Buston's Styford (103), — by Bolingbroke (86).

  (43404)   JUPITER,
Red, calved October 2, 1878, bred by Sir G. R. Philips, Bart., Weston Park; got by Grand Duke 29th (38372), dam (Jasmine) by Third Duke of Geneva (21592), g. d. (Desdemona) by Colonel Dan (21445), gr. g. d. (Jessica) by Seventh Duke of York (17754), — (Jardine) by Lord Warden (7167), — by Fawsley (6004), — by Warden (5595), — by Javelin (4093), — by Blyth (797), — by Wellington (684), — by Phenomenon (491), — by Favourite (252), — by Favourite

(252), — by Favourite (252), — by Hubback (319), — by Snowdon's Bull (612), — by Waistell's Bull (669), — by Masterman's Bull (422), — by the Studley Bull (626).

(43405) JUPITER,

Roan, calved March 22, 1879, bred by Mr. F. Garth, Crackpot; got by David (36425), dam (Windsor's Gem) by Junius (31454), g. d. (Annetta Windsor) by Windsor Fitz-Windsor (25458), gr. g. d. (Annetta Bumper) by Bumper (19371), — (Lady Annetta) by Consolation (14316), — by Whittington (12299), — by Noble (4578), — by Newton (2367), — by Emperor (3716), — by Satellite (1420), — by Cato (119), — by Jupiter (342), — by George (273), — by Chilton (136), — by Irishman (329), — by B. (45).

(43406) JUPITER,

Roan, calved April 7, 1879, bred by Mr. J. A. Gordon, Udale, the property of Mr. G. Duff Dunbar, Ackergill; got by Balnagowan (41007), dam (Rosa Undine) by Rosario (35315), g. d. (Undine 13th) by Balliemore (30412), gr. g. d. (Undine 5th) by Young Talpa (20928), — (Undine 4th) by Hiawatha (14705), — by Lord Bathurst (13173), — by Young Fourth Duke (9037), — by Duke of Richmond (7996), — by Sir Thomas Fairfax 2nd (6493), — by Darius (3560), — by Sir Walter (2639), — by Young Star (5319), — by Roseberry (567), — by Constitution (166).

(43407) KAFFIR,

Red, calved January 23, 1879, bred by Mr. W. Faulkner, Rothersthorpe, the property of Mr. C. T. Riley, Milton Reynes; got by Fair Thane (31127), dam (Katherine 3rd) by Prince Regent (35169), g. d. (Katherine 2nd) by Stormy Petrel (37538), gr. g. d. (Katherine) by Royal Hamlet (18769), — (Sybilla) by King Arthur (13110), — by Royal Buck (10750), — by Hamlet (8126), — by Buckingham (3239).

(43408) KESTREL,

Red and white, calved March 10, 1879, bred by Mr. H. Fawcett, Old Bramhope; got by Rufus (37413), dam (Queen of St. Albans) by Hamlet (31334), g. d. (Primula) by Barleycorn (17348), gr. g. d. (Pride of Bushey) by Cock of the Walk (15782), — (Khirkee 5th) by Master Butterfly 2nd (14918), — by Young Fourth Duke (9037), — by Duke of Richmond (7996), — by Sir Walter (2639), — by Young Jerry (8177), — by Roseberry (567), — by Roseberry (567), — by Constellation (163), — by Hastings (293), — by Hastings (293), — by Leopold (372).

(43409) KILDARE DUKE,

Red and white, calved March 30, 1879, bred by Mr. J. H. Casswell, Laughton; got by Duke of Barrington 5th (33575), dam (Duchess of Kent 2nd) by Cambridge Duke 5th (30644), g. d. (Duchess of Kent) by Grand Duke of Kent 2nd (28759), gr. g. d. (Lady Walton 2nd) by Earl of Glo'ster (21644), — (Lady Walton) by Third Earl of Walton (19673), — by Lord Thoresby (14856), — by Horrox (11591), — by Second Duke of Oxford (9046), — by Cleveland Lad (3407), — by Red Rose Bull (2493), — by Rex (1375), — bred by Mr. Richardson, of Hart.

(43410) KILDONNAN,

Roan, calved May 27, 1879, bred by Mr. J. A. Gordon, Udale; got by Rosario

(35315), dam (Luxury) by Heir of Windsor (26364), g. d. (Lemon) by Havelock
of Lucknow (16242), gr. g. d. (Legacy) by La Blache (10387), — (Lovely Kate)
by Hosills (14720), — by Young Wynyard (15524), — by Young Lenton (4206), —
by Burleigh (3244), — by Woodville (2856), — by George (1068), — by Constellation (919), — by Young Favourite (254), — by Midas (436), — by Major (397).

(43411)   KILLIECRANKIE,
Red, calved April 21, 1877, bred by Mr. A. Longmore, Rettie, the property of
Mr. Watt, Lintmill, Boyndie; got by Wallace (32787), dam (La Superbe) by
Lord Forth (26649), g. d. (Dame Durden) by Warrior (21068), gr. g. d. (Virago)
by Inheritor (13065), — (Bloomer) by Seafield (9616), — by Sultan (5349), —
by Plenipo (4725), — by Abbot (2899).

(43412)   KILMUIR,
Red and white, calved April 22, 1879, bred by Mr. J. Cran, Kirkton; got by
Bridegroom (33201), dam (Lady Elma 2nd) by Baronet (25564), g. d. (Lady
Elma) by Lord Elgin (20170), gr. g. d. (Ruth) by Dipple (14401), — (Duchess)
by General (19835).

(43413)   KINELLAR,
Roan, calved January 2, 1879, bred by Mr. S. Campbell, Kinellar, the property
of Mr. J. Gough, Mains of Drum ; got by Golden Prince (38363), dam (Claret
2nd) by Novelist (34929), g. d. (Claret 1st) by Duke (28342), gr. g. d. (Claret)
by Scarlet Velvet (16916), — (Barbara) by Unrivalled (13926), — by The Pacha
(7612), — by Second Duke of Northumberland (3646), — by Sillery (5131),
— by Sillery (5131), — by Young Western Comet (1575), — by Diamond (205),
— by Favourite (256), — by Mr. Charge's Red Bull (1810).

(43414)   KING BERRY,
White, calved May 13, 1879, bred by Mr. J. Stanford, Haxted Mills; got by
King Harry (36841), dam (Sloeberry) by King of Britain (31498), g. d.
(Strawberry) by Fitz-Hopewell (23953), gr. g. d. (Cloudberry) by Cæsarian
(15717), — (Dewberry) by Douglas (12714), — by Rivers (10716), — by Diamond
(5918), — by Albert (2950), — by Stirling (5330), — by Commodore (1858), —
by Tathwell Studley (5401), — by Blyth Comet (85).

(43415)   KING BRILLIANT 2ND,
Red, calved June 7, 1879, bred by Mr. A. Garfit, Scothern ; got by Grand Duke
25th (34065), dam (Blanche Rosette) by Ninth Duke of Geneva (28391), g. d.
(Brilliant Rose 3rd) by Second Wharfdale Oxford (30298), gr. g. d. (Brilliant
Rose) by General Napier (24023), — (Brilliant) by May Duke (13320), — by
Antinous (12401); — by Diamond (5918), — by Norfolk (2377), — by Belvedere
(1706), — by Belvedere (1706), — by Lancaster (360), — by Petrarch (488), —
by Major (397), — by Chapman's Son of Punch (122), — by Dickson's Grandson
of Punch (213), — by Checks (132), — by R. Grimston's Bull (282), — by J.
Coates's Bull (148).

(43416)   KING CHARLES,
Roan, calved May 29, 1879, bred by Mr. J. H. Braikenridge, The Rookery; got
by Pioneer (38868), dam (Kathleen Mavourneen) by Monarch (31930), g.d. (Katherine 6th) by Prince Regent (35169), gr. g. d. (Katherine 2nd) by Stormy Petrel
(37538), — (Katherine) by Royal Hamlet (18769), — by King Arthur (13113),
— by Royal Buck (10750), — by Hamlet (8126), — by Buckingham (3239).

(43417) **KING DAVID,**
Roan, calved December 15, 1879, bred by the Executors of Mr. T. C. Booth, Warlaby; got by King James (28971), dam (Marchioness) by Royal Benedict (27348), g. d. (Margaret) by Commander-in-Chief (21451), gr. g. d. (Maggie) by Lord of the Hills (18267), — (Melina) by Crown Prince (10087), — by Fitz-Leonard (7010), — by Buckingham (3239),— by Roseberry (5011), — by Priam (2452), — by Jerry (4097), — by Young Pilot (4702).

(43418) **KING EDWARD,**
Roan, calved April 10, 1876, bred by Mr. J. Wood, Midtown, the property of Mr. W. Henderson, Mossfield; got by Phœnix (35034), dam (Venus 15th) by Vanguard (30204), g. d. (Venus 12th) by Speculator (13775), gr. g. d. (Venus 4th) by Red Knight (11976), — (Venus 2nd) by Grand Duke (10284), — by Kelly 2nd (9265), — by Leander (4199), — by Son of Matchem (2281), — by Sir Henry (1446), — by Young Neswick (1268), — by Son of Rose's Red Bull (5009), — by Southampton, — by Prince (521).

(43419) **KING MALCOLM,**
White, calved November 3, 1879, bred by the Executors of Mr. T. C. Booth, Warlaby; got by King James (28971), dam (Meta) by Royal Benedict (27348), g. d. (Margaret) by Commander-in-Chief (21451), gr. g. d. (Maggie) by Lord of the Hills (18267), — (Melina) by Crown Prince (10087), — by Fitz-Leonard (7010), — by Buckingham (3239), — by Roseberry (5011), — by Priam (2452), — by Jerry (4097), — by Young Pilot (4702).

(43420) **KING OF BALLIOL,**
Red and white, calved June 17, 1878, bred by Mr. C. Stephenson, Balliol College Farm, the property of Mr. C. A. Cantlie, Keithmore; got by Mayor of Windsor (31897), dam (Lady April) by Sir Roger Gwynne (27496), g. d. (Lady Augusta) by Oxford 2nd (18507), gr. g. d. (Lady Richmond) by Oxford 2nd (18507), — (Mrs. Richmond) by Duke of Richmond (7996), — by Cotherstone (6903), — by Yellow Boy (5694), — by Red Robin (2492), — by Burley (1766), — by Son of Young Albion (15), — by Sir Marton (1453), — by Palmsun (7311), — by a Bull of Mr. Parrington's.

(43421) **KING OF DENMARK,**
Roan, calved November 21, 1879, bred by Mr. W. Little, Littleport; got by King Hamlet (34318), dam (Queen of the Moors) by Western Abbot (35967), g. d. (Queen of the Abbey) by Lord Abbot (29052), gr. g. d. (Queen of Bridport) by Bridport (25668), — (Queen of the Valley) by Bridport (25668), — by The Friar (23037), — by Royal Prince (18776).

(43422) **KING OF ENGLAND,**
Red, calved February 6, 1876, bred by Mr. R. Welsted, Ballywalter, the property of Mr. T. Robertson, Narraghmore; got by England's Glory (23889), dam (Queen Victoria) by Sir James (16980), g. d. (Elfin Queen) by Elfin King (17796), gr. g. d. (British Queen) by British Prince (14197), — (Primrose) by Orson (13432), — by Duke of Cornwall (5947), — by Mowbray (4516), — by Fergus (3782), — by Woodford (2854), — by Sir Walter (2637), — by Hotspur (1117), — by Coxcomb (928), — by Midas (435), — by Comet (155), — by R. Colling's Son of Favourite (252), — by same Son of Favourite (252), — by Hubback (319).

(43423) **KING RUFUS 4TH,**

Red and white, calved May 30, 1879, bred by Mr. E. H. Cheney, Gaddesby Hall, the property of Mrs. C. H. Stopford-Sackville, Drayton House; got by Third Duke of Glo'ster (33653), dam (Seraphina 24th) by Royal Cumberland (27358), g. d. (Seraphina 20th) by Grand Duke 6th (19876), gr. g. d. (Seraphina 19th) by Imperial Oxford (18084), — (Seraphina 11th) by May Duke (13320), — by Duke of Sussex (12772), — by Sweet William (7571), — by Earl of Essex (6955), — by Stratton (5336), — by Fanatic (1996), — by Red Rover (4902), — by Rufus (2576), — by Emperor (1014), — by the Rev. R. Pointer's Old Darlington.

(43424) **KINGSLEY,**

Roan, calved March 19, 1879, bred by Mr. H. Fawcett, Old Bramhope; got by Rufus (37413), dam (Katherine 6th) by Prince Regent (35169), g. d. (Katherine 2nd) by Stormy Petrel (37538), gr. g. d. (Katherine) by Royal Hamlet (18769), — (Sybilla) by King Arthur (13110), — by Royal Buck (10750), — by Hamlet (8126), — by Buckingham (3239).

(43425) **KINGSWOOD,**

Red and white, calved December 1, 1878, bred by Mr. E. T. Wright, The Laurels; got by Earl of Chester (33784), dam (Jantja Geneva) by Third Duke of Geneva (21592), g. d. (Jantja 3rd) by Euxine (12845), gr. g. d. (Jantja) by Lycurgus (7180), — (Jovenini) by Sweet William (5368), — by Javelin (4093), — by Blyth (797), — by Wellington (684), — by Phenomenon (491), — by Favourite (252), — by Favourite (252), — by Favourite (252), — by Hubback (319), — by Snowdon's Bull (612), — by Waistell's Bull (669), — by Masterman's Bull (422), — by the Studley Bull (626).

(43426) **KING WILLIAM 2ND,**

Red and white, calved February 27, 1873, bred by Mr. W. Johnson, Prumplestown House, the property of Mr. P. Hanlon, Grangeford; got by King William (31523), dam (Amy) by Koh-i-noor Diamond (36881), g. d. (Lucia) by First Fruits (16049), gr. g. d. (Laurestina) by Berrington Boy (11173), — (Laurel Leaf) by Claret (10057), — by Woodranger (15522), — by Son of Cupid, — by Exotic (16012), — by Duke (1934), — by Isaac (1129), — by Blucher (6793), — by Cecil (120).

(43427) **KINSMAN,**

Red, calved October 29, 1879, bred by Mr. W. Hawkes, Thenford, the property of Mr. F. Golby, Middleton Cheney; got by Sir Swithin (40722), dam (Kate) by Abbot of Windsor (32903), g. d. (Katherine 3rd) by Prince Regent (35169), gr. g. d. (Katherine 2nd) by Stormy Petrel (37538), — (Katherine) by Royal Hamlet (18769), — by King Arthur (13110), — by Royal Buck (10750), — by Hamlet (8126), — by Buckingham (3239).

(43428) **KIRKLEVINGTON EMPEROR,**

Roan, calved November 3, 1879, bred by Lord Fitzhardinge, Berkeley Castle; got by Duke of Oxford 45th (39775), dam (Kirklevington Empress 3rd) by Duke of Connaught (33604), g. d. (Kirklevington Empress) by Second Duke of Tregunter (26022), gr. g. d. (Siddington 7th) by Seventh Duke of York (17754), — (Siddington 3rd) by Seventh Duke of York (17754), — by Earl of Derby (10177), — by Earl of Liverpool (9061), — by Duke of Northumberland (1940), — by Belvedere (1706), — by Son of Second Hubback (2683), — a Cow of Mr. Bates's, descended from the stock of Mr. Maynard, of Eryholme.

(43429)   KIRKLEVINGTON PRINCE,
Roan, calved September 22, 1879, bred by Lord Fitzhardinge, Berkeley Castle;
got by Duke of Connaught (33604), dam (Kirklevington Princess) by Grand
Duke of Geneva (28756), g. d. (Siddington) by Fourth Duke of Oxford (11387),
gr. g. d. (Kirklevington 7th) by Earl of Derby (10177), — (Kirklevington 4th)
by Earl of Liverpool (9061), — by Duke of Northumberland (1940), — by
Belvedere (1706), — by Son of Second Hubback (2683), — a Cow of Mr. Bates's,
descended from the stock of Mr. Maynard, of Eryholme.

(43430)   KIRKLEVINGTON PRINCE,
Roan, calved February 16, 1879, bred by Mr. R. Lodge, The Rookery; got by
Earl of Horton 10th (36587), dam (Kirklevington Duchess 13th) by Second
Duke of Tregunter (26022), g. d. (Kirklevington Duchess 4th) by Thirteenth
Duke of Oxford (21604), gr. g. d. (Kirklevington 14th) by Fourth Duke of Oxford
(11387), — (Kirklevington 7th) by Earl of Derby (10177), — by Earl of Liverpool (9061), — by Duke of Northumberland (1940), — by Belvedere (1706),
— by Son of Second Hubback (2683), — a Cow of Mr. Bates's, descended from
the stock of Mr. Maynard, of Eryholme.

(43431)   KIRKLEVINGTON PRINCE 2ND,
Roan, calved November 23, 1879, bred by Mr. R. Lodge, The Rookery; got by
Oxford's King (34997), dam (Kirklevington 19th) by Seventh Duke of York
(17754), g. d. (Kirklevington 10th) by Delhi (15865), gr. g. d. (Kirklevington
8th) by General Canrobert (12926), — (Kirklevington 7th) by Earl of Derby
(10177), — by Earl of Liverpool (9061), — by Duke of Northumberland (1940),
— by Belvedere (1706), — by Son of Second Hubback (2683), — a Cow of Mr.
Bates's, descended from the stock of Mr. Maynard, of Eryholme.

(43432)   KNIGHT HAROLD,
White, calved April 23, 1879, bred by Mr. W. Mitchell, Cleasby; got by
King Harold (40053), dam (Queen of the Roses) by Squire Booth (30049), g. d.
(Queen of the Day) by Sir Christopher (22895), gr. g. d. (Queen Rose) by Windsor
(14013), — (Queen Bess 4th) by King Arthur (13110), — by Wellington (13989),
— by Hamlet (8126), — by Pam (4643), — by Young Matchem (2282), — by
Jack Tar (1133), — by Pilot (496), — by Young Albion (15).

(43433)   KNIGHTLY,
White, calved October 12, 1877, bred by Mr. J. Spencer, Murrah Hall, the
property of Mr. J. Richardson, Row House; got by Knight of England (38513),
dam (Beatrice Windsor) by Grand Duke of Lightburne 2nd (26291), g. d.
(Alexandra Windsor) by Prince of Wales (24851), gr. g. d. (Ormolu Windsor)
by Imperial Windsor (18086), — (Ormolu Gwynne) by Master Hopewell
(14929), — by General Sale (8099), — by Lord Warden (7167), — by Orontes
(4623), — by Guardian (3947), — by Mercury (2301), — by Monarch (2324),
— by St. Albans (2584), — by Jupiter (342), — by Sir Oliver (605), — by
Trunnell (659), — by Favourite (252), — by Favourite (252), — by Dalton
Duke (188), —by R. Alcock's Bull (19), — by J. Smith's Bull (608), — by
Jolly's Bull (337).

(43434)   KNIGHT OF DARLINGTON.
Red and white, calved November 30, 1879, bred by Mr H. Allsopp, Hindlip
Hall; got by Third Duke of Hillhurst (30975), dam (Diamond) by Second Duke

of Claro (21576), g. d. (Diadem) by Marmaduke (14897), gr. g. d. (Darlington 5th) by Fourth Duke of Oxford (11387), — (Darlington 2nd) by Percy (9472), — by Thomas (5471), — by Eryholme (3736), — by Reformer (4914), — by Young Favourite (3770), — by Wellington (2825).

(43435)   KNIGHT OF EBOR,
Red, calved April 25, 1878, bred by Mr. W. P. Horne, Moulton; got by Ryedale Duke (35435), dam (Marvel of Peru) by Frederick First Fruits (23989), g. d. (Marvel) by Bumper (19371), gr. g. d. (Marvelmore) by Bumper (19371), — (Marigold) by Sunbeam (15360).

(43436)   KNIGHT OF FARNLEY,
Roan, calved January 14, 1875, bred by Mr. T. Harris, Stoneylane House, the property of Mr. J. Avery, Bentley Heath; got by Winterfold (36020), dam (Miss Fawkes) by Golden Duke 2nd (21837), g. d. (Lady Ada) by Lord Cobham (20164), gr. g. d. (Ada) by Nabob (11834), — (Lady Anna) by Zadig (8796), — by Lord Durham (9311), — by St. Helena (5055), — by Emperor (3716), — by Satellite (1420), — by Cato (119), — by Jupiter (342), — by George (273), — by Chilton (136), — by Irishman (329), — by B. (45).

(43437)   KNIGHT OF GAZELLE 2ND,
Roan, calved January 10, 1879, bred by Mr. B. Hoddinott, Moor Court, the property of Mr. W. Hoddinott, Lighthorne; got by Knight of Fame (34379), dam (Sidonia 3rd) by Fifth Duke of Wharfdale (26033), g. d. (Sidonia) by John O'Gaunt (16322), gr. g. d. (Siddons) by Grand Sultan (16189), — (Duchess of Siddington) by Tortworth Duke (13892), — by Annuity (9892), — by Leo (4208), — by Henwood (2114), — by Sir Stephen (1456), — by Prince of Waterloo (528), — by Mayflower (425), — by a Bull of Mr. Nicholson's, descended from the stock of Mr. J. Brown, of Aldborough.

(43438)   KNIGHT OF GWYNNE,
Roan, calved May 23, 1879, bred by Mr. L. Rawstorne, Hutton Hall; got by Duke of Underley (33745), dam (Christmas Gwynne) by The Christmas Duke (32663), g. d. (Fauna Gwynne) by Grand Duke 5th (19875), gr. g. d. (Faustina Gwynne) by May Duke (13320), — (Flora Gwynne) by Young Benedict (15641), — by Sir Thomas (10777), — by Prime Minister (2456), — by Marmion (406), — by Merlin (430), — by Layton (366), — by Phenomenon (491), — by Favourite (252), — by Favourite (252), — by Hubback (319), — by Snowdon's Bull (612), — by Waistell's Bull (669), — by Masterman's Bull (422), — by the Studley Bull (626).

(43439)   KNIGHT OF MOOR COURT,
Roan, calved April 28, 1879, bred by Mr. B. Hoddinott, Moor Court, the property of Mr. G. Downer, Runcton; got by Knight of Stratton Roses (40086), dam (Rosalie 2nd) by Knight of Fame (34379), g. d. (Wild Rose) by Miracle (24602), gr. g. d. (Yellow Rose) by St. George (20773), — (Rosemary) by Knight of the Lagan (20083), — by Waterloo (11025), — by The Red Duke (8694), — by Hero of the West (8150), — by Lottery (4280), — by Phœnix (6290).

(43440)   KNIGHT OF OXFORD 2ND,
Roan, calved March 3, 1879, bred by Mr. R. P. Davies, Horton, the property of Mr. H. Allsopp, Hindlip Hall; got by Duke of Hillhurst (28401), dam (Marchioness of Oxford) by Fourth Duke of Geneva (30958), g. d. (Eighth

Maid of Oxford) by Second Duke of Geneva (23752), gr. g. d. (Second Maid of Oxford) by Grand Duke of Oxford (16184), — (Oxford 20th) by Marquis of Carrabas (11789), — by Duke of Northumberland (1940), — by Short Tail (2621), — by Matchem (2281), — by Young Wynyard (2859).

(43441)   KNIGHT OF OXFORD 3RD,

Roan, calved March 9, 1879, bred by Mr. H. Allsopp, Hindlip Hall; got by Third Duke of Collingham (38134), dam (Grand Duchess of Morecambe) by Second Duke of Tregunter (26022), g. d. (Grand Duchess of Oxford 18th) by Baron Oxford 4th (25580), gr. g. d. (Grand Duchess of Oxford 11th) by Grand Duke 10th (21848), — (Grand Duchess of Oxford 5th) by Priam (18567), — by Earl of Warwick (11412), — by Fourth Duke of York (10167), — by Second Duke of Northumberland (3646), — by Short Tail (2621), — by Matchem (2281), — by Young Wynyard (2859).

(43442)   KNIGHT OF PROBUS,

Roan, calved October 31, 1878, bred by Mr. W. Trethewy, Tregoose, the property of Messrs. Lawry and Co., Gorran; got by M. C. (31898), dam (Ruth 78th) by Crœsus (30820), g. d. (Ruth 60th) by Lord Montgomery (26686), gr. g. d. (Ruth 25th) by Duke of Manchester (33690), — (Ruth 9th) by Vandumper (23114), — by Lord Fingal (11716), — by Frantick (8088), — by Harold (8131), — by Cedric (3311), — by Nimrod (4571), — by Grandson of Blyth Comet (85), — by Crispin (174), — by Meteor (431), — by Meteor (431).

(43443)   KNIGHT OF WARCOP,

Roan, calved December 31, 1879, bred by Captain T. Chamley, Warcop House; got by Grand Duke of Morecambe (36722), dam (Czarinina) by Earl of Strafford (31084), g. d. (Duchess of Oxford) by Barrington Oxford (25607), gr. g. d. (Dulcimer) by Oxford (20450), — (Czarina) by Grand Duke 2nd (12961), — by Grand Duke (10284), — by Humber (7102), — by Zenith (5702), — by Roman (2561), — by Childers (1824), — by Richard (1376), — by Jupiter (342), — by Charles (127), — by Windsor (698), — by Chilton (136), — by Colonel (152).

(43444)   KNIGHT OF WINDSOR,

Roan, calved December 4, 1879, bred by the Duke of Northumberland, Alnwick Castle; got by Sir Raymond (40716), dam (Maid of Windsor) by Mayor of Windsor (31897), g. d. (Village Maid) by Jeweller (26460), gr. g. d. (Harriet) by Vice-President (23125), — (Buttercup) by George 3rd (16147), — by Cleveland Lad (3407), — Cherry, bred by the Duke of Northumberland.

(43445)   KNIGHT TEMPLAR,

White, calved August 10, 1878, bred by the Rev. J. J. Moutray, Favour Royal; got by Gladiator (36698), dam (Governess 2nd) by Prince of the Woods (32185), g. d. (Governess) by The Governor (32681), gr. g. d. (Norma) by Knight of the Grand Cross (31558), — (Nosegay) by Prince Duke 2nd (16731), — by White Jack (36089), — by Collingwood (8964), — by Brilliant (1740), — by Monarch (2324), — by Cato (119), — by George (273), — by Chilton (136), — by Irishman (329), — by B. (45).

(43446)   KNOLLSMAN,

Red and white, calved December 31, 1879, bred by Sir W. C. Worsley, Bart., Hovingham; got by Sir Arthur Ingram (32490), dam (Irwin's Hope) by Lord Irwin (29123), g. d. (Emily) by Earl Torquil (23854), gr. g. d. (Josephine) by Earl of

Windsor (17788), — (Mint) by Magnus Troil (14880), — by Third Duke of Athol (12734), — by Bates (12451), — by Ingram (9236), — by Liberator (7140), — by Prince Albert (4791), — by a descendant of Mars (1199).

(43447) LAIRD O'COCKPEN,

Roan, calved December 31, 1878, bred by the Earl of Tankerville, Chillingham Castle; got by Blair Athol (37866), dam (Costly's Pride) by Ben Brace (30524), g. d. (Costly 4th) by Victor (25370), gr. g. d. (Costly 2nd) by Young Brilliant (15689), — (Costly) by Abraham Parker (9856), — by Bumper (10005), — by Young Hastings (3988), — by Wallace (5588), — by President (4752), — by Wellington (683), — by Sir Harry (5155), — by Traveller (655), — by Colonel (152), — by Son of Hubback (319).

(43448) LALLY HILLHURST SWEETHEART,

Roan, calved February 10, 1879, bred by Mr. W. Blackler, Wottons; got by Lally' Hillhursts Duke (38538), dam (Sweetheart 23rd) by Second Duke of Wellington (28465), g.d. (Sweetheart 21st) by The Corsair (25291), gr.g.d. (Sweetheart 9th) by The Baron (13833), — (Sweetheart 3rd) by Daybreak (11338), — by Accordion (5708), — by Little John (4232), — by Caliph (1774), — by Sir Walter (2637), — by Hotspur (1117), — by Coxcomb (928), — by Midas (435), — by Comet (155), — by R. Colling's Son of Favourite (252), — by same Son of Favourite (252), — by Hubback (319).

(43449) LALLY'S GRAND DUKE,

Red, calved April 4, 1877, bred by Mrs. Fawcett, Scaleby Castle; got by Grand Duke of Kirklevington (34071), dam (Lally 9th) by Seventh Duke of York (17754), g. d. (Lally 2nd) by Malachite (18313), gr. g. d. (Lally) by Earl of Derby (10177), — (Olive Leaf 3rd) by Earl of Liverpool (9061), — by Second Duke of Cambridge (3638), — by Belvedere (1706), — by Son of Herdsman (304), — by Wonderful (700), — by Alfred (23), — by Young Favourite (6994).

(43450) LAMMAS BOY,

Roan, calved August 1, 1879, bred by Mr. J. Jones, Ballyloughan; got by White Boy (40908), dam (Heroine of Lothian) by Heir of Lothian (28841), g.d. (Leo) by Earl of Cleveland (23828), gr.g.d. (Lily) by Prince Arthur (13497), — (Miss Topham) by Doctor Eck (14406), — by Marquis of Rockingham (10506), — by Prince Albert (4781), — by Emperor 2nd (3724), — by Emperor (1978), — by Ivanhoe (1131), — bred by Mr. Leighton.

(43451) LAMWATH PLUTO,

Red, calved March 1, 1877, bred by Mr. H. F. Smith, Lamwath House, the property of Mr. G. Allen, Sleaford; got by Pluto (35050), dam (Countess Stephenson) by Robert Stephenson (32313), g. d. (Countess Royal) by British Crown (21322), gr. g. d. (Countess of Panton) by Royal Buckingham (20718), — (Countess of Wragby) by Sir Roger (16991), — by Lambton (9273), — by General Washington (6036), — by Plenipo (4724), — by Ratify (2481), — by Wellington (680), — by Favourite (252), — by Punch (531).

(43452) YOUNG LANCASTER,

Roan, calved March 7, 1879, bred by Mr. T. Nash, Featherstone; got by Duke of Goldsmith (41386), dam (Queen of May) by Duke of Lancaster (28411), g. d. (Queen of Trumps 3rd) by Manager (24521), gr. g. d. (Queen of Trumps 2nd) by Romeo (22754), — (Queen of Trumps 1st) by Grand Duke of York

(12966), — by Brownie (8909), — by Lord Adolphus Fairfax (4249), — by
Young Favourite (3770), — by Waterloo (2816), — by Lawnsleeves (365), —
by Phenomenon (491), — by Favourite (252), — by Favourite (252), — by
Favourite (252), — by Hubback (319), — by Snowdon's Bull (612), — by
Waistell's Bull (669), — by Masterman's Bull (422), — by the Studley Bull
(626).

(43453)   LANCASTRIAN,
Roan, calved December 13, 1878, bred by Mr. A. F. Hurt, Alderwasley ; got by
Paul Gwynne (37181), dam (Lancaster 66th) by Lancaster Star (22073), g. d.
(Lancaster 30th) by Third Duke of Norfolk (21601), gr. g. d. (Lancaster 21st)
by Chilton Hero (17564), — (Lancaster 37th) by Lord George (13191), — by
George 3rd (7038), — by Spectator (2688), — by Albion (1619), — by Lan-
caster (360), — by Son of Windsor (698), — by Comet (155).

(43454)   LANDLORD,
Red, calved April 2, 1879, bred by Mr. J. P. Tynte, Tynte Park ; got by British
Peer (33224), dam (Elder Sister) by Bright Spur (25671), g. d. (Hebe) by Ho-
henlohe (18074), gr. g. d. (Crinoline) by Comrade (14307), — (Barmaid) by
Barley Sugar (11140), — by Romulus (6408), — by California (17487), — by
Sockburn (6509), — by Monarch (2324), — by Planet (2432).

(43455)   LAST NOT LEAST,
Red, calved March 8, 1879, bred by Earl Fitzwilliam, Coollattin Park ; got by
Robert Burns (29795), dam (Lady Lilias 3rd) by Lord Stanley (24466), g. d.
(Lady Lilias 2nd) by Fusilier (24000), gr. g. d. (Lady Lilias) by Windsor
Augustus (19157) — (Lilias) by Valasco (15443), — by Prince Imperial (15095),
— by Earl of Derby (10177), — by Earl of Liverpool (9061), — by Second
Duke of Cambridge (3638), — by Belvedere (1706), — by Son of Herdsman
(304), — by Wonderful (700), — by Alfred (23), — by Young Favourite
(6994).

(43456)   LAST OF THE K. C. B.'s,
Red and white, calved December 20, 1879, bred by Mr. J. B. Booth, Killerby ;
got by K. C. B. (26492), dam (Princess Brigantine) by Royal Benedict (27348),
g. d. (Queen of the Brigantes) by Brigade Major (21312), gr. g. d. (Queen of
Trumps) by Welcome Guest (15497), — (Hecuba) by Hopewell (10332), — by
Hamlet (8126), — by Leonard (4210).

(43457)   LAWERS VICEROY,
Roan, calved December 16, 1879, bred by Colonel Williamson, House of Lawers ;
got by Viceroy (32764), dam (Lavender 22nd) by Royal Duke of Glo'ster (29864),
g. d. (Lavender 13th) by Count Bickerstaffe 2nd (25838), gr. g. d. (Lavender
9th) by Brian Boru (17440), — (Lavender 7th) by Friar Tuck (14578), — by
Eclipse (10186), — by Queen's Roan (7389), — by Will Honeycomb (5660), —
by Spectator (2688), — by Albion (1619), — by Lancaster (360), — by Son of
Windsor (698), — by Comet (155).

(43458)   LEANDER,
Roan, calved April 5, 1877, bred by Mr. G. Mann, Scawsby ; got by Zenith
(39344), dam by Iron Duke (26440), g. d. by Lord of Roslin (29169), gr. g. d.
by Ambrose (12392), — by Lord Thoresby (14856), — by Master Thankful
(10514), — by Vanish (5546), — by Wilkinson's Queen's Roan.

(43459) LEBE WOHL,

Red and white, calved December 10, 1879, bred by Mr. F. J. S. Foljambe, Osberton Hall; got by Titan (35805), dam (Farewell's Rose) by Sweet-Pea (35708), g. d. (June Rose) by Cambridge Duke 4th (25706), gr. g. d. (Clematis) by Sir John (12084), — (Clementina) by Clementi (3399), — by Young Matchem (4422), — by Isaac (1129), — by Young Pilot (4702), — by Pilot (496), — by Julius Cæsar (1143).

(43460) LEINSTER,

Red, calved July 15, 1879, bred by Mr. J. P. Tynte, Tynte Park ; got by British Peer (33224), dam (Flash) by Royal Fern (29865), g. d. (Surprise) by Bright Spur (25671), gr. g. d. (Wonder) by Knight of Windsor (16350), — (Echo) by Emperor (12835), — by Barley Sugar (11140), — by Son of Juan (6106), — by Chieftain (11274), — by Glendale (3902), — by Constitution (11307), — by Snowball (1464), — by Vulcan (667).

(43461) LEONARD,

Red and white, calved May 4, 1877, bred by Mr. E. Hodgkinson, Topcliffe, the property of Mr. J. Wildon, Knapton Grange ; got by Cherry Crown (36346), dam (Red Raspberry) by Sandringham (32452), g. d. (Bramble) by Bramble Berry (28068), gr. g. d. (Lucy) by Conqueror (17607), — (Lydia) by Baron Barton (14123), — by Earl of Derby (12810), — by Gay Lad (12922), — by Belted Will (6780), — by Son of Newton (2367), — by The Colonel (5428), — by Dinsdale.

(43462) LEONIDAS,

Red and white, calved January 10, 1878, bred by Mr. T. Morris, Maisemore Court, the property of Mr. S. Hoddinott, Timsbury ; got by Baron Tregunter (36208), dam (Levity 10th) by Duke of Devonshire (23738), g. d. (Levity 3rd) by Macdonald (13268), gr. g. d. (Levity) by Shamrock (10803), — (Larkspur) by Belus (8879), — by Selim (10792), — by Sackville (2583), — by White's Bull.

(43463) LEOPOLD,

Red and white, calved December 18, 1878, bred by Mr. M. Boyes, Wandale House ; got by Duke of Tregunter 5th (33743), dam (Victoria 4th) by Twentieth Duke of Oxford (28432), g. d. (Victoria 3rd) by Orestes (22443), gr. g. d. (Victoria) by Louis 18th (18284), — (Violet) by Cavalier (14250), — by Bletsoe (9970), — by Viscount Winkle (9788), — by Sultan (7566), — by Norfolk (2377), — by Burley (1766), — by Pilot (496), — by Warlaby (672), — by Albion (14), — by Lame Bull (359), — by Shipton (587), — by Son of Suworrow (636), — by Son of Twin Brother to Ben (88), — by Twin Brother to Ben (660).

(43464) LESLIE CHIEF,

Roan, calved December 24, 1879, bred by Mr. J. Tweedie, Deuchrie, the property of the Hon. George Waldegrave Leslie, Leslie House ; got by Earl (38208), dam (Rose of Deuchrie) by British Grenadier (33217), g. d. (Rose of Warlaby) by Prince of Warlaby (20593), gr. g. d. (Roseleaf) by Le Moor (22090), — (Rosebush) by Oxford (15035), — by Redmond (11978), — by Cotherstone (6903), — by Young Matchem (4425), — by Fairfax (1023), — by Son of Marske (418), — by Palmsun (7311), — by a Bull of Mr. Maynard's, — by Son of Favourite.

(43465) **LIBERATOR,**
Red, calved March 29, 1872, bred by Mr. A. Cruickshank, Sittyton, the property
of Colonel Balfour, Balfour Castle; got by Breadalbane (28073), dam (Lovely
13th) by Cæsar Augustus (25704), g. d. (Lovely 9th) by Windsor Augustus
(19157), gr. g. d. (Lovely 8th) by Bosquet (14183), — (Lovely) by Kelly 2nd
(9265), — by Robin O'Day (4973), — by Favourite (9116), — by Antony
(1640), — by Antony (1640), — by Edgcott (1953), — by Son of Merlin (6522),
— by Acton (1607).

(43466) **LIEUTENANT,**
Roan, calved November 12, 1876, bred by Mr. R. Chaloner, King's Fort, the property of Lord Rathdonnell, Lisnavagh; got by Lieutenant General (31600), dam
(Pretty Lucy) by King James (28971), g. d. (Louisa 7th) by Ravenspur (20628),
gr. g. d. (Louisa 6th) by Dr. McHale (15887), — (Louisa 2nd) by Monk (11824),
— by Sir John Sinclair (5168), — by Collard (3419), — by Matchem (2281), —
by Cato (119), — by Jupiter (342), — by George (273), — by Chilton (136), —
by Irishman (329), — by B. (45).

(43467) **LIFEBOAT,**
Red, calved January 6, 1879, bred by Mr. R. Stratton, The Duffryn; got by
Pearl Diver (37182), dam (Rosa) by Lamp of Lothian (16356), g. d. (Michael-
mas) by Hermit (14697), gr. g. d. (Young Moss Rose) by Lottery (4280), —
(Moss Rose) by Phœnix (6290).

(43468) **LIGHTBURNE'S DUKE,**
White, calved November 16, 1879, bred by Mr. W. G. Shelton, Wergs; got by
Duke of Goldsmith (41386), dam (Clara 3rd) by Fourth Duke of Claro (30940),
g. d. (Clara 2nd) by Lord Lyon (24417), gr. g. d. (Clara) by Sultan (15355), —
(Queen Anne) by Hobgoblin (13035), — by Musician (13361), — by Fame
(10221), — by Lord Adolphus Fairfax (4249), — by Barmpton Grazier (3091),
— by Rex (4942), — by Wellington (680), — by Favourite (252).

(43469) **LILLESHALL,**
Roan, calved June 11, 1879, bred by Mr. G. T. Phillips, Sheriff Hales Manor;
got by Ruby Duke (40654), dam (Honeysuckle 2nd) by Marshal MacMahon
(34792), g. d. (Charming Princess) by Prince Charming (27130), gr. g. d.
(Honeysuckle) by Duke of Manchester (15934), — (Honey Flower) by Sultan
(15355), — by Musician (13361), — by Fame (10221), — by Lord Adolphus
Fairfax (4249), — by Barmpton Grazier (3091), — by Rex (4942), — by
Wellington (680), — by Favourite (252).

(43470) **LINDLEY,**
Red and white, calved January 17, 1879, bred by Mr. T. M. Hopkins,
Lower Wick Farm; got by Scarlatti (39085), dam (Lady Lucretia) by Grand
Duke of Clarence (28750), g. d. (Letitia) by Ancient Druid (21178), gr. g. d.
(Laura) by Duke of Cambridge (15921), — (Loyalty) by The Corsair (15378),
— by Usurer (9763), — by Dan O'Connell (3557), — by Brutus (1752), — by
Frederick (1060), — by Cato (1794), — by Son of Wellington (679), — bred by
Mr. Robertson, of Ladykirk.

(43471) **LIQUIDATOR,**
Roan, calved November 6, 1878, bred by Mr. G. Nelson, Great Salkeld; got by
Brilliant Butterfly (36270), dam (Belle of Burrell Green) by Ignoramus (28887),

g. d. (Bellisima) by Borderer (15676), gr. g. d. (Bella) by Omer Pacha (13429), — (Belinda) by Silk Velvet (12070), — by Buchan Hero (3238), — by Dan O'Connell (3557), — by Brutus (1752).

(43472) **LITTLE JIMMY,**

Roan, calved January 11, 1879, bred by Mr. J. Kerfoot, Faenol Bach; got by Favourite (33895), dam (Charlotte 4th) by Oxford Wild Eyes 2nd (32038), g. d. (Charlotte 3rd) by New York (22412), gr. g. d. (Charlotte) by Duke of Wellington (12776), — (Bona) by Monk (11824), — by Dan O'Connell (9011), — by Lord Adolphus Fairfax (4249), — by Young Favourite (3770), — by Waterloo (2816), — by Lawnsleeves (365), — by Phenomenon (491), — by Favourite (252), — by Favourite (252), — by Favourite (252), — by Hubback (319), — by Snowdon's Bull (612), — by Waistell's Bull (669), — by Masterman's Bull (422), — by the Studley Bull (626).

(43473) **LORD ACOMB,**

Roan, calved January 31, 1879, bred by Mr. E. Bowly, Siddington House; got by Duke of Hillhurst (28401), dam (Astrea) by The Bursar (35742), g. d. (Asenath) by Duke of Brailes (23724), gr. g. d. (Acacia) by Count de Gourcy (17632), — (Asia) by Grand Duke 2nd (12961), — by Fusilier (11499), — by Third Duke of York (10166), — by Second Cleveland Lad (3408), — by Duke of Cleveland (1937), — by Belvedere (1706), — a Cow of Mr. Bates's, of Kirklevington.

(43474) **LORD ACOMB 2ND,**

Roan, calved March 27, 1878, bred by Mr. G. Allen, Knightley Hall, the property of Mr. G. Bolton, 16 Merrion Square North, Dublin; got by Duke of Clarence 5th (36479), dam (Duchess) by Second Duke of Wetherby (21618), g. d. (Annette 3rd) by Lord Oxford 2nd (20215), gr. g. d. (Annette) by Horrox (11591), — (Active) by Fourth Duke of York (10167), — by Duke of Northumberland (1940), — by Short Tail (2621), — by Belvedere (1706), — a Cow of Mr. Bates's, of Kirklevington.

(43475) **LORD AILESBURY 4TH,**

Roan, calved May 28, 1879, bred by Mr. F. G. Dalgety, Lockerley Hall; got by Parnassus (38850), dam (Windsor) by Shuttlecock 3rd (35521), g. d. (Lilian) by Star Prince (18925), gr. g. d. by Savernake (35475), — (Violet) by Mazeppa (34831).

(43476) **LORD ALEXANDER 2ND,**

White, calved March 27, 1879, bred by Mr. C. Ellis, Meldreth; got by Meldreth Duke (34832), dam (Charlotte) by General McNab (34012), g. d. (Faustula) by Victorious (25378), gr. g. d. (Fair Spots) by Newland (22411), — (Fair Water) by Plato (18552), — by Pioneer (11904), — by The Captain (5422), — by Hero (4021), — by Percy (1314), — by Lord Grantham's Son of Comet (155).

(43477) **LORD ALINGTON,**

White, calved October 10, 1877, bred by Mr. J. G Attwater, Britford, the property of Mr. J. B. Workman, Upton-on-Severn; got by Duke of Ozleworth (31008), dam (Pearl 2nd) by Sir Michael (29999), g. d. (Pearl) by Lord Raleigh (29200), gr. g. d. (White Duchess) by Fourth Duke of Thorndale (17750), — (Imogene) by Neptune (11847), — by Albert (8816), — by Broughton Hero (6811), — by Gainford (2044), — by Waterloo (2816), — by Pyramid (4852), — by Grandson of Cupid (177), — by Son of Cupid (177), — by Cupid (177).

(43478) **LORD ALVERTON,**
White, calved July 31, 1877, bred by Messrs. N. Russell and Son, Northallerton; got by King Harry (36841), dam (Virginia) by Clarion (33393), g. d.
(Verbena 8th) by Sir Richard (20843), gr. g. d. (Verbena 4th) by Red Duke
(18676), — (Verbena 3rd) by Marc Antony (14895), — by Walburn Buchan
(9797), — by Hurricane (4060), — by Luck's All (2230), — by Bedford (68),
— by Barrister (776), — by Whitworth (1584), — by Young Jupiter (1148), —
by White Comet (1582), — by Cattley's Grey Bull (1798).

(43479) **LORD APRIL,**
Roan, calved July 10, 1879, bred by Mr. C. Stephenson, Balliol College Farm,
the property of Mr. C. A. Cantlie, Keithmore; got by Cheveley (36357), dam
(Lady April) by Sir Roger Gwynne (27496), g. d. (Lady Augusta) by Oxford
2nd (18507), gr. g. d. (Lady Richmond) by Oxford 2nd (18507), — (Mrs.
Richmond) by Duke of Richmond (7996), — by Cotherstone (6903), — by
Yellow Boy (5694), — by Red Robin (2492), — by Burley (1766), — by Son of
Young Albion (15), — by Sir Marton (1453), — by Palmsun (7311), — by a
Bull of Mr. Parrington's.

(43480) **LORD AUDLEY 2ND,**
Roan, calved June 1, 1879, bred by Lord Braybrooke, Audley End, the property
of the Rev. G. M. Wilson, Great Canfield; got by Duke of Rosedale 3rd
(33723), dam (Grand Duchess of Oxford 7th) by Duke of Rosedale (33721), g. d.
(Grand Duchess of Oxford 5th) by Heydon Duke 2nd (31370), gr. g. d. (Grand
Duchess of Oxford) by Grand Duke 7th (19877), — (Duchess of Oxford) by
Rembrandt (13587), — by Jasper (11609), — by Second Duke of Oxford
(9046), — by Second Duke of Northumberland (3646), — by Belvedere
(1706), — by Son of Second Hubback (2683), — a Cow of Mr. Bates's, of
Kirklevington.

(43481) **LORD BEACONSFIELD,**
Red and white, calved November 2, 1879, bred by Mr. P. Cadman, Union Mills;
got by Quick Hope (40558), dam (Sweet May) by Farnley Royal (28573), g. d.
(Sweet Willow) by Sir Sam (25171), gr. g. d. (Sweetbrier) by White Hamlet
(15518), — (Butterfly) by General Fairfax (14594), — by Brawith Boy (7846),
— by Sir Walter (2639), — by Sir Walter (2639), — by Roseberry (567), — by
Roseberry (567), — by Constitution (166), — by Hastings (293), — by Constellation (163).

(43482) **LORD BENNET,**
White, calved April 22, 1879, bred by the Earl of Tankerville, Chillingham
Castle, the property of the Duke of Northumberland, Alnwick Castle; got by
Sir Raymond (40716), dam (Well-Born) by Knight of the Shire (26552), g. d.
(Weal Royal) by Booth Royal (15673), gr. g. d. (Weal Princess) by British
Prince (14197), — (Water Coryphée) by Crown Prince (10087), — by Fourth
Duke of Northumberland (3649), — by Norfolk (2377), — by Waterloo (2816),
— by Waterloo (2816).

(43483) **LORD BERTHWAITE,**
Roan, calved February 12, 1879, bred by the Earl of Bective, Underley Hall;
got by Duke of Underley (33745), dam (Duchess Gwynne 2nd) by Third Duke
of Clarence (23727), g. d. (Duchess Gwynne) by Duke of Wetherby (17753),

gr. g. d. (Polly Gwynne) by Flying Dutchman (10235), — (Young Dowager Gwynne) by St. Thomas (10777), — by Prime Minister (2456), — by Wallace (5586), — by Marmion (406), — by Merlin (430), — by Layton (366), — by Phenomenon (491), — by Favourite (252), — by Favourite (252), — by Hubback (319), — by Snowdon's Bull (612), — by Waistell's Bull (669), — by Masterman's Bull (422), — by the Studley Bull (626).

(43484)   LORD BEVERLEY 4TH,

Red, calved November 19, 1878, bred by Sir W. C. Trevelyan, Bart., Wallington, the property of Mr. J. Yorke, Bewerley Hall; got by Oxford Beau 4th (34964), dam (Miss Beverley 2nd) by Lieutenant Oxford (24336), g. d. (Miss Beverley 18th) by Royal Butterfly 16th (20724), gr. g. d. (Miss Beverley 6th) by Ivanhoe (14735), — (Miss Beverley 3rd) by Voltigeur (13964), — by Disraeli (10125), — by Lord George Bentinck (9317), — by Second Earl of Beverley (5963), — — by Second Cleveland Lad (3408), — Red Darlington.

(43485)   LORD BEVERLEY 5TH,

Roan, calved December 10, 1878, bred by Sir W. C. Trevelyan, Bart., Wallington, the property of Mr. J. Swann, Bedlington; got by Duke of Oxford 27th (33709), dam (Miss Beverley 3rd) by Duke of Tynedale (28462), g. d. (Miss Beverley 2nd) by Lieutenant Oxford (24336), gr. g. d. (Miss Beverley 18th) by Royal Butterfly 16th (20724), — (Miss Beverley 6th) by Ivanhoe (14735), — by Voltigeur (13964), — by Disraeli (10125), — by Lord George Bentinck (9317), — by Second Earl of Beverley (5963), — by Second Cleveland Lad (3408), — Red Darlington.

(43486)   LORD BLANCHE,

Red and white, calved March 20, 1878, bred by Mr. J. C. Musters, Annesley Park; got by Lord Chaworth (36924), dam (Blanche 10th) by Fifth Lord Oxford (31738), g. d. (Blanche Rose 5th) by Baron Waterloo (27977), gr. g. d. (Blanche Kale) by Knightley Grand Duke (24268), — (Olga) by Old Buck (15017), — by May Duke (13319), — by Selim (6454), — by Rex (6385), — by Norfolk (2377), — by Belvedere (1706), — by Belvedere (1706), — by Lancaster (360), — by Petrarch (488), — by Major (397), — by Chapman's Son of Punch (122), — by Dickson's Grandson of Punch (213), — by Checks (132), — by R. Grimston's Bull (282), — by J. Coates's Bull (148).

(43487)   LORD BUGLER,

Red, calved March 1, 1879, bred by Mr. H. F. Smith, Lamwath House; got by Lamwath Pluto (43451), dam (Lamwath Music) by Bismarck Baron (30545), g. d. (Harmonium) by Knight of the Whistle (26558), gr. g. d. (Concertina) by May Duke (16553), — (Seraphine) by Monarch (13347), — by Lord of Brawith (10465), — by Pestalozzi (10603), — by Lamplighter (8204), — by Zenith (5702), — by Warlock (5599), — by Alabaster (1616), — by Monarch (2324), — by Satellite (1420), — by Cato (119), — by Jupiter (342), — by George (273), — by Chilton (136), — by Irishman (329), — by B. (45).

(43488)   LORD CAWDOR,

Red and white, calved October 9, 1878, bred by Mr. C. A. Barnes, Solesbridge, the property of Mr. W. Hyslop, Church Stretton; got by Lord of Lochaber 2nd (36983), dam (Lady Celia 6th) by Lord Eglintoun (31652), g. d. (Lady Celia 3rd) by Royal Duke (32375), gr. g. d. (Lady Celia) by Lord Wallace (24473), — (Colleen Bawn) by First Fruits (19751), — by Young Duke of Cambridge

(14433), — by The Minstrel (8687), — by Percy (6283), — by Third Duke of Northumberland (3647), — by Bashaw (1692), — by Helmsman (2109), — by Columella (904), — by Regent (544), — by Palatine (478), — by Palmflower (480), — by Patriot](486), — by Driffield (223), — by C. Holmes's Bull (314).

(43489)   LORD CHELMSFORD,

Roan, calved April 27,]1879, bred by Mrs. Severne, Thenford House; got by Petrarch (42136), dam (Lavender Kid) by Filho da Puta (31159), g. d. (Lavinia) by Hailstorm (28807), gr. g. d. (Lady Lavender) by Superintendent (25254), — (Lavender 7th) by Sir Sam (16992), — by Hero of the West (8150), — by Barrister (7815), — by Marcellus (6181), — by Major (4338), — by Eclipse (1949), — by Son of Dishley Ketton (982), — by Cossack (925), — by Patriot (486).

(43490)   LORD CHELMSFORD,

Roan, calved October 21, 1879, bred by Mr. R. H. Crabb, Baddow Place; got by Oxford's Waterloo (40433), dam (Oxford's Minnie) by Baron Oxford 3rd (25579), g. d. (Miss Minnie 5th) by Cambridge Duke 3rd (23503), gr. g. d. (Miss Minnie 4th) by Grand Duke of York (24071), — (Miss Minnie) by Leonidas (14791), — by Arthur (14106), — by Earl of Dublin (10178), — by Gustavus the Hero (7058), — by Tathwell (5399), — by Belvedere 2nd (3127), — by Bellerophon (3119), — by Barmpton (5774), — by Kitt (7127), — by Kitt (7127), — by Page's Bull (6269), — by Middleton's Bull (438).

(43491)   LORD CHESTERFIELD,

Roan, calved February 20, 1879, bred by Mr. R. Welsted, Ballywalter, the property of Mr. R. McDougall, Arundel, Keilor, Melbourne, Australia; got by Master of Arts (34816), dam (Ladylike 7th) by England's Glory (23889), g. d. (Ladylike 5th) by Prince Christian (22581), gr. g. d. (Ladylike 2nd) by Prince Regent (18637), — (Ladylike) by Stars and Stripes (12148), — by Sir Henry (10824), — by Buckingham (3239), — by Marcus (2262), — by Matchem (2281), — by Alderman (1622), — by Pilot (496), — by Remus (550), — by Sir Charles (592), — by R. Colling's Son of Favourite (252), — by Son of Favourite (252), — Strawberry.

(43492)   LORD CLONMEL,

Roan, calved March 29, 1879, bred by Mr. R. J. M. Gumbleton, Glanatore; got by Royal Fitz-Rose (37390), dam (Buttercup 6th) by King Richard (26523), g. d. (Buttercup 2nd) by Young Windsor (19156), gr. g. d. (Buttercup) by Famous (12857), — (Blanche 3rd) by Baron Wild Eyes (8869), — by Auld Robin Grey (6753), — by James Chrisp's Bull, — by Burley (1766), — by Isaac (1129), — by Pilot (496), — by Albion (14), — by Lame Bull (359), — by Shipton (587), — by Son of Suworrow (636), — by Son of Twin Brother to Ben (88), — by Twin Brother to Ben (660).

(43493)   LORD CLWYD,

White, calved January 25, 1879, bred by Mr. R. Blezard, Pool Park; got by Duke of Siddington 3rd (38183), dam (Lady Ruthin) by Baron Oxford 4th (25580), g. d. (Cleopatra 2nd) by Wild Duke (27808), gr. g. d. (Cleopatra) by Grand Duke 3rd (16182), — (Duchess of Cambridge 2nd) by Second Duke of Cambridge (12743), — by Weathercock (9815), — by Second Cleveland Lad (3408), — by Fourth Duke of Northumberland (3649), — by Short Tail (2621), — by Belvedere (1706), — by Son of Young Wynyard (2859), — descended from J. Brown's Red Bull (97).

(43494) LORD COBHAM,

Red and white, calved March 23, 1879, bred by Captain D. H. Mytton, Garth ;
got by Constantine 2nd (33439), dam (Early Rose) by Richmond (35269), g. d.
(Eglantine) by Manager (24521), gr. g. d. (White Rose 2nd) by Victor (19061),
— (White Rose) by Cronstadt (14352), — by Autocrat (19249), — by Warrior
(12287), — by Sweetmeat (8645), — Old Strawberry (winner of First Prize,
County Show, Shrewsbury, 1844), bred by Dr. Rowley.

(43495) LORD CORISANDE,

Red, calved October 28, 1879, bred by Mr. Y. R. Graham, Yardley Stud Farm ;
got by Baron Fantail (37790), dam (Corisande 2nd) by Coronation (30796), g. d.
(Corisande) by The Earl (27624), gr. g. d. (British Belle) by Second Duke of
Thorndale (17748), — (British Beauty) by British Prince (14197), — by Cardinal (11246), — by Sheldon (8557), — by Benjamin (1710), — by Malbro'
(1189), — by Ebor (997), — by Regent (546), — by North Star (459), — by
R. Colling's White Bull (151), — bred by Mr. R. Colling.

(43496) LORD CRAGGS,

Roan, calved June 2, 1878, bred by Mr. A. P. Clear, Maldon, and exported to
South America ; got by Cherry Grand Duke 6th (33348), dam (Janet) by
Caractacus (28141), g. d. (Johanna Southcott) by John O'Gaunt (16322),
gr. g. d. (Duchess of Sussex) by Duke of Sussex (12772), — (Countess of
Beverley) by Lord Foppington (10437), — by Second Cleveland Lad (3408), —
by Duke of Northumberland (1940), — by Son of Second Hubback (2683), — a
Cow of Mr. Bates's, descended from the stock of Mr. Maynard, of Eryholme.

(43497) LORD DACRE,

Red and white, calved January 31, 1879, bred by Mr. R. E. Oliver, Sholebroke
Lodge ; got by Grand Duke 30th (38373), dam (Daphne) by Caractacus (28141),
g. d. (Diamond) by Second Duke of Claro (21576), gr. g. d. (Diadem) by Marmaduke (14897), — (Darlington 5th) by Fourth Duke of Oxford (11387), — by
Percy (9472), — by Thomas (5471), — by Eryholme (3736), — by Reformer
(4914), — by Young Favourite (3770), — by Wellington (2825).

(43498) LORD DE CLIFFORD,

Roan, calved September 7, 1877, bred by Mr. W. A. Nicholson, Pembroke
House ; got by Hubback Junior (31395), dam (Lady De Clifford) by Edwin
(21671), g. d. (General's Daughter) by General Haynau (11520), gr. g. d. (Red
Rose 6th) by Gainford 2nd (10255), — (Red Rose 5th) by Emperor (9084), —
by Sir Thomas Newton (6495), — by General (3867).

(43499) LORD DENISON,

Roan, calved October 21, 1877, bred by Mr. D. Arkell, Butlers Court, the property of Mr. E. W. Harcourt, Nuneham Park ; got by Lord Fitzclarence 12th
(34551), dam (Jane Gray) by Oxford's Gwynne (32033), g. d. (Jenny Lind) by
Lancaster (24302), gr. g. d. by Royal Butterfly 14th (20722), — by Uncle Tom
(13912), — by Cedric (3311), — by Fisher's Son of Favourite (1028), — by
Coelebs (897), — by Rose's Old Red Bull (5000), — by Fisher's Old Bull
(3799).

(43500) LORD DE ROSS,

Roan, calved November 22, 1878, bred by Mr. J. Strong, Culgaith, the property
of Mr. T. Brown, Little Bramton ; got by Earl of Doune (36579), dam (Golden

Lady Eliza) by Pure Gold (32226), g. d. (Lady Eliza) by Lord of Nunwick (26702), gr. g. d. (Eliza) by Brigadier (33203), — by General Haynau (11520).

(43501) **LORD DINORBEN,**

Red and white, calved October 23, 1878, bred by Mr. J. Kerfoot, Faenol Bach; got by Favourite (33895), dam (Bona 2nd) by Oxford Wild Eyes 2nd (32038), g. d. (Charlotte 3rd) by New York (22412), gr. g. d. (Charlotte) by Duke of Wellington (12776), — (Bona) by Monk (11824), — by Dan O'Connell (9011), — by Lord Adolphus Fairfax (4249), — by Young Favourite (3770), — by Waterloo (2816), — by Lawnsleeves (365), — by Phenomenon (491), — by Favourite (252), — by Favourite (252), — by Favourite (252), — by Hubback (319), — by Snowdon's Bull (612), — by Waistell's Bull (669), — by Masterman's Bull (422), — by the Studley Bull (626).

(43502) **LORD DOUNE,**

Roan, calved December 6, 1878, bred by Mr. J. Strong, Culgaith; got by Earl of Doune (36579), dam (Princess Louise) by Lord Lieutenant (29133), g. d. (Royal Annie) by Edgar (19680), gr. g. d. (Gilbanks 6th) by Speculation (12135), — (Gilbanks 5th) by Teddington (12174), — by Son of Eden (3689), — by Lord de Roos (4254), — by Rose Castle Bull, — by Whitefield Bull.

(43503) **LORD DUKE SPENCER,**

Red, calved May 1, 1879, bred by Mr. J. Lynn, Church Farm, the property of Mr. Hack, Buckminster; got by Cambridge Duke 8th (33274), dam (Lady Rose Spencer) by Cambridge Duke 4th (25706), g. d. (Lady Sylvia Spencer) by Grand Duke of Lightburne (26290), gr. g. d. (Lady Susan Spencer) by Second Duke of Airdrie (19600), — (Lady Selina Spencer) by Second Duke of Cambridge (12743), — by Usurer (9763), — by Shamrock (7488), — by Monarch (2324), — by Jupiter (342), — by Sir Oliver (605), — by Trunnell (659), — by Favourite (252), — by Favourite (252), — by Dalton Duke (188), — by R. Alcock's Bull (19), — by J. Smith's Bull (608), — by Jolly's Bull (337).

(43504) **LORD EDGCOTT 3RD,**

Red and white, calved March 10, 1879, bred by Lieut.-Colonel Webb, Elford House; got by Grand Gwynne 2nd (38384), dam (Lady 3rd) by Senator (29950), g. d. (Lady 2nd) by Oxford Comet (22479), gr. g. d. (Lady) by Marmaduke (14897), — (La Cantatrice) by King Arthur (13110), — by The Minstrel (7242), — by Mercury (4459), — by Bagdad (3075), — by Anthony (1640), — by Anthony (1640), — by Edgcott (1953), — by Son of Merlin (6522), — by Acton (1607).

(43505) **LORD EDWARD.**

Roan, calved January 23, 1879, bred by Mr. W. Little, Littleport; got by King Hamlet (34318), dam (Lady Knowlmere) by Lord Abbot (29052), g. d. (Lady Lowndes) by Imperial Windsor (18086), gr. g. d. (Welcome Mary) by Prince Alfred (13494), — (Welcome Lady) by Welcome Guest (15497), — by Vanguard (10994), — by Hamlet (8126).

(43506) **LORD ELLENHALL,**

Roan, calved December 23, 1878, bred by Mr. S. T. Addison, Ellenhall; got by Duke of Darlington 5th (41377), dam (Laura 6th) by Roan Prince 2nd (29791), g. d. (Laura 3rd) by Liverpool (36899), gr. g. d. (Laura 1st) by Ninason 4th (37129), — (Walton Lady 1st) by Laurason (18177), — by Sebastopol (15252),

— by Walton (12281), — by Lark (8211), — by Rathreagh (6366), — by Belvedere 2nd (3126), — by Young Waterloo (8757), — by Waterloo (2816), — by Kitt (7127), — by Kitt (7127), — by Page's Bull (6269), — by Middleton's Bull (438).

(43507)   LORD EXETER,
Red and white, calved November 9, 1876, bred by Mr. W. Carr, Low Gate, Balne; got by Telemachus 7th (35726), dam (Lady Farnley) by Gladstone (31251), g. d. (First Lady) by Lord Darlington (26633), gr. g. d. (Valerian) by Royal Oak (16873), — (Vanilla) by Robinson Crusoe (13610), — by Borrowby Boy (9980), — by Laudable (9282), — by Sir Thomas Fairfax (5196), — by Colossus (1847), — by Burley (1766), — by Pilot (496), — by Warlaby (672), — by Albion (14), — by Lame Bull (359), — by Shipton (587), — by Son of Suworrow (636), — by Son of Twin Brother to Ben (88), — by Twin Brother to Ben (660).

(43508)   LORD FLUTE,
Red, calved May 1, 1878, bred by Mr. H. F. Smith, Lamwath House; got by Lord Violet 3rd (34692), dam (Harmonium) by Knight of the Whistle (26558), g. d. (Concertina) by May Duke (16553), gr. g. d. (Seraphine) by Monarch (13347), — (Seraph) by Lord of Brawith (10465), — by Pestalozzi (10603), — by Lamplighter (8204), — by Zenith (5702), — by Warlock (5599), — by Alabaster (1616), — by Monarch (2324), — by Satellite (1420), — by Cato (119), — by Jupiter (342), — by George (273), — by Chilton (136), — by Irishman (329), — by B. (45).

(43509)   LORD FOGGATHORPE,
Red, calved February 26, 1877, bred by Mr. W. T. Brackenbury, Thorpe Hall, the property of Mr. E. Durrant, Wimbotsham; got by Knightley Foggathorpe (34372), dam (Lady Spot) by Prince Imperial (22594), g. d. (The Lady) by El Sol (15986), gr. g. d. (Duchess) by El Sol (15986), — by Plato (18552), — by Richard (8482), — by Chance (3333), — by The Captain (5422).

(43510)   LORD FORDHAM,
Red and white, calved September 18, 1879, bred by Mr. J. P. Clark, North Ferriby; got by Fordham Duke of Oxford 4th (41569), dam (Guelder Rose 10th) by Duke of Thorndale (33740), g. d. (Guelder Rose 7th) by Earl of Collingham (28498), gr. g. d. (Guelder Rose 2nd) by Seventh Duke of York (17754), — (Geranium) by Cupid (14359), — by Valentine (13932), — by Berkeley (7830), — by Son of Holkar (4041), — bred by Mr. G. Angus, Byfield.

(43511)   LORD GARLAND 7TH,
Roan, calved September 9, 1878, bred by Mr. H. J. Sheldon, Brailes House; got by Duke of Rothesay (36534), dam (Lady Florence 7th) by Duke of Rosedale (33721), g. d. (Lady Florence) by Duke of Brailes (23724), gr. g. d. (Countess) by Duke of Cambridge (12742), — (Chrysalis) by Earl of Dublin (10178), — by Grey Friar (9172), — by Fawsley (6004), — by Marcellus (2260), — by Rufus (2576), — by Wellington (683), — by Windsor (698), — by Windsor (698), — by Own Brother to North Star (459).

(43512)   LORD GEORGE,
Red and white, calved February 8, 1877, bred by Mr. C. Bruce, Broadland, the property of Mr. D. C. Bruce, Broadland; got by King George (28968), dam

(Lady March 2nd) by Jacob (28904), g. d. (Lady March) by Dipple (14401), gr. g. d. (Young Jenny Lind) by Orbliston (24691), — (Jenny Lind) by Jacob (6101), — by Proselyte (4842), — by Othello (4631), — by Juniper (1145), — by Mentor (426), — by Lancaster (360), — by Ketton (346), — by George (274).

(43513)   LORD GEORGE HAMILTON,

Roan, calved August 19, 1879, bred by Mr. B. St. John Ackers, Prinknash Park ; got by Lord Prinknash 2nd (38653), dam (Princess Georgie) by County Member (28268), g. d. (Georgie's Queen) by Brigade Major (21312), gr. g. d. (Georgie) by Prince George (13510), — (Hopeful) by Hopewell (10332), — by Warrior (12287).

(43514)   LORD GIFFORD 3rd,

Roan, calved March 1, 1879, bred by Lord Fitzhardinge, Berkeley Castle, the property of Mr. J. C. Croome, Bagendon House ; got by Duke of Connaught (33604), dam (Lady Gifford) by Oxford's Tony (35000), g. d. (Gazelle 20th) by Seventh Duke of York (17754), gr. g. d. (Gazelle 4th) by Seventh Duke of York (17754), — (Selina) by Fourth Duke of Oxford (11387), — by Snowstorm (12119), — by Hampden (8129), — by Leo (4208), — by Henwood (2114), — by Sir Stephen (1456), — by Prince of Waterloo (528), — by Mayflower (425), — by a Bull of Mr. Nicholson's, descended from the stock of Mr. J. Brown, of Aldborough.

(43515)   LORD GIFFORD 4th,

Roan, calved October 26, 1879, bred by Lord Fitzhardinge, Berkeley Castle ; got by Duke of Oxford 45th (39775), dam (Gazelle 37th) by Beau of Oxford 2nd (33129), g. d. (Gazelle 30th) by Third Duke of Clarence (23727), gr. g. d. (Gazelle 21st) by Second Duke of Tregunter (26022), — (Gazelle 11th) by Seventh Duke of York (17754), — by Earl of Walton (17787), — by Fourth Duke of Oxford (11387), — by Snowstorm (12119), — by Hampden (8129), — by Leo (4208), — by Henwood (2114), — by Sir Stephen (1456), — by Prince of Waterloo (528), — by Mayflower (425), — by a Bull of Mr. Nicholson's, descended from the stock of Mr. J. Brown, of Aldborough.

(43516)   LORD GORDON,

White, calved October 10, 1878, bred by Mr. J. A. Dickinson, Brough Sowerby ; got by British Sovereign (36285), dam (Clarissa) by Count of the Realm (23640), g. d. (Clara) by Royal Guest (18768), gr. g. d. (Lady Caroline Gordon) by Lord Raglan (13244), — (Lady Kenmure) by Baron of Ravensworth (7811), — by Adam (2920), — by Sir Thomas (5194), — by Noble Henry (2374), — by Abraham (2905), — by Mustachios (4527), — by Simon (5134), — by Young George (3885), — by George (276).

(43517)   LORD GRANVILLE 5th,

Red, calved March 20, 1878, bred by Mr. F. Simmers, Cullerlie ; got by Grand Duke (41640), dam (Young Flavia) by Lord Granville (24395), g. d. (Flavia) by Marmaduke (20284), gr. g. d. (Mary Anne) by Garibalbi (17916), — (Mary) by Moss Trooper (11827), — by Robin O'Day (4973), — by Holkar (4041), — by Jopp's Bull (9256), — Kate of Darlington.

(43518)   LORD GREY,

Roan, calved May 20, 1878, bred by Mr. H. A. Brassey, Preston Hall, the property of Mr. W. Kirby, London Land's Farm, Reigate; got by Grand Duke 24th

(34064), dam (Graceful 2nd) by Airdrie Geneva (32920), g. d. (Graceful) by Duke of Albany (25931), gr. g. d. (Lucinda) by Priam (18567), — (Coral) by Cardinal (11246), — by Earl of Dublin (10178), — by Little John (4232), — by Caliph (1774), — by Sir Walter (2637), — by Hotspur (1117), — by Coxcomb (928), — by Midas (435), — by Comet (155), — by R. Colling's Son of Favourite (252), — by same Son of Favourite (252), — by Hubback (319).

(43519) LORD HARLEY,

Roan, calved August 3, 1879, bred by Mr. T. Brocklebank, Springwood ; got by Oxford's Heir (37172), dam (Florentia 29th) by Grand Duke of Kirklevington (34071), g. d. (Florentia 25th) by Second Duke of Collingham (23730), gr. g. d. (Florentia 13th) by Seventh Duke of York (17754), — (Fairy Queen) by Duke of Ulster (12774), — by Koh-i-noor (11642), — by Usurer (9763), — by Zenith (5702), — by Sweet William (5368), — by Roman (2561), — by Monarch (2324), — by Jupiter (342), — by Sir Oliver (605), — by Trunnell (659), — by Favourite (252), — by Favourite (252), — by Dalton Duke (188), — by R. Alcock's Bull (19), — by J. Smith's Bull (608), — by Jolly's Bull (337).

(43520) LORD HAROLD,

Roan, calved March 23, 1879, bred by Mr. W. Mitchell, Cleasby ; got by King Harold (40053), dam (Queen of the Palace) by Squire Booth (30049), g. d. (Queen of the Day) by Sir Christopher (22895), gr. g. d. (Queen Rose) by Windsor (14013), — (Queen Bess 4th) by King Arthur (13110), — by Wellington (13989), — by Hamlet (8126), — by Pam (4643), — by Young Matchem (2282), — by Jack Tar (1133), — by Pilot (496), — by Young Albion (15).

(43521) LORD HAWTHORN,

Roan, calved July 4, 1879, bred by Mr. R. Taylor, Sigglesthorne Manor, the property of the Marquis of Bute, Cardiff Castle ; got by Wild Oxford (40926), dam (Penelope 2nd) by Waltron (30255), g. d. (Penelope) by Jerry (28910), gr. g. d. (Pearl) by Next of Kin (20405), — (White Thorn) by Bullion (15706), — by Highthorn (13028), — by Ulysses (7654), — by The Lord of Gilling (6587), — by Belshazzar (1703), — by Camden (1776), — by Simon (5134), — by Son of Symmetry (641).

(43522) LORD HIGHTHORN,

Red, calved January 28, 1879, bred by Mr. E. Heinemann, Ratton Park ; got by Earl of Sheffield (33812), dam (Lady James) by Sir James (35566), g. d. (British Queen) by British Crown (21322), gr. g. d. (Lady Booth) by The Sutler (23061), — (Lady Alfred) by Prince Alfred (13494), — by Fitz-Clarence (14552), — by Highthorn (13028), — by Banker (11136).

(43523) LORD HOLKER,

Roan, calved April 12, 1879, bred by Mr. E. Bowly, Siddington House ; got by Duke of Holker 2nd (39749), dam (Gertrude) by Duke of Killhow (30980), g. d. (Gazelle 10th) by Seventh Duke of York (17754), gr. g. d. (Sibyl) by Union (19031), — (Selina) by Fourth Duke of Oxford (11387), — by Snowstorm (12119), — by Hampden (8129), — by Leo (4208), — by Henwood (2114), — by Sir Stephen (1456), — by Prince of Waterloo (528), — by Mayflower (425), — by a Bull of Mr. Nicholson's, descended from the stock of Mr. J. Brown, of Aldborough.

(43524)   **LORD HOLKER 2ND,**

White, calved September 8, 1879, bred by Mr. E. Bowly, Siddington House, the property of Mr. A. B. Gregory, Wroxall; got by Duke of Holker 2nd (39749), dam (Florida) by Twenty-second Duke of Oxford (31000), g. d. (Fortuna) by Seventeenth Duke of Oxford (25994), gr. g. d. (Fairy Belle) by Seventh Duke of York (17754), — (Fata Morgana) by Cupid (14359), — by Fourth Duke of Oxford (11387), — by Percy (9472), — by Gainford (2044), — by Monitor (2331), — by Snowdrop (2653), — by Barmpton (1677), — by Waverley (2819).

(43525)   **LORD HOLKER 3RD,**

White, calved December 7, 1879, bred by Mr. E. Bowly, Siddington House; got by Duke of Holker 2nd (39749), dam (Gazelle 36th) by Beau of Oxford 2nd (33129), g. d. (Gazelle 4th) by Seventh Duke of York (17754), gr. g. d. (Selina) by Fourth Duke of Oxford (11387), — (Buttercup) by Snowstorm (12119), — by Hampden (8129), — by Leo (4208), — by Henwood (2114), — by Sir Stephen (1456), — by Prince of Waterloo (528), — by Mayflower (425), — by a Bull of Mr. Nicholson's, descended from the stock of Mr. J. Brown, of Aldborough.

(43526)   **LORD HOWE,**

Roan, calved March 30, 1878, bred by Mr. T. G. Lofthouse, Boroughbridge, the property of Messrs. C. and J. Mason, Dishforth; got by Knight of St. Patrick (38520), dam (Platina) by Great Hope (24082), g. d. (Paulina) by Heir of Windsor (26364), gr. g. d. (Pauline 4th) by Hopewell (19973), — (Pauline) by British Boy (11206), — by Hopewell (10332), — by Hamlet (8126), — a Cow bred by Mr. Booth.

(43527)   **LORD HUDSON 3RD,**

Red and white, calved April 4, 1879, bred by Messrs. F. Leney and Sons, Wateringbury; got by Sixth Duke of Oneida (30997), dam (Lady Hudson's Duchess 2nd) by Grand Duke 15th (21852), g. d. (Lady Hudson) by Fourth Duke of Oxford (11387), gr. g. d. (Hudson 3rd) by General Canrobert (12927), — (Hudson 2nd) by Horrox (11591), — by The Duke (8676).

(43528)   **LORD HUDSON 4TH,**

Red and white, calved April 6, 1879, bred by Messrs. F. Leney and Sons, Wateringbury; got by Sixth Duke of Oneida (30997), dam (Lady Hudson's Duchess) by Grand Duke 15th (21852), g. d. (Lady Hudson) by Fourth Duke of Oxford (11387), gr. g. d. (Hudson 3rd) by General Canrobert (12927), — (Hudson 2nd) by Horrox (11591), — by The Duke (8676).

(43529)   **LORD JAMES,**

Red and white, calved March 25, 1879, bred by Mr. W. A. Mitchell, Auchnagathle, the property of Colonel Hay, Leith Hall; got by Duke of Chamburgh (36052), dam (Lady Jane) by Lord Lincoln (34583), g. d. (Bessie) by Old England (24681), gr. g. d. (Strawberry 2nd) by Master Gunner (22316), — (Strawberry) by Canterbury (33280).

(43530)   **LORD JOHN,**

Roan, calved December 26, 1879, bred by Messrs. F. Leney and Sons, Wateringbury; got by Sixth Duke of Oneida (30997), dam (Jenny 15th) by Eighth Duke of Geneva (28390), g. d. (Jenny 12th) by Clifford (21437), gr. g. d. (Jenny 4th) by Tippoo 2nd (23075), — (Jenny 3rd) by Tippoo (23074), — by Sultan

(22989), — by Edrick (10188), — by Hengist (10315), — by Partisan (6276), — by Highflyer (2122), — by Montgomery (2334), — by Forester (1055), — by Eclipse (238), — by White Comet (1582), — by Western Comet (689), — by Favourite (252), — descended from the Studley White Bull (627).

(43531)　　　　　　　LORD KILGOBBIN,

Roan, calved August 23, 1878, bred by Mr. T. Crompton, Lowthorpe; got by Mantalini Chief (34753), dam (Danthorpe Lady 18th) by Genuine Prince (24031), g. d. (Danthorpe Lady 7th) by Lord Greta (20174), gr. g. d. (Danthorpe Lady) by Crusade (7938), — (Angelique Borel) by Sliding Scale (9669), — by Nicholas Nickleby (9440), — by Invincible (4077), — by Son of Patriot (486).

(43532)　　　　　　　LORD KIMBOLTON,

Roan, calved June 4, 1877, bred by Mr. F. Sartoris, Rushden Hall, the property of the Hon. C. W. Fitzwilliam, Alwalton; got] by Grand Duke of Waterloo (28766), dam (Duchess of Waterloo) by Duke of Kingscote (25981), g. d. (Wellingtonia 8th) by Baron Oxford (23375), gr. g. d. |(Wellingtonia 4th) by Grand Duke 4th (19874), — (Waterloo 16th) by Duke of Bolton (12738), — by Second Cleveland Lad (3408), — by Duke of Northumberland (1940), — by Norfolk (2377), — by Waterloo (2816), — by Waterloo (2816).

(43533)　　　　　　　LORD KIRKBY,

Roan, calved November 29, 1879, bred by Mr. G. Ashburner, Low Hall; got by Duke of Oxford 41st (38174), dam (Victoria Melita) by Duke of Oxford 27th (33709), g. d. (Grand Duchess Marie) by Grand Duke of Oxford (28764), gr. g. d. (Blanche 5th) by Grand Duke 10th (21848), — (Sylph) by Glo'ster (14619), — by Childers (10052), — by Selim (6454), — by Rex (6385), — by Norfolk (2377), — by Belvedere (1706), — by Belvedere (1706), — by Lancaster (360), by Petrarch (488), — by Major (397), — by Chapman's Son of Punch (122), — by Dickson's Grandson of Punch (213), — by Checks (132), — by R. Grimston's Bull (282), — by J. Coates's Bull (148).

(43534)　　　　　　　LORD LINCOLN,

Roan, calved November 22, 1879, bred by Mr. J. M. Frudd, Bloxholm Moor; got by Oliver Cromwell (37145), dam (Lady Ashby) by Royal Brilliant (32361), g. d. (Prize Rose) by Cambridge Duke 4th (25706), gr. g. d. (Peep o' Day) by Great Comet (16192), — (Pomp) by Sir John (12084), — by The Bonus (10922), — by Hercules (7077), — by Richmond (4947), — by Belvoir (1708), — by Sir Roger de Coverley (5187), — by Albion (1624), — by Son of Favourite (252).

(43535)　　　　　　　LORD LLAY,

Roan, calved February 9, 1879, bred by Mr. T. Beakbane, Llay Place, Gresford; got by Lord Fulbeck (38608), dam (Penelope's Pride) by Earl Warwick (31092), g. d. (Penelope's Princess) by General Napier (24023), gr. g. d. (Penelope) by Lord Raglan (14849), — (Prosperity) by Captain Edwards (8929), — by Almacks (7779), — by Dan O'Connell (3557), — by Young Grazier (3928), — by Milton (2314), — by Policy (1329), — by Barmpton (54).

(43536)　　　　　　　LORD LOCKET,

Roan, calved May 6, 1879, bred by Messrs. Horsley and Son, Colton Manor House; got by Woodpecker (42629), dam (Locket 3rd) by Fire King 2nd (33921), g. d. (Locket 2nd) by Fire King (28598), gr. g. d. (Locket) by Lord Foppington (26647), — (Luna) by Lord Rosliston (22210).

(43537)   LORD LOFTUS,

Red, calved August 15, 1878, bred by Mr. G. Graham, The Oaklands, the property of Mr. C. Bayes, Kettering; got by King Christmas (36837) or Ambassador (39372), dam (Lady Clara 2nd) by The Cardinal (27612), g. d. (Lady Clara) by Fourth Duke of Thorndale (17750), gr. g. d. (Cordelia) by Mameluke (13289), — (Charmer 2nd) by White Friar (9827), — by Little John (4232), — by Caliph (1774), — by Sir Walter (2637), — by Hotspur (1117), — by Coxcomb (928), — by Midas (435), — by Comet (155), — by R. Colling's Son of Favourite (252), — by same Son of Favourite (252), — by Hubback (319).

(43538)   LORD LORN,

Roan, calved March 29, 1879, bred by Mr. G. Ashburner, Low Hall, the property of Mr. C. Stephenson, Crosby Thwaite; got by Duke of Oxford 41st (38174), dam (Lady Lorn) by Duke of Oxford (31004), g. d. (Bride of Lorn) by Sockburn Lad (30024), gr. g. d. (Lady Oxford) by Tenth Duke of Oxford (17739), — (Lily) by Hope (13042), — by Duke of Richmond (8000), — by Bachelor (5770), — by Romulus (6405), — by Sillery (5131), — by Young Western Comet (1575), — by Western Comet (689), — by Son of Favourite (252).

(43539)   LORD MAIDSTONE,

Roan, calved June 16, 1878, bred by Mr. E. Leney, Hadlow Place, the property of Mr. J. K. Pain, Eridge; got by Lord Barrington (36914), dam (Rubina 4th) by Duke of Maidstone (30989), g. d. (Rubina 3rd) by Duke of Kirklevington (25982), gr. g. d. (Rubina 2nd) by Grand Duke 16th (24063), — (Rubina) by Sixth Duke of Airdrie (19602), — by Parade (18518), — by Lord Ruby (14851), — by Preston (20514), — by Monzani (6222), — by Firby (1040), — by Arbutus (1649).

(43540)   LORD MANFRED,

Roan, calved March 23, 1879, bred by Mr. W. Mitchell, Cleasby; got by King Manfred (38491), dam (Ladyship) by Major (26790), g. d. (Lady Ann) by Sir Christopher (22895), gr. g. d. (Lady Bird) by Parmesan (20471), — (Royal Maid) by Good Friday (14635), — by Neptune (11847), — by The Dandy (10926), — by Hamlet (8126), — by Sir Roger (7515), — by Studley Royal (5342).

(43541)   LORD MAR 2ND,

Roan, calved March 25, 1878, bred by Mrs. Fawcett, Scaleby Castle, the property of Mr. J. Grainger, Fenton; got by Grand Duke of Kirklevington 2nd (38379), dam (Lady Mary 2nd) by Earl of Glo'ster (21644), g. d. (Mary Jane 2nd) by Lord Ravensworth (20222), gr. g. d. (Mary Jane) by General Canrobert (12927), — (Mary) by Earl of Derby (10177), — a Cow of Mr. Thompson's, of Kirklevington, bred from Mr. Bates's Bulls.

(43542)   LORD MUNCASTER 2ND,

Roan, calved April 23, 1879, bred by Mr. J. Harward, Winterfold, the property of the Executors of Mr. J. Harward; got by Baron Barrington 6th (33008), dam (Dido 10th) by Waterloo Cherry (27763), g. d. (Dido 9th) by Sir Windsor (22927), gr. g. d. (Dido 7th) by Knight of Distington (18158), — (Dido 3rd) by Joscelin (16326), — by Carlisle (14244), — by Flying Dutchman (10235), — by Muncaster (20387), — by Broad Oak, — by Young Western Comet (1575), — by Diamond (205), — by a Bull of Mr. Charge's.

(43543) **LORD MUNSTER,**
Roan, calved March 1, 1879, bred by Lord Fitzhardinge, Berkeley Castle, the property of Mr. S. Leonard, Berkeley ; got by Duke of Connaught (33604), dam (Minstrel) by Second Duke of Tregunter (26022), g. d. (Musical 14th) by Seventh Duke of York (17754), gr. g. d. (Musical 6th) by Seventh Duke of York (17754), — (Chorus) by Fourth Duke of Oxford (11387), — by Alchemist (11097), — by Monzani (6222), — by The Prince (7615), — by Fitzroy (3808), — by Morpeth (2339), — by Roman (2559), — by Admiral (5), — by Son of Blyth Comet (85), — by Foljambe's Bull (1051).

(43544) **LORD NELSON,**
Roan, calved May 14, 1879, bred by Mr. C. Ellis, Meldreth ; got by Anoop Sing (37734), dam (Lady Braybrooke) by Englishman (19701), g. d. (Miriam) by Snowball (15309), gr. g. d. (Red Rose) by Horatio (10335), — (Maria) by Third Duke of Northumberland (3647), — by Velocipede (5552), — by Sir Thomas (2636), — by Marske (418), — by Comet (155), — by Tom (652), — by Favourite (1033), — by Hutton's Bull (323), — by Barningham (56).

(43545) **LORD NELSON,**
Red and white, calved June 21, 1879, bred by Mr. T. Horsfall, Limefield House, the property of Messrs. S. Whitehead and Son, Bradely Hall, Nelson ; got by Oxford 16th (38822), dam (Wild Flower) by Baron Bolderston (30422), g. d. (Wild Brier) by Royal Scotforth (25042), gr. g. d. (Brier Bud) by Napoleon (20395), — (Miss Brier) by Petteril (20488), — by Splendour (10869), — by Son of Young Bedlamite (6775).

(43546) **LORD NONSUCH,**
Red and white, calved February 20, 1879, bred by Captain Gandy, Castle Bank ; got by Lord Bowness (40139), dam (Nonsuch Princess) by Crisis (33474) g. d. (Nonsuch) by Baron Fennel (27937), gr. g. d. (Nancy 7th) by Storrs (25238), — (Nancy 5th) by Master Gwynne (16539), — by General Havelock (14598), — by Despot (11348), — by Balco (9918), — by Homer (2134), — by Lord Marlboro' (7166), — by Rockingham (2550), — by Bulmer (1760), — by Don Juan (1923), — by a Bull of Lord Feversham's.

(43547) **LORD OF ATHENS,**
Red, calved March 25, 1878, bred by Mr. T. Barber, Sproatley Rise, the property of Mr. T. Bainton, Arram Hall ; got by Baron Wild Eyes (33100), dam (Lady of Athens) by Oxford's Baronet (29499), g. d. (Lady Cherry) by Cherry Duke (25753), gr. g. d. (Purity) by Duke of Wellington (12776), — (Lady Clementina 2nd) by Master Charlie (13312), — by Selim (6454), — by Rex (6385), — by Sir Thomas Fairfax (5196), — by Ambo (1636), — by Memnon (2295), — by Pilot (496), — by Agamemnon (9), — by Burrell's Bull of Burdon.

(43548) **LORD OF EWOOD 4TH,**
Roan, calved June 21, 1879, bred by Mr. T. Riley, Ewood Hall ; got by Grandee (36726), dam (Pearl of Beauty) by Junior Lord (31453), g. d. (Countess Granville) by Lord Belmore (31617), gr. g. d. (Lady Rollings) by Knowsley (24289), — by Lord Stanley (16452), — by Rifleman (15163), — by Young Hopewell (14719), — by Fitz-William (14555). — from Sir W. Lawson's stock.

(43549) **LORD OF KILLIGRAY,**
Red, calved September 7, 1879, bred by Mr. F. Barchard, Horsted ; got by

Marquis of Oxford 2nd (37055), dam (Red Rose of Killigray) by Fourth Duke of Geneva (30958), g. d. (Duchess 3rd) by Dandy Duke (30855), gr. g. d. (Duchess 2nd) by Pilot (32066), — (Duchess) by Buena Vista (30623), — by Prince Charles 2nd (32113), — by Shakespeare (12062), — by Reformer (2505), — by Belvedere (1706), — by Second Hubback (1423), — by His Grace (311), — by Yarborough (705), — by Favourite (252), — by Punch (531), — by Foljambe (263), — by Hubback (319).

(43550) LORD OF LUNE,

Roan, calved January 27, 1879, bred by Mr. A. Graham, Yanwath Hall; got by The Colonel (35747), dam (Lunesdale Rose 5th) by Master Dragon Fly (34807), g. d. (Lunesdale Rose 4th) by Sir Charles Deans (32496), gr. g. d. (Lunesdale Rose 2nd) by Tommy Bates (23079), — (Lunesdale Rose) by Duke of Norfolk (25991), — by Orphan Duke (20442), — by Young Eclipse (14485), — by Pilot (13473), — by Major (4344), — by Champion (3323), — by Young Rockingham (7440).

(43551) LORD OF ROWFANT,

Red and white, calved December 30, 1878, bred by Sir Curtis M. Lampson, Bart., Rowfant; got by Duke of Underley 2nd (36551), dam (Cordelia 8th) by Fourth Duke of Geneva (30958), g. d. (Cordelia 2nd) by Airdrie (30365), gr. g. d. (Cordelia 1st) by Dandy Duke (30855), — (Duchess 2nd) by Pilot (32066), — by Buena Vista (30623), — by Prince Charles 2nd (32113), — by Shakespeare (12062), — by Reformer (2505), — by Belvedere (1706), — by Second Hubback (1423), — by His Grace (311), — by Yarborough (705), — by Favourite (252), — by Punch (531), — by Foljambe (263), — by Hubback (319).

(43552) LORD OF STRATHEARNE,

Red and white, calved January 14, 1879, bred by the Earl of Dunmore, Dunmore, the property of the Earl of Zetland, Aske Hall; got by Marquis of Oxford 2nd (37055), dam (Red Rose of Strathearne) by Sixth Duke of Geneva (30959), g. d. (Red Rose of Strathtay) by Joe Johnson (31440), gr. g. d. (Duchess 4th) by Airdrie (30365), — (Duchess 2nd) by Pilot (32066), — by Buena Vista (30623), — by Prince Charles 2nd (32113), — by Shakespeare (12062), — by Reformer (2505), — by Belvedere (1706), — by Second Hubback (1423), — by His Grace (311), — by Yarborough (705), — by Favourite (252), — by Punch (531), — by Foljambe (263), — by Hubback (319).

(43553) LORD OF THE GRANGE,

Roan, calved November 1, 1878, bred by Mr. T. Nichols, The Grange; got by Grand Duke of Morecambe (36722), dam (Lady Brailes) by Duke of Brailes (23724), g. d. (Formosa) by Duke of Brailes (23724), gr. g. d. (Gionetta) by Sarawak (15238), — (Smock Frock) by Earl of Dublin (10178), — by Janizary (8175), — by Snowball (8602), — by Caliph (1774), — by Norman (2379), — by White Boy (1580), — by Wyville's Bull, — by a Bull bred by Mr. Charge.

(43554) LORD OF THE NORTH,

Roan, calved September 23, 1878, bred by Mr. J. Falder, Winskill, the property of Mr. J. Longrigg, The Demesne, Kirkoswald; got by Baron Underley (37824), dam (Portrait 3rd) by Interest (34210), g. d. (Portrait 2nd) by Sulyman (12157), gr. g. d. (Portrait 1st) by Patron (7315), — by Archibald (9902), — by Gainford (10254).

(43555) **LORD OF THE NORTH,**
Roan, calved January 30, 1879, bred by Mr. R. Lodge, The Rookery ; got by
Duke of Oxford 36th (38170), dam (Ida 7th) by Baron Lightburne 3rd (36192),
g. d. (Ida 6th) by Grand Duke of Lightburne 3rd (28761), gr. g. d. (Ida 5th) by
Lord of the Valley (31732), — (Ida 4th) by Red Windsor (20653), — (Ida 3rd)
by Benedict (17403), — (Ida 2nd) by a Bull of Mr. Birchall's, — (Ida) by Half
Brother to Ben (14156).

(43556) **LORD OF THE RHINE,**
Red and white, calved September 1, 1879, bred by Major O'Reilly, Knock Abbey;
got by Red Rhine (38976), dam (Lady Sarai) by Cherry Prince (23555), g. d.
(Lady Adah) by Killerby Monk (20053), gr. g. d. (Lady Eve) by Doctor McHale
(15887), — (Lady Mary Bountiful) by Baron Warlaby (7813), — by Usurer
(9763), — by Ranunculus (2479), — by Sir Walter (2637), — by Hotspur
(1117), — by Coxcomb (928), — by Midas (435), — by Comet (155), — by R.
Colling's Son of Favourite (252), — by same Son of Favourite (252), — by
Hubback (319).

(43557) **LORD OF THE ROSES,**
Roan, calved July 22, 1879, bred by Mr. J. P. Clark, North Ferriby ; got by
Baron Turncroft Oxford 3rd (36209), dam (Guelder Rose 8th) by Nineteenth
Duke of Oxford (28431), g. d. (Guelder Rose 6th) by Fourth Duke of Thorndale
(17750), gr. g. d. (Guelder Rose 2nd) by Seventh Duke of York (17754), —
(Geranium) by Cupid (14359), — by Valentine (13932), — by Berkeley (7830),
— by Son of Holkar (4041), — bred by Mr. G. Angus, Byfield.

(43558) **LORD OF THE TWEED 2ND,**
Red, calved April 7, 1879, bred by Mr. H. Allsopp, Hindlip Hall ; got by
Duke of Collingham 3rd (38134), dam (Red Rose of Tweeddale) by Fourth Duke
of Geneva (30958), g. d. (Rosebud 6th) by Airdrie (30365), gr. g. d. (Rosebud
1st) by General [Winfield] Scott (12936), — (White Rose) by Young Paragon
(11886), — by Prince Charles 2nd (32113), — by Shakespeare (12062), — by
Reformer (2505), — by Belvedere (1706), — by Second Hubback (1423), — by
His Grace (311), — by Yarborough (705), — by Favourite (252), — by Punch
(531), — by Foljambe (263), — by Hubback (319).

(43559) **LORD OF THORNDALE 5TH,**
Roan, calved January 8, 1879, bred by Mr. C. H. Cock, Bridgefoot ; got by
Young England (31110), dam (Kate Thorndale) by Costa (21487), g. d. (Lady
Lucy Thorndale) by Third Duke of Thorndale (17749), gr. g. d. (Lady Bird) by
Count de Gourcy (17632), — (Luxury) by Cheltenham (12588), — by Horatio
(10335), — by Third Duke of Northumberland (3647), — by Velocipede (5552),
— by Sir Thomas (2636), — by Marske (418), — by Comet (155), — by Tom
(652), — by Favourite (1033), — by Hutton's Bull (323), — by Barningham (56).

(43560) **LORD OF WHALE,**
Roan, calved August 22, 1879, bred by Mr. T. Dargue, Whale Farm ; got by
Lord of the Mere (40223), dam (Windsor's Duchess 2nd) by Shirley Fitz Windsor
(27451), g. d. (Duchess Hamilton) by Duke of Hamilton (19618), gr. g. d.
by Master Hopewell (14929), — by Benedict (7828).

(43561) **LORD OXFORD,**
Red, calved June 15, 1879, bred by Mr. R. Thompson, Mythop Lodge ; got by
Lord of the Forth 2nd (38639), dam (My Darling) by Baron Oxford 6th

(33075), g. d. (My Pet) by Royal Cambridge (25009), gr. g. d. (My Lass) by Glo'ster's Grand Duke (12949), — (Meretrix) by Field Marshal (14545), — by Lord Milton (10461), — by Locksley (4240), — by Stanhope (5315), — by Hercules (21919), — a Cow bought at Hexham in 1834.

(43562) LORD PARAMOUNT,

Red and white, calved October 8, 1879, bred by Mr. R. Jefferson, Preston Hows; got by Lord Prinknash (34655), dam (Souvenir) by Duke of Aosta (28356), g. d. (Primrose) by Prowler (22662), gr. g. d. (Rose 2nd) by Peak (24733), — (Napier Rosebud) by Lord Napier (14832), — by Sam Glen (10780), — by Peter Plough (10606), — by Eildon (9076), — by Ethelred (5990), — by Emperor (1974), — by Barmpton (1677), — by St. Albans (2584), — by Simon (590), — by Pope (514).

(43563) LORD PENRHYN,

Roan, calved August 15, 1877, bred by Mr. R. Loder, Whittlebury, the property of Mr. G. Chapman, Brook Farm; got by Grand Duke 22nd (34062), dam (Ruby Lips 2nd) by Grand Duke of Kent (26289), g. d. (Ruby Lips) by Sir Charles Knightley (27466), gr. g. d. (Ruby) by Lord of the Harem (16430), — (Cornelian) by Mameluke (13289), — by Cardinal (11246), — by Earl of Dublin (10178), — by Little John (4232), — by Caliph (1774), — by Sir Walter (2637), — by Hotspur (1117), — by Coxcomb (928), — by Midas (435), — by Comet (155), — by R. Colling's Son of Favourite (252), — by same Son of Favourite (252), — by Hubback (319).

(43564) LORD PRIVY SEAL 2ND,

Red, calved March 30, 1866, bred by Mr. J. Innes, Wogle, late the property of Mr. J. Hepburn, Overtown of Keithfield; got by Lord Balco (22123), dam (Croma) by Lord Privy Seal (16444), g. d. (Maital) by Goldsmith (14632), gr. g. d. (Lady Lothian) by New Year's Gift (10564), — (Dew Lass) by Belted Will (6780), — by Roger (13615), — by Favourite (2008), — by Romulus (1398), — by Charles 2nd (128), — by Alfred (23).

(43565) LORD PROTECTOR,

Roan, calved December 5, 1879, bred by the Executors of Mr. T. C. Booth, Warlaby; got by Royal Stuart (40646), dam (Lady Cheerful) by Royal Benedict (27348), g. d. (Lady Mirthful) by British Crown (21322), gr. g. d. (Lady Mirth) by Sir Samuel (15302), — (Lady Blithe) by Windsor (14013), — by Hopewell (10332), — by Leonard (4210), — by Young Matchem (2282), — by Jerry (4097), — by Young Pilot (4702), — by Pilot (496), — by a Son of Apollo (36).

(43566) LORD PYTHUS,

Red, calved April 4, 1877, bred by Mr. G. Brown, Willow Hill; got by Janizary (34243), dam (Pythia 10th) by Wellesley (25421), g. d. (Pythia 7th) by Cambridge Barrington 1st (14223), gr. g. d. (Pythia 4th) by Master Thankful (10514), — (Pythia 3rd) by Governor (10280), — by Friar Tuck (3848), — by Algernon (1631), — by Emperor (1014), — by Meteor (1226).

(43567) LORD RAGLAN,

Red, calved March 16, 1861, bred by Mr. R. W. Saunders, Nunwick Hall, late the property of Mr. W. Longrigg, Sceugh Dyke; got by Nunwick (16635), dam (Lovely) by Eugene Sing (14519), g. d. by Eden (3689), gr. g. d. by Buston's Styford (103), — by Catterick, — by Barmpton (54).

(43568) **LORD RAUNDS,**
Red, calved November 4, 1879, bred by Mr. T. Nichols, The Grange; got by
Lord Clarence Waterloo (36926), dam (Lady Edith 2nd) by Fourth Duke of
Barrington (30924), g. d. (Lady Edith) by Duke of Brailes (23724), gr. g. d.
(Edith of Fawsley) by Prince Christian (22582), — (Sylphide) by Sarawak
(15238), — by Janizary (8175), — by Caliph (1774), — by Scipio (1421), — by
Billy (787), — by Western Comet (689), — by Favourite (252), — descended
from the Studley White Bull (627).

(43569) **LORD REDLIGHT,**
Red, calved January 30, 1879, bred by Mr. H. Bettridge, East Hanney; got by
Lord Lamplight (40186), dam (Belinda) by Second Duke of Tregunter (26022),
g. d. (Purity) by Miracle (24602), gr. g. d. (Filigree) by Warwick (19120), —
(Embroidery) by Clarendon (12605), — by The Red Duke (8694), — a Cow
bought in the North.

(43570) **LORD ROCHFORD,**
Roan, calved April 24, 1878, bred by Mr. H. Cross, Barling; got by Champion's
Oxford (37963), dam (Grisi Thorndale) by Thorndale Butterfly (23508), g. d.
(Grisi) by Musician (18432), gr. g. d. (Delhi) by Marquis (11786), — (Anna
Maria) by Vandyke (7669), — by Albus (2975), — by Belshazzar (1704), — by
Sultan (1485), — by Tarrare (1501), — by Midas (435), — by Grandson of
Phenomenon (491), — by Punch (531).

(43571) **LORD ROSEBERY,**
Red and white, calved March 7, 1878, bred by Mr. J. Thompson, Newseat of
Dumbreck, the property of Mr. A. Milne, Corse of Kinmoir; got by Blucher
(33170), dam (Florette) by Albert (30370), g. d. (Flora) by Allan (21172),
gr. g. d. (Roseblush) by Lord Sackville (13249), — (Rosa) by Procurator
(10657), — by Fourth Duke of Northumberland (3649), — by Red Highflyer
(2488), — by Pyramid (4852), — by Harry Lorrequer (3985), — by Blucher
(84), — by Magnum Bonum (4322), — by Buston's Styford (103), — by Son of
Wetherell's Bull (690).

(43572) **LORD ROSEBERY,**
Roan, calved June 12, 1879, bred by Mr. A. Fawkes, Farnley Hall; got by
Baron Winsome 6th (33111), dam (Lady-le-Grand) by Oxford-le-Grand (29496),
g. d. (Lady Vane) by Lord Cobham (20164), gr. g. d. (Vanity) by Reformer
(18687), — (Vanille) by Robinson Crusoe (13610), — by Borrowby Boy (9980),
— by Laudable (9282), — by Sir Thomas Fairfax (5196), — by Colossus (1847),
— by Burley (1766), — by Pilot (496), — by Warlaby (672), — by Albion
(14), — by Lame Bull (359), — by Shipton (587), — by Son of Suworrow
(636), — by Son of Twin Brother to Ben (88), — by Twin Brother to
Ben (660).

(43573) **LORD ROSEDALE,**
Roan, calved December 1, 1874, bred by Mr. R. Welsted, Ballywalter, the
property of Mr. H. B. Mitchell, Lleiniog Castle; got by England's Glory
(23889), dam (Rosabel) by Sir James (16980), g. d. (Elfin Rose) by Elfin King
(17796), gr. g. d. (Carolina Rose) by Uncle Tom (13913), — (Rosa) by Crusade
(7938), — by Cotherstone (6903), — by Young Matchem (4425), — by Fairfax
(1023), — by Son of Marske (418), — by Palmsun (7311), — by a Bull of Mr.
Maynard's, — by Son of Favourite.

(43574) **LORD RYEDALE,**
Roan, calved August 9, 1879, bred by H.R.H. the Prince of Wales, Sandringham;
got by Baron Ryedale (37813), dam (Wild Musical) by Wildfire (32866), g. d.
(Musical 15th) by Grand Duke 13th (21850), gr. g. d. (Musical 7th) by Seventh
Duke of York (17754), — (Harpsichord) by Earl of Walton (17787), — by
Fourth Duke of Oxford (11387), — by Snowball (8602), — by Monzani (6222),
— by The Prince (7615), — by Fitzroy (3808), — by Morpeth (2339), — by
Roman (2559), — by Admiral (5), — by Son of Blyth Comet (85), — by
Foljambe's Bull (1051).

(43575) **LORD SEATON,**
Roan, calved June 18, 1879, bred by Mr. T. Purkis, West Wratting Grange; got
by Knightley York (38511), dam (Moss Rose) by Cambridge Duke 5th (30644),
g. d. (Wild Rose) by Telemachus (27603), gr. g. d. (Wild Sage) by Nestor
(24648), — (Mulleen) by Fourth Duke of Thorndale (17750), — by Achilles
(14056), — by Earl of Dublin (10178), — by Janizary (8175), — by Snowball
(8602), — by Caliph (1774), — by Norman (2379), — by White Boy (1580),
— by Wyville's Bull, — by a Bull bred by Mr. Charge.

(43576) **LORD SHENDISH FAWSLEY,**
Red, calved October 2, 1877, bred by Mr. A. H. Longman, Shendish; got by
Oxford Duke (34977), dam (Fawsley 5th) by Second Duke of Collingham
(23730), g. d. (Fawsley 4th) by Grand Duke 4th (19874), gr. g. d. (Archduchess
of Cambridge) by Archduke 2nd (15588), — (Coquelicot) by Duke of Cambridge
(12742), — by Earl of Dublin (10178), — by Janizary (8175), — by Caliph
(1774), — by Rob Roy (557), — by Satellite (1420), — by Sir Dimple (594), —
by Styford (629).

(43577) **LORD SHIPSTON,**
Red, calved August 12, 1878, bred by Mr. T. Nichols, The Grange; got by Duke
of Wellington 9th (33754), dam (Lady Brailes 3rd) by Wild Eyes Duke (36007),
g. d. (Lady Brailes) by Duke of Brailes (23724), gr. g. d. (Formosa) by Duke
of Brailes (23724), — (Gionetta) by Sarawak (15238), — by Earl of Dublin
(10178), — by Janizary (8175), — by Snowball (8602), — by Caliph (1774), —
by Norman (2379), — by White Boy (1580), — by Wyville's Bull, — by a Bull
bred by Mr. Charge.

(43578) **LORD SOMERVILLE,**
Red, calved June 4, 1877, bred by Mr. R. Bell, Linton, Kelso; got by Baron
Oxford 5th (27958), dam (Rosebud) by Duke of Norfolk (33697), g. d. (The
Nun) by Clifton (23578), gr. g. d. (Vanity) by Heir-at-Law (13005), — (Hoddam)
by Baron of Ravensworth (7811), — by Hudibras (10339), — by Baronet
(1686), — by Spectator (2688), — by Fitz-Remus (2025), — by Whitworth (695),
— a Cow bought by Mr. Mason, of Chilton.

(43579) **LORD SURREY,**
Roan, calved February 3, 1879, bred by Mr. J. R. Singleton, Great Givendale;
got by Duke of Oxford 37th (38171), dam (Ate) by Chatsworth (23546), g. d.
(Alert) by Surrey (17067), gr. g. d. (Arabella) by Apollo (9899), — (Viscountess)
by Fourth Duke of York (10167), — by Second Cleveland Lad (3408), — by
Velocipede (5552), — by Colonel Cradock's Son of Rob Roy (557), — bred by
Mr. Kitchen, from the stock of Messrs. C. and R. Colling.

(43580)   LORD SWEET PEA,

Roan, calved April 26, 1877, bred by Mr. W. Sheraton, Broom House, the property of Mr. Booth, Haslington; got by Lilleshall (34456), dam (Sweet Pea 2nd) by May Duke (26885), g. d. (Sweet Pea) by Don Windsor 2nd (21550), gr. g. d. (Miss Louisa Heald) by Silk Velvet (12070), — (Miss Heald)' by Tom of Lincoln (8714), — by Buchan Hero (3238), — by Lord John (4257), — by Berryman (3143), — by Lenny (2197), — by Cappy (1782), — by Western Comet (689).

(43581)   LORD TANKERVILLE,

Roan, calved March 22, 1879, bred by Captain D. H. Mytton, Garth; got by Constantine 2nd (33439), dam (Hopbine) by Manager (24521), g. d. (Mistletoe Berry) by St. Helena (35450), gr. g. d. (Mistletoe) by Sir Charles (16946), — (Nectarine) by Cronstadt (14352), — by Young Pompey (13480), — by Son of David (1912), — bred by Mr. Slye.

(43582)   LORD TEHIDY,

Roan, calved October 23, 1878, bred by Mr. W. Trethewy, Tregoose, the property of Mr. Willoughby, Tuckingmill; got by M. C. (31898), dam (Ruth 74th) by Crœsus (30820), g. d. (Ruth 45th) by Lord Montgomery (26686), gr. g. d. (Ruth 26th) by Duke of Manchester (33690), — (Ruth 20th) by Rufus (35423), — by Earl Ducie (12797), — by Lord Fingal (11716), — by Frantick (8088), — by Harold (8131), — by Cedric (3311), — by Nimrod (4571), — by Grandson of Blyth Comet (85), — by Crispin (174), — by Meteor (431), — by Meteor (431).

(43583)   LORD THIRLMERE,

Red, calved July 2, 1878, bred by the Earl of Bective, Underley Hall, the property of Mr. G. Blackwell, Hazlecote; got by Lord of the Isles (34631), dam (Red Rose of Nithsdale) by Fourth Duke of Geneva (30958), g. d. (Leonora) by Airdrie (30365), gr. g. d. (Nora 3rd) by Anacreon Moore (37731), — (Nora 1st) by Renick (32290), — by Ashland (11122), — by Prince Charles 2nd (32113), — by Shakespeare (12062), — by Reformer (2505), — by Belvedere (1706), — by Second Hubback (1423), — by His Grace (311), — by Yarborough (705), — by Favourite (252), — by Punch (531), — by Foljambe (263), — by Hubback (319).

(43584)   LORD THORNDALE 10TH,

Roan, calved March 23, 1879, bred by Mr. J. Allan, Billie Mains; got by Lord Thorndale (37010), dam (Grand Duchess of Essex 10th) by Earl of Studley (31085), g. d. (Grand Duchess of Essex 8th) by Glory of the Mountains (28717), gr. g. d. (Grand Duchess of Essex 2nd) by Confederate (21457), — (Grand Duchess of Essex) by Grand Duke of Essex (17995), — by Dreadnought (15900), — by Hendon (14688), — by Lamplighter (8204), — by Zenith (5702), — by Warlock (5599), — by Alabaster (1616), — by Monarch (2324), — by Satellite (1420), — by Cato (119), — by Jupiter (342), — by George (273), — by Chilton (136), — by Irishman (329), — by B. (45).

(43585)   LORD THORNDALE 11TH,

Red and white, calved April 14, 1879, bred by Mr. J. Allan, Billie Mains; got by Lord Thorndale (37010), dam (Grand Duchess of Essex 9th) by Earl of Studley (31085), g. d. (Grand Duchess of Essex 2nd) by Confederate (21457), gr. g. d. (Grand Duchess of Essex) by Grand Duke of Essex (17995), — (Silkworm) by Dreadnought (15900), — by Hendon (14688), — by Lamplighter (8204), — by Zenith (5702), — by Warlock (5599), — by Alabaster (1616), — by Monarch (2324), — by Satellite (1420), — by Cato (119), — by Jupiter (342), — by George (273), — by Chilton (136), — by Irishman (329), — by B. (45).

(43586)   LORD TREGUNTER,

Roan, calved January 21, 1879, bred by Mr. R. Loder, Whittlebury; got by Duke of Tregunter 7th (38194), dam (Grand Duchess of Fawsley 2nd) by Grand Duke 22nd (34062), g. d. (Oxford Fawsley 3rd) by Lord Oxford 2nd (20215), gr. g. d. (Fawsley 3rd) by Grand Duke 4th (19874), — (Coquelicot) by Duke of Cambridge (12742), — by Earl of Dublin (10178), — by Janizary (8175), — by Caliph (1774), — by Rob Roy (557), — by Satellite (1420), — by Sir Dimple (594), — by Styford (629).

(43587)   LORD TYRLEY,

Roan, calved January 23, 1878, bred by Mr. P. Buchanan, Hales Hall, the property of Mr. G. Taylor, The Old Springs; got by Oneida (37146), dam (Duchess 9th) by Grand Duke of Brockton (26284), g. d. (Duchess 3rd) by Grand Duke of York (12966), gr. g. d. (Duchess 1st) by Duke of Oxford (11386), — (Honeysuckle) by Romulus (12016), — Lady Rosa.

(43588)   LORD TYRLEY 2ND,

Roan, calved June 12, 1878, bred by Mr. P. Buchanan, Hales Hall; got by Oneida (37146), dam (Duchess 17th) by Second Duke of Wetherby (21618), g. d. (Duchess 9th) by Grand Duke of Brockton (26284), gr. g. d. (Duchess 3rd) by Grand Duke of York (12966), — (Duchess 1st) by Duke of Oxford (11386), — by Romulus (12016), — Lady Rosa.

(43589)   LORD WALLACE 3RD,

Roan, calved July 9, 1879, bred by Mr. T. W. Cadman, Ballifield Hall, the property of Mr. G. Revitt, Myrtle Bank; got by Duke of Sandale 2nd (38180) dam (Lady Wallace 2nd) by Iron Duke 2nd (31421), g. d. (Lady Wallace) by Lord Cobham (20164), gr. g. d. (Wild Rose) by Sir Walter Scott (13753), — (Moss Rose 3rd) by Sir Walter (2639), — by Belvedere 2nd (3127), — by Waterloo (2816), — by Barmpton (5774), — by Kitt (7127), — by Kitt (7127), — by Page's Bull (6269), — by Middleton's Bull (438).

(43590)   LORD WATERLOO 2ND,

White, calved October 8, 1879, bred by Mr. R. Harrett, Kirkwhelpington; got by Prince of Waterloo 2nd (38944), dam (Duchess Charlotte Florentia 3rd) by Oxford Beau 4th (34964), g. d. (Duchess Charlotte Florentia) by Stanwick (22965), gr. g. d. (Moss Rose) by Moss Trooper (13357), — (Flowery) by Redesdale (22697), — by Tom Boy (9749), — by Pilot (7333).

(43591)   LORD WELCOME FAME,

Roan, calved October 13, 1879, bred by Mr. R. J. M. Gumbleton, Glanatore; got by Captain Cook (39548), dam (Lady Welcome) by Vain Hope (23102), g. d (Castanet 2nd) by Ravenspur (20628), gr. g. d. (Castanet) by Prince Arthur (13497), — (Coquette) by Comet (11298), — by Druid (10140), — by Patriarch (7329), — by Second Duke of York (5959), — by Raspberry (4875), — by Young Matchem (4422), — by Isaac (1129), — by Young Pilot (4702), — by Pilot (496), — by Julius Cæsar (1143).

(43592) **LORD WILD EYES,**
Roan, calved September 6, 1878, bred by Mr. J. Woodhouse, Scale Hall ; got by
Fennel Duke (36639), dam (Minstrel 3rd) by Fifth Lord Wild Eyes (26762),
g. d. (Musical 3rd) by Seventh Duke of York (17754), gr. g. d. (Minstrel) by
Nundi (9453), — (La Polka) by Monzani (6222), — by The Prince (7615), —
by Fitzroy (3808), — by Morpeth (2339), — by Roman (2559), — by Admiral
(5), — by Son of Blyth Comet (85), — by Foljambe's Bull (1051).

(43593) **LORD WILD EYES 2ND,**
Roan, calved May 17, 1878, bred by Mr. E. H. Cheney, Gaddesby Hall, the
property of Mr. E. C. Tisdall, Holland Park Farm ; got by Third Duke of
Glo'ster (33653), dam (Wild Oxford) by Lord Oxford 2nd (20215), g. d. (Wild
Eyes 24th) by Lord Barrington 3rd (16382), gr. g. d. (Wild Eyes 16th) by
Second Duke of Oxford (9046), — (Wild Eyes 15th) by Fourth Duke of
Northumberland (3649), — by Duke of Northumberland (1940), — by Belvedere
(1706), — by Emperor (1975), — by Wonderful (700), — by Cleveland (145),
— by Butterfly (104), — by Hollon's Bull (313), — by Mowbray's Bull (2342),
— by Masterman's Bull (422), — descended from M. Dobison's stock.

(43594) **LORD WILTON,**
White, calved March 30, 1879, bred by Mr. A. Fawkes, Farnley Hall ; got by
Baron Winsome 6th (33111), dam (Sixth Lady) by Lord Darlington (26633),
g. d. (Lady Maggie) by Lord Cobham (20164), gr. g. d. (Magnolia) by Royal
Oak (16873), — (Magenta) by Don Giovanni (15893), — by Lord Marquis
(10459), — by Petrarch (7329), — by Sir Thomas Fairfax (5196), — by Garton
(2052), — by Harold (291), — by Count (170), — by Badsworth (47), — by
Driffield (223), — bred by Sir G. Strickland.

(43595) **LORD WOLFERTON,**
Red and white, calved March 23, 1879, bred by H.R.H. the Prince of Wales,
Sandringham ; got by Homer (34170), dam (Marie) by High Sheriff (26392),
g. d. (Satinet) by Duke of Edinburgh (23741), gr. g. d. (Seamstress) by Skipper
(20854), — (Seraph) by Baron Albany (11151), — by The Captain (5422), —
by The Captain (5422), — by Percy (1314), — by Lord Grantham's Son of
Comet (155).

(43596) **LORD ZETLAND,**
Roan, calved April 12, 1879, bred by the Earl of Zetland, Aske Hall, the property
of Mr. J. Outhwaite, Bainesse ; got by Royal Windsor (29890), dam (Florella)
by George Peabody (28710), g. d. (Floss) by Windsor Augustus (19157), gr. g. d.
(Flirt) by Cobham (14287), — (Woodnymph) by Ravensworth (10681), — by
Postmaster (9487), — by Ranunculus (2479) — by William (2840), — by Firby
(1040), — by Rodney (1392), — by Tyrant (1537).

(43597) **LORENZO,**
Roan, calved November 14, 1879, bred by Mr. W. H. Wodehouse, Woolmers Park ;
got by Lord Janicula (41873), dam (Lode Star) by Woolmers Duke (32890),
g. d. (Lausanne) by Vulcan (27739), gr. g. d. (Lucerne) by Fawsley Baronet
(23920), — (Lucy) by Wolfsbane (15518), — by Dukedom (12729), — by
Honeycomb (10330), — by Bower's Bull (19332), — by May Duke (424).

(43598) **LOTHAIR,**
Roan, calved June 24, 1879, bred by Messrs. A. and A. Mitchell, Alloa ; got by

Brocklesby (36288), dam (Corisande) by Lord Plymouth (24455), g. d. (Coral) by Valasco (15443), gr. g. d. (Carmine) by Prince Arthur (13497), — (Carnation) by Newton (2367), — by Young Magog (2247), — by Margrave (2263), — by Sir Charles (593), — by Sir Dimple (594), — by St. Albans (2584), — by Layton (366), — by Charge's Grey Bull (872), — by J. Brown's Red Bull (97).

(43599)   LOVER,

Roan, calved August 14, 1879, bred by Mr. H. Smith, Mountmellick; got by Earl of Aylesby (38215), dam (Lovesome 5th) by Royal Duke (25014), g. d. (Lovesome 3rd) by Leviathan (20120), gr. g. d. (Lovesome 2nd) by Favourite (21728), — (Lovesome) by Great Mogul (14651), — by Lord Foppington (10437), — by Mambrino (7196), — by Wiseton (2848), — by Firby (1040), — by Ivanhoe (1131), — by Son of Irishman (329), — by Son of Favourite (252).

(43600)   LUCIFER,

Roan, calved June 5, 1879, bred by the Duke of Northumberland, Alnwick Castle; got by Sir Raymond (40716), dam (Lucretia 5th) by Duke of Tyne (33744), g. d. (Lucretia 2nd) by Royal Butterfly 23rd (27355), gr. g. d. (Lucretia) by Knight of the Grand Cross 2nd (26551), — (Bianca) by Majestic (16492), — by Ambo (12391), — by Lord George (10439), — by Vanguard (10994), — by Hamlet (8126), — by Sir Roger (7515), — by Studley Royal (5342).

(43601)   LYONS,

Red and white, calved July 6, 1878, bred by Captain C. R. Conwy, Bodrhyddan, the property of Mr. R. C. Welsby, Criccin; got by Sir Frederick Gwynne 3rd (39117), dam (Caper) by Favourite (33895), g. d. (Columbine 5th) by Oxford Wild Eyes 2nd (32038), gr. g. d. (Columbine) by New York (22412), — (Charlotte) by Duke of Wellington (12776), — by Monk (11824), — by Dan O'Connell (9011), — by Lord Adolphus Fairfax (4249), — by Young Favourite (3770), — by Waterloo (2816), — by Lawnsleeves (365), — by Phenomenon (491), — by Favourite (252), — by Favourite (252), — by Favourite (252), — by Hubback (319), — by Snowdon's Bull (612), — by Waistell's Bull (669), — by Masterman's Bull (422), — by the Studley Bull (626).

(43602)   MACHAON,

Red and white, calved January 4, 1879, bred by Mr. T. Pears, Hackthorne; got by Earl of Ashfield (36571), dam (Mantalini Breeze) by Whiff (30299), g. d. (Maid of Connington) by Quis (20618), gr. g. d. (Maid of the Mist) by Prince Imperial (16740), — (Maid of the Valley) by Dumbarton (14462), — by Ivanhoe (14735), — by Heir-at-Law (13005), — by Sir Henry (10824), — by Buckingham (3239), — by Marcus (2262), — by Matchem (2281), — by Alderman (1622), — by Pilot (496), — by Remus (550), — by Sir Charles (592), — by Son of Favourite (252), — by Son of Favourite (252), — Strawberry.

(43603)   MAGNATE,

Roan, calved May 14, 1878, bred by Mr. T. Dargue, Whale Farm; got by Knight of the Forest (31556), dam (Magdalen 7th) by Knight of Warlaby (31571), g. d. (Magdalen 6th) by Royal Duke Hamilton (29863), gr. g. d. (Filagree) by Hauberk (21905), — (Frantic) by Facsimile (17821), — by Red Rover (11982), — by Isaac (9645), — by Victory (5566), — by Miracle (2321), — from the stock of Sir George Crewe.

(43604) **MAGNET,**

Red, calved June 6, 1879, bred by Earl Beauchamp, Madresfield Court; got by Magnum Bonum (41954), dam (Moonbeam) by Gay Boy (31222), g. d. (Sunbeam) by Archduke (21185), gr. g. d. (Sunshine) by Ortolan (18490), — (Gay Lass) by Columbus (17591), — by Duke William (11400), — by Cotgrave (6901), — by Progress (4839).

(43605) **MAGOR BUTTERFLY,**

Roan, calved February 28, 1878, bred by Mr. W. Murdoch, Old What; got by Emperor (39840), dam (Miss Butterfly 5th) by Bismarck (33154), g. d. (Miss Butterfly) by Symmetrical (23006), gr. g. d. (Exquisite Butterfly) by Master Butterfly 5th (14918), — (Rosa) by Procurator (10657), — by Fourth Duke of Northumberland (3649), — by Red Highflyer (2488), — by Pyramid (4852), — by Harry Lorrequer (3985), — by Blucher (84), — by Magnum Bonum (4322), — by Buston's Styford (103), — by Son of T. Wetherell's Bull (690).

(43606) **MAHOMED,**

Roan, calved June 22, 1877, bred by Mr. S. Campbell, Kinellar, the property of Mr. J. Wilson, Mains of Scotstown; got by Novelist (34929), dam (Bessie 6th) by Foljambe (33950), g. d. (Bessie 3rd) by British Prince (23468), gr. g. d. (Bessie 2nd) by Rory O'More (24991), — (Bessy) by Frolic (16086), — by Red Rover (20651), — by Robin Hood (8494), — by Mahomed 2nd (10492), — by Billy (3151), — by Belshazzar (1703), — by Abraham (2905), — by Simon (5134).

(43607) **MAJOR,**

Roan, calved March 29, 1877, bred by Lord Polwarth, Mertoun House; got by Major Errant (34732), dam (Mertoun Ruby) by Coronet (33448), g. d. (Missie 19th) by Pretender (29579), gr. g. d. (Missie 12th) by Master Gunner (22316), — (Missie 4th) by Clarendon (14280), — by Son of Duke 3rd (17697), — by The Pacha (7612), — by Mahomed (6170), — by Plenipo (4725), — by Abbot (2899).

(43608) **MAJOR BLANCHE,**

Red, calved May 1, 1879, bred by Mr. T. R. Hulbert, North Cerney; got by Viscount Blanche (39269), dam (Short) by Shuttlecock 3rd (35521), g. d. by Clint (33399), gr. g. d. by Star Prince (18925), — by Savernake (35475), — (Violet) by Mazeppa (34831).

(43609) **MAJOR DOMO,**

White, calved May 19, 1879, bred by Mr. J. B. Booth, Killerby; got by King Richard 2nd (28984), dam (My Partner) by K. C. B. (26492), g. d. (Lady Patroness) by Brigade Major (21312), gr. g. d. (Lady of the Lake) by Knight Errant (18154), — (Forest Queen) by Royal Buck (10750), — by Hopewell (10332) — by Hamlet (8126), — by Leonard (4210).

(43610) **MAJOR TODDLES,**

Roan, calved August 20, 1878, bred by Mr. R. Arklay, Ethiebeaton; got by Master Toddles (40331), dam (Border Girl) by Master Booth (34801), g. d. (Border Woman) by Craigrossie (30815), gr. g. d. (Border Lady) by Keir Butterfly 1st (24235), — (Border Lass) by Knight of the Border (18161), — by Hiawatha (14705), — by Sir Charles (13705), — by Strathmore (6547), — by Playfellow (6297), — by Sir William (12102), — by Togston (5487), — by Emperor (1974).

(43611) MAJOR WINDSOR,

Roan, calved August 1, 1879, bred by Messrs. J. and D. D. Lazonby, Calthwaite House ; got by Baron Helbeck (33053), dam (Alexandra Windsor) by Prince of Wales (24851), g. d. (Ormolu Windsor) by Imperial Windsor (18086), gr. g. d. (Ormolu Gwynne) by Master Hopewell (14929), — (Ormolu) by General Sale (8099), — by Lord Warden (7167), — by Orontes (4623), — by Guardian (3947), — by Mercury (2301), — by Monarch (2324), — by St. Albans (2584), — by Jupiter (342), — by Sir Oliver (605), — by Trunnell (659), — by Favourite (252), — by Favourite (252), — by Dalton Duke (188), — by R. Alcock's Bull (19), — by J. Smith's Bull (608), — by Jolly's Bull (337).

(43612) MANFRED,

Roan, calved August 31, 1879, bred by Mr. T. Dargue, Whale Farm ; got by Knight of the Forest (31556), dam (Magdalen 7th) by Knight of Warlaby (31571), g. d. (Magdalen 6th) by Royal Duke Hamilton (29863), gr. g. d. (Fila-gree) by Hauberk (21905), — (Frantic) by Facsimile (17821), — by Red Rover (11982), — by Isaac (9645), — by Victory (5566), — by Miracle (2321), — from the stock of Sir George Crewe.

(43613) MAPLE,

Red, calved July 25, 1879, bred by Mr. T. Purkis, West Wratting Grange ; got by Wharfdale Oxford (27786), dam (Harmonia) by Grand Duke of Essex 4th (24068), g. d. (Harmony) by Cherry Duke 3rd (15763), gr. g. d. (Floret) by Douglas (12714), — (Florimel) by Duke of Cambridge (12742), — by Grey Friar (9172), — by Allan-a-Dale (7778), — by Little John (4232), — by Mar-cellus (2260), — by Caliph (1774), — by Swing (2721), — by Argus (759), — by Defender (194), — by Petrarch (488), — by Own Brother to R. Colling's White Heifer, — by Butterfly (104), — by Globe (278).

(43614) MARCUS,

Red and white, calved August 21, 1879, bred by Mr. J. Downing, Ashfield ; got by Robert Stephenson (32313), dam (Brilliant Butterfly) by Great Hope (24082), g. d. (Broughton Butterfly) by Victorious (25378), gr. g. d. (Alice Butterfly) by Master Butterfly (13311), — (Alice 2nd) by Duke of Athol (10150), — by Marcus (2262), — by Matchem (2281), — by Pilot (496), — by Young Albion (15), — by Albion (14), — by Suworrow (636), — by Son of Twin Brother to Ben (88), — by Twin Brother to Ben (660).

(43615) MARKHAM,

Red, calved March 4, 1878, bred by Mr. P. Stevenson, Rainton, the property of Mr. C. Wells, Fawdington ; got by Duke of Edlingham 3rd (36489), dam (Copper Queen) by Prince of Thorndale (32187) g. d. (Copper Plate) by Lord Lally 3rd (24408), gr. g. d. (Cordelia) by Red Duke (18676), — (Cowslip 3rd) by Earl of Derby (10177), — by Second Cleveland Lad (3408), — by Brougham (1746), — by Wallace (5588), — by Mr. Corner's Bull, — by Short Legs (5124), — by Adonis (2933), — by Barmpton (54), — by Hector (2103), or Leopold (2199), — by Sir Harry (1444), — by Hollon's Bull (313), — by Traveller (655), — by Bolingbroke (86).

(43616) MARKSMAN,

Red, calved October 10, 1878, bred by Mr. G. Britten, Overstone Farm, the pro-perty of Mr. T. Perkins, Kingsthorpe ; got by Fearless (36637), dam (Milkmaid

2nd) by Lord York Fawsley (34709), g. d. (Milkmaid 1st) by General Hesse (34008), gr. g. d. (Nightfall) by Twelfth Duke of Oxford (19633), — (Nightingale) by The Hero (20958), — by General Hopewell (17953), — by Albus (14075), — by Prince George (13510), — by Son of Ranunculus (2479), — by Remus (4932), — by Prince Comet (1342).

(43617)   MARMADUKE,
Red and white, calved March 1,'1876, bred by Major Ramsay, Straloch, the property of Mr. J. Cassie, Westertown; got by Ferdinand (36640), dam (Matilda) by Baron Oxford (25581), g. d. (Lady Mary 3rd) by Champion of England (17526), gr. g. d. (Emily) by Lord Sackville (13249), — (Eliza) by Brutus (10000), — by Duplicate Duke (6952), — by Sir Walter (2639), — by Son of Emperor (2671), — by Wonderful (700), — by Cardinal (841), — by Baronet (61), — by Cleveland (145), — by Butterfly (104), — by Globe (278).

(43618)   MARQUIS,
Red and white, calved November 24, 1878, bred by Lord Lovat, Beaufort Farm; got by Duke of Beaufort (38122), dam (Young Julia 3rd) by Bachelor of Arts (32982), g. d. (Young Julia) by King's Seal (26525), gr. g. d. (Julia) by Allan (21172), — (Josephine) by Matadore (11800), — by Fairfax Royal (6987), — by Premier (6308), — by Saturn (5089), — by Favourite (6997), — by Grindon (3942), — bred by Mr. Rennie, of Phantassie.

(43619)   MARQUIS BARRINGTON,
Roan, calved February 24, 1879, bred by Mr. J. Harward, Winterfold, the property of the Executors of Mr. J. Harward; got by Baron Barrington 6th (33008), dam (Marchioness) by Earl Mortimer (33780), g. d. (Kelstern's Marchioness 2nd) by Second Duke of Wetherby (21618), gr. g. d. (Kelstern's Marchioness) by Marquis of Oxford (18339), — (Kelstern Lady) by Viceroy (13945), — by Monzani (6222), — by Studley (628), — by Young Major (8266), — by Radical (4858).

(43620)   MARQUIS MONTHERMER 4TH,
Roan, calved May 14, 1879, bred by the Duke of Manchester, Kimbolton Castle; got by Duke of Underley 3rd (38196), dam (Lady Montagu) by Duke of Hillhurst (28401), g. d. (Lady Worcester 10th) by Eighth Duke of Geneva (28390), gr. g. d. (Lady Worcester 2nd) by Charleston (21400), — (Clear Star) by Marton Duke (22307), — by Red Duke (18676), — by Third Duke of York (10166), — by Second Cleveland Lad (3408), — by Duke of Northumberland (1940), — by Belvedere (1706), — by Emperor (1975), — by Wonderful (700), — by Cleveland (145), — by Butterfly (104), — by Hollon's Bull (313), — by Mowbray's Bull (2342), — by Masterman's Bull (422), — descended from M. Dobison's stock.

(43621)   MARQUIS OF GLO'STER,
Roan, calved July 18, 1878, bred by Mr. J. Abell, Higham; got by Fifth Duke of Glo'ster (36494), dam (May Lass 6th) by Marquis of York (34791), g. d. (May Lass) by May Duke 2nd (18372), gr. g. d. (Music) by Vocalist (13960), — (Moss Rose) by Lord of Brawith (10465), — by Laudable (9282), — by Grouchy (6051), — by Sir Thomas Fairfax (5196), — by Young Colling (1843), — by Red Bull (2838), — by Son of Hollings (2131), — by Partner (2409), — by R. Alcock's Bull (19).

(43622)   MARQUIS OF TELLURIA,
Roan, calved March 31, 1878, bred by Mr. J. Spencer, Murrah Hall, the property
of Mr. W. Graham, Hawksdale ; got by Marquis of Lorne (31847), dam (Telluria 4th) by Beau of Oxford (21254), g. d. (Telluria 3rd) by Romulus (15185),
gr. g. d. (Telluria 2nd) by Horatio (10335), — (Taglioni) by Lord George (10439),
— by Orontes (4623), — by Guardian (3947), — by Mercury (2301), — by
Monarch (2324), — by St. Albans (2584), — by Jupiter (342), — by Sir Oliver
(605), — by Trunnell (659), — by Favourite (252), — by Favourite (252), —
by Dalton Duke (188), — by R. Alcock's Bull (19), — by J. Smith's Bull (608),
— by Jolly's Bull (337).

(43623)   MARQUIS OF WORCESTER 8TH,
Red, calved February 6, 1879, bred by the Earl of Dunmore, Dunmore, the property of Mr. J. Darling, Beau Desert ; got by Marquis of Oxford 2nd (37055),
dam (Lady Worcester 18th) by Duke of Connaught (33604), g. d. Lady Worcester) by Charleston (21400), gr. g. d. (Clear Star) by Marton Duke (22307),
— (Bright Star) by Red Duke (18676), — by Third Duke of York (10166), —
by Second Cleveland Lad (3408), — by Duke of Northumberland (1940), — by
Belvedere (1706), — by Emperor (1975), — by Wonderful (700), — by Cleveland (145), — by Butterfly (104), — by Hollon's Bull (313), — by Mowbray's
Bull (2342), — by Masterman's Bull (422), — descended from M. Dobison's
stock.

(43624)   MARQUIS OF WORCESTER 9TH,
Red and white, calved May 22, 1879, bred by the Earl of Dunmore, Dunmore,
the property of Mr. J. Darling, Beau Desert ; got by Marquis of Oxford 2nd
(37055), dam (Lady Worcester 2nd) by Charleston (21400), g. d. (Clear Star)
by Marton Duke (22307), gr. g. d. (Bright Star) by Red Duke 18676), —
(Bright Eyes) by Third Duke of York (10166), — by Second Cleveland Lad
(3408), — by Duke of Northumberland (1940), — by Belvedere (1706), — by
Emperor (1975), — by Wonderful (700), — by Cleveland (145), — by Butterfly (104), — by Hollon's Bull (313), — by Mowbray's Bull (2342), — by
Masterman's Bull (422), — descended from M. Dobison's stock.

(43625)   MARS,
Red and white, calved April 6, 1879, bred by Mr. J. Whyte, Aldborough ; got
by Gunpowder (28801), dam (Duchess of Knightley 4th) by Valentine Vox
(32752), g. d. (Damsel) by Lord Hopewell (18239), gr. g. d. (Datura) by Fitz-
Clarence (14552), — (Duchess) by Duke of Cambridge (12742), — by Earl of
Dublin (10178), — by Grey Friar (9172), — by Fawsley (6004), — by Little
John (4232), — by Marcellus (2260), — by Caliph (1774), — by Swing (2721),
— by Argus (759), — by Defender (194), — by Petrarch (488), — by Own
Brother to R. Colling's White Heifer, — by Butterfly (104), — by Globe (278).

(43626)   MARSHAL BENEDICT,
Red and white, calved February 26, 1879, bred by Mr. E. Heinemann, Ratton
Park, the property of Mr. J. Ramsbotham, Crowborough Warren ; got by Prince
Benedict (38893), dam (Marshal's Gift) by Marshal Booth (31852), g. d. (New
Year's Gift) by Thorntree (25312), gr. g. d. (Dewy Morn) by Le Moor (22090),
— (Rosy Morn) by Redmond (11978), — by Cotherstone (6903), — by Young
Matchem (4425), — by Fairfax (1023), — by Son of Marske (418), — by Palm-
sun (7311), — by a Bull of Mr. Maynard's, — by Son of Favourite.

(43627)   **MARSHAL NIEL,**

Roan, calved October 20, 1879, bred by Mr. G. Graham, The Oaklands; got by Duke of Oaklands 2nd (39765), dam (Cheerful Rose 6th) by Lord Oxford Rosy (38647), g. d. (Cheerful Rose) by Cambridge Duke 4th (25706), gr. g. d. (Cheerful) by Mystic (20391), — (Coquette) by Mameluke (13289), — by Garrick (11506), — by Earl of Dublin (10178), — by Little John (4232), — by Caliph (1774), — by Sir Walter (2637), — by Hotspur (1117), — by Coxcomb (928), — by Midas (435), — by Comet (155), — by R. Colling's Son of Favourite (252), — by same Son of Favourite (252), — by Hubback (319).

(43628)   **MARSHAL OXFORD,**

Roan, calved July 1, 1879, bred by Dr. Mackay, Marshall Green; got by Duke of Oxford 27th (33709), dam (Duchess of Wallington 3rd) by Third Lord Wharfdale (26759), g. d. (Duchess of Wallington) by Second Duke of Wharfdale (19649), gr. g. d. (Amity) by Seventh Duke of York (17754), — (Amy) by Sixth Duke of Oxford (12765), — by Handel (14663), — by Mozart (11830), — by Diamond (5918), — by Albert (2950), — by Stirling (5330), — by Commodore (1858), — by Tathwell Studley (5401), — by Blyth Comet (85).

(43629)   **MARY'S CHERRY DUKE,**

Red, calved April 26, 1879, bred by Mr. J. Singleton, Teresa Cottage; got by Thorndale Cherry Lad (40818), dam (Mary 12th) by Grand Duke of Lightburne 2nd (26291), g. d. (Mary 10th) by Waterloo Chief (23184), gr. g. d. (Mary 8th) by General Jackson 2nd (17954), — (Mary 6th) by The Baronet (10918), — by Burgundy (7861), — by Sizergh (5205), — by Chancellor (1809), — by Umpire (5530), — by Northern Light (4586), — by Young Alfred (2993), — by Lemon Horns (2196), — by Witham's White Bull (5670).

(43630)   **MASTER BUTTERFLY,**

Roan, calved November 28, 1879, bred by Colonel R. Loyd Lindsay, Lockinge Park; got by Earl of Horton 11th (36588), dam (Bella Donna) by Lord Napier (26691), g. d. (Burlesque) by Fawsley Baronet (23920), gr. g. d. (Britannia) by Master Coleshill (18344), — (Blossom) by Sultan (15358), — by Neptune (11847), — by Fanatic (8054), — by Earl of Durham (5965), — by Pedestrian (7321), — by Edrom (1956), — by Scipio (1421), — by Hector (1104), — by Midas (435), — by Marquis (407), — by Chilton (136), — by Ben (70).

(43631)   **MASTER CAWOOD,**

Roan, calved January 5, 1879, bred by Mr. W. Handley, Green Head, the property of Mr. N. Eckersley, Standish Hall; got by Lord St. Vincent (40239), dam (Cawood's Rose 6th) by Lord Chancellor (26622), g. d. (Cawood's Rose) by Third Lord Cawood (24368), gr. g. d. (White Cow) by Golden Eclipse (14625), — (Roan Cow) by Reindeer (15150), — (Red Cow) by Horton Boy (13050), — a Roan Heifer bought at Underley Hall.

(43632)   **MASTER DUDLEY,**

Red and white, calved June 11, 1879, bred by Mr. J. Christy, Boynton Hall; got by Oxford's Baron (32030), dam (Miss Dudley) by Baron Oxford 3rd (25579), g. d. (Lady Dudley 7th) by Third Duke of Claro (23729), gr. g. d. (Lady Dudley) by Third Duke of Lancaster (19624), — (Soprano) by Fitzroy (16058), — by Duke of Ulster (12774), — by Belshazzar 2nd (14154), — by Napoleon

(10552), — by Plenipo (4724), — by Sir John (16985), — by Rex (1375), — by Baron (58), — by a Bull of Mr. Mason's, — by Falstaff (250), — by Irishman (329).

(43633) MASTER EGLINTON,
Red, calved March 23, 1879, bred by Mr. W. Handley, Green Head, the property of Mr. H. Dennison, Whitwell Folds ; got by Alfred the Great (36121), dam (Lady Eglinton) by Earl of Eglinton (23832), g. d. (Quickly) by Marquis of Cobham (22299), gr.g.d. (Sprightly) by Tweedside (12246), — (Rose) by Strathmore (6547), — by Sir William (12102), — by Emperor (1974).

(43634) MASTER INGLEWOOD 2ND,
Red and white, calved October 13, 1879, bred by Mr. R. Thompson, Inglewood Bank ; got by Brilliant Butterfly (36270), dam (Moss Rose 3rd) by Oxford (20450), g. d. (Moss Rose) by Lord of Brawith (10465), gr. g. d. (Lady Millicent) by Laudable (9282), — (Millicent) by Grouchy (6051), — by Sir Thomas Fairfax (5196), — by Young Colling (1843), — by Red Bull (2838), — by Son of Hollings (2131), — by Partner (2409), — by Hollings (2131), — by R. Alcock's Bull (19).

(43635) MASTER NOBODY,
Roan, calved August 24, 1879, bred by Lord Rathdonnell, Lisnavagh ; got by Anchor (32947), or Lieutenant (43466), dam (James's Nancy) by King James (28971), g. d. (Red Nancy) by Sovereign (27538), gr. g. d. (Nancy) by Baron Warlaby (7813), — (Red Rose) by Hamlet (8126), — by Prince Ernest (7366), — by Vanquish (2793), — by Monarch (4495).

(43636) MASTER OF ESKDALE,
Red and white, calved May 18, 1878, bred by the Earl of Bective, Underley Hall, the property of Miss Taunton, Ashley ; got by Duke of Underley (33745), dam (Red Rose of Eskdale) by Airdrie 3rd (32919), g. d. (Poppy 8th) by Joe Johnson (31440), gr. g. d. (Poppy 3rd) by Airdrie (30365), — (Poppy 2nd) by Duke of Airdrie (12730), — by Ashland (11122), — by Prince Charles 2nd (32113), — by Shakespeare (12062), — by Reformer (2505), — by Belvedere (1706), — by Second Hubback (1423), — by His Grace (311), — by Yarborough (705), — by Favourite (252), — by Punch (531), — by Foljambe (263), — by Hubback (319).

(43637) MASTER OXFORD,
Red and white, calved in 1874, bred by Mr. H. Mousell, Tuffleigh Court ; got by Peri's Duke of Collingham (43748), dam (Baroness Oxford 5th) by Fifth Lord Wild Eyes (26762), g. d. (Baroness Oxford 4th) by Baron Oxford 2nd (23376), gr. g. d. (Nell Gwynne) by Lord Jersey (20185), — (Niagara 2nd) by Harry of Glo'ster (14674), — by Chinaman (12599), — by Master Shaftoe (10513), — by Duke of St. Albans (6941), — by Will Honeycomb (5660), — by Spectator (2688), — by Son of Sir Roger de Coverley (5187), — by a Son of Albion (1619), — by Alexander (1624).

(43638) MASTER QUICKLY 2ND,
White, calved October 30, 1879, bred by Mr. R. C. Morton, Lane House ; got by Baron Barrington 4th (33006), dam (Quickly 4th) by Young Knightley (31529), g. d. (Wine) by Wolfsbane (15518), gr. g. d. (Erigone) by Artichoke (14107), — (Mimosa) by Scimitar (10788), — by Allan-a-Dale (7778), — by

Little John (4232), — by Caliph (1774), — by Swing (2721), — by Argus (759), — by Defender (194), — by Petrarch (488), — by Own Brother to R. Colling's White Heifer, — by Butterfly (104), — by Globe (278).

(43639) MASTER ROSY,
Roan, calved May 15, 1879, bred by Mr. M. Savidge, Sarsden Lodge Farm, the property of Mr. J. B. White, Street End House ; got by Rag Merchant (43865), dam (Lady Rosy 2nd) by Lord Rosy (34661), g. d. (Fresh Start) by Duke of Towneley (21615), gr. g. d. (Silver Crescent) by Archduke 2nd (15588), — (Champion) by Vagabond (9765), — (Silver) by Locksley (4240), — by Stanhope (5315), — by Marquis (22292), — bred from the stock of Mr. Jobling, of Styford.

(43640) MASTER SURMISE 2ND,
Red and white, calved March 26, 1879, bred by Messrs. F. Leney and Sons, Wateringbury ; got by Sixth Duke of Oneida (30997), dam (Tacita) by Duke of Rutland (19641), g. d. (Surmise 3rd) by May Duke (13320), gr. g. d. (Surmise) by Duke of Glo'ster (11382), — (Silence) by Earl of Derby (10177), — by Duke of Sutherland (6945), — by Locomotive (4242), — by Short Tail (2621), — by Gambier (2046), — by Young Wynyard (2859), — by Bulls of Messrs. C. and R. Colling's.

(43641) MASTER TENTERDEN 3RD,
Roan, calved October 3, 1879, bred by Mr. H. Stonham, Thurnham ; got by Tenterden (42479), dam (Miss Thorndale) by Royal Thorndale (42336), g. d. (Miss Wrotham) by Duke of Wrotham (23803), gr. g. d. (Cashmere) by Noble Arthur (16621), — (Cassandra) by Norfolk (9442), — by Rex (6385), — by Sir Thomas Fairfax (5196), — by Ambo (1636), — by Memnon (2295), — by Pilot (496), — by Agamemnon (9), — by Burrell's Bull, of Burdon.

(43642) MASTER WYNNE,
Red, calved June 3, 1877, bred by Mr. W. Talbot Crosbie, Ardfert Abbey, the property of Mr. N. Mac Vicar, Kirmond ; got by England's Glory (23889), dam (Royal Florentine) by Regal Booth (27262), g. d. (Light of Florence) by Northern Light (24670), gr. g. d. (Florentine Grove) by Castle Grove (19408), — (Florentine) by Lamp of Lothian (16356), — by Desmond (10112), — by Old Bull (8368), — by Young Spectator (8619), — by Phantassie (8389), — by Young Rockingham (8498).

(43643) MATCHWORK,
Red, calved May 16, 1879, bred by Earl Beauchamp, Madresfield Court ; got by Magnum Bonum (41954), dam (Sunshine 3rd) by Gay Boy (31222), g. d. (Sunshine 2nd) by Festival (26147), gr. g. d. (Sunbeam) by Archduke (21185), — (Sunshine) by Ortolan (18496), — by Columbus (17591), — by Cotgrave (6901), — by Progress (4839).

(43644) MAURICE,
Roan, calved May 30, 1878, bred by Mr. J. Angus, Bearl ; got by Captain Brace (37938), dam (May Morn) by Royal Frederick (35366), g. d. (May Day) by Roan Chief (27294), gr. g. d. (May) by Bloomfield (23430), — (May Flower) by Richard Cœur-de-Lion (13590), — by Red Prince (13576), — by Brilliant (7851), — by Young Hector (7074), — by Wallace (5588), — descended from C. Colling's Son of Hubback (319).

(43645)   MAY BARON,

Roan, calved April 7, 1879, bred by Mr. B. Hale, Holly Hill; got by Baron Lampson (39419), dam (May Duchess 4th) by Grand Duke 15th (21852), g. d. (May Queen) by May Duke (13320), gr. g. d. (Lady Foppington) by Lord Foppington (10437), — (Dowager Queen) by Sir Frederick Fairfax (5152), — by Albert (727), — by Pilot (496), — by Albion (14), — by Lame Bull (359), — by Shipton (587), — by Son of Suworrow (636), — by Son of Twin Brother to Ben (88), — by Twin Brother to Ben (660).

(43646)   MAZURKA LAD,

Red, calved February 8, 1879, bred by Mr. G. Fox, Elmhurst Hall; got by Duke of Oxford 39th (38173), dam (Mazurka 2nd of Oakdale) by Malcolm (41963), g. d. (Mazurka 8th) by Albion (19209), gr. g. d. (Mazurka 5th) by Duke of Airdrie (12730), — (Mazurka 2nd) by Orontes 2nd (11877), — by Harbinger (10297), — by Baron of Ravensworth (7811), — by Mariner (7204), — by Mina (2316), — by Commodore (1858), — by Rival (553).

(43647)   MENTOR,

Red, calved February 9, 1879, bred by Mr. W. Bolton, The Island; got by Albion (36112), dam (Mina Gwynne) by Lieutenant General (31600), g. d. (Margery Gwynne) by Grey Gauntlet (19908), gr. g. d. (Rose Gwynne) by Equinox (17810), — (Modesty 5th) by Young Usurer (10985), — by Sir Thomas Fairfax (5196), — by Wallace (5586), — by Wellington (2824), — by Marmion (406), — by Merlin (430), — by Layton (366), — by Phenomenon (491), — by Favourite (252), — by Favourite (252), — by Hubback (319), — by Snowdon's Bull (612), — by Waistell's Bull (669), — by Masterman's Bull (422), — by the Studley Bull (626).

(43648)   MEROPE,

Red and white, calved January 11, 1879, bred by Mr. T. Pears, Hackthorne; got by Lichtenstein (41801), dam (Alpine Mantalini) by Warlaby (32792), g. d. (Lady Mowbray) by Whiff (30299), gr. g. d. (Maid of Connington) by Quis (20618), — (Maid of the Mist) by Prince Imperial (16740), — by Dumbarton (14462), — by Ivanhoe (14735), — by Heir-at-Law (13005), — by Sir Henry (10824), — by Buckingham (3239), — by Marcus (2262), — by Matchem (2281), — by Alderman (1622), — by Pilot (496), — by Remus (550), — by Sir Charles (592), — by Son of Favourite (252), — by Son of Favourite (252), — Strawberry.

(43649)   MERRY BARON,

Roan, calved in June 1875, bred by Mr. N. Reid, Danestown, the property of Mr. W. Innes, Meikle, Clinterty; got by Duke of Lancaster (30986), dam (Mary Anne of Lancaster) by Baronet (30448), g. d. (Anne of Lancaster) by Lord Raglan (13244), gr. g. d. (Lancaster 25th) by Matadore (11800), — (Lancaster 16th) by The Marquis (10938), — by Will Honeycomb (5660), — by George 3rd (7038), — by Spectator (2688), — by Albion (1619), — by Lancaster (360), — by Son of Windsor (698), — by Comet (155).

(43650)   METEOR,

Roan, calved August 3, 1879, bred by Mr. J. B. Booth, Killerby; got by Moonstone (37107), dam (Maggie) by K. C. B. (26492), g. d. (Mabel) by Hailstone (21891), gr. g. d. (Lady Percy) by Percival (20486), — (Lady Day) by Man Friday (13290), — by The Irishman (5446), — by Sir Richard (5175), — by Son of Booth's White Bull, — by Seymour's Red Bull.

(43651) **MICHAELMAS BOY,**
Roan, calved November 17,.1879, bred by Mr. R. Blezard, Pool Park; got by
Viscount Dursley (37636), dam (Michaelmas Daisy) by Viceroy (27712), g. d.
(Daisy Flower) by Lord Waterloo (24475), gr. g. d. (Daisy) by Daisy Bull
(30841), — (Amy) by Abbot 3rd (17270), — by Alexander (5741).

(43652) **MICHIGAN 2ND,**
Roan, calved March 3, 1868, bred by the Duke of Richmond and Gordon, Gordon
Castle, the property of Mr. J. Fletcher, Rosehaugh ; got by Michigan (24594),
dam (Distraction) by Whipper In (19139), g. d. (Flirt) by Magnum Bonum
(13277), gr. g. d. (Romp) by Bloomsbury (9972), — (Queen) by The Pacha
(7612), — by Second Duke of Northumberland (3646), — by Mahomed (6170),
— by Sillery (5131), — by Carleton (843), — by Diamond (205), — by Diamond
(205).

(43653) **MIDAS,**
Red, calved May 27, 1879, bred by Earl Fitzwilliam, Coollattin Park, sold to
the Provincial Government of Liege, Belgium ; got by Lord Lilias (31690), dam
(May Duchess) by Robert Burns (29795), g. d. (May Queen) by Lord Lyons
(26677), gr. g. d. (May Duchess) by May Duke (16553), — (Blanche) by Monarch
(13347), — by Lord of Brawith (10465), — by Pestalozzi (10603), — by Lamplighter (8204), — by Zenith (5702), — by Warlock (5599), — by Alabaster
(1616), — by Monarch (2324), — by Satellite (1420), — by Cato (119), — by
Jupiter (342), — by George (273), — by Chilton (136), — by Irishman (329),
— by B. (45).

(43654) **MIDAS,**
Roan, calved May 29, 1879, bred by Mr. J. Stratton, Alton Priors ; got by
Royal James (35387), dam (Mabel) by James 1st (24202), g. d. (Miranda) by
Knight of the Lagan (20083), gr. g. d. (Moss Rose 4th) by Hickory (14706),
— (Moss Rose) by Phœnix (6290).

(43655) **MIDLOTHIAN,**
Red, calved April 6, 1879, bred by Mr. W. Scott, Glendronach ; got by Ivanhoe
(36796), dam (Pansy 3rd) by Rosethorn (32345), g. d. (Pansy 2nd) by Prince of
Worcester (20597), gr. g. d. (Pansy) by Signet Seal (18824), — (Rosa) by Lennox
(31593), — by Corporal (6899), — by Commander (8978), — by Young Duke.

(43656) **MILCOTE GALLEON,**
Red, calved April 27, 1879, bred by Mr. J. C. Adkins, Milcote ; got by Hotspur
(38440), dam (Oxford Galleon) by Duke of Oxford 32nd (36527), g. d. (Galleon
2nd) by Cherry Fawsley (30711), gr. g. d. (Guinevere) by Fourth Duke of
Geneva (25964), — (Galleon) by Morocco (16590), — by Earl of Dublin (10178),
— by Janizary (8175), — by Snowball (8602), — by Little John (4232), — by
Caliph (1774), — by Rob Roy (557), — by Satellite (1420), — by Sir Dimple
(594), — by Styford (629).

(43657) **MILTON LORD,**
Roan, calved March 4, 1879, bred by Mr. J. Strong, Culgaith ; got by Earl of
Doune (36579), dam (White Baroness 2nd) by Pure Gold (32226), g. d. (Baroness)
by Edwin (21671), gr. g. d. by Young Baronet (33038), — by The Rejected
(15399), — by Son of Eden (3689).

(43658) **MILTON REFORMER,**
Red and white, calved November 1, 1877, bred by Mr. B. James, Halls Hannery

the property of Mr. R. Baker, Milton Damrell; got by Iron Duke (31420), dam (Lady Danby) by Royal Windsor (29890), g. d. (Miss Danby) by Emmanuel (31105), gr. g. d. by Champion (23529), — by Baron Warlaby (7813), — by Young Thornton (13885), — by California (10019), — by Gainford (7029), — by Son of Roland.

(43659) MINARD,

Roan, calved May 29, 1879, bred by Mr. J. Ganly, 18 Usher's Quay, Dublin; got by Prince Albert (37211), dam (Nanny 2nd) by Prince of Lothian (35152), g. d. (Nanny) by Little Wonder (40121), gr. g. d. (Fair Maid) by The Squire (13875), — (Florence 2nd) by Prince Ernest (7366), — by Marquis of Chandos (6190), — by Prince Paul (4827), — by Monarch (2324), — by Dr. Syntax (220), — by Charles (127), — by St. John (572), — by White Bull (421), — by Favourite (252), — by Dalton Duke (188), — by R. Alcock's Bull (19), — by J. Smith's Bull (608), — by Jolly's Bull (337).

(43660) MINO TAURUS,

Red and white, calved March 3, 1879, bred by Mr. T. Pears, Hackthorne; got by Earl of Ashfield (36571), dam (Summer Morn) by Valentine Vox (32752), g. d. (Rosy Morn) by Brother Windsor (25690), gr. g. d. (Early Dawn) by Fashion (21724), — (New Year's Morn) by Baltic (12431), — by Master Charlie (13312), — by Rex (6385), — by Sir Thomas Fairfax (5196), — by Ambo (1636), — by Memnon (2295), — by Pilot (496), — by Agamemnon (9), — by Burrell's Bull of Burdon.

(43661) MINSTREL,

Roan, calved March 2, 1878, bred by Mr. C. Lyall, Old Montrose; got by Leopold (38561), dam (Mysie 12th) by Brother Booth (25689), g. d. (White Mysie 11th) by Jolly Prince (24222), gr. g. d. (Mysie 3rd) by Barnaby Rudge (11142), — (Mysie) by Casham (14248), — by Carlos (1788), — by Sultan (1485), — by Albion (731), — by Diomed (974), — by Barmpton (54), — by Wellington (679), — by Sultan (631), — by Son of Punch (531), — by Hubback (319), — by Dalton Duke (188).

(43662) MINSTREL BOY,

Roan, calved May 3, 1879, bred by Mr. C. Hobbs, Maisey Hampton; got by Duke of Hazlecote 55th (39746), dam (Chorus 5th) by Duke of Hazlecote 36th (33671), g. d. (Chorus 3rd) by Bates Tertius (21249), gr. g. d. (Chorus) by Fourth Duke of Oxford (11387), — (Opera) by Alchemist (11097), — by Monzani (6222), — by The Prince (7615), — by Fitzroy (3808), — by Morpeth (2339), — by Roman (2559), — by Admiral (5), — by Son of Blyth Comet (85), — by Foljambe's Bull (1051).

(43663) MINSTREL BOY 2ND,

Roan, calved June 18, 1879, bred by Mr. C. Hobbs, Maisey Hampton; got by Duke of Hazlecote 36th (33671), dam (Musical) by Lord Hastings (29115), g. d. (Musical 2nd) by Earl of Walton (17787), gr. g. d. (Songstress) by Mountaineer (14966), — (Minstrel) by Nundi (9453), — by Monzani (6222), — by The Prince (7615), — by Fitzroy (3808), — by Morpeth (2339), — by Roman (2559), — by Admiral (5), — by Son of Blyth Comet (85), — by Foljambe's Bull (1051).

(43664) MINSTREL BOY 3RD,

Red, calved June 18, 1879, bred by Mr. C. Hobbs, Maisey Hampton; got by

Duke of Hazlecote 55th (39746), dam (Musical 3rd) by Duke of Hazlecote 36th
(33671), g. d. (Musical 2nd) by Lord Hastings (29115), gr. g. d. (Musical 2nd)
by Earl of Walton (17787), — (Songstress) by Mountaineer (14966), — by
Nundi (9453), — by Monzani (6222), — by The Prince (7615), — by Fitzroy
(3808), — by Morpeth (2339), — by Roman (2559), — by Admiral (5), — by
Son of Blyth Comet (85), — by Foljambe's Bull (1051).

(43665)   MINSTREL PRINCE,
Red and white, calved July 24, 1879, bred by Mr. J. Martin, Hawkshead Hall;
got by Duke of Oxford 34th (36529), dam (Minstrel 6th) by Sixth Duke of
Kirklevington (30982), g. d. (Minstrel 5th) by Hematite 2nd (24129), gr. g. d
(Minstrel 4th) by Tenth Duke of Oxford (17739), — (Minstrel 2nd) by Prince
of Glo'ster (13517), — by Count Conrad (3510), — by Wallace (5586), — by
Wellington (2824), — by Marmion (406), — by Merlin (430), — by Layton
(366), — by Phenomenon (491), — by Favourite (252), — by Favourite (252),
— by Hubback (319), — by Snowdon's Bull (612), — by Waistell's Bull (669),
— by Masterman's Bull (422), — by the Studley Bull (626).

(43666)   MINUTE GUN,
Roan, calved July 6, 1879, bred by Mr. M. Matthews, Fifield; got by Cherry
Walnut (42934), dam (Mineral) by Seraphina's Grand Duke (37430), g. d.
(Miniver) by Lord Lytton (34596), gr. g. d. (Minn) by Liddington (24334), —
(Minnie) by Greville (19907), — by Rebec (15132), — by Ivanhoe (9240), — by
Othello (4631), — by Abbot (2899), — by Juniper (1145), — by Colling (900),
— by Lancaster (360), — by Neswick (1266).

(43667) ·   MISCHIEF MAKER,
White, calved February 21, 1879, bred by Mr. L. C. Chrisp, Hawkhill, the
property of Mr. J. Nicholson, Murton; got by Muscovite (38773), dam
(Primrose 3rd) by Fitz-Roland (33936), g. d. (Rose 2nd) by Peak (24733),
gr. g. d. (Napier Rosebud) by Lord Napier (14832), — (Primrose) by Tam
Glen (10780), — by Peter Plough (10606), — by Eildon (9076), — by Ethelred
(5990), — by Emperor (1974), — by Barmpton (1677), — by St. Albans (2584),
— by Simon (590), — by Pope (514).

(43668)   MR. WINKLE,
Roan, calved May 23, 1879, bred by Mr. A. Hamond, Westacre; got by Heir
Apparent (36760), dam (Whisper) by Hotspur (34190), g. d. (Wretham) by
Prince Louis (29643), gr. g. d. (Shepherdess) by Ant Eater (19228), — (Shepherd)
by Windham (15514), — by Gay Lad (11510), — by Newham (4563), — by
Dulverton (3959), — by Juniper (2165), — by The Squire (1512), — by Midas
(435), — by Boughton (90).

(43669)   MONOMEATH,
Roan, calved May 4, 1879, bred by Mr. J. Mickle, Stoneshiel; got by Lord
Thorndale (37010), dam (Rosebud 10th) by Red Knight (24916), g. d. (Rosebud
7th) by Snowflake (18888), gr. g. d. (Rosebud 2nd) by Duke of Buckingham
(14429), — (Rose de Meaux) by Ravensworth (7400), — by Son of Regent
(2517), — by Togston (5487), — by Son of Lawnsleeves (365), — by Son of
Bolingbroke (86).

(43670)   MONTENEGRO,
Roan, calved January 1, 1879, bred by Mr. J. A. Gordon, Udale, the property

of Mr. J. Reid, Inchberry; got by Rosario (35315), dam (Missie) by Baron Killerby (27949), g. d. (Mysie 12th) by Baron Laurie 2nd (25570), gr. g. d. (Mysie 9th) by Whipper In (19139), — (Mysie 8th) by Young Grand Duke (12964), — by The Hero (10934), — by Kelly 2nd (9265), — by The Pacha (7612), — by Mahomed (6170), — by Sillery (5131), — by Carleton (843), — by Diamond (205), — by Diamond (205).

(43671) MONUMENT,

Roan, calved January 28, 1879, bred by Mr. F. C. Matthews, Easterfield House; got by Ignoramus (28887), dam (Lady Agatha) by Duke of Towneley (21615), g. d. (Selina) by Photograph (20492), gr. g. d. (Soprano) by Vice Chancellor (17180), — (Symphony) by Jock O'Hazledean (13085), — by Rodolph (9568), — by Sir Launcelot (5166), — by Snowball (2651), — by Pirate (2430), — by Pioneer (1321), — by Candour (107), — by Cardinal (111), — by Lawnsleeves (365), — by C. Colling's White Bull (150).

(43672) MOON RAKER,

Roan, calved April 15, 1879, bred by Mr. J. H. Braikenridge, The Rookery; got by Pioneer (38868), dam (Lunaire) by Prince William (32217), g. d. (Luna) by Prince George (13510), gr. g. d. (Aurora) by King Arthur (13110), — (Aster) by Sylvan King (13819), — by The Silkey Laddie (10947), — by Morning Star (6223), — by Roland (2556), — by Priam (2452), — by Matchem (2281), — by Son of Peter (487).

(43673) MORRIS DANCER,

White, calved December 9, 1875, bred by Lord Chesham, Latimer House, the property of Mr. W. Nicholson, Basing Park; got by Duke of Oxford 28th (33710), dam (Countess Morisco) by Baron Wastwater (30492), g. d. (Morisco's Blanche) by Morisco (18421), gr. g. d. (May Fly) by Duke of Moscow (14447), — (Constance) by Selim (6454), — by Rex (6385), — by Norfolk (2377), — by Belvedere (1706), — by Belvedere (1706), — by Lancaster (360), — by Petrarch (488), — by Major (397), — by Chapman's Son of Punch (122), — by Dickson's Grandson of Punch (213), — by Checks (132), — by R. Grimston's Bull (282), — by J. Coates's Bull (148).

(43674) MOSSTROOPER,

Roan, calved October 27, 1879, bred by Mrs. Pery, Coolcronan House; got by Don Diego (33539), dam (Blooming Butterfly) by King Victor (28986), g. d. (Blithe Butterfly) by Lord Blithe (22126), gr. g. d. (Alice Butterfly) by Master Butterfly (13311), — (Alice 2nd) by Duke of Athol (10150), — by Marcus (2262), — by Matchem (2281), — by Pilot (496), — by Young Albion (15), — by Albion (14), — by Suworrow (636), — by Son of Twin Brother to Ben (88), — by Twin Brother to Ben (660).

(43675) MOUNTAIN CHIEF,

Roan, calved April 28, 1879, bred by Mr. R. Parker, Moss End; got by Bates Duke (36223), dam (Pink) by Golden Duke (26266), g. d. (Moss Rose) by Regent (20658), gr. g. d. (Lady Love) by Constantine (14318), — (Necklace) by Prince of Wales (4833), — by General Chasse (3874), — by Archibald (1652).

(43676) MOUNTAINEER,

Roan, calved July 6, 1878, bred by Mr. J. H. Braikenridge, The Rookery, the property of Mr. G. Edwards, Congresbury; got by Pioneer (38868), dam (Rock

Rose 65th) by Rosary Monk (35316), g. d. (Rock Rose 48th) by Genuine Prince
(24031), gr. g. d. (Rock Rose 37th) by Lord Greta (21074), — (Rock Rose 28th)
by Actor (17274), — by Earl of Wentworth (14481), — by Blood Royal
(13638), — by Cavaignac (10033), — by Boccaccio (7838), — by Belshazzar
(1703), — by Velocipede (5552), — by Matchem (2281), — by Sir Stephen
(1456), — by Newton (1271), — by Son of Comet (155), — by Western Comet
(689), — by Charge's Grey Bull (872), — by Favourite (252).

(43677) MOUNTAIN HERO,
Roan, calved February 20, 1878, bred by Mr. J. Shaw, Tillyching, the property
of Mr. J. Rust, Bow Butts; got by Duke of Chamburgh (36052), dam (Mysie
34th) by Bullion (36294), g. d. (Mysie 30th) by Denmark (33515), gr. g. d.
(Mysie 20th) by Lee (20109), — (Mysie 15th) by Cherry Duke 2nd (14265),
— by Red Knight (11976), — by The Hero (10934), — by Kelly 2nd (9265), —
by The Pacha (7612), — by Mahomed (6170), — by Sillery (5131), — by
Carleton (843), — by Diamond (205), — by Diamond (205).

(43678) MOUNTAIN KING,
Roan, calved May 11, 1879, bred by Mr. E. W. Meade-Waldo, Stonewall Park;
got by King Harry (36841), dam (Mountain Dell) by Waltron (30255), g. d.
(Mountain Vale) by Blinkhoolie (23428), gr. g. d. (Mountain Mist) by Booth
Royal (15673), — (Mountain Maid) by Vanguard (10994), — by Vanguard
(10994), — by Baron Warlaby (7813), — by Leonard (4210), — by Prince
Comet (1342), — by Constellation (163), — by Prince of Waterloo (528), — by
Young Favourite (255).

(43679) MUSICIAN,
Roan, calved April 16, 1879, bred by Messrs. J. G. W. and W. R. Lyle, Donaghmore House; got by Lord of the Manor (38641), dam (Flora's Choice) by Sailor
Prince (35442), g. d. (Flora) by Alabama (30366), gr. g. d. (Clara) by Fugleman
(14580), — (Rose of Tullyree) by Defender (12687), — by Gold Dust (11536),
— by Nimrod (7279), — by Regent (2517), — by Dinning's Bull (973), — by
Young Lancaster (361), — by Comet (155).

(43680) NATAL,
Red, calved March 22, 1879, bred by Mr. W. Hewer, Sevenhampton; got by
British Prince (37907), dam (Nymph) by Executor (26117), g. d. (Nancy) by
Experience (23900), gr. g. d. (Norah) by Viscount Killerby (19081), — (Norna)
by Neptune (11847), — by Albinus (9879), — by Pioneer (4711), — by Lofty
(2217), — by Cleveland (143), — by Windsor (698).

(43681) NEASHAM DUKE,
Red and white, calved August 13, 1879, bred by Mr. S. R. C. Ward, Neasham
Hill; got by Grand Duke 25th (34065), dam (Red and White Duchess) by
Prince Imperial (27150), g. d. (Red Roan Duchess) by Duke of Wharfdale
(19648), gr. g. d. (Another Roan Duchess) by Master Frederick (18348), —
(Violante) by Valiant (10980), — by Frederick (11409), — by Whittington
(12299), — by Second Cleveland Lad (3408), — by Duke of Northumberland
(1940), — by Norfolk (2377), — by Belvedere (1706), — by Belvedere (1706),
— by Lancaster (360), — by Petrarch (488), — by Major (397), — by
Chapman's Son of Punch (122), — by Dickson's Grandson of Punch (213), —
by Checks (132), — by R. Grimston's Bull (282), — by J. Coates's Bull (148).

(43682) NEPTUNE,
Red and white, calved June 14, 1878, bred by Mr. L. Hodgson, Highthorn ; got
by Standard (40750), dam (Nelly 29th) by Newburgh 3rd (29434), g. d. (Nelly
16th) by Statesman (22974), gr. g. d. (Nelly 11th) by Cheam (21403), — (Nelly
4th) by Dandy (14367), — by Crusader (12668), — by Third Duke of York
(10166), — by Nimrod (7279), — by Counsellor (1882), — by Regent (2517),
— by Young Lancaster (361), — by Comet (155), — by Washington (674).

(43683) NEPTUNE,
Roan, calved September 5, 1879, bred by Mr. C. E. Lyon, Johnson Hall, the property
of Mr. C. W. Falcon, Nuthurst House, Hockley Heath ; got by Saturn (40673),
dam (Leda) by Vandyke (35848), g. d. (Duchess of Padworth 2nd) by Duke of
Padworth (31009), gr. g. d. (Duchess of Padworth) by Duke of Kennet (30977),
— (Lelia 8th) by Mainprize (22260), — by Final Hope (17848), — by Meteor
(13334), — by Killjoy (14759), — by Third Duke of Northumberland (3647),
— by Bashaw (1692), — by Columella (904), — by Palatine (478), — by
Palmflower (480), — by Patriot (486), — by Driffield (223), — by C. Holmes's
Bull (314).

(43684) NIGEL,
Roan, calved May 10, 1879, bred by Mr. J. Stratton, Alton Priors ; got by
Primus (38886), dam (Niobe) by Royal James (35387), g. d. (Nancy) by
Brilliant (28084), gr. g. d. (Novice) by Young Windsor (17241), — (Red
Marigold) by Hermit (14697), — by Clarendon (12605), — by Hero of the West
(8150), — by Mr. Grant's Son of Phœnix (6290).

(43685) NIL DESPERANDUM,
White, calved December 12, 1879, bred by Mr. W. Handley, Green Head ; got
by Master Harbinger (40324), dam (Earl's Portrait) by Earl of Derwent (28503),
g. d. (Venus) by Sir Walter Trevelyan (25179), gr. g. d. (Old Venus) by General
Garibaldi (21813), — by Tenant Farmer (13828).

(43686) NIMROD,
Roan, calved February 16, 1879, bred by Mr. W. Curtis, Fernham ; got by Count
of Hanney (42980), dam (Nancy 4th) by Oxford's Gwynne (32033), g. d. (Nancy
2nd) by Duke of Flamborough (25960), gr. g. d. (Nancy) by Lord Walton (22227),
— by Midsummer (8314), — descended from the stock of Mr. Tucker, of Bourton.

(43687) NIMROD,
Roan, calved June 3, 1879, bred by Mr. F. J. S. Foljambe, Osberton Hall ; got by
Titan (35805), dam (November Rose) by M.P. (29398), g. d. (China Rose) by
Cambridge Duke 4th (25706), gr. g. d. (Clematis) by Sir John (12084), —
(Clementina) by Clementi (3399), — by Young Matchem (4422), — by Isaac
(1129), — by Young Pilot (4702), — by Pilot (496), — by Julius Cæsar (1143).

(43688) NIMROD 2ND,
Roan, calved May 2, 1878, bred by Mr. T. Walker, Stowell Park ; got by Nimrod
(37128), dam (Bertha) by Lord Lyon (24424), g. d. (Britannia) by Viscount
Killerby (19081), gr. g. d. (Barbara) by Napier (13368), — (Brunette) by
Protector (10665), — by Rob Roy (7434), — by Raffler (7391).

(43689) NOBLE GENERAL,
Red and white, calved July 7, 1878, bred by Mr. J. A. M. Cope, Drummilly, the
property of Messrs. J. G. W. and W. R. Lyle, Donaghmore House; got by

Gloster's Gwynne (38360), dam (Peggy Gwynne) by King Richard (26523), g. d. (Phœbe Gwynne) by Killerby Lad (20052), gr. g. d. (Princess Gwynne) by Paul Potter (16688), — (Empress Gwynne) by Emperor (14500), — by Cadet (12521), — by Young Usurer (10985), — by Sir Thomas Fairfax (5196), — by Wallace (5586), — by Wellington (2824), — by Marmion (406), — by Merlin (430), — by Layton (366), — by Phenomenon (491), — by Favourite (252), — by Favourite (252), — by Hubback (319), — by Snowdon's Bull (612), — by Waistell's Bull (669), — by Masterman's Bull (422), — by the Studley Bull (626).

(43690)   NOBLE GIFT,

Red and white, calved October 15, 1879, bred by Mr. A. Metcalfe, Park House, the property of Mr. J. Barker, Gunnerthwaite; got by Earl of Sheffield (33812), dam (Golden Gift) by Peer of the Realm (27057), g. d. (Gift 8th) by King Richard (26523), gr. g. d. (Gift 5th) by The Druid (18981), — (Gift 3rd) by Prince Ernest (7366), — by Albion (7771), — by Hamlet (8126), — by Second Comet (5101), — by Lucifer (4293), — by Prince George (2464), — by Frederick (7023), — by Kearney's Bull (4144).

(43691)   NOBLEMAN,

Roan, calved March 12, 1879, bred by Major E. G. S. Hornby, Dalton Hall; got by Royal Ensign (37388), dam (Lady Nancy 10th) by Oxford Cherry Duke 2nd (34971), g. d. (Lady Nancy 6th) by General Napier (26238), gr. g. d. (Lady Nancy 2nd) by Viscount Oxford (25386), — (Lady Nancy 1st) by Golden Major (19861), — (Lady Nancy) by John Scott's Bull.

(43692)   NORMAN,

Roan, calved April 21, 1879, bred by Mr. G. Marr, Cairnbrogie, the property of Mr. J. Smith, Balmain; got by Cossack (39033), dam (Averne 2nd) by Ben Nevis (39462), g. d. (Averne) by Red Prince (32267), gr. g. d. (Countess) by Hydra (19995), — (Matchless) by Emperor (19689), — by Picotee (15063), — by Report (10704), — by Duke 3rd (17697), — by Bucephalus (6784), — by Crusader (934), — by Sultan (1485), — by Mars (411), — by North Star (458).

(43693)   NORMAN DAISY BULL,

Roan, calved October 4, 1879, bred by Mr. J. Garsed, The Moorlands, the property of Mr. D. J. Jenkins, Lancadle; got by Kentish Charmer (38482), dam (Notable) by Fatherland (28574), g. d. (Namely) by Duke of York (23804), gr. g. d. (Nanny) by Fortunatus (19773), — (Nance) by Lord Mayor (14828), — by Nicety 13386), — by Lord Alexander Buchan 3rd (11706), — by The Miller (5451), — by Lisbon (1172), — by Norman (1276), — by Acklam (713), — by Meteor (431), — by Duke (226), — by Favourite (252), — by Punch (531), — by Hubback (319).

(43694)   NORSEMAN 2ND,

Red, calved March 6, 1878, bred by Mr. W. Scott, Glendronach; got by Norseman (34924), dam (Bridesmaid) by Medallion (20331), g. d. (Bride) by Lord Elcho (22150), gr. g. d. (Queen Mary) by Red Luggs (24919), — (Mary Gray) by Red Orbliston (24920), — by Major (24514), — by Jock O'Darlington, — from the stock of Mr. Grant Duff.

(43695)   · NORTHAMPTON,

Roan, calved April 6, 1879, bred by Mr. T. Nichols, The Grange; got by Lord Clarence Waterloo (36926), dam (Lady Brailes 4th) by Grand Duke 21st (34061)

' g. d. (Lady Brailes) by Duke of Brailes (23724), gr. g. d. (Formosa) by Duke of Brailes (23724), — (Gionetta) by Sarawak (15238), — by Earl of Dublin (10178), — by Janizary (8175), — by Snowball (8602), — by Caliph (1774), — by Norman (2379), — by White Boy (1580), — by Wyville's Bull, — by a Bull bred by Mr. Charge.

(43696)   **NORTHERN DUKE 3RD,**

Roan, calved January 14, 1879, bred by Mr. R. Harrett, Kirkwhelpington ; got by Prince of Waterloo 2nd (38944), dam (Lady Wharfdale) by Third Lord of Wharfdale (26759), g. d. (Duchess of Oxford) by Second Earl of Oxford (23843), gr. g. d. (Ruby) by Rubens (20756), — (Lucy) by Pantaloon (13450), — by Pilot (7333), — by Exmouth (6983), — by Young Sir Harry Liddell (7503), — by Captain (6830), — by Sir Harry Liddell (5157), — by Lord Prudhoe (8251), — by Sir Harry (5155), — by Hollon's Bull (313).

(43697)   **NORTHERN LIGHT 2ND,**

Roan, calved September 3, 1866, bred by Mr. T. Bell, Brockton Hall, late the property of Mr. R. H. Frank, Ashbourne Hall ; got by Second Baron Westbury (19288), dam (Peach Blossom 2nd) by Baron Westbury (19287), g. d. (Peach) by General Canrobert (12927), gr. g. d. (Poppy) by Second Cleveland Lad (3408), — (Place) by Second Earl of Darlington (1945), — by Son of Second Hubback (2683), — a Cow of Mr. Bates's, of Kirklevington.

(43698)   **NORTH NORTHUMBERLAND,**

White, calved November 29, 1879, bred by Mr. Y. R. Graham, Yardley Stud Farm ; got by Duke of Yardley (36556), dam (Lorelei) by Touchstone (20986), g. d. (Lady of the Lake) by Second Duke of Bolton (12739), gr. g. d. (Lady Foggathorpe) by Third Duke of Oxford (9047), — (Foggathorpe 4th) by Duke of Northumberland (1940), — by Malbro' (1189), — by Ebor (997), — by Regent (546), — by North Star (459), — by R. Colling's White Bull (151), — bred by Mr. R. Colling.

(43699)   **NORTH STAR,**

White, calved February 28, 1879, bred by Mr. G. W. Lambart, Beau Parc ; got by Jupiter (38477), dam (Evening Star) by Rupert (29902), g. d. (Fair Star) by British Sailor (23472), gr. g. d. (Flag) by British Flag (21323), — (Woman in White) by Volunteer (15476), — by Prince Ernest (7366), — by Solway (7530), — by General (9146), — by Narcissus (4540), — by Monarch (2324), — by Planet (1325).

(43700)   **NORTH STAR,**

White, calved May 17, 1879, bred by Mr. J. Bousfield, Soulby ; got by Lord of the Manor (38642), dam (Blooming Maid) by Prince of Paris (29657), g. d. (Bright Eyes 4th) by Archdeacon (21184), gr. g. d. (Bright Eyes 3rd) by Fennel (19737), — (Bright Eyes 2nd) by Betony (30533), — by Windsor (21117), — by China (19447), — by Vanguard (21010), — by Oswald (20445).

(43701)   **OBSTRUCTIONIST,**

White, calved July 14, 1879, bred by Mr. F. B. Greenwood, Swarcliffe ; got by Knight of the Heather (38524), dam (Lady Rochester) by King James (28971), g. d. (Revival) by Reformer (18687), gr. g. d. (Valerian) by Royal Oak (16873), — (Vanilla) by Robinson Crusoe (13610), — by Borrowby Boy (9980), — by Laudable (9282), — by Sir Thomas Fairfax (5196), — by Colossus (1847), —

by Beverley (1766), — by Pilot (496), — by Warlaby (672), — by Albion (14), — by Lame Bull (359), — by Shipton (587), — by Son of Suworrow (636), — by Son of Twin Brother to Ben (88), — by Twin Brother to Ben (660).

(43702) OLD DAISY BULL 3RD,
Roan, calved March 12, 1877, bred by the Rev. W. Holt Beever, Pencraig Court, the property of Mr. C. Williams, Pilton House; got by Clarence of that Ilk (33391), dam (Astonishment) by Mountain Dew (34879), g. d. (Ancienne) by The Baron (25277), gr. g. d. (Isola) by Lord Raglan (13222), — (Donna) by Patriot (10594), — by Sir Frederick (8577), — by Marquis (4386), — by Mynheer (1255), — by Enchanter (244), — by Major (398), — by Windsor (698), — by Favourite (252), — by Punch (531), — by Hubback (319).

(43703) OLD DAISY BULL 8TH,
Roan, calved July 3, 1879, bred by the Rev. W. Holt Beever, Pencraig Court; got by Twentieth Duke of Oxford (28432), dam (Quietude) by Royal Cumberland (27358), g. d. (Ireni) by The Bully (23019), gr. g. d. (Isola) by Lord Raglan (13222), — (Donna) by Patriot (10594), — by Sir Frederick (8577), — — by Marquis (4386), — by Mynheer (1255), — by Enchanter (244), — by Major (398), — by Windsor (698), — by Favourite (252), — by Punch (531), — by Hubback (319).

(43704) OLD DAISY BULL 9TH,
Roan, calved August 12, 1879, bred by the Rev. W. Holt Beever, Pencraig Court; got by Twentieth Duke of Oxford (28432), dam (Ireni) by The Bully (23019), g. d. (Isola) by Lord Raglan (13222), gr. g. d. (Donna) by Patriot (10594), — (Duchess) by Sir Frederick (8577), — by Marquis (4386), — by Mynheer (1255), — by Enchanter (244), — by Major (398), — by Windsor (698), — by Favourite (252), — by Punch (531), — by Hubback (319).

(43705) OLDFIELD FRIAR,
Roan, calved May 9, 1879, bred by Mr. James Fry, Lacock, the property of Messrs. J. C. and J. H. Fry, Oldfield; got by Grand Duke of Waterloo (34077), dam (Brown 7th) by Lord Hastings (29115), g. d. (Brown 4th) by Bates Tertius (21249), gr. g. d. (Brown) by Sidon (20803), — (Brown) by Romulus (16853), — by Alchemist (11097).

(43706) OLIVER CROMWELL,
Roan, calved January 9, 1876, bred by Mr. F. J. S. Foljambe, Osberton Hall, the property of Mr. Evan Baillie, Dochfour; got by Sweet Pea (35708), dam (May Flower) by Knight of the Bath (26546), g. d. (May Duchess) by May Duke (16553), gr. g. d. (Blanche) by Monarch (13347), — (Seraph) by Lord of Brawith (10465), — by Pestalozzi (10603), — by Lamplighter (8204), — by Zenith (5702), — by Warlock (5599), — by Alabaster (1616), — by Monarch (2324), — by Satellite (1420), — by Cato (119), — by Jupiter (342), — by George (273), — by Chilton (136), — by Irishman (329), — by B. (45).

(43707) OLIVER CROMWELL,
Roan, calved December 10, 1878, bred by the Earl of Suffolk and Berkshire, Charlton Park; got by Protector (42241), dam (Lady 2nd) by Viscount Walton (23154), g. d. (Lady 1st) by Heir of Oxford (19940), gr. g. d. by Longfellow (18206), — by Jupiter (8186).

(43708)   ONEIDA DUKE,

Roan, calved May 13, 1877, bred by Mr. R. Taylor, Sigglesthorne Manor, the property of Mr. T. Bainton, Arram Hall; got by Oneida Prince (34948), dam (Cherry Countess 2nd) by Netta (31970), g. d. (Cherry Countess) by Cherry Royal (23556), gr. g. d. (Strawberry) by Star 2nd (13786), — (Flora) by Sir Charles (13706), — Danthorpe Belle, from the herd of Mr. Colling, Danthorpe.

(43709)   ORANGE,

Roan, calved November 13, 1879, bred by Mr. A. Garfit, Scothern; got by Scothern Butterfly 3rd (42364), dam (Orange Lass) by Grand Duke 24th (34064), g. d. (Orange Girl) by Grand Duke 19th (28746), gr. g. d. (Orange Leaf) by Grand Duke 7th (19877), — (Orange Fruit) by First Fruits (16048), — by Tambour (15366), — by Concord (11302), — by Sir Launcelot (5166), — by Orville (4625), — by Tomboy (2765), — by Vesper (1547).

(43710)   ORANGE LORD,

White, calved April 28, 1879, bred by Mr. A. F. Hurt, Alderwasley; got by Orange Duke (37148), dam (Lady 11th) by Lancaster Comet 2nd (34420), g. d. (Lady 7th) by Lancaster Star (22073), gr. g. d. (Lady 6th) by Third Duke of Norfolk (21601), — (Lady 5th) by Wonderful (14022), — by Priam (15079), — by Earl of Chester (14472), — by Young Major (6175), — by Norfolk (2377), — by Warrior (673), — by Blyth Comet (85), — by Neswick (1266), — by Son of Favourite (252).

(43711)   ORION,

Roan, calved January 29, 1879, bred by Mr. R. P. Maxwell, Finnebrogue, the property of Mr. R. Maxwell, Kilmore Hill; got by Woodranger (39340), dam (Sister to Stella) by Half Sovereign (34104), g. d. (Southern Lass) by Prince Victor (20606), gr. g. d. (Gem of the South's Butterfly) by Royal Butterfly 5th (18756), — (Gem of the South) by Great Mogul (14651), — by Young Fourth Duke (9037), — by Duke of Richmond (7996), — by Sir Walter (2639), — by Young Jerry (8177), — by Roseberry (567), — by Roseberry (567), — by Constellation (163), — by Hastings (293), — by Hastings (293), — by Leopold (372).

(43712)   OSIRIS,

Roan, calved March 4, 1878, bred by Mr. J. M. Graham, Battleby, the property of Mr. Fenwick, Leadketty; got by Bellerus (37848), dam (Nyanza) by Knight of the Rock (24279), g. d. (Lucy 5th) by Lord of the Meadow (22197), gr. g. d. (Lucy) by Hawksworth (14681), — (Maud) by Mainfred (11767), — by Phœnix (10608), — by Corporal Trim (7932), — by Brougham (1746), — by St. Albans 2nd (5048), — by St. Albans (2584), — by Lawnsleeves (365).

(43713)   OSMAN,

Roan, calved May 1, 1876, bred by the Earl of Caledon, Caledon; got by Sir Booth Gwynne (35542), dam (Mabella) by White Rock (42602), g. d. (Mab) by Earl of Erne (17773), gr. g. d. (Mayflower) by Pro Bono Publico (13528), — (Sultana) by Pat (13456), — by Fairfax, — by Sir John Sinclair (5165), — by Coriander (6895), — by Cato (119).

(43714)   OSMAN,

Red and white, calved March 19, 1877, bred by Mr. W. R. Meade, Ballymartle, the property of the Representatives of Mr. T. Hungerford, The Island; got by Lord Percy (34653), dam (Peony) by Danny Man (21527), g. d. (Pansy) by St.

George (22829), gr. g. d. (Polyanthus) by Australian (12414), — (Primrose) by
Rollo 2nd (13618), — by President (8407), — by Hymen (6088), — by Marton
Comet (4409), — by Cyrus (3538), — by Bedford Junior (1701), — by Northern
Light (1281), — by Snowdrop (614), — by Son of Kitt (7127).

(43715)   OUT OF TUNE,
Roan, calved April 14, 1879, bred by Mr. H. F. Smith, Lamwath House; got
by Lord Violet 3rd (34692), dam (Harmonium) by Knight of the Whistle
(26558), g. d. (Concertina) by May Duke (16553), gr. g. d. (Seraphine) by
Monarch (13347), — (Seraph) by Lord of Brawith (10465), — by Pestalozzi
(10603), — by Lamplighter (8204), — by Zenith (5702), — by Warlock (5599),
— by Alabaster (1616), — by Monarch (2324), — by Satellite (1420), — by
Cato (119), — by Jupiter (342), — by George (273), — by Chilton (136), —
by Irishman (329), — by B. (45).

(43716)   OXFORD 22ND,
Red and white, calved August 14, 1877, bred by Mr. R. Botterill, Wauldby, the
property of Mr. J. Cobo, Buenos Ayres; got by Oxford-le-Grand (29496), dam
(Vanity Fair) by Nineteenth Duke of Oxford (28431), g. d. (Vanity) by Thorndale Lad (23066), gr. g. d. (Lady Vane) by Lord Cobham (20164), — (Vanity)
by Reformer (18687), — by Robinson Crusoe (13610), — by Borrowby Boy
(9980), — by Laudable (9282), — by Sir Thomas Fairfax (5196), — by Colossus
(1847), — by Burley (1766), — by Pilot (496), — by Warlaby (672), — by
Albion (14), — by Lame Bull (359), — by Shipton (587), — by Son of Suworrow (636), — by Son of Twin Brother to Ben (88), — by Twin Brother to
Ben (660).

(43717)   OXFORD 25TH,
Red, calved March 11, 1879, bred by Mr. R. Botterill, Wauldby, the property
of Mr. J. S. Egginton, Kirk Ella; got by Oxford-le-Grand (29496), dam (Lady
Fortunate 3rd) by Nineteenth Duke of Oxford (28431), g. d. (Lady Fortunate
2nd) by Volunteer (30239), gr. g. d. (Lady Fortunate) by Second Duke of
Wharfdale (19649), — (Lady Flora) by Gainford 5th (12913), — by Bumper
(10005), — by Royal Buck (10750), — by Rousseau (10746), — by Harlsonio
(6055), — by Belvedere 2nd (3127), — by Waterloo (2816), — by Kitt (7127),
— by Kitt (7127), — by Page's Bull (6269), — by Middleton's Bull (438).

(43718)   OXFORD BOY,
Red and white, calved April 2, 1879, bred by Mr. W. Lavender, Biddenham, the
property of Mr. J. Fyson, Wickhambrook; got by Duke of Oxford 32nd (36527),
dam (Louisa 15th) by Duke Valentine (33765), g. d. (Louisa 14th) by The
Worcester Knight (30156), gr. g. d. (Louisa 11th) by Nestor (24648), — (Louisa
8th) by Lord Augustus (20147), — by Baron Farnley (14129), — by Third Duke
of York (10166), — by Hamlet (8126), — by Dulcimer (3658), — by Sol (2655),
— by Maximus (2284), — by Matchem (2281), — by Scipio (1421), — by Sir
Stephen (1456), — by Western Comet (689), — by Charge's Grey Bull (872),
— by Favourite (252), — by Bartle (777), — descended from the Studley White
Bull (627).

(43719)   OXFORD BOY 2ND,
Roan, calved April 19, 1877, bred by Earl Howe, Gopsall Hall, the property of
Mr. J. Gardner, Twycross; got by Oxford Boy (34966), dam (May Rose 23rd)

by Burleigh (25697), g. d. (May Rose 14th) by Royal Arch 8th (22771), gr. g. d. (May Rose 3rd) by Guardsman (14656), — (May Rose) by Second Duke of Oxford (9046), — by Robin Hood (8491), — by Simon (5135), — by Crispin (174), — by Fisher's Old Bull (3799), — by Grandson of Favourite (252), — by Grandson of Favourite (252), — descended from the stock of Mr. Cornforth, of Barforth.

(43720)   OXFORD DUKE OF KILLHOW 2ND,
Roan, calved February 6, 1879, bred by Mr. S. P. Foster, Killhow ; got by Duke of Ormskirk (36526), dam (Grand Duchess of Oxford 18th) by Baron Oxford 4th (25580), g. d. (Grand Duchess of Oxford 11th) by Grand Duke 10th (21848), gr. g. d. (Grand Duchess of Oxford 5th) by Priam (18567), — (Countess of Oxford) by Earl of Warwick (11412), — by Fourth Duke of York (10167), — by Second Duke of Northumberland (3646), — by Short Tail (2621), — by Matchem (2281), — by Young Wynyard (2859).

(43721)   OXFORD LAD 1ST,
Roan, calved June 23, 1874, bred by Messrs. J. Horswell and Sons, Burns Hall, the property of Mr. T. Chope, Hartland; got by Oxford Duke 3rd (32019), dam (Cometilla 3rd) by General Barrington (21810), g. d. (Ceres 3rd) by Duke (15908), gr. g. d. (Ceres) by Mistor (13343), — (Cometilla 6th) by Duke of Devonshire (7990), — by Scaleby (6448), — by Smeatonian (5212), — by Don Quixote (987), — by Regent (1366), — by Young Comet (156), — by Jupiter (343), — by Grandson of Favourite (252), — by Barningham (56).

(43722)   OXFORD PRINCE 2ND,
Red and white, calved February 13, 1879, bred by Mr. G. Fox, Elmhurst Hall; got by Duke of Oxford 39th (38173), dam (Rosalie) by Saladin (35461), g. d. (Rosette) by Oxford Lad (24713), gr. g. d. (Red Rose 10th) by Third Duke of Cambridge (5941), — (Red Rose 4th) by Earl of Chatham (10176), — by Napier (6238), — by South Durham (5281), — by Bellerophon (3119), — by Belvedere (1706), — by Waterloo (2816), — by Baron (58), — by Phenomenon (491), — by Favourite (252), — by Favourite (252), — by Favourite (252), — by Hubback (319), — by Snowdon's Bull (612), — by Waistell's Bull (669), — by Masterman's Bull (422), — by the Studley Bull (626).

(43723)   OXFORD SURMISE,
Roan, calved April 25, 1879, bred by Mr. H. Lovatt, Low Hill; got by Duke of Oxford 36th (38170), dam (Tacita 5th) by Oxford Beau 2nd (32012), g. d. (Tacita 3rd) by Third Duke of Claro (23729), gr. g. d. (Surmise 3rd) by May Duke (13320), — (Surmise) by Duke of Glo'ster (11382), — by Earl of Derby (10177), — by Duke of Sutherland (6945), — by Locomotive (4242), — by Short Tail (2621), — by Gambier (2046), — by Young Wynyard (2859), — by Bulls of Messrs. C. and R. Colling's.

(43724)   OXFORD SWELL 5TH,
Red, calved March 6, 1879, bred by Mr. E. Bowly, Siddington House ; got by Beau of Oxford 2nd (33129), dam (Melody) by Second Duke of Collingham (23730), g. d. (Musical 9th) by Seventh Duke of York (17754), gr. g. d. (Songstress) by Mountaineer (14966), — (Minstrel) by Nundi (9453), — by Monzani (6222), — by The Prince (7615), — by Fitzroy (3808), — by Morpeth (2339), — by Roman (2559), — by Admiral (5), — by Son of Blyth Comet (85), — by Foljambe's Bull (1051).

(43725) **OXFORD SWELL 6TH,**
Red and white, calved April 12, 1879, bred by Mr. E. Bowly, Siddington House; got by Beau of Oxford 2nd (33129), dam (Gazelle 23rd) by Second Duke of Tregunter (26022), g. d. (Gazelle 8th) by Seventh Duke of York (17754), gr. g. d. (Gazelle 2nd) by Earl of Walton (17787), — (Selina) by Fourth Duke of Oxford (11387), — by Snowstorm (12119), — by Hampden (8129), — by Leo (4208), — by Henwood (2114), — by Sir Stephen (1456), — by Prince of Waterloo (528), — by Mayflower (425), — by a Bull of Mr. Nicholson's, descended from the stock of Mr. J. Brown, of Aldborough.

(43726) **OXFORD SWELL 7TH,**
Roan, calved May 14, 1879, bred by Mr. E. Bowly, Siddington House; got by Beau of Oxford 2nd (33129), dam (Dewlap) by Third Duke of Geneva (21592), g. d. (Diadem) by Touchstone (20986), gr. g. d. (Dewdrop) by Earl of Derby (21637), — (Dora) by Amiens (14095), — by Rubens (15209), — by Meteor (10526), — by Phœnix (6290), — by Shakespeare (5108), — by Rival (2534), — by Darlington (956), — by Denton (198), — by Rockingham (560), — by Jobling's Son of Phenomenon (491), — by Phenomenon (491), — by Colonel (152), — by Styford (629).

(43727) **OXFORD'S WATERLOO 2ND,**
Red and white, calved May 7, 1879, bred by Mr. R. H. Crabb, Baddow Place; got by Duke of Oxford (39770), dam (Waterloo 42nd) by Oxford's Baron (32030), g. d. (Waterloo 36th) by Grand Duke 11th (21849), gr. g. d. (Waterloo 26th) by Duke of Geneva (19614), — (Waterloo 24th) by Cherry Duke 2nd (14265), — by Red Knight (11976), — by Grand Duke (10284), — by Third Duke of Oxford (9047), — by Second Cleveland Lad (3408), — by Duke of Northumberland (1940), — by Norfolk (2377), — by Waterloo (2816), — by Waterloo (2816).

(43728) **OXFORD WILD EYES,**
Red, calved March 7, 1879, bred by Mr. R. H. Crabb, Baddow Place; got by Oxford's Baron (32030), dam (Wild Eyes A 1) by Baron Oxford 3rd (25579), g. d. (Wild Eyes 24th) by Fourth Duke of Oxford (11387), gr. g. d. (Wild Eyes 22nd) by Wild Duke (19148), — (Wild Eyes 20th) by Lord Barrington 1st (13170), — by Second Duke of Oxford (9046), — by Fourth Duke of Northumberland (3649), — by Duke of Northumberland (1940), — by Belvedere (1706), — by Emperor (1975), — by Wonderful (700), — by Cleveland (145), — by Butterfly (104), — by Hollon's Bull (313), — by Mowbray's Bull (2342), — by Masterman's Bull (422), — descended from M. Dobison's stock.

(43729) **OXFORD WILD EYES 3RD,**
White, calved June 29, 1879, bred by Lord Penrhyn, Penrhyn Castle; got by Grand Duke of Oxford (31293), dam (Wild Duchess 4th) by Sixth Duke of Oneida (30997), g. d. (Wild Duchess 3rd) by Lord Oxford 2nd (20215), gr. g. d. (Wild Duchess) by Duke of Wetherby (17753), — (Wild Eyes 24th) by Lord Barrington 3rd (16382), — by Second Duke of Oxford (9046), — by Fourth Duke of Northumberland (3649), — by Duke of Northumberland (1940), — by Belvedere (1706), — by Emperor (1975), — by Wonderful (700), — by Cleveland (145), — by Butterfly (104), — by Hollon's Bull (313), — by Mowbray's Bull (2342), — by Masterman's Bull (422), — descended from M. Dobison's stock.

(43730) PAGANINI,'

Roan, calved May 19, 1879, bred by Mr. J. Mason, Dishforth, the property of Mr. C. Stephenson, Newcastle-on-Tyne; got by Merryman (42016), dam (Platina) by Great Hope (24082), g. d. (Paulina) by Heir of Windsor (26364), gr. g. d. (Pauline 4th) by Hopewell (19973), — (Pauline) by British Boy (11206), — by Hopewell (10332), — by Hamlet (8126), — a Cow bred by Mr. Booth.

(43731) PANTALOON,

Red, calved July 11, 1879, bred by Mr. T. Purkis, West Wratting Grange; got by Sweet William (40778), dam (Lady Furbelow 4th) by Wharfdale Oxford (27786), g. d. (Harmonia) by Grand Duke of Essex 4th (24068), gr. g. d. (Harmony) by Cherry Duke 3rd (15763), — (Floret) by Douglas (12714), — by Duke of Cambridge (12742), — by Grey Friar (9172), — by Allan-a-Dale (7778), — by Little John (4232), — by Marcellus (2260), — by Caliph (1774), — by Swing (2721), — by Argus (759), — by Defender (194), — by Petrarch (488), — by Own Brother to R. Colling's White Heifer, — by Butterfly (104), — by Globe (278).

(43732) PANTON ROBIN,

Roan, calved February 17, 1876, bred by Messrs. Dudding, Panton House, the property of Mr. P. Cadman, Union Mills; got by Robert Stephenson (32313), dam (Robin's Rose) by Robin (24968), g. d. (Ruby Rose) by Lord Panton (22204), gr. g. d. (Royal Rose) by Prince Alfred (13494), — (Blooming Rose) by Baron Warlaby (7813).

(43733) PARK RUBY LAD 4TH,

Roan, calved October 17, 1878, bred by Mr. E. Pease, Greencroft West; got by Foster Brother (36661), dam (Red Ruby) by Manfred (26801), g. d. (Red Gem) by Prince Patrick (16760), gr. g. d. (Ruby 6th) by General Havelock (16108), — (Ruby) by Duke (9032), — by Harlsonio (6055), — by Planet (4718), — by Sol (2655), — by Harlsey (2091), — by Regent (2514), — by St. Albans (2584), — by North Star (459).

(43734) PARK RUBY LAD 5TH,

Roan, calved November 1, 1879, bred by Mr. E. Pease, Greencroft West; got by Foster Brother (36661), dam (Red Ruby) by Manfred (26801), g. d. (Red Gem) by Prince Patrick (16760), gr. g. d. (Ruby 6th) by General Havelock (16108), — (Ruby) by Duke (9032), — by Harlsonio (6055), — by Planet (4718), — by Sol (2655), — by Harlsey (2091), — by Regent (2514), — by St. Albans (2584), — by North Star (459).

(43735) PARTISAN,

Red and white, calved November 18, 1879, bred by Mr. J. How, Broughton; got by Regal Booth (35252), dam (Pauline 17th) by King Victor (28986), g. d. (Pauline 10th) by Prince of the Realm (22627), gr. g. d. (Pauline 6th) by Heir of Windsor (26364), — (Pauline 3rd) by Ravenspur (20628), — by British Boy (11206), — by Hopewell (10332), — by Hamlet (8126), — a Cow bred by Mr. Booth.

(43736) PATER,

Roan, calved July 23, 1879, bred by Mr. A. Hamond, Westacre; got by Hotspur (34190), dam (Pansy) by Prince Louis (29643), g. d. (Princess) by Prince Leopold (20557), gr. g. d. (Shepherdess) by Ant Eater (19228), — (Shepherd)

by Windham (15514), — by Gay Lad (11510), — by Newham (4563), — by
Dulverton (3959), — by Juniper (2165), — by The Squire (1512), — by Midas
(435), — by Boughton (90).

(43737) PATRIOT,
Red, calved January 17, 1878, bred by Mr. J. A. M. Cope, Drummilly; got by
Knight of Raby (34393), dam (Pink Blossom) by Baron Eccleswall (33037),
g. d. (Pink 9th) by Royal Benedict (27348), gr. g. d. (Patience) by Royal Oak
(16870), — (Pink) by Marchmont (9367), — by Young Consul (6893), — by
Newnham (2365), — by Satellite (1420), — by Jupiter (342), — by Sir Oliver
(605), — by Trunnell (659), — by Favourite (252), — by Favourite (252), —
by Dalton Duke (188), — by R. Alcock's Bull (19), — by J. Smith's Bull (608),
— by Jolly's Bull (337).

(43738) PATRIOTIC,
Roan, calved March 10, 1879, bred by Mr. S. C. Pilgrim, The Outwoods; got
by John of Gaunt (34265), dam (Jumper) by Leamington (22084), g. d. by
Marengo (13294), gr. g. d. by Plato (11907), — by Geddington (8102).

(43739) PAUL PRY,
Roan, calved May 25, 1879, bred by Mr. J. Thompson, Anlaby; got by Beverley
Oxford 2nd (39468), dam (Peony) by Baron Ruth (30482), g. d. (Pry) by Sir Tatton
(27497), gr. g. d. (Princess) by Earl of Derby (23831), — (Peeress) by Humphrey
(13054), — by Brocado (11210), — by Brocado (11210), — by Elevator (9080),
— by Highlander (9216), — by Amateur (3007), — by Sir Walter (2639).

(43740) PAXTON,
Red, calved January 30, 1879, bred by Mr. T. Harris, Stonylane House; got by
Baron Winsome 3rd (33108), dam (Niobe 18th) by North Star (31988), g. d.
(Niobe 4th) by Oxartes (22471), gr. g. d. (Niobe) by Sultan (15358), —
(Brownie) by Neptune (11847), — by Essex Hero (9096), — by Duke of Marl-
borough (3645), — by Harlequin (3975), — by Captain (3273), — by Emperor
(1974), — by Snowball (2647), — by Son of St. Albans (2584), — by Colonel
Trotter's Son of Lawnsleeves (365), — by Barnaby (1678).

(43741) PEACH STONE,
Roan, calved May 5, 1879, bred by Mr. J. Taber, Rivenhall; got by Osman
Pasha (42078), dam (Peach Bud) by Heydon Duke 2nd (31370), g. d. (White
Peach Blossom) by Costa (21487), gr. g. d. (Rich Peach Blossom) by Clumber
2nd (21438), — (Peach Blossom) by Young Baron Godolphin (17366), — by
Cheltenham (12588), — by Scrivener (10791), — by St. Martin (8525), — by
Prince Albert (7358), — by Favourite (3769), — by Oliver (2386), — by Albany
(13), — by Son of George (273), — by Son of Mason's White Bull (421).

(43742) PEACOCK BUTTERFLY,
Roan, calved July 13, 1879, bred by Mr. J. A. M. Cope, Drummilly; got by
Knight of Raby (34393), dam (Towneley Butterfly 2nd) by Earl of Thorndale
(28521), g. d. (Towneley Butterfly) by Count of Windsor (21498), gr. g. d.
(White Butterfly) by Butterfly's Nephew (15714), — (Paris Butterfly) by
Master Butterfly (13311), — by Gavazzi (11508), — by Baron of Ravensworth
(7811), — by Raree Show (4874), — by Thick Hock (6601), — by Expectation
(1988), — by Belzoni (1709), — by Comus (1861), — by Denton (198).

(43743) PEARL DIVER,
Roan, calved January 30, 1879, bred by the Duke of Northumberland, Alnwick Castle; got by Sir Raymond (40716), dam (Windsor's Pearl) by Mayor of Windsor (31897), g. d. (Dewdrop) by President (20510), gr. g. d. (Maid of Aln) by Melsonby (18380), — (Young Jessy) by George 3rd (16147), — by Shaftoe (5107), — by Richardi (4944), — Cowslip, bought of Mr. Patterson, of Wood Houses.

(43744) PEARL FINDER,
Red, calved January 8, 1879, bred by Mr. R. Stratton, The Duffryn, the property of Mr. H. Hussey Vivian, Park Wern; got by Pearl Diver (37182), dam (Glitter) by Eighth Duke of York (23808), g. d. (Brilliance) by Lamp of Lothian (16356), gr. g. d. (Duchess of Glo'ster 8th) by His Highness (14708), — (Duchess of Glo'ster 4th) by Clarendon (12605), — by The Red Duke (8694), — by Lottery (4280), — by Lottery (4280), — by Phœnix (6290).

(43745) PEARL SETTER,
Red, calved January 8, 1879, bred by Mr. R. Stratton, The Duffryn, the property of Mr. T. R. Hulbert, North Cerney; got by Pearl Diver (37182), dam (Glitter) by Eighth Duke of York (23808), g. d. (Brilliance) by Lamp of Lothian (16356), gr. g. d. (Duchess of Glo'ster 8th) by His Highness (14708), — (Duchess of Glo'ster 4th) by Clarendon (12605), — by The Red Duke (8694), — by Lottery (4280), — by Lottery (4280), — by Phœnix (6290).

(43746) PEREGRINE,
Roan, calved March 16, 1879, bred by Mr. S. C. Pilgrim, The Outwoods; got by John of Gaunt (34265), dam (Perry) by Lord of the Lilacs 5th (26712), g. d. by Leamington (22084), gr. g. d. by Marengo (13294), — by Plato (11907), — by Geddington (8102).

(43747) PERICLES,
White, calved February 28, 1879, bred by Mr. J. Gamble, Shouldham Thorpe; got by Havelock (36749), dam (Plastic) by Zealot (25480), g. d. (Prize) by Forester (19767), gr. g. d. (Pansy) by Plato (18552), — (Precious) by Baron Albany (11151), — by Orontes (4623), — by Roman (2561), — by Mercury (2301), — by Monarch (2324), — by St. Albans (2584), — by Jupiter (342), — by Sir Oliver (605), — by Trunnell (659), — by Favourite (252), — by Favourite (252), — by Dalton Duke (188), — by R. Alcock's Bull (19), — by J. Smith's Bull (608), — by Jolly's Bull (337).

(43748) PERI'S DUKE OF COLLINGHAM,
White, calved in 1869, bred by Mr. H. Mousell, Tufflcigh Court; got by Second Duke of Collingham (23730), dam (Peri) by Seventh Duke of York (17754), g. d. (Palmyra) by Marquis of Oxford (18339), gr. g. d. (Persia) by Cupid (14359), — (Phantom) by Sixth Duke of Oxford (12765), — by Phœbus (9475), — by Tom of Lincoln (8714), — by Daniel (3554), — by Danby (1900), — by Grazier (3927), — by Wilkinson's Red Bull (5658), — by Sedbury (1424), — by Smurthwaite's Bull (5219), — by Nailer (2347), — by Danby (190).

(43749) PERSEUS,
Roan, calved December 1, 1879, bred by Colonel R. Loyd Lindsay, Lockinge Park; got by Earl of Horton 11th (36588), dam (Pearl) by Marquis of Sockburn (34787), g. d. (Peach Flower) by Young Duke of Jamaica (30976), gr. g. d.

(Peach) by Lord Stanley (22217), — (Lily) by Sarsden Clipper (20787), — by
Norman (16628), — by Fitz-Hardinge (8073), — by Harold (8131), — by Lord
John (4257), — by Newnham (2365), — by Satellite (1420), — by Jupiter
(342), — by Sir Oliver (605), — by Trunnell (659), — by Favourite (252), —
by Favourite (252), — by Dalton Duke (188), — by R. Alcock's Bull (19), —
by J. Smith's Bull (608), — by Jolly's Bull (337).

(43750)   PETER,
Red, calved February 8, 1879, bred by Sir Wilfrid Lawson, Bart., Brayton; got
by Waterloo Lad (40894), dam (Princess Royal 2nd) by Baron Oxford 6th
(33075), g. d. (Princess Royal) by Twenty-second Duke of Oxford (31000),
gr. g. d. (Royal Gertrude) by Royal Cambridge (25009), — (Gertrude) by Sir
John (16983), — by General Cavaignac (11517), — by Emperor (11436), — by
Magician 2nd (11764), — by Regent (1366), — by Wellesley (1571), — by
Layton (366), — by Mac George (1183).

(43751)   PETERSBURG,
White, calved September 19, 1877, bred by Lord Polwarth, Mertoun House, the
property of Mr. J. C. Toppin, Musgrave Hall; got by Rapid Rhone (35205),
dam (Pauline 7th) by King Charles (24240), g. d. (Pauline 5th) by British Hope
(21324), gr. g. d. (Pauline 3rd) by Ravenspur (20628), — (Pauline) by British
Boy (11206), — by Hopewell (10332), — by Hamlet (8126), — a Cow bred by
Mr. Booth.

(43752)   PHILOLOGIST,
Roan, calved July 7, 1879, bred by Lord Bolton, Bolton Hall; got by St.
Swithin's Star Drop (40667), dam (Prestonia 5th) by Heir-at-Law (34124), g. d.
(Prestonia 2nd) by Dandelion (30849), gr. g. d. (Prestonia 1st) by Marmion
(26821), — (Hunter) by Mineralist (20359), — by Day Star (12686), — by Maro
(14901).

(43753)   PILGRIM'S PROGRESS,
Roan, calved March 26, 1879, bred by Mr. S. C. Pilgrim, The Outwoods; got
by John of Gaunt (34265), dam (Maggie) by Knight of Wetherby (24283),
g. d. by Leamington (22084), gr. g. d. by Marengo (13294), — by Plato (11907),
— by Geddington (8102).

(43754)   PILOT,
Roan, calved January 13, 1879, bred by Mr. Hugh Aylmer, West Dereham
Abbey, the property of Mr. B. Hoddinott, Moor Court Farm; got by Royal
Commander (29857), dam (Phillis 15th) by Royal Broughton (27352), g. d.
(Phillis 10th) by General Hopewell 2nd (24021), gr. g. d. (Phillis 5th) by Hil-
debrand (18068), — (Phillis 2nd) by Red Knight (16809), — by Homer (14714),
— by Cardigan (12556), — by Young Rufus (13649), — by Constitution
(12634), — by Young Comet (1853), — descended from Jolly's Bull (4115).

(43755)   PILOT,
Roan, calved March 9, 1879, bred by Mr. W. Bolton, The Island; got by Albion
(36112), dam (Princess Gwynne) by British Sailor (23471), g. d. (Phœbe
Gwynne) by Killerby Lad (20052), gr. g. d. (Princess Gwynne) by Paul Potter
(16688), — (Empress Gwynne) by Emperor (14500), — by Cadet (12521), —
by Young Usurer (10985), — by Sir Thomas Fairfax (5196), — by Wallace
(5586), — by Wellington (2824), — by Marmion (406), — by Merlin (430), — by

Layton (366), — by Phenomenon (491), — by Favourite (252), — by Favourite (252), — by Hubback (319), — by Snowdon's Bull (612), — by Waistell's Bull (669), — by Masterman's Bull (422), — by the Studley Bull (626).

(43756)   PIRATE,
Roan, calved May 17, 1878, bred by Mr. D. G. Briggs, Kelstern Grange; got by Homer (34170), dam (Pancake) by Victorious (25378), g. d. (Pansy) by Plato (18552), gr. g. d. (Precious) by Baron Albany (11151), — (Prudence) by Orontes (4623), — by Roman (2561), — by Mercury (2301), — by Monarch (2324), — by St. Albans (2584), — by Jupiter (342), — by Sir Oliver (605), — by Trunnell (659), — by Favourite (252), — by Favourite (252), — by Dalton Duke (188), — by R. Alcock's Bull (19), — by J. Smith's Bull (608), — by Jolly's Bull (337).

(43757)   PISCATOR,
Roan, calved March 28, 1879, bred by Mr. S. C. Pilgrim, The Outwoods; got by John of Gaunt (34265), dam (Penelope Lorne) by Duke of Lorne 3rd (30987), g. d. (Penelope) by Lord of the Lilacs 5th (26712), gr. g. d. by Knight of Wetherby (24283), — by Field Marshal (19742), — by Marengo (13294), — by Plato (11907), — by Geddington (8102).

(43758)   PLANET,
Roan, calved November 17, 1878, bred by Mr. J. Gamble, Shouldham Thorpe; got by Framemaker (33963), dam (Pleasant) by Lord Cyril (29094), g. d. (Pun) by Zealot (25480), gr. g. d. (Pungent) by Plato (18552), — (Pride) by The Saxon (12215), — by Orontes (4623), — by Roman (2561), — by Mercury (2301), — by Monarch (2324), — by St. Albans (2584), — by Jupiter (342), — by Sir Oliver (605), — by Trunnell (659), — by Favourite (252), — by Favourite (252), — by Dalton Duke (188), — by R. Alcock's Bull (19), — by J. Smith's Bull (608), — by Jolly's Bull (337).

(43759)   PLATO,
Roan, calved July 5, 1877, bred by Mr. A. Baird, Robeston Hall, the property of Mr. W. H. Shield, Gilfach; got by Alfred (36119), dam (Patty) by Earl of Horton 3rd (33796), g. d. (Polly) by Sea-Serpent (25107), gr. g. d. (Peggy) by Woolsack (23243), — (Aberglasney) by Koh-i-noor (13126), — by Union (12249), — by Oregon (8371), — by Bristol (7852).

(43760)   PLATO,
Roan, calved January 20, 1879, bred by Mr. F. Garth, Crackpot; got by David (36425), dam (Annabelle) by Frederick First Fruits (23989), g. d. (Annette Bumper) by Bumper (19371), gr. g. d. (Lady Annetta) by Consolation (14316), — (Lady Annabella) by Whittington (12299), — by Noble (4578), — by Newton (2367), — by Emperor (3716), — by Satellite (1420), — by Cato (119), — by Jupiter (342), — by George (273), — by Chilton (136), — by Irishman (329), — by B. (45).

(43761)   PLENIPOTENTIARY,
Roan, calved January 10, 1879, bred by Mr. J. Jardine, Dryfeholm; got by Frankland (39895), dam (Wild Agnes) by Wild Boy (25447), g. d. (Dahlia) by Pizarro (20497), gr. g. d. (Venus) by Master Annandale (14916), — (Sprightly) by Young Earl (14467), — by Snowy Down (8607), — by Prince Albert (4778), — by Son of Cumberland (5256), — by Exmouth (3747), — by Prince (4765), — by Leopold (2199).

(43762)   PLOUGHBOY,
Red, calved May 1, 1878, bred by Mr. R. Bruce, Manor House Farm, the property
of Mr. T. M. Tod, West Brackly; got by Lord of the Forth (40215), dam (Lady
Elma 2nd) by Baronet (25564), g. d. (Lady Elma) by Lord Elgin (20170),
gr. g. d. (Ruth) by Dipple (14401), — (Duchess) by General (19835).

(43763)   PLUTARCH,
Roan, calved April 11, 1879, bred by Mr. J. B. Booth, Killerby; got by Heart
of Oak (39982), dam (Proserpine) by K. C. B. (26492), g. d. (Hecate) by Knight
Errant (18154), gr. g. d. (Hecuba) by Hopewell (10332), — (Helen) by Hamlet
(8126), — by Leonard (4210).

(43764)   PLUTO,
Roan, calved April 3, 1879, bred by Mr. J. Tyacke, Merthen; got by Monarch
Gwynne (37103), dam (Peony) by Orlando (32000), g. d. (Pet) by Don Pedro
(25910), gr. g. d. (Polly Perkins) by Favourite (19730), — (Phœbe) by Sir
Roger de Coverley (12095), — by Hero of the West (8150), — by Lottery
(4280), — by Phœnix (6290).

(43765)   PLUTUS,
Roan, calved March 14, 1879, bred by Mr. J. Tyacke, Merthen; got by Monarch
Gwynne (37103), dam (Penelope 2nd) by Don Pedro (25910), g. d. (Penelope) by
Hector (24117), gr. g. d. (Polly Perkins) by Favourite (19730), — (Phœbe) by
Sir Roger de Coverley (12095), — by Hero of the West (8150), — by Lottery
(4280), — by Phœnix (6290).

(43766)   POLE STAR,
Red, calved May 29, 1879, bred by Earl Fitzwilliam, Coollattin Park, sold to
the Provincial Government of Liege, Belgium; got by Robert Burns (29795),
dam (Portia) by Chief Justice (28188), g. d. (Portia) by The Bushel (17092),
gr. g. d. (Poplin) by Tom Boy (10961), — (Polly) by Volcano (21044), — by
Municipal Bill (2344), — by Bonaparte (19327), — Phœnix, bought of Mr. De
Renzi, and descended from the stock of Mr. R. Holmes.

(43767)   POLYPUS,
Red and white, calved May 23, 1879, bred by Sir G. R. Philips, Bart., Weston
Park; got by Grand Duke 29th (38372), dam (Polycharm) by Duke Polycherry
(33763), g. d. (Polyanthus) by Third Duke of Geneva (21592), gr. g. d. (Polly)
by Stepping Stone (22978), — (Polytint) by Earl of Dublin (10178), —
by Janizary (8175), — by Snowball (8602), — by Little John (4232), —
by Caliph (1774), — by Rob Roy (557), — by Satellite (1420), — by Sir
Dimple (594), — by Styford (629).

(43768)   POPE JOAN,
Red, calved June 4, 1879, bred by Earl Fitzwilliam, Coollattin Park; got
by Lord Lilias (31690), dam (Patience) by Robert Burns (29795), g. d. (Prudence) by Sir Charles Napier (13708), gr. g. d. (Phœbe) by Soldier (20864), —
(Peggy) by Volcano (21014), — by Municipal Bill (2344), — by Bonaparte
(19327), — Phœnix, bought of Mr. De Renzi, and descended from the stock of
Mr. R. Holmes.

(43769)   POSTULANT,
Roan, calved November 15, 1879, bred by the Rev. T. Staniforth, Storrs; got by
Royal Stuart (40646), dam (Probation) by High Sheriff (26392), g. d. (Noviciate)

by Juryman (20043), gr. g. d. (Novice) by Monk (11824), — (Duchess of Norfolk) by Norfolk (9442), — by Laudable (9282), — by The Stuart (7623), — by Norfolk (2377), — by Ambo (1636), — by Memnon (2295), — by Pilot (496), — by Agamemnon (9), — by Burrell's Bull, of Burdon.

(43770) PREMIER,
Red, calved May 3, 1879, bred by Mr. W. Bolton, The Island ; got by Albion (36112), dam (Pansy Gwynne) by King Richard (26523), g. d. (Polly Gwynne) by Killerby Lad (20052), gr. g. d. (Pauline Gwynne) by Paul Potter (16688), — (White Moll Gwynne) by Cadet (12521), — by Young Usurer (10985), — by Sir Thomas Fairfax (5196), — by Wallace (5586), — by Wellington (2824), — by Marmion (406), — by Merlin (430), — by Layton (366), — by Phenomenon (491), — by Favourite (252), — by Favourite (252), — by Hubback (319), — by Snowdon's Bull (612), — by Waistell's Bull (669), — by Masterman's Bull (422), — by the Studley Bull (626).

(43771) PREMIUM,
Roan, calved May 12, 1879, bred by Mr. E. Lythall, Radford Hall ; got by Virginus (40873), dam (Snowdrop) by Fitz-Killerby (26166), g. d. (Lady of the Lake) by Lord of the Isles (20206), gr. g. d. (Lavender) by Monk (13353), — (Red Lucy 2nd) by Broomborough (9994), — by The Minstrel (8687), — by Chorister (8952), — by Whitaker (5640), — by Rebel (4882), — by Guy (3956), — by Woodford (5684), — by Pumpkin (4847), — by Son of Minor (441).

(43772) YOUNG PRETENDER,
Roan, calved December 19, 1878, bred by the Duke of Northumberland, Alnwick Castle ; got by Sir Raymond (40716), dam (Windsor's Rosebud) by Mayor of Windsor (31897), g. d. (Rosebud 11th) by Brigand (28080), gr. g. d. (Rosebud 7th) by Snowflake (18888), — (Rosebud 2nd) by Duke of Buckingham (14429), — by Ravensworth (7400), — by Son of Regent (2517), — by Togston (5487), — by Son of Lawnsleeves (365), — by Son of Bolingbroke (86).

(43773) PRIAM,
Roan, calved March 19, 1878, bred by Mr. J. Tyacke, Merthen, the property of Mr. H. Benny, Panalguy ; got by Monarch Gwynne (37103), dam (Peony) by Orlando (32000), g. d. (Pet) by Don Pedro (25910), gr. g. d. (Polly Perkins) by Favourite (19730), — (Phœbe) by Sir Roger de Coverley (12095), — by Hero of the West (8150), — by Lottery (4280), — by Phœnix (6290).

(43774) PRIDE OF DEVON,
Red and white, calved January 12, 1878, bred by Mr. J. Cruse, Cleave House ; got by Oxford Duke 9th (34979), dam (Cora 7th) by Duke of Gaddesby (30956), g. d. (Cora 4th) by First Earl Ducie (23814), gr. g. d. (Cora 3rd) by Lord Barrington (18213), — (Cora) by Waterloo (11025), — by John (4108), — by Liberty (7141), — by Martin (2279), — by Vertumnus (1546), — by Sir Stephen (1456), — by Prince of Waterloo (528), — by Mayflower (425), — by a Bull of Mr. Nicholson's, descended from the stock of Mr. J. Brown, of Aldborough.

(43775) PRIDE OF ERIN,
Roan, calved January 5, 1878, bred by Mr. W. Talbot Crosbie, Ardfert Abbey, the property of Earl Fitzwilliam, Coollattin Park ; got by England's Glory

(23889), dam (Pride of Adare) by Northern Light (24670), g. d. (Maid of Adare) by Lamp of Lothian (16356), gr. g. d. (Lady Dunraven) by Volunteer (15478), — (Lady Camilla) by Norfolk (9442), — by Rex (6385), — by Sir Thomas Fairfax (5196), — by Ambo (1636), — by Memnon (2295), — by Pilot (496), — by Agamemnon (9), — by Burrell's Bull, of Burdon.

(43776)   PRIDE OF OSMASTON,

Red and white, calved March 28, 1879, bred by Mr. J. Osmaston, Osmaston Manor; got by The Druid (35753), dam (Fancy) by Count Bickerstaffe (23630), g. d. (Lady) by The General (13856), gr. g. d. (Isabella 5th) by Daybreak (11338), — (Isabella 2nd) by Alfred (6732), — by Accordion (5708), — by Despot (1915), — by Napoleon (4531), — by Rival (2534), — by Darlington (956), — by Denton (198), — by Rockingham (560), — by Jobling's Son of Phenomenon (491), — by Phenomenon (491), — by Colonel (152), — by Styford (629).

(43777)   PRIME MINISTER 2ND,

White, calved July 6, 1879, bred by Mr. R. Taylor, New House; got by Major Irwin (34735), dam (Primula) by Knight of Killerby (29000), g. d. (Prairie Flower) by Cavalier Gwynne (28148), gr. g. d. (Princess) by Count of the Realm (23640), — (Playful) by Inkermann (14730), — by Lord Barrington (9308), — by Sir Robert (5180), — by Sir Robert (5180), — by Son of Farmer (2001), — by a Son of Mentor (2300), — by Mars.

(43778)   PRINCE,

Red and white, calved March 6, 1876, bred by Mr. F. W. Low, Kilshane; got by Visigoth (35907), dam (Virginia) by Warlaby Knight (35931), g. d. (Victoria) by Lamp of Florence (26564), gr. g. d. (Rosette) by Adjutant (30360), — (Rosalia) by Emperor (33842), — by The Baron (13833), — by Prince Edward Fairfax (9506), — by Premier (6308), — by Sovereign (7539), — by Favourite (6997), — by Saturn (5089), — bred by Mr. Rennie, of Phantassie.

(43779)   PRINCE ALBERT,

Roan, calved October 8, 1879, bred by Mr. W. Parker, Great Stanney Hall; got by Hard Times (38402), dam (Princess) by Golden Duke (26266), g. d. (Empress) by Majestic (20264), gr. g. d. (Regina) by Constantine (14318), — (Lucy Long) by The Duke of Lancaster (10929), — by Gainford 2nd (6030), — by Wharton (2833), — by Count (1883), — by Baronet (1686), — by Young Rockingham (2549), — by Wellington (2824), — by Northumberland (464), — by Buston's Styford (103), — by Lame Bull (358), — by Bolingbroke (86).

(43780)   PRINCE ALEXANDER,

Roan, calved September 21, 1879, bred by Mr. R. Crowe, Speeton; got by Monarch (40362), dam (Princess Alexandra 2nd) by Knight of Osberton (29003), g. d. (Princess Alexandra) by Knight Errant (18154), gr. g. d. (Lady Amy) by Cardigan (12556), — (Lady Alicia) by Prince Arthur (13497), — by Noble (4578), — by Newton (2367), — by Emperor (3716), — by Satellite (1420), — by Cato (119), — by Jupiter (342), — by George (273), — by Chilton (136), — by Irishman (329), — by B. (45).

(43781)   PRINCE ALFRED,

Roan, calved March 29, 1877, bred by the Trustees of the late Mr. J. Dickson, Cambushinnie, the property of Mr. C. S. H. D. Moray, Abercairny; got by Prince of Schleswig (35158), dam (Lady Mary 1st) by Duke of Glasgow (33649),

g. d. (Lady Mary) by Windsor (37683), gr. g. d. (Blythsome) by Major (4344), — (Jenny Lind) by Ethelred (5990), — by Thorpe, — by Rennie's Romulus.

(43782) PRINCE ALFRED,

Roan, calved March 14, 1879, bred by Mr. R. Laycock, Winlaton; got by Lion of Flanders (34460), dam (Florence) by Prince of Wales (32190), g. d. (Louise) by Grand Duke 13th (21850), gr. g. d. (Waterwitch 15th) by Field Marshal (26150), — (Waterwitch 8th) by Havelock (26351), — by Young Duke of Oxford (12767), — by Zadig (8796), — by Wizard (6688), — by Hecatomb (2102), — a Cow of Earl Spencer's.

(43783) PRINCE ANGELUS,

Roan, calved February 1, 1879, bred by the Earl of Bective, Underley Hall; got by Grand Duke 31st (38374), dam (Lady Angelina) by Ninth Duke of Geneva (28391), g. d. (Tube Rose 44th) by Saladin (35461), gr. g. d. (Tube Rose 40th) by Zanoni (37700), — (Tube Rose 4th) by Wolviston (21125), — by Third Duke of Cambridge (5941), — by Earl of Antrim (10174), — by South Durham (5281), — by Bellerophon (3119), — by Belvedere (1706), — by Waterloo (2816), — by Baron (58), — by Phenomenon (491), — by Favourite (252), — by Favourite (252), — by Favourite (252), — by Hubback (319), — by Snowdon's Bull (612), — by Waistell's Bull (669), — by Masterman's Bull (422), — by the Studley Bull (626).

(43784) PRINCE ARTHUR,

Red, calved January 9, 1879, bred by Mr. J. Bromley, Forton, the property of Mr. R. Richardson, Croston Barns; got by Duke of Richmond 2nd (38175), dam (Duchess 10th) by Bretwalda (33198), g. d. (Duchess 4th) by Favourite (23916), gr. g. d. (Duchess 2nd) by Prince Albert (20529), — (Duchess) by Duke of Cambridge (19610), — by Grand Duke 2nd (12961), — by Brocardo (9992), — by Duke (3633), — by Reformer (2510), — by Snowball (2648), — by Wellington (678).

(43785) PRINCE ARTHUR,

Red and white, calved February 7, 1879, bred by Mr. A. Graham, Yanwath Hall; got by The Colonel (35747), dam (Primula) by Royal Scotforth (25042), g. d. (Sanicle) by Napoleon (20395), gr. g. d. (Sundew) by White Hamlet (15508), — (Sunflower) by General Fairfax (14594), — by Symmetry (12168), — by Young Frederick (3836), — by Commodore (1858), — by Tathwell Studley (5401), — by Blyth Comet (85), — a Turnell Cow.

(43786) PRINCE ARTHUR,

Roan, calved April 4, 1879, bred by Mr. F. Fowler, Henlow; got by Cecrops (37958), dam (Princess Ellen) by Regal Booth (35252), g. d. (Princess Maud) by Prince Regent (29677), gr. g. d. (Princess Pearl) by Prince Pearl (29674), — (Lettuce Hopewell) by Prince Hopewell (22592), — by Prince George (13510), — by Clarence (14279), — by Justice (14753), — by Young Rufus (13649), — by Constitution (12634), — by Young Comet (1853), — descended from Jolly's Bull (4115).

(43787) PRINCE ARTHUR,

Red, calved September 4, 1879, bred by Mr. A. Raine, The Grove, Bow Bank; got by Whiff (30299), dam (Necklace) by British Crown (21322), g. d. (Polly) by Sir Samuel (15302), gr. g. d. (Mary) by Majestic (13279), — (Bella) by Harbinger (10297), — by Buckingham (3239).

(43788) PRINCE BEAUTY,

White, calved September 14, 1879, bred by Mr. C. Cradock, Hartforth; got by Crown Prince (38061), dam (Beauty) by Guy (28803), g. d. (Blameless) by Strawberry Prince (25240), gr. g. d. (Eanfleda) by Enterprise (16002), — (Sultana) by Sultan (15354), — by Cardinal Wiseman (12560), — by Son of Clementi (3399), — by Magnum Bonum (2243), — by Thorp (2757), — by Pirate (2430), — by Houghton (318), — by Marshal Blucher (416), — from the stock of Messrs. Wright and Charge.

(43789) PRINCE BUTTERFLY,

Red and white, calved February 20, 1873, bred by the Earl of Dartrey, Dartrey House, the property of the Earl of Caledon, Caledon; got by Warlaby Knight (35931), dam (Lady Butterfly) by Czar (23670), g. d. (Butterfly) by Master Butterfly 2nd (14918), gr. g. d. (Campanula) by Lord Wiseton (13256), — (Carolina) by Hudson (9228), — by Fairfax Royal (6987), — by The Toucher (6596), — by George (2057), — by Togston (5487), — bred by Mr. Laing, of Longhoughton.

(43790) PRINCE BUTTERFLY,

Red and white, calved January 8, 1878, bred by Mr. J. A. M. Cope, Drummilly, the property of Mr. Douglas Strain, Market Hill; got by Knight of Raby (34393), dam (Butterfly Princess) by Baron Eccleswall (33037), g. d. (Queen Butterfly) by Rose Butterfly (24993), gr. g. d. (Lady Butterfly) by Great Mogul (14651), — (Red Butterfly) by Master Butterfly (13311), — by Valiant (10989), — by Tom of Lincoln (8714), — by Bellerophon (3119), — by Renown (2525), — by Tartar (2738), — by Colton (1849), — by Pioneer (1321), — by Marshal Beresford (415), — by Cecil (120), — by Favourite (252), — by Cupid (177), — by Grandson of Bolingbroke (280), — by Foljambe (263), — by R. Alcock's Bull (19), — by J. Smith's Bull (608), — by Jolly's Bull (337).

(43791) PRINCE CHARLEY,

White, calved April 18, 1879, bred by Mr. R. H. Gould, Didmarton; got by Prince Arthur (38892), dam (Charmer 2nd) by Grand Duke of York (24071), g. d. (Charmer) by Monarch (16575), gr. g. d. (Cherry) by West Australian (17223), — (Clarissa) by Victor (13951), — by Berkeley (7830), — by Son of Old Strickland (11870).

(43792) PRINCE CHARLIE,

Red, calved January 8, 1879, bred by Messrs. Annandale and Sons, Lintzford; got by Blair Athole (37866), dam (Princess) by Beverley Butterfly (28021), g. d. (Lady) by Benjamin (19303), gr. g. d. (Queen Elizabeth) by Lord Nelson (13207), — (Wealthy) by Son of Fitz-Maurice (36084).

(43793) PRINCE CHARLIE,

Roan, calved July 8, 1879, bred by Mr. C. Ashton, Delrow; got by Lord of Lochaber 2nd (36983), dam (Queen Maude) by Young England (31110), g. d. (Queen Eleanor) by Cherry Duke (25752), gr. g. d. (Queen Anne) by Second Duke of Geneva (21591), — (Queen of Airdrie) by Second Duke of Airdrie (19600), — by May Duke (13320), — by Third Duke of Oxford (9047), — by Freeman (10244), — by Hamlet (8128), — by Short Tail (2621), — by Emperor (1974), — by Young Lancaster (361), — by St. Albans (2584), — by Lawn-sleeves (365), — bred by Messrs. James, of Stamford.

(43794) **PRINCE EUGENE,**
Roan, calved September 14, 1878, bred by the Earl of Suffolk and Berkshire, Charlton Park ; got by Cockadilly (36372), dam (Eunice 3rd) by Duke of Hazlecote 17th (30965), g. d. (Eunice 2nd) by Honeysuckle Wetherby 6th (24155), gr. g. d. (Eunice 1st) by Heir of Walton (24125), — (Eunice) by Duke of Brompton (19607), — by Longfellow (18206), — by The Minstrel (8688).

(43795) **PRINCE FRAGRANT,**
White, calved April 11, 1878, bred by Mr. W. T. Crosbie, Ardfert Abbey, the property of the Rev. H. Beckwith, Eaton Constantine ; got by Foreign Prince (36656), dam (Fragrance) by Regal Booth (27262), g. d. (Oriental) by Castle Grove (19408), gr. g. d. (Queen of the East) by Nobleman (18457), — (Circassia) by Lamp of Lothian (16356), — by Matadore (11800), — by Lord of Brawith (10465), — by Amateur (3007), — by Belshazzar (1703), — by Abraham (2905), — by Simon (5134), — by Young George (3885), — by George (276).

(43796) **PRINCE GEORGE,**
Red and white, calved February 21, 1879, bred by Mr. W. Charley, Seymour Hill ; got by Lord Seafield (40240), dam (Princess Louisa) by Red Knight (32261), g. d. (Princess Elizabeth) by Ulysses (25346), gr. g. d. (Princess Maude) by Prince of Warlaby (15107), or Fawsley Prince (17837), — (Princess Alice) by Prince of Warlaby (15107), — by Defiance (10107), — by Sir John Sinclair (5165), — by Coriander (6895), — by Escape (11446), — by Carter's Falstaff, — by Whisker (2835), — by Hero, — Vesta.

(43797) **PRINCE GEORGE,**
Red and white, calved August 5, 1879, bred by Mr. J. C. Musters, Annesley Park ; got by Duke of Glo'ster 6th (39734), dam (Princess Kelly) by Oxonian (38840), g. d. (Princess of Windsor 4th) by Lord Lorne (34587), gr. g. d. (Princess of Windsor 2nd) by Duke of Burghley (36472), — (Princess of Windsor) by Don Windsor (19585), — by British Prince (14197), — by Vanguard (10994), — by Baron Warlaby (7813), — by Fourth Duke of Northumberland (3649), — by Norfolk (2377), — by Waterloo (2816), — by Waterloo (2816).

(43798) **PRINCE GEORGE 2ND,**
White, calved February 18, 1871, bred by Mr. R. Welsted, Ballywalter, the property of Mr. M. Whelan, Co. Kerry ; got by Prince Christian (22581), dam (Little Gem) by Roan Oxford (16841), g. d. (Little Dorrit) by Lord Raglan (13240), gr. g.d (Donna Isabella) by Master Charlie (13312), — (Lady Grace) by Laudable (9282), — by The Stuart (7623), — by Norfolk (2377), — by Ambo (1636), — by Memnon (2295), — by Pilot (496), — by Agamemnon (9), — by Burrell's Bull, of Burdon.

(43799) **PRINCE GWYNNE 3RD,**
Red, calved August 28, 1878, bred by Mr. H. Sharpley, Acthorpe ; got by Grand Duke 27th (34067), dam (Duchesse Gwynne) by Grand Duke 21st (34061), g. d. (Favourite Gwynne) by Grand Duke of Lightburne (26290), gr. g. d. (Fairy Gwynne) by Grand Duke 5th (19875), — (Fortuna Gwynne) by Duke of Leinster (17724), — by Captain Hardinge (10023), — by St. Thomas (10777), — by Prime Minister (2456), — by Marmion (406), — by Merlin (430), — by Layton (366), — by Phenomenon (491), — by Favourite (252),

— by Favourite (252), — by Hubback (319), — by Snowdon's Bull (612), — by Waistell's Bull (669), — by Masterman's Bull (422), — by the Studley Bull (626).

(43800)   PRINCE HECTOR,

Roan, calved February 7, 1879, bred by Mr. E. Cazalet, Fairlawn ; got by Charming Prince 2nd (33334), dam (Princess 2nd) by Eighth Duke of Geneva (28390), g. d. (Princess Alice) by British Prince (14197), gr. g. d. (Duchess) by Duke of Cambridge (12742), — (Cold Cream) by Earl of Dublin (10178), — by Grey Friar (9172), — by Fawsley (6004), — by Little John (4232), — by Marcellus (2260), — by Caliph (1774), — by Swing (2721), — by Argus (759), — by Defender (194), — by Petrarch (488), — by Own Brother to R. Colling's White Heifer, — by Butterfly (104), — by Globe (278).

(43801)   PRINCE HUMBERT,

Roan, calved April 5, 1877, bred by the Rev. J. J. Moutray, Favour Royal, the property of Mr. R. D. Harrison, Hollywood ; got by Gladiator (36698), dam (Governess 2nd) by Prince of the Woods (32185), g. d. (Governess) by The Governor (32681), gr. g. d. (Norma) by Knight of the Grand Cross (31558), — (Nosegay) by Prince Duke 2nd (16731), — by White Jack (36089), — by Collingwood (8964), — by Brilliant (1740), — by Monarch (2324), — by Cato (119), — by George (273), — by Chilton (136), — by Irishman (329), — by B. (45).

(43802)   PRINCE IMPERIAL,

Roan, calved December 6, 1877, bred by Mr. F. Fowler, Henlow ; got by Prince Leopold (37239), dam (Laurestina) by Prince Royal (29680), g. d. (Laurel) by Prince James (20554), gr. g. d. (Lettuce) by Prince George (13510), — (Lucretia) by Clarence (14279), — by Justice (14753), — by Young Rufus (13649), — by Constitution (12634), — by Young Comet (1853), — descended from Jolly's Bull (4115).

(43803)   PRINCE IMPERIAL,

Roan, calved January 12, 1878, bred by Messrs. Annandale and Sons, Lintzford ; got by Ben Brace (30524), dam (Princess) by Beverley Butterfly (28021), g. d. (Lady) by Benjamin (19303), gr. g. d. (Queen Elizabeth) by Lord Nelson (13207), — (Wealthy) by Son of Fitz-Maurice (36084).

(43804)   PRINCE IMPERIAL,

Red and white, calved October 2, 1878, bred by the Executors of Mr. J. Morton, Skelsmergh Hall, the property of Mr. Leach, Northsceugh ; got by Duke of Barrington 2nd (36463), dam (Princess 4th) by Oxford's Victor (35001), g. d. (Oxford's Princess) by Atherton's Oxford (21195), gr. g. d. (Princess 2nd) by Atherton's Oxford (21195), — (Princess) by Rob Roy (18725), — by Prelate (11919), — by Tomboy (5494), — by Rokeby (8502), — by Eclipse (8021), — by Eclipse (8021), — by Chance (7890).

(43805)   PRINCE IMPERIAL,

White, calved October 22, 1878, bred by Messrs. W. and H. Morley, East Gate Farm, the property of Mr. Dewar, Aberdeen ; got by Broomley Victor (37917), dam (Lady Flora 4th) by British Baron (33213), g. d. (Lady Flora 2nd) by Master Blithe (29313), gr. g. d. (Lady Flora) by Protector (20609), — (Cleasby Daisy) by King Consort (16335), — by Emperor (14498), — by a Son of Belleville (6778).

(43806)   PRINCE IMPERIAL,

Red, calved February 21, 1879, bred by Mrs. Mace, Sherborne ; got by Duke of Wellington 7th (33753), dam (Rose of Summer) by Count Glo'ster (23637), g. d. (Rose of Spring) by Lord Elgin (18230), gr. g. d. (Rosebud) by General Pelissier (14605), — (Moss Rose) by Marchmont (9367), — by Fitz-Hardinge (8073), — by Augustus (6751), — by Son of Anthony (1640), — by a Bull of Mr. Champion's, of Blyth.

(43807)   PRINCE IMPERIAL,

Red and white, calved September 15, 1879, bred by Mr. T. W. Cadman, Balli-field Hall, the property of Mr. G. Appleyard, Canklow; got by Duke of Sandale 2nd (38180), dam (Lady Joan 2nd) by Iron Duke 4th (34220), g. d. (Lady Joan) by Knight of Derby (31535), gr. g. d. (Lady Julian) by Prince Imperial (27145), — (Joan 5th) by Henry 5th (19944), — by Scimitar (16917), — by Reo Bulla (10703), — by Diamond (5918), — by Second Duke of Northumberland (3646), — by Young Sea Gull (5100), — by Blucher (795), — by Sultan (1485), — by Charles (6852), — by Yarborough (705).

(43808)   PRINCE JAMES,

Roan, calved October 27, 1877, bred by Mr. B. Dickson, Gilford House ; got by Fitz-James (36647), dam (Jane 3rd) by Gallant Knight (26214), g. d. (Jane 2nd) by Mountebank (24626), gr. g. d. (Jane) by Downshire (15898), — (Beauty) by Musician (13362), — by Hopewell (10332), — by Leonard (4210),

(43809)   PRINCE JAMES,

Red and white, calved April 12, 1879, bred by Messrs. J. G. W. and W. R. Lyle, Donaghmore House ; got by Sir James (37466), dam (Hoops) by Silk and Scarlet (20806), g. d. (Heroine) by Comet (11298), gr. g. d. (Jewel 3rd) by Essex Hero (10207), — (Jewel) by Wiseton (2848), — by Firby (1040), — by Ivanhoe (1131), — by Regent (544), — by Blyth Comet (85).

(43810)   PRINCE JOHN,

White, calved April 3, 1877, bred by Mr. W. Robinson, Edge Hill, the property of Mr. J. Holmes, Mansriggs ; got by Prince Alfred (29593), dam (Flower Child) by Cherry Prince (23555), g. d. (Flower Damsel) by Breast Plate (19337), gr. g. d. (Flower Maid) by Vanguard (10994), — (Flower Girl) by Londesboro' (6142), — by Rinaldo (4949), — by Sir Thomas (2636), — by Sir Alexander (591), — by Marske (418), — by North Star (459), — by Wellington (680), — by Favourite (252), — by Favourite (252), — by Ben (70), — by Hubback (319), — by Snowdon's Bull (612), — by Sir J. Pennyman's Bull (601).

(43811)   PRINCE LEOPOLD,

White, calved June 28, 1879, bred by Mr. J. Hunter, The Palms; got by Prince Regent (29676), dam (Ringlet 3rd) by Prince Charlie (29606), g. d. (Ringlet) by Merry Monarch (22349), gr. g. d. (Blossom) by Alderman (17291), — (Young Matchless) by Duke of Tyne (12773), — by Young Hector (7074), — by Young Hastings (3988), — by Short Legs (5124), — by Ajax (723), — by Hector (2103), — by Surly (2715), — by Sir Harry (5155), — by Colonel (152), — by Son of Hubback (319).

(43812)   PRINCE LEOPOLD GWYNNE,

Roan, calved March 1, 1878, bred by Mr. C. Howard, Biddenham ; got by Earl of Leicester 2nd (33803), dam (Lady Gwynne) by Grand Duke 15th (21852),

g. d. (Kentish Gwynne 2nd) by Fourth Duke of Thorndale (17750), gr. g. d. (Patty Gwynne) by Young Benedict (15641), — (Pricky Gwynne) by St. Thomas (10777), — by Conservative (3472), — by Chorister (3378), — by Wellington (2824), — by Marmion (406), — by Merlin (430), — by Layton (366), — by Phenomenon (491), — by Favourite (252), — by Favourite (252), — by Hubback (319), — by Snowdon's Bull (612), — by Waistell's Bull (669), — by Masterman's Bull (422), — by the Studley Bull (626).

(43813)　　　　PRINCE LICHTENSTEIN,
Roan, calved January 3, 1879, bred by Mr. J. W. Humphrys, Ballyhaise House; got by Earl of Derby (33786), dam (Blossom) by Abercorn (25484), g. d. (Moss) by Prince of Lurg (22617), gr. g. d. (Moss) by Ancient Briton (19225), — (Ada) by Trory (13901), — by Drury (10140), — by Belted Will (6780), — by Regent (2517), — by Borderer (3191), — by Eclipse (1949), — by Togston (5487), — by Bolingbroke (3184), — by Son of Midas (435), — by Twin Brother to Ben (660).

(43814)　　　　PRINCE LOUIS,
Roan, calved December 19, 1878, bred by Mr. G. Taylor, Stanton Prior; got by Waterloo Prince (35949), dam (Belinda) by Winterfold (36020), g. d. (Lavender 12th) by Lord Red Eyes 2nd (24460), gr. g. d. (Lavender 7th) by Sir Sam (16992), — (Lavender 4th) by Hero of the West (8150), — by Barrister (7815), — by Marcellus (6181), — by Major (4338), — by Eclipse (1949), — by Son of Dishley Ketton (982), — by Cossack (925), — by Patriot (486).

(43815)　　　　PRINCE LUD,
Roan, calved October 2, 1878, bred by Major O'Reilly, Knock Abbey; got by Grand Prince (34085), dam (Lady Sarai) by Cherry Prince (23555), g. d. (Lady Adah) by Killerby Monk (20053), gr. g. d. (Lady Eve) by Doctor McHale (15887), — (Lady Mary Bountiful) by Baron Warlaby (7813), — by Usurer (9763), — by Ranunculus (2479), — by Sir Walter (2637), — by Hotspur (1117), — by Coxcomb (928), — by Midas (435), — by Comet (155), — by R. Colling's Son of Favourite (252), — by same Son of Favourite (252), — by Hubback (319).

(43816)　　　　PRINCELY,
White, calved March 2, 1879, bred by Mr. R. B. Brockbank, Crosby; got by Borderer (33183), dam (Princess Helena) by Baron Charleston (27922), g. d. (Princess Beatrice) by Kildonan (20051), gr. g. d. (Princess Alexandra) by Eighth Duke of Oxford (15939), — (Catchit) by Earl of Dublin (10178), — by Grey Friar (9172), — by Fawsley (6004), — by Little John (4232), — by Marcellus (2260), — by Caliph (1774), — by Swing (2721), — by Argus (759), — by Defender (194), — by Petrarch (488), — by Own Brother to R. Colling's White Heifer, — by Butterfly (104), — by Globe (278).

(43817)　　　　PRINCE MALCOLM 3RD,
Red and white, calved November 2, 1879, bred by Messrs. Dalton, Cummersdale; got by Duke of Siddington (38182), dam (Princess Maude 3rd) by Earl of York (33824), g. d. (Princess Maude 2nd) by Oxford Isis (32024), gr. g. d. (Princess Maude) by Royal Cumberland (27358), — (Maude) by Fourteenth Duke of Oxford (21605), — by Palmerston (15041), — by Stapleton (15336), — by Bracebridge (14187), — by Mingo (7239), — by Snowball (1465), — by Pizarro (1323).

(43818) **PRINCE MARMION,**
Roan, calved December 25, 1878, bred by Mr. J. Cattley, Stearsby ; got by
Sergeant (37434), dam (Princess Maud) by Chancellor (30679), g. d. (Princess
Mary) by Trifolium (25330), gr. g. d. (Princess Charlotte) by Blood Royal
(17423), — (White Princess) by Earl of Athole (14471), — by De Grey (11346),
— by Prince Alfred (10638), — by Prince of Wales (6348), — by Cyrus
(3538), — by Bedford Junior (1701), — by Romulus (1403), — by Isaac (1129),
— by Northern Light (1281), — by White Comet (1582), — by Cattley's Grey
Bull (1798).

(43819) **PRINCE OF AUSTRIA,**
White, calved February 6, 1879, bred by.Mr. W. Charley, Seymour Hill ; got by
Lord Seafield (40240), dam (Princess Anna) by Red Knight (32261), g. d.
(Princess Elizabeth) by Ulysses (25346), gr. g. d. (Princess Maude) by Prince
of Warlaby (15107), or Fawsley Prince (17837), — (Princess Alice) by Prince of
Warlaby (15107), — by Defiance (10107), — by Sir John Sinclair (5165), — by
Coriander (6895), — by Escape (11446), — by Carter's Falstaff, — by Whisker
(2835), — by Hero, — Vesta.

(43820) **PRINCE OF BRAILES 2ND,**
Red and white, calved June 25, 1878, bred by Mr. H. J. Sheldon, Brailes House;
got by Duke of Rothesay (36534), dam (Princess of the Valley 2nd) by Baron
Morley 2nd (41047), g. d. (Damask 2nd) by Millbrook (34851), gr. g. d. (Damask)
by Mosstrooper (34877), — (Double Rose) by Double Duke (41333), — by
Napier (6238), — by South Durham (5281), — by Bellerophon (3119), — by
Belvedere (1706), — by Waterloo (2816), — by Baron (58), — by Phenomenon
(491), — by Favourite (252), — by Favourite (252), — by Favourite (252), —
by Hubback (319), — by Snowdon's Bull (612), — by Waistell's Bull (669), —
by Masterman's Bull (422), — by the Studley Bull (626).

(43821) **PRINCE OF BRAILES 3RD,**
White, calved April 6, 1879, bred by Mr. H. J. Sheldon, Brailes House ; got by
Duke of Rothesay (36534), dam (Princess of Geneva 2nd) by Ninth Duke of
Geneva (28391), g. d. (Princess) by General Havelock (17952), gr. g. d.
(Countess) by Earl of Chatham (10176), — (Clara) by Napier (6238), — by
Mameluke (2258), — by Waterloo (2816), — by Baron (58), — by Phenomencn
(491), — by Favourite (252), — by Favourite (252), — by Favourite (252), —·
by Hubback (319), — by Snowdon's Bull (612), — by Waistell's Bull (669), —
by Masterman's Bull (422), — by the Studley Bull (626).

(43822) **PRINCE OF FALKLAND,**
Roan, calved March 26, 1876, bred by Mr. A. H. Tyndall Bruce, House of
Falkland; got by Work of Art (36033), dam (Oxford Policy 2nd) by Fourteenth
Duke of Oxford (21605), g. d. (Policy) by Lord Raglan (14849), gr. g. d.
(Prosperity) by Captain Edwards (8929), — (Panic) by Almacks (7779), — by
Dan O'Connell (3557), — by Young Grazier (3928), — by Milton (2314), — by
Policy (1329), — by Barmpton (54).

(43823) **PRINCE OF GENEVA 21ST,**
Red, calved May 13, 1879, bred by Mr. G. Garne, Churchill Heath, the property
of Count Renard, Castle Strehlitz, Upper Silesia ; got by Grand Duke of Geneva
2nd (31288), dam (Portrait 9th) by Cherry Grand Duke 5th (30712), g. d.

(Peahen) by Bashaw (12449), gr. g. d. (Peace) by Valiant (7662), — (Young Portrait) by Fitz-Hardinge (8073), — by Lord John (4259), — by Young Consul (6893), — by Newnham (2365), — by Satellite (1420), — by Jupiter (342), — by Sir Oliver (605), — by Trunnell (659), — by Favourite (252), — by Favourite (252), — by Dalton Duke (188), — by R. Alcock's Bull (19), — by J. Smith's Bull (608), — by Jolly's Bull (337).

(43824)   PRINCE OF GENEVA 22ND,
Roan, calved July 7, 1879, bred by Mr. G. Garne, Churchill Heath; got by Grand Duke of Geneva 2nd (31288), dam (Premium 2nd) by Grand Duke of Geneva 2nd (31288), g. d. (Prima Donna) by Royal Butterfly 20th (25007), gr. g. d. (Paradox) by Cynric (19542), — (Prejudice) by Telegraph (15370), — by Enterprise (11443), — by Young Consul (6893), — by Newnham (2365), — by Satellite (1420), — by Jupiter (342), — by Sir Oliver (605), — by Trunnell (659), — by Favourite (252), — by Favourite (252), — by Dalton Duke (188), — by R. Alcock's Bull (19), — by J. Smith's Bull (608), — by Jolly's Bull (337).

(43825)   PRINCE OF GENEVA 23RD,
White, calved September 27, 1879, bred by Mr. G. Garne, Churchill Heath; got by Grand Duke of Geneva 2nd (31288), dam (Portrait 8th) by Grand Duke of Geneva 2nd (31288), g. d. (Pride of the Heath) by Cynric (19542), gr. g. d. (Peach) by Havelock (14676), — (Peace) by Valiant (7662), — by Fitz-Hardinge (8073), — by Lord John (4259), — by Young Consul (6893), — by Newnham (2365), — by Satellite (1420), — by Jupiter (342), — by Sir Oliver (605), — by Trunnell (659), — by Favourite (252), — by Favourite (252), — by Dalton Duke (188), — by R. Alcock's Bull (19), — by J. Smith's Bull (608), — by Jolly's Bull (337).

(43826)   PRINCE OF GENEVA 24TH,
Roan, calved December 23, 1879, bred by Mr. G. Garne, Churchill Heath; got by Grand Duke of Geneva 2nd (31288), dam (Premium) by Royal Cambridge 2nd (25010), g. d. (Precedent) by Cynric (19542), gr. g. d. (Precedence) by Progression (16770), — (Prejudice) by Telegraph (15370), — by Enterprise (11443), — by Young Consul (6893), — by Newnham (2365), — by Satellite (1420), — by Jupiter (342), — by Sir Oliver (605), — by Trunnell (659), — by Favourite (252), — by Favourite (252), — by Dalton Duke (188), — by R. Alcock's Bull (19), — by J. Smith's Bull (608), — by Jolly's Bull (337).

(43827)   PRINCE OF GOLDCLIFF,
Red and white, calved May 12, 1876, bred by Mr. W. Price, Goldcliff; got by Cromwell (33480), dam (Prairie Flower) by Master Broomstick (34803), g. d. (Promised Land 2nd) by Prince of Gwent (24830), gr. g. d. (Promised Land) by Star of Gwent (25227), — (Lady Bird 2nd) by Prince of the Empire (20578), — by Prince Imperial (15095) or Chief (18976), — by Sir Thomas (13749), — by Lord George (10439), — by Fitz-Leonard (7010), — by Velocipede (5552), — by Young Matchem (4422), — by Jack Tar (1133), — by Pilot (496), — by Young Albion (15).

(43828)   PRINCE OF KERRY,
Roan, calved May 1, 1879, bred by Mr. B. McKenna, Lea Grange, Blackley; got by Foreign Glory (39890), dam (Rose of Ardfert) by England's Glory (23889), g. d. (Rose of Cashmere) by Irish Baron (31417), gr. g. d. (Miriam) by

Northern Light (24670), — (Oriental) by Castle Grove (19408), — by Nobleman (18457), — by Lamp of Lothian (16356), — by Matadore (11800), — by Lord of Brawith (10465), — by Amateur (3007), — by Belshazzar (1703), — by Abraham (2905), — by Simon (5134), — by Young George (3885), — by George (276).

(43829)   PRINCE OF KIRKLEVINGTON,
Roan, calved September 20, 1879, bred by Messrs. Dalton, Cummersdale ; got by Duke of Siddington (38182), dam (Princess of Kirklevington) by Grand Duke of Kirklevington (34071), g. d. (Princess of York) by Eighth Duke of York (28480), gr. g. d. (Princely) by Warrior (23178), — (Princess of Cambridge) by Duke of Cambridge (12742), — by Earl of Dublin (10178), — by Grey Friar (9172), — by Allan-a-dale (7778), — by Little John (4232), — by Marcellus (2260), — by Caliph (1774), — by Swing (2721), — by Argus (759), — by Defender (194), — by Petrarch (488), — by Own Brother to R. Colling's White Heifer, — by Butterfly (104), — by Globe (278).

(43830)   PRINCE OF MAPLEWELL 2ND,
Red, calved May 9, 1879, bred by Sir W. H. Salt, Bart., Maplewell, the property of Mrs. Barrs, Odstone Hall; got by Fifth Duke of Glo'ster (36494), dam (Princess 7th) by Second Duke of Tregunter (26022), g. d. (Princess 4th) by Baron Oxford 3rd (25579), gr. g. d. (Princess 3rd) by Fawsley Baronet (23920), — (Princess 2nd) by Third Duke of Thorndale (17749), — by Old Rowley (15020), — by Jethro Tull (11616), — by General Sale (8099), — by Napier (6238), — by Mameluke (2258), — by Waterloo (2816), — by Baron (58), — by Phenomenon (491), — by Favourite (252), — by Favourite (252), — by Favourite (252), — by Hubback (319), — by Snowdon's Bull (612), — by Waistell's Bull (669), — by Masterman's Bull (422), — by the Studley Bull (626).

(43831)   PRINCE OF MARLBORO' 3RD,
Red, calved May 28, 1879, bred by Mr. G. Garne, Churchill Heath ; got by Grand Duke of Marlboro' (38380), dam (Polly Geneva) by Grand Duke of Geneva 2nd (31288), g. d. (Penelope) by Duke of Towneley (21615), gr. g. d. (Pearl) by Luxury (18293), — (Project) by Homer (16277), — by Lord George (9314), — by Manager (8271), — by Newnham (2365), — by Satellite (1420), — by Jupiter (342), — by Sir Oliver (605), — by Trunnell (659), — by Favourite (252), — by Favourite (252), — by Dalton Duke (188), — by R. Alcock's Bull (19), — by J. Smith's Bull (608), — by Jolly's Bull (337).

(43832)   PRINCE OF MARTLEY,
Roan, calved October 23, 1878, bred by Mr. W. Robinson, Edge Hill, the property of Mr. J. Robinson, Rodge Hill; got by Prince Alfred (29593), dam (Wild Spray) by Cherry Grand Duke 3rd (28174), g. d. (Wave Rise) by Blinkhoolie (23428), gr. g. d. (Wave Ripple) by Killerby Monk (20053), — (Wave Princess) by British Prince (14197), — by Vanguard (10994), — by Baron Warlaby (7813), — by Fourth Duke of Northumberland (3649), — by Norfolk (2377), — by Waterloo (2816), — by Waterloo (2816).

(43833)   PRINCE OF MENTMORE,
Red and white, calved December 10, 1879, bred by Mr. E. Freeman, Chilton ; got by King Alfred (36833), dam Chauntress) by Charming Lad (30699), g. d.

(Lady Superior 4th)·by Darlington (23679), gr. g. d. (Lady Superior 3rd) by Prince Leopold (20558), — (Lady Superior) by British Baronet (15691), — by Cardinal (11246), — by Marquis of Rockingham (10560), — by Sheldon (8557), — by Benjamin (1710), — by Malbro' (1189), — by Ebor (997), — by Regent (546), — by North Star (459), — by R. Colling's White Bull (151), — bred by Mr. R. Colling.

(43834) ***PRINCE OF ORANGE,**
White, calved February 15, 1852, bred by Mr. D. C. Collins, High Cliff Farm, Duchess Co., New ‚York, late the property of Mr. T. Cowles, Farmington, Connecticut, U.S.A.; got by Third Duke of Cambridge (5941), dam (Phœnix) by Hero (4020), g. d. (Princess) by Washington (1566), gr. g. d. (Pansy) by Blaize (76), — (Primrose) by Charles (127), — by Blyth Comet (85), — by Prince (521), — by Patriot (486).

(43835) ***PRINCE OF ORANGE 2ND,**
Roan, calved May 24, 1856, bred by Mr. P. Lathrop, Hadley Falls, Massachusetts, late the property of Mr. A. M. Winslow, Putney, Vermont, U.S.A.; got by Prince of Orange (43834), dam (Novice) by Coxcomb (12655), g. d. (Nina) by Sancho (2597), gr. g. d. (Nannette) by Patriot (2412), — (Nonpareil) by Young Denton (963), — by North Star (460), — by Comet (155), — by Henry (301), — by Danby (190).

(43836) **PRINCE OF POWYS,**
Roan, calved January 26, 1879, bred by Captain D. H. Mytton, Garth; got by Constantine 2nd (33439), dam (Janet) by Vespasian (32759), g. d. (Hebe) by Berwick (23411), gr. g. d. (Hecuba) by Hector (14685), — (Cornflower) by Columbus (12616), — by Red Rover (11982), — by Earl of Durham 2nd (9060), — by Raree Show (4874), — by Premier (2449), — by Albion (731), — by Sirius (598), — by Wellington (679), — by Sultan (631), — by Punch (531), — by Ladykirk (355).

(43837) **PRINCE OF PRUSSIA,**
Roan, calved August 13, 1879, bred by Mrs. Mace, Sherborne; got by Geneva's Duke (38349), dam (Paradox 4th) by Duke of Wellington 7th (33753), g. d. (Paradox 2nd) by Royal Cambridge 2nd (25010), gr. g. d. (Paradox) by Cynric (19542), — (Prejudice) by Telegraph (15370), — by Enterprise (11443), — by Young Consul (6893), — by Newnham (2365), — by Satellite (1420), — by Jupiter (342), — by Sir Oliver (605), — by Trunnell (659), — by Favourite (252), — by Favourite (252), — by Dalton Duke (188), — by R. Alcock's Bull (19), — by J. Smith's Bull (608), — by Jolly's Bull (337).

(43838) **PRINCE OF RYEDALE,**
Roan, calved August 20, 1879, bred by Mr. J. Snarry, Marramatt Farm; got by Lord of Ryedale (40214), dam (Eastern Princess) by Ignoramus (28887), g. d. (Princess 3rd) by Duke of Towneley (21615), gr. g. d. (Red Princess) by Gipsy Prince (17965), — (Brandy) by Codrington (14290), — by Starlight (12146), — by Prince Edward (6334), — by Traveller (6617), — by Charles (3343), — by Commodore (5874), — by Cupid (5900), — by Parson (1306), — by St. John (572), — by Pope (514), — by Chilton (136).

(43839) **PRINCE OF THE MIDLANDS,**
Roan, calved November 3, 1879, bred by Mr. E. Freeman, Chilton; got by Kin

Alfred (36833), dam (Barrington Foggathorpe) by The Bursar (35742), g. d. (Grace) by Oxford Butterfly (29488), gr. g. d. (Lady Superior 3rd) by Prince Leopold (20558), — (Lady Superior) by British Baronet (15691), — by Cardinal (11246), — by Marquis of Rockingham (10560), — by Sheldon (8557), — by Benjamin (1710), — by Malbro' (1189), — by Ebor (997), — by Regent (546), — by North Star (459), — by R. Colling's White Bull (151), — bred by Mr. R. Colling.

(43840)			PRINCE OF THE PINKS 2ND,
Red, calved November 10, 1878, bred by Mr. G. Hewer, Ley Gore House; got by Grand Duke of Geneva 2nd (31288), dam (Pink 23rd) by Ranger (35203), g. d. (Pink 10th) by Royal Benedict (27348), gr. g. d. (Postscript) by Cynric (19542), — (Young Pink) by General Pelissier (14605), — by Marchmont (9367), — by Young Consul (6893), — by Newnham (2365), — by Satellite (1420), — by Jupiter (342), — by Sir Oliver (605), — by Trunnell (659), — by Favourite (252), — by Favourite (252), — by Dalton Duke (188), — by R. Alcock's Bull (19), — by J. Smith's Bull (608), — by Jolly's Bull (337).

(43841)			PRINCE OF WALES,
Red, calved June 21, 1879, bred by Mr. W. Bliss, Chipping Norton; got by Elmhurst Prince (41503), dam (Elmhurst Mazurka) by Duke of Barrington 4th (30924), g. d. (Mazurka 2nd of Oakdale) by Malcolm (41963), gr. g. d. (Mazurka 8th) by Albion (19209), — (Mazurka 5th) by Duke of Airdrie (12730), — by Orontes 2nd (11877), — by Harbinger (10297), — by Baron of Ravensworth (7812), — by Mariner (7204), — by Mina (2316), — by Commodore (1858), — by Rival (553).

(43842)			PRINCE OF WARLABY,
Roan, calved October 24, 1877, bred by Mr. J. Close, Holmescales, the property of Mr. Case, Mireside; got by Knight of the Bath (38521), dam (Familiar Warlaby) by Knight of Warlaby (31571), g. d. (Familiar Windsor) by Roan Windsor (24967), gr. g. d. (Familiar Hopewell Gwynne) by Hopewell Gwynne (19976), — (Familiar 4th) by Red Duke (13571), — by Master Belleville (11795), — by Fitz-Leonard (7010), — by Velocipede (5552), — by Young Matchem (2282), — by Jack Tar (1133), — by Pilot (496), — by Young Albion (15).

(43843)			PRINCE OF WITLEY,
Roan, calved April 13, 1879, bred by Mr. W. Robinson, Edge Hill, the property of Mr. J. Mason, Hillhampton; got by Prince Waterloo (40546), dam (Rosabella 2nd) by Hyperion (21966), g. d. (Princess) by Warwick (15486), gr. g. d. (Victoria) by Lablache (11656), — (Empress) by Spearman (5288), — (Countess) by Prime Minister (2456), — by Hetherington's Bull (4029), — by Julius Cæsar (1143), — by Pilot (496), — by Rockingham (559), — by Lame Bull (359), — by Shipton (587), — by Son of Suworrow (636), — by Son of Twin Brother to Ben (88), — by Twin Brother to Ben (660).

(43844)			PRINCE OF YORK,
Roan, calved November 28, 1879, bred by Messrs. Dalton, Cummersdale; got by Duke of Siddington (38182), dam (Princess of York) by Eighth Duke of York (28480), g. d. (Princely) by Warrior (23178), gr. g. d. (Princess of Cambridge) by Duke of Cambridge (12742), — (Chemisette) by Earl of Dublin (10178), —

by Grey Friar (9172), — by Allan-a-dale (7778), — by Little John (4232), — by Marcellus (2260), — by Caliph (1774), — by Swing (2721), — by Argus (759), — by Defender (194), — by Petrarch (488), — by Own Brother to R. Colling's White Heifer, — by Butterfly (104),— by Globe (278).

(43845) PRINCE OSCAR,

White, calved October 18, 1878, bred by Mr. R. Marsh, Little Offley, the property of Mr. T. Y. Sindall, Fulney Hall; got by King Offa (38494), dam (Rose of Pride) by Mantalini Prince (29273), g. d. (Rose of Eden) by Baron Warlaby (23381), gr. g. d. (Rose of Warlaby) by British Flag (19351), — (Rose of Hope) by Prince Alfred (13494), — by Heir-at-Law (13005), — by Sir Henry (10824), — by Buckingham (3239), — by Marcus (2262), — by Matchem (2281), — by Alderman (1622), — by Pilot (496), — by Remus (550), — by Sir Charles (592), — by R. Colling's Son of Favourite (252), — by Son of Favourite (252), — Old Strawberry.

(43846) PRINCE PAN,

Roan, calved February 27, 1879, bred by Mrs. Collingwood, Clayworth, the property of Mr. C. Greenwood, Clayworth; got by Jack of the Grange (40026), dam (Princess Pansy) by Pan (35005), g. d. (Royal Selene) by Double First (23700), gr. g. d. (Selina 7th) by Royal Gwynne (22784), — (Selina 2nd) by Royal Duke (16865), — by Guy Faux (12980), — by Young Pompey (13480), — by Broomborough (9994), — by The Minstrel (8687), — by Bloomsbury (3171), — by Red Rover (4906), — by Wharfdale (1578), — by Palemon (479), — by Meteor (432), — by Western Comet (689), — by Favourite (252), — by Cupid (177), — by Grandson of Bolingbroke (280), — by Foljambe (263), — by R. Alcock's Bull (19), — by J. Smith's Bull (608), — by Jolly's Bull (337).

(43847) PRINCE PATRICK,

Roan, calved January 23, 1879, bred by Mr. J. A. M. Cope, Drummilly; got by Pilgrim Father (35037), dam (Princess) by Lord Spencer (26738), g. d. (Little Queen) by Red Rover (24923), gr. g. d. (Pickle) by Highland Laddie (13026), — (Pauline 2nd) by Hamlet (8126), — by Prince of Northumberland (4826), — by Mason's Son of Matchem (2678), — by Falstaff (1993), — by Richard (1376), — by Jupiter (342), — by Rufus (570), — by Pope (514).

(43848) PRINCE PAUL,

Roan, calved October 7, 1873, bred by Mr. W. Sisman, Buckworth; got by Paris (27039), dam (Khirkee 3rd) by Sir Edward (20819), g. d. (Keepsake) by Great Mogul (14651), gr. g. d. (Khirkee 4th) by Marmaduke (14897), — (Khirkee 2nd) by Melbourne (13327), — by Young Fourth Duke (9037), — by Duke of Richmond (7996), — by Sir Walter (2639), — by Young Jerry (8177), — by Roseberry (567), — by Constellation (163), — by Hastings (293), — by Hastings (293), — by Leopold (372).

(43849) YOUNG PRINCE REGENT,

Red, calved March 27, 1878, bred by Mr. W. Langhorn, East Mill Hills; got by Prince Regent (29676), dam (Dinah 5th) by Lucky Lad (34714), g. d. (Diadem 1st) by Wild Boy (25447), gr. g. d. (Dinah 4th) by Pizarro (20497), — (Dinah 3rd) by Master Annandale (14916), — by The Earl (13850), — by Matchem (10517).

(43850) **PRINCE RUPERT,**
Roan, calved January 16, 1876, bred by Mr. B. J. Greene, Lecarrow; got by
Rupert (35430), dam (Lucy 17th) by Terry (30106), g. d. (Lucy 7th) by Anthony
(21180), gr. g. d. (Lucy 2nd) by Orphan Boy (13429), — (Lucy) by Jeweller
(10354), — by Royal Prince (5024), — by Eolus (3733), — by Belvedere 2nd
(3126), — by Alive 'O (2995), — by Eclipse (236), — by Charge's Grey Bull
(872), — by Paddock Bull (477), — by J. Brown's Old Red Bull (97).

(43851) **PRINCE RUPERT,**
Red, calved November 5, 1879, bred by Mrs. Mace, Sherborne; got by Geneva's
Duke (38349), dam (Oxford Belle) by Oxford's Tony (35000), g. d. (Roan
Tucker 2nd) by Royal Cambridge 2nd (25010), gr. g. d. (Red Tucker) by Orangeman (18485), — (Miss Nightingale) by Chieftain (12595), — by Morpeth (8325),
— by The Prince (7615), — by Colling (902), — by a Son of Alexander (1624),
— by a Grandson of Favourite (252).

(43852) **PRINCE SALADIN 3RD,**
Red, calved August 8, 1879, bred by the Earl of Bective, Underley Hall; got
by Duke of Underley (33745), dam (Lady Sale of Putney) by Ninth Duke of
Thorndale (31023), g. d. (Lady Sale 6th) by Red Knight (35232), gr. g. d. (Lady
Sale 3rd) by Third Duke of Cambridge (5941), — (Lady Sale 2nd) by Earl of
Chatham (10176), — by General Sale (8099), — by Napier (6238), — by
Mameluke (2258), — by Waterloo (2816), — by Baron (58), — by Phenomenon
(491), — by Favourite (252), — by Favourite (252), — by Favourite (252), —
by Hubback (319), — by Snowdon's Bull (612), — by Waistell's Bull (669), —
by Masterman's Bull (422), — by the Studley Bull (626).

(43853) **PRINCE VICTOR,**
Red and white, calved January 20, 1879, bred by Mr. W. T. Wakefield, Fletchampstead Hall; got by Thorndale Grand Duke 2nd (39218), dam (Vine Leaf)
by Duke of Brailes 2nd (30930), g. d. (Victoria) by Grand Duke 9th (19879),
gr. g. d. (Verity) by White Prince (21102), — (Vine 6th) by Sir Sam (16992),
— by Third Duke of Glo'ster (12754), — by Prince (6322), — by Major (4338)
— by Young Rockingham (2547), — by Young Major (2254), — by Monarch
(22372), — by Cossack (925).

(43854) **PRINCE WATERLOO 2ND,** ●
Red and white, calved January 13, 1879, bred by Sir John Swinburne, Bart.,
Capheaton; got by Twenty-seventh Duke of Oxford (33709), dam (Lady Waterloo 23rd) by Seventh Duke of York (17754), g. d. (Lady Waterloo 11th) by Patriot
(20475), gr. g. d. (Lady Waterloo 4th) by Ferdinand (12871), — (Lady Waterloo)
by Third Duke of Oxford (9047), — by Cleveland Lad (3407), — by Norfolk
(2377), — by Waterloo (2816), — by Waterloo (2816).

(43855) **PRINCIPAL,**
Roan, calved March 12, 1878, bred by Mr. W. Peacey, Chedglow; got by Lord
Fitzclarence 11th (34550), dam (Patience) by Crown Prince (30829), g. d. (Passive) by Cynric (19542), gr. g. d. (Patience) by Royal Oak (16870), — (Pink)
by Marchmont (9367), — by Young Consul (6893), — by Newnham (2365),
— by Satellite (1420), — by Jupiter (342), — by Sir Oliver (605), — by Trunnell (659), — by Favourite (252), — by Favourite (252), — by Dalton Duke
(188), — by R. Alcock's Bull (19), — by J. Smith's Bull (608), — by Jolly's
Bull (337).

(43856) PROCTOR,
Roan, calved March 14, 1879, bred by Mr. S. C. Pilgrim, The Outwoods; got
by John of Gaunt (34265), dam (Flora) by Duke of Lorne 3rd (30987), g. d.
(Flower Girl) by Lord of the Lilacs 5th (26712), gr. g. d. (Market Woman) by
Knight of Wetherby (24283), — by Marengo (13294), — by Plato (11907), —
by Geddington (8102).

(43857) PROCYON,
Red and white, calved December 15, 1877, bred by Mr. J. C. Daubuz, Killiow;
got by Vanguard (35850), dam (Moss Rose) by Young Hector (28830), g. d.
(White Rose) by Knight of Branches (20076), gr. g. d. (Dairy Maid) by Scimitar
(16917), — (Deborah) by Protection (11956), — by Hector (9200), — by Nelson
(4547), — by Milton (8315), — by Merlin (2302), — by Midas (435), — by
Denton (198).

(43858) PROFESSOR,
Red, calved May 12, 1879, bred by Lieut.-Colonel Milligan, Caldwell Hall; got
by Rob Roy (37353), dam (Problem 2nd) by Colonel Tregunter 4th (30766),
g. d. (Problem) by Blue Gown (25644), gr. g. d. (Conundrum) by Mountebank
(24627), — (Riddle) by Freddy (19780), — by Lenton Favourite (14789), — by
Lancaster Comet (11663), — by Queen's Roan (7389), — by Roman (2561),
— by Mercury (2301), — by Monarch (2324), — by St. Albans (2584), — by
Jupiter (342), — by Sir Oliver (605), — by Trunnell (659), — by Favourite
(252), — by Favourite (252), — by Dalton Duke (188), — by R. Alcock's Bull
(19), — by J. Smith's Bull (608), — by Jolly's Bull (337).

(43859) PROMOTION,
Roan, calved March 20, 1879, bred by Mr. S. C. Pilgrim, The Outwoods; got
by John of Gaunt (34265), dam (Marjory) by Lord of the Lilacs 5th (26712),
g. d. by Field Marshal (19742), gr. g. d. by Marengo (13294), — by Plato
(11907), — by Geddington (8102).

(43860) PUNJAUB,
Roan, calved May 2, 1878, bred by Mr. R. P. Maxwell, Finnebrogue; got by
Czarowitz 4th (33497), dam (Maid of the Mountain) by Prince Victor (20606),
g. d. (Mountain Sylph 2nd) by Prince Andrew (18588), gr. g. d. (Mountain
Sylph) by Beau Bill (9940), — (Collina) by Ballinasloe (12429), — by Buchan
Hero (3238), — by Sir Thomas Fairfax (5196), — by Hubback (2142), — by
Son of Young Warlaby (2812), — by Imperial (2151), — supposed by Young
Comet (905).

(43861) Q. L.,
Red and white, calved August 11, 1879, bred by Mr. J. P. Tynte, Tynte Park;
got by British Peer (33224), dam (Mountain Dew) by Mac Callum Mohr (31789),
g. d. (Wine Cup) by Knight of Windsor (16350), gr. g. d. (Ceres) by Coronet
(15818), — (Corn Cup) by Vanguard (10994), — by Helmsman (8141), — by
Tarrare (2735), — by Mameluke (2257), — by Diamond (209), — by George
(1067), — by Major (398), — by Major (397), — by Brown's Bull of Aldborough (820).

(43862) QUETTAH,
Red and white, calved June 5, 1879, bred by Mr. R. P. Maxwell, Finnebrogue;
got by Czarowitz 4th (33497), dam (Maid of the Mountain) by Prince Victor

0606), g. d. (Mountain Sylph 2nd) by Prince Andrew (18588), gr. g. d. (Mountain Sylph) by Beau Bill (9940), — (Collina) by Ballinasloe (12429), — by Buchan Hero (3238), — by Sir Thomas Fairfax (5196), — by Hubback (2142), — by Son of Young Warlaby (2812), — by Imperial (2151), — supposed by Young Comet (905).

(43863)　　　　　QUICKSILVER,
Red, calved December 16, 1879, bred by Mr. H. Allsopp, Hindlip Hall; got by Third Duke of Hillhurst (30975), dam (Quiet 3rd) by Windrush (32873), g. d. (Quiet 2nd) by Marnhull Duke (29283), gr. g. d. (Quiet) by Lord Waterloo 2nd (26755), — (Quietly) by Festival (26147), — by Sir John (25154), — Lady, bought from the herd of Lady Mary Vyner, Newby Hall, Yorkshire.

(43864)　　　　　YOUNG RABY,
Red and white, calved September 5, 1879, bred by Mr. T. Wynne, Lislea; got by Knight of Raby (34393), dam (Pink Blossom) by Baron Eccleswall (33037), g. d. (Pink 9th) by Royal Benedict (27348), gr. g. d. (Patience) by Royal Oak (16870), — (Pink) by Marchmont (9367), — by Young Consul (6893), — by Newnham (2365), — by Satellite (1420), — by Jupiter (342), — by Sir Oliver (605), — by Trunnell (659), — by Favourite (252), — by Favourite (252), — by Dalton Duke (188), — by R. Alcock's Bull (19), — by J. Smith's Bull (608), — by Jolly's Bull (337).

(43865)　　　　　RAG MERCHANT,
White, calved December 31, 1876, bred by Mr. M. Savidge, Sarsden Lodge Farm; got by Ragman (35198), dam (Young Wheedle) by Grand Marshal (21865), g. d. (Wedlock) by Grand Duke of Oxford 2nd (17996), gr. g. d. (Wheedle) by Earl of Dublin (10178), — (Lobelia) by Grey Friar (9172), — by Little John (4232), — by Young Norman (4584), — by White Boy (1580), — by Wyville's Bull, — by a Bull bred by Mr. Charge.

(43866)　　　　　RAMSEY,
Red and white, calved January 19, 1879, bred by Mr. A. H. Browne, Callaly Castle, the property of Mr. W. Langhorn, East Mill Hills; got by Muscovite (38773), dam (Mona's Princess) by Windsor's Prince of Mona (32883), g. d. (Duchess of Abbotsford) by Prince Regent (29676), gr. g. d. (Abbotsford Princess) by Abbotsford (23256), — (Wolviston Princess) by Wolviston (19163), — by Ettrick (14518), — by Sam Glen (10780), — by Willie (9835), — by Bachelor (1666), — by Snowball (2647), — by St. Albans (2584), — by Lawnsleeves (365), — bred by Messrs. James, of Stamford.

(43867)　　　　　RANGER,
Roan, calved April 22, 1879, bred by Mr. W. Curtis, Fernham; got by Count of Hanney (42980), dam (Jane Dennison) by Marquis of Sockburn (34787), g. d. (Jenny Dennison 6th) by Royal Butterfly 14th (20722), gr. g. d. (Jenny Dennison 3rd) by General Pelissier (14605), — (Jenny Dennison) by Cedric (3311), — by Fisher's Son of Favourite (1028), — by Cœlebs (897), — by Rose's Old Red Bull (5009), — by Fisher's Old Bull (3799).

(43868)　　　　　RAPID FOGGATHORPE,
Roan, calved November 3, 1877, bred by Lord Polwarth, Mertoun House; got by Rapid Rhone (35205), dam (Riby Foggathorpe) by Breast Plate (19337), g. d. (Georgie) by Elfin King (17796), gr. g. d. (Geraldine) by Brideman (12493),

— (Gertrude) by Vanguard (10994), — by Royal Prince (5024), — by Ganthorpe (2049), — by Malbro' (1189), — by Ebor (997), — by Regent (546),
— by North Star (459), — by R. Colling's White Bull (151), — bred by Mr. R.
Colling.

(43869)   RASSELAS,
Red and white, calved February 18, 1879, bred by Lord Bolton, Bolton Hall,
the property of Colonel the Hon. A. M. Cathcart, Mowbray House ; got by Heir-
at-Law (34124), dam (Rose of the Hills) by Manfred (26801), g. d. (Rosetta)
by Lord of the Hills (18267), gr. g. d. (Rose) by Baron Barton (14123), —
(White Rose) by Baron Farnley (14129), — by Banker (11136), — by Ravensworth (9532), — by Nelson (4549), — by Newton (4567), — by Wonderful
(700), — by Cleveland (145), — by Butterfly (104), — by Hollon's Bull (313),
— from the stock of Sir J. Pennyman.

(43870)   RATHBARRY,
Roan, calved March 2, 1878, bred by Lord Carbery, Castle Freke ; got by
Gwynne Fitz-Blithe (39970), dam (Cowslip 24th) by England's Glory (23889),
g. d. (Cowslip 16th) by Prince Christian (22581), gr. g. d. (Cowslip 9th) by
Uncle Tom (13913), — (Cowslip 8th) by Consolation (14316), — by Gay Lad
(12922), — by Earl of Antrim (10174), — by Duke of Norfolk (5952), — by
Waterloo (2816), — by Kitt (7127), — by Kitt (7127), — by Page's Bull
(6269), — by Middleton's Bull (438).

(43871)   RAVENSWORTH,
Red, calved November 23, 1878, bred by Earl Beauchamp, Madresfield Court ;
got by Marquis of Blandford 2nd (34779), dam (Fairy Belle) by Festival
(26147), g. d. (Favourite Lass) by Archduke (21185), gr. g. d. (Favourite) by
Castlereagh (19409), — (Fashion) by Ortolan (18496), — by Columbus (17591),
— by Progress (4839), — by Progress (4839), — by Progress (4839).

(43872)   RAYMOND,
Roan, calved September 26, 1879, bred by Mr. J. Stratton, Alton Priors ; got
by Primus (38886), dam (Rosette) by James 1st (24202), g. d. (May Rose 2nd)
by Bude Light (21342), gr. g. d. (May Rose) by Young Windsor (17241), —
(Essence of Roses) by His Highness (14708), — by Waterloo (11025), — by
Lottery (4280), — by Lottery (4280), — by Phœnix (6290).

(43873)   RECRUIT,
Red and white, calved July 3, 1879, bred by Mr. H. Smith, Mountmellick ; got
by Earl of Aylesby (38215), dam (Red Breast 3rd) by Conqueror (30786), g. d.
(Red Breast) by Sheet Anchor 2nd (25121), gr. g. d. (Goodluck 2nd) by Gipsy
Boy (14617), — (Goodluck) by Emperor (12835), — by Prince Ernest (7366),
— by Booth (11195), — by Linton (4227), — by Streamer (624).

(43874)   RED BERRY,
Red and white, calved October 4, 1879, bred by Mr. R. S. Bruere, Braithwaite
Hall ; got by St. Swithin's Star Drop (40667), dam (Spring Flower) by Booth
Satellite (30564), g. d. (Scintilla Flower) by Booth's Kinsman (25658), gr. g. d.
(Sunflower) by Prince George (13510), — (Marigold Flower) by King Arthur
(13110), — by Gainford (7029), — by Rouge (5012), — by Cleveland (3404),
— by Burton (3250).

(43875)   **RED CAPTAIN,**
Red, calved February 16, 1879, bred by Mr. A. Stobo, Porterstown; got by
Captain Hardy (39549), dam (Baroness 2nd) by Marquis (26829), g. d. (Baroness)
by Baron of Kidsdale (11156), gr. g. d. (Duchess) by Whitaker's Comet (8771),
— (Meg) by Redkirk (11975), — by Playfellow (6297), — by Tryo (6620), —
by Memnon (4452), — by Duke of Wellington (231), — by a Bull from the
stock of Mr. Mason.

(43876)   **RED DUKE,**
Red and white, calved February 11, 1875, bred by Mr. T. Mace, Sherborne, the
property of Mr. R. H. Frank, Ashbourne Hall; got by Wild Duke (32864), dam
(Louisa) by Lord of the Hills (26705), g. d. (Sally) by Butter Boy (21349),
gr. g. d. (Louisa) by Royal Butterfly 14th (20722), — (Lucy Long) by Orange-
man (18485), — by General Pelissier (14605), — by Chieftain (12595), — by
Morpeth (8325), — by The Prince (7615), — by Colling (902), — by a Son of
Alexander (1624), — by a Grandson of Favourite (252).

(43877)   **RED DUKE 2ND,**
Red, calved December 3, 1879, bred by Mr. J. Todd, Mereside; got by Oxford
Waterloo (40434), dam (Red Duchess 9th) by Favonius (31143), g. d. (Red
Duchess 8th) by Grand Duke of Lancaster (19883), gr. g. d. (Red Duchess 6th)
by Oxford (20450), — (Red Duchess 5th) by Tortworth (10966), — by Harbinger (9183), — by Bachelor (1666), — by Wellington (683), — by Admiral
(4), — by Sir Harry (5155), — by Colonel (152), — by Grandson of Hubback
(319), — by Son of Hubback (319).

(43878)   **RED KING,**
Red, calved February 1, 1878, bred by Mr. A. Cameron, Artafallie; got by
King of Kessock (38495), dam (Golden Gauntlet) by Flower of the Forest
(33948), g. d. (Golden Rose) by Scotch Rose (25099), gr. g. d. (Golden Reed) by
Baronet (15614), — (Mint 5th) by Lord Raglan (13244), — by Duplicate Duke
(6952), — by Robin O'Day (4973), — by Sir Walter (2639), — by Jerry (4097),
— by Roseberry (567), — by Roseberry (567), — by Constellation (163), — by
Hastings (293), — by Hastings (293), — by Leopold (372).

(43879)   **RED KING,**
Roan, calved September 3, 1879, bred by Sir Pryse Pryse, Bart., Gogerddan;
got by Prince Rufus (40540), dam (Julia 4th) by Blair Athol (23426), g. d.
(Julia 3rd) by Fitz Sir James (19761), gr. g. d. (Julia Joan) by Cavalier (14250),
— (Joan) by Diamond (5918), — by Second Duke of Northumberland (3646),
— by Young Seagull (5100), — by Blucher (795), — by Sultan (1485), — by
Charles (6852), — by Yarborough (705).

(43880)   **RED KNIGHT,**
Red, calved April 28, 1879, bred by Mr. R. P. Maxwell, Finnebrogue; got by
Czarowitz 4th (33497), dam (Birthday 2nd) by Prince Victor (20606), g. d.
(Birthday) by Priam (18568), gr. g. d. (Honeysuckle) by Champion (12577), —
(Hybla) by Master Fairfax (13314), — by Sir Thomas Fairfax (5196), — by
Hubback (2142), — by Son of Young Warlaby (2812), — by Imperial (2151),
— supposed by Young Comet (905).

(43881)   **RED PRINCE,**
Red, calved March 5, 1878, bred by Mr. C. Mason, Dishforth; got by Cardinal

Windsor (37949), dam (Laura 10th) by Master Fred (31866), g. d. (Laura 4th)
by Prince Edward (32129), gr. g. d. (Laura 3rd) by Marc]Antony (14895), —
(Laura) by Maunby (7223), — by Lot (2223), — by Mark (2266), — by Meteor
(431).

(43882)   RED PRINCE,
Red, calved March 29, 1878, bred by Messrs. W. and H. Morley, East Gate
Farm; got by British Prince (36284), dam (Roan Princess) by White Duke
(32849), g. d. (Princess Patrick) by Prince Patrick (16760), gr. g. d. (Lilac) by
Abraham Parker (9856), — (Wellborn) by Bumper (10005), — by Tom Boy
(9750), — by Magician (7185), — by Vivian (5575), — by Belvedere 2nd
(3126), — by Alive 'O (2995), — by Eclipse (236), — by Charge's Grey Bull
(872), — by Paddock Bull (477), — by J. Brown's Red Bull (97).

(43883)   RED RIBY,
Red, calved March 19, 1877, bred by Mr. G. J. Day, Horsford Hall; got by
Young Robert Peel (37348), dam (Fanny Fern) by Lord Blithesome (29067),
g. d. (Duchess Windsor) by Windsor Lad (27823), gr. g. d. (Queen of the Mist)
by Volunteer (19087), — (Maid of the Mist) by Admiral (14064), — by Hero
(18055), — by Rubens (5027), — by Young Matchem (4422), — by Isaac
(1129), — by Young Pilot (4702), — by Pilot (496), — by Julius Cæsar
(1143).

(43884)   RED ROVER,
Red, calved November 1, 1879, bred by Mr. W. Sheraton, Broom House, the
property of Mr. T. J. Provis, The Grange, Ellesmere; got by The Sultan (40811),
dam (Miss Nelly Beetham) by Darius (30860), g. d. (Miss Beetham 3rd) by
Antonia's Oxford (32952), gr. g. d. (Miss Beetham 2nd) by Union Jack (27691),
— (Miss Beetham) by Banncret (19257), — by War Knight (27748), — by
Sinbad (27460), — by Harold (14670), — by Grandson of Emperor 2nd (11438),
— Harlsonia.

(43885)   REDWING,
Red, calved November 16, 1879, bred by Mr. J. How, Broughton; got by Regal
Booth (35252), dam (Fairy Princess) by Fairy Prince (36627), g. d. (Queen
Victoria) by King Victor (28986), gr. g. d. (Queen of the Valley) by Quis
(20618), — (May Queen 2nd) by Claxton (21433), — by Master Rembrandt
(16545), — by Horatio (10335), — by Lord George (10439), — by Orontes
(4623), — by Guardian (3947), — by Mercury (2301), — by Monarch (2324),
— by St. Albans (2584), — by Jupiter (342), — by Sir Oliver (605), — by
Trunnell (659), — by Favourite (252), — by Favourite (252), — by Dalton
Duke (188), — by R. Alcock's Bull (19), — by J. Smith's Bull (608), — by
Jolly's Bull (337).

(43886)   REFORMATION,
Roan, calved June 24, 1879, bred by Mr. G. J. Bell, The Nook, the property of
Mr. Winter, High Berries; got by Captain (37936), dam (Pride) by Keir But-
terfly 8th (28951), g. d. (Purity) by Rifleman (32304), gr. g. d. (Miss Pigot) by
Baron Cherry (19268), — (Lady Pigot) by Mac Turk (14872), — by Heir-at-
Law (13005), — by Baron of Ravensworth (7811), — by Hudibras (10339),
— by Baronet (1686), — by Spectator (2688), — by Fitz-Remus (2025), — by
Whitworth (695), — a Cow bought by Mr. Mason, of Chilton.

(43887) REGAL,

Red and white, calved March 11, 1879, bred by Mr. R. W. Gaussen, Brookmans Park; got by Duke of Worcester 5th (39797), dam (Ursula 25th) by Lord Lally (22161), g. d. (Ursula 15th) by Third Duke of Lancaster (19624), gr. g. d. (Ursula Booth) by Richard (15161), — (Ursula) by Usurer (9763), — by Prince Ernest (4818), — by A-la-Mode (725), — by Ratify (2481), — by Wellington (680), — by Favourite (252), — by Punch (531).

(43888) REGAL CHERRY,

Red and white, calved December 27, 1879, bred by Mr. G. Murton Tracy, Redlands; got by Hillhurst Cherry Duke (39999), dam (Cherry Queen 2nd) by Cherry Duke of Glo'ster (36348), g. d. (Cherry Queen) by Horsa (34188), gr. g. d. (Cherry Blanche) by Dairy Prince (17655), — (Sultana) by Sultan (15354), — by Cardinal Wiseman (12560), — by Son of Clementi (3399), — by Magnum Bonum (2243), — by Thorp (2757), — by Pirate (2430), — by Houghton (318), — by Marshal Blucher (416), — from the stock of Messrs. Wright and Charge.

(43889) REGAL CROWN,

Red and white, calved December 31, 1877, bred (by Lord Polwarth, Mertoun House; got by Royal Commander (29857), dam (Countess 2nd) by British Crown (21322), g. d. (Countess) by Sir Sam (25171), gr. g. d. (Calendula) by Majestic (13279), — (Calomel) by Hamlet (8126), — by Leonard (4210), — by Bucking ham (3239), — from the stock of Sir M. W. Ridley, Bart.

(43890) REGAL LIND,

Roan, calved February 2, 1879, bred by Mr. B. Hannan, Riverstown; got by St. Ronan (35458), dam (Jenny Lind 15th) by Abercorn (25484), g. d. (Jenny Lind 7th) by Prince of Lurg (22617), gr. g. d. (Jenny Lind 4th) by Hohenlohe (18074), — (Jenny Lind 2nd) by Napoleon (18438), — by Druid (10140), — by Belted Will (6780), — by Regent (2517), — by Borderer (3191), — by Eclipse (1949), — by Togston (5487), — by Bolingbroke (3184), — by Son of Midas (435), — by Twin Brother to Ben (660).

(43891) REGAL ROAN,

Roan, calved June 1, 1878, bred by Miss Milne, Otterburn, the property of Mr. J. W. Annett, Houndalee; got by Vice Regal Booth (39258), dam (Prince Regent's Rosebud) by Prince Regent (29676), g. d. (Prowler Rosebud) by Prowler (22662), gr. g. d. (Stanley Rosebud) by Lord Stanley (18277), — (Primrose) by Sam Glen (10780), — by Peter Plough (10606), — by Eildon (9076), — by Ethelred (5990), — by Emperor (1974), — by Barmpton (1677), — by St. Albans (2584), — by Simon (590), — by Pope (514).

(43892) REGAL RUFUS,

Red, calved June 16, 1878, bred by Mr. H. P. Baxter, Southall Green Farm; got by King Rufus (34351), dam (Diana) by Spring Buck (32585), g. d. (Diadem) by Prince of Saxe Coburg (20576), gr. g. d. (Dignity) by Prince Alfred (13494), — (Dairymaid) by Lord Foppington (10437), — by Goldsmith (10277), — by Brilliant (8905).

(43893) REGULATOR,

Red and white, calved February 18, 1879, bred by Mr. J. Brigham, Slingsby; got by Bastion (37836), dam (Lady Mason 1st) by Brigadier (28079), g. d. (Veronica) by Lord Waterloo (24475), gr. g. d. (Verona) by Veteran (13491), —

(Venus) by Lord Mayor (14828), — by Forerunner (12891), — by Usurer (9763), — by Zenith (5702), — by Sweet William (5368), — by Roman (2561), — by Monarch (2324), — by Jupiter (342), — by Sir Oliver (605), — by Trunnell (659), — by Favourite (252), — by Favourite (252), — by Dalton Duke (188), — by R. Alcock's Bull (19), — by J. Smith's Bull (608), — by Jolly's Bull (337).

(43894)　　　　　REIGNING MONARCH,
Roan, calved December 4, 1879, bred by Mr. J. B. Booth, Killerby; got by King James (28971), dam (Regalia) by K. C. B. (26492), g. d. (Crown Diamond) by Royal Sovereign (22802), gr. g. d. (Creusa) by Valasco (15443), — (Hecuba) by Hopewell (10332), — by Hamlet (8126), — by Leonard (4210).

(43895)　　　　　REINDEER,
Roan, calved October 22, 1878, bred by Mr. J. G. Attwater, Britford, the property of Archdeacon Sanctuary, Powerstock; got by Lord Louis (38620), dam (Wild Chance) by Duke of Ozleworth (31008), g. d. (Chance 4th) by Prince Pearl 3rd (20599), gr. g. d. (Chance 2nd) by Dreadnought (12718), — (Chance) by Cramer 2nd (10084), — by Duke of Richmond (10163), — by Anthony (1640), — by Sportsman (5304), — by Anthony (1640), — by Edgcott (1953), — by Audley (3055), — by Son of Merlin (6522).

(43896)　　　　　REMUS,
Red, calved October 22, 1878, bred by Messrs. T. and R. Theakston, Masham; got by Star Regent (35697), dam (Rosebud) by Sandringham (32452), g. d. (Rosetta) by Lord of the Hills (18267), gr. g. d. (Rose) by Baron Barton (14123), — (White Rose) by Baron Farnley (14129), — by Banker (11136), — by Ravensworth (9532), — by Nelson (4549), — by Newton (4567), — by Wonderful (700), — by Cleveland (145), — by Butterfly (104), — by Hollon's Bull (313), — from the stock of Sir J. Pennyman.

(43897)　　　　　REMUS 3RD,
White, calved February 19, 1879, bred by Mr. E. Robinson, Nafferton; got by Halloo (34105), dam (Gertrude 2nd) by Rose Duke 2nd (32341), g. d. (Lady Fanny) by Remus (18693), gr. g. d. (Gem) by Prince Tom (15110), — (Lady Griselda) by Braithwaite Laddie (12488), — by Monarch (9406), — by Young Bellerophon (7824), — by Belshazzar (1703), — by Bachelor (1666), — by Adonis (2933), — by Leopold (2199), or Hector (2103), — by Sir Harry (5155).

(43898)　　　　　RENOWN,
Roan, calved March 9, 1879, bred by Mr. G. Low, Birtown; got by Great Fame (34087), dam (Prude) by Escape (28556), g. d. (Phœbe) by Lord Stanley (24466), gr. g. d. (Patty) by The Bushel (17092), — (Patricia) by Sir Charles Napier (13708), — by Tom Boy (10961), — by Volcano (21044), — by Municipal Bill (2344), — by Bonaparte (19327), — bought of Mr. De Renzi, and descended from the stock of Mr. R. Holmes.

(43899)　　　　　RETRIEVER,
Red and white, calved November 14, 1874, bred by Sir John Leslie, Bart., Glaslough, the property of Mr. W. Bradford, Donaghmore; got by Dunbar (33766), dam (Gipsy 2nd) by Second Earl of Rosedale (26073), g. d. (Gipsy) by Best Hope (23413), gr. g. d. (Banshee) by Lumley (16478), — (Baroness) by Master Goldsmidt (13315), — by Vandyke (7669), — by Carcase (3285), — by Tyro

(2781), — by Falstaff (1993), — by Dr. Syntax (220), — by Charles (127), — by Henry (301), — by Favourite (252), — by White Bull (421), — by Bolingbroke (86), — by Foljambe (263), — by Hubback (319), — bred by Mr. Maynard.

(43900) RIBY EARL,
Roan, calved April 15, 1878, bred by Messrs. J. W. and E. Cruickshank, Lethenty; got by Knight of St. Patrick (38520), dam (Riby Princess) by Knight of the Shire (26552), g. d. (Riby Empress) by Duke of York (23804), gr. g. d. (Riby Peeress) by Breast Plate (19337), — (Riby Queen) by Booth Royal (15673), — by Vanguard (10994), — by Fanatic (8054), — by Auld Robin Grey (6753), — by Scrip (2604), — by Burley (1766), — by Isaac (1129), — by Pilot (496), — by Albion (14), — by Lame Bull (359), — by Shipton (587), — by Son of Suworrow (636), — by Son of Twin Brother to Ben (88), — by Twin Brother to Ben (660).

(43901) RIGEL,
Red, calved November 22, 1878, bred by Mr. J. C. Daubuz, Killiow; got by M.C. (31898), dam (Moss Rose) by Young Hector (28830), g. d. (White Rose) by Knight of Branches (20076), gr. g. d. (Dairy Maid) by Scimitar (16917), — (Deborah) by Protection (11956), — by Hector (9200), — by Nelson (4547), — by Milton (8315), — by Merlin (2302), — by Midas (435), — by Denton (198).

(43902) RINGLEADER,
Red and white, calved December 30, 1879, bred by Mr. T. Purkis, West Wratting Grange; got by Cambridge Duke 5th (30644), dam (Foggathorpe 2nd) by Liston (22105), g. d. (Lady Lind) by Baron Lind (17374), gr. g. d. (First Love) by Norman (13394), — (Lady Marguerite) by Third Duke of Oxford (9047), — by Sheldon (8557), — by Benjamin (1710), — by Malbro' (1189), — by Ebor (997), — by Regent (546), — by North Star (459), — by R. Colling's White Bull (151), — bred by Mr. R. Colling.

(43903) RINGWOOD,
Roan, calved March 10, 1879, bred by Sir R. J. Paul, Bart., Ballyglan; got by Red-Wood (42265), dam (Satchel) by Troubadour (32736), g. d. (Sable) by Glory (24043), gr. g. d. (Sarcenet) by Mainsheet (20262), — (Satinet) by Little Pat (18203), — by Acrobat (14058), — by Sovereign (15323), — by Sir Charles (7501), — Silkworm.

(43904) ROAN ROBIN,
Roan, calved August 12, 1878, bred by Mr. C. Mason, Dishforth; got by Lord Mirth (34602), dam (Lady Pink) by Master Frank (31865), g. d. (Lady White) by Marc Antony (14895), gr. g. d. (Miss Frances Smith) by Aaron Smith (12331), — (Frances) by Hercules (11572), — by Marton (11794), — by Pioneer (4711), — by Clark's Son of Norfolk (2377), — by Red Simon (2499), — by Imperial (2151), — by Young Favourite (2007), — by Snowball (2648).

(43905) ROAN SEAL,
Roan, calved February 6, 1879, bred by Mr. H. P. Baxter, Southall Green Farm, the property of Mr. W. G. Garne, Broadmoor; got by Rose Seal (39023), dam (Lady Rose) by Ranger (35023), g. d. (Rose of Clitheroe) by Cynric (19542), gr. g. d. (Rosebud) by General Pelissier (14605), — (Moss Rose) by Marchmont (9367), — by Fitz-Hardinge (8073), — by Augustus (6751), — by Son of Anthony (1640), — by a Bull of Mr. Champion's, of Blyth.

(43906)   ROBERT BRUCE,
Red and white, calved February 2, 1879, bred by Mr. C. W. Schroeter, Tedfold;
got by Prince Alfred (37215), dam (Lady Belle) by Baron Killerby (27949),
g. d. (Lady Britomart) by Britomart (23476), gr. g. d. (Golden Red Rose) by Sir
Christopher (22895), — (Golden Rose) by Wyville (14033), — by Newby (13380),
— by Lord Cardigan (7158), — by Tom Fairfax (6610), — by Clementi (3399),
— by Rockingham (2551), — by Priam (2452).

(43907)   ROBESPIERRE,
Red and white, calved April 5, 1879, bred by Mr. H. J. Marshall, Poulton Priory;
got by The Barrister (35740), dam (Red Lass) by Marksman (26814), g. d. (Red
Lady) by Experience (23900), gr. g. d. (Red Bess) by Economist (15977), —
(Red Rose) by Napier (13368), — by Norval (9450), — by Rob Roy (7434), —
by Gazer (7030).

(43908)   ROBIN,
Red, calved March 7, 1878, bred by Mr. W. Ray, Sunbank; got by Baron Outh-
waite (36197), dam (Duchess 18th) by Baron Colling (25560), g. d. (Duchess 6th)
by Michigan (24594), gr. g. d. (Duchess 3rd) by Prince Arthur (16723), —
(Duchess 2nd) by Magnum Bonum (13277), — by Bloomsbury (9972), — by
Mons. Vestris (6220), — by Second Duke of Northumberland (3646), — by
Sultan Selim (2710), — by Young Wynyard (2859).

(43909)   ROBIN,
Red, calved April 23, 1878, bred by Mr. W. Hewer, Sevenhampton; got by
British Prince (37907), dam (Red Lass) by Marksman (26814), g. d. (Red Lady)
by Experience (23900), gr. g. d. (Red Bess) by Economist (15977), — (Red Rose)
by Napier (13368), — by Norval (9450), — by Rob Roy (7434), — by Gazer
(7030).

(43910)   ROBIN ADAIR,
Roan, calved March 28, 1878, bred by Mr. G. Stewart, Craigniston, the property
of Mr. J. Smith, Balmain; got by The Scotsman (39211), dam (Rosedale) by
Honeycomb (28866), g. d. (Rosette) by Allan (21172), gr. g. d. (Ring Dove) by
Lord Raglan (13244), — (Roseberry) by Somerset (10858), — by Guy Faux
(7062), — by Pedestrian (7321), — by Lucien (2288), — by Raby (2473).

(43911)   ROBIN ADAIR,
Roan, calved January 1, 1879, bred by the Exors. of Mr. J. Morton, Skelsmergh
Hall, the property of Mr. Dove, Fisher Beck; got by Duke of Barrington 2nd
(36463), dam (Ruth) by Atherton's Oxford (21195), g. d. (Robina) by Coxcomb
(19527), gr. g. d. (Rowena) by Duke of Moscow (14447), — (Ruby) by Red
Rover (11982), — by Sir Isaac (9645), — by Victory (5566), — by Matchem
3rd (4420), — by Young Eryholme (1981), — by Wellington (2825), — by Alfred
(23), — by Windsor (698), — by Cupid (177), — by Barker's Bull of Layton
(53).

(43912)   ROBIN ADAIR,
Roan, calved January 26, 1879, bred by Mr. R. Thompson, Inglewood Bank;
got by Brilliant Butterfly (36270), dam (Lady Locksly) by Grand Duke of
Fawsley 3rd (31286), g. d. (Lady Eglinton) by Earl of Eglinton (23832), gr. g. d.
(Lively) by Master Annandale (14916), — (Sprightly) by Tweedside (12246),
— by Strathmore (6547), — by Sir William (12102), — by Emperor (1974).

(43913) ROB ROY,
Roan, calved November 19, 1878, bred by Mr. W. Parkin, Sen., Southwaite Mill, the property of Mr. W. Parkin, Jun., Southwaite; got by Glenluce (36700), dam (Iole 5th) by Flash (33943), g. d. (Iole 2nd) by Grand Duke of Lightburne 2nd (26291), gr. g. d. (Iole) by Hogarth (13036), — (White Rose) by Avalanche (12419), — by Tybalt (19024), — by Stentor (9691), — by Charles (17535), — by Sorcerer (13771), — by Waverley (677), — by Western Comet (689), — by Harlequin (289), — by Sidney (640), — by Traveller (655), — by Bolingbroke (86).

(43914) ROCKET,
Red and white, calved September 25, 1878, bred by Messrs. T. and J. McElderry, Ballymoney, the property of Mr. John McElderry, Ballymoney; got by Robert Burns (29795), dam (Pelerine) by British Hero (30604), g. d. (Elfleda) by Blood Royal (23429), gr. g. d. (Mantalini) by Master Rembrandt (16545), — (Elfleda) by Gainford 4th (11501), — by Stars and Stripes (12148), — by Sir Henry (10824), — by Buckingham (3239), — by Marcus (2262), — by Matchem (2281), — by Alderman (1622), — by Pilot (496), — by Remus (550), — by Sir Charles (592), — by Son of Favourite (252), — by Son of Favourite (252), — Strawberry.

(43915) ROLAND,
Roan, calved December 8, 1878, bred by Mr. J. Turner, Ulceby; got by Grand Duke of Glo'ster (36721), dam (Rosette 3rd) by Grand Duke 6th (19876), g. d. (Rosette 2nd) by Royal Butterfly 3rd (18754), gr. g. d. (Rosette) by Agamemnon (17279), — (Rose of Summer) by Rembrandt (13587), — by Valiant (10989), — by Duke of Cornwall (5947), — by Fergus (3782), — by Dandy (1902), — by Blyth (797), — by Midas (435), — by Boughton (90), — by Windsor (698), — by R. Colling's Son of Favourite (252).

(43916) ROMPING JACK,
White, calved February 8, 1879, bred by Mr. J. Kerfoot, Faenol Bach; got by Edward Waterloo (38246), dam (Romping Girl) by Hotspur (28878), g. d. (Reckless) by Mandarin (26799), gr. g. d. (Rispah) by Duke of Grafton (21594), — (Rivulet) by Carolus (14246), — by King Arthur (13110), — by Preston (8408), — by South Star (7538), — by Van Amburgh (5543), — by Plenipo (4724), — by Young Matchem (4425), — by Sir Thomas (2636), — by Marske (418), — by Comet (155), — by Tom (652), — by Favourite (1033), — by Hutton's Bull (323), — by Barningham (56).

(43917) ROSAMOND'S BEAU,
Roan, calved March 27, 1878, bred by Mr. W. G. Garne, Broadmoor, the property of Mr. P. H. Couper, Bulwell Hall; got by Skylark (37489), dam (Rosamond Beauty) by Buccaneer (25693), g. d. (Rosamond) by A. A. (23253), gr. g. d. (Rosette) by Duke of Towneley (21615), — (Rosary) by Royal Oak (16870), — by Bashaw (12449), — by Marchmont (9367), — by Fitz-Hardinge (8073), — by Augustus (6751), — by Son of Anthony (1640), — by a Bull of Mr. Champion's, of Blyth.

(43918) ROSARIO,
Roan, calved April 4, 1879, bred by Mr. C. Cradock, Hartforth; got by Rose King (39019), dam (Bella) by Wanderer (32790), g. d. (Donna Bella) by Straw-

berry Prince (25240), gr. g. d. (Donna) by Valasco (15443), — (Roaned Ciss) by
Rifleman (15163), — by Young Hopewell (14719), — by Bellemont (11164), —
by Dan O'Connell (3557), — by Lord Lieutenant (4260), — by Outhwaite's
White Bull (9825), — by North Star (459), — by Young Comet (157), — by
Easby (232).

(43919) ROSEBERRY ROAN,
Roan, calved January 30, 1873, bred by Mr. T. Walker, Berkswell Hall, the
property of Mr. J. Woollaston, Berkswell; got by Cremorne (30818), dam
(Charmer 4th) by Grand Duke 9th (19879), g. d. (Charmer 3rd) by Grand Duke
4th (19874), gr. g. d. (Charmer) by Mainstay (16490), — (Sunset) by Mame-
luke (13289), — by Daybreak (11338), — by Accordion (5708), — by Little
John (4232), — by Caliph (1774), — by Sir Walter (2637), — by Hotspur
(1117), — by Coxcomb (928), — by Midas (435), — by Comet (155), — by R.
Colling's Son of Favourite (252), — by same Son of Favourite (252), —
Hubback (319).

(43920) ROSE DUKE 9TH,
Roan, calved July 6, 1878, bred by Lord Mowbray and Stourton, Stourton, the
property of Mr. J. Key, Musley Bank; got by Lord Oxford Bright Eyes (34648),
dam (Juno 21st) by Sandown (35470), g. d. (Juno 16th) by Golden Horn (31267),
gr. g. d. (Juno 12th) by Rose Duke (22760), — (Juno 11th) by Hesperus
(18062), — by Abbot (14054), — by Sting (15345), — by Sparta (13773), — by
Calcetto Bianco (10014), — by Chancellor (1809), — by Umpire (5530), — by
Northern Light (4586), — by Young Alfred (2993), — by Lemon Horns (2196),
— by Witham's Bull.

(43921) ROSEWOOD,
Red, calved February 28, 1879, bred by Sir R. J. Paul, Bart., Ballyglan; got by
Red-Wood (42265), dam (Reprisal) by Pirate (37191), g. d. (Reine) by Trouba-
dour (32736), gr. g. d. (Royalty) by Glory (24043), — (Regalia) by Lord Clyde
(18219), — by Prince Ernest 2nd (10644), — by Forrest (10240), — by Second
Comet (5101), — by Prince Ernest (7366), — by Prince George (2464), — by
Madcap (7183).

(43922) ROSSINI,
Red and white, calved February 20, 1879, bred by Mr. T. M. Hopkins, Lower
Wick Farm; got by Scarlatti (39085), dam (Rock Rose 56th) by Roan Prince
(32310), g. d. (Rock Rose 49th) by Lord Greta (20174), gr. g. d. (Rock Rose
26th) by Earl of Wentworth (14481), — (Rock Rose 7th) by Duke of Kent
(12756), — by General Bem (11516), — by Boccaccio (7838), — by Belshazzar
(1703), — by Velocipede (5552), — by Matchem (2281), — by Sir Stephen (1456),
— by Newton (1271), — by Son of Comet (155), — by Western Comet (689), —
by Charge's Grey Bull (872), — by Favourite (252).

(43923) ROSY'S GRAND DUKE,
Roan, calved July 18, 1878, bred by Sir G. R. Philips, Bart., Weston Park; got
by Grand Duke 29th (08372), dam (Polyanthus) by Third Duke of Geneva
(21592), g. d. (Polly) by Stepping Stone (22978), gr. g. d. (Polytint) by Earl of
Dublin (10178), — (Cornbind) by Janizary (8175), — by Snowball (8602), —
by Little John (4232), — by Caliph (1774), — by Rob Roy (557), — by Satellite
(1420), — by Sir Dimple (594), — by Styford (629).

(43924)   **ROVER,**

Roan, calved May 12, 1879, bred by Mr. J. Stratton, Alton Priors; got by Proteus (40552), dam (Rosebud) by Royal (35331), g. d. (May Rose 2nd) by Bude Light (21342), gr. g. d. (May Rose) by Young Windsor (17241), — (Essence of Roses) by His Highness (14708), — by Waterloo (11025), — by Lottery (4280), — by Lottery (4280), — by Phœnix (6290).

(43925)   **ROVER,**

Roan, calved June 13, 1879, bred by Mr. W. Bolton, The Island; got by Albion (36112), dam (Roseleaf) by Lieutenant General (31600), g. d. (Rosebud) by Manrico (26805), gr. g. d. (Rosanna) by Sir James (16980), — (Rosalie) by Uncle Ned (19026), — by Crusade (7938), — by Cotherstone (6903), — by Young Matchem (4425), — by Fairfax (1023), — by Son of Marske (418), — by Palmsun (7311), — by a Bull of Mr. Maynard's, — by Son of Favourite.

(43926)   **ROWFANT DUKE OF OXFORD,**

Red and white, calved January 16, 1879, bred by Sir Curtis M. Lampson, Bart., Rowfant, the property of Mr. J. J. D. Jefferson; got by Duke of Underley 2nd (36551), dam (Grand Duchess of Oxford 11th) by Grand Duke 10th (21848), g. d. (Grand Duchess of Oxford 5th) by Priam (18567), gr. g. d. (Countess of Oxford) by Earl of Warwick (11412), — (Oxford 15th) by Fourth Duke of York (10167), — by Second Duke of Northumberland (3646), — by Short Tail (2621), — by Matchem (2281), — by Young Wynyard (2859).

(43927)   **ROWFANT DUKE OF OXFORD 2ND,**

Red, calved May 25, 1879, bred by Sir Curtis M. Lampson, Bart., Rowfant; got by Duke of Underley 2nd (36551), dam (Rowfant Oxford) by Third Duke of Hillhurst (30975), g. d. (Grand Duchess of Oxford 11th) by Grand Duke 10th (21848), gr. g. d. (Grand Duchess of Oxford 5th) by Priam (18567), — (Countess of Oxford) by Earl of Warwick (11412), — by Fourth Duke of York (10167), — by Second Duke of Northumberland (3646), — by Short Tail (2621), — by Matchem (2281), — by Young Wynyard (2859).

(43928)   **ROWLAND,**

Roan, calved September 18, 1879, bred by Colonel R. Loyd Lindsay, Lockinge Park; got by Earl of Horton 11th (36588), dam (Ronda) by Rob Roy (29806), g. d. (Rosetta) by Costa (21487), gr. g. d. (Rosette) by Prince of Prussia (16752), — (Red Rose) by Horatio (10335), — by Third Duke of Northumberland (3647), — by Velocipede (5552), — by Sir Thomas (2636), — by Marske (418), — by Comet (155), — by Tom (652), — by Favourite (1033), — by Hutton's Bull (323), — by Barningham (56).

(43929)   **ROYAL,**

Roan, calved February 12, 1879, bred by Mr. R. Reynell, Killynon; got by Royal Baron (40617), dam (Woodbine) by Lieutenant General (31600), g. d. (Hawthorn) by Royal Prince (27384), gr. g. d. (Sweetbrier) by Nimrod (13388) — (Charlotte) by Selim (6454), — by Rex (6385), — by Sir Thomas Fairfax (5196), — by Ambo (1636), — by Memnon (2295), — by Pilot (496), — by Agamemnon (9), — by Burrell's Bull, of Burdon.

(43930)   **ROYAL ALPHONSO,**

Roan, calved August 23, 1878, bred by Mr. R. Chaloner, King's Fort, the property of Mr. H. Bruen, Oak Park; got by Royal Arthur (37372), dam

(Irish Sweetheart) by Patrician (24728), g. d. (Sweetheart 16th) by Duke
of Albemarle (21560), gr. g. d. (Sweetheart 8th) by The Baron (13833), —
(Sweetheart 6th) by Mameluke (13289), — by Daybreak (11338), — by
Accordion (5708), — by Little John (4232), — by Caliph (1774), — by Sir
Walter (2637), — by Hotspur (1117), — by Coxcomb (928), — by Midas (435),
— by Comet (155), — by R. Colling's Son of Favourite (252), — by same Son
of Favourite (252), — by Hubback (319).

(43931)   ROYAL ALPINE,
Roan, calved March 12, 1879, bred by Mr. B. St. John Ackers, Prinknash Park;
got by King William (34358), dam (Alpine 2nd) by Lieutenant General (31600),
g. d. (Alpine) by British Prince (14197), gr. g. d. (Sylph) by Hopewell (10332),
— (Ophelia) by Hamlet (8126), — by Albion (7771), — by Lord Stanley (4269),
— by Marcus (2262), — by Matchem (2281), — by Alderman (1622), — by
Pilot (496), — by Remus (550), — by Sir Charles (592), — by R. Colling's Son
f Favourite (252), — by Son of Favourite (252), — Strawberry.

(43932)   ROYAL ARCH,
ed and white, calved July 12, 1879, bred by Mrs. Pery, Coolcronan House;
got by Royal Crown (40628), dam (Lady Love) by The Earl (27623), g. d. (Lady
Sarah) by Best Hope (23413), gr. g. d. (Marion) by Duke of Leinster (17724),
— (Violet) by Baron Warlaby (7813), — by Burley Fairfax (6822), — by
Monarch (4495), — by Invalid (4076), — by Satellite (1420), — by Cato (119),
— by Jupiter (342), — by George (273), — by Chilton (136), — by Irishman
(329), — by B. (45).

(43933)   ROYAL BALMORAL,
Roan, calved February 19, 1879, bred by Sir R. Sutton, Bart., Benham Park;
got by Lord Balmoral (38576), dam (Flutter) by Gang Forward (33988), g. d.
(Flavia) by Fashion (21724), gr. g. d. (Lady Elma) by Lord Elgin (20170), —
(Ruth) by Dipple (14401), — by General (19835).

(43934)   ROYAL BENEDICTION,
Red, calved July 22, 1878, bred by Mr. J. F. Bomford, Drumlargan; got by
Field Marshal (39873), dam (Red Breast) by Royal Benedict (27348), g. d.
(Rosette) by Duke of Towneley (21615), gr. g. d. (Rosary) by Royal Oak
(16870), — (Rosy) by Bashaw (12449), — by Marchmont (9367), — by
Fitz-Hardinge (8073), — by Augustus (6751), — by Son of Anthony (1640),
— by a Bull of Mr. Champion's, of Blyth.

(43935)   ROYAL BOY,
Red and white, calved May 4, 1879, bred by Mr. G. W. Lambart, Beau Parc;
got by Jupiter (38477), dam (Royal Maid) by Rupert (29902), g. d. (Red Flag)
by Royal Standard (40644), gr. g. d. (Heiress) by Heir of Windsor (16255), —
(Lady Ranelagh) by Tom Jones (40825), — by Cato (12566), — by Narcissus
(4540).

(43936)   ROYAL CARLISLE,
White, calved July 26, 1879, bred by Mr. J. Wright, Green Gill Head; got by
Game Bird (39912), dam (Countess) by Luck Estate (34713), g. d. (Ada) by
Edwin (21671), gr. g. d. (General's Daughter) by General Haynau (11520), —
(Red Rose 6th) by Gainford 2nd (10255), — by Emperor (9084), — by Sir
Thomas Newton (6495), — by General (3867).

(43937)   ROYAL CHERRY,
White, calved May 18, 1879, bred by Mr. G. Murton Tracy, Redlands; got by
King of the Weald (38499), dam (Cherry Empress) by Cherry Prince 4th (25765),
g. d. (Cherry Blanche) by Dairy Prince (17655), gr. g. d. (Sultana) by Sultan
(15354), — (Peggy) by Cardinal Wiseman (12560), — by Son of Clementi
(3399), — by Magnum Bonum (2243), — by Thorp (2757), — by Pirate
(2430), — by Houghton (318), — by Marshal Blucher (416), — from the stock
of Messrs. Wright and Charge.

(43938)   ROYAL CLARENCE,
Roan, calved March 15, 1879, bred by Mr. A. T. Matthews, Church Hanborough;
got by Lord Fitzclarence 16th (36943), dam (Regalia) by Pompey (35059), g. d.
(Red Empress) by Second Duke of Wateringbury (26024), gr. g. d. (Revenge) by
Second Baron Wetherby (21243), — (Revision) by Fairleigh (19714), — by
Dukedom (12729), — by Monarch (13347), — by Rodolph (9569), — by
Beckingham (6772), — by Jenner (4095), — by Son of Ormsby (4621).

(43939)   ROYAL COMMISSIONER,
Roan, calved January 21, 1879, bred by the Duke of Northumberland, Alnwick
Castle; got by Sir Raymond (40716), dam (Rose of Tyne) by Duke of Tyne
(33744), g. d. (Marion) by Vanguard (23115), gr. g. d. (Gipsy) by President
(20510), — (Garland) by Highflyer (11576), — by Phœnix (10608), — by
Timothy (12225), — by Young Alive 'O (2996), — by Belvedere 2nd (3126), —
by Alive 'O (2995).

(43940)   ROYAL DUKE,
Roan, calved March 3, 1878, bred by Mr. G. Marr, Cairnbrogie, the property of
Mr. Gavin Catto, Mains of Gight; got by Sweet-Pea (35708), dam (Dainty 8th)
by Young Hero (26385), g. d. (Dainty 7th) by Grand Prince (26308), gr. g. d.
(Dainty 4th) by Baron Sebastopol (21241), — (Dainty 3rd) by Fourth Duke of
Bolton (15915), — by Fairfax Hero (9106), — by Second Duke of Northumber-
land (3646), — by Emperor (3716), — by Invalid (4076), — by Magnet (392),
— by Palmflower (480).

(43941)   ROYAL DUKE,
Roan, calved April 17, 1878, bred by Mr. W. Johnson, Little Kelk; got by
Prince Butterfly (42169), dam (Gem 9th) by Neville (34903), g. d. (Gem 8th)
by Soldier Laddie (22941), gr. g. d. (Gem 7th) by Effinel (14489), — (Gem 6th)
by Orthodox 5th (13435), — by Young Lord Adolphus (7153), — by Brigadier
(5807), — by Orthodox 1st (4624), — by Vulpes (1558), — by Edmund (1954).

(43942)   ROYAL DUKE,
Roan, calved July 18, 1878, bred by Mr. W. Lambe, Aubourn; got by Duke of
Charming Land 2nd (36477), dam (Romance) by Lord of the Manor (29178),
g. d. (Rose of York) by Sir Christopher (22895), gr. g. d. by Prince of Delhi
(22611), — by Marmion (20285), — by North Star (15008), — by White
Knight (14001), — by Frederick (10243), — by Prince Comet (9505), — by
Sovereign (18902), — by Young Favourite (17836), — by Crispin (174).

(43943)   ROYAL DUKE,
Roan, calved February 28, 1879, bred by Lord Fitzhardinge, Berkeley Castle,
the property of Mr. F. E. Pope, Great Toller; got by Duke of Connaught
(33604), dam (Rose Niblett) by Grand Duke of Geneva 2nd (31288), g. d.

(Rugia Niblett) by Royal Butterfly 20th (25007), gr. g. d. (Ruth Niblett) by Second Duke of Jamaica (25977), — (Rebecca Niblett) by Cynric (19542), — by Nauplius (16607), — by British Boy (11206), — by Actæon (14059), — by Dædalus (3539), — by Chilo (1825),. — by Nero (1265), — by Harold Junior (1096), — by Adonis (1612), — by R. Colling's Son of Phenomenon (491), — by Punch (531).

(43944)   **ROYAL DUKE 7TH,**
Red and white, calved April 13, 1878, bred by Mr. C. A. Barnes, Solesbridge ; got by Lord of Lochaber 2nd (36983), dam (Royal Duchess 5th) by Barrington Duke (27985), g. d. (Royal Duchess) by Bodenbach (23433), gr. g. d. (Lady Love) by Cock of the Walk (15782), — (Looey) by Marmaduke (14897), — by Count Glo'ster (12650), — by Fourth Duke of York (10167), — by Cramer (6907), — by Cato (6836), — by Helicon (2107), — by Matchem (2281), — by Sir Alexander (591), — by Sir Stephen (1456), — by Western Comet (689), — by Charge's Grey Bull (872), — by Favourite (252), — by Bartle (777), — descended from the Studley White Bull (627).

(43945)   **ROYAL DUKE 8TH,**
Roan, calved April 19, 1879, bred by Mr. C. A. Barnes, Solesbridge ; got by Lord of Lochaber 2nd (36983), dam (Royal Duchess 5th) by Barrington Duke (27985), g. d. (Royal Duchess) by Bodenbach (23433), gr. g..d. (Lady Love) by Cock of the Walk (15782), — (Looey) by Marmaduke (14897), — by Count Glo'ster (12650), — by Fourth Duke of York (10167), — by Cramer (6907), — by Cato (6836), — by Helicon (2107), — by Matchem (2281), — by Sir Alexander (591), — by Sir Stephen (1456), — by Western Comet (689), — by Charge's Grey Bull (872), — by Favourite (252), — by Bartle (777), — descended from the Studley White Bull (627).

(43946)   **ROYAL FARMER,**
Red and white, calved July 22, 1879, bred by Mr. T. H. Miller, Singleton Park ; got by Frederick the Great (36665), dam (Ringlet 4th) by White Duke (32849), g. d. (Ringlet 2nd) by Bywell Victor (21353), gr. g. d. (Ringlet) by Lord of the Valley (14837), — (Rose Duchess) by Red Duke (13571), — by Vanguard (10994), — by Diamond (5918), — by Young Matchem (2282), — by Jack Tar (1133), — by Pilot (496), — by Young Albion (15).

(43947)   **ROYAL FRANK,**
Red and white, calved April 28, 1879, bred by Mr. H. Chandos-Pole-Gell, Hopton Hall ; got by Royal Saxon (39057), dam (Princess Franziska) by Prince Christian (22581), g. d. (Leopoldine 4th) by King Charles (24240), gr. g. d. (Leopoldine 3rd) by Killerby Lad (20052), — (Leopoldine 2nd) by Prince Patrick (16760), — by Bumper.(10005), — by Second Colonel (8544), — by Sir Walter (2639), — by Marquis (2270), — by Wellington (2824), — by Leopold (2199), — by Dundas (1943), — by Cupid (177), — by Simon (590), — by Punch (531), — by Bolingbroke (86).

(43948)   **ROYAL GENERAL,**
Red and white, calved May 7, 1879, bred by the Duke of Richmond and Gordon, Gordon Castle ; got by Royal Hope (32392), dam (Queen Esther) by Squire Booth (30049), g. d. (Queen of the Valley) by Sir Christopher (22895), gr. g. d. (Queen Mary) by Parmesan (20471), — (Queen Bess 4th) by King

Arthur (13110), — by Wellington (13989), — by Hamlet (8126), — by Pam (4643), — by Young Matchem (2282), — by Jack Tar (1133), — by Pilot (496), — by Young Albion (15).

(43949) ROYAL GEORGE,

Roan, calved July 16, 1879, bred by Mr. J. Close, Holmescales ; got by Royal Victor (35414), dam (Vacillation) by K. C. B. (26492), g. d. (Vexation) by Knight Errant (18154), gr. g. d. (Vanity) by Valasco (15443), — (Hopeful) by Hopewell (10332), — by Warrior (12287).

(43950) ROYAL HOPE,

Roan, calved December 20, 1878, bred by Mr. M. Savidge, Sarsden Lodge Farm ; got by Royal Blithesome (40620), dam (Faith) by British Hope (21324), g. d. (Charity) by Lord of the Harem (16430), gr. g. d. (Cherry Pie) by Fitz-Milton (14554), — (Red Cherry) by Prince of Wales (8432), — by Bucephalus (6816), — by Locksley (4240), — by Stanhope (5315), — a Cow bought in the North.

(43951) ROYAL HUBBACK,

Roan, calved February 20, 1879, bred by Mr. J. Woodhouse, Scale Hall ; got by Fennel Duke (36639), dam (Marie Stuart) by Second Hubback (28880), g. d. (Fancy 16th) by Festus 3rd (26148), gr. g. d. (Fancy 13th) by Festus 2nd (21738), — (Fancy 9th) by Festus (16040), — by Coriander (10072), — by Biscay (7834), — by Austerlitz (3063), — by Norfolk (2377), — by Bright (1739), — by Shylock (2622), — by Percy (1314), — by Blucher (84), — by Jupiter (344), — by Marshal Beresford (415), — by Crocus (932), — by a Bull of Mr. R. Colling's.

(43952) ROYALIST,

Red and little white, calved May 30, 1879, bred by Mr. R. Reynell, Killynon ; got by Royal Baron (40617), dam (Sweetbrier) by King James (28971), g. d. (Hawthorn) by Royal Prince (27384), gr. g. d. (Sweetbrier) by Nimrod (13388), — (Charlotte) by Selim (6454), — by Rex (6385), — by Sir Thomas Fairfax (5196), — by Ambo (1636), — by Memnon (2295), — by Pilot (496), — by Agamemnon (9), — by Burrell's Bull, of Burdon.

(43953) ROYAL JOHN,

Roan, calved February 8, 1877, bred by Mr. J. N. Beasley, Brampton House, the property of Lord Tredegar, Tredegar Park ; got by John of Oxford (34267), dam (Royal Janette) by Juvenile (22021), g. d. (Queen Janette) by Royal Butterfly 5th (18756), gr. g. d. (Countess Janette) by Lilyvick (10421), — (Young Jocund) by Japetus (10350), — by Monarch (7249), — by Mac Ivor (2237), — by Northern Light (1280), — by Wellington (684), — by Phenomenon (491), — by Favourite (252), — by Favourite (252), — by Favourite (252), — by Hubback (319), — by Snowdon's Bull (612), — by Waistell's Bull (669), — by Masterman's Bull (422), — by the Studley Bull (626).

(43954) ROYAL LANCASTER,

Roan, calved June 25, 1878, bred by Mr. H. Fawcett, Old Bramhope ; got by Rufus (37413), dam (Bramhope Darling) by Telemachus (27603), g. d. (Royal Dora) by Royal Cambridge (25009), gr. g. d. (Turk's Darling) by Royal Turk (16875), — (Delightful) by Field Marshal (14545), — by Lord Milton (10461), — by Prince of Wales (8432), — by Bucephalus (6816), — by Stanhope (5315), — by Blyth Favourite (801), — by Son of Wellington (683).

(43955) ROYAL LEO,
Red and white, calved March 6, 1879, bred by Mr. H. Chandos-Pole-Gell, Hopton Hall, sold to Monsieur de Sainte Marie for the French Government; got by Royal Saxon (39057), dam (Leopoldine 4th) by King Charles (24240), g. d. (Leopoldine 3rd) by Killerby Lad (20052), gr. g. d. (Leopoldine 2nd) by Prince Patrick (16760), — (Sally) by Bumper (10005), — by Second Colonel (8544), — by Sir Walter (2639), — by Marquis (2270), — by Wellington (2824), — by Leopold (2199), — by Dundas (1943), — by Cupid (177), — by Simon (590), — by Punch (531), — by Bolingbroke (86).

(43956) ROYAL MANFRED,
Roan, calved June 10, 1879, bred by Mr. W. Mitchell, Cleasby; got by King Manfred (38491), dam (Lady Bloom) by Squire Booth (30049), g. d. (Ladyship) by Major (26790), gr. g. d. (Queen Ann) by Sir Christopher (22895), — (Lady Bird) by Parmesan (20471), — by Good Friday (14635), — by Neptune (11847), — by The Dandy (10926), — by Hamlet (8126), — by Sir Roger (7515), — by Studley Royal (5342).

(43957) ROYAL OXFORD 2ND,
Red, calved March 21, 1879, bred by Messrs. W. Hosken and Son, Hayle; got by Duke of Oxford 33rd (36528), dam (Miss Maggie) by Second Earl of Oxford (23844), g. d. (Miss Garne) by Garibaldi (17925), gr. g. d. (Princess Ellen) by A 1 (15538), — (Princess Charlotte) by Colchicum (8963), — by Fitz-Hardinge (8073), — by Elevator (6969), — by Raffler (7391), — by Second Fairfax (8050).

(43958) ROYAL OXFORD 3RD,
White, calved April 15, 1879, bred by Messrs. W. Hosken and Son, Hayle; got by Duke of Oxford 33rd (36528), dam (Gertrude 2nd) by Duke of Oxford (31005), g. d. (Gertrude) by Second Earl of Oxford (23844), gr. g. d. (Grateful) by Thorndale Mason (23067), — (Graceful) by Prince Frederick (16734), — by Sir John Barleycorn (12085), — by Critic (9000).

(43959) ROYAL OXFORD 4TH,
Red and white, calved July 22, 1879, bred by Messrs. W. Hosken and Son, Hayle; got by Duke of Oxford 33rd (36528), dam (Rose of Oxford) by Fifth Earl of Oxford (28515), g. d. (White Rose) by Thorndale Mason (23067), gr. g. d. (Moss Rose) by Prince Frederick (16734), — (Fancy 2nd) by Sir John Barleycorn (12085), — by Critic (9000).

(43960) ROYAL OXFORD 5TH,
Roan, calved July 28, 1879, bred by Messrs. W. Hosken and Son, Hayle; got by Duke of Oxford 33rd (36528), dam (Countess of Oxford 6th) by Second Baron Wild Eyes (30497), g. d. (Countess of Oxford 3rd) by Towneley Oxford (30170), gr. g. d. (Countess) by Prince Frederick (16734), — (Jocund) by Brigadier (14193), — by Usurper (13929), — by Fawsley (6004), — by Javelin (4093), — by Blyth (797), — by Wellington (684), — by Phenomenon (491), — by Favourite (252), — by Favourite (252), — by Favourite (252), — by Hubback (319), — by Snowdon's Bull (612), — by Waistell's Bull (669), — by Masterman's Bull (422), — by the Studley Bull (626).

(43961) ROYAL OXFORD 6TH,
Roan, calved August 1, 1879, bred by Messrs. W. Hosken and Son, Hayle; got

by Duke of Oxford 33rd (36528), dam (Countess of Oxford 7th) by Second Baron Wild Eyes (30497), g. d. (Countess of Oxford 3rd) by Towneley Oxford (30170), gr. g. d. (Countess) by Prince Frederick (16734), — (Jocund) by Brigadier (14193), — by Usurper (13929), — by Fawsley (6004), — by Javelin (4093), — by Blyth (797), — by Wellington (684), — by Phenomenon (491), — by Favourite (252), — by Favourite (252), — by Favourite (252), — by Hubback (319), — by Snowdon's Bull (612), — by Waistell's Bull (669), — by Masterman's Bull (422), — by the Studley Bull (626).

(43962) ROYAL OXFORD 7TH,

Roan, calved November 22, 1879, bred by Messrs. W. Hosken and Son, Hayle; got by Duke of Oxford 33rd (36528), dam (Red Rose) by Second Earl of Oxford (23844), g. d. (Rosebud) by Prince Frederick (16734), gr. g. d. (Kitchen Maid) by Sir Richard (15298), — (Kate) by Romeo (13621), — by Guizot (9174), — by Recovery (7403), — by Baronet.

(43963) ROYAL PATRON,

Roan, calved July 4, 1879, bred by Mr. J. Angus, Bearl; got by Hawthorn (36751), dam (Christmas Snowdrop) by King Christmas (36836), g. d. (Snowdrop 3rd) by Manfred (26801), gr. g. d. (Snowdrop) by Brigade Major (21312), — (Lily) by Lord Albert (20143), — by Frederick (14571), — by Red Prince (13576), — by Brilliant (7851), — by Sir Robert (5180), — by Bachelor (1665), — by Short Legs (5124), — by Ajax (723), — by Hector (2103), — by Surly (2715), — by Sir Harry (5155), — by Colonel (152), — by Son of Hubback (319).

(43964) ROYAL PRINCE,

Roan, calved June 13, 1878, bred by Mr. J. Close, Holmescales, the property of Mr. J. Relph, Meaburn Hall; got by Royal Victor (35414), dam (Maid of the Mist) by Golden Duke (31265), g. d. (Shamrock) by Golden Dick (24047), gr. g. d. (Sarah) by General Wilson (16143), — (Susan) by Young Count (14338), — by Abraham Parker (9856), — by Sir Francis (9635), — by Petteril (4685), — by Young Brian.

(43965) ROYAL PRINCE,

Roan, calved September 30, 1879, bred by Mrs. Mace, Sherborne; got by Geneva's Duke (38349), dam (Sarah) by Don Pedro (30887), g. d. (Sally) by Butter Boy (21349), gr. g. d. (Louisa) by Royal Butterfly 14th (20722), — (Lucy Long) by Orangeman (18485), — by General Pelissier (14605), — by Chieftain (12595), — by Morpeth (8325), — by The Prince (7615), — by Colling (902), — by Son of Alexander (1624), — by Grandson of Favourite (252).

(43966) ROYAL ROSEBUD,

Roan, calved April 16, 1879, bred by Mr. J. Moffat, Ballyhyland, sold to the Provincial Government of Brabant, Belgium; got by Royalist (37396), dam (Rosalind 3rd) by Dey of Algiers (25892), g. d. (Rosalind 2nd) by Grey Gauntlet (19908), gr. g. d. (Rosalind) by Syntax (13821), — (Splendour 6th) by Hamlet (8126), — by Baronet (6763), — by Guy Fawkes (7062), — by Pedestrian (7321), — by Edrom (1956), — by Scipio (1421), — by Hector (1104), — by Midas (435), — by Marquis (407), — by Chilton (136), — by Ben (70).

(43967) ROYAL SCEPTRE,

White, calved April 24, 1879, bred by the Exors. of Mr. T. C. Booth, Warlaby; got by King Harold (40053), dam (Bright Jewel) by Knight of the Shire (26552), g. d. (Bright Spangle) by Prince of Warlaby (15107), gr. g. d. (Bright Dew) by British Prince (14197), — (Bright Morn) by Vanguard (10994), — by Crown Prince (10087), — by Zadig (8796), — by Auld Robin Grey (6753), — by James· Chrisp's Bull, — by Burley (1766), — by Isaac (1129), — by Pilot (496), — by Albion (14), — by Lame Bull (359), — by Shipton (587), — by Son of Suworrow (636), — by Son of Twin Brother to Ben (88), — by Twin Brother to Ben (660).

(43968) ROYAL SPLENDOUR,

Red, calved February 10, 1879, bred by Mr. J. Oastler, Loxwood House; got by Royal Bell (39030), dam (Splendour 26th) by Keithmore (34297), g. d. (Splendour 16th) by Remus (20665), gr. g. d. (Splendour 12th) by Orphan Boy (13429), — (Splendour 10th) by General Lax (12933), — by Marquis of Rockingham (10506), — by Baronet (6763), — by Guy Faux (7062), — by Pedestrian (321), — by Edrom (1956), — by Scipio (1421), — by Hector (1104), — by Midas (435), — by Marquis (407), — by Chilton (136), — by Ben (70).

(43969) ROYAL STANDARD,

Red and white, calved March 1, 1879, bred by Mr. C. W. Schroeter, Tedfold; got by Sir Wilfrid (37484), dam (Phillis 11th) by Royal Broughton (27352), g. d. (Phillis 7th) by Norfolk Thorndale Duke (24666), gr. g. d. (Phillis 2nd) by Red Knight (16809), — (Phillis) by Homer (14714), — by Cardigan (12556), — by Young Rufus (13649), — by Constitution (12634), — by Young Comet (1853), — descended from Jolly's Bull (4115).

(43970) ROYAL STRANGER,

Roan, calved April 17, 1879, bred by Messrs. J. G. W. and W. R. Lyle, Donaghmore House; got by Lord of the Manor (38641), dam (Bessy) by Lord Francis (24393), g. d. (Grisi) by Duke of Montrose (21599), gr. g. d. (De Beriot) by Lumley (16478), — (Malibran) by Master Goldschmidt (13315), — by Marquis (11786), — by The Lord of Gilling (6587), — by Buckingham (3239), — by Son of Chance (868), — by Chance (868), — by Grazier (1085), — by Plutarch (1327), — by Midas (435), — by Grandson of Simon (590).

(43971) YOUNG ROYALTY,

Roan, calved March 16, 1879, bred by Mr. T. Taylor, Hall Garth; got by Royalty (39059), dam (Lady Deans) by Hubback Junior (31395), g. d. (Effy Deans) by Edgar (19680), gr. g. d. (Jenny Deans) by Great Mogul (14651), — (Young Daisy) by Zadig (8796), — by Duke of Richmond (7996), — by Royal Prince (5024), — by Lisbon (1172), — by Norman (1276), — by Acklam (713), — by Meteor (431), — by Duke (226), — by Favourite (252), — by Punch (531), — by Hubback (319).

(43792) ROYAL VICTOR,

Roan, calved November 30, 1878, bred by Mr. A. Cruickshank, Sittyton; got by Pride of the Isles (35072), dam (Victoria 48th) by Lord Lancaster (26666), g. d. (Victoria 39th) by Champion of England (17526), gr. g. d. (Victoria 29th) by Red Knight (11976), — (Victoria 19th) by Lord John (11731), — by Prince Albert (11933), — by Belzoni (783), — by Satellite (1420), — by Cato (119),

— by Pope (514), — by Favourite (252), — by White Bull (421), — by Favourite (252), — by Dalton Duke (188), — by R. Alcock's Bull (19), — by J. Smith's Bull (608), — by Jolly's Bull (337).

(43973) **ROYAL WINDSOR,**
Roan, calved October 29, 1878, bred by Messrs. J. and D. D. Lazonby, Calthwaite House ; got by Bright Duke (37893), dam (Lady Windsor) by Royal Fantail (32383), g. d. (Alexandra Windsor) by Prince of Wales (24851), gr. g. d. (Ormolu Windsor) by Imperial Windsor (18086), — (Ormolu Gwynne) by Master Hopewell (14929), — by General Sale (8099), — by Lord Warden (7167), — by Orontes (4623), — by Guardian (3947), — by Mercury (2301), — by Monarch (2324), — by St. Albans (2584), — by Jupiter (342), — by Sir Oliver (605), — by Trunnell (659), — by Favourite (252), — by Favourite (252), — by Dalton Duke (188), — by R. Alcock's Bull (19), — by J. Smith's Bull (608), — by Jolly's Bull (337).

(43974) **ROYAL WINDSOR,**
Roan, calved April 5, 1879, bred by Mr. J. Close, Holmescales, the property of Mr. J. Tander, Berrier End ; got by Royal Victor (35414), dam (Familiar Windsor) by Roan Windsor (24967), g. d. (Familiar Hopewell Gwynne) by Hopewell Gwynne (19976), gr. g. d. (Familiar 4th) by Red Duke (13571), — (Familiar 2nd) by Master Belleville (11795), — by Fitz-Leonard (7010), — by Velocipede (5552), — by Young Matchem (2282), — by Jack Tar (1133), — by Pilot (496), — by Young Albion (15).

(43975) **RUBICON,**
Roan, calved February 6, 1879, bred by Colonel R. Loyd Lindsay, Lockinge Park ; got by Earl of Horton 11th (36588), dam (Ruby) by Rob Roy (29806), g.d. (Rosetta) by Costa (21487), gr.g.d. (Rosette) by Prince of Prussia (16752), — (Red Rose) by Horatio (10335), — by Third Duke of Northumberland (3647), — by Velocipede (5552), — by Sir Thomas (2636), — by Marske (418), — by Comet (155), — by Tom (652), — by Favourite (1033), — by Hutton's Bull (323), — by Barningham (56).

(43976) **RUBY'S GRAND DUKE,**
Roan, calved September 7, 1878, bred by Sir G. R. Philips, Bart., Weston Park ; got by Grand Duke 29th (38372), dam (Fawsley Cherry) by Third Duke of Geneva (21592), g. d. (Fawsley) by Third Duke of Thorndale (17749), gr. g. d. (Fawsley Garland) by Earl of Dublin (10178), — (Garland) by Grey Friar (9172), — by Fawsley (6004), — by Marcellus (2260), — by Rufus (2576), — by Wellington (683), — by Windsor (698), — by Windsor (698), — by Own Brother to North Star (459).

(43977) **RUBY'S GRAND DUKE 2ND,**
Roan, calved February 12, 1879, bred by Sir G. R. Philips, Bart., Weston Park ; got by Grand Duke 29th (38372), dam (Fawsley Cherry 2nd) by Third Cherry Duke (28171), g.d. (Fawsley Cherry) by Third Duke of Geneva (21592), gr.g.d. (Fawsley) by Third Duke of Thorndale (17749), — (Fawsley Garland) by Earl of Dublin (10178), — by Grey Friar (9172), — by Fawsley (6004), — by Marcellus (2260), — by Rufus (2576), — by Wellington (683), — by Windsor (698), — by Windsor (698), — by Own Brother to North Star (459).

(43978) RUFUS,
Red, calved March 22, 1879, bred by Mrs. Mace, Sherborne; got by Geneva's
Duke (38349), dam (Rose of Summer 2nd) by Duke Geneva (30911), g.d. (Rose
of Summer) by Count Glo'ster (23637), gr.g.d. (Rose of Spring) by Lord Elgin
(18230), — (Rosebud) by General Pelissier (14605), — by Marchmont (9367),
— by Fitz-Hardinge (8073), — by Augustus (6751), — by Son of Anthony
(1640), — by a Bull of Mr. Champion's, of Blyth.

(43979) RUFUS,
Red, calved April 7, 1879, bred by the Duke of Northumberland, Alnwick
Castle; got by Sir Raymond (40716), dam (Rosebud 12th) by Hotspur (28876),
g. d. (Rosebud 11th) by Brigand (28080), gr. g. d. (Rosebud 7th) by Snowflake
(18888), — (Rosebud 2nd) by Duke of Buckingham (14429), — by Ravensworth (7400), — by Son of Regent (2517), — by Togston (5487), — by Son of
Lawnsleeves (365), — by Son of Bolingbroke (86).

(43980) RUFUS,
Red, calved April 22, 1879, bred by Sir Victor Brooke, Bart., Colebrooke; got
by Royal Baron (40617), dam (The Sweetest Heart) by Lieutenant General
(31600), g. d. (James's Sweetheart) by King James (28971), gr. g. d. (Irish
Sweetheart) by Patrician (24728), — (Sweetheart 16th) by Duke of Albemarle
(21560), — by The Baron (13833), — by Mameluke (13289), — by Daybreak
(11338), — by Accordion (5708), — by Little John (4232), — by Caliph (1774),
— by Sir Walter (2637), — by Hotspur (1117), — by Coxcomb (928), — by
Midas (435), — by Comet (155), — by R. Colling's Son of Favourite (252), —
by same Son of Favourite (252), — by Hubback (319).

(43981) RUSHDEN,
Red, calved July 26, 1879, bred by Mr. T. Nichols, The Grange; got by Lord
Clarence Waterloo (36926), dam (Lady Brailes 2nd) by Baron Oxford 5th (27958),
g. d. (Lady Brailes) by Duke of Brailes (23724), gr. g. d. (Formosa) by Duke
of Brailes (23724), — (Gionetta) by Sarawak (15238), — by Earl of Dublin
(10178), — by Janizary (8175), — by Snowball (8602), — by Caliph (1774),
— by Norman (2379), — by White Boy (1580), — by Wyville's Bull, — by a
Bull bred by Mr. Charge.

(43982) RUSTIC,
Red, calved May 30, 1879, bred by Earl Beauchamp, Madresfield Court; got by
Magnum Bonum (41954), dam (Lavinia 2nd) by Festival (26147), g.d. (Lavinia)
by Diplomatist (19571), gr. g. d. (Bess) by Impatience (14723), — (Berangeria)
by Hero (8146), — by Major (4338), — by Ormsby (4621), — by Cossack (925),
— by Alpha (3004), — by Captain (3271), — by Captain (108).

(43983) SAILOR PRINCE,
Red and little white, calved August 4, 1879, bred by Mr. A. Graham, Yanwath
Hall; got by The Colonel (35747), dam (Stamper 3rd) by Hackthorpe (21889),
g. d. (Stamper 1st) by Baronet (19274), gr. g. d. (Selah) by Albert Edward
(6723), — (Young Beatrice) by Mountaineer (6228), — by Exmouth (3747), —
by Bolivar (1730).

(43984) ST. AUGUSTIN,
White, calved December 16, 1879, bred by Mr. R. C. Morton, Lane House; got
by Duke of Holker (38153), dam (Monica) by Baron Barrington 4th (33006),

g. d. (Royal Rose) by Royal George (32387), gr. g. d. (Todd 3rd) by Sir George (27471), — (Todd 2nd) by Charley (42901), — Todd, bought of Mr. J. Todd, Ladyford, Kendal.

(43985) ST. CUTHBERT,

Roan, calved October 7, 1878, bred by Mr. J. Lumsden, Castle Heaton; got by Duke of Bamburgh (39711), dam (Empress) by The Wizard (35796), g. d. (Mulberry) by Faldonside (28564), gr. g. d. (Mistake) by Shaftoe (27447), — (Mistletoe) by Enterprise (14514), — by Rob Roy (12007), — by Romulus (10741), — by Zadig (8796), — by Young Hastings (3988), — by Bachelor (1666), — by Reformer (2502), — by Shortlegs (5124), — by Leopold (2199), — by Sir Harry (5155), — by Traveller (655), — by Colonel (152), — by Colling's Son of Hubback (319).

(43986) ST. SWITHIN,

Roan, calved July 15, 1879, bred by Mr. A. H. Browne, Callaly Castle, the property of Mr. W. Langhorn, East Mill Hills; got by Lord Prinknash (34655), dam (Prune) by Windsor's Prince of Mona (32883), g. d. (Prunette) by Lord Plymouth (24455), gr. g. d. (Prunella) by Knight Errant (18154), — (Prude) by Valasco (15443), — by Prince Arthur (13497), — by Noble (4578), — by Master Charley (7215), — by Baronet (1686), — by Young Magog (2247), — by Reformer (2502), — by Margrave (2263), — by Leopold (2199), — by Hector (2103), — by Surly (2715), — by Traveller (655), — by Colonel (152).

(43987) ST. SWITHIN,

Red and white, calved July 15, 1879, bred by Mr. S. R. C. Ward, Neasham Hill; got by Lord Ripon (38656), dam (Flora 8th) by Windsor Booth (32875), g. d. (Flora) by Squire of Bushey (20889), gr. g. d. (Tulip) by Richard Cœur-de-Lion (13590), — (Opulent) by Abraham Parker (9856), — by Tom Boy (9750), — by Magician (7185), — by Vivian (5575), — by Belvedere 2nd (3126), — by Alive 'O (2995), — by Young Alive 'O (2996), — by Eclipse (236), — by Charge's Grey Bull (872), — by Paddock Bull (477).

(43988) SALAR JUNG 2ND,

Roan, calved October 11, 1878, bred by Mr. R. S. Brundell, Leicester House, Doncaster, the property of Mr. J. Elmhirst, Thorne; got by Salar Jung (40669), dam (Miss Maxwell 4th) by Belvedere (33138), g. d. (Miss Maxwell) by Maxwell (40334), gr. g. d. by Victor (21025), — by Festus (16040).

(43989) SAMSON,

Roan, calved March 11, 1879, bred by Mr. A. Fawkes, Farnley Hall; got by Baron Winsome 6th (33111), dam (Sycamore 7th) by Oxford-le-Grand (29496), g. d. (Salvia) by Thorndale Lad (23066), gr. g. d. (Lady Salome) by Lord Cobham (20164), — (Sycamore) by Royal Oak (16873), — by Sir Edmund Lyons (15284), — by Lamartine (11660), — by Matchless (4428), — by True Blue (5522), — by Baronet (1688), — by A-la-mode (725), — by Mameluke (2257).

(43990) SANCHO PANZA,

Roan, calved May 27, 1876, bred by Mr. J. P. Tynte, Tynte Park, the property of Mr. P. Hanlon, Grangeforth; got by British Peer (33224), dam (Buss) by British Boy (25675), g. d. (Secret) by Bright Spur (25671), gr. g. d. (Hush) by Hohenlohe (18074), — (Whisper) by Knight of Windsor (16350), — by Emperor (12835), — by Barley Sugar (11140), — by Prince Ernest (7366), — by Captain (11241), — by Doctor (2748), — by Brough (3228), — by Pilot (496).

(43991) SAP,
Roan, calved March 3, 1879, bred by Messrs. Pender, Budockvean; got by
Fawsley (41531), dam (Sappho 2nd) by Sailor Boy (37419), g. d. (Sappho) by
Orlando (32000), gr. g. d. (Stately) by Red Duke (16803), — (Sylph) by Cæsar
(36300), — by The Red Duke (8694), — by Phœnix (6290), — Wallflower,
bought of Mr. Wall.

(43992) SCARLET,
Red, calved August 27, 1879, bred by Mr. H. Burtt, Fulbeck Grange; got by
Favonius (39865), dam (Fancy Pompeii) by Pompeii (35058), g. d. (Fancy 7th)
by Comus (25821), gr. g. d. (Fancy) by Fitz-York 2nd (16060), — (Fancy) by
Exhibition (11450), — by Biscay (7834), — by Austerlitz (3063), — by Norfolk
(2377), — by Bright (1739), — by Shylock (2622), — by Percy (1314), — by
Blucher (84), — by Jupiter (344), — by Marshal Beresford (415), — by Crocus
(932), — by a Bull of Mr. R. Colling's.

(43993) SCOTLAND,
Roan, calved September 14, 1878, bred by Mr. J. A. Gordon, Udale; got by
Rosario (35315), dam (Miss Danby 2nd) by Royal Windsor (29890), g. d. (Miss
Danby) by Emmanuel (31105), gr. g. d. by Champion (23529), — by Baron
Warlaby (7813), — by Young Thornton (13885), — by California (10019), —
by Gainford (7029), — by Son of Roland.

(43994) SCOTS FUSILIER 3RD,
Roan, calved January 22, 1879, bred by the Earl of Dunmore, Dunmore, the
property of Mr. J. A. M. Cope, Drummilly; got by Marquis of Oxford 2nd
(37055), dam (Fuchsia 15th) by Duke of Connaught (33604), g. d. (Fuchsia
11th) by Second Duke of Collingham (23730), gr. g. d. (Fuchsia 6th) by Earl
of Glo'ster (21644), — (Cambridge Fuchsia) by Second Duke of Cambridge
(12743), — by Kirklevington 3rd (13120), — by Second Cleveland Lad (3408),
— by Fourth Duke of Northumberland (3649), — by Short Tail (2621), — by
Belvedere (1706), — by Son of Young Wynyard (2859), — descended from J.
Brown's Old Red Bull (97).

(43995) SEDGWICK LAD 3RD,
Roan, calved March 31, 1879, bred by Mr. W. H. Wakefield, Sedgwick; got by
Duke of Holker (38153), dam (Sedgwick Lass) by Dunrobin (28486), g. d.
(Banks 3rd) by Frederick Warlaby (23990), gr. g. d. (Banks 1st) by Sir Colin
(16957), — (Lancaster) by Leap Year (11677).

(43996) SERAPIS,
Red, calved August 3, 1879, bred by Mr. E. Byron, Coulsdon Court; got by
Silenus (37444), dam (Priscilla 6th) by Lord St. Leonards (29202), g. d. (Priscilla 3rd) by Prince Arthur (16723), gr. g. d. (Priscilla) by Water King (13980),
— (Peggy) by Dannecker (7949), — by The Pacha (7612), — by Second Duke
of Northumberland (3646), — by Sir Henry (1446), — by Juniper (1145), —
by Lancaster (360), — by Neswick (1200), — Yellow Cow.

(43997) SERGEANT GARNETT,
Roan, calved April 7, 1879, bred by Mr. T. Isherwood, Fryton; got by Lord
Oxford Bright Eyes 2nd (36998), dam (Dorothy) by General Monk (34013),
g. d. (Dora) by Sir Roger (35590), gr. g. d. (Martha) by Friar Tuck (33976), —
(Young Olivia) by Problem (18644), — by Snyders (7525), — by Bulmer

(1760), — by Harry (3981), — by Navigator (1260), — by Percy (1312), — by Ketton (346), — by Expectation (247), — by Magnum Bonum (2882), — by H. Chapman's Bull, — by R. Grimston's Bull, — by a Son of Dalton Duke (188)

(43998)   SEVERUS,
Roan, calved June 12, 1879, bred by Mr. H. Allsopp, Hindlip Hall; got by Duke of Collingham 3rd (38134), dam (Red Rose of Severn) by Twenty-fourth Duke of Airdrie (36460), g. d. (Grace Rose 4th) by Airdrie 3rd (32919), gr. g. d. (Mayflower of Stony Point) by Airdrie (30365), — (Dorothy) by Prince Charles 2nd (32113), — by Shakespeare (12062), — by Reformer (2505), — by Belvedere (1706), — by Second Hubback (1423), — by His Grace (311), — by Yarborough (705), — by Favourite (252), — by Punch (531), — by Foljambe (263), — by Hubback (319).

(43999)   SHAKESPEARE,
Red and white, calved June 19, 1879, bred by Mr. C. Phelps, Brays Court; got by Duke of Connaught (33604), dam (Prima Donna) by Royal Butterfly 20th (25007), g. d. (Paradox) by Cynric (19542), gr. g. d. (Prejudice) by Telegraph (15370), — (Pledge) by Enterprise (11443), — by Young Consul (6893), — by Newnham (2365), — by Satellite (1420), — by Jupiter (342), — by Sir Oliver (605), — by Trunnell (659), — by Favourite (252), — by Favourite (252), — by Dalton Duke (188), — by R. Alcock's Bull (19), — by J. Smith's Bull (608), — by Jolly's Bull (337).

(44000)   SHARON DUKE,
Roan, calved May 17, 1879, bred by Mr. G. Fox, Elmhurst Hall; got by Twenty-fourth Duke of Airdrie (36460), dam (Julia's Rose) by Fourth Duke of Geneva (30958), g. d. (Poppy's Julia) by Airdrie Duke (40959), gr. g. d. (Poppy 5th) by Thirteenth Duke of Airdrie (36459), — (Poppy 4th) by Airdrie (30365), — by Duke of Airdrie (12730), — by Ashland (11122), — by Prince Charles 2nd (32113), — by Shakspeare (12062), — by Reformer (2505), — by Belvedere (1706), — by Second Hubback (1423), — by His Grace (311), — by Yarborough (705), — by Favourite (252), — by Punch (531), — by Foljambe (263), — by Hubback (319).

(44001)   SHEPHERD BOY,
Roan, calved February 7, 1879, bred by Mr. J. Harris, Langham; got by Lord Hardwicke (41866), dam (Shepherdess) by Manton (24525), g. d. (Spencer 2nd) by Bumper (14211), gr. g. d. (Spencer) by Father Matthew (9112), — (Moss Rose) by Romulus (15184), — by Rasper (13562), — by Roderick Random (4984).

(44002)   SHERE ALI,
Red and white, calved November 9, 1878, bred by Mr. W. S. Marr, Uppermill, the property of Mr. W. Duthie, Collynie; got by Aldrick (40970), dam (Missie 49th) by Heir of Englishman (24122), g. d. (Missie 20th) by Gold Digger (24044), gr. g. d. (Missie 9th) by Lord Surrey (20230), — (Missie 4th) by Clarendon (14280), — by Son of Duke 3rd (17697), — by The Pacha (7612), — by Mahomed (16170), — by Plenipo (4725), — by Abbot (2899).

(44003)   SHERE ALI,
Red and white, calved January 4, 1879, bred by Sir P. Miles, Bart., Leigh Court; got by The Prior (42493), dam (Careless) by Meteor (29364), g. d.

(Crocus) by Lygon (24494), gr. g. d (Carmine) by Young Duke of Cambridge
(17708), — (Constance) by Ottoman (13442), — by Ruber (13644), — by
Brownie (8909), — by Ranunculus (2479), — by William (2840), — by Childers
(1824), — by Richard (1376), — by Jupiter (342), — by Charles (127), — by
Windsor (698), — by Chilton (136), — by Colonel (152).

(44004)　　SHERWOOD CHIEFTAIN 2ND,
Red, calved August 16, 1879, bred by Viscountess Ossington, Ossington; got
by Sherwood Chieftain (39098), dam (Pretty Maid 5th) by Twenty-first Duke
of Oxford (30999), g. d. (Pretty Maid) by Robin (24968), gr. g. d. (Pride of
Panton) by Lord Panton (22204), — (Soldier's Bride) by Superior (15362), —
by Baron Warlaby (7813), — by Leonard (4210), — by Counsellor (6904), —
by Snowball (8605).

(44005)　　SHIPMATE,
Red and white, calved December 29, 1879, bred by Lord Bolton, Bolton Hall;
got by Vice Admiral (39257), dam (Sprightliness) by Heir-at-Law (34124), g. d.
(Singleness) by Dandelion (30849), gr. g. d. (Simpleness) by Marmion (26821),
— (Shepherdess) by Stonegrave (27575), — by Homer (14714), — by Starling
(15338), — by Baronet (19273), — by Lord Cardigan (7158), — by Clementi
(3399).

(44006)　　SIDNEY,
Red, calved April 20, 1879, bred by Earl Beauchamp, Madresfield Court; got by
Magnum Bonum (41954), dam (Lady Bird 4th) by Gay Boy (31222), g. d.
(Lady Bird 2nd) by Diplomatist (19571), gr. g. d. (Lady Bird) by Ortolan
(18496), — (Stella) by Columbus (17591), — by Snitterfield (27513), — by
Balkan (14120), — by Jolly's Bull (4116).

(44007)　　SIGNET,
Roan, calved March 25, 1878, bred by Mr. G. Wilson, Netherton Clatt; got by
Braithwaite (28066), dam (Butterfly 33rd) by Cæsar Augustus (25704), g. d.
(Butterfly 14th) by Baronet (15614), gr. g. d. (Butterfly 8th) by Champion of
England (17526), — (Butterfly 2nd) by John Bull (11618), — by Matadore
(11800), — by Report (10704), — by The Pacha (7612), — by Second Duke of
Northumberland (3646), — by Mahomed (6170), — by Sillery (5131), — by
Carleton (843), — by Diamond (205), — by Diamond (205).

(44008)　　SILENT DUKE,
Red and white, calved January 9, 1879, bred by Mr. A. T. Matthews, Church
Hanborough; got by Grand Duke of Geneva 2nd (31288), dam (Silence 9th)
by Senator (32470), g. d. (Silence 5th) by Conte de Vermandois (19507), gr. g. d.
(Silence 2nd) by Sir Colin Campbell (13718), — (Silence) by Earl of Derby
(10177), — by Duke of Sutherland (6945), — by Locomotive (4242), — by
Short Tail (2621), — by Gambier (2046), — by Young Wynyard (2859), — by
Bulls of Messrs. C. and R. Colling's.

(44009)　　SILVER CROWN,
White, calved February 20, 1879, bred by Mrs. Pery, Coolcronan House, the
property of Mr. Utred A. Knox, Mount Falcon; got by Royal Crown (40628),
dam (Mabel) by Lord Broughton (31626), g. d. (Memnonia) by King Richard
(26523), gr. g. d. (Minerva) by Duke of York (23804), — (Medora) by Dr.
McHale (15887), — by Nimrod (13388), — by Selim (6454), — by Rex (6385),

— by Sir Thomas Fairfax (5196), — by Ambo (1636), — by Memnon (2295), — by Pilot (496), — by Agamemnon (9), — by Burrell's Bull, of Burdon.

(44010)   SILVIO,

Red and little white, calved January 2, 1879, bred by Mr. A. Graham, Yanwath Hall; got by The Colonel (35747), dam (Sunshine) by Dean (30063), g. d. (Stamper 3rd) by Hackthorpe (21889), gr. g. d. (Stamper 1st) by Baronet (19274), — (Selah) by Albert Edward (6723), — by Mountaineer (6228), — by Exmouth (3747), — by Bolivar (1730).

(44011)   SIR ALFRED,

White, calved July 20, 1877, bred by Mr. H. D. de Vitre, Charlton House, the property of Mr. G. T. Wright, Stokes Farm; got by King Alfred (36833), dam (Towneley Butterfly 3rd) by Lord Thorndale (31756), g. d. (Towneley Butterfly) by Count of Windsor (21498), gr. g. d. (White Butterfly) by Butterfly's Nephew (15714), — (Paris Butterfly) by Master Butterfly (13311), — by Gavazzi (11508), — by Baron of Ravensworth (7811), — by Raree Show (4874), — by Thick Hock (6601), — by Expectation (1988), — by Belzoni (1709), — by Comus (1861), — by Denton (198).

(44012)   SIR ALLEN,

Roan, calved September 28, 1879, bred by Mr. Hugh Aylmer, West Dereham Abbey; got by Sir Wilfrid (37484), dam (Eastthorpe Rose) by High Sheriff (26392), g. d. (Eastthorpe Strawberry 6th) by Royal Broughton (27352), gr. g. d. (Eastthorpe Strawberry 4th) by Ravenspur (20628), — (Eastthorpe Strawberry) by Prince Leopold (20557), — by Red Knight (16809), — by Sir Samuel (15302), — by Lemnos (13146), — by General Washington (6036), — by Woldsman (2851), — by Malbro' (1189), — by Harold (291), — by Count (170), — by Badsworth (47), — by Driffield (223), — bred by Sir George Strickland.

(44013)   SIR ARTHUR,

White, calved February 28, 1875, bred by the Earl of Lichfield, Shugborough; got by Sir Henry (35559), dam (Shugborough) by Knight Templar 3rd (34412), g. d. (Orange Blossom) by Knight Gwynne (34368), gr. g. d. (Lady Olivia) by Sir James the Rose (15290), — (Little Red Rose) by Petrarch (7329), — by Second Duke of York (5959), — by Raspberry (4875), — by Young Matchem (4422), — by Isaac (1129), — by Young Pilot (4702), — by Pilot (496), — by Julius Cæsar (1143).

(44014)   SIR ARTHUR,

Roan, calved June 6, 1879, bred by Mr. Hugh Aylmer, West Dereham Abbey; got by Sir Wilfrid (37484), dam (Mistress Mary) by Royal Broughton (27352), g. d. (Mistress May) by Prince of Rosedale (24837), gr. g. d. (Mistress Margaret) by Paterfamilias (18521), — (Modred) by Valasco (15443), — by Baron Warlaby (7813), — by Royal Buck (10750), — by Hopewell (10332), — by Hamlet (8126).

(44015)   SIR ARTHUR,

Roan, calved October 9, 1879, bred by the Duke of Northumberland, Alnwick Castle; got by Sir Raymond (40716), dam (Cherry Bud) by Fitz-Roland (33936), g. d. (Red Blossom) by Rifleman (27283), gr. g. d. (Maid of Aln) by Melsonby (18380), — (Young Jessy) by George 3rd (16147), — by Shaftoe

(5107), — by Richardi (4944), — Cowslip, bought of Mr. Patterson, of Wood Houses.

(44016)   SIR ARTHUR IRWIN,
Roan, calved August 17, 1879, bred by Mr. W. Linton, Sheriff Hutton, the property of Mr. R. Taylor, New House; got by Sir Arthur Ingram (32490), dam (Irwin Branch) by Sergeant Major (29957), g. d. (Home Beauty) by Mountain Chief (20383), gr. g. d. (Hand Maid) by May Day (20323), — (White Rose) by Magnus Troil (14880), — by Magnus Troil (14880), — by Bates (12451), — by Lord Warden (20233), — by Snyders (7525), — by Duke of York (9049).

(44017)   SIR BARTLE,
Roan, calved July 7, 1879, bred by Mr. G. Ashburner, Low Hall; got by Duke of Oxford 41st (38174), dam (Polly's Duchess) by Grand Duke 9th (19879), g. d. (Polly) by Oxford 4th (24706), gr. g. d. (Moss Rose) by His Royal Highness (18072), — (Short Tail) by Columbus (14299).

(44018)   SIR BEVIS,
Roan, calved May 28, 1879, bred by Mr. S. L. Horton, Park House; got by Marquis of Blandford 6th (41983), dam (Waterloo Cherry 7th) by Abacot (32900), g. d. (Waterloo Cherry 4th) by Idsal (31404), gr. g. d. (Waterloo Cherry) by Kirbythore Waterloo (24263), — (Aurora 2nd) by Brutus (17469), — by Constantine (14318), — by Prince of Wales (4833), — by Hector (16250), — by Prince Ernest (4818), — by Sir William (2640), — by Young Rockingham (2549), — by Wellington (2824), — by Northumberland (464), — by Buston's Styford (103), — by Lame Bull (358), — by Bolingbroke (86).

(44019)   SIR CHARLES,
Red, calved May 21, 1878, bred by Sir W. C. Worsley, Bart., Hovingham, the property of Mr. J. Sharples, Liverpool; got by Doctor (36433), dam (Robin's Rose) by Robin (24968), g. d. (Ruby Rose) by Lord Panton (22204), gr. g. d. (Royal Rose) by Prince Alfred (13494), — (Blooming Rose) by Baron Warlaby (7813).

(44020)   SIR CHARLES,
Red and little white, calved September 4, 1878, bred by Mr. Hugh Aylmer, West Dereham Abbey, the property of Mr. D. Pugh, Manoravon; got by Sir Wilfrid (37484), dam (Lady Mayoress) by High Sheriff (26392), g. d. (Lady Leonora 2nd) by Emperor of the North (23888), gr. g. d. (Lady Leonore) by Sir Samuel (15302), — (Lady Leonora) by Prince Alfred (13494), — by Homer (14714), — by Justice (14753), — by Young Rufus (13649), — by Constitution (12634), — by Young Comet (1853), — descended from Jolly's Bull (4115).

(44021)   SIR CHARLIE,
Red and white, calved March 16, 1879, bred by Mr. H. D. de Vitre, Charlton House, the property of Messrs. J. and E. Belcher, Bradfield Farm; got by Blushing Duke (37869), dam (Lady Culshaw 3rd) by Lord Thorndale (31756), g. d. (Lady Culshaw) by Royal Butterfly 17th (22774), gr. g. d. (Rose of Lancashire) by Master Butterfly 4th (14920), — (Young Barmpton Rose) by Richard Cœur-de-Lion (13590), — by Lord John (11731), — by Baron of Ravensworth (7811), — by Raree Show (4874), — by Thick Hock (6601), — by Expectation (1988), — by Belzoni (1709), — by Comus (1861), — by Denton (198).

(44022)   SIR CHARLTON,
Roan, calved March 7, 1879, bred by Captain D. H. Mytton, Garth; got by
Constantine 2nd (33439), dam (Etiquette) by Duke of York (31038), g. d.
(Eglantine) by Manager (24521), gr. g. d. (White Rose 2nd) by Victor (19061),
— (White Rose) by Cronstadt (14352), — by Autocrat (19249), — by Warrior
(12287), — by Sweetmeat (8645), — Old Strawberry (winner of First Prize,
County Show, Shrewsbury, 1844), bred by Dr. Rowley.

(44023)   SIR COLIN,
Red, calved December 22, 1879, bred by Mr. Hugh Aylmer, West Dereham
Abbey; got by Sir Wilfrid (37484), dam (Strawberry Duchess 5th) by High
Sheriff (26392), g. d. (Strawberry Duchess 3rd) by Royal Broughton (27352),
gr. g. d. (Strawberry Duchess 2nd) by Prince Christian (22581), — (Windsor
Strawberry) by Imperial Buckingham (19998), — by Duke of Buckingham
(14428), — by Duke of Cambridge (12747), — by St. Thomas (10777), —
by Chorister (3378), — by Tom Gwynne (5498), — by Wellington (2824), —
by Marmion (406), — by Ossian (476).

(44024)   SIR DAVID DEANS,
Roan, calved March 14, 1879, bred by Mr. J. Graham, Calthwaite Hall; got by
Earl of York 2nd (36605), dam (Effy Deans) by Edgar (19680), g. d. (Jenny
Deans) by Great Mogul (14651), gr. g. d. (Young Daisy) by Zadig (8796), —
(Wild Flower) by Duke of Richmond (7996), — by Royal Prince (5024), — by
Lisbon (1172), — by Norman (1276), — by Acklam (713), — by Meteor (431),
— by Duke (226), — by Favourite (252), — by Punch (531), — by Hubback
(319).

(44025)   SIR DUDLEY,
Roan, calved April 21, 1879, bred by Mr. J. Christy, Boynton Hall; got by
Oxford's Dudley (42101), dam (Lady Dudley 7th) by Third Duke of Claro
(23729), g. d. (Lady Dudley) by Third Duke of Lancaster (19624), gr. g. d.
(Soprano) by Fitzroy (16058), — (Symmetry) by Duke of Ulster (12774), —
by Belshazzar 2nd (14154), — by Napoleon (10552), — by Plenipo (4724), —
by Sir John (16985), — by Rex (1375), — by Baron (58), — by a Bull of Mr.
Mason's, — by Falstaff (250), — by Irishman (329).

(44026)   SIR FRANCIS,
White, calved May 19, 1879, bred by Mr. Hugh Aylmer, West Dereham Abbey;
got by Sir Wilfrid (37484), dam (Foreign Beauty) by Knight of the Shire
(26552), g. d. (Foreign Empress) by Fitz Royal (26167), gr. g. d. (Foreign
Princess) by Prince of Warlaby (15107, — (Flower of Denmark) by Fitz-
Clarence (14552), — by Vanguard (10004), — by Londesboro' (0142), — by
Rinaldo (4949), — by Sir Thomas (2636), — by Sir Alexander (591), — by
Marske (418), — by North Star (459), — by Wellington (680), — by Favourite
(252), — by Favourite (252), — by Ben (70), — by Hubback (319), — by
Snowdon's Bull (612), — by Sir J. Pennyman's Bull (601).

(44027)   SIR FRANK,
Roan, calved February 11, 1878, bred by Mr. J. Lamb, Glasson; got by Lord
Ashton Wild Eyes 2nd (36912), dam (Lady Alice) by Windsor Duke (23230),
g. d. (Trinket) by Constantine (14318), gr. g. d. (Bracelet) by Prince of Wales
(4833), — (Young Chance) by General Chasse (3874), — by Archibald (1652).

(44028)   SIR FREDERICK,

Roan, calved April 6, 1879, bred by Mr. J. P. Tynte, Tynte Park; got by Royal Fern (29865), dam (Beeswing) by British Boy (25675), g. d. (Wine Cup) by Knight of Windsor (16350), gr. g. d. (Ceres) by Coronet (15818), — (Corn Cup) by Vanguard (10994), — by Helmsman (8141), — by Tarrare (2735), — by Mameluke (2257), — by Diamond (209), — by George (1067), — by Major (398), — by Major (397), — by Brown's Bull of Aldborough (820).

(44029)   SIR GARNET,

Roan, calved July 6, 1878, bred by Mr. H. Lees, Pickhill Hall; got by Edward Waterloo (38246), dam (Cachucha) by Favourite (33895), g. d. (Columbine 2nd) by Churchwarden (23572), gr. g. d. (Columbine) by New York (22412), — (Charlotte) by Duke of Wellington (12776), — by Monk (11824), — by Dan O'Connell (9011), — by Lord Adolphus Fairfax (4249), — by Young Favourite (3770), — by Waterloo (2816) — by Lawnsleeves (365), — by Phenomenon (491), — by Favourite (252), — by Favourite (252), — by Favourite (252), — by Hubback (319), — by Snowdon's Bull (612), — by Waistell's Bull (669), — by Masterman's Bull (422), — by the Studley Bull (626).

(44030)   SIR GARNET,

Red, calved March 4, 1879, bred by Mr. Hugh Aylmer, West Dereham Abbey; got by Sir Wilfrid (37484), dam (Lady Abbess) by Heirloom (34125), g. d. (Lady Lilian) by Royal Broughton (27352), gr. g. d. (Lady Leonora 2nd) by Emperor of the North (23888), — (Lady Leonore) by Sir Samuel (15302), — by Prince Alfred (13494), — by Homer (14714), — by Justice (14753), — by Young Rufus (13649), — by Constitution (12634), — by Young Comet (1853), — descended from Jolly's Bull (4115).

(44031)   SIR GARNET,

Roan, calved March 6, 1879, bred by Mr. R. Stratton, The Duffryn; got by Pearl Diver (37182), dam (Gertrude 3rd) by Protector (32221), g. d. (Miss Glanville 3rd) by Buckingham (15700), gr. g. d. (Miss Glanville 2nd) by Waterloo (11025), — (Rosanne) by Hero of the West (8150), — by Lottery (4280), — by Phœnix (6290).

(44032)   SIR GEORGE,

Roan, calved November 19, 1879, bred by Mr. T. Stamper, Highfield House; got by Duke of Nawton 3rd (39764), dam (New Year's Gift 2nd) by Newburgh 5th (34904), g. d. (New Year's Gift) by Star of Brightness (32604), gr. g. d. (Clarice) by Lord Abbot (20140), — (Claret 2nd) by Warlaby (21063), — by Constituent (17614), — by Third Duke of Athol (12734), — by Bates (12451), — by Ingram (9236), — by Liberator (7140), — by Prince Albert (4791), — by a descendant of Mars (1199).

(44033)   SIR GEORGE FREDERICK,

Roan, calved June 2, 1879, bred by Mr. W. P. Horne, Moulton; got by Mount Blanc (40373), dam (Marvel of Peru) by Frederick First Fruits (23989), g. d. (Marvel) by Bumper (19371), gr. g. d. (Marvelmore) by Bumper (19371), — (Marigold) by Sunbeam (15360).

(44034)   SIR GLO'STER BARRINGTON,

Roan, calved January 16, 1879, bred by the Earl of Dunmore, Dunmore, the property of Mr. J. P. Clark, North Ferriby; got by Duke of Glo'ster 7th (39735), dam (Countess of Barrington 7th) by Baron Barrington 4th (33006), g. d. (Lally 15th) by Eighth Duke of Geneva (28390), gr. g. d. (Lally 8th) by Seventh Duke of York (17754), — (Lally 3rd) by Fourth Duke of Oxford (11387), — by Earl of Derby (10177), — by Earl of Liverpool (9061), — by Second Duke of Cambridge (3638), — by Belvedere (1706), — by Son of Herdsman (304), — by Wonderful (700), — by Alfred (23), — by Young Favourite (6994).

(44035)   SIR GRIFFITH,
Roan, calved July 19, 1879, bred by Captain D. H. Mytton, Garth; got by Blanche's Oxford (42796), dam (Hebe) by Berwick (23411), g. d. (Hecuba) by Hector (14685), gr. g. d. (Cornflower) by Columbus (12616), — (Oriflamme 2nd) by Red Rover (11982), — by Earl of Durham 2nd (9060), — by Raree Show (4874), — by Premier (2449), — by Albion (731), — by Sirius (598), — by Wellington (679), — by Sultan (631), — by Punch (531), — by Ladykirk (355).

(44036)   SIR GWYON,
Roan, calved April 15, 1877, bred by Mr. W. T. Crosbie, Ardfert Abbey, the property of Mr. R. S. Drought, Passage West; got by England's Glory (23889), dam (Fairy Queen) by Irish Baron (31417), g. d. (Sprite of Lothian) by Fairy King (21716), gr. g. d. (Chastity) by Lord of Lothian (18260), — (Prim) by Volunteer (15478), — by Falstaff (14529), — by Marquis of Rockingham (10506), — by Baronet (6763), — by Prince Albert (4781), — by Madcap (4308), — by Major (398), — by Cossack (925), — by Captain (108).

(44037)   SIR HENRY,
Roan, calved March 9, 1879, bred by Mr. J. Bousfield, Soulby; got by Lord of the Manor (38642), dam (Lady Alice) by Prince of Paris (29657), g. d. (Belle of Borrenthwaite 4th) by Royal Lord (29871), gr. g. d. (Belle of Borrenthwaite 3rd) by Archdeacon (21184), — (Belle of Borrenthwaite 2nd) by Fennel (19737), — by Betony (30533), — by Windsor (21117), — by China (19447), — by Vanguard (21010), — by Oswald (20445).

(44038)   SIR HERCULES,
Roan, calved September 16, 1878, bred by Mr. W. F. Beaven, Woodborough; got by Hudibras (38441), dam (Lady Selina 2nd) by Roman (35308), g. d. (Lady Selina) by Prince Ulric (29687), gr. g. d. (Village Lass) by Orion (18488), — (Alice) by Apollo (17308), — by Sir Hugh (12082), — by Son of Percy (9472), — by Son of Admiral (8806).

(44039)   SIR HUGH ROSE,
Roan, calved April 7, 1875, bred by Mr. R. Chaloner, King's Fort, the property of the Hon. Mrs. Maxwell, Fortland; got by King James (28971), dam (Lady Rose) by King Richard (26523), g. d. (Nancy) by Baron Warlaby (7813), gr. g. d. (Red Rose) by Hamlet (8126), — (Spectacles 2nd) by Prince Ernest (7366), — by Vanquish (2793), — by Monarch (4495).

(44040)   SIR JAMES,
Red, calved April 5, 1879, bred by Mr. Hugh Aylmer, West Dereham Abbey; got by Sir Wilfrid (37484), dam (Lady Lilian) by Royal Broughton (27352), g. d. (Lady Leonora 2nd) by Emperor of the North (23888), gr. g. d. (Lady Leonore) by Sir Samuel (15302), — (Lady Leonora) by Prince Alfred (13494), — by Homer (14714), — by Justice (14753), — by Young Rufus (13649), —

by Constitution (12634), — by Young Comet (1853), — descended from Jolly's Bull (4115).

(44041) SIR JOSEPH,
White, calved March 31, 1879, bred by Captain Beak, Somerford; got by Pompey (35059), dam (Japonica) by Purple Emperor (27222), g. d. (Josephine) by Viscount Killerby (19081), gr. g. d. (Juno) by Experiment (16014), — (Julia) by Pluralist (10620), — by Albert (12364), — by Augustus (6751), — by Raffler (7391).

(44042) SIR KENNETH,
Roan, calved November 18, 1878, bred by Mr. J. Cran, Kirkton; got by Bridegroom (33201), dam (Snowdrop 5th) by Regal Windsor (29765), g. d. (Snowdrop 2nd) by Scottish Chief (25102), gr. g. d. (Snowdrop) by Red Luggs (32263), — (Young Lousea) by Major (24514), — by Jock O'Darlington, — Mary of Eden, from the stock of Mr. Grant Duff.

(44043) SIR LOUIS,
Roan, calved March 9, 1879, bred by Mr. A. Mackenzie Lyle, Donaghmore House, the property of Messrs. J. G. W. and W. R. Lyle, Donaghmore House; got by Lord of the Manor (38641), dam (Flora's Pride) by Lord Wodehouse (29224), g. d. (Flora) by Alabama (30366), gr. g. d. (Clara) by Fugleman (14580), — (Rose of Tullyree) by Defender (12687), — by Gold Dust (11536), — by Nimrod (7279), — by Regent (2517), — by Dinning's Bull (973), — by Young Lancaster (361), — by Comet (155).

(44044) SIR MICHAEL,
Roan, calved July 22, 1879, bred by Mrs. Mace, Sherborne; got by Geneva's Duke (38349), dam (Beauty) by Lord of the Hills (26705), g. d. (Sally) by Butter Boy (21349), gr. g. d. (Louisa) by Royal Butterfly 14th (20722), — (Lucy Long) by Orangeman (18485), — by General Pelissier (14605), — by Chieftain (12595), — by Morpeth (8325), — by The Prince (7615), — by Colling (902), — by Son of Alexander (1624), — by Grandson of Favourite (252).

(44045) SIR MICHAEL,
Red and white, calved August 6, 1879, bred by Mr. Hugh Aylmer, West Dereham Abbey; got by Sir Wilfrid (37484), dam (Mistress Maud) by Royal Booth (29842), g. d. (Mistress Mary) by Royal Broughton (27352), gr. g. d. (Mistress May) by Prince of Rosedale (24837), — (Mistress Margaret) by Paterfamilias (18521), — by Valasco (15443), — by Baron Warlaby (7813), — by Royal Buck (10750), — by Hopewell (10332), — by Hamlet (8126).

(44046) SIR NATTY,
Red and white, calved February 5, 1879, bred by Mr. E. Freeman, Chilton; got by King Alfred (36833), dam (Chauntress) by Charming Lad (30699), g. d. (Lady Superior 4th) by Darlington (23679), gr. g. d. (Lady Superior 3rd) by Prince Leopold (20558), — (Lady Superior) by British Baronet (15691), — by Cardinal (11246), — by Marquis of Rockingham (10560), — by Sheldon (8557), — by Benjamin (1710), — by Malbro' (1189), — by Ebor (997), — by Regent (546), — by North Star (459), — by R. Colling's White Bull (151), — bred by Mr. R. Colling.

(44047)   SIR RICHARD,

Red, calved June 19, 1878, bred by Mr. H. Aylmer, West Dereham Abbey, the property of Lord Polwarth, Mertoun House; got by Sir Wilfrid (37484), dam (Cleopatra) by High Sheriff (26392), g. d. (Cinderella) by Prince Christian (22581), gr. g. d. (Calendula) by Majestic (13279), — (Calomel) by Hamlet (8126), — by Leonard (4210), — by Buckingham (3239), — from the stock of Sir M. W. Ridley, Bart.

(44048)   SIR ROBIN,

Roan, calved February 17, 1879, bred by Mr. J. Wright, Green Gill Head; got by Game Bird (39912), dam (Strawberry 2nd) by Shafto (32476), g. d. (Strawberry) by Wild Boy (25447), gr. g. d. (Lady Elvino 2nd) by Marplot (22291), — (Lady Elvino) by Orphan Doctor (13430), — by Leopold (9293), — by Duke of Norfolk (12759), — by Preston's Son of Gainford (2044), — by Chorister (3378), — by Noble's Son of Gainford (2044).

(44049)   SIR RODERIC,

Red and white, calved February 22, 1879, bred by Mr. Hugh Aylmer, West Dereham Abbey, the property of Mr. R. McDougall, Arundel, Keilor, Australia; got by Sir Wilfrid (37484), dam (Castanet 4th) by High Sheriff (26392), g. d. (Castanet 3rd) by Royal Knight (25032), gr. g. d. (Castanet 2nd) by Ravenspur (20628), — (Castanet) by Prince Arthur (13497), — by Comet (11298), — by Druid (10140), — by Petrarch (7329) — by Second Duke of York (5959), — by Raspberry (4875), — by Young Matchem (4422), — by Isaac (1129), — by Young Pilot (4702), — by Pilot (496), — by Julius Cæsar (1143).

(44050)   SIR ROGER,

Red, calved March 3, 1879, bred by Mrs. Mace, Sherborne; got by Geneva's Duke (38349), dam (Fame 2nd) by Wild Duke (32864), g. d. (Fame) by Mr. Peabody (24606), gr. g. d. (Fairy Flower) by The Druid (20948), — (Fanny Fairfax) by Homer (16277), — by Chieftain (12595), — by Magnet (10488), — by Fitz-Hardinge (8073), — by Manager (8271), — by Consul (1868), — by Gazer (7030), — by Second Fairfax (8050), — from the stock of Mr. Champion, of Blyth.

(44051)   SIR ROSE,

Roan, calved June 24, 1879, bred by Mr. L. C. Chrisp, Hawkhill, the property of Dr. Marshall, Chatton Park; got by Sir Raymond (40716), dam (Primrose 2nd) by Duke of Aosta (28356), g. d. (Rose 2nd) by Peak (24733), gr. g. d. (Napier Rosebud) by Lord Napier (14832), — (Primrose) by Sam Glen (10780), — by Peter Plough (10606), — by Eildon (9076), — by Ethelred (5990), — by Emperor (1974), — by Barmpton (1677), — by St. Albans (2584), — by Simon (590), — by Pope (514).

(44052)   SIR ROWLAND,

Roan, calved December 24, 1879, bred by Mr. Hugh Aylmer, West Dereham Abbey; got by Sir Wilfrid (37484), dam (Golden Hope) by Royal Broughton (27352), g. d. (Golden Drop) by Prince Christian (22581), gr. g. d. (Golden Chain) by Fitz-Windsor (17860), — (Gilt Hope) by Hopewell (10332), — by Vanguard (10994), — by Leonard (4210), — by Remus (4932), — by Prince Comet (1342), — by Count (170), — by Prince of Waterloo (528), — by Young Favourite (255).

(44053) SIR SIDNEY NEWPORT,

Roan, calved January 7, 1878, bred by Mr. W. G. Garne, Broadmoor, the property of Mr. Fisher, South America; got by Skylark (37489), dam (Neatness) by Brigadier (23456), g. d. (Nemophila) by Cynric (19542), gr. g. d. (Nectarine) by General Pelissier (14605), — (Necklace) by Uncle Tom (13912), — by Fitz-Hardinge (8073), — by Raffler (7391), — by Consul (1868), — by Son of Speculation (1472).

(44054) SIR THOMAS,

Red, calved March 10, 1878, bred by Mr. H. Aylmer, West Dereham Abbey, the property of Mr. A. Hamond, Westacre; got by Sir Wilfrid (37484), dam (Lady Abbess) by Heirloom (34125), g. d. (Lady Lilian) by Royal Broughton (27352), gr. g. d. (Lady Leonora 2nd) by Emperor of the North (23888), — (Lady Leonore) by Sir Samuel (15302), — by Prince Alfred (13494), — by Homer (14714), — by Justice (14753), — by Young Rufus (13649), — by Constitution (12634), — by Young Comet (1853), — descended from Jolly's Bull (4115).

(44055) SIR TRISTRAM,

Red and white, calved June 6, 1879, bred by Mr. J. Turner, Ulceby; got by Grand Duke of Glo'ster (36721), dam (Una) by Marquis of Thorndale (29304), g. d. (Ursula 32nd) by Second Duke of Waterloo (23800), gr. g. d. (Ursula 10th) by Archduke (17316), — (Ursula Booth) by Richard (15161), — by Usurer (9763), — by Prince Ernest (4818), — by A-la-Mode (725), — by Ratify (2481), — by Wellington (680), — by Favourite (252), — by Punch (531).

(44056) SIR ULSHAW,

White, calved January 14, 1879, bred by Mr. J. Topham, Middleham House, the property of Mr. F. Robinson, Ulshaw Farm; got by Heir-at-Law (34124), dam (Cowslip) by Booth's Royal Signet (28061), g. d. (Middleham Lassie) by Yorkshireman (17263), gr. g. d. (Roan Lady) by Middleham Laddie (13338), — (Lady) by The Silkey Laddie (10947), — by Bloomsbury (3173), — by Cleveland (3404), — by Danby (3550), — by Studley Grange (1483), — by Duke (225).

(44057) SIR WALTER,

Roan, calved October 16, 1878, bred by Mr. J. A. Dickinson, Brough Sowerby; got by Earl of Sheffield (33812), dam (Violet) by Whiff (30299), g. d. (Lady of the Valley) by Lord of the Valley (26722), gr. g. d. (Nancy) by Prince Patrick (16760), — (Violet) by Majestic (16492), — by Mickleton (18392), — by Len (14788), — by Voltaire (13963), — by Lord George (10439), — by Vanguard (10994), — by Hamlet (8126), — by Sir Roger (7515), — by Studley Royal (5342).

(44058) SIR WALTER OXFORD 2ND,

Roan, calved March 5, 1873, bred by Mr. R. H. Frank, Ashbourne Hall; got by Sir Walter Oxford (30012), dam (Rosina) by Royal Scotforth (25042), g. d. (Wild Rose) by Napoleon (20395), gr. g. d. (Miss Brier) by Petteril (20488), — (Sweet Brier) by Splendour (10869), — (Beauty) by Son of Young Bedlamite (6775).

(44059) SIR WILLIAM,

Roan, calved March 26, 1879, bred by Mr. W. Lett, Rushock; got by Waterloo Prince (35949), dam (Herbena) by Lord of the Lilacs 5th (26712), g. d. (Helen) by Field Marshal (19742), gr. g. d. by Marengo (13294), — by Plato (11907), — by Geddington (8102).

(44060)   SIR WILLIAM,

Roan, calved April 1, 1879, bred by Mr. S. Shaw, Brooklands ; got by Lord
Oxford Sockburn 2nd (38648), dam (Pearl) by Lord Danby (36930), g. d. (The
Snow Queen) by Royal Windsor (29890), gr. g. d. (Snowdrop) by British Hope
(21324), — (Sunflower) by Son of Apollo (9899), — by Chieftain (10048), — by
Postmaster (9487), — by Albert (7767), — by Noble (4579), — by Corelli (3485),
— by Wauldby (2818), — by Son of Wonderful (700).

(44061)   SIR WILLIAM,

Roan, calved October 13, 1879, bred by the Duke of Northumberland, Alnwick
Castle ; got by Sir Raymond (40716), dam (Lucretia 6th) by Fitz-Roland (33936),
g. d. (Lucretia 4th) by Mayor of Windsor (31897), gr. g. d. (Lucretia 2nd) by
Royal Butterfly 23rd (27355), — (Lucretia) by Knight of the Grand Cross 2nd
(26551), — by Majestic (16492), — by Ambo (12391), — by Lord George
(10439), — by Vanguard (10994), — by Hamlet (8126), — by Sir Roger (7515),
— by Studley Royal (5342).

(44062)   SKELSMERGH JACK,

White, calved May 2, 1879, bred by the Exors. of Mr. Morton, Skelsmergh
Hall, the property of Mr. R. Martindale, Benson Hall ; got by Duke of Barrington 2nd (36462), dam (Jemima 2nd) by Atherton's Oxford (21195), g. d.
(Jemima) by Freedom (17884), gr. g. d. (Jennet 2nd) by Duke of Westmoreland
(12780), — (Jessy 5th) by Pantaloon (9467), — by Tommy Lad (6611), — by
Young Rockingham (2549), — by Monitor (2331), — by Young Rockingham
(2549), — by Major (2255), — by a Bull of Mr. Newby's, — by Northumberland
(464), — by Buston's Styford (103), — by Bolingbroke (86).

(44063)   SKYLARK,

Red, calved May 13, 1879, bred by Mr. J. Singleton, Teresa Cottage ; got by
Duke of Oxford 37th (38171) dam (Countess of Oxford 2nd) by Telemachus
(27603), g. d. (Lady Oxford) by Imperial Oxford (18084), gr. g. d. (Sandpiper)
by The Briar (15376), — (Water Wagtail) by Francisco (12893), — by Columbus
(10063), — by Prince of Wales (6345), — by Edward (3696), — by Topper
(2768), — by Favourite (3768), — by Windsor (698), — by Windsor (698).

(44064)   SNOWBALL,

White, calved February 14, 1879, bred by Messrs. W. Hosken and Son, Hayle.
the property of Mr. R. Humphreys, Brew, Sennen, Cornwall ; got by Duke of
Oxford 33rd (36528), dam (Miss Ada 5th) by Towneley Oxford (30170), g. d.
(Miss Ada 3rd) by Second Earl of Oxford (23844), gr. g. d. (Miss Ada) by
Royal Oak (22793), — (Duchess) by Count Cavour (19523), — by Brigadier
(14193), — by Usurper (13929), — by Fawsley (6004), — by Javelin (4093),
— by Blyth (797), — by Wellington (684), — by Phenomenon (491), — by
Favourite (252), — by Favourite (252), — by Favourite (252), — by Hubback
(319), — by Snowdon's Bull (612), — by Waistell's Bull (669), — by Masterman's Bull (422), — by the Studley Bull (626).

(44065)   SNOWDON,

White, calved May 11, 1879, bred by Mrs. Healey, Morris Grange ; got by
Mount Blanc (40373), dam (Kitty) by K. C. B. (26492), g. d. (Susan) by British
Hope (21324), gr. g. d. (Sunflower) by Son of Apollo (9899), — (Sally) by
Chieftain (10048), — by Postmaster (9487), — by Albert (7767), — by Noble

(4579), — by Corelli (3485), — by Wauldby (2818), — by Son of Wonderful (700).

(44066) SNOWDON,

White, calved May 11, 1879, bred by Mr. J. Tyacke, Merthen; got by Plantagenet (38871), dam (Syringa) by Monarch Gwynne (37103), g. d. (Syren 2nd) by Don Pedro (25910), gr. g. d. (Syren) by Hector (24117), — (Sultana 3rd) by Red Duke (16803), — by Sir Roger de Coverley (12095), — by The Red Duke (8694), — by Phœnix (6290), — Wallflower, bought of Mr. Wall.

(44067) SNOWFLAKE,

White, calved May 18, 1879, bred by Mr. J. Moffat, Ballyhyland, sold to the Provincial Government of Brabant, Belgium; got by Royalist (37396), dam (Snowball) by Lord Mayor (29153), g. d. (Saucy 2nd) by Grandee (26298), gr. g. d. (Saucy) by Sugar Candy (22987), — (Saffron) by Berrington Boy (11173), — by Claret (10057), — by Sir Charles (7501), — by Son of Cupid (177), — by Exotic (16012).

(44068) SNOWFLAKE,

White, calved December 3, 1879, bred by Messrs. N. Russell and Son, Northallerton; got by Sir Andrew (42387), dam (Virginia) by Clarion (33393), g. d. (Verbena 8th) by Sir Richard (20843), gr. g. d. (Verbena 4th) by Red Duke (18676), — (Verbena 3rd) by Marc Antony (14895), — by Walburn Buchan (9797), — by Hurricane (4060), — by Luck's All (2230), — by Bedford (68), — by Barrister (776), — by Whitworth (1584), — by Young Jupiter (1148), — by White Comet (1582), — by Cattley's Grey Bull (1798).

(44069) SNOW MAN,

White, calved February 20, 1879, bred by Mr. C. Cradock, Hartforth; got by Chaser (37970), dam (Verona) by Wanderer (32790), g. d. (Valencia) by Valasco (15443), gr. g. d. (Lady Celia) by Rifleman (15163), — (Ciss) by Young Hopewell (14719), — by Bellemont (11164), — by Dan O'Connell (3557), — by Lord Lieutenant (4260), — by Outhwaite's White Bull (9825), — by North Star (459), — by Young Comet (157), — by Easby (232).

(44070) SOCKBURN GWYNNE,

White, calved March 6, 1879, bred by Mr. S. Shaw, Brooklands; got by Lord Oxford Sockburn 2nd (38648), dam (Silky Gwynne 2nd) by Duke of Waterloo (28464), g. d. (Silky Gwynne) by Sir Windsor (22927), gr. g. d. (Susan Gwynne) by Duke of York (14461), — (Sally Gwynne) by Lablache (11656), — by Sir Thomas (10777), — by Prime Minister (2456), — by Wallace (5586), — by Marmion (406), — by Merlin (430), — by Layton (366), — by Phenomenon (491), — by Favourite (252), — by Favourite (252), — by Hubback (319), — by Snowdon's Bull (612), — by Waistell's Bull (669), — by Masterman's Bull (422), — by the Studley Bull (626).

(44071) SOLWAY,

Red and white, calved February 28, 1878, bred by Lord Polwarth, Mertoun House, the property of the Hon. R. Baillie Hamilton, Langton; got by Rapid Rhone (35205), dam (Silver Chain) by Coronet (33448), g. d. (Sincere) by Scotland's Pride (25100), gr. g. d. (Superb) by The Czar (20947), — (Splendour) by Lord Sackville (13249), — by Duke of Athol (10150), — by Earl of Derby (10177), — by Duke of Sutherland (6945), — by Locomotive (4242), — by Short Tail (2621), — by Gambier (2046), — by Young Wynyard (2859), — by Bulls of Messrs. C. and R. Colling's.

(44072)   SONSIE BOY,

Red, calved October 14, 1879, bred by Mr. L. Rawstorne, Hutton Hall; got by Duke of Wellington 12th (38199), dam (Sonsie Belle) by Baron Oxford 6th (33075), g. d. (Sonsie 18th) by Royal Cambridge (25009), gr. g. d. (Sonsie 13th) by Perth (22513), — (Sonsie 11th) by Lord Durham (16410), — by Comus (12626), — by Hunt (7103), — by Mons. Vestris (6220), — by Sultan Selim (2710), — by Prince Edward (2462), — by Sultan (1485), — by Son of Trunnell (659), — by Middleton (1235), — by Son of Ben (70), or Punch (531).

(44073)   SONSIE LAD 2ND,

Roan, calved September 3, 1878, bred by Sir Wilfrid Lawson, Bart., Brayton; got by Wild Eyes Duke (36007), dam (Sonsie 22nd) by Royal Cambridge (25009), g. d. (Sonsie 12th) by Brayton 3rd (19335), gr. g. d. (Sonsie 11th) by Lord Durham (16410), — (Sonsie Lass) by Comus (12626), — by Hunt (7103), — by Mons. Vestris (6220), — by Sultan Selim (2710), — by Prince Edward (2462), — by Sultan (1485), — by Son of Trunnell (659), — by Middleton (1235), — by Son of Ben (70), or Punch (531).

(44074)   SOVEREIGN,

Roan, calved February 9, 1878, bred by Mr. G. Wilson, Netherton Clatt; got by Braithwaite (28066), dam (Lady Dorothy 6th) by King of the Isles (31506), g. d. (Lady Dorothy 3rd) by Master Butterfly 13th (16531), gr. g. d. (Lady Dorothy 2nd) by Champion (14255), — (Lady Dorothy) by Guy Fawkes (12981), — by The Pacha (7612), — by Second Duke of Northumberland (3646), — by Sillery (5131), — by Carleton (843), — by Diamond (205), — by Diamond (205).

(44075)   SOVEREIGN 10TH,

Red and white, calved July 10, 1878, bred by Mr. T. Morris, Maisemore Court; got by Duke of Siddington 3rd (38183), dam (Stranger) by Charleston (21400), g. d. (Stranger) by Kent Oxford (20047), gr. g. d. (Stranger) by Samson (12045), — (Splendour) by Shakespeare (10801), — by Vincent (5567), — by Duke (1934), — by Acton (716), — bred by Mr. Strickland.

(44076)   SPANIARD,

Roan, calved February 4, 1879, bred by Mr. G. W. Lambart, Beau Parc; got by Jupiter (38477), dam (Armada) by Rupert (29902), g. d. (Queen Elizabeth) by British Sailor (23472), gr. g. d. (Her Royal Highness) by Royal Standard (40644), — (Her Majesty) by Prince Ernest (7366), — by Solway (7530), — by General (9146), — by Narcissus (4540), — by Monarch (2324), — by Planet (1325).

(44077)   SPECK,

Roan, calved March 16, 1879, bred by Sir P. Miles, Bart., Leigh Court; got by The Prior (42493), dam (Spot) by Proud Youth (32224), g. d. (Pride) by Hopeful (26410), gr. g. d. (Jilt) by Maynard (24569), — (Coquette) by Bladrid (14165), — by Franklin (14568), — by Latimer (9280), — by Son of Zoophite (15537), — by Magnum Bonum (2343), — Chance, from the stock of Colonel Cradock, of Hartforth.

(44078)   SPECULATION,

Roan, calved May 20, 1877, bred by Mr. M. Boyes, Wandale House; got by Jocelyn's Tregunter (36808), dam (Special Notice) by Duke Royal 2nd (26038),

g. d. (Seven Stars) by Sensation (22869), gr. g. d. (Strawberry) by Neal Dow (16608), — (Kora) by Uncle Tom (13915), — by Grimaldi (7054), — by Antonio (3019), — by Victory (5565), — by Pyramid (4852), — by Harry Lorrequer (3985), — by Blucher (84), — by Magnum Bonum (4322), — by Buston's Styford (103), — by Son of T. Wetherell's Bull (690).

(44079)   **SPENSER,**

Roan, calved January 17, 1878, bred by Mr. G. Murton Tracy, Redlands, the property of Mr. F. R. Moser, Carbery; got by Prince of the Blood (37259), dam (Faery Queene) by Cherry Emperor (33347), g. d. (Fanny) by Enterprise (16002), gr. g. d. (Emma 3rd) by Richmond (13592), — (Emma 1st) by Free Trader (10246), — by Gainford (2044), — by Magnum Bonum (2243), — by Rob Roy (557), — by a Son of Houghton (318), — by Sir Stephen (1456), — by Sedbury (1424).

(44080)   **SPLENDID BUTTERFLY,**

Roan, calved August 9, 1878, bred by Lord Polwarth, Mertoun House; got by Rapid Rhone (35205), dam (Broughton Butterfly) by Victorious (25378), g. d. (Alice Butterfly) by Master Butterfly (13311), gr. g. d. (Alice 2nd) by Duke of Athol (10150), — (Madeline) by Marcus (2262), — by Matchem (2281), — by Pilot (496), — by Young Albion (15), — by Albion (14), — by Suworrow (636), — by Son of Twin Brother to Ben (88), — by Twin Brother to Ben (660).

(44081)   **SPLENDOUR ROYAL,**

Roan, calved March 30, 1879, bred by Mr. J. Moffat, Ballyhyland, sold to the Provincial Government of Brabant, Belgium; got by Royalist (37396), dam (Splendour) by Lord of the Manor (29181), g. d. (Rosalind 3rd) by Dey of Algiers (25892), gr. g. d. (Rosalind 2nd) by Grey Gauntlet (19908), — (Rosalind) by Syntax (13821), — by Hamlet (8126), — by Baronet (6763), — by Guy Fawkes (7062), — by Pedestrian (7321), — by Edrom (1956), — by Scipio (1421), — by Hector (1104), — by Midas (435), — by Marquis (407), — by Chilton (136), — by Ben (70).

(44082)   **SPRIGHTLY VICTOR,**

Roan, calved March 16, 1879, bred by Messrs. W. and H. Morley, East Gate Farm; got by Broomley Victor (37917), dam (Sprightly Duchess) by White Duke (32849), g. d. (Sprightly Maid) by Royal Arthur (29840), gr. g. d. (Sprightly 3rd) by Richard Cœur-de-Lion (13590), — (Sprightly) by Abraham Parker (9856), — by Bumper (10005), — by Speculation (8621), — by Young Hastings (3988), — by Bachelor (1666), — by Sir Harry Liddell (5157), — by Leopold (2199), — by Sir Harry (5155), — by Surly (2715), — by Traveller (655), — by Colonel (152).

(44083)   **SQUIRE GARNETT,**

White, calved November 14, 1878, bred by Mr. W. H. Wakefield, Sedgwick, the property of Mr. J. Bentham, Blease Hall; got by Baron Barrington 4th (33006), dam (Garnett 7th) by Dunrobin (28486), g. d. (Garnett 3rd) by Frederick Warlaby (23990), gr. g. d. (Garnett 1st) by General Garibaldi (21813), — (Lancaster) by Faust (16033).

(44084)   **SQUIRE ROSEDALE,**

Roan, calved February 23, 1879, bred by Mr. R. Darling, Plawsworth; got by Titanus (40821), dam (Rosalind) by Sirius (39121), g. d. (Rosedale 3rd) by Royal Buckingham (20718), gr. g. d. (Rosy) by Master Belleville (11795), —

(Red Rose) by Vanguard (10994), — by Diamond (5918), — by Young Matchem (2282), — by Jack Tar (1133), — by Pilot (496), — by Young Albion (15).

(44085)　　　　STANDARD,
Roan, calved January 5, 1879, bred by Messrs. Christy Brothers, Fort Union, the property of Lady Jane Moore, Moore Park ; got by Lord Broughton (31626), dam (Ringlet) by Fairy King (21716), g. d. (Choral Ruby) by Chorister (17565), gr. g. d. (Ruby) by Abacus (15542), — (Limerick Lass 2nd) by Lord Spencer (13251), — by Young Shaftoe (9625), — by Young Zealot (8797), — by Vandyke (2791), — by Despatch (983), — by Remus (1369), — by Stamford (1476).

(44086)　　　　STANDARD BEARER,
Red, calved June 1, 1879, bred by Sir D. Baird, Bart., Newbyth ; got by Skirmisher (40729), dam (Jezebel) by George 1st (19848), g. d. (Jura) by Tenedos (20936), gr. g. d. (Jane) by The Sheriff (12216), — (Janette) by Triumph (8717), — by Eclipse (3688), — by Coxcomb (3514), — by Juniper (2165), — by Cedric (1801), — by The Squire (1512), — by Midas (435), — by Boughton (90).

(44087)　　　　STANLEY,
Roan, calved May 25, 1879, bred by Mr. R. B. Brockbank, Crosby ; got by Borderer (33183), dam (Iris Flower) by Third Duke of Oxford (23783), g. d. (Alfred's Flower) by Alfred Fitz-Clarence (19215), gr. g. d. (Miss Nicety) by Veteran (13941), — (Nina) by Waterloo Hero (13981), — by Duke of Richmond (7996), — by Lord Stanley (4269), — by Velocipede (5552), — by Priam (2452), — by Jerry (4097), — descended from the stock of Mr. Booth.

(44088)　　　　STANLEY ROSE'S EARL,
Roan, calved January 26, 1879, bred by Mr. H. Smith, Mountmellick ; got by Earl of Aylesby (38215), dam (Stanley Rose) by Lord Stanley (24466), g. d. (Alpine Rose) by Prince Bertram (27119), gr. g. d. (Rosette) by Duke of York (23804), — (Young Moss Rose) by Dr. McHale (15887), — by Dorrington (17691), — by Hopewell (10332), — by Baron Warlaby (7813), — by Pilgrim (4701), — by Shakespeare (2614), — by Sheridan (2616), — by Normanby (1278), — by a Bull of Mr. W. Smith's, West Rasen.

(44089)　　　　STAR 3RD,
White, calved June 26, 1879, bred by Sir W. G. Armstrong, Cragside ; got by Duke of Oxford 27th (33709), dam (Oxford Rose 5th) by Oxford Beau 4th (34964), g. d. (Wharfdale Rose) by Third Lord Wharfdale (26759), gr. g. d. (Yetholm Rose) by Prince of Yetholm (22639), — (Clarence Rose) by Duke of Clarence (19611), — by Sixth Duke of Oxford (12765), — by Ravensworth (9532), — by Freebooter (7025), — by Ganthorp (2049), — by Belshazzar (1704), — by Don Juan (1923), — by Shylock (2622), — by Muggeen's Bull.

(44090)　　　　STAR KING,
White, calved August 29, 1879, bred by Mr. A. E. W. Darby, Adcote ; got by Magnate Star (37029), dam (Vesper Queen) by Royal Commander (29857), g. d. (Vernal Star) by The Sutler (23061), gr. g. d. (Venus Star) by Prince George (13510), — (Vesper) by King Arthur (13110), — by Morning Star (6223), — by Roland (2556), — by Priam (2452), — by Matchem (2281), — by Son of Peter (487).

(44091) STAR OF GOLDCLIFF,

Red and white, calved June 26, 1879, bred by Mr. W. Price, Goldcliff; got by Prince of Goldcliff (43827), dam (Promised Land 5th) by James 2nd (24203), g. d. (Promised Land) by Star of Gwent (25227), gr. g. d. (Lady Bird 2nd) by Prince of the Empire (20578), — (Lady Olivia) by Prince Imperial (15095), or Chief (18976), — by Sir Thomas (13749), — by Lord George (10439), — by Fitz-Leonard (7010), — by Velocipede (5552), — by Young Matchem (4422), — by Jack Tar (1133), — by Pilot (496), — by Young Albion (15).

(44092) STAR OF HALNABY,

Red, calved March 20, 1879, bred by Mr. W. T. Crosbie, Ardfert Abbey; got by Royal Halnaby (39041), dam (Vesper Star) by Sir Windsor Broughton (27507), g. d. (Star Queen) by The Sutler (23061), gr. g. d. (Star of Windsor) by Windsor (14013), — (Vesper) by King Arthur (13110), — by Morning Star (6223), — by Roland (2556), — by Priam (2452), — by Matchem (2281), — by Son of Peter (487).

(44093) STAR OF THE BORDER,

Roan, calved May 21, 1878, bred by Mr. A. Bain, Legars, Stichel, the property of Mr. W. Duthie, Collynie; got by Lancer (41786), dam (Annabel Lee) by Lord Charles (34503), g. d. (Grahamslaw 3rd) by Free Stanley (26204), gr. g. d. (Grahamslaw 1st) by Redmond 2nd (20645), — by Gold Dust (11536), — by Chieftain (11273), — by Lilyson (6131), — by Bowmont (3200), — by Young Exmouth (9101), — by Mr. Robson's Bull (9562), — by Prince (4765), — by Wellington (2824).

(44094) STAR OF WORCESTER,

Red, calved March 7, 1879, bred by Mr. H. Webb, Streetly Hall; got by Viscount Worcester (40881), dam (Barrington Star) by Duke of Barrington 5th (33575), g. d. (Star of the Evening) by Lord Chancellor (20160), gr. g. d. (Saturn) by Sambo (20781), — (Shining Light) by Henry 5th (19944), — by Reo Bulla (10703), — by Preston (8408), — by Pioneer (9480), — Strawberry, bought of Mr. Lyon.

(44095) STATESMAN,

Red, calved February 21, 1864, bred by Mr. W. S. Marr, Uppermill; got by Master Gunner (22316), dam (Salvia) by The Baron (13833), g. d. (Stephania) by Procurator (10657), gr. g. d. (Sympathy) by Prince Edward Fairfax (9506), — (Fancy) by Billy (3151), — by Sovereign (7539), — by Satellite (1420), — by Baronet (60), — by Cleveland (144), — by Symmetry (641).

(44096) STATESMAN 1st,

Red, calved in July, 1878, bred by Mr. S. Campbell, Kinellar, the property of Mr. J. Isaac, Bowmanton, Ontario, Canada; got by Golden Prince (38363), dam (Nonpareil 30th) by Royal Duke (35356), g. d. (Nonpareil 28th) by Sir Christopher (22895), gr. g. d. (Nonpareil 24th) by Lord Sackville (13249), — (Nonpareil 23rd) by The Baron (13833), — by Matadore (11800), — by Prince Edward Fairfax (9506), — by Diamond (5918), — by Young Frederick (3836), — by Commodore (1858), — by Tathwell Studley (5401), — by Blyth Comet (85).

(44097) STATIRA DUKE 10th,

Red, calved July 26, 1879, bred by Mr. H. Lovatt, Low Hill; got by Baron Turncroft Oxford 4th (37822), dam (Statira Duchess 2nd) by Fourth Duke of

Grafton (28396), g. d. (Statira 8th) by Twelfth Duke of Oxford (19633), gr. g. d. (Statira 6th) by Britannicus 2nd (19349), — (Statira) by Duke of Glo'ster (11382), — by Balco (9918), — by Sir Launcelot (5166), — by Major (4345), — by Ganthorpe (2049), — by Don Juan (1923), — by Shylock (2622).

(44098) STOIC,
Roan, calved July 6, 1879, bred by Mr. J. Moffat, Ballyhyland, sold to the Provincial Government of Brabant, Belgium; got by Royalist (37396), dam (Roan Soho) by Dey of Algiers (25892), g. d. (Soho) by White Chieftain (21096), gr. g. d. (Impudence) by Sir Colin (16960), — (Soho) by Lord St. Leonards (13250), — by Juan (6106), — by Gainford (2044), — by Telemachus (5412), — by King Ernest, — by Fergus (3783).

(44099) STRAWBERRY CHIEF,
Roan, calved March 12, 1879, bred by Sir Wilfrid Lawson, Bart., Brayton ; got by Wild Eyes Duke (36007), dam (Strawberry 2nd) by Wellington (32825), g. d. (Strawberry) by Grand Vizier (26313), gr. g. d. (Spotless) by Perth (22513), — (Spot) by Duke (14419), — by Captain Hardinge (14232), — by Evan Gwynne (14521), — by a Son of Wallace (5586), — by Cupid (1893), — by Charlie (130), — by Western Comet (689).

(44100) STRAWBERRY DUKE,
Roan, calved April 17, 1879, bred by Mr. J. Owen, Middleton; got by White Prince (40914), dam (Miss Helena) by Prince Charlie (27129), g. d. (Strawberry) by Grand Monarch 2nd (19887), gr. g. d. (Buttercup 2nd) by Don Pedro (19584), — (Young Buttercup) by Alma (14087), — by Concord (11301), — by Second Duke of Lancaster (5951), — by Crichton (3516), — by Ploughboy (4726).

(44101) STREPHON,
Red, calved July 29, 1879, bred by Lord Bolton, Bolton Hall; got by St. Swithin's Star Drop (40667), dam (Simplicity 2nd) by Heir-at-Law (34124), g. d. (Simplicity) by Marmion (26821), gr. g. d. (Shepherdess) by Stonegrave (27575), — (Strawberry) by Homer (14714), — by Starling (15338), — by Baronet (19273), — by Lord Cardigan (7158), — by Clementi (3399).

(44102) STUDENT,
Roan, calved March 3, 1879, bred by Miss Graham, Yardley Stud Farm ; got by Baron Fantail (37790), dam (Oxford Mixture 2nd) by Lord Oxford Rosy (38647), g. d. (Oxford Mixture) by Gräf Renard (31278), gr. g. d. (Red Daisy 2nd) by The Earl (27624), — (Red Daisy) by Cæsar (23499), — by Barleycorn (17348), — Red Cow, purchased from Mr. Bates, Kingsheath.

(44103) STUDGILL BANK,
Roan, calved January 23, 1879, bred by Mr. T. Blenkarn, Keisley; got by Kirkbythore Wild Eyes 4th (36865), dam (Bright Eyes 4th) by Earl of Carlisle (23826), g. d. (Young Bright Eyes) by Lord Newby (26695), gr. g. d. (Bright Eyes) by Garibaldi (24008), — (Lady White) by Benedict 2nd (25624), — by Hopeful (9222), — by Loyal (6154).

(44104) STUDLEY,
Roan, calved June 12, 1879, bred by the Rev. T. Sheepshanks, Arthington Hall; got by Earl Roseberry (33825), dam (Villette) by Baron Winsome 6th (33111), g. d. (Lady Vane) by Lord Cobham (20164), gr. g. d. (Vanity) by Reformer (18687), — (Vanilla) by Robinson Crusoe (13610), — by Borrowby Boy (9980),

— by Laudable (9282), — by Sir Thomas Fairfax (5196), — by Colossus (1847), — by Burley (1766), — by Pilot (496), — by Warlaby (672), — by Albion (14), — by Lame Bull (359), — by Shipton (587), — by Son of Suworrow (636), — by Son of Twin Brother to Ben (88), — by Twin Brother to Ben (660).

(44105)   **SUGAR BARLEY,**
Red and white, calved March 15, 1879, bred by Mr. J. P. Tynte, Tynte Park, the property of Mr. H. Smith, Mountmellick ; got by Royal Fern (29865), dam (Sugar Candy) by Bright Spur (25671), g. d. (Date) by British Duke (19350), gr. g. d. (Cinnamon) by Comrade (14307), — (Honeysuckle) by High Sheriff (13027), — by Barley Sugar (11140), — by Prince Ernest (7366), — by Captain (11241), — by The Doctor (2748), — by Brough (3228), — by Pilot (496).

(44106)   **SULTAN 2ND,**
Roan, calved March 22, 1877, bred by Mr. W. Arkell, Dudgrove ; got by Fifteenth Baron Wetherby (25596), dam (Satellite) by Sixteenth Baron Wetherby (25597), g. d. (Snowdrop) by Brazenose (15686), gr. g. d. (Snow Blossom) by Napoleon 2nd (14975), — (Lily) by Protector (13537), — by Homer (2134), — by Reformer (4915), — by Son of Sir Roger de Coverley (2634), — by Young York (1592), — by Sultan (5344), — by Burgess's Red Bull (3139), — Old Countess.

(44107)   **SUNRISE,**
Roan, calved July 22, 1879, bred by Mr. E. Randall, East Down Lodge, the property of Mr. E. Martin, East Down Lodge ; got by Fred Vokes (43248), dam (Sunbeam) by Spectator (37508), g. d. (Sunflower) by Brazenose (15686), gr. g. d. (Flash) by Napoleon 2nd (14975), — (Cowslip) by Atlas (8845), — by Colling (902), — by Son of Alexander (1624), — by Grandson of Favourite (252).

(44108)   **SURMISE DUKE 7TH,**
Red, calved August 4, 1878, bred by Sir Curtis M. Lampson, Bart., Rowfant ; got by Duke of Underley 2nd (36551), dam (Surmise Duchess 11th) by Grand Duke of Geneva (28756), g. d. (Fancy Duchess) by Ninth Duke of Geneva (28391), gr. g. d. (Fancy 2nd) by Rowfant 1st (22767), — (Fancy) by Fourth Duke of Thorndale (17750), — by Duke of Glo'ster (11382), — by Earl of Derby (10177), — by Duke of Sutherland (6945), — by Locomotive (4242), — by Short Tail (2621), — by Gambier (2046), — by Young Wynyard (2859), — by Bulls of Messrs. C. and R. Colling's.

(44109)   **SURVEYOR,**
Red and white, calved February 13, 1879, bred by Mr. J. Brigham, Slingsby ; got by Mowbray (40374), dam (Lady Maynard 19th) by Brigadier (28079), g. d. (Lady Maynard 8th) by Third May Duke (24567), gr. g. d. (Lady Maynard 3rd) by Euclid (17815), — (Charade) by Syntax (13821), — by Marquis of Rockingham (10506), — by Wizard (6688), — by Ranunculus (2479), — by Monarch (2324), — by Jupiter (342), — by Sir Oliver (605), — by Trunnell (659), — by Favourite (252), — by Favourite (252), — by Dalton Duke (188), — by R. Alcock's Bull (19), — by J. Smith's Bull (608), — by Jolly's Bull (337).

(44110)   **SUTHERLAND,**
Roan, calved May 24, 1879, bred by Mr. G. T. Phillips, Sheriff Hales Manor ; got by Viscount Cherry (37635), dam (Charming Princess) by Prince Charming

(27130), g. d. (Honeysuckle) by Duke of Manchester (15934), gr. g. d. (Honey Flower) by Sultan (15355), — (Caroline) by Musician (13361), — by Fame (10221), — by Lord Adolphus Fairfax (4249), — by Barmpton Grazier (3091), — by Rex (4942), — by Wellington (680), — by Favourite (252).

(44111)　　　　　　SWEETBREAD,
Red and white, calved October 20, 1879, bred by Mr. F. J. S. Foljambe, Osberton Hall; got by Titan (35805), dam (Sweetheart 31st) by Sweet-Pea (35708), g. d. (Sweetheart 28th) by Count Leinster (23638), gr. g. d. (Sweetheart 11th) by The Baron (13833), — (Sweetheart 3rd) by Daybreak (11338), — by Accordion (5708), by Little John (4232), — by Caliph (1774), — by Sir Walter (2637), — by Hotspur (1117), — by Coxcomb (928), — by Midas (435), — by Comet (155), — by R. Colling's Son of Favourite (252), — by same Son of Favourite (252), — by Hubback (319).

(44112)　　　　　　SWEET SAUCE,
Red, calved May 30, 1879, bred by Mr. F. J. S. Foljambe, Osberton Hall; got by Titan (35805), dam (Sweetheart 32nd) by Sweet-Pea (35708), g. d. (Sweetheart 29th) by Knight of the Bath (26546), gr. g. d. (Sweetheart 28th) by Count Leinster (23638), — (Sweetheart 11th) by The Baron (13833), — by Daybreak (11338), — by Accordion (5708), — by Little John (4232), — by Caliph (1774), — by Sir Walter (2637), — by Hotspur (1117), — by Coxcomb (928), — by Midas (435), — by Comet (155), — by R. Colling's Son of Favourite (252), — by same Son of Favourite (252), — by Hubback (319).

(44113)　　　　　　SWEET WILLIAM,
Roan, calved May 22, 1879, bred by Mr. R. P. Maxwell, Finnebrogue; got by Czarowitz 4th (33497), dam (Honey Dew) by Prince Victor (20606), g. d. (Honeysuckle 3rd) by Dillon (17680), gr. g. d. (Honeysuckle) by Champion (12577), — (Hybla) by Master Fairfax (13314), — by Sir Thomas Fairfax (5196), — by Hubback (2142), — by Son of Young Warlaby (2812), — by Imperial (2151), — supposed by Young Comet (905).

(44114)　　　　　　SYDNOPE DAISY BULL,
Roan, calved August 3, 1879, bred by Mr. B. Langdale Barrow, Sydnope Hall, the property of Major Pountain, Barrow Hall; got by Twentieth Duke of Oxford (28432), dam (Royal Bijou) by Royal Cumberland (27358), g. d. (Bijou 2nd) by The Baron (25277), gr. g. d. (Bijou) by The Bully (23019), — (Jessie) by Lord Raglan (13222), — by Patriot (10594), — by Sir Frederick (8577), — by Marquis (4386), — by Mynheer (1255), — by Enchanter (244), — by Major (398), — by Windsor (698), — by Favourite (252), — by Punch (531), — by Hubback (319).

(44115)　　　　　　SYLVAN,
Roan, calved October 12, 1879, bred by Messrs. Pender, Budockvean; got by Fawsley (41531), dam (Silene) by Sailor Boy (37419), g. d. (Syren 2nd) by Don Pedro (25910), gr. g. d. (Syren) by Hector (24117), — (Sultana 2nd) by Red Duke (16803), — by Sir Roger de Coverley (12095), — by The Red Duke (8694), — by Phœnix (6290), — Wallflower, bought of Mr. Wall.

(44116)　　　　　　TACITURN 3RD,
Red and white, calved July 27, 1879, bred by Messrs. F. Leney and Sons, Water-ingbury; got by Sixth Duke of Oneida (30997), dam (Tacita 4th) by Fifth Duke

of Wharfdale (26033), g. d. (Tacita 2nd) by Didmarton Duke (21546), gr. g. d. (Surmise 3rd) by May Duke (13320), — (Surmise) by Duke of Glo'ster (11382), — by Earl of Derby (10177), — by Duke of Sutherland (6945), — by Locomotive (4242), — by Short Tail (2621), — by Gambier (2046), — by Young Wynyard (2859), — by Bulls of Messrs. C. and R. Colling's.

(44117)   TACITUS,
Red, calved October 10, 1878, bred by Captain D. H. Mytton, Garth; got by Constantine 2nd (33439), dam (Agnes) by Vespasian (32759), g. d. (Gem) by Berwick (23411), gr. g. d. (Glossary 2nd) by Hardware (19919), — (Glossy) by Hazel (26352), — by Harlequin (26340), — by Emperor (11438), — by Priam (15079).

(44118)   TAURUS,
Red, calved June 11, 1879, bred by Mr. J. F. Armistead, Cob Wall House; got by Brierfield (39489), dam (Jemima) by Volunteer (30239), g. d. (Jennet) by Orion 2nd (22445), gr. g. d. (Red Sal) by Sportsman (18909), — (Young Sal) by May Duke (13320), — by Tom of Lincoln (8714), — by Daniel (3554), — by Danby (1900), — by Grazier (3927), — by Red Bull (5658), — by Sedbury (1424), — by Smurthwaite's Bull (5219), — by Nailer (2347), — by Danby (190).

(44119)   TAXUS,
Red, calved February 17, 1879, bred by Mr. A. Garfit, Scothern, the property of Mr. J. L. Newham; got by Grand Duke 25th (34065), dam (Falernian) by Red Cross Knight (35218), g. d. (Epernay) by Rose Butterfly (24993), gr. g. d. (Cœcuba) by Champagne (17522), — (Signora) by Grand Signor (14644), — by Artichoke (14107), — by Scimitar (10788), — by Allan-a-Dale (7778), — by Little John (4232), — by Caliph (1774), — by Swing (2721), — by Argus (759), — by Defender (194), — by Petrarch (488), — by Own Brother to R. Colling's White Heifer, — by Butterfly (104), — by Globe (278).

(44120)   TENNIS BALL,
Roan, calved June 26, 1879, bred by Mr. C. Cradock, Hartforth; got by Chaser (37970), dam (Racket) by Pilgrim (35036), g. d. (Strawberry Drop) by Strawberry Prince (25240), gr. g. d. (Raindrop) by Dairy Prince (17655), — (Eardrop) by Apollo (9899), — by Richmond (13592), — by Free-Trader (10246), — by Gainford (2044), — by Magnum Bonum (2243), — by Rob Roy (557), — by Son of Houghton (318), — by Sir Stephen (1456), — by Sedbury (1424).

(44121)   THAKOOR,
Red and white, calved May 30, 1878, bred by Mr. R. S. Brundell, Leicester House, Doncaster, the property of Mr. J. Elmhirst, Thorne; got by Shazadah (37440), dam (Amelia) by Barbarossa (30415), g. d. (Amy) by Earl of Kerry (23839), gr. g. d. (Amaryllis) by Prince Rupert (20605), — (Crocus) by Norfolk (14999), — by Kingston (13118), — by Jack Andrews (9243), — by Lane's Bull (8208), — by Aisthorpe Bull (6708), — by Herdsman (6076), — by Young Quaternion (8447), — by Young Warrior (8755), — by Waddingworth (668).

(44122)   THAMES WHARFDALE,
Red and white, calved December 11, 1879, bred by Mr. F. Dodd, Rush Court got by Lord Darlington 14th (40149), dam (Annie Wharfdale) by His Lordship

(26398), g. d. (Agnes Wharfdale) by Duke of Wharfdale (19648), gr. g. d.
(Abigail) by Marquis of Oxford (18339), — (Arbelle) by Union (12249), — by
Contract (10071), — by Bristol (7852), — by Brunel (7857), — by Leo (4208),
— by Ovis (4635), — by Rival (2534), — by Darlington (956), — by Denton
(198), — by Rockingham (560), — by Jobling's Son of Phenomenon (491), —
by Phenomenon (491), —by Colonel (152), — by Styford (629).

(44123)   THE ABBOT,
Red and white, calved March 21, 1879, bred by Sir P. Miles, Bart., Leigh Court ;
got by The Prior (42493), dam (Coquette 2nd) by Proud Youth (32224), g. d.
(Pride) by Hopeful (26410), gr. g. d. (Jilt) by Maynard (24569), — (Coquette)
by Bladrid (14165), — by Franklin (14568), — by Latimer (9280), — by Son of
Zoophite (15537), — by Magnum Bonum (2243), — Chance, from the stock of
Colonel Cradock, of Hartforth.

(44124)   THE ANCHORITE,
Roan, calved April 25, 1879, bred by Mr. G. D. Beresford, The Palace, Armagh ;
got by Pilgrim Father (35037), dam (Sheelagh 9th) by Red Robin (32274), g. d.
(Sheelagh 4th) by Loadstone (31605), gr. g. d. (Sheelagh 1st) by Jupiter (11625),
— (Ernestine) by Prince Ernest 2nd (10644), — by Beau of Killerby (7821),
— by King of the Valley (9270), — by Monarch.

(44125)   THE BARON,
Red, calved January 19, 1878, bred by Mr. A. P. Clear, Maldon, and exported to
South America ; got by Oxford's Baron (32030), dam (Queen Oxford) by Oxford's
Baron (32030), g. d. (Cranberry) by Second Duke of Claro (21576), gr. g. d.
(Yewberry) by Seventh Duke of York (17754), — (Hollyberry) by Douglas (12714),
— by Rivers (10716), — by Diamond (5918), — by Albert (2950), — by Stirling
(5330), — by Commodore (1858), — by Tathwell Studley (5401), — by Blyth
Comet (85).

(44126)   THE BARON,
Roan, calved January 22, 1878, bred by Mr. W. Fox, St. Bees Abbey, the
property of Mr. J. Ostle, Mowbray Cote ; got by British Boy (30597), dam
(Rowena Gwynne) by Waterloo Cherry (27763), g. d. (Robina Gwynne) by
Hematite (21917), gr. g. d. (Ruby Gwynne) by Garibaldi (19820), — (Ruth
Gwynne) by Exquisite (14524), — by St. Thomas (10777), — by Prime
Minister (2456), — by Wallace (5586), — by Marmion (406), — by Merlin
(430), — by Layton (366), — by Phenomenon (491), — by Favourite (252),
— by Favourite (252), — by Hubback (319), — by Snowdon's Bull (612),
— by Waistell's Bull (669), — by Masterman's Bull (422), — by the Studley
Bull (626).

(44127)   THE BARON,
Roan, calved February 15, 1879, bred by Mr. W. H. Wakefield, Sedgwick ; got
by Baron Barrington 4th (33006), dam (Well Heads Rose 2nd) by Sir Arthur
Windsor (35541), g. d. (Well Heads Rose) by Dunrobin (28486), gr. g. d.
(Rosebud 2nd) by Albert Victor (23293), — (Rosebud 1st) by Squire Stuart
(20891), — by Precedent (11918), — by Abraham Parker (9856).

(44128)   THE BARON,
Roan, calved February 27, 1879, bred by Mr. T. Harris, Stonylane House, the
property of Mr. Dixon, Tardebig ; got by Waterloo Prince (35949), dam

(Flower of the Link) by Festival (26147), g. d. (Belle of Braithwaite) by Prince George (13510), gr. g. d. (Bessie) by The Silkey Laddie (10947), — (Mary) by Fitz-Leonard (7010), — by Bloomsbury (3173), — by Coldstorms (3418), — by Remus (550), — by Kay's Grey Bull (4140).

(44129)   THE CHEESE,
Roan, calved February 20, 1879, bred by Mr. G. W. Lambart, Beau Parc; got by Jupiter (38477), dam (Dairy Maid) by Rupert (29902), g. d. (Petrel) by British Sailor (23472), gr. g. d. (Royal Lady) by British Flag (21323), — (Heiress) by Heir of Windsor (16255), — by Tom Jones (40825), — by Cato (12566), — by Narcissus (4540).

(44130)   THE CID,
Red, calved June 21, 1879, bred by Colonel R. Loyd Lindsay, Lockinge Park; got by Don Carlos (39693), dam (Red Streak) by Rob Roy (29806), g. d. (Red Rose) by Fairleigh (19714), gr. g. d. (Rosabel) by Ivanhoe (18096), — (Reflection) by Hercules (14692), — by Clarendon (12605), — by Admiral (9861), — by Will Honeycomb (5660), — by Spectator (2688), — by Albion (1619), — by Lancaster (360), — by Son of Windsor (698), — by Comet (155).

(44131)   THE CLOWN,
Roan, calved April 20, 1873, bred by Miss Calvert, Sandysike, the property of Mr. J. P. Law, Irthington; got by Merryman (37090), dam (Annie) by Faldonside (21718), g. d. by Rufus (22811), gr. g. d. by Prince Charlie (32114), — by Prince (8413), — by Alexander (1627), — by Brandon (3206), — by Lonsdale (379), — by Meteor (431), — by Butterfly (104).

(44132)   THE CZAR,
Roan, calved December 28, 1879, bred by Messrs. Annandale and Sons, Lintzford; got by Prince Imperial (43803), dam (Little Emily) by Beverley Butterfly 2nd (28022), g. d. (Lady) by Benjamin (19303), gr. g. d. (Queen Elizabeth) by Lord Nelson (13207), — (Wealthy) by Son of Fitz-Maurice (36084).

(44133)   THE DOCTOR,
Red and white, calved November 3, 1877, bred by Mr. W. Fox, St. Bees Abbey, the property of Mr. H. Walker, Sellafield; got by British Boy (30597), dam (Louisa 7th) by Hæmatite (31332), g. d. (Louisa 6th) by Comet (21449), gr. g. d. (Louisa 5th) by Baron Farnley (14129), — (Louisa 2nd) by Third Duke of York (10166), — by Hamlet (8126), — by Dulcimer (3658), — by Sol (2655), — by Maximus (2284), — by Matchem (2281), — by Scipio (1421), — by Sir Stephen (1456), — by Western Comet (689), — by Charge's Grey Bull (872), — by Favourite (252), — by Bartle (777), — descended from the Studley White Bull (627).

(44134)   THE DUKE,
Roan, calved February 1, 1878, bred by Earl Howe, Gopsall Hall, the property of Mr. R. Low, Ashby; got by Duke of Hazlecote 38th (36498), dam (May Rose 24th) by Burleigh (25697), g. d. (May Rose 17th) by Admiral (21146), gr. g. d. (May Rose 2nd) by Lord Moreton (14831), — (May Rose) by Second Duke of Oxford (9046), — by Robin Hood (8491), — by Simon (5135), — by Crispin (174), — by Fisher's Old Bull (3799), — by Grandson of Favourite (252), — by Grandson of Favourite (252), — descended from the stock of Mr. Cornforth, of Barforth.

(44135) THE DUKE,
Red, calved November 14, 1879, bred by Messrs. F. Leney and Sons, Wateringbury; got by Duke of Oxford 44th (39774), dam (Grand Duchess of Geneva 6th) by Sixth Duke of Oneida (30997), g.d. (Grand Duchess of Geneva) by Grand Duke 15th (21852), gr. g. d. (Duchess of Geneva 7th) by Third Lord Oxford (22200), — (Duchess of Geneva 3rd) by Oxford Lad (24713), — by Grand Duke 2nd (12961), — by Duke of Glo'ster (11382), — by Fourth Duke of York (10167), — by Fourth Duke of Northumberland (3649), — by Norfolk (2377), — by Belvedere (1706), — by Second Hubback (1423), — by The Earl (646), — by Ketton 2nd (710), — by Comet (155), — by Favourite (252), — by Daisy Bull (186), — by Favourite (252), — by Hubback (319), — by J. Brown's Red Bull (97).

(44136) THE EARL,
White, calved October 20, 1878, bred by Mr. A. Graham, Yanwath Hall; got by The Colonel (35747), dam (Topsy) by Dean (30863), g. d. (Tulip) by Midsummer (22358), gr. g. d. (Spotted Tremble) by Baronet (19274), — by Duke (14421), — by General Haynau (11520).

(44137) THE EARL,
Roan, calved February 10, 1879, bred by Mr. W. Duthie, Collynie, the property of Mr. Paterson,' Oldwhat; got by Earl of Derby 2nd (31061), dam (Evangeline 2nd) by Diphthong (17681), g. d. (Evangeline) by Hotspur (21960), gr. g. d. (Pride of the Dairy) by Guy Fawkes (12981), — (Bashful) by Young Ury (10984), — by The Pacha (7612), — by Second Duke of Northumberland (3646), — by Sillery (5131), — by Carleton (843), — by Diamond (205), — by Diamond (205).

(44138) THE EXPRESS,
Red and white, calved August 2, 1879, bred by Mr. A. Ashworth, Egerton Hall; got by Star Regent (35679), dam (Rosa Marguerita) by Booth's Royal Signet (28061), g. d. (Rosa Sabina) by Royal Booth (22772), gr. g. d. (Rose Wreath) by Windsor (14013), — (Rose Garland) by Baron Warlaby (7813), — by The Silkey Laddie (10947), — by Rouge (5012), — by Shipton (2620), — by Cleveland (3404), — by Danby (3550), — by Son of Studley Grange (1483), — by Son of Duke (225), — by Midas (4470), — by Nailer (4528), — by Ambo (746).

(44139) THE GEM,
Red, calved February 28, 1879, bred by Mr. E. Bowly, Siddington House; got by Lord Fitzclarence 15th (36942), dam (Ruby 5th) by Second Duke of Tregunter (26022), g. d. (Ruby 2nd) by Seventh Duke of York (17754), gr. g. d. (Ruby) by Lord Raglan (13225), — (Gem) by Annuity (9892), — by The Prince (7615), — by Colling (902), — by Son of Alexander (1624), — by Grandson of Favourite (252).

(44140) THE GRIFFIN,
Roan, calved March 23, 1878, bred by Mr. C. W. Griffin, Werrington, the property of Mr. C. W. Brierley, Drinkwater Park; got by Telemachus 10th (35728), dam (Miss Sicily) by Red Duke (29735), g. d. (Sicily) by Louis (14861), gr. g. d. (Young Snail Horn) by Haydn (11566), — (Snail Horn) by Young Frederick (3836), — by Favourite (3768), — by Commodore (1858), — by Favourite (3768), — by Tathwell Studley (5401), — a Turnell Cow.

(44141) THE MISER,
White, calved May 13, 1879, bred by H.R.H. the Prince of Wales, Sandringham; got by Admiral (39353), dam (Mirabel) by Homer (34170), g. d. (Satinet) by Duke of Edinburgh (23741), gr. g. d. (Seamstress) by Skipper (20854), — (Seraph) by Baron Albany (11151), — by The Captain (5422), — by The Captain (5422), — by Percy (1314), — by Lord Grantham's Son of Comet (155).

(44142) THE MONK,
White, calved November 15, 1878, bred by Mr. J. B. Aylmer, Fincham Hall; got by Prince of Knowlmere (37257), dam (Sweet Abbess) by Royal Abbot (35332), g. d. (Sweet Lady) by Lord Abbot (29052), gr. g. d. (Sweet Dane) by War Dane (23169), — (Sweet Duchess) by Duke Homfrey (19599), — by Prince George (13510), — by Vatican (12260), — by Baron Foggathorpe (9931), — by Quality (8446), — by Sweet William (5368), — by Homer (2134), — by Sirius (5161), — by Lubin (4291), — by Sultan (5344), — by Burgess's Bull (3139), — Old Countess.

(44143) THE PREMIER,
Roan, calved March 29, 1879, bred by Mr. W. Handley, Green Head; got by Alfred the Great (36121), dam (Palm Flower 3rd) by Melrose (29357), g. d. (Palm Flower 2nd) by Cecil (25725), gr. g. d. (Palm Leaf) by Royal Charlie (25012), — (Palmetto) by Cavendish (15745), — by Sir Charles (16949), — by Zadig (8796), — by Sir Harry (5156), — by Premier (2449), — by Emperor (1974), — by Barmpton (1677), — by St. Albans (2584), — by Simon (590), — by Pope (514).

(44144) THE SQUIRE,
Roan, calved May 27, 1879, bred by Mr. J. C. Musters, Annesley Park; got by Lord Chaworth (36924), dam (Prize Leaf 2nd) by Oxonian (38840), g. d. (Prize Leaf) by Lord Chancellor (20160), gr. g. d. (Oak Leaf) by Lucifer (13261), — (Oak Apple) by Reo Bulla (10703), — by Daring Ranger (9012), — by Monarch (7249), — by Eclipse (3684), — by Magnum Bonum (2243), — by Spectre (2689), — by Mac Gregor (2235), — by Goldfinder (2064), — by Star (2699), — by Broughton (90).

(44145) THE SULTAN,
Red and white, calved May 24, 1878, bred by Mr. D. Hartley, Westerdale; got by Captain Cook (37939), dam (Beauty) by Sixth Duke of Wetherby (28474), g. d. (Rose Bud) by Gerald (24034), gr. g. d. (Rachel) by Prince of I. (35149), — (Rose of Shamrock) by Sepoy (35499), — by Hero (18055).

(44146) THORALBY,
Red and white, calved March 21, 1879, bred by Mr. J. Topham, Middleham House; got by Heir-at-Law (34124), dam (Una) by Booth's Royal Signet (28061), g. d. (Jessie) by Royal Booth (22772), gr. g. d. (Middleham Lassie) by Yorkshireman (17263), — (Roan Lady) by Middleham Laddie (13338), — by The Silkey Laddie (10947), — by Bloomsbury (3173), — by Cleveland (3404), — by Danby (3550), — by Studley Grange (1483), — by Duke (225).

(44147) THORNDALE GENEVA 3RD,
Roan, calved December 27, 1879, bred by Mr. G. Garne, Churchill Heath; got by Grand Duke of Geneva 2nd (31288), dam (Thorndale Blanche 2nd) by Sir Rainald (27485), g. d. (Thorndale Blanche) by Twelfth Duke of Thorndale

(26020), gr. g. d. (Blanche 8th) by Clifford (21437), — (Blanchette) by Magistrate (13274), — by Antinous (12401), — by Killjoy (14759), — by Diamond (5918), — by Norfolk (2377), — by Belvedere (1706), — by Belvedere (1706), — by Lancaster (360), — by Petrarch (488), — by Major (397), — by Chapman's Son of Punch (122), — by Dickson's Grandson of Punch (213), — by Checks (132), — by R. Grimston's Bull (282), — by J. Coates's Bull (148).

(44148) THORNDALE KING,
Roan, calved April 11, 1879, bred by Mr. H. Allsopp, Hindlip Hall, the property of Mr. R. H. Godard, Evelench, Tibberton ; got by Oxford's King (34997) dam (Thorndale Maid 2nd) by Butterfly (33255), g. d. (Thorndale Maid) by Third Duke of Thorndale (17749), gr. g. d. (Passion Flower 3rd) by Fourth Duke of Thorndale (17750), — (Passion Flower) by Homer (16277), — by Bashaw (12449), — by Enterprise (11443), — by Young Consul (6893), — by Newnham (2365), — by Satellite (1420), — by Jupiter (342), — by Sir Oliver (605),— by Trunnell (659), — by Favourite (252), — by Favourite (252), — by Dalton Duke (188), — by R. Alcock's Bull (19), — by J. Smith's Bull (608), — by Jolly's Bull (337).

(44149) THORNDALE RUTH,
Red and white, calved June 14, 1879, bred by Mr. J. R. Singleton, Great Givendale; got by Thorndale Cherry Lad (40818), dam (Mary Ruth 2nd) by Beau Siddington (30515), g. d. (Mary Ruth) by Bobby (25648), gr. g. d. (Maude) by Patriot (20475), — (Graceful) by Ferdinand (12871), — by Surplice (10901), — by Remus (9545), — by Snowball (5227), — by Red Highflyer (2488), — by Old Snowball (2648), — descended from the stock of Mr. Newby, of Thorpe.

(44150) TICHBORNE,
Roan, calved May 5, 1873, bred by Mr. T. Burgess, Burleydam ; got by Jerome (31433), dam (Charlotte Corday) by Hawser (18038), g. d. (Lucy Long) by Sir James (16980), gr. g. d. (Lucy Locket) by Usurer (9763), — (Lavender) by Dan O'Connell (3557), — by Brutus (1752), — by Frederick (1060), — by Cato (1794), — by Son of Wellington (679), — bred by Mr. Robertson, of Ladykirk.

(44151) TINCROFT,
Red and white, calved February 12, 1879, bred by Mr. W. Trethewy, Tregoose, the property of Mr. S. T. Tregaskis, St. Issey ; got by M. C. (31898), dam (Ruth 76th) by Crœsus (30820), g. d. (Ruth 50th) by Lord Montgomery (26686), gr. g. d. (Ruth 19th) by Sir Roger (18863), — (Ruth 11th) by Prince Frederick (16734), — by Frantick (8088), — by Harold (8131), — by Cedric (3311), — by Nimrod (4571), — by Grandson of Blyth Comet (85), — by Crispin (174), — by Meteor (431), by Meteor (431).

(44152) TOLLENDAL,
White, calved January 10, 1879, bred by the Duke of Manchester, Kimbolton Castle ; got by Duke of Underley 3rd (38196), dam (Lally 16th) by Third Duke of Claro (23729), g. d. (Lally 9th) by Seventh Duke of York (17754), gr. g. d. (Lally 2nd) by Malachite (18313), — (Lally) by Earl of Derby (10177), — by Earl of Liverpool (9061), — by Second Duke of Cambridge (3638), — by Belvedere (1706), — by Son of Herdsman (304), — by Wonderful (700), — by Alfred (23), — by Young Favourite (6994).

(44153)   TOM CHRISP,

Red, calved March 29, 1879, bred by Miss Milne, Otterburn; got by Vice Regal
Booth (39258), dam (Abbotsford Princess) by Abbotsford (23256), g. d. (Wolviston
Princess) by Wolviston (19163), gr. g. d. (Ettrick Princess) by Ettrick (14518),
— (Red and White Princess) by Sam Glen (10780), — by Willie (9835), — by
Bachelor (1666), — by Snowball (2647), — by St. Albans (2584), — by Lawn-
sleeves (365), — bred by Messrs. James, of Stamford.

(44154)   TOM TIT,

Roan, calved October 16, 1878, bred by Messrs. T. Jennings and Son, Cockfield
Hall; got by Lord Marcellus (40202), dam (Violet) by Monarque (37104), g. d.
(White Duchess) by Prolific 3rd (27214), gr. g. d. (Archduchess) by Archduke
(23318), — (Welcome Lass 2nd) by Sir Walter Raleigh (15303), — by Paris
(7314), — by King of Trumps (4156), — by Vanguard (5545), — by Red Rover
(4903), — by Anticipation (750), — by Emperor (1014), — by Young Windsor
(699), — by Windsor (698).

(44155)   TOWN COUNCILLOR,

Roan, calved September 14, 1875, bred by Mr. A. Thompson, The Cross; got by
Borough Member (33186), dam (Cressida 4th) by Cupid (25861), g. d. (Cressida
3rd) by Cerdic (19415), gr. g. d. (Polyxena) by Duke of Buckingham (14428),
— (Hecuba) by Florian (12887), — by Duke of Athol (10150), — by Cossack
(1880), — by Miracle (2320), — by Matchem (2281), — by Fitz-Remus (2025),
— by Cato (119), — by Whitworth (695).

(44156)   TRAFALGAR,

Red, calved May 18, 1879, bred by Earl Beauchamp, Madresfield Court; got by
Magnum Bonum (41954), dam (Lady Kirk) by Gay Boy (31222), g. d. (Kathleen) by Battersea Gwynne (25617), gr. g. d. (Christmas Rose) by Castlereagh
(19409), — (Bess) by Impatience (14723), — by Hero (8146), — by Major
(4338), — by Ormsby (4621), — by Cossack (925), — by Alpha (3004), — by
Captain (3271), — by Captain (108).

(44157)   TRAPPIST 3RD,

Roan, calved February 14, 1879, bred by the Duke of Manchester, Kimbolton
Castle; got by Duke of Underley 3rd (38196), dam (Princess Surmise) by Wild
Eyes Duke (36007), g. d. (Princess) by May Duke (13320), gr. g. d. (Surmise)
by Duke of Glo'ster (11382), — (Silence) by Earl of Derby (10177), — by Duke
of Sutherland (6945), — by Locomotive (4242), — by Short Tail (2621), — by
Gambier (2046), — by Young Wynyard (2859), — by Bulls of Messrs. C. and
R. Colling's.

(44158)   TRAPPIST 4TH,

Roan, calved October 12, 1879, bred by the Duke of Manchester, Kimbolton
Castle; got by Duke of Underley 3rd (38196), dam (Princess Silence) by Wild
Eyes Duke (36007), g. d. (Princess) by May Duke (13320), gr. g. d. (Surmise)
by Duke of Glo'ster (11382), — (Silence) by Earl of Derby (10177), — by Duke
of Sutherland (6945), — by Locomotive (4242), — by Short Tail (2621), — by
Gambier (2046), — by Young Wynyard (2859), — by Bulls of Messrs. C. and
R. Colling's.

(44159)   TRAVELLER,

Roan, calved March 10, 1879, bred by Mr. J. Bousfield, Soulby; got by Lord of the Manor (38642), dam (Jessie) by Duke of Gainford (33644), g. d. (Jeanette) by General Windsor (28701), gr. g. d. (Nelly) by Roan Duke (20685), — by Bloody Douglas (12478).

(44160)   TREGUNTER CHARMING DUKE,
Red, calved April 12, 1879, bred by Mr. L. H. Wraith, Newfield ; got by Duke of Tregunter 7th (38194), dam (Red Duchess) by Seventeenth Duke of Oxford (25994), g. d. (Duchess) by Grand Duke 15th (21852), gr. g. d. (Countess) by Knightley Grand Duke (24268), — (Chorus) by Fourth Duke of Thorndale (17750), — by Mameluke (13289), — by Cardinal (11246), — by White Friar (9827), — by Little John (4232), — by Caliph (1774), — by Sir Walter (2637). — by Hotspur (1117), — by Coxcomb (928), — by Midas (435), — by Comet (155), — by R. Colling's Son of Favourite (252), — by same Son of Favourite (252), — by Hubback (319).

(44161)   TREGUNTER KNIGHTLEY,
Red, calved March 6, 1879, bred by the Earl of Feversham, Duncombe Park ; got by Duke of Tregunter 5th (33743), dam (Gwendoline) by Second Duke of Collingham (23730), g. d. (Lady Geneva) by Duke of Geneva (19614), gr. g. d. (Rosette) by Marmaduke (14897), — (Rosalie) by Vanguard (10994), — by Earl of Dublin (10178), — by Grey Friar (9172), — by Little John (4232), — by Young Norman (4584), — by White Boy (1580), — by Wyville's Bull, — by a Bull bred by Mr. Charge.

(44162)   TROUBADOUR,
Red and white, calved June 24, 1879, bred by Messrs. N. Russell and Son, Northallerton; got by King Harold (40053), dam (Truthful) by Lord Lally 3rd (24408), g. d. (Truth) by Oxford 2nd (18507), gr. g. d. (Faith) by Favour (9114), — (Annabella) by Young Cheshireman (17555), — by Valentine Jun. (6630), — by Sir William (5203), — by Noble (4577), — by Imperial (2151), — by Fairfax (1023), — by Young Warlaby (5598), — by Young Dimple (971), — by Snowball (2648), — by Layton (2190).

(44163)   TRUANT,
Roan, calved March 4, 1879, bred by Mr. H. Allsopp, Hindlip Hall; got by Oxford's King (34997), dam (Tiny 4th) by Marnhull Duke (29283), g. d. (Tiny 3rd) by Butterfly (33255), gr. g. d. (Tiny 2nd) by Mameluke (24520), — (Tiny) by Sir John (25154), — by Prince (18575).

(44164)   TRUE BRITON,
Red, calved May 1, 1879, bred by Mr. J. Strong, Culgaith; got by Earl of Doune (36579), dam (Gamblesby 2nd) by Duke of Hazlecote 29th (33667), g. d. (Gamblesby) by Edgar (19680), gr. g. d. (Gamblesy) by Marksman (18329), — (Gamblesby) by Pugnator (18656), — by Solon (13766), — by Leopold (4216), — by Premier (6307), — by Prime Minister (2456), — by Monarch (4494).

(44165)   TRUE EARL,
Roan, calved March 10, 1879, bred by Mr. H. Smith, Mountmellick; got by Earl of Aylesby (38215), dam (True Love 8th) by Red Cross (32247), g. d. (True Love 2nd) by Dr. McHale (15887), gr. g. d. (True Love) by Diamond (15878), — (Trinket) by Villikins (15465), — by Hopewell (10332), — by Hamlet (8126), — by Second Comet (5101), — by Prince of Northumberland (4826), —

by Mason's Son of Matchem (2678), — by Falstaff (1993), — by Richard (1376),
— by Jupiter (342), — by Rufus (570), — by Pope (514).

(44166) TRUMPETER,
Roan, calved August 22, 1879, bred by Mr. C. H. Cock, Bridgefoot; got by
Duke of Worcester 5th (39797), dam (Tuneful) by Young England (31110), g. d.
(Madrigal) by Prince Llewellyn (32154), gr. g. d. (Songstress) by Grand Duke
16th (24063), — (Seraphina 3rd) by Royal Essex (18767), — by Fitz Clarence
(11477), — by Sweet William (7571), — by Earl of Essex (6955), — by Stratton
(5336), — by Fanatic (1996), — by Red Rover (4902), — by Rufus (2576), —
by Emperor (1014), — the Rev. R. Pointer's Old Darlington.

(44167) TURVEY,
Roan, calved May 17, 1879, bred by Mr. P. R. Norton, Turvey House; got by
Grand Lama (41656), dam (Cybele 3rd) by Duke of Cranagh (39726), g. d.
(Cybele) by Prince Arthur (29598), gr. g. d. (Keopolina) by Professor Miller
(18649), — (Cherry Ripe) by Cœur-de-Lion (15783), — by The Templar 2nd
(13879), — by South Durham (9676), — by Pilot (7333), — by Young Sir Harry
Liddell (7503), — by Mickley (7234), — by Adonis (2933), — by Lord Prudhoe
(8251), — by Hollon's Bull (313).

(44168) ULSTERMAN
Roan, calved January 23, 1879, bred by Mr. R. P. Maxwell, Finnebrogue; got
by Woodranger (39340), dam (Rosamond) by Prince Victor (20606), g. d.
(Rosabell) by Royal Dane (18762), gr. g. d. (Moss Rose) by Tipton Slasher
(13888), — (Rose of Westmoreland) by Brewster (7847), — by Rokeby (8502),
— by Petteril (4685), — by Duke (1933), — by Northumberland (464), — by
Buston's Styford (103), — by Lame Bull (358), — by Bolingbroke (86).

(44169) UMPIRE,
Red, calved August 12, 1879, bred by Messrs. J. G. W. and W. R. Lyle, Donaghmore House; got by Sir James (37466), dam (Lady of the Vale) by Sailor Prince
(35442), g. d. (Lady of the Manor) by Earl of Donaghmore (31064), gr. g. d.
(Perfume 2nd) by Best Hope (23414), — (Perfume) by Fugleman (14580), —
by White Jack (36089), — by Collingwood (8964), — by Brilliant (1740), — by
Monarch (2324), — by Cato (119), — by George (273), — by Chilton (136), —
by Irishman (329), — by B. (45).

(44170) UNCAS,
Red and white, calved March 10, 1879, bred by Mr. W. Davidson, Esker,
Timahoe, the property of Mr. W. Johnson, Prumplestown House; got by Frank
(43246), dam (Indian Queen) by Indian Chief (34205), g. d. (Lady St. Patrick)
by The Red Duke (27643), gr. g. d. (Mary Fitz Patrick) by Chancellor Hopewell
(19422), — by Count de Grey (23633), — by Lord John Russell (16417).

(44171) UNION ROSE 2ND,
Roan, calved October 30, 1875, bred by Mr. W. Tippler, Roxwell; got by Union
Rose (35839), dam (Rosalind Butterfly 5th) by Royal Baron Butterfly (25001),
g. d. (White Rosalind 2nd) by Sultan (15358), gr. g. d. (Isabel) by Kalafat
(13101), — (Rocket) by Fanatic (8054), — by Earl of Durham (5965), — by
Pedestrian (7321), — by Edrom (1956), — by Scipio (1421), — by Hector (1104),
— by Midas (435), — by Marquis (407), — by Chilton (136), — by Ben (70).

(44172) **UNIQUE,**
Roan, calved March 20, 1878, bred by Mr. J. Turner, Ulceby, the property of Mr.
Nash, Buenos Ayres; got by Geneva's Oxford (34025), dam (Ursula 32nd) by
Second Duke of Waterloo (23800), g. d. (Ursula 10th) by Archduke (17316),
gr. g. d. (Ursula Booth) by Richard (15161), — (Ursula) by Usurer (9763), —
by Prince Ernest (4818), — by A-la-Mode (725), — by Ratify (2481), — by
Wellington (680), — by Favourite (252), — by Punch (531).

(44173) **UPLEATHAM,**
Roan, calved April 22, 1876, bred by the Earl of Zetland, Upleatham; got by
Brisk (33212), dam (Maple) by Chatsworth (23546), g. d. (Marcia) by King of
the Roses (22043), gr. g. d. (Music) by Sir James the Rose (15290), — (Timbrel)
by Apollo (9899), — by Second Duke of Northumberland (3646), — by Clarion
(3392), — by Sultan (1485), — by Blucher (795), — by Wellington (680), —
by Phenomenon (491), — from the stock of Mr. Hill, of Blackwell.

(44174) **VALENTINE,**
Red and white, calved February 14, 1879, bred by Mr. B. Greenhill, Rowde, the
property of Mr. E. G. Greenhill, Inmarch Farm; got by Butterfly's Relative
2nd (36298), dam (Sweetheart 2nd) by Warrior (32801), g. d. (Sweetheart) by
Fugleman (28660), gr. g. d. (Margaret) by Colonel (28221), — (Moggy Lauder)
by Cambridge Lad (28121), — by Romeo (29819), — by Fitz George (28606),
— bred by Mr. E. Clarke.

(44175) **VALENTINE DUKE,**
Red, calved February 14, 1879, bred by Sir R. C. Musgrave, Bart., Eden Hall;
got by Duke of Underley (33745), dam (Wild Eyes 30th) by Fourteenth Duke
of Oxford (21605), g. d. (Wild Eyes 21st) by Lord Stanley (14854), gr. g. d.
(Wild Eyes 19th) by Solon (13766), — (Wild Eyes 16th) by Second Duke of
Oxford (9046), — by Fourth Duke of Northumberland (3649), — by Duke of
Northumberland (1940), — by Belvedere (1706), — by Emperor (1975), — by
Wonderful (700), — by Cleveland (145), — by Butterfly (104), — by Hollon's
Bull (313), — by Mowbray's Bull (2342), — by Masterman's Bull (422), —
descended from M. Dobison's stock.

(44176) **VANDERVELDE,**
White, calved May 21, 1879, bred by Mr. A. Fawkes, Farnley Hall; got by
Baron Winsome 6th (33111), dam (Lady Verena) by The Baron (32654), g. d.
(Lady Valentine) by Lord Cobham (20164), gr. g. d. (Vanity) by Reformer
(18687), — (Vanilla) by Robinson Crusoe (13610), — by Borrowby Boy (9980),
— by Laudable (9282), — by Sir Thomas Fairfax (5196), — by Colossus (1847),
— by Burley (1766), — by Pilot (496), — by Warlaby (672), — by Albion
(14), — by Lame Bull (359), — by Shipton (587), — by Son of Suworrow (636),
— by Son of Twin Brother to Ben (88), — by Twin Brother to Ben (660).

(44177) **VANGUARD,**
Roan, calved April 22, 1879, bred by Mr. C. Bayes, Kettering; got by Grand
Duke of Darlington (39948), dam (Vinery) by Prince Royal (24859), g. d.
(Vintage) by Ernest (17812), gr. g. d. (Vineleaf) by Hamlet (8126), — (Vaux-
hall) by Duke of Cambridge (7987), — by Zohrab (5705), — by Commodore
(3445), — by Ormsby (4621), — by Barmpton Major (3092), — by Cossack
(925), — by Captain (3271).

(44178)   **VASSAL,**

Roan, calved March 17, 1879, bred by Lord Bolton, Bolton Hall, the property of Mr. Graves, Selby; got by Heir-at-Law (34124), dam (Vensela) by King Charles (24240), g. d. (Vignette) by Marmion (26821), gr. g. d. (Viscountess) by Valasco (15443), — by Highflyer (11576), — by Maximus (14939).

(44179)   **VEDETTE,**

Roan, calved May 5, 1878, bred by Sir D. Baird, Bart, Newbyth; got by Skirmisher (40729), dam (Jezebel) by George 1st (19848), g. d. (Jura) by Tenedos (20936), gr. g. d. (Jane) by The Sheriff (12216), — (Jannette) by Triumph (8717), — by Eclipse (3688), — by Coxcomb (3514), — by Juniper (2165), — by Cedric (1801), — by The Squire (1512), — by Midas (435), — by Boughton (90).

(44180)   **VENTRILOQUIST,**

White, calved February 7, 1879, bred by Mr. A. Cruickshank, Sittyton, the property of Mr. J. Black, Bartol Chapel, Tarves; got by Roan Gauntlet (35284), dam (Victoria 41st) by Lord Privy Seal (16444), g. d. (Victoria 39th) by Champion of England (17526), gr. g. d. (Victoria 29th) by Red Knight (11976), — (Victoria 19th) by Lord John (11731), — by Prince Albert (11933), — by Belzoni (783), — by Satellite (1420), — by Cato (119), — by Pope (514), — by Favourite (252), — by White Bull (421), — by Favourite (252), — by Dalton Duke (188), — by R. Alcock's Bull (19), — by J. Smith's Bull (608), — by Jolly's Bull (337).

(44181)   **VETERAN,**

Red and white, calved March 10, 1879, bred by Earl Fitzwilliam, Coollattin Park, sold to the Provincial Government of Liege, Belgium; got by Lord Lilias (31690), dam (Venus 22nd) by Robert Burns (29795), g. d. (Venus 17th) by Fusilier (24000), gr. g. d. (Venus 11th) by Clydesdale(15779), — (Venus 10th) by Cavaignac (10033), — by Splendid (12136), — by Marquis of Rockingham (10506), — by Baronet (6763), — by Prince Albert (4781), — by Madcap (4308), — by Major (398), — by Cossack (925), — by Captain (108).

(44182)   **VETERAN,**

White, calved November 25, 1879, bred by Mr. W. Handley, Green Head; got by Alfred the Great (36121), dam (Hopeful Flora) by Sir Arthur Windsor (35541), g. d. (Earl's Flora) by Earl of Eglinton (23832), gr. g. d. (Flora Cobham) by Marquis of Cobham (22299), — (Flower of Fitz Clarence) by Alfred Fitz Clarence (19215), — by Veteran (13941), — by Waterloo Hero (13981), — by Duke of Richmond (7996), — by Lord Stanley (4269), — by Velocipede (5552), — by Priam (2452), — by Jerry (4097), — descended from the stock of Mr. Booth.

(44183)   **VICE-ADMIRAL WINDSOR,**

Roan, calved January 16, 1879, bred by Mr. T. Willis, Carperby; got by Vice-Admiral (39257), dam (Rose Fitz Clarence) by Major Windsor (34739), g. d. (Rose Clarence) by Fitz Clarence (14552), gr. g. d. (Moss Rose) by Gavazzi (11508), — (Moss Rosebud) by Wilberforce (9830), — by Swintonian (9702), — by Young Symmetry (7577), — by Shipton (2620), — by Fitz-Champion (7007), — by Young Snowball (7521).

(44184)   **VICEROY,**

Roan, calved February 11, 1877, bred by Mr. S. P. Savage, Lays Farm, the property of Messrs. T. and J. Young, Earthcott; got by Fulgens (33980), dam (Virtue) by Duke of Cambridge (15921), g. d. (Violet 4th) by Royal Harbinger (18770), gr. g. d. (Violet 3rd) by Burbage (10006), — (Violet) by White Comet (11041), — by Monzani (6222), — by Studley (628), — by Young Major (8266), — bred by Mr. Leighton, of Willingham.

(44185)   **VICEROY,**

Red, calved August 15, 1878, bred by Mr. A. Mackenzie Lyle, Donaghmore House, the property of Messrs. J. G. W. and W. R. Lyle, Donaghmore House; got by Sir James (37466), dam (Lady of the Vale) by Sailor Prince (35442), g. d. (Lady of the Manor) by Earl of Donaghmore (31064), gr. g. d. (Perfume 2nd) by Best Hope (23414), — (Perfume) by Fugleman (14580), — by White Jack (36089), — by Collingwood (8964), — by Brilliant (1740), — by Monarch (2324), — by Cato (119), — by George (273), — by Chilton (136), — by Irishman (329), — by B. (45).

(44186)   **VICEROY,**

Red and white, calved April 1, 1879, bred by Earl Fitzwilliam, Coollattin Park; got by Robert Burns (29795), dam (Venus 18th) by Lord Stanley (24466), g. d. (Venus 11th) by Clydesdale (15779), gr. g. d. (Venus 10th) by Cavaignac (10033), — (Venus 9th) by Splendid (12136), — by Marquis of Rockingham (10506), — by Baronet (6763), — by Prince Albert (4781), — by Madcap (4308), — by Major (398), — by Cossack (925), — by Captain (108).

(44187)   **VICKS,**

White, calved February 9, 1879, bred by Mr. J. Cran, Kirkton; got by Bridegroom (33201), dam (Victoria Rose) by Frederick Fitz Windsor (31196), g. d. (Victoria) by Duke of Cornwall (23732), gr. g. d. (Countess) by Earl of Windsor (15968), — (Bella) by Roger de Norton (12012), — by Marmion (6379), — by Champion, — by Barmpton, — by Wellington (680), — by Favourite (252).

(44188)   **VICTOR,**

Roan, calved May 11, 1873, bred by Mr. R. Coulson, Coastley; got by Earl of Derwent (28503), dam (Cherry) by Lofty (26596), g. d. (Victoria 4th) by Cherry Prince (17554), gr. g. d. (Victorine) by Ivanhoe (14735), — (Victoria 3rd) by Wentworth (11034), — by Anick (12399), — by Speculation (8621), — by Guardian (3948), — by a Son of Miracle (2320).

(44189)   **VICTOR,**

Red, calved March 24, 1874, bred by the Earl of Caledon, Caledon, the property of Mr. M. C. Close, Drumbanagher Castle; got by Harlequin (31340), dam (Venus) by Best Hope (23414), g. d. (Vesper) by Dr. McHale (15887), gr. g. d. (Douglas Beatrice) by Gold Dust (11536), — (Beatrice) by Benedict (7828), — by Brilliant (7851), — by Young Hastings (3988), — by Wallace (5588), — by Ivanhoe (1130), — by Leopold (2199), — by Hector (2103), — by Sir Harry (5155), — by Surly (2715), — by Burrell's Bull (3248).

(44190)   **VICTOR,**

Roan, calved in March, 1878, bred by Mr. J. White, Cowleigh Gate; got by Lord Darlington 13th (38597), dam (Victory) by Duke of Brailes 2nd (30930),

g. d. (Victoria) by Grand Duke 9th (19879), gr. g. d. (Verity) by White Prince (21102), — (Vine 6th) by Sir Sam (16992), — by Third Duke of Glo'ster (12754), — by Prince (6322), — by Major (4338), — by Young Rockingham (2547), — by Young Major (2254), — by Monarch (22372), — by Cossack (925).

(44191) VICTOR,

Red, calved November 18, 1878, bred by Mr. A. Geekie, Baldowrie; got by Neptune (38788), dam (Alma) by Mosstrooper (29395), g. d. (Juanita 3rd) by Fancy Boy (26129), gr. g. d. (Juanita 2nd) by Too Late (25319), — (Juanita) by Young John O'Groat (16324), — by John O'Groat (13090), — by Lovemore (10476), — by Hector (9200), — by Eclipse (3684), — by Boughton (7841), — by Young Merlin (6204), — by Midas (435), — by Denton (198).

(44192) VICTOR CHIEF,

Roan, calved January 28, 1879, bred by Mr. A. Cruickshank, Sittyton, the property of Mr. A. Geekie, Baldowrie; got by Lord of the Isles (40218), dam (Victoria 53rd) by Royal Duke of Gloster (29864), g. d. (Victoria 46th) by Cæsar Augustus (25704), gr. g. d. (Victoria 43rd) by Champion of England (17526), — (Victoria 36th) by Baronet (15614), — by Master Butterfly 2nd (14918), — by Red Knight (11976), — by Lord John (11731), — by Prince Albert (11933), — by Belzoni (783), — by Satellite (1420), — by Cato (119),— by Pope (514), — by Favourite (252), — by White Bull (421), — by Favourite (252), — by Dalton Duke (188), — by R. Alcock's Bull (19), — by J. Smith's Bull (608), — by Jolly's Bull (337).

(44193) VICTOR GWYNNE,

Roan, calved January 12, 1878, bred by Mr. W. T. Crosbie, Ardfert Abbey, the property of Mr. J. Jameson, St. Marnocks; got by England's Glory (23889), dam (Bright Gwynne) by Northern Light (24670), g. d. (Gwynne Grove) by Castle Grove (19408), gr. g. d. (Sweet Poll Gwynne) by Duke of Cambridge (12747), — (Peg Gwynne) by Young Benedict (15641), — by Sir Harry (10819), — by St. Thomas (10777), — by Conservative (3472), — by Chorister (3378), — by Wellington (2824), — by Marmion (406), — by Merlin (430), — by Layton (366), — by Phenomenon (491), — by Favourite (252), — by Favourite (252), — by Hubback (319), — by Snowdon's Bull (612), — by Waistell's Bull (669), — by Masterman's Bull (422), — by the Studley Bull (626).

(44194) VICTORIOUS,

White, calved March 10, 1879, bred by Mr. C. Bayes, Kettering; got by Grand Duke of Darlington (39948), dam (Victoria) by Robin Adair (35290), g. d. (Paulina) by Prince Royal (24859), gr. g. d. (Hecate) by Denver (21534), — (Daisy) by Hero of Kars (19956), — by Testator (20938), — by Pompey (10622), — by Auckland (5766), — by Duncan (1942), — by Trinculo (2775), — by Mac Gregor (2235), — by Goldfinder (2064), — by Boughton (90).

(44195) VILLAGE PRINCE,

Roan, calved April 28, 1879, bred by Mr. W. Graham, Eden Grove; got by Earl of Doune (36579), dam (Village Queen) by Monarch (31930), g. d. (Village Rose) by Flag of Truce (31172), gr. g. d. (Village Joy) by First Favourite (21747), — (Village Belle) by Blondin (21289), — by The Squire (23060), — by Donation (5927), — by Young Sea Gull (5100), — by Crusader (934), — by Sultan (1485) — by Mars (411), — by North Star (458).

(44196) **VILLIERS WATERLOO,**

Roan, calved May 26, 1879, bred by Lord Skelmersdale, Lathom House; got by Duke of Rosedale 6th (38176), dam (Maud Waterloo) by Baron Oxford 4th (25580), g. d. (Waterloo 33rd) by Grand Duke 11th (21849), gr. g. d. (Waterloo 25th) by Duke of Geneva (19614), — (Waterloo 17th) by Red Knight (11976), — by Grand Duke (10284), — by Third Duke of Oxford (9047), — by Second Cleveland Lad (3408), — by Duke of Northumberland (1940), — by Norfolk (2377), — by Waterloo (2816), — by Waterloo (2816).

(44197) **VIOLET'S OXFORD,**

Red, calved June 27, 1879, bred by Mr. J. Singleton, Teresa Cottage; got by Duke of Oxford 37th (38171), dam (Violet 6th) by Third Duke of Glo'ster (33653), g. d. (Violet 5th) by Sixth Earl of Walton (26078), gr. g. d. (Violet) by Seventh Duke of York (17754), — (Ptarmigan) by Oxford Grand Duke (16674), — by Siddington Duke (15263), — by Harold (10299), — by Leo (4208), — by Henwood (2114), — by Wharfdale (1578), — by Young Marske (419), — by Meteor (432), — by Western Comet (689), — by Favourite (252), — by Cupid (177), — by Grandson of Bolingbroke (280), — by Foljambe (263), — by R. Alcock's Bull (19), — by J. Smith's Bull (608), — by Jolly's Bull (337).

(44198) **VIOLINIST,**

Red, calved May 10, 1876, bred by Mr. A. Cruickshank, Sittyton, the property of Mr. J. B. Manson, Kilblean; got by Barmpton Prince (32995), dam (Violet Flower) by Windsor Augustus (19157), g. d. (Sweet Violet) by Lord Stanley (16454), gr. g. d. (Violet) by Lord Bathurst (13173), — (Roseate) by Matadore (11800), — by Hudson (9228), — by Fairfax Royal (6987), — by Inkhorn (6091), — by Grazier (1085), — by Sampson, — by Wallace (1560).

(44199) **VIRGIL,**

Red and white, calved September 27, 1858, bred by the Marquis of Exeter, Burghley Park, late the property of Mrs. T. H. Vivian, Singleton; got by Madman (14873), dam (Teal) by Master Thankful (10514), g. d. (Duck) by Prince of Wales (6345), gr. g. d. (Darling) by Edward (3696), — (Pet) by Topper (2768), — by Favourite (3768), — by Windsor (698), — by Windsor (698).

(44200) **VIRGIL,**

Roan, calved March 18, 1879, bred by Mr. E. Lythall, Radford Hall; got by Virginus (40873), dam (Alicia) by Duke of Lancaster (28411), g. d. (Alice Swinnerton) by Lord of the Isles (20206), gr. g. d. (Alice) by Alma (14085), — (Venus) by Sultan (13801), — by Planet (10616), — by The Hero (6584), — by Freeman (2042), — by Grazier (3030), by Ebor (007), — by a Bull of Mr. Howard's.

(44201) **VIRGINIUS,**

White, calved April 29, 1879, bred by Mr. W. Crickmore, Seething; got by Lamarque (38540), dam (Virginia) by Victoria's Brian (23141), g. d. (Snowdrop) by Virginius (23147), gr. g. d. (Roan Cow) by Garibaldi (21792), — (Loddon Prize Cow) by Merman (13330).

(44202) **VISCOUNT,**

Roan, calved May 28, 1878, bred by Major Barton, Straffan House, the property of Mr. W. T. Trench, Loughton, Roscrea; got by Valorous (27701), dam (Tulip

2nd) by Iron Duke (31420), g. d. (Tulip) by Sovereign (27538), gr. g. d. (Luna) by Red Knight (18681), — (Young Moonshade) by Dandy (11326), — by Anticipation (6747), — by Hero (7079), — by Oliver (2386), — by Blyth (797), — by Midas (435), — by Broughton (90), — by Windsor (698), — by R. Colling's Son of Favourite (252).

(44203)   VISCOUNT COLLINGHAM,
Red and white, calved December 31, 1878, bred by Mr. J. Turner, Ulceby; got by Grand Duke of Glo'ster (36721), dam (Belle of Collingham) by Second Duke of Collingham (23730), g. d. (Syrian Belle) by Lord Lally (22161), gr. g. d. (Queen of Tyre) by Archduke (17316), — (Syria) by Duke of Ulster (12774), — by His Highness (11580), — by Gainford 2nd (6030), — by Wharton (2833), — by Count (1883), — by Sir William (2640), — by Young Rockingham (2549), — by Wellington (2824), — by Northumberland (464), — by Buston's Styford (103), — by Bolingbroke (86).

(44204)   VISCOUNT CORWEN,
Roan, calved April 17, 1879, bred by Mr. R. Blezard, Pool Park; got by Duke of Siddington 3rd (38183), dam (Matchless 19th) by Eighth Duke of York (28480), g. d. (Matchless 18th) by Royal Cumberland (27358), gr. g. d. (Matchless 7th) by Palmerston (15041), — (Matchless 4th) by Stapleton (15336), — by Grisdale (14654), — by Mingo (7239), — by Snowball (1465), — by Pizarro (1323).

(44205)   VISCOUNT DURSLEY 2ND,
Roan, calved November 6, 1879, bred by Lord Fitzhardinge, Berkeley Castle, the property of Mr. J. Bennett, Downhouse Farm, Dursley; got by Duke of Oxford 45th (39775), dam (Dowager) by Duke of Connaught (33604), g. d. (Dorothea) by Grand Duke of Waterloo (28766), gr. g. d. (Dora) by Second Duke of Airdrie (19600), — (Darling) by Fourth Duke of Oxford (11387), — by Sir Hugh (12082), — by Percy (9472), — by Thomas (5471), — by Eryholme (3736), — by Reformer (4914), — by Young Favourite (3770), — by Wellington (2825).

(44206)   VISCOUNT FROGMORE,
Roan, calved March 24, 1879, bred by Mr. W. G. Garne, Broadmoor; got by Sir Robert Frogmore (40719), dam (Dido 4th) by Royal Butterfly 20th (25007), g. d. (Duchess 3rd) by Bibury Butterfly (25631), gr. g. d. (Duchess) by Oberon (15016), — (Darling) by Meteor (10526), — by Phœnix (6290), — by Shakespeare (5108), — by Rival (2534), — by Darlington (956), — by Denton (198), — by Rockingham (560), — by Jobling's Son of Phenomenon (491), — by Phenomenon (491), — by Colonel (152), — by Styford (629).

(44207)   VISCOUNT LINCOLN,
Roan, calved April 19, 1879, bred by Mr. F. J. Moss, Stainfield Hall, the property of Mr. R. Moss, Whisby; got by Snowstorm (40732), dam (Viscountess) by Dunbar (31045), g. d. (Her Ladyship) by Lord of the Dales (26703), gr. g. d. (Heiress) by Duke of Clarence (19611), — (Rosemary) by Duke Arthur (14422), — by Cardigan (12556), — by Lord Mayor (18248), — by Duke of Cornwall (5947), — by Fergus (3782), — by Woodford (2854), — by Sir Walter (2637), — by Hotspur (1117), — by Coxcomb (928), — by Midas (435), — by Comet (155), — by R. Colling's Son of Favourite (252), — by same son of Favourite (252), — by Hubback (319).

(44208) VISCOUNT OXFORD 3RD,

Red, calved January 5, 1879, bred by Mr. T. Holford, Papillon Hall; got by Grand Duke 23rd (34063), dam (Baroness Oxford 3rd) by Duke of Hillhurst (28401), g. d. (Baroness Oxford) by Second Duke of Claro (21576), gr. g. d. (Lady Oxford 5th) by Third Duke of Thorndale (17749), — (Lady Oxford 4th) by Grand Duke 2nd (12961), — by The Lord of Eryholme (12205), — by Third Duke of York (10166), — by Duke of Northumberland (1940), — by Short Tail (2621), — by Matchem (2281), — by Young Wynyard (2859).

(44209) VISCOUNT OXFORD 4TH,

Roan, calved July 1, 1879, bred by Mr. T. Holford, Papillon Hall; got by Duke of Rosedale 6th (38176), dam (Viscountess Oxford) by Grand Duke 23rd (34063), g. d. (Baroness Oxford 3rd) by Duke of Hillhurst (28401), gr. g. d. (Baroness Oxford) by Second Duke of Claro (21576), — (Lady Oxford 5th) by Third Duke of Thorndale (17749), — by Grand Duke 2nd (12961), — by The Lord of Eryholme (12205), — by Third Duke of York (10166), — by Duke of Northumberland (1940), — by Short Tail (2621), — by Matchem (2281), — by Young Wynyard (2859).

(44210) VLADIMER'S LING-CROPPER,

Red and white, calved December 25, 1879, bred by Mr. J. Waind, Ankness; got by Count Vladimer (33461), dam (Gipsy Girl) by Duke of Rubies (33724), g. d. (Gipsy Maid) by Snowstorm (35616), gr. g. d. (Countess) by Wathstone's Hero (25417), — (Cygnet) by Captain (14229), — by Captain (14229), — by Chevy Chase (7897), — by Monarch (9407), — by Gainford (2044), — by Samson (5080), — by Young Sovereign (5286), — by White Walton (5652).

(44211) VOLIS HILLHURST,

Red, calved March 18, 1879, bred by Messrs. W. and G. Bird, Volis; got by Lally's Hillhurst Duke (38538), dam (Lydia) by Cherry Duke 6th (30705), g. d. (Laura) by Wellington (32830), gr. g. d. (Venilia 2nd) by Botanist (30568), — (Venilia) by Cherry Duke 3rd (15763), — by Mistor (13343), — by Third Duke of Cambridge (5941), — by Scaleby (6448), — by Smeatonian (5213), — by Don Quixote (987), — by Regent (1366), — by Young Comet (156), — by Jupiter (343), — by Grandson of Favourite (252), — by Barningham (56).

(44212) VULCAN,

Roan, calved December 19, 1879, bred by Mr. M. Boyes, Wandale House; got by Lord Oxford Bright Eyes 2nd (36998), dam (Victoria 4th) by Twentieth Duke of Oxford (28432), g. d. (Victoria 3rd) by Orestes (22443), gr. g. d. (Victoria) by Louis 18th (18284), — (Violet) by Cavalier (14250), — by Bletsoe (9970), — by Viscount Winkle (9788), — by Sultan (7566), — by Norfolk (2377), — by Burley (1766), — by Pilot (496), — by Warlaby (672), — by Albion (14), — by Lame Bull (359), — by Shipton (587), — by Son of Suworrow (636), — by Son of Twin Brother to Ben (88), — by Twin Brother to Ben (660).

(44213) WALESBY LAD,

Red and white, calved July 20, 1879, bred by Mr. J. Moffat, Ballyhyland, sold to the Provincial Government of Brabant, Belgium; got by Royalist (37396), dam (Walesby Lass) by Protector (35188), g. d. (Lady Walesby) by Lord of the Manor (29181), gr. g. d. (Walesby Rose) by Dey of Algiers (25892), — (Walesby Twin 15th) by Refiner (20655), — by Lord John (11731), — by Splendid (12136), — by Shamrock (7488), — by Belvedere 4th (3130), — by Raby (2474), — by Blyth Eclipse (1728), — bred by Mr. Walesby, of Ranby.

(44214) **WALNUT BARRINGTON,**
Red and white, calved April 19, 1879, bred by Mr. C. C. Garner, The Wolds;
got by Duke of Barrington 6th (33576), dam (Symphony) by Knightley
Wellington (31533), g. d. (Gossamer 1st) by Cambridge (25705), gr. g. d.
(Midnight) by Economist (21669), — (Nelly O'Brien) by Cromwell (19528),
— by Old Buck (15017), — by Earl of Dublin (10178), — by Janizary (8175), —
by Snowball (8602), — by Caliph (1774), — by Norman (2379), — by White
Boy (1580), — by Wyville's Bull, — by a Bull bred by Mr. Charge.

(44215) **WAMBA,**
Red, calved November 20, 1879, bred by Mr. R. Pinder, Whitwell; got by
Prince William (37283), dam (Rebecca) by Fiery Star (33914), g. d. (Rowena)
by The Stuart (27650), gr. g. d. (Cleodora) by Falstaff (21720), — (Cranmore
Belle) by Sir Simon (18867), — by Sugar Plum (10894), — by Nero (4557), —
by Thorp (2757), — by Beaufort (1696), — by Czar (495), — by Meteor (1226),
— by Marske (418), — by Freeman (269), — by Danby (190), — by Son of
Danby (190), — by Danby (190).

(44216) **WARLABY WARRIOR,**
Roan, calved March 27, 1876, bred by Mr. J. Beattie, Newbie House; got by
Red Cross Knight (35219), dam (Warrior's Gem) by British Crown (21322),
g. d. (Warrior's Plume) by Breast Plate (19337), gr. g. d. (Warrior's Pride) by
Dr. McHale (15887), — (Warrior's Bride) by Brideman (12493), — by Baron
Warlaby (7813), — by Fourth Duke of Northumberland (3649), — by Norfolk
(2377), — by Waterloo (2816), — by Waterloo (2816).

(44217) **WARREN,**
Red and white, calved in March 1876, bred by the Marquis of Ailesbury,
Savernake, the property of Mr. R. S. Lyne, Barton; got by Æolus (27861), dam
(Pert) by Clint (33399), g. d. (Pretty) by Star Prince (18925), gr. g. d. by
Savernake (35475), — (Violet) by Mazeppa (34831).

(44218) **WARRIOR,**
Roan, calved August 30, 1879, bred by Mr. R. C. Morton, Lane House; got by
Viscount Oxford (40876), dam (Walnut 6th) by Duke of Havering (33664),
g. d. (Walnut) by Third Duke of Geneva (23753), gr. g. d. (Auricula) by Protector (22660), — (Wheedle) by Earl of Dublin (10178), — by Grey Friar
(9172), — by Little John (4232), — by Young Norman (4584), — by White
Boy (1580), — by Wyville's Bull, — by a Bull bred by Mr. Charge.

(44219) **WARRIOR BRAVE,**
White, calved July 26, 1877, bred by Lord Polwarth, Mertoun House; got by
Rapid Rhone (35205), dam (Wave Surf) by Knight of Knowlmere 2nd (31542),
g. d. (Wave Foam) by Manfred (26801), gr. g. d. (Wave Breast) by Breast
Plate (19337), — (Wave Princess) by British Prince (14197), — by Vanguard
(10994), — by Baron Warlaby (7813), — by Fourth Duke of Northumberland
(3649), — by Norfolk (2377), — by Waterloo (2816), — by Waterloo (2816).

(44220) **WASTWATER,**
Red and white, calved October 5, 1876, bred by Messrs. I. and J. Gaitskell, Hall
Santon, the property of Mr. J. Highfield, Frandley House; got by Sultan
(30088), dam (Butterfly) by Mentana (26893), g. d. (Pearl) by Joseph (20033),
gr. g. d. (Rose) by Codrington (14290), — (Alma) by Earl of Scarborough

(11410), — by Prince Edward (6334), — by Traveller (6617), — by Charles (3343), — by Commodore (5874), — by Cupid (5900), — by Parson (1306), — by St. John (572), — by Pope (514), — by Chilton (136).

(44221) WATERLOO DUKE,
Red and white, calved August 29, 1879, bred by Mr. G. Fox, Elmhurst Hall; got by Twenty-fourth Duke of Airdrie (36460), dam (Water Lass 2nd) by General Napier (24023), g. d. (Water Lass) by Oxford 2nd (18507), gr. g. d. (Water Maid) by Marc Antony (14895), — (Water Witch) by Squire Blanche (12139), — by Fourth Duke of Northumberland (3649), — by Second Cleveland Lad (3408), — by Duke of Northumberland (1940), — by Norfolk (2377), — by Waterloo (2816), — by Waterloo (2816).

(44222) WATERLOO FLOWER,
Roan, calved March 16, 1878, bred by Mr. W. Chapman, Kingshaugh House; got by Flower Prince (38307), dam (Water Belle) by Bright Prince (28083), g. d. (Water Lily) by Lord Napier (26688), gr. g. d. (Water Fay) by Regal Dane (22710), — (Water Fairy) by Booth Royal (15673), — by Vanguard (10994), — by Crown Prince (10087), — by Fourth Duke of Northumberland (3649), — by Norfolk (2377), — by Waterloo (2816), — by Waterloo (2816).

(44223) WATERLOO HERO,
Red and white, calved December 11, 1879, bred by Mr. T. Stamper, Highfield House; got by Duke of Nawton 3rd (39764), dam (Sweet Rosa) by Duke of Waterloo (21616), g. d. (Mountain Rose) by Hartforth (16224), gr. g. d. (Rose Bay) by Sir Colin Campbell (15279), — (White Violet) by Esholt (10206), — by Walter Buchan (7692), — by Marmion (4381), — by Charles (1815), — by Remus (550), — by Sedbury (1424), — by Hollings (2131), — by Partner (2409), — by Hutton's Bull (2154), — by Marske (418), — by R. Alcock's Bull (19).

(44224) WATERLOO LORD,
Roan, calved February 28, 1877, bred by Mr. T. Barber, Sproatley Rise, the property of Mr. J. Brown, Park Cottage, Burton Constable; got by Oxford's Baronet (29499), dam (Water Duchess) by Grand Duke 6th (19876), g. d. (Water Lass) by Oxford 2nd (18007), gr. g. d. (Water Maid) by Marc Antony (14895), — (Water Witch) by Squire Blanche (12139), — by Fourth Duke of Northumberland (3649), — by Second Cleveland Lad (3408), — by Duke of Northumberland (1940), — by Norfolk (2377), — by Waterloo (2816), — by Waterloo (2816).

(44225) WATERLOO RANGER,
Roan, calved April 6, 1877, bred by Mr. W. Chapman, Kingshaugh House, the property of Mr. W. Oxley, Broughton; got by Ranger (35203), dam (Water Lily) by Lord Napier (26688), g. d. (Water Fay) by Regal Dane (22710), gr. g. d. (Water Fairy) by Booth Royal (15673), — (Water Elf) by Vanguard (10994), — by Crown Prince (10087), — by Fourth Duke of Northumberland (3649), — by Norfolk (2377), — by Waterloo (2816), — by Waterloo (2816).

(44226) WEAL EARL,
Roan, calved August 21, 1879, bred by Mr. H. Smith, Mountmellick; got by Earl of Aylesby (38215), dam (Weal Trust) by Lord Cain (31630), g. d. (Weal Hope) by Royal Prince (27384), gr. g. d. (Weal Royal) by Booth Royal (15673),

— (Weal Princess) by British Prince (14197), — by Crown Prince (10087), — by Fourth Duke of Northumberland (3649), — by Norfolk (2377), — by Waterloo (2816), — by Waterloo (2816).

(44227) WELCOME WORCESTER,

Red and white, calved April 7, 1879, bred by Mr. H. Webb, Streetly Hall; got by Viscount Worcester (40881), dam (Welcome Lady) by Hairy Boy (34102), g. d. (Welcome Lassie 3rd) by Costa King (28254), gr. g. d. (Welcome Lassie) by Costa (21487), — (Welcome Lass 4th) by The Plumber (18993), — by Paris (7314), — by King of Trumps (4156), — by Vanguard (5545), — by Red Rover (4903), — by Anticipation (750), — by Emperor (1014), — by Young Windsor (699), — by Windsor (698).

(44228) WELFORD,

Roan, calved February 23, 1879, bred by Earl Beauchamp, Madresfield Court; got by Magnum Bonum (41954), dam (Lady Blanche) by Festival (26147), g. d. (Butterworth) by Castlereagh (19409), gr. g. d. (Bess) by Impatience (14723), — (Berangeria) by Hero (8146), — by Major (4338), — by Ormsby (4621), — by Cossack (925), — by Alpha (3004), — by Captain (3271), — by Captain (108).

(44229) WELLESLEY,

Red and white, calved February 22, 1879, bred by Mr. C. Hobbs, Maisey Hampton; got by Grand Duke of Waterloo (34077), dam (Fanny 5th) by Colonel Tregunter 7th (30769), g. d. (Fanny 2nd) by King of Barton (28976), gr. g. d. (Fanny) by Captain Wetherby (21371), — (Fanny) by Romulus (16853), — by Lord Milton (10461), — by Belton (9954), — by Cedric (3311), — by a Grandson of Blyth Comet (85), — by Cœlebs (897), — by Rose's Old Red Bull (5009), — by Neswick (1266), — by Fisher's Old Bull (3799).

(44230) WELLESLEY 2ND,

Red and white, calved March 18, 1879, bred by Mr. C. Hobbs, Maisey Hampton, the property of Mr. W. H. Poynder, Hartham Park; got by Grand Duke of Waterloo (34077), dam (Melody 5th) by Prince Charlie (32116), g. d. (Melody) by Grand Duke 13th (21850), gr. g. d. (Musical 4th) by Northern Duke (22431), — (Musical) by Fourth Duke of Oxford (11387), — by Mountaineer (14966), — by Nundi (9453), — by Monzani (6222), — by The Prince (7615), — by Fitzroy (3808), — by Morpeth (2339), — by Roman (2559, — by Admiral (5), — by Son of Blyth Comet (85), — by Foljambe's Bull (1051).

(44231) WELLESLEY 2ND,

* Red and white, calved September 25, 1879, bred by the Duke of Manchester, Kimbolton Castle; got by Duke of Underley 3rd (38196), dam (Waterloo 33rd) by Duke of Hillhurst (28401), g. d. (Waterloo 32nd) by Seventh Duke of York (17754), gr. g. d. (Waterloo 28th) by Grand Duke 3rd (16182), — (Waterloo 13th) by Third Duke of Oxford (9047), — by Second Cleveland Lad (3408), — by Duke of Northumberland (1940), — by Norfolk (2377), — by Waterloo (2816), — by Waterloo (2816).

(44232) WELLINGTON,

White, calved July 10, 1879, bred by the Duke of Northumberland, Alnwick Castle; got by Waterloo (40892), dam (Snowflake 2nd) by Hawthorn (36751), g. d. (Snowflake) by Ben Brace (30524), gr. g. d. (Snowdrop) by Wellington

Hero (32832), — (Daisy 2nd) by Merry Monarch (22349), — by Alderman (17291), — by Duke of Tyne (12773), — by Young Hector (7074), — by Young Hastings (3988), — by Short Legs (5124), — by Ajax (723), — by Hector (2103), — by Surly (2715), — by Sir Harry (5155), — by Colonel (152), — by Son of Hubback (319).

(44233)   WELLINGTON DUKE 3RD,

White, calved August 22, 1879, bred by Mr. R. Lodge, The Rookery ; got by Baron Siddington 2nd (37816), dam (Duchess of Wellington 6th) by Baron Barrington 5th (33007), g. d. (Duchess of Wellington 5th) by Duke of Lorn (25985), gr. g. d. (Duchess of Wellington 2nd) by Garibaldi 2nd (26220), — (Wellingtonia 5th) by Grand Duke 4th (19874), — by Third Duke of Thorndale (17749), — by Duke of Bolton (12738), — by Second Cleveland Lad (3408), — by Duke of Northumberland (1940), — by Norfolk (2377), — by Waterloo (2816), — by Waterloo (2816).

(44234)   WELLINGTON DUKE 4TH,

Red and white, calved October 27, 1879, bred by Mr. R. Lodge, The Rookery ; got by Baron Siddington 2nd (37816), dam (Duchess of Wellington 5th) by Duke of Lorn (25985), g. d. (Duchess of Wellington 2nd) by Garibaldi 2nd (26220), gr. g. d. (Wellingtonia 5th) by Grand Duke 4th (19874), — (Wellingtonia 2nd) by Third Duke of Thorndale (17749), — by Duke of Bolton (12738), — by Second Cleveland Lad (3408), — by Duke of Northumberland (1940), — by Norfolk (2377), — by Waterloo (2816), — by Waterloo (2816).

(44235)   WERRINGTON 2ND,

Red and white, calved March 6, 1879, bred by Mr. C. W. Griffin, Werrington ; got by Telemachus 10th (35728), dam (Lady Blanche 2nd) by Nestor (24648), or Sugar Cane (32625), g. d. (Lady Blanche) by Red Butterfly (22690), gr. g. d. (Bella) by Louis (14861), — (Lady Blanche) by Rivers (10716), — by Mozart (11830), — by Stirling (5330), — by Young Frederick (3836), — by Favourite (3768), — by Tathwell Studley (5401), — by Son of Waddingworth (668).

(44236)   WESTBOURNE,

Roan, calved August 7, 1879, bred by Miss Graham, Yardley Stud Farm ; got by Baron Fantail (37790), dam (Lady Jocelyn 2nd) by Janitor (24204), g. d. (Lady Jocelyn) by Seventh Duke of York (17754), gr. g. d. (Lady Jane), by Duke of Glo'ster (11382), — (Jardine) by Lord Warden (7167), — by Fawsley (6004), — by Warden (5595), — by Javelin (4093), — by Blyth (797), — by Wellington (684), — by Phenomenon (491), — by Favourite (252), — by Favourite (252), — by Favourite (252), — by Hubback (319), — by Snowdon's Bull (612), — by Waistell's Bull (669), — by Masterman's Bull (422), — by the Studley Bull (626).

(44237)   WETHERBY,

White, calved June 5, 1876, bred by Colonel Gunter, Wetherby Grange, the property of Mr. Maude, Blawith ; got by Duke of Tregunter 5th (33743), dam (Lady's Maid) by Duke Ferdinand 3rd (28354), g. d. (Milkmaid) by Marquis (20290), gr. g. d. (Dairymaid) by Le Moor (22090), — (Barmaid) by Oxford (15035), — by Red Richmond (13578), — by Earl Stanhope (5966), — by Lord Stanley (4269), — by Young Matchem (4425), — by Fairfax (1023), — by Son of Marske (418), — by Palmsun (7311), — by a Bull of Mr. Maynard's — by Son of Favourite.

(44238)   **WHIMPER,**

Roan, calved December 8, 1876, bred by the Rev. H. Beckwith, Eaton Constantine, the property of Mr. G. T. Wright, Stokes Farm; got by Lord of Wicken (34644), dam (Essie) by Leander (22085), g. d. (Bonnie Bess) by Cronkhill (19530), gr. g. d. (Bonne Bouche) by Sir Roger (16991), — (Bona Fide) by May Duke (13320), — by Brawith Boy (7846), — by Sir Walter (2639), — by Roseberry (567), — by Roseberry (567), — by Constitution (166), — by Hastings (293), — by Constellation (163).

(44239)   **WHIRLBONE,**

Red and white, calved June 7, 1878, bred by Mr. B. Hale, Holly Hill; got by Duke of Anglesea (38118), dam (Whirlwind) by Cherry Grand Duke 3rd (28174), g. d. (Wave Elf) by Cherry Prince (23555), gr. g. d. (Wave Naiad) by Breast Plate (19337), — (Wave Witch) by Fitz-Clarence (14552), — by British Prince (14197), — by Vanguard (10994), — by Baron Warlaby (7813), — by Fourth Duke of Northumberland (3649), — by Norfolk (2377), — by Waterloo (2816), — by Waterloo (2816).

(44240)   **WHITE BAIT,**

White, calved November 21, 1879, bred by Mr. J. Morton, Fences Farm; got by Hesperus (39994), dam (Datura) by Broughton Royal (33236), g. d. (Daphne) by Hogarth 2nd (24148), gr. g. d. (Dahlia 7th) by Butley Butterfly (21348), — (Dahlia 4th) by Conte de Vermandois (19507), — by Lumley (16478), — by Grand Duke (16183), — by Harper (16223), — by Brennus (8902), — by Earl Fitzwilliam (8016), — by Duke of Norfolk (5952), — by Cleveland Lad (3407), — by Red Highflyer (2488), — by Sir Charles (5146), — by Harry Lorrequer (3985), — by Blucher (84), — by Magnum Bonum (4322), — by Buston's Styford (103), — by Son of T. Wetherell's Bull (690).

(44241)   **WHITE BUTTERFLY,**

White, calved April 6, 1879, bred by Mr. J. H. Braikenridge, The Rookery; got by Pioneer (38868), dam (Queen's Butterfly) by King James (28971), g. d. (Duke's Butterfly) by Duke of York (23804), gr. g. d. (Baron's Butterfly) by Baron Hopewell (14134), — (White Butterfly) by Butterfly's Nephew (15714), — by Master Butterfly (13311), — by Gavazzi (11508), — by Baron of Ravensworth (7811), — by Raree Show (4874), — by Thick Hock (6601), — by Expectation (1988), — by Belzoni (1709), — by Comus (1861), — by Denton (198),

(44242)   **WHITE CHIEF,**

White, calved January 15, 1879, bred by Mr. J. Jardine, Dryfeholm; got by Frankland (39895), dam (Convolvulus) by Sir William Fox (40726), g. d. (Dahlia) by Pizarro (20497), gr. g. d. (Venus) by Master Annandale (14916), — (Sprightly) by Young Earl (14467), — by Snowy Down (8607), — by Prince Albert (4778), — by Son of Cumberland (5256), — by Exmouth (3747), — by Prince (4765), — by Leopold (2199).

(44243)   **WHITE CHIEF,**

White, calved April 22, 1879, bred by Mr. R. B. Brockbank, Crosby; got by Borderer (33183), dam (Miss Brayton) by Oxford Don (29489), g. d. (Lady Lawson) by Kildonan (20051), gr. g. d. (Pansy) by Sir Colin (16951), — (Prosperity) by The Baron (13833), — by Prince Edward Fairfax (9506), —. by Fourth Duke of Northumberland (3649), — by Red Highflyer (2488), — by

Pyramid (4852), — by Sir Harry Lorrequer (3985), — by Blucher (84), — by Magnum Bonum (4322), — by Buston's Styford (103), — by Son of T. Wetherell's Bull (690).

(44244)　　　　　WHITE HAWTHORN,
White, calved April 10, 1879, bred by Mr. N. Milne, Dryhope, the property of Mr. J. Muir, Dryhope; got by Earl of Kelso (38228), dam (Lady Hawthorn) by Thorndale Duke 2nd (32712), g. d. (Lady Knightley) by Grand Knight (26302), gr. g. d. (Lady Mary) by Redmond (20646), — (Young Sophy) by MacTurk (14872), — by Heir-at-Law (13005), — by Hudibras (10339), — by Baron of Ravensworth (7811), — by Baronet (1686), — by Spectator (2688), — by Fitz-Remus (2025), — by Whitworth (695).

(44245)　　　　　WHITE KNIGHT,
White, calved May 13, 1879, bred by Mr. E. Lythall, Radford Hall; got by Tablet (37558), dam (Miss Knightley) by Knightley Wellington (31533), g. d. (Gillyflower) by Standard Bearer (25221), gr. g. d. (Gwendoline) by Old Buck (15017), — (Gundreda) by Earl of Dublin (10178), — by Janizary (8175), — by Snowball (8602), — by Caliph (1774), — by Norman (2379), — by White Boy (1580), — by Wyville's Bull — by a Bull bred by Mr. Charge.

(44246)　　　　　WHITE PRINCE,
White, calved September 28, 1878, bred by Mr. F. Fowler, Henlow, the property of Mr. T. White, Wootten, Kempston; got by Prince Leopold (37239), dam (Princess Mary) by Prince Regent (29677), g. d. (Lettuce Hopewell) by Prince Hopewell (22592), gr. g. d. (Lettuce) by Prince George (13510), — (Lucretia) by Clarence (14279), — by Justice (14753), — by Young Rufus (13649), — by Constitution (12634), — by Young Comet (1853), — descended from Jolly's Bull (4115).

(44247)　　　　　WHITE RAYMOND,
White, calved December 11, 1878, bred by Mr. J. Bruce, Barmoor Castle; got by Sir Raymond (40716), dam (Wave Queen) by Knight of the Shire (26552), g. d. (Wave Naiad) by Breast Plate (19337), gr. g. d. (Wave Witch) by Fitz-Clarence (14552), — (Wave Princess) by British Prince (14197), — by Vanguard (10994), — by Baron Warlaby (7813), — by Fourth Duke of Northumberland (3649), — by Norfolk (2377), — by Waterloo (2816), — by Waterloo (2816).

(44248)　　　　　WHITE SIRE,
White, calved December 29, 1878, bred by Mr. J. Cran, Kirkton; got by Bridegroom (33201), dam (Dewdrop 1st) by Baronet (25504), g. d. (Dewdrop) by Scottish Chief (25102), gr. g. d. (Snowdrop) by Red Luggs (32263), — (Young Lousea) by Major (24514), — by Jock O'Darlington, — Mary of Eden, from the stock of Mr. Grant Duff.

(44249)　　　　　WIDEAWAKE,
Red and white, calved February 15, 1879, bred by Mr. H. Holborow, Willsley; got by Emperor (41508), dam (Wallfruit) by Northfleet (34926), g. d. (Wallflower 11th) by Second Earl of Walton (19672), gr. g. d. (Wallflower 5th) by Chaffcutter (12572), — (Wallflower 3rd) by Oregon (8371), — by Bristol (7852), — Wallflower, a Cow of Colonel Kingscote's.

(44250) WIDEAWAKE 2ND,
Roan, calved January 4, 1877, bred by Mr. F. W. Low, Kilshane; got by
Wideawake (27805), dam (Limerick Lass 2nd) by Star of the West (22972),
g. d. (Lady Clarina) by Volunteer (15478), gr. g. d. (Lady Camilla) by Norfolk
(9442), — (Florence) by Rex (6385), — by Sir Thomas Fairfax (5196), — by
Ambo (1636), — by Memnon (2295), — by Pilot (496), — by Agamemnon (9),
— by Burrell's Bull, of Burdon.

(44251) WILD CHERRY DUKE 2ND,
Roan, calved February 8, 1877, bred by Mr. J. J. Stone, Scyborwen, the
property of the Exors. of Mr. J. J. Stone; got by Oxford Duke (32022),
dam (Wild Cherry Duchess) by Oxford Duke (32022), g. d. (Wild Cherry)
by Cherry Grand Duke 2nd (25758), gr. g. d. (Wild Duchess) by Grand
Duke 3rd (16182), — (Wild Eyes 25th) by Lablache (16353), — by Vocalist (13960), — by Solon (13766), — by Second Duke of Oxford (9046),
— by Fourth Duke of Northumberland (3649), — by Duke of Northumberland
(1940), — by Belvedere (1706), — by Emperor (1975), — by Wonderful (700),
— by Cleveland (145), — by Butterfly (104), — by Hollon's Bull (313), — by
Mowbray's Bull (2342), — by Masterman's Bull (422), — descended from M.
Dobison's stock.

(44252) WILD CLEOPATRA,
Roan, calved August 13, 1879, bred by Mr. H. P. Baxter, Southall Green
Farm; got by Wild Duke of Connaught (40921), dam (Lady Margaret) by
Second Duke of Collingham (23730), g. d. (Cleopatra 6th) by Ninth Duke of
Oxford (17738), gr. g. d. (Cleopatra 4th) by Duke of Buckingham (14428), —
(Cleopatra) by Earl of Dublin (10178), — by Standish (7550), — by Bellerophon
(3119), — by Alderman (2976), — by Waterloo (2816), — by Young Wynyard
(2859), — by Son of Simon (590), — by Buston's Styford (103).

(44253) WILD COLONEL,
Red and little white, calved December 29, 1878, bred by Mr. R. Lodge, The
Rookery, the property of Mr. Rawlinson, Cartmel; got by Lightburne's Duke of
Oxford (36895), dam (Wild Eyes E.) by Eighteenth Duke of Oxford (25995),
g. d. (Wild Eyes D.) by Diamond Eyes (28318), gr. g. d. (Wild Eyes) by Lord
Lally 3rd (24408), — (Wild Eyes 22nd) by Wild Duke (19148), — by Lord
Barrington 1st (13170), — by Second Duke of Oxford (9046), — by Fourth
Duke of Northumberland (3649), — by Duke of Northumberland (1940, — by
Belvedere (1706), — by Emperor (1975), — by Wonderful (700), — by Cleveland
(145), — by Butterfly (104), — by Hollon's Bull (313), — by Mowbray's Bull
(2342), — by Masterman's Bull (422), — descended from M. Dobison's stock.

(44254) WILD DUKE 3RD,
Red, calved October 25, 1879, bred by Messrs. F. Leney and Sons, Wateringbury; got by Duke of Oxford 44th (39774), dam (Wild Duchess 5th) by Sixth
Duke of Oneida (30997), g. d. (Wild Duchess 3rd) by Lord Oxford 2nd (20215),
gr. g. d. (Wild Duchess) by Duke of Wetherby (17753), — (Wild Eyes 24th)
by Lord Barrington 3rd (16382), — by Second Duke of Oxford (9046), — by
Fourth Duke of Northumberland (3649), — by Duke of Northumberland (1940),
— by Belvedere (1706), — by Emperor (1975), — by Wonderful (700), — by
Cleveland (145), — by Butterfly (104), — by Hollon's Bull (313), — by Mow-

bray's Bull (2342), — by Masterman's Bull (422), — descended from M. Dobison's stock.

(44255)   WILD DUKE 4TH,

Roan, calved April 1, 1879, bred by Lord Fitzhardinge, Berkeley Castle; got by Duke of Connaught (33604), dam (Lady Wild Eyes 4th) by Cherry Grand Duke 2nd (25758), g. d. (Lady Wild Eyes 3rd) by Cherry Grand Duke (23554), gr. g. d. (Lady Wild Eyes 2nd) by Touchstone (20986), — (Lady Wild Eyes) by Weathercock (9815), — by Second Cleveland Lad (3408), — by Second Duke of Northumberland (3646), — by Short Tail (2621), — by Emperor (1975), — by Wonderful (700), — by Cleveland (145), — by Butterfly (104), — by Hollon's Bull (313), — by Mowbray's Bull (2342), — by Masterman's Bull (422), — descended from M. Dobison's stock.

(44256)   WILD DUKE 5TH,

Roan, calved September 2, 1879, bred by Lord Fitzhardinge, Berkeley Castle; got by Duke of Connaught (33604), dam (Lady Wild Eyes 3rd) by Cherry Grand Duke (23554), g. d. (Lady Wild Eyes 2nd) by Touchstone (20986), gr. g. d. (Lady Wild Eyes) by Weathercock (9815), — (Wild Eyes 27th) by Second Cleveland Lad (3408), — by Second Duke of Northumberland (3646), — by Short Tail (2621), — by Emperor (1975), — by Wonderful (700), — by Cleveland (145), — by Butterfly (104), — by Hollon's Bull (313), — by Mowbray's Bull (2342), — by Masterman's Bull (422), — descended from M. Dobison's stock.

(44257)   WILD DUKE OF COLLINGHAM,

Roan, calved June 2, 1879, bred by Major J. Chaffey, Prince Hill; got by Duke of Collingham 3rd (38134), dam (Wild Duchess of Geneva 5th) by Ninth Duke of Geneva (28391), g. d. (Wild Duchess of York) by Seventh Duke of York (17754), gr. g. d. (Wild Oxford) by Lord Oxford 2nd (20215), — (Wild Eyes 24th) by Lord Barrington 3rd (16382), — by Second Duke of Oxford (9046), — by Fourth Duke of Northumberland (3649), — by Duke of Northumberland (1940), — by Belvedere (1706), — by Emperor (1975), — by Wonderful (700), — by Cleveland (145), — by Butterfly (104), — by Hollon's Bull (313), — by Mowbray's Bull (2342), — by Masterman's Bull (422), — descended from M. Dobison's stock.

(44258)   WILD EARL,

Roan, calved July 13, 1879, bred by Mr. H. Lovatt, Low Hill; got by Baron Turncroft Oxford 4th (37822), dam (Wild Eyes 37th) by Lord Darlington 8th (34520), g. d. (Wild Eyes 30th) by Don John (19583), gr. g. d. (Wild Eyes 27th) by Gainford 5th (12913), — (Wild Eyes 26th) by Second Cleveland Lad (3408), — by Short Tail (2621), — by Emperor (1975), — by Wonderful (700), — by Cleveland (145), — by Butterfly (104), — by Hollon's Bull (313), — by Mowbray's Bull (2342), — by Masterman's Bull (422), — descended from M. Dobison's stock.

(44259)   WILD MAID'S DUKE,

Red, calved December 29, 1879, bred by Sir R. C. Musgrave, Bart., Eden Hall; got by Duke of Underley (33745), dam (Wild Maid) by Seventeenth Duke of Oxford (25994), g. d. (Grand Duchess 6th) by Duke of Geneva (19614), gr. g. d. (Grand Duchess) by Cherry Duke 2nd (14265), — (Wildair) by Santiago (12047), — by Duke of Northumberland (1940), — by Belvedere (1706), — by Emperor

(1975), — by Wonderful (700), — by Cleveland (145), — by Butterfly (104),
— by Hollon's Bull (313), — by Mowbray's Bull (2342), — by Masterman's
Bull (422), — descended from M. Dobison's stock.

(44260)　　　　　　WILD PRINCE 7TH,
Roan, calved March 24, 1879, bred by Lord Skelmersdale, Lathom House, the
property of Mr. H. Rawcliffe, Euxton Hall ; got by Baron Oxford 4th (25580),
dam (Lady Wild Eyes) by Eighth Duke of Geneva (28390), g. d. (Wild Eyes
24th) by Fourth Duke of Oxford (11387), gr. g. d. (Wild Eyes 22nd) by Wild
Duke (19148), — (Wild Eyes 20th) by Lord Barrington 1st (13170), — by
Second Duke of Oxford (9046), — by Fourth Duke of Northumberland (3649),
— by Duke of Northumberland (1940), — by Belvedere (1706), — by Emperor
(1975), — by Wonderful (700), — by Cleveland (145), — by Butterfly (104),
— by Hollon's Bull (313), — by Mowbray's Bull (2342), — by Masterman's
Bull (422), — descended from M. Dobison's stock.

(44261)　　　　　　WILD PRINCE 8TH,
Red, calved April 9, 1879, bred by Lord Skelmersdale, Lathom House, the
property of Mr. A. J. Scott, Rotherfield Park ; got by Fifth Duke of Wetherby
(31033), dam (Lady Bright Eyes 3rd) by Seventh Duke of York (17754), g. d.
(Bright Eyes 2nd) by Royal Butterfly 3rd (18754), gr. g. d. (Bonny) by Oxford
Duke (15036), — (Beauty) by Crusade (7938), — by Third Duke of York
(10166), — by Second Cleveland Lad (3408), — by Duke of Northumberland
(1940), — by Belvedere (1706), — by Emperor (1975), — by Wonderful (700),
— by Cleveland (145), — by Butterfly (104), — by Hollon's Bull (313), — by
Mowbray's Bull (2342), — by Masterman's Bull (422), — descended from M.
Dobison's stock.

(44262)　　　　　　WILD PRINCE 9TH,
White, calved June 21, 1879, bred by Lord Skelmersdale, Lathom House ; got
by Baron Oxford 4th (25580), dam (Bright Eyes 5th) by Grand Duke 6th
(19876), g. d. (Bonny) by Oxford Duke (15036), gr. g. d. (Beauty) by Crusade
(7938), — (Bright Eyes) by Third Duke of York (10166), — by Second Cleveland Lad (3408), — by Duke of Northumberland (1940), — by Belvedere (1706),
— by Emperor (1975), — by Wonderful (700), — by Cleveland (145), — by
Butterfly (104), — by Hollon's Bull (313), — by Mowbray's Bull (2342), —
by Masterman's Bull (422), — descended from M. Dobison's stock.

(44263)　　　　　　WILD PRINCE 10TH,
White, calved October 28, 1879, bred by Lord Skelmersdale, Lathom House ; got
by Duke of Rosedale 6th (38176), dam (Winsome 18th) by Baron Oxford 4th
(25580), g. d. (Winsome 4th) by Grand Duke 10th (21848), gr. g. d. (Winsome)
by Oxford 2nd (18507), — (Beauty) by Crusade (7938), — by Third Duke of
York (10166), — by Second Cleveland Lad (3408), — by Duke of Northumberland (1940), — by Belvedere (1706), — by Emperor (1975), — by Wonderful
(700), — by Cleveland (145), — by Butterfly (104), — by Hollon's Bull (313),
— by Mowbray's Bull (2342), — by Masterman's Bull (422), — descended from
M. Dobison's stock.

(44264)　　　　　　WILD STRAWBERRY,
White, calved July 5, 1878, bred by Sir Wilfrid Lawson, Bart., Brayton ; got by
Wild Eyes Duke (36007), dam (Strawberry) by Grand Vizier (26313), g. d.
(Spotless) by Perth (22513), gr. g. d. (Spot) by Duke (14419), — (Spot) by

Captain Hardinge (14232), — by Ewan Gwynne (14521), — by a Son of Wallace (5586), — by Cupid (1893), — by Charlie (130), — by Western Comet (689).

(44265)　　　　　　WILD TREGUNTER,

Red and white, calved October 24, 1879, bred by the Earl of Feversham, Duncombe Park; got by Duke of Tregunter 5th (33743), dam (Sparkling Eyes) by Sixth Duke of Geneva (30959), g. d. (Wild Eyes Duchess) by Grand Duke 9th (19879), gr. g. d. (Wild Eyes 19th) by Lablache (16353), — (Wild Eyes 18th) by Solon (13766), — by Second Duke of Oxford (9046), — by Fourth Duke of Northumberland (3649), — by Duke of Northumberland (1940) — by Belvedere (1706), — by Emperor (1975), — by Wonderful (700), — by Cleveland (145), — by Butterfly (104), — by Hollon's Bull (313), — by Mowbray's Bull (2342), — by Masterman's Bull (422), — descended from M. Dobison's stock.

(44266)　　　　　　WILD WATTIE,

Roan, calved June 10, 1879, bred by Mr. J. Kerfoot, Faenol Bach; got by Edward Waterloo (38246), dam (Agnes 2nd) by Norman (31980), g. d. (Agnes) by Charles Edward (25743), gr. g. d. (Annette) by Mac Turk (14872), — (Seclusion) by Heir-at-Law (13005), — by Baron of Ravensworth (7811), — by Hudibras (10339), — by Baronet (1686), — by Spectator (2688), — by Fitz-Remus (2025), — by Whitworth (695), — a Cow bought by Mr. Mason, of Chilton.

(44267)　　　　　　WILEY DUKE 3rd,

Red and white, calved June 23, 1879, bred by Mr. G. Fox, Elmhurst Hall; got by Duke of Oxford 39th (38173), dam (Wiley Duchess 2nd) by Geneva Lad (41608), g. d. (Duchess Paulina) by Oxford Lad (24713), gr. g. d. (Pride of the Springs) by Duke of Glo'ster (11382), — (Miss Wiley 7th) by Baron Martin (12444), — by Prince Royal (8428), — by Hermes (8145), — by Carcase (3285), — by Tyro (2781), — by Falstaff (1993), — by Dr. Syntax (220), — by Charles (127), — by Henry (301), — by Favourite (252), — by Mason's White Bull (421), — by Bolingbroke (86), — by Foljambe (263), — by Hubback (319), — bred by Mr. Maynard.

(44268)　　　　　　WILLIE GWYNNE,

Red, calved December 9, 1879, bred by Mr. G. Fox, Elmhurst Hall; got by Twenty-fourth Duke of Airdrie (36460), dam (Polly Gwynne 8th) by Oxford 4th (29481), g. d. (Polly Gwynne 2nd) by Wild Duke 4th (21107), gr. g. d. (Polly Gwynne) by Flying Dutchman (10235), — (Young Dowager Gwynne) by St. Thomas (10777), — by Prime Minister (2456), — by Wallace (5586), — by Marmion (406), — by Merlin (430), — by Layton (366), — by Phenomenon (491), — by Favourite (252), — by Favourite (252), — by Hubback (319), — by Snowdon's Bull (612), — by Waistell's Bull (669), — by Masterman's Bull (422), — by the Studley Bull (626).

(44269)　　　　　　WINDSOR CHERRY,

Roan, calved March 3, 1869, bred by Lady Pigot, Branches Park, the property of Mr. J. Y. Burgess, Parkanaur; got by Roan Windsor (24967), dam (Cherry) by Gipsy King (21830), g. d. (Violet) by Majestic (16492), gr. g. d. (Vivandière) by Mickleton (18392), — (Necklace) by Len (14788), — by Voltaire (13963), — by Lord George (10439), — by Vanguard (10994), — by Hamlet (8126), — by Sir Roger (7515), — by Studley Royal (5342).

(44270)   WINDSOR'S ACKWORTH,

Red, calved July 31, 1878, bred by Mr. W. Brown, The Villa, Ackworth, the property of Mr. H. Burtt, Fulbeck Grange; got by Windsor's Royal Seal (36016), dam (Sweetbrier) by Farnley Royal (28573), g. d. (Sweetbroom) by Twin Athol (27684), gr. g. d. (Sweet Willow) by Sir Sam (25171), — (Sweetbrier) by White Hamlet (15518), — by General Fairfax (14594), — by Brawith Boy (7846), — by Sir Walter (2639), — by Sir Walter (2639), — by Roseberry (567), — by Roseberry (567), — by Constitution (166), — by Hastings (293), — by Constellation (163).

(44271)   WINDSOR'S FAME,

Roan, calved November 25, 1879, bred by Mr. R. Lawson, Elswick; got by Warrior's Fame (40890), dam (Rose of Windsor) by Water Wizard (37657), g. d. (Rose of Summer) by Sir Windsor Broughton (27507), gr. g. d. (Diamond Rose) by Sir David (25135), — (Rose Blush) by King Arthur (13110), — by The Silkey Laddie (10947), — by Rouge (5012), — by Chance (3329), — by Cleveland (3404), — by Danby (3550), — by Son of Studley Grange (1483), — by Son of Duke (225), — by Midas (4470), — by Nailer (4528), — by Ambo (746).

(44272)   WINDSOR'S PRINCE,

Roan, calved April 4, 1879, bred by Mr. J. Close, Holmescales; got by Royal Victor (35414), dam (Windsor's Duchess 3rd) by Flag of Windsor (28614), g. d. (Windsor's Duchess 2nd) by Shirley Fitz-Windsor (27451), gr. g. d. (Duchess) by Duke of Hamilton (19618), — by Master Hopewell (14929), — by Benedict (7828).

(44273)   WINDSOR VAIL,

Roan, calved March 21, 1879, bred by Mr. C. A. Cantlie, Keithmore, the property of Mr. W. Duthie, Collynie; got by Sir Windsor Broughton (27507), dam (Vail 7th) by Watchman 2nd (27756), g. d. (Vail 2nd) by Lord Jersey (26661), gr. g. d. (Vail) by Old England (24681), — (Rosabella) by Master Gunner (22316), — by California (12529), — by Bang (17346), — by Sir Thomas Fairfax (5196), — by Duke (3630), — by Reveller (2528), — by Grazier (1085), — by Cato (857), — by Atlas (42), — by Favourite (257), — by Mr. Robinson's Bull (4974), — by Badsworth (47).

(44274)   WINSOME DUKE 3RD,

White, calved August 11, 1879, bred by the Earl of Feversham, Duncombe Park; got by Duke of Tregunter 5th (33743), dam (Wild Winsome 2nd) by Twentieth Duke of Oxford (28432), g. d. (Winsome 11th) by Baron Oxford 4th (25580), gr. g. d. (Winsome 5th) by Grand Duke 10th (21848), — (Winsome 2nd) by Lord Oxford (20214), — by Oxford 2nd (18507), — by Crusade (7938), — by Third Duke of York (10166), — by Second Cleveland Lad (3408), — by Duke of Northumberland (1940), — by Belvedere (1706), — by Emperor (1975), — by Wonderful (700), — by Cleveland (145), — by Butterfly (104), — by Hollon's Bull (313), — by Mowbray's Bull (2342), — by Masterman's Bull (422), — descended from M. Dobison's stock.

(44275)   WISEMAN,

Roan, calved May 23, 1879, bred by Mr. H. W. Goulder, Wimbotsham; got by Holman Hunt (36777), dam (Dainty) by Royal Highness (40635), g. d. (Daisy 2nd) by Master Hopewell (24555), gr. g. d. (Daisy 1st) by Viscount Windsor (19082), — (Dora) by Oliver (40407).

(44276)   WODEN,

Red and little white, calved March 4, 1879, bred by Sir Wilfrid Lawson, Bart., Brayton; got by Doctor (39687), dam (Waterloo 38th) by Baron Oxford 6th (33075), g. d. (Waterloo 35th) by Waterloo Chief (23184), gr. g. d. (Waterloo 20th) by Cherry Duke 2nd (14265), — (Waterloo 16th) by Red Knight (11976), — by Grand Duke (10284), — by Third Duke of Oxford (9047), — by Second Cleveland Lad (3408), — by Duke of Northumberland (1940), — by Norfolk (2377), — by Waterloo (2816), — by Waterloo (2816).

(44277)   WOODMAN,

Roan, calved May 8, 1879, bred by Mr. E. W. Meade-Waldo, Stonewall Park; got by King Harry (36841), dam (Wave Spirit) by Knight of the Shire (26552), g. d. (Wave Naiad) by Breast Plate (19337), gr. g. d. (Wave Witch) by Fitz-Clarence (14552), — (Wave Princess) by British Prince (14197), — by Vanguard (10994), — by Baron Warlaby (7813), — by Fourth Duke of Northumberland (3649), — by Norfolk (2377), — by Waterloo (2816), — by Waterloo (2816).

(44278)   WOODREEVE,

Roan, calved July 4, 1879, bred by Mr. E. W. Meade-Waldo, Stonewall Park; got by King Harry (36841), dam (Wave Drift) by King Rufus (34351), g. d. (Wave Ebb) by Blinkhoolie (23428), gr. g. d. (Wave Foam) by Breast Plate (19337), — (Wave Princess) by British Prince (14197), — by Vanguard (10994), — by Baron Warlaby (7813), — by Fourth Duke of Northumberland (3649), — by Norfolk (2377), — by Waterloo (2816), — by Waterloo (2816).

(44279)   WOODSTOCK,

Roan, calved July 11, 1879, bred by Mr. H. Chandos-Pole-Gell, Hopton Hall; got by Favourite (31145), dam (Rosamond 2nd) by Gunpowder (28801), g. d. (Ringdove 2nd) by King Charles (24240), gr. g. d. (Ringdove) by Warlaby Knight (25403), — (Ringlet 2nd) by Bywell Victor (21353), — by Lord of the Valley (14837), — by Red Duke (13571), — by Vanguard (10994), — by Diamond (5918), — by Young Matchem (2282), — by Jack Tar (1133), — by Pilot (496), — by Young Albion (15).

(44280)   WOODYTUFT,

Red, calved January 24, 1879, bred by Mr. H. Bruen, Oak Park, the property of Mr. J. Scanlon, Bohermore; got by Wodehouse Gwynne (37690), dam (Candytuft) by Fiddler (33911), g. d. (Crinoline 2nd) by Glory (24043), gr. g. d. (Crinoline) by Fawsley Lad (41536), — (Croker 4th) by Garibaldi (41589), — by Paddy (15037), — by Shabbaknk (42372), — by Warwick, — by Barley Sugar.

(44281)   YAKOOB KHAN,

Red, calved January 9, 1878, bred by Mr. P. Taaffe, Foxborough, the property of Mr. R. A. Burris, Ballintemple; got by Bates-Booth (33121), dam (Precious 2nd) by Earl of Mayo (21511), g. d. (Precious) by British Flag (19351), gr. g. d. (Cora) by Gerard Dow (19849), — (Ellen) by Minos (11816), — by Premier (11920), — by Narcissus (4540), — by Linton (4227).

(44282)   YORKSHIREMAN,

Roan, calved June 26, 1878, bred by Mr. F. Barroby, Dishforth, the property of Mr. W. Mitchell, Cleasby; got by Cardinal Windsor (37949), dam (Lady Stanley 10th) by Clarion (33393), g. d. (Lady Stanley) by Lord Stanley (16452), gr. g. d. by Rifleman (15163), — by Young Hopewell (14719), — by Fitz-William (14555).

(44283) ZENOBIUS,
Roan, calved June 2, 1879, bred by Mr. T. Willis, Carperby; got by Major Windsor (34739), dam (Zenobia) by K. C. B. (26492), g. d. (Zone) by Windsor Fitz-Windsor (25458), gr. g. d. (Venus) by Lord Frederick (22156), — (Gentle Breeze) by King Alfred (16334), — by Wilberforce (9830), — by Wharfdale Hero (9821), — by Symmetry (5384), — by Fitz-Champion (7007), — by Young Snowball (7521).

(44284) ZERO,
Roan, calved March 3, 1879, bred by Mr. F. Morice, Springfield; got by Undergraduate (35835), dam (Duchess of Thorndale 5th) by Hero of Thorndale (18061), g. d. (Duchess of Thorndale 3rd) by Hero of Thorndale (18061), gr. g. d. (Gold Lace) by Soubadar (18901), — (Spangle) by Western Wonder (17225), — by Duke of Beauford (11377), — by Humber (7102), — by Guardian (3947), — by Firby (1040), — by Hector (1104), — by Regent (544), — Old York.

(44285) ZINGAREE,
Roan, calved July 9, 1879, bred by Mr. T. Purkis, West Wratting Grange; got by Cambridge Duke 5th (30644), dam (Zenobia 12th) by Wharfdale Oxford (27786), g. d. (Zenobia 6th) by Baron Killian (21226), gr. g. d. (Zenobia 3rd) by Belleville (6778), — (Zenobia) by Ernest (3735), — by Marton (4408), — by Uptaker (5534), — by Wonderful (700), — by Mars (412), — by Trunnell (659), — by George (273), — by Dash (191).

(44286) ZULU,
Red and white, calved January 8, 1878, bred by Mr. P. Taaffe, Foxborough, the property of the Earl of Courtown, Courtown House; got by Bates-Booth (33121), dam (Violet) by Leviathan (20120), g. d. (Welcome) by May Boy (16552), gr. g. d. (Leopardy 3rd) by Voltigeur (12274), — (Leopardy) by Burgundy (7861), — by Prince Albert (4801), — by Brutus (1751), — by Daniel O'Connell (1905), — by Son of Sir Kenneth (1450), — by Pilot (1319).

(44287) ZULU,
Red and white, calved March 11, 1879, bred by Mr. C. H. Cock, Bridgefoot; got by Young England (31110), dam (Melody) by Prince Llewellyn (32154), g. d. (Songstress) by Grand Duke 16th (24063), gr. g. d. (Seraphina 3rd) by Royal Essex (18767), — (Spangle) by Fitz-Clarence (11477), — by Sweet William (7571), — by Earl of Essex (6955), — by Stratton (5336), — by Fanatic (1996), — by Red Rover (4902), — by Rufus (2576), — by Emperor (1014), — the Rev. R. Pointer's Old Darlington.

(44288) ZULU,
Roan, calved April 3, 1879, bred by Mr. H. W. Goulder, Wimbotsham; got by Holman Hunt (36777), dam (Lily) by Royal Highness (40035), g. d. (Violet 2nd) by Master Hopewell (24555), gr. g. d. (Violet 1st) by Viscount Windsor (19082), — (Primrose) by Oliver (40407).

COWS

AND THEIR PRODUCE.

HER MAJESTY THE QUEEN,
Windsor Park.

ALIX 6TH, roan, calved March 24, 1873, Vol. xxiii. p. 301. Bred by Her Majesty the Queen; got by Royal Benedict (27348), dam (Alix 3rd) by Waterloo Duke (21077), &c.

Produce in	Names, &c.	By what Bull.	By whom bred.
1879, Aug. 2, white, C.C.	Alix 7th	PrinceAlbertVictor,40479	H. M. the Queen

BARONESS HOPEWELL, roan, calved October 23, 1870, Vols. xxi., xxiv., and xxv. pp. 843, 485, 305. Bred by Mr. H. W. Martin, Upton House; got by King Hopewell (26506), dam (Baroness of Windsor) by Count of Windsor (21498), &c.

Produce in	Names, &c.	By what Bull.	By whom bred.
1879, May 11, white, C.C.	Baroness Hope	PrinceAlbertVictor,40479	H. M. the Queen

BENEDICTA, roan, calved February 1, 1870, Vols. xx., xxiv., and xxv. pp. 408, 285, 305. Bred by Messrs. T. Garne and Son, Broadmoor; got by Royal Benedict (27348), dam (Blue Bell) by Cynric (19542), &c.

Produce in	Names, &c.	By what Bull.	By whom bred.
1879, Aug. 6, roan, C.C.	Beauty	PrinceAlbertVictor, 40479	H. M. the Queen

CAWLINA 5TH, roan, calved May 10, 1874, Vols. xxiii. and xxv. pp. 301, 305. Bred by Her Majesty the Queen; got by King Tom (31521), dam (Cawlina 2nd) by Prince of Saxe Coburg (20576), &c.

Produce in	Names, &c.	By what Bull.	By whom bred.
1879, Aug. 7, r. & w., D.C.	Prince Leopold	PrinceAlbertVictor,40479	H. M. the Queen

CAWLINA 9TH, roan, calved November 18, 1876. Bred by Her Majesty the Queen; got by Manrico (26805), dam (Cawlina 5th) by King Tom (31521), &c. See Vol. xxiii. p. 301.

Produce in	Names, &c.	By what Bull.	By whom bred.
1879, Feb. 20, roan, B.C.	The Red Prince	PrinceAlbertVictor, 40479	H. M. the Queen

The Red Prince, sold to Mrs. A. C. Pullein, Clifton Castle, Bedale, Yorkshire.

CAWLINA 9TH A, roan, calved July 3, 1877. Bred by Her Majesty the Queen; got by Manrico (26805), dam (Cawlina 2nd) by Prince of Saxe Coburg (20576), &c. See Vol. xxiv. p. 285.

Produce in	Names, &c.	By what Bull.	By whom bred.
1879, Aug. 3, white, B.C.	Duke of Connaught	PrinceAlbertVictor,40479	H. M. the Queen

DAME EUROPA, roan, calved March 4, 1871, Vol. xxi. p. 919. Bred by Mr. W. C. Scott, Thorpe; got by England's Glory (23889), dam (Spring) by Duke of Cambridge (21574), &c.

Produce in	Names, &c.	By what Bull.	By whom bred.
1879, Feb. 11, white, C.C.	Empress of India	Captain Walton, 36314	H. M. the Queen

ESTHER, roan, calved September 27, 1872, Vols. xxiv. and xxv. pp. 286, 306. Bred by Her Majesty the Queen; got by Royal Benedict (27348), dam (Eva) by Prince Louis (22603), &c.

1879, Feb. 23, r. & w., C.C.	Eliza	PrinceAlbertVictor,40479	H. M. the Queen

FAWSLEY 10TH, red and white, calved June 24, 1873, Vols. xxiii., xxiv., and xxv. pp. 303, 286, 306. Bred by Her Majesty the Queen; got by Royal Benedict (27348), dam (Fawsley 8th) by Knightley (22051), &c.

1879, Mar. 1, roan, B.C.	(dead)	PrinceAlbertVictor,40479	H. M. the Queen

FAWSLEY 12TH, red and white, calved March 27, 1877. Bred by Her Majesty the Queen; got by Manrico (26805), dam (Fawsley 10th) by Royal Benedict (27348), &c.

1879, Jan. 25, roan, C.C.	Fawsley 13th	PrinceAlbertVictor,40479	H. M. the Queen

NELLY, roan, calved May 12, 1873, Vol. xxv. p. 306. Bred by Her Majesty the Queen; got by Prince Albert 2nd (29588), dam (Nosegay) by Prince of Orange (22618), &c.

1879, July 31, r. & w., B.C.	(Steer)	PrinceAlbertVictor,40479	H. M. the Queen

PASTIME, red and white, calved February 8, 1872, Vols. xxi., xxii., xxiv., and xxv. pp. 531, 286, 286, 306. Bred by Her Majesty the Queen; got by My Lord of Dublin (29411), dam (Pink) by Prince Louis (22603), &c.

1879, Sept. 2, r. & w., B.C.	Prince George	PrinceAlbertVictor,40479	H. M. the Queen

ROSETTE, red and white, calved December 26, 1871, Vols. xxiv. and xxv. pp. 287, 306. Bred by Her Majesty the Queen; got by My Lord of Dublin (29411), dam (Rosebud) by Rajah (22670), &c.

1879, Aug. 18, roan, C.C.	Rosebud	PrinceAlbertVictor,40479	H. M. the Queen

ROYAL ROSE, roan, calved October 2, 1870, Vols. xxi., xxiii., xxiv., and xxv. pp. 857, 605, 614, 306. Bred by Mr. W. Morley, Sweet Wells; got by Royal Arthur (29840), dam (Rose of Beauty) by Bywell Victor (21353), &c.

1879, Sept. 7, roan, B.C.	Royal Albert	PrinceAlbertVictor,40479	H. M. the Queen

WHARFDALE COUNTESS, roan, calved June 3, 1874, Vol. xxv. p. 307. Bred by Mr. E. J. Coleman, Stoke Park; got by Wharfdale Oxford (27786), dam (Juniper Berry) by George 1st (19848), &c.

1879, June 22, r. & w., B.C.	Warlaby Chief	PrinceAlbertVictor,40479	H. M. the Queen

H.R.H. THE PRINCE OF WALES,
Sandringham, King's Lynn, Norfolk.

BLYTHESOME EYES, red, calved December 22, 1874. Bred by the Earl of Dunmore, Dunmore; got by Third Duke of Hillhurst (30975), dam (Wild Eyes Duchess) by Grand Duke 9th (19879), &c. See "Babingly Duke," p. 8.

1879, Nov. 26, r. & w., B.C.	Babingly Duke	M. of Oxford 2nd, 37055	Prince of Wales

DIADEM 2ND, roan, calved August 19, 1875, Vol. xxv. p. 307. Bred by the Rev. J. N. Micklethwait, Taverham Hall; got by Royal Dublin (35354), dam (Diadem) by Fawsley Prince (31150), &c.

Produce in		Names, &c.	By what Bull.	By whom bred.
1879, Dec. 3, white,	C.C.	Diadem 8th	Admiral, 39353	Prince of Wales

DIADEM 3RD, red and white, calved December 1, 1876, Vol. xxv. p. 307. Bred by the Rev. J. N. Micklethwait, Taverham Hall; got by Royal Dublin (35354), dam (Diadem) by Fawsley Prince (31150), &c.

1879, Dec. 4, roan,	C.C.	Diadem 9th	Admiral, 39353	Prince of Wales

ERIGONE 3RD, red, calved November 13, 1874. Bred by Mr. G. Murton Tracy, Redlands; got by Red Belvedere (29724), dam (Erigone 2nd) by Cock of the Midden (23585), &c. See Vol. xxi. p. 966.

1879, June 14, roan,	C.C.	Erigone 5th	Baron Ryedale, 37813	Prince of Wales

FESTIVE, white, calved February 9, 1877. Bred by H.R.H. the Prince of Wales; got by Homer (34170), dam (Feather) by Theodorus (27639), &c. See "Fusilier," p. 104.

1879, Jan. 27, roan,	B.C.	Fusilier	Frederic, 38317	Prince of Wales

FLASH, red, calved August 6, 1871, Vol. xxi. p. 532. Bred by the Rev. J. N. Micklethwait, Taverham Hall; got by Robert Peel (29796), dam (Flush) by Zaratan (21134), &c.

Produce in		Names, &c.	By what Bull.	By whom bred.
1875, Dec. 6,	{ roan, B.C. { roan, B.C.	Don Carlos (Steer)	} General McNab, 34012	Prince of Wales
1877, Mar. 26, roan,	B.C.	Santander	Homer, 34170	do.
1878, April 21, roan,	C.C.	Fleecy	do.	do.
1879, May 31, roan,	C.C.	Fleecy 2nd	do.	do.

GEM, roan, calved July 3, 1874, Vol. xxiii. p. 303. Bred by Her Majesty the Queen, Windsor Park; got by King Tom (31521), dam (Gertrude) by Prince Albert (22560), &c.

1878, July 3 (slipped calf)				Prince of Wales
1879, Dec. 18, white,	C.C.	Geraldine	Admiral, 39353	do.

GRAND DUCHESS ROSY, red and little white, calved February 20, 1877. Bred by Mr. F. Sartoris, Rushden Hall; got by Grand Duke 23rd (34063), dam (Lady Geneva Rosy) by Ninth Duke of Geneva (28391), &c. See Vol. xxii. p. 551.

1879, Apr. 17, r. & w.,	C.C.	Gr.Duchess Rosy 2nd	Baron Ryedale, 37813	Prince of Wales

LADY KINGSCOTE ROSY, roan, calved November 6, 1875, Vol. xxv. p. 307. Bred by Mr. F. Sartoris, Rushden Hall; got by Lord Kingscote Rosy (34576), dam (Lady Geneva Rosy) by Ninth Duke of Geneva (28391), &c.

1879, Dec. 24, roan,	C.C.	Ly Kingscote Rosy 2d	Baron Ryedale, 37813	Prince of Wales

MARIE, roan, calved February 1, 1874, Vol. xxv. p. 307. Bred by H.R.H. the Prince of Wales; got by High Sheriff (26392), dam (Satinet) by Duke of Edinburgh (23741), &c.

1879, Mar. 23, r. & w.,	B.C.	Lord Wolferton	Homer, 34170	Prince of Wales

MIRABEL, roan, calved April 27, 1877. Bred by H.R.H. the Prince of Wales; got by Homer (34170), dam (Satinet) by Duke of Edinburgh (23741), &c. See "The Miser," p. 250.

1879, May 13, white,	B.C.	The Miser	Admiral, 39353	Prince of Wales

WILD MUSICAL, red and white, calved August 30, 1874. Bred by Mr. J. D'A.
Samuda, Chillies; got by Wildfire (32866), dam (Musical 15th) by Grand
Duke 13th (21850), &c. See "Lord Ryedale," p. 157.

Produce in		Names, &c.	By what Bull.	By whom bred.
1878, May 25, roan,	B.C.	Musical Oxford	Marquis of Oxford, 34786	Prince of Wales
1879, Aug. 9, roan,	B.C.	Lord Ryedale	Baron Ryedale, 37813	do.

ABELL, John,
Valley Farm, Higham, Hinckley.

MAY LASS, roan, calved June 8, 1863, Vol. xix. p. 624. Bred by Mr. F.
Wythes, Ravensden House; got by May Duke 2nd (18372), dam (Music) by
Vocalist (13960), &c.

1871, Jan. 14, roan,	C.C.	May Lass 5th	General Napier, 24023	Mr. Cheney
1874, June 12, roan,	C.C.	May Lass 6th	Marquis of York, 34791	Mr. Abell

ACKERS, B. St. John,
Prinknash Park, Painswick.

ALPINE 2ND, roan, calved April 18, 1873, Vols. xxii. and xxiii. pp. 328, 341.
Bred by Mr. W. Bolton, The Island; got by Lieutenant General (31600),
dam (Alpine) by British Prince (14197), &c.

1878, Jan. 22, red,	C.C.	Alpine 4th	King William, 34358	Mr. Bolton
1879, Mar. 12, roan,	B.C.	Royal Alpine	do.	Mr. Ackers

ALPINE 3RD, red and white, calved August 19, 1876. Bred by Mr. W. Bolton,
The Island; got by King James (28971), dam (Alpine 2nd) by Lieutenant
General (31600), &c.

1878, Aug. 6, roan,	C.C.	Alpine 5th	Regal Buck, 38982	Mr. Bolton
1879, Oct. 18, red,	C.C.	Alpine 6th	King Harold, 40053	Mr. Ackers

ANNA COMNENA, roan, calved December 13, 1873, Vol. xxv. p. 371. Bred
by the Hon. M. H. Cochrane, Hillhurst, Canada; got by Royal Commander
(29857), dam (Bright Lady) by Lord Blithe (22126), &c.

1879, Nov. 15; white, C.C.	Anna Mia	Lord Prinknash, 34655	Mr. Ackers

BRIGHT DOWAGER, red, calved November 12, 1873, Vols. xxiii., xxiv., and
xxv. pp. 304, 287, 308. Bred by Mr. W. Torr, Aylesby Manor; got by Duke
of York (23804), dam (Bright Queen) by Fitz-Clarence (14552), &c.

1879, Oct. 8, red,	B.C.	Earl of Aylesby 3rd	King Harold, 40053	Mr. Ackers

DIADEM, red, calved March 31, 1871, Vols. xxii. and xxiv. pp. 288, 287. Bred
by Mr. T. E. Pawlett, Beeston; got by Royal Booth (27350), dam (Ringlet)
by Fitz-Killerby (26166), &c.

1879, Feb. 2, roan,	C.C.	Fame's Diadem	Lord Prinknash 2nd, 38653	Mr. Ackers

EMMA FITZ-WINDSOR, roan, calved September 25, 1871, Vols. xxi. and xxiii.
pp. 632, 304. Bred by Mr. J. Close, Lingstubbs; got by Shirley Fitz-Windsor
(27451), dam (Familiar Windsor) by Roan Windsor (24967), &c.

1879, Nov. 23, red,	C.C.	Lady Fitz-Windsor	King Harold, 40053	Mr. Ackers

FAIR SAXON, red and white, calved March 11, 1869, Vols. xx., xxi., and xxv.
pp. 509, 962, 308. Bred by Mr. W. Torr, Aylesby Manor; got by Breast
Plate (19337), dam (Fair Dane) by Fitz-Clarence (14552), &c.

1879, April 10, red,	C.C.	Fair Celt	K. C. B., 26492	Mr. Ackers

FLOWER, white, calved March 11, 1873, Vol. xxiv. p. 287. Bred by Mr. B. St. John Ackers ; got by County Member (28268), dam (Ringlet) by Fitz-Killerby (26166), &c.

Produce in	Names, &c.	By what Bull.	By whom bred.
1879, May 30, roan, C.C.	Fairy Fame	Lord Prinknash 2nd, 38653	Mr. Ackers

LADY BLITHE, red, calved September 21, 1872, Vols. xxii. and xxv. pp. 316, 308. Bred by Mr. F. Heugh, Broomfield House; got by Lord Blithe (22126), dam (Lady Booth) by The Sutler (23061), &c.

Produce in	Names, &c.	By what Bull.	By whom bred.
1879, Nov. 30, roan, C.C.	Lady Elizabeth	King Harold, 40053	Mr. Ackers

LADY CAREW 2ND, roan, calved September 22, 1875. Bred by Mr. B. St. John Ackers; got by County Member (28268), dam (Lady Jane) by Baron Killerby (23364), &c. See Vol. xxii. p. 288.

Produce in	Names, &c.	By what Bull.	By whom bred.
1879, Mar. 14, white, C.C.	Lady Carew 6th	Lord Prinknash 2nd, 38653	Mr. Ackers

LADY CAREW 3RD, roan, calved October 5, 1876. Bred by Mr. B. St. John Ackers; got by County Member (28268), dam (Lady Jane) by Baron Killerby (23364), &c. See Vol. xxiii. p. 304.

Produce in	Names, &c.	By what Bull.	By whom bred.
1879, Jan. 6, roan, C.C.	Lady Carew 4th	Lord Prinknash 2nd, 38653	Mr. Ackers

LADY ELIMA, roan, calved December 17, 1874. Bred by Mr. J. Beattie, Newbie House; got by Knight of Knowlmere 2nd (31542), dam (Lady Booth) by The Sutler (23061), &c. See Vol. xx. p. 587.

Produce in	Names, &c.	By what Bull.	By whom bred.
1879, May 12, white, B.C.	Manxman	Lord Prinknash, 34655	Mr. Ackers

LADY GAY, roan, calved May 29, 1869, Vols. xx., xxi., and xxv. pp. 596, 533, 308. Bred by Mr. J. B. Booth, Killerby Hall; got by Brigade Major (21312), dam (Lady Georgina) by Knight Errant (18154), &c.

Produce in	Names, &c.	By what Bull.	By whom bred.
1879, Nov. 21, roan, C.C.	Lady Geraldine 2nd	King Harold, 40053	Mr. Ackers

LADY GEORGINA TURBITT, roan, calved March 16, 1875. Bred by Mr. B. St. John Ackers; got by County Member (28268), dam (Patience Heatherstone) by British Crown (21322), &c. See Vol. xxii. p. 288.

Produce in	Names, &c.	By what Bull.	By whom bred.
1879, July 29, roan, C.C.	Ly G'rgina Newcomb	Lord Prinknash 2nd, 38653	Mr. Ackers

LADY JANE, red, calved May 28, 1869, Vols. xx., xxi., xxii., xxiii., and xxiv. pp. 599, 533, 288, 304, 287. Bred by Mr. T. E. Pawlett, Beeston; got by Baron Killerby (23364), dam (Miracle) by Prince James (20554), &c.

Produce in	Names, &c.	By what Bull.	By whom bred.
1879, Jan. 24, roan, C.C.	Lady Carew 5th	Lord Prinknash 2nd, 38653	Mr. Ackers

LOWLAND FLOWER, roan, calved April 12, 1871, Vols. xxi., xxiii., and xxv. pp. 962, 305, 308. Bred by Mr. W. Torr, Aylesby Manor; got by Manfred (26801), dam (Clarence Flower) by Fitz-Clarence (14552), &c.

Produce in	Names, &c.	By what Bull.	By whom bred.
1879, April 19, roan, C.C.	Fl. of Prinknash 3rd	K. C. B., 26492	Mr. Ackers

MAID OF GLO'STER, roan, calved July 23, 1873, Vol. xxiv. p. 287. Bred by Mr. B. St. John Ackers; got by County Member (28268), dam (Maid Marion) by Ravenspur (20628), &c.

Produce in	Names, &c.	By what Bull.	By whom bred.
1879, Mar. 4, { roan, C.C. r. & w., C.C.	Maid of Glo'ster 2d Maid of Glo'ster 3d }	K. C. B., 26492	Mr. Ackers

PATIENCE HEATHERSTONE, red and white, calved April 18, 1869, Vols. xx., xxii., xxiii., and xxv. pp. 682, 288, 305, 308. Bred by Mr. J. B. Booth, Killerby Hall; got by British Crown (21322), dam (Virtue) by Valasco (15443), &c.

Produce in	Names, &c.	By what Bull.	By whom bred.
1879, Jan. 27, roan, C.C.	Patience Beverley	K. C. B., 26492	Mr. Ackers

PRINCESS GEORGIE, roan, calved August 1, 1874, Vol. xxiii. p. 305. Bred
by Mr. B. St. John Ackers; got by County Member (28268), dam (Georgie's
Queen) by Brigade Major (21312), &c.

Produce in		Names, &c.		By what Bull.		By whom bred.
1879, Aug. 19, roan,	B.C.	Lord Geo. Hamilton	Lord Prinknash 2nd, 38653			Mr. Ackers

ADKINS, John C.,
Milcote, Stratford-on-Avon.

FLOSS, roan, calved July 1, 1874, Vol. xxiv. p. 328. Bred by Sir G. R. Philips,
Bart., Weston Park; got by Cherry Grand Duke 5th (30712), dam (Farewell)
by Eighteenth Duke of Oxford (25995), &c.

1879, Mar. 30, roan,	C.C.	Milcote Flame	D. of Oxford 32nd, 36527	Mr. Adkins

PRIMROSE 6TH, red and white, calved October 21, 1876. Bred by the Rev. J.
Storer, Hellidon; got by Young Knightley (31529), dam (Primrose 3rd) by
Prince of Fawsley (32171), &c. See Vol. xxiv. p. 421.

1879, Jan. 11, red,	C.C.	Milcote Primrose	Hotspur, 38440	Mr. Adkins

AITCHISON, Peter,
West Garleton, Haddington, N.B.

GRAND LADY, red and white, calved February 28, 1877. Bred by Mr. P.
Aitchison; got by Sir John (37470), dam (Grand Maid) by Grand Knight
(26302), &c. See Vol. xxi. p. 534.

1879, Apr. 21, white,	B.C.	Smeaton Grand Duke	Peter the Great, 38863	Mr. Aitchison

Smeaton Grand Duke, sold to Sir T. B. Hepburn, Bart., Smeaton, Prestonkirk.

ALEXANDER, S. M.,
Roe Park, Limavady, Ireland.

MYRTLE 22ND, roan, calved February 9, 1872, Vols. xxii. and xxiv. pp. 291,
288. Bred by Mr. J. Alexander, Limavady; got by King Oberon (26510),
dam (Myrtle 18th) by Ancient Briton (19225), &c.

1878, July 2, white,	B.C.	(dead)	Abab, 27850	Mr. S.M. Alexander
1879, June 4, roan,	C.C.	Myrtle 31st	Fascinator, 33890	do.

MYRTLE 23RD, roan, calved January 7, 1873, Vol. xxiv. p. 288. Bred by Mr.
S. M. Alexander; got by King Oberon (26510), dam (Myrtle 18th) by Ancient
Briton (19225), &c.

1878, May 26, roan,	C.C.	Myrtle 29th	Fascinator, 33890	Mr. Alexander
1879, Apr. 10, r. & w.,	C.C.	Myrtle 30th	do.	do.

MYRTLE 25TH, roan, calved June 21, 1875. Bred by Mr. S. M. Alexander;
got by King Oberon (26510), dam (Myrtle 18th) by Ancient Briton (19225),
&c.

1878, Nov. 5, white,	B.C.	(Steer)	Flag of Ireland, 28613	Mr. Alexander

MYRTLE 26TH, roan, calved June 28, 1875. Bred by Mr. S. M. Alexander;
got by King Oberon (26510), dam (Myrtle 17th) by Lord Winton (18282),
&c. See Vol. xxii. p. 290.

1878, May 23, white,	B.C.	Flag of Truce	Flag of Ireland, 28613	Mr. Alexander
1879, June 19, roan,	B.C.	Flaxman	Fascinator, 33890	do.

Flag of Truce, sold to Mr. R. A. Gilfillan, Lismacarrol, Londonderry.

ROSE OF WARLABY, red and white, calved January 3, 1870, Vols. xx. and xxiv. pp. 744, 289. Bred by Messrs. Garne and Son, Broadmoor; got by Royal Benedict (27348), dam (Rose of Clitheroe) by Cynric (19542), &c.

Produce in	Names, &c.	By what Bull.	By whom bred.
1879, Jan. 31, r. & w., C.C.	Rose of Carrichue	Fascinator, 33890	Mr. Alexander

VESTAL, roan, calved in 1868, Vol. xxiv. p. 289. Bred by Mr. J. W. E. Macartney, Clogher Park; got by Ravenspur (20628), dam (Virginia) by Dr. McHale (15887), &c.

Produce in	Names, &c.	By what Bull.	By whom bred.
1878, Nov. 12, roan, C.C.	Violette	Fascinator, 33890	Mr. Alexander
1879, Oct. 28, r. & w., B.C.	(dead)	do.	do.

ALLAN, John,
Billie-mains, Ayton, N.B.

GRAND DUCHESS OF ESSEX 8TH, red, calved April 14, 1873, Vols. xxiii. and xxiv. pp. 306, 290. Bred by Mr. J. Allan; got by Glory of the Mountains (28717), dam (Grand Duchess of Essex 2nd) by Confederate (21457), &c.

Produce in	Names, &c.	By what Bull.	By whom bred.
1879, Mar. 11, red, C.C.	G. D'ss of Essex 12th	Lord Thorndale, 37010	Mr. Allan

GRAND DUCHESS OF ESSEX 9TH, roan, calved April 17, 1875. Bred by Mr. J. Allan; got by Earl of Studley (31085), dam (Grand Duchess of Essex 2nd) by Confederate (21457), &c. See "Lord Thorndale 11th," p. 158.

Produce in	Names, &c.	By what Bull.	By whom bred.
1879, Apr. 14, r. & w., B.C.	Lord Thorndale 11th	Lord Thorndale, 37010	Mr. Allan

GRAND DUCHESS OF ESSEX 10TH, roan, calved March 3, 1876. Bred by Mr. J. Allan; got by Earl of Studley (31085), dam (Grand Duchess of Essex 8th) by Glory of the Mountains (28717), &c. See "Lord Thorndale 10th," p. 158.

Produce in	Names, &c.	By what Bull.	By whom bred.
1879, Mar. 23, roan, B.C.	Lord Thorndale 10th	Lord Thorndale, 37010	Mr. Allan

GRAND DUCHESS OF ESSEX 11TH, red, calved March 23, 1877. Bred by Mr. J. Allan; got by Lord Thorndale (37010), dam (Grand Duchess of Essex 8th) by Glory of the Mountains (28717), &c.

Produce in	Names, &c.	By what Bull.	By whom bred.
1879, Mar. 31, roan, C.C.	G. D'ss of Essex 13th	Fitz-Rose, 41555	Mr. Allan

GRAND DUCHESS OF WETHERBY 6TH, roan, calved August 11, 1874, Vol. xxv. p. 309. Bred by Mr. J. Allan; got by Glory of the Mountains (28717), dam (Grand Duchess of Wetherby 4th) by Red Knight (24916), &c.

Produce in	Names, &c.	By what Bull.	By whom bred.
1879, June 15, roan, C.C.	G.D. of Weth'rby 11th	Lord Thorndale, 37010	Mr. Allan

LADY RAINE 6TH, red, calved March 15, 1876. Bred by Mr. J. Allan; got by Earl of Studley (31085), dam (Lady Raine 3rd) by Glory of the Mountains (28717), &c. See Vol. xxiii. p. 306.

Produce in	Names, &c.	By what Bull.	By whom bred.
1879, March 16, red, C.C.	Lady Raine 8th	Lord Thorndale, 37010	Mr. Allan

ALLAN, John,
Crieffvechter, Crieff, N.B.

GRAND DUCHESS MARIE, roan, calved December 18, 1872, Vol. xxii. p. 292. Bred by Mr. J. McQueen, Divers Wells; got by Hopeful (34175), dam (Constancy) by Alfred (32934), &c.

Produce in	Names, &c.	By what Bull.	By whom bred.
1877, Jan. 10, roan, C.C.	Gd D'ss Marie 2nd	Calculator, 33266	Mr. Allan
1878, Jan. 12, roan, C.C.	Gd D'ss Marie 3rd	Bellerus, 37848	do.
1879, Jan. 2, roan, B.C.	Reformer	do.	do.

GRAND DUCHESS MARIE 2ND, roan, calved January 10, 1877. Bred by
Mr. J. Allan; got by Calculator (33266), dam (Grand Duchess Marie) by
Hopeful (34175), &c.

Produce in		Names, &c.	By what Bull.	By whom bred.
1879, Jan. 20, red,	C.C.	Gd D'ss Marie 4th	British Ensign, 41147	Mr. Allan

ALLEN, B. Haigh,
Clifford Priory, Hereford.

BEETROOT, red and white, calved December 29, 1873, Vol. xxiv. p. 291. Bred
by Mr. B. H. Allen ; got by Macgregor (29241), dam (Blithesome) by Barleycorn the Younger (21209), &c.

Produce in		Names, &c.	By what Bull.	By whom bred.
1878, Aug. 16, roan,	B.C.	Brecon	King Rufus, 38502	Mr. Allen
1879, Dec. 8, roan,	C.C.	Beetroot 2nd	do.	do.

BELLA 5TH, roan, calved February 7, 1873, Vols. xxiii. and xxiv. pp. 307, 291.
Bred by Mr. B. H. Allen; got by Satellite (29930), dam (Arabella 7th) by
Knight of the Shire (20085), &c.

Produce in		Names, &c.	By what Bull.	By whom bred.
1878, Aug. 1, roan,	C.C.	Bella 7th	King Rufus, 38502	Mr. Allen
1879, Sept. 1, white,	B.C.	White Chief	do.	do.

BELLA 6TH, red, calved July 2, 1876. Bred by Mr. B. H. Allen ; got by
Wetherby Lad (32838), dam (Bella 5th) by Satellite (29930), &c.

Produce in		Names, &c.	By what Bull.	By whom bred.
1879, May 13, red,	C.C.	Bella 8th	King Rufus, 38502	Mr. Allen

CHARITY 3RD, red and white, calved January 21, 1874, Vol. xxiv. p. 291.
Bred by Mr. B. H. Allen ; got by Lord Tortworth (31757), dam (Charity 2nd)
by Madresfield (22254), &c.

Produce in		Names, &c.	By what Bull.	By whom bred.
1878, May 14, roan,	C.C.	Charity 7th	George. 34028	Mr. Allen
1879, June 6, roan,	B.C.	Cardinal	King Rufus, 38502	do.

Charity 7th, sold to Mr. B. L. Barrow, Sydnope Hall, Matlock.

CHARITY 4TH, white, calved February 19, 1876. Bred by Mr. B. H. Allen ;
got by Wetherby Lad (32838), dam (Charity 2nd) by Madresfield (22254), &c.

Produce in		Names, &c.	By what Bull.	By whom bred.
1879, Mar. 12, roan,	B.C.	Cupid	King Rufus, 38502	Mr. Allen

GRENADINE 2ND, roan, calved January 28, 1866, Vols. xix., xx., xxi., xxii.,
xxiii., and xxiv. pp. 539, 550, 536, 292, 307, 291. Bred by Mr. W. Wood,
Holly Bank ; got by Lord Wild Eyes 2nd (22234), dam (Grenadine) by Rob
Roy (20694), &c.

Produce in		Names, &c.	By what Bull.	By whom bred.
1878, June 8, roan,	C.C.	Grenade	King Rufus, 38502	Mr. Allen

GRENADINE 3RD, white, calved March 15, 1872, Vols. xxiii. and xxiv. pp.
307, 291. Bred by Mr. B. H. Allen ; got by Catton (25721), dam (Grenadine
2nd) by Lord Wild Eyes 2nd (22234), &c.

Produce in		Names, &c.	By what Bull.	By whom bred.
1878, Oct. 26, roan,	C.C.	Ada	King Rufus, 38502	Mr. Allen
1879, Dec. 2, white,	B.C.	(Steer)	do.	do.

PYRENEE 2ND, red, calved April 23, 1874, Vol. xxiv. p. 292. Bred by Mr.
B. H. Allen ; got by Wetherby Lad (32838), dam (Pyrenee) by Etoile du
Nord (21710), &c.

Produce in		Names, &c.	By what Bull.	By whom bred.
1878, June 22, r. & w.,	C.C.	(dead)	King Rufus, 38502	Mr. Allen
1879, Sept. 23, roan,	B.C.	Plato	do.	do.

WALLFLOWER, red, calved January 27, 1873, Vol. xxiii. p. 307. Bred by
Mr. B. H. Allen ; got by Majestic (29256), dam (Water Lily) by Marmaduke
(22287), &c.

Produce in		Names, &c.	By what Bull.	By whom bred.
1878, Feb. 2, roan,	B.C.	Wanderer	George, 34028	Mr. Allen
1879, July 13, roan,	B.C.	William Rufus	King Rufus, 38502	do.

Wanderer, sold to Mr. Farmer, New Zealand.

WELCOME, red, calved May 1, 1874, Vol. xxiv. p. 292. Bred by Mr. B.
H. Allen ; got by Macgregor (29241), dam (Woodlass 2nd) by Friar Tuck
(31204), &c.

Produce in		Names, &c.	By what Bull.	By whom bred.
1878, May 24, roan,	C.C.	Welcome 3rd	George, 34028	Mr. Allen
1879, Aug. 26, roan,	C.C.	Welcome 4th	King Rufus, 38502	do.

WOODBINE, roan, calved February 4, 1872, Vols. xxiii. and xxiv. pp. 307, 292.
Bred by Mr. B. H. Allen ; got by Catton (25721), dam (Woodlass 2nd) by
Friar Tuck (31204), &c.

Produce in		Names, &c.	By what Bull.	By whom bred.
1878, Mar. 23, white,	B.C.	Woodpecker	George, 34028	Mr. Allen
1879, Mar. 25, roan,	B.C.	Watchman	King Rufus, 38502	do.

Woodpecker, sold to Mr. Hutton, Compton Verney, Warwick.

ALLEN, George,
Unicarville, Comber, Co. Down.

ACHIEVEMENT, roan, calved January 29, 1866, Vols. xix., xx., xxi., xxii.,
xxiii., and xxv. pp. 384, 383, 538, 294, 310, 312. Bred by Mr. G. Allen ;
got by Prince of Warlaby (15107), dam (Heather Bell) by Harbinger
(10297), &c.

Produce in		Names, &c.	By what Bull.	By whom bred.
1879, Feb. 18, roan,	C.C.	Achievement 2nd	Irish Hero, 36791	Mr. Allen

ADA 5TH, red, calved February 26, 1874, Vols. xxii. and xxiii. pp. 294, 310.
Bred by Mr. N. M. Archdall, Crock-na-Crieve ; got by Abercorn (25484), dam
(Ada 2nd) by Prince of Lurg (22617), &c.

Produce in		Names, &c.	By what Bull.	By whom bred.
1879, Jan. 25, red,	C.C.	Irish Ada	Irish Hero, 36791	Mr. Allen

DARLING, red and white, calved May 6, 1875. Bred by Mr. G. Allen ; got
by British Hero (30604), dam (Honey Dew) by Great Hope (24082), &c. See
Vol. xxii. p. 294.

Produce in		Names, &c.	By what Bull.	By whom bred.
1879, Feb. 18, roan,	C.C.	Bright Rose	Bright Baronet, 37891	Mr. Allen

FANNY 11TH, roan, calved May 5, 1875. Bred by Mr. G. Allen ; got by
British Hero (30604), dam (Fanny 9th) by Great Hope (24082), &c. See
Vol. xxii. p. 294.

Produce in		Names, &c.	By what Bull.	By whom bred.
1879, Feb. 18, roan,	C.C.	Fanny 13th	Bright Baronet, 37891	Mr. Allen

HEATHER BELL, roan, calved March 17, 1871, Vols. xxi. and xxv. pp. 538,
313. Bred by Mr. G. Allen ; got by Prince Victor (20606), dam (Achievement) by Prince of Warlaby (15107), &c.

Produce in		Names, &c.	By what Bull.	By whom bred.
1879, June 14, roan,	C.C.	Rosebud	Bright Baronet, 37891	Mr. Allen

LADYLIKE, red and white, calved March 30, 1867, Vols. xx., xxi., xxii., xxiii.,
and xxv. pp. 604, 538, 295, 310, 313. Bred by Mr. G. Allen ; got by Prince
Victor (20606), dam (Mantalini) by Master Rembrandt (16545), &c.

Produce in		Names, &c.	By what Bull.	By whom bred.
1879, Apr. 19, roan,	B.C.	Bright Viscount	Bright Baronet, 37891	Mr. Allen

MAIDEN, red and white, calved April 20, 1873, Vols. xxii. and xxv. pp. 295, 313.
Bred by Mr. G. Allen; got by Great Hope (24082), dam (Elfrida) by Blood
Royal (23429), &c.

Produce in		Names, &c.	By what Bull.	By whom bred.
1879, May 5, roan,	C.C.	Bright Maiden	Bright Baronet, 37891	Mr. Allen

MAY QUEEN, red and white, calved June 13, 1875. Bred by Mr. G. Allen;
got by British Hero (30604), dam (May Bloom) by Prince Victor (20606), &c.
See Vol. xxii. p. 295.

Produce in		Names, &c.	By what Bull.	By whom bred.
1879, Jan. 12, r.&w.,	C.C.	Heroine	Irish Hero, 36791	Mr. Allen

MISS WARLABY, red and white, calved July 26, 1875, Vol. xxv. p. 313. Bred
by Mr. G. Allen; got by British Hero (30604), dam (Bonny Belle) by Great
Hope (24082), &c.

Produce in		Names, &c.	By what Bull.	By whom bred.
1879, May 16, roan,	B.C.	Bright Warlaby	Bright Baronet, 37891	Mr. Allen

MOYA, roan, calved March 27, 1876, Vol. xxv. p. 313. Bred by Mr. G. Allen;
got by British Hero (30604), dam (Colleen Rua) by Prince of Orange (20574),
&c.

Produce in		Names, &c.	By what Bull.	By whom bred.
1879, May 25, roan,	C.C.	Kathleen	Bright Baronet, 37891	Mr. Allen

PERFECTION, white, calved March 22, 1876. Bred by Mr. W. S. Garnett,
Williamston; got by Lieutenant-General (31600), dam (Pattern) by Prince
Bertram (27119), &c. See Vol. xxiv. p. 289.

Produce in		Names, &c.	By what Bull.	By whom bred.
1879, Oct. 9, white,	C.C.	Bright Perfection	Bright Baronet, 37891	Mr. Allen

QUEEN BUTTERFLY, red and white, calved February 6, 1877. Bred by
Mr. R. Jefferson, Preston Hows; got by British King (36279), dam (Proud
Butterfly) by Gay Cavalier (31223), &c. See Vol. xxiv. p. 294.

Produce in		Names, &c.	By what Bull.	By whom bred.
1879, June 13, roan,	C.C.	Bright Madeline	Bright Baronet, 37891	Mr. Allen

ALLEN, Stephen H.,
Eastover, Andover.

BEAUTY, roan, calved in 1869. Bred by the Marquis of Ailesbury, Savernake;
got by Clint (33399), dam (Pretty) by Star Prince (18925), g. d. by Savernake
(35475), gr. g. d. (Violet) by Mazeppa (34831).

Produce in		Names, &c.	By what Bull.	By whom bred.
1873, June 24, r.&w.,	B.C.	Waxwork	Shuttlecock 3rd, 35521	Mqs. of Ailesbury
1874, May, roan,	B.C.	The Beau	Æolus, 27861	do.
1875, May 3, roan,	B.C.	Ailesbury	do.	do.
1876, Sept. 25, r.&w.,	C.C.	Bluebell	J. P., 41731	Mr. Allen
1877, Oct. 15, r.&w.,	C.C.	Bonny Belle	do.	do.
1878, Nov. 4, roan,	C.C.	Bijou	do.	do.
1879, Nov. 18, roan,	B.C.	(dead)	J. P. D. 43402	do.

Ailesbury, sold to Mr. Hoddinott, Dinton, Salisbury.

CLARA, red and white, calved in 1874. Bred by the Marquis of Ailesbury,
Savernake; got by Æolus (27861), dam (Curley) by Shuttlecock 3rd (35521),
g. d. (Cherry) by Clint (33399), gr. g. d. (Old Cherry) by Clint (33399), — by
Star Prince (18925), — by Savernake (35475), — (Violet) by Mazeppa (34831).

Produce in		Names, &c.	By what Bull.	By whom bred.
1876, Aug. 15, r.&w.,	C.C.	(dead)	J. P., 41731	Mr. Allen
1877, Sept. 4,	r.&w.,B.C.	(Steer)	do.	do.
	r.&w.,C.C.	(slaughtered)		
1878, Oct. 19,	r.&w.,C.C.	Charlotte	do.	do.
	r.&w.,C.C.	Charity		
1879, Nov. 4,	red, B.C.	Counsellor	J. P. D., 43402	do.
	red, C.C.	(dead)		

DATURA, red and white, calved in 1874. Bred by the Marquis of Ailesbury Savernake ; got by Æolus (27861), dam (Dorcas) by Clint (33399), g. d. (Dodo) by Star Prince (18925), gr. g. d. by Savernake (35475), — (Violet) by Mazeppa (34831).

Produce in	Names, &c.	By what Bull.	By whom bred.
1876, Dec. 24, r. & w., B.C.	J. P. D.	J. P., 41731	Mr. Allen
1877, Oct. 23, { r.&w.,B.C. / r.&w.,C.C.	(Steer) / (slaughtered) }	do.	do.
1878, Oct. 22, r. & w., C.C.	Dido	do.	.do.
1879, Sept. 12, { red, B.C. / red, C.C.	Bencher / (dead) }	Barrister, 42761	do.

ALLSOPP, Henry,
Hindlip Hall, Worcester.

AIRDRIE DUCHESS 3RD, roan, calved June 20, 1873, Vol. xxv. p. 314. Bred by Mr. G. Murray, Slausondale, Wisconsin, U.S.A.; got by Duke of Geneva 11th (41385), dam (Airdrie Duchess) by Fourteenth Duke of Thorndale (28459), &c.

1879, June 29, roan,	C.C.	D'ss of Hindlip 2nd	3rd D. of Hillhurst, 30975	Mr. H. Allsopp

ARIEL COUNTESS 2ND, red, calved April 5, 1876. Bred by Lord Penrhyn, Penrhyn Castle; got by Grand Duke 20th (31281), dam (Ariel Countess) by Third Duke of Clarence (23727), &c. See Vol. xxiii. p. 596.

1879, Apr. 16, red,	B.C.	(Steer)	D. of Collingham 3d, 38134	Mr. H. Allsopp

BARBERRY, red, calved June 21, 1874, Vol. xxiv. p. 294. Bred by Mr. H. Allsopp; got by Windrush (32873), dam (Bella) by Mameluke (24520), &c.

1879, May 27, red,	C.C.	Bernice	D. of Collingham 3d, 38134	Mr. H. Allsopp

BELLE OF OXFORD 6TH, roan, calved October 25, 1876. Bred by Lord Penrhyn, Penrhyn Castle ; got by Grand Duke of Oxford (31293), dam (Belle of Oxford 3rd) by Marmaduke (14897), &c. See Vol. xxiii. p. 597.

1879, Mar. 9, r. & w.,	C.C.	Hindlip Belle of Ox'd	D. of Collingham 3d, 38134	Mr. H. Allsopp

BELLONA, roan, calved November 28, 1875. Bred by Mr. H. Allsopp; got by Windrush (32873), dam (Bride) by Butterfly (33255), g. d. (Beauty) by Mameluke (24520), gr. g. d. (Blush) by Sir John (25154), — (Bloomer) by Benedict (39457), — (Blythe) by Prince (18575).

1879, Jan. 23, white, B.C.	Brutus	Oxford's King, 34997	Mr. H. Allsopp

Brutus, sold to Mr. Harris, Zouch Farm, Culham, Oxon.

BRIGHT EYES 6TH, white, calved January 30, 1876. Bred by Mr. E. Musgrove, Aughton Old Hall; got by Baron Oxford 4th (25580), dam (Bright Eyes 4th) by Third Duke of Wharfdale (21619), &c. See Vol. xx. p. 421.

1879, Mar. 19, roan,	C.C.	(dead)	D. of Collingham 3d, 38134	Mr. H. Allsopp

CHESNUT, red, calved October 11, 1874, Vol. xxv. p. 314. Bred by Mr. H. Allsopp; got by Windrush (32873), dam (Cactus) by Lord Wharfdale (26761), &c.

1879, Dec. 16, r. & w., C.C.	Charity	3rd D. of Hillhurst, 30975	Mr. H. Allsopp

DARLING, roan, calved January 20, 1875, Vol. xxiv. p. 295. Bred by Mr. A. J. Robarts, Lillingstone Dayrell; got by Caractacus (28141), dam (Diamond) by Second Duke of Claro (21576), &c.

1879, May 1, r. & w.,	C.C.	Darlington Belle 2nd	3rd D. of Hillhurst, 30975	Mr. H. Allsopp

DIAMOND, red and white, calved April 22, 1869, Vols. xxii., xxiii., xxiv., and xxv. pp. 545, 620, 627, 314. Bred by Mr. A. J. Robarts, Lillingstone Dayrell; got by Second Duke of Claro (21576), dam (Diadem) by Marmaduke (14897), &c.

Produce in	Names, &c.	By what Bull.	By whom bred.
1879, Nov. 30, r.&w., B.C.	Kn't of Darlington	3rd D. of Hillhurst, 30975	Mr. H. Allsopp

DRUSILLA, roan, calved October 31, 1871, Vol. xxiv. p. 295. Bred by Mr. H. Allsopp; got by Lord Worcester (29225), dam (Diana) by Mameluke (24520), &c.

1879, July 9, roan, C.C.	Delilah	D. of Collingham 3d, 38134	Mr. H. Allsopp

DUCHESS 114TH, red and white, calved November 15, 1875, Vol. xxv. p. 433. Bred by the Earl of Dunmore, Dunmore, the property of Mr. H. Allsopp; got by Sixth Duke of Geneva (30959), dam (Duchess 97th) by Third Duke of Wharfdale (21619), &c.

1879, Mar. 12, r.&w., B.C.	D. of Cornwall 2nd	Mq. of Oxford 2nd, 37055	Earl of Dunmore

Duke of Cornwall 2nd, sold to Sir Curtis M. Lampson, Bart., Rowfant, Crawley.

GARLAND, red, calved May 8, 1875. Bred by Mr. H. Allsopp; got by Windrush (32873), dam (Geranium) by Butterfly (33255), g. d. (Grace) by Lord Wharfdale (26761), &c. See " Grimaldi," p. 116.

1879, Mar. 9, roan, C.C.	Georgiana	Oxford's King, 34997	Mr. H. Allsopp

GRACEFUL, red, calved June 24, 1873, Vol. xxiv. p. 296. Bred by Mr. H. Allsopp; got by Marnhull Duke (29283), dam (Grace) by Lord Wharfdale (26761), &c.

1879, Sept. 22, r.&w., B.C.	Grimaldi	3rd D. of Hillhurst, 30975	Mr. H. Allsopp

GRACE ROSE 4TH, red, calved August 3, 1873, Vol. xxiv. p. 296. Bred by Mr. G. Morrow, Bourbon County, Kentucky, U.S.A; got by Airdrie 3rd (32919), dam (Mayflower of Stony Point) by Airdrie (30365), &c.

1879, May 30, r. & w., C.C.	R.Rose of Hindlip 2d	D. of Collingham 3d, 38134	Mr. H. Allsopp

GRAND DUCHESS 29TH, red, calved September 12, 1875, Vol. xxv. p. 315. Bred by Mr. R. E. Oliver, Sholebroke Lodge; got by Grand Duke 22nd (34062), dam (Grand Duchess 24th) by Third Duke of Clarence (23727), &c.

1879, Oct. 24, red, B.C.	G. D. of Worcester 2d	3rd D. of Hillhurst, 30975	Mr. H. Allsopp

GRAND DUCHESS OF MORECAMBE, roan, calved October 27, 1876. Bred by the Earl of Bective, Underley Hall; got by Second Duke of Tregunter (26022), dam (Grand Duchess of Oxford 18th) by Baron Oxford 4th (25580), &c. See " Knight of Oxford 3rd," p. 135.

1879, Mar. 9, roan, B.C.	Kn't of Oxford 3rd	D. of Collingham 3d, 38134	Mr. H. Allsopp

KIRKLEVINGTON PRINCESS 4TH, red and white, calved October 18, 1875, Vol. xxv. p. 315. Bred by Mr. J. W. Larking, Ashdown House; got by Third Duke of Glo'ster (33653), dam (Kirklevington Duchess 9th) by Grand Duke of Clarence (28750), &c.

1879, Aug. 17, r.&w., B.C.	Baron Kirklevington	Duke of Underley, 33745	Mr. H. Allsopp

LADY ANGELINA 2ND, red and white, calved March 12, 1877. Bred by Mr. E. H. Cheney, Gaddesby Hall; got by Third Duke of Glo'ster (33653), dam (Tube Rose 43rd) by Saladin (35461), &c. See Vol. xxii. p. 358.

1879, Aug. 2, r. & w., C.C.	(dead)	D. of Collingham 3d, 38134	Mr. H. Allsopp

LADY OF OXFORD 13TH, roan, calved January 21, 1871, Vols. xxii. and xxiv.
pp. 358, 296. Bred by Messrs. Walcott and Campbell, New York Mills,
U.S.A.; got by Baron of Oxford (23371), dam (Lady of Oxford 7th) by Sixth
Duke of Thorndale (23794), &c.

Produce in	Names, &c.	By what Bull.	By whom bred.
1879, Mar. 10, r. & w., B.C.	(dead)	3rd Duke of Glo'ster, 33653	Mr. H. Allsopp

LADY WORCESTER 18TH, roan, calved May 2, 1876. Bred by the Earl of
Dunmore, Dunmore, the property of Mr. H. Allsopp; got by Duke of
Connaught (33004), dam (Lady Worcester) by Charleston (21400), &c. See
"Marquis of Worcester 8th," p. 165.

1879, Feb. 6, red, B.C.	M. of Worcester 8th	Mq. of Oxford 2nd, 37055	Earl of Dunmore

Marquis of Worcester 8th, sold to Mr. J. Darling, Beau Desert, Rugeley.

MARCHIONESS 11TH, red and white, calved July 25, 1876. Bred by the
Earl of Bective, Underley Hall; got by Duke of Underley (33745), dam
(Marchioness 6th) by Second Duke of Collingham (23730), &c. See Vol.
xxiii. p. 330.

1879, Jan. 19, red, B.C.	(dead)	D. of Collingham 3d, 38134	Mr. H. Allsopp

MARCHIONESS OF OXFORD, roan, calved January 21, 1871, Vols. xxi.,
xxii., xxiv., and xxv. pp. 666, 394, 407, 420. Bred by Mr. J. O. Sheldon,
Geneva, New York, U.S.A., the property of Mr. H. Allsopp; got by Fourth
Duke of Geneva (30958), dam (Maid of Oxford 8th) by Second Duke of Geneva
(23752), &c.

1879, Mar. 3, roan, B.C.	Knight of Oxford 2d	Duke of Hillhurst, 28401	Mr. Davies

Knight of Oxford 2nd, sold to Sir H. Allsopp, Bart.

PEGGY, roan, calved February 1, 1871, Vol. xxiv. p. 297. Bred by Mr. H.
Allsopp; got by Butterfly (33255), dam (Patience) by Sir John (25154), &c.

1879, June 19, r. & w., C.C.	Prude	D. of Collingham 3d, 38134	Mr. H. Allsopp

PRESTON BUTTERFLY 4TH, roan, calved February 6, 1873, Vol. xxv.
p. 316. Bred by Mr. H. Allsopp; got by Marnhull Duke (29283), dam
(Preston Butterfly 2nd) by Grand Duke 4th (19874), &c.

1879, Dec. 5, red, C.C.	Prston Butterfly 11th	3rd Dk. of Hillhurst, 30975	Mr. H. Allsopp

QUIET 3RD, red and white, calved March 10, 1874. Bred by Mr. H. Allsopp;
got by Windrush (32873), dam (Quiet 2nd) by Marnhull Duke (29283), &c.
See "Quicksilver," p. 205.

1879, Dec. 16, red, B.C.	Quicksilver	3rd Dk. of Hillhurst, 30975	Mr. H. Allsopp

RED ROSE OF ANNANDALE 2ND, white, calved March 7, 1877. Bred by
Mr. E. J. Coleman, Stoke Park; got by Second Duke of Tregunter (26022),
dam (Red Rose of Annandale) by Fourth Duke of Geneva (30958), &c. See
Vol. xxiv. p. 298.

1879, June 12, roan, C.C.	R. Rose of Hindlip 3d	D. of Collingham 3d, 38134	Mr. H. Allsopp

RED ROSE OF SEVERN, roan, calved January 17, 1877. Bred by Mr. G.
Fox, Elmhurst Hall; got by Twenty-fourth Duke of Airdrie (36460), dam
(Grace Rose 4th) by Airdrie 3rd (32919), &c. See "Severus," p. 227.

1879, June 12, roan, B.C.	Severus	D. of Collingham 3d, 38134	Mr. H. Allsopp

RED ROSE OF TWEEDDALE, red, calved March 6, 1873, Vols. xxiii. and
xxv. pp. 330, 316.   Bred by Mr. A. Renick, Clark County, Kentucky, U.S.A.;
got by Fourth Duke of Geneva (30958), dam (Rosebud 6th) by Airdrie
(30365), &c.

Produce in	Names, &c.	By what Bull.	By whom bred.
1879, April 7, red,	B.C.	Lord of the Tweed 2d	D. of Collingham 3d, 38134	Mr. H. Allsopp

THORNDALE MAID 2ND, red, calved October 9, 1871, Vol. xxiv. p. 299.
Bred by Mr. H. Allsopp; got by Butterfly (33255), dam (Thorndale Maid)
by Third Duke of Thorndale (17749), &c.

1879, April 11, roan, B.C.|Thorndale King	Oxford's King, 34997	Mr. H. Allsopp

Thorndale King, sold to Mr. R. H. Godard, Tibberton, Droitwich.

THORNDALE MAID 3RD, roan, calved November 30, 1873, Vol. xxiv. p. 299.
Bred by Mr. H. Allsopp; got by Marnhull Duke (29283), dam (Thorndale
Maid) by Third Duke of Thorndale (17749), &c.

1879, Feb. 28, red, B.C.|Heir of Thorndale	D. of Collingham,3d,38134|Mr. H. Allsopp

THORNDALE ROSE 7TH, red, calved June 20, 1875, Vols. xxiv. and xxv.
pp. 345, 364.   Bred by Lord Braybrooke, Audley End; got by Sixth Duke of
Oneida (30997), dam (Thorndale Rose 3rd) by Third Duke of Geneva (23753),
&c.

1879, Aug. 2, roan, C.C. |(dead)	D. of Underley 3rd, 38196|Mr. H. Allsopp

TINY 4TH, roan, calved February 22, 1873.   Bred by Mr. H. Allsopp; got by
Marnhull Duke (29283), dam (Tiny 3rd) by Butterfly (33255), &c.   See
"Truant," p. 253.

1879, Mar. 4, roan, B.C.|Truant	Oxford's King, 34997	Mr. H. Allsopp

WATERLOO 24TH, roan, calved March 22, 1876, Vol. xxv. p. 316.   Bred by
Mr. J. P. Foster, Killhow; got by Twenty-second Duke of Oxford (31000),
dam (Waterloo 22nd) by Speculator (13775), &c.

1879, Dec. 23, roan, C.C.	Waterloo Belle 4th	Grand D. of Airdrie, 43310	Mr. H. Allsopp

WATERLOO 37TH, roan, calved May 31, 1872, Vol. xxii. p. 551.   Bred by
Lord Penrhyn, Penrhyn Castle; got by Oxford Beau (29485), dam (Waterloo
30th) by Third Duke of Wharfdale (21619), &c.

1879, Feb. 11, roan, B.C.	Count Waterloo	Marquis of Oxford, 34786	Mr. H. Allsopp

WATERLOO DUCHESS 5TH, roan, calved July 23, 1874, Vols. xxiii. and xxv.
pp. 537, 316.   Bred by Mr. R. Lodge, The Rookery; got by Fifth Duke of
Wetherby (31033), dam (Waterloo Duchess 4th) by Thirteenth Duke of Oxford
(21604), &c.

1879, Dec. 8, roan, C.C.|Waterloo Belle 3rd	3rd Dk. of Hillhurst, 30975|Mr. H. Allsopp

ANDREWS, James,
Carnesune, Comber, Co. Down.

COUNTESS, red and white, calved January 2, 1871, Vols. xxi. and xxv. pp. 541,
317.   Bred by Mr. W. Humphrys, Ballyhaise House; got by Napoleon 3rd
(29417), dam (Nautilus) by Chieftain (30732), &c.

1879, May 6, roan, B.C.|Europe	Bright Baronet, 37891	Mr. Andrews

DAISY, red and white, calved March 8, 1877. Bred by Mr. J. Andrews; got
by British Mantalini (36281), dam (Ruby) by British Hero (30604), &c. See
Vol. xxv. p. 317.

Produce in	Names, &c.	By what Bull.	By whom bred.
1879, June 26, roan, C.C.	Briony	Bright Baronet, 37891	Mr. Andrews

RUBY, red and white, calved September 12, 1874, Vol. xxv. p. 317. Bred by
Mr. J. Andrews; got by British Hero (30604), dam (Countess) by Napoleon
3rd (29417), &c.

1879, Mar. 5, roan, C.C.	Woodbine	Bright Baronet, 37891	Mr. Andrews

ANDREWS, Thomas,
Ardara, Comber, Co. Down.

PRIMROSE, red and white, calved January 17, 1877. Bred by Mr. J. Andrews,
Carnesune; got by British Mantalini (36281), dam (Countess) by Napoleon
3rd (29417), &c. See Vol. xxv. p. 317.

1879, June 7, r. & w. B.C.	Master John	Bright Baronet, 37891	Mr. T. Andrews

ANGAS, John Howard,
Collingrove, Angaston, South Australia.

DUCHESS OF WORTLEY, roan, calved October 5, 1875. Bred by Mr. F.
Cartwright, The Grove, the property of Mr. J. H. Angas; got by Duke of
Wortley (33759), dam (Statira 16th) by Thirteenth Duke of Oxford (21604),
&c. See "Duke of Wellington 2nd," p. 85.

1879, July 13, red, B.C.	D. of Wellington 2nd	B.T'ncroftOxfd 4th,37822	Mr. Bliss

GAZELLE 20TH, white, calved July 13, 1870, Vol. xxi. p. 710. Bred by Mr.
E. Bowly, Siddington House, the property of Lord Fitzhardinge, Berkeley
Castle; got by Seventh Duke of York (17754), dam (Gazelle 4th) by Seventh
Duke of York (17754), &c.

1878, Feb. 22, roan, C.C.	Lady Gracious	Duke of Connaught, 33604	Lord Fitzhardinge

Lady Gracious, sold to Mr. J. H. Angas.

GRAND DUCHESS OF OXFORD 8TH, roan, calved March 2, 1877. Bred
by Lord Braybrooke, Audley End; got by Duke of Rosedale (33721), dam
(Grand Duchess of Oxford 4th) by Claro's Rose (25784), &c. See Vol. xxiv.
p. 344.

1879, July 30, white, C.C.	Gr. D'ss of Oxfd 10th	Dk. of Underley 3d, 38196	Mr. Angas

LADY WILD EYES 8TH, roan, calved October 5, 1873, Vol. xxiii. p. 438.
Bred by Lord Fitzhardinge, Berkeley Castle; got by Grand Duke of Waterloo
(28766), dam (Lady Wild Eyes 3rd) by Cherry Grand Duke (23554), &c.

1877, April 7, roan, C.C.	Lady Wild Eyes 12th	Duke of Connaught, 33604	Lord Fitzhardinge

Lady Wild Eyes 12th, sold to Mr. J. H. Angas.

MUSICAL 18TH, red, calved March 6, 1877. Bred by Mr. E. H. Cheney, Gaddesby Hall; got by Third Duke of Glo'ster (33653), dam (Musical 17th) by
Third Duke of Claro (23729), g. d. (Musical 13th) by Seventh Duke of York
(17754), &c. See Vol. xxii. p. 296.

1879, July 18, red, B.C.	Earl of Berkeley	Duke of Oxford 45th, 39775	Mr. Angas

ANGUS, John,
Bearl, Stocksfield-on-Tyne.

CHRISTMAS SNOWDROP, roan, calved January 15, 1877. Bred by Mr. J.
  Angus; got by King Christmas (36836), dam (Snowdrop 3rd) by Manfred
  (26801), &c. See "Royal Patron," p. 221.

Produce in	Names, &c.	By what Bull.	By whom bred.
1879, July 4, roan, · B.C.	Royal Patron	Hawthorn, 36751	Mr. Angus

COSTLY'S HOPE, roan, calved July 15, 1877. Bred by the Exors. of Mr. G.
  Angus, Broomley; got by Ben Brace (30524), dam (Costly 5th) by King
  Charles (26500), &c. See Vol. xxiv. p. 299.

1879, Dec. 8, roan, B.C.	(dead)	Waterloo, 40892	Mr. J. Angus

DAIRY MAID, roan, calved February 20, 1877. Bred by Mr. J. Angus; got
  by Ben Brace (30524), dam (Milkmaid) by Roan Chief (27294), &c. See
  Vol. xxi. p. 542.

1879, July 25, roan, C.C.	Dairy Girl	Hawthorn, 36751	Mr. Angus

DAISY GIRL, roan, calved February 25, 1868, Vol. xxii. p. 298. Bred by Mr.
  G. Angus, Broomley; ,got by Merry Monarch (22349), dam (Daisy) by
  Alderman (17291), &c.

1877, June 11, roan, C.C.	Daisy Flower	Ben Brace, 30524	Mr. J. Angus
1878, July 12, roan, C.C.	Daisy of Styford	do.	do.

MAY MORN, roan, calved January 20, 1876. Bred by Mr. J. Angus; got by
  Royal Frederick (35366), dam (May Day) by Roan Chief (27294), &c. See
  "Maurice," p. 168.

1878, May 30, roan, B.C.	Maurice	Captain Brace, 37938	Mr. Angus
1879, June 28, roan, B.C.	Horace	Hawthorn, 36751	do.

ANNETT, John W.,
Houndalee, Acklington.

LILYWHITE, white, calved January 3, 1875. Bred by Mr. R. Burdon, Castle
  Eden; got by Mountain Hero (31944), dam (Lady Lilias) by Emperor
  Maximilian (26100), &c. See "Sir Matthew," Vol. xxv. p. 263.

1877, May 12, white, B.C.	(dead)	Knight of Murrah, 34391	Mr. Annett
1878, April 24, roan, B.C.	Sir Matthew	do.	do.
1879, April 12, white, C.C.	Cream of Tartar	do.	do.

Sir Matthew, sold to Mr. J. White, Yorkshire.

WHITE ROSE 2ND, white, calved January 22, 1875. Bred by Mr. R. Burdon,
  Castle Eden; got by Mountain Hero (31944), dam (Burnett Rose) by Young
  Freedom (21777), &c. See "Euclid," Vol. xxv. p. 104.

1877, July 4, roan, B.C.	Zoroaster	Knight of Murrah, 34391	Mr. Annett
1878, June 13, roan, B.C.	Euclid	do.	do.
1879, May 7, roan, C.C.	Rose Cream	do.	do.

Zoroaster, sold to Mr. R. C. Rouse, Learmouth, Coldstream.

ARCHDALE, E. M.,
Crock-na-Crieve, Ballinamallard, Ireland.

LADY GEORGINA 6TH, white, calved April 26, 1875. Bred by Major
  Hamilton, Brownhall; got by Baron Underley (30490), dam (Lady Georgina
  2nd) by Duke of Montrose (21599), &c. See "Mob Law," Vol. xx. p. 238.

1879, Mar 1, roan, B.C.	Baron Ballintra	King Frederick, 41756	Mr. Archdale

PRINCESS OF LURG, red and white, calved May 28, 1875. Bred by Mr. N.
M. Archdale, Crock-na-Crieve; got by Abercorn (25484), dam (Jenny Lind
7th) by Prince of Lurg (22617), &c. See " Victor Lind," Vol. xxv. p. 288.

Produce in		Names, &c.		By what Bull.		By whom bred.
1879, Oct. 23, red,	B.C.	Squire of Lurg		Vivian Grey, 42561		Mr. E. M. Archdale

RHODA 2ND, white, calved February 28, 1875. Bred by Mr. W. Humphrys,
Ballyhaise House; got by Prince of Lothian (35152), dam (Rhoda) by
Napoleon 3rd (29417), g. d. (Victoria 6th) by Loadstone (31605), gr. g. d.
(Victoria 5th) by Sir John Sinclair (5165), &c. See " Emerald," Vol. xxii. p. 94.

Produce in		Names, &c.	By what Bull.	By whom bred.
1878, Mar. 25, roan,	B.C.	(dead)	King Frederick, 41756	Mr. Archdale
1879, May 1, roan,	B.C.	Squire of Lothian	do.	do.

ARKELL, Daniel,
Butler's Court, Lechlade.

CLOUDY, roan, calved April 5, 1876. Bred by Mr. D. Arkell; got by Viceroy
(30216), dam (Crystal) by Research (27270), g. d. (Clodge) by Oxford Don
(20451), gr. g. d. (Pigeon) by The Slave (27647).

Produce in		Names, &c.	By what Bull.	By whom bred.
1879, April 20, roan,	B.C.	(Steer)	Baron Lee 4th, 37799	Mr. Arkell

FILLPAIL 3RD, red and white, calved April 6, 1874. Bred by Mr. D. Arkell;
got by Factory Boy (33870), dam (Fairmaid) by Royal George (29866), g. d.
(Fillpail 2nd) by Mercury (22342), gr. g. d. (Fillpail) by Oxford Don (20451).

Produce in		Names, &c.	By what Bull.	By whom bred.
1879, Mar. 9, r. & w.,	B.C.	Baron Lechlade	Baron Lee 4th, 37799	Mr. Arkell

FILLPAIL 7TH, red and white, calved May 7, 1877. Bred by Mr. D. Arkell; got
by Brilliant (37896), dam (Fillpail 4th) by Seventh Baron Wetherby (23384),
g. d. (Fillpail 2nd) by Mercury (22342), gr. g. d. (Fillpail) by Oxford Don
(20451).

Produce in		Names, &c.	By what Bull.	By whom bred.
1879, Oct. 3, r. & w.,	C.C.	Fillpail 9th	Baron Lee 4th, 37799	Mr. Arkell

FLOWER OF SPRING, roan, calved March 24, 1875. Bred by Mr. D. Arkell;
got by Seventh Baron Wetherby (23384), dam (Flower Girl) by Mercury
(22342), g. d. (Flourish) by The Slave (27647), &c. See Vol. xviii. p. 500.

Produce in		Names, &c.	By what Bull.	By whom bred.
1879, Mar. 20, roan,	B.C.	Baron Lechdale 2nd	Baron Lee 4th, 37799	Mr. Arkell

FRISKY, red and white, calved March 4, 1871. Bred by Mr. W. Arkell,
Dudgrove; got by Fitz Walton (28609), dam (Flower Girl) by Mercury
(22342), &c.

Produce in		Names, &c.	By what Bull.	By whom bred.
1877, April 2, roan,	C.C.	Fuchsia	Viceroy, 30216	Mr. D. Arkell
1878, Feb. 18, roan,	C.C.	(dead)	do.	do.
1879, Mar. 2, r. & w.,	C.C.	Festive	Baron Lee 4th, 37799	do.

JEMIMA, roan, calved January 5, 1875. Bred by Mr. J. B. Jenkins, Kingston
House; got by Oxford's Gwynne (32033), dam (Jenny Lind) by Lancaster
(24302), &c. See Vol. xxi. p. 789.

Produce in		Names, &c.	By what Bull.	By whom bred.
1878, May 10, r. & w.,	C.C.	Lady Dennison 3rd	Remus, 32286	Mr. Arkell
1879, Sept., roan,	B.C.	(dead)	Pompey, 35059	do.

LADY DENNISON 2ND, white, calved January 19, 1876. Bred by Mr. D.
Arkell; got by Marquis of Sockburn (34787), dam (Lady Dennison) by Second
Lord Waterloo (29219), g. d. (Jenny Dennison 6th) by Royal Butterfly 14th
(20722), &c. See " Jenny Butterfly," Vol. xxi. p. 894.

Produce in		Names, &c.	By what Bull.	By whom bred.
1879, April 22, roan,	C.C.	Lady Dennison 4th	Baron Lee 4th, 37799	Mr. Arkell

LIZZIE LEE 2ND, roan, calved May 5, 1877. Bred by Mr. W. Arkell, Dudgrove; got by Pompey (35059), dam (Lizzie Lee) by Fitz Walton (28609), g. d. (Rosa) by Young Cobham (17580), &c. See Vol. xviii. p. 694.

Produce in	Names, &c.	By what Bull.	By whom bred.
1879, Sept. 25, r. & w., B.C.	(Steer)	Sultan 2nd, 44106	Mr. D. Arkell

SATELLITE, roan, calved December 16, 1870. Bred by Mr. W. Arkell, Dudgrove; got by Sixteenth Baron Wetherby (25597), dam (Snowdrop) by Brazenose (15686), &c. See "Sultan 2nd," p. 244.

Produce in	Names, &c.	By what Bull.	By whom bred.
1879, Mar. 8, roan, B.C.	Statesman	Pompey, 35059	Mr. D. Arkell

SOPHIE LEE, red and white, calved January 9, 1871. Bred by Mr. W. Arkell, Dudgrove; got by Sixteenth Baron Wetherby (25597), dam (Rosa) by Young Cobham (17580), &c. See Vol. xviii. p. 694.

Produce in	Names, &c.	By what Bull.	By whom bred.
1879, Jan. 11, roan, C.C.	Alice Lee	Pompey, 35059	Mr. D. Arkell

ARKELL, William,
Dudgrove, Fairford, Gloucestershire.

ROMP, roan, calved April 3, 1874. Bred by Mr. W. Arkell, the property of Mr. F. R. Moser, Carbery; got by Viceroy (30216), dam (Royal Lady) by Royal George (29866), g. d. (Friendling) by Lord of the Vale (29183), gr. g. d. (Forester) by Oxford Don (20451).

Produce in	Names, &c.	By what Bull.	By whom bred.
1879, Mar. 21, roan, B.C.	Roving Lad	L. Fitzclarence 16th, 36943	Mr. Arkell

ARKELL, William, jun.,
Hatherop, Fairford, Gloucestershire.

DAMSEL 8TH, roan, calved February 27, 1876. Bred by Mr. W. Arkell, jun.; got by Wandering Minstrel (35924), dam (Damsel 2nd) by Freemason (28654), g. d. (Damsel) by Coleshill (23590), gr. g. d. (Daffodil) by Royal Arch 3rd (27346).

Produce in	Names, &c.	By what Bull.	By whom bred.
1879, April 5, roan, B.C.	(dead)	Fitz-Wetherby, 36650	Mr. Arkell, jun.

FACTORY MAID, roan, calved February 13, 1876. Bred by Mr. W. Arkell, sen., Dudgrove; got by Fifteenth Baron Wetherby (25596), dam (Factory Girl) by Mercury (22342), &c. See Vol. xxv. p. 692.

Produce in	Names, &c.	By what Bull.	By whom bred.
1878, Dec. 24, white, B.C.	Fitz-William 2nd	L. Fitzclarence 16th, 36943	Mr. Arkell, jun.

Fitz-William 2nd, sold to Mr. F. Larkworthy, Carnarvon Estate, Wanawater, New Zealand.

FASTING GIRL, roan, calved February 11, 1877. Bred by Mr. W. Arkell, sen., Dudgrove; got by Pompey (35059), dam (Factory Girl) by Mercury (22342), &c.

Produce in	Names, &c.	By what Bull.	By whom bred.
1879, May 5, r. & w., C.C.	(dead)	Sultan 2nd, 44106	Mr. Arkell, jun.

JANETTE 4TH, red and white, calved March 18, 1876. Bred by Mr. W. Arkell, jun; got by Wandering Minstrel (35924), dam (Janette) by Bothwell (25661), g. d. (Judy) by Lord Lyon (24424), gr. g. d. (Jewel) by Viscount Killerby (19081), &c. See Vol. xviii. p. 542.

Produce in	Names, &c.	By what Bull.	By whom bred.
1879, Mar. 10, roan, C.C.	Janette 6th	Fitz-Wetherby, 36650	Mr. Arkell, jun.

MERMAID, roan, calved May 19, 1876. Bred by Mr. W. Arkell, sen., Dudgrove, the property of Mr. W. Arkell, jun.; got by Baron Bounty (33012), dam (Merry Lass) by Viceroy (30216), g. d. (Mulberry 2nd) by Lord of the Vale (29183), gr. g. d. (Mulberry) by Prince Albert (29589).

Produce in	Names, &c.	By what Bull.	By whom bred.
1879, Mar. 7, red,	B.C. Merry Monarch 2nd	Sultan 2nd, 44106	Mr. W. Arkell, sen.
Merry Monarch 2nd, sold to Mr. W. Arkell, jun.			

SWEET BRIAR 3RD, roan, calved March 14, 1874. Bred by Mr. W. Arkell, jun.; got by Remus (32286), dam (Sweet Briar) by Iceberg (26426), g. d. (Dainty) by Coleraine (25793), gr. g. d. (Pretty) by Royal (27343).

Produce in	Names, &c.	By what Bull.	By whom bred.
1877, July 15, roan,	C.C. Sweet Briar 5th	D. of Hazlecote 36th, 33671	Mr. Arkell, jun.
1878, May 29, red,	C.C. Sweet Briar 6th	do.	do.

URSULINA 4TH, white, calved July 11, 1873. Bred by Mr. T. Arkell, Draycott; got by Baron Draycott 3rd (33035), dam (Ursula 12th) by General Canrobert (12927), &c. See Vol. xvi. p. 720.

Produce in	Names, &c.	By what Bull.	By whom bred.
1876, Oct. 17, roan,	C.C. Ursulina 5th	D. of Hazlecote 36th, 33671	Mr. W. Arkell, jun.
1878, Sept. 25, roan,	B.C. Umpire	do.	do.

ARKLAY, Robert,
Ethiebeaton, Dundee, N.B.

ANNANDALE MISS TODDLES, roan, calved November 15, 1873. Bred by Mr. R. Arklay; got by Annandale (30386), dam (Miss Toddles 4th) by Red Friar 2nd (27247), g. d. (Miss Toddles) by Andy (23312), &c. See " Master Toddles," Vol. xxiv. p. 176.

Produce in	Names, &c.	By what Bull.	By whom bred.
1876, Aug. 6, roan,	B.C. Master Toddles	Monarch of Lorn, 34869	Mr. Arklay
1878, April 16, red,	C.C. Red Miss Toddles 3rd	King Hope, 36843	do.

ARMISTEAD, J. F.,
Cob Wall House, Blackburn.

NEWFIELD FAVOURITE 2ND, roan, calved May 27, 1877. Bred by Mr. L. H. Wraith, Newfield; got by Baron Turncroft Bates (33086), dam (Newfield Favourite) by Wharfdale's Duke (35974), &c. See Vol. xxiv. p. 716.

Produce in	Names, &c.	By what Bull.	By whom bred.
1879, Oct. 21, roan,	C.C. Sweetbriar	Brierfield, 39489	Mr. Armistead

ROSE OF LANCASTER, red, calved October 16, 1872, Vol. xxiv. p. 301. Bred by Mr. R. Whittam, Brierfield; got by Earl of Thorndale (28521), dam (Fancy 16th) by Festus 3rd (26148), &c.

Produce in	Names, &c.	By what Bull.	By whom bred.
1878, July 31, roan,	C.C. Ophelia	White Hamlet, 37669	Mr. Armistead

ARMSTRONG, Samuel,
Gally House, Enniscorthy, Co. Wexford.

ANNIE GWYNNE, roan, calved March 7, 1873, Vols. xxiii., xxiv., and xxv. pp. 341, 301, 318. Bred by Mr. W. Bolton, The Island; got by King Richard (26523), dam (Ally Gwynne) by Grey Gauntlet (19908), &c.

Produce in	Names, &c.	By what Bull.	By whom bred.
1879, July 27, r. & w.,	C.C. Adela Gwynne	British Lad, 41151	Mr. Armstrong

GOLDEN BUD, roan, calved July 24, 1871, Vols xxi., xxiii., and xxiv. pp. 544, 312, 301. Bred by Mr. S. Armstrong; got by King Richard (26523), dam (Golden Rose) by Blood Royal (14169), &c.

Produce in	Names, &c.	By what Bull.	By whom bred.
1879, Jan. 13, roan, C.C.	(slaughtered)	British Lad, 41151	Mr. Armstrong

GOLDEN CHAIN, roan, calved February 23, 1874, Vol. xxiv. p. 301. Bred by Mr. S. Armstrong; got by St. Ringan (27417), dam (Golden Drop) by Royal Sovereign (22802), &c.

1879, Jan. 20, r.&w., B.C.	Golden Banner	British Lad, 41151	Mr. Armstrong

GOLDEN LILY, white, calved July 7, 1871, Vols. xxi., xxiii., and xxiv. pp. 545, 312, 302. Bred by Mr. S. Armstrong, the property of Mr. W. Bolton, The Island; got by King Richard (26523), dam (Golden Pin) by Duke of Marlboro' (23768), &c.

1879, Jan. 27, { roan, C.C.	Golden Gem	} British Lad, 41151	Mr. Armstrong
roan, C.C.	Golden Joy		

LOUISA 27TH, roan, calved April 27, 1876. Bred by Mr. W. S. Garnett, Williamston; got by Lieutenant General (31600), dam (Louisa 8th) by Ravenspur (20628), g. d. (Louisa 4th) by Dr. McHale (15887), &c. See Vol. xviii. p. 594.

1879, Jan. 16, white, B.C.	(dead)	Albion, 36112	Mr. Armstrong

LUNA, roan, calved July 9, 1875. Bred by Mr. A. W. Connon, Cremore; got by Lieutenant General (31600), dam (Roan Lass) by Main Royal (24505), &c. See Vol. xxii. p. 371.

1878, Jan. 3, red, B.C.	(dead)	British Hero, 36278	Mr. Armstrong
1879, Mar. 23, r.&w., C.C.	Letitia	British Lad, 41151	do.

Letitia, sold to Mr. G. H. Lett, Millpark, Enniscorthy.

ARMSTRONG, Sir W. G.,
Cragside, Morpeth.

OXFORD ROSE 5TH, roan, calved February 18, 1875, Vol xxv. p. 319. Bred by Sir W. G. Armstrong; got by Oxford Beau 4th (34964), dam (Wharfdale Rose) by Third Lord Wharfdale (26759), &c.

1879, June 26, white, B.C.	Star 3rd	Duke of Oxford 27th, 33709	Sir W. Armstrong

WILD DUCHESS OF GENEVA, roan, calved January 28, 1872, Vols. xxi., xxii., and xxiv. pp. 547, 299, 304. Bred by Mr. E. H. Cheney, Gaddesby Hall; got by Ninth Duke of Geneva (28391), dam (Wild Oxford) by Lord Oxford 2nd (20215), &c.

1879, April 27, roan, C.C.	W'd D. of Geneva 10th	Duke of Oxford 27th, 33709	Sir W. Armstrong

WILD DUCHESS OF GENEVA 2ND, red, calved January 1, 1873, Vols. xxii., xxiii., and xxiv. pp. 299, 314, 304. Bred by Mr. E. H. Cheney, Gaddesby Hall; got by Ninth Duke of Geneva (28391), dam (Wild Oxford) by Lord Oxford 2nd (20215), &c.

Jan. 5, roan, C.C.	W'd D. of Geneva 9th	Duke of Oxford 27th, 33709	Sir W. Armstrong

ASHBURNER, George,
Low Hall, Kirkby Ireleth, Carnforth,

BRIDE OF LORN, white, calved May 29, 1872, Vol. xxiv. p. 304. Bred by Mr. G. Ashburner; got by Sockburn Lad (30024), dam (Lady Oxford) by Tenth Duke of Oxford (17739), &c.

Produce in		Names, &c.	By what Bull.	By whom bred.
1878, Oct. 23, roan,	C.C.	Duchess of Lorn	Duke of Oxford, 31004	Mr. Ashburner
1879, Nov. 9, roan,	B.C.	General Lorn	Duke of Oxford 41st, 38174	do.

FANTAIL 6TH, roan, calved January 3, 1871, Vols. xxi., xxiv., and xxv. pp. 855, 304, 320. Bred by Mr. G. Moore, Whitehall; got by Seventeenth Duke of Oxford (25994), dam (Fantail 5th) by Royal Cambridge (25009), &c.

Produce in		Names, &c.	By what Bull.	By whom bred.
1879, July 23, roan,	C.C.	Fantail 10th	Duke of Oxford 41st, 38174	Mr. Ashburner

GRAND DUCHESS MARIE, roan, calved March 23, 1874, Vol. xxiv. p. 304. Bred by Mr. G. Ashburner; got by Grand Duke of Oxford (28764), dam (Blanche 5th) by Grand Duke 10th (21848), &c.

Produce in		Names, &c.	By what Bull.	By whom bred.
1878, May 29, r. & w.,	B.C.	Royal Oak	2nd Duke of Glo'ster, 28392	Mr. Ashburner
1879, June 4, red,	C.C.	Blanche 6th	Duke of Oxford 41st. 38174	do.

Royal Oak, sold to Mr. J. B. Benson, Kirkby Hall, Ulverston.

HOLKER DARLING, red and white, calved January 4, 1875. Bred by Mr. R. Postlethwaite, The Hollins; got by Fifth Duke of Wetherby (31033), dam (Severn Girl) by Third Duke of Clarence (23727), &c. See Vol. xxii. p. 478.

Produce in		Names, &c.	By what Bull.	By whom bred.
1878, Feb. 5, r. & w.,	B.C.	(dead)	Baron Tregunter, 36208	Mr. Ashburner
1879, Apr. 22, r. & w.,	C.C.	Seraphina	Duke of Oxford 41st, 38174	do.

JEWEL, roan, calved March 3, 1875. Bred by Mr. G. Ashburner; got by Grand Duke of Lightburne 3rd (28761), dam (Oxford's Gem) by Oxford 4th (24706), &c. See "Jeweller," Vol. xxv. p. 141.

Produce in		Names, &c.	By what Bull.	By whom bred.
1877, Nov. 14, roan,	C.C.	Trinket	G. D. of Lightb'rne 3d, 28761	Mr. Ashburner
1878, Oct. 20, r. & w.,	B.C.	Jeweller	Duke of Oxford, 31004	do.
1879, Sept. 11, red,	C.C.	Diamond	Duke of Oxford 41st, 38174	do.

Trinket, sold to Mr. W. Ashburner, Conishead Grange, Ulverston; Jeweller, to Mr. A. Brogden, Grange over Sands, Carnforth.

KIRKLEVINGTON 27TH, roan, calved May 26, 1875. Bred by Mr. W. Ashburner, Conishead Grange; got by Grand Duke of Kent 2nd (28759), dam (Kirklevington Duchess 7th) by Duke of Kirklevington (25982), &c. See Vol. xxii. p. 301.

Produce in		Names, &c.	By what Bull.	By whom bred.
1877, Dec. 18, red,	B.C.	(dead)	Baron Oxford 7th, 36199	Mr. G. Ashburner
1879, May 1, red,	C.C.	Kirklevington 28th	Duke of Glo'ster 7th, 39735	do.

MAY QUEEN, roan, calved May 6, 1874. Bred by Mr. G. Ashburner; got by Grand Duke of Lightburne 3rd (28761), dam (Queen Maria) by Erl King (26109), &c. See Vol. xxiii. p. 315.

Produce in		Names, &c.	By what Bull.	By whom bred.
1877, Oct. 12, roan,	C.C.	Autumn Queen	Duke of Oxford, 31004	Mr. Ashburner
1879, Jan. 15, r. & w.,	B.C.	Ameer	Duke of Oxford 41st, 38174	do.

OXFORD'S BLANCHE, red and white, calved August 19, 1875. Bred by Mr. G. Ashburner; got by Duke of Oxford (31004), dam (Blanche 5th) by Grand Duke 10th (21848), &c. See "Duke of Millom," p. 80.

Produce in		Names, &c.	By what Bull.	By whom bred.
1878, Apr. 16, r. & w.,	B.C.	Berlin	Wild Boy of the Dale, 39315	Mr. Ashburner
1879, May 21, red,	B.C.	Duke of Millom	Duke of Glo'ster 7th, 39735	do.

Berlin, sold to Mr. W. Inman, Coniston Hall, Coniston; Duke of Millom, to Mr. J. Harker, Salthouse, Millom.

OXFORD'S DUCHESS, red and white, calved October 4, 1873, Vol. xxiv.
p. 305. Bred by Mr. G. Ashburner; got by Grand Duke 9th (19879), dam
(Christmas Oxford) by Fifteenth Duke of Oxford (23776), &c.

Produce in		Names, &c.	By what Bull.	By whom bred.
1879, Jan. 13, red,	C.C.	Oxford Lass	Duke of Oxford 41st, 38174	Mr. Ashburner
1879, Dec. 29, roan,	B.C.	Cherry Oxford	do.	do.

OXFORD'S GEM, roan, calved in March 1869, Vol. xxii. p. 300. Bred by
Mr. R. W. Ashburner, Low Hall; got by Oxford 4th (24706), dam (Double
Oxford) by Oxford (20449), &c.

Produce in		Names, &c.	By what Bull.	By whom bred.
1876, May 2, r. & w.,	B.C.	Precious Metal	Duke of Oxford, 31004	Mr. G. Ashburner
1877, May 1, roan,	B.C.	Jasper	Ch'y D. Lightburne, 36349	do.
1878, May 23, roan,	B.C.	Emerald	do.	do.
1879, April 22, roan,	B.C.	Goldsmith	Duke of Oxford 41st, 38174	do.

Precious Metal, sold to Mr. E. Brown, Grasmere; Jasper, to Mr. Elwood, Staveley, Kendal;
Emerald, to Mr. W. Stephenson, Nettleslack, Ulverston; Goldsmith, to Mr. J. Newby,
Loweswater.

OXFORD'S GWYNNE, red and white, calved May 27, 1874. Bred by Mr.
G. Ashburner; got by Duke of Oxford (31004), dam (Wild Eyes Gwynne
2nd) by Sir Windsor (22927), g. d. (Wild Eyes Gwynne) by Baron Wild Eyes
(19290), &c. See Vol. xx. p. 823.

Produce in		Names, &c.	By what Bull.	By whom bred.
1878, Oct. 21, roan,	C.C.	Cherry Gwynne	Ch'y D. Lightburne, 36349	Mr. Ashburner
1879. Oct. 14, r. & w.,	B.C.	(dead)	Duke of Oxford 41st, 38174	do.

PARK MINSTREL, red and white, calved March 16, 1867, Vols. xix. and xxii.
pp. 660, 300. Bred by Mr. W. Slater, Dalton in Furness; got by Hematite
(21917), dam (Minstrel 4th) by Tenth Duke of Oxford (17739). &c.

Produce in		Names, &c.	By what Bull.	By whom bred.
1877, April 4, roan,	B.C.	Cherry Minstrel	Ch'y D. Lightburne, 36349	Mr. Ashburner
1878, April 2, red,	C.C.	Cambridge Minstrel	G.D.of Lightb'rne 3d, 28761	do.
1879, April 3, red	C.C.	Oxford Minstrel 3rd	Duke of Oxford 41st, 38174	do.

Cherry Minstrel, sold to Mr. I. Bargh, Scaleber, Kirkby Lonsdale.

POLLY'S DUCHESS, white, calved in June 1873. Bred by Mr. G. Ashburner;
got by Grand Duke 9th (19879), dam (Polly) by Oxford 4th (24706), &c.
See " Sir Bartle," p. 230.

Produce in		Names, &c.	By what Bull.	By whom bred.
1876, June 29, roan,	B.C.	Duke of Witherslack	Duke of Oxford, 31004	Mr. Ashburner
1878, July 25, roan,	B.C.	Zulu	G.D.of Lightb'rne 3d, 28761	do.
1879, July 7, roan,	B.C.	Sir Bartle	Duke of Oxford 41st, 38174	do.

Duke of Witherslack, sold to Mr. F. Clarke, Tarn Green, Cartmel; Zulu, to Mr. Whittaker,
Denham Hall, Chorley.

PRIMROSE, roan, calved April 21, 1873. Bred by Mr. G. Ashburner; got by
Grand Duke 9th (19879), dam (Pink) by Temperance Star (23012), g. d.
(Pansy) by Garibaldi (19815), gr. g. d. (Poplar) by Hopeful (14716).

Produce in		Names, &c.	By what Bull.	By whom bred.
1876, Dec. 23, roan,	B.C.	Duke of Windermere	Duke of Oxford, 31004	Mr. Ashburner
1878, Feb. 26, r. & w.,	B.C.	Ottoman	G.D.of Lightb'rne 3d, 28761	do.
1879, Jan. 24, r. & w.,	C.C.	Peony	Duke of Oxford 41st, 38174	do.
1879, Dec. 31, roan,	C.C.	Primula	do.	do.

Duke of Windermere, sold to Mr. R. Postlethwaite, Newby Bridge, Ulverston; Ottoman, to
Mr. R. Atkinson, Oxenpark, Ulverston.

PRINCESS OF OXFORD, red and white, calved in December 1869, Vol. xxiii.
p. 315. Bred by Mr. G. Ashburner, the property of Mr. E. Brown, Grasmere;
got by Baron Fennel (27937), dam (Countess of Oxford) by Tenth Duke of
Oxford (17739), &c.

Produce in		Names, &c.	By what Bull.	By whom bred.
1877, July 5, roan,	C.C.	Princess Cherry	Ch'y D. Lightburne, 36349	Mr. Ashburner

ASHBURNER, William,
Conishead Grange, Ulverston.

BRIGHT EYES 5TH, roan, calved December 16, 1872, Vol. xxiii. p. 315. Bred by Mr. E. Musgrove, Aughton Old Hall; got by Royal Lancaster (29870), dam (Bright Eyes 4th) by Third Duke of Wharfdale (21619), &c.

Produce in	Names, &c.	By what Bull.	By whom bred.
1877, June 10, white, C.C.	Conishead Wild Eyes	24th Duke of Airdrie, 36460	Mr. Ashburner
1878, April 29, white, B.C.	Conishead Duke 3rd	2nd Duke of Glo'ster, 28392	do.
1879, Mar. 29, roan, C.C.	Conish'd W'd Eyes 2d	do.	do.

Conishead Wild Eyes, sold to Mr. H. Allsopp, Hindlip Hall, Worcester; Conishead Duke 3rd and Conishead Wild Eyes 2nd, to Mr. H. Lovatt, Low Hill, Bushbury.

DAYLIGHT, roan, calved June 1, 1876. Bred by Archdeacon Holbech, Farnborough Hall; got by Earl of Chatham (28495), dam (Dahlia) by Second Earl of Cleveland (26054), &c. See Vol. xxi. p. 768.

1879, Aug. 16, roan, B.C.	Daisy Prince	Earl of Horton 10th, 36587	Mr. Ashburner

DOWAGER GWYNNE 3RD, roan, calved July 24, 1875, Vol. xxv. p. 321. Bred by Mr. W. Ashburner; got by Second Duke of Glo'ster (28392), dam (Double Gwynne) by Rufus (27397), &c.

1879, Feb. 8, roan, C.C.	Dowager Gwynne 4th	2nd Duke of Glo'ster, 28392	Mr. Ashburner

DUCHESS NANCY 2ND, white, calved July 20, 1875. Bred by Lord Penrhyn, Penrhyn Castle, the property of Mr. J. J. Hetherington, Middle Farm; got by Grand Duke of Oxford (31293), dam (Grand Duchess of Oxford 3rd) by Thorndale Duke (27661), &c. See Vol. xxii. p. 529.

1878, July 2, white, B.C.	Duke Nancy	Beau of Oxford 3rd, 33130	Mr. Ashburner
1879, June 10, roan, C.C.	Conish'd D'ss Nancy	Duke of Oxford 34th, 36529	do.

WILD EYES 36TH, red and white, calved May 18, 1875. Bred by Mr. J. Robinson, Fell Side; got by Oliver Gwynne (31998), dam (Wild Eyes 29th) by Knight of the Harem (24278), &c. See Vol. xxii. p. 547.

1878, April 8, r. & w., C.C.	Wild Eyes 39th	2nd Duke of Glo'ster, 28392	Mr. Ashburner
1879, Mar. 10, r. & w., C.C.	Wild Eyes 40th	do.	do.

ASHBY, Captain G. A.,
Naseby Woolleys, Rugby.

DESDEMONA, roan, calved October 10, 1871, Vols. xxi., xxiii., xxiv., and xxv. pp. 549, 316, 306, 322. Bred by Mr. S. Canning, Snitterfield; got by Grand Duke of Wateringbury (26296), dam (Daphne) by Comet (19484), &c.

1879, July 7, roan, C.C.	Desd'mona of Naseby	Earl of Geneva, 33794	Capt. Ashby

DOROTHY, red, calved July 15, 1876. Bred by Captain Ashby; got by Telemachus 3rd (32650), dam (Desdemona) by Grand Duke of Wateringbury (26296), &c. See " Dorothy's Geneva," p. 68.

1879, Feb. 11, roan, B.C.	Dorothy's Geneva	Earl of Geneva, 33794	Capt. Ashby

FERONIA, roan, calved March 27, 1874, Vol. xxv. p. 322. Bred by Mr. J. P. Foster, Killhow; got by Twenty-second Duke of Oxford (31000), dam (False Fanny) by Edgar (19680), &c.

1879, Dec. 17, roan, C.C.	Feronia of Naseby	Earl of Geneva, 33794	Capt. Ashby

FLIRT, roan, calved December 1, 1876. Bred by Captain Ashby; got by Fawsley Knight (33903), dam (Miss Fawsley) by Stanley (27560), &c. See Vol. xxiii. p. 316.

Produce in	Names, &c.	By what Bull.	By whom bred.
1879, June 19, roan, B.C.	Fop	E.Le'ster's Furbelow,41478	Capt. Ashby

GRACE COSTA 3RD, roan, calved November 5, 1874, Vols. xxiv. and xxv. pp. 505, 322. Bred by Sir G. R. Philips, Bart., Weston Park; got by Cherry Fawsley (30711), dam (Grace Costa) by Costa (21487), &c.

Produce in	Names, &c.	By what Bull.	By whom bred.
1879, Dec. 19. roan, B.C.	Costa of Naseby	Earl of Geneva, 33794	Capt. Ashby

INNOCENCE, roan, calved May 12, 1876. Bred by Captain Ashby; got by Telemachus 3rd (32650), dam (Inquiry) by Third Duke of Geneva (21592), &c. See Vol. xxiii. p. 316.

Produce in	Names, &c.	By what Bull.	By whom bred.
1879, April 29, roan, B.C.	(dead)	E.Le'ster's Furbelow,41478	Capt. Ashby

INQUIRY, red, calved June 13, 1869, Vols. xxi., xxiii., and xxiv. pp. 885, 316, 307. Bred by Sir G. R. Philips, Bart., Weston Park; got by Third Duke of Geneva (21592), dam (Invoice) by Pan (18516), &c.

Produce in	Names, &c.	By what Bull.	By whom bred.
1879, Oct. 1, roan, B.C.	Brother Innocence	Earl of Geneva, 33794	Capt. Ashby

IRENE 2ND, red, calved February 28, 1873, Vols. xxiii. and xxv. pp. 316, 322. Bred by Sir G. R. Philips, Bart., Weston Park; got by Third Cherry Duke (28171), dam (Irene) by Barleycorn the Younger (21209), &c.

Produce in	Names, &c.	By what Bull.	By whom bred.
1879, May 19, roan, B.C.	Imp	Earl of Geneva, 33794	Capt. Ashby

LADY FURBELOW, roan, calved January 11, 1875, Vols. xxiv. and xxv. pp. 307, 322. Bred by Mr. C. M. Hamer, Snitterfield; got by Knightley Wellington (31533), dam (Harmonia) by Grand Duke of Essex 4th (24068), &c.

Produce in	Names, &c.	By what Bull.	By whom bred.
1879, Aug. 8, roan, C.C.	(dead)	Earl of Geneva, 33794	Capt. Ashby

LADY KNIGHTLEY 2ND, roan, calved February 10, 1867, Vol. xxv. p. 322. Bred by Mr. E. L. Betts, Preston Hall; got by Grand Duke 4th (19874), dam (Maidenhair) by Mocassin (18406), &c.

Produce in	Names, &c.	By what Bull.	By whom bred.
1879, Dec. 3, white, C.C.	Rose K.of Naseby 2d	Earl of Geneva, 33794	Capt. Ashby

MABEL, roan, calved December 12, 1875, Vol. xxv. p. 322. Bred by Captain Ashby; got by Telemachus 3rd (32650), dam (Miranda) by Magician (26778). &c.

Produce in	Names, &c.	By what Bull.	By whom bred.
1879, July 16, roan, B.C.	Mainstay	Earl of Geneva, 33794	Capt. Ashby

MARY OF N., roan, calved September 20, 1874, Vol. xxiv. p. 307. Bred by Captain Ashby; got by Third Duke of Barrington (30923), dam (Miranda) by Magician (26778), &c.

Produce in	Names, &c.	By what Bull.	By whom bred.
1879, Jan. 23, roan, B.C.	G. of N.	Earl of Geneva, 33794	Capt. Ashby

PRIESTESS, red and white, calved January 22, 1868, Vols. xix. and xxi. pp. 673, 549. Bred by Mr. C. W. Goode, Pewsey; got by Florist (23962), dam (Poppy) by Cambridge Grand Duke (15722), &c.

Produce in	Names, &c.	By what Bull.	By whom bred.
1879, Feb. 15, roan, B.C.	Geneva Priest	Earl of Geneva, 33794	Capt. Ashby

Geneva Priest, sold to Mr. A. H. Pearson. Chilwell. Notts.

REBECCA, roan, calved March 29, 1876. Bred by Captain Ashby; got by Telemachus 3rd (32650), dam (Ridlington Lass) by Ridlington (22726), &c. See Vol. xxiii. p. 316.

Produce in	Names, &c.	By what Bull.	By whom bred.
1879, Mar. 7. roan, C.C.	Rebecca 2nd	Earl of Geneva, 33794	Capt. Ashby

ROAN DUCHESS, roan, calved May 16, 1875, Vol. xxv. p. 507.   Bred by Mr.
G. Moore, Whitehall; got by Seventeenth Duke of Oxford (25994), dam
(Duchess) by Grand Duke 15th (21852), &c.

Produce in	Names, &c.	By what Bull.	By whom bred.
1879, Jan. 26, r. & w., C.C.	Double Duchess	Grand Duke 23rd, 34063	Capt. Ashby

ASHWORTH, Alfred,
Egerton Hall, Bolton-le-Moors.

ROSA MARGUERITA, roan, calved July 28, 1873, Vols. xxiv. and xxv.
pp. 308, 323.   Bred by Mr. A. Ashworth, Egerton Hall; got by Booth's
Royal Signet (28061), dam (Rosa Sabina) by Royal Booth (22772), &c.

1879, Aug. 2, r. & w., B.C.	The Express	Star Regent, 35679	Mr. Ashworth

SERENE STAR, roan, calved March 7, 1874, Vols. xxiii. and xxiv. pp. 317,
308.   Bred by Mr. R. S. Bruere, Braithwaite Hall; got by Booth's Royal
Signet (28061), dam (Vestal Star) by Sir Windsor Broughton (27507), &c.

1879, Jan. 11, white, B.C.	Twilight	Heir-at-Law, 34124	Mr. Ashworth
1879, Dec. 5, roan, C.C.		S.Swthn'sStarDrop,40667	do.

SWEET PRIMROSE, roan, calved May 15, 1872, Vols. xxii., xxiv., and xxv.
pp. 303, 308, 323.   Bred by Mr. R. S. Bruere, Braithwaite Hall; got by
Booth's Royal Signet (28061), dam (Sweet Brier Perfume) by Prince of the
Realm (22627), &c.

1879, Dec. 12, red, C.C.	(dead)	Star Regent, 35679	Mr. Ashworth

ATTWATER, J. G.,
Britford, Salisbury.

BLUSH 4TH, red and white, calved February 4, 1876.   Bred by Mr. R. E.
Oliver, Sholebroke Lodge; got by Grand Duke 21st (34061), dam (Blush
2nd) by Volunteer (30238), &c.   See Vol. xxiii. p. 587.

1878, May 19, roan, B.C.	Spot	Grand Duke 29th, 38372	Mr. Attwater
1879, Mar. 11, r. & w., B.C.	(dead)	Duke Furbelow, 38116	do.
Spot, sold to Mr. Sargent, Amesbury, Wilts.			

CHANCE 12TH, roan, calved February 5, 1877.   Bred by Mr. T. Hands,
Canley; got by Cæsarewitch 4th (33263), dam (Chance 8th) by Marmaduke
(29280), g. d. (Chance 6th) by Prince Pearl 5th (24855), gr. g. d. (Chance 4th)
by Prince Pearl 3rd (20599), &c.   See "Reindeer," p. 210.

1879, Mar. 18, roan, B.C.	Canley	D. Charng. L'd 8th, 41367	Mr. Attwater

CHANCE 13TH, roan, calved February 5, 1877.   Bred by Mr. T. Hands,
Canley; got by Cæsarewitch 4th (33263), dam (Chance 8th) by Marmaduke
(29280), &c.

1879, May 14, red, C.C.	Chance 15th	Duke Furbelow, 38116	Mr. Attwater

CORYPHEE 1ST, red and white, calved August 21, 1874.   Bred by Mr. H.
Harding, West Stour; got by Lord Lancaster (31682), dam (Coryphee) by
Count Bickerstaffe 2nd (25838), &c.   See "Earl Beaconsfield," Vol. xxiii.
p. 94.

1877, Mar. 13, red, B.C.	(dead)	Earl Fantail, 33775	Mr. Attwater
1878, Feb. 12, roan, C.C.	Coryphee 3rd	Duke of Ozleworth, 31008	do.
1879, Jan. 16, red, B.C.	(dead)	Lord Louis, 38620	do.

CORYPHEE 2ND, roan, calved July 24, 1875. Bred by Mr. H. Harding, West Stour; got by Lord Lancaster (31682), dam (Coryphee) by Count Bickerstaffe 2nd (25838), &c.

Produce in	Names, &c.	By what Bull.	By whom bred.
1878, Mar. 6, r. & w., C.C.	Coryphee 4th	Lord Louis, 38620	Mr. Attwater
1879, Feb. 6, r. & w., B.C.	(dead)	do.	do.

FLORENTIA, red and white, calved February 1, 1875. Bred by Mr. J. G. Attwater; got by Duke of Ozleworth (31008), dam (Roan Florentia) by Fifth Lord Wild Eyes (26762), &c. See Vol. xxiv. p. 311.

1877, Feb. 20, r. & w., B.C.	(dead)	Irregular, 38451	Mr. Attwater
1878, Feb. 15, roan, B.C.	(slaughtered)	Duke of Ozleworth, 31008	do.
1879, Apr. 16, r. & w., C.C.	Charming Florentia	D. Charmg. L'd 8th,41367	do.

LADY BEAUJOLAIS 4TH, red, calved October 1, 1876. Bred by Mr. R. E. Oliver, Sholebroke Lodge; got by Grand Duke 21st (34061), dam (Lady Beaujolais 2nd) by Lord Darlington (26633), &c. See Vol. xxiii. p. 588.

1879, Dec. 17, r. & w., C.C.	Lady Beaujolais 6th	Lord Garland 6th, 41864	Mr. Attwater

LAVENDER 10TH, roan, calved April 7, 1876. Bred by Mr. T. Hands, Canley; got by Cæsarewitch 4th (33263), dam (Lavender 6th) by Canley Duke (23506), &c. See Vol. xxiii. p. 483.

1878, Nov. 18, r. & w., C.C.	Lavender 13th	Lord Louis, 38620	Mr. Attwater
1879, Oct. 29, roan, C.C.	Lavender 14th	Duke Furbelow, 38116	do.

MISS OXFORD 4TH, roan, calved April 12, 1876. Bred by Mr. J. G. Attwater; got by Augustus (32971), dam (Miss Oxford 2nd) by Don Fawsley (28329), g. d. (Miss Oxford) by Beau of Oxford (21254), &c. See Vol. xxi. p. 596.

1879, Mar. 9, r. & w., C.C.	Miss Oxford 6th	D. Charmg. L'd 8th, 41367	Mr. Attwater

MISS PEARL 18TH, roan, calved in May 1872. Bred by Mr. T. Hands, Canley; got by Second Duke of Darlington (28378), dam (Miss Pearl 3rd) by White Prince (21102), &c. See Vol. xxiii. p. 483.

1878, Feb. 15, r. & w., C.C.	Miss Pearl 35th	Cæsarewitch 4th, 33263	Mr. Attwater
1879, April 13, roan, C.C.	Miss Pearl 36th,	Lord Louis, 38620	do.

MOSS ROSE 2ND, roan, calved February 6, 1874. Bred by Mr. J. G. Attwater; got by Duke of Ozleworth (31008), dam (Moss Rose) by James 1st (24202), g. d. (Ada) by Speculation (20882), &c. See Vol. xix. p. 385.

1876, Mar. 4, roan, B.C.	(dead)	Duke of Ozleworth, 31008	Mr. Attwater
1877, April 1, white, B.C.	(Steer	do.	do.
1878, Mar. 31, white, C.C.	Moss Rose 4th	Lord Louis, 38620	do.
1879, April 24, roan, C.C.	Moss Rose 6th	do.	do.

MOSS ROSE 3RD, roan, calved January 22, 1875. Bred by Mr. J. G. Attwater; got by Duke of Ozleworth (31008), dam (Moss Rose) by James 1st (24202), &c.

1877, Mar. 22, roan, B.C.	(killed)	Irregular, 38451	Mr. Attwater
1878, Feb. 25, roan, B.C.	Avon	Duke of Ozleworth, 31008	do.
1879, Feb. 7, red, C.C.	Moss Rose 5th	Duke Furbelow, 38116	do.

Avon, sold to Mr. J. Allsop. jun., West Welton, Romsey.

PHILLIS 4TH, red and white, calved August 5, 1876, Vol. xxv. p. 325. Bred by Mr. J. W. Larking, Ashdown House; got by Boston 4th (37876), dam (Phillis) by Field Marshal (26149), &c.

1879, Nov. 10, r. & w., C.C.	Phillis 6th	D. Charmg. L'd 8th, 41367	Mr. Attwater

POPPY 5TH, red, calved March 7, 1877. Bred by Mr. J. G. Attwater; got by Duke of Ozleworth (31008), dam (Poppy 2nd) by Florist (23962), &c. See Vol. xxiv. p. 311.

Produce in	Names, &c.	By what Bull.	By whom bred.
1879, Dec. 19, r.&w., C.C.	Poppy 7th	Duke Furbelow, 38116	Mr. Attwater

ROYAL ROSE 2ND, red, calved February 21, 1876. Bred by Mr. J. G. Attwater; got by Duke of Ozleworth (31008), dam (Royal Rose) by King Charles (24240), g. d. (Cherry Rose 2nd) by Royal Hamlet (18769), &c. See Vol. xix. p. 443.

1878, April 7, r.&w., B.C.	Royalist	Lord Louis, 38620	Mr. Attwater
1879, Mar. 24, red, C.C.	Royal Rose 4th	Duke Furbelow, 38116	do.

Royalist, sold to Mr. J. Allsop, sen., Welton, Romsey.

SONGSTRESS 4TH, red and white, calved April 25, 1876. Bred by Mr. J. G. Attwater; got by Duke of Ozleworth (31008), dam (Songstress 3rd) by Duke of Ozleworth (31008), &c. See Vol. xxiv. p. 311.

1878, April 1, roan, C.C.	Songstress 6th	Lord Louis, 38620	Mr. Attwater
1879, Mar. 18, r.&w., C.C.	Songstress 7th	Duke Furbelow, 38116	do.

AYLMER, Hugh,
West Dereham Abbey, Stoke Ferry, Norfolk.

BALMFUL, red and white, calved February 16, 1874, Vols. xxiv. and xxv. pp. 312, 325. Bred by Mr. H. Aylmer; got by High Sheriff (26392), dam (Banter) by Cerdic (19415), &c.

1879, Sept. 13, red, C.C.	Bijou	Sir Wilfrid, 37484	Mr. Aylmer

BEAUTIFUL STAR, roan, calved March 26, 1877. Bred by Mr. H. Aylmer; got by Hyperion (34196), dam (Balmful) by High Sheriff (26392), &c.

1879, Sept. 9, red, C.C.	Beta	Sir Wilfrid, 37484	Mr. Aylmer

CASSANDRA, red, calved March 3, 1875, Vol. xxiv. p. 312. Bred by Mr. H. Aylmer; got by Royal Monk (35392), dam (Celeste) by Royal Broughton (27352), &c.

1879, Feb. 2, red, C.C.	Ceres	Sir Wilfrid, 37484	Mr. Aylmer

CASTANET 4TH, roan, calved March 17, 1874. Bred by Mr. H. Aylmer; got by High Sheriff (26392), dam (Castanet 3rd) by Royal Knight (25032), &c. See "Sir Roderic," p. 235.

1879, Feb. 22, r.&w., B.C.	Sir Roderic	Sir Wilfrid, 37484	Mr. Aylmer

Sir Roderic, sold to Mr. R. McDougall, Arundel, Keilor, Australia.

CHRISTINA, red and white, calved January 19, 1868, Vols. xx., xxiv., and xxv. pp. 446, 312, 326. Bred by Mr. H. Aylmer; got by Prince Christian (22581), dam (Calendula) by Majestic (13279), &c.

1879, Sept. 3, r.&w., C.C.	Cheerful	Sir Wilfrid, 37484	Mr. Aylmer

CLEMATIS, red, calved September 26, 1877. Bred by Mr. H. Aylmer; got by Sir Wilfred (37484), dam (Christina) by Prince Christian (22581), &c. See Vol. xxiv. p. 312.

1879, Nov. 9, r.&w., C.C.	Cyclamen	Sir Simeon, 42412	Mr. Aylmer

CLEOPATRA, red, calved February 7, 1876, Vol. xxv. p. 326. Bred by Mr. H. Aylmer; got by High Sheriff (26392), dam (Cinderella) by Prince Christian (22581), &c.

Produce in		Names, &c.	By what Bull.	By whom bred.
1879, Aug. 8, red,	C.C.	Chloe	Sir Wilfrid, 37484	Mr. Aylmer

EASTTHORPE STRAWBERRY 9TH, roan, calved June 19, 1875, Vol. xxv. p. 326. Bred by Mr. H. Aylmer; got by Royal Magnate (32398), dam (Eastthorpe Strawberry 5th) by British Crown (21322), &c.

Produce in		Names, &c.	By what Bull.	By whom bred.
1879, Oct. 31, roan,	C.C.	East'rp.Strwbry12th	Sir Wilfrid, 37484	Mr. Aylmer

FLORA BELL, roan, calved November 28, 1876. Bred by Mr. H. Aylmer; got by Royal Commander (29857), dam (Fleurette) by Royal Prince (32404), &c. See Vol. xxiii. p. 319.

Produce in		Names, &c.	By what Bull.	By whom bred.
1879, June 7, red,	C.C.	Fides	Sir Wilfrid, 37484	Mr. Aylmer

GAZELLE, roan, calved November 3, 1875. Bred by Mr. H. Aylmer; got by High Sheriff (26392), dam (Graceful 2nd) by Ravenspur (20628), &c. See Vol. xxii. p. 304.

Produce in		Names, &c.	By what Bull.	By whom bred.
1879, Feb. 28, roan,	C.C.	Greta	Sir Wilfrid, 37484	Mr. Aylmer

GERTRUDE, red, calved November 20, 1876. Bred by Mr. H. Aylmer; got by Royal Commander (29857), dam (Graceful 2nd) by Ravenspur (20628), &c.

Produce in		Names, &c.	By what Bull.	By whom bred.
1879, Sept. 4, roan,	C.C.	Gaiety	Sir Wilfrid, 37484	Mr. Aylmer

GOLDEN FEATHER, roan, calved January 2, 1873, Vol. xxiii. p. 319. Bred by Mr. H. Aylmer; got by Royal Prince (32404), dam (Golden Hope) by Royal Broughton (27352), &c.

Produce in		Names, &c.	By what Bull.	By whom bred.
1879, Mar. 1, red,	C.C.	Golden Belle	Sir Wilfrid, 37484	Mr. Aylmer

GOLDEN HOPE, white, calved October 10, 1870, Vols. xxi., xxii., and xxiv. pp. 553, 304, 312. Bred by Mr. H. Aylmer; got by Royal Broughton (27352), dam (Golden Drop) by Prince Christian (22581), &c.

Produce in		Names, &c.	By what Bull.	By whom bred.
1879, Dec. 24, roan,	B.C.	Sir Rowland	Sir Wilfrid, 37484	Mr. Aylmer

KILLERBY QUEEN 5TH, roan, calved October 20, 1873, Vols. xxiii. and xxv. pp. 319, 326. Bred by Mr. H. Aylmer; got by Royal Broughton (27352), dam (Killerby Queen 4th) by Royal Buckingham (20718), &c.

Produce in		Names, &c.	By what Bull.	By whom bred.
1879, Mar. 6, r. & w.,	C.C.	Killerby Maid	Sir Wilfrid, 37484	Mr. Aylmer

LADY LEIGH, roan, calved March 12, 1872, Vols. xxiv. and xxv. pp. 313, 326. Bred by Mr. H. Aylmer; got by Royal Broughton (27352), dam (Lady Leonore) by Sir Samuel (15302), &c.

Produce in		Names, &c.	By what Bull.	By whom bred.
1879, Feb. 10, white,	C.C.	Lady Victoria	Royal Commander, 29857	Mr. Aylmer

MISTRESS MAGGIE, roan, calved May 12, 1877. Bred by Mr. H. Aylmer; got by Royal Commander (29857), dam (Mistress Mary) by Royal Broughton (27352), &c. See "Sir Michael," p. 234.

Produce in		Names, &c.	By what Bull.	By whom bred.
1879, Aug. 30, red.	C.C.	Mistress Muriel	Sir Wilfrid, 37484	Mr. Aylmer

ROSELEAF 7TH, red and white, calved August 19, 1875, Vol. xxiv. p. 313. Bred by Mr. H. Aylmer; got by High Sheriff (26392), dam (Roseleaf 4th) by Prince Christian (22581), &c.

Produce in		Names, &c.	By what Bull.	By whom bred.
1879, June 19, roan,	C.C.	Roseleaf 9th	Heir of Dereham, 38417	Mr. Aylmer

STRAWBERRY DUCHESS, red and white, calved November 10, 1867, Vols. xix., xxi., and xxv. pp. 740, 554, 327. Bred by Mr. H. Aylmer; got by Prince Christian (22581), dam (Windsor Strawberry) by Imperial Buckingham (19998), &c.

Produce in		Names, &c.		By what Bull.		By whom bred.
1879, Aug. 17, red,	C.C.	Strawberry D'ss 12th	Sir Wilfrid, 37484			Mr. Aylmer

STRAWBERRY DUCHESS 5TH, red, calved March 22, 1875, Vol. xxv. p. 327. Bred by Mr. H. Aylmer; got by High Sheriff (26392), dam (Strawberry Duchess 3rd) by Royal Broughton (27352), &c.

Produce in		Names, &c.		By what Bull.		By whom bred.
1879, Dec. 22, red,	B.C.	Sir Colin	Sir Wilfrid, 37484			Mr. Aylmer

STRAWBERRY DUCHESS 7TH, roan, calved February 27, 1876. Bred by Mr. H. Aylmer; got by High Sheriff (26392), dam (Strawberry Duchess 3rd) by Royal Broughton (27352), &c.

Produce in		Names, &c.		By what Bull.		By whom bred.
1878, July 17, red,	B.C.	(dead)	Sir Wilfrid, 37484			Mr. Aylmer
1879, July 25, roan,	C.C.	Strawberry D'ss 11th	do.			do.

AYLMER, J. B.,
Fincham Hall, Downham, Norfolk.

FAME, roan, calved June 23, 1874. Bred by Mr. J. B. Aylmer; got by Zeno (32897), dam (Fatima) by Skipper (20854), &c. See Vol. xxiii. p. 320.

Produce in		Names, &c.		By what Bull.		By whom bred.
1877, Oct. 15, roan,	C.C.	Floss	Comus, 33423			Mr. Aylmer
1878, Dec. 4, white,	C.C.	Fleecy	Prince of Kno'lmere, 37257			do.

FATIMA, red and white, calved April 9, 1865, Vols. xviii. and xxiii. pp. 490, 320. Bred by Mr. J. B. Aylmer; got by Skipper (20854), dam (Floss) by Newton (20403), &c.

Produce in		Names, &c.		By what Bull.		By whom bred.
1878, Feb. 15, r. & w.,	C.C.	Frill	Prince of Kno'lmere, 37257			Mr. Aylmer

FLORAMOUR, white, calved September 10, 1869. Bred by Mr. J. Gamble, Shouldham Thorpe; got by Zealot (25480), dam (Fantail) by Felix (19734), g. d. (Fame) by Second Duke of Thorndale (17748), &c. See Vol. xviii. p. 484.

Produce in		Names, &c.		By what Bull.		By whom bred.
1876, April 21, roan,	C.C.	Coquette	Comus, 33423			Mr. Aylmer

FLOWER-DE-LUCE, white, calved March 30, 1865, Vols. xviii. and xxiii. pp. 501, 321. Bred by Mr. J. B. Aylmer; got by Skipper (20854), dam (Florence) by Plato (18552), &c.

Produce in		Names, &c.		By what Bull.		By whom bred.
1877, Mar. 10, roan,	C.C.	Folly	Prince of Kno'lmere, 37257			Mr. Aylmer

FREDERICA, red and white, calved August 29, 1872. Bred by Mr. J. B. Aylmer; got by Counsellor (28256), dam (Feodorowna) by Zaratan (21134), &c. See Vol. xxiii. p. 321.

Produce in		Names, &c.		By what Bull.		By whom bred.
1877, Jan. 1, r. & w.,	C.C.	Flutter	Comus, 33423			Mr. Aylmer

SATIN, red and white, calved July 25, 1874. Bred by Mr. J. B. Aylmer; got by Zeno (32897), dam (Seraphin) by Zaratan (21134), &c. See Vol. xxiii. p. 321.

Produce in		Names, &c.		By what Bull.		By whom bred.
1877, Sept. 1, roan,	C.C.	Superb	Prince of Kno'lmere, 37257			Mr. Aylmer
1878, Aug. 11, roan,	C.C.	Sukey	do.			do.

SEWING GIRL, roan, calved September 1, 1867, Vol. xxiii. p. 321. Bred by Mr. J. B. Aylmer; got by Zaratan (21134), dam (Seamstress) by Skipper (20854), &c.

Produce in		Names, &c.		By what Bull.		By whom bred.
1877, Dec. 24, roan,	C.C.	Spinster	Prince of Kno'lmere, 37257			Mr. Aylmer

SINCERITY, roan, calved January 6, 1869, Vol. xxi. p. 554. Bred by Mr. J.
B. Aylmer; got by Duke of Edinburgh (23741), dam (Snowdrop) by Plato
(18552), &c.

Produce in	Names, &c.	By what Bull.	By whom bred.
1878, Mar. 15, roan, C.C.	Simplicity	Prince of Kno'lmere, 37257	Mr. Aylmer

SWEET ABBESS, white, calved December 23, 1875. Bred by Mr. H. W.
Martin, Littleport; got by Royal Abbot (35332), dam (Sweet Lady) by Lord
Abbot (29052), &c. See Vol. xxii. p. 502.

Produce in	Names, &c.	By what Bull.	By whom bred.
1878, Nov. 15, white, B.C.	The Monk	Prince of Kno'lmere, 37257	Mr. Aylmer
1879, Oct. 16, white, C.C.	Sweet Alice	do.	do.

BAILEY, Stephen,
Hornshay, Nynehead, Wellington, Somerset.

BABY 3RD, roan, calved January 14, 1876, Vol. xxv. p. 327. Bred by Mr. S.
Bailey; got by Cardinal (28144), dam (Baby) by Augustus Windsor (19248),
&c.

Produce in	Names, &c.	By what Bull.	By whom bred.
1879, Dec. 9, r. & w., C.C.	Baby 7th	D. of Wel'ngton 10th, 36554	Mr. Bailey

BRIDE ELECT, roan, calved March 23, 1873, Vols. xxiii., xxiv., and xxv.
pp. 322, 314, 328. Bred by Mr. J. S. Bult, Dodhill House; got by Cardinal
(28144), dam (Bride) by Conqueror (21466), &c.

Produce in	Names, &c.	By what Bull.	By whom bred.
1879, Jan. 22, r. & w., B.C.	(Steer)	Birthday, 36245	Mr. Bailey
1879, Dec. 26, r. & w., C.C.	Hornshay D'ss 7th	D. of Wel'ngton 10th, 36554	do.

DUCHESS OF LANCASTER 16TH, roan, calved January 20, 1876. Bred by
Mr. S. Bailey; got by The Claimant (36085), dam (Duchess of Lancaster 11th)
by Duke of Geneva (19614), &c. See Vol. xxi. p. 555.

Produce in	Names, &c.	By what Bull.	By whom bred.
1879, May 9, r. & w., C.C.	D'ss of L'ncaster 18th	D. of Wel'ngton 10th, 36554	Mr. Bailey

HOLOFERNES BETHULIA, red and white, calved March 31, 1875, Vol. xxv.
p. 328. Bred by Mr. S. Bailey; got by Cardinal (28144), dam (Judith) by
Cherry Duke (25752), &c.

Produce in	Names, &c.	By what Bull.	By whom bred.
1879, May 20, r. & w., C.C.	Judith 4th	D. of Wel'ngton 10th, 36554	Mr. Bailey

HORNSHAY DUCHESS 2ND, roan, calved January 24, 1877. Bred by Mr. S.
Bailey; got by Third Duke of Waterloo (23801), dam (Bride Elect) by
Cardinal (28144), &c. See Vol. xxiv. p. 314.

Produce in	Names, &c.	By what Bull.	By whom bred.
1879, Oct. 16, r. & w., C.C.	Hornshay D'ss 6th	D. of Wel'ngton 10th, 36554	Mr. Bailey

JUDITH, red, calved July 9, 1869, Vols. xx., xxii., and xxiv. pp. 574, 307, 314.
Bred by Lord Penrhyn, Wicken Park; got by Cherry Duke (25752), dam
(Jewess) by Second Duke of Geneva (21591), &c.

Produce in	Names, &c.	By what Bull.	By whom bred.
1879, May 3, r. & w., C.C.	Judith 3rd	D. of Wel'ngton 10th, 36554	Mr. Bailey

BAILLIE, Evan,
Dochfour, Inverness, N.B.

GOLDEN LINK, red and white, calved June 14, 1875. Bred by Mr. E. Baillie;
got by Flower of the Forest (33948), dam (Golden Locket) by No Mistake
(31918), g. d. (Golden Rose) by Scotch Rose (25099), &c. See Vol. xxi. p. 555.

Produce in	Names, &c.	By what Bull.	By whom bred.
1878, Aug. 16, roan, C.C.	Golden Lace	Abbot of Windsor, 32903	Mr. Baillie

GOLDEN LOCKET, roan, calved March 10, 1873. Bred by Mr. E. Baillie; got by No Mistake (34918), dam (Golden Rose) by Scotch Rose (25099), &c. See Vol. xxi. p. 555.

Produce in	Names, &c.	By what Bull.	By whom bred.
1875, June 14,r. & w., C.C.	Golden Link	Flower of the Forest, 33948	Mr. Baillie
1876, Mar. 29, roan, B.C.	Golden Hope	do.	do.
1878, Mar. 7, white, B.C.	Snow Wreath	Abbot of Windsor, 32903	do.
1879, April 14, roan, C.C.	Golden Lily	do.	do.

Golden Hope, sold to Mr. McLean, Groam, Inverness; Snow Wreath, to Mr. J. Haggart, Leys, Inverness.

GOLDEN ROSE, red and white, calved April 1, 1869, Vol. xxi. p. 555. Bred by Mr. A. Cruickshank, Sittyton; got by Scotch Rose (25099), dam (Golden Reed) by Baronet (15614), &c.

1878, May 18, red,	C.C.	Golden Gazelle	Abbot of Windsor, 32903	Mr. Baillie
1879, April 13, red,	C.C.	Golden Windsor	do.	do.

LADY MARGARET, red, calved August 30, 1876. Bred by Mr. A. Robotham, Drayton Bassett; got by Red Prince (35240), dam (Annie 2nd) by Count Bickerstaffe (23630), &c. See Vol. xxi. p. 911.

1879, Mar. 5, roan,	C.C.	Belle of Albion	Abbot of Windsor, 32903	Mr. Baillie

MARIGOLD 13TH, red and white, calved March 26, 1872, Vol. xxi. p. 555. Bred by Mr. W. S. Marr, Uppermill; got by Gold Digger (24044), dam (Marigold 6th) by Young Pacha (24057), &c.

1878, Mar. 31, red,	C.C.	Marigold 20th	Abbot of Windsor, 32903	Mr. Baillie
1879, April 18, red,	C.C.	Windsor Gold	do.	do.

NECTAR 22ND, roan, calved April 14, 1875. Bred by Mr. W. S. Marr, Uppermill; got by Livingstone (34465), dam (Nectar 12th) by Prince Louis (27158), g. d. (Nectar) by Jemmy (11611), &c. See Vol. xviii. p. 638.

1878, Mar. 20, red,	B.C.	Shareholder	Abbot of Windsor, 32903	Mr. Baillie
1879, April 7, roan,	C.C.	Sugar Plum	do.	do.

PRETTY GOLD, red and white, calved December 28, 1875. Bred by Mr. E. Baillie; got by Flower of the Forest (33948), dam (Marigold 13th) by Gold Digger (24044), &c.

1878, Jan. 13, r. & w.,	C.C.	Sweet Pea	Oliver Cromwell, 43706	Mr. Baillie
1879, April 10, roan,	B.C.	Dochfour	Abbot of Windsor, 32903	do.

RED ROCKET, red, calved April 20, 1875. Bred by Mr. E. Baillie; got by Golden Glove (34047), dam (Rocket) by Macduff (26773), &c. See Vol. xxi. p. 555.

1877, April 8, roan,	C.C.	Game Rocket	Game Bird, 38326	Mr. Baillie
1878, April 3, red,	C.C.	Sweet Rocket	Abbot of Windsor, 32903	do.
1879, April 17, red,	C.C.	Lady Montague	do.	do.

BAINTON, Thomas,
Arram Hall, Hull.

EVA 2ND, white, calved August 1, 1871, Vols. xxi., xxii., and xxiv. pp. 951, 587, 671. Bred by Mr. W. Clapham, Hatfield Magna; got by Royal Prince (29874), dam (Emma) by Second Prince of Waterloo (22636), &c.

1878, Mar. 13, white, C.C.	Lily	Oneida Prince, 34948	Mr. Bainton
1879, June 11, white, C.C.	Edith	Oneida Duke, 43708	do.

LADY CLEVELAND 2ND, red, calved September 24, 1876. Bred by Mr. T.
Feetenby, Stonegrave; got by Major (34729), dam (Lady Cleveland) by Duke
of Cleveland (33599), &c. See Vol. xxiv. p. 638.

Produce in		Names, &c.	By what Bull.	By whom bred.
1879, April 16, red,	B.C.	Rufus	Oneida Duke, 43708	Mr. Bainton

OXFORD'S PRIDE, white, calved August 2, 1876. Bred by Mr. J. Crust,
Catwick, the property of Mr. T. Bainton; got by Aughton's Oxford 2nd
(32969), dam (Waterloo's Pride) by Netta (31970), &c. See Vol. xxiii. p. 411.

Produce in		Names, &c.	By what Bull.	By whom bred.
1879, Mar. 30, white,	C.C.	Snowdrop	Wild Oxford, 40926	Mr. Crust

ROSEMARY, roan, calved November 2, 1876. Bred by Mr. T. Feetenby,
Stonegrave; got by Major (34729), dam (Rosebud) by Surplice (32636), g. d.
(Moss Rose 2nd) by Pride of England (32083), &c. See Vol. xxi. p. 703.

Produce in		Names, &c.	By what Bull.	By whom bred.
1879, Mar. 19, roan,	B.C.	Leo	Oneida Duke, 43708	Mr. Bainton

STRAWBERRY 3RD, roan, calved February 3, 1877. Bred by Mr. W. H.
Carter, Brockholme; got by Snowstorm 2nd (35619), dam (Strawberry 2nd)
by Netta (31970), g. d. (Strawberry) by Star 2nd (13786), &c. See "Cherry
Countess 2nd," Vol. xxiii. p. 667.

Produce in		Names, &c.	By what Bull.	By whom bred.
1879, May 20, roan,	B.C.	Wellington	Oneida Duke, 43708	Mr. Bainton

BAIRD, Alexander,
Robeston Hall, Milford Haven.

ADA, red, calved August 16, 1871, Vol. xxiii. p. 323. Bred by Mr. R. P.
Davies, Horton; got by Duke of Kirklevington (25982), dam (Adeliza 2nd)
by Second Duke of Wetherby (21618), &c.

Produce in		Names, &c.	By what Bull.	By whom bred.
1877, Mar. 8, roan,	C.C.	Adeliza 4th	Earl of Horton 3rd, 33796	Mr. Baird
1878, Mar. 25, roan,	B.C.	Ameer of Afghan	do.	do.
1879, April 4, r. & w.,	B.C.	Earl of Pembroke	Earl of Horton 5th, 33797	do.

Ameer of Afghan, sold to Mr. T. Joseph, Broomhill, Taibach.

HONEY FLOWER 5TH, white, calved May 26, 1875. Bred by Mr. E.
Vaughan, Fernhill; got by Earl of Robeston 2nd (33811), dam (Honey
Flower 4th) by Second Duke of Claro (21576), g. d. (Honey Flower) by
Grand Duke of Oxford (16184), &c. See Vol. xix. p. 547.

Produce in		Names, &c.	By what Bull.	By whom bred.
1879, Dec. 29, roan,	C.C.	Honey Flower 6th	Earl of Horton 5th, 33797	Mr. Baird

PATTY, red, calved May 13, 1875. Bred by Mr. A. Baird, the property of
Mr. C. E. G. Philipps, Picton Castle; got by Earl of Horton 3rd (33796),
dam (Polly) by Sea Serpent (25107), &c. See "Plato," p. 187.

Produce in		Names, &c.	By what Bull.	By whom bred.
1877, July 5, roan,	B.C.	Plato	Alfred, 36119	Mr. Baird
1878, May 29, red,	C.C.	Phœbe	Earl of Horton 5th, 33797	do.

Plato, sold to Mr. W. H. Shield, Gilfach, Narberth; Phœbe, to Mr. R. Ward, Sodstone,
Narberth.

BAIRD, Sir David, Bart.,
Newbyth, Prestonkirk, N.B.

LIVELY, red, calved April 16, 1875. Bred by Sir D. Baird, Bart.; got by
Baron Laurie 3rd (30467), dam (Lace) by Hardicanute (26338), g. d. (Lovely)
by George 1st (19848), gr. g. d. (Luna) by Tenedos (20936), — (Lavender)
by California (10018), — (Lucy) by The Sheriff (12216), — (Larkspur) by Sir
John, — (Lucy) by Hambletonian (9179), &c. See Vol. xi. p. 555.

Produce in		Names, &c.	By what Bull.	By whom bred.
1879, May 20, red,	B.C.	Forlorn Hope	Skirmisher, 40729	Sir D. Baird, Bart.

RONA, red, calved April 18, 1875. Bred by Sir D. Baird, Bart. ; got by Baron
Laurie 3rd (30467), dam (Rosebud) by Hardicanute (26338), g. d. (Reckless)
by George 1st (19848), gr. g. d. (Racket) by Tenedos (20936), &c. See
"Rhoda," Vol. xxi. p. 826.

Produce in		Names, &c.	By what Bull.	By whom bred.
1878, Feb. 11, roan,	B.C.	Rearguard	Skirmisher, 40729	Sir D. Baird, Bart.
1879, Mar. 12, red,	B.C.	Grenadier	do.	do.

Rearguard, sold to Mr. A. Smith, Stevenson Mains, Haddington.

BAKER, Thomas,
Harpole, Weedon.

EIGHTH DUCHESS OF OXFORD, red and white, calved August 12, 1869,
Vols. xx., xxii., and xxiii. pp. 492, 548, 323. Bred by Mr. T. Robinson,
Burton-on-Trent; got by Garibaldi 4th (24009), dam (Seventh Duchess of
Oxford) by Royal Master Butterfly (22789), &c.

Produce in		Names	By what Bull	By whom bred
1878, Jan. 9, r. & w.,	B.C.	Duke of Oxford 2nd	Ducal Knight, 33552	Mr. Baker
1879, Apr. 28, r. & w.,	C.C.	Duch's of Oxford 12th	Duke of Oxford, 39769	do.

Duke of Oxford 2nd, sold to Viscount Clifden, Naseby Farm, Rugby.

GERANIUM, red and white, calved April 30, 1869, Vol. xxi. p. 617. Bred by
Mr. S. Canning, Snitterfield; got by Standard Bearer (25221), dam (Gwendoline) by Old Buck (15017), &c.

Produce in		Names	By what Bull	By whom bred
1879, April 19, red,	B.C.	Sir B'gton Knightley	D. of Barrington 6th, 33576	Mr. Baker

TRUTH, red and white, calved April 18, 1871, Vol. xxi. p. 769. Bred by Archdeacon Holbech, Farnborough Hall; got by Second Earl of Cleveland (26054),
dam (Goodness) by Lord Hardinge (13193), &c.

Produce in		Names	By what Bull	By whom bred
1879, Mar. 10, r. & w.,	C.C.	Alexandra	Duke of Oxford 32nd, 36527	Mr. Baker

BALFOUR, Colonel,
Balfour Castle, Kirkwall, N.B.

EFFIE DEANS, roan, calved April 19, 1868, Vols. xx., xxii., and xxiv. pp. 497,
308, 316. Bred by Colonel Balfour; got by Sir Burdett Coutts (25131), dam
(Christie Steele) by Artizan (15593), &c.

Produce in		Names	By what Bull	By whom bred
1878, Mar. 6, roan,	C.C.	Old Mortality	Laudable, 31587	Colonel Balfour
1879, Feb. 14, roan,	C.C.	Peggy	do.	do.

FENELLA, white, calved February 22, 1869, Vols. xx. and xxii. pp. 519, 309.
Bred by Colonel Balfour; got by Sir Burdett Coutts (25131), dam (Beenie)
by Artizan (15593), &c.

Produce in		Names	By what Bull	By whom bred
1879, Mar. 28, roan,	C.C.	Oyesse	Shuttleworth, 35523	Colonel Balfour

GENUBATH, roan, calved April 20, 1870, Vols. xxi., xxii., and xxiv. pp. 556,
309, 316. Bred by Colonel Balfour; got by Sir Burdett Coutts (25131), dam
(Miriam) by Aaron (15540), &c.

Produce in		Names	By what Bull	By whom bred
1878, Mar. 20, red,	C.C.	Omicron	Shuttleworth, 35523	Colonel Balfour

GINATH, red and white, calved April 1, 1870, Vols. xxii. and xxiv. pp. 309, 316.
Bred by Colonel Balfour; got by Sir Burdett Coutts (25131), dam (Rachel)
by Artizan (15593), &c.

Produce in		Names	By what Bull	By whom bred
1878, Feb. 27, red,	C.C.	Orpah	Laudable, 31587	Colonel Balfour
1879, Jan. 11, red,	B.C.	Pilate	do.	do.

GRIZEL OLDBUCK, red, calved April 14, 1870, Vols. xxi. and xxiv. pp. 557, 317. Bred by Colonel Balfour; got by Sir Burdett Coutts (25131), dam (Christie Steele) by Artizan (15593), &c.

Produce in		Names, &c.	By what Bull.	By whom bred.
1878, April 10, red,	C.C.	Orra	Laudable, 31587	Colonel Balfour
1879, Mar. 13, red,	B.C.	(Steer)	do.	do.

HANNAH, white, calved February 26, 1871, Vols. xxi., xxii., and xxiv. pp. 557, 309, 317. Bred by Colonel Balfour; got by Sir Burdett Coutts (25131), dam (Mary) by Prince Frederick (18603), &c.

1878, Mar. 20, red,	C.C.	Omega	Shuttleworth, 35523	Colonel Balfour
1879, Mar. 13, roan,	B.C.	(Steer)	Laudable, 31587	do.

HENA, red and white, calved February 26, 1871, Vols. xxi. and xxiv. pp. 557, 317. Bred by Colonel Balfour; got by Sir Burdett Coutts (25131), dam (Mary) by Prince Frederick (18603), &c.

1878, May 12, roan,	C.C.	Ocina	Laudable, 31587	Colonel Balfour
1879, Feb. 10, red,	C.C.	Prisca	do.	do.

JOANNA, red, calved March 15, 1873, Vol. xxiv. p. 317. Bred by Colonel Balfour; got by Remus (29768), dam (Anna) by Artizan (15593), &c.

1878, Mar. 23, red,	C.C.	Ophir	Shuttleworth, 35523	Colonel Balfour
1879, Feb. 24, red,	C.C.	Phœbe	Liberator, 43465	do.

KEZIA, roan, calved February 20, 1874. Bred by Colonel Balfour; got by Laudable (31587), dam (Hannah) by Sir Burdett Coutts (25131), &c. See Vol. xxi. p. 557.

1877, May 5, roan,	B.C.	Nabal	Shuttleworth, 35523	Colonel Balfour
1878, Mar. 6, roan,	C.C.	Olympia	do.	do.

KILLIAN, red, calved March 9, 1874, Vol. xxiv. p. 318. Bred by Colonel Balfour; got by Laudable (31587), dam (Flora McIvor) by Sir Burdett Coutts (25131), &c.

1878, Mar. 22, red,	C.C.	Orhonarie	Shuttleworth, 35523	Colonel Balfour

KITTY, red, calved March 19, 1874, Vol. xxiv. p. 318. Bred by Colonel Balfour; got by Laudable (31587), dam (Grizel Oldbuck) by Sir Burdett Coutts (25131), &c.

1878, April 1, red,	C.C.	Oxeye	Shuttleworth, 35523	Colonel Balfour

LOUISE, roan, calved April 22, 1875. Bred by Colonel Balfour; got by Shuttleworth (35523), dam (Fenella) by Sir Burdett Coutts (25131), &c. See Vol. xxii. p. 309.

1878, Feb. 20, roan,	C.C.	Orkney	Laudable, 31587	Colonel Balfour
1879, Jan. 9, red,	C.C.	Penelope	Liberator, 43465	do.

LUCETTA, red, calved May 1, 1875. Bred by Colonel Balfour; got by Laudable (31587), dam (Coral) by Artizan (15593), &c. See Vol. xxii. p. 308.

1878, Jan. 13, red,	C.C.	Ophelia	Shuttleworth, 35523	Colonel Balfour
1879, Jan. 19, red,	B.C.	(Steer)	Liberator, 43465	do.

BARBER, Thomas,
Sproatley Rise, Hull.

BLANCHE 3RD, roan, calved March 22, 1873, Vols. xxiii. and xxiv. pp. 628, 318. Bred by Sir W. H. Salt, Bart., Maplewell; got by Marquis of York (34791), dam (Bantling) by Britannicus (17452), &c.

1878, April 15, red,	C.C.	Blanche 4th	Baron Wild Eyes, 33100	Mr. Barber

DUCHESS OF CLARENCE 11TH, roan, calved January 22, 1874, Vol. xxiii.
p. 324. Bred by Mr. T. Barber; got by Oxford's Baronet (29499), dam
(Duchess of Clarence 2nd) by Squire of Clarence (25214), &c.

Produce in		Names, &c.	By what Bull.	By whom bred.
1878, Oct. 18, red,	B.C.	Red Duke	Baron Wild Eyes, 33100	Mr. Barber
1879, Sept. 17, roan,	C.C.	D'ss of Clarence 15th	do.	do.

LADY OF LORNE, roan, calved June 4, 1870, Vols. xxi. and xxiii. pp. 559,
324. Bred by Mr. F. H. Fawkes, Farnley Hall; got by Lord Cobham (20164),
dam (Blue Belle) by Reformer (18687), &c.

Produce in		Names, &c.	By what Bull.	By whom bred.
1877, May 28, roan,	C.C.	Queen of Lorne	Baron Wild Eyes, 33100	Mr. Barber
1878, May 3, roan,	C.C.	(dead)	do.	do.
1879, Sept. 14, { roan,	C.C.	Maid of Lorne 2nd }	do.	do.
roan,	C.C.	Maid of Lorne 3rd }		

MAY DUKE'S ROSE, roan, calved December 1, 1871, Vols. xxii. and xxiv. pp.
312, 319. Bred by Mr. T. Barber; got by May Duke (31892), dam (Grand
Duke's Rose) by Grand Duke 6th (19876), &c.

Produce in		Names, &c.	By what Bull.	By whom bred.
1878, July 25, { roan,	C.C.	Wild Rose 1st }	Baron Wild Eyes, 33100	Mr. Barber
roan,	C.C.	Wild Rose 2nd }		

WATER DUCHESS 2ND, red, calved September 27, 1870, Vols. xxi. and xxiii.
pp. 560, 324. Bred by Mr. T. Barber; got by Grand Duke 13th (21850),
dam (Water Duchess) by Grand Duke 6th (19876), &c.

Produce in		Names, &c.	By what Bull.	By whom bred.
1877, May 13, red,	C.C.	Water Duchess 5th	Baron Wild Eyes, 33100	Mr. Barber
1879, Sept. 24, red,	B.C.	Waterloo Baron	do.	do.

BARCHARD, Francis,
Horsted, Uckfield, Sussex.

GRISELDA, red and white, calved September 27, 1874. Bred by Mr. J. W.
Philips, Heybridge; got by Duke of Hillhurst (28401), dam (Graceful) by Duke
of Albany (25931), &c. See Vol. xxi. p. 886.

1878, April 25, roan,	C.C.	Grisette	20th Duke of Oxford, 28432	Earl of Feversham
1879, July 15, r. & w.,	C.C.	Coral Charm	G. Sultan of Turkey, 41657	Mr. Barchard

RED ROSE OF KILLIECRANKIE, red and white, calved April 30, 1875.
Bred by the Earl of Dunmore, Dunmore, the property of Mr. R. Loder,
Whittlebury; got by Airdrie Geneva (32920), dam (Red Rose of Athole) by
Airdrie (30365), &c. See Vol. xxii. p. 405.

1879, June 19, r. & w.,	C.C.	Red Rose of Pitlochrie	Marq. of Oxford 2nd, 37055	Earl of Dunmore

Red Rose of Pitlochrie, sold to Mr. F. Barchard, Horsted.

*RED ROSE OF KILLIGRAY, red and white, calved May 25, 1873. Bred
by Mr. A. Renick, Kentucky, U.S.A.; got by Fourth Duke of Geneva
(30958), dam (Duchess 3rd) by Dandy Duke (30855), &c. See "Lord of
Killigray," p. 152.

1879, Sept. 7, red,	B.C.	Lord of Killigray	Marq. of Oxford 2nd, 37055	Mr. Barchard

BARNETT, The Misses,
Costislost, Bodmin, Cornwall.

ALEXANDRA 2ND, red, calved September 26, 1873, Vols. xxiii. and xxiv.
pp. 625, 631. Bred by Mr. J. A. Rolls, The Hendre, the property of the
Misses Barnett; got by Towneley Wild Eyes (27674), dam (Alexandra) by
Prince Henry (18608), &c.

1879, Oct. 16, r. & w.,	C.C.	Alexandra Daisy 3rd	G. Duke of Clarence, 28750	Mr. Rolls

DUCHESS OF NORTHUMBERLAND 3RD, red and white, calved December 22, 1871, Vols. xxi. and xxv. pp. 772, 330.   Bred by Messrs. J. Horswell and Sons, Burns Hall; got by Duke of Oxford and Glo'ster (28436), dam (Duchess of Northumberland 2nd) by First Earl Ducie (23814), &c.

Produce in	Names, &c.	By what Bull.	By whom bred.
1879, July 27, roan, C.C.	D'ss of Bar'ngton 4th	Tregunter, 42516	Misses Barnett

LADY FLORA, roan, calved June 19, 1874, Vols. xxiv. and xxv. pp. 649, 330. Bred by Mr. J. Singleton, Teresa Cottage; got by Beau Siddington (30515), dam (Lady April) by Sir Roger Gwynne (27496), &c.

1879, Dec. 18, roan, C.C.	Lady Flora 5th	Tregunter, 42516	Misses Barnett

BARROW, B. Langdale,
Sydnope Hall, Matlock.

CHARMING BLUSH, red and white, calved April 20, 1873, Vol. xxiii. p. 330. Bred by the Rev. W. Holt Beever, Pencraig Court, the property of Lord Fitzhardinge, Berkeley Castle; got by Sir John (35572), dam (Blush) by Prince Henry (18608), &c.

1879, Jan. 21, r. & w., C.C.	Clarence Blush	G. Duke of Clarence, 28750	Rev. W. H. Beever

Clarence Blush, sold to Mr. B. L. Barrow.

EARLY MORN, red and white, calved September 30, 1874.   Bred by the Rev. W. Holt Beever, Pencraig Court; got by Mountain Dew (34879), dam (Jessica) by The Baron (25277), &c.   See Vol. xxi. p. 568.

1879, Aug. 7, white, C.C.	Dewy Morn	20th Duke of Oxford, 28432	Mr. Barrow

VIOLET, roan, calved February 14, 1877.   Bred by Archdeacon Holbech, Farnborough Hall; got by The Bursar (35742), dam (Daffodil) by Eleventh Duke of Oxford (19632), &c.   See Vol. xix. p. 465.

1879, Dec. 14, roan, C.C.	Glo'ster's Daisy	Duke of Glo'ster 6th, 39734	Mr. Barrow

BARTON, Major,
Straffan House, Straffan Station, Kildare, Ireland.

BUTTERCUP, red and white, calved May 20, 1873, Vols. xxiii. and xxv. pp. 325, 331.   Bred by Major Barton; got by Iron Duke (31420), dam (Primrose) by Flag of Kildare (23956), &c.

1879, Jan. 7, r. & w., C.C.	Daisy 2nd	Valorous, 27701	Major Barton

BUTTERCUP 2ND, roan, calved February 2, 1874.   Bred by Major Barton; got by Iron Duke (31420), dam (Primrose) by Flag of Kildare (23956), &c. See Vol. xxii. p. 314.

1879, Jan. 12, r. & w., B.C.	Leander	Valorous, 27701	Major Barton

FAWN, white, calved February 27, 1872, Vols. xxiii. and xxv. pp. 325, 331. Bred by Major Barton; got by Iron Duke (31420), dam (Wreath) by British Flag (19351), &c.

1879, June 7, roan, C.C.	Queen 3rd	Valorous, 27701	Major Barton

GERTY 2ND, roan, calved February 24, 1874.   Bred by Major Barton; got by Iron Duke (31420), dam (Princess of Erin) by Prince of Rosedale (24837), &c.   See Vol. xxi. p. 561.

1879, Feb. 2, roan, C.C.	Sweet Pea	Valorous, 27701	Major Barton

GERTY 3RD, roan, calved August 8, 1876. Bred by Major Barton; got by
King Charles (24240), dam (Princess of Erin) by Prince of Rosedale (24837),
&c. See Vol. xxi. p. 561.

Produce in		Names, &c.		By what Bull.		By whom bred.
1879, Mar. 25, r. & w., B.C.		Basil		Ben Ledi, 37853		Major Barton

JEMIMA, red and white, calved February 27, 1872, Vols. xxiii. and xxv. pp. 325,
331. Bred by Major Barton; got by Iron Duke (31420), dam (Princess
Alexandra) by Red Knight (18681), &c.

1879, April 8, roan,	B.C.	Jasper		Valorous, 27701		Major Barton

LILIAN, red and white, calved November 17, 1865, Vols. xix., xxii., and xxv.
pp. 601, 314, 331. Bred by Mr. N. Barton, Straffan; got by The Friar
(20954), dam (Lily of the Valley) by Master Goldsmidt (14926), &c.

1879, Jan. 1, r. & w.,	B.C.	Vesuvius		Valorous, 27701		Major Barton

LILY 2ND, roan, calved February 18, 1874, Vol. xxv. p. 332. Bred by Major
Barton; got by Iron Duke (31420), dam (Diana of Geneva) by Duke of
Geneva (19614), &c.

1879, May 19, white, C.C.		Lilywhite		Ben Ledi, 37853		Major Barton

MAGGIE, red, calved January 14, 1873, Vols. xxiii. and xxv. pp. 325, 332.
Bred by Major Barton; got by Iron Duke (31420), dam (Lizzy) by Cumberland 1st (23662), &c.

1879, April 9, red,	C.C.	Meta		Valorous, 27701		Major Barton

MAGGIE 2ND, red, calved April 9, 1874, Vol. xxiii. p. 325. Bred by Major
Barton; got by Iron Duke (31420), dam (Lizzy) by Cumberland 1st (23662),
&c.

1879, Feb. 13, red,	C.C.	Rosa		Ben Ledi, 37853		Major Barton

MILKMAID, red, calved April 24, 1873, Vol. xxv. p. 332. Bred by Major
Barton; got by First Rate (31164), dam (Moonlight) by Prince of Rosedale
(24837), &c.

1879, Mar. 21, r. & w., C.C.		Dairymaid		Valorous, 27701		Major Barton

MYSIE 21ST, white, calved February 8, 1874, Vol. xxv. p. 332. Bred by
Major Barton; got by Iron Duke (31420), dam (Mysie 20th) by Cumberland
1st (23662), &c.

1879, Mar. 20, roan,	C.C.	Mysie 24th		Red Cross, 42254		Major Barton

NORTHERN LASS, roan, calved January 9, 1873. Bred by Major Barton;
got by Iron Duke (31420), dam (Empress of the North) by Royal Duke
(25014), &c. See Vol. xxi. p. 560.

1879, Jan. 19, roan,	B.C.	Barbarossa		Valorous, 27701		Major Barton

PRINCESS OF ERIN, roan, calved March 16, 1870, Vol. xxi. p. 561. Bred
by Major Barton; got by Prince of Rosedale (24837), dam (Princess
Alexandra) by Red Knight (18681), &c.

1876, Aug. 8, roan,	C.C.	Gerty 3rd		King Charles, 24240		Major Barton
1879, Feb. 13, red,	C.C.	Gerty 4th		Valorous, 27001		do.

QUEEN, roan, calved February 20, 1875. Bred by Major Barton; got by
Lord Rose (34659), dam (Fawn) by Iron Duke (31420), &c. See Vol. xxiii.
p. 325.

1879, Mar. 26, roan,	C.C.	Royal Wreath		Ben Ledi, 37853		Major Barton

ROSE OF APRIL, roan, calved April 8, 1866, Vols. xix., xxii., and xxv.
pp. 709, 315, 332.   Bred by Mr. N. Barton, Straffan ; got by British Flag
(19351), dam (Rose of May) by Red Knight (18681), &c.

Produce in	Names, &c.	By what Bull.	By whom bred.
1879, April 24, roan, B.C.	Rainbow	Valorous, 27701	Major Barton

SNOWDROP 2ND, white, calved March 20, 1874.   Bred by Major Barton ; got
by Iron Duke (31420), dam (Duchess of Rosedale) by Prince of Rosedale
(24837), &c.   See Vol. xxii. p. 314.

1879, April 5, roan, B.C.	Opechee	Red Cross, 42254	Major Barton

STELLA 3RD, red and white, calved June 29, 1876.   Bred by Major Barton ;
got by King Charles (24240), dam (Stella) by Prince of Rosedale (24837),
&c.   See Vol. xxv. p. 333.

1879, Mar. 20, roan, C.C.	Twilight	Ben Ledi, 37853	Major Barton

TULIP 2ND, roan, calved March 20, 1874, Vol. xxv. p. 333.   Bred by Major
Barton ; got by Iron Duke (31420), dam (Tulip) by Sovereign (27538), &c.

1879, April 17, roan, C.C.	Ranunculus	Valorous, 27701	Major Barton

VICTORIA 43RD, white, calved February 11, 1872, Vols. xxi. and xxv.
pp. 561, 333.   Bred by Major Barton ; got by Iron Duke (31420), dam
(Victoria 42nd) by Flag of Kildare (23956), &c.

1879, May 30, roan, B.C.	Virgil	Valorous, 27701	Major Barton

BARTON, Richard,
Caldy Manor, Birkenhead.

NETTA 3RD, roan, calved November 12, 1875.   Bred by Mr. W. G. Garne,
Broadmoor ; got by Prince of the Heath (35161), dam (Netta) by Buccaneer (25693), &c.   See Vol. xxiii. p. 451.

1879, May 2. roan, B.C.	Gazetteer	Gazette, 39918	Mr. Barton

PICOTEE, white, calved February 22, 1869, Vols. xxii., xxiv., and xxv. pp. 315,
319, 334.   Bred by Mr. R. Barton ; got by Sir Samuel Butterfly (25172); dam
(Carnation 2nd) by Baron Belvedere (19265), &c.

1879, April 18, roan, C.C.	Picotee 6th	Gazette, 39918	Mr. Barton

PICOTEE 2ND, roan, calved February 17, 1873, Vol. xxv. p. 334.   Bred by Mr.
R. Barton ; got by Salamander (32445), dam (Picotee) by Sir Samuel Butterfly
(25172), &c.

1879, June 30, roan, C.C.	Patience	Gazette, 39918	Mr. Barton

PICOTEE 3RD, roan, calved March 15, 1875, Vol. xxv. p. 334.   Bred by Mr.
R. Barton ; got by Salamander (32445), dam (Picotee) by Sir Samuel Butterfly
(25172), &c.

1879, Sept. 1, r. & w., C.C.	(slaughtered)	Gazette, 39918	Mr. Barton

PICOTEE 4TH, roan, calved February 21, 1876.   Bred by Mr. R. Barton ; got by
Proud Preston (37286), dam (Picotee) by Sir Samuel Butterfly (25172), &c.
See Vol. xxiv. p. 319.

1879, Feb. 3, roan, B.C.	Gem of Caldy	Gazette, 39918	Mr. Barton
n of Caldy, sold to Mrs. J. Davies, Bidston Hill, Birkenhead.			

PICOTEE 5TH, roan, calved February 12, 1877. Bred by Mr. R. Barton; got
by Proud Preston (37286), dam (Picotee) by Sir Samuel Butterfly (25172),
&c. See Vol. xxiv. p. 319.

Produce in	Names, &c.	By what Bull.	By whom bred.
1879, Aug. 4, white, C.C.	(slaughtered)	Gazette, 39918	Mr. Barton

BARWISE, John,
Edderside Hall, Maryport.

LADY GWYNNE 9TH, red and white, calved May 12, 1876. Bred by Mr. A.
Thompson, The Cross; got by Romance (32328), dam (Lady Gwynne) by
Captain Gwynne (21365), &c. See "Reality," Vol. xxiv. p. 217.

1879, Mar. 6, r. & w., C.C.	L'dy Benson Gwynne	Baron Benson 4th, 36157	Mr. Barwise

BAXTER, H. P.,
Southall Green Farm, Southall, Middlesex.

CATHLEEN, roan, calved August 21, 1875, Vol. xxiv. p. 527. Bred by Mr. J.
W. Larking, Ashdown House; got by Grand Duke of Geneva (28756), dam
(Crystal 2nd) by Second Duke of Waterloo (23800), &c.

1879, Mar. 30, roan, C.C.	Cathleen of Geneva	W'd D. of C'nnaught, 40921	Mr. Baxter

DIANA, red and white, calved March 24, 1875. Bred by Her Majesty the
Queen; got by Spring Buck (32585), dam (Diadem) by Prince of Saxe Coburg
(20576), &c. See "Regal Rufus," p. 209.

1878, June 16, red, B.C.	Regal Rufus	King Rufus, 34351	Mr. Baxter
1879, Aug. 12, roan, C.C.	Rose Blossom	Rose Seal, 39023	do.

LADY CLEOPATRA, red and white, calved February 3, 1877. Bred by Mr.
H. P. Baxter; got by Duke of Rosedale 2nd (33722), dam (Lady Margaret)
by Second Duke of Collingham (23730), &c. See "Wild Cleopatra," p. 268.

1879, April 16, r. & w., C.C.	Lady Cleopatra 2nd	W'd D. of C'nnaught, 40921	Mr. Baxter

LADY MARGARET, roan, calved May 21, 1872. Bred by the Earl of Dunmore, Dunmore; got by Second Duke of Collingham (23730), dam (Cleopatra
6th) by Ninth Duke of Oxford (17738), &c. See "Wild Cleopatra," p. 268.

1877, Feb. 3, r. & w., C.C.	Lady Cleopatra	D. of Rosedale 2nd, 33722	Mr. Baxter
1878, Aug. 10, r. & w., C.C.	Cherry Cleopatra	Cherry D. of Glo'ster, 36348	do.
1879, Aug. 13, roan, B.C.	Wild Cleopatra	W'd D. of C'nnaught, 40921	do.

LOVELY, roan, calved February 20, 1873. Bred by Mr. N. Catchpole, Bramford; got by Oxford Prize (29498), dam (Cherry) by Victor 4th (30220), &c.
See "Connaught Oxford." p. 55.

1879, Mar. 30, roan, B.C.	Connaught Oxford	W'd D. of C'nnaught, 40921	Mr. Baxter

PRINCESS TECK, red, calved August 25, 1874. Bred by Mr. H. P. Baxter; got
by Duke of Kennet (30977), dam (Queen of Cambridge) by First Duke of
Oxford and Glo'ster (19637), &c. See Vol. xx. p. 712.

1877, Oct. 19, roan, C.C.	(dead)	Duke of Connaught, 33604	Mr. Baxter
1879, May 13, red, C.C.	Princess Teck 2nd	W'd D. of C'nnaught, 40921	do.

ROAN DUCHESS 2ND, roan, calved October 12, 1876. Bred by Mr. W. Ash-
burner, Conishead Grange; got by Second Duke of Glo'ster (28392), dam
(Roan Duchess) by Barrington Oxford (25607), g. d. (Red Duchess 6th) by
Oxford (20450), &c. See Vol. xix. p. 691.

1879, April 2, r. & w., B.C.	The Red Duke	W'd D. of C'nnaught, 40921	Mr. Baxter

WILD EYES 34TH, red and white, calved June 25, 1873. Bred by Mr. J. Robinson, Fell Side; got by Bramble 2nd (36259), dam (Wild Eyes 30th) by Don John (19583), &c. See "Connaught Wild Eyes," p. 55.

Produce in		Names, &c.	By what Bull.	By whom bred.
1877, Jan. 16, roan,	C.C.	Derwent Wild Eyes	Derwent Duke, 38089	Mr. Baxter
1879, Mar. 5, roan,	B.C.	C'nnaught W'd Eyes	W'd D.of C'nnaught,40921	do.

BEAK, Captain W. E.,
The Manor, Somerford, Chippenham.

BRITISH LASS, roan, calved May 18, 1877. Bred by Mr. W. Arkell, Dudgrove; got by Pompey (35059), dam (Brilliance) by Research (27270), &c. See "British Lion," p. 36.

1879, Nov. 14, white,	B.C.	British Tar	Sultan 2nd, 44106	Captain Beak

PRUDENTIA, red and white, calved February 26, 1877. Bred by Mr. H. Bettridge, East Hanney; got by Bismarck (33159), dam (Prudence) by Baron Booth 1st (27915), &c. See Vol. xxiv. p. 322.

1879, Dec. 18, r. & w.,	B.C.	Primogeniture	Burgundy, 37926	Capt. Beak

ROSE BLOSSOM 2ND, red and white, calved May 22, 1877. Bred by Mr. W. Arkell, Dudgrove; got by Roderick (35300), dam- (Rose Blossom) by Whist (30301), g. d. (Roseleaf) by Oxford Don (20451), gr. g. d. (Rosebud) by Proctor (27210), &c. See Vol. xviii. p. 701.

1879, Oct. 21, roan,	C.C.	Roseberry	Baron Lee 5th, 39420	Capt. Beak

BEATTIE, James,
Newbie House, Annan, N.B.

COMONA, roan, calved November 8, 1875. Bred by Mr. R. Jefferson, Preston Hows; got by England's Fame (31111), dam (Cressida 6th) by Cerdic (19415), &c. See "King Dick," Vol. xxi. p. 235.

1878, Feb. 1, white,	B.C.	(dead)	Warlaby Warrior, 44216	Mr. Beattie

LADY HELEN, red, calved March 6, 1876. Bred by Mr. J. Beattie; got by Red Cross Knight (35219), dam (Lady Booth) by The Sutler (23061), &c. See Vol. xx. p. 587.

1878, July 5, r. & w.,	C.C.	Lady Mary	Warlaby Warrior, 44216	Mr. Beattie
1879, Sept. 13, red,	C.C.	Lady Jane	do.	do.

LILY OF THE VALE, white, calved March 30, 1875. Bred by Mr. J. Beattie; got by Knight of Knowlmere 2nd (31542), dam (Eliza) by Mac Turk (14872), &c. See Vol. xxiv. p. 464.

1878, Aug. 15, white,	C.C.	(dead)	Warlaby Warrior, 44216	Mr. Beattie
1879, Oct. 3, white,	B.C.	Imperial	do.	do.

BEAUCHAMP, Earl,
Madresfield Court, Malvern.

LADY ADONIS, red, calved February 1, 1875. Bred by Earl Beauchamp; got by Gay Boy (31222), dam (Lady Adair) by Festival (26147), &c. See Vol. xxii. p. 317.

1878, April 26, red,	C.C.	Lady Adeline	M'quis Blandford 2d, 34779	Earl Beauchamp
1879, April 24, red,	C.C.	Lady Alice	Magnum Bonum, 41954	do.

LADY LAVENDER, roan, calved June 16, 1875. Bred by Earl Beauchamp; got by Gay Boy (31222), dam (Lavinia) by Diplomatist (19571), &c. See "Jock," p. 127.

Produce in		Names, &c.	By what Bull.	By whom bred.
1879, Aug. 29, roan,	B.C.	Hornbeam	Magnum Bonum, 41954	Earl Beauchamp

MADRESFIELD ROSE, roan, calved June 12, 1874. Bred by Earl Beauchamp; got by Festival (26147), dam (Miss Valentine) by Second Duke of Airdrie (19600), &c. See "Matchfield," Vol. xxv. p. 192.

1877, Dec. 3, red,	B.C.	Matchfield	M'quis Blandford 2d, 34779	Earl Beauchamp
1878, Nov. 21, roan,	C.C.	Madresfield Maid	do.	do.
1879, Nov. 2, red,	B.C.	Madresfield	Magnum Bonum, 41954	do.

MISS KITTY, roan, calved November 14, 1874. Bred by Earl Beauchamp; got by Gay Boy (31222), dam (Kathleen) by Battersea Gwynne (25617), &c. See "Trafalgar," p. 252.

1878, April 9, roan,	C.C.	Kitty Dale	M'quis Blandford 2d, 34779	Earl Beauchamp
1879, April 17, roan,	C.C.	Kathleen 2nd	Magnum Bonum, 41954	do.

NINEVET, red, calved January 20, 1873. Bred by Earl Beauchamp; got by Festival (26147), dam (Ninette) by Diplomatist (19571), &c. See "Nettleworth," Vol. xxiii. p. 200.

1877, June 19, red,	C.C.	Numedia 4th	M'quis Blandford 2d, 34779	Earl Beauchamp
1878, May 16, roan,	C.C.	Ninevet 2nd	do.	do.
1879, June 25, red,	C.C.	Ninette 2nd	Magnum Bonum, 41954	do.

SUNFLOWER, roan, calved March 4, 1873. Bred by Earl Beauchamp; got by Festival (26147), dam (Sunbeam) by Archduke (21185), &c. See "Clarence," p. 51.

1877, Dec. 23, roan,	C.C.	May Flower	M'quis Blandford 2d, 34779	Earl Beauchamp
1878, Nov. 21, red,	C.C.	Sunshine	do.	do.
1879, Sept. 18, red,	C.C.	Sunshade	Magnum Bonum, 41954	do.

BEAUCHAMP, E. B.,
Trevince, Scorrier, Cornwall.

DORCAS, red, calved June 2, 1867. Bred by Messrs. W. Hosken and Son, Hayle; got by Royal Oak (22793), dam (Princess Ellen) by A 1 (15538), &c. See Vol. xvii. p. 683.

1877, Mar. 3, r. & w.,	C.C.	Dorcas 2nd	L. Ashton Wild Eyes, 34485	Mr. Beauchamp
1879, June 30, r. & w.,	C.C.	Dorcas 3rd	Alexander, 37725	do.

BECKWITH, Rev. H.,
Eaton Constantine, Iron Bridge, Salop.

BELLA, red and white, calved February 20, 1874. Bred by the Rev. H. Beckwith; got by Turco (32740), dam (Lydia Bell) by Duke of Flanders (21590), &c. See Vol. xxi. p. 565.

1876, Oct. 2, roan,	B.C.	Forester	Lord of Wicken, 34644	Rev. H. Beckwith
1877, Oct. 27, red,	B.C.	(dead)	do.	do.
1878, Sept. 30, r. & w.,	B.C.	Consul	do.	do.

CHAMOIS, red, calved June 30, 1865, Vols. xx., xxi., and xxii. pp. 435, 564, 317. Bred by Mr. C. Stubbs, Preston Hill; got by Thorndale's Grand Duke (20976), dam (Cassida) by Mac Turk (14872), &c.

1876, Sept. 25, r. & w.,	B.C.	The Kid	Lord of Wicken, 34644	Rev. H. Beckwith
1878, Mar. 13, roan,	B.C.	Gregory	do.	do.

CONSTANCE, red, calved July 27, 1876. Bred by the Rev. H. Beckwith ; got
by Lord of Wicken (34644), dam (Bonnie Bess) by Cronkhill (19530), &c.
See Vol. xxiii. p. 328.

Produce in	Names, &c.	By what Bull.	By whom bred.
1879, May 24, roan, C.C.	Gussey Constantine	Paul Gwynne, 37181	Rev. H. Beckwith

CONSTANTIA, red, calved July 27, 1876. Bred by the Rev. H. Beckwith ;
got by Lord of Wicken (34644), dam (Bonnie Bess) by Cronkhill (19530), &c.
See Vol. xxiii. p. 328.

1879, May 23, red, C.C.	Polly Constantine	Paul Gwynne, 37181	Rev. H. Beckwith

DUCHESS MARIE, white, calved March 7, 1874, Vol. xxv. p. 335. Bred by
the Rev. H. Beckwith ; got by Turco (32740), dam (Peeress) by Castlereagh
(19409), &c.

1879, Nov. 20, roan, C.C.	Jemima Constantine	Tacitus, 39184	Rev. H. Beckwith

FORTUNE 2ND, roan, calved January 23, 1876. Bred by the Rev. H. Beckwith ;
got by Lord of Wicken (34144), dam (Fortune) by Drummer (25919), &c.
See Vol. xxiv. p. 319.

1878, Sept. 24, roan, B.C.	Fortunatus 3rd	The Kid, 40804	Rev. H. Beckwith

FORTUNE 3RD, red, calved January 10, 1877. Bred by the Rev. H. Beckwith ;
got by Lord of Wicken (34644), dam (Fortune) by Drummer (25919), &c.
See Vol. xxiv. p. 319.

1879, Nov. 29, roan, C.C.	Fortune 4th	Tacitus, 39184	Rev. H. Beckwith

ISARD, roan, calved September 8, 1875. Bred by the Rev. H. Beckwith ; got by
Lord of Wicken (34644), dam (Chamois) by Thorndale's Grand Duke (20976),
&c. See Vol. xxii. p. 317.

1878, Mar. 23, white, B.C.	Chamouny	Geneva Prince, 36691	Rev. H. Beckwith
1879, Mar. 23, white, C.C.	Minna Constantine	Tacitus, 39184	do.

LADY LOO, roan, calved March 2, 1874. Bred by the Rev. H. Beckwith ; got
by Turco (32740), dam (Bonnie Bess) by Cronkhill (19530), &c. See Vol. xxi.
p. 564.

1876, Dec. 19, roan, B.C.	Ambassador	Lord of Wicken, 34644	Rev. H. Beckwith
1878, Mar. 23, roan, B.C.	Envoy	do.	do.
1879, April 7, roan, C.C.	Frances Constantine	Tacitus, 39184	do.

MAY QUEEN, roan, calved March 2, 1873. Bred by the Rev. H. Beckwith ;
got by Turco (32740), dam (Bonnie Bess) by Cronkhill (19530), &c. See
Vol. xxi. p. 564.

1875, Aug. 27, white, C.C.	(dead)	Lord of Wicken, 34644	Rev. H. Beckwith
1876, Sept. 14, red, B.C.	Jim	do.	do.
1877, Oct. 7, red, B.C.	Dick	do.	do.
1878, Sept. 23, roan, B.C.	Oliver	do.	do.

MYRTLE LEAF, roan, calved November 1, 1876. Bred by Mr. D. A. Green,
East Donyland ; got by Earl of Horton 7th (33799), dam (Myrtle Flower)
by Heydon Duke 2nd (31370), &c. See Vol. xxiii. p. 470.

1879, Jan. 21, roan, C.C.	Martha Constantine	Lord Oxford 7th, 38645	Rev. H. Beckwith

BECTIVE, Earl of,
Underley Hall, Carnforth.

AMETHYST, roan, calved August 12, 1871, Vol. xxv. p. 335. Bred by Lord Skelmersdale, Lathom House; got by Ninth Duke of Geneva (28391), dam (Anemone) by Duke of Kent (19619), &c.

Produce in		Names, &c.	By what Bull.	By whom bred.
1879, Feb. 8, roan,	C.C.	Lady Appleby	Duke of Underley, 33745	Earl of Bective

BELLE OF RYEDALE, roan, calved March 1, 1870, Vols. xxii. and xxiv. pp. 404, 422. Bred by Mr. Sigsworth, Stiltons; got by Manchester (26798), dam (Blaze) by Artillery (21190), &c.

Produce in		Names, &c.	By what Bull.	By whom bred.
1879, Jan. 31, roan,	C.C.	Duncombe Belle	D. of Tregunter 5th, 33743	Earl of Bective

BUTTERFLY DUCHESS, roan, calved May 16, 1876. Bred by Mr. J. P. Foster, Killhow; got by Duke of Hillhurst (28401), dam (Butterfly Princess 21st) by Royal Cumberland (27358), g. d. (Butterfly Princess 15th) by Fourteenth Duke of Oxford (21605), gr. g. d. (Lady Butterfly Princess 13th) by Richard (16834), &c. See " Firefly 2nd," p. 99.

Produce in		Names, &c.	By what Bull.	By whom bred.
1879, Feb. 17, roan,	C.C.	Butterfly Queen	Grand Duke 31st, 38374	Earl of Bective

BUTTERFLY PRINCESS 22ND, roan, calved December 22, 1871, Vol. xxiii. p. 441. Bred by Mr. J. Fawcett, Scaleby Castle; got by Royal Cumberland (27358), dam (Lady Butterfly Princess 13th) by Richard (16834), &c.

Produce in		Names, &c.	By what Bull.	By whom bred.
1878, Mar. 25, roan,	B.C.	Firefly 2nd	24th D. of Airdrie, 36400	Earl of Bective
1879, Mar. 3, roan,	C.C.	Butterfly Queen 2nd	Turcoman, 39238	do.

Firefly 2nd, sold to Mr. Holliday, Silloth, Cumberland.

CHERRY CHEEK, roan, calved February 27, 1873, Vol. xxiv. p. 320. Bred by Mr. J. W. Philips, Heybridge; got by Duke of Albany (25931), dam (Cherry) by Monarch (16575), &c.

Produce in		Names, &c.	By what Bull.	By whom bred.
1878, Mar. 28, roan,	C.C.	Rosy Cheek	Lord of the Isles, 34631	Earl of Bective
1879, Mar. 1, red,	B.C.	Ceytawayo	do.	do.

CYRENE, red, calved May 19, 1876. Bred by the Earl of Bective; got by Duke of Underley (33745), dam (Cherry Cheek) by Duke of Albany (25931), &c. See Vol. xxiv. p. 320.

Produce in		Names, &c.	By what Bull.	By whom bred.
1879, Jan. 24, roan,	C.C.	Cypris	Grand Duke 31st, 38374	Earl of Bective

DORA GWYNNE, red and white, calved January 14, 1876. Bred by Messrs. I. and J. Gaitskell, Hall Santon; got by Duke of Oxford (31004), dam (Dorothy Gwynne) by Oxford le Valiant (27021), &c. See Vol. xxiii. p. 447.

Produce in		Names, &c.	By what Bull.	By whom bred.
1879, Oct. 21, r. & w.,	B.C.	General Gough	L'tb'ne's D.of Oxf'd, 30895	Earl of Bective

DUCHESS GWYNNE 2ND, white, calved November 7, 1869, Vols. xxi., xxiii., and xxiv. pp. 565, 329, 320. Bred by the Earl of Bective; got by Third Duke of Clarence (23727), dam (Duchess Gwynne) by Duke of Wetherby (17753), &c.

Produce in		Names, &c.	By what Bull.	By whom bred.
1879, Feb. 12, roan,	B.C.	Lord Berthwaite	Duke of Underley, 33745	Earl of Bective

ELIA, red, calved November 19, 1870, Vols. xxi. and xxii. pp. 873, 524. Bred by Lord Southampton, Whittlebury Lodge; got by Sam of Oxford (25084), dam (Laura) by Marmaduke (14897), &c.

Produce in		Names, &c.	By what Bull.	By whom bred.
1878, July 22, red,	C.C.	Elia 2nd	Duke of Underley, 33745	Earl of Bective
1879, Aug. 9, r. & w.,	C.C.	Elia 3rd	do.	do.

GAZELLE 3RD, white, calved December 31, 1865, Vols. xviii., xx., xxi., xxii., and xxv. pp. 509, 635, 844, 502, 577. Bred by Mr. E. Bowly, Siddington House ; got by Seventh Duke of York (17754), dam (British Lass) by Union (29031), &c.

Produce in	Names, &c.	By what Bull.	By whom bred.
1879, Jan. 11, white, B.C.	Gay Gallant	2nd D. of Glo'ster, 28392	Earl of Bective

GAZELLE OF CHAMOUNIX, roan, calved March 24, 1875. Bred by Mr. J. D'A. Samuda, Chillies ; got by Eighth Duke of Geneva (28390), dam (Gazelle 28th) by Second Duke of Tregunter (26022), &c. See Vol. xxii. p. 551.

1879, Jan. 28, roan, C.C.	Gazelle of Castra	Marquis of Oxford, 34786	Earl of Bective

GRAND DUCHESS 23RD, roan, calved August 17, 1869, Vols. xx., xxii., and xxiii. pp. 547, 524, 587. Bred by Mr. R. E. Oliver, Sholebroke Lodge ; got by Grand Duke 7th (19877), dam (Grand Duchess 17th) by Imperial Oxford (18084), &c.

1878, Mar. 17, red, B.C.	Grand Duke 34th	Duke of Underley, 33745	Earl of Bective
1879, May 23, roan, B.C.	Grand Duke 37th	do.	do.

Grand Duke 34th, sold to Messrs. Hosken and Son, Loggans Mill, Hayle, Cornwall.

GRAND PRINCESS OF LIGHTBURNE 3RD, roan, calved October 17, 1876. Bred by Mr. A. Brogden, Lightburne Park ; got by Second Duke of Glo'ster (28392), dam (Grand Princess of Lightburne) by Grand Duke 10th (21848), &c. See Vol. xxiii. p. 354.

1879, April 20, roan, C.C.	Underley Princess	L'tb'ne'sD.ofOxf.2d,38564	Earl of Bective

HARMONY, red and white, calved October 9, 1872, Vol. xxii. p. 417. Bred by the Duke of Devonshire, Holker Hall ; got by Prince of Lightburne (29653), dam (Music) by Juryman (20043), &c.

1877, Oct. 10, red, C.C.	(dead)	Duke of Underley, 33745	Earl of Bective

JUNO 2ND, roan, calved April 23, 1871, Vols. xxi. and xxiii. pp. 637, 567. Bred by Mr. Caddy, Rougholm ; got by Waterloo Cherry (27763), dam (Juno) by Muchland (24629), &c.

1878, Nov. 11, roan, B.C.	Mars	Duke of Underley, 33745	Earl of Bective
1879, Dec. 16, r. & w., B.C.	Jingo	do.	do.

LADY SALE OF PUTNEY, red, calved December 12, 1869, Vols. xxi., xxii., xxiii., and xxiv. pp. 566, 318, 329, 320. Bred by Messrs. A. M. Winslow and Sons, Putney, Vermont, U.S.A. ; got by Ninth Duke of Thorndale (31023), dam (Lady Sale 6th) by Red Knight (35232), &c.

1878, Aug. 16, r. & w., B.C.	Prince Saladin 2nd	Lord of the Isles, 34631	Earl of Bective
1879, Aug. 8, red, B.C.	Prince Saladin 3rd	Duke of Underley, 33745	do.

*LADY WISETON, red, calved May 19, 1874. Bred by Mr. G. M. Bedford, Stonor Farm, Bourbon County, Kentucky, U.S.A. ; got by Duke of Geneva 11th (41385), dam (Duchess of Goodness 25th) by Fourteenth Duke of Thorndale (28459), g. d. (Duchess of Goodness 2nd) by Duke of Airdrie (12730), gr. g. d. (Goodness 3rd) by Senator 2nd (13687), — (Goodness) by Orontes (4623), &c. See Vol. x. p. 387.

1879, Mar. 10, r. & w., C.C.	Lady Wiseton 2nd	Duke of Underley, 33745	Earl of Bective

MARCHIONESS 6TH, roan, calved April 6, 1872, Vols. xxii., xxiii., and xxv. pp. 318, 330, 336. Bred by the Earl of Bective; got by Second Duke of Collingham (23730), dam (Siddington 4th) by Seventh Duke of York (17754), &c.

Produce in		Names, &c.	By what Bull.	By whom bred.
1879, Mar. 27, roan,	C.C.	Marchioness 13th	Duke of Underley, 33745	Earl of Bective

RED ROSE OF ESKDALE, red, calved August 2, 1873, Vols. xxiii. and xxiv. pp. 330, 321. Bred by Mr. A. Renick, Kentucky, U.S.A.; got by Airdrie 3rd (32919), dam (Poppy 8th) by Joe Johnson (31440), &c.

Produce in		Names, &c.	By what Bull.	By whom bred.
1878, May 18, r. & w.,	B.C.	Master of Eskdale	Duke of Underley, 33745	Earl of Bective
1879, June 20, r. & w.,	B.C.	Master of Eskdale 2d	do.	do.

Master of Eskdale, sold to Miss Taunton, Ashley, Stockbridge.

RED ROSE OF ETTRICK, red and white, calved June 25, 1876. Bred by the Earl of Bective; got by Duke of Underley (33745), dam (Red Rose of Eskdale) by Airdrie 3rd (32919), &c. See Vol. xxiii. p. 330.

Produce in		Names, &c.	By what Bull.	By whom bred.
1878, Dec. 17, r. & w.,	C.C.	R. Rose of Elderbeck	Grand Duke 31st. 38374	Earl of Bective

RED ROSE OF MARYLAND, roan, calved March 29, 1877. Bred by the Earl of Dunmore, Dunmore; got by Sixth Duke of Geneva (30959), dam (Red Rose of Ohio) by Imperial Starlight (36786), &c. See Vol. xxiv. p. 423.

Produce in		Names, &c.	By what Bull.	By whom bred.
1879, July 21, roan,	B.C.	Umbelini	Grand Duke 31st, 38374	Earl of Bective

RED ROSE OF NITHSDALE, red, calved May 30, 1873, Vol. xxiv. p. 321. Bred by Mr. A. Renick, Kentucky, U.S.A.; got by Fourth Duke of Geneva (30958), dam (Leonora) by Airdrie (30365), &c.

Produce in		Names, &c.	By what Bull.	By whom bred.
1878, July 2, red,	B.C.	Lord Thirlmere	Lord of the Isles, 34631	Earl of Bective
1879, June 18, r. & w.,	C.C.	Rd. Rose of Netherby	Grand Duke 31st, 38374	do.

Lord Thirlmere, sold to Mr. G. Blackwell, Hazlecote, Kingscote, Wotton-under-Edge.

REVELRY 12TH, roan, calved October 18, 1873, Vol. xxiv. p. 321. Bred by the Earl of Dunmore, Dunmore; got by Marquis 3rd (31826), dam (Ruby) by Cherry Duke (25752), &c.

Produce in		Names, &c.	By what Bull.	By whom bred.
1878, Dec. 24, roan,	C.C.	Rydal	Duke of Underley, 33745	Earl of Bective

REVELRY 17TH, red and white, calved June 1, 1877. Bred by the Earl of Dunmore, Dunmore; got by Sixth Duke of Geneva (30959), dam (Revelry 11th) by Baron Oxford 5th (27958), &c. See Vol. xxiv. p. 424.

Produce in		Names, &c.	By what Bull.	By whom bred.
1879, Dec. 13, {	red, C.C.	Lady Rydal	} Marquis of Oxford 2d, 37055	Earl of Bective
	red, C.C.	Lady Rydal 2nd		

ROSE OF UNDERLEY, roan, calved October 31, 1875. Bred by the Earl of Bective; got by Captain Tregunter (28136), dam (Cawood's Rose 7th) by Lord Chancellor (26622), &c. See Vol. xxii. p. 318.

Produce in		Names, &c.	By what Bull.	By whom bred.
1879, May 10, r. & w.,	B.C.	Zulu	Grand Duke 31st, 38374	Earl of Bective

UNDERLEY DARLING, red and white, calved November 15, 1868, Vols. xx., xxi., and xxiv. pp. 795, 567, 321. Bred by the Earl of Bective; got by Royal Cambridge (25009), dam (Turk's Darling) by Royal Turk (16875), &c.

Produce in		Names, &c.	By what Bull.	By whom bred.
1878, Jan. 11, {	red, B.C.	Turcoman 3rd	} Duke of Underley, 33745	Earl of Bective
	red, C.C.	Underley Darl'g 4th		
1879, May 23, r. & w.,	C.C.	Underley Darling 6th	do.	do.

Turcoman 3rd, sold to Mr. J. Woodhouse, Scale Hall, Lancaster.

UNDERLEY DARLING 2ND, roan, calved August 29, 1873, Vols. xxiv. and
xxv. pp. 321, 336. Bred by the Earl of Bective; got by Grand Duke of
Kent 2nd (28759), dam (Underley Darling) by Royal Cambridge (25009), &c.

Produce in	Names, &c.	By what Bull.	By whom bred.
1879, Nov. 25, roan, C.C.	Underley Darling 7th	Duke of Underley, 33745	Earl of Bective

UNDERLEY DARLING 3RD, roan, calved January 7, 1877. Bred by the
Earl of Bective; got by Second Duke of Tregunter (26022), dam (Underley
Darling) by Royal Cambridge (25009), &c. See Vol. xxiv. p. 321.

1879, Dec. 15, roan, B.C.	Turcoman 4th	Grand Duke 31st, 38374	Earl of Bective

WATER LASS 4TH, roan, calved February 8, 1872. Bred by Mr. E. H.
Cheney, Gaddesby Hall; got by Ninth Duke of Geneva (28391), dam
(Water Lass) by Oxford 2nd (18507), &c. See Vol. xx. p. 813.

1879, Feb. 20, roan, B.C.	Water Wizard	L'tb'ne's D. of Oxf. 2d, 38564	Earl of Bective

WATERLOO 25TH, roan, calved July 26, 1867, Vol. xxiv. p. 701. Bred by
Mr. J. P. Kearney, Miltown House; got by Dr. McHale (15887), dam
(Waterloo 22nd) by Speculator (13775), &c.

1879, Aug. 3, roan, C.C.	Windermere	Duke of Underley, 33745	Earl of Bective

BEEVER, Rev. W. Holt,
Pencraig Court, Ross, Herefordshire.

BIJOU 2ND, red and white, calved April 4, 1870, Vols. xxiii. and xxv.
pp. 330, 336. Bred by Mr. S. O. Priestley, Trefan; got by The Baron
(25277), dam (Bijou) by The Bully (23019), &c.

1879, Mar. 16, r. & w., B.C.	(Steer)	G. Dk. of Clarence, 28750	Rev. W. H. Beever

BLUE BLOOD, roan, calved December 10, 1875. Bred by the Rev. W. Holt
Beever; got by Royal Cumberland (27358), dam (Ancienne) by The Baron
(25277), &c. See Vol. xxii. p. 319.

1879, Feb. 14, r. & w., C.C.	Clarence Isola	G. Dk. of Clarence, 28750	Rev. W. H. Beever

BLUSH, red, calved February 9, 1862, Vols. xvii., xviii., xix., xxi., and xxii.
pp. 386, 398, 414, 567, 319. Bred by Mr. S. O. Priestley, Trefan; got by
Prince Henry (18608), dam (Donna) by Patriot (10594), &c.

1879, May 4, r. & w., B.C.	Blushing Duke 5th	G. Dk. of Clarence, 28750	Rev. W. H. Beever

CARMINE, red and white, calved September 2, 1873, Vol. xxv. p. 337. Bred
by the Rev. W. Holt Beever; got by Mountain Dew (34879), dam (Court
Blush) by The Baron (25277), &c.

1879, Dec. 7, roan, B.C.	Blushing Duke 7th	20th D. of Oxford, 28432	Rev. W. H. Beever

DAISY QUEEN 2ND, red and white, calved June 20, 1877. Bred by the Rev.
W. Holt Beever; got by Royal Cumberland (27358), dam (Hecuba) by Sir
Knight (25157), g. d. (Ireni) by The Bully (23019), &c. "See "Old Daisy
Bull 9th," p. 178.

1879, Nov. 15, r. & w., B.C.	Daisy King	Old Daisy Bull 6th, 42065	Rev. W. H. Beever

EARLY BLUSH, red and white, calved April 3, 1874, Vol. xxv. p. 337. Bred
by the Rev. W. Holt Beever; got by Mountain Dew (34879), dam (Blush)
by Prince Henry (18608), &c.

1879, Sept. 8, roan, B.C.	Blushing Duke 6th	20th D. of Oxford, 28432	Rev. W. H. Beever

HELEN, red and white, calved March 28, 1875, Vol. xxv. p. 337. Bred by
the Rev. W. Holt Beever; got by Mountain Dew (34879), dam (Hecuba) by
Sir Knight (25157), &c.

Produce in	Names, &c.	By what Bull.	By whom bred.
1879, Sept. 5, roan, C.C.	Sweet Ruin	Old Daisy Bull 6th, 42065	Rev. W. H. Beever

IRENI, red and white, calved April 2, 1866, Vols. xxi., xxiii., and xxv. pp. 568,
331, 337. Bred by Mr. S. O. Priestley, Trefan; got by The Bully (23019),
dam (Isola) by Lord Raglan (13222), &c.

1879, Aug. 12, roan, B.C.	Old Daisy Bull 9th	20th D. of Oxford, 28432	Rev. W. H. Beever

PRINCESS MAUD, red, calved January 16, 1870, Vols. xxii. and xxiii. pp. 320,
331. Bred by Mr. S. O. Priestley, Trefan; got by Sir Knight (25157), dam
(Alexandra) by Prince Henry (18608), &c.

1879, May 20, r. & w., C.C.	Clarence Maud	G. Dk. of Clarence, 28750	Rev. W. H. Beever

PRINCESS MAUVE, roan, calved October 3, 1875, Vol. xxv. p. 338. Bred
by the Rev. W. Holt Beever; got by Royal Cumberland (27358), dam
(Princess Maud) by Sir Knight (25157), &c.

1879, May 26, roan, C.C.	Clarence Mauve	G. Dk. of Clarence, 28750	Rev. W. H. Beever

QUIETUDE, red, calved October 18, 1876. Bred by the Rev. W. Holt Beever;
got by Royal Cumberland (27358), dam (Ireni) by The Bully (23019), &c.
See "Old Daisy Bull 8th," p. 178.

1879, July 3, roan, B.C.	Old Daisy Bull 8th	20th D. of Oxford, 28432	Rev. W. H. Beever

ROYAL BIJOU 2ND, red, calved August 2, 1876. Bred by the Rev. W. Holt
Beever; got by Royal Cumberland (27358), dam (Bijou 2nd) by The Baron
(25277), &c. See Vol. xxiii. p. 330.

1879, Feb. 23, r. & w., C.C.	Clarence Bijou	G. Dk. of Clarence, 28750	Rev. W. H. Beever

ROYAL BLUSH, roan, calved May 28, 1875. Bred by the Rev. W. Holt
Beever; got by Royal Cumberland (27358), dam (Court Blush) by The Baron
(25277), &c. See "Blushing Duke 4th," p. 29.

1879, Mar. 14, roan, B.C.	Blushing Duke 4th	G. Dk. of Clarence, 28750	Rev. W. H. Beever

ZOE, red and white, calved October 6, 1873. Bred by the Rev. W. Holt
Beever; got by Mountain Dew (34879), dam (Zoraid) by The Bully (23019),
&c. See Vol. xxi. p. 568.

1879, Oct. 1, r. & w., B.C.	Double Daisy Bull	Old Daisy Bull 6th, 42065	Rev. W. H. Beever

ZORAID, roan, calved April 22, 1868, Vols. xxi., xxii., and xxiii. pp. 568, 320,
331. Bred by Mr. S. O. Priestley, Trefan; got by The Bully (23019), dam
(Jessie) by Lord Raglan (13222), &c.

1879, Aug. 30, white, C.C.	Zenith	20th D. of Oxford, 28432	Rev. W. H. Beever

BELL, Alexander,
Barrowford, Burnley, Lancashire.

VENETIA'S BRIDE, roan, calved December 31, 1875. Bred by Mr. T.
Strickland, Thirsk Junction, the property of Mr. A. Bell; got by Bride-
groom (30584), dam (Venetia) by Ravenswood (22682), &c. See "Bell
Oxford," p. 25.

1878, Sept. 16, red, B.C.	Bell Oxford	Earl of Waterloo 3d, 36601	Mr. Robinson

Bell Oxford, sold to Mr. A. Bell, Barrowford.

BELL, William,
Spittal, Rhuddlan.

PRINCESS ROSE, roan, calved October 19, 1876. Bred by Mr. S. Canning, Snitterfield, the property of Mr. W. Bell; got by Lord Kingscote Rosy (34576), dam (Patricia) by Patrician (24728), &c. See "Duke of Hadley," p. 78.

Produce in	Names, &c.	By what Bull.	By whom bred.
1879, Oct. 14, r.&w., B.C.	Duke of Hadley	2d Duke of Rowley, 28441	Mr. Allen

Duke of Hadley, sold to Mr. G. Henshall, Castle Mead, Hadley, Wellington, Salop.

BENTON, W.,
Cattie, Keig, Whitehouse, N.B.

LAMBERTEE 2ND, red, calved August 1, 1875. Bred by Mr. W. Benton; got by Glo'ster Royal (34042), dam (Lambertee) by Old England (24681), &c. See "Miss Maria," Vol. xxii. p. 321.

1878, May, red,	B.C.	(dead)	Baron Booth, 36162	Mr. Benton
1879, May 28, red,	B.C.	Duke of Connaught	Oxford Duke, 38829	do.

BETTRIDGE, Henry,
East Hanney, Wantage, Berkshire.

JUVENILE, red and white, calved April 4, 1875. Bred by Mr. H. Bettridge; got by Marquis of Sockburn (34787), dam (Judy 2nd) by Baron Oxford 2nd (27962), &c. See Vol. xxii. p. 322.

1878, Jan. 28, r.&w., C.C.	Judy Rock	Rockville 2nd, 37356	Mr. Bettridge
1879, Jan. 29, red, C.C.	Young Judy	Burgundy, 37926	do.

Judy Rock, sold to Mr. W. Curtis, Fernham, Faringdon.

PROPRIETY, red, calved February 4, 1875. Bred by Mr. H. Bettridge; got by Prince Rupert (35178), dam (Prude) by Miracle (24602), &c. See Vol. xxiii. p. 334.

1878, Jan. 15, red,	C.C.	Prudent Jane	Rockville 2nd, 37356	Mr. Bettridge

ROSE OF POUGHLEY, red and white, calved January 4, 1872, Vol. xxv. p. 340. Bred by Mr. H. Bettridge; got by Baron Booth 1st (27915), dam (Redheart Rose) by Artemus Ward (23326), &c.

1879, Jan. 9, roan,	C.C.	Glorie Dijon	Earl of Horton 11th, 36588	Mr. Bettridge
1879, Dec. 31, red,	C.C.	China Rose	Burgundy, 37926	do.

BIRD, W. and G.,
Volis, Kingston, Taunton.

BONBON, roan, calved August 18, 1875. Bred by Mr. W. Woodward, Northway House; got by Owen Glendwr (32009), dam (Snowball) by Majestic (29256), &c. See Vol. xxii. p. 615.

1879, Oct. 22, white,	C.C.	White Lily	Glo'ster Gwynne, 39938	Messrs. Bird

KISS 8TH, white, calved November 18, 1874. Bred by Messrs. W. and G. Bird; got by Cardinal (28144), dam (Kiss 7th) by Earl of Fife (23835), &c. See Vol. xxi. p. 573.

1879, April 27, roan,	C.C.	Kiss 9th	Pretender, 35069	Messrs. Bird

MATLN, red, calved February 16, 1876. Bred by Mr. W. Woodward, Northway
House; got by Prince Butterfly (35093), dam (Lady Sarah) by Duke of Wellington (26026), &c. See Vol. xxiii. p. 706.

Produce in	Names, &c.	By what Bull.	By whom bred.
1879, Dec. 2, roan, C.C.	Elphie	Glo'ster Gwynne, 39938	Messrs. Bird

VENTURE, roan, calved March 28, 1876. Bred by Mr. D. R. Scratton,
Ogwell; got by Cherry Duke 6th (30705), dam (Vigel) by Lord Derby (29098),
&c. See Vol. xxiii. p. 638.

1879, May 20, roan, C.C.	Vocal	L. Rothesay S'w'th'rt, 40208	Messrs. Bird

BIRKETT, James,
Bolton-le-Sands, Lancaster.

DUCHESS OF KNIGHTLEY 2ND, red, calved May 12, 1872. Bred by Mr.
D. Fisher, Pitlochrie; got by Fawsley Prince (31150), dam (Damsel) by Lord
Hopewell (18239), &c. See Vol. xx. p. 473.

1878, May 1, r. & w., C.C.	D'ss of Knightley 4th	G.D. of Kirklev'ton, 34071	Mr. Birkett

BLACKWELL, George,
Hazlecote, Kingscote, Wotton-under-Edge.

SPRING FLOWER 3RD, white, calved April 10, 1874. Bred by Mr. A. T.
Playne, Minchinhampton; got by Severn Boy (32473), dam (Wallflower 17th)
by Bleeding Heart 5th (25639), &c. See "Albion," p. 4.

1877, June 20, white, B.C.	Albion	Emp. of Gipsies 2nd, 33851	Mr. Blackwell
1878, May (dead calf)		Lord Calcot, 34500	do.
1879, Dec. 28, roan, C.C.	Spring Flower 4th	D. of Hazlecote 52nd, 43104	do.

Albion, sold to Mr. J. Kendal, Trotshill, Worcester.

BLANTERN, G. G.,
Haston Hadnall, Shropshire.

ECONOMY, red, calved December 23, 1868, Vols. xx., xxi., xxii., xxiii., xxiv., and
xxv. pp. 496, 969, 600, 682, 686, 697. Bred by Mr. P. Stevenson, Rainton;
got by Lord Lally 3rd (24408), dam (Emilia) by Duke (17700), &c.

1879, April 15, red, C.C.	Emma 14th	Oxford Beau 4th, 34964	Mr. Blantern

BLENKARN, Thomas,
Keisley, Appleby, Westmoreland.

BRIGHT EYES 6TH, roan, calved September 15, 1876. Bred by Mr. T. Blenkarn; got by Sir John (32507), dam (Bright Eyes 3rd) by Earl of Waterloo
(23849), &c. See "Bright Eyes 5th," Vol. xxv. p. 341.

1879, Dec. 11, white, B.C.	Studgill	K'bythore W. E. 4th, 36865	Mr. Blenkarn

BRIGHT EYES 7TH, red, calved April 4, 1877. Bred by Mr. T. Blenkarn;
got by Kirkbythore Wild Eyes 2nd (34362), dam (Bright Eyes 4th) by Earl
of Carlisle (23826), &c. See "Studgill Bank," p. 243.

1879, Nov. 19, roan, C.C.	Bright Eyes 9th	K'bythore W. E. 4th, 36865	Mr. Blenkarn

BLEZARD, Robert,
Pool Park, Ruthin, North Wales.

BLANCHE 10TH, roan, calved December 3, 1870, Vols. xxii. and xxiv. pp. 295, 326. Bred by the Duke of Devonshire, Holker Hall; got by Grand Duke 10th (21848), dam (Blanche 4th) by Lord Oxford (20214), &c.

Produce in		Names, &c.	By what Bull.	By whom bred.
1879, Aug. 4, roan,	C.C.	Lady Blanche 16th	D. of Siddington 3rd, 38183	Mr. Blezard

BLANCHE 14TH, roan, calved January 11, 1877. Bred by Mr. R. Blezard; got by Duke of Rosedale 2nd (33722), dam (Blanche 10th) by Grand Duke 10th (21848), &c. See Vol. xxiv. p. 326.

Produce in		Names, &c.	By what Bull.	By whom bred.
1879, Feb. 23, white,	C.C.	Lady Blanche 15th	D. of Siddington 3rd, 38183	Mr. Blezard

CZARINA, roan, calved April 19, 1873, Vol. xxiii. p. 337. Bred by Mr. J. A. Mumford, Brill House; got by Duke John (30913), dam (Consolation) by Earl of Lancaster (21647), &c.

Produce in		Names, &c.	By what Bull.	By whom bred.
1879, Oct. 31, white,	C.C.	Scylla 2nd	Viscount Dursley, 37636	Mr. Blezard

HONEYSUCKLE, red, calved March 8, 1872, Vol. xxi. p. 931. Bred by Lord Skelmersdale, Lathom House; got by Satyr (29933), dam (Honeymoon) by Viscount Curly (25382), &c.

Produce in		Names, &c.	By what Bull.	By whom bred.
1877, Feb. 24, r. & w.,	B.C.	Astor 2nd	Astor, 32962	Mr. Blezard
1879, Nov. 23, red,	C.C.	Lady Honeysuckle	Viscount Dursley, 37636	do.

KATHLEEN, red, calved April 15, 1870, Vol. xxi. p. 864. Bred by Mr. J. A. Mumford. Brill House; got by Earl of Lancaster (21647), dam (Spotted Kate) by First Lord (14550), &c.

Produce in		Names, &c.	By what Bull.	By whom bred.
1876, Jan. 10, roan,	B.C.	Boscobel	Notley, 31991	Mr. Blezard
1877, April 17, roan,	B.C.	Astor 3rd	Astor, 32962	do.
1878, July 20, red,	B.C.	(Steer)	do.	do.
1879, May 28, { red,	C.C.	Margharita	} D. of Siddington 3rd, 38183	do.
{ roan,	C.C.	Margharita 2nd		

Boscobel, sold to Mr. H. R. Sandbach, Abergele.

LADY ALICE, roan, calved September 20, 1875. Bred by Mr. R. E. Oliver, Sholebroke Lodge; got by Grand Duke 21st (34061), dam (Lady Adelaide) by Northern Duke (22431), &c. See Vol. xxii. p. 525.

Produce in		Names, &c.	By what Bull.	By whom bred.
1879, April 10, white,	C.C.	Lady Alice 2nd	D. of Siddington 3rd, 38183	Mr. Blezard

MOUNTAIN DAISY 4TH, roan, calved March 18, 1876. Bred by Mr. R. Blezard; got by Astor (32962), dam (Mountain Daisy) by Duke of Walton (28463), g. d. (Michaelmas Daisy) by Viceroy (27712), &c. See " Michaelmas Boy," p. 170.

Produce in		Names, &c.	By what Bull.	By whom bred.
1879, June 14, roan,	C.C.	Mountain Daisy 7th	D. of Siddington 3rd, 38183	Mr. Blezard

NARCISSUS, roan, calved March 9, 1869, Vols. xxii. and xxv. pp. 324, 341. Bred by Mr. J. A. Mumford, Brill House; got by Earl of Lancaster (21647), dam (Nancy 6th) by Prince of Paris (16748), &c.

Produce in		Names, &c.	By what Bull.	By whom bred.
1879, Dec. 15, white,	C.C.	Narcissus 3rd	Viscount Dursley, 37636	Mr. Blezard

WATER DUCHESS 3RD, red and white, calved January 26, 1874, Vol. xxiii. p. 337. Bred by Mr. T. Barber, Sproatley Rise; got by Oxford's Baronet (29499), dam (Water Duchess) by Grand Duke 6th (19876), &c.

Produce in		Names, &c.	By what Bull.	By whom bred.
1878, Sept. 28, roan,	B.C.		D. of Siddington 3rd, 38183	Mr. Blezard
1879, Nov. 23, red,	C.C.	Lady Clwyd 2nd	D. of Barrington 7th, 39715	do.

YORK'S CERTAINTY DUCHESS, roan, calved December 1, 1872, Vol. xxii.
p. 324. Bred by Mr. J. Fawcett, Scaleby Castle; got by Eighth Duke of
York (28480), dam (Certainty Duchess) by Royal Cumberland (27358), &c.

Produce in		Names, &c.	By what Bull.	By whom bred.
1879, Oct. 27, red,	C.C.	Certainty Duchess 2d	Viscount Dursley, 37636	Mr. Blezard

BLISS, William,
Chipping Norton, Oxon.

EARL'S FLOWER, red and white, calved September 11, 1871, Vol. xxiii. p. 442.
Bred by Mr. R. R. Brockbank, Burgh-by-Sands; got by Earl of Eglinton
(23832), dam (Marquis's Flower) by Marquis of Cobham (22299), &c.

Produce in		Names, &c.	By what Bull.	By whom bred.
1879, Nov. 27, red,	C.C.	Earl's Flower 2nd	Elmhurst Prince, 41503	Mr. Bliss

Earl's Flower 2nd, sold to Count Renard, Castle Strehlitz, Upper Silesia.

GENEVA VERITY, red, calved July 22, 1874. Bred by Mr. J. W. Wilson,
Broadway; got by Grand Duke of Geneva 2nd (31288), dam (Verity 3rd) by
Paragon (24722), &c. See Vol. xxi. p. 998.

Produce in		Names, &c.	By what Bull.	By whom bred.
1879, Jan. 29, red,	C.C.	Geneva Verity 2nd	G. D. of Marlboro', 38380	Mr. Bliss

JUNIPER BERRY, roan, calved May 26, 1868, Vols. xxi. and xxiii. pp. 637,
338. Bred by Mr. W. B. Stopford-Sackville, Drayton House; got by George 1st
(19848), dam (Juniper) by Firebrand (19747), &c.

Produce in		Names, &c.	By what Bull.	By whom bred.
1879, April 14, red,	C.C.	Juniper Berry 2nd	Elmhurst Prince, 41503	Mr. Bliss

KENTISH NONSUCH, red, calved October 5, 1872. Bred by Mr. J. K.
Fowler, Prebendal Farm; got by Grand Duke of Kent (26289), dam (Nonsuch 5th) by Charleston (21400), &c. See "Duke of Glos'ter," p. 77.

Produce in		Names, &c.	By what Bull.	By whom bred.
1878, May 16, r. & w.,	C.C.	Countess Nonsuch	Duke of Connaught, 33604	Lord Fitzhardinge
1879, Mar. 27, roan,	B.C.	Duke of Glo'ster	do.	Mr. Bliss

MAID OF OXFORD 10TH, red, calved August 22, 1874. Bred by Mr. T. G.
Curtler, Bevere House; got by Grand Duke of Clarence (28750), dam (Maid
of Oxford 6th) by Lord Waterloo 2nd (26755), &c. See Vol. xxi. p. 662.

Produce in		Names, &c.	By what Bull.	By whom bred.
1879, Mar. 28, red,	C.C.	Maid of Oxford 11th	Elmhurst Prince, 41503	Mr. Bliss

Maid of Oxford 11th, sold to Mr. J. Garne, Rissington, Burford, Oxon.

OAKLAND MAZURKA, red, calved December 16, 1874. Bred by Mr. G.
Fox, Elmhurst Hall; got by Duke of Hillhurst 2nd (39748), dam (Mazurka
of Lyndale) by Duke of Airdrie 17th (41349), &c. See Vol. xxv. p. 454.

Produce in		Names, &c.	By what Bull.	By whom bred.
1879, Aug. 18, red,	C.C.	Oakland Mazurka 2d	Elmhurst Prince, 41503	Mr. Bliss

BLUNDELL, J. H.,
Woodside, Luton, Bedfordshire.

FRAULEIN GWYNNE, roan, calved June 19, 1875, Vols. xxiv. and xxv. pp.
328, 343. Bred by Mr. J. H. Blundell; got by Sir Walter (37482), dam
(Frisky Gwynne) by Thorndale Knightley (23065), &c.

Produce in		Names, &c.	By what Bull.	By whom bred.
1879, June 6, roan,	C.C.	Oxford Gwynne	D. of Oxford 32nd, 36527	Mr. Blundell

LACTEA ONEIDA, roan, calved August 6, 1874, Vol. xxiv. p. 329. Bred by
Mr. J. H. Blundell; got by Sixth Duke of Oneida (30997), dam (Lactea
Oxonensis) by Imperial Oxford (18084), &c.

Produce in		Names, &c.	By what Bull.	By whom bred.
1878, Mar. 15, red,	B.C.	(dead)	Duke of Oxford 32nd, 36527	Mr. Blundell
1879, April 7, { roan,	B.C.	Baron Lactea }	do.	do.
{ red,	B.C.	Lord Lactea 2nd }		

LADY GENEVA ROSY 2ND, roan, calved August 4, 1876. Bred by Mr. F.
N. Sartoris, Rushden Hall; got by Ninth Duke of Geneva (28391), dam
(Polythorn) by Fourth Duke of Thorndale (17750), &c. See Vol. xxi. p. 917.

Produce in	Names, &c.	By what Bull.	By whom bred.
1879, Mar. 28, r. & w., B.C.	Lord Oxford Geneva	D. of Oxford 32nd, 36527	Mr. Blundell

ROSE OF FARNDISH, roan, calved November 22, 1876. Bred by Mr. F. N.
Sartoris, Rushden Hall; got by Lord York Fawsley (34709), dam (Lady
Knightley 2nd) by Grand Duke 4th (19874), &c. See Vol. xxv. p. 322.

1879, Mar. 18, red,	C.C.	Rose of Woodside	Duke of Oxford 32nd, 36527	Mr. Blundell

SIDDINGTON 16TH, roan, calved August 9, 1874, Vols. xxiv. and xxv. pp.
528, 343. Bred by Mr. E. Bowly, Siddington House; got by Third Duke of
Clarence (23727), dam (Siddington 11th) by Second Duke of Tregunter (26022),
&c.

1879, July 15, red,	C.C.	Oxford Siddington	Duke of Oxford 32nd, 36527	Mr. Blundell

SURPRISE DUCHESS 2ND, red and white, calved November 12, 1876. Bred
by Mr. J. H. Blundell; got by Fifth Duke of Wetherby (31033), dam (Surprise) by Eighth Duke of Geneva (28390), &c. See Vol. xxiii. p. 339.

1879, Jan. 29, red,	B.C.	Surprise Duke 2nd	Duke of Oxford 32nd, 36527	Mr. Blundell

BLYTH, R. Burn,
Woolhampton House, Reading.

CLARA, roan, calved April 6, 1870, Vols. xxi. and xxiv. pp. 579, 329. Bred
by Mr. J. Blyth, Woolhampton; got by Fra Diavola (26190), dam (British
Lass) by Union (19031), &c.

1879, Jan. 30, r.&w., B.C.	Boreas	Alphonso, 32943	Mr. R. B. Blyth

Boreas, sold to Mr. R. Attenborough, Whitley Grove, Reading.

DUCHESS MARIE, roan, calved August 23, 1873, Vols. xxiv. and xxv. pp.
329, 344. Bred by Mr. J. Blyth, Woolhampton; got by Duke of Kennet
(30977), dam (Valentine) by Tam O'Shanter (20930), &c.

1879, May 2, roan,	B.C.	Bend Or	Lord Balmoral, 38576	Mr. R. B. Blyth

Bend Or, sold to Mr. G. Fox, Adbury House, Newbury.

PARIS, roan, calved August 20, 1870, Vols. xxi., xxiv., and xxv. pp. 579,
330, 344. Bred by Mr. J. Blyth, Woolhampton; got by Snowball (30020),
dam (Crummie) by Tam O'Shanter (20930), &c.

1879, Mar. 12, white, C.C.	Barbara	Lord Balmoral, 38576	Mr. R. B. Blyth

PITTERI, roan, calved April 18, 1875, Vol. xxv. p. 345. Bred by Mr. R.
Burn Blyth; got by Knight of Geneva (31539), dam (Fair Lady) by Earl
of Walton (17787), &c.

1879, July 27, roan,	B.C.	Beaudesert	Lord Balmoral, 38576	Mr. R. B. Blyth

SYMPHONY, roan, calved November 30, 1869, Vols. xxi., xxiii., xxiv., and
xxv. pp. 579, 340, 330, 345. Bred by Lord Southampton, Whittlebury Lodge;
got by Sam of Oxford (25084), dam (Lady Seraphina) by Imperial Oxford
(18084), &c.

1879, Dec. 1, white,	C.C.	Symphony 3rd	Lord Balmoral, 38576	Mr. R. B. Blyth

BOLTON, Lord,
Bolton Hall, Bedale.

BLANDITIA, white, calved October 13, 1870, Vols. xxi., xxiii., xxiv., and
xxv. pp. 579, 340, 330, 345. Bred by Lord Bolton; got by Marmion (26821),
dam (Amorous) by Strawberry Prince (25240), &c.

Produce in	Names, &c.	By what Bull.	By whom bred.
1879, Nov. 3, roan, B.C.	Bassoon	Heir-at-Law, 34124	Lord Bolton

COCO 1ST, roan, calved September 10, 1870, Vols. xxi., xxii., xxiii., xxiv., and
xxv. pp. 580, 327, 340, 330, 345. Bred by Lord Bolton; got by Marmion
(26821), dam (Graceful) by Dairy Prince (17655), &c.

1879, Mar. 27, roan, B.C.	Clown	Heir-at-Law, 34124	Lord Bolton

PRESTONIA 4TH, calved December 19, 1875, Vol. xxv. p. 345. Bred by Lord
Bolton; got by Admiral Windsor (32912), dam (Prestonia 2nd) by Dandelion
(30849), &c.

1879, Oct. 18, { r.&w.,C.C. / red, C.C.	Prestonia 6th / Prestonia 7th	Heir-at-Law, 34124	Lord Bolton

PRESTONIA 5TH, white, calved December 1, 1876. Bred by Lord Bolton;
got by Heir-at-Law (34124), dam (Prestonia 2nd) by Dandelion (30849), &c.
See " Philologist," p. 186.

1879, July 7, roan, B.C.	Philologist	S.Swthn's Star Drop, 40667	Lord Bolton

ROSE OF THE HILLS, red and white, calved July 6, 1869, Vol. xxiv. p. 331.
Bred by Mr. W. White, Burrill; got by Manfred (26801), dam (Rosetta) by
Lord of the Hills (18267), &c.

1878, Feb. 27, roan, B.C.	Reversioner	Heir-at-Law, 34124	Lord Bolton
1879, Feb. 18, r. & w., B.C.	Rasselas	do.	do.

Reversioner, sold to Mr. John Willis, Carperby, Bedale; Rasselas, to Colonel the Hon. A.
M. Cathcart, Mowbray House. Ripon.

SIMPLICITY, red and white, calved March 10, 1871, Vols. xxi., xxii., xxiv.,
and xxv. pp. 581, 327, 331, 346. Bred by Lord Bolton; got by Marmion
(26821), dam (Shepherdess) by Stonegrave (27575), &c.

1879, Mar. 14, red, C.C.	Simplicity 3rd	Heir-at-Law, 34124	Lord Bolton

SIMPLICITY 1ST, white, calved April 9, 1875. Bred by Lord Bolton; got by
Dandelion (30849), dam (Simplicity) by Marmion (26821), &c. See " Strephon," p. 243.

1878, July 11, white, B.C.	(Steer)	Heir-at-Law, 34124	Lord Bolton
1879, Oct. 20, white, C.C.	Simplicity 4th	do.	do.

SIMPLICITY 2ND, red and white, calved March 9, 1877. Bred by Lord Bolton; got by Heir-at-Law (34124), dam (Simplicity) by Marmion (26821), &c.
See " Strephon," p. 243.

1879, July 29, red, B.C.	Strephon	S.Swthn's Star Drop, 40667	Lord Bolton

SPRIGHTLINESS, roan, calved February 25, 1877. Bred by Lord Bolton;
got by Heir-at-Law (34124), dam (Singleness) by Dandelion (30849), &c. See
" Shipmate," p. 228.

1879, Dec. 29, r. & w., B.C.	Shipmate	Vice-Admiral, 39257	Lord Bolton

VENSELA, roan, calved September 22, 1873. Bred by Lord Bolton ; got by
King Charles (24240), dam (Vignette) by Marmion (26821), &c. See " Vassal," p. 256.

Produce in	Names, &c.	By what Bull.	By whom bred.
1877, June 19, white, B.C.	(Steer)	Heir-at-Law, 34124	Lord Bolton
1879, Mar. 17, roan, B.C.	Vassal	do.	do.
Vassal, sold to Mr. Graves, Selby.			

BOLTON, William,
The Island, Oulart, Ireland.

BUTTERCUP 5TH, red and white, calved January 1, 1872, Vol. xxiv. p. 331.
Bred by Mr. W. Bolton ; got by King Richard (26523), dam (Buttercup 2nd)
by Young Windsor (19156), &c.

1879, Feb. 8, roan, C.C.	Buttercup 9th	Albion, 36112	Mr. Bolton

EVEY GWYNNE, roan, calved April 1, 1871, Vols. xxi. and xxv. pp. 582, 346.
Bred by Mr. W. Bolton ; got by King Richard (26523), dam (Eveline Gwynne)
by Grey Gauntlet (19908), or Royal Duke (20727), &c.

1879, Oct. 12. r. & w., C.C.	Ethel Gwynne	Albion, 36112	Mr. Bolton

GAME HEN 5TH, white, calved July 8, 1873, Vols. xxiv. and xxv. pp. 331,
347. Bred by Mr. W. Bolton ; got by Lieutenant-General (31600), dam
(Game Hen 3rd) by Mercury (22342), &c.

1879, May 26, roan, C.C.	Game Hen 8th	Albion, 36112	Mr. Bolton

GOLDEN RING, red and white, calved July 21, 1870, Vols. xxi. and xxiv.
pp. 545, 332. Bred by Mr. W. Bolton ; got by King Richard (26523), dam
(Golden Rose) by Blood Royal (14169), &c.

1879, May 6, roan, C.C.	Golden Chance	Albion, 36112	Mr. Bolton

PHILLIS GWYNNE, white, calved March 8, 1876. Bred by Mr. W. Bolton ;
got by King James (28971), dam (Pansy Gwynne) by King Richard (26523),
&c. See " Premier," p. 189.

1879, Aug. 5, white, C.C.	Posy Gwynne	Albion, 36112	Mr. Bolton

SUSAN GWYNNE, red and white, calved July 14, 1873, Vol. xxv. p. 347.
Bred by Mr. W. Bolton; got by Lieutenant-General (31600), dam (Sall Gwynne)
by Grey Gauntlet (19908), &c.

1879, Feb. 27, red, C.C.	Satanella Gwynne	Albion, 36112	Mr. Bolton

WILLIE'S GLOSSY, red, calved March 30, 1870, Vol. xxi. p. 585. Bred by
Mr. W. Bolton ; got by Manrico (26805), dam (Island Glossy) by Grey Gauntlet (19908), &c.

1879, Mar. 27, r. & w., C.C.	Anna Glossy	Albion, 36112	Mr. Bolton

BOND, G. M.,
Alrewas House, Ashbourne, Derbyshire.

DUCHESS 28TH, red and white, calved April 24, 1876. Bred by Mr. E. T.
Tunnicliffe, Bromley Hall ; got by Duke of Bromley 2nd (36050), dam
(Duchess 6th) by Earl of Oxford (21651), &c. See " Duchess 10th," Vol. xxiii.
p. 363.

1879, July 1, r. & w., C.C.	Clifton Duchess 1st	Ct. Bickerstaffe 5th, 36400	Mr. Bond

BOOTH, J. B.,
Killerby Hall, Catterick.

BRENHILDA, roan, calved May 7, 1874, Vols. xxiv. and xxv. pp. 351, 371.
Bred by the Hon. M. H. Cochrane, Hillhurst, Canada, the property of Mr. J.
B. Booth; got by Royal Commander (29857), dam (Welcome Lady) by Banner
Bearer (27907), &c.

Produce in	Names, &c.	By what Bull.	By whom bred.
1879, Feb. 28, white, C.C.	Fair Ellen	Muscovite, 38773	Mr. A. H. Browne

CLARA DEA, roan, calved in December 1869, Vol. xxv. p. 348. Bred by Mr.
J. B. Booth; got by K. C. B. (26492), dam (Clara) by Fitz-Clarence (14552),
&c.

1879, Aug. 16, roan, C.C.	Clarissa	Bezique, 33148	Mr. Booth

GIPSIES' DORCAS, red, calved March 30, 1874. Bred by Lord Kinnaird,
Rossie Priory; got by Prince of the Gipsies (27179), dam (Hope's Dorcas)
by Great Hope (24082), &c. See " Brian Fitzalan," p. 32.

1879, May 28, roan, B.C.	Brian Fitzalan	King Brian, 34308	Mr. Booth

HECATE, red and white, calved September 27, 1866, Vols. xix., xxiv., and xxv.
pp. 545, 333, 348. Bred by Mr. J. B. Booth; got by Knight Errant (18154),
dam (Hecuba) by Hopewell (10332), &c.

1879, July 3, r. & w., C.C.	Heroine	K. C. B., 26492	Mr. Booth

MADGE, roan, calved in August 1869, Vols. xxi. and xxiii. pp. 585, 341. Bred
by Mr. J. B. Booth; got by K. C. B. (26492), dam (Mabel) by Hailstone
(21891), &c.

1878, Mar. 2, r. & w., B.C.	Mercury	Janus, 34245	Mr. Booth
1879, May 23, roan, C.C.	Minnie	Bezique, 33148	do.

MAGGIE, red, calved in August 1870, Vols. xxi., xxii., xxiv., and xxv. pp. 585,
328, 333, 348. Bred by Mr. J. B. Booth; got by K. C. B. (26492), dam
(Mabel) by Hailstone (21891), &c.

1879, Aug. 3, roan, B.C.	Meteor	Moonstone, 37107	Mr. Booth

MY PARTNER, roan, calved March 29, 1875. Bred by Mr. J. B. Booth; got
by K. C. B. (26492), dam (Lady Patroness) by Brigade Major (21312), &c.
See " Major Domo," p. 162.

1879, May 19, white, B.C.	Major Domo	King Richard 2nd, 28984	Mr. Booth

PRINCESS BRIGANTINE, red and white, calved November 28, 1874. Bred
by Mr. J. B. Booth; got by Royal Benedict (27348), dam (Queen of the Brigantes) by Brigade Major (21312), &c. See " Last of the K. C. B.'s," p. 137.

1879, Dec. 20, r. & w., B.C.	Last of the K.C. B.'s	K. C. B., 26492	Mr. Booth

PROSERPINE, red and white, calved June 2, 1870, Vols. xxii. and xxiv. pp. 329,
333. Bred by Mr. J. B. Booth; got by K. C. B. (26492), dam (Hecate) by
Knight Errant (18154), &c.

1879, Apr. 11, roan, B.C.	Plutarch	Heart of Oak, 39982	Mr. Booth

QUEEN MARGARET, red, calved September 3, 1873, Vol. xxv. p. 348. Bred
by Mr. J. B. Booth; got by King Richard 2nd (28984), dam (Maggie) by
K. C. B. (26492), &c.

1879, Sept. 3, white, C.C.	Queen Maud	Bezique, 33148	Mr. Booth

QUEEN OF HEARTS, roan, calved November 24, 1871, Vol. xxv. p. 348.
Bred by Mr. J. B. Booth; got by Merry Monarch (22349), dam (Queen of
Trumps) by Welcome Guest (15497), &c.

Produce in	Names, &c.	By what Bull.	By whom bred.
1879, May 28, white, C.C.	Queen Editha	King Harold, 40053	Mr. Booth

REGALIA, roan, calved July 30, 1869, Vols. xxi., xxii., xxiv., and xxv. pp. 586,
329, 333, 348. Bred by Mr. J. B. Booth; got by K. C. B. (26492), dam
(Crown Diamond) by Royal Sovereign (22802), &c.

Produce in	Names, &c.	By what Bull.	By whom bred.
1879, Dec. 4, roan, B.C.	Reigning Monarch	King James, 28971	Mr. Booth

REGINA, roan, calved November 22, 1875. Bred by Mr. J. B. Booth; got by
Royal Benedict (27348), dam (Regalia) by K. C. B. (26492), &c. See Vol. xxii.
p. 329.

Produce in	Names, &c.	By what Bull.	By whom bred.
1879, Dec. 29, white, C.C.	Roy's Wife	King James, 28971	Mr. Booth

WELCOME LADY, roan, calved December 30, 1870, Vol. xxiv. p. 382. Bred by
the Hon. M. H. Cochrane, Hillhurst, Canada; got by Banner Bearer (27907),
dam (Lady of the Lake) by Knight Errant (18154), &c.

Produce in	Names, &c.	By what Bull.	By whom bred.
1879, Apr. 15, r. & w., C.C.	Welcome Lassie	King of Trumps, 31512	Mr. Booth

BOOTH, Miss,
Bagstones, Danebridge, Macclesfield.

DUCHESS 27TH, roan, calved July 2, 1875. Bred by Mr. J. Brett, Burton
Joyce; got by Athenian (30400), dam (Duchess 18th) by Baron Hartforth 4th
(27943), g. d. (Duchess 12th) by Royal Butterfly 12th (20720), &c. See
Vol. xx. p. 485.

Produce in	Names, &c.	By what Bull.	By whom bred.
1879, June 30, roan, C.C.	Fair Saxon	Royal Saxon, 39057	Miss Booth

RUBY 25TH, red and white, calved February 3, 1874, Vol. xxv. p. 349. Bred
by Mr. J. Brett, Burton Joyce; got by Borderer (28064), dam (Ruby 11th)
by Jupiter (18124), &c.

Produce in	Names, &c.	By what Bull.	By whom bred.
1878, Mar. 20, r. & w., C.C.	Zubdeh	Sir John Ridd, 35577	Miss Booth

BOOTH, T. C., Executors of,
Warlaby, Northallerton.

BRIDAL MORN, red and white, calved September 20, 1870, Vols. xxi., xxii.,
xxiii., xxiv., and xxv. pp. 586, 329, 342, 333, 349. Bred by Mr. T. C. Booth;
got by Commander-in-Chief (21451), dam (Blooming Bride) by Prince of Battersea (20567), &c.

Produce in	Names, &c.	By what Bull.	By whom bred.
1879, Dec. 3, roan, C.C.	Bridal Queen	King James, 28971	Exors. Mr. Booth

BRIDAL VEIL, roan, calved May 23, 1875. Bred by Mr. T. C. Booth; got
by Royal Benedict (27348), dam (Bridal Morn) by Commander-in-Chief
(21451), &c. See Vol. xxii. p. 329.

Produce in	Names, &c.	By what Bull.	By whom bred.
1879, Mar. 22, white, C.C.	Bridal Favour	King Harold, 40053	Exors. Mr. Booth

BRIGHT DESIGN, roan, calved February 7, 1875, Vols. xxiv. and xxv. pp.
334, 349. Bred by the Executors of Mr. W. Torr, Aylesby Manor; got by
Knight of the Shire (26552), dam (Bright Spangle) by Prince of Warlaby
(15107), &c.

Produce in	Names, &c.	By what Bull.	By whom bred.
1879, June 6, r. & w., C.C.	Royal Cypher	King Harold, 40053	Exors. Mr. Booth

BRIGHT JEWEL, roan, calved February 1, 1874, Vol. xxv. p. 349. Bred by
Mr. W. Torr, Aylesby Manor; got by Knight of the Shire (26552), dam
(Bright Spangle) by Prince of Warlaby (15107), &c.

Produce in	Names, &c.	By what Bull.	By whom bred.
1879, April 24, white, B.C.	Royal Sceptre	King Harold, 40053	Exors. Mr. Booth

BRIGHT MARCHIONESS, white, calved July 20, 1871, Vols. xxiii., xxiv.,
and xxv. pp. 342, 334, 349. Bred by Mr. W. Torr, Aylesby Manor; got by
Lord Napier (26688), dam (Bright Countess) by Breast Plate (19337), &c.

1879, Sept. 1, white, B.C.	(dead)	Royal Stuart, 40646	Exors. Mr. Booth

BRIGHT SAXON, roan, calved February 22, 1872, Vols. xxi., xxiii., and xxv.
pp. 961, 342, 349. Bred by Mr. W. Torr, Aylesby Manor; got by Royal
Prince (27384), dam (Bright Spangle) by Prince of Warlaby (15107), &c.

1879, Jan. 4, r. & w., C.C.	Saxon Queen	Royal Halnaby, 39041	Exors. Mr. Booth

HOMESPUN, roan, calved October 2, 1875, Vol. xxv. p. 350. Bred by
Mr. T. C. Booth; got by Royal Benedict (27348), dam by King James
(28971), &c.

1879, Mar. 29, roan, C.C.	(dead)	King Harold, 40053	Exors. Mr. Booth

LADY CHEERFUL, white, calved March 18, 1876, Vol. xxv. p. 350. Bred
by Mr. T. C. Booth; got by Royal Benedict (27348), dam (Lady Mirthful) by
British Crown (21322), &c.

1879, Dec. 5, roan, B.C.	Lord Protector	Royal Stuart, 40646	Exors. Mr. Booth

MARCHIONESS, red and white, calved August 12, 1874, Vols. xxiv. and xxv.
pp. 334, 350. Bred by Mr. T. C. Booth; got by Royal Benedict (27348), dam
(Margaret) by Commander-in-Chief (21451), &c.

1879, Dec. 15, roan, B.C.	King David	King James, 28971	Exors. Mr. Booth

MARCIA, white, calved March 16, 1873, Vols. xxii. and xxiv. pp. 329, 334.
Bred by Mr. T. C. Booth; got by King James (28971), dam (Melissa) by
General Hopewell (17953), &c.

1878, Jan. 27, roan, B.C.	Sir Andrew	Royal Benedict, 27348	Mr. Booth
1879, Jan. 11, roan, C.C.	Mortimer	Royal Halnaby, 39041	Exors. Mr. Booth

META, roan, calved January 20, 1877. Bred by Mr. T. C. Booth; got by Royal
Benedict (27348), dam (Margaret) by Commander-in-Chief (21451), &c. See
" King Malcolm," p. 131.

1879, Nov. 3, white, B.C.	King Malcolm	King James, 28971	Exors. Mr. Booth

MIRIAM, red and white, calved March 11, 1876, Vol. xxv. p. 350. Bred by
Mr. T. C. Booth; got by Royal Benedict (27348), dam (Martha) by King
James (28971), &c.

1879, May 20, r. & w., C.C.	Mavourneen	King Harold, 40053	Exors. Mr. Booth

MURIEL, roan, calved January 15, 1874, Vols. xxiii. and xxv. pp. 343, 351.
Bred by Mr. T. C. Booth; got by King Richard (26523), dam (Margery) by
Commander-in-Chief (21451), &c.

1879, Jan. 16, roan, B.C.	(dead)	Royal Halnaby, 39041	Exors. Mr. Booth

MYRA, white, calved September 13, 1875, Vol. xxv. p. 351. Bred by Mr. T. C.
Booth; got by Royal Benedict (27348), dam (Lady Maude) by Lord Blithe
(22126), &c.

1879, Aug. 14, white, C.C.	Mary Stuart	Royal Stuart, 40646	Exors. Mr. Booth

ROYAL LADY, red and white, calved August 23, 1877. Bred by Mr. T. C.
Booth; got by Royal Benedict (27348), dam (Riby Lassie) by Blinkhoolie
(23428), &c. See Vol. xxiv. p. 335.

Produce in	Names, &c.	By what Bull.	By whom bred.
1879, Dec. 2, roan,	C.C. Royal Damsel	King James, 28971	Exors. Mr. Booth

BOTTERILL, R.,
Wauldby, Brough, Yorkshire.

BEVERLEY DUCHESS, roan, calved October 25, 1871, Vols. xxi., xxiii., and
xxv. pp. 587, 344, 351. Bred by Mrs. Dawson, Weston Hall; got by Grand
Duke of Weston (28768), dam (Miss Beverley 2nd) by Thorndale Lad (23066),
&c.

1879, Nov. 22, white, B.C.	Beverley Oxford 9th	Oxford-le-Grand, 29496	Mr. Botterill

BEVERLEY DUCHESS 3RD, red and white, calved December 7, 1872, Vols.
xxii., xxiii., xxiv., and xxv. pp. 330, 344, 335, 352. Bred by Mrs. Dawson,
Weston Hall; got by Grand Duke of Weston (28768), dam (Miss Beverley
2nd) by Thorndale Lad (23066), &c.

1879, May 19, roan, B.C.	Beverley Oxford 8th	Oxford-le-Grand, 29496	Mr. Botterill

BEVERLEY DUCHESS 7TH, red and white, calved December 1, 1874, Vols.
xxiv. and xxv. pp. 336, 352. Bred by Mr. R. Botterill; got by Oxford-le-
Grand (29496), dam (Beverley Duchess 2nd) by Grand Duke of Weston
(28768), &c.

1879, Apr. 25, r. & w., C.C.	Bev'rley Duchess 17th	B.T'croft Oxford 3rd, 36209	Mr. Botterill

BEVERLEY DUCHESS 11TH, red and white, calved April 18, 1877. Bred
by Mr. R. Botterill; got by Baron Turncroft Oxford 3rd (36209), dam
(Beverley Duchess 7th) by Oxford-le-Grand (29496), &c. See Vol. xxiv.
p. 336.

1879, Oct. 16, roan, C.C.	Bev'rley Duchess 20th	Oxford-le-Grand, 29496	Mr. Botterill

BEVERLEY DUCHESS 12TH, red, calved May 23, 1877. Bred by Mr. R.
Botterill; got by Oxford-le-Grand (29496), dam (Beverley Duchess 3rd) by
Grand Duke of Weston (28768), &c. See Vol. xxiv. p. 335.

1879, Sept. 19, red, C.C.	Bev'rley Duchess 19th	Fordh'm D.Oxf'd 4th, 41569	Mr. Botterill

BLOSSOM 6TH, roan, calved January 7, 1876, Vol. xxv. p. 352. Bred by Mr.
G. Allen, Knightley Hall; got by Duke of Wetherby 6th (33756), dam
(Blossom 2nd) by Eighth Duke of York (28480), &c.

1879, Sept. 1, r. & w., B.C.	Baron Peach 2nd	Oxford-le-Grand, 29496	Mr. Botterill

BLOSSOM 11TH, roan, calved April 19, 1877. Bred by Mr. G. Allen, Knightley Hall; got by Duke of Wetherby 6th (33756), dam (Blossom 5th) by
Second Duke of Wetherby (21618), &c. See Vol. xxiv. p. 292.

1879, Oct. 22, roan, B.C.	Baron Peach 3rd	Oxford-le-Grand, 29496	Mr. Botterill

BRAMBLE, roan, calved May 27, 1873, Vol. xxiv. p. 336. Bred by the Earl
of Zetland, Upleatham; got by Grand Monarch (28774), dam (Bellona) by
Musician (26936), &c.

1879, Feb. 23, roan, B.C.	Prince Gwynne 3rd	Oxford-le-Grand, 29496	Mr. Botterill

CHERRY OXFORD 4TH, roan, calved September 18, 1875, Vol. xxv. p. 352.
Bred by Mr. W. Ashburner, Conishead Grange; got by Second Duke of
Glo'ster (28392), dam (Cherry Oxford) by Thirteenth Duke of Oxford
(21604), &c.

Produce in	Names, &c.	By what Bull.	By whom bred.
1879, July 24, red,	C.C.|Cherry Oxford 6th	Fordh'mD.Oxf'd4th,41569	Mr. Botterill

COWSLIP 25TH, roan, calved August 21, 1874, Vol. xxiii. p. 713. Bred by
Mr. T. Wilson, Shotley Hall, the property of Mr. J. R. Singleton, Givendale;
got by Waterloo Duke (32815), dam (Cowslip 12th) by Beverley Butterfly
(28021), &c.

1878, Jan. 6, r. & w., B.C.		Duke of Oxford 31st, 33713	Mr. Wilson
1879, Jan. 17, r. & w., C.C.	Oxford Cowslip	do.	Mr. Botterill

Oxford Cowslip, sold to Mr. W. Parkin, Blaithwaite.

DAYLIGHT, roan, calved February 25, 1874, Vol. xxv. p. 353. Bred by the
Earl of Zetland, Upleatham, the property of Mr. J. S. Spence, Skirlaugh;
got by Grand Monarch (28774), dam (Dance) by King of the Roses (22043),
&c.

1879, April 8, red,	B.C.|Prince Gwynne 5th	Oxford-le-Grand, 29496	Mr. Botterill

DESTINY, roan, calved March 12, 1870, Vols. xxi., xxii., xxiii., and xxiv. pp.
588, 330, 344, 336. Bred by the Earl of Zetland, Upleatham, the property
of Mr. J. H. Stephenson, Sancton Grange; got by Doubtful (25914), dam
(Duenna) by Grand Duke of Essex 2nd (21860), &c.

1879, Mar. 10, white, B.C.	Prince Gwynne 4th	Oxford-le-Grand, 29496	Mr. Botterill

Prince Gwynne 4th, sold to Mr. A. Botterill, Garton, Driffield, Yorks.

DUCHESS VANE, white, calved May 1, 1872, Vols. xxii., xxiii., xxiv., and
xxv. pp. 330, 344, 336, 353. Bred by Mr. R. Botterill; got by Duke of
Clarence (19611), dam (Vanity) by Thorndale Lad (23066), &c.

1879, Mar. 3, roan,	B.C.|Oxford 24th	Oxford-le-Grand, 29496	Mr. Botterill

Oxford 24th, sold to Mr. C. S. Holgate, Lincolnshire.

FIDELITY 2ND, roan, calved January 8, 1870, Vol. xxv. p. 353. Bred by Mr.
J. W. Wadsworth, Geneseo, U.S.A.; got by Millbrook (34851), dam (Frantic
4th) by Duke of Oxford (33703), &c.

1879, April 9, roan,	C.C.|Lady Fidget Bates 3rd|Oxford-le-Grand, 29496	Mr. Botterill

FIDELITY 3RD, roan, calved May 31, 1871, Vols. xxiv. and xxv. pp. 372, 353.
Bred by Mr. J. W. Wadsworth, Geneseo, U.S.A.; got by Millbrook (34851),
dam (Frantic 4th) by Duke of Oxford (33703), &c.

1879, July 24, roan,	C.C.|Lady Fidget Bates 4th|Oxford-le-Grand, 29496	Mr. Botterill

GWYNNE PRINCESS 2ND, roan, calved March 25, 1873, Vols. xxii., xxiv.,
and xxv. pp. 331, 336, 353. Bred by the Earl of Zetland, Upleatham; got
by Grand Monarch (28774), dam (Dance) by King of the Roses (22043), &c.

1879, Feb. 2, white,	B.C.|Prince Gwynne 2nd	Oxford-le-Grand, 29496	Mr. Botterill

GWYNNE PRINCESS 9TH, red and white, calved March 31, 1877. Bred by
Mr. R. Botterill; got by Grand Duke 26th (34066), dam (Dance) by King of
the Roses (22043), &c. See Vol. xxiv. p. 336.

1879, Sept. 11, r. & w., B.C.	Prince Gwynne 6th	Oxford-le-Grand, 29496	Mr. Botterill

HEYDON ROSE 3RD, white, calved May 8, 1873, Vols. xxiii. and xxv. pp. 353, 363. Bred by Lord Braybrooke, Audley End; got by Grand Duke 17th (24064), dam (Heydon Rose) by Englishman (19701), &c.

Produce in	Names, &c.	By what Bull.	By whom bred.
1879, July 13, roan, C.C.	Cambridge Rose 7th	Duke of Rosedale 3rd, 33723	Mr. Botterill

LADY ACOMB, roan, calved February 11, 1877. Bred by Mr. R. Botterill; got by Oxford-le-Grand (29496), dam (Abundance) by General Napier (24023), &c. See Vol. xxiv. p. 335.

1879, Sept. 23, roan, C.C.	Lady Acomb 3rd	Fordh'm D.Oxf'd 4th, 41569	Mr. Botterill

LADY FORTUNATE 2ND, roan, calved February 19, 1871, Vols. xxi., xxii., xxiii., and xxiv. pp. 588, 331, 345, 337. Bred by Mr. R. Botterill, the property of Mr. J. H. Angas, Collingrove, Adelaide, South Australia; got by Volunteer (30239), dam (Lady Fortunate) by Second Duke of Wharfdale (19649), &c.

1879, May 21, roan, C.C.	Lady Fortunate 18th	Oxford-le-Grand, 29496	Mr. Botterill

LADY FORTUNATE 3RD, roan, calved June 19, 1873, Vol. xxii. p. 331. Bred by Mr. R. Botterill, the property of Mr. W. Parkin, Blaithwaite; got by Nineteenth Duke of Oxford (28431), dam (Lady Fortunate 2nd) by Volunteer (30239), &c.

1879, Mar. 11, red, B.C.	Oxford 25th	Oxford-le-Grand, 29496	Mr. Botterill

Oxford 25th, sold to Mr. J. S. Egginton, Kirk Ella, Hull.

LADY FORTUNATE 4TH, roan, calved October 12, 1873, Vols. xxiii., xxiv., and xxv. pp. 345, 337, 353. Bred by Mr. R. Botterill, the property of Mr. J. H. Angas, Collingrove, Adelaide, South Australia; got by Nineteenth Duke of Oxford (28431), dam (Lady Fortunate) by Second Duke of Wharfdale (19649) &c.

1879, Feb. 23, roan, C.C.	Lady Fortunate 16th	Oxford-le-Grand, 29496	Mr. Botterill

LADY FORTUNATE 5TH, roan, calved September 30, 1874, Vols. xxiv. and xxv. pp. 337, 354. Bred by Mr. R. Botterill; got by Oxford-le-Grand (29496), dam (Lady Fortunate 2nd) by Volunteer (30239), &c.

1879, April 14, red, C.C.	Lady Fortunate 17th	L'd of Dalmally 2nd, 40209	Mr. Botterill

LADY FORTUNATE 7TH, white, calved August 18, 1875, Vols. xxiv. and xxv. pp. 337, 354. Bred by Mr. R. Botterill; got by Oxford-le-Grand (29496), dam (Lady Fortunate 2nd) by Volunteer (30239), &c.

1879, April 15, r. & w., B.C.	Oxford 26th	L'd of Dalmally 2nd, 40209	Mr. Botterill

LADY FORTUNATE 9TH, roan, calved April 7, 1876, Vol. xxv. p. 354. Bred by Mr. R. Botterill; got by Oxford-le-Grand (29496), dam (Lady Fortunate 4th) by Nineteenth Duke of Oxford (28431), &c.

1879, Dec. 22, roan, C.C.	Lady Fortunate 19th	Fordh'm D.Ox'd 4th, 41569	Mr. Botterill

LADY HILDA 7TH, white, calved April 25, 1875, Vol. xxiv. p. 293. Bred by Mr. G. Allen, Knightley Hall; got by Duke of Wetherby 6th (33756), dam (Lady Hilda 2nd) by Eighth Duke of York (28480), &c.

1879, Sept. 6, roan, B.C.	Oxford 28th	Oxford-le-Grand, 29496	Mr. Botterill

LADY HILDA 10TH, white, calved March 27, 1876. Bred by Mr. G. Allen, Knightley Hall; got by Duke of Wetherby 6th (33756), dam (Lady Hilda) by Lord Liverpool (22168), &c. See Vol. xxiii. p. 309.

1879, Mar. 23, roan, C.C.	Lady Hudson Bates	2nd Duke of Rowley, 28441	Mr. Botterill

LADY JANE, roan, calved December 1, 1872. Bred by Mr. F. N. Sartoris, Rushden Hall, the property of Mr. J. Webb, Melton Ross; got by Duke of Kingscote (25981), dam (Oxford Jantja 10th) by Twelfth Duke of Oxford (19633), &c. See "Janus," Vol. xxv. p. 139.

Produce in	Names, &c.	By what Bull.	By whom bred.
1878, Feb. 25, roan, B.C.	Janus	Earl of Leicester 7th, 33806	Mr. Sartoris
1879, Feb. 7, roan, C.C.	J. Princess 3rd	L. Clarence Waterloo, 36926	Mr. Botterill

Janus, sold to Mr. C. Ashton, Delrow House, Watford; J. Princess 3rd, to Mr. J. Webb, Melton Ross, Ulceby.

LADY OF LYONS 2ND, white, calved July 30, 1871, Vols. xxi., xxii., xxiii., xxiv., and xxv. pp. 589, 332, 345, 337, 354. Bred by Mr. R. Botterill, the property of Mr. M. Bust, West Halton, Brigg; got by Van Thol (32755), dam (Lady of Lyons) by Lord Cobham (20164), &c.

1879, Mar. 13, roan, C.C.	Lady of Lyons 11th	Oxford-le-Grand, 29496	Mr. Botterill

LADY WATERLOO 14TH, red and white, calved February 14, 1867, Vols. xxi., xxiii., and xxiv. pp. 589, 345, 337. Bred by Mr. J. R. Singleton, Givendale; got by Second Lord of Waterloo (22198), dam (Lady Waterloo 11th) by Patriot (20475), &c.

1879, July 26, r.&w., B.C.	F M. W'lington 2nd	Oxford-le-Grand, 29496	Mr. Botterill

LADY WATERLOO 27TH, roan, calved May 13, 1873, Vols. xxiii., xxiv., and xxv. pp. 682, 687, 698. Bred by Mr. E. H. Cheney, Gaddesby Hall; got by Twentieth Duke of Oxford (28432), dam (Lady Waterloo 11th) by Patriot (20475), &c.

1879, May 19, roan, C.C.	D'ss of Wellington 2d	Duke of Oxford 27th, 33709	Mr. Botterill

LADY WATERLOO 30TH, roan, calved May 12, 1876. Bred by Sir W. C. Trevelyan, Bart., Wallington; got by Oxford Beau 4th (34964), dam (Lady Waterloo 27th) by Twentieth Duke of Oxford (28432), &c. See Vol. xxiii. p. 682.

1879, May 19, roan, C.C.	D'ss of Wellington 3d	Duke of Oxford 27th, 33709	Mr. Botterill

MINSTREL PRINCESS, red and white, calved September 14, 1877. Bred by Mr. J. Martin, Bardsea; got by Duke of Oxford 34th (36529), dam (Minstrel 6th) by Sixth Duke of Kirklevington (30982), &c. See Vol. xxiv. p. 565.

1879, Dec. 8, r. & w., C.C.	Gwynne Pr'cess 16th	B'n Siddington 2nd, 37816	Mr. Botterill

MISS BEVERLEY 3RD, roan, calved May 21, 1870, Vols. xxi., xxiv., and xxv. pp. 589, 338, 354. Bred by Mrs. Dawson, Weston Hall; got by Thorndale Lad (23066), dam (Miss Beverley 22nd) by Royal Butterfly 16th (20724), &c.

1879, June 6, white, C.C.	Bev'ley Duchess 18th	Oxford-le-Grand, 29496	Mr. Botterill

OXFORD'S DOWAGER, roan, calved March 10, 1876. Bred by Mr. R. P. Davies, Horton; got by Oxford's King (34997), dam (Dowager Duchess 2nd) by Grand Duke 11th (21849), &c. See Vol. xxiii. p. 414.

1879, Nov. 3, roan, C.C.	Nancy Duchess 1st	L. T'croft Oxford 2nd, 38668	Mr. Botterill

WESTON DUCHESS 4TH, roan, calved June 25, 1877. Bred by Mr. R. Botterill, the property of Mr. W. Parkin, Blaithwaite; got by Oxford-le-Grand (29496), dam (Weston Duchess) by Grand Duke of Weston (28768), &c. See Vol. xxiv. p. 338.

1879, June 13, roan, B.C.	Oxford 27th	Oxford 22nd, 43716	Mr. Botterill

BOURKE, W. C.,
Ballynahown, Fermoy, Ireland.

ALMA 5TH, roan, calved September 27, 1874, Vol. xxiv. p. 362. Bred by Mr. A. J. Campbell, Ballynahown; got by Vain Hope (23102), dam (Alma 4th) by Roan Robin (27306), &c.

Produce in	Names, &c.	By what Bull.	By whom bred.
1879, Jan. 20, r. & w., C.C.	Alma 8th	Prince of Killerby, 37256	Mr. Bourke

LOVESOME 8TH, red and white, calved June 3, 1875, Vol. xxv. p: 355. Bred by Mr. A. J. Campbell, Ballynahown; got by Vain Hope (23102), dam (Lovesome 3rd) by Leviathan (20120), &c.

1879, Jan. 23, r. & w., C.C.	Lovesome 11th	Prince of Killerby, 37256	Mr. Bourke

ROSE OF THE VALLEY, red, calved February 14, 1875. Bred by Mr. A. J. Campbell, Ballynahown; got by Vain Hope (23102), dam (Young Moss Rose) by Dr. McHale (15887), &c. See Vol. xxii. p. 351.

1879, Mar. 5, roan, C.C.	Rosamond	Robert Stephenson, 32313	Mr. Bourke

VENUS 5TH, roan, calved June 6, 1869, Vols. xxi., xxiii., and xxiv. pp. 616, 370, 363. Bred by Mr. R. Smith, Blossomfort; got by Prairie King (24768), dam (Venus) by Western Wonder (17225), &c.

1879, Feb. 22, red, C.C.	Venus 6th	Prince of Killerby, 37256	Mr. Bourke

BOWERS, William,
Harewood Park, Cheadle, Stoke-on-Trent.

HOPBINE, roan, calved September 27, 1871, Vols. xxi., xxiii., xxiv., and xxv. pp. 897, 346, 338, 356. Bred by Mr. T. Purkis, West Wratting Grange; got by Sir Rainald (27485), dam (Foggathorpe 2nd) by Liston (22105), &c.

1879, May 1, roan, C.C.	The Gem	Harewood, 36741	Mr. Bowers

ZENOBIA 8TH, roan, calved November 18, 1868, Vols. xx., xxi., xxiii., xxiv. and xxv. pp. 829, 898, 347, 339, 356. Bred by Mr. T. Purkis, West Wratting Grange; got by Liston (22105), dam (Zenobia 5th) by Mason (20299), &c.

1879, May 28, roan, C.C.	Zenobia 14th	Harewood, 36741	Mr. Bowers

ZENOBIA 10TH, roan, calved February 4, 1874. Bred by Mr. W. Bowers; got by Zachariah (32894), dam (Zenobia 8th) by Liston (22105), &c. See Vol. xxi. p. 898.

1879, Mar. 2, roan, C.C.	Edith	Zachariah 5th, 40943	Mr. Bowers

BOWLY, Edward,
Siddington House, Cirencester.

ASTREA, roan, calved July 2, 1876. Bred by Archdeacon Holbech, Farnborough Hall; got by The Bursar (35742), dam (Asenath) by Duke of Brailes (23724), &c. See "Lord Acomb," p. 140.

1879, Jan. 31, roan, B.C.	Lord Acomb	Duke of Hillhurst, 28401	Mr. Bowly

DEWLAP, roan, calved September 17, 1871, Vols. xxii., xxiii., xxiv., and xxv. pp. 531, 347, 339, 356. Bred by Sir G. R. Philips, Bart., Weston Park; got by Third Duke of Geneva (21592), dam (Diadem) by Touchstone (20986), &c.

1879, May 14, roan, B.C.	Oxford Swell 7th	Beau of Oxford 2nd, 33129	Mr. Bowly

FLORIDA, white, calved July 27, 1875, Vol. xxv. p. 357. Bred by Mr. J. P. Foster, Killhow; got by Twenty-second Duke of Oxford (31000), dam (Fortuna) by Seventeenth Duke of Oxford (25994), &c.

Produce in	Names, &c.	By what Bull.	By whom bred.
1879, Sept. 8, white, B.C.	Lord Holker 2nd	Duke of Holker 2nd, 39749	Mr. Bowly

Lord Holker 2nd, sold to Mr. A. B. Gregory, Wraxall, Bristol.

GAZELLE 23RD, red, calved December 9, 1870, Vols. xxi., xxiii., xxiv., and xxv. pp. 592, 347, 340, 357. Bred by Mr. E. Bowly; got by Second Duke of Tregunter (26022), dam (Gazelle 8th) by Seventh Duke of York (17754), &c.

Produce in	Names, &c.	By what Bull.	By whom bred.
1879, April 12, r. & w., B.C.	Oxford Swell 6th	Beau of Oxford 2nd, 33129	Mr. Bowly

GAZELLE 28TH, roan, calved October 19, 1872, Vols. xxi., xxiii., xxiv., and xxv. pp. 592, 347, 340, 357. Bred by Mr. E. Bowly; got by Second Duke of Tregunter (26022), dam (Gazelle 5th) by Seventh Duke of York (17754), &c.

Produce in	Names, &c.	By what Bull.	By whom bred.
1879, Oct. 10, roan, C.C.	Gazelle 41st	Beau of Oxford 2nd, 33129	Mr. Bowly

GAZELLE 30TH, roan, calved April 26, 1873, Vols. xxiii. and xxiv. pp. 347, 340. Bred by Mr. E. Bowly; got by Third Duke of Clarence (23727), dam (Gazelle 21st) by Second Duke of Tregunter (26022), &c.

Produce in	Names, &c.	By what Bull.	By whom bred.
1879, Oct. 30, white, C.C.	Gazelle 42nd	Duke of Holker 2nd, 39749	Mr. Bowly

GAZELLE 36TH, roan, calved September 18, 1877. Bred by Mr. E. Bowly; got by Beau of Oxford 2nd (33129), dam (Gazelle 4th) by Seventh Duke of York (17754), &c. See "Lord Holker 3rd," p. 149.

Produce in	Names, &c.	By what Bull.	By whom bred.
1879, Dec. 7, white, B.C.	Lord Holker 3rd	Duke of Holker 2nd, 39749	Mr. Bowly

GERTRUDE, roan, calved November 16, 1874, Vols. xxiv. and xxv. pp. 340, 357. Bred by Mr. J. P. Foster, Killhow; got by Duke of Killhow (30980), dam (Gazelle 10th) by Seventh Duke of York (17754), &c.

Produce in	Names, &c.	By what Bull.	By whom bred.
1879, April 12, roan, B.C.	Lord Holker	Duke of Holker 2nd, 39749	Mr. Bowly

LILY GAZELLE, white, calved September 17, 1875, Vol. xxiv. p. 528. Bred by Mr. J. W. Larking, Ashdown House; got by Third Duke of Clarence (23727), dam (Gazelle 29th) by Second Duke of Tregunter (26022), &c.

Produce in	Names, &c.	By what Bull.	By whom bred.
1879, Jan. 29, roan, C.C.	Gazelle 40th	3rd D. of Hillhurst, 30975	Mr. Bowly

MELODY, red and white, calved March 10, 1870, Vols. xxi., xxiii., xxiv., and xxv. pp. 887, 348, 340, 357. Bred by the Earl of Dunmore, Dunmore; got by Second Duke of Collingham (23730), dam (Musical 9th) by Seventh Duke of York (17754), &c.

Produce in	Names, &c.	By what Bull.	By whom bred.
1879, Mar. 6, red, B.C.	Oxford Swell 5th	Beau of Oxford 2nd, 33129	Mr. Bowly

MINSTREL 2ND, white, calved March 4, 1871, Vols. xxi., xxiii., and xxv. pp. 592, 348, 357. Bred by Mr. E. Bowly; got by Second Duke of Tregunter (26022), dam (Musical 7th) by Seventh Duke of York (17754), &c.

Produce in	Names, &c.	By what Bull.	By whom bred.
1879, Mar. 19, white, C.C.	Minstrel 13th	Beau of Oxford 2nd, 33129	Mr. Bowly

MINSTREL 6TH, white, calved October 26, 1875, Vol. xxv. p. 357. Bred by Mr. E. Bowly; got by Third Duke of Clarence (23727), dam (Minstrel 3rd) by Second Duke of Tregunter (26022), &c.

Produce in	Names, &c.	By what Bull.	By whom bred.
1879, Mar. 25, roan, C.C.	Minstrel 14th	Duke of Holker 2nd, 39749	Mr. Bowly

MINSTREL 9TH, roan, calved March 14, 1877. Bred by Mr. E. Bowly; got by Third Duke of Clarence (23727), dam (Melody) by Second Duke of Collingham (23730), &c. See "Oxford Swell 5th," p. 181.

Produce in	Names, &c.	By what Bull.	By whom bred.
1879, May 30, r. & w., C.C.	Minstrel 15th	Duke of Holker 2nd, 39749	Mr. Bowly

MUSICAL 7TH, roan, calved February 12, 1866, Vols. xix., xx., xxi., xxiii., xxiv., and xxv. pp. 642, 666, 592, 348, 340, 357. Bred by Mr. E. Bowly; got by Seventh Duke of York (17754), dam (Harpsichord) by Earl of Walton (17787), &c.

Produce in	Names, &c.	By what Bull.	By whom bred.
1879, Oct. 24, roan, C.C.	Minstrel 16th	Duke of Holker 2nd, 39749	Mr. Bowly

QUEEN OF WESTON 3RD, red, calved November 10, 1873, Vols. xxiii., xxiv., and xxv. pp. 348, 341, 358. Bred by Sir G. R. Philips, Bart., Weston Park; got by Cherry Fawsley (30711), dam (Queen of Weston 2nd) by Duke of Kent (25979), &c.

| 1879, May 27, roan, C.C. | Lady Weston 4th | Duke of Holker 2nd, 39749 | Mr. Bowly |

RUBY 5TH, red and white, calved February 3, 1872, Vols. xxi. and xxiii. pp. 593, 348. Bred by Mr. E. Bowly; got by Second Duke of Tregunter (26022), dam (Ruby 2nd) by Seventh Duke of York (17754), &c.

| 1879, Feb. 28, red, B.C. | The Gem | L. Fitzclarence 15th, 36942 | Mr. Bowly |

BOWNESS, Moses,
Low Wray, Ambleside, Westmoreland.

BELLE MAHON, white, calved December 4, 1876. Bred by Mr. R. Taylor, New House; got by Duke of Kirkby (33682), dam (Belle of Greengill) by Marplot's Farewell (34767), &c. See Vol. xxiv. p. 670.

| 1879, Dec. 4, white, B.C. | Sir Henry Tufton | Major Irwin, 34735 | Mr. Bowness |

CHERRY RUBY 2ND, roan, calved February 27, 1877. Bred by Mr. G. Ashburner, Low Hall; got by Cherry Duke of Lightburne (36349), dam (Ruby 9th) by Oxford (20449), &c. See Vol. xxv. p. 320.

| 1879, Dec. 27, roan, C.C. | (dead) | Duke of Wray, 43150 | Mr. Bowness |

BOYES, Matthew,
Wandale House, Slingsby, York.

SPECIAL NOTICE, red and white, calved October 29, 1872, Vol. xxii. p. 562. Bred by Mr. J. Stephenson, Wheldrake; got by Duke Royal 2nd (26038), dam (Seven Stars) by Sensation (22869), &c.

| 1877, May 20, roan, B.C. | Speculation | Jocelyn's Tregunter, 36808 | Mr. Boyes |
| 1878, June 30, r. & w., C.C. | Sophia | L. Oxf'd Br't Eyes 2d, 36998 | do. |

BRACKENBURY, W. T.,
Thorpe Hall, Downham, Norfolk.

BEAUTY OF THORPE, roan, calved July 31, 1877. Bred by Mr. W. T. Brackenbury; got by Knight of Foggathorpe (34372), dam (Beauty) by El Sol (15986), &c. See "Thorpe Knightley," Vol. xxii. p. 266.

| 1879, Oct. 17, roan, C.C. | Beauty of Fog'thorpe | Lord Foggathorpe, 43509 | Mr. Brackenbury |

JAUNTY, red, calved January 28, 1872, Vol. xxiii. p. 349. Bred by Mr. W. T. Brackenbury; got by Jason (26453), dam (Smart 3rd) by Prince Imperial (22594), &c.

| 1879, Jan. 23, r. & w., C.C. | Joy | Havelock, 36749 | Mr. Brackenbury |

MARION, red, calved June 16, 1874, Vol. xxiii. p. 349. Bred by Mr. W. T. Brackenbury; got by Jason (26453), dam (Marchioness) by Prince Imperial (22594), &c.

Produce in	Names, &c.	By what Bull.	By whom bred.
1879, Mar. 22, roan, C.C.	Minnehaha	Havelock, 36749	Mr. Brackenbury

WATERLOO ROSEBUD, roan, calved January 25, 1876, Vol. xxv. p. 358. Bred by the Rev. J. N. Micklethwait, Taverham Hall; got by Third Duke of Geneva (23753), dam (Waterloo Rose 3rd) by Third Duke of Geneva (23753), &c.

1879, June 23, roan, C.C.	Wat'loo Rosebud 3rd	Havelock, 36749	Mr. Brackenbury

BRAIKENRIDGE, J. H.,
The Rookery, Chew Magna, Somerset.

ANNA 7TH, red, calved January 22, 1873, Vols. xxii., xxiv., and xxv. pp. 534, 341, 359. Bred by the Rev. J. Storer, Hellidon; got by Rosedale Favourite (29831), dam (Anna 3rd) by Mantalini Prince (22276), &c.

1879, July 3, { roan, C.C. Bright Eyes / roan, C.C. Anna 13th }	Pioneer, 38868	Mr. Braikenridge

BEESWING 2ND, roan, calved April 12, 1869, Vol. xxiv. p. 341. Bred by Mr. A. B. Winnall, The Hawthorns; got by Cantab (23508), dam (Beeswing) by Wellington (21090), &c.

1879, May 24, roan, C.C.	Beeswing 7th	Pioneer, 38868	Mr. Braikenridge

COWSLIP 17TH, white, calved March 29, 1871, Vols. xxii., xxiv., and xxv. pp. 334, 342, 359. Bred by Mr. R. Welsted, Ballywalter; got by Prince Christian (22581), dam (Cowslip 13th) by Sir James (16980), &c.

1879, May 12, white, C.C.	Cowslip 21st	Pioneer, 38868	Mr. Braikenridge

GOLDEN FLEECE, roan, calved March 5, 1877. Bred by Mr. R. Jefferson, Preston Hows; got by Banner Bearer (27907), dam (Golden Fruit) by King Richard (26523), &c. See "Grey Somerset," p. 116.

1879, Sept. 6, roan, C.C.	Golden Pet	Pioneer, 38868	Mr. Braikenridge

GOLDEN FRUIT, roan, calved August 6, 1871, Vols. xxi. and xxv. pp. 544, 359. Bred by Mr. S. Armstrong, Gally House; got by King Richard (26523), dam (Golden Locket) by Grey Gauntlet (19908), &c.

1879, Aug. 10, roan, B.C.	Grey Somerset	Lord Prinknash 2nd, 38653	Mr. Braikenridge

ISABELLA, roan, calved April 25, 1875. Bred by Messrs. R. Fisher and Son, Leconfield; got by Rosary Monk (35316), dam (Isabel) by Lord Greta (20174), g. d. (Red Rose) by Falstaff (14530), &c. See "Red Rose 2nd," Vol. xx. p. 721.

1879, Nov. 9, r. & w., C.C.	Isabel	Pride of the Isle, 42153	Mr. Braikenridge

KATHLEEN MAVOURNEEN, roan, calved November 7, 1874, Vols. xxiv. and xxv. pp. 342, 359. Bred by Mr. J. C. Bowstead, Hackthorpe Hall; got by Monarch (31930), dam (Katherine 6th) by Prince Regent (35169), &c.

1879, May 29, roan, B.C.	King Charles	Pioneer, 38868	Mr. Braikenridge

LUNAIRE, red and white, calved May 18, 1874.   Bred by Mr. H. Pickersgill, Middleton Quernhow; got by Prince William (32217), dam (Luna) by Prince George (13510), &c.   See "Moonraker," p. 173.

Produce in		Names, &c.	By what Bull.	By whom bred.
1879, April 15, roan,	B.C.	Moonraker	Pioneer, 38868	Mr. Braikenridge

OLIVE BRANCH, white, calved April 24, 1873, Vol. xxiv. p. 342.   Bred by Mr. J. H. Casswell, Laughton, the property of Mr. W. Cox, Badbury, Swindon; got by Baron York (30500), dam (Olive Leaf) by Fitz Sir James (19761), &c.

1879, Jan. 12, white,	C.C.	White Dahlia	Pioneer, 38868	Mr. Braikenridge

PENANCE, roan, calved February 2, 1870, Vols. xxi. and xxiv. pp. 637, 342. Bred by Messrs. T. Garne and Son, Broadmoor; got by Royal Benedict (27348), dam (Penelope) by Cynric (19542), &c.

1879, May 12, white,	C.C.	Penitence	Pioneer, 38868	Mr. Braikenridge

PORTRAIT 6TH, roan, calved February 26, 1874.   Bred by Mr. G. Garne, Churchill Heath; got by Buccaneer (25693), dam (Penelope) by Cynric (19542), &c.   See Vol. xxi. p. 728.

1879, April 8, r. & w.,	C.C.	Pretty Star	Star Regent, 35679	Mr. Braikenridge

QUEEN'S BUTTERFLY, red, calved March 12, 1874, Vols. xxiv. and xxv. pp. 343, 359.   Bred by Mr. W. A. Barnes, Westland; got by King James (28971), dam (Duke's Butterfly) by Duke of York (23804), &c.

1879, April 6, white,	B.C.	White Butterfly	Pioneer, 38868	Mr. Braikenridge

ROCK ROSE 65TH, roan, calved January 9, 1875.   Bred by Messrs. R. Fisher and Son, Leconfield; got by Rosary Monk (35316), dam (Rock Rose 48th) by Genuine Prince (24031), &c.   See "Mountaineer," p. 173.

1878, July 6, roan,	B.C.	Mountaineer	Pioneer, 38868	Mr. Braikenridge
1879, July 10, white,	C.C.	White Rose	do.	do.

Mountaineer, sold to Mr. G. Edwards, Congresbury, Somerset.

BRAYBROOKE, Lord,
Audley End, Saffron Walden, Essex.

CHARM, red and white, calved July 7, 1876, Vol. xxv. p. 362.   Bred by Lord Braybrooke; got by General Knightley (28695), dam (Caress) by Thorndale Duke (27661), &c.

1879, Sept. 9, r. & w.,	C.C.	Amulet	Knight Grand Cross, 40071	Lord Braybrooke

CHRISTMAS ROSE 3RD, roan, calved September 10, 1876.   Bred by Lord Braybrooke, the property of Messrs. F. Leney and Sons, Wateringbury; got by Duke of Rosedale 3rd (33723), dam (Christmas Rose 2nd) by Grand Duke 17th (24064), &c.   See "Christmas Duke 4th," p. 50.

1879, April 8, roan,	B.C.	Christmas Duke 4th	D. of Underley 3rd, 38196	Lord Braybrooke

DREAM, red and white, calved November 1, 1876.   Bred by Lord Braybrooke; got by Duke of Rosedale (33721), dam (Fancy) by Thorndale Duke (27661), &c.   See Vol. xxiii. p. 352.

1879, Mar. 12, r. & w.,	C.C.	Slumber	Knight Grand Cross, 40071	Lord Braybrooke

GRAND DUCHESS OF OXFORD 6TH, roan, calved October 21, 1875, Vol.
xxv. p. 362. Bred by Lord Braybrooke, the property of Mr. S. P. Foster,
Killhow; got by Duke of Rosedale (33721), dam (Grand Duchess of Oxford
4th) by Claro's Rose (25784), &c.

Produce in	Names, &c.	By what Bull.	By whom bred.
1879, May 30, roan, C.C.	Lady Audley 3rd	D. of Rosedale 3rd, 33723	Lord Braybrooke

Lady Audley 3rd, sold to Mr. T. Baker, Harpole, Weedon.

GRAND DUCHESS OF OXFORD 7TH, white, calved April 12, 1876, Vol.
xxv. p. 362. Bred by Lord Braybrooke, the property of Mr. A. P. Clear,
Maldon; got by Duke of Rosedale (33721), dam (Grand Duchess of Oxford
5th) by Heydon Duke 2nd (31370), &c.

Produce in	Names, &c.	By what Bull.	By whom bred.
1879, June 1, roan, B.C.	Lord Audley 2nd	D. of Rosedale 3rd, 33723	Lord Braybrooke

Lord Audley 2nd, sold to the Rev. G. Maryon Wilson, Great Canfield, Essex.

HEYDON ROSE 2ND, white, calved March 1, 1870, Vols. xxi., xxiv., and xxv.
pp. 596, 344, 363. Bred by Lord Braybrooke, the property of Mr. F. Barchard, Horsted; got by Third Duke of Geneva (23753), dam (Heydon Rose)
by Englishman (19701), &c.

Produce in	Names, &c.	By what Bull.	By whom bred.
1879, May 3, roan, B.C.	Heydon Duke 10th	D. of Rosedale 3rd, 33723	Lord Braybrooke

HEYDON ROSE 5TH, roan, calved October 20, 1876. Bred by Lord Braybrooke, the property of the Australian Agricultural Company; got by Duke
of Rosedale 3rd (33723), dam (Heydon Rose 3rd) by Grand Duke 17th(24064),
g. d. (Heydon Rose) by Englishman (19701), &c. See "Heydon Duke 10th,"
p. 120.

Produce in	Names, &c.	By what Bull.	By whom bred.
1879, Jan. 5, r. & w., C.C.	Heydon Rose 8th	D. of Underley 3rd, 38196	Lord Braybrooke

SHADOW, red and white, calved December 5, 1875, Vol. xxv. p. 363. Bred by
Lord Braybrooke, the property of Mr. J. R. Cowell, Ashdon; got by Duke
of Rosedale (33721), dam (Fancy) by Thorndale Duke (27661), &c.

Produce in	Names, &c.	By what Bull.	By whom bred.
1879, Mar. 17, r. & w., B.C.	Shade	Knight Grand Cross, 40071	Lord Braybrooke

Shade, sold to the Rev. W. F. Thursby, Burgh Apton, Norwich.

THORNDALE ROSE 2ND, red, calved February 4, 1867, Vols. xix., xx., xxiii.,
xxiv., and xxv. pp. 752, 790, 353, 345, 363. Bred by Mr. E. L. Betts, Preston
Hall; got by Grand Duke 4th (19874), dam (Thorndale Rose) by Fourth
Duke of Thorndale (17750), &c.

Produce in	Names, &c.	By what Bull.	By whom bred.
1879, June 1, roan, C.C.	Thorndale Rose 16th	D. of Underley 3rd, 38196	Lord Braybrooke

THORNDALE ROSE 5TH, roan, calved December 18, 1872, Vols. xxii., xxiii.,
xxiv., and xxv. pp. 337, 353, 345, 363. Bred by Lord Braybrooke; got by
Grand Duke 17th (24064), dam (Thorndale Rose 2nd) by Grand Duke 4th
(19874), &c.

Produce in	Names, &c.	By what Bull.	By whom bred.
1879, Aug. 12, white, C.C.	Thorndale Rose 17th	D. of Underley 3rd, 38196	Lord Braybrooke

THORNDALE ROSE 6TH, roan, calved June 6, 1874, Vol. xxv. p. 363. Bred
by Lord Braybrooke; got by Eighth Duke of Geneva (28390), dam (Thorndale Rose 3rd) by Third Duke of Geneva (23753), &c.

Produce in	Names, &c.	By what Bull.	By whom bred.
1879, Jan. 13, r. & w., B.C.	(Steer)	Duke of Underley, 33745	Lord Braybrooke

THORNDALE ROSE 8TH, roan, calved September 4, 1876. Bred by Lord
Braybrooke; got by Sixth Duke of Oneida (30997), dam (Thorndale Rose
3rd) by Third Duke of Geneva (23753), &c. See Vol. xxiii. p. 353.

Produce in	Names, &c.	By what Bull.	By whom bred.
1879, Jan. 24, white, C.C.	Thorndale Rose 15th	D. of Underley 3rd, 38196	Lord Braybrooke

BRIGGS, D. Grant,
Kelstern Grange, Louth.

PEONY, roan, calved August 29, 1873, Vol. xxiii. p. 304. Bred by Mr. H. D. Barclay, Eastwick Park; got by Albert Victor (30371), dam (Pancake) by Victorious (25378), &c.

Produce in		Names, &c.		By what Bull.		By whom bred.
1878, April 6, r. & w.,	C.C.	Primula		The General, 35768		Mr. Briggs
1879, Sept. 27, roan,	C.C.	(dead)		Ld. Lightburne 4th, 36967		do.

PRIDE OF THE GLEN, red, calved December 16, 1876. Bred by Messrs. Dudding, Panton House, the property of Mr. D. Grant Briggs; got by Pluto (35050), dam (Pride of the Rose) by Robert Stephenson (32313), &c. See Vol. xxv. p. 432.

1879, April 27, r. & w., C.C.	Pride of the Dale	Sluggish Arve, 39136	Messrs. Dudding

WEE WEE, white, calved August 9, 1874. Bred by H.R.H. the Prince of Wales, Sandringham; got by Albert Victor (30371), dam (Pancake) by Victorious (25378), &c.

1878, Jan. 25, white,	C.C.	Poppy	The General, 35768	Mr. Briggs
1879, May, roan,	B.C.	(Steer)	Ravenlock, 38968	do.

BRIGHAM, James,
Slingsby, York.

DAISY 9TH, roan, calved May 8, 1871, Vols. xxiv. and xxv. pp. 345, 364. Bred by Mr. J. Brigham; got by Volunteer (27737), dam (Daisy 3rd) by Prince of Wales (20589), &c.

1879, April 3, roan,	C.C.	Daisy 18th	Mowbray, 40374	Mr. Brigham

LADY MAYNARD 17TH, roan, calved April 4, 1875. Bred by Mr. J. Brigham; got by Second Earl of Durham (33790), dam (Lady Maynard 12th) by Volunteer (27737), &c. See Vol. xxii. p. 338.

1879, Jan. 9, roan,	C.C.	Lady Maynard 22nd	Bastion, 37836	Mr. Brigham

MOSS ROSE 1ST, red, calved March 19, 1872, Vol. xxi. p. 598. Bred by Mr. J. Brigham; got by Brigadier (28097), dam (Brenda) by Twentieth Duke of Oxford (23778), &c.

1879, Feb. 10, red,	C.C.	Moss Rose 5th	Mowbray, 40374	Mr. Brigham

BRISCOE, John,
Hill Croome, Worcester.

BETTY, roan, calved February 18, 1870, Vol. xxv. p. 365. Bred by Mr. J. Briscoe; got by Mars (29307), dam (Lady Love) by Gold Nugget (16176). &c.

1879, Jan. 14, red,	B.C.	Beacon	D. of St. Johns 2nd, 36539	Mr. Briscoe
1879, Nov. 30, roan,	C.C.	Bess	do.	do.
Beacon, sold to Mr. Wilson, Stettin.				

CAROLINA, roan, calved March 24, 1873. Bred by Mr. J. Briscoe; got by Cremorne (30818), dam (Caroline) by Grand Duke 9th (19879), &c. See Vol. xxiv. p. 345.

1877, Dec. 30, roan,	C.C.	(dead)	D. of St. Johns 2nd, 36539	Mr. Briscoe
1878, Nov. 22, roan,	C.C.	Carry	do.	do.

LADY VALENTINE, roan, calved February 14, 1875. Bred by Mr. J.
Briscoe; got by Third Duke of Waterloo (23801), dam (Lady Love) by Gold
Nugget (16176), &c. See "Croome Diamond," p. 60.

Produce in	Names, &c.	By what Bull.	By whom bred.
1877, Dec. 28, r. & w., B.C.	Duke of Waterloo	D. of St. Johns 2nd, 36539	Mr. Briscoe
1878, Dec. 11, roan, C.C.	Lady Agnes	do.	do.
1879, Nov. 28, white, B.C.	Croome Diamond	do.	do.

PRINCESS VALENTINE, roan, calved February 14, 1876. Bred by Mr. J.
Briscoe; got by Darlingtonis Walnut (33509), dam (Princess Alice) by Grand
Duke 9th (19879), &c. See Vol. xxv. p. 366.

1878, Nov. 14, r. & w., C.C.	Princess of Orange	D. of St. Johns 2nd, 36539	Mr. Briscoe

BRITTEN, George,
Overstone, Northampton.

MILKMAID 1st, red and white, calved November 15, 1871, Vol. xxiv. p. 346.
Bred by Mr. G. Britten; got by General Hesse (34008), dam (Nightfall) by
Twelfth Duke of Oxford (19633), &c.

1878, Mar. 19, roan, C.C.	Milkmaid 6th	Fair Thane, 31127	Mr. Britten
1879, April 11, r. & w., C.C.	Milkmaid 8th	Wallace, 40885	do.

MILKMAID 2nd, red and white, calved December 24, 1874, Vol. xxiv. p. 346.
Bred by Mr. G. Britten; got by Lord York Fawsley (34709), dam (Milkmaid
1st) by General Hesse (34008), &c.

1879, Sept. 22, roan, C.C.	Milkmaid 9th	Fearless, 36637	Mr. Britten

MILKMAID 3rd, roan, calved February 6, 1876. Bred by Mr. G. Britten;
got by Southern (35643), dam (Milkmaid 1st) by General Hesse (34008), &c.
See "Marksman," p. 163.

1878, Nov. 20, roan, C.C.	Milkmaid 7th	Fearless, 36637	Mr. Britten

ROSA 2nd, roan, calved July 26, 1876. Bred by Mr. G. Britten; got by
Southern (35643), dam (Rosa 1st) by Regent (32283), &c. See Vol. xxiv.
p. 347.

1879, May 6, roan, C.C.	Rosa 4th	Wallace, 40885	Mr. Britten

TRINKET, roan, calved October 10, 1873. Bred by Mr. L. B. Bagshaw,
Newton; got by Robin Adair (35290), dam (Nymph) by Merlin (24581),
g. d. (Jessica) by Lord Dundreary (20169), &c. See Vol. xviii. p. 541.

1879, Feb. 25, roan, C.C.	Newton 1st	Fearless, 36637	Mr. Britten

VESPER, roan, calved June 6, 1874. Bred by Mr. L. B. Bagshaw, Newton;
got by Penrhyn Castle (32059), dam (Rachel) by Satan (27430), &c. See
Vol. xxii. p. 306.

1879, Mar. 9, roan, C.C.	Newton 2nd	Fearless, 36637	Mr. Britten

VINE, roan, calved July 13, 1874. Bred by Mr. L. B. Bagshaw, Newton; got
by Satan (27430), dam (Grape) by Lord Dundreary (20169), &c. See Vol.
xviii. p. 519.

1878, Dec. 11, r. & w., C.C.	Vine Leaf	Fearless, 36637	Mr. Britten

BROCKBANK, R. B.,
Crosby, Maryport.

BENSON 3RD, roan, calved in November 1872. Bred by Mr. J. Littleton, Arkleby Hall; got by Oxford's Pride (27023), dam (Benson 1st) by Kildonan (20051), g. d. (Old Fanny Benson) by Duke (14419), &c. See "Benson 17th," Vol. xxii. p. 479.

Produce in		Names, &c.	By what Bull.	By whom bred.
1879, Oct. 26, roan,	C.C.	Crosby Benson	Farmer's Friend, 38275	Mr. Brockbank

COUNTESS BIANCA, red and white, calved February 25, 1875. Bred by Mr. R. B. Brockbank; got by Wild Earl (32865), dam (Bianca) by Royal Duke (25015), &c. See "Count Bianca," Vol. xxv. p. 60.

1878, May 4, roan,	B.C.	Count Bianca	Borderer, 33183	Mr. Brockbank
1879, May 7, roan.	C.C.	Border Countess	do.	do.

Count Bianca, sold to Mr. A. Stockdale, Bromfield, Aspatria.

DORA 2ND, roan, calved in 1872. Bred by Miss Calvert, Sandysike, the property of Mr. R. B. Brockbank; got by Snowdrift (32558), dam (Dora) by Prince Charlie (32114), &c. See "Lowry Calvert," Vol. xxv. p. 180.

1875, Aug. 14, white,	B.C.	White Knight	Kt. of the Parsonage, 31561	Mr. Todd
1876, Oct. 22, roan,	C.C.	Dora 3rd	G.D. of Lightb'ne 2d, 26291	do.
1878, Feb. 11, roan,	B.C.	Lowry Calvert	Grandeur, 36727	do.
1879, July 4, roan.	C.C.	Cantab's Dora	Cantab, 39544	do.

White Knight, sold to Mr. R. Jefferson, Preston Hows, Whitehaven; Lowry Calvert, to Mr. J. Barnes, Langrigg, Aspatria; Cantab's Dora, to Mr. R. B. Brockbank.

DUCHESS WILD EYES, roan, calved March 29, 1870, Vols. xxi. and xxiii. pp. 599, 354. Bred by Mr. J. Beswick, Raby Cote; got by Third Duke of Oxford (23783), dam (Wild Eyes 2nd) by General Havelock (16118), &c.

1877, April 1, roan,	C.C.	(dead)	Borderer, 33183	Mr. Brockbank
1878, Mar. 26, roan,	B.C.	Crosby Duke	do.	do.
1879, May 28, roan,	C.C.	Bright Eyes	do.	do.

Crosby Duke, sold to Mr. J. Gibson, Parkhouse.

IRIS FLOWER, roan, calved June 29, 1873. Bred by Mr. J. Todd, Mereside; got by Third Duke of Oxford (23783), dam (Alfred's Flower) by Alfred Fitz Clarence (19215), &c. See "Stanley," p. 241.

1878, April 27, white,	B.C.	(slaughtered)	Borderer, 33183	Mr. Brockbank
1879, May 25, roan,	B.C.	Stanley	do.	do.

MISS BRAYTON, roan, calved January 31, 1872, Vol. xxv. p. 366. Bred by Mr. W. Thompson, Aigle Gill; got by Oxford Don (29489), dam (Lady Lawson) by Kildonan (20051), &c.

1879, April 22, white,	B.C.	White Chief	Borderer, 33183	Mr. Brockbank

PHANTOM 5TH, roan, calved May 17, 1876. Bred by Mr. J. Postlethwaite, The Hollins; got by Baron Tregunter (33085), dam (Phantom 4th) by King James (28972), g. d. (Phantom 2nd) by His Lordship (21939), &c. See Vol. xx. p. 688.

1879, April 17, white,	B.C.	(Steer)	Borderer, 33183	Mr. Brockbank

PRINCESS HELENA, roan, calved July 20, 1872. Bred by Mr. R. B. Hetherington, Park Head ; got by Baron Charleston (27922), dam (Princess Beatrice) by Kildonan (20051), &c. See " Princely," p. 196.

Produce in		Names, &c.	By what Bull.	By whom bred.
1875, May 6, roan,	B.C.	Border Prince	Borderer, 33183	Mr. Brockbank
1876, April 12, roan,	B.C.	Crosby	do.	do.
1877, April 29, roan,	C.C.	(dead)	do.	do.
1878, April 8, roan,	B.C.	(dead)	do.	do.
1879, Mar. 2, white,	B.C.	Princely	do.	do.

Crosby, sold to Mr. Waugh, Allerby Mill, Maryport.

QUEEN 11TH, red, calved August 16, 1872, Vol. xxv. p. 366. Bred by Mr. J. Fawcett, Scaleby Castle ; got by Royal Cumberland (27358), dam (Queen 7th) by Fourteenth Duke of Oxford (21605), &c.

1879, July 11, roan,	C.C.	Border Queen	Borderer, 33183	Mr. Brockbank

TWIN ROSE, roan, calved April 18, 1876. Bred by Mr. J. Todd, Mereside, the property of Mr. R. B. Brockbank ; got by Grand Duke of Lightburne 2nd (26291), dam (Cressida Benson) by Pammon (22486), &c. See Vol. xxiii. p. 674.

1879, July 10, white, C.C.	Miss Todd	Gwynne Duke, 34098	Mr. Todd

Miss Todd, sold to Mr. W. Walker, Edderside.

WILD LILY, white, calved April 16, 1876. Bred by Mr. R. B. Brockbank ; got by Borderer (33183), dam (Duchess Wild Eyes) by Third Duke of Oxford (23783), &c. See Vol. xxiii. p. 354.

1879, May 13, roan, C.C.	Crosby Lily	Crosby, 38055	Mr. Brockbank

BROCKHOLES, Mrs.,
Clifton Hill, Garstang, Lancashire.

MARGERY, red and white, calved May 16, 1875. Bred by the Rev. J. Swarbrick, Thurnham ; got by Oscar (32006), dam (York's Fairie) by Eighth Duke of York (28480), &c. See Vol. xxii. p. 584.

1878, Feb. 18, r. & w., B.C.	(dead)	Earl Blanche, 36566	Mrs. Brockholes
1879, May 13, r. & w., C.C.	Norah	G.D. of Th'ndale 3rd, 36723	do.

YORK'S FAIRIE, roan, calved October 8, 1872, Vols. xxii. and xxiii. pp. 584, 663. Bred by Mr. J. Fawcett, Scaleby Castle ; got by Eighth Duke of York (28480), dam (Lady Chance 9th) by Cato (21380), &c.

1877, May 20, r. & w., B.C.	Young Ketton	Bn T'croft Oxf'd 2d, 33087	Rev. J. Swarbrick
1878, May 15, r. & w., C.C.	Fay	Earl Blanche, 36566	Mrs. Brockholes
1879, April 11, red, C.C.	Elfin	do.	do.

Young Ketton, sold to Mr. H. Mackereth, Thurnham.

BROCKLEBANK, T.,
Springwood, Woolton, Liverpool.

BUTTERFLY PRINCESS 27TH, roan, calved March 5, 1876, Vol. xxv. p. 366. Bred by the Executors of Mr. J. Fawcett, Scaleby Castle ; got by Eighth Duke of York (28480), dam (Butterfly Princess 20th) by Royal Cumberland (27358), &c.

1879, May 25, roan, C.C.	B'terfly Princess 28th	Oxford's Heir, 37172	Mr. Brocklebank

FLORENTIA 29TH, white, calved September 25, 1876. Bred by the Executors of Mr. J. Fawcett, Scaleby Castle; got by Grand Duke of Kirklevington (34071), dam (Florentia 25th) by Second Duke of Collingham (23730), &c. See "Lord Harley," p. 148.

Produce in		Names, &c.	By what Bull.	By whom bred.
1879, Aug. 3, roan,	B.C.	Lord Harley	Oxford's Heir, 37172	Mr. Brocklebank

LADY HILDA 14TH, roan, calved June 20, 1877. Bred by Mr. G. Allen, Knightley Hall; got by Duke of Clarence 5th (36479), dam (Lady Hilda 7th) by Duke of Wetherby 6th (33756), &c. See Vol. xxiv. p. 293.

1879, Aug. 23, red,	C.C.	Lady Hilda 15th	Oxford's Heir, 37172	Mr. Brocklebank

MODESTY, roan, calved December 10, 1865, Vol. xxv. p. 366. Bred by Mr. J. Fawcett, Scaleby Castle; got by Fourteenth Duke of Oxford (21605), dam (Matilda) by Richard (16834), &c.

1879, Mar. 29, red,	C.C.	Bashful 2nd	Oxford's Heir, 37172	Mr. Brocklebank

SHAMROCK 26TH, red, calved April 8, 1871, Vol. xxiv. p. 347. Bred by Mr. W. Harland, Blois Hall; got by Victorious (30226), dam (Shamrock 18th) by Perseverance (24738), &c.

1879, May 2, r. & w.,	C.C.	Shamrock 31st	Oxford's Heir, 37172	Mr. Brocklebank

VICTORIA 6TH, roan, calved May 20, 1875, Vols. xxiv. and xxv. pp. 347, 367. Bred by Mr. W. Harland, Blois Hall; got by Victor 4th (35878), dam (Faith) by The Hero (25296), &c.

1879, April 28, roan,	C.C.	Victoria 8th	Oxford's Heir, 37172	Mr. Brocklebank

BROMET, W. R.,
Cocksford, Tadcaster.

AMY BROOK, red and white, calved February 13, 1877. Bred by Mr. W. R. Bromet; got by Second Cherry Duke (28170), dam (Lizzie 3rd) by Duke Ferdinand 3rd (28354), g. d. (Lizzie 2nd) by Third Duke of Flanders (23750), gr. g. d. (Lizzie) by Duke of Clarence (19611), — a Cow bought of Colonel Gunter.

1879, Sept. 21, r. & w.,	C.C.	Amy Brook 3rd	D. of Flanders 6th, 38144	Mr. Bromet

BARONESS BEVERLEY, roan, calved February 3, 1873, Vols. xxiii., xxiv., and xxv. pp. 354, 347, 367. Bred by Mr. W. R. Bromet; got by Duke of Clarence (19611), dam (Wharfdale Beverley) by Third Duke of Wharfdale (21619), &c.

1879, July 12, roan,	B.C.	Baron Beverley 4th	D. of Clarence 5th, 38133	Mr. Bromet

BARONESS BEVERLEY 2ND, roan, calved July 22, 1873, Vols. xxiii., xxiv., and xxv. pp. 355, 347, 367. Bred by Mr. W. R. Bromet; got by Third Duke of Tregunter (31026), dam (Clarence Beverley) by Duke of Clarence (19611), &c.

1879, Aug. 11, roan,	B.C.	Baron Beverley 5th	D. of Clarence 5th, 38133	Mr. Bromet

BARONESS BEVERLEY 3RD, roan, calved February 21, 1874, Vols. xxiii. and xxiv. pp. 355, 347. Bred by Mr. W. R. Bromet; got by Fifth Duke of Wetherby (31033), dam (Wharfdale Beverley) by Third Duke of Wharfdale (21619), &c.

1879, Mar. 4, roan,	C.C.	B'ness Beverley 14th	Oxf'd Cherry D. 2nd, 34972	Mr. Bromet

BARONESS BEVERLEY 5TH, red and white, calved April 30, 1876. Bred by Mr. W. R. Bromet; got by Duke of Tregunter 5th (33743), dam (Baroness Beverley 2nd) by Third Duke of Tregunter (31026), &c. See "Baron Beverley 5th," p. 12.

Produce in		Names, &c.		By what Bull.		By whom bred.
1879, Jan. 5, r. & w.,	B.C.	(dead)		D. of Clarence 5th, 38133		Mr. Bromet

BARONESS BEVERLEY 6TH, roan, calved May 22, 1876. Bred by Mr. W. R. Bromet; got by Baron Turncroft Oxford 2nd (33087), dam (Baroness Beverley) by Duke of Clarence (19611), &c. See "Baron Beverley 3rd," p. 11.

Produce in		Names, &c.		By what Bull.		By whom bred.
1879, April 8, roan,	B.C.	Baron Beverley 3rd		D. of Clarence 5th, 38133		Mr. Bromet

COUNTESS OF FLANDERS, roan, calved July 24, 1869, Vols. xxi., xxiii., and xxv. pp. 599, 355, 367. Bred by Mr. W. R. Bromet; got by Second Duke of Wetherby (21618), dam (Fame) by Fourth Duke of Oxford (11387), &c.

Produce in		Names, &c.		By what Bull.		By whom bred.
1879, April 5, roan,	B.C.	Duke of Flanders 8th	18th D. of Oxford, 25995			Mr. Bromet

COUNTESS OF FLANDERS 2ND, roan, calved October 8, 1870, Vols. xxi., xxiii., xxiv., and xxv. pp. 600, 355, 348, 367. Bred by Mr. W. R. Bromet; got by Third Duke of Wharfdale (21619), dam (Fame) by Fourth Duke of Oxford (11387), &c.

Produce in		Names, &c.		By what Bull.		By whom bred.
1879, Mar. 30, roan,	B.C.	Duke of Flanders 7th	18th D. of Oxford, 25995			Mr. Bromet

COUNTESS OF FLANDERS 6TH, roan, calved August 14, 1876. Bred by Mr. W. R. Bromet; got by Duke of Tregunter 5th (33743), dam (Countess of Flanders) by Second Duke of Wetherby (21618), &c. See "Duke of Flanders 9th," p. 77.

Produce in		Names, &c.		By what Bull.		By whom bred.
1879, May 30, r. & w.,	B.C.	Duke of Flanders 9th	D. of Clarence 5th, 38133			Mr. Bromet

COUNTESS OF FLANDERS 7TH, roan, calved February 2, 1877. Bred by Mr. W. R. Bromet; got by Duke of Tregunter 5th (33743), dam (Countess of Flanders 5th) by Fifth Duke of Wetherby (31033), g. d. (Countess of Flanders) by Second Duke of Wetherby (21618), &c. See "Duke of Flanders 8th," p. 77.

Produce in		Names, &c.		By what Bull.		By whom bred.
1879, Nov. 17, roan,	C.C.	C'tess of Fl'nders 11th	D. of Clarence 5th, 38133			Mr. Bromet

COUNTESS OF FLANDERS 8TH, roan, calved March 8, 1877. Bred by Mr. W. R. Bromet; got by Duke of Tregunter 5th (33743), dam (Countess of Flanders 2nd) by Third Duke of Wharfdale (21619), &c. See "Duke of Flanders 10th," p. 77.

Produce in		Names, &c.		By what Bull.		By whom bred.
1879, Dec. 22, roan,	B.C.	D. of Flanders 10th	D. of Clarence 5th, 38133			Mr. Bromet

BROMLEY, James,
Forton, Garstang, Lancashire.

BOLINA, white, calved December 23, 1876. Bred by Mr. J. Bromley; got by The Squire (35791), dam (Strawberry 2nd) by Duke of Lancaster 3rd (28409), &c. See Vol. xxiii. p. 357.

Produce in		Names, &c.		By what Bull.		By whom bred.
1879, Nov. 7, roan,	C.C.	Bolina 3rd		Knight of Forton, 40073		Mr. Bromley

DUCHESS 9TH, roan, calved February 24, 1875. Bred by Mr. J. Bromley; got by Duke of Lancaster 3rd (28409), dam (Duchess 4th) by Favourite (23916), &c. See "Prince Arthur," p. 191.

Produce in	Names, &c.	By what Bull.	By whom bred.
1879, Dec. 23, roan, B.C.	Baron Forton	Knight of Forton, 40073	Mr. Bromley

STRAWBERRY 3RD, roan, calved February 3, 1875, Vol. xxiv. p. 348. Bred by Mr. J. Bromley; got by Duke of Lancaster 3rd (28409), dam (Bridget 4th) by Prince of Wales (22629), &c.

1879, Nov. 15, white, C.C.	Bolina 4th	Knight of Forton, 40073	Mr. Bromley

BROMLEY, Richard,
Felton Butler, Salop.

MAID MARIAN 2ND, roan, calved March 3, 1877. Bred by the Rev. H. O. Wilson, Church Stretton; got by Coventry (33463), dam (Maid Marian) by Lord Warwick (26753), &c. See Vol. xxi. p. 996.

1879, Dec. 18, red, C.C.	Maid Marian 3rd	Duke of Stretton, 43128	Mr. Bromley

BROOKE, Sir Victor, Bart.,
Colebrooke, Brookeboro', Co. Fermanagh.

GRANA UILE, roan, calved January 2, 1876. Bred by Mr. R. Chaloner, King's Fort; got by King James (28971), dam (Ladies' Pet) by Irish Baron (31417), &c. See Vol. xxiii. p. 573.

1879, Jan. 11, r. & w., C.C.	Lady's Garland	Adjutant General, 40952	Sir V. Brooke

LADY HESTER, roan, calved March 10, 1875. Bred by Mr. R. Chaloner, King's Fort; got by King James (28971), dam (Lady Harriette) by Sovereign (27538), &c. See "Royal Hector," Vol. xxv. p. 249.

1878, April 16, r. & w., B.C.	Royal Hector	Royal Arthur, 37372	Sir V. Brooke
1879, Mar. 1, white, C.C.	Lady Helen	Adjutant-General, 40952	do.

QUEEN'S PRIZE, roan, calved March 19, 1876. Bred by Mr. R. Chaloner, King's Fort; got by Lieutenant General (31600), dam (Lady's Prize) by Regal Dane (22710), &c. See "King John," Vol. xxii. p. 135.

1879, Mar. 3, roan, C.C.	Lady's Pearl	Royal Baron, 40617	Sir V. Brooke

BROWN, George,
Willow Hill, Morhanger, Sandy, Beds.

BABRAHAM LILY 2ND, white, calved September 10, 1876. Bred by Mr. G. Brown; got by Cambridge Walnut (33278), dam (Babraham Lily) by Englishman (19701), &c. See Vol. xxiii. p. 357.

1879, Mar. 28, roan, C.C.	Babraham Lily 3rd	Jupiter, 38476	Mr. Brown

CELIA BURDETT, red and white, calved May 16, 1873, Vol. xxiii. p. 357. Bred by Mr. G. Brown; got by Heydon Duke 2nd (31370), dam (Lady Florence Burdett) by Duke of Kent (19619) &c.

1879, Dec. 28, r. & w., C.C.	Celia Burdett 2nd	Coronet 2nd, 39629	Mr. Brown

JUNO, red and white, calved in February 1868, Vol. xxiv. p. 349. Bred by Mr. Stopford-Sackville, Drayton House; got by George 1st (19848), dam (Lady Jocelyn) by Rasselas (29714), &c.

Produce in		Names, &c.	By what Bull.	By whom bred.
1879, Feb. 4, r. & w.,	C.C.	Juniper	Ld. of the Border 3rd, 38637	Mr. Brown

JUNO 2ND, red and white, calved February 20, 1877. Bred by Mr. G. Brown; got by Cambridge Walnut (33278), dam (Juno) by George 1st (19848), &c. See Vol. xxiv. p. 349.

1879, Oct. 29, red.	C.C.	Juno 3rd	Lord Pythius, 43566	Mr. Brown

TRIFOLIUM 7TH, red, calved January 27, 1876. Bred by Mr. C. Barnett, Stratton Park; got by Red Clover (35216), dam (Marie Christine) by Mason (20299), &c. See Vol. xxiii. p. 358.

1879, June 21, r. & w., C.C.	Christine 2nd	Jupiter, 38476	Mr. Brown

TWELFTH TWIN CELIA, red and white, calved December 11, 1876. Bred by Mr. G. Brown; got by Cambridge Walnut (33278), dam (Lady Florence Burdett) by Duke of Kent (19619), &c. See Vol. xxiii. p. 358.

1879, Sept. 20, r. & w., C.C.	Coral	Janizary, 34243	Mr. Brown

BROWN, John,
Park Cottage, Burton Constable, Hull.

CARNATION, red, calved March 26, 1872, Vols. xxi. and xxiv. pp. 603, 349. Bred by Lord Skelmersdale, Lathom House; got by Cherry Grand Duke 2nd (25758), dam (Cherry) by Mufti (20385), &c.

1878, Sept. 21, roan, C.C.	Carnation 2nd	Oneida Prince, 34948	Mr. Brown

CHERRY BELLE, red and white, calved February 29, 1872, Vols. xxi. and xxiv. pp. 603, 349. Bred by Lord Skelmersdale, Lathom House; got by Cherry Grand Duke 2nd (25758), dam (Belle of Collingham) by Second Duke of Collingham (23730), &c.

1879, May 12, roan, B.C.	Cottage Boy	Cherry D. of L'burne, 36349	Mr. Brown

CHRISTINE 3RD, roan, calved June 17, 1875. Bred by Mr. J. Brown; got by Nineteenth Duke of Oxford (28431), dam (Christine 2nd) by Battersea First Fruits (21250), &c. See Vol. xxii. p. 340.

1878, April 8, red,	C.C.	Christine 6th	Prince of Geneva, 37246	Mr. Brown
1879, Nov. 16, r. & w.,	B.C.	Prince Christine	Prince of Glo'ster, 40517	do.

Christine 6th, sold to Mr. Steffingham, Germany.

WHITE FROST, roan, calved September 20, 1868, Vols. xx. and xxiv. pp. 820, 350. Bred by the Rev. W. Holt Beever, Pencraig Court; got by Royal Butterfly 17th (22774), dam (Chill) by Champagne (17522), &c.

1878, April 17, roan,	C.C.	Winter Rose 3rd	Prince of Geneva, 37246	Mr. Brown
1879, Nov. 17, r. & w.,	C.C.	Winter Rose 4th	Waterloo Prince, 37655	do.

Winter Rose 3rd, sold to Mr. Steffingham, Germany.

WINTER ROSE 2ND, roan calved February 27, 1877. Bred by Mr. J. Brown; got by Duke of Hazlecote 23rd (30970), dam (White Frost) by Royal Butterfly 17th (22774), &c. See Vol. xxiv. p. 350.

1879, Aug. 8, roan,	B.C.	Champagne Charlie	Waterloo Lord, 44224	Mr. Brown

BROWNLOW, Earl,
Belton, Grantham.

ARCHDUCHESS, red and white, calved February 28, 1876. Bred by Earl Brownlow; got by Cambridge Duke 4th (25706), dam (Almack's Queen) by Robin (24968), &c. See Vol. xxiv. p. 352.

Produce in		Names, &c.	By what Bull.	By whom bred.
1879, May 8, r. & w.,	C.C.	Archeress	Malcolm, 34740	Earl Brownlow

FLOURISH, red, calved June 7, 1873, Vol. xxv. p. 372. Bred by Mr. G. Bland, Coleby Hall; got by Waterloo Prince (30279), dam (Florence) by Macduff (24500), &c.

1879, Jan. 10, red,	B.C.	Matchlock	Malcolm, 34740	Earl Brownlow
1879, Nov. 26, red,	C.C.	Flush	do.	do.

PANSY, red, calved March 15, 1876. Bred by Earl Brownlow; got by Brigand (39493), dam (Periwinkle) by Prince Excellent (35108), &c. See Vol. xxv. p. 372.

1879, Jan. 2, red,	C.C.	Peony	Malcolm, 34740	Earl Brownlow
1879, Dec. 21, red,	B.C.	Malmsey	do.	do.

PERIWINKLE, red, calved March 10, 1873, Vol. xxv. p. 372. Bred by Earl Brownlow; got by Prince Excellent (35108), dam (Polyanthus) by Chieftain (21421), &c.

1879, Mar. 6, red,	B.C.	Maroyas	Malcolm, 34740	Earl Brownlow

TIMBREL, red and white, calved May 10, 1875. Bred by Earl Brownlow; got by Duncan (33767), dam (Tyrol) by Macbeth (26772), &c. See Vol. xxiv. p. 353.

1879, Dec. 16, r. & w.,	C.C.	Tambourine	Confucius, 39615	Earl Brownlow

TYROL, red, calved April 15, 1870, Vols. xxiv. and xxv. pp. 353, 372. Bred by Mr. G. Bland, Coleby Hall; got by Macbeth (26772), dam (Tunic) by Grand Duke 6th (19876), &c.

1879, June 4, red,	C.C.	Tyranness	Malcolm, 34740	Earl Brownlow

BRUCE, D. C.,
Broadland, Huntly, N.B.

MINNIE 3RD, red, calved April 13, 1877. Bred by Mr. C. Bruce, Broadland; got by Baron Fitz Killerby (36181), dam (Minnie 2nd) by Trochu (32734), &c. See Vol. xxv. p. 373.

1879, Dec. 27, red,	C.C.	Minnie 5th	Ben Lomond, 41098	Mr. D. C. Bruce

BRUCE, Robert,
Manor House Farm, Great Smeaton, Northallerton.

FLOWER GIRL, roan, calved May 8, 1870, Vols. xxi. and xxv. pp. 875, 374. Bred by Mr. J. Outhwaite, Bainesse; got by Baron Killerby (27949), dam (Sylvia) by Champion (23529), &c.

1879, Mar. 2, r. & w.,	B.C.	Hymen	Hyperion, 34196	Mr. Bruce

SYLVIA 2ND, roan, calved October 8, 1875, Vol. xxv. p. 374. Bred by Mr. J. Outhwaite, Bainesse; got by Royal Windsor (29890), dam (Flower Girl) by Baron Killerby (27949), &c.

Produce in	Names, &c.	By what Bull.	By whom bred.
1879, Mar. 9, white, B.C.	Hylas	Hyperion, 34196	Mr. Bruce

WHITE ROSE, white, calved October 13, 1876. Bred by Mr. J. Thom, Chorley; got by Aylesby (32978), dam (Royal Rose) by Royal Duke (25014), &c. See Vol. xxiii. p. 361.

| 1879, Oct. 4, roan, B.C. | Plato | Pluto, 35050 | Mr. Bruce |

BRUEN, Henry,
Oak Park, Carlow, Ireland.

CANDYTUFT, red, calved March 4, 1875. Bred by Mr. H. Bruen; got by Fiddler (33911), dam (Crinoline 2nd) by Glory (24043), &c. See "Woody-tuft," p. 273. ·

| 1877, June 20, r. & w., C.C. | Candy | King William 2nd, 43426 | Mr. Bruen |
| 1879, Jan. 24, red, B.C. | Woodytuft | Woodhouse G'ynne, 37690 | do. |

Woodytuft, sold to Mr. J. Scanlon, Bohermore, Bagnalstown.

BRUERE, Raymond S.,
Braithwaite Hall, Middleham, Yorkshire.

AUTUMNAL STAR, red, calved October 3, 1870, Vols. xxi., xxii., xxiv., and xxv., pp. 609, 344, 353, 375. Bred by Mr. R. S. Bruere; got by Regal Booth (27262), dam (Vernal Star) by The Sutler (23061), &c.

| 1879, May 30, red, B.C. | (dead) | S.Swthin'sStarDrop,40667 | Mr. Bruere |

CROWN MEDALLION, red, calved September 3, 1876. Bred by Mr. R. S. Bruere; got by Star Regent (35679), dam (Crown Jewel) by Booth's Royal Signet (28061), &c. See Vol. xxiii. p. 361.

| 1879, Aug. 9, red, C.C. | Crown Garnet | S.Swthin'sStarDrop,40667 | Mr. Bruere |

HEATHER BELL FLOWER, roan, calved June 5, 1875, Vol. xxiv. p. 354. Bred by Mr. R. S. Bruere; got by Booth Satellite (30564), dam (Heather Flower) by Booth's Royal Signet (28061), &c.

| 1879, Jan. 22, roan, C.C. | Wild Heath Flower | Star Regent, 35679 | Mr. Bruere |
| 1879, Dec. 26, r. & w. C.C. | Christmas Flower | S.Swthin'sStarDrop,40667 | do. |

ROSA LAVINIA, white, calved July 11, 1872, Vols. xxiii., xxiv., and xxv. pp. 362, 354, 375. Bred by Mr. R. S. Bruere; got by Booth's Royal Signet (28061), dam (Rosa Lætitia) by Booth's Kinsman (25658), &c.

| 1879, Oct. 14, roan, C.C. | Rosa Lita | S.Swthin'sStarDrop,40667 | Mr. Bruere |

SCINTILLA FLOWER, roan, calved March 4, 1869, Vols. xx., xxi., xxii., xxiv., and xxv. pp. 758, 609, 345, 355, 376. Bred by Mr. R. S. Bruere; got by Booth's Kinsman (25658), dam (Sunflower) by Prince George (13510), &c.

| 1879, April 3, roan, C.C. | Lenten Flower | Star Regent. 35679 | Mr. Bruere |

SILVER LILY FLOWER, white, calved September 3, 1875, Vol. xxv. p. 376. Bred by Mr. R. S. Bruere; got by Booth Satellite (30564), dam (Pearl Flower) by Booth's Royal Signet (28061), &c.

| 1879, June 10, roan, B.C. | Goshawk | S.Swthin'sStarDrop,40667 | Mr. Bruere |

SPANGLED STAR, red and white, calved July 2, 1875, Vol. xxv. p. 376.
Bred by Mr. R. S. Bruere; got by Booth Satellite (30564), dam (Autumnal
Star) by Regal Booth (27262), &c.

Produce in	Names, &c.	By what Bull.	By whom bred.
1879, April 20, r. & w., B.C.	(dead)	S.Swthin'sStarDrop,40667	Mr. Bruere

SPRING FLOWER, roan, calved April 30, 1877. Bred by Mr. R. S. Bruere;
got by Booth Satellite (30564), dam (Scintilla Flower) by Booth's Kinsman
(25658), &c. See "Red Berry," p. 206.

1879, Oct. 4, r. & w., B.C.	Red Berry	S.Swthin'sStarDrop,40667	Mr. Bruere

VESPER REGINA, roan, calved March 26, 1875, Vol. xxv. p. 376. Bred by
Mr. R. S. Bruere; got by Royal Benedict (27348), dam (Vestal Star) by Sir
Windsor Broughton (27507), &c.

1879, Aug. 19, red, C.C.	Vesper Soverina	S.Swthin'sStarDrop,40667	Mr. Bruere

VESTAL STAR, white, calved February 21, 1870, Vols. xx., xxi., xxii., and
xxv. pp. 803, 609, 345, 376. Bred by Mr. R. S. Bruere; got by Sir Windsor
Broughton (27507), dam (Morning Star) by Prince George (13510), &c.

1879, Jan. 2, roan, B.C.	(dead)	Heir-at-Law, 34124	Mr. Bruere

BRUNDELL, R. S.,
Leicester House, Doncaster.

AMELIA, red and white, calved November 14, 1874. Bred by Mr. R. H.
Wrightson, Warmsworth Hall; got by Barbarossa (30415), dam (Amy) by
Earl of Kerry (23839), &c. See "Thakoor," p. 246.

1878, May 30, r. & w., B.C.	Thakoor	Shazadah, 37440	Mr. Brundell
1879, April 10, r. & w., C.C.	Amelia 3rd	Salar Jung, 40669	do.
Thakoor, sold to Mr. J. Elmhurst, Thorne, Yorkshire.			

RANEE, red and white, calved July 29, 1877. Bred by Mr. R. S. Brundell;
got by Shazadah (37440), dam (Cow Lady) by Pretender (29578), &c. See
Vol. xxii. p. 346.

1879, June 12, r. & w., C.C.	Bhaien	Bob Cherry, 33172	Mr. Brundell

BRUNYEE, S. C.,
Sand Hall, Crowle, Lincolnshire.

CLEMATIS, roan, calved September 20, 1869, Vols. xx. and xxiii. pp. 450, 362.
Bred by Mr. R. H. Wrightson, Warmsworth Hall; got by Iron Duke (26440),
dam (Candy Tuft) by Guy Fawkes (16211), &c.

1879, Mar. 1, roan, C.C.	Snowdrop	Isle Axholme 2nd, 40024	Mr. Brunyee

DOURA, red and white, calved July 23, 1876. Bred by Mr. S. C. Brunyee;
got by Clovis (38007), dam (Clematis) by Iron Duke (26440), &c. See Vol.
xxiii. p. 362.

1879, Dec. 20, white, C.C.	Doura 2nd	Isle Axholme 2nd, 40024	Mr. Brunyee

BUCHANAN, Phillips,
Hales Hall, Market Drayton.

DUCHESS 17TH, roan, calved in December 1873, Vols. xxiv. and xxv. pp. 356, 377. Bred by Mr. E. T. Tunnicliffe, Bromley Hall; got by Second Duke of Wetherby (21618), dam (Duchess 9th) by Grand Duke of Brockton (26284), &c.

Produce in	Names, &c.	By what Bull.	By whom bred.
1879, April 23, r. & w., C.C.	Lady Tyrley 7th	Oneida, 37146	Mr. Buchanan

BUDDS, W. F.,
Courtstown, Freshford, Co. Kilkenny.

EUGENA 13TH, red, calved February 16, 1875. Bred by Mr. J. T. Riddell, Grange House; got by Lord of the South (31730), dam (Eugena 3rd) by Beautiful Star (21256), g. d. (Eugena) by Professor Miller (18649), &c. See Vol. xxv. p. 636.

Produce in	Names, &c.	By what Bull.	By whom bred.
1878, Jan. 7, r. & w., C.C.	Ellie	True Hope, 35824	Mr. Budds
1879, Feb. 10, red, C.C.	Eumenes	Duke of Cranagh, 39726	do.

BULT, J. S.,
Dodhill House, Kingston, Taunton.

ANEMONE 8TH, roan, calved February 5, 1869, Vol. xx. p. 392. Bred by Mr. J. S. Bult; got by Earl of Fife (23835), dam (Anemone 2nd) by Duke of Cambridge (12742), &c.

Produce in	Names, &c.	By what Bull.	By whom bred.
1873, Aug. 17, roan, B.C.	Archer	Cardinal, 28144	Mr. Bult
1874, Aug. 11, r. & w., C.C.	Annie	do.	do.
1876, June 5, r. & w., C.C.	Anemone 11th	do.	do.
1877, May 27, roan, B.C.	Anthony	Gallant Gay, 33983	do.
1878, Mar. 27, r. & w., C.C.	Anemone 12th	do.	do.

Archer, sold to Mr. J. Rood, Cannington, Bridgewater; Anthony, to Mr. J. King, Steart, Dunster, Taunton.

ANNIE, red and white, calved August 11, 1874. Bred by Mr. J. S. Bult; got by Cardinal (28144), dam (Anemone 8th) by Earl of Fife (23835), &c.

Produce in	Names, &c.	By what Bull.	By whom bred.
1877, Oct. 3, roan, B.C.	Antonio	Gallant Gay, 33983	Mr. Bult

BRIDE CAKE, roan, calved December 4, 1869, Vol. xxi. p. 611. Bred by Mr. J. S. Bult; got by Earl of Fife (23835), dam (Bride) by Conqueror (21466), &c.

Produce in	Names, &c.	By what Bull.	By whom bred.
1876, June 26, white, B.C.	(Steer)	Cardinal, 28144	Mr. Bult
1877, June 7, white, B.C.	(Steer)	Gallant Gay, 33983	do.
1878, April 30, r. & w., C.C.	Bride Cake 2nd	do.	do.
1879, Mar. 15, roan, B.C.	Best Man	D. of Hazlecote 40th, 00743	do.

Best Man, sold to Mr. W. S. Gibbs, Manor House, Cothelestone, Taunton.

BURDON, Rev. John,
The Castle, Castle Eden, Co. Durham.

LADY ROSABEL, roan, calved July 7, 1871, Vol. xxv. p. 379. Bred by Mr. R. Burdon, Castle Eden; got by Red Errant (24912), dam (Lady Blanche) by Statesman (18927), &c.

Produce in	Names, &c.	By what Bull.	By whom bred.
1879, June 20, white, C.C.	Lady of Eden	Prince of Georgia, 37251	Rev. J. Burdon

MAYFLOWER, white, calved May 7, 1876, Vol. xxv. p. 379. Bred by the Rev.
J. Burdon ; got by Mountain Hero (31944), dam (Sweetbriar) by Red Errant
(24912), &c.

Produce in	Names, &c.	By what Bull.	By whom bred.
1879, Dec. 7, roan,	C.C.	Queen of the May 2nd	Prince of Georgia, 37251	Rev. J. Burdon

SWEETBRIAR, roan, calved June 20, 1871, Vol. xxv. p. 380. Bred by Mr. R.
Burdon, Castle Eden ; got by Red Errant (24912), dam (Burnett Rose) by
Young Freedom (21777), &c.

1879, Sept. 7, r. & w.,	B.C.|Forest Briar	Lord of the Forest, 38638|Rev. J. Burdon

BURNABY, J. D. A.,
Asfordby, Melton Mowbray.

DEWDROP, red and white, calved October 12, 1873, Vols. xxiii. and xxv.
pp. 664, 381. Bred by Mr. T. Swingler, Langham ; got by Union Jack
(32746), dam (Kate) by Manton (24525), &c.

1879, Nov. 23, r. & w.,	C.C.|Florence	3rd Duke of Glo'ster, 33653|Mr. Burnaby

BURTON, George,
Thorpe Willoughby, Selby.

FEODORE, red, calved April 26, 1873, Vols. xxiii., xxiv., and xxv. pp. 511, 518,
381. Bred by Mr. T. Lister, Groby Park ; got by Baron York (30500), dam
(Fragrance) by Baron Panton (23377), &c.

1879, Sept. 28, r. & w.,	C.C.|Dora	Kalamazoo, 40039	Mr. Burton

LADY LAUREL, red, calved June 26, 1876, Vol. xxv. p. 381. Bred by Mr. E.
Holden, Laurel Mount ; got by Oxford Cherry Duke 2nd (34972), dam
(Constance) by Lord Lally 3rd (24408), &c.

1879, Aug. 4, red,	B.C.|Prince Willoughby	Kalamazoo, 40039	Mr. Burton

BURTT, Henry,
Fulbeck Grange, Grantham.

CRIMSON KNIGHTLEY 2ND, red, calved January 10, 1875. Bred by Mr.
H. Burtt ; got by Earl Warwick 2nd (33829), dam (Crimson Knightley) by
Don Fawsley (28329), &c. See Vol. xxiv. p. 358.

1877, Sept. 13, red,	C.C.|Crimson Bride	Red Robin, 37326	Mr. Burtt
1878, Sept. 17, red,	C.C.|Crimson K'ghtley 3rd|Favonius, 39865	do.
1879, Aug. 16, red,	C.C.|Crimson K'ghtley 4th|Francolino, 39893	do.
Crimson Bride, sold to Mr. T. Beakbane, Llay Place, Wrexham.			

FANCY POMPEII, red, calved October 10, 1875. Bred by Mr. H. Burtt ; got
by Pompeii (35058), dam (Fancy 7th) by Comus (25821), &c. See " Scarlet,"
p. 226.

1878, Sept. 5, roan,	C.C.|Fancy	Favonius, 39865	Mr. Burtt
1879, Aug. 27, red,	B.C.|Scarlet	do.	do.

BUST, Martin,
West Halton, Brigg, Lincolnshire.

EDITH OXFORD, red and white, calved May 28, 1873, Vols. xxii., xxiii., and
xxv. pp. 331, 344, 353. Bred by Mr. R. Botterill, Wauldby; got by Nineteenth Duke of Oxford (28431), dam (Lady Edith) by Duke of Clarence
(19611), &c.

Produce in		Names, &c.	By what Bull.	By whom bred.
1879, July 5, roan,	C.C.	Oxf'dStrawberry2nd	Oxford-le-Grand, 29406	Mr. Bust

BYRON, Edmund,
Coulsdon Court, Surrey.

NELLIE 3RD, white, calved December 30, 1875, Vol. xxv. p. 383. Bred by Mr.
E. Byron; got by Highland Chief (34159), dam (Nellie 2nd) by The Count
(35749), &c.

1879, Mar. 26, r. & w., B.C.	(Steer)	Silenus, 37444	Mr. Byron

PRISCILLA 7TH, roan, calved November 2, 1875, Vol. xxv. p. 383. Bred by
Mr. E. Byron; got by Highland Chief (34159), dam (Priscilla 6th) by Lord
St. Leonards (29202), &c.

1879, Mar. 15, r. & w., B.C.	(Steer)	Silenus, 37444	Mr. Byron

PRISCILLA 8TH, roan, calved October 9, 1876. Bred by Mr. E. Byron; got
by Highland Chief (34159), dam (Priscilla 6th) by Lord St. Leonards (29202),
&c. See "Serapis," p. 226.

1879, Mar. 23, r. & w., C.C.	Priestess	Silenus, 37444	Mr. Byron

PRISCILLA 9TH, red and white, calved August 31, 1877. Bred by Mr. E. Byron;
got by Hartley (36746), dam (Priscilla 6th) by Lord St. Leonards (29202), &c.
See "Serapis," p. 226.

1879, Dec. 31, red,	B.C.	Spartan	Silenus, 37444	Mr. Byron

CADDY, Henry,
Rougholm, Bootle, Cumberland.

CASTANET 4TH, red and white, calved May 2, 1870. Bred by Mr. J. B.
Chirnside, Newham; got by Royal Knight (25032), dam (Castanet 2nd) by
Ravenspur (20628), &c. See "Coningsby," p. 54.

1876, Aug. 24, r. & w., C.C.	Clochette	Royal Fame, 35364	Mr. Caddy
1878, June 30, { red, B.C.	(dead)	do.	do.
{ red, B.C.	(dead)		

ROSE OF THE VALLEY, red and white, calved May 25, 1871. Bred by Mr.
J. Downing, Ashfield; got by Sir Egbert (27468), dam (Rosette) by Duke of
York (23804), &c. See Vol. xx. p. 746.

1877, July 22, red, B.C.	(dead)	British King, 36279	Mr. Caddy
1878, Aug. 28, r. & w., C.C.	Rose of Rougholm	British Boy, 30597	do.

CADMAN, Peter,
Ballahutchin, Union Mills, Isle of Man.

SWEET MAY, red, calved May 30, 1874, Vol. xxiv. p. 350. Bred by Mr. W. Brown, Ackworth; got by Farnley Royal (28573), dam (Sweet Willow) by Sir Sam (25171), &c.

Produce in		Names, &c.	By what Bull.	By whom bred.
1878, Jan. 10, red,	C.C.	Ruby	W'dsor's Royal Seal, 36016	Mr. Brown
1879, Nov. 2, r. & w.,	B.C.	Lord Beaconsfield	Quick Hope, 40558	Mr. Cadman

CADMAN, T. W.,
Ballifield Hall, Sheffield.

ADA 2ND, roan, calved April 20, 1876. Bred by Mr. T. W. Cadman; got by Baron Havering 9th (33051), dam (Ada 8th) by Cavalier (23524), &c. See Vol. xxiii. p. 368.

1879, Dec. 9, roan,	B.C.	Adrian	Duke of Sandale 2nd, 38180	Mr. Cadman

ADA 8TH, red and white, calved January 2, 1870, Vols. xx., xxi., xxii., xxiii., xxiv., and xxv. pp. 384, 763, 453, 368, 359, 383. Bred by Mr. R. Hemming, Bentley Manor; got by Cavalier (23524), dam (Ada) by Orthodox 28th (18493), &c.

1879, Nov. 6, r. & w.,	B.C.	Ada's Prince	Duke of Sandale 2nd, 38180	Mr. Cadman

FAIRY 2ND, red, calved April 5, 1875, Vol. xxv. p. 383. Bred by Mr. T. W. Cadman; got by Knight of Lothian (31545), dam (Fairy) by Baron Grantham (27940), &c.

1879, Dec. 12, r. & w.,	B.C.	Fairy Prince	Duke of Sandale 2nd, 38180	Mr. Cadman

FAME, red and white, calved December 30, 1869, Vols. xx., xxi., xxii., xxiii., and xxiv. pp. 512, 620, 349, 368, 359. Bred by Mr. J. H. Casswell, Laughton; got by Lord Chancellor (20160), dam (Flame) by Henry 5th (19944), &c.

1879, Feb. 2, r. & w.,	B.C.	(Steer)	Duke of Sandale 2nd, 38180	Mr. Cadman
1879, Dec. 18,	r.&w., C.C.	Fame 5th	do.	do.
	red, C.C.	Fame 6th		

FAME 3RD, red and white, calved March 2, 1876, Vol. xxv. p. 383. Bred by Mr. T. W. Cadman; got by Knight of Lothian (31545), dam (Fame) by Lord Chancellor (20160), &c.

1879, Sept. 21, r. & w.,	C.C.	Fame Church	Duke of Sandale 2nd, 38180	Mr. Cadman

FILLPAIL, roan, calved March 15, 1876. Bred by Mr. T. W. Cadman; got by Knight of Lothian (31545), dam (Lady Mary) by Cambridge Duke 4th (25706), &c. See "Earl Sandale," p. 92.

1879, Dec. 7, r. & w.,	B.C.	Earl Sandale	Duke of Sandale 2nd, 38180	Mr. Cadman

Earl Sandale, sold to Mr. G. J. Young, Claxby House, Market Rasen.

FLORENCE, white, calved May 13, 1872, Vols. xxi. and xxv. pp. 804, 365. Bred by Mr. J. W. Larking, Ashdown House; got by Grand Duke of Geneva (28756), dam (Florentia 6th) by Third Duke of Lancaster (19624), &c.

1879, March 8, white,	C.C.	Florence 2nd	D. of St. Johns 2nd, 36539	Mr. Cadman

LADY GRANADA, roan, calved February 10, 1875, Vols. xxiv. and xxv. pp. 359, 384. Bred by Mr. H. Pickersgill, Middleton Quernhow; got by Marshal Booth (31852), dam (Princess of Granada) by Prince Boabdil (27120), &c.

Produce in	Names, &c.	By what Bull.	By whom bred.
1879, Sept. 25, roan, C.C.	Lady Granada 3rd	Duke of Sandale 2nd, 38180	Mr. Cadman

LADY JOAN, roan, calved March 28, 1874, Vol. xxiv. p. 359. Bred by Mr. T. Adwick, Staythorpe; got by Knight of Derby (31535), dam (Lady Julian) by Prince Imperial (27145), &c.

1879, Jan. 20, r. & w., C.C.	Lady Joan 3rd	Duke of Sandale 2nd, 38180	Mr. Cadman

LADY JOAN 2ND, roan, calved February 21, 1877. Bred by Mr. T. W. Cadman, the property of Mr. G. Appleyard, Canklow; got by Iron Duke 4th (34220), dam (Lady Joan) by Knight of Derby (31535), &c.

1879, Sept. 15, r. & w., B.C.	Prince Imperial	Duke of Sandale 2nd, 38180	Mr. Cadman

Prince Imperial, sold to Mr. G. Appleyard, Canklow, Rotherham.

LADY MARY, roan, calved February 8, 1871, Vols. xxi., xxiii., and xxv. pp. 534, 368, 384. Bred by Mr. T. Adwick, Staythorpe; got by Cambridge Duke 4th (25706), dam (Lady Julia) by Julius Cæsar (22008), &c.

1879, May 19, roan, B.C.	Duke of Rother	Duke of Sandale 2nd, 38180	Mr. Cadman

Duke of Rother, sold to Mr. G. Appleyard, Canklow, Rotherham.

LADY SEYMOUR 2ND, roan, calved August 11, 1875, Vol. xxv. p. 384. Bred by Mr. T. W. Cadman; got by Knight of Derby (31535), dam (Lady Seymour) by General Seymour (28697), &c. See Vol. xxii. p. 349.

1879, April 15, r. & w., B.C.	General Seymour 2nd	Duke of Sandale 2nd, 38180	Mr. Cadman

LADY WALLACE 2ND, roan, calved February 12, 1873, Vols. xxii., xxiii., xxiv., and xxv. pp. 349, 368, 360, 384. Bred by Mr. T. W. Cadman; got by Iron Duke 2nd (31421), dam (Lady Wallace) by Lord Cobham (20164), &c.

1879, July 9, roan, B.C.	Lord Wallace 3rd	Duke of Sandale 2nd, 38180	Mr. Cadman

Lord Wallace 3rd, sold to Mr. G. Revitt, Myrtle Bank, Handsworth, Sheffield.

LADY WALLACE 3RD, roan, calved November 8, 1875, Vol. xxv. p. 384. Bred by Mr. T. W. Cadman; got by Knight of Lothian (31545), dam (Lady Wallace 2nd) by Iron Duke 2nd (31421), &c.

1879, Oct. 20, roan, B.C.	Lord Wallace 4th	Duke of Sandale 2nd, 38180	Mr. Cadman

LADY WALLACE 4TH, roan, calved October 23, 1876. Bred by Mr. T. W. Cadman; got by Knight of Lothian (31545), dam (Lady Wallace 2nd) by Iron Duke 2nd (31421), &c.

1879, Jan. 15, roan, C.C.	Lady Wallace 5th	Duke of Sandale 2nd, 38180	Mr. Cadman

LADY WILD ROSE, roan, calved October 17, 1871, Vols. xxiv. and xxv. pp. 360, 384. Bred by Mr. C. L. Whalley, Richmond House; got by Royal Scotforth (25042), dam (Wild Rose) by Napoleon (20395), &c.

1879, Oct. 25, roan, B.C.	Wild Boy	Duke of Sandale 2nd, 38180	Mr. Cadman

LUNETTE, roan, calved September 20, 1873, Vols. xxiv. and xxv. pp. 360, 384. Bred by Mr. T. Whiteside, Hesketh End; got by Oxford Cheerboy (32014), dam (Luna 4th) by Lord Oxford (29193), &c.

1879, Oct. 4, r. & w., C.C.	Lunette 4th	Duke of Sandale 2nd, 38180	Mr. Cadman

LUNETTE 2ND, roan, calved February 13, 1877. Bred by Mr. T. W. Cadman;
got by Knight of Lothian (31545), dam (Lunette) by Oxford Cheerboy
(32014), &c.

Produce in		Names, &c.	By what Bull.	By whom bred.
1879, Nov. 29, red,	B.C.	Moon Light	Duke of Sandale 2nd, 38180	Mr. Cadman

THE ABBESS, red and white, calved February 6, 1876. Bred by Mr. F.
Lythall, Offchurch; got by Telemachus 5th (35724), dam (Lodge Oxford) by
Brockley Oxford (25688), &c. See Vol. xxiii. p. 494.

1879, May 13, r. & w.,	C.C.	Abbess 2nd	Duke of Sandale 2nd, 38180	Mr. Cadman

CALTHORPE, Lord,
Elvetham Park, Winchfield.

LADY FAWSLEY 4TH, roan, calved April 4, 1872, Vols. xxii. and xxiv.
pp. 559, 361. Bred by Mr. H. J. Sheldon, Brailes House; got by Eighteenth
Duke of Oxford (25995), dam (Lady Fawsley) by Duke of Darlington
(21586), &c.

1879, Mar. 26, white,	C.C.	Farina	D. of Tregunter 5th, 33743	Lord Calthorpe

CAMERON, Archibald,
Artafallie, Inverness, N.B.

SNOWDROP 10TH, roan, calved February 21, 1876. Bred by Mr. J. Cran,
Kirkton; got by Bridegroom (33201), dam (Snowdrop 3rd) by Baronet
(25564), &c. See Vol. xxi. p. 649.

1878, May 7, red,	C.C.	Rosebud	King of Kessock, 38495	Mr. Cameron
1879, May 17, roan,	B.C.	Roan Prince	do.	do.

CAMPBELL, S.,
Kinellar, Blackburn, N.B.

CLARET 2ND, roan, calved July 2, 1875. Bred by Mr. S. Campbell; got by
Novelist (34929), dam (Claret 1st) by Duke (28342), &c. See "Kinellar,"
p. 130.

1878, Mar. 4, roan,	B.C.	(dead)	Emperor, 39841	Mr. Campbell
1879, Jan. 2, roan,	B.C.	Kinellar	Golden Prince, 38363	do.

Kinellar, sold to Mr. J. Gough, Mains of Drum, Drumoak.

ROSEBUD 2ND, red, calved May 4, 1870. Bred by Mr. S. Campbell; got by
Prince of Worcester (20597), dam (Rosebud) by Scarlet Velvet (16916),
&c. See "British Statesman," p. 36.

1879, April 5, roan,	B.C.	Forest King	Luminary, 34715	Mr. Campbell

CANTLIE, Charles A.,
Keithmore, Dufftown, N.B.

MYRTLE 25TH, roan, calved January 7, 1875. Bred by Mr. W. Cantlie, Keithmore; got by Benvolio (28017), dam (Myrtle 21st) by Boanerges (25647), &c.
See "Irwin Myrtle," p. 125.

1878, May 3, roan,	C.C.	Myrtle 31st	Lord Irwin. 29123	Mr. C. A. Cantlie

MYRTLE 29TH, white, calved March 12, 1877.   Bred by Mr. W. Cantlie, Keithmore; got by Baron's Heir (36205), dam (Myrtle 21st) by Boanerges (25647),
   &c.   See "Irwin Myrtle," p. 125.

Produce in		Names, &c.	By what Bull.	By whom bred.
1879, Oct. 20, red,	B.C.	Irwin Myrtle	Lord Irwin, 29123	Mr. C. A. Cantlie

CARBERY, Lord,
Castle Freke, Clonakilty, Co. Cork.

BLITHESOME GEORGIA, red and white, calved March 1, 1876.   Bred by
   Mr. W. T. Crosbie, Ardfert Abbey; got by Lord Blithesome (29067), dam
   (Georgia) by Castle Grove (19408), &c.   See Vol. xxiv. p. 393.

1878, April 19, roan,	B.C.	Prince of Georgia	Foreign Prince, 36656	Mr. Crosbie
1879, April 29, red,	B.C.	Fitz-Georgia	Gwynne Fitz Blithe, 39970	Lord Carbery

COWSLIP 24TH, roan, calved March 31, 1874, Vol. xxiv. p. 363.   Bred by Mr.
   R. Welsted, Ballywalter; got by England's Glory (23889), dam (Cowslip 16th)
   by Prince Christian (22581), &c.

1878, Mar. 2, roan,	B.C.	Rathbarry	Gwynne Fitz Blithe, 39970	Lord Carbery
1879, Mar. 7, roan,	C.C.	Cowslip 25th	do.	do.

LITTLE SPARKLE, roan, calved July 2, 1874, Vol. xxiv. p. 363.   Bred by
   Mr. W. T. Crosbie, Ardfert Abbey; got by Regal Booth (27262), dam
   (Frontlet) by Northern Light (24670), &c.

1878, May 28, red,	C.C.	Ruby Sparkle	Master Primrose, 37070	Lord Carbery
1879, May 17, roan,	C.C.	Sea Sparkle	do.	do.

PATIENCE, red and white, calved June 21, 1869, Vols. xx. and xxiii. pp. 681,
   363.   Bred by Mr. R. Welsted, Ballywalter; got by Uncle Ned (19026), dam
   (Prudence) by Roan Oxford (16841), &c.

1878, Mar. 29, roan,	C.C.	Patience 2nd	Master Primrose, 37070	Lord Carbery

CARR, W.,
Lowgate, Balne, near Selby.

LADY FARNLEY, roan, calved April 30, 1873.   Bred by Mr. W. Carr; got
   by Gladstone (31251), dam (First Lady) by Lord Darlington (26633), &c.   See
   "Lord Exeter," p. 146.

1876, Nov. 9, r. & w.,	B.C.	Lord Exeter	Telemachus 7th, 35726	Mr. Carr
1877, Nov. 10, roan,	C.C.	Lady Fawkes	British Hero, 36276	do.
1878, Nov. 20, roan,	C.C.	Lady Farewell	Yarburgh, 37698	do.
1870, Oct. 31, roan,	B.C.	Roan Duke	Duke of Heck, 38152	do.

CARTER, George,
Sandhill, Bedale.

SAUCY JANEY, red and white, calved November 2, 1872, Vols. xxiii. and xxiv.
   pp. 469, 472.   Bred by Mr. W. White, Burrill; got by Star of the Valley
   (32605), dam (Saucy Jane) by Manfred (26801), &c.

1878, Sept. 6, roan,	C.C.	Mary Jane	Baron Waterloo, 39444	Mr. Carter

CASSWELL, J. H.,
Laughton, Folkingham.

BLANCHE ARTENAY, roan, calved December 31, 1872, Vol. xxiii. p. 375.
Bred by Mr. C. O. Eaton, Tixover Hall; got by Duke of Artenay (30918),
dam (Lady Blanche) by Costa (21487), &c.

Produce in	Names, &c.	By what Bull.	By whom bred.
1879, Jan. 21, roan, C.C.	Bessie Artenay	D. of Barrington 5th, 33575	Mr. Casswell

CLEOPATRA LUNESDALE, roan, calved May 9, 1877. Bred by Mr. J. H.
Casswell; got by Lord Lunesdale Bates (34592), dam (Cleopatra 10th) by
Lord Oxford (20214), &c. See Vol. xxiv. p. 365.

Produce in	Names, &c.	By what Bull.	By whom bred.
1879, Aug. 27, roan, C.C.	Cleopatra	Duke of Elmhurst, 39731	Mr. Casswell

DUCHESS OF KENT, roan, calved March 29, 1872, Vols. xxii. and xxiv.
pp. 353, 365. Bred by Mr. W. W. Slye, Beaumont Grange; got by Grand
Duke of Kent 2nd (28759), dam (Lady Walton 2nd) by Earl of Glo'ster
(21644), &c.

Produce in	Names, &c.	By what Bull.	By whom bred.
1879, Oct. 21, white, C.C.	Duchess of Kent 4th	Duke of Elmhurst, 39731	Mr. Casswell

DULCIBELLA, red, calved April 7, 1873, Vols. xxii. and xxiv. pp. 353, 365.
Bred by Mr. W. W. Slye, Beaumont Grange; got by Grand Duke of Thorndale (31297), dam (Dulcimer) by Oxford (20450), &c.

Produce in	Names, &c.	By what Bull.	By whom bred.
1879, June 28, red, C.C.	Dairy Maid	D. of Barrington 5th, 33575	Mr. Casswell

GRACE KNIGHTLEY, red and white, calved October 12, 1869, Vols. xxi.,
xxiii., and xxv. pp. 839, 375, 389. Bred by Mr. J. Clayden, Littlebury; got by
Captain Knightley (25716), dam (Grace Costa) by Costa (21487), &c.

Produce in	Names, &c.	By what Bull.	By whom bred.
1879, April 11, r. & w., C.C.	Grace Darling 2nd	D. of Barrington 5th, 33575	Mr. Casswell

KIRKLEVINGTON DUCHESS 12TH, red, calved November 6, 1872, Vol.
xxv. p. 389. Bred by Mr. R. P. Davies, Horton; got by Second Duke of
Glo'ster (28392), dam (Kirklevington Rose) by Earl of Glo'ster (21644), &c.

Produce in	Names, &c.	By what Bull.	By whom bred.
1879, Sept. 2, red, C.C.	D'ss of Laughton 2nd	D. of Barrington 5th, 33575	Mr. Casswell

PATTIE LUNESDALE, roan, calved September 19, 1875. Bred by Mr. D.
Mackinder, Sempringham; got by Lord Lunesdale Bates (34592), dam
(Purple Jar) by Fourth Duke of Thorndale (17750), &c. See " Prince
Leopold," Vol. xxv. p. 224.

Produce in	Names, &c.	By what Bull.	By whom bred.
1879, Dec. 20, r. & w., C.C.	Pattie Lunesdale 2nd	Earl of Geneva 2nd, 38224	Mr. Casswell

PRIMROSE, roan, calved February 28, 1873, Vol. xxv. p. 390. Bred by Mr.
J. H. Casswell; got by Baron York (30500), dam (Perfection) by Royalist
(27370), &c.

Produce in	Names, &c.	By what Bull.	By whom bred.
1879, May 15, white, C.C.	Priscilla	Earl of Geneva 2nd, 28224	Mr. Casswell

SPARTAN QUEEN, red and white, calved May 1, 1875. Bred by Mr. J. H.
Casswell; got by Earl of Barrington 2nd (28494), dam (Spicy Lightburne) by
Grand Duke of Lightburne (26290), &c. See " Baron Lincoln," Vol. xxiv. p. 15.

Produce in	Names, &c.	By what Bull.	By whom bred.
1879, May 16, roan, C.C.	Spartan Beauty	Earl of Geneva 2nd, 38224	Mr. Casswell

SPICY LIGHTBURNE, roan, calved August 16, 1869, Vols. xxi., xxii., xxiv.,
and xxv. pp. 620, 354, 366, 390. Bred by Mr. C. Howard, Biddenham; got
by Grand Duke of Lightburne (26290), dam (The Sort) by Grand Duke 5th
(19875), &c.

Produce in	Names, &c.	By what Bull.	By whom bred.
1879, Feb. 18, r. & w., C.C.	Spicy Barrington	D. of Barrington 5th, 33575	Mr. Casswell

TREGUNTER GWYNNE, red and white, calved October 5, 1872, Vol. xxv. p. 390. Bred by Mr. W. W. Slye, Beaumont Grange; got by Barrington Duke (27985), dam (Orphan Gwynne) by Duke of Glo'ster (11382), &c.

Produce in		Names, &c.	By what Bull.	By whom bred.
1879, Mar. 2,	r.&w., C.C.	Tulip Gwynne	D.ofBarrington5th,33575	Mr. Casswell
	r.&w., C.C.	Tiny Gwynne		

URSULINE BATES, red and white, calved November 5, 1876. Bred by Mr. J. H. Casswell; got by Lord Lunesdale Bates (34592), dam (Ursuline) by Second Earl of Oxford (33809), &c. See Vol. xxiii. p. 376.

1879, Apr. 25, r. & w., C.C.	Ursuline Bates 2nd	Duke of Elmhurst, 39731	Mr. Casswell

CATHER, George,
Carrichue, Londonderry, Ireland.

FLEDA 5TH, roan, calved March 21, 1871, Vol. xxiii. p. 376. Bred by Mr. T. Cather, Lisnakilly; got by King Oberon (26510), dam (Fleda 4th) by Oberon (22438), &c.

		Names	By what Bull	By whom bred
1877, June 11, white,	C.C.	Fleda 10th	Flag of Ireland, 28613	Mr. G. Cather
1878, May 9, roan,	B.C.	Flagstaff	do.	do.
1879, April 26, white,	B.C.	White Flag	do.	do.

FLEDA 6TH, roan, calved June 12, 1874, Vol. xxiii. p. 376. Bred by Mr. T. Cather, Lisnakilly; got by Abab (27850), dam (Fleda 4th) by Oberon (22438), &c.

		Names	By what Bull	By whom bred
1877, May 28, roan,	C.C.	Fleda 9th	Fascinator, 33890	Mr. G. Cather
1878, April 28, roan,	B.C.	Forester	do.	do.

Forester, sold to Mr. Macdonnell, Widgion Lodge, Limavady.

JEANETTE, roan, calved March 12, 1873, Vol. xxiii. p. 377. Bred by Mr. N. M. Archdall, Crock-na-Crieve; got by Abercorn (25484), dam (Jenny Lind 11th) by Napoleon 3rd (29417), &c.

		Names	By what Bull	By whom bred
1877, Jan. 7, roan,	B.C.	(dead)	Abab, 27850	Mr. Cather
1878, Feb. 16, r. &w.,	B.C.	Fee-Simple	Fascinator, 33890	do.
1879, Mar. 19, roan,	B.C.	Flash	do.	do.

Fee-Simple, sold to Mr. T. Wilson, Drumeroon, Coleraine.

JOLLITY 5TH, roan, calved April 29, 1866, Vols. xx. and xxii. pp. 574, 354. Bred by Mr. G. Cather; got by Sir Roderick (22914), dam (Jollity) by Carlysle (14245), &c.

		Names	By what Bull	By whom bred
1876, Mar. 19, roan,	C.C.	Jollity 24th	Abab, 27850	Mr. Cather
1877, Apr. 11, r. & w.,	B.C.	Factotum	Fascinator, 33890	do.
1878, June 5, roan,	B.C.	Factor	do.	do.

Factotum, sold to Sir H. H. Bruce, Bart., Downhill, Coleraine; Factor, to Mr. Joseph Irwin, Ballyarton, Killaloo, Londonderry.

JOLLITY 7TH, roan, calved February 28, 1869, Vol. xxii. p. 354. Bred by Mr. G. Cather; got by King Oberon (26510), dam (Jollity 4th) by Sir Roderick (22914), &c.

		Names	By what Bull	By whom bred
1876, April 24, roan,	B.C.	Agricola	Abab, 27850	Mr. Cather
1877, April 10, white,	C.C.	Jollity 29th	do.	do.
1878, June 19, roan,	B.C.	(dead)	do.	do.
1879, July 5, roan,	C.C.	Jollity 42nd	Fascinator, 33890	do.

Agricola, sold to Mr. McCutcheon, Tirbracken, Eglinton, Londonderry.

JOLLITY 11TH, roan, calved February 8, 1872, Vol. xxii. p. 355.　Bred by Mr. G. Cather; got by King Oberon (26510), dam (Jollity) by Carlysle (14245), &c.

Produce in		Names, &c.	By what Bull.	By whom bred.
1876, Mar. 19, roan,	B.C.	Apropos	Abab, 27850	Mr. Cather
1877, June 25, roan,	C.C.	Jollity 31st	Fascinator, 33890	do.
1878, June 18, r. & w.,	B.C.	Allspice	Abab, 27850	do.
1879, July 7, { red,	B.C.	Fox-Glove	} Fascinator, 33890	do.
{ r. & w.,	C.C.	Jollity 43rd		

Apropos, sold to Mr. James Cherry, Thorn Hill, Bally Kelly; Allspice, to Mr. Henderson, Barna Kelly, Carrichue.

JOLLITY 13TH, roan, calved April 19, 1872, Vol. xxiii. p. 377.　Bred by Mr. G. Cather; got by King Oberon (26510), dam (Jollity 5th) by Sir Roderick (22914), &c.

Produce in		Names, &c.	By what Bull.	By whom bred.
1877, Feb. 26, white,	C.C.	Jollity 28th	Fascinator, 33890	Mr. Cather
1878, Mar. 13, r. & w.,	B.C.	Freeholder	do.	do.
1879, Mar. 14, roan,	C.C.	Jollity 40th	do.	do.

Freeholder, sold to Mr. W. Stewart, Coleraine.

JOLLITY 15TH roan, calved June 5, 1873.　Bred by Mr. G. Cather; got by Abab (27850), dam (Jollity 5th) by Sir Roderick (22914), &c.　See Vol. xxii. p. 354.

Produce in		Names, &c.	By what Bull.	By whom bred.
1876, May 30, white,	C.C.	Jollity 25th	Artizan, 36137	Mr. Cather
1877, April 28, roan,	C.C.	Jollity 30th	Fascinator, 33890	do.
1878, April 15, roan,	B.C.	Field Marshal	do.	do.
1879, Mar. 13, roan,	C.C.	Jollity 39th	do.	do.

Field Marshal, sold to Mr. Rankin, New Park, Garvagh.

JOLLITY 17TH, red and white, calved May 25, 1874.　Bred by Mr. G. Cather; got by Abab (27850), dam (Jollity 11th) by King Oberon (26510), &c.　See Vol. xxii. p. 355.

Produce in		Names, &c.	By what Bull.	By whom bred.
1877, Jan. 3, r. & w.,	C.C.	Jollity 27th	Fascinator, 33890	Mr. Cather
1878, April 13, roan,	C.C.	Jollity 34th	do.	do.
1879, Mar. 11, roan,	C.C.	Jollity 38th	do.	do.

JOLLITY 18TH, roan, calved September 27, 1874.　Bred by Mr. G. Cather; got by Abab (27850), dam (Jollity 13th) by King Oberon (26510), &c.　See Vol. xxiii. p. 377.

Produce in		Names, &c.	By what Bull.	By whom bred.
1877, May 27, r. & w.,	B.C.	Figaro	Fascinator, 33890	Mr. Cather
1878, Apr. 16, r. & w.,	C.C.	Jollity 35th	do.	do.
1879, Mar. 26, white,	C.C.	Jollity 41st	do.	do.

Figaro, sold to Mr. R. Cochrane, Spring Hill, Co. Donegal.

JOLLITY 19TH, roan, calved October 19, 1874.　Bred by Mr. G. Cather; got by Abab (27850), dam (Jollity 8th) by King Oberon (26510), &c.　See Vol. xxii. p. 354.

Produce in		Names, &c.	By what Bull.	By whom bred.
1877, Oct. 27, roan,	C.C.	Jollity 33rd	Fascinator, 33890	Mr. Cather
1878, Oct. 25, roan,	B.C.	Fustian	do.	do.

JOLLITY 21ST, white, calved April 13, 1875.　Bred by Mr. G. Cather; got by Abab (27850), dam (Jollity 5th) by Sir Roderick (22914), &c.　See Vol. xxii. p. 354.

Produce in		Names, &c.	By what Bull.	By whom bred.
1877, Aug. 29, white,	C.C.	Jollity 32nd	Fascinator, 33890	Mr. Cather
1879, Jan. 24, white,	C.C.	Jollity 37th	do.	do.

JOSEPHINE, red and white, calved May 15, 1873, Vol. xxiii. p. 377. Bred by Mr. G. Cather; got by King Oberon (26510), dam (July) by Sir Roderick (22914), &c.

Produce in		Names, &c.	By what Bull.	By whom bred.
1877, May 13, roan,	B.C.	Ad-Cap-Tandum	Abab, 27850	Mr. Cather
1878, May 26, red,	C.C.	Josephine 2nd	do.	do.
1879, May 21, red,	C.C.	Josephine 3rd	Fascinator, 33890	do.

Ad-Cap-Tandum, sold to Mr. W. J. Hanna, White House, Carrigans, Co. Donegal.

JULY, red and white, calved July 16, 1866, Vols. xx. and xxii. pp. 576, 355. Bred by Mr. G. Cather; got by Sir Roderick (22914), dam (Juniper) by Prince Arthur (13497), &c.

1876, Sept. 9, roan,	C.C.	Juicy	Abab, 27850	Mr. Cather
1877, Oct. 11, r. & w.,	C.C.	Jill	do.	do.

LIZZIE, roan, calved October 28, 1873, Vol. xxiii. p. 377. Bred by Mr. N. M. Archdall, Crock-na-Crieve; got by Meteor (38754), dam (Virginia) by Dr. McHale (15887), &c.

1877, May 11, { roan,	C.C.	Lovely	} Abab, 27850	Mr. Cather
{ roan,	C.C.	Lively		
1878, May 16, roan,	C.C.	Lucy	do.	do.
1879, April 26, red,	C.C.	Letty	Fascinator, 33890	do.

SELINA 13TH, roan, calved April 3, 1869, Vol. xxii. p. 355. Bred by Mr. G. Cather; got by King Oberon (26510), dam (Selina 2nd) by Prince Arthur (13497), &c.

1876, Feb. 22, red,	C.C.	Selina 34th	Abab, 27850	Mr. Cather
1877, Mar. 14, roan,	C.C.	Selina 36th (dead)	do.	do.
1878, April 1, roan,	B.C.	Favourite	Fascinator, 33890	do.
1879, Mar. 10, white,	B.C.	Fusilier	do.	do.

Favourite, sold to Mr. McCutcheon, Tirbracken, Eglinton, Londonderry.

SELINA 19TH, red, calved March 14, 1871, Vol. xxii. p. 355. Bred by Mr. G. Cather; got by King Oberon (26510), dam (Selina 9th) by Oberon (22438), &c.

1876, Mar. 27, r. & w.,	B.C.	Abdul	Abab, 27850	Mr. Cather
1877, Mar. 18, r. & w.,	C.C.	Selina 37th (dead)	do.	do.
1878, April 4, r. & w.,	C.C.	Selina 38th	Fascinator, 33890	do.
1879, May 7, roan,	C.C.	Selina 41st	do.	do.

Abdul, sold to Mr. D. McClelland, Balteagh, Limavady.

SELINA 28TH, roan, calved July 28, 1874. Bred by Mr. G. Cather; got by Abab (27850), dam (Selina 15th) by King Oberon (26510), &c. See Vol. xxii. p. 355.

1877, June 17, roan,	B.C.	Freebooter	Fascinator, 33890	Mr. Cather
1878, April 26, white,	C.C.	Selina 30th	do.	do.

Freebooter, sold to Mr. C. McElroy, Aghadovey, Coleraine.

SELINA 33RD, red, calved August 24, 1875. Bred by Mr. G. Cather; got by Abab (27850), dam (Selina 23rd) by King Oberon (26510), g. d. (Selina 10th) by Sir Richard (27487), gr. g. d. (Selina 5th) by Elfin King (17796), &c. See "Ali Baba," Vol. xxii. p. 6.

1878, Apr. 21, r. & w.,	B.C.	Flambeau	Fascinator, 33890	Mr. Cather
1879, June 24, roan,	C.C.	Selina 42nd	do.	do.

Flambeau, sold to Mr. Gamble, Newton Stuart, Co. Tyrone.

SELINA 34TH, red, calved February 22, 1876. Bred by Mr. G. Cather; got
by Abab (27850), dam (Selina 13th) by King Oberon (26510), &c. See Vol.
xxii. p. 355.

Produce in	Names, &c.	By what Bull.	By whom bred.
1878, July 12, r. & w., C.C.	Selina 40th	Fascinator, 33890	Mr. Cather
1879, June 30, roan, B.C.	Fac-simile	do.	do.

CATTLEY, John,
Stearsby, Easingwold, Yorkshire.

RED NANCY, roan, calved June 10, 1873, Vol. xxiii. p. 378. Bred by Mr. J.
Cattley; got by Lord Stanley (31752), dam (Red Nell) by The General
(27632), &c.

1878, Jan. 28, roan,	B.C.	Reform		Sergeant, 37434	Mr. Cattley
1879, Jan. 10, { roan,	B.C.	(Steer)	}	May Prince, 38746	do.
{ roan,	C.C.	Roan Nancy			

Reform, sold to Mr. Farrah, Craike, Easingwold.

RINGLET, roan, calved November 26, 1875. Bred by Mr. J. Cattley; got by
Bismarck (28040), dam (Red Nancy) by Lord Stanley (31752), &c.

1878, Oct. 12, roan,	C.C.	Ringdove	May Prince, 38746	Mr. Cattley
1879, Sept. 10, roan,	C.C.	Regina	Reformer, 42266	do.

CATTO, Gavin,
Mains of Gight, Fyvie, Aberdeenshire.

QUEEN ANNE, roan, calved April 12, 1871. Bred by Mr. W. Mackie, Petty
Fyvie; got by Boanerges (25647), dam (Queen) by Commander (25811), g. d.
(Daisy 2nd) by Guy Fawkes (12981), &c. See "Louise," Vol. xix. p. 605.

1875, April 12, roan,	B.C.	Charles	Attorney General, 30402	Mr. Catto
1876, Mar. 28, red,	B.C.	General	do.	do.
1877, April 8, roan,	C.C.	Lizzie	Richard, 40594	do.
1878, April 20, roan.	C.C.	Jeannie	do.	do.

Charles and General, sold to Mr. John Catto, New Maud, Aberdeenshire.

CAZALET, Edward,
Fairlawn, Tonbridge, Kent.

BEAUTY, roan, calved March 11, 1872, Vol. xxiii. p. 378. Bred by Messrs.
F. Leney and Sons, Wateringbury; got by Grand Duke of Oxford (28764),
dam (Sultana 2nd) by Man in the Moon (18320), &c.

1879, Feb. 3, roan,	C.C.	Beatrice	Charming Prince 2d, 33334	Mr. Cazalet

CHAFFEY, Major John,
Prince Hill, Worton, Devizes.

CLEOPATRA'S DUCHESS 4TH, roan, calved March 1, 1875. Bred by Mr.
E. H. Cheney, Gaddesby Hall; got by Ninth Duke of Geneva (28391), dam
(Cleopatra 12th) by Grand Duke of Essex 4th (24068), &c. See Vol. xxiv.
p. 368.

1879, April 20, roan,	C.C.	Cleop's D. of Glo'ster	3rd Duke of Glo'ster, 33653	Major Chaffey

GIRDLE, white, calved October 14, 1872.   Bred by Mr. J. Blandy Jenkins,
Kingston House; got by Duke of Lancaster (28411), dam (Victoria 8th) by
Manager (24521), &c.   See "Orion," Vol. xxiv. p. 190.

Produce in		Names, &c.	By what Bull.	By whom bred.
1878, Mar. 19, roan,	C.C.	Victoria of Worton	But'fly'sRelative2d, 36298	Major Chaffey
1879, Mar. 20, roan,	C.C.	Victoria of Worton 2d	do.	do.

LILY OF TOWNELEY, red and white, calved June 20, 1875.   Bred by Colonel
Towneley, Towneley Park; got by Second Hubback (28880), dam (Lily of
Rossendale 1st) by Pride of Rossendale (29584), &c.   See Vol. xxii. p. 595.

Produce in		Names, &c.	By what Bull.	By whom bred.
1878, Apr. 26, r. & w.,	C.C.	Lily of Towneley 2nd	But'fly'sRelative2d, 36298	Major Chaffey
1879, Apr. 1, r. & w.,	C.C.	Lily of Towncley 3rd	do.	do.

SERAPHINA 23RD, red, calved December 25, 1870, Vol. xxiii. p. 527.   Bred
by Mr. J. Fawcett, Scaleby Castle; got by Royal Cumberland (27358), dam
(Seraphina 19th) by Imperial Oxford (18084), &c.

Produce in		Names, &c.	By what Bull.	By whom bred.
1879, Sept. 15, red,	C.C.	Seraphina's D'ss 4th	Duke of Glo'ster 6th, 39734	Mr. Cheney

Seraphina's Duchess 4th, sold to Major Chaffey, Prince Hill.

SONSIE GIRL, roan, calved December 25, 1875.   Bred by Mr. C. H. Cock,
Bridgefoot; got by Master Fragrant (34810), dam (Daisy) by Lord Claret
(24376), &c.   See Vol. xx. p. 472.

Produce in		Names, &c.	By what Bull.	By whom bred.
1878, April 13, roan,	B.C.	Sir Wilfrid	D. of Worcester 3rd, 39796	Major Chaffey
1879, July 24, r. & w.,	B.C.	(Steer)	But'fly'sRelative2d, 36298	do.

Sir Wilfrid, sold to Mr. G. H. Woods, East Bradley, Trowbridge.

WILD DUCHESS OF GENEVA 4TH, red and white, calved September 20,
1874.   Bred by Mr. E. H. Cheney, Gaddesby Hall, the property of Mr. C.
Magniac, Colworth House; got by Ninth Duke of Geneva (28391), dam (Wild
Duchess of York) by Seventh Duke of York (17754), &c.   See "Wild Duke
of Collingham," p. 269.

Produce in		Names, &c.	By what Bull.	By whom bred.
1878, Dec. 3, red,	C.C.	W'dD'sofGlo'ster 2d	Duke of Glo'ster 6th, 39734	Mr. E. H. Cheney

Wild Duchess of Glo'ster 2nd, sold to Major Chaffey, Prince Hill.

WILD DUCHESS OF GENEVA 5TH, roan, calved December 14, 1876.   Bred
by Mr. E. H. Cheney, Gaddesby Hall; got by Ninth Duke of Geneva (28391),
dam (Wild Duchess of York) by Seventh Duke of York (17754), &c.

Produce in		Names, &c.	By what Bull.	By whom bred.
1879, June 2, roan,	B.C.	W'dD.ofCollingham	D. of Collingham 3d, 38134	Major Chaffey

WINDSOR'S LEAF, red and white, calved August 10, 1874.   Bred by Mr. R.
S. Bruere, Braithwaite Hall; got by Booth's Royal Signet (28061), dam
(Tulip Leaf) by Sir Windsor Broughton (27507), &c.   See "Sir Windsor
Towneley," Vol. xxv. p. 265.

Produce in		Names, &c.	By what Bull.	By whom bred.
1879, Feb. 23, r. & w.,	C.C.	Leaf of Towneley	But'fly'sRelative2d, 36298	Major Chaffey

CHALK, Thomas,
Linton, Cambridgeshire.

BATTERSEA BATTERSEA, roan, calved February 13, 1867, Vols. xxi., xxiv.,
and xxv. pp. 983, 368, 391.   Bred by Mr. Z. Walker, Acock's Green; got by
Battersea First Fruits (21250), dam (Miss Battersea) by Earl of Hardwicke
(14476), &c.

Produce in		Names, &c.	By what Bull.	By whom bred.
1879, Sept. 19, r. & w.,	C.C.	Lady Battersea 3rd	Baron Aston, 37770	Mr. Chalk

DUCHESS OF GRAFTON, roan, calved August 9, 1874. Bred by Mr. H.
Webb, Streetly Hall; got by Prolific 3rd (27214), dam (Babraham Duchess
2nd) by Duke of Grafton (21594), &c. See Vol. xxi. p. 983.

Produce in	Names, &c.	By what Bull.	By whom bred.
1879, May 15, red, C.C.	Lady Grafton	Cambridge Duke 5th, 30644	Mr. Chalk

FLOWER 2ND, roan, calved October 7, 1873, Vols. xxii. and xxv. pp. 501, 391.
Bred by Mr. H. W. Martin, Upton House, Littleport; got by Hamlet (31335),
dam (Flower) by Duke Homfrey (19599), &c.

Produce in	Names, &c.	By what Bull.	By whom bred.
1879, April 28, roan, B.C.	Flower of Day	Viscount Worcester, 40881	Mr. Chalk

LILY LASSIE, white, calved January 6, 1878. Bred by Mr. T. Chalk; got
by Young Heydon Duke (34156), dam (Welcome Lassie) by Costa (21487),
&c. See Vol. xxv. p. 391.

Produce in	Names, &c.	By what Bull.	By whom bred.
1879, Dec. 27, white, C.C.	Lily Lassie 2nd	Baron Aston, 37770	Mr. Chalk

WELCOME LASSIE, roan, calved March 12, 1869, Vols. xx., xxi., xxiv., and
xxv. pp. 818, 984, 699, 391. Bred by Mr. H. Webb, Streetly Hall; got by
Costa (21487), dam (Welcome Lass 4th) by The Plumber (18993), &c.

Produce in	Names, &c.	By what Bull.	By whom bred.
1879, Jan. 20, roan, C.C.	Welcome Lassie 9th	Cambridge Duke 5th, 30644	Mr. Chalk

WELCOME LASSIE 6TH, red, calved March 7, 1877. Bred by Mr. H. Webb,
Streetly Hall; got by Hairy Boy (34102), dam (Welcome Lassie) by Costa
(21487), &c. See Vol. xxiv. p. 699.

Produce in	Names, &c.	By what Bull.	By whom bred.
1879, April 1, r. & w., C.C.	Linton Lassie	Viscount Worcester, 40881	Mr. Chalk

CHAMLEY, Captain T.,
Warcop House, Penrith.

CZARINA 2ND, roan, calved February 16, 1875. Bred by Mr. T. Whiteside,
Hesketh End; got by Baron of Raby 2nd (33071), dam (Duchess of Oxford)
by Barrington Oxford (25607), &c.

Produce in	Names, &c.	By what Bull.	By whom bred.
1877, May 11, roan, C.C.	(dead)	Lord of Sleddale, 36985	Mr. Whiteside
1879, April 24, white, B.C.	(Steer)	Sir O. Barrington 3d, 39126	Capt. Chamley

CZARININA, roan, calved March 17, 1874. Bred by Mr. T. Whiteside, Hesketh
End; got by Earl of Strafford (31084), dam (Duchess of Oxford) by Barrington Oxford (25607), &c. See " Knight of Warcop," p. 135.

Produce in	Names, &c.	By what Bull.	By whom bred.
1877, Nov. 10, roan, C.C.	(dead)	Lord of Sleddale, 36985	Capt. Chamley
1879, Dec. 31, roan, B.C.	Knight of Warcop	Gd. D. of Morecambe, 36722	do.

DUCHESS OF OXFORD, red, calved June 7, 1871, Vols. xxi. and xxv.
pp. 990, 392. Bred by Mr. W. W. Slye, Beaumont Grange; got by Barrington Oxford (25607), dam (Dulcimer) by Oxford (20450), &c.

Produce in	Names, &c.	By what Bull.	By whom bred.
1879, April 2, roan, C.C.	Czarina 4th	Sir O. Barrington 3d, 39126	Capt. Chamley

GOLDEN DUCHESS 2ND, roan, calved May 29, 1869, Vols. xx., xxi., xxii.,
and xxv. pp. 542, 636, 369, 392. Bred by Mr. W. W. Slye, Beaumont
Grange; got by Barrington Oxford (25607), dam (Golden Duchess) by Golden
Duke (19860), &c.

Produce in	Names, &c.	By what Bull.	By whom bred.
1879, Sept. 3, white, C.C.	Golden Duchess 8th	Sir O. Barrington 3d, 39126	Capt. Chamley

GOLDEN DUCHESS 4TH, red and white, calved August 11, 1875. Bred by
Mr. E. J. Coleman, Stoke Park ; got by Third Duke of Glo'ster (33653), dam
(Golden Duchess 2nd) by Barrington Oxford (25607), &c.

Produce in	Names, &c.	By what Bull.	By whom bred.
1877, Oct. 30, r. & w., C.C.	Golden Duchess 6th	Lord of Lochaber 2d, 36983	Mr. Barnes
1879, Oct. 22, r. & w., C.C.	Golden Duchess 9th	Gd.D.of Morecambe, 36722	Capt. Chamley

Golden Duchess 6th, sold to Captain Chamley, Warcop House.

GOLDEN DUCHESS 5TH, red and white, calved August 5, 1876. Bred by
Mr. C. A. Barnes, Charleywood ; got by Third Duke of Glo'ster (33653),
dam (Golden Duchess 2nd) by Barrington Oxford (25607), &c.

Produce in	Names, &c.	By what Bull.	By whom bred.
1879, July 5, roan,	C.C.	Golden Duchess 7th	Sir O. Barrington 3d,39126	Capt. Chamley

CHANDOS-POLE-GELL, H.,
Hopton Hall, Wirksworth, Derbyshire.

*BRIGHT LADY OF THE REALM, red and white, calved December 28,
1874. Bred by the Hon. M. H. Cochrane, Hillhurst, Compton, Canada ; got
by Star of the Realm (40758), dam (Bright Lady) by Lord Blithe (22126),
g. d. (Bright Countess) by Breast Plate (19337), &c. See Vol. xviii. p. 404.

Produce in	Names, &c.	By what Bull.	By whom bred.
1879, Feb. 22, r. & w., C.C.	Bright Lodestar	Lord Lamech, 34578	Mr. H.C. Pole-Gell

LEONORA, red and white, calved March 18, 1876, Vol. xxv. p. 392. Bred by
Mr. H. Chandos-Pole-Gell ; got by Favourite (31145), dam (Leopoldine 4th)
by King Charles (24240), &c.

Produce in	Names, &c.	By what Bull.	By whom bred.
1879, Nov. 1, r. & w., C.C.	Rebecca	Royal Saxon, 39057	Mr. H.C. Pole-Gell

LEOPOLDINE 4TH, red and white, calved July 18, 1871, Vols. xxi., xxii.. xxiii.,
and xxiv. pp. 623, 356, 380, 369. Bred by Messrs. Atkinson, Peepy ; got by
King Charles (24240), dam (Leopoldine 3rd) by Killerby Lad (20052), &c.

Produce in	Names, &c.	By what Bull.	By whom bred.
1879, Mar. 6, r. & w., B.C.	Royal Leo	Royal Saxon, 39057	Mr. H.C. Pole-Gell

Royal Leo, sold to Monsieur de Saint Marie, for the French Government.

PRINCESS FRANZISKA, red and white, calved March 6, 1875. Bred by
Mr. H. Chandos-Pole-Gell ; got by Prince Christian (22581), dam (Leopoldine
4th) by King Charles (24240), &c.

Produce in	Names, &c.	By what Bull.	By whom bred.
1878, May 10, roan,	B.C.	Freebooter	County Member, 28268	Mr. H.C. Pole-Gell
1879, Apr. 28, r. & w., B.C.	Royal Frank	Royal Saxon, 39057	do.

Freebooter, sold to Mr. W. C. Booth, Oran, Catterick.

ROSAMOND 2ND, roan, calved April 7, 1876. Bred by Mr. Donald Fisher,
Pitlochrie ; got by Gunpowder (28801), dam (Ringdove 2nd) by King Charles
(24240), &c. See "Woodstock," p. 273.

Produce in	Names, &c.	By what Bull.	By whom bred.
1878, July 17, r. & w., C.C.	Fair Rosamond	Favourite, 31145	Mr. Fisher
1879, July 11, roan, B.C.	Woodstock	do.	Mr. H.C. Pole-Gell

WEAL CHARITY, white, calved November 20, 1876. Bred by Mr. Donald
Fisher, Pitlochrie ; got by King Brian (34308), dam (Weal Hope) by Royal
Prince (27384), &c. See Vol. xxiii. p. 438.

Produce in	Names, &c.	By what Bull.	By whom bred.
1879, June 5, white, C.C.	(dead)	Favourite, 31145	Mr. H.C. Pole-Gell

CHESTER, Thomas W.,
Park House, Crawley, Sussex.

LETITIA 5TH, roan, calved March 19, 1874. Bred by Major McCraith, Lough-loher; got by Richard Gwynne (32299), dam (Letitia 4th) by Vavasour (35852), g. d. (Letitia) by Little Pat (18203), gr. g. d. (Lizzie Lindsay) by Lord Spencer (14853), &c. See "Patrician," Vol. xvi. p. 210.

Produce in		Names, &c.	By what Bull.	By whom bred.
1878, May 3, r. & w.,	C.C.	Lady Gay	Great Fame, 34087	Mr. Chester

LETITIA 6TH, red, calved April 7, 1875. Bred by Major McCraith, Lough-loher; got by Richard Gwynne (32299), dam (Letitia 4th) by Vavasour (35852), &c.

1878, Apr. 12, r. & w., C.C.			Great Fame, 34087	Mr. Chester

MAGICAL, roan, calved May 1, 1875. Bred by Major McCraith, Loughloher; got by Richard Gwynne (32299), dam (Magic) by Vavasour (35852), g. d. (Witchcraft) by Master Harbinger (18352), gr. g. d. (Wizard's Daughter) by Wizard (19162), — by Childe Harold (30733), — by Duke of Argyll (15912), — by Royalty (27392), — by Scamp (13674).

1879, Feb. 6, r. & w.,	B.C.	Romulus	Prince of Allerton, 38920	Mr. Chester

ROSY QUEEN, red, calved March 10, 1875. Bred by Major McCraith, Loughloher; got by Richard Gwynne (32299), dam (Fairy Queen) by Master Harbinger (18352), g. d. by Narcissus (26948), &c. See "Oberon," Vol. xviii. p. 234.

1878, Feb. 2, roan,	C.C.	Buttercup	Great Fame, 34087	Mr. Chester
1879, Jan. 28, red,	C.C.	Rose Red	Prince of Allerton, 38920	do.

CHOPE, Thomas,
Farford Farm, Hartland, North Devon.

LADY GLENBURNIE, roan, calved August 27, 1876. Bred by Mr. J. B. James, Hallsanney, Bideford; got by Lord Godolphin (36065), dam (Lady Danby) by Royal Windsor (29890), g. d. (Miss Danby) by Emmanuel (31105), &c. See Vol. xxi. p. 875.

1879, May 31, roan,	C.C.	Lady Farford	Marquis of Oxford, 40311	Mr. Chope

CHRISTY, James,
Boynton Hall, Chelmsford.

BARONESS DUDLEY, red, calved December 12, 1874, Vol. xxv. p. 393. Bred by Mr. J. Christy; got by Oxford's Baron (32030), dam (Lady Dudley 7th) by Third Duke of Claro (23729), &c.

1879, Oct. 21, red,	C.C.	Boynton Dudley	Dudley's Oxford, 38110	Mr. Christy

LOU, red, calved January 6, 1877. Bred by Mr. J. Christy; got by Kidderminster (41747), dam (Princess Louise) by Manhatten (26802), &c. See Vol. xxv. p. 394.

1879, Sept. 28, red,	C.C.	Lucky	Dudley's Oxford, 38110	Mr. Christy

LOVELY ROSE, roan, calved January 22, 1877. Bred by Mr. J. Christy; got by Oxford's Baron (32030), dam (Geneva Rose) by Third Duke of Geneva (23753), &c. See Vol. xxv. p. 393.

Produce in		Names, &c.	By what Bull.	By whom bred.
1879, July 16, r. & w., C.C.		Sweet Rose	Duke of Oxford, 39770	Mr. Christy

MISS LIVERPOOL, roan, calved June 7, 1876, Vol. xxv. p. 393. Bred by Mr. J. Christy; got by Baron Dudley (33036), dam (Lady Liverpool) by Duke of Liverpool (33688), &c.

1879, July 12, red,	C.C.	Luna	Secundus, 42367	Mr. Christy

PEARLY, red and white, calved April 29, 1877. Bred by Mr. J. Christy; got by Canterbury (41181), dam (Pearl) by Baron Dudley (33036), &c. See Vol. xxv. p. 393.

1879, Oct. 16, r. & w., C.C.		Pimpernel	Dudley's Oxford, 38110	Mr. Christy

CLANCARTY, Earl of,
Garbally, Ballinasloe, Co. Galway.

CAROLINE, roan, calved August 5, 1876. Bred by the Earl of Clancarty, Garbally; got by Roan Sovereign (35289), dam (Snowdrop 2nd) by Gallant Knight (26214), &c. See Vol. xxii. p. 365.

1879, May 1, roan,	C.C.	Virtue	Czarowitz 5th, 43016	Earl of Clancarty

DAISY 10TH, roan, calved June 8, 1872, Vol. xxii. p. 364. Bred by Mr. R. F. Dunlop, Monasterboice House; got by Sovereign (27538), dam (Daisy 6th) by Royal Duke (25014), &c.

1876, July 29, roan,	B.C.	Sovereignty	Roan Sovereign, 35289	Earl of Clancarty
1877, Sept. 28, roan,	B.C.	Knight of the West	do.	do.
1878, Aug. 10, roan,	B.C.	King Coffey	do.	do.
1879, June 23, roan,	B.C.	Barometer	Czarowitz 5th, 43016	do.

Knight of the West, sold to Mr. F. K. Pilkington, Fort Tyrrel's Pass; King Coffey, to Mr. G. Rutherford, Turin Castle, Co. Mayo.

SNOWDROP 2ND, roan, calved June 1, 1872, Vol. xxii. p. 365. Bred by Mr. W. D. Dunlop, Monasterboice House; got by Gallant Knight (26214), dam (Snowdrop) by Mountebank (24626), &c.

1876, Aug. 5, roan,	C.C.	Caroline	Roan Sovereign, 35289	Earl of Clancarty
1877, Aug. 27, roan,	B.C.	Baron Galway	do.	do.
1878, Aug. 1, white,	B.C.	Mohammed	do.	do.
1879, July 21, roan,	B.C.	Hard Times	Czarowitz 5th, 43016	do.

Baron Galway, sold to Mr. T. Dawson, Tullyyard, Newbliss, Co. Monaghan; Mohammed, to Mr. M. B. Joyce, Rosina, Clonbur.

CLARK, J. P.,
North Ferriby, Brough, Yorkshire.

COUNTESS OF WETHERBY 2ND, roan, calved January 10, 1877. Bred by Mr. J. P. Clark; got by Duke of Thorndale (33740), dam (Countess of Wetherby) by Fifth Duke of Wetherby (31033), &c. See Vol. xxiv. p. 376.

1879, Oct. 11, r. & w., B.C.		(Steer)	B'n. T'croft Oxf'd 3d, 36209	Mr. Clark

DUCHESS GWYNNE, red and white, calved February 10, 1876. Bred by
Mr. W. W. Slye, Beaumont Grange; got by Grand Duke of Thorndale 2nd
(31298), dam (Tregunter Gwynne) by Barrington Duke (27985), g. d. (Orphan
Gwynne) by Duke of Glo'ster (11382), &c. See Vol. xx. p. 676.

Produce in	Names, &c.	By what Bull.	By whom bred.
1879, Jan. 23, r. & w., C.C.	Airdrie's Gwynne	24th D. of Airdrie, 36460	Mr. Clark

GAZELLE 32ND, roan, calved September 12, 1873, Vol. xxiv. p. 635. Bred
by Mr. J. D'A. Samuda, Chillies; got by Wildfire (32866), dam (Gazelle
25th) by Second Duke of Tregunter (26022), &c.

1879, Feb. 18, white, B.C.	Duke of Sussex	Marquis of Oxford, 34786	Mr. Clark

GUELDER ROSE 10TH, red and white, calved March 28, 1876, Vol. xxv.
p. 394. Bred by Mr. J. P. Clark; got by Duke of Thorndale (33740), dam
(Guelder Rose 7th) by Earl of Collingham (28498), &c.

1879, Sept. 18, r. & w., B.C.	Lord Fordham	Fordham D.Oxf'd 4th, 41569	Mr. Clark

LADY JULIA 2ND, red, calved June 10, 1876. Bred by Mr. G. Drewry,
Holker House; got by Duke of Oxford 28th (33710), dam (Lady Julia) by
Second Duke of Collingham (23730), &c. See Vol. xxiii. p. 426.

1879, Feb. 19, r. & w., C.C.	Lady Julia 3rd	Baron Oxford 7th, 36199	Mr. Clark

LADY URSULA 2ND, roan, calved August 12, 1874, Vol. xxv. p. 395. Bred
by Mr. J. P. Clark; got by Nineteenth Duke of Oxford (28431), dam (Ursula
33rd) by Third Duke of Wharfdale (21619), &c.

1879, Oct. 25, roan, B.C.	(Steer)	B'n T'croft Oxf'd 3d, 36209	Mr. Clark

MUSICAL 15TH, red and white, calved January 19, 1869, Vols. xx., xxi., and
xxiv. pp. 666, 916, 636. Bred by Mr. E. Bowly, Siddington House; got by
Grand Duke 13th (21850), dam (Musical 7th) by Seventh Duke of York
(17754), &c.

1879, Jan. 21, roan, B.C.	Music Master	Marquis of Oxford, 34786	Mr. Clark

Music Master, sold to Monsieur Michiels van Kessenick, Limbourg, Holland.

OXFORD'S BLANCHE, roan, calved May 6, 1873, Vol. xxiii. p. 518. Bred
by Mr. M. Kennedy, Stone Cross; got by Twenty-third Duke of Oxford
(31001), dam (Blanche Rose 3rd) by General Napier (24023), &c.

1879, Feb. 16, roan, B.C.	England's Oxford	Dk. of Oxford 34th, 36529	Mr. Clark

England's Oxford, sold to Monsieur J. Loge, Namur, Belgium.

PRINCESS VICTORIA 7TH, roan, calved November 17, 1868, Vols. xxi., xxiii.,
and xxiv. pp. 639, 393, 377. Bred by Mr. T. Atherton, Chapel House; got
by Thirteenth Duke of Oxford (21604), dam (Princess Victoria) by Fourth
Duke of Oxford (11387), &c.

1879, Mar. 8, r. & w., C.C.	Oxford's Princess	B'n T'croft Oxf'd 3d, 36209	Mr. Clark

ROSE OF ONEIDA, red, calved September 9, 1875, Vol. xxv. p. 395. Bred
by Mr. J. P. Clark; got by Sixth Duke of Oneida (30997), dam (Guelder
Rose 6th) by Fourth Duke of Thorndale (17750), &c.

1879, Mar. 11, r. & w., C.C.	Oxf'd's Rose of Oneida	B'n T'croft Oxf'd 3d, 36209	Mr. Clark

THORNDALE ROSETTE, red and white, calved March 21, 1877. Bred by
Mr. J. P. Clark; got by Duke of Thorndale (33740), dam (Oxford's Rosette)
by Nineteenth Duke of Oxford (28431), &c. See Vol. xxiv. p. 377.

1879, Sept. 9, r. & w., B.C.	(Steer)	Baron Ursula, 41073	Mr. Clark

WATERLOO 42ND, red, calved July 17, 1875. Bred by Lord Penrhyn, Penrhyn Castle; got by Grand Duke 20th (31281), dam (Waterloo 31st) by Grand Duke 11th (21849), &c. See Vol. xxii. p. 530.

Produce in		Names, &c.	By what Bull.	By whom bred.
1879, Aug. 23, red,	C.C.	Duchess Waterloo	FordhamD.Ox'd4th,41569	Mr. Clark

CLEAR, Albert P.,
Maldon, Essex.

OXFORD BLANCHE 2ND, roan, calved October 23, 1876. Bred by Mr. A. P. Clear; got by Oxford's Baron (32030), dam (Oxford Blanche) by Baron Oxford 3rd (25579), &c. See Vol. xxiii. p. 387.

1879, Mar. 24, r. & w.,	B.C.	Baron Blanche 4th	Oxford's Baron, 32030	Mr. Clear

QUEEN OXFORD, roan, calved May 17, 1875. Bred by Mr. A. P. Clear; got by Oxford's Baron (32030), dam (Cranberry) by Second Duke of Claro (21576), &c. See "The Baron," p. 247.

1878, Jan. 19, red,	B.C.	The Baron	Oxford's Baron, 32030	Mr. Clear
1879, Mar. 29, red,	C.C.	Empress of Oxford	do.	do.

The Baron, exported to South America.

CLEASBY, William,
Eden Place, Kirkby Stephen, Westmoreland.

MOSS ROSE 2ND, roan, calved March 9, 1876. Bred by Mr. W. Cleasby; got by True Briton (35823), dam (Moss Rose) by Gamester (31216), g. d. (Red and White Rose) by Panton Squire (29520), gr. g. d. (Rose) by Conservative (25827), — Princess.

1879, May 22, roan,	C.C.	Westmoreland Rose	Earl of Sheffield, 33812	Mr. Cleasby

CLOSE, James,
Holmescales, Milnthorpe.

BEADED MANTLE, roan, calved December 26, 1872, Vols. xxi. and xxiv. pp. 632, 378. Bred by Mr. E. W. Meade-Waldo, Stonewall Park; got by Beadsman (27998), dam (Mantle) by British Flag (25678), &c.

1878, Dec. 28, red,	C.C.	(dead)	Knight of the Bath. 38521	Mr. Close
1879, Nov. 19, white,	B.C.	White Bull	Royal Victor, 35414	do.

BELLADONNA, red, calved May 9, 1871, Vols. xxi. and xxiv. pp. 632, 378. Bred by Mr. A. Raine, Bow Bank; got by Whiff (30299), dam (Nancy) by Prince Patrick (16760), &c.

1878, Sept. 24, r. & w.,	B.C.	Royal Heir	Royal Victor, 35414	Mr. Close
1879, Nov. 1, red,	B.C.	Royal Guest	do.	do.

Royal Heir, sold to Mr. F. Williamson, Bell Mount, Penrith; Royal Guest, to Mr. Whitaker, Dale Barnes, Lancaster.

BLOOMING GIRL, roan, calved June 3, 1872, Vols. xxi. and xxiv. pp. 632, 378. Bred by Mr. J. Davidson, Greengill; got by Errant Boy (31117), dam (Gipsy Girl) by Earl of Eglinton (23832), &c.

1878, Mar. 31, red,	C.C.	Dainty Girl	Royal Victor, 35414	Mr. Close
1879, Feb. 26, roan,	C.C.	Dandy Girl	do.	do.

BRIDAL ROBE, red, calved February 27, 1875.   Bred by Mr. J. Close ; got
by Knight of Warlaby (31571), dam (Bridal Rose) by Sir David (25135), &c.
See Vol. xxv. p. 395.

Produce in		Names, &c.	By what Bull.	By whom bred.
1878, May 27, red,	C.C.	Bridal Ring	Knight of the Bath, 38521	Mr. Close
1879, July 10, roan,	B.C.	Royal Bridesman	Royal Victor, 35414	do.

EMILY, roan, calved March 13, 1875.   Bred by Mr. J. Close ; got by Knight of
Warlaby (31571), dam (Evaline) by Whiff (30299), &c.   See Vol. xxiv. p. 378.

Produce in		Names, &c.	By what Bull.	By whom bred.
1878, Feb. 25, r. & w.,	B.C.	(Steer)	Knight of the Bath, 38521	Mr. Close
1879, March 1, roan,	C.C.	Emmeline	Royal Victor, 35414	do.

EVA, roan, calved March 18, 1874, Vol. xxiii. p. 388.   Bred by Mr. J. Close ;
got by Royal Duke Hamilton (29863), dam (Evaline) by Whiff (30299), &c.

Produce in		Names, &c.	By what Bull.	By whom bred.
1877, Aug. 10, roan,	B.C.	Ebenezer	Knight of Warlaby, 31571	Mr. Close
1878, July 15, roan,	B.C.	Ethelred	Knight of the Bath. 38521	do.
1879, July 13, red,	B.C.	Royal Edict	Royal Victor, 35414	do.

Ebenezer, sold to Mr. Birket, Middleton Hall.

EVALINE, red, calved July 6, 1871, Vols. xxi. and xxiv. pp. 633, 378.   Bred
by Mr. A. Raine, Bow Bank ; got by Whiff (30299), dam (Lady Fragrant) by
Roan Windsor (24967), &c.

Produce in		Names, &c.	By what Bull.	By whom bred.
1878, July 8, red,	C.C.	Emma	Royal Victor, 35414	Mr. Close
1879, July 14, red,	C.C.	Emerald	do.	do.

FAMILIAR WINDSOR, red, calved December 30, 1866, Vols. xix., xx., xxi.,
and xxiv. pp. 507, 513, 633, 379.   Bred by Mr. J. Unthank, Netherscales ;
got by Roan Windsor (24967), dam (Familiar Hopewell Gwynne) by Hopewell Gwynne (19976), &c.

Produce in		Names, &c.	By what Bull.	By whom bred.
1878, May 14, roan,	C.C.	Familiar Victor	Royal Victor, 35414	Mr. Close
1879, April 5, roan,	B.C.	Royal Windsor	do.	do.

Royal Windsor, sold to Mr. J. Tander, Berrier End, Penrith.

LADY MANFRED, roan, calved June 29, 1874, Vol. xxiv. p. 379.   Bred by
Mr. A. Raine, Bow Bank ; got by Manfred (26801), dam (Lady Grateful) by
The Sutler (23061), &c.

Produce in		Names, &c.	By what Bull.	By whom bred.
1878, April 9, r. & w.,	C.C.	Royal Lady	Royal Victor, 35414	Mr. Close
1879, May 5, roan,	C.C.	Lady Victor	do.	do.

Royal Lady, sold to Mr. J. C. Toppin, Musgrave Hall, Skelton, Penrith.

LILY OF THE VALE, white, calved February 27, 1875.   Bred by Mr. J.
Close ; got by Knight of Warlaby (31571), dam (Columbia) by Hail Columbia
(24095), &c.   See Vol. xxii. p. 366.

Produce in		Names, &c.	By what Bull.	By whom bred.
1877, Nov. 10, roan,	C.C.	Roan Lily	Knight of the Bath, 38521	Mr. Close
1879, Feb. 26, roan,	B.C.	Knight of the Valley	do.	do.

VERITY, white, calved February 27, 1875, Vol. xxiv. p. 379.   Bred by Mr. J.
Close ; got by County Member (28268), dam (Vacillation) by K. C. B. (26492),
&c.

Produce in		Names, &c.	By what Bull.	By whom bred.
1878, Nov. 1, roan,	C.C.	Veracity	Baron Aylesby, 39397	Mr. Close
1879, Oct. 8, roan,	C.C.	Vivacity	Royal Victor, 35414	do.

Veracity, sold to Mr. J. C. Toppin, Musgrave Hall, Skelton, Penrith.

WINDSOR'S DUCHESS 3RD, roan, calved October 21, 1873, Vol. xxiii. p. 389.
Bred by Mr. T. Dargue, Whale Farm ; got by Flag of Windsor (28614), dam
(Windsor's Duchess 2nd) by Shirley Fitz-Windsor (27451), &c.

Produce in		Names, &c.	By what Bull.	By whom bred.
1878, March 11, red,	C.C.	Windsor's Rose	Knight of the Bath, 38521	Mr. Close
1879, April 4, roan,	B.C.	Windsor's Prince	Royal Victor, 35414	do.

CLOWES, Charles,
Stockbridge House, Stockbridge, Hants.

DARLING, roan, calved May 21, 1877. Bred by Mr. W. Smith, Hatch Farm, the property of Mr. C. Clowes; got by Duke Cambridge (39703), dam (Duchess Wetherby) by Third Duke of Wetherby (26030), &c. See Vol. xxiv. p. 670.

Produce in		Names, &c.	By what Bull.	By whom bred.
1879, July 5, red,	C.C.	Little Darling	Holcus, 40004	Miss Taunton

COATES, C.,
Cleasby, Darlington.

AMPTHILL ROSE, roan, calved September 29, 1876. Bred by Mr. C. Coates; got by Prince of Osborne (35155), dam (Amber Rose) by King Charles (26500), &c. See Vol. xxiii. p. 389.

Produce in		Names, &c.	By what Bull.	By whom bred.
1878, Nov. 21, roan,	C.C.	Annie Rose	Manfred's Prince, 40291	Mr. Coates

ARTFUL ROSE, white, calved December 26, 1877. Bred by Mr. C. Coates; got by Manfred Booth (34749), dam (Amber Rose) by King Charles (26500), &c.

Produce in		Names, &c.	By what Bull.	By whom bred.
1879, Sept. 27, white,	C.C.	Artless Rose	King Manfred, 38491	Mr. Coates

PRIMROSE 4TH, roan, calved May 29, 1874, Vol. xxiv. p. 380. Bred by Mr. C. Coates; got by Squire Booth (30049), dam (Primrose 3rd) by King Charles (26500), &c.

Produce in		Names, &c.	By what Bull.	By whom bred.
1879, Jan. 21, roan,	C.C.	Primrose 5th	King Manfred, 38491	Mr. Coates

ROSE OF WINTER, white, calved February 11, 1876. Bred by Mr. C. Coates; got by Prince of Osborne (35155), dam (Rose of Summer) by Squire Booth (30049), &c. See Vol. xxiii. p. 389.

Produce in		Names, &c.	By what Bull.	By whom bred.
1879, Mar. 27, white,	C.C.	Rose of Spring	King Manfred, 38491	Mr. Coates

VICTORINE 2ND, roan, calved April 7, 1873, Vols. xxiii. and xxiv. pp. 389, 380. Bred by Mr. C. Coates; got by Squire Booth (30049), dam (Victorine) by General (19836), &c.

Produce in		Names, &c.	By what Bull.	By whom bred.
1879, May 10, white,	B.C.	Vic	King Manfred, 38491	Mr. Coates

WHITE ROSE, white, calved October 23, 1875. Bred by Mr. C. Coates; got by Prince of Osborne (35155), dam (Wild Rose) by Squire Booth (30049), &c. See Vol. xxii. p. 367.

Produce in		Names, &c.	By what Bull.	By whom bred.
1878, July 3, roan,	B.C.	Turk	Manfred Booth, 34749	Mr. Coates
1879, June 24, white,	B.C.	Wilful	King Manfred, 38491	do.
Turk, sold to Mr. Jordison, Dalton Percy.				

WILD ROSE, roan, calved March 8, 1873, Vols. xxii. and xxiv. pp. 367, 380. Bred by Mr. C. Coates; got by Squire Booth (30049), dam (Garnet Rose) by King Charles (26500), &c.

Produce in		Names, &c.	By what Bull.	By whom bred.
1878, July 7, roan,	B.C.	Sam	Manfred Booth, 34749	Mr. Coates
1879, July 8, { roan,	B.C.	Tim	} King Manfred, 38491	do.
white,	C.C.	West Rose		

COCHRANE, Major,
Aldwark Manor, Easingwold, Yorkshire.

BETSY, roan, calved February 20, 1874. Bred by Major Cochrane; got by
Squire Booth (30049), dam (Blossom) by Double Magnus (23702), g. d.
(Buttercup) by Blood Royal (17423), gr. g. d. (Miss Butterfly) by Wild Man
(14008), — (Irish Lass) by De Grey (11346).

Produce in	Names, &c.	By what Bull.	By whom bred.
1876, Sept. 24, r. & w., C.C.	Butterscotch	Master Booth, 38724	Major Cochrane

BUTTERSCOTCH, red and white, calved September 24, 1876. Bred by Major
Cochrane; got by Master Booth (38724), dam (Betsy) by Squire Booth
(30049), &c.

1879, Dec. 4, r. & w., C.C.	Butterpat	Star of Windsor, 39166	Major Cochrane

DUCHESS OF WINDSOR, roan, calved May 11, 1867, Vols. xxi. and xxiii.
pp. 635, 390. Bred by Mr. J. Unthank, Netherscales; got by Roan Windsor
(24967), dam (Duchess) by Duke of Hamilton (19618), &c.

1879, Sept. 4, roan, C.C.	Victoria	Manfred Booth, 34749	Major Cochrane

FAMILIAR BENVENUTA, roan, calved January 21, 1874, Vol. xxv. p. 396.
Bred by Major Cochrane; got by Squire Booth (30049), dam (Familiar
Hamilton) by Duke of Hamilton (19618), &c.

1879, May 26, roan, C.C.	Familiar Langton	Star of Windsor, 39166	Major Cochrane

COCK, Charles H.,
Bridgefoot, Barnet, Herts.

CONISHEAD WATERLOO, roan, calved April 25, 1877. Bred by Mr. W.
Ashburner, Conishead Grange; got by Second Duke of Glo'ster (28392), dam
(Waterloo 29th) by Grand Duke of Lightburne 2nd (26291), &c. See Vol.
xxiv. p. 306.

1879, Nov. 29, roan, C.C.	Waterloo Gem	D. of Worcester 5th, 39797	Mr. Cock

SEVERN GEM, red and white, calved August 13, 1876. Bred by Mr. C. H.
Cock; got by Minstrel Boy (34855), dam (Severn Lady) by Second Duke of
Wetherby (21618), &c. See "Farley," Vol. xxiv. p. 95.

1879, June 6, roan, C.C.	Severn Jewel	D. of Worcester 5th, 39797	Mr. Cock

TUNEFUL, red and little white, calved February 19, 1877. Bred by Mr. C. H.
Cock; got by Young England (31110), dam (Madrigal) by Prince Llewellyn
(32154), &c. See "Trumpeter," p. 254.

1879, Aug. 22, roan, B.C.	Trumpeter	D. of Worcester 5th, 39797	Mr. Cock

COCK, J. and SONS,
Coat Green Farm, Burton, Westmoreland.

ELIZA NAPIER 2nd, red and white, calved August 18, 1871, Vol. xxii. p. 368.
Bred by Messrs. J. Cock and Sons; got by Sir Robert Napier (30005), dam
(Eliza 4th) by Second Duke of Buckingham (17706), &c.

1878, May 20, roan, C.C.	Eliza 10th	Royal Ensign, 37388	M'ssrs. Cock & Sons
1879, Sept. 5, { r. & w., C.C.	Eliza 11th	} Warlaby Squire, 37646	do.
r. & w., C.C.	Eliza 12th		

ISABELLA 3RD, white, calved March 18, 1876. Bred by Messrs. J. Cock and
Sons ; got by Arch Hero (27893), dam (Rachel Napier) by Sir Robert Napier
(30005), g. d. (Reggi) by Constitution (30793), &c. See "British Pilot,"
p. 36.

Produce in		Names, &c.	By what Bull.	By whom bred.
1879, Feb. 2, roan,	C.C.	Isabella 6th	British Gwynne, 39503	M'ssrs.Cock&Sons

RACHEL NAPIER, roan, calved April 16, 1870, Vol. xxii. p. 368. Bred by
Messrs. J. Cock and Sons ; got by Sir Robert Napier (30005), dam (Reggi)
by Constitution (30793), &c.

1876, Mar. 18, white,	C.C.	Isabella 3rd	Arch Hero, 27893	M'ssrs.Cock&Sons

REGGI, roan, calved in 1867, Vols. xx. and xxii. pp. 724, 368. Bred by Mr. F.
C. Ellison, Rawsons ; got by Constitution (30793), dam (Regina) by King
Otho (18148), &c.

1877, April 16, roan,	C.C.	Isabella 4th	Warlaby Squire, 37646	M'ssrs.Cock&Sons
1878, Oct. 28, roan,	C.C.	Isabella 5th	British Gwynne, 39503	do.

COLLARD, Charles,
Little Barton, Canterbury.

BARTON SWEETHEART, roan, calved November 24, 1876. Bred by Mr. C.
Collard ; got by Wild Duke 2nd (36002), dam (Sweetheart 25th) by Count
Leinster (23638), &c. See Vol. xxiii. p. 391.

1879, May 15, roan,	B.C.	Cherry D. of Barton	Ch'ry D. of Glo'ster, 36348	Mr. Collard

COUNTESS OF BARTON, white, calved June 6, 1877. Bred by Mr. C. Collard ; got by Wild Duke 2nd (36002), dam (Countess of Kent) by Grand Duke
of Oxford (28764), &c. See Vol. xxiii. p. 391.

1879, Dec. 31, roan,	B.C.	(Steer)	Ch'ry D. of Glo'ster, 36348	Mr. Collard

FLORENCE KNIGHTLEY, red, calved March 30, 1875, Vol. xxiv. p. 383.
Bred by Mr. C. Collard ; got by Duke of Maidstone (30989), dam (Florentia
6th) by Third Duke of Lancaster (19624), &c.

1878, Sept. 4, roan,	C.C.	Florentia Wild Eyes	Wild Duke 2nd, 36002	Mr. Collard
1879, Aug. 7, red,	B.C.	Lord Fordwich	Ch'ry D. of Glo'ster, 36348	do.

HOLLY, roan, calved December 25, 1875. Bred by Mr. C. Collard ; got by
Wild Duke 2nd (36002), dam (Laurel) by Baron Torr (23380), &c. See
Vol. xxii. p. 370.

1878, Nov. 5, roan,	B.C.	Guy Fawkes	Kentish Cherry, 40042	Mr. Collard
1879, Oct. 20, r. & w.,	C.C.	Laurestina	Ch'ry D. of Glo'ster, 36348	do.
Guy Fawkes, sold to Mr. Wythe.				

KENT LILY, roan, calved May 25, 1875. Bred by Mr. C. Collard ; got by
Heydon Duke 2nd (31370), dam (Roan Lily) by Costa King (28254), &c. See
Vol. xxii. p. 370.

1878, Feb. 7, roan,	C.C.	Barton Lily	Wild Duke 2nd, 36002	Mr. Collard
1879, Mar. 9, r. & w.,	C.C.	Canterbury Lily	Ch'ry D. of Glo'ster, 36348	do.

KENT SWEETHEART, white, calved March 26, 1875. Bred by Mr. C. Collard ; got by Boston 2nd (30567), dam (Sweetheart 25th) by Count Leinster
(23638), &c. See Vol. xxii. p. 370.

1879, Jan. 6, roan,	C.C.	Charming Sweeth'rt	Kentish Cherry, 40042	Mr. Collard
1879, Dec. 27, roan,	C.C.	(dead)	Ch'ry D. of Glo'ster, 36348	do.

LADY OXFORD, red, calved February 28, 1872, Vols. xxi. and xxiii. pp. 637, 391. Bred by Mr. C. Collard; got by Comet (28230), dam (Wild Oxford) by Oxford (20450), &c.

Produce in		Names, &c.	By what Bull.	By whom bred.
1878, July 29, roan,	C.C.	Duchess of Oxford	Wild Duke 2nd, 36002	Mr. Collard

LIZZY, white, calved December 16, 1870, Vols. xxii. and xxiii. pp. 370, 391. Bred by Mr. J. Webb, Horseheath; got by Costa King (28254), dam (Lucy) by Prolific 2nd (27213), &c.

Produce in		Names, &c.	By what Bull.	By whom bred.
1877, June 13, roan,	C.C.	Lottie	Cherry Oxford 2nd, 36353	Mr. Collard
1879, April 28, roan,	C.C.	Lydia	Ch'ry D. of Glo'ster, 36348	do.

LOUISE, roan, calved May 24, 1876. Bred by Mr. C. Collard; got by Lord Cambridge (38588), dam (Lizzy) by Costa King (28254), &c. See Vol. xxiii. p. 391.

Produce in		Names, &c.	By what Bull.	By whom bred.
1879, March 13, roan,	C.C.	Laura	Ch'ry D. of Glo'ster, 36348	Mr. Collard

MAY, roan, calved September 30, 1874. Bred by Mr. G. Mount, Nackington, the property of Captain Knight, Bilting House; got by Bloomfield (33169), dam (Maude) by Claxton (21433), &c. See Vol. xx. p. 646.

Produce in		Names, &c.	By what Bull.	By whom bred.
1877, April 18, roan,	C.C.	(dead)	Wild Duke 2nd, 36002	Mr. Collard
1878, March 26, roan,	C.C.	Mabel	do.	do.

MAY LILY, roan, calved May 1, 1874, Vol. xxiv. p. 383. Bred by Mr. D. A. Green, East Donyland; got by Heydon Duke 2nd (31370), dam (Roan Lily) by Costa King (28254), &c.

Produce in		Names, &c.	By what Bull.	By whom bred.
1878, April 9, roan,	C.C.	(slaughtered)	Wild Duke 2nd, 36002	Mr. Collard
1879, March 21, red,	C.C.	Red Lily	Ch'ry D. of Glo'ster, 36348	do.

NASTURTIUM, red and white, calved June 27, 1873, Vol. xxiii. p. 391. Bred by Mr. J. A. Mumford, Brill House; got by Duke John (30913), dam (Natural) by Prince of Geneva (24829), &c.

Produce in		Names, &c.	By what Bull.	By whom bred.
1877, Dec. 16, roan,	C.C.	Nancy	Wild Duke 2nd, 36002	Mr. Collard
1879, Feb. 26, red,	B.C.	Lord Luddenham	Ch'ry D. of Glo'ster, 36348	do.

Lord Luddenham, sold to Mr. W. Carter, Luddenham, Faversham, Kent.

NELLIE, roan, calved November 30, 1876. Bred by Mr. C. Collard; got by Wild Duke 2nd (36002), dam (Nasturtium) by Duke John (30913), &c. See Vol. xxiii. p. 391.

Produce in		Names, &c.	By what Bull.	By whom bred.
1879, May 7, red,	B.C.	Noisy Boy	Ch'ry D. of Glo'ster, 36348	Mr. Collard

Noisy Boy, sold to Mr. A. Amos, Boughton Court, Ashford, Kent.

PRINCESS EUGENIE, roan, calved December 13, 1873, Vol. xxiii. p. 391. Bred by Mr. R. H. Crabb, Baddow Place; got by Baron Oxford 3rd (25579), dam (Princess of Denmark) by Baron Roxwell (21240), &c.

Produce in		Names, &c.	By what Bull.	By whom bred.
1878, Sept. 11, white,	C.C.	Princess Eleanor	Wild Duke 2nd, 36002	Mr. Collard
1879, Nov. 9, red,	C.C.	Princess Ellen	Ch'ry D. of Glo'ster, 36348	do.

PRINCESS OF WALES, white, calved May 3, 1876. Bred by Mr. C. Collard; got by Wild Duke 2nd (36002), dam (Princess Eugenie) by Baron Oxford 3rd (25579), &c. See Vol. xxiii. p. 391.

Produce in		Names, &c.	By what Bull.	By whom bred.
1879, April 1, roan,	C.C.	Princess of Glo'ster	Ch'ry D. of Glo'ster, 36348	Mr. Collard

ROAN LILY, roan, calved March 13, 1871, Vols. xxii. and xxiii. pp. 370, 391. Bred by Mr. J. Webb, Horseheath; got by Costa King (28254), dam (Lily) by Clumber 2nd (21438), &c.

Produce in		Names, &c.	By what Bull.	By whom bred.
1878, Jan. 1, white,	C.C.	White Lily	Wild Duke 2nd, 36002	Mr. Collard
1879, April 14, red,	B.C.	(Steer)	Ch'ry D. of Glo'ster, 36348	do.

ROXWELL ROSE 1st, roan, calved July 8, 1875. Bred by Mr. W. Tippler, Roxwell; got by Heydon Duke 2nd (31370), dam (Lizzy) by Costa King (28254), &c. See Vol. xxii. p. 370.

Produce in	Names, &c.	By what Bull.	By whom bred.
1878, Feb. 15, white, B.C.	Duke William	Wild Duke 2nd, 36002	Mr. Collard
1879, Mar. 17, roan, C.C.	Barton Rose	Ch'ry D. of Glo'ster, 36348	do.

Duke William, sold to Mrs. Collard, Ickham Court, Wingham, Kent.

ROXWELL ROSE 2nd, roan, calved July 8, 1875. Bred by Mr. W. Tippler, Roxwell; got by Heydon Duke 2nd (31370), dam (Lizzy) by Costa King (28254), &c. See Vol. xxii. p. 370.

Produce in	Names, &c.	By what Bull.	By whom bred.
1878, Mar. 5, white, C.C.	Wild Rose	Wild Duke 2nd, 36002	Mr. Collard
1879, May 20, r. & w., C.C.	Rose of Glo'ster	Ch'ry D. of Glo'ster, 36348	do.

SWEETHEART 25th, roan, calved March 5, 1868, Vols. xxii. and xxiii. pp. 370, 391. Bred by Mr. G. Murton Tracy, Redlands; got by Count Leinster (23638), dam (Sweetheart 3rd) by Daybreak (11338), &c.

Produce in	Names, &c.	By what Bull.	By whom bred.
1878, Feb. 15, roan, C.C.	Canterb'ry Sweeth'rt	Wild Duke 2nd, 36002	Mr. Collard
1879, Mar. 5, r. & w., C.C.	Cherry Sweetheart	Ch'ry D. of Glo'ster, 36348	do.

WILD DUCHESS OF OXFORD, roan, calved January 13, 1876. Bred by Mr. C. Collard; got by Wild Duke 2nd (36002), dam (Lady Oxford) by Comet (28230), &c. See Vol. xxiii. p. 391.

Produce in	Names, &c.	By what Bull.	By whom bred.
1879, Nov. 6, r. & w., B.C.	Lord Westgate	Ch'ry D. of Glo'ster, 36348	Mr. Collard

WILD OXFORD, red and white, calved August 10, 1868, Vols. xx. and xxiii. pp. 823, 391. Bred by Mr. J. T. Noakes, Brockley House; got by Oxford (20450), dam (Wild Cherry) by Cherry Duke 3rd (15763), &c.

Produce in	Names, &c.	By what Bull.	By whom bred.
1878, March 6, roan, B.C.	Lord Cowper	Wild Duke 2nd, 36002	Mr. Collard
1879, Dec. 18, r. & w., B.C.	Earl of Glo'ster	Ch'ry D. of Glo'ster, 36348	do.

Lord Cowper, sold to Mr. J. Robinson, Dambridge, Wingham, Kent.

COLTHURST, John,
Chew Court, Chew Magna, Somerset.

LANCASTER 35th, red, calved February 3, 1868, Vols. xx. and xxv. pp. 620, 397. Bred by Mr. E. Hall, Shallcross Hall; got by Third Duke of Norfolk (21601), dam (Lancaster 20th) by Chilton Hero (17564), &c.

Produce in	Names, &c.	By what Bull.	By whom bred.
1879, Feb. 1, red, C.C.	Lancaster D'ss 2nd	Adjutant, 36097	Mr. Colthurst

LANCASTER 51st, red and white, calved May 1, 1871, Vols. xxii. and xxiv. pp. 444, 384. Bred by Mr. E. Hall, Shallcross Hall; got by Lancaster Star (22073), dam (Lancaster 20th) by Chilton Hero (17564), &c.

Produce in	Names, &c.	By what Bull.	By whom bred.
1878, Mar. 20, { red, B.C.	(dead)	Adjutant, 36097	Mr. Colthurst
{ white, B.C.	(dead)		
1879, Mar. 15, roan, C.C.	Lancaster Pet	do.	do.

LANCASTER 57th, roan, calved February 11, 1872, Vols. xxiv. and xxv. pp. 384, 398. Bred by Mr. E. Hall, Shallcross Hall; got by Lancaster Star (22073), dam (Lancaster 30th) by Third Duke of Norfolk (21601), &c.

Produce in	Names, &c.	By what Bull.	By whom bred.
1879, Aug. 12, roan, C.C.	Lancaster Beauty	Lancaster Duke, 40097	Mr. Colthurst

OLIVE BUD, white, calved August 1, 1876. Bred by Mr. J. H. Braikenridge, The Rookery; got by Earl of Killerby (33802), dam (Olive Branch) by Baron York (30500), &c. See Vol. xxiv. p. 342.

Produce in	Names, &c.	By what Bull.	By whom bred.
1879, Mar. 11, white, C.C.	Olive Blossom	Adjutant, 36097	Mr. Colthurst

CONWY, Captain C. R.,
Bodrhyddan, Rhyl, North Wales.

HORTENSIA, roan, calved February 14, 1874, Vol. xxiv. p. 488. Bred by Mr. T. Harris, Stonylane House; got by Winterfold (36020), dam (Helena) by Lord Red Eyes 2nd (24460), &c.

Produce in		Names, &c.	By what Bull.	By whom bred.
1879, Feb. 16, r. & w.,	B.C.	Cefn Prince	Waterloo Prince, 35949	Capt. Conwy

WATERLOO BELLE, red, calved April 28, 1876, Vol. xxv. p. 553. Bred by Mr. T. Wilson, Shotley Hall; got by Oxford Beau 3rd (32013), dam (Duchess of Waterloo) by Earl of Jersey (23838), &c.

1879, Dec. 8, r. & w.,	C.C.	Waterloo Belle 2nd	Prince Saladin, 40542	Capt. Conwy

COOKE, Charles,
Whiteley Green, Adlington, Cheshire.

LANCASTER 36TH, red and white, calved March 9, 1868, Vol. xxii. p. 443. Bred by Mr. E. Hall, Shallcross Hall; got by Lenton Hero (22092), dam (Lancaster 26th) by Third Duke of Norfolk (21601), &c.

1876, May 14, red,	B.C.	Duke of Lorn 8th	Duke of Lorn, 25985	Mr. Hall
1877, July 9, r. & w.,	C.C.	Lancaster 92nd	Paul Gwynne, 37181	do.
1878, Oct. 9, r. & w.,	B.C.	Banner Bearer	do.	Mr. Cooke
1879, Dec. 29, roan,	C.C.	Duchess of Lorn	Duke of Lorn 6th, 36518	do.

COOPER, Percy H.,
Bulwell Hall, Nottingham.

LADY FAREWELL, roan, calved December 24, 1873. Bred by Mr. R. Tennant, Scarcroft Lodge; got by Duke of Clarence (19611), dam (Miss Farewell) by Second Duke of Wharfdale (19649), &c. See Vol. xxiv. p. 673.

1876, June 20, roan,	C.C.	Miss Frederick	Dk. of Tregunter 5th, 33743	Mr. Tennant
1879, July 21, red,	C.C.	Tregunter's Farewell	do.	Mr. Cooper

COPE, J. A. M.,
Drummilly, Loughgall, Co. Armagh.

BARONESS OF RABY, roan, calved February 12, 1875. Bred by Mr. J. A. M. Cope; got by Baron of Raby 3rd (33072), dam (Genuine) by Polyphemus (27079), &c. See Vol. xxii. p. 373.

1878, July 22, r. & w.,	C.C.	Baroness of Raby 2nd	Gloster's Gwynne, 88360	Mr. Cope
1879, Dec. 26, r. & w.,	B.C.	Goshawk	Burghley, 42855	do.

Baroness of Raby 2nd, sold to Mr. G. Gray, Glenapp, Co. Armagh.

DUCHESS 16TH, red and white, calved November 6, 1874. Bred by Mr. J. A. M. Cope, the property of Mr. G. Gray, Glenapp; got by Baron of Raby 3rd (33072), dam (Duchess 13th) by Baron Oxford (23375), &c. See Vol. xxi. p. 643.

1879, June 9, roan,	B.C.	Young Duke	Knight of Raby, 34393	Mr. Cope

FLOSS 2ND, roan, calved April 27, 1877. Bred by Mr. J. H. Blundell, Woodside; got by Earl of Strafford (31084), dam (Floss) by Cherry Grand Duke 5th (30712), &c. See Vol. xxiv. p. 328.

1879, May 28, r. & w.,	C.C.	Flossy	Dk. of Oxford 32nd, 36537	Mr. Cope

HONEYBELL 2ND, roan, calved March 24, 1870, Vol. xxiv. p. 387. Bred by
Colonel Kingscote, Kingscote; got by Third Duke of Clarence (23727), dam
(Honeybell) by Grand Duke of Oxford (16184), &c.

Produce in		Names, &c.	By what Bull.	By whom bred.
1878, June 9, r. & w.,	C.C.	Honeybell 4th	Knight of Raby, 34393	Mr. Cope
1879, July 30, roan,	C.C.	Honeybell 5th	do.	do.

LOUISA 15TH, red, calved May 1, 1874. Bred by the Marquis of Exeter,
Burghley Park; got by Lord Chancellor (20160), dam (Louisa 10th) by St.
Matthew (22830), g. d. (Louisa 8th) by Lord Augustus (20147), &c. See Vol.
xviii. p. 593.

1878, May 21, r. & w.,	B.C.	Lochiel	Glo'ster's Gwynne, 38360	Mr. Cope
1879, June 9, red,	B.C.	Lothair	Knight of Raby, 34393	do.

MAID OF THE VALE, red and white, calved September 25, 1874. Bred by
Mr. J. A. M. Cope; got by Baron Eccleswall (33037), dam (Lady of the Vale
2nd) by Garibaldi 2nd (26220), &c. See Vol. xxi. p. 643.

1878, June 13, r. & w.,	C.C.	Maid of the Vale 3rd	Glo'ster's Gwynne, 38360	Mr. Cope
1879, June 9, red,	C.C.	Maid of the Vale 4th	Knight of Raby, 34393	do.

MELODY 2ND, red and white, calved January 10, 1870, Vol. xxii. p. 373. Bred
by Mr. J. Blyth, Woolhampton House; got by Fra Diavolo (26190), dam
(Musical 4th) by Northern Duke (22431), &c.

1878, Feb. 13, roan,	C.C.	Melody 5th	Knight of Raby, 34393	Mr. Cope
1879, Feb. 3, red,	C.C.	Melody 6th	do.	do.

RABY'S LAUNDRESS, roan, calved November 30, 1874. Bred by Mr. J. A.
M. Cope, the property of Mr. Thompson, Clandeboye; got by Baron of Raby
3rd (33072), dam (Laundress 3rd) by Lord Darlington 2nd (29096), &c. See
Vol. xxi. p. 643.

1878, Nov. 14, roan,	C.C.	Laundress 4th	Knight of Raby, 34393	Mr. Cope

Laundress 4th, sold to Mr. G. Gray, Glenapp, Co. Armagh.

REVELRY 15TH, red, calved in April 1871. Bred by Mr. J. Longland,
Grendon; got by Duke of Wicken (26035), dam (Duchess of Oxford 3rd) by
Twelfth Duke of Oxford (19633), g. d. (Red Hart) by May Duke (13320), &c.
See Vol. xvii. p. 699.

1879, April 11, roan,	C.C.	Frolic	Knight of Raby, 34393	Mr. Cope

TOWNELEY BUTTERFLY 2ND, roan, calved December 13, 1872. Bred by
Colonel Towneley, Towneley Park; got by Earl of Thorndale (28521), dam
(Towneley Butterfly) by Count of Windsor (21498), &c. See "Peacock But-
terfly," p. 184.

1876, Feb. 26, red,	B.C.	Towneley Butterfly	Lord Stafford, 34680	Mr. Cope
1877, Feb. 27, roan,	B.C.	Butterfly	Knight of Raby, 34393	do.
1878, April 16, white,	C.C.	White Butterfly	do.	do.
1879, July 13, roan,	B.C.	Peacock Butterfly	do.	do.

WOODBINE 14TH, red, calved March 2, 1873, Vol. xxiii. p. 396. Bred by Mr.
W. Bolton, The Island, the property of Mr. McLellan; got by Lieutenant
General (31600), dam (Woodbine 6th) by Duke of Marlborough (23768), &c.

1878, Jan. 9, roan,	B.C.	Woodranger	Pilgrim Father, 35037	Mr. Cope
1879, Jan. 25, r. & w.,	C.C.	Woodbine 18th	do.	do.

Woodbine 18th, sold to Mr. G. Gray, Glenapp, Co. Armagh.

COSTELLO, Michael,
Ardbash, Edenderry, King's Co., Ireland.

WANDERING BEAUTY, roan, calved April 25, 1876. Bred by Messrs. Christy Brothers, Fort Union, the property of Mr. M. Costello; got by Lord Broughton (31626), dam (Golden Wonder) by Goldsmith (26276), &c. See Vol. xxiii. p. 384.

Produce in	Names, &c.	By what Bull.	By whom bred.
1879, April 18, roan, C.C.	Limerick Lass	Verger, 35860	Messrs. Christy

COUPLAND, William,
Bolton House, Yeadon, Leeds.

MARGERY PERCY, roan, calved May 12, 1876. Bred by Mr. W. Gomersall, Otterburn, the property of Mr. W. Coupland; got by Baron Percy (30477), dam (Lady Margery) by Duke of Otterburn (30998), &c. See Vol. xxiv. p. 426.

1878, Nov. 27, roan, C.C.	Dame Margery	Ligulf, 41805	Mr. Gomersall
1879, Dec. 12, roan, B.C.	Percy Don Yeadon	do.	do.
Dame Margery, sold to Mr. M. Lamb, Caley Hall Farm, Otley.			

COURTOWN, Earl of,
Courtown House, Gorey, Co. Wexford.

BRIDE, roan, calved January 12, 1876. Bred by the Earl of Courtown; got by Champagne Charley (33314), dam (Bess) by Prince Arthur (35086), &c. See Vol. xxii. p. 374.

1879, Mar. 4, roan, B.C.	(dead)	Prince Milan, 38916	Earl of Courtown

BRIDESMAID, red and white, calved February 24, 1877. Bred by the Earl of Courtown; got by New Year's Gift (37127), dam (Bess) by Prince Arthur (35086), &c. See Vol. xxii. p. 374.

1879, Feb. 24, r. & w., B.C.	Courtown Borderer	Trooper, 42520	Earl of Courtown

CATTY, red, calved January 2, 1877. Bred by the Earl of Courtown; got by Champagne Charley (33314), dam (Queen Cathleen) by King Richard (26523), &c. See Vol. xxiii. p. 396.

1879, Feb. 28, roan, C.C.	Cleopatra	Trooper, 42520	Earl of Courtown

COWELL, J. R.,
Ashdon, Essex.

SILENCE, red and white, calved September 2, 1876, Vol. xxv. p. 363. Bred by Lord Braybrooke, Audley End; got by Duke of Rosedale 3rd (33723), dam (Secret) by Heydon Duke 2nd (31370), &c.

1879, Nov. 7, r. & w., B.C.	Silent Knight	Kt. Grand Cross, 40071	Mr. Cowell

CRABB, R. H.,
Baddow Place, Chelmsford.

GENEVA'S WILD ROSE, roan, calved July 23, 1876. Bred by Mr. R. H. Crabb; got by Third Duke of Geneva (23753), dam (Wharfdale's Wild Rose) by Third Duke of Wharfdale (21619), &c. See Vol. xxiii. p. 397.

1879, July 19, roan, B.C.	(dead)	Duke of Oxford, 39770	Mr. Crabb

WATERLOO 41st, roan, calved April 18, 1875, Vol. xxiv. p. 387. Bred by Mr. R. H. Crabb; got by Third Duke of Geneva (23753), dam (Waterloo 36th) by Grand Duke 11th (21849), &c.

Produce in	Names, &c.	By what Bull.	By whom bred.
1879, Jan. 29, r. & w., C.C.	Waterloo 44th	Oxford's Baron, 32030	Mr. Crabb

CRADOCK, C.,
Hartforth, Richmond, Yorkshire.

ACONITE, white, calved August 31, 1873, Vols. xxiii. and xxv. pp. 398, 400. Bred by Mr. C. Cradock; got by Gamester (33985), dam (Primrose) by Guy (28803), &c.

Produce in	Names, &c.	By what Bull.	By whom bred.
1879, June 13, white, C.C.	Aconite 3rd	Chaser, 37970	Mr. Cradock

BEAUTY, roan, calved April 10, 1870, Vols. xx., xxi., xxiii., xxiv., and xxv. pp. 404, 645, 398, 388, 400. Bred by Mr. C. Cradock; got by Guy (28803), dam (Blameless) by Strawberry Prince (25240), &c.

Produce in	Names, &c.	By what Bull.	By whom bred.
1879, Sept. 14, { roan, B.C. / white, B.C.	Crown Beauty / Prince Beauty }	Crown Prince, 38061	Mr. Cradock

Crown Beauty, sold to Mr. J. Outhwaite, Bainesse, Catterick.

BELLA, roan, calved March 23, 1874. Bred by Mr. C. Cradock; got by Wanderer (32790), dam (Donna Bella) by Strawberry Prince (25240), &c. See "Rosario," p. 213.

Produce in	Names, &c.	By what Bull.	By whom bred.
1879, April 4, roan, B.C.	Rosario	Rose King, 39019	Mr. Cradock

COMELY, roan, calved October 2, 1875, Vol. xxv. p. 400. Bred by Mr. C. Cradock; got by Pilgrim (35036), dam (Valentine) by The Laird (32685), &c.

Produce in	Names, &c.	By what Bull.	By whom bred.
1879, Oct. 10, r. & w., C.C.	Comfort	Lythe, 40277	Mr. Cradock

DRIP DROP, red, calved May 20, 1873, Vols. xxiii. and xxiv. pp. 398, 388. Bred by Mr. C. Cradock; got by The Don (35752), dam (Strawberry Drop) by Strawberry Prince (25240), &c.

Produce in	Names, &c.	By what Bull.	By whom bred.
1879, April 17, roan, B.C.	Deepdale	Chaser, 37970	Mr. Cradock

DRIPPING, roan, calved March 8, 1876. Bred by Mr. C. Cradock; got by His Grace (36774), dam (Drip Drop) by The Don (35752), &c. See "Deepdale," p. 66.

Produce in	Names, &c.	By what Bull.	By whom bred.
1879, Sept. 16, r. & w., C.C.	Lard	Lythe, 40277	Mr. Cradock

EMILY, roan, calved April 2, 1872, Vols. xxii., xxiii., xxiv., and xxv. pp. 375, 399, 388, 401. Bred by Mr. C. Cradock; got by Wanderer (32790), dam (Princess Emma) by Dairy Prince (17655), &c.

Produce in	Names, &c.	By what Bull.	By whom bred.
1879, Sep. 1, { r. & w., C.C. / white, C.C.	Empress / Emerald }	Chaser, 37970	Mr. Cradock

FAITHLESS, roan, calved June 25, 1875. Bred by Mr. C. Cradock; got by Pilgrim (35036), dam (Faithful) by Guy (28803), &c. See Vol. xxii. p. 375.

Produce in	Names, &c.	By what Bull.	By whom bred.
1879, Jan. 27, roan, C.C.	Fickle	Chaser, 37970	Mr. Cradock

FATIMA, red and white, calved August 4, 1874, Vols. xxiv. and xxv. pp. 388, 401. Bred by Mr. C. Cradock; got by Painter (32040), dam (Faithful) by Guy (28803), &c.

Produce in	Names, &c.	By what Bull.	By whom bred.
1879, June 9, roan, C.C.	Fancy	Chaser, 37970	Mr. Cradock

HELP, roan, calved January 16, 1875.   Bred by Mr. C. Cradock ; got by Pilgrim
(35036), dam (Gravity) by Vandyke (30202), &c.   See Vol. xxii. p. 375.

Produce in	Names, &c.	By what Bull.	By whom bred.
1879, Sept. 14, white, C.C.	Helpmate	Chaser, 37970	Mr. Cradock

MARY, roan, calved July 9, 1868, Vols. xxi., xxiv., and xxv. pp. 1015, 388, 401.
Bred by the Earl of Zetland, Upleatham ; got by Chatsworth (23546), dam
(Marigold) by Maximilian (20322), &c.

1879, Dec. 2, white, B.C.	Duke Hilda	Chaser, 37970	Mr. Cradock

MIDNIGHT, red, calved October 28, 1870, Vols. xxi. and xxiv. pp. 705, 435.
Bred by Mr. C. O. Eaton, Tolethorpe ; got by Lord Nelson (26693), dam
(Alexandra) by Cromwell (19528), &c.

1879, Feb. 10, roan, C.C.	Midnight 2nd	20th Dk. of Oxford, 28432	Mr. Cradock

PRIMROSE, roan, calved September 21, 1870, Vols. xxi. and xxiii. pp. 646, 399.
Bred by Mr. C. Cradock ; got by Guy (28803), dam (Lily Grey) by Warlaby
(21063), &c.

1879, Aug. 7, roan, C.C.	Primula	Chaser, 37970	Mr. Cradock

PRINCESS GRACE 2ND, roan, calved April 29, 1875.   Bred by Mr. C.
Cradock ; got by Pilgrim (35036), dam (Princess Grace) by Dairy Prince
(17655), &c.   See Vol. xxv. p. 401.

1878, April 20, roan, C.C.	(dead)	Chaser, 37970	Mr. Cradock
1879, June 9, roan, C.C.	Princess Grace 3rd	do.	do.

RACKET, roan, calved April 15, 1876.   Bred by Mr. C. Cradock ; got by
Pilgrim (35036), dam (Strawberry Drop) by Strawberry Prince (25240). &c.
See " Tennis Ball," p. 246.

1879, June 26, roan, B.C.	Tennis Ball	Chaser, 37970	Mr. Cradock

RECEPTION, roan, calved February 14, 1875, Vol. xxv. p. 401.   Bred by Mr.
C. Cradock ; got by Wanderer (32790), dam (Welcome) by Dairy Prince
(17655), &c.

1879, Feb. 8, roan, C.C.	Reception 3rd	Chaser, 37970	Mr. Cradock

STRAWBERRY DROP 2ND, red and white, calved December 29, 1871, Vols.
xxii. and xxv. pp. 375, 402.   Bred by Mr. C. Cradock ; got by Wanderer
(32790), dam (Strawberry Drop) by Strawberry Prince (25240), &c.

1879, Apr. 14, { r.&w.,B.C. / roan, C.C.	(dead) / F. M.	Chaser, 37970	Mr. Cradock

VALENTINE, red and white, calved February 14, 1872, Vols. xxii., xxiii., xxiv.,
and xxv. pp. 375, 399, 389, 402.   Bred by Mr. C. Cradock ; got by The Laird
(32685), dam (Beauty) by Guy (28803), &c.

1879, Dec. 18, roan, C.C.	Clara	Chaser, 37970	Mr. Cradock

VERONA, roan, calved October 14, 1872, Vols. xxii. and xxv. pp. 376, 402.
· Bred by Mr. C. Cradock ; got by Wanderer (32790), dam (Valencia) by Valasco
(15443), &c.

1879, Feb. 20, white, B.C.	Snow Man	Chaser, 37970	Mr. Cradock

VESTA, white, calved January 22, 1875, Vol. xxv. p. 402.   Bred by Mr. C.
Cradock ; got by Wanderer (32790), dam (Valencia) by Valasco (15443), &c.

1879, April 14, roan, C.C.	Vestal	Chaser, 37970	Mr. Cradock

VIOLET, white, calved November 24, 1876. Bred by Mr. C. Cradock; got by
Vagrant (35843), dam (Primrose) by Guy (28803), &c. See Vol. xxiii. p. 399.

Produce in	Names, &c.	By what Bull.	By whom bred.
1879, Dec. 26, roan, C.C.	Lavender	Lythe, 40277	Mr. Cradock

CRADOCK, Captain R. W.,
Derrycalaghan, Roscrea, Ireland.

DIAGRAM, red, calved March 14, 1873. Bred by Mr. J. H. Blundell, Woodside; got by Third Cherry Duke (28171), dam (Deodora) by Third Duke of
Wetherby (26030), &c. See Vol. xxi. p. 643.

1879, Mar. 31, r. & w., C.C.	Cherry Gwynne	Gloster's Gwynne, 38360	Capt. Cradock

CRAGG, W. S.,
Arkholme, Carnforth.

CAWOOD'S HONEYSUCKLE, white, calved May 9, 1877. Bred by Mr. W.
S. Cragg; got by Captain Tregunter (28136), dam (Honeysuckle) by Double
Freedom (23701), &c. See Vol. xxiv. p. 389.

1879, Sept. 26, white, B.C.	Duke Honeysuckle	Earl Ashton, 39808	Mr. Cragg

CAWOOD'S HOPE 2ND, white, calved December 20, 1875, Vol. xxv. p. 402.
Bred by Mr. W. S. Cragg; got by Captain Tregunter (28136), dam (Double
Cawood) by Third Lord Cawood (24368), &c.

1879, Oct. 23, white, B.C.	Earl Cawood	Earl Ashton, 39808	Mr. Cragg

CHARMER, roan, calved May 29, 1874, Vol. xxiii. p. 400. Bred by Mr. H.
Fawcett, Old Bramhope; got by Fifth Lord Wild Eyes (26762), dam (Cordelia) by Mars (24543), &c.

1877, Dec. 26, roan, B.C.	Captain	Captain Tregunter, 28136	Mr. Cragg
1879, April 17, roan, C.C.	Cawood's Charmer	Earl Ashton, 39808	do.

Captain, sold to Mr. J. Ridding, Westhouse, Ingleton, Yorkshire.

COWSLIP 26TH, white, calved February 28, 1875, Vol. xxiv. p. 711. Bred by
Mr. T. Wilson, Shotley Hall, the property of Mr. W. S. Cragg; got by Waterloo Duke (32815), dam (Cowslip 21st) by Lord of Nunwick (26702), &c.

1879, Feb. 9, white, C.C.	Miss Cliburn	Duke of Oxford 31st, 33713	Earl of Lonsdale

Miss Cliburn, sold to Mr. W. S. Cragg, Arkholme.

GAZELLE'S DUCHESS, roan, calved March 17, 1876, Vol. xxv. p. 403. Bred
by Mr. H. Fawcett, Old Bramhope; got by Grand Duke of Weston (28768),
dam (Gazelle 17th) by Sixth Earl of Walton (26078), &c.

1879, Dec. 9, red, B.C.	Gazelle's Earl	Earl Ashton, 39808	Mr. Cragg

LADY CAWOOD 13TH, roan, calved November 3, 1875. Bred by Mr. W. S.
Cragg; got by Captain Tregunter (28136), dam (Lady Cawood 7th) by Lord
Chancellor (26622), &c. See Vol. xxii. p. 377.

1878, Sept. 18, r. & w., C.C.	(dead)	L. Ashton W. Eyes 2d, 36912	Mr. Cragg
1879, Aug. 6, white, C.C.	Lady Cawood 14th	Earl Ashton, 39808	do.

LUCRETIA 4TH, roan, calved December 1, 1868, Vols. xxi. and xxiii. pp. 648,
400. Bred by Mr. W. S. Cragg; got by Third Lord Cawood (24368), dam
(Lucretia 2nd) by Duke of Kirklevington (19622), &c.

1877, Oct. 18, roan, C.C.	Lucretia Cawood 2nd	Captain Tregunter, 28136	Mr. Cragg

LUCRETIA 6TH, roan, calved January 25, 1874, Vol. xxiii. p. 401. Bred by Mr. W. S. Cragg; got by Captain Tregunter (28136), dam (Lucretia 2nd) by Duke of Kirklevington (19622), &c.

Produce in		Names, &c.	By what Bull.	By whom bred.
1878, Jan. 16, roan,	B.C.	(slaughtered)	Captain Tregunter, 28136	Mr. Cragg
1879, Jan. 4, roan,	C.C.	Lucretia 7th	Earl Ashton, 39808	do.
1879, Dec. 2, roan,	C.C.	Lucretia 8th	do.	do.

ROSALIE DAISY, red and white, calved January 4, 1877. Bred by Mr. J. A. Rolls, The Hendre; got by Grand Duke of Clarence (28750), dam (Rosalie) by The Bully (23019), &c. See Vol. xxiv. p. 633.

1879, Sept. 12, r. & w., B.C.	Daisy Duke	D.of Siddington 2nd,33732	Mr. Cragg

ROYAL THORNDALE CHARMER, red and white, calved April 5, 1873, Vol. xxiii. p. 401. Bred by Mr. W. W. Slye, Beaumont Grange; got by Grand Duke of Thorndale (31297), dam (Royal Charmer 4th) by Barrington Oxford (25607), &c.

1877, Nov. 18, roan,	B.C.	(slaughtered)	Captain Tregunter, 28136	Mr. Cragg
1879, Jan. 3, roan,	B.C.	(slaughtered)	Earl Ashton, 39808	do.
1879, Dec. 18, roan,	C.C.	Cawood's R'l Ch'mer	do.	do.

CRAN, John,
Kirkton, Bunchrew Station, Highland Railway.

CHERRY IRWIN, roan, calved February 21, 1875. Bred by Mr. R. Bruce, Newton of Struthers, the property of Mr. J. Cran; got by Lord Irwin (29123), dam (Cherry Ripe) by Royal Errant (22780), &c. See Vol. xxii. p. 343.

1879, Feb. 28. r. & w., C.C.	Cherry Rose	Imperial Windsor, 36787	Mr. Fraser

CROWN PRINCESS 7TH, red and white, calved March 3, 1876. Bred by Mr. J. Cran; got by Bridegroom (33201), dam (Crown Princess 6th) by Baron Laurie 2nd (25570), &c. See Vol. xxi. p. 648.

1878, Nov. 17, red,	C.C.	Bridal Princess	Bridegroom, 33201	Mr. Cran

DAINTY 8TH, red and white, calved January 23, 1873. Bred by Mr. G. Marr, Cairnbrogie, the property of Mr. J. Cran; got by Young Hero (26385), dam (Dainty 7th) by Grand Prince (26308), &c. See "Ben Wyvis," p. 26.

1879, Feb. 20, roan,	B.C.	Ben Wyvis	Bromley, 36289	Mr. Marr

Ben Wyvis, sold to Mr. J. Cran, Kirkton.

MAY, red, calved April 14, 1876. Bred by Mr. R. Bruce, Newton of Struthers; got by Baron Killerby (27949), dam (Mysie 13th) by Baron Colling (25560), &c. See "Maxwell," Vol. xxii. p. 176.

1879, Jan. 8, red,	C.C.	May 1st	Commandant, 39610	Mr. Cran

PRINCESS 1ST, roan, calved November 6, 1876. Bred by Mr. J. Cran; got by Earl of Derby 2nd (31061), dam (Princess) by Prince of Worcester (20597), &c. See Vol. xxiv. p. 390.

1879, June 1, roan,	B.C.	Cadboll	Commandant, 39610	Mr. Cran

PRINCESS WINDSOR, red and white, calved February 1, 1871, Vols. xxi. and xxiv. pp. 738, 463. Bred by Her Majesty the Queen; got by England's Glory (23889), dam (Daybreak) by Rajah (22670), &c.

1879, Feb. 22, r. & w., C.C.	Sweet Windsor	Sir W'sor Broughton,27507	Mr. Cran

STAMFORD 12TH, red, calved April 15, 1873. Bred by Mr. W. S. Marr, Uppermill; got by Heir of Englishman (24122), dam (Stamford 6th) by Prince Louis (27158), &c. See " Stanley," Vol. xxi. p. 461.

Produce in		Names, &c.	By what Bull.	By whom bred.
1877, Feb. 2, red,	C.C.	Stamford Rose	Fred. Fitz-Windsor, 31196	Mr. Cran
1878, Nov. 19, roan,	C.C.	Stamford Bride	Bridegroom, 33201	do.
1879, Dec. 11, red,	C.C.	Stamford Belle	Commandant, 39610	do.

STAR OF HOPE, roan, calved April 8, 1875, Vol. xxv. p. 456. Bred by Mr. W. A. Fraser, Brackla, the property of Mr. J. Cran; got by Baron Havering (33043), dam (Star of Peace) by Lord Forth (26649), &c.

1879, Mar. 22, roan,	C.C.	Star of Brackla	Imperial Windsor, 36787	Mr. Fraser

VICTORIA ROSE, white, calved April 9, 1875. Bred by Mr. J. Cran; got by Frederick Fitz-Windsor (31196), dam (Victoria) by Duke of Cornwall (23732), &c. See " Vicks," p. 257.

1878, Jan. 24, white,	C.C.	Victoria Rosebud	Bridegroom, 33201	Mr. Cran
1879, Feb. 9, white,	B.C.	Vicks	do.	do.

CRICKMORE, William,
Seething, Brooke, Norfolk.

VANITY, red, calved May 8, 1876, Vol. xxv. p. 404. Bred by Mr. W. Crickmore; got by John Hopper (34264), dam (Venus) by Bonbon (30560), &c.

1879, April 21, red,	C.C.		Romulus, 39006	Mr. Crickmore

VENUS, roan, calved October 20, 1873, Vols. xxiii. and xxiv. pp. 402, 391. Bred by Mr. W. Crickmore; got by Bonbon (30560), dam (Virginia) by Victoria's Brian (23141), &c.

1878, Mar. 15, roan,	B.C.	(Steer)	John Hopper, 34264	Mr. Crickmore
1879, Mar. 4, roan,	B.C.	Cupid	Romulus, 39006	do.

VERBENA, red, calved May 11, 1875, Vol. xxiv. p. 391. Bred by Mr. W. Crickmore; got by Bonbon (30560), dam (Violet) by Duke Robert (31039), &c.

1878, May 17, roan,	B.C.	Lord Clyde	John Hopper, 34264	Mr. Crickmore
1879, April 23, roan,	B.C.	Valentine	Romulus, 39006	do.

VESTA, red, calved April 11, 1876, Vol. xxiv. p. 391. Bred by Mr. W. Crickmore; got by Bonbon (30560), dam (Valeria) by Lord of the Isles (24443), &c.

1879, April 27, roan.	C.C.		Romulus, 39006	Mr. Crickmore

VIGNETTE, red, calved April 1, 1874, Vols. xxiii. and xxiv. pp. 403, 391. Bred by Mr. W. Crickmore; got by Bonbon (30560), dam (Valeria) by Lord of the Isles (24443), &c.

1878, Mar. 23, red,	B.C.	(Steer)	John Hopper, 34264	Mr. Crickmore
1879, Mar. 27, roan,	C.C.	Victoria	Romulus, 39006	do.

VIRGINIA, roan, calved in 1868, Vols. xxi., xxiv., and xxv. pp. 650, 391, 404. Bred by Mr. J. Batchelder, Claxton Abbey; got by Victoria's Brian (23141), dam (Snowdrop) by Virginius (23147), &c.

1879, Apr. 29, white,	B.C.	Virginius	Lamarque, 38540	Mr. Crickmore

CROMPTON, Thomas,
Lowthorpe, Hull.

DANTHORPE LADY 18TH, roan, calved August 8, 1875, Vol. xxv. p. 404.
Bred by Mr. R. Fisher, Leconfield; got by Genuine Prince (24031), dam
(Danthorpe Lady 7th) by Lord Greta (20174), &c.

Produce in		Names, &c.	By what Bull.	By whom bred.
1879, Oct. 3, roan,	C.C.	Adela	Baronet, 36175	Mr. Crompton

CROOME, J. Capel,
Bagendon House, Cirencester.

CROCEA 3RD, roan, calved November 19, 1876.   Bred by Mr. J. Capel Croome;
got by Siddington Orange (42380), dam (Crocea) by Lord Tregunter (31758),
&c. See Vol. xxv. p. 404.

1879, Mar. 15, r.&w., B.C.	Cherry Stone	Viscount Blanche, 39269	Mr. Croome

DAMSEL 6TH, red, calved December 14, 1876.   Bred by Mr. L. Little, Harnhill;
got by Prince of Clarence 5th (35139), dam (Damsel 5th) by Lord Napier
(29156), g. d. (Damsel 4th) by General Havelock (16120), gr. g. d. (Darlington 3rd) by Sir Hugh (12082), &c. See Vol. xii. p. 344.

1879, April 7, roan, B.C.	Ranger	Connaught, 39618	Mr. Croome

LAUREL 3RD, roan, calved November 26, 1876.   Bred by Mr. J. Capel Croome;
got by Siddington Orange (42380), dam (Laurel) by Duke of Albany (25931),
&c. See Vol. xxv. p. 404.

1879, March 15, roan, B.C.	Portugal	Viscount Blanche, 39269	Mr. Croome

SWEETBRIER 5TH, roan, calved March 8, 1877.   Bred by Mr. J. Capel
Croome; got by Siddington Orange (42380), dam (Sweetbrier 4th) by Duke
of Albany (25931), &c. See Vol. xxv. p. 405.

1879, Aug. 7, red,	C.C.	Sweetbrier 6th	Viscount Blanche, 39269	Mr. Croome

CROSBIE, W. Talbot,
Ardfert Abbey, Ardfert, Ireland.

BLANCHE VENUS, white, calved December 21, 1876.   Bred by Mr. W. T.
Crosbie; got by England's Glory (23889), dam (Regal Venus) by Regal Booth
(27262), &c. See Vol. xxiii. p. 404.

1879, April 18, roan, B.C.	Royal Ardfert	Royal Halnaby, 39041	Mr. Crosbie

BLITHESOME BRENDA, roan, calved March 29, 1876.   Bred by Mr. W. T.
Crosbie; got by Lord Blithsome (29067), dam (Royal Brenda) by Royal
Sovereign (22802), &c. See Vol. xxiv. p. 395.

1879, June 3, white, C.C.	Blithesome Glory	Foreign Glory, 39890	Mr. Crosbie

BLITHESOME FLORENTINE, roan, calved March 31, 1875, Vol. xxv. p. 405.
Bred by Mr. W. T. Crosbie; got by Lord Blithesome (29067), dam (Florentine Grove) by Castle Grove (19408), &c.

1879, May 15, red,	C.C.	Florentine Queen	Royal Halnaby, 39041	Mr. Crosbie

BLITHESOME GWYNNE, red, calved November 22, 1876, Vol. xxv. p. 405.
Bred by Mr. W. T. Crosbie; got by Lord Blithesome (29067), dam (Gwynne
White) by Castle Grove (19408), &c.

Produce in		Names, &c.	By what Bull.	By whom bred.
1879, April 3, roan,	C.C.	Halnaby Gwynne	Royal Halnaby, 39041	Mr. Crosbie

BRENDA'S ROSE, red and white, calved March 20, 1877. Bred by Mr. W.
T. Crosbie; got by Royal Fitz-Rose (37390), dam (Brendina) by Regal Booth
(27262), &c. See Vol. xxiv. p. 392.

1879, May 24, r. & w.,	B.C.	Brenda's King	Royal Halnaby, 39041	Mr. Crosbie

BRIGHT GWYNNE, white, calved March 24, 1869, Vol. xx. p. 422. Bred by
Mr. W. T. Crosbie; got by Northern Light (24670), dam (Gwynne Grove) by
Castle Grove (19408), &c.

1873, Feb. 1, white,	B.C.	Irish Gwynne	Irish Baron, 31417	Mr. Crosbie
1874, Jan. 5, white,	B.C.	Western Gwynne	Regal Booth, 27262	do.
1875, May 20, white,	B.C.	Blithe Gwynne	Lord Blithesome, 29067	do.
1876, Mar. 25, white,	B.C.	Sir Pawlett Gwynne	do.	do.
1877, Jan. 22, roan,	B.C.	Gwynne Briton	England's Glory, 23889	do.
1878, Jan. 12, roan,	B.C.	Victor Gwynne	do.	do.
1879, Feb. 28, white,	C.C.	Aylesby Gwynne	Foreign Glory, 39890	do.

Victor Gwynne, sold to Mr. J. Jameson, Malahide, Dublin.

BRITANNIA HOWARD, roan, calved August 11, 1877. Bred by Mr. W. T.
Crosbie; got by England's Glory (23889), dam (Duchess of Norfolk) by Castle
Grove (19408), &c. See Vol. xxiv. p. 392.

1879, Nov. 25, roan,	B.C.	Prince of Norfolk	Riby Prince, 40593	Mr. Crosbie

BRITISH BELLE, roan, calved February 26, 1877. Bred by Mr. W. T. Crosbie; got by England's Glory (23889), dam (Lothian Belle) by Regal Booth
(27262), &c. See Vol. xxiv. p. 394.

1879, Apr. 21, r. & w.,	C.C.	Halnaby Belle	Royal Halnaby, 39041	Mr. Crosbie

BRITISH DAISY, white, calved June 21, 1877. Bred by Mr. W. T. Crosbie;
got by England's Glory (23889), dam (Miriam) by Northern Light (24670),
&c. See Vol. xxiv. p. 395.

1879, Dec. 13, roan,	B.C.	Knight of Halnaby	Royal Halnaby, 39041	Mr. Crosbie

BRITISH NYMPH, red, calved March 17, 1877. Bred by Mr. W. T. Crosbie;
got by England's Glory (23889), dam (Sprite of Lothian) by Fairy King
(21716), &c. See Vol. xxiv. p. 396.

1879, June 26, roan,	B.C.	Linhurst	Foreign Glory, 39890	Mr. Crosbie

BRITISH VENUS, red and white, calved August 5, 1877. Bred by Mr. W.
T. Crosbie; got by England's Glory (23889), dam (Royal Eunice) by Royal
Sovereign (22802), &c. See Vol. xxiv. p. 396.

1879, Nov. 22, r. & w.,	C.C.	Venus Queen	Royal Halnaby, 39041	Mr. Crosbie

CORONA, white, calved January 28, 1872, Vols. xxii. and xxiv. pp. 379, 392.
Bred by Mr. W. T. Crosbie; got by Regal Booth (27262), dam (Crown of
Light) by Northern Light (24670), &c.

1878, Feb. 19, white,	B.C.	Crown of Britain	England's Glory, 23889	Mr. Crosbie
1879, May 21. roan,	C.C.	Crown of Halnaby	Royal Halnaby, 39041	do.

Crown of Britain, sold to Mr. W. R. Meade, Ballymartle, Kinsale.

FAIRY QUEEN, roan, calved February 15, 1873, Vols. xxiv. and xxv. pp. 393,
406.   Bred by Mr. W. T. Crosbie; got by Irish Baron (31417), dam (Sprite
of Lothian) by Fairy King (21716), &c.

Produce in		Names, &c.		By what Bull.		By whom bred.
1879, May 23, roan,	C.C.	Royal Nymph		Royal Halnaby, 39041		Mr. Crosbie

FOREST FLOWER, red, calved January 27, 1877.   Bred by Mr. A. H. Browne,
Doxford; got by Royal Benedict (27348), dam (Forget-me-Not) by Lieutenant-
General (31600), &c.   See Vol. xxiv. p. 351.

1879, Oct. 26, roan,	B.C.	(dead)		Lord Prinknash, 34655		Mr. Crosbie

LADY GWYNNE BOOTH, roan, calved June 6, 1871, Vols. xxi., xxii., and
xxiv. pp. 651, 381, 394.   Bred by Mr. W. T. Crosbie; got by Regal Booth
(27262), dam (Gwynne Grove) by Castle Grove (19408), &c.

1878, Mar. 11, white, B.C.	Sir Bruere Gwynne	Royal Fitz-Rose, 37390	Mr. Crosbie
1879, Feb. 14, white, C.C.	White Rose Gwynne	do.	do.

Sir Bruere Gwynne, sold to Mr. P. Fitzgerald, Corkbeg House, Whitegate, Cloyne, Co. Cork.

LOUISA 19TH, roan, calved February 15, 1873, Vols. xxiii. and xxv. pp. 404,
406.   Bred by Mr. W. S. Garnett, Williamston; got by Royal Prince (27384),
dam (Louisa 5th) by Dr. McHale (15887), &c.

1879, May 3, r. & w., C.C.	Royal Portia	Royal Halnaby, 39041	Mr. Crosbie

MAID OF GEORGIA, roan, calved February 26, 1870, Vols. xxi., xxii., and
xxiv. pp. 652, 382, 394.   Bred by Mr. W. T. Crosbie; got by Northern Light
(24670), dam (Georgia) by Castle Grove (19408), &c.

1878, March 5, red,	C.C.	(dead)		England's Glory, 23889	Mr. Crosbie
1879, Feb. 5, { white,	C.C.	Twin Rose	}	Royal Fitz-Rose, 37390	do.
{ roan,	C.C.	Twin Georgia			

MAID OF MEDORA, red, calved February 19, 1876, Vol. xxv. p. 406.   Bred
by Mr. W. T. Crosbie; got by Lord Blithesome (29067), dam (Lady Augusta)
by Northern Light (24670), &c.

1879, April 5, red.	C.C.	Royal Medora	Royal Halnaby, 39041	Mr. Crosbie

MAID OF NORFORK, red, calved May 30, 1875, Vol. xxiv. p. 394.   Bred by
Mr. W. T. Crosbie; got by Lord Blithesome (29067), dam (Maid of Booth) by
Regal Booth (27262), &c.

1879, Feb. 9, red,	C.C.	Rosa Howard	Royal Fitz-Rose, 37390	Mr. Crosbie

MAY DAISY, roan, calved May 7, 1876.   Bred by Mr. W. T. Crosbie; got by
Lord Blithesome (29067), dam (Oriental) by Castle Grove (19408), &c.   See
Vol. xxiv. p. 395.

1879, May 8, r. & w., B.C.	Daisy King	Royal Halnaby, 39041	Mr. Crosbie

MERIDIAN, roan, calved February 17, 1872, Vols. xxii. and xxiv. pp. 382, 394.
Bred by Mr. W. T. Crosbie; got by Regal Booth (27262), dam (Miriam) by
Northern Light (24670), &c.

1879, Jan. 10, roan,	C.C.	Meridian Rose	Royal Fitz-Rose, 37390	Mr. Crosbie

PERI, roan, calved April 19, 1869, Vols. xx. and xxiv. pp. 687, 395.   Bred by
Mr. W. T. Crosbie; got by Northern Light (24670), dam (Circassia) by Lamp
of Lothian (16356), &c.

1878, Feb. 24, roan,	B.C.	Gautama	Royal Fitz-Rose, 37390	Mr. Crosbie
1879, Feb. 24, roan,	C.C.	Aylesby Daisy	Foreign Glory, 39890	do.

Gautama, sold to Mr. M. King, Strangemore, Londonderry.

RED DAISY, red, calved April 8, 1876. Bred by Mr. W. T. Crosbie; got by
Lord Blithesome (29067), dam (Khiva) by Irish Baron (31417), &c. See
Vol. xxiii. p. 404.

Produce in		Names, &c.	By what Bull.	By whom bred.
1879, June 7, red,	C.C.	Royal Daisy	Royal Halnaby, 39041	Mr. Crosbie

RIBY MARCHIONESS, roan, calved March 31, 1875, Vols. xxiv. and xxv. pp.
395, 407. Bred by Mr. W. Torr, Aylesby Manor; got by Knight of the Shire
(26552), dam (Riby Peeress) by Breast Plate (19337), &c.

1879, March 13, roan, C.C.	Riby Anna	Royal Halnaby, 39041	Mr. Crosbie

ROSA GWYNNE, roan, calved April 15, 1877. Bred by Mr. W. T. Crosbie;
got by Royal Fitz-Rose (37390), dam (Lady Gwynne Booth) by Regal Booth
(27262), &c. See Vol. xxiv. p. 394.

1879, April 16, roan, C.C.	Rosetta Gwynne	Foreign Glory, 39890	Mr. Crosbie

ROYAL ADARE, red and white, calved March 28, 1867, Vols. xx., xxii., and
xxiv. pp. 748, 383, 395. Bred by Mr. W. T. Crosbie; got by Royal Sovereign
(22802), dam (Maid of Adare) by Lamp of Lothian (16356), &c.

1879, Mar. 23, r. & w., C.C.	Queen of Halnaby	Royal Halnaby, 39041	Mr. Crosbie

ROYAL FLORENTINE, red, calved January 7, 1872, Vols. xxii. and xxiv.
pp. 383, 396. Bred by Mr. W. T. Crosbie; got by Regal Booth (27262), dam
(Light of Florence) by Northern Light (24670), &c.

1878, June 29, roan, B.C.	Gilmerton	Royal Fitz-Rose, 37390	Mr. Crosbie
1879, Aug. 6, r. & w., C.C.	Halnaby Florentine	Royal Halnaby, 39041	do.

Gilmerton, sold to Mr. Cruickshank, Listowel, Co. Kerry.

SAXON MAID, roan, calved June 6, 1877. Bred by Mr. W. T. Crosbie; got
by England's Glory (23889), dam (Maid of Booth) by Regal Booth (27262),
&c. See Vol. xxiv. p. 394.

1879, Dec. 1, roan, B.C.	Saxon King	Royal Halnaby, 39041	Mr. Crosbie

CROSS, J.,
Chiswell Hall, Edenbridge, Kent.

DIDO 4TH, red and white, calved February 20, 1877. Bred by Mr. J. Cross;
got by Duke of Underley 2nd (36551), dam (Delight) by Langley (26567),
&c. See "Cherry Dido," Vol. xxv. p. 48.

1879, Nov. 6, r. & w., B.C.	Dido Boy	Lord Knightley, 40184	Mr. Cross

DIDO 5TH, roan, calved March 9, 1877. Bred by Mr. J. Cross; got by Milcote
Duke (40355), dam (Diadem) by Monitor (24615), &c. See Vol. xxi. p. 652.

1879, Sept. 9, roan, B.C.	(Steer)	Lord Knightley, 40184	Mr. Cross

KNIGHTLEY QUEEN 2ND, red, calved June 18, 1874. Bred by Mr. F. N.
Sartoris, Rushden Hall, the property of Mr. J. Cross; got by Baron Oxford
5th (27958), dam (Knightley Queen) by Hardicanute (26338), g. d. (Pipalee)
by Bull's Run (19368), &c. See Vol. xx. p. 690.

1878, Dec. 26, red,	C.C.	Knightley Princess	Earl of Leic'ster 7th, 33806	Mr. Griffin

NELLY KNIGHTLEY, red and white, calved July 7, 1875. Bred by Mr. G.
Murton Tracy, Redlands; got by Squire Knightley (35663), dam (Nelly
O'Brien) by Cromwell (19528), &c. See Vol. xxii. p. 597.

1879, July 2, r. & w., B.C.	Knightley Squire	Lord Knightley, 40184	Mr. Cross

ROSE OF HILLHURST, red and white, calved April 22, 1877. Bred by Mr.
J. W. Larking, Cansiron Farm; got by Third Duke of Hillhurst (30975), dam
(Rose of Summer) by Third Duke of Waterloo (23801), &c. See Vol. xxiv.
p. 528.

Produce in	Names, &c.	By what Bull.	By whom bred.
1879, Sept. 26, r. & w., C.C.	Knightley Rose	Lord Knightley, 40184	Mr. Cross

CROUDSON, J.,
Urswick, Ulverstone.

BRIGHT EYES, roan, calved March 28, 1874, Vol. xxv. p. 407. Bred by Mr.
J. Croudson; got by Tichborne (32719), dam (Oxford Wild Eyes) by Lord
Bright Eyes (29070), &c.

Produce in	Names, &c.	By what Bull.	By whom bred.
1879, Feb. 11, roan, C.C.	Bright Lady	Viscount Barrington,39268	Mr. Croudson

COMMEMORATION 4TH, roan, calved September 12, 1874, Vols. xxiv. and
xxv. pp. 397, 408. Bred by Mr. W. Croudson, Greenslack; got by Duke of
Oxford (31004), dam (Commemoration 2nd) by Tenth Duke of Oxford (17739),
&c.

Produce in	Names, &c.	By what Bull.	By whom bred.
1879, Feb. 13, roan, B.C.	Colonel Gunter	Viscount Barrington,39268	Mr. J. Croudson

OLGA 2ND, white, calved February 29, 1872, Vols. xxii. and xxiv. pp. 384, 397.
Bred by Mr. J. Croudson; got by Grand Duke of Cambridge 2nd (26285),
dam (Olga) by Old Buck (15017), &c.

Produce in	Names, &c.	By what Bull.	By whom bred.
1878, April 12, roan, C.C.	Olga's Rose	G. Duke Wild Eyes, 34081	Mr. Croudson
1879, March 8, white, B.C.	Blanche Boy	Viscount Barrington,39268	do.

Olga's Rose, sold to Mr. Leece, Melling, Lancaster.

OLGA'S KIRKLEVINGTON 2ND, red and white, calved May 4, 1874,
Vol. xxiv. p. 398. Bred by Mr. J. Croudson; got by Sixth Duke of Kirklevington (30982), dam (Olga) by Old Buck (15017), &c.

Produce in	Names, &c.	By what Bull.	By whom bred.
1878, June 17, white, C.C.	Sockburn Sall 2nd	G. Duke Wild Eyes, 34081	Mr. Croudson
1879, June, roan, C.C.	Sockburn Sall 3rd	Viscount Barrington,39268	do.

CROUDSON, W.,
Greenslack, Broughton-in-Furness.

DUCHESS OF BRAILES, roan, calved October 9, 1873. Bred by Mr. W.
Croudson; got by Duke of Southcott (28449), dam (Commemoration 2nd) by
Tenth Duke of Oxford (17739), &c. See Vol. xxii. p. 384.

Produce in	Names, &c.	By what Bull.	By whom bred.
1879, March 28, roan, C.C.	D'ss of Brailes 2nd	G.D.Kirklev'ton 3rd,34073	Mr. Croudson

ROSE OF DUDDON, roan, calved July 27, 1872, Vol. xxii. p. 384. Bred by
Mr. W. Croudson; got by Baron Fennel (27937), dam (Commemoration 2nd)
by Tenth Duke of Oxford (17739), &c.

Produce in	Names, &c.	By what Bull.	By whom bred.
1876, Feb. 14, r. & w., C.C.	Rose of Duddon 2nd	Duke of Oxford, 31004	Mr. Croudson
1878, April 19, roan, C.C.	Rose of Duddon 4th	G.D.Kirklev'ton3rd,34073	do.
1879, April 12, roan, B.C.	Lord of Duddon	do.	do.

ROSE OF DUDDON 2ND, red and white, calved February 14, 1876. Bred
by Mr. W. Croudson; got by Duke of Oxford (31004), dam (Rose of Duddon)
by Baron Fennel (27937), &c.

Produce in	Names, &c.	By what Bull.	By whom bred.
1879, Sept. 9, r. & w., B.C.	Squire of Duddon	G.D.Kirklev'ton 3rd,34073	Mr. Croudson

CROWE, Robert,
Speeton, Bempton, Hull.

LADY ALBERTINE, roan, calved February 3, 1874.   Bred by Mr. R. Crowe; got by Prince Charlie (32117), dam (Lady Adelina) by Baron Booth (23354), &c.   See Vol. xxi. p. 653.

Produce in		Names, &c.	By what Bull.	By whom bred.
1878, Sept. 14, roan,	B.C.	(Steer)	Bn.Goldschmidt 3d, 36182	Mr. Crowe
1879, Aug. 28, red,	C.C.	Lady Annette	Monarch, 40362	do.

LADY FLORENCE, roan, calved January 11, 1872, Vol. xxi. p. 654.   Bred by Mr. T. Stamper, Highfield House; got by Grindelwald (26323), dam (Lady Wiske Monk) by Forerunner (12891), &c.

Produce in		Names, &c.	By what Bull.	By whom bred.
1876, Sept. 18, red,	B.C.	(Steer)	Bn. Goldschmidt 2d, 33040	Mr. Crowe
1877, Aug. 22, white,	B.C.	(Steer)	Bn. Goldschmidt 3d, 36182	do.
1878, Sept. 14, roan	B.C.	(Steer)	do.	do.
1879, Oct. 28, roan,	C.C.	Lady Fanny	Monarch, 40362	do.

MADAME GOLDSCHMIDT 2ND, white, calved August 30, 1868, Vols. xx., xxi., and xxii. pp. 635, 654, 385.   Bred by Mr. R. Crowe; got by Baron Booth (23354), dam (Madame Goldschmidt) by Master Goldschmidt (20305), &c.

Produce in		Names, &c.	By what Bull.	By whom bred.
1878, April 17, roan,	B.C.	Bn. Goldschmidt 5th	Sir Gregory Gwynne, 37465	Mr. Crowe
1879, Aug. 26, white,	C.C.	MdeGoldschmidt7th	Monarch, 40362	do.

Baron Goldschmidt 5th, sold to Mr. W. Crowe, Flamborough, Yorkshire.

MADAME GOLDSCHMIDT 5TH, white, calved January 7, 1874.   Bred by Mr. R. Crowe; got by Prince Charlie (32117), dam (Madame Goldschmidt 2nd) by Baron Booth (23354), &c.   See "Baron Goldschmidt 5th," p. 14.

Produce in		Names, &c.	By what Bull.	By whom bred.
1878, Aug. 18, roan,	B.C.	(Steer)	Prince Alfonso, 37214	Mr. Crowe
1879, Aug. 17, { white,	C.C.	MdeGoldschmidt8th }	Monarch, 40362	do.
{ white,	B.C.	(Steer)		

PRINCESS ALEXANDRINA, red and white, calved February 11, 1873, Vol. xxv. p. 408.   Bred by Mr. R. Crowe; got by Pretender (29581), dam (Princess Alexandra) by Knight Errant (18154), &c.

Produce in		Names, &c.	By what Bull.	By whom bred.
1879, Aug. 15, roan,	C.C.	Pr'ssAlexandrina3rd	Monarch, 40362	Mr. Crowe

CRUICKSHANK, A.,
Sittyton, Aberdeen, N.B.

AMARYLLIS, roan, calved January 1, 1877.   Bred by Mr. A. Cruickshank; got by Lord Lancaster (26666), dam (Azalea) by Cæsar Augustus (25704), &c.   See Vol. xxv. p. 409.

Produce in		Names, &c.	By what Bull.	By whom bred.
1879, April 14, red,	C.C.	Amaranth	Barmpton, 37763	Mr. Cruickshank

CIRCASSIA, red, calved March 31, 1869, Vols. xx. and xxiv. pp. 447, 398.   Bred by Mr. A. Cruickshank; got by Champion of England (17526), dam (Cicely) by Lancaster Royal (18167), &c.

Produce in		Names, &c.	By what Bull.	By whom bred.
1878, Dec. 26, roan,	C.C.	Celandine	Pride of the Isles, 35072	Mr. Cruickshank
1879, Nov. 25, { red,	C.C.	Cineraria }	Viceroy, 32764	do.
{ roan,	C.C.	Convolvulus		

COSTUME, roan, calved April 22, 1877.   Bred by Mr. A. Cruickshank; got by Bridesman (30586), dam (Cactus) by Champion of England (17526), &c.   See Vol. xxv. p. 409.

Produce in		Names, &c.	By what Bull.	By whom bred.
1879, Sept. 22, white,	B.C.	Chillingham	Lord of the Isles, 40218	Mr. Cruickshank

DUCHESS OF GLO'STER 21st, roan, calved February 7, 1877. Bred by
Mr. A. Cruickshank; got by Barmpton Prince (32995), dam (Duchess of
Glo'ster 13th) by Grand Duke of Glo'ster (26288), &c. See Vol. xxi. p. 655.

Produce in	Names, &c.	By what Bull.	By whom bred.
1879, Nov. 4, roan,	C.C.,D's of Glo'ster 24th	Lord of the Isles, 40218	Mr. Cruickshank

GERANIUM, red and white, calved February 18, 1877. Bred by Mr. A. Cruickshank; got by Pride of the Isles (35072), dam (Garland) by Scotland's Pride
(25100), &c. See Vol. xxi. p. 655.

1879, Nov. 13, r. & w., C.C.,Gloxinia		Lord of the Isles, 40218	Mr. Cruickshank

JULIET, red, calved March 2, 1877. Bred by Mr. A. Cruickshank; got by
Barmpton Prince (32995), dam (Joyful) by Master of Arts (26867), g. d.
(Jealousy) by Champion of England (17526), &c. See Vol. xx. p. 566.

1879, Oct. 16, r. & w., C.C.,Jessamine		Lord of the Isles, 40218	Mr. Cruickshank

ORANGE BLOSSOM 8th, red, calved February 23, 1867, Vol. xix. p. 655.
Bred by Mr. A. Cruickshank; got by Sir Walter Scott (22922), dam (Orange
Blossom) by Dr. Buckingham (14405), &c.

1878, April 5, roan,	B.C. Ostensible	Pride of the Isles, 35072	Mr. Cruickshank
1879, Mar. 29, r. & w., C.C. Orange Blossom 30th		do.	do.

ROSE OF KNOWLMERE, red, calved April 20, 1876. Bred by Mr. A. Cruickshank; got by Knight of Knowlmere (22055), dam (Red Violet) by Allan
(21172), &c. See Vol. xxiii. p. 407.

1878, Sept. 28, roan,	B.C. Ravenshaw	Lord of the Isles, 40218	Mr. Cruickshank
1879, Nov. 17. roan,	C.C. Violet Perfume	do.	do.

Ravenshaw, sold to Mr. Haigh, Fife.

SYBIL 16th, roan, calved February 28, 1876. Bred by Mr. A. Cruickshank;
got by Bridesman (30586), dam (Sybil 11th) by Scotch Rose (25099), &c. See
Vol. xxiv. p. 400.

1879, Apr. 22, white, C.C. Sybil 20th		Roan Gauntlet, 35284	Mr. Cruickshank

SYCAMORE, roan, calved November 23, 1876. Bred by Mr. A. Cruickshank;
got by Count Robert (30812), dam (Surmise) by Champion of England (17526),
&c. See Vol. xxiii. p. 408.

1879, April 3. roan, C.C. Semptress		Barmpton, 37763	Mr. Cruickshank

VICTORIA 45th, roan, calved November 28, 1871. Bred by Mr. A. Cruickshank; got by Cæsar Augustus (25704), dam (Victoria 42nd) by Forth
(17866), &c. See Vol. xxi. p. 658.

1878, May 8, roan,	C.C. Victoria 63rd	Pride of the Isles, 35072	Mr. Cruickshank
1879, Oct. 7, roan,	C.C. Victoria 65th	Roan Gauntlet, 35284	do.

VICTORIA 53rd, red, calved February 10, 1876. Bred by Mr. A. Cruickshank; got by Royal Duke of Glo'ster (29864), dam (Victoria 46th) by Cæsar
Augustus (25704), &c. See "Victor Chief," p. 258.

1879, Jan. 28, roan,	B.C. Victor Chief	Lord of the Isles, 40218	Mr. Cruickshank

Victor Chief, sold to Mr. Geekie, Baldowrie, Forfarshire.

VICTORIA 54th, roan, calved March 18, 1876. Bred by Mr. A. Cruickshank;
got by Royal Duke of Glo'ster (29864), dam (Victoria 39th) by Champion of
England (17526), &c. See "Ventriloquist," p. 256.

1878. Nov. 16, red,	B.C. Vandal	Lord of the Isles, 40218	Mr. Cruickshank

VICTORIA 57TH, roan, calved February 22, 1877. Bred by Mr. A. Cruickshank; got by Barmpton Prince (32995), dam (Victoria 41st) by Lord Privy
Seal (16444), &c. See "Ventriloquist," p. 256.

Produce in	Names, &c.	By what Bull.	By whom bred.
1879, Apr. 15, r. & w., C.C.	Victoria 64th	Royal Violet, 40649	Mr. Cruickshank

VICTORIA 58TH, red, calved March 30, 1877. Bred by Mr. A. Cruickshank;
got by Pride of the Isles (35072), dam (Victoria 43rd) by Champion of England
(17526), &c. See Vol. xxii. p. 387.

1879, Sept. 15, roan, B.C.	Vittoria	Lord of the Isles, 40218	Mr. Cruickshank

VIOLET'S ROSE, white, calved July 12, 1876. Bred by Mr. A. Cruickshank ;
got by Pride of the Isles (35072), dam (Violet's Pride) by Scotland's Pride
(25100), &c. See Vol. xx. p. 809.

1879, Feb. 24, roan, C.C.	Violet Bud	Barmpton, 37763	Mr. Cruickshank

CRUICKSHANK, J. W. and E.,
Lethenty, Inverurie, N.B.

BRIGHT GEM, red, calved March 9, 1874, Vols. xxiii. and xxv. pp. 408, 411.
Bred by Mr. A. W. Connon, Cremore; got by Lieutenant General (31600),
dam (Red Rose) by King Richard (26523), &c.

1879, May 27, roan, C.C.	Bright Casquet	Kt. of St. Patrick, 38520	Msrs.Cruickshank

BRIGHT GLEAM, roan, calved November 17, 1875, Vol. xxv. p. 411. Bred
by Messrs. J. W. and E. Cruickshank; got by Third Duke of Geneva (23753),
dam (Rose of Cashmere) by Royal Sovereign (22802), &c.

1879, July 27, roan, C.C.	Bright Ripple	Kt. of St. Patrick, 38520	Msrs.Cruickshank

BRIGHT HOPE, roan, calved August 22, 1876, Vol. xxv. p. 411. Bred by
Messrs. J. W. and E. Cruickshank; got by King James (28971), dam (Bright
Gem) by Lieutenant General (31600), &c.

1879, Nov. 31, white, C.C.	Bright Pearl	Kt. of St. Patrick, 38520	Msrs.Cruickshank

COMFREY, white, calved June 19, 1877. Bred by Messrs. J. W. and E.
Cruickshank; got by Sir Windsor Broughton (27507), dam (Cream) by
Athelstane (23331), &c. See Vol. xxiv. p. 400.

1879, July 27, white, C.C.	Curfew	Kt. of St. Patrick, 38520	Msrs.Cruickshank

COQUETTE, roan, calved May 15, 1876. Bred by Messrs. J. W. and E.
Cruickshank; got by King Richard 2nd (28984), dam (Cream) by Athelstane
(23331), &c. See Vol. xxiii. p. 408.

1879, Aug. 17, roan, C.C.	Coral	Kt. of St. Patrick, 38520	Msrs.Cruickshank

CREAM, roan, calved January 31, 1873, Vols. xxiii., xxiv., and xxv. pp. 408,
400, 411. Bred by Mr. E. A. Fawcett, Childwick Hall; got by Athelstane
(23331), dam (Cowslip) by Cambridge Barrington 1st (14223), &c.

1879, Aug. 27, roan, C.C.	Cystus	Kt. of St. Patrick, 38520	Msrs.Cruickshank

DIAMOND BRACELET, roan, calved January 25, 1870, Vols. xxi., xxiii., and
xxv. pp. 658, 409, 411. Bred by Mr. C. L. Ellison, Loughglyn; got by
Trumpeter (35825), dam (Bridal) by Buckingham (11219), &c.

1879, Feb. 14, roan, B.C.	Ben Cruachan	Kt. of St. Patrick, 38520	Msrs.Cruickshank

FANNY 16TH, red, calved February 25, 1875, Vol. xxiv. p. 400.	Bred by Messrs. J. W. and E. Cruickshank; got by Lord of the Manor (29181), dam (Fanny 15th) by Duke of York (23804), &c.

Produce in	Names, &c.	By what Bull.	By whom bred.
1879, Nov. 30, roan, C.C.	Feodora	Kt. of St. Patrick, 38520	Msrs.Cruickshank

HAZLE BRACELET, red, calved May 12, 1873, Vols. xxiii., xxiv., and xxv. pp. 409, 401, 412.	Bred by The O'Connor Don, Clonalis; got by Duke of Hazlecote 18th (30966), dam (Diamond Bracelet) by Trumpeter (35825), &c.

Produce in	Names, &c.	By what Bull.	By whom bred.
1879, April 14, roan, C.C.	Birthday Bracelet	Kt. of St. Patrick, 38520	Msrs.Cruickshank

JEWELLED BRACELET, roan, calved May 12, 1875, Vol. xxv. p. 412.	Bred by Messrs. J. W. and E. Cruickshank; got by Larry the Lad (31584), dam (Bracelet 3rd) by Lord of Rocklands (22183), &c.

Produce in	Names, &c.	By what Bull.	By whom bred.
1879, Oct. 27, white, B.C.	Ben Lawers	Kt. of St. Patrick, 38520	Mrs.Cruickshank

LADY OF THE VALLEY, red and white, calved April 27, 1872, Vols. xxi., xxiv., and xxv. pp. 659, 401, 412.	Bred by Mr. G. Allen, Unicarville; got by Great Hope (24082), dam (Ladylike) by Prince Victor (20606), &c.

Produce in	Names, &c.	By what Bull.	By whom bred.
1879, Oct. 15, roan, C.C.	Rose of the Forest	Kt. of St. Patrick, 38520	Msrs.Cruickshank

OLIVE BRACELET, red, calved June 19, 1877.	Bred by Messrs. J. W. and E. Cruickshank; got by Manfred (26801), dam (Oak Bracelet) by Lord of the Isles (26706), &c. See Vol. xxiv. p. 401.

Produce in	Names, &c.	By what Bull.	By whom bred.
1879, Aug. 18, roan, B.C.	Ben Vorlich	Kt. of St. Patrick, 38520	Msrs.Cruickshank

QUEEN OF THE CITY, roan, calved April 13, 1873, Vols. xxii., xxiv., and xxv. pp. 389, 401, 412.	Bred by Mr. W. Mitchell, Cleasby; got by Squire Booth (30049), dam (Queen of the Throne) by Major (26790), &c.

Produce in	Names, &c.	By what Bull.	By whom bred.
1879, June 17, white, C.C.	Queen of the East	Kt. of St. Patrick, 38520	Msrs.Cruickshank

RIBY PRINCESS, roan, calved October 15, 1875, Vol. xxv. p. 412.	Bred by Messrs. J. W. and E. Cruickshank; got by Knight of the Shire (26552), dam (Riby Empress) by Duke of York (23804), &c.

Produce in	Names, &c.	By what Bull.	By whom bred.
1879, Mar. 5, roan, B.C.	Riby Baron	Kt. of St. Patrick, 38520	Msrs.Cruickshank

ROSE OF EASTER, red, calved April 15, 1876, Vol. xxv. p. 412.	Bred by Messrs. J. W. and E. Cruickshank; got by King Richard 2nd (28984), dam (Rose of Erin) by Great Hope (24082), &c.

Produce in	Names, &c.	By what Bull.	By whom bred.
1879, Aug. 13, roan, C.C.	Rose Alpine	Kt. of St. Patrick, 38520	Msrs.Cruickshank

ROSE OF ERIN, red, calved June 10, 1873, Vols. xxiii. and xxv. pp. 409, 412.	Bred by Mr. G. Allen, Unicarville; got by Great Hope (24082), dam (Blooming Rose) by Prince Victor (20606), &c.

Produce in	Names, &c.	By what Bull.	By whom bred.
1879, Aug. 18, roan, C.C.	Rose of Nithsdale	Kt. of St. Patrick, 38520	Msrs.Cruickshank

VAIN FANCY, white, calved June 6, 1873, Vol. xxiv. p. 402.	Bred by Mr. J. B. Booth, Killerby Hall; got by King Richard 2nd (28984), dam (Vain Woman) by Merry Monarch (22349), &c.

Produce in	Names, &c.	By what Bull.	By whom bred.
1879, Jan. 31, white, C.C.	Vain Maiden	Balmoral, 36151	Msrs.Cruickshank

VAIN PEERESS, roan, calved October 27, 1877.	Bred by Messrs. J. W. and E. Cruickshank; got by Sir Windsor Broughton (27507), dam (Vain Fancy) by King Richard 2nd (28984), g. d. (Vain Woman) by Merry Monarch (22349), &c. See "Glenfalloch," p. 109.

Produce in	Names, &c.	By what Bull.	By whom bred.
1879, Dec. 20, white, C.C.	Vain Countess	Kt. of St. Patrick, 38520	Msrs.Cruickshank

VAIN PRINCESS, red, calved June 5, 1877. Bred by Messrs. J. W. and E.
Cruickshank; got by King Richard 2nd (28984), dam (Vain Woman) by
Merry Monarch (22349), &c. See "Glenfalloch," p. 109.

Produce in		Names, &c.		By what Bull.		By whom bred.
1879, July 24, roan,	C.C.	Vain Beauty		Kt. of St. Patrick, 38520		Msrs.Cruickshank

VAIN QUEEN, roan, calved May 25, 1876. Bred by Messrs. J. W. and E.
Cruickshank; got by King Richard 2nd (28984), dam (Vain Woman) by
Merry Monarch (22349), &c. See "Glenfalloch," p. 109.

1879, Aug. 9, roan,	B.C.	Glenlyon		Kt. of St. Patrick, 38520		Msrs.Cruickshank

VICTORIA AURICOMA, red and white, calved March 1, 1876, Vol. xxv. p. 413.
Bred by Messrs. J. W. and E. Cruickshank; got by Royal Benedict (27348),
dam (Victoria Aurifera) by Bythis (25700), &c.

1879, Aug. 26, roan,	B.C.	Royal Harbinger		Kt. of St. Patrick, 38520		Msrs.Cruickshank

WINDSOR BRACELET, red and white, calved May 13, 1877. Bred by Messrs.
J. W. and E. Cruickshank; got by Sir Windsor Broughton (27507), dam
(Hazle Bracelet) by Duke of Hazlecote 18th (30966), &c. See Vol. xxiv. p. 401.

1879, Nov. 10, roan,	C.C.	Crown Bracelet		Kt. of St. Patrick, 38520		Msrs.Cruickshank

CRUSE, Jabez,
Cleave House, Bulkworthy, Brandiscorner, North Devon.

CORA 7TH, red and white, calved January 27, 1875. Bred by Messrs. Horswell
and Sons, Burns Hall; got by Duke of Gaddesby (30956), dam (Cora 4th) by
First Earl Ducie (23814), &c. See "Pride of Devon," p. 189.

1878, Jan. 12, r. & w.,	B.C.	Pride of Devon	Oxford Duke 9th, 34979	Mr. Cruse
1879, Jan. 4, r. & w.,	B.C.	Cetewayo	Oxford Duke 10th, 38830	do.

LADY DANBY, roan, calved August 4, 1874. Bred by Mr. J. Outhwaite,
Bainesse; got by Royal Windsor (29890), dam (Miss Danby) by Emmanuel
(31105), &c. See "Milton Reformer," p. 170.

1877, Nov. 1, r. & w.,	B.C.	Milton Reformer	Iron Duke, 31420	Mr. James
1878, Sept. 15, roan,	B.C.	ParagonOxfordDuke	Oxford Duke 10th, 38830	Mr. Cruse

Milton Reformer, sold to Mr. R. Baker, Mansion House, Milton Damrell, Brandiscorner.

LADY FLOSSY, white, calved November 5. 1877. Bred by Mr. J. B. James,
Halls Annery; got by Iron Duke (31420), dam (Charmer 2nd) by Royal
Windsor (29890), g. d. (Charmer) by Cistercian (28202), &c. See "Pre-
tender," Vol. xxi. p. 359.

1879. Oct. 10. roan,	B.C.	Parag'n Oxf'd D. 2nd	Pride of Devon, 43774	Mr. Cruse

DALGETY, F. G.,
Lockerley Hall, Romsev, Hants.

STRAWBERRY 2ND, roan, calved January 6, 1875. Bred by Mr. F. G.
Dalgety; got by Æolus (27861), dam (Strawberry) by Shuttlecock 3rd (35521),
&c. See Vol. xxii. p. 391.

1877, Oct., roan,	B.C.	(dead)	Lord Ailesbury, 36908	Mr. Dalgety
1878, Oct. 20, roan,	B.C.	(Steer)	do.	do.
1879, Oct. 11, red,	C.C.	Strawberry 3rd	Parnassus, 38850	do.

DALTON, Messrs.,
Cummersdale, Carlisle.

BUTTERFLY PRINCESS 23RD, roan, calved August 10, 1872. Bred by Mr. J. Fawcett, Scaleby Castle; got by Eighth Duke of York (28480), dam (Butterfly Princess 16th) by Fourteenth Duke of Oxford (21605), &c. See "Butterfly Prince 2nd," Vol. xxv. p. 41.

Produce in	Names, &c.	By what Bull.	By whom bred.
1874, Dec. 29, r. & w., B.C.	York's But'fly Prince	Baron Deepdale, 30438	Messrs. Dalton.
1876, Mar. 7, white, B.C.	Oxford Prince	Oxford Isis, 32024	do.
1877, April 8, roan, C.C.	(dead)	Earl of York, 33824	do.
1878, April 18, red, B.C.	Butterfly Prince 2nd	Duke of Siddington, 38182	do.
1879, April 1, r. & w., C.C.	Butterfly Prn'ss 25th	do.	do.

York's Butterfly Prince, sold to Mr. Gill, Scuggar House, Carlisle; Oxford Prince, to Mr. Weatherston, Greysouthen, Cockermouth.

CHERRY RIPE 3RD, white, calved March 12, 1874. Bred by Messrs. Dalton; got by Oxford Isis (32024), dam (Cherry Ripe) by Royal Duke (25015), g. d. (Cheritta) by Fourteenth Duke of Oxford (21605), &c. See "Cherry Duke 7th," p. 47.

1878, Feb. 5, roan, C.C.	Cherry Ripe 5th	Duke of Siddington, 38182	Messrs. Dalton
1879, Dec. 20, roan, C.C.	Cherry Ripe 6th	do.	do.

JELLY FLOWER 4TH, roan, calved August 12, 1867, Vol. xxi. p. 663. Bred by Mr. J. P. Foster, Killhow; got by Thirteenth Duke of Oxford (21604), dam (Jelly Flower 3rd) by Specimen (17026), &c.

1875, Feb. 17, roan, C.C.	Jelly Flower 6th	Oxford Isis, 32024	Messrs. Dalton
1875, Dec. 30, white, C.C.	Jelly Flower 7th	do.	do.
1878, Jan. 4, roan, C.C.	Jelly Flower 8th	Duke of Siddington, 38182	do.
1878, Dec. 3, r. & w., C.C.	Jelly Flower 9th	do.	do.
1879, Oct. 20, r. & w., C.C.	Jelly Flower 10th	do.	do.

DARBY, A. E. W.,
Adcote, Little Ness, Shrewsbury.

CHERRY ROSE, red, calved March 24, 1877. Bred by Mr. R. Chaloner, King's Fort; got by Lieutenant General (31600), dam (Rosabelle) by King James (28971), g. d. (Lady Rose) by King Richard (26523), &c. See Vol. xxv. p. 426.

1879, July 12, r. & w., B.C.	(Steer)	Braithwaite Booth, 33192	Mr. Darby

CRIMSON FLOWER, red and little white, calved December 16, 1873, Vols. xxiii. and xxv. pp. 412, 416. Bred by Mr. R. S. Bruere, Braithwaite Hall; got by Booth's Royal Signet (28061), dam (Lilla Flower) by Booth's Kinsman (25658), &c.

1879, July 16, roan, B.C.	(Steer)	Magnate Star, 37029	Mr. Darby

GILLY FLOWER, red and white, calved March 13, 1876, Vol. xxv. p. 417. Bred by Mr. A. E. W. Darby; got by Braithwaite Booth (33192), dam (Crimson Flower) by Booth's Royal Signet (28061), &c.

1879, May 31, r. & w., C.C.	Jonquil Flower	Magnate Star, 37029	Mr. Darby

ROSABELLE, roan, calved May 7, 1874. Bred by Mr. R. Chaloner, King's Fort; got by King James (28971), dam (Lady Rose) by King Richard (26523), &c. See Vol. xxv. p. 426.

1879, May 21, roan, B.C.	Baron Kells	Royal Baron, 40617	Mr. Darby

ROSE CELESTIAL, red and white, calved May 15, 1874, Vols. xxiii. and xxiv. pp. 579, 589. Bred by Mr. J. P. Haslam, Heaton; got by Sidus (29969), dam (Rosa's Rose) by Algernon (27876), &c.

Produce in	Names, &c.	By what Bull.	By whom bred.
1879, Sept. 6, r. & w., C.C.	Rosa Rebecca	Magnate Star, 37029	Mr. Darby

VESPER QUEEN, roan, calved March 7, 1876. Bred by the Hon. M. H. Cochrane, Hillhurst, Compton, Canada; got by Royal Commander (29857), dam (Vernal Star) by The Sutler (23061), &c. See "Star King," p. 241.

1879, Aug. 29, white, B.C.	Star King	Magnate Star, 37029	Mr. Darby

DARLING, John,
Beau Desert, Rugeley.

DAPHNE 2ND, red and white, calved May 9, 1874, Vol. xxiv. p. 351. Bred by Mr. W. H. Brown, Stepple Hall; got by Duke of Rowley (28440), dam (Daphne) by Bull's Bay (28490), &c.

1878, Feb. 10, roan,	B.C.	Baron Stanley	Bn. Barrington 6th, 33008	Mr. Brown
1879, Dec. 29, roan,	C.C.	Daphne 3rd	do.	Mr. Darling

Baron Stanley, sold to Mr. W. Bentham, Crosthwaite.

EMMA 7TH, red, calved January 23, 1875, Vols. xxiv. and xxv. pp. 686, 697. Bred by Sir W. C. Trevelyan, Bart., Wallington; got by Oxford Beau 4th (34964), dam (Energy) by Lord Lally 3rd (24408), &c.

1879, Nov. 6, red,	C.C.	Emma 12th	Oxford's Prince, 34998	Mr. Darling

MISS CRADOCK, roan, calved in May 1876. Bred by Mr. H. Lovatt, Low Hill; got by Duke of Winterfold (33757), dam (Lady Cradock) by Kirbythore Waterloo (24263), &c. See Vol. xxiv. p. 620.

1879, July 6, red,	C.C.	Miss Cradock 2nd	Duke of Oxford 36th, 38170	Mr. Darling

STANTON BUTTERFLY, roan, calved January 31, 1876, Vol. xxv. p. 708. Bred by Mr. H. Wardle, Highfield; got by Royal Victor (35414), dam (Windsor Butterfly) by Rose Butterfly (24993), &c.

1879, Aug. 23, roan,	C.C.	Stanton Butterfly 2nd	Lord Ringlet, 38655	Mr. Darling

DARLING, Robert,
Plawsworth, Chester-le-Street, Durham.

ROSALIND, red and white, calved September 14, 1876. Bred by the Hon. M. H. Cochrane, Hillhurst, Compton, Canada; got by Sirius (39121), dam (Rosedale 3rd) by Royal Buckingham (20718), &c. See "Squire Rosedale," p. 240.

1879, Feb. 23, roan,	B.C.	Squire Rosedale	Titanus, 40821	Mr. Darling

DAUBUZ, J. C.,
Killiow, Truro, Cornwall.

MOSS ROSE, roan, calved June 4, 1872. Bred by Mr. W. Faulkner, Rothersthorpe; got by Young Hector (28830), dam (White Rose) by Knight of Branches (20076), &c. See "Procyon," p. 204.

1876, Mar. 26, roan,	B.C.	Orion	Savoy, 35476	Mr. Daubuz
1877, Dec. 15, r. & w.,	B.C.	Procyon	Vanguard, 35850	do.
1878, Nov. 22, red,	B.C.	Rigel	M.C., 31898	do.

DAVIDSON, A.,
Mains of Cairnbrogie, Old Meldrum, N.B.

DAINTY 9TH, roan, calved February 19, 1873. Bred by Mr. G. Marr, Cairn-brogie, the property of Mr. A. Davidson ; got by Young Hero (26385), dam (Dainty 4th) by Baron Sebastopol (21241), &c. See "Young Prince," Vol. xxii. p. 198.

Produce in	Names, &c.	By what Bull.	By whom bred.
1879, Mar. 5, { roan, B.C. / white,C.C.	Dauntless	} Bromley, 36289	Mr. Marr.

Dauntless, sold to Mr. A. Davidson, Mains of Cairnbrogie.

DAISY 10TH, roan, calved in 1872. Bred by Mr. G. Marr, Cairnbrogie, the property of Mr. A. Davidson ; got by King John (31494), dam (Daisy 4th) by Prince (16716), &c. See "Albert," Vol. xxi. p. 5.

Produce in	Names, &c.	By what Bull.	By whom bred.
1879, Mar. 11, red, B.C.	Drum Major	Bromley, 36289	Mr. Marr

Drum Major, sold to Mr. A. Davidson, Mains of Cairnbrogie.

FLORA 80TH, roan, calved May 5, 1877. Bred by Mr. G. Marr, Cairnbrogie, the property of Mr. A. Davidson ; got by Freemason (33973), dam (Flora 33rd) by Valiant (23108), &c. See "Czar," Vol. xxiv. p. 58.

Produce in	Names, &c.	By what Bull.	By whom bred.
1879, Feb. 8, roan, B.C.	Formartine Hero	Brabazon, 37881	Mr. Marr

Formartine Hero, sold to Mr. A. Davidson, Mains of Cairnbrogie.

MARIAN, red, calved January 4, 1876. Bred by Dr. Reid, Hillhead ; got by Standard Bearer (35665), dam (Maud Mary) by Forth 3rd (33958), g. d. (Mary) by Prince Arthur 2nd (22571), &c. See "Lord Lorne," Vol. xx. p. 206.

Produce in	Names, &c.	By what Bull.	By whom bred.
1879, Jan. 24, r. & w., B.C.	Mazzaroth	Mountain Rose, 38770	Mr. Davidson

MARY ANNE 34TH, red, calved April 2, 1877. Bred by Mr. G. Marr, Cairn-brogie, the property of Mr. A. Davidson ; got by Bromley (36289), dam (Mary Anne 24th) by Scotsman 3rd (32465), g. d. (Mary Anne 13th) by Grand Prince (26308), &c. See "Count Andrassy," Vol. xxiv. p. 53.

Produce in	Names, &c.	By what Bull.	By whom bred.
1879, Mar. 23, red, C.C.	Mary Anne 35th	Brabazon, 37881	Mr. Marr

Mary Anne 35th, sold to Mr. A. Davidson, Mains of Cairnbrogie.

M·ERRYMAID 4TH, roan, calved February 1, 1876. Bred by Mr. A. Davidson ; got by Baron Cecil (27921), dam (Merrymaid 3rd) by Baron 4th (27938), g. d. (Merrymaid 2nd) by Liberty (24332), gr. g. d. (Merrymaid) by Merryman (22348), — (Garnet 4th) by Goldfinder (14629), — (Garnet 2nd) by Corre-spondent (10076), — (Garnet) by Eildon (9076).

Produce in	Names, &c.	By what Bull.	By whom bred.
1878, Apr. 24, white, C.C.	Merrymaid Alba	Mountain Rose, 38770	Mr. Davidson
1879, July 22, { roan, C.C. / roan, C.C.	Merry Twin Prima / Merry Twin Secunda	} Titus, 40822	do.

ROSEMARY, roan, calved March 28, 1872, Vol. xxii. p. 393. Bred by Mr. R. Burdon, Castle Eden ; got by Emperor Maximilian (26100), dam (Moss Rose) by Young Freedom (21777), &c.

Produce in	Names, &c.	By what Bull.	By whom bred.
1878, Mar. 11, roan, C.C.	Rose of Corioli	Coriolanus, 33448	Mr. Davidson
1879, Feb. 14, roan, C.C.	Rose of Balmoral	Balmoral, 36151	do.

VESTA 2ND, roan, calved February 9, 1875. Bred by Mr. Henderson, Moss-field ; got by Corduroy (39625), dam (Vesta) by Golden Eagle (26267), g. d. (Victoria) by Sir Thomas Stanley (25176), gr. g. d. (Duchess) by Marquis of Bute (18336), &c. See "Duchess," Vol. xvi. p. 426.

Produce in	Names, &c.	By what Bull.	By whom bred.
1877, Dec. 10, roan, B.C.	Viscount	Cassius Booth, 33302	Mr. Davidson

DAVIDSON, James,
Bank House, Acklington.

BRIDAL DAY, roan, calved in September 1873.   Bred by Mr. W. Vickers, Howl John; got by Warlaby Manfred (30260), dam (Bridal Morn) by Sovereign (27537), g. d. (Belle of the Valley) by Knight of Richard Cœur-de-Lion (20080), &c.   See Vol. xix. p. 408.

Produce in		Names, &c.	By what Bull.	By whom bred.
1878, Sept. 26, roan,	C.C.	Bride	Roan Duke, 35283	Mr. Davidson

MOSS ROSE, red, calved February 26, 1870, Vols. xxiii., xxiv., and xxv. pp. 585, 593, 605.   Bred by the Duke of Northumberland, Alnwick Castle; got by Briton (25686), dam (Mulberry) by President (20510), &c.

1879, July 21, red,	B.C.	Knight of Guyzance	Knight of Murrah, 34391	Mr. Davidson

PRIMROSE 3RD, roan, calved September 28, 1876.   Bred by Mr. L. C. Chrisp, Hawkhill, the property of Mr. J. Davidson; got by Fitz-Roland (33936), dam (Rose 2nd) by Peak (24733), &c.   See "Mischief Maker," p. 172.

1879, Feb. 21, white,	B.C.	Mischief Maker	Muscovite, 38773	Mr. Chrisp

Mischief Maker, sold to Mr. J. Nicholson, Murton, Berwick-on-Tweed.

WHARFDALE ROSE 2ND, roan, calved November 27, 1870, Vols. xxi., xxiv., and xxv. pp. 970, 689, 699.   Bred by Sir W. C. Trevelyan, Bart., Wallington; got by Third Lord Wharfdale (26759), dam (Yetholme Rose) by Prince of Yetholme (22639), &c.

1879, Aug. 12, red,	C.C.	Warkworth Rose	Oxford Beau 4th, 34964	Mr. Davidson

DAY, G. J.,
Horsford Hall, Norwich.

DAFFY GWYNNE 8TH, roan, calved October 23, 1875.   Bred by Mr. H. Aylmer, West Dereham Abbey; got by High Sheriff (26392), dam (Daffy Gwynne 6th) by Ravenspur (20628), &c.   See Vol. xxii. p. 536.

1878, Sept. 1,	roan, C.C.	Bertha Gwynne	Sir Campbell Gwynne 2nd, 37455	Mr. Day
	roan, C.C.	Ruth Gwynne		

TRIFOLIUM 5TH, roan, calved November 14, 1875.   Bred by Mr. C. Barnett, Stratton Park; got by Red Clover (35216), dam (Haughty) by Paris (29523), &c.   See Vol. xxii. p. 313.

1878, Sept. 23, roan,	C.C.	Breda	Young Robert Peel, 37348	Mr. Day

DENT, A. C.,
Wharton Hall, Kirkby Stephen.

BRITISH QUEEN, red, calved in September 1871, Vols. xxi., xxiv., and xxv. pp. 668, 408, 421.   Bred by Mr. F. Heugh, Broomfield House; got by British Crown (21322), dam (Lady Booth) by The Sutler (23061). &c.

1879, Dec. 29, red,	B.C.	Lord Wharton	Prince Regent, 29076	Mr. Dent

MANTLE'S BEAUTY 2ND, white, calved April 21, 1877. Bred by Mr. A. C. Dent; got by Majestic (34725), dam (Mantle) by British Flag (25678), &c. See Vol. xxiv. p. 409.

Produce in	Names, &c.	By what Bull.	By whom bred.
1879, Oct. 23, roan, B.C.	Prince Victor	Prince of Roses, 35157	Mr. Dent

MISS FLORENCE, roan, calved January 23, 1876, Vol. xxv. p. 421. Bred by Mr. A. C. Dent; got by Sir James (35566), dam (Miss Nightingale) by Baron Mantalini (27951), &c.

1879, Sept. 5, red, B.C.	British Prince	British King, 37904	Mr. Dent

VIOLET 4TH, white, calved December 20, 1875, Vol. xxv. p. 421. Bred by Mr. A. C. Dent; got by Sir James (35566), dam (Violet 3rd) by Whiff (30299) or Baron Mantalini (27951), &c.

1879, Oct. 3, white, C.C.	British Maid	British King, 37904	Mr. Dent

DENT, William,
Kaber Fold, Brough, Westmoreland.

BIRTHDAY, roan, calved December 11, 1874, Vol. xxiv. p. 409. Bred by Mr. D. Nesham, Gainford Hall; got by Boaz (30552), dam (Sweetmeat) by King Richard (26523), &c.

1878, Oct. 21, r. & w., B.C.	King David	Earl of Sheffield, 33812	Mr. Dent
1879, June 14, B.C.	(premature, dead)	do.	do.

King David, sold to Mr. H. Lawson, Manor House, Water Fulford, York.

HANNAH, red and white, calved April 22, 1875, Vol. xxv. p. 421. Bred by Mr. W. Dent; got by Warlaby (32792), dam (Susan) by Third Duke of Athol (12734), &c.

1879, Mar. 10, roan, B.C.	Liberator	Earl of Sheffield, 33812	Mr. Dent

LADY BLUSH, roan, calved November 23, 1873, Vols. xxiii. and xxiv. pp. 417, 409. Bred by Mr. W. Mitchell, Cleasby; got by Squire Booth (30049), dam (Ladyship) by Major (26790), &c.

1878, Apr. 24, r. & w., B.C.	Baron de Cleasby	Earl of Sheffield, 33812	Mr. Dent
1879, Mar. 17, white, B.C.	Village Squire	do.	do.

Baron de Cleasby, sold to Mr. J. Hudson, Wakefield.

LADY CRISPIN, red and white, calved March 25, 1876. Bred by Mr. P. H. Rowlandson, Eden Bank; got by Warlaby (32792), dam (Lady Booth 3rd) by Squire Booth (30049), &c. See "Earl Crispin," p. 89.

1879, Mar. 28, roan, B.C.	Earl Crispin	Earl of Sheffield, 33812	Mr. Dent

LADY MANFRED, white, calved June 19, 1876. Bred by Mr. W. Dent; got by Manfred Booth (34749), dam (Lady Blush) by Squire Booth (30049), &c. See Vol. xxiii. p. 417.

1879, Apr. 15, white, C.C.	Lady Sheffield	Earl of Sheffield, 33812	Mr. Dent

SUSANNAH, red and white, calved April 13, 1872, Vol. xxii. p. 397. Bred by Mr. W. Dent; got by General Windsor (28701), dam (Susan) by Third Duke of Athol (12734), &c.

1877, May 31, r. & w., B.C.	Royal Warlaby	Warlaby, 30702	Mr. Dent
1878, May 6, r. & w., B.C.	Financier	Earl of Sheffield, 33812	do.
1879, Apr. 12, roan, C.C.	Susette	do.	do.

Royal Warlaby, sold to Mrs. Ewbanke, Borrenthwaite; Financier, to Mr. Thomas Simpson, Loading, Brough, Westmoreland.

WARLABY'S ROSE, roan, calved March 18, 1874. Bred by Mr. W. Dent; got by Warlaby (32792), dam (Susannah) by General Windsor (28701), g. d. (Susan) by Third Duke of Athol (12734), &c. See "British Earl," p. 35.

Produce in		Names, &c.	By what Bull.	By whom bred.
1877, Mar. 13, roan,	B.C.	General Manfred	Manfred Booth, 34749	Mr. Dent
1878, Mar. 31, roan,	B.C.	Earl De Kaber	Earl of Sheffield, 33812	do.
1879, Mar. 1, roan,	B.C.	Whisker	do.	do.

General Manfred, sold to Mr. J. G. Taylor, Hambridge, Taunton ; Earl De Kaber, to Mr. J. Bentley, Wythen, Coventry.

WELCOME, roan, calved April 1, 1874, Vols. xxiv. and xxv. pp. 409, 422. Bred by Mr. W. Dent; got by Warlaby (32792), dam (Susan) by Third Duke of Athol (12734), &c.

1879, Feb. 27, red,	B.C.	British Earl	Earl of Sheffield, 33812	Mr. Dent

British Earl, sold to Mr. W. Hutchinson, Rookby Scarth, Brough, Westmoreland.

De VITRE, Rev. G. E. D.,
Keep Hatch, Wokingham, Berks.

ROGUERY 10TH, roan, calved April 22, 1870, Vol. xxi. p. 927. Bred by Mr. A. F. M. Druce, Burghfield; got by Fra Diavolo (26190), dam (Roguery 9th) by Thornhill Spencer (23072), &c.

1878, Mar. 27, red,	C.C.	Roguery 13th	Janus, 34244	Rev. G. De Vitre
1879, Mar. 3, roan,	C.C.	Roguery 14th	Whimper, 44238	do.

ROGUERY 12TH, roan, calved March 22, 1876. Bred by Mr. T. Simonds, Arborfield; got by Janus (34244), dam (Roguery 10th) by Fra Diavolo (26190), &c.

1878, Aug. 30, roan,	B.C.		Caractacus, 36315	Rev. G. De Vitre
1879, July 16, white,	C.C.	Roguery 15th	Whimper, 44238	do.

1878 Bull calf, sold to Mr. J. Walter, Bearwood, Wokingham.

de VITRE, H. Denis,
Charlton House, Wantage.

BERKSHIRE BUTTERFLY, roan, calved June 5, 1876. Bred by Mr. H. Denis de Vitre; got by Grand Duke of Kent 2nd (28759), dam (Double Butterfly) by Royal Butterfly (16862), &c. See Vol. xix. p. 480.

1879, Feb. 5, red,	B.C.	Butterfly Boy	Blushing Duke, 37869	Mr. de Vitre

Butterfly Boy, sold to Mr. R. Aldworth, East Hagbourne, Berkshire.

KENTISH BUTTERFLY, white, calved August 5, 1875. Bred by Mr. H. Denis de Vitre; got by Grand Duke of Kent 2nd (28759), dam (Grand Duke's Butterfly) by Grand Duke 4th (19874), &c. See Vol. xx. p. 549.

1878, May 6, roan,	C.C.	(dead)	Third D. of Clarence, 23727	Mr. de Vitre
1879, May 25, roan,	C.C.	Blanche's Blush	Blushing Duke, 37869	do.

LADY JESSICA, roan, calved January 10, 1875. Bred by Mr. H. Denis de Vitre; got by Lord Thorndale (31756), dam (Not for Joe) by Hush (28883), &c. See "Blushing Joe," p. 30.

1878, Apr. 28, white,	C.C.	Lady Jessica 2nd	Grand D. of Kent 2nd, 28759	Mr. de Vitre
1879, Mar. 22, red,	C.C.	Lady Janet	Blushing Duke, 37869	do.

LANCASHIRE BEAUTY, roan, calved June 21, 1876. Bred by Mr. H. Denis de Vitre; got by Grand Duke of Kent 2nd (28759), dam (Lancashire Witch) by Royal Butterfly 17th·(22774), g. d. (Rose of Lancashire) by Master Butterfly 4th (14920), &c. See "Blushing Joe," p. 30.

Produce in	Names, &c.	By what Bull.	By whom bred.
1879, Feb. 19, roan, C.C.	Barmpton Belle	Blushing Duke, 37869	Mr. de Vitre

ROAN BLANCHE, roan, calved July 27, 1876. Bred by Mr. H. Denis de Vitre; got by Grand Duke of Kent 2nd (28759), dam (Grand Duke's Butterfly) by Grand Duke 4th (19874), &c. See Vol. xx. p. 549.

| 1879, Mar. 18, roan, C.C. | (dead) | Blushing Duke, 37869 | Mr. de Vitre |

DEVONSHIRE, Duke of,
Holker Hall, Carke in Cartmel, Carnforth.

BARONESS OXFORD, roan, calved April 30, 1871, Vols. xxi., xxii., xxiii., and xxiv. pp. 674, 398, 418, 411. Bred by the Duke of Devonshire; got by Second Duke of Claro (21576), dam (Lady Oxford 5th) by Third Duke of Thorndale (17749), &c.

| 1879, Jan. 7, r. & w., B.C. | Baron Oxford 9th | Fifth D. of Wetherby, 31033 | D. of Devonshire |

BARONESS OXFORD 4TH, red, calved October 26, 1874, Vols. xxiv. and xxv. pp. 411, 423. Bred by the Duke of Devonshire; got by Third Duke of Hillhurst (30975), dam (Baroness Oxford) by Second Duke of Claro (21576), &c.

| 1879, Sept. 12, red, C.C. | Baroness Oxford 9th | Duke of Glo'ster 7th, 39735 | D. of Devonshire |

BARONESS OXFORD 6TH, roan, calved December 10, 1876. Bred by the Duke of Devonshire; got by Fifth Duke of Wetherby (31033), dam (Baroness Oxford) by Second Duke of Claro (21576), &c. See "Baron Oxford 10th," p. 18.

| 1879, June 1, roan, B.C. | Baron Oxford 10th | D. of Barrington 4th, 39712 | D. of Devonshire |

CHERRY DUCHESS OF HILLHURST, roan, calved March 18, 1877. Bred by Mr. J. W. Larking, Cansiron Farm; got by Third Duke of Hillhurst (30975), dam (Cherry Queen) by Baron Oxford 5th (27958), &c. See "Cherry Duke of Holker," p. 48.

| 1879, Dec. 9, roan, B.C. | Cherry D. of Holker | Duke of Glo'ster 7th, 39735 | D. of Devonshire |

COUNTESS OF BARRINGTON 5TH, roan, calved February 25, 1869, Vols. xx., xxiii., xxiv., and xxv. pp. 461, 418, 411, 423. Bred by the Duke of Devonshire; got by Grand Duke 10th (21848), dam (Countess of Barrington 4th) by Lord Oxford (20214), &c.

| 1879, Apr. 6, roan, B.C. | D. of Barrington 10th | Fifth D. of Wetherby, 31033 | D. of Devonshire |

COUNTESS OF BARRINGTON 8TH, roan, calved June 9, 1875, Vols. xxiv. and xxv. pp. 411, 423. Bred by the Duke of Devonshire; got by Second Duke of Tregunter (26022), dam (Lady Laura Barrington) by Baron Oxford 4th (25580), &c.

| 1879, Oct. 17, r. & w., C.C. | C'ss of Bar'gton 10th | D. of Glo'ster 7th, 39735 | D. of Devonshire |

GRAND DUCHESS OF OXFORD 14TH, red and white, calved May 10, 1869, Vols. xx., xxiii., xxiv., and xxv. pp. 548, 419, 411, 423. Bred by the Duke of Devonshire; got by Grand Duke 10th (21848), dam (Grand Duchess of Oxford 7th) by Lord Oxford (20214), &c.

| 1879, Apr. 4, r. & w., B.C. | Duke of Oxford 51st | 5th D. of Wetherby, 31033 | D. of Devonshire |

Duke of Oxford 51st, sold to Lord Skelmersdale, Latham House.

GRAND DUCHESS OF OXFORD 26TH, roan, calved May 14, 1873, Vols.
xxiii. and xxiv. pp. 420, 412. Bred by the Duke of Devonshire; got by Baron
Oxford 4th (25580), dam (Grand Duchess of Oxford 11th) by Grand Duke 10th
(21848), &c.

Produce in	Names, &c.	By what Bull.	By whom bred.
1879, Feb. 6, roan,	C.C.|Gd.D'ss of Oxf'd 46th	D. of Glo'ster 7th, 39735	D. of Devonshire

GRAND DUCHESS OF OXFORD 27TH, roan, calved November 15, 1873,
Vol. xxiv. p. 412. Bred by the Duke of Devonshire; got by Baron Oxford
4th (25580), dam (Grand Duchess of Oxford 6th) by Imperial Oxford (18084),
&c.

Produce in	Names, &c.	By what Bull.	By whom bred.
1879, Sept. 5, roan,	B.C.|(dead)	D. of Glo'ster 7th, 39735	D. of Devonshire

GRAND DUCHESS OF OXFORD 32ND, red and white, calved September 8,
1875, Vol. xxv. p. 424. Bred by the Duke of Devonshire; got by Fifth Duke
of Wetherby (31033), dam (Grand Duchess of Oxford 22nd) by Baron Oxford
4th (25580), &c.

Produce in	Names, &c.	By what Bull.	By whom bred.
1879, Feb. 19, r. & w.,	B.C.|Duke of Oxford 50th	D. of Glo'ster 7th, 39735	D. of Devonshire

GRAND DUCHESS OF OXFORD 33RD, red, calved May 23, 1876, Vol. xxv.
p. 424. Bred by the Duke of Devonshire; got by Fifth Duke of Wetherby
(31033), dam (Grand Duchess of Oxford 21st) by Baron Oxford 4th (25580),
&c.

Produce in	Names, &c.	By what Bull.	By whom bred.
1879, Oct. 29, r. & w.,	C.C.|Gd.D'ss of Oxf'd 48th	D. of Glo'ster 7th, 39735	D. of Devonshire

GRAND DUCHESS OF OXFORD 34TH, red and white, calved July 24, 1876.
Bred by the Duke of Devonshire; got by Duke of Oxford 27th (33709), dam
(Grand Duchess of Oxford 26th) by Baron Oxford 4th (25580), &c. See
" Duke of Oxford 53rd," p. 82.

Produce in	Names, &c.	By what Bull.	By whom bred.
1879, Nov. 24, roan,	B.C.|Duke of Oxford 53rd	D. of Oxford 43rd, 39773	D. of Devonshire

GRAND DUCHESS OF OXFORD 35TH, red, calved August 9, 1876. Bred
by the Duke of Devonshire; got by Fifth Duke of Wetherby (31033), dam
(Grand Duchess of Oxford 23rd) by Baron Oxford 4th (25580), g. d. (Grand
Duchess of Oxford 14th) by Grand Duke 10th (21848), &c. See " Duke of
Oxford 51st," p. 82.

Produce in	Names, &c.	By what Bull.	By whom bred.
1879, Aug. 6, red,	C.C.|Gd.D'ss of Oxf'd 47th	D. of Glo'ster 7th, 39735	D. of Devonshire

GRAND DUCHESS OF OXFORD 39TH, roan, calved April 18, 1877. Bred
by the Duke of Devonshire; got by Baron Oxford 7th (36199), dam (Grand
Duchess of Oxford 27th) by Baron Oxford 4th (25580), &c. See " Duke of
Oxford 54th," p. 82.

Produce in	Names, &c.	By what Bull.	By whom bred.
1879, Nov. 29, roan,	B.C.|Duke of Oxford 54th	D. of Oxford 43rd, 39773	D. of Devonshire

LADY BRIGHT EYES 2ND, red, calved May 27, 1870, Vols. xxi. and xxv.
pp. 705, 446. Bred by Mr. E. H. Cheney, Gaddesby Hall; got by General
Napier (24023), dam (Bright Eyes 2nd) by Royal Butterfly 3rd (18754), &c.

Produce in	Names, &c.	By what Bull.	By whom bred.
1879, Aug. 1, red,	C.C.|Winsome 23rd	D. of Oxford 43rd, 39773	D. of Devonshire

LADY CHISHOLM, roan, calved March 20, 1876. Bred by Mr. J. P. Foster,
Killhow; got by Twenty-second Duke of Oxford (31000), dam (Lady Cressida)
by Seventeenth Duke of Oxford (25994), &c. See " Cato," p. 43.

Produce in	Names, &c.	By what Bull.	By whom bred.
1879, Dec. 9, r. & w.,	B.C.|Cato	We'rby Winsome 2d, 40905	D. of Devonshire

LADY CRESSIDA, roan, calved April 21, 1872, Vols. xxii., xxiii., and xxiv. pp. 418, 420, 412. Bred by Mr. J. P. Foster, Killhow; got by Seventeenth Duke of Oxford (25994), dam (Andromache) by Hogarth (13036), &c.

Produce in		Names, &c.	By what Bull.	By whom bred.
1879, May 30, red,	C.C.	Lady Cressida 4th	D. of Glo'ster 7th, 39735	D. of Devonshire

LALLY 15TH, roan, calved July 19, 1872, Vols. xxii., xxiii., xxiv., and xxv. pp. 398, 420, 413, 424. Bred by Mr. J. Harward, Winterfold; got by Eighth Duke of Geneva (28390), dam (Lally 8th) by Seventh Duke of York (17754), &c.

Produce in		Names, &c.	By what Bull.	By whom bred.
1879, Aug. 21, r. & w.,	B.C.	(Steer)	D. of Oxford 43rd, 39773	D. of Devonshire

MARCHIONESS OF WORCESTER, roan, calved April 18, 1873, Vols. xxiv. and xxv. pp. 528, 424. Bred by Mr. J. W. Larking, Cansiron Farm; got by Eighth Duke of Geneva (28390), dam (Lady Worcester 6th) by Third Duke of Claro (23729), &c.

Produce in		Names, &c.	By what Bull.	By whom bred.
1879. Sept. 29, red,	C.C.	(dead)	D. of Glo'ster 7th, 39735	D. of Devonshire

WINSOME 4TH, red, calved December 2, 1866, Vols. xix., xx., and xxi. pp. 789, 826, 676. Bred by the Duke of Devonshire; got by Grand Duke 10th (21848), dam (Winsome) by Oxford 2nd (18507), &c.

Produce in		Names, &c.	By what Bull.	By whom bred.
1879, March 15, red,	C.C.	Winsome 22nd	D. of Glo'ster 7th, 39735	D. of Devonshire

WINSOME BEAUTY 4TH, roan, calved April 23, 1877. Bred by Lord Skelmersdale, Lathom House; got by Baron Oxford 4th (25580), dam (Bright Eyes 5th) by Grand Duke 6th (19876), &c. See Vol. xxiv. p. 651.

Produce in		Names, &c.	By what Bull.	By whom bred.
1879, Nov. 15, roan,	C.C.	Winsome 25th	D. of Rosedale 6th, 38176	D. of Devonshire

WINSOMEDALE 2ND, roan, calved June 4, 1874, Vol. xxiv. p. 529. Bred by the Earl of Bective, Underley Hall; got by Second Duke of Tregunter (26022), dam (Winsome 9th) by Grand Duke 17th (24064). &c.

Produce in		Names, &c.	By what Bull.	By whom bred.
1879, Dec. 4, red,	B.C.	Baron Winsome 7th	We'rby Winsome 2d, 40905	D. of Devonshire

DICKINSON, J. A.,
Brough Sowerby, Brough, Penrith.

CHARITY, red and white, calved in 1872. Bred by Mr. W. Marshall, Patterdale Hall; got by Lord Ullin (29211), dam (Charity) by Emperor (23883), &c. See Vol. xxii. p. 399.

Produce in		Names, &c.	By what Bull.	By whom bred.
1875, May 30, r. & w.,	B.C.	Oxford Charity 2nd	Y'ng Oxfrd Gwynne, 29492	Mr. Dickinson
1876, Sept. 29, r. & w.,	B.C.	(slaughtered)	Oxford Hill, 34984	do.
1878, Apr. 7, r. & w.,	B.C.	British Charity	British Sovereign, 36285	do.
1879, Apr. 23, roan,	C.C.	Charity 3rd	Lord Clare, 41842	do.

Oxford Charity 2nd, sold to Messrs. Littlefair, Hartley Fold, Kirkby Stephen; British Charity, to Mr. J. Thompson, Intack Bottom, Kirkby Stephen.

CHARITY, red and white, calved May 6, 1874. Bred by Mr. J. A. Dickinson, the property of Mr. T. James, Kirkby Stephen; got by Warlaby (32792), dam (Charity) by Emperor (23883), &c. See Vol. xxii. p. 399.

Produce in		Names, &c.	By what Bull.	By whom bred.
1877, April 28, roan,	B.C.	(dead)	True Briton, 35823	Mr. Dickinson
1878, Mar. 31, r. & w.,	C.C.	Charity 2nd	British Sovereign, 36285	do.

CLARISSA, white, calved February 23, 1872, Vol. xxiv. p. 414. Bred by Mr. J. C. Toppin, Musgrave Hall; got by Count of the Realm (20040), dam (Clara) by Royal Guest (18768), &c.

Produce in		Names, &c.	By what Bull.	By whom bred.
1878, Oct. 10, white,	B.C.	Lord Gordon	British Sovereign, 36285	Mr. Dickinson
1879, Oct. 3, white,	C.C.	Lady Emily	British King, 37904	do.

FAMILIARITY, roan, calved February 12, 1875. Bred by Mr. J. C. Toppin, Musgrave Hall; got by King Tom (34356), dam (Familiar 10th) by Count of the Realm (23640), &c. See Vol. xxii. p. 618.

Produce in		Names, &c.		By what Bull.		By whom bred.
1879, June 25, white,	C.C.	Familiar		Marmaduke, 40297		Mr. Dickinson

VIOLET, red, calved August 9, 1875. Bred by Mr. A. Raine, Bow Bank; got by Whiff (30299), dam (Lady of the Valley) by Lord of the Valley (26722), &c. See "Sir Walter," p. 236.

1878, Oct. 16, roan,	B.C.	Sir Walter	Earl of Sheffield, 33812	Mr. Dickinson
1879, Nov. 29, roan,	B.C.	Sir Evelyn	Lord Clare, 41842	do.

DICKSON, B.,
Gilford House, Gilford, Ireland.

ALINE, red and white, calved February 24, 1875, Vol. xxv. p. 425. Bred by Mr. J. Downing, Ashfield; got by Vain Hope (23102), dam (Alma 3rd) by Clear the Way (21434), &c.

1879, Feb. 20, roan,	B.C.	(dead)	Robert Stephenson, 32313	Mr. Dickson

BELLE OF LOTHIAN, roan, calved January 7, 1877. Bred by Mr. B. Dickson, the property of Mr. H. Dickson, Elimfield; got by Royal Oxford (35395), dam (Countess of Lothian) by Czarowitz (30837), &c. See Vol. xxiv. p. 415.

1879, April 2, roan,	C.C.	Lady Lothian	M. of the Ceremonies, 40328	Mr. Dickson

COUNTESS OF LOTHIAN, roan, calved March 20, 1873, Vols. xxiv. and xxv. pp. 415, 425. Bred by Mr. S. Purves, Lissanerin; got by Czarowitz (30837), dam (Maid of Lothian) by Crown of Lothian (19533), &c.

1879, Feb. 10, r. & w.,	B.C.	Mid Lothian	M. of the Ceremonies, 40328	Mr. Dickson

DAISY, red and white, calved January 3, 1872, Vols. xxii., xxiv., and xxv. pp. 399, 415, 425. Bred by the Rev. W. Moutray, Favour Royal; got by The Governor (32681), dam (Delight) by Knight of the Grand Cross (31558), &c.

1879, Feb. 20, r. & w.,	C.C.	(dead)	M. of the Ceremonies, 40328	Mr. Dickson

DAISY WREATH, roan, calved February 21, 1869, Vols. xxiv. and xxv. pp. 415, 426. Bred by Mr. R. Chaloner, King's Fort; got by Sovereign (27538), dam (Daisy Chain 2nd) by Tally Ho (20927), &c.

1879, March 12, roan,	C.C.	(dead)	M. of the Ceremonies, 40328	Mr. Dickson

HEARTSEASE, roan, calved February 13, 1877. Bred by Mr. B. Dickson; got by Royal Oxford (35395), dam (Pansy) by Napoleon 3rd (29417), &c. See Vol. xxiv. p. 415.

1879, April 4, roan,	B.C.	Benedict	M. of the Ceremonies, 40328	Mr. Dickson

JANE 3RD, roan, calved June 1, 1871, Vol. xxiv. p. 415. Bred by Mr. W. D. Dunlop, Monasterboice House; got by Gallant Knight (26214), dam (Jane 2nd) by Mountebank (24626), &c.

1879, Feb. 5, roan,	B.C.	Aid-de-Camp	M. of the Ceremonies, 40328	Mr. Dickson

MAID OF LOTHIAN 2ND, white, calved June 25, 1875. Bred by Mr. B. Dickson; got by Royal Duke (37384), dam (Countess of Lothian) by Czarowitz (30837), &c. See Vol. xxiv. p. 415.

1879, April 10, roan,	C.C.	Maid of Lothian 3rd	M. of the Ceremonies, 40328	Mr. Dickson

MISS GWYNNE 3RD, red, calved June 26, 1873, Vols. xxii., xxiv., and xxv.
pp. 399, 415, 426. Bred by Mr. R. F. Dunlop, Monasterboice House; got by
Guinea Cock (31326), dam (Miss Gwynne) by Mountebank (24626), &c.

Produce in	Names, &c.	By what Bull.	By whom bred.
1879, June 10, r. & w., B.C.	Master Gwynne	M. of the Ceremonies, 40328	Mr. Dickson

PRINCESS, roan, calved December 27, 1870, Vol. xxv. p. 426. Bred by the
Rev. W. Moutray, Favour Royal; got by The Governor (32681), dam (Norma)
by Knight of the Grand Cross (31558), &c.

1879, Feb. 13, roan, B.C.	Royal Master	M. of the Ceremonies, 40328	Mr. Dickson

ROYAL NANCY, roan, calved August 25, 1874, Vols. xxiv. and xxv. pp. 369,
426. Bred by Mr. R. Chaloner, King's Fort; got by King James (28971),
dam (Roan Nancy) by Frederick Fitz Booth (26195), &c.

1879, June 28, roan, B.C. (dead)		M. of the Ceremonies, 40328	Mr. Dickson

STRAWBERRY, roan, calved June 8, 1873, Vol. xxiv. p. 415. Bred by Miss
Sanderson, Cloverhill; got by Bright Duke (30592), dam (Princess Helena)
by Valasco (15443), &c.

1879, April 13, red, C.C.	Strawberry 4th	M. of the Ceremonies, 40328	Mr. Dickson

DODD Francis,
Rush Court, Wallingford, Berks.

ADA WHARFDALE, white, calved November 15, 1872, Vol. xxiv. p. 417.
Bred by Mr. F. Dodd; got by Baron Colby (27924), dam (Agnes Wharfdale)
by Duke of Wharfdale (19648), &c.

1879, Jan. 26, roan, C.C.	Lina Wharfdale	L. Darlington 14th, 40149	Mr. Dodd

ALICE WHARFDALE, red, calved February 13, 1875. Bred by Mr. F.
Dodd; got by Grand Prince of Claro (28781), dam (Annie Wharfdale) by
His Lordship (26398), &c. See "Thames Wharfdale," p. 246.

1879, Apr. 16, r. & w., C.C.	Maggie Wharfdale	L. Darlington 14th, 40149	Mr. Dodd

FLORENTINA, red and white, calved April 10, 1875. Bred by Mr. F. Dodd;
got by Grand Prince of Claro (28781), dam (Flower Girl) by Duke of Albany
(25931), &c. See "Flavius 2nd," p. 101.

1878, Jan. 31, roan, B.C.	Flavius	Duke of Argyle, 36462	Mr. Dodd
1879, Apr. 13, r. & w., C.C.	Florentina 4th	L. Darlington 14th, 40149	do.

FLOWER GIRL, roan, calved January 22, 1872, Vols. xxii. and xxiii. pp. 400,
421. Bred by Captain Blathwayt, Dyrham Park; got by Duke of Albany
(25931), dam (Florentia 9th) by Seventh Duke of York (17754), &c.

1879, Jan. 12, roan, C.C.	Florentina 3rd	Gd. D. of Geneva 2d, 31288	Mr. Dodd

LADY LUCY 7TH, red and white, calved August 8, 1871, Vols. xxi. and xxv.
pp. 677, 428. Bred by Mr. F. Dodd; got by Baron Colby (27924), dam (Lucy
Neal) by Royal Arch (18749), &c.

1879, July 18, roan, C.C.	Lady Lucy 12th	L. Darlington 14th, 40149	Mr. Dodd

LADY LUCY 8TH, roan, calved March 22, 1875. Bred by Mr. F. Dodd; got
by Grand Prince of Claro (28781), dam (Lady Lucy 6th) by Lord Red Eyes
(24459), &c. See Vol. xxii. p. 400.

1879, Jan. 3, roan, C.C.	Lady Lucy 11th	L. Darlington 14th, 40149	Mr. Dodd
1879, Dec. 25, roan, B.C.	Lucas 2nd	do.	do.

LADY LUCY 9TH, roan, calved March 30, 1876. Bred by Mr. F. Dodd; got
by Grand Prince of Claro (28781), dam (Lady Lucy 6th) by Lord Red Eyes
(24459), &c. See Vol. xxiii. p. 421.

Produce in	Names, &c.	By what Bull.	By whom bred.
1879, Dec. 30, r. & w., C.C.	Lady Lucy 13th	L. Darlington 14th, 40140	Mr. Dodd

DORMER, C. U. Cottrell,
Rousham, Oxford.

ADA, red, calved December 27, 1875. Bred by Mrs. Strickland, Cokethorpe
Park; got by North Star (31988), dam (Ariel) by Earl of Verulam (26077),
g. d. (Amy) by Oxarles (22471), gr. g. d. (Abbess) by Freeholder (19792), —
(Adeline) by Duke of Chester (17710), — (Nancy) by Prosperson (18654),
&c. See Vol. xiv. p. 617.

1879, Dec. 27, r. & w., B.C.	Abraham	Earl of Leicester 3d, 33804	Mr. Dormer

COUNTESS 2ND, red and white, calved October 30, 1871. Bred by Mr. W.
Strickland, Cokethorpe Park; got by Earl of Verulam (26077), dam (Patience)
by Waterloo Duke (21077), &c. See " Protector 3rd," Vol. xxiv. p. 215.

1879, Jan. 26, roan, C.C.	Patience	Baron Winsome 3d, 33108	Mr. Dormer

COUNTESS 4TH, roan, calved November 16, 1873. Bred by Mrs. Strickland,
Cokethorpe Park; got by North Star (31988), dam (Patience) by Waterloo
Duke (21077), &c. See Vol. xxi. p. 946.

1879. Aug. 25, white, C.C.	Countess 5th	Earl of Leicester 3d, 33804	Mr. Dormer

FAVOURITE GWYNNE, red and white, calved September 2, 1873, Vols. xxii.
and xxv. pp. 400, 428. Bred by Mr. C. U. C. Dormer; got by Second Cherry
Duke (28170), dam (Faustina Gwynne) by May Duke (13320), &c.

1879, April 29, roan, C.C.	Favourita Gwynne	Earl of Leicester 3d, 33804	Mr. Dormer

FLEDA GWYNNE, red and white, calved December 31, 1875. Bred by Mr.
C. U. C. Dormer; got by Earl of Leicester 3rd (33804), dam (Favourite
Gwynne) by Second Cherry Duke (28170), &c. See Vol. xxii. p. 400.

1879, Mar. 2, r. & w., B.C.	Frederick Gwynne	Peach Stone, 42127	Mr. Dormer

FLORENCE GWYNNE, roan, calved October 24, 1875. Bred by Mr. C. U. C.
Dormer; got by Second Cherry Duke (28170), dam (Faustina Gwynne) by
May Duke (13320), &c. See Vol. xxii. p. 400.

1879, May 17, roan, B.C.	Francis Gwynne	Earl of Leicester 3d, 33804	Mr. Dormer

GEORGIANA GWYNNE, red and white, calved July 22, 1875. Bred by Mr.
C. U. C. Dormer; got by Second Cherry Duke (28170), dam (Goody Gwynne)
by Grand Duke 5th (19875), &c. See Vol. xxii. p. 400.

1879, Mar. 18, roan, B.C.	Gregory Gwynne	Peach Stone, 42127	Mr. Dormer

LADY CHERRY SPENCER, red and white, calved October 10, 1874, Vol.
xxv. p. 428. Bred by Mr. C. U. C. Dormer; got by Second Cherry Duke
(28170), dam (Lady Oxford Spencer) by Twelfth Duke of Oxford (19633), &c.

1879, Dec. 28, r. & w., C.C.	L'y Charl'tte Spencer	Earl of Leicester 3d, 33804	Mr. Dormer

MERMAID, roan, calved February 4, 1870, Vols. xxi. and xxii. pp. 679, 401.
Bred by Mr. C. U. C. Dormer; got by Crown Prince (21510), dam (Motley)
by May Duke (13320), &c.

1876, July 6, white, B.C.	Merman	Earl of Leicester 3d, 33804	Mr. Dormer
1878, Feb. 25, roan, C.C.	Mignonette	do.	do.
1879, Mar. 28, roan, C.C.	Magnolia	do.	do.

MILDRED, roan, calved June 24, 1875. Bred by Mr. C. U. C. Dormer; got by Second Cherry Duke (28170), dam (Mermaid) by Crown Prince (21510), &c. See Vol. xxii. p. 401.

Produce in	Names, &c.	By what Bull.	By whom bred.
1879, Feb. 21, r. & w., B.C.	Moonshine	Earl of Leicester 3d, 33804	Mr. Dormer

MILKMAID 3RD, red and white, calved June 19, 1873. Bred by Mr. C. U. C. Dormer; got by Second Cherry Duke (28170), dam (Milkmaid 2nd) by Juvenile (22021), &c. See Vol. xxi. p. 679.

1876, Mar. 17, roan,	B.C.	Mystery	Earl of Leicester 3d, 33804	Mr. Dormer
1878, May 7, roan,	C.C.	Milkmaid 5th	do.	do.
1879, Apr. 23, r. & w.,	C.C.	Milkmaid 6th	do.	do.

MILKMAID 4TH, red and white, calved June 7, 1874, Vol. xxv. p. 429. Bred by Mr. C. U. C. Dormer; got by Second Cherry Duke (28170), dam (Milkmaid 2nd) by Juvenile (22021), &c.

| 1879, Jan. 29, roan, | C.C. | Milkmaid 7th | Earl of Leicester 3d, 33804 | Mr. Dormer |

MOSS ROSE 2ND, white, calved November 22, 1874. Bred by Mrs. Strickland, Cokethorpe Park; got by North Star (31988), dam (Moss Rose) by Earl of Verulam (26077), g. d. (Rosalie 2nd) by Mentor (22341), &c. See Vol. xviii. p. 696.

| 1877, Dec. 31, roan, | C.C. | Moss Rose 3rd | Proctor, 38956 | Mrs. Strickland |

Moss Rose 3rd, sold to Mr. C. U. C. Dormer.

NIOBE 13TH, red, calved March 15, 1873. Bred by Mrs. Strickland, Cokethorpe Park; got by North Star (31988), dam (Niobe 7th) by Napier (26944), g. d. (Niobe 2nd) by Freeholder (19792), &c. See "Napier," Vol. xxi. p. 330.

| 1879, July 12, r. & w., | C.C. | Nebula | Earl of Leicester 3d, 33804 | Mr. Dormer |

SWEETHEART 33RD, roan, calved July 12, 1874, Vol. xxv. p. 429. Bred by Mr. C. U. C. Dormer; got by Second Cherry Duke (28170), dam (Sweetheart 32nd) by Patrician (24728), &c.

| 1879, Jan. 1, roan, | C.C. | Sweetheart 36th | Earl of Leicester 3d, 33804 | Mr. Dormer |

DOWNING, John,
Ashfield, Fermoy, Ireland.

BRACELET, roan, calved March 13, 1869, Vols. xxi. and xxiv. pp. 679, 418. Bred by Mr. T. Barnes, Westland; got by Royal Duke (25014), dam (Grand Duchess) by Grand Duke 3rd (16182), &c.

| 1878, Mar. 29, roan, | C.C. | Mantle | Robert Stephenson, 32313 | Mr. Downing |
| 1879, Mar. 2, white, | C.C. | Maiden | do. | do. |

Mantle, sold to Mr. J. Watt, Claragh, Ramelton, Co. Donegal.

BREASTPLATE, roan, calved August 28, 1869, Vols. xxi., xxii., xxiii., and xxiv. pp. 679, 401, 422, 418. Bred by Sir R. J. Paul, Bart., Ballyglan; got by Knight of the Empire (22061), dam (Bracelet 2nd) by Vanguard (21009). &c.

| 1878, Apr. 12, { | roan, C.C. | Brilliant Jewel 2nd } | Robert Stephenson, 32313 | Mr. Downing |
| | roan, B.C. | Brave Chieftain } | | |

Brave Chieftain, sold to Mr. A. Moutray, Favour Royal.

BRILLIANT BUTTERFLY, red and white, calved December 3, 1873, Vol. xxiii. p. 422. Bred by Mr. J. How, Broughton; got by Great Hope (24082), dam (Broughton Butterfly) by Victorious (25378), &c.

Produce in	Names, &c.	By what Bull.	By whom bred.
1878, Sept. 9, r. & w., B.C.	Marcellus	Robert Stephenson, 32313	Mr. Downing
1879, Aug. 21, r. & w., B.C.	Marcus	do.	do.

Marcellus, sold to the Earl of Listowel, Convamore, Mallow.

BRILLIANT NECKLACE, red and white, calved March 28, 1876. Bred by Mr. J. Downing; got by Vain Hope (23102), dam (Birthright) by Duke of Hazlecote 18th (30966), &c. See Vol. xxiii. p. 422.

1870, Mar. 1, r. & w., B.C.	British Warrior	Robert Stephenson, 32313	Mr. Downing

British Warrior, sold to Mr. D. F. Slattery, Coolnagour, Dungarvan.

BROOCH, roan, calved July 1, 1870, Vols. xxi., xxiii., and xxiv. pp. 679, 422, 418. Bred by Sir R. J. Paul, Bart., Ballyglan; got by Knight of the Empire (22061), dam (Bracelet 2nd) by Vanguard (21009), &c.

1878, Mar. 29, roan, B.C.	Baron Stephenson	Robert Stephenson, 32313	Mr. Downing
1879, Feb. 23, r. & w., B.C.	Brigadi'r Stephenson	do.	do.

Baron Stephenson, sold to the Earl of Erne, Crom Castle, Newtownbutler.

COUNTESS, roan, calved May 23, 1869, Vols. xxi., xxii., and xxiii. pp. 680, 401, 422. Bred by Mr. J. G. Grove, Castle Grove; got by The Sutler (23061), dam (Coquette) by Comet (11298), &c.

1878, April 11. roan, B.C.	Count Stephenson	Robert Stephenson, 32313	Mr. Downing
1879, May 6, roan, B.C.	Commandant	do.	do.

Count Stephenson, sold to the Duke of Devonshire, Lismore Castle, Lismore.

FAREWELL 3RD, red and white, calved April 8, 1873, Vols. xxiii. and xxiv. pp. 422, 418. Bred by Mr. J. Downing; got by Lord Stanley (24466), dam (Norman Lady) by Sir Roger (16991), &c.

1878, Apr. 28, r. & w., B.C.	Norman Baron	Robert Stephenson, 32313	Mr. Downing
1879, June 20. r. & w., C.C.	Farewell 8th	do.	do.

Norman Baron, sold to Messrs. T. and J. McElderry, Cameron Place, Ballymoney, Co. Antrim.

GOLDEN PIN, roan, calved August 14, 1868, Vols. xx., xxi., and xxiii. pp. 543, 850, 423. Bred by Mr. W. Bolton, The Island, the property of Mr. M. C. Cramer, Rathmore; got by Duke of Marlborough (23768), dam (Golden Rose) by Blood Royal (14169), &c.

1878, Feb. 5, { r. & w., C.C.	Guiding Star	Robert Stephenson, 32313	Mr. Downing .
{ r. & w., B.C.	Gauntlet		

Gauntlet, sold to The Commissioners of National Education, Ireland.

VICTRIX, roan, calved June 11, 1871, Vols. xxi. and xxiii. pp. 680, 423. Bred by Mr. J. Downing; got by Sir Egbert (27468), dam (Vestal Queen) by Hero of Thorndale (18061), &c.

1877, April 4, red, B.C.	Viscount	Vain Hope, 23102	Mr. Downing
1879, April 1, roan, C.C.	Vivacity	Robert Stephenson, 32313	do.

Viscount, sold to Sir Charles Coote, Bart., Ballyfinn, Mountrath.

DRUMMOND, John,
Blackruthven, Perth, N.B.

BEAUTY, red and white, calved February 6, 1876, Vol. xxv. p, 431. Bred by Mr. J. Drummond; got by Baron of the Isles (36196), dam (Christina) by Mozart (26933), &c.

1879, Aug. 7, red, C.C.	Rose of Sharon	Lord of Alabama, 38630	Mr. Drummond

CATHLEEN 2ND, red and white, calved April 6, 1877. Bred by Mr. J.
Drummond; got by Baron of the Isles (36196), dam (Cathleen) by Victor
Royal (21028), &c. See Vol. xxiv. p. 421.

Produce in	Names, &c.	By what Bull.	By whom bred.
1879, July 15, r. & w.,	B.C. Star Baron	Star Crest, 39164	Mr. Drummond

CHRISTINA, roan, calved February 18, 1872, Vols. xxii., xxiii., and xxiv.
pp. 402, 426, 421. Bred by Mr. J. Drummond; got by Mozart (26933), dam
(Cathleen) by Victor Royal (21028), &c.

Produce in	Names, &c.	By what Bull.	By whom bred.
1879, Mar. 7, r. & w.,	B.C. Vesper Knight	Star Crest, 39164	Mr. Drummond

COLD CREAM 3RD, roan, calved in May 1869, Vol. xx. p. 453. Bred by Mr.
D. Fisher, Pitlochrie; got by England's Glory (23889), dam (Cold Cream
2nd) by Prince of Saxe Coburg (20576), &c.

Produce in	Names, &c.	By what Bull.	By whom bred.
1879, May 29, red,	B.C. Knight of the Shire	Lord of Alabama, 38630	Mr. Drummond

DANDY, red and white, calved March 6, 1875, Vols. xxiii. and xxv. pp. 426,
431. Bred by Mr. J. Drummond; got by Prince Frederick Charles (32135),
dam (Norma) by Knight of the Rock (24279), &c.

Produce in	Names, &c.	By what Bull.	By whom bred.
1879, Mar. 1, roan,	B.C. Star Crest 3rd	Star Crest, 39164	Mr. Drummond

DUCHESS OF CONNAUGHT, white, calved January 29, 1876, Vol. xxv.
p. 431. Bred by Mr. J. Gordon, Cluny Castle; got by Duke of Connaught
(33604), dam (Lady Mary Burdett) by Thorndale Duke (27661), &c.

Produce in	Names, &c.	By what Bull.	By whom bred.
1879, June 15, roan,	C.C. D's of Connaught 2d	Lord of Alabama, 38630	Mr. Drummond

ELVIRA'S ROSE, red, calved April 12, 1875, Vol. xxv. p. 431. Bred by Mr.
J. Lynn, Stroxton; got by Cambridge Duke 4th (25706), dam (Elvira 8th) by
Grand Duke 17th (24064), &c.

Produce in	Names, &c.	By what Bull.	By whom bred.
1879, April 11, red,	B.C. Duke of Elvira	Lord of Alabama, 38630	Mr. Drummond

LILY, white, calved June 6, 1877. Bred by Colonel Williamson, Lawers
House; got by King Brian (34308), dam (Louisa's Gipsy) by Prince of the
Gipsies (27179), g. d. (Louisa Hope) by Great Hope (24082), &c. See "King
Hope," Vol. xxii. p. 134.

Produce in	Names, &c.	By what Bull.	By whom bred.
1879, June 30, roan,	C.C. Lily 2nd	Lord of Alabama, 38630	Mr. Drummond

LITTLE GREEK DOT, red and white, calved February 26, 1873, Vol. xxv.
p. 431. Bred by Mr. G. H. M. Binning Home, Argaty; got by Duke of
Athens (28360), dam (Little Dot) by Royal Butterfly 3rd (18754), &c.

Produce in	Names, &c.	By what Bull.	By whom bred.
1879, May 21, red,	B.C. Lord of the Manor	Lord of Alabama, 38630	Mr. Drummond

ORANGE JELLY 2ND, red and white, calved January 10, 1876, Vol. xxv.
p. 431. Bred by Mr. J. Drummond; got by Baron of the Isles (36196), dam
(Orange Jelly) by Lord Lansdowne (29128), &c.

Produce in	Names, &c.	By what Bull.	By whom bred.
1879, July 21, red,	B.C. Vesper Star 2nd	Star Crest, 39164	Mr. Drummond

STRAWBERRY 12TH, red, calved February 19, 1877. Bred by Mr. A.
Bethune, Blebo; got by Rovin Boy (35329), dam (Strawberry 3rd) by Master
Blithe (29314), &c. See Vol. xxiii. p. 333.

Produce in	Names, &c.	By what Bull.	By whom bred.
1879, July 10, red,	B.C. Barrister	Lord of Alabama, 38630	Mr. Drummond

SUNSHINE, red and white, calved March 30, 1870, Vols. xix., xxi., xxii., xxiii.,
xxiv., and xxv. pp. 711, 683, 403, 427, 421, 432. Bred by the Earl of Dunmore, Dunmore; got by Second Duke of Collingham (23730), dam (Rose of
Oxford 2nd) by Fourth Duke of Oxford (11387), &c.

Produce in	Names, &c.	By what Bull.	By whom bred.
1879, July 21, red,	B.C. Duke of Perth	Lord of Alabama, 38630	Mr. Drummond

SUNSHINE 2ND, red and white, calved April 20, 1877. Bred by Mr. J.
Drummond; got by Baron of the Isles (36196), dam (Sunshine) by Second
Duke of Collingham (23730), &c. See Vol. xxiv. p. 421.

Produce in	Names, &c.	By what Bull.	By whom bred.
1879, April 4, red,	B.C. Lord of Alabama 2nd	Lord of Alabama, 38630	Mr. Drummond

SUSAN, red and white, calved March 30, 1876, Vol. xxv. p. 432. Bred by Mr.
J. Drummond; got by Baron of the Isles (36196), dam (Norah) by Criffel
(23653), &c.

1879, July 2, red,	B.C. Lord of the Isles	Lord of Alabama, 38630	Mr. Drummond

DUNCOMBE, Hon. Cecil,
Nawton Grange, York.

DUCHESS OF WATERLOO, roan, calved September 2, 1870, Vol. xxi. p. 917.
Bred by Mr. F. N. Sartoris, Rushden Hall; got by Duke of Kingscote (25981),
dam (Wellingtonia 8th) by Baron Oxford (23375), &c.

1879, July 28, white,	C.C. D'ss of Waterloo 2nd	D. of Tregunter 5th, 33743	Hon. C. Duncombe

DUNNING, J. H.,
Court Barton, Creech St. Michael, Taunton.

MELINDA, red and white, calved November 11, 1874. Bred by Mr. J. H.
Dunning; got by Baron (42694), dam (Ma Belle) by Henchman (21918), &c.
See Vol. xix. p. 611.

1878, Mar. 1, r. & w.,	B.C. Neptune	Birthright, 33152	Mr. Dunning

EDMONDS, W. J.,
Southrop House, Lechlade.

BIANCA OXFORD, roan, calved October 29, 1872, Vols. xxii. and xxv. pp. 312,
434. Bred by Mr. C. Barnett, Stratton Park; got by Brockley Oxford
(25688), dam (Balm) by Britannicus (17452), &c.

1879, May 8, roan,	B.C. Earl of Southrop 5th	Oxford Swell, 38838	Mr. Edmonds

BROWN 8TH, roan, calved December 24, 1873, Vol. xxv. p. 434. Bred by Mr.
C. Hobbs, Maisey Hampton; got by Bates Tertius (21249), dam (Brown) by
Sidon (20803), &c.

1879, Oct. 3, roan,	B.C. Southrop Lad	Oxford Swell, 38838	Mr. Edmonds

CLOUDY MORN 2ND, red and white, calved February 28, 1872, Vol. xxv.
p. 434. Bred by Mr. C. Hobbs, Maisey Hampton; got by Lord Hastings
(29115), dam (Cloudy) by Captain Wetherby (21371), &c.

1879, Mar. 12, roan,	B.C. Earl of Southrop 2nd	Oxford Swell, 38838	Mr. Edmonds

CLOUDY MORN 3RD, roan, calved March 4, 1877. Bred by Mr. W. J.
Edmonds; got by Colonel Tregunter 7th (30769), dam (Cloudy Morn 2nd)
by Lord Hastings (29115), &c.

1879, June 11, roan,	B.C. Earl of Southrop 7th	Oxford Swell, 38838	Mr. Edmonds

DIDO 7TH, red and white, calved March 8, 1874, Vol. xxv. p. 434. Bred by
Mr. D. R. Scratton, Ogwell; got by Cherry Duke 6th (30705), dam (Domina
2nd) by Red Richard (29749), &c.

1879, Feb. 6, roan,	B.C. Earl of Southrop	Oxford Swell, 38838	Mr. Edmonds

DULCE 2ND, red, calved March 31, 1873, Vol. xxv. p. 435. Bred by Mr. W.
J. Edmonds; got by Second Earl of Walton (19672), dam (Dulce) by The
Baronet (17088), &c.

Produce in		Names, &c.	By what Bull.	By whom bred.
1879, May 21, roan,	C.C.	Dulce 4th	Oxford Swell, 38838	Mr. Edmonds

FAIRY 3RD, roan, calved September 4, 1868, Vols. xxi., xxii., and xxv. pp. 689,
406, 435. Bred by Mr. W. J. Edmonds; got by Fifth Baron Wetherby
(23382), dam (Fairy) by Fourth Duke of Oxford (11387), &c.

1879, Feb. 9, white,	C.C.	Fairy 18th	Oxford Swell, 38838	Mr. Edmonds

FAIRY 4TH, roan, calved in 1871, Vols. xxi. and xxv. pp. 689, 435. Bred by
Mr. W. J. Edmonds; got by Third Lord Waterloo (29220), dam (Fairy 3rd)
by Fifth Baron Wetherby (23382), &c.

1879, May 26, roan,	B.C.	Earl of Southrop 6th	Oxford Swell, 38838	Mr. Edmonds

FAIRY 6TH, roan, calved June 30, 1873, Vol. xxv. p. 435. Bred by Mr. W. J.
Edmonds; got by Second Earl of Walton (19672), dam (Fairy 3rd) by Fifth
Baron Wetherby (23382), &c.

1879, Mar. 19, white,	C.C.	Fairy 20th	Oxford Swell, 38838	Mr. Edmonds

FAIRY 7TH, white, calved February 4, 1874. Bred by Mr. W. J. Edmonds; got
by Dalkeith (30847), dam (Fairy 4th) by Third Lord Waterloo (29220), &c.

1878, March, roan,	B.C.	(dead)	Col. Tregunter 9th, 83406	Mr. Edmonds
1879, April 8, white,	C.C.	Fairy 21st	Oxford Swell, 38838	do.

FAIRY 9TH, roan, calved October 21, 1875, Vol. xxv. p. 435. Bred by Mr. W.
J. Edmonds; got by Colonel Tregunter 9th (33406), dam (Fairy 3rd) by
Fifth Baron Wetherby (23382), &c.

1879, Mar. 15, red,	C.C.	Fairy 19th	Oxford Swell, 38838	Mr. Edmonds

IMPUDENCE, red and white, calved March 25, 1875. Bred by Sir G. R.
Philips, Bart., Weston Park; got by Cherry Fawsley (30711), dam (Irene) by
Barleycorn the Younger (21209), &c. See Vol. xx. p. 563.

1879, Feb. 1, red,	B.C.	Homeleaze Lad	Oxford Swell, 38838	Mr. Edmonds
1879, Dec. 21, r. & w.,	B.C.	Homeleaze Lad 2nd	do.	do.

ROSE OF FRANCE 5TH, roan, calved March 7, 1870, Vols. xxii. and xxv.
pp. 406, 436. Bred by Mr. W. J. Edmonds; got by Fifth Baron Wetherby
(23382), dam (Rose of France 3rd) by Second Northern Duke (22432), &c.

1879, April 3, red,	B.C.	Earl of Southrop 3rd	Oxford Swell, 38838	Mr. Edmonds

ROSE OF FRANCE 7TH, roan, calved November 2, 1872, Vol. xxv. p. 436.
Bred by Mr. W. J. Edmonds; got by Second Earl of Walton (19672), dam
(Rose of France 3rd) by Second Northern Duke (22432), &c.

1879, March 24, red,	C.C.	Rose of France 15th	Oxford Swell, 38838	Mr. Edmonds

ROSE OF FRANCE 8TH, roan, calved March 24, 1874, Vol. xxv. p. 436. Bred
by Mr. W. J. Edmonds; got by Dalkeith (30847), dam (Rose of France 6th)
by Fifth Baron Wetherby (23382), &c.

1879, May 8, r. & w.,	B.C.	Earl of Southrop 4th	Oxford Swell, 38838	Mr. Edmonds

ROSE OF FRANCE 9TH, roan, calved March 19, 1875, Vol. xxv. p. 436. Bred
by Mr. W. J. Edmonds; got by Colonel Tregunter 9th (33406), dam (Rose of
France 5th) by Fifth Baron Wetherby (23382), &c.

1879, Apr. 28, r. & w.,	B.C.	Baron Homeleaze	Oxford Swell, 38838	Mr. Edmonds

ROSE OF FRANCE 11TH, roan, calved February 2, 1877. Bred by Mr. W. J. Edmonds; got by Colonel Tregunter 9th (33406), dam (Rose of France 8th) by Dalkeith (30847), &c.

Produce in		Names, &c.	By what Bull.	By whom bred.
1879, June 16, red,	B.C.	Earl of Southrop 8th	Oxford Swell, 38838	Mr. Edmonds

ROSE OF OXFORD 3RD, roan, calved in 1871, Vols. xxi. and xxv. pp. 690, 436. Bred by Mr. W. J. Edmonds; got by Fifth Baron Wetherby (23382), dam (Rose of Oxford) by Fourth Duke of Oxford (11387), &c.

Produce in		Names, &c.	By what Bull.	By whom bred.
1879, March 12, red,	C.C.	Rose of Oxford 9th	Oxford Swell, 38838	Mr. Edmonds

ROSE OF OXFORD 4TH, roan, calved April 20, 1874, Vol. xxv. p. 436. Bred by Mr. W. J. Edmonds; got by Lord Walton (31765), dam (Rose of Oxford 2nd) by Son of Virago (27533), &c.

Produce in		Names, &c.	By what Bull.	By whom bred.
1879, March 17, roan,	C.C.	Rose of Oxford 10th	Oxford Swell, 38838	Mr. Edmonds

ROSE OF OXFORD 6TH, white, calved February 9, 1877. Bred by Mr. W. J. Edmonds; got by Colonel Tregunter 9th (33406), dam (Rose of Oxford 3rd) by Fifth Baron Wetherby (23382), &c.

Produce in		Names, &c.	By what Bull.	By whom bred.
1879, Oct. 8, roan,	C.C.	Rose of Oxford 11th	Oxford Swell, 38838	Mr. Edmonds

RUBY 8TH, red and little white, calved March 12, 1877. Bred by Mr. E. Bowly, Siddington House; got by Third Duke of Clarence (23727), dam (Ruby 2nd) by Seventh Duke of York (17754), &c. See Vol. xxiv. p. 341.

Produce in		Names, &c.	By what Bull.	By whom bred.
1879, Sept. 4, r. & w.,	B.C.	Baron Homeleaze 2nd	Oxford Swell, 38838	Mr. Edmonds

TULIP, red and white, calved December 17, 1871, Vols. xxi., xxiii., and xxv. pp. 922, 638, 437. Bred by the Marquis of Exeter, Burghley Park; got by Telemachus (27603), dam (Poppy) by Royal Oxford (27380), &c.

Produce in		Names, &c.	By what Bull.	By whom bred.
1879, May 9, { roan,	C.C.	Tulip 3rd	} Oxford Swell, 38838	Mr. Edmonds
red,	B.C.	Nero		

VIOLET 5TH, red, calved May 14, 1877. Bred by Mr. W. J. Edmonds; got by Duke of Hazlecote 36th (33671), dam (Violet 4th) by Lord Hastings (29115), &c. See Vol. xxv. p. 437.

Produce in		Names, &c.	By what Bull.	By whom bred.
1879, Nov. 3, roan,	C.C.	Violet 6th	Oxford Swell, 38838	Mr. Edmonds

EDMONDSON, Bernard,
Lower Hood House, Burnley.

LUNA 4TH, roan, calved June 1, 1871. Bred by Mr. T. Whiteside, Hesketh End; got by Lord Oxford (29193), dam (Luna 2nd) by Sir Charles (22891), &c. See "Lunette," Vol. xxiv. p. 360.

Produce in		Names, &c.	By what Bull.	By whom bred.
1879, Jan. 0, r. & w.,	C.C.	Luna 5th	Braithwaite, 30258	Mr. Edmondson

ELLESMERE, Earl of,
Worsley Hall, Manchester.

DUCHESS OF WORSLEY, roan, calved March 18, 1874, Vol. xxiv. p. 426. Bred by the Earl of Ellesmere; got by Duke of Lorne (25985), dam (Duchess of Wellington 2nd) by Garibaldi 2nd (26220), &c.

Produce in		Names, &c.	By what Bull.	By whom bred.
1879, June 13, r. & w.,	B.C.	Baron Worsley	D. of Glo'ster 7th, 39735	Earl of Ellesmere

DUCHESS OF WORSLEY 2ND, roan, calved March 16, 1875, Vol. xxv. p. 437. Bred by the Earl of Ellesmere; got by Earl of Thorndale (28521), dam (Duchess of Wellington 2nd) by Garibaldi 2nd (26220), &c.

Produce in	Names, &c.	By what Bull.	By whom bred.
1879, Jan. 20, r. & w., C.C.	D'ss of Worsley 6th	Kentucky Airdrie, 41746	Earl of Ellesmere

DUCHESS OF WORSLEY 4TH, roan, calved February 16, 1877. Bred by the Earl of Ellesmere; got by Attractive Lord (32968), dam (Duchess of Wellington 2nd) by Garibaldi 2nd (26220), &c. See Vol. xxiv. p. 426.

1879, Nov. 27, roan, C.C.	D'ss of Worsley 7th	Kentucky Airdrie, 41746	Earl of Ellesmere

HARMONY, roan, calved November 25, 1873, Vol. xxiv. p. 426. Bred by the Earl of Ellesmere; got by Nicholas (31974), dam (Sympathy) by Photograph (20492), &c.

1879, Aug. 24, roan, C.C.	Melody 2nd	Kentucky Airdrie, 41746	Earl of Ellesmere

ELLIS, Charles,
Oak Villa, Meldreth, Royston.

BARONESS BRAYBROOKE, red and white, calved August 5, 1875, Vol. xxv. p. 437. Bred by Mr. C. Ellis; got by Balmoral (32987), dam (Lady Braybrooke) by Englishman (19701), &c.

1879, June 2, roan, C.C.	Princess Louise	Anoop Sing, 37734	Mr. Ellis

BARONESS HOPE, red and white, calved November 5, 1875, Vol. xxv. p. 437. Bred by Mr. C. Ellis; got by Baron Hope (30460), dam (Florentine) by King Oxford (26521), &c.

1879, May 16, roan, C.C.	Oxford Lassie	Anoop Sing, 37734	Mr. Ellis

COUNTESS, red and white, calved May 23, 1872, Vol. xxv. p. 386. Bred by Sir A. de Rothschild, Bart., Aston Clinton; got by Prince Alexander (35082), dam (Country Girl) by Duke of York (23804), &c.

1879, Aug. 8, r. & w., C.C.	Duchess of York	Meldreth Duke, 34832	Mr. Ellis

FLORENTINE 2ND, red and white, calved November 13, 1873. Bred by Mr. C. Ellis; got by Baron Hope (30460), dam (Florentine) by King Oxford (26521), &c. See "Wellington," Vol. xxv. p. 295.

1879, Sept. 15, roan, C.C.	Fanny	Anoop Sing, 37734	Mr. Ellis

ICKLETON DUCHESS, red and white, calved April 3, 1875, Vol. xxv. p. 438. Bred by Mr. S. A. Wilson, Ickleton Grange; got by Heydon Duke 3rd (31371), dam (Xmas Rose) by Xmas Duke (27839), &c.

1879, Nov. 20, roan, C.C.	Heydon Rose	Meldreth Duke, 34832	Mr. Ellis

PAVILION, white, calved July 6, 1875, Vol. xxv. p. 538. Bred by Mr. C. Barnett, Stratton Park; got by Paris (29523), dam (Pipe Light) by Provost (27878), &c.

1879, May 13, roan, C.C.	Water Lily	Meldreth Duke, 34832	Mr. Ellis

ROSE OF MELDRETH, red and white, calved March 23, 1876, Vol. xxv. p. 438. Bred by Mr. C. Ellis; got by Baron Hope (30460), dam (Rose of Spring) by Lord Oxford 2nd (20215), &c.

1879, Dec. 17, r. & w., C.C.	Moss Rose	Meldreth Duke, 34832	Mr. Ellis

ELLIS, Colonel J. J.,
Ellistown, Leicester.

STAR OF NIGHT, red and white, calved October 30. 1874. Bred by Mr. J. W. Wilson, Broadway; got by Grand Duke of Waterloo (28766), dam (Star of Eve) by Seventh Duke of York (17754), &c. See " Prince of Wales," Vol. xxv. p. 231.

Produce in		Names, &c.	By what Bull.	By whom bred.
1879, Dec. 26, roan,	C.C.	Princess Mary	Duke of Connaught, 33004	Colonel Ellis

ELSTON, W.,
Bank House, Selby.

ROSE OF RYEDALE 7TH, white, calved March 31, 1875. Bred by the Earl of Feversham, Duncombe Park; got by Twentieth Duke of Oxford (28432), dam (Convolvulus) by Orestes (22443), &c. See Vol. xxii. p. 415.

Produce in		Names, &c.	By what Bull.	By whom bred.
1879, June 1, r. & w.,	C.C.	Lady Ryedale Rose	D. of Tregunter 5th, 33743	Mr. Elston

ELVIDGE, Henry,
Long Riston, Hull.

LADY OF LYONS 9TH, roan, calved April 29, 1877. Bred by Mr. R. Botterill, Wauldby, the property of Mr. H. Elvidge; got by Oxford-le-Grand (29496), dam (Laurel) by Thorndale Lad (23066), &c. See Vol. xxiv. p. 337.

Produce in		Names, &c.	By what Bull.	By whom bred.
1879, Mar. 28, white,	B.C.	Prince of Lyons	Prince of Oxford, 42210	Mr. Crust

ERRINGTON, Robson,
Scotby Farm, Carlisle.

BLOSSOM 14TH, roan, calved May 29, 1876. Bred by Mr. J. Todd, Mereside; got by Grand Duke of Lightburne 2nd (26291), dam (Blossom 10th) by Brayton 4th (23450), &c. See Vol. xxi. p. 959.

Produce in		Names, &c.	By what Bull.	By whom bred.
1879, July 18, roan,	C.C.	Scotby Blossom	Gwynne Duke, 34098	Mr. Errington

TULIP 2ND, roan, calved April 1, 1873. Bred by Miss Calvert, Sandysike; got by Merryman (37090), dam (Young Tulip) by Snowdrift (32558), g. d. (Tulip) by Rufus (22811), &c. See " Tarquin," Vol. xxii. p. 260.

Produce in		Names, &c.	By what Bull.	By whom bred.
1879, Nov. 13, white,	C.C.	Scotby Tulip	Oxford Waterloo, 40434	Mr. Errington

EVANS, John,
Uffington, Shrewsbury.

BERTHA, red and white, calved March 6, 1872, Vols. xxii. and xxiv. pp. 407, 427. Bred by Messrs. Perry, Acton Pigott; got by Duke of Lancaster (28411), dam (Bride) by Manager (24521), &c.

Produce in		Names, &c.	By what Bull.	By whom bred.
1878, Dec. 20, roan,	C.C.	Bouquet	Geneva Prince, 36691	Mr. Evans
1879, Dec. 8, roan,	B.C.	Buckskin	The Buck, 40795	do.

BONNIE BRIDE, roan, calved January 19, 1877. Bred by Mr. J. Evans; got by Geneva Prince (36691), dam (Bride) by Manager (24521), &c. See Vol. xxiv. p. 427.

Produce in		Names, &c.	By what Bull.	By whom bred.
1879, Oct. 4, roan,	C.C.	Bonnie Belle	Prince Rudolf, 40539	Mr. Evans

BRIDAL RING, roan, calved February 16, 1877. Bred by Mr. J. Evans; got by Geneva Prince (36691), dam (Bertha) by Duke of Lancaster (28411), &c. See Vol. xxiv. p. 427.

Produce in	Names, &c.	By what Bull.	By whom bred.
1879, Sept. 22, roan, B.C.	(Steer)	The Buck, 40795	Mr. Evans

BRIDE, red, calved March 24, 1868, Vols. xxii. and xxiv. pp. 407, 427. Bred by Messrs. Perry, Acton Pigott; got by Manager (24521), dam (Bridesmaid) by Romeo (22754), &c.

1878, Dec. 15, r. & w., B.C.	Bruce	Geneva Prince, 36691	Mr. Evans
1879, Dec. 11, roan, B.C.	Brutus	Prince Rudolf, 40539	do.

BRIGHT GEM, roan, calved March 17, 1877. Bred by Mr. J. Evans; got by Geneva Prince (36691), dam (Ruby 3rd) by Manager (24521), &c. See Vol. xxiv. p. 527.

1879, April 11, red, C.C.	Ruby 12th	The Buck, 40795	Mr. Evans

CHARMER 3RD, roan, calved December 2, 1865. Bred by Mr. D. McIntosh, Havering Park; got by Grand Duke 4th (19874), dam (Charmer) by Mainstay (16490), &c. See Vol. xvii. p. 410.

1879, Mar. 16, red, C.C.	Ch'ming Sweetheart	The Buck, 40795	Mr. Evans

CLARA, roan, calved May 2, 1876, Vol. xxv. p. 439. Bred by Mr. J. Evans; got by Richmond (35269), dam (Charm) by St. Helena (35450), &c.

1879, April 28, white, C.C.	Camilla	The Buck, 40795	Mr. Evans

COLLEEN BAWN, roan, calved April 10, 1876, Vol. xxv. p. 439. Bred by Mr. J. Evans; got by Richmond (35269), dam (Dear Shamrock) by Manager (24521), &c.

1879, April 9, roan, B.C.	Fusilier	The Buck, 40795	Mr. Evans

CONSTANCE, red, calved April 20, 1874, Vol. xxv. p. 439. Bred by Mr. J. Evans; got by Manager (24521), dam (Charm) by St. Helena (35450), &c.

1879, Aug. 6, red, B.C.	Baronet	The Buck, 40795	Mr. Evans

DEAR SHAMROCK, roan, calved May 12, 1874. Bred by Mr. J. Evans; got by Manager (24521), dam (Erin) by Shamrock (20797), &c. See "Colleen Bawn," Vol. xxv. p. 439.

1878, April 20, roan, C.C.	Rose Bell	D. of Kirklevington, 38158	Mr. Evans
1879, July 15. roan, B.C.	Tyrant	The Buck. 40795	do.

ELSIE CRAGGS, roan, calved February 14, 1877. Bred by Lord Moreton, Tortworth Court; got by Oxford's Prince (34998), dam (Elsie Stuart) by Standard Bearer (25221), &c. See Vol. xxv. p. 439.

1879, Apr. 7, roan, C.C.	(dead)	The Buck, 40795	Mr. Evans

ELSIE HILLHURST, roan, calved March 13, 1875, Vol. xxv. p. 439. Bred by Colonel Kingscote, Kingscote Park; got by Duke of Hillhurst (28401), dam (Elsie Stuart) by Standard Bearer (25221), &c.

1879, April 5, roan, B.C.	Baron Oxford Craggs	Bn. T'croft Oxf'd 4th, 37822	Mr. Evans

ELSIE STUART, roan, calved October 19, 1872, Vols. xxii., xxiii., and xxv. pp. 513, 573, 439. Bred by Mr. J. Canning, Snitterfield; got by Standard Bearer (25221), dam (Lady Stuart) by John O'Groat (18115), &c.

1879, June 2, roan, C.C.	Lady Gazelle Craggs	The Buck, 40795	Mr. Evans

EMPRESS, red, calved December 12, 1876. Bred by Mr. J. Evans; got by
Richmond (35269), dam (Evergreen) by Manager (24521), &c. See Vol. xxiii.
p. 432.

Produce in		Names, &c.		By what Bull.	By whom bred.
1879, Apr. 29, {	red, B.C.	Romulus	}	The Buck, 40795	Mr. Evans
	red, B.C.	Remus			

ERIN LILY, white, calved May 7, 1877. Bred by Mr. J. Evans; got by
Geneva Prince (36691), dam (Dear Shamrock) by Manager (24521), &c. See
"Colleen Bawn," Vol. xxv. p. 439.

1879, April 2, white, C.C.	Dear Erin	The Buck, 40795	Mr. Evans

ERIN ROSE, red, calved March 15, 1876. Bred by Mr. J. Evans; got by
Richmond (35269), dam (Norah) by Manager (24521), &c. See Vol. xxiii.
p. 432.

1879, Mar. 4, roan, B.C.	(dead)	The Buck, 40795	Mr. Evans

EVERGREEN, roan, calved May 11, 1871, Vols. xxii. and xxiii. pp. 408, 432.
Bred by Mr. J. Evans; got by Manager (24521), dam (Emerald) by Victor
(19061), &c.

1878, Dec. 20, roan, C.C.	Eugenie	Geneva Prince, 36691	Mr. Evans
1879, Dec. 1, roan, B.C.	Macgregor	The Buck, 40795	do.

HONEYDEW, red, calved April 27, 1877. Bred by Mr. J. Evans; got by
Richmond (35269), dam (Honeysuckle) by St. Helena (35450), g. d. (Mistletoe)
by Sir Charles (16946), &c. See "Christmas Rose," Vol. xxii. p. 408.

1879, July 24, roan, B.C.	Havelock	The Buck, 40795	Mr. Evans

HONEYSUCKLE, roan, calved March 20, 1865, Bred by Mr. J. Evans; got
by St. Helena (35450), dam (Mistletoe) by Sir Charles (16946), &c.

1879, April 3, roan, B.C.	Broadside	The Buck, 40795	Mr. Evans

JESSIE 3RD, roan, calved July 2, 1874, Vol. xxv. p. 439. Bred by Mr. J. Evans;
got by Manager (24521), dam (Jessie 2nd) by Shamrock (20797), &c.

1879, Mar. 19, white, B.C.	Jester	The Buck, 40795	Mr. Evans

JUNO 5TH, roan, calved September 15, 1872. Bred by Mr. W. Bradburn, Wednesfield, the property of Mr. J. Evans; got by Wednesfield (30281), dam (Juno
3rd) by Thorndale Grand Duke (20976), &c. See "Prince of Wales,"
Vol. xxii. p. 209.

1879, Sept. 20, roan, C.C.	Juno 7th	Magnate Star, 37029	Mr. D. H. Owen

KATHLEEN, roan, calved March 25, 1876. Bred by Mr. J. Evans; got by
Richmond (35269), dam (Emerald) by Victor (19061), &c. See "Regent,"
Vol. xxi. p. 391.

1879, April 6, roan, B.C.	Rory O'More	The Buck, 40795	Mr. Evans

LADY WHITE FLANK, red and white, calved March 20, 1872, Vol. xxv.
p. 439. Bred by Mr. J. Evans; got by Manager (24521), dam (Erin) by
Shamrock (20797), &c.

1879, May 20, red, B.C.	Red Robin	The Buck, 40795	Mr. Evans

LAURA, red, calved March 7, 1875. Bred by Mr. J. Evans; got by Richmond
(35269), dam (Lucy) by Manager (24521), &c. See Vol. xxiii. p. 432.

1878, Mar. 10, roan, B.C.	(Steer)	Geneva Prince, 36691	Mr. Evans
1879, Mar. 24, red, B.C.	Admiral	The Buck, 40795	do.

LAUREL, roan, calved March 2, 1877. Bred by Mr. J. Evans; got by Geneva Prince (36691), dam (Lady Bird) by Volunteer (30238), &c. See Vol. xxiv. p. 427.

Produce in	Names, &c.	By what Bull.	By whom bred.
1879, May 20, white, B.C.	Lanercost	The Buck, 40795	Mr. Evans

LUFRA, red and white, calved March 20, 1877. Bred by Mr. J. Evans; got by Geneva Prince (36691), dam (Lucy) by Manager (24521), &c. See Vol. xxiii. p. 432.

1879, April 2, white, B.C.	(Steer)	The Buck, 40795	Mr. Evans

MARCHIONESS 3RD, roan, calved January 3, 1876, Vol. xxv. p. 440. Bred by Mr. G. Juckes, Beslow Hall; got by Stretton (35691), dam (Marchioness 2nd) by Prince Patrick (22642), &c.

1879, April 14, roan, B.C.	(Steer)	The Buck, 40795	Mr. Evans

PRINCESS ERIN, roan, calved March 12, 1877. Bred by Mr. J. Evans; got by Geneva Prince (36691), dam (Norah) by Manager (24521), &c. See Vol. xxiii. p. 432.

1879, Oct. 25, roan, B.C.	(Steer)	The Buck, 40795	Mr. Evans

QUEEN OF GENEVA, roan, calved November 12, 1876. Bred by Mr. G. Garne, Churchill Heath; got by Grand Duke of Geneva 2nd (31288), dam (Princess Royal) by Grand Duke of Lightburne 2nd (26291), &c. See Vol. xxiii. p. 449.

1879, Mar. 20, roan, C.C.	Queen Furbelow	The Buck, 40795	Mr. Evans

ROSA BELLA, red, calved April 10, 1877. Bred by Mr. J. Evans; got by Richmond (35269), dam (Ruby Bright) by Prince Patrick (22642), &c. See Vol. xxiv. p. 427.

1879, April 25, red, B.C.	Donald	The Buck, 40795	Mr. Evans

RUBY 1ST, red, calved March 1, 1866, Vol. xxii. p. 408. Bred by Mr. J. Evans; got by Victor (19061), dam (Charm) by St. Helena (35450), &c.

1878, April 27, roan, C.C.	Ruby 7th	Geneva Prince, 36691	Mr. Evans
1879, April 10, roan, C.C.	Ruby 11th	The Buck, 40795	do.

RUBY 2ND, red, calved April 2, 1872, Vol. xxii. p. 408. Bred by Mr. J. Evans; got by Manager (24521), dam (Ruby 1st) by Victor (19061), &c.

1878, Mar. 7, roan, C.C.	Ruby 6th	Geneva Prince, 36691	Mr. Evans
1879, April 5, red, C.C.	Ruby 10th	The Buck, 40795	do.

RUBY 3RD, red, calved March 22, 1874, Vols. xxiv. and xxv. pp. 427, 440. Bred by Mr. J. Evans; got by Manager (24521), dam (Ruby 1st) by Victor (19061), &c.

1879, Mar. 20, roan, B.C.	Hercules	The Buck, 40795	Mr. Evans

RUBY BRIGHT, red, calved April 4, 1868, Vol. xxiv. p. 427. Bred by Mr. J. Evans; got by Prince Patrick (22642), dam (Ruby 1st) by Victor (19061), &c.

1878, Mar. 6, roan, C.C.	Ruby 5th	Geneva Prince, 36691	Mr. Evans
1879, Mar. 9, roan, B.C.	Sunbeam	The Buck, 40795	do.

RUBY PRINCESS, roan, calved April 20, 1877. Bred by Mr. J. Evans; got by Geneva Prince (36691), dam (Ruby 1st) by Victor (19061), &c.

1879, April 4, r. & w., C.C.	Ruby 9th	The Buck, 40795	Mr. Evans

RUBY QUEEN, red, calved March 7, 1876, Vol. xxv. p. 440. Bred by Mr. J.
Evans; got by Richmond (35269), dam (Ruby 2nd) by Manager (25421),
&c.

Produce in		Names, &c.		By what Bull.		By whom bred.
1879, Apr. 2, roan,	C.C.	Ruby 8th		The Buck, 40795		Mr. Evans

SAPPHIRE, roan, calved March 2, 1875, Vol. xxv. p. 440. Bred by Mr. J.
Evans; got by Richmond (35269), dam (Emerald) by Victor (19061), &c.

1879, April 20, white, B.C.	(Steer)		The Buck, 40795		Mr. Evans

WILD DUCHESS 3RD, white, calved June 11, 1876. Bred by Lieut.-Colonel
Webb, Elford House; got by Marquis 6th (34777), dam (Wild Duchess 2nd)
by Caractacus (28141), &c. See Vol. xxiii. p. 698.

1879, Dec. 19, white, B.C.	(dead)		The Buck, 40795		Mr. Evans

FARRER, John,
Thornyholme, Burnley.

LADY PADIHAM, red, calved April 7, 1875, Vol. xxv. p. 440. Bred by Mr.
J. Farrer; got by Baron Padiham (36200), dam (Lady Athelstane) by Duke
of Athelstane (21562), &c.

1879, May 2, red,	C.C.	Thornyholme Rose	Oxford 16th, 38822		Mr. Farrer

LILY ATHELSTANE, white, calved January 16, 1873, Vols. xxii. and xxv.
pp. 410, 440. Bred by Mr. J. Farrer; got by Pamerose (32042), dam (Lady
Athelstane) by Duke of Athelstane (21562), &c.

1879, Mar. 13, roan,	C.C.	Thornyholme Belle	Oxford 16th, 38822		Mr. Farrer

MAY FLOWER, white, calved May 10, 1872. Bred by Mr. J. Farrer; got by
Pamelax (29516), dam (Butterflower) by Duke of Kent (23760), g. d. (Miss
Butterfield) by Royal Butterfly 8th (18759), gr. g. d. (White Cow) by Gipsy
Boy (11530), — Big Heifer, a Shorthorn Cow bred by Mr. Bolton, Barrow-
ford.

1876, June 27, roan,	B.C.	(dead)	Double Athelstane, 33544	Mr. Farrer
1878, Mar. 27, roan,	C.C.	(dead)	Major Hubback, 37035	do.
1879, Aug. 14, roan,	C.C.	August Rose	Oxford 16th, 38822	do.

FAULKNER, W.,
Rothersthorpe, Northampton.

DAHLIA 11TH, roan, calved March 18, 1872, Vols. xxi., xxii., xxiii., and xxiv.
pp. 694, 410, 433, 428. Bred by Mr. W. Faulkner; got by Athelstane (23331),
dam (Dahlia 7th) by Saracen (25091), &c.

1879, Mar. 3, roan,	C.C.	Dahlia 22nd	Fair Thane, 31127		Mr. Faulkner

DAHLIA 12TH, roan, calved July 7, 1874, Vols. xxiv. and xxv. pp. 428, 441.
Bred by Mr. W. Faulkner; got by Royalty (35412), dam (Dahlia 11th) by
Athelstane (23331), &c.

1879, Aug. 29, roan,	C.C.	Dahlia 23rd	Fair Thane, 31127		Mr. Faulkner

DAHLIA 14TH, white, calved November 18, 1875. Bred by Mr. W. Faulkner;
got by Royalty (35412), dam (Dahlia 11th) by Athelstane (23331), &c.

1879, Dec. 19, roan,	B.C.	Dreamer	Rupee, 39069		Mr. Faulkner

DAHLIA 15TH, red, calved September 22, 1876.   Bred by Mr. W. Faulkner ;
got by Abbot of Windsor (32903), dam (Dahlia 11th) by Athelstane (23331),
&c.

Produce in		Names, &c.	By what Bull.	By whom bred.
1879, July 10, red,	B.C.	Dudley	Rupee, 39069	Mr. Faulkner

DAHLIA 16TH, roan, calved October 6, 1876.   Bred by Mr. W. Faulkner ; got
by Abbot of Windsor (32903), dam (White Rose) by Knight of Branches
(20076), &c.   See Vol. xxiii. p. 434.

1879, Aug. 10, roan,	B.C.	Duncan	Firebrand, 39878	Mr. Faulkner

FAIR LADY, roan, calved February 18, 1876.   Bred by Mr. W. Faulkner ;
got by Royalty (35412), dam (Fairy) by Athelstane (23331), &c.   See Vol.
xxiii. p. 433.

1878, July 7, red,	B.C.	Figaro	Rupee, 39069	Mr. Faulkner
1879, Aug. 4, red,	C.C.	Fair Maid	do.	do.

FLAME, red, calved January 1, 1875, Vols. xxiv. and xxv. pp. 429, 441.   Bred
by Mr. J. Gurney, Winslow ; got by Rosehill (29833), dam (Faith) by Romeo
(24985), &c.

1879, July 10, r. & w.,	C.C.	Fern	Fair Thane, 31127	Mr. Faulkner

FORMOSA, roan, calved February 8, 1871, Vol. xxi. p. 695.   Bred by Mr. W.
Faulkner ; got by Rosehill (29833), dam (Fancy) by Knight of Branches
(20076), &c.

1879, Aug. 7, roan,	B.C.	Fair Day	Fair Thane, 31127	Mr. Faulkner

Fair Day, sold to Mr. Hutchinson, Bozeat, Wellingboro'.

FRAGRANCE, roan, calved March 14, 1872, Vols. xxi. and xxiv. pp. 695, 429.
Bred by Mr. W. Faulkner ; got by Athelstane (23331), dam (Fancy) by Knight
of Branches (20076), &c.

1878, July 29, roan,	C.C.	(dead)	Fair Thane, 31127	Mr. Faulkner
1879, June 21, roan,	C.C.	Filagree	do.	do.

KATHERINE 3RD, roan, calved January 22, 1869, Vols. xxii. and xxiii. pp. 411,
433.   Bred by Mr. G. Pell, Heyford ; got by Prince Regent (35169), dam
(Katherine 2nd) by Stormy Petrel (37538), &c.

1879, Jan. 23, red,	B.C.	Kaffer	Fair Thane. 31127	Mr. Faulkner
1879, Dec. 26, white,	C.C.	Kitty	Rupee, 39069	do.

Kaffer, sold to Mr. C. T. Riley, Milton Reynes, Newport Pagnell.

MATCHLESS, roan, calved in May 1867, Vols. xxi., xxii., and xxiv. pp. 695,
411, 429.   Bred by Mr. J. Faulkner, The Woodlands ; got by Knight of
Branches (20076), dam (Mulberry) by August (12412), &c.

1878, Jan. 20, white,	B.C.	Mariner	King of the Roses, 36853	Mr. Faulkner
1879, May 1, roan,	C.C.	Rink	Fair Thane, 31127	do.

Mariner, sold to Mr. Robinson, Courteen Hall, Northampton.

RED ROSE, red, calved November 1, 1875.   Bred by Mr. W. Faulkner ; got
by Royalty (35412), dam (Matchless) by Knight of Branches (20076), &c.

1879, Mar. 4, roan,	B.C.	Rayleigh	Fair Thane, 31127	Mr. Faulkner

ROSALIA, red, calved July 14, 1871, Vols. xxi., xxii., and xxv. pp. 695, 411,
441.   Bred by Mr. W. Faulkner ; got by Athelstane (23331), dam (Rhoda)
by Saracen (25091), &c.

1879, Mar. 18, {	red, B.C.	Roarer	} Fair Thane, 31127	Mr. Faulkner
	roan, B.C.	Rumpus		do.

ROSALIND, roan, calved November 1, 1875. Bred by Mr. W. Faulkner; got by Prince Rufus (35177), dam (Rosebud) by Athelstane (23331), &c. See Vol. xxii. p. 411.

Produce in		Names, &c.	By what Bull.	By whom bred.
1879, May 19, red,	B.C.	Rocket	Fair Thane, 31127	Mr. Faulkner

ROSE 16TH, roan, calved March 15, 1872, Vols. xxi., xxii., xxiv., and xxv. pp. 695, 411, 429, 441. Bred by Mr. W. Faulkner; got by Athelstane (23331), dam (Rose 6th) by Knight of Branches (20076), &c.

Produce in		Names, &c.	By what Bull.	By whom bred.
1879, June 27, roan,	C.C.	Rose 27th	Fair Thane, 31127	Mr. Faulkner

ROSE 22ND, roan, calved October 8, 1875. Bred by Mr. W. Faulkner; got by Royalty (35412), dam (Rose 16th) by Athelstane (23331), &c.

Produce in		Names, &c.	By what Bull.	By whom bred.
1879, July 28, r. & w.,	C.C.	Rose 28th	Firebrand, 39878	Mr. Faulkner

ROSE BUD, white, calved March 2, 1872, Vols. xxii. and xxiii. pp. 411, 433. Bred by Mr. W. Faulkner; got by Athelstane (23331), dam (Rosamond) by Saracen (25091), &c.

Produce in		Names, &c.	By what Bull.	By whom bred.
1877, Aug. 20, roan,	B.C.	(dead)	Abbot of Windsor, 32903	Mr. Faulkner
1879, Mar. 1, roan,	B.C.	Rodney	Fair Thane, 31127	do.

SWEETBRIER 9TH, roan, calved July 14, 1863, Vols. xix., xx., xxi., and xxv. pp. 745, 780, 696, 441. Bred by Mr. W. Faulkner; got by Knight of Branches (20076), dam (Spicy) by Dreadnought (12719), &c.

Produce in		Names, &c.	By what Bull.	By whom bred.
1879, July 4, roan,	C.C.	Sweetbrier 83rd	Fair Thane, 31127	Mr. Faulkner

SWEETBRIER 26TH, roan, calved September 22, 1867, Vols. xix., xxii., xxiii., xxiv., and xxv. pp. 746, 412, 433, 429, 442. Bred by Mr. W. Faulkner; got by General Burnside (24016), dam (Sweetbrier 7th) by Sardinia (16908), &c.

Produce in		Names, &c.	By what Bull.	By whom bred.
1879, May 25, white,	C.C.	Sweetbrier 81st	Fair Thane, 31127	Mr. Faulkner

SWEETBRIER 48TH, roan, calved March 10, 1872. Bred by Mr. W. Faulkner; got by Athelstane (23331), dam (Sweetbrier 9th) by Knight of Branches (20076), &c. See "Sir Frederick," Vol. xxv. p. 261.

Produce in		Names, &c.	By what Bull.	By whom bred.
1878, Aug. 6, roan,	B.C.	Sir Frederick	King of the Roses, 36853	Mr. Faulkner
1879, July 8, r. & w.,	C.C.	Sweetbrier 84th	Fair Thane, 31127	do.

Sir Frederick, sold to Mr. Higgins, Stoke Goldington, Newport Pagnell.

SWEETBRIER 62ND, red, calved April 12, 1874. Bred by Mr. W. Faulkner; got by Rosehill (29883), dam (Sweetbrier 24th) by General Burnside (24016), &c. See "Lord Salisbury," Vol. xxv. p. 176.

Produce in		Names, &c.	By what Bull.	By whom bred.
1878, May 14, red,	B.C.	Lord Salisbury	Fair Thane, 31127	Mr. Faulkner
1879, Apr. 3, red,	C.C.	Sweetbrier 79th	do.	do.

Lord Salisbury, sold to Mr. Walker, Newnham, Daventry.

SWEETBRIER 70TH, roan, calved September 25, 1875. Bred by Mr. W. Faulkner; got by Rosehill (29833), dam (Spangle) by Trojan (35817), &c. See Vol. xxii. p. 411.

Produce in		Names, &c.	By what Bull.	By whom bred.
1878, Nov. 1, roan,	B.C.	Stalworth	Rupee, 39069	Mr. Faulkner
1879, Sept. 20, red,	B.C.	Strapper	do.	do.

Stalworth, sold to Mr. M. George, Gayton, Northampton.

SWEETBRIER 71ST, roan, calved November 18, 1875, Vol. xxv. p. 442. Bred by Mr. W. Faulkner; got by Royalty (35412), dam (Sweetbrier 28th) by Saracen (25091), &c.

Produce in		Names, &c.	By what Bull.	By whom bred.
79, Dec. 13, roan,	B.C.	Stamboul	Fair Thane, 31127	Mr. Faulkner

SWEETBRIER 72ND, red, calved March 1, 1876. Bred by Mr. W. Faulkner;
got by Prince Rufus (35177), dam (Sweetbrier 56th) by Athelstane (23331),
&c. See Vol. xxiii. p. 434.

Produce in		Names, &c.		By what Bull.		By whom bred.
1879, May 16, red,	C.C.	Sweetbrier 80th		Rambler, 40561		Mr. Faulkner

SWEETBRIER 73RD, red, calved May 23, 1876. Bred by Mr. W. Faulkner;
got by Rosehill (29833), dam (Sweetbrier 26th) by General Burnside (24016),
&c. See "Standish," Vol. xxv. p. 269.

1878, Sept. 10, roan,	B.C.	Standish	King of the Roses, 36853	Mr. Faulkner
1879, Aug. 17, red,	B.C.	Sir Hector	Fair Thane, 31127	do.

Standish, sold to M. de Biolley, Province de Liege, Belgium.

SWEETHEART, roan, calved in February 1874. Bred by Mr. J. Faulkner,
The Woodlands; got by Sirloin (32510), dam (Spring) by Trojan (35817),
g. d. (Sweetbread) by General (17947), gr. g. d. (Sunshine) by Protection
(11956), &c. See Vol. xiv. p. 711.

1879, June 1, red,	C.C.	Sweetbrier 82nd	Rupee, 39069	Mr. W. Faulkner

WESTERN, roan, calved November 6, 1868, Vols. xx., xxi., xxiii., and xxiv.
pp. 818, 697, 434, 430. Bred by Mr. W. Faulkner; got by Romeo (24985),
dam (Woman in White) by Marmaduke (14897), &c.

1878, Aug. 7, r. & w.,	B.C.	Webster	Fair Thane, 31127	Mr. Faulkner
1879, Aug. 28, white,	C.C.	Westeria	do.	do.

Webster, sold to Mr. J. Faulkner, The Woodlands, Stowe.

WHITE ROSE, white, calved in May 1867, Vols. xxi., xxiii., xxiv., and xxv.
pp. 698, 434, 430, 442. Bred by Mr. J. Faulkner, The Woodlands; got by
Knight of Branches (20076), dam (Dairymaid) by Scimitar (16917), &c.

1879, Dec. 4, white,	B.C.	(Steer)	Rupee, 39069	Mr. W. Faulkner

FAWCETT, Henry,
Old Bramhope, Otley, Yorkshire.

BALMY GALE, red, calved February 22, 1876. Bred by Mr. H. Chandos-
Pole-Gell, Hopton Hall; got by Favourite (31145), dam (Zephyr 3rd) by
Wolfran (25469), &c. See Vol. xxiii. p. 381.

1879, Mar. 6, roan,	C.C.	March Wind	Royal Saxon, 39057	Mr. Fawcett

KATHERINE 6TH, roan, calved March 17, 1870, Vols. xxii., xxiii., and xxiv.
pp. 412, 434, 432. Bred by Mr. E. A. Fawcett, Childwick Hall; got by
Prince Regent (35169), dam (Katherine 2nd) by Stormy Petrel (37538) &c.

1879, Mar. 19, roan,	B.C.	Kingsley	Rufus, 37413	Mr. Fawcett

KATHERINE 8TH, roan, calved December 21, 1875. Bred by Mr. H. Fawcett;
got by Monarch (31930), dam (Katherine 6th) by Prince Regent (35169),
&c. See "Kingsley," p. 132.

1878, Mar. 24, roan,	C.C.	Katherine 12th	Rufus, 37413	Mr. Fawcett
1879, Mar. 11, roan,	B.C.	Kaffir	do.	do.

KHIRKEE, roan, calved June 11, 1875. Bred by Mr. H. Fawcett; got by
Merry Lord (34843), dam (Queen of St. Albans) by Hamlet (31334), &c.
See "Kestrel," p. 129.

1879, Jan. 17, red,	C.C.	Khirkee 3rd	Rufus, 37413	Mr. Fawcett
1879, Dec. 20, roan,	B.C.	Kirkman	Peter the Great, 38863	do.

LADY BRIGHT EYES 2ND, roan, calved January 3, 1870, Vols. xx. and xxiv. pp. 587, 432. Bred by Mr. C. L. Whalley, Richmond House; got by Royal Scotforth (25042), dam (Bright Eyes) by Crown Duke (17643), &c.

Produce in		Names, &c.	By what Bull.	By whom bred.
1878, Mar. 3, white,	B.C.	(dead)	Prince Imperial, 32142	Mr. Fawcett
1879, Oct. 1, r. & w.,	C.C.	Lady Bright Eyes 4th	Peter the Great, 38863	do.

MAGGIE MILDRED, red, calved September 4, 1874, Vol. xxiv. p. 432. Bred by Messrs. Dudding, Panton House; got by Robert Stephenson (32313), dam (Minnie Mildred) by Duke of Lorn (25985), &c.

1878, Oct. 5, roan,	B.C.	(dead)	Prince Imperial, 32142	Mr. Fawcett
1879, Sept. 21, roan,	B.C.	The Czar	Peter the Great, 38863	do.

MERRY ROSE, roan, calved November 18, 1875. Bred by Mr. H. Fawcett; got by Merry Lord (34843), dam (Satin Rose) by White Satin (27800), &c. See Vol. xxii. p. 413.

1878, Feb. 8, roan,	B.C.	Tudor	Richmond, 37340	Mr. A. Fawkes
1879, Jan. 16, white,	B.C.		do.	do.
1879, Dec. 14, white,	C.C.	White Rose	Baron Winsome 6th, 33111	Mr. Fawcett

MILLY GWYNNE, white, calved April 24, 1876. Bred by Lady Pigot, West Hall; got by Damon (33503), dam (Matchless Gwynne) by Sir Windsor (22927), &c. See Vol. xxiii. p. 604.

1879, Jan. 20, roan,	C.C.	Nell Gwynne	Rufus, 37413	Mr. Fawcett

FAWCETT, Mrs.,
Scaleby Castle, Carlisle.

DUCHESS LALLY 2ND, roan, calved April 4, 1877. Bred by Mrs. and Mr. M. Fawcett; got by Grand Duke of Kirklevington (34071), dam (Lally 9th) by Seventh Duke of York (17754), &c. See "Lally's Grand Duke," p. 136.

1879, July 8, r. & w.,	C.C.	Duchess Lally 4th	G. D. of K'lev'ton 2d, 38379	Mrs. Fawcett

Duchess Lally 4th, sold to Mr. W. Ashburner, Conishead Grange.

GEORGIANA OF YORK 2ND, roan, calved June 12, 1876. Bred by Mrs. Fawcett, the property of Mr. W. Ashburner, Conishead Grange; got by Eighth Duke of York (28480), dam (Georgiana 5th) by General Canrobert (12926), &c. See Vol. xvii. p. 507.

1879, Dec. 26, roan,	C.C.	G'rgiana of York 4th	G.D. of K'lev'ton 2d, 38379	Mrs. Fawcett

GRAND DUCHESS URSULA, red, calved November 11, 1870. Bred by Mr. J. P. Foster, Killhow; got by Grand Duke 10th (21848), dam (Ursula 19th) by Seventh Duke of York (17754), g. d. (Ursula 12th) by General Canrobert (12927), &c.. See "Duke of Ulster 2nd," p. 83.

1879, June 5, red,	C.C.	G. Duchess Ursula 2d	Wild Eyes Duke, 36007	Mrs. Fawcett

LADY CLARENCE BATES, red and white, calved January 27, 1874, Vol. xxiv. p. 433. Bred by Mr. W. W. Slye, Beaumont Grange; got by Grand Duke of Thorndale (31297), dam (Lady Thorndale Bates) by Fourth Duke of Thorndale (17750), &c.

1879, Feb. 28, red,	C.C.	Lady Thorndale and Oxford Bates	B.T'croft Oxf'd 4th, 37822	Mrs. Fawcett

LADY YORK AND THORNDALE BATES 2ND, white, calved December 28, 1874, Vol. xxiv. p. 432. Bred by Mrs. Fawcett; got by Eighth Duke of York (28480), dam (Lady Tregunter Bates) by Second Duke of Tregunter (26022), &c.

Produce in	Names, &c.	/	By what Bull.	By whom bred.
1879, Feb. 23, roan, C.C.	{ Lady York and Oxford Bates }		B.T'croft Oxf'd 4th, 37822	Mrs. Fawcett

LADY YORK AND THORNDALE BATES 3RD, white, calved February 4, 1876. Bred by Mrs. and Mr. M. Fawcett; got by Eighth Duke of York (28480), dam (Lady Tregunter Bates) by Second Duke of Tregunter (26022), &c. See Vol. xxi. p. 701.

Produce in	Names, &c.	By what Bull.	By whom bred.
1878, Sept. 8, roan, B.C.	(dead)	B.T'croft Oxf'd 4th, 37822	{ Mrs. and Mr. M. Fawcett
1879, Nov. 30, roan, C.C.	{ Lady York and Thr'ndale Bates 4th }	} Grand Duke of Kirk- levington 2nd, 38379	} do.

PEACH BLOSSOM 15TH, white, calved March 18, 1877. Bred by Mrs. Fawcett; got by Grand Duke of Kirklevington (34071), dam (Peach Blossom 6th) by Earl of Glo'ster (21644), &c. See Vol. xxi. p. 701.

Produce in	Names, &c.	By what Bull.	By whom bred.
1879, Oct. 9, roan, C.C.	Peach Blossom 16th	Lally's Grand Duke, 43449	Mrs. Fawcett

YORK'S DUCHESS GWYNNE, roan, calved July 26, 1875. Bred by Mrs. Fawcett; got by Eighth Duke of York (28480), dam (Amy Gwynne) by Earl of Eglinton (23832), &c. See Vol. xxiii. p. 525.

Produce in	Names, &c.	By what Bull.	By whom bred.
1878, June 4. roan, B.C.	(dead)	G. D. of Kirklev'ton, 34071	Mrs. Fawcett
1879, July 21. roan, C.C.	York's D'ss Gwynne 2d	do.	do.

FAWKES, Ayscough,
Farnley Hall, Otley.

LADY IDA, roan, calved July 31, 1876. Bred by Mr. A. Fawkes; got by Baron Winsome 6th (33111), dam (Lady Imogene) by The Baron (32654), &c. See "Irving," p. 125.

Produce in	Names, &c.	By what Bull.	By whom bred.
1879, Oct. 18, roan, C.C.	Lady Rosebery	Richmond, 37340	Mr. Fawkes

LADY LE GRAND, roan, calved January 23, 1875. Bred by Mr. A. Fawkes; got by Oxford le Grand (29496), dam (Lady Vane) by Lord Cobham (20164), &c. See "Lord Rosebery," p. 156.

Produce in	Names, &c.	By what Bull.	By whom bred.
1877, July 14, roan, B.C.	Rothschild	Richmond, 37340	Mr. Fawkes
1878, June 10, roan, B.C.	(dead)	do.	do.
1879, June 12, roan, B.C.	Lord Rosebery	Baron Winsome 6th, 33111	do.

ROYAL DUCHESS, roan, calved March 21, 1875. Bred by Mr. A. Fawkes; got by Royal Crown (32372), dam (First Duchess) by Reformer (24930), g. d. (Lady Valentine) by Lord Cobham (20164), &c. See "Vandervelde." p. 255.

Produce in	Names, &c.	By what Bull.	By whom bred.
1877, Oct. 17, roan, C.C.	Rosalind	Richmond, 37340	Mr. Fawkes
1878, Nov. 10, white, B.C.	Royalist	do.	do.
1879, Oct. 19, roan, C.C.	Rosamund	Baron Winsome 6th, 33111	do.

Rosalind, sold to Mr. Fawcett, Old Bramhope, Otley.

ROYAL LADY, white, calved September 10, 1874. Bred by Mr. A. Fawkes; got by Royal Crown (32372), dam (Fourth Lady) by Lord Darlington (26633), &c. See "Rubens," Vol. xxv. p. 252.

Produce in	Names, &c.	By what Bull.	By whom bred.
1877, May 14, white, C.C.	White Lady	Baron Winsome 6th, 33111	Mr. Fawkes
1878, May 26, roan, B.C.	Rubens	Richmond, 37340	do.
1879, May 27, roan, C.C.	Mad'selle Croisette	Baron Winsome 6th, 33111	do.

White Lady, sold to Mr. J. W. Hartley, Ashfield House, Otley; Rubens, to Mr. R. J. Ramsden, Worksop.

FERGUSSON, John,
Brettenham Manor, Thetford.

DUCHESS 2ND, red and white, calved June 1, 1874, Vol. xxiv. p. 434. Bred by Mr. J. Fergusson; got by Hogarth 3rd (34167), dam (Wharfdale Duchess) by Royal Wharfdale (22805), &c.

Produce in	Names, &c.	By what Bull.	By whom bred.
1878, Nov. 30 (dead twins)		Lord Waterloo 2nd, 37016	Mr. Fergusson

GREY LADY, roan, calved June 1, 1875, Vol. xxiv. p. 434. Bred by Mr. J. Fergusson; got by Hogarth 3rd (34167), dam (Lady de Grey) by De Grey (25883), &c.

Produce in	Names, &c.	By what Bull.	By whom bred.
1878, Oct. 18, white, C.C.	White Lady	Lord Waterloo 2nd, 37016	Mr. Fergusson
1879, Aug. 10, white, B.C.	(Steer)	do.	do.

LADY DE GREY, roan, calved January 9, 1870, Vol. xxii. p. 414. Bred by Mr. J. Fergusson; got by De Grey (25883), dam (Cherry Blossom) by Hayman (16245), &c.

Produce in	Names, &c.	By what Bull.	By whom bred.
1876, April 29, roan, C.C.	Lady De Grey 2nd	Hogarth 3rd, 34167	Mr. Fergusson
1877, May 1 (dead twins)		Lord Waterloo 2nd, 37016	do.

LADY DE GREY 2ND, roan, calved April 29, 1876. Bred by Mr. J. Fergusson; got by Hogarth 3rd (34167), dam (Lady de Grey) by De Grey (25883), &c.

Produce in	Names, &c.	By what Bull.	By whom bred.
1879, July 20, roan, C.C.	Lady de Grey 3rd	Lord Waterloo 2nd, 37016	Mr. Fergusson

THORNDALE LASS, roan, calved December 31, 1868, Vols. xxi., xxii., and xxiv. pp. 703, 415, 434. Bred by Mr. J. Fergusson; got by Norfolk Thorndale Duke (24666), dam (Cherry Blossom) by Hayman (16245), &c.

Produce in	Names, &c.	By what Bull.	By whom bred.
1878, May 10, white, C.C.	Thorn	Lord Waterloo 2nd, 37016	Mr. Fergusson
1879, Mar. 17, roan, C.C.	Thorny	do.	do.

WHARFDALE DUCHESS, red and white, calved August 14, 1867, Vols. xxi., xxii., and xxiv. pp. 704, 415, 434. Bred by Mr. J. Fergusson; got by Royal Wharfdale (22805), dam (Cherry Blossom) by Hayman (16245), &c.

Produce in	Names, &c.	By what Bull.	By whom bred.
1878, Mar. 6, roan, C.C.	Duchess 5th	Lord Waterloo 2nd, 37016	Mr. Fergusson

FEVERSHAM, Earl of,
Duncombe Park, Helmsley.

BEESWING, red and white, calved July 30, 1867, Vols. xx., xxi., xxii., and xxiv. pp. 405, 704, 415, 434. Bred by the Earl of Feversham; got by Orestes (22443), dam (Barefoot) by Chanticleer (17530), &c.

Produce in	Names, &c.	By what Bull.	By whom bred.
1879, June 10, roan, B.C.	Baron Ryedale 4th	Conn'ht Knightley, 41267	Earl of Feversham

BELLE OF RYEDALE 3RD, red and white, calved March 6, 1873, Vols. xxiv. and xxv. pp. 434, 445. Bred by the Earl of Bective, Underley Hall; got by Second Duke of Tregunter (26022), dam (Princess of Battersea) by Mandarin (18317), &c.

Produce in	Names, &c.	By what Bull.	By whom bred.
1879, Dec. 30, r. & w., C.C.	Belle of Ryedale 6th	D. of Tregunter 5th, 33743	Earl of Feversham

BLOOM OF RYEDALE 4TH, roan, calved September 3, 1875, Vol. xxv. p. 445. Bred by the Earl of Feversham; got by Twentieth Duke of Oxford (28432), dam (Beeswing) by Orestes (22443), &c.

Produce in	Names, &c.	By what Bull.	By whom bred.
1879, Oct. 26, white, B.C.	Baron Ryedale 5th	D. of Tregunter 5th, 33743	Earl of Feversham

CORA, roan, calved November 16, 1872, Vol. xxiii. p. 715. Bred by Sir G. O.
Wombwell, Bart., Newburgh Park; got by Twentieth Duke of Oxford
(28432), dam (Cerito) by Vesuvius (21017), &c.

Produce in	Names, &c.	By what Bull.	By whom bred.
1879, Feb 8, roan,	C.C.|Cashmere	|Waterloo Grand D., 39291|Earl of Feversham

GWENDOLINE, roan, calved July 16, 1873, Vols. xxiii. and xxiv. pp. 435, 435.
Bred by the Earl of Dunmore, Dunmore; got by Second Duke of Collingham
(23730), dam (Lady Geneva) by Duke of Geneva (19614), &c.

1879, Mar. 6, red,	B.C.	Tregunter Knightley	D.of Tregunter 5th, 33743	Earl of Feversham

LADY CLEVELAND 3RD, red and white, calved May 6, 1875, Vol. xxv.
p. 446. Bred by the Earl of Feversham; got by Duke of Ryedale 2nd
(33727). dam (Lady Cleveland) by Tisamines (27670), &c.

1879, Feb. 3, roan,	C.C.	Lady Cleveland 4th	D.of Tregunter 5th, 33743	Earl of Feversham

LADY OF RYEDALE 7TH, roan, calved November 27, 1874. Bred by the
Earl of Feversham; got by Cleveland (33396), dam (Lady of Ryedale 2nd) by
Cedric (30673), &c. See Vol. xxi. p. 705.

1879, Nov. 2, roan,	B.C.|Ryedale Lad 6th	|D.of Tregunter 5th, 33743|Earl of Feversham

LADY OF RYEDALE 8TH, white, calved May 18, 1876. Bred by the Earl
of Feversham; got by Twentieth Duke of Oxford (28432), dam (Lady Bird)
by Mandarin (18317), &c. See Vol. xxiii. p. 436.

1879, Aug. 5, white,	B.C.|Ryedale Lad 4th	|D.of Tregunter 5th, 33743|Earl of Feversham

LADY OF RYEDALE 9TH, roan, calved July 29, 1876. Bred by the Earl of
Feversham; got by Winsome Duke (36019), dam (Lady Oxford Ryedale 2nd)
by Twentieth Duke of Oxford (28432), &c. See Vol. xxiii. p. 436.

1879, April 19, roan,	C.C.	Lady of Ryedale 11th	D.of Tregunter 5th. 33743	Earl of Feversham

LADY OXFORD RYEDALE, white, calved December 14, 1872, Vols. xxii.,
xxiii., and xxv. pp. 415, 436, 446. Bred by the Earl of Feversham; got
by Twentieth Duke of Oxford (28432), dam (Lioness) by Orestes (22443), &c.

1879, April 22, roan,	C.C.	Lady of Ryedale 12th	D.of Tregunter 5th, 33743	Earl of Feversham

LILIAN, red, calved June 14, 1865, Vols. xxi. and xxv. pp. 684, 433. Bred by
the Earl of Feversham; got by Photograph (20492), dam (Leonora) by
Splendour (15329), &c.

1879, Oct. 30, roan,	B.C.|Ryedale Lad 5th	|D.of Tregunter 5th, 33743|Earl of Feversham

LILY OF ISIS, roan, calved May 8, 1873, Vols. xxii., xxiii., and xxiv. pp. 404,
428, 422. Bred by the Hon. Cecil Duncombe, Nawton Grange; got by
Twentieth Duke of Oxford (28432), dam (Lilian) by Photograph (20492), &c.

1879, Oct. 27, roan,	C.C.	Lady of Ryedale 13th	D.of Tregunter 5th, 33743	Earl of Feversham

ROSE OF RYEDALE 8TH, roan, calved June 13, 1877. Bred by the Earl of
Feversham; got by Winsome Duke (36019), dam (Convolvulus) by Orestes
(22443), &c. See Vol. xxiv. p. 435.

1879, Dec. 14. roan,	C.C.	Rose of Ryedale 11th	D.of Tregunter 5th, 33743	Earl of Feversham

SPARKLING EYES, red and white, calved November 18, 1873, Vols. xxiv.
and xxv. pp. 435, 440. Died by the Earl of Dunmore, Dunmore; got by
Sixth Duke of Geneva (30959), dam (Wild Eyes Duchess) by Grand Duke 9th
(19879), &c.

1879, Oct. 24, r. & w.,	B.C.|Wild Tregunter	|D.of Tregunter 5th, 33743|Earl of Feversham

WILD WINSOME, roan, calved March 18, 1876, Vol. xxv. p. 446. Bred by
the Earl of Feversham; got by Twentieth Duke of Oxford (28432), dam
(Winsome 5th) by Grand Duke 10th (21848), &c.

Produce in	Names, &c.	By what Bull.	By whom bred.
1879, Oct. 26, roan, C.C.	Wild Winsome 5th	D.of Tregunter 5th, 33743	Earl of Feversham

WILD WINSOME 2ND, white, calved September 5, 1876. Bred by the Earl
of Feversham: got by Twentieth Duke of Oxford (28432), dam (Winsome
11th) by Baron Oxford 4th (25580), &c. See "Winsome Duke 3rd," p. 272.

1879, Aug. 11, white, B.C.	Winsome Duke 3rd	D.of Tregunter 5th, 33743	Earl of Feversham

WINSOME 11TH, roan, calved January 5, 1871, Vols. xxi., xxiii., xxiv., and
xxv. pp. 706, 436, 435, 447. Bred by the Duke of Devonshire, Holker Hall;
got by Baron Oxford 4th (25580), dam (Winsome 5th) by Grand Duke 10th
(21848), &c.

1879, Dec. 30, white, C.C.	Wild Winsome 6th	D.of Tregunter 5th, 33743	Earl of Feversham

WINSOME EYES 3RD, roan, calved February 11, 1869, Vols. xxi. and xxv.
pp. 687, 447. Bred by Mr. W. R. Bromet, Cocksford; got by Fifth Duke
of Wharfdale (26033), dam (Winsome Eyes) by Rose Duke (22760), &c.

1879, June 1, r. & w., C.C.	Winsome Winnie	D.of Tregunter 5th, 33743	Earl of Feversham

FIELDEN, John,
Grimston Park, Tadcaster.

QUEEN OF THE HILLS, roan, calved May 29, 1871, Vols. xxii. and xxiv
pp. 416, 436. Bred by Mr. J. B. Booth, Killerby Hall; got by Merry
Monarch (22349), dam (Queen of the Glen) by Valasco (15443), &c.

1879, April 17, roan, C.C.	Nymph	Janus, 34245	Mr. Fielden

FINDLATER, James,
Milltack, King Edward, N.B.

GOLDEN LOCKS 3RD, red, calved March 21, 1876. Bred by Mr. J. Findlater,
the property of Mr. Duff Dunbar, Ackergill Tower; got by Lord Forbes 2nd
(31658), dam (Golden Locks 2nd) by Lad O'Deveron (36882), g. d. (Golden
Locks) by Prince Imperial (22595), gr. g. d. (Golden Duchess) by Windsor
Augustus (19157), &c. See "Golden Drop," Vol. xxi. p. 536.

1878, Dec. 15, r. & w., C.C.	Golden Locks 4th	Meridian, 38748	Mr. Findlater
1879, Dec. 29, red, B.C.	North Star	do.	Mr. Duff Dunbar

FITZHARDINGE, Lord,
Berkeley Castle, Gloucestershire.

BLANCHE ROSE 5TH, red, calved October 24, 1873, Vol. xxii. p. 416. Bred
by Lord Fitzhardinge; got by Grand Duke of Waterloo (28766), dam
(Blanche Rose 4th) by General Napier (24023), &c.

1878, Mar. 8, roan, C.C.	Blanche Rose 6th	Duke of Connaught, 33604	Lord Fitzhardinge
1879, Mar. 5, red, C.C.	Blanche Rose 7th	do.	do.

Blanche Rose 6th, sold to Mr. J. H. Angas, Collingrove, Angaston, South Australia.

DOWAGER, white, calved October 5, 1877. Bred by Lord Fitzhardinge; got by Duke of Connaught (33604), dam (Dorothea) by Grand Duke of Waterloo (28766), &c. See "Viscount Dursley 2nd," p. 260.

Produce in	Names, &c.	By what Bull.	By whom bred.
1879, Nov. 6, roan,	B.C.|Viscount Dursley2nd|D.of Oxford 45th, 39775	Lord Fitzhardinge

Viscount Dursley 2nd, sold to Mr. J. Bennett, Downhouse Farm, Dursley.

GAZELLE 37TH, roan, calved November 1, 1877. Bred by Mr. E. Bowly, Siddington House; got by Beau of Oxford 2nd (33129), dam (Gazelle 30th) by Third Duke of Clarence (23727), &c. See "Lord Gifford 4th," p. 147.

Produce in	Names, &c.	By what Bull.	By whom bred.
1879, Oct. 26, roan,	B.C.|Lord Gifford 4th	D. of Oxford 45th, 39775	Lord Fitzhardinge

KIRKLEVINGTON EMPRESS 3RD, roan, calved August 18, 1877. Bred by Lord Fitzhardinge; got by Duke of Connaught (33604), dam (Kirklevington Empress) by Second Duke of Tregunter (26022), &c. See "Kirklevington Emperor," p. 132.

Produce in	Names, &c.	By what Bull.	By whom bred.
1879, Nov. 3, roan,	B.C.|Kirklev'ton Emperor|D. of Oxford 45th, 39775	Lord Fitzhardinge

KIRKLEVINGTON PRINCESS, roan, calved August 25, 1873. Bred by Mr. J. W. Larking, Ashdown House; got by Grand Duke of Geneva (28756), dam (Siddington) by Fourth Duke of Oxford (11387), &c. See "Kirklevington Prince," p. 133.

Produce in	Names, &c.	By what Bull.	By whom bred.
1879, Sept. 22, roan,	B.C.	Kirklev'gton Prince	Duke of Connaught, 33604	Lord Fitzhardinge

MINSTREL, roan, calved February 18, 1871, Vol. xxi. p. 710. Bred by Lord Fitzhardinge; got by Second Duke of Tregunter (26022), dam (Musical 14th) by Seventh Duke of York (17754), &c.

Produce in	Names, &c.	By what Bull.	By whom bred.
1878, Jan. 6, roan,	C.C.|Minstrel 8th	Duke of Connaught, 33604	Lord Fitzhardinge
1879, Mar. 1, roan,	B.C.|Lord Munster	do.	do.

Minstrel 8th, sold to Mr. J. H. Angas, Collingrove, Angaston, South Australia; Lord Munster, to Mr. S. Leonard, Fishing House, Berkeley.

OXFORD DUCHESS 2ND, roan, calved June 20, 1872, Vols. xxiii. and xxiv. pp. 438, 436. Bred by the Earl of Dunmore, Dunmore; got by Second Duke of Collingham (23730), dam (Eleventh Lady of Oxford) by Baron of Oxford (23371), &c.

Produce in	Names, &c.	By what Bull.	By whom bred.
1879, April 29, roan,	B.C.	Berkeley D.of Oxford	Duke of Connaught, 33604	Lord Fitzhardinge

Berkeley Duke of Oxford, sold to Mr. W. McCulloch, Glenroy, Melbourne.

WISDOM, roan, calved May 18, 1875, Vol. xxv. p. 448. Bred by Lord Fitzhardinge; got by Third Duke of Glo'ster (33653), dam (Winsomedale) by Baron Oxford 4th (25580), &c.

Produce in	Names, &c.	By what Bull.	By whom bred.
1879, June 2, roan,	C.C.|Wisdom 2nd	Duke of Connaught, 33604|Lord Fitzhardinge

FITZWILLIAM, Earl,
Coollattin Park, Shillelagh, Co. Wicklow.

APPLE BLOSSOM, red and white, calved April 25, 1874. Bred by Mr. W. Bolton, The Island; got by King Charles (24240), dam (Nectarine Blossom) by Doctor Collins (25905), &c. See Vol. xxi. p. 583.

Produce in	Names, &c.	By what Bull.	By whom bred.
1877, Nov. 20, red,	C.C.|Appeal	King Lud, 34327	Earl Fitzwilliam
1879, Mar. 24, r.& w.,	C.C.|Apricot	Robert Burns, 29795	do.

Appeal, sold to Monsieur F. R. de la Trehonnais, Château du Saron, sur Seine (Marne), France.

IMOGENE, roan, calved March 15, 1875.   Bred by Earl Fitzwilliam; got by
Robert Burns (29795), dam (Iris) by Lord Stanley (24466), &c.   See "In-
Esse," p. 123.

Produce in	Names, &c.	By what Bull.	By whom bred.
1878, May 22, white, C.C.	Infant	King Lud, 34327	Earl Fitzwilliam
1879, July 21, roan, C.C.	Ianthe	do.	do.

LADY LILIAS 9TH, red, calved January 6, 1874, Vol. xxiv. p. 437.   Bred by
Earl Fitzwilliam; got by Robert Burns (29795), dam (Lady Lilias) by Windsor
Augustus (19157), &c.

1879, May 3, red, C.C.	Lady Lilias 16th	King Lud, 34327	Earl Fitzwilliam

Lady Lilias 16th, sold to the Provincial Government of Liege, Belgium.

RASPBERRY, red, calved May 4, 1876.   Bred by Earl Fitzwilliam; got by
Robert Burns (29795), dam (Rachel) by Fusileer (24000), &c.   See Vol. xxi.
p. 713.

1879, Oct. 14, red, C.C.	Regina	King Lud, 34327	Earl Fitzwilliam

VENUS 27TH, red, calved March 20, 1876.   Bred by Earl Fitzwilliam; got by
Robert Burns (29795), dam (Venus 15th) by Ben Lawers (23408), g. d. (Venus
11th) by Clydesdale (15779), &c.   See "Viceroy," p. 257.

1879, Feb. 16, roan, C.C.	Venus 31st	King Lud, 34327	Earl Fitzwilliam

FOLJAMBE, F. J. S.,
Osberton Hall, Worksop.

BRIGHT DUCHESS, white, calved June 2, 1871, Vol. xxii. p. 403.   Bred by
Mr. H. A. Brassey, Preston Hall; got by Grand Duke 15th (21852), dam
(Bright Halo) by Breast Plate (19337), &c.

1879, Mar. 21, roan, C.C.	Bright Titanius	Titan, 35805	Mr. Foljambe

CRESSIDA 4TH A 3, roan, calved June 10, 1875.   Bred by the Rev. T. Staniforth, Storrs; got by Judge of Assize (34280), dam (Cressida 4th A) by
Grenadier (21876), &c.   See Vol. xxii. p. 572.

1879, Dec. 28, roan, C.C.	Cassandra	Titan, 35805	Mr. Foljambe

FORTESCUE, A. I.,
Kingcausie, Aberdeen.

LADY ELEANOR, red, calved March 10, 1876.   Bred by Mr. A. I.
Fortescue; got by Jacobite (34236), dam (Ranger) by Prince Frederick
(18603), &c.   See "Knight Errant," Vol. xxv. p. 150.

1879, Jan. 25, red, B.C.	Aneroid	Ben Nevis, 39462	Mr. Fortescue

NERISSA 5TH, roan, calved March 20, 1876.   Bred by Mr. A. I. Fortescue;
got by Jacobite (34236), dam (Nerissa) by Red Rover (42261), &c.   See
"Knight Errant," Vol. xxv. p. 150.

1879, Jan. 24, white, C.C.	Narcissus	Ben Nevis, 39462	Mr Fortescue

SITTYTON BUTTERFLY, red, calved March 8, 1874.   Bred by Mr. A. I.
Fortescue; got by Masterpiece (31882), dam (Butterfly 7th) by President
Lincoln (20513), g. d. (Butterfly 2nd) by John Bull (11618), gr. g. d.
(Butterfly) by Matadore (11800), &c.   See Vol. xv. p. 400.

1878, June 15, roan, B.C.	Coronet	Ben Nevis, 39462	Mr. Fortescue
1879, May 10, roan, C.C.	Beeswing	do.	do.

FOSTER, S. P.,
Killhow, Mealsgate, Carlisle.

CONCORD, roan, calved June 7, 1873, Vols. xxiii. and xxiv. pp. 440, 439. Bred by Mr. J. P. Foster, Killhow; got by Ravensbourne (29716), dam (Charmer 14th) by Third Duke of Geneva (23753), &c.

Produce in	Names, &c.	By what Bull.	By whom bred.
1879, Mar. 4, roan, B.C.	Cabul	Duke of Ormskirk, 36526	Mr. S. P. Foster

DAINTY 2ND, red, calved February 5, 1876. Bred by Mr. J. P. Foster, Killhow; got by Duke of Hillhurst (28401), dam (Dutiful) by Third Duke of Wharfdale (21619), &c. See " Dainty Duke 5th," p. 63.

Produce in	Names, &c.	By what Bull.	By whom bred.
1879, Feb. 12, red, B.C.	Dainty Duke 5th	D. of Glo'ster 7th, 39735	Mr. S. P. Foster

DUTIFUL, red, calved January 16, 1869, Vols. xx., xxi., xxiii., and xxiv. pp. 495, 715, 440, 440. Bred by Lord Penrhyn, Penrhyn Castle; got by Third Duke of Wharfdale (21619), dam (Duenna) by Grand Duke 11th (21849), &c. .

Produce in	Names, &c.	By what Bull.	By whom bred.
1879, Dec. 1, roan, C.C.	Dainty 4th	Sir Ormsk'k Gw'ne, 42405	Mr. Foster

GAZELLE 27TH, red, calved December 26, 1871, Vols. xxi. and xxiv. pp. 997, 440. Bred by Mr. E. Bowly, Siddington House; got by Second Duke of Tregunter (26022), dam (Gazelle 13th) by Grand Duke 13th (21850), &c.

Produce in	Names, &c.	By what Bull.	By whom bred.
1879, April 15, roan, C.C.	Grace Gazelle	22nd D. of Oxford, 31000	Mr. Foster

GRAND DUCHESS OF OXFORD 18TH, roan, calved September 27, 1870, Vols. xxi. and xxiii. pp. 566, 329. Bred by the Duke of Devonshire, Holker Hall; got by Baron Oxford 4th (25580), dam (Grand Duchess of Oxford 11th) by Grand Duke 10th (21848), &c.

Produce in	Names, &c.	By what Bull.	By whom bred.
1879, Feb. 6, roan, B.C.	Oxf'd D. of K'how 2d	Duke of Ormskirk, 36526	Mr. Foster

GRAND DUCHESS OF OXFORD 31ST, red and white, calved June 20, 1875, Vols. xxiv. and xxv. pp. 440, 451. Bred by Mr. G. Moore, Whitehall; got by Fifth Duke of Wetherby (31033), dam (Grand Duchess of Oxford 11th) by Grand Duke 10th (21848), &c.

Produce in	Names, &c.	By what Bull.	By whom bred.
1879, Dec. 29, roan, C.C.	Ox'd.D's. of K'how 3d	Duke of Ormskirk, 36526	Mr. Foster

OPAL GWYNNE, roan, calved May 15, 1874, Vol. xxiv. p. 440. Bred by Mr. J. P. Foster, Killhow; got by Baron Oxford 4th (25580), dam (Ora Gwynne) by Grand Duke of Lightburne (26290), &c.

Produce in	Names, &c.	By what Bull.	By whom bred.
1879, Jan. 8, roan, B.C.	Gerad Gwynne	Duke of Underley, 33745	Mr. S. P. Foster

ORA GWYNNE, red and white, calved July 9, 1869, Vols. xx., xxi., and xxii. pp. 675, 716, 418. Bred by Mr. C. Howard, Biddenham; got by Grand Duke of Lightburne (26290), dam (Orange Gwynne) by Grand Duke 5th (19875), &c.

Produce in	Names, &c.	By what Bull.	By whom bred.
1879, Jan. 24, r. & w., B.C.	(dead)	Duke of Ormskirk, 36526	Mr. Foster

ROSE GWYNNE 2ND, red and white, calved in August, 1870, Vols. xxii. and xxiii. pp. 418, 440. Bred by Mr. J. P. Foster, Killhow; got by Royal Cambridge (25009), dam (Polly Gwynne) by Flying Dutchman (10235), &c.

Produce in	Names, &c.	By what Bull.	By whom bred.
1879, April 7, red, C.C.	Royal Gwynne 3rd	Duke of Ormskirk, 36526	Mr. S. P. Foster

ROYAL GWYNNE 2ND, roan, calved March 21, 1876. Bred by Mr. J. P.
Foster, Killhow; got by Twenty-second Duke of Oxford (31000), dam (Ross
Gwynne 2nd) by Royal Cambridge (25009), &c.

Produce in	Names, &c.	By what Bull.	By whom bred.
1879, Oct. 14, white, C.C.	Royal Gwynne 4th	Duke of Ormskirk, 36526	Mr. S. P. Foster

SILENCE 9TH, red, calved June 25, 1874, Vol. xxiv. p. 440. Bred by Mr. A. H.
Longman, Shendish; got by Senator (32470), dam (Silence 6th) by Etoile du
Nord (21710), &c.

Produce in	Names, &c.	By what Bull.	By whom bred.
1879, Dec. 1, r. & w., B.C.	Silent Frank	Sir Ormsk'k Gw'ne, 42405	Mr. Foster

SURMISE DUCHESS 12TH, red, calved April 23, 1876. Bred by Mr. J. P.
Foster, Killhow; got by Duke of Hillhurst (28401), dam (Surmise 2nd) by
May Duke (13320), &c. See Vol. xxiii. p. 440.

Produce in	Names, &c.	By what Bull.	By whom bred.
1879, Feb. 16, roan, C.C.	Surmise Duch's 19th	Duke of Ormskirk, 36526	Mr. S. P. Foster

WILD EYES LASSIE 2ND, roan, calved May 12, 1875, Vol. xxiv. p. 441.
Bred by Mr. J. P. Foster, Killhow; got by Twenty-second Duke of Oxford
(31000), dam (Wild Eyes Lassie) by Third Duke of Claro (23729), &c.

Produce in	Names, &c.	By what Bull.	By whom bred.
1879, Mar. 24, roan, C.C.	Wild Eyes Lassie 4th	Duke of Ormskirk, 36526	Mr. S. P. Foster

WINSOME BEAUTY, roan, calved May 29, 1876. Bred by Lord Skelmersdale, Lathom House; got by Baron Oxford 4th (25580), dam (Bright
Eyes 5th) by Grand Duke 6th (19876), &c. See Vol. xxiii. p. 645.

Produce in	Names, &c.	By what Bull.	By whom bred.
1879, Nov. 9, roan. B.C.	Baron Waver	D. of Rosedale 6th, 38176	Mr. Foster

FOWLER, Francis,
Henlow, Biggleswade.

GRATITUDE, roan, calved September 23, 1875. Bred by Mr. F. Fowler; got
by Royal Warrior (35415), dam (Governess) by Prince Royal (29680), &c.
See Vol. xxiii. p. 441.

Produce in	Names, &c.	By what Bull.	By whom bred.
1878, Jan. 23, roan, C.C.	Grateful	Prince Leopold, 37239	Mr. Fowler
1879, April 12, roan, B.C.	Governor	Cecrops, 37958	do.

HEBE, roan, calved November 26, 1866, Vols. xx. and xxiii. pp. 555, 441. Bred
by Mr. T. E. Pawlett, Beeston; got by Prince Hopewell (22592), dam
(Heather Bell) by Hero (18055), &c.

Produce in	Names, &c.	By what Bull.	By whom bred.
1878, July 29, white, C.C.	Princess Helena	Prince Leopold, 37239	Mr. Fowler
1879, Dec. 5, roan, B.C.	Prince Henry	Cecrops, 37958	do.

LAURESTINA, roan, calved August 26, 1872, Vols. xxi. and xxiii. pp. 717, 441.
Bred by Mr. F. Fowler; got by Prince Royal (29680), dam (Laurel) by Prince
James (20554), &c.

Produce in	Names, &c.	By what Bull.	By whom bred.
1877, Dec. 6, roan, B.C.	Prince Imperial	Prince Leopold, 37239	Mr. Fowler
1878, Dec. 4, roan, C.C.	Laurel	do.	do.
1879, Oct. 28, roan, B.C.	Prince Napoleon	do.	do.

PRINCESS LOUISE, roan, calved September 17, 1876. Bred by Mr. F.
Fowler; got by Prince Regent (29677), dam (Hebe) by Prince Hopewell
(22592), &c. See Vol. xxiii. p. 441.

Produce in	Names, &c.	By what Bull.	By whom bred.
1879, Feb. 25, roan, C.C.	Marchioness of Lorne	Cecrops, 37958	Mr. Fowler

PRINCESS MARY, roan, calved May 29, 1874, Vol. xxiv. p. 441. Bred by
Mr. F. Fowler; got by Prince Regent (29677), dam (Lettuce Hopewell) by
Prince Hopewell (22592), &c.

Produce in	Names, &c.	By what Bull.	By whom bred.
1879, Dec. 30, roan, C.C.	Princess Agnes	Cecrops, 37958	Mr. Fowler

PRINCESS MAUD, roan, calved May 27, 1874, Vol. xxiv. p. 441. Bred by
Mr. F. Fowler; got by Prince Regent (29677), dam (Princess Pearl) by Prince
Pearl (29674), &c.

Produce in		Names, &c.		By what Bull.		By whom bred.
1878, Jan. 10, white,	C.C.	White Princess		Prince Leopold, 37239		Mr. Fowler
1879, Jan. 6, roan,	C.C.	Princess Marion		Cecrops, 37958		do.

UNA 4TH, roan, calved August 21, 1874, Vol. xxiv. p. 441. Bred by Mr. F.
Fowler; got by Hercules (34139), dam (Una 3rd) by Prince Royal (29680),
&c.

1878, Sept. 10, roan,	C.C.	Una 6th		Prince Leopold, 37239		Mr. Fowler
1879, Sept. 2, white,	C.C.	Una 7th		do.		do.

FOX, George,
Elmhurst Hall, Lichfield.

AUSTRALIA 13TH, roan, calved October 9, 1871, Vol. xxv. p. 451. Bred by
J. W. Wadsworth, Geneseo, New York, U.S.A.; got by Grand Duke of
Oxford 4th (34075), dam (Australia 4th) by Reynolds (42277), &c.

1879, April 14, roan,	C.C.	Airdrie's Aust'lia 1st	24th D. of Airdrie, 36460	Mr. Fox

CHERRY DUCHESS 25TH, red, calved April 5, 1874, Vols. xxiv. and xxv.
pp. 442, 452. Bred by Lord Penrhyn, Penrhyn Castle; got by Grand Duke
20th (31281), dam (Cherry Duchess 17th) by Third Duke of Wharfdale
(21619), &c.

1879, Dec. 7, red,	C.C.	Ch'y D's of Elmhs't 3d	5th D. of Wetherby, 31033	Mr. Fox

DAMASK 2ND, roan, calved March 16, 1870, Vol. xxv. p. 452. Bred by Mr.
J. W. Wadsworth, Geneseo, New York, U.S.A.; got by Millbrook (34851),
dam (Damask) by Mosstrooper (34877), &c.

1879, Mar. 1, r. & w.,	C.C.	Elmhurst Princess 2d	24th D. of Airdrie, 36460	Mr. Fox

DEEPDALE, roan, calved March 2, 1873, Vols. xxii., xxiv., and xxv. pp. 419,
442, 452. Bred by the Earl of Bective, Underley Hall; got by Second Duke
of Tregunter (26022), dam (Darlington 17th) by Grand Duke 11th (21849),
&c.

1879, May 19, roan,	B.C.	Darlington Duke 3rd	24th D. of Airdrie, 36460	Mr. Fox

DOWAGER DUCHESS 5TH, roan, calved July 23, 1873, Vols. xxiii. and xxv.
pp. 442, 452. Bred by Lord Penrhyn, Penrhyn Castle; got by Oxford Beau
(29485), dam (Dowager Duchess) by Third Duke of Wharfdale (21619), &c.

1879, May 22, red,	C.C.	Oxford's Dowager D's	D. of Oxford 39th, 38173	Mr. Fox

*DUCHESS 19TH, red, calved March 14, 1874. Bred by Mr. A. Renick, Clark
Co., Kentucky, U.S.A.; got by Fourth Duke of Geneva (30958), dam
(Duchess 4th) by Airdrie (30365), &c. See "Red Rose of Strathtay," Vol.
xxii. p. 406.

1879, June 13, red,	C.C.	Red Rose of Thames	24th D. of Airdrie, 36460	Mr. Fox

DUCHESS OF AIRDRIE 20TH, roan, calved January 9, 1874, Vols. xxiii.
and xxv. pp. 442, 452. Bred by Mr. A. J. Alexander, Kentucky, U.S.A.;
got by Tenth Duke of Thorndale (28458), dam (Duchess of Airdrie 9th) by
Royal Oxford (18774), &c.

1879, June 21, red,	B.C.	Duke of Elmhurst 2d	D. of Oxford 39th, 38173	Mr. Fox

DUCHESS OF KIRKLEVINGTON 3rd, roan, calved April 25, 1873, Vols. xxiii., xxiv., and xxv. pp. 442, 442, 452. Bred by Mr. J. W. Wadsworth, Geneseo, New York, U.S.A.; got by Second Duke of Oneida (33702), dam (Kirklevington 13th) by St. Valentine (35459), &c.

Produce in	Names, &c.	By what Bull.	By whom bred.
1879, Sept. 9, red,	C.C.|Wth'byKirkl'tonD's|5th D. of Wetherby, 31033	Mr. Fox

JULIA'S ROSE, red and white, calved November 23, 1873, Vol. xxv. p. 453. Bred by Mr. B. F. Vanmeter, Stock Place, Clark Co., Kentucky, U.S.A.; got by Fourth Duke of Geneva (30958), dam (Poppy's Julia) by Airdrie Duke (40959), &c.

Produce in	Names, &c.	By what Bull.	By whom bred.
1879, May 17, roan,	B.C.|Sharon Duke	24th D. of Airdrie, 36480	Mr. Fox

JULIA'S 2nd ROSE, roan, calved August 10, 1874, Vol. xxv. p. 453. Bred by Mr. B. F. Vanmeter, Stock Place, Clark Co., Kentucky, U.S.A.; got by Fourth Duke of Geneva (30958), dam (Poppy's 2nd Julia) by Airdrie Duke (40959), &c.

Produce in	Names, &c.	By what Bull.	By whom bred.
1879, June 5, red,	C.C.|Red Rose of Trent 2d|D. of Oxford 39th, 38173	Mr. Fox

KIRKLEVINGTON DUCHESS 6th, roan, calved June 23, 1870, Vols. xxiii. and xxv. pp. 443, 453. Bred by Mr. R. P. Davies, Horton; got by Second Duke of Claro (21576), dam (Kirklevington 18th) by Third Lord Oxford (22200), &c.

Produce in	Names, &c.	By what Bull.	By whom bred.
1879, June 9, roan,	B.C.|Airdr'sKirkl't'nD.3d|24th D. of Airdrie, 36460	Mr. Fox

*MAY ROSE 6th, roan, calved October 17, 1876. Bred by Mr. G. Fox at Vinewood, Kentucky, U.S.A.; got by Duke of Airdrie 14th (41348), dam (May Rose 2nd) by Miss Butterfly's Son (34860), &c. See Vol. xxv. p. 454.

Produce in	Names, &c.	By what Bull.	By whom bred.
1879, Oct. 20, r. & w.,	C.C.|Red Rose of Clwyd	D. of Oxford 39th, 38173	Mr. Fox

MAZURKA 2nd OF OAKDALE, roan, calved March 10, 1873, Vol. xxv. p. 454. Bred by Mr. G. V. Hoyle, Champlin, Clinton County, U.S.A.; got by Malcolm (41963), dam (Mazurka 8th) by Albion (19209), &c.

Produce in	Names, &c.	By what Bull.	By whom bred.
1879, Feb. 8, red,	B.C.|Mazurka Lad	D. of Oxford 39th, 38173	Mr. Fox

NORA 8th, roan, calved April 24, 1874, Vols. xxiv. and xxv. pp. 443, 454. Bred by Mr. A. Renick, Clark Co., Kentucky, U.S.A.; got by Fourth Duke of Geneva (30958), dam (Nora 6th) by Airdrie 2nd (37717), &c.

Produce in	Names, &c.	By what Bull.	By whom bred.
1879, Dec. 1, roan,	C.C.|Red Rose of Wye	5th D. of Wetherby, 31033	Mr. Fox

ONEIDA'S ROSE, roan, calved December 15, 1874, Vol. xxv. p. 454. Bred by Mr. J. G. Kinnard, Fayette Co., Kentucky, U.S.A.; got by Duke of Oneida 2nd (33702), dam (Norma) by Airdrie (30365), &c.

Produce in	Names, &c.	By what Bull.	By whom bred.
1879, June 26, roan,	C.C.|Oxford's Rose	D. of Oxford 39th, 38173	Mr. Fox

POLLY GWYNNE 8th, red, calved April 16, 1873, Vols. xxii. and xxiv. pp. 410, 443. Bred by Mr. J. J. Hetherington, Middle Farm; got by Oxford 4th (29481), dam (Polly Gwynne 2nd) by Wild Duke 4th (21107), &c.

Produce in	Names, &c.	By what Bull.	By whom bred.
1879, Dec. 9, red,	B.C.|Willie Gwynne	24th D. of Airdrie, 36460	Mr. Fox

*PRINCESS OF THORNDALE 3rd, roan, calved September 25, 1874. Bred by Colonel King, Minneapolis, Minnesota, U.S.A.; got by Duke of Hillhurst 2nd (39748), dam (Princess of Thorndale) by Sixth Duke of Thorndale (23794), g. d. (Lady Sale 9th) by Comet (42955), gr. g. d. (Lady Sale 6th) by Red Knight (35232), &c. See "Prince Saladin 3rd," p. 203.

Produce in	Names, &c.	By what Bull.	By whom bred.
1879, June 23, roan,	C.C.|Princess Airdrie	24th D. of Airdrie, 36460	Mr. Fox

ROSALIE, red and white, calved February 23, 1874, Vols. xxiii. and xxv.
pp. 444, 455. Bred by Mr. E. H. Cheney, Gaddesby Hall; got by Saladin
(35461), dam (Rosette) by Oxford Lad (24713), &c.

Produce in	Names, &c.	By what Bull.	By whom bred.
1879, Feb. 13, r. & w., B.C.	Oxford Prince 2nd	D. of Oxford 39th, 38173	Mr. Fox

*ROSE OF MASON 8TH, roan, calved November 16, 1876. Bred by Mr. G.
Fox, at Vinewood, Kentucky, U.S.A.; got by Duke of Airdrie 14th (41348),
dam (Rose of Mason 6th) by Double Duke (41334), &c. See Vol. xxv. p. 455.

1879, Aug. 14, roan, C.C.	Red Rose of Dove	D. of Oxford 39th, 38173	Mr. Fox

*SHARON LADY, roan, calved March 30, 1876. Bred by Mr. G. Fox, at
Vinewood, Kentucky, U.S.A.; got by Duke of Airdrie 14th (41348), dam
(Poppy's 2nd Julia) by Airdrie Duke (40959), g. d. (Poppy 5th) by Thirteenth
Duke of Airdrie (36459), &c. See "Sharon Duke," p. 227.

1879, June 4, roan, B.C.	(Steer)	Airdrie's Kir'ton'D., 40961	Mr. Fox

*SHARON'S ROSE, red, calved May 25, 1876. Bred by Mr. G. Fox, at Vine-
wood, Kentucky, U.S.A.; got by Duke of Vinewood 3rd (41439), dam (Belle
Rose of Duchess 2nd) by 13th Duke of Airdrie (36459), g. d. (Belle Rose
of Duchess) by Airdrie (30365), gr. g. d. (Mayflower 2nd) by Anacreon Moore
(37731), — (Mayflower) by General [Winfield] Scott (12936), &c. See
"Duke of Liverpool," Vol. xxv. p. 84.

1879, Aug. 23, red, C.C.	Red Rose of Thame	Prince Saladin, 40542	Mr. Fox

WATER LASS 2ND, red and white, calved February 19, 1870, Vols. xxi. and
xxiv. pp. 721, 443. Bred by Mr. E. H. Cheney, Gaddesby Hall; got by
General Napier (24023), dam (Water Lass) by Oxford 2nd (18507), &c.

1879, Aug. 29, r. & w., B.C.	Waterloo Duke	24th D. of Airdrie, 36460	Mr. Fox

WILEY DUCHESS 2ND, red and white, calved March 16, 1874, Vol. xxv.
p. 455. Bred by Messrs. Bush and Hampton, Clark Co., Kentucky, U.S.A.;
got by Geneva Lad (41608), dam (Duchess Paulina) by Oxford Lad (24713),
&c.

1879, June 23, { r. & w., B.C. / r. & w., B.C.	Wiley Duke 3rd / (dead) }	D. of Oxford 39th, 38173	Mr. Fox

FRANK, R. Hayston,
Ashbourne Hall, Derbyshire.

DAISY 3RD, red, calved March 31, 1875. Bred by Mr. R. H. Frank; got by
Sir Walter Oxford 2nd (44058), dam (Dewdrop 2nd) by Sir Walter Oxford
(30012), g. d. (Daisy 2nd) by Northern Light 2nd (43697), gr. g. d. (Daisy 1st)
by Sir Henry Fitz-Herbert (20827), — (Dewdrop 1st) by Lablache (16353),
— Old Daisy, bought at Kirklevington.

1877, Mar. 2, red, C.C.	Daisy 4th	Red Duke, 43876	Mr. Frank

ELEANOR 2ND, roan, calved February 12, 1873. Bred by Mr. A. Robotham,
Drayton Bassett, the property of Mr. R. H. Frank; got by Baron Butterfly
(27920), dam (Eleanor) by Castlereagh (19409), &c. See Vol. xxi. p. 911.

79. Mar. 5, roan, B.C.		Guardsman, 43333	Mr. Coath

LADY STANHOPE 4TH, roan, calved October 28, 1873. Bred by Mr. R. H.
Frank; got by Sir Walter Oxford (30012), dam (Lady Stanhope 3rd) by Northern Light 2nd (43697), g. d. (Lady Stanhope 2nd) by Sir Henry Fitz-Herbert
(20827), gr. g. d. (Sultana) by Lablache (16353), — (Lady Stanhope) by
Horrox (11591), &c. See Vol. xiv. p. 548.

Produce in	Names, &c.	By what Bull.	By whom bred.
1876. Mar. 25, white, B.C.	Ashbourne Oxford	Sir Walter Oxford, 30012	Mr. Frank

Ashbourne Oxford, sold to Mr. J. Osmaston, Osmaston Manor, Derby.

ROSINA, roan, calved March 26, 1870. Bred by Mr. C. L. Whalley, Richmond
House; got by Royal Scotforth (25042), dam (Wild Rose) by Napoleon
(20395), &c. See "Sir Walter Oxford 2nd," p. 236.

1873, Mar. 5, roan,	B.C. Sir Walter Oxford 2d	Sir Walter Oxford, 30012	Mr. Frank

RUBY 3RD, red, calved May 13, 1878. Bred by Messrs. Horsley and Son,
Colton Manor House; got by Fire King 2nd (33921), dam (Miss Field) by
Wednesfield (30281), &c. See Vol. xxv. p. 509.

1879, Dec. 27, red,	B.C. Ruben	Lord Wetherby, 40262	Mr. Frank

SUNBEAM 2ND, red and white, calved January 10, 1872. Bred by Mr. R. H.
Frank; got by Sir Walter Oxford (30012), dam (Sunbeam) by Rifleman
(20678), g. d. (Sultana) by Lablache (16353), gr. g. d. (Lady Stanhope) by
Horrox (11591), &c. See Vol. xiv. p. 548.

1879. April 16, red,	C.C. Sunbeam 3rd	Bradley Duke, 42820	Mr. Frank

VIRGINIA 5TH, roan, calved April 7, 1875. Bred by Mr. W. H. Brown,
Stepple Hall; got by Second Duke of Rowley (28441), dam (Virginia 4th)
by Eighth Duke of Geneva (28390), &c. See Vol. xxiii. p. 574.

1879, Dec. 17, white, B.C.	Paul Siddington	B. T'croft Sidd'gton, 37823	Mr. Frank

FROST, F.,
Bowforth, Kirby Moorside, York.

DAISY 3RD, white, calved April 2, 1874, Vol. xxv. p. 457. Bred by Mr. F.
Frost; got by Snowball (30018), dam (Daisy 2nd) by Feversham (23932),
&c.

1879, May 25, white, B.C.	Bowforth Duke	L.Oxf. Bri't Eyes 3rd, 38646	Mr. Frost

FANNY, red and white, calved November 25, 1874. Bred by Mr. F. Frost;
got by Snowball (30018), dam (Lady Flora) by Duke of Pickering (26006),
&c. See Vol. xxi. p. 722.

1879, Dec. 15, r. & w., C.C.	Fanny 2nd	L.Oxf. Bri't Eyes 3rd 38646	Mr. Frost

LADY FLORA, red and white, calved July 11, 1871, Vols. xxi., xxiv., and xxv.
pp. 722, 444, 457. Bred by Mr. F. Frost; got by Duke of Pickering (26006),
dam (Lucky Lass) by Lord Nelson (34606), &c.

1879, Nov. 12, red,	B.C. (Steer)	L.Oxf. Bri't Eyes 3rd, 38646	Mr. Frost

LADY FLORA 2ND, red and white, calved September 14, 1877. Bred by Mr.
F. Frost; got by Bowforth (30572), dam (Lady Flora) by Duke of Pickering
(26006), &c. See Vol. xxiv. p. 444.

1879, July 4, red,	C.C. Lady Flora 3rd	L.Oxf. Bri't Eyes 3rd, 38646	Mr. Frost

LILY 2ND, roan, calved May 17, 1875, Vol. xxv. p. 458. Bred by Mr. F. Frost; got by Bowforth (30572), dam (Lily) by Snowball (30018), &c.

Produce in		Names, &c.	By what Bull.	By whom bred.
1879, Oct. 15, red,	B.C.	Lastingham	L.Oxf.Bri't Eyes3rd,38646	Mr. Frost

MAUDE, roan, calved July 17, 1875. Bred by Mr. F. Frost; got by Bowforth (30572), dam (Lady Mary) by True Blue 2nd (25337), &c. See Vol. xxii. p. 422.

Produce in		Names, &c.	By what Bull.	By whom bred.
1878, Jan. 8, white,	B.C.	(dead)	Bowforth, 30572	Mr. Frost
1879, Aug. 8, roan,	C.C.	Maude 2nd	L.Oxf. Bri'tEyes3rd,38646	do.

MAY, roan, calved May 13, 1876. Bred by Mr. F. Frost; got by Bowforth (30572), dam (Lady Mary) by True Blue 2nd (25337), &c. See Vol. xxiii. p. 446.

Produce in		Names, &c.	By what Bull.	By whom bred.
1879, May 21, red,	B.C.	May Gosling	L.Oxf. Bri't Eyes3rd.38646	Mr. Frost

MOLLY, white, calved July 14, 1874, Vol. xxv. p. 458. Bred by Mr. F. Frost; got by Snowball (30018), dam (Lady Mary) by True Blue 2nd (25337), &c.

Produce in		Names, &c.	By what Bull.	By whom bred.
1879, May 18, white,	B.C.	Lord Howard	L.Oxf. Bri't Eyes3rd,38646	Mr. Frost

MOSS ROSE 2ND, roan, calved September 5, 1877. Bred by Mr. F. Frost; got by Bowforth (30572), dam (Moss Rose) by Snowball (30018), g. d. (Red Rose 2nd) by Feversham (23932), gr. g. d. (Red Rose) by True Blue 2nd (25337).

Produce in		Names, &c.	By what Bull.	By whom bred.
1879, July 5, roan,	C.C.	Moss Rose 3rd	L.Oxf. Bri't Eyes3rd,38646	Mr. Frost

WHITE ROSE, white, calved April 4, 1875, Vol. xxv. p. 458. Bred by Mr. F. Frost; got by Bowforth (30572), dam (Rose 3rd) by Snowball (30018), &c.

Produce in		Names, &c.	By what Bull.	By whom bred.
1879, June 8, white,	C.C.	White Rosette	L.Oxf. Bri't Eyes3rd,38646	Mr. Frost

FRY, James,
Lacock, Chippenham.

BROWN 7TH, red, calved March 7, 1873. Bred by Mr. C. Hobbs, Maisey Hampton; got by Lord Hastings (29115), dam (Brown 4th) by Bates Tertius (21249), &c. See "Oldfield Friar," p. 178.

Produce in		Names, &c.	By what Bull.	By whom bred.
1876, Apr. 1, r. & w.,	B.C.		Col. Tregunter 7th, 30769	Mr. Hobbs
1877, Mar. 15, r. & w.,	C.C.	Brown 16th	do.	do.
1878, Mar. 18, r. & w.,	B.C.	Cherry Brandy	do.	do.
1879, May 9, roan,	B.C.	Oldfield Friar	Gr'ndD.of Waterloo,34077	Mr. Fry

1876 Bull Calf, sold to Mr. Papon, Coate; Brown 16th, to Mr. T. Walker, Lechlade; Oldfield Friar, to Messrs. J. C. and J. H. Fry, Oldfield, Marshfield.

FRY, J. C. and J. H.,
Oldfield, Marshfield, Chippenham.

CLYTIE, roan, calved May 15, 1869. Bred by Mr. J. Thompson, Badminton; got by Duke of Hazlecote (25969), dam (Caroline) by Young Favourite (19729), &c. See Vol. xix. p. 430.

Produce in		Names, &c.	By what Bull.	By whom bred.
1878, April 10, roan,	C.C.	Oldfield Rose	Duke of Hinton, 43105	Messrs. Fry
1879, May 3, r. & w.,	C.C.	Oldfield Rose 2nd	Lord Tregunter, 31758	do.

FYSON, John,
Gesyns, Wickhambrook.

CARESS, roan, calved May 24, 1871, Vols. xxii., xxiii., and xxv. pp. 337, 352, 362. Bred by Lord Braybrooke, Audley End; got by Thorndale Duke (27661), dam (Memento) by Old Buck (15017), &c.

Produce in	Names, &c.	By what Bull.	By whom bred.
1879, Dec. 18, r. & w., C.C.	Fanciful	D. of Rosedale 3rd, 33723	Mr. Fyson

GAMBLE, John,
Shouldham Thorpe, Downham, Norfolk.

FAN, roan, calved March 18, 1876. Bred by Mr. J. Gamble; got by Ferdinand (36641), dam (Faultless) by Zealot (25480), &c. See Vol. xxiii. p. 447.

1879, Feb. 23, roan, C.C.	Fairy	Havelock, 36749	Mr. Gamble

FAN, white, calved December 24, 1875, Vol. xxv. p. 460. Bred by the Rev. J. N. Micklethwait, Taverham Hall; got by Royal Dublin (35354), dam (Heather Bell) by General Wetherby (24026), &c.

1879, Feb. 24, roan, C.C.	Flippant	Havelock, 36749	Mr. Gamble

FAULTLESS, red and white, calved August 10, 1869, Vols. xxi., xxii., xxiii., and xxiv. pp. 726, 424, 447, 445. Bred by Mr. J. Gamble; got by Zealot (25480), dam (Fairwater) by Plato (18552), &c.

1879, Feb. 14, roan, C.C.	Fama	Havelock, 36749	Mr. Gamble

FIERY, red, calved in February 1874, Vol. xxiv. p. 446. Bred by Mr. J. Gamble; got by Lord Cyril (29094), dam (Faultless) by Zealot (25480) &c.

1879, Feb. 22, r. & w., C.C.	Flighty	Havelock, 36749	Mr. Gamble

FRUITFUL, red, calved in December 1874, Vol. xxv. p. 460. Bred by Mr. J. Gamble; got by First Lord (33923), dam (Faultless) by Zealot (25480), &c.

1879, Oct. 29, r. & w., C.C.	Fruit Blossom	Framemaker, 33963	Mr. Gamble

HOPEFUL, roan, calved October 26, 1875, Vol. xxv. p. 460. Bred by the Rev. J. N. Micklethwait, Taverham Hall; got by Royal Dublin (35354), dam (Hearty) by Robert Peel (29796), &c.

1879, July 10, roan, C.C.	Hoiden	Framemaker, 33963	Mr. Gamble

PANDORA, roan, calved March 16, 1876. Bred by Mr. J. Gamble; got by Ferdinand (36641), dam (Pun) by Zealot (25480), &c. See "Planet," p. 187.

1879, Jan. 4, white, B.C.	(Steer)	Pr. of Knowlmere, 37257	Mr. Gamble

PLASTIC, white, calved March 28, 1868, Vols. xxi., xxiii., and xxiv. pp. 726, 447, 446. Bred by Mr. J. Gamble; got by Zealot (25480), dam (Prize) by Forester (19767), &c.

1879, Feb. 29, white, B.C.	Pericles	Havelock, 36749	Mr. Gamble

PUN, red, calved March 2, 1869, Vols. xxi., xxiii., xxiv., and xxv. pp. 726, 447, 446, 460. Bred by Mr. J. Gamble; got by Zealot (25480), dam (Pungent) by Plato (18552), &c.

1879, June 1, roan, C.C.	Pensive	Havelock, 36749	Mr. Gamble

GANDY, Captain H.,
Castle Bank, Appleby.

GRACEFUL, roan, calved September 8, 1876. Bred by Captain Gandy; got by Marquis 3rd (31826), dam (Garnet) by Grenadier (21876), g. d. (Grace) by Coxcomb (19527), &c. See Vol. xix. p. 535.

Produce in		Names, &c.	By what Bull.	By whom bred.
1879, Sept. 2, red,	B.C.	Lord Garnet	Lord Bowness, 40139	Capt. Gandy

MERRY CARLISLE, red and white, calved March 24, 1870, Vol. xxiv. p. 447. Bred by Captain Gandy; got by Earl of Carlisle (23826), dam (Charity) by Duke of Lancaster (21597), &c.

1878, July 22, red,	C.C.	Red Cowslip 5th	Marquis 3rd, 31826	Capt. Gandy
1879, Oct. 8, r. & w.,	B.C.	Lord Carlisle	Lord Bowness, 40139	do.

RED COWSLIP 2ND, red and white, calved May 28, 1875, Vol. xxv. p. 461. Bred by Captain Gandy; got by Red Duke (29734), dam (Merry Carlisle) by Earl of Carlisle (23826), &c.

1879, July 17, r. & w.,	B.C.	Gr. Duke of Cowslip	Gr. D. of Morecambe, 36722	Capt. Gandy

ROAN COWSLIP, roan, calved June 4, 1870, Vol. xxv. p. 461. Bred by Captain Gandy; got by Earl of Carlisle (23826), dam (Campanula) by Earl of Oxford (15966), &c.

1879, Mar. 20, roan,	C.C.	R'n Cowslip's F'well	Gr. D. of Morecambe, 36722	Capt. Gandy

GARBUTT, Mrs.,
Brag House, Farndale, Kirby Moorside.

LADY MARY, red and white, calved May 3, 1872, Vol. xxiv. p. 447. Bred by Mr. I. Garbutt, Sinnington; got by Handel (26335), dam (Mary) by Salesman (25081), &c.

1878, April 21, red,	C.C.	Marion	Nonpareil, 40388	Mr. I. Garbutt
1879, May 5, r. & w.,	C.C.	Marigold	do.	Mrs. Garbutt

MOSS ROSE 4TH, roan, calved April 12, 1872, Vol. xxiv. p. 447. Bred by Mr. C. Smith, Westerdale; got by Handel (26335), dam (Moss Rose 3rd) by Salesman (25081), &c.

1878, April 25, red,	B.C.	(Steer)	Nonpareil, 40388	Mr. I. Garbutt
1879, April 6, roan,	B.C.	Roseberry	do.	Mrs. Garbutt

GARFIT, Arthur,
Scothern, Lincoln.

ALICE HAWTHORN, roan, calved June 7, 1875. Bred by Mr. A. Garfit; got by Heydon Duke 3rd (31371), dam (Alonzo Premium) by Alonzo (19219), &c. See "Alec," p. 4.

1878, Mar. 20, white,	C.C.	Amalie	Lord of Scothern, 34826	Mr. Garfit
1879, Aug. 12, white,	B.C.	Alec	Grand Duke 25th, 34065	do.

Amalie, sold to Mr. G. Appleyard, Canklow Mills.

BRILLIANT ROSE 3RD, red, calved September 27, 1872, Vol. xxiii. p. 447. Bred by Mr. A. Garfit; got by Second Wharfdale Oxford (30298), dam (Brilliant Rose) by General Napier (24023), &c.

1877, Aug. 28, red,	C.C.	Blanche Rosette 4th	Lord of Scothern, 34826	Mr. Garfit
1878, Sept. 20, red,	B.C.	Lord Brilliant 4th	Grand Duke 25th, 34065	do.
1879, Aug. 16, roan,	C.C.	Blanche Rosette 5th	do.	do.

Lord Brilliant 4th, sold to Mr. C. R. Fieldsend, Kirmond, Market Rasen.

RANBY LADY, red and white, calved November 23, 1874. Bred by Mr. A. Garfit; got by Second Wharfdale Oxford (30298), dam (Reginella) by Young Cock of the Walk (23586), g. d. (Riven Dairymaid) by Welford (23196), &c. See "Recruit," Vol. xx. p. 286.

Produce in	Names, &c.	By what Bull.	By whom bred.
1877, Aug. 17, r. & w., C.C.	Rhotee	Lord of Scothern, 34626	Mr. Garfit
1878, July 7, r. & w., B.C.	Rouble	do.	do.
1879, May 13, roan, C.C.	Rozzie	Grand Duke 25th, 34065	do.

Rhotee, sold to Mr. Peacock.

GARNE, George,
Churchill Heath, Chipping Norton.

CREAM OF THE VALLEY, red and white, calved in April 1876. Bred by Mr. W. Taylor, Churchill; got by Lord Brougham (31625), dam (Cream) by President Lincoln (42151), g. d. (Clara 2nd) by Young Weathercock (15495), gr. g. d. (Clara) by Lord Milton (10461), — (Roan Crummy) by Sherborne (10805), &c. See Vol. xii. p. 577.

1879, Feb. 5, r. & w., C.C.	Cream of the Vale	Gr. D. of Marlboro', 38380	Mr. Garne

GENEVA'S SILENCE, roan, calved January 10, 1877. Bred by Mr. G. Garne; got by Grand Duke of Geneva 2nd (31288), dam (Silence 9th) by Senator (32470), &c. See Vol. xxiv. p. 450.

1879, Sept. 19, r. & w., C.C.	Semele	Gr. D. of Marlboro', 38380	Mr. Garne

J. GRAND DUCHESS, red and white, calved August 27, 1873. Bred by Mr. G. Garne; got by Grand Duke 11th (21849), dam (Japonica) by Cherry Duke (25752), &c. See Vol. xxi. p. 728.

1878, June 3, red, C.C.	(dead)	Gr. D. of Geneva 2nd, 31288	Mr. Garne
1879, April 10, red, C.C.	Jemima Geneva	do.	do.

JURA, roan, calved March 25, 1869, Vols. xxi. and xxiv. pp. 886, 327 Bred by Mr. G. C. Adkins, The Lightwoods; got by Streamer (25241), dam (Lady Julia) by Lodowick (20136), &c.

1879, June 23, roan, C.C.	Jura 2nd	Elmhurst Prince, 41503	Mr. Garne

LADY FLORA, roan, calved March 21, 1873, Vols. xxiv. and xxv. pp. 667, 428. Bred by Mrs. Strickland, Cokethorpe Park; got by Earl of Verulam (26077), dam (Lady Augusta) by Oxford 2nd (18507), &c.

1879, Mar. 3, white, C.C.	Lady Fanny	Baron Winsome 3rd, 33108	Mr. Garne

LADY GENEVA BURDETT, roan, calved March 29, 1877. Bred by Mr. G. Garne; got by Grand Duke of Geneva 2nd (31288), dam (Lady Rose Burdett) by Claro's Rose (25784), &c. See Vol. xxiv. p. 448.

1879, Dec. 30, roan, C.C.	L'y Gwynne Burdett	Gr. D. of Marlboro', 38380	Mr. Garne

MUSICAL 3RD, roan, calved June 4, 1873, Vols. xxiv. and xxv. pp. 449, 463. Bred by Mr. R. B. Hetherington, Park Head; got by Grand Duke of Lightburne 2nd (26291), dam (Musical) by Grenadier (21876), &c.

1879, Feb. 7, red, C.C.	Geneva's Gwynne 3rd	Gr. D. of Geneva 2nd, 31288	Mr. Garne

PHANTOM 10TH, roan, calved October 4, 1876. Bred by Mr. G. Garne; got by Lord Chief Justice (34507), dam (Phantom 6th) by Royal Cambridge 2nd (25010), &c. See Vol. xxiii. p. 449.

1879, Dec. 8, red, C.C.	Phantom 12th	Gr. D. of Marlboro', 38380	Mr. Garne

PINK 35TH, red and white, calved April 14, 1876.   Bred by Mr. G. Garne; got
by Grand Duke of Geneva 2nd (31288), dam (Pink 25th) by Royal Butterfly
20th (25007), &c.   See Vol. xxiii. p. 449.

Produce in	Names, &c.	By what Bull.	By whom bred.
1878, Nov. 19, r. & w., B.C.	Prince of Marlboro'	Gr. D. of Marlboro', 38380	Mr. Garne
1879, Dec. 24, r. & w., C.C.	Pink 43rd	do.	do.

THORNDALE BLANCHE, white, calved October 27, 1868, Vols. xix., xxii., and
xxiv. pp. 752, 426, 450.   Bred by Messrs. F. Leney and Sons, Wateringbury;
got by Twelfth Duke of Thorndale (26020), dam (Blanche 8th) by Clifford
(21437), &c.

1878, Feb. 25, roan, B.C.	Th'ndale Geneva 2nd	Gr. D. of Geneva 2nd, 31288	Mr. Garne
1879, Feb. 1, white, C.C.	Blanche Geneva 5th	do.	do.

Thorndale Geneva 2nd, sold to Mr. J. S. Foster, Louth.

THORNDALE BLANCHE 2ND, white, calved July 19, 1870, Vols. xxi., xxii.,
and xxiii. pp. 898, 426, 449.   Bred by Mr. T. Purkis, West Wratting Grange;
got by Sir Rainald (27485), dam (Thorndale Blanche) by Twelfth Duke of
Thorndale (26020), &c.

1877, Dec. 24, white, B.C.	Baron Geneva	Gr. D. of Geneva 2nd, 31288	Mr. Garne
1879, Jan. 7, white, C.C	Blanche Geneva 4th	do.	do.
1879, Dec. 27, roan, B.C.	Th'ndale Geneva 3rd	do.	do.

Baron Geneva, sold to Mons. A. De Woncke, Belgium.

GARNE, John,
Rissington, Burford, Oxon.

EPITAPH 27TH, roan, calved July 26, 1876.   Bred by Mr. W. Hewer, Sevenhampton; got by Numa (34931), dam (Epitaph 25th) by Colesbourne (30757),
&c.   See "Episode," Vol. xxiv. p. 93.

1878, Oct. 12, roan, C.C.	Epithet	British Prince, 37907	Mr. Garne

PARADE 2ND, roan, calved October 26, 1875.   Bred by Mr. E. J. Coleman,
Stoke Park; got by Third Duke of Glo'ster (33653), dam (Parade) by Count
Glo'ster (23637), g. d. (Parasol) by Royal Oak (16870), &c.   See Vol. xviii.
p. 649.

1879, Aug. 11, roan, B.C.	Paragon Prince	Elmhurst Prince, 41503	Mr. Garne

ROCKING GIRL, roan, calved August 6, 1875.   Bred by Mr. W. Hewer,
Sevenhampton; got by Duke of Cambridge (33587), dam (Romping Girl) by
Hotspur (28878), &c.   See Vol. xxiv. p. 386.

1878, Feb. 14, roan, C.C.	Racket	British Prince, 37907	Mr. Hewer
1879, Jan. 27, red, C.C.	Rissington Girl	do.	Mr. Garne

Racket, sold to Mr. J. Garne, Rissington.

GARNE, W. G.,
Broadmoor, Northleach, Gloucestershire.

DIDO 4TH, red and white, calved August 30, 1873, Vol. xxv. p. 464.   Bred by Mr.
G. Garne, Churchill Heath, the property of Mr. C. Walker Didcot; got by Royal
Butterfly 20th (25007), dam (Duchess 3rd) by Bibury Butterfly (25631), &c.

1879, Mar. 24, roan, B.C.	Viscount Frogmore	Sir Robert Frogmore, 40719	Mr. W. G. Garne

DUCHESS OF WARWICK, roan, calved January 8, 1873, Vol. xxiii. p. 450.
Bred by Mr. G. Garne, Churchill Heath; got by Earl of Warwickshire 3rd
(28524), dam (Butterfly's Duchess) by Royal Butterfly 20th (25007), &c.

Produce in		Names, &c.	By what Bull.	By whom bred.
1877, May 16, r. & w.,	C.C.	D'ss of Warwick 4th	Skylark, 37489	Mr. W. G. Garne
1878, May 8, roan,	B.C.	Duke of Warwick	Gr. D. of Geneva 2nd, 31288	do.
1879, April 18, r. & w.,	B.C.	Duke of Warwick 3rd	Sir Robert Frogmore, 40719	do.

Duke of Warwick, sold to M. E. von Gronow, Prussia.

NEATNESS, red and white, calved in February 1867, Vols. xix. and xx. pp. 647,
670. Bred by Messrs. T. Garne and Son, Broadmoor, the property of Mr. W.
T. Carrington, Croxden Abbey; got by Brigadier (23456), dam (Nemophila)
by Cynric (19542), &c.

Produce in		Names, &c.	By what Bull.	By whom bred.
1878, Jan. 7, roan,	B.C.	Sir Sidney Newport	Skylark, 37489	Mr. Garne
1879, Feb. 10, red,	C.C.	Young Neatness	Sir Robert Frogmore, 40719	do.

Sir Sidney Newport, sold to Mr. Fisher, South America.

NIGHT QUEEN, roan, calved December 7, 1875. Bred by Mr. A. T. Matthews,
Church Hanborough; got by Riby Royal (35265), dam (Nightfall) by Royal
Cambridge 2nd (25010), &c. See Vol. xxii. p. 503.

Produce in		Names, &c.	By what Bull.	By whom bred.
1879, July 31, roan,	B.C.	Night Prince	Scotland's Glory, 40679	Mr. Garne

NILSSON, roan, calved April 20, 1872, Vol. xxiv. p. 451. Bred by Messrs. T.
Garne and Son, Broadmoor; got by Buccaneer (25693), dam (Naomi) by
Prince Consort (22583), &c.

Produce in		Names, &c.	By what Bull.	By whom bred.
1878, July 22, roan,	C.C.	New Empress	Emperor, 41508	Mr. Garne
1879, May 27, red,	C.C.	Nancy	Sir Robert Frogmore, 40719	do.

New Empress, sold to Mr. M. Matthews, Fifield, Burford.

NONPAREIL, roan, calved in November 1867, Vols. xix. and xxi. pp. 650, 662.
Bred by Messrs. T. Garne and Son, Broadmoor, the property of Mr. J.
Widdows, Standlake; got by Prince Consort (22583), dam (New Novel) by
Royal Oak (16870), &c.

Produce in		Names, &c.	By what Bull.	By whom bred.
1877, May, white,	B.C.	(Steer)	Ragman, 35198	Mr. Garne
1879, Feb. 12, red,	C.C.	Young Nonpareil	Skylark, 37489	do.

PINK 13TH, roan, calved in March 1869, Vol. xxi. p. 730. Bred by Messrs. T.
Garne and Son, Broadmoor; got by Masterpiece (24561), dam (Young Pink)
by General Pelissier (14605), &c.

Produce in		Names, &c.	By what Bull.	By whom bred.
1876, Jan. 30, roan,	B.C.	Prince-de-Paris	Prince-le-Grand, 35127	Mr. Garne
1877, June 26, roan,	B.C.	Blooming Prince	Baron Bloom, 41014	do.
1878, June 30, roan,	B.C.	Rough Prince	Prince Arthur, 37217	do.
1879, July 16, red,	C.C.	Frogmore Pink	Sir Robert Frogmore, 40719	do.

Blooming Prince, sold to Mr. Ash, Cherington, Shipston-on-Stour; Rough Prince, to Mr. T.
Hyde, Caldicote House, Abingdon.

ROSE OF WARLABY 3RD, white, calved August 4, 1876. Bred by Mr. W.
G. Garne; got by General Primo (34018), dam (Rose of Warlaby 2nd) by
Buccaneer (25693), &c. See Vol. xxiii. p. 452.

Produce in		Names, &c.	By what Bull.	By whom bred.
1879, Jan. 3, roan,	C.C.	Rose of Warlaby 4th	Sir Robert Frogmore, 40719	Mr. Garne
1879, Nov. 24, white,	C.C.	Rose of Warlaby 5th	Sir Sidney Newport, 44063	do.

SOVERINA, roan, calved January 9, 1873, Vols. xxiii. and xxiv. pp. 405, 396.
Bred by Mr. W. T. Crosbie, Ardfert Abbey, the property of Mr. J. Harbage,
Little Compton House; got by Irish Baron (31417), dam (Sovereign's Beauty)
by Royal Sovereign (22802), &c.

Produce in		Names, &c.	By what Bull.	By whom bred.
1878, May 10, roan,	B.C.	Irish Sovereign	Royal Fitz-Rose, 37390	Mr. Garne
1879, April 21, white,	B.C.	Irish Light	Scotland's Glory, 40679	do.

Irish Sovereign, sold to Mons. A. de Woncke, Belgium.

TIMBREL, white, calved January 23, 1874, Vol. xxv. p. 465. Bred by Mr. W. T. Crosbie, Ardfert Abbey, the property of Mr. W. Mace, Milton; got by Regal Booth (27262), dam (Miriam) by Northern Light (24670), &c.

Produce in	Names, &c.	By what Bull.	By whom bred.
1879, Mar. 4, white, C.C.	Timbrel 3rd	Scotland's Glory, 40679	Mr. Garne

TULIP BLOSSOM, roan, calved March 11, 1877. Bred by Mr. W. G. Garne; got by Colonel Tregunter 7th (30769), dam (Tulip) by Duke of Lancaster (25983), g. d. (Tulip) by Captain Wetherby (21371), gr. g. d. (Blossom) by Romulus (16853), — (Flower) by Alchemist (11097).

Produce in	Names, &c.	By what Bull.	By whom bred.
1879, July 25, roan, C.C.	Tulip Flower	Prince Rupert, 40541	Mr. Garne

WINDSOR BEAUTY, red and white, calved September 18, 1876. Bred by Mr. W. G. Garne; got by Lord Chief Justice (34507), dam (Windsor Butterfly) by Royal Butterfly 20th (25007), &c. See Vol. xxiii. p. 452.

Produce in	Names, &c.	By what Bull.	By whom bred.
1878, Nov. 11, r. & w., B.C.	Young Emperor	Emperor, 41508	Mr. Garne
1879, Sept. 13, r. & w., B.C.	Lord Brett	Sir Robert Frogmore, 40719	do.

GARNER, Charles C.,
The Wolds, Snitterfield, Stratford-on-Avon.

CELESTE 3RD, red and white, calved April 28, 1868, Vols. xx. and xxii. pp. 434, 582. Bred by Lord Sudeley, Toddington; got by Lord Wild Eyes 3rd (22235), dam (Seraphina 15th) by John O'Gaunt (16322), &c.

Produce in	Names, &c.	By what Bull.	By whom bred.
1878, June 16, roan, B.C.	(dead)	Baron Winsome 3rd, 33108	Mr. Garner
1879, Sept. 29, r. & w., B.C.	(dead)	Kalafat, 40038	do.

CHLORIS, white, calved November 16, 1874, Vol. xxiv. p. 451. Bred by Mr. J. W. Philips, Heybridge; got by Lord Tregunter (31758), dam (Cherry Ripe) by Cardinal York (25719), &c.

Produce in	Names, &c.	By what Bull.	By whom bred.
1878, June 26, roan, B.C.	(Steer)	D. of Barrington 6th, 33576	Mr. Garner
1879, May 5, roan, B.C.	Cherry Barrington	do.	do.

FLORA KNIGHTLEY, roan, calved November 29, 1874. Bred by Mr. C. M. Hamer, Snitterfield; got by Knightley Wellington (31533), dam (Ruth) by Friponnier (26208), &c. See Vol. xxi. p. 753.

Produce in	Names, &c.	By what Bull.	By whom bred.
1879, Jan. 14, roan, B.C.	(dead)	Gr. D. of Oxford 2nd, 38381	Mr. Garner

GAZETTE, roan, calved March 24, 1874, Vol. xxiv. p. 451. Bred by Mr. J. W. Wilson, Broadway; got by Duke Wild Eyes (31042), dam (Gazelle 27th) by Second Duke of Tregunter (26022), &c.

Produce in	Names, &c.	By what Bull.	By whom bred.
1879, Oct. 22, roan, C.C.	Gazette 3rd	Grand Duke 29th, 38372	Mr. Garner

LAVINIA, roan, calved May 31, 1872, Vols. xxiii. and xxiv. pp. 452, 452. Bred by Mr. J. W. Wilson, Broadway; got by Eighteenth Duke of Oxford (25995), dam (Lucretia) by Barleycorn (17348), &c.

Produce in	Names, &c.	By what Bull.	By whom bred.
1879, Mar. 4, r. & w., C.C.	Leah	D. of Barrington 6th, 33576	Mr. Garner

LEONORA, roan, calved May 19, 1870, Vols. xxiii. and xxiv. pp. 453, 452. Bred by Mr. H. J. Sheldon, Brailes House; got by Duke of Brailes (23724), dam (Lucretia) by Barleycorn (17348), &c.

Produce in	Names, &c.	By what Bull.	By whom bred.
1878, Aug. 8, r. & w., C.C.	Letty	D. of Barrington 6th, 33576	Mr. Garner

LOTTIE, red and white, calved February 18, 1874, Vol. xxiv. p. 452.   Bred by
Mr. J. W. Wilson, Broadway; got by Severn Lad (29959), dam (Lucretia) by
Barleycorn (17348), &c.

Produce in	Names, &c.	By what Bull.	By whom bred.
1879, Oct. 9, r. & w., C.C.	Lottie Lumley	Kalafat, 40038	Mr. Garner

QUEEN ISABELLA, red, calved September 22, 1871.   Bred by Mr. J. C.
Adkins, Milcote; got by Nobility (26965), dam (Queen) by Grand Monarch
2nd (19887), g. d. (Isabella 5th) by Daybreak (11338), &c.   See "Red Knight
3rd," Vol. xxii. p. 219.

1878, Aug. 22, r. & w., B.C.	(dead)	Lord KingscoteRosy,34576	Mr. Garner

SYMBOL, roan, calved August 29, 1876.   Bred by Mr. C. C. Garner; got by
Duke of Sockburn (33734), dam (Symphony) by Knightley Wellington
(31533), &c.   See Vol. xxiii. p. 453.

1879, Jan. 25, roan, C.C.	Symbol 3rd	Kalafat, 40038	Mr. Garner

SYMPHONY, roan, calved September 4, 1873, Vols. xxiii. and xxiv. pp. 453,
452.   Bred by Mr. C. M. Hamer, Snitterfield; got by Knightley Wellington
(31533), dam (Gossamer 1st) by Cambridge (25705), &c.

1878, May 16, r. & w., B.C.	(Steer)	D. of Barrington 6th, 33576	Mr. Garner
1879, April 19, r. & w., B.C.	Walnut Barrington	do.	do.

GARSED, John,
The Moorlands, Cowbridge.

BLONDE, red and white, calved September 16, 1869, Vols. xx., xxiii., and xxv.
pp. 412, 453, 466.   Bred by Mr. J. Garsed; got by Lord of the Lake (26710),
dam (Blanche) by Duffryn (19592), &c.

1879, April 14, roan, B.C.	Boanerges	Kentish Charmer, 38482	Mr. Garsed

NAMELY, roan, calved October 6, 1869, Vols. xxi., xxii., and xxv. pp. 671, 397,
467.   Bred by Sir A. de Rothschild, Bart., Aston Clinton; got by Duke of
York (23804), dam (Nanny) by Fortunatus (19773), &c.

1879, Feb. 13, roan, C.C.	Nelly Daisy	Kentish Charmer, 38482	Mr. Garsed

NAUGHTY, red and white, calved November 4, 1874, Vol. xxv. p. 467.   Bred
by Sir A. de Rothschild, Bart., Aston Clinton; got by Ranger (35203), dam
(Namely) by Duke of York (23804), &c.

1879, April 13, roan, C.C.	Nina Daisy	Kentish Charmer, 38482	Mr. Garsed

NEATNESS, red and white, calved November 9, 1875.   Bred by Sir A. de
Rothschild, Bart., Aston Clinton; got by Fatherland (28574), dam (Lady
Nancy) by Duke of York (23804), g. d. (Nanny) by Fortunatus (19773), &c.
See "Norman Daisy Bull," p. 176.

1879, Jan. 17, roan, C.C.	Nymph Daisy	Kentish Charmer, 38482	Mr. Garsed

NOTABLE, roan, calved March 16, 1873, Vol. xxv. p. 467.   Bred by Sir A. de
Rothschild, Bart., Aston Clinton; got by Fatherland (28574), dam (Namely)
by Duke of York (23804), &c.

1879, Oct. 4, roan, B.C.	Norman Daisy Bull	Kentish Charmer, 38482	Mr. Garsed

Norman Daisy Bull, sold to Mr. D. J. Jenkins, Llancadle, Cowbridge.

GARTH, Francis,
Crackpot, Reeth, Yorkshire.

CARRIE ANNETTA, roan, calved February 4, 1876. Bred by Mr. F. Garth;
got by Marmion (36069), dam (Annabelle) by Frederick First Fruits (23989),
&c. See "Plato," p. 187.

Produce in		Names, &c.	By what Bull.	By whom bred.
1879, Feb. 23, red,	B.C.	(dead)	David, 36425	Mr. Garth

ROSABELLE 2ND, roan, calved February 15, 1876. Bred by Mr. F. Garth;
got by Marmion (36069), dam (Rosabelle) by Frederick First Fruits (23989),
&c. See Vol. xxi. p. 732.

1879, Dec. 13, roan,	B.C.		Provost, 40563	Mr. Garth

GAUSSEN, R. W.,
Brookmans Park, Hatfield.

JUNE ROSE, roan, calved January 20, 1874, Vol. xxv. p. 468. Bred by Mr.
R. W. Gaussen; got by Ginx (28708), dam (May Rose) by Victoria's Chip
(25377), &c.

1879, June 14, roan,	C.C.	Rosy	Royal Blood, 35340	Mr. Gaussen

MAY ROSE, roan, calved in May 1871, Vols. xxi., xxiii., xxiv., and xxv. pp. 732,
454, 454, 468. Bred by Mr. R. W. Gaussen; got by Victoria's Chip (25377),
dam (Mary Anne) by Master Wilson (34823), &c.

1879, Nov. 2, r. & w.,	C.C.	May	Royal Blood, 35340	Mr. Gaussen

MINNIE, roan, calved March 10, 1876. Bred by Mr. R. W. Gaussen; got by
St. Albans (35443), dam (Polly 2nd) by Victoria's Chip (25371), &c. See Vol.
xxiii. p. 454.

1879, Nov. 19, roan,	C.C.	Polly 3rd	Royal Blood, 35340	Mr. Gaussen

MOSS ROSE 3RD, red, calved March 7, 1873, Vol. xxv. p. 468. Bred by Mr.
R. W. Gaussen; got by Ginx (28708), dam (Moss Rose 2nd) by Master Wilson
(34823), &c.

1879, Jan. 18, red,	B.C.	(Steer)	Royal Blood, 35340	Mr. Gaussen
1879, Dec. 27, roan,	C.C.	Noisette	do.	do.

URGENT, roan, calved March 21, 1875. Bred by Mr. R. W. Gaussen; got by
Young England (31110), dam (Ursula 25th) by Lord Lally (22161), &c. See
"Regal," p. 209.

1879, Jan. 18, red,	C.C.	Active	Royal Blood, 35340	Mr. Gaussen

URSULA 25TH, roan, calved February 25, 1867, Vols. xix., xxi., xxiii., and xxv.
pp. 759, 733, 454, 468. Bred by Mr. S. Rich, Didmarton; got by Lord Lally
(22161), dam (Ursula 15th) by Third Duke of Lancaster (19624), &c.

1879, Mar. 11, r. & w.,	B.C.	Regal	Dk. of Worcester 5th, 39797	Mr. Gaussen

URSULINE, red and white, calved March 27, 1873, Vol. xxiv. p. 455. Bred by
Mr. R. W. Gaussen; got by Ginx (28708), dam (Ursula 25th) by Lord Lally
(22161), &c.

1879, Mar. 21, roan,	C.C.	The Nun	Royal Blood, 35340	Mr. Gaussen

WHITE BUTTERFLY, white, calved July 12, 1872, Vol. xxv. p. 469. Bred
by Mr. R. W. Gaussen; got by Grand Duke of Geneva (28756), dam (Pride of
Aylesford 2nd) by Grand Duke 16th (24063), &c.

Produce in	Names, &c.	By what Bull.	By whom bred.
1879, Mar. 4, white, B.C.	Hero	Dk. of Worcester 5th, 39797	Mr. Gaussen

GEEKIE, A.,
Baldowrie, Cupar Angus, N.B.

ABBESS, red and white, calved February 6, 1877. Bred by Mr. A. Geekie; got
by Belmore (33135), dam (Bessie 1st) by Mosstrooper (29395), g. d. (Lady
Gray) by Windsor Castle (27819), gr. g. d. (Bessie) by Panmure 5th (20464),
&c. See Vol. xix. p. 409.

1879, Oct. 26, r. & w., B.C.	Leopold	Neptune, 38788	Mr. Geekie

CECILY 2nd, red and white, calved February 6, 1877. Bred by Mr. A. Geekie;
got by Belmore (33135), dam (Cecily) by Windsor Castle (27819), g. d. (Lady
Betty) by Charles (17536), &c. See " Ruby," Vol. xvii. p. 730.

1879, Nov. 28, r. & w., C.C.	Cecily 3rd	Neptune, 38788	Mr. Geekie

EMMELINE 4th, red and white, calved October 20, 1876. Bred by Mr. A.
Geekie; got by Belmore (33135), dam (Emmeline 3rd) by Mosstrooper (29395),
&c. See Vol. xxiii. p. 455.

1879, Oct. 10, r. & w., C.C.	Emmeline 5th	Neptune, 38788	Mr. Geekie

ROSA, red and white, calved November 20, 1876. Bred by Mr. A. Geekie; got
by Belmore (33135), dam (Vesta 3rd) by Golden Star (31272), g. d. (Vesta) by
Windsor Castle (27819), &c. See Vol. xxii. p. 428.

1879, Dec. 5, r. & w., C.C.	Rosa 1st	Neptune, 38788	Mr. Geekie

ROSE, red and white, calved November 20, 1876. Bred by Mr. A. Geekie; got
by Belmore (33135), dam (Vesta 3rd) by Golden Star (31272), &c.

1879, Dec. 2, r. & w., B.C.	Hector	Neptune, 38788	Mr. Geekie

GEEKIE, Robert,
Rosemount, Blairgowrie, N.B.

BELLA 2nd, roan, calved January 20, 1876. Bred by Mr. R. Geekie; got by
Remus (35259), dam (Bella) by Jack (26448), &c. See Vol. xxiii. p. 455.

1878, Sept. 10, r. & w., C.C.	Bella 3rd	Coomassie, 38029	Mr. Geekie

SNOWDROP, white, calved April 6, 1874. Bred by Mr. R. Geekie; got by
Bismarck (30539), dam (Charlotte) by Jack (26448), &c. See Vol. xx. p. 437.

1877, May 10, roan, C.C.	Snowdrop 2nd	Remus, 35259	Mr Geekie
1878, April 16, roan, C.C.	Agnes	Belmore, 33135	do.

GILBERT, Rev. George,
Claxton Grange, Norwich.

RUGA, roan, calved April 20, 1877. Bred by the Rev. G. Gilbert; got by
Monarch (31930), dam (La Reine) by Victim (32767), &c. See Vol. xxiv.
p. 457.

1879, May 23, white, C.C.	Madame Vidot	Admiral, 39353	Rev. G. Gilbert

GODSON, J. S.,
Saundby, Gainsborough.

VELVET, red, calved November 12, 1868. Bred by Mr. W. Godson, Normanby
 by Stow ; got by Dulcimer (41457), dam by Lord Mar (16424), &c. See
 "Moleskin," Vol. xxv. p. 471.

Produce in		Names, &c.	By what Bull.	By whom bred.
1872, Jan. 1, red,	C.C.	Young Velvet	Ingleby Prince, 43374	Mr. J. S. Godson
1873, Mar. 5, red,	C.C.	Sealskin	Waterloo Prince, 30279	do.
1874, Mar. 10, red,	C.C.	Moleskin	D. of Liverpool, 33688	do.

YOUNG VELVET, red, calved January 1, 1872. Bred by Mr. J. S. Godson ;
 got by Ingleby Prince (43374), dam (Velvet) by Dulcimer (41457), &c.

1874, Dec. 13, roan,	C.C.	Viscountess	Royal Brilliant, 32361	Mr. Godson

VISCOUNTESS, roan, calved December 13, 1874. Bred by Mr. J. S. Godson ;
 got by Royal Brilliant (32361), dam (Young Velvet) by Ingleby Prince
 (43374), &c.

1878, Jan. 9, r. & w.,	C.C.	Vanity	Prince of the Pasture, 42214	Mr. Godson

GODSON, J. W.,
Edge Hill House, Banbury, Oxon.

TIDY, red and white, calved June 16, 1875. Bred by Mr. J. W. Godson ; got
 by Paragon (32045), dam (Trotty) by Lord Valiant (29213), &c. See Vol.
 xxiii. p. 457.

1878, May 26, roan,	C.C.	Tulip	Pasha, 38851	Mr. Godson
1879, May 22, r. & w.,	B.C.	(Steer)	do.	do.

TRISTE, roan, calved November 8, 1876. Bred by Mr. J. W. Godson ; got by
 Joe (31439), dam (Trinket) by Second Earl of Cleveland (26054), &c. See
 Vol. xxiii. p. 457.

1879, Oct. 31, r. & w.,	C.C.	Tragedy	Pasha, 38851	Mr. Godson

TROTTY, roan, calved January 27, 1871, Vols. xxi. and xxiii. pp. 735, 457.
 Bred by Mr. J. W. Godson ; got by Lord Valiant (29213), dam (Trousseau)
 by Eleventh Duke of Oxford (19632), &c.

1878, Apr. 25, r. & w.,	C.C.	Trifle	Pasha, 38851	Mr. Godson
1879, Feb. 1, white,	C.C.	Telephone (dead)	do.	do.

GODWIN, J. S. S.,
Hazlewood, Hadlow, Tonbridge

HECATE, red, calved August 22, 1876. Bred by Mr. J. S. S. Godwin ; got by
 Duke of Kirklevington 2nd (36509), dam (Ruthless) by Cambridge Duke 3rd
 (23503), &c. See Vol. xxiii. p. 457.

1879, Feb. 3, red,	C.C.	Hecate 2nd	Wild Geneva, 39320	Mr. Godwin

MARY GWYNNE, red, calved January 10, 1877. Bred by Mr. J. S. S. Godwin ;
 got by Duke of Kirklevington 2nd (36509), dam (Flora Gwynne) by Oxford
 Gwynne (24711), &c. See Vol. xxiv. p. 458.

1879, Apr. 14, r. & w.,	C.C.	Kathleen Gwynne	Wild Geneva, 39320	Mr. Godwin

SARAH 2ND, red and white, calved January 10, 1876. Bred by Mr. J. S. S. Godwin ; got by Duke of Kirklevington 2nd (36509), dam (Sarah) by Cambridge Duke 3rd (23503), &c. See Vol. xxiii. p. 458.

Produce in	Names, &c.	By what Bull.	By whom bred.
1878, Sept.20, r.&w., C.C.	Sarah 3rd	Wild Geneva, 39320	Mr. Godwin

SYLPH, red and white, calved May 30, 1875, Vol. xxv. p. 472. Bred by Mr. J. S. S. Godwin ; got by Sir Walter (37482), dam (Sweetheart 21st) by The Corsair (25291), &c.

Produce in	Names, &c.	By what Bull.	By whom bred.
1879, July 7,	roan, C.C. Sylph 3rd roan, C.C. Sylph 4th	Duke Carolus, 43067	Mr. Godwin

GORDON, James A.,
Udale, Invergordon, N.B.

ANNIE A., red and white, calved January 26, 1876. Bred by Mr. J. A. Gordon ; got by Royal Eden (35363), dam (Ada 19th) by Quintus (29705), g. d. (Ada 2nd) by Prince Carl (20538), &c. See Vol. xxi. p. 763.

Produce in	Names, &c.	By what Bull.	By whom bred.
1878, Jan. 8, red,	B.C. Albert	Rosario, 35315	Mr. Gordon
1879, Apr. 10, red,	C.C. (dead)	do.	do.

Albert, sold to Mr. Paterson, Cullaird, Inverness.

CHERRY ROSE, roan, calved January 31, 1877. Bred by Mr. J. A. Gordon ; got by Rosario (35315), dam (Cherry Ripe 3rd) by Balliemore (30412), &c. See Vol. xxiv. p. 459.

Produce in	Names, &c.	By what Bull.	By whom bred.
1879, Apr. 24, white, B.C.	Lord Chelmsford	Rosario's Heir, 42300	Mr. Gordon

Lord Chelmsford, sold to Mr. Davie, Ardgay, Ross-shire.

CLARINDA, roan, calved January 16, 1878. Bred by Mr. J. A. Gordon ; got by Rosario (35315), dam (Rhua Eden) by Royal Eden (35363), g. d. (Clara) by Argus (27894), gr. g. d. (Daisy) by Fashion (21724), — Sunflower.

Produce in	Names, &c.	By what Bull.	By whom bred.
1879, Dec. 20, roan, C.C.	Clarinda 2nd	Rosario 2nd, 42299	Mr. Gordon

CROMA 5TH, red and white, calved April 1, 1877. Bred by Mr. J. Gordon Cluny Castle ; got by Sir Windsor Broughton (27507), dam (Croma 4th) by K. C. B. (26492), &c. See Vol. xxiv. p. 461.

Produce in	Names, &c.	By what Bull.	By whom bred.
1879, Sept.22, r.&w., B.C.	Cromarty	Lord Mayor, 38626	Mr. J. A. Gordon

HIGHLAND CHERRY, roan, calved September 15, 1875. Bred by Mr. J. A. Gordon ; got by Duke of Cerisia 2nd (33595), dam (Chloris) by Cambridge Duke 4th (25706), &c. See Vol. xxii. p. 431.

Produce in	Names, &c.	By what Bull.	By whom bred.
1877, June 6, white, C.C.	(dead)	Rosario, 35315	Mr. Gordon
1879, April 1, roan, B.C.	Highlander	Highland Chief, 39996	do.

Highlander, sold to Mr. Stirling, Fairburn, Beauly.

LADY CLARA, roan, calved January 13, 1876. Bred by Mr. R. Bruce, Great Smeaton ; got by Baron Killerby (27949), dam (Lady Cecil) by Baron Cecil (27021), g. d. (Lady Hay 1st) by Aberdeen (21142), &c. See "Red Lady," Vol. xxiv. p. 460.

Produce in	Names, &c.	By what Bull.	By whom bred.
1878, July 5, red,	B.C. Strathcarron	Rosario, 35315	Mr. Gordon
1879, Oct. 18, red,	C.C. Lady Helen	do.	do.

Strathcarron, sold to Mr. W. Mitchell, Pullrossie, Sutherlandshire.

LIZZIE, red and white, calved January 8, 1877. Bred by Mr. J. A. Gordon got by Rosario (35315), dam (Luxury) by Heir of Windsor (26364), &c. See Vol. xxiii. p. 458.

Produce in	Names, &c.	By what Bull.	By whom bred.
1879, Nov. 5, r.&w., B.C.	Gladstone	Highland Chief, 39996	Mr. Gordon

MISSIE, roan, calved March 12, 1876. Bred by Mr. R. Bruce, Great Smeaton ; got by Baron Killerby (27949), dam (Mysie 12th) by Baron Laurie 2nd (25570), &c. See " Montenegro," p. 172.

Produce in		Names, &c.	By what Bull.	By whom bred.
1879, Jan. 1, roan,	B.C.	Montenegro	Rosario, 35315	Mr. Gordon

Montenegro, sold to Mr. Reid, Inchberry, Inverness.

MISS MOLLY, roan, calved March 30, 1876. Bred by Mr. R. Bruce, Great Smeaton ; got by Baron Killerby (27949), dam (Mysie 14th) by Baron Colling (25560), &c. See Vol. xxii. p. 343.

1878, June 4, white,	B.C.	Marmion	Rosario, 35315	Mr. Gordon
1879, Nov. 25, roan,	C.C.	Myrtle	do.	do.

Marmion, sold to Mr. Grant, Pollo, Invergordon.

WILLIANNA, roan, calved January 16, 1876. Bred by Mr. J. A. Gordon ; got by Royal Eden (35363), dam (Miss Castro) by Tichborne (32719), &c. See Vol. xxiii. p. 459.

1878, Jan. 9, roan,	B.C.	William	Rosario, 35315	Mr. Gordon
1879, June 2, r. & w.,	C.C.	Welcome	do.	do.

William, sold to Mr. Fleming, Ardullie, Dingwall.

GORDON, Mrs.,
Cluny Castle, Aberdeenshire.

FLORA 2ND, red and white, calved July 20, 1868, Vols. xxi., xxii., and xxiv. pp. 737, 432, 461. Bred by Mr. J. Gordon, Cluny Castle; got by Masterman (26864), dam (Flora) by Squire of Bushey (20889), &c.

1878, April 23, roan,	C.C.	Flora 18th	Baron of Knowlmere, 30474	Mrs. Gordon
1879, April 16, r. & w.,	B.C.	Elcho Cluny	Friar of Knowlmere, 33975	do.

FLORA 8TH, roan, calved February 8, 1873, Vols. xxii. and xxiv. pp. 432, 462. Bred by Mr. J. Gordon, Cluny Castle ; the property of Mr. S. R. C. Ward, Neasham Hill ; got by Windsor Booth (32875), dam (Flora) by Squire of Bushey (20889), &c.

1878, Mar. 7, white,	C.C.	Flora 16th	Sir W'dsor Br'ghton, 27507	Mr. Gordon

FLORA 10TH, red, calved June 29, 1873, Vols. xxii. and xxiv. pp. 432, 462. Bred by Mr. J. Gordon, Cluny Castle; got by Watchman 2nd (27756), dam (Flora 3rd) by Masterman (26864), &c.

1878, Mar. 13, r. & w.,	C.C.	Flora 17th	Lollius Booth, 34470	Mr. Gordon
1879, Mar. 20, r. & w.,	B.C.	Erskine Cluny	do.	Mrs. Gordon

FLORA 12TH, roan, calved November 23, 1875. Bred by Mr. J. Gordon, Cluny Castle ; got by Baron of Knowlmere (30474), dam (Flora 2nd) by Masterman (26864), &c. See " Elphinstone Cluny," p. 94.

1878, June 28, roan,	C.C.	Flora 20th	Lollius Booth, 34470	Mrs. Gordon
1879, May 26, roan,	B.C.	Elphinstone Cluny	Sir W'dsor Br'ghton, 27507	do.

MISS GOLDIE, red and white, calved April 30, 1872, Vols. xxiii. and xxiv. pp. 461, 463. Bred by Mr. J. Cochrane, Little Haddo ; got by Golden Era (31266), dam (Miss Ayre) by British Hope (21324), &c.

1878, Mar. 18, roan,	C.C.	Miss Goldie 4th	Sir W'dsor Br'ghton, 27507	Mr. Gordon
1879, April 27, r. & w.,	C.C.	Miss Goldie 5th	do.	Mrs. Gordon

PETULANCE 2ND, roan, calved March 8, 1876. Bred by Mr. J. Gordon, Cluny Castle; got by Baron of Knowlmere (30474), dam (Petulance) by Baron Colling (25560), &c. See Vol. xxiii. p. 461.

Produce in	Names, &c.	By what Bull.	By whom bred.
1879, May 18, r. & w., C.C.	Petulance 4th	Brabazon, 37881	Mrs. Gordon

VAIL 2ND, red, calved May 23, 1866, Vols. xx., xxi., xxii., xxiii., and xxiv. pp. 796, 738, 433, 462, 464. Bred by Mr. J. Gordon, Cluny Castle; got by Lord Jersey (26661), dam (Vail) by Old England (24681), &c.

| 1878, April 25, roan, C.C. | Vail 20th | Sir W'dsor Br'ghton, 27507 | Mrs. Gordon |

VAIL 12TH, roan, calved March 22, 1875. Bred by Mr. J. Gordon, Cluny Castle; got by Friar of Knowlmere (33975), dam (Vail 8th) by Watchman 2nd (27756), &c. See Vol. xxii. p. 433.

| 1878, April 26, roan, B.C. | Deas | Sir W'dsor Br'ghton, 27507 | Mr. Gordon |
| 1879, April 17, roan, C.C. | Vail 22nd | do. | Mrs. Gordon |

GORRINGE, Hugh,
Kingston-by-Sea, Brighton, Sussex.

ANEMONE, red and white, calved January 15, 1875. Bred by Mr. H. Gorringe got by Shannon (35514), dam (Amy) by Wrestler (30344), &c. See Vol. xxii. p. 434.

| 1879, Feb. 6, red, B.C. | (Steer) | Alfonso, 36116 | Mr. Gorringe |

AZALEA, roan, calved October 13, 1877. Bred by Mr. H. Gorringe; got by Shannon (35514), dam (Amy) by Wrestler (30344), &c. See Vol. xxiv. p. 464.

| 1879, Dec. 17, red, C.C. | Aloe | Sir Walter, 40725 | Mr. Gorringe |

HOPE, red, calved March 4, 1874. Bred by Mr. J. Blyth, Woolhampton; got by Knight of Geneva (31539), dam (Robsart) by Fra Diavolo (26190), &c. See Vol. xxii. p. 434.

1876, Nov. 20, roan, B.C.	(dead)	Shannon, 35314	Mr. Gorringe
1878, Jan. 6, roan, C.C.	(dead)	do.	do.
1879, Jan. 7, roan, B.C.	(Steer)	Alfonso, 36116	do.

IONE, white, calved November 8, 1874. Bred by H.R.H. the Prince of Wales Sandringham; got by General McNab (34012), dam (Flutter) by Robert Peel (29796), &c. See Vol. xxi. p. 532.

1877, June 30, roan, C.C.	Irene	The General, 35768	Prince of Wales
1878, April 14, roan, B.C.	(dead)	do.	Mr. Gorringe
1879, Mar. 30, white, C.C.	Iris	Alfonso, 36116	do.
Irene, sold to Mr. Jarvis.			

LOUISE, red, calved April 1, 1874, Vols. xxiv. and xxv. pp. 464, 473. Bred by Mr. C. Collard, Little Barton; got by Hotspur (28878), dam (Lady Louisa) by Duke of Grafton (21594), &c.

| 1879, Aug. 17, roan, C.C. | Laurel | Alfonso, 36116 | Mr. Gorringe |

MAYFLOWER, red, calved September 27, 1873. Bred by Mr. T. G. Curtler, Bevere House; got by Grand Duke of Clarence (28750), dam (Meadow Flower) by Lord Waterloo 2nd (26755), g. d. (Mermaid) by Blenheim (21288), &c. See Vol. xix. p. 627.

1876, Aug. 10, roan, B.C.	(Steer)	Bn. Barrington 6th, 33008	Mr. Curtler
1878, Jan. 17, roan, B.C.	(Steer)	Vespasian, 37621	Mr. Gorringe
1879, Mar. 6, roan, C.C.	March Flower	Alfonso, 36116	do.

YOUNG PINE APPLE 1st, roan, calved March 3, 1873, Vols. xxiii. and xxiv. pp. 482, 484. Bred by Mr. W. Hampton, Applesham; got by Wrestler (30344), dam (Pine Apple) by Harbinger (18023), &c.

Produce in		Names, &c.	By what Bull.	By whom bred.
1878, Oct. 10, roan,	C.C.	Young PineApple4th	Young Marchmont, 38704	Mr. Hampton
1879, Oct. 6, roan,	C.C.	Kingston Pine	do.	Mr. Gorringe
Young Pine Apple 4th, sold to Mr. J. Robinson.				

ROSALIND, red, calved April 2, 1877. Bred by Mr. H. Gorringe; got by The Knight (37582), dam (Ruby) by Rosicrucian (27337), g. d. (Rosy) by Rustic Prince (18789), gr. g. d. (Ruby) by Shine (13697), — bred by Mr. Harrison, Leven Hall Garth.

1879, Dec. 24, roan,	B.C.	(Steer)	Alfonso, 36116	Mr. Gorringe

WARRIOR'S HELMET, red and white, calved November 17, 1875. Bred by the Rev. J. N. Micklethwait, Taverham Hall; got by Royal Dublin (35354), dam (Wreath) by Cherry Grand Duke 3rd (28174), &c. See Vol. xxii. p. 510.

1878, Oct. 21, roan,	B.C.	(Steer)	Alfonso, 36116	Mr. Gorringe
1879, Dec. 2, roan,	B.C.	(Steer)	do.	do.

WASSAIL, roan, calved February 16, 1872, Vol. xxv. p. 474. Bred by Mr. H. A. Brassey, Preston Hall; got by Cherry Grand Duke 3rd (28174), dam (Water Snowdrop) by Prince of Warlaby (15107), &c.

1879, Feb. 5, red,	C.C.	Wassail Cup	Alfonso, 36116	Mr. Gorringe

GOULD, R. H.,
Didmarton, Chippenham.

CLARENCE BADMINTON, red and white, calved January 3, 1872, Vols. xxiv. and xxv. pp. 466, 474. Bred by Mr. W. Fox Beavan, Woodborough; got by Heir of Clarence 4th (28833), dam (Badminton Maid) by Grand Duke of York (24071), &c.

1879, April 22, roan,	C.C.	Princess of Clarence	Lowlander, 37022	Mr. Gould

COMELY 2nd, red, calved April 3, 1873, Vols. xxiii. and xxiv. pp. 463, 466. Bred by Mr. R. H. Gould; got by General Clarence (28689), dam (Comely) by Juggernaut (20035), &c.

1879, Mar. 30, red,	C.C.	Cowslip	Prince Arthur, 38892	Mr. Gould

COUNTESS, roan, calved February 12, 1876. Bred by Mr. R. H. Gould; got by Wild Duke (36001), dam (Countess of York) by Imperial Duke of York (26431), &c. See Vol. xxiii. p. 464.

1879, April 23, red,	C.C.	Countless	Prince Arthur, 38892	Mr. Gould

GLO'STER ROSE, roan, calved September 29, 1869, Vols. xx. and xxiv. pp. 541, 466. Bred by Mr. R. H. Gould; got by Second Duke of Collingham (23730), dam (Guelder Rose 3rd) by Lord Lally (22161), &c.

1879, Jan. 30, roan,	C.C.	Glo'ster Favourite	Prince Arthur, 38892	Mr. Gould

RHODA NIBLETT, roan, calved March 6, 1872, Vol. xxiv. p. 466. Bred by Mr. G. Garne, Churchill Heath; got by Third Earl of Warwickshire (28524), dam (Rebecca Niblett) by Cynric (19542), &c.

1879, April 30, roan,	C.C.	Low-Niblett	Lowlander, 37022	Mr. Gould

SATIN 2ND, red, calved May 22, 1873, Vol. xxiv. p. 466. Bred by Mr. R. H. Gould; got by General Clarence (28689), dam (Satin) by Imperial Duke of York (26431), &c.

Produce in		Names, &c.	By what Bull.	By whom bred.
1879, Feb. 21, roan,	C.C.	Saucy	Prince Arthur, 38892	Mr. Gould

GOULDER, Herbert W.,
Wimbotsham, Downham Market.

FEATHER, roan, calved October 14, 1870, Vol. xxi. p. 532. Bred by the Rev. J. N. Micklethwait, Taverham Hall; got by Theodorus (27639), dam (Fledge) by Zealot (25480), &c.

| 1878, Jan. 18, white, | C.C. | Feather White | Homer, 34170 | Mr. Goulder |
| 1879, July 30, red, | C.C. | Feather Red | Holman Hunt, 36777 | do. |

GRACE, red, calved May 20, 1872, Vol. xxii. p. 334. Bred by Mr. W. T. Brackenbury, Thorpe Hall; got by Jason (26453), dam (Graceless) by Prince Imperial (22594), &c.

| 1878, May 29, red, | C.C. | Grace 2nd | Kn'tleyFogg'thorpe,34372 | Mr. Goulder |
| 1879, June 4, roan, | C.C. | Grace 3rd | Holman Hunt, 36777 | do. |

LADY AUGUSTA, roan, calved June 13, 1876. Bred by the Rev. J. N. Micklethwait, Taverham Hall; got by Royal Dublin (35354), dam (Lady Agony) by Lord Cyril (29094), &c. See Vol. xxiii. p. 565.

| 1878, June 3, roan, | C.C. | Lady Alice | Balmoral, 32987 | Mr. Goulder |
| 1879, Dec. 24, roan, | C.C. | Lady Anna | Holman Hunt, 36777 | do. |

LADY FOGGATHORPE, roan, calved February 8, 1876. Bred by Mr. W. T. Brackenbury, Thorpe Hall; got by Knightley Foggathorpe (34372), dam (Lady Spot) by Prince Imperial (22594), &c. See Vol. xxiii. p. 349.

| 1878, July 25, red, | C.C. | Lady Fun | Holman Hunt, 36777 | Mr. Goulder |
| 1879. July 8, roan, | C.C. | Lady Fair | do. | do. |

GOW, Thomas,
Cambo, Newcastle-on-Tyne.

ANNETTE, red and white, calved June 24, 1870, Vols. xxi., xxiv. and xxv. pp. 741, 467, 474. Bred by Sir J. Rolt, Ozleworth Park; got by Duke of Brailes (23724), dam (Amy) by Duke of Darlington (21586), &c.

| 1879, July 3, roan, | C.C. | Lady Acomb 8th | D. of Oxford 27th, 33709 | Mr. Gow |

GRAHAM, Arthur,
Yanwath Hall, Penrith.

BRIGHT EYES 7TH, roan, calved October 21, 1876. Bred by Mr. A. Graham; got by Major Sandown (37037), dam (Bright Eyes 6th) by Waterloo Commander (32811), g. d. (Bright Eyes 5th) by British Hero (30606), &c. See "British Yeoman," Vol. xxiii. p. 39.

| 1879, Nov. 16, roan, | C.C. | Bright Eyes 8th | Farnley Prince, 39864 | Mr. Graham |

DUCHESS OF KNIGHTLEY 3RD, roan, calved April 3, 1877. Bred by Mrs. Fawcett, Scaleby Castle; got by Grand Duke of Kirklevington (34071), dam (Duchess of Knightley 2nd) by Fawsley Prince (31150), g. d. (Damsel) by Lord Hopewell (18239), &c. See Vol. xx. p. 473.

| 1879, Dec. 26, roan, | C.C. | D'ss of Knightley 4th | Master Smartly 2d, 40330 | Mr. Graham |

FLOWER BUD, roan, calved April 29, 1876. Bred by Mr. A. Graham ; got
by Major Sandown (37037), dam (Fairmaid 4th) by Waterloo Commander
(32811), g. d. (Fairy Queen) by Hackthorpe (21889), &c. See "Flag of Westmoreland," p. 100.

Produce in	Names, &c.	By what Bull.	By whom bred.
1879, Nov. 29, r. & w., C.C.	Floriline	Farnley Prince, 39864	Mr. Graham

GRAHAM, George,
The Oaklands, Birmingham.

ARETHUSA 2ND, red and white, calved May 8, 1872, Vols. xxi. and xxiv.
pp. 741, 468. Bred by Mr. H. J. Sheldon, Brailes House ; got by Third Earl
of Fawsley (28506), dam (Arethusa) by Duke of Brailes (23724), &c.

1878, Feb. 17, red, C.C.	Arethusa 4th	Duke of Yardley, 36556	Mr. Graham

ARETHUSA 3RD, roan, calved October 27, 1874. Bred by Mr. G. Graham ; got
by Coronation (30796), dam (Arethusa 2nd) by Third Earl of Fawsley (28506),
&c.

1879, Jan. 30, roan, C.C.	Arethusa 5th	Duke of Yardley, 36556	Mr. Graham

CAMBRIDGE ROSE 4TH, roan, calved October 13, 1875. Bred by Mr. G.
Graham ; got by Coronation (30796), dam (Cheerful Rose) by Cambridge
Duke 4th (25706), &c. See Vol. xxii. p. 436.

1878, April 17, white, C.C.	Cheerful Rose 8th	Sir Salar Jung, 39130	Mr. Graham

CAROLINE 4TH, roan, calved March 14, 1874, Vol. xxiv. p. 468. Bred by Mr.
W. Angerstein, Weeting Hall ; got by Third Duke of Claro (23729), dam
(Caroline) by Grand Duke 10th (21848), &c.

1879, Jan. 22, red, C.C.	Caroline 7th	Duke of Yardley, 36556	Mr. Graham

CHEERFUL ROSE 2ND, red and white, calved September 2, 1872, Vols. xxiii.
and xxiv. pp. 465, 468. Bred by Mr. G. Graham ; got by Lord Thorndale
(29210), dam (Cheerful Rose) by Cambridge Duke 4th (25706), &c.

1879, Jan. 20, red, B.C.	Baron Rothschild	Baron Fantail, 37790	Mr. Graham

CHEERFUL ROSE 6TH, red, calved October 8, 1876. Bred by Mr. G. Graham ; got by Lord Oxford Rosy (38647), dam (Cheerful Rose) by Cambridge
Duke 4th (25706), &c.

1879, Oct. 20, roan, B.C.	Marshal Niel	D. of Oaklands 2nd, 39765	Mr. Graham

DUCHESS 7TH, red, calved February 16, 1875, Vol. xxiv. p. 468. Bred by
Messrs. F. Leney and Sons, Wateringbury ; got by Sixth Duke of Oneida
(30997), dam (Duchess 5th) by Grand Duke of Kent (26289). &c.

1878, Aug. 1, red, B.C.	Duke of Oaklands 3d	Baron Fantail, 37790	Mr. Graham

FANTAIL 10TH, roan, calved August 5, 1874, Vol. xxiv. p. 468. Bred by Mr.
W. Angerstein, Weeting Hall ; got by Third Duke of Claro (23729), dam
(Fantail 5th) by Royal Cambridge (25009), &c.

1878, Nov. 23, roan, C.C.	Fantail 12th	King Christmas, 36837	Mr. Graham

GRÄFIN FOGGATHORPE 9TH, roan, calved April 3, 1874, Vol. xxiv. p. 468.
Bred by Mr. G. Graham ; got by Lord Thorndale (29210), dam (Gräfin Foggathorpe 4th) by Touchstone (20986), &c.

1878, June 14, white, C.C.	Gräfin F'thorpe 15th	Duke of Yardley, 36556	Mr. Graham
1879, June 1, white, C.C.	Gräfin F'thorpe 17th	do.	do.

GRÄFIN FOGGATHORPE 10th, roan, calved April 9, 1875, Vol. xxiv. p. 468. Bred by Mr. G. Graham; got by Coronation (30796), dam (Gräfin Foggathorpe 4th) by Touchstone (20986), &c.

Produce in		Names, &c.	By what Bull.	By whom bred.
1878, Dec. 24, roan,	C.C.	Gräfin F'thorpe 16th	Duke of Yardley, 36556	Mr. Graham

IDALIA 2nd, red, calved September 9, 1871, Vol. xxiii. p. 465. Bred by Mr. H. J. Sheldon, Brailes House; got by Eighteenth Duke of Oxford (25995), dam (Idalia) by Duke of Darlington (21586), &c.

Produce in		Names, &c.	By what Bull.	By whom bred.
1878, Mar. 11, roan,	C.C.	Idalia 3rd	Duke of Yardley, 36556	Mr. Graham
1879, Dec. 1, red,	C.C.	Idalia 4th	do.	do.

JESSICA 3rd, white, calved May 5, 1869. Bred by Mr. H. Adkins, Edgbaston; got by Janitor (24204), dam (Jessica) by Lodowick (20136), &c. See "Jesuit," p. 127.

Produce in		Names, &c.	By what Bull.	By whom bred.
1872, Oct. 10, white,	C.C.	Jessica 5th	Lord Thorndale, 29210	Mr. Graham
1873, Oct. 29, roan,	C.C.	Jessica 6th	do.	do.
1874, Aug. 30, roan,	B.C.	Jackdaw	do.	do.
1875, Sept. 30, white,	B.C.	Snow King	Coronation, 30796	do.
1878, Aug. 22, roan,	B.C.	Joskin	King Christmas, 36837	do.
1879, Oct. 17, white,	C.C.	Jessica 15th	D. of Oaklands 2nd, 39765	do.

JESSICA 4th, roan, calved September 27, 1872, Vol. xxiv. p. 469. Bred by Mr. G. Graham; got by Lord Thorndale (29210), dam (Jessica) by Lodowick (20136), &c.

Produce in		Names, &c.	By what Bull.	By whom bred.
1878, Feb. 17, roan,	C.C.	Jessica 11th	Duke of Yardley, 36556	Mr. Graham
1879, Jan. 28, roan,	B.C.	Jesuit	do.	do.

JESSICA 9th, white, calved April 13, 1876. Bred by Mr. G. Graham; got by Lord Oxford Rosy (38647), dam (Jessica 6th) by Lord Thorndale (29210), g. d. (Jessica 3rd) by Janitor (24204), &c.

Produce in		Names, &c.	By what Bull.	By whom bred.
1878, Oct. 30, roan,	C.C.	Jessica 13th	King Christmas, 36837	Mr. Graham

LADY CLARA 2nd, red and white, calved August 1, 1872, Vols. xxiii. and xxiv. pp. 465, 469. Bred by Mr. G. Graham; got by The Cardinal (27612) dam (Lady Clara) by Fourth Duke of Thorndale (17750), &c.

Produce in		Names, &c.	By what Bull.	By whom bred.
1878, Aug. 15, red,	B.C.	Lord Loftus	{ King Christmas, 36837, or Ambassador, 39372 }	Mr. Graham

Lord Loftus, sold to Mr. C. Bayes, Kettering.

LADY JOCELYN 11th, red, calved October 4, 1874, Vol. xxiv. p. 469. Bred by Mr. G. Graham; got by Duke of Oaklands (30995), dam (Lady Jocelyn 2nd) by Janitor (24204), &c.

Produce in		Names, &c.	By what Bull.	By whom bred.
1879, June 18, r. & w.,	C.C.	Lady Jocelyn 20th	D. of Oaklands 2nd, 39765	Mr. Graham

LADY JOCELYN 13th, red, calved December 15, 1874, Vol. xxiv. p. 469. Bred by Mr. G. Graham; got by Coronation (30796), dam (Lady Jocelyn 6th) by The Earl (27624), &c.

Produce in		Names, &c.	By what Bull.	By whom bred.
1878, Oct. 31, roan,	C.C.	Lady Jocelyn 18th	King Christmas, 36837	Mr. Graham

LADY JOCELYN 16th, white, calved May 13, 1876. Bred by Mr. G. Graham; got by Lord Oxford Rosy (38647), dam (Lady Jocelyn 10th) by Lord Thorndale (29210), &c. See Vol. xxiii. p. 465.

Produce in		Names, &c.	By what Bull.	By whom bred.
1878, Nov. 21, roan,	C.C.	Lady Jocelyn 19th	Ambassador, 39372	Mr. Graham

LADY JOCELYN 17th, red, calved April 18, 1877. Bred by Mr. G. Graham; got by Ironstone (36792), dam (Lady Jocelyn 13th) by Coronation (30796), &c.

Produce in		Names, &c.	By what Bull.	By whom bred.
1879, Nov. 1, red,	C.C.	Lady Jocelyn 21st	D. of Oaklands 2nd, 39765	Mr. Graham

WESTON'S COLUMBIA, roan, calved April 7, 1876. Bred by Mr. G. Fox, Elmhurst Hall; got by Grand Duke of Weston (34079), dam (Columbia's Duchess 4th) by Eighth Duke of Geneva (28390), &c. See Vol. xxiii. p. 442.

Produce in		Names, &c.	By what Bull.	By whom bred.
1878, June 4, red,	C.C.	West'n's Col'mbia 2d	Baron Fantail, 37790	Mr. Graham
1879, July 17, white,	C.C.	West'n's Col'mbia 3d	D. of Oaklands 2nd, 39765	do.

GRAHAM, Miss Isabella,
Yardley Stud Farm, Birmingham.

LADY JOCELYN 6TH, red, calved June 10, 1871, Vols. xxi., xxiii., and xxv. pp. 742, 465, 475. Bred by Mr. G. Graham, The Oaklands; got by The Earl (27624), dam (Lady Jocelyn 3rd) by Janitor (24204), &c.

1879, Jan. 9, roan,	C.C.	Bella Jocelyn 2nd	Duke of Yardley, 36556	Miss Graham
1879, Dec. 8, roan,	C.C.	Bella Jocelyn 3rd	do.	do.

LADY JOCELYN 15TH, roan, calved January 23, 1876. Bred by Mr. G. Graham, The Oaklands; got by Coronation (30796), dam (Lady Jocelyn 6th) by The Earl (27624), &c. See Vol. xxiii. p. 465.

1879, July 20, roan,	C.C.	Bella Jocelyn 4th	Baron Fantail, 37790	Miss Graham

MARIE STUART 2ND, red, calved February 9, 1876. Bred by Miss I. Graham; got by Lord Oxford Rosy (38647), dam (Marie Stuart) by Coronation (30796), &c. See Vol. xxiii. p. 467.

1879, Mar. 21, red,	C.C.	Marie Stuart 3rd	Baron Fantail, 37790	Miss Graham

GRAHAM, J. Maxtone,
Battleby, Redgorton, N.B.

VERITY, roan, calved March 12, 1875. Bred by Mr. J. M. Graham; got by Aide-de-Camp (30364), dam (Vanity) by Forth (17856), &c. See Vol. xx. p. 797.

1878, Feb. 15, roan,	B.C.	(dead)	Prince Rupert, 38953	Mr. Graham
1879, April 14, red,	C.C.		Knightley Friar, 41773	do.

VOLATILE, roan, calved March 15, 1876. Bred by Mr. J. M. Graham; got by Prince Rupert (38953), dam (Vanity) by Forth (17856), &c.

1879, May 9, white,	B.C.		Knightley Friar, 41773	Mr. Graham

WATER LADY, roan, calved March 22, 1876. Bred by Mr. D. Fisher, Pitlochrie, the property of Mr. J. M. Graham; got by Lieutenant General (31600), dam (Waterloo 27th) by Prince Bertram (27119), g. d. (Waterloo 24th) by Doctor McHale (15887), gr. g. d. (Waterloo 22nd) by Speculator (13775), &c. See Vol. xviii. p. 775.

1878, Sept. 30. r. & w.,	C.C.	Water Witch	Prince Thomas, 35183	Mr. Fisher

GRAHAM, Rev. P.,
Turncroft, Over Darwen, Lancashire.

BELLE OF OXFORD 5TH, roan, calved March 14, 1873, Vols. xxii., xxiii., xxiv., and xxv. pp. 437, 467, 469, 475. Bred by Lord Penrhyn, Penrhyn Castle; got by Oxford Beau (29485), dam (Belle of Oxford 3rd) by Marmaduke (14897), &c.

1879, Dec. 5, roan,	C.C.	T'e'ft Belle of O'fd 2d	D. of Tregunter 7th. 38194	Rev. P. Graham

CHERRY DUCHESS 24TH, red, calved February 14, 1874, Vols. xxiv. and
xxv. pp. 470, 475. Bred by Lord Penrhyn, Penrhyn Castle; got by Grand
Duke 20th (31281), dam (Cherry Duchess 11th) by Duke of Geneva
(19614), &c.

Produce in	Names, &c.	By what Bull.	By whom bred.
1879, April 20, red, C.C.	T'c'ft Cherry D'ss 2d	D. of Tregunter 7th, 38194	Rev. P. Graham

FANTAIL 4TH, roan, calved May 21, 1867, Vols. xxi., xxiii., xxiv., and xxv.
pp. 700, 468, 470, 475. Bred by Mr. J. P. Foster, Killhow; got by Touchstone (20986), dam (Fantail) by Barleycorn (17348), &c.

1879, Sept. 16, red, B.C.	B. T'croft Fantail 3d	D. of Tregunter 7th, 38194	Rev. P. Graham

GRAND DUCHESS OF OXFORD 16TH, red, calved November 14, 1869,
Vols. xx., xxiii., and xxiv. pp. 549, 468, 470. Bred by the Duke of Devonshire, Holker Hall; got by Grand Duke 17th (24064), dam (Grand Duchess
of Oxford 6th) by Imperial Oxford (18084), &c.

1879, Feb. 11, red, C.C.	Ly. T'croft Oxford 2d	D. of Tregunter 7th, 38194	Rev. P. Graham

LADY ASHTON WILD EYES 2ND, red, calved November 8, 1874, Vols.
xxiv. and xxv. pp. 661, 476. Bred by Mrs. C. Starkie, Ashton Hall; got by
Grand Duke of Thorndale 2nd (31298), dam (Winsome 10th) by Eighteenth
Duke of Oxford (25995), &c.

1879, Nov. 29, red, C.C.	Ly. T'c'ft W'd Eyes 2d	D. of Tregunter 7th, 38194	Rev. P. Graham

LADY THORNDALE BATES, red and white, calved October 10, 1868. Vols.
xx., xxiv., and xxv. pp. 615, 636, 476. Bred by Mr. W. W. Slye, Beaumont
Grange; got by Fourth Duke of Thorndale (17750), dam (Lady Bates 3rd) by
Fourth Duke of Oxford (11387), &c.

1879, Dec. 30, r. & w., B.C.	B. T'croft Bates 6th	D. of Tregunter 7th, 38194	Rev. P. Graham

LADY TURNCROFT BATES, white, calved August 7, 1872, Vols. xxiii. and
xxiv. pp. 468, 470. Bred by the Rev. P. Graham; got by Third Duke of
Wharfdale (26119), dam (Lady Calcaria Bates) by Fourth Duke of Thorndale
(17750), &c.

1879, Jan. 23, roan, C.C.	Ly. T'croft Bates 4th	B. T'croft Oxf'd 2nd, 33087	Rev. P. Graham

LADY TURNCROFT BATES 2ND, roan, calved September 20, 1875, Vol xxv.
p. 476. Bred by the Rev. P. Graham; got by Sixth Duke of Oneida (30997),
dam (Lady Calcaria Bates) by Fourth Duke of Thorndale (17750), &c.

1879, July 4, red, C.C.	Ly. T'croft Bates 6th	D. of Tregunter 7th, 38194	Rev. P. Graham

LADY TURNCROFT BATES 3RD, roan, calved May 6, 1876, Vol. xxv. p. 476.
Bred by the Rev. P. Graham; got by Baron Turncroft Oxford 2nd (33087),
dam (Lady Turncroft Bates) by Third Duke of Wharfdale (21619), &c.

1879, Nov. 11, roan, C.C.	(dead)	D. of Tregunter 7th, 38194	Rev. P. Graham

LADY TURNCROFT KIRKLEVINGTON, red, calved October 29, 1870.
Bred by the Rev. P. Graham; got by Baron Turncroft Oxford 2nd (33087),
dam (Kirklevington 7th) by Northern Duke (22431), &c. See Vol. xxiii. p. 468.

1879, Jan. 30, red, C.C.	Ly. T'c'ft Kirk'ton 2d	D. of Tregunter 7th, 38194	Rev. P. Graham

MARCHIONESS 5TH, red and white, calved June 3, 1871, Vols. xxiii. and
xxv. pp. 468, 476. Bred by the Earl of Dunmore, Dunmore; got by Second
Duke of Collingham (23730), dam (Siddington) by Fourth Duke of Oxford
(11387), &c.

1879, Aug. 12, red, C.C.	M'ch'ness of T'croft 3d	D. of Tregunter 7th, 38194	Rev. P. Graham

PRINCESS TURNCROFT GWYNNE, roan, calved November 23, 1875. Bred
by the Rev. P. Graham; got by Baron Turncroft Oxford 2nd (33087), dam
(Princess Gwynne 2nd) by Royal Cambridge (25009), &c.   See Vol. xxii.
p. 437.

Produce in		Names, &c.		By what Bull.		By whom bred.
1879, Mar. 16, roan,	B.C.	Pr. T'croft Gwyn'e 2d	D. of Tregunter 7th, 38194	Rev. P. Graham

SATYRA, roan, calved March 13, 1875, Vols. xxiv. and xxv. pp. 471, 477.
Bred by Lord Skelmersdale, Lathom House; got by Prince Victor (32211),
dam (Satanella) by Grand Duke 7th (19877), &c.

1879, May 29, red,	C.C.	Ly. T'croftCh'mer 2d	D. of Tregunter 7th, 38194	Rev. P. Graham

GRAHAM, William,
Eden Grove, Bolton, Penrith.

BELLE OF THE HALL, roan, calved December 26, 1874, Vol. xxiv. p. 471.
Bred by Mr. J. Lamb, Burrell Green; got by Hubback Junior (31395), dam
(Belle of Burrell Green) by Ignoramus (28887), &c.

1878, Dec. 5, roan,	C.C.	Brilliant Belle	Brilliant Butterfly, 36270	Mr. Graham

LAURESTINA 2ND, red, calved December 28, 1871, Vols. xxi. and xxiv.
pp. 799, 471. Bred by Mr. J. Lamb, Burrell Green; got by Indian Chief
(26434), dam (Laurestina) by Edgar (19680), &c.

1879, Mar. 14, r. & w., B.C.	(dead)	Baron Oxford 5th, 27958	Mr. Graham

LAURESTINA 4TH, red and white, calved January 14, 1875, Vol. xxiv. p. 471.
Bred by Mr. J. Lamb, Burrell Green; got by Hubback Junior (31395), dam
(Laurestina 2nd) by Indian Chief (26434), &c.

1879, Jan. 18, red	C.C.	Laurestina 10th	Brilliant Butterfly, 36270	Mr. Graham

LAURESTINA 7TH, red and white, calved October 31, 1876. Bred by Mr. J.
Lamb, Burrell Green; got by The Colonel (35747), dam (Laurestina 3rd) by
Hubback Junior (31395), &c.   See Vol. xxiv. p. 471.

1879, Oct. 21, roan,	B.C.	Lord Bolton 2nd	L. Ormskirk Gw'ne, 41905	Mr. Graham

LAURESTINA 8TH, roan, calved February 5, 1877. Bred by Mr. J. Lamb,
Burrell Green; got by Baron Winsome (30498), dam (Laurestina 2nd) by
Indian Chief (26434), &c.   See Vol. xxiv. p. 471.

1879, Oct. 5, roan,	B.C.	Lord Bolton 1st	L. Ormskirk Gw'ne, 41905	Mr. Graham

GRAHAM, Colonel,
Graham's Hill, Ecclefechan, N.B.

AVICE NO. 6, roan, calved July 1, 1871, Vols. xxi. and xxiv. pp. 742, 467.
Bred by Colonel Graham; got by Prelate (29570), dam (Avice) by Mac Turk
(14872), &c.

1878, Feb. 9, roan,	C.C.	Avice No. 7	Sweetheart's D. 2nd, 37552	Colonel Graham
1879, Feb. 3, roan,	B.C.	Borderer	do.	do.

BRENDA, roan, calved February 3, 1875. Bred by Colonel Graham; got by
Prelate (29570), dam (Princess Dagmar) by Fourteenth Duke of Oxford
(21605), &c.   See Vol. xxiii. p. 466.

1879, April 17, r. & w., C.C.	Min na	Sweetheart's D. 2nd, 37552	Colonel Graham

COWSLIP, roan, calved March 14, 1875. Bred by Colonel Graham ; got by
   Prelate (29570), dam (Cara Mia) by Warrior (21067), &c.   See Vol.
   xxiii. p. 466.

Produce in	Names, &c.	By what Bull.	By whom bred.
1879, Mar. 17, r. & w., C.C.	Tulip	Sweetheart's D. 2nd, 37552	Colonel Graham

PRINCESS DAGMAR, roan, calved April 20, 1866, Vols. xxi. and xxiii.
   pp. 743, 466. Bred by Mr. J. Fawcett, Scaleby Castle ; got by Fourteenth
   Duke of Oxford (21605), dam (Matchless 9th) by Richard (16834), &c.

1878, Jan. 1, roan, C.C.	Dagmar 2nd	Sweetheart'sD. 2nd, 37552	Colonel Graham

GRAHAM, Y. R.,
Yardley Stud Farm, Birmingham.

DUCHESS OF UIST, roan, calved March 16, 1876, Vol. xxv. p. 477. Bred by
   Mr. Y. R. Graham ; got by Coronation (30796), dam (Countess 4th) by Touchstone (20986), &c.

1879, Aug. 8, red, C.C.	Lady Uist 2nd	Baron Fantail, 37790	Mr. Graham

FANTAIL 4TH, roan, calved December 2, 1876. Bred by Mr. Y. R. Graham ;
   got by Coronation (30796), dam (Fantail 3rd) by Lord Thorndale (29210), &c.
   See " Beaudesert," p. 23.

1879, Feb. 20, red, C.C.	Fantail 5th	Royal Cadet, 40621	Mr. Graham

LADY MARGARET, white, calved May 29, 1877. Bred by Mr. Y. R.
   Graham ; got by Duke of Yardley (36556), dam (Lorelei 2nd) by Coronation
   (30796), g. d. (Lorelei) by Touchstone (20986), &c. See " North Northumberland," p. 177.

1879, Oct. 16, roan, C.C.	Lady Margaret 4th	Royal Cadet, 40621	Mr. Graham

LORELEI 2ND, white, calved March 12, 1875, Vols. xxiv. and xxv. pp. 472,
   477. Bred by Mr. Y. R. Graham ; got by Coronation (30796), dam (Lorelei)
   by Touchstone (20986), &c.

1879, April 30, roan, C.C.	Lady Margaret 3rd	Baron Fantail, 37790	Mr. Graham

LORELEI 3RD, roan, calved March 27, 1876, Vol. xxv. p. 478. Bred by Mr.
   Y. R. Graham ; got by Coronation (30796), dam (Lorelei) by Touchstone
   (20986), &c.

1879, April 24, roan, C.C.	Merry Duchess 2nd	Baron Fantail, 37790	Mr. Graham

VIOLETTA, roan, calved May 17, 1871, Vol. xxiv. p. 473. Bred by Mr. Y. R.
   Graham ; got by The Earl (27624), dam (British Belle) by Second Duke of
   Thorndale (17748), &c.

1879, April 28, roan, C.C.	Violetta 3rd	Baron Fantail, 37790	Mr. Graham

GRAHAME, Thomas,
20 Chiswick Street, Carlisle.

PERFECTION, roan, calved April 10, 1875. Bred by Mr. J. Johnstone,
   Halleaths ; got by Magician (34720), dam (Purity) by Rifleman (32304), &c.
   See Vol. xxiii. p. 515.

1879, May 30, roan, C.C.	Perfection 2nd	Captain, 37936	Mr. Grahame

GRAVE, Stephen,
Mirkholme, Keswick.

COLUMBINE, roan, calved September 24, 1869. Bred by Mr. W. Marshall, Patterdale Hall; got by Emperor (23883), dam (White Columbine) by Helvellyn (34136), &c. See "Columbus," p. 53.

Produce in		Names, &c.	By what Bull.	By whom bred.
1877, Jan. 13, roan,	C.C.	Columbine Charmer	Consol, 38024	Mr. Grave
1878, Jan. 9, roan,	C.C.	Dent's Columbine	Sweethearts Duke, 39179	do.
1879, Mar. 16, roan,	B.C.	Columbus	Young Catesby, 41192	do.

GRAY, George,
Glenapp, Co. Armagh.

DUCHESS 13TH, red, calved January 28, 1872, Vol. xxi. p. 643. Bred by Colonel Towneley, Towneley Park; got by Baron Oxford (23375), dam (Duchess 7th) by Grand Duke of Lancaster (19883), &c.

1879, Sept. 10, roan,	C.C.	Duchess 19th	Knight of Raby, 34393	Mr. Gray

GREEN, D. A.,
East Donyland, Colchester.

ETTY GWYNNE, roan, calved March 3, 1873, Vol. xxiv. p. 474. Bred by Captain Townshend, Caldicote; got by Lord Versailles (29214), dam (Flavia Gwynne) by Oxford Gwynne (24711), &c.

1878, Aug. 6, roan,	B.C.	Gwynne Boy	Lord Oxford 7th, 38645	Mr. Green
1879, June 28, roan,	C.C.	Edith Gwynne	do.	do.

FLUFFY GWYNNE, roan, calved February 26, 1877. Bred by Mr. D. A. Green; got by Duke of Glo'ster 5th (36494), dam (Etty Gwynne) by Lord Versailles (29214), &c. See Vol. xxiv. p. 474.

1879, May 21, roan,	C.C.	Fancy Gwynne	Duke of Oxford, 39770	Mr. Green

GERTRUDE, red and little white, calved September 12, 1877. Bred by Mr. D. A. Green; got by Lord Oxford 7th (38645), dam (Cambridge Geraldine 3rd) by Lord Cambridge (38588), &c. See Vol. xxiv. p. 474.

1879, Oct. 26, roan,	C.C.	Gaiety	Prince of the Roses, 42216	Mr. Green

KIRKLEVINGTON DUCHESS 22ND, white, calved August 6, 1875. Bred by Mr. R. P. Davies, Horton; got by Oxford's King (34997), dam (Kirklevington Duchess 10th) by Second Duke of Tregunter (26022), &c. See Vol. xxii. p. 394.

1879, Jan. 7, roan,	C.C.	Kirklevington Lady	Lord Oxford 7th, 38645	Mr. Green

GREENWOOD, F. B.,
Swarcliffe, Ripley, Yorkshire.

CALAIS, roan, calved March 8, 1873, Vols. xxii., xxiv., and xxv. pp. 439, 475, 479. Bred by Mr. J. Greenwood, Swarcliffe; got by Tenth Lord (29050), dam (Crecy) by Royal Buck (25004), &c.

1879, July 1, r. & w.,	C.C.	Calicia	Kt. of Windermere, 38533	Mr. F. Greenwood

CHATELAINE, roan, calved February 21, 1874, Vols. xxiv. and xxv. pp. 475, 479. Bred by Mr. F B Greenwood; got by High Constable (34158), dam (Crecy) by Royal Buck (25004), &c.

Produce in	Names, &c.	By what Bull.	By whom bred.
1879, Dec. 5, white, C.C.	Cashmere	Kt. of Knowlmere, 38533	Mr. F. Greenwood

CHICAGO, white, calved September 28, 1876. Bred by Mr. F. B. Greenwood; got by Peer of the Realm (27057), dam (Corisande) by Grindelwald (26323), &c. See Vol. xxiii. p. 471.

Produce in	Names, &c.	By what Bull.	By whom bred.
1879, Aug. 12, white, C.C.	Cheyenne	Kt. of Windermere, 38533	Mr. F. Greenwood

CONFERENCE, roan, calved February 22, 1877. Bred by Mr. F. B. Greenwood; got by Hautboy (36748), dam (Chatelaine) by High Constable (34158), &c.

Produce in	Names, &c.	By what Bull.	By whom bred.
1879, June 8, roan, B.C.	Lord Beaconsfield	Kt. of Windermere, 38533	Mr. F. Greenwood

Lord Beaconsfield, sold to Mr. J. Marshall, Hayshaw Moor, Dacre, Yorks.

FERULA, red, calved December 21, 1873, Vols. xxiii. and xxv. pp. 472, 479. Bred by Mr. J. Greenwood, Swarcliffe; got by Tenth Lord (29050), dam (Flora) by Grindelwald (26323), &c.

Produce in	Names, &c.	By what Bull.	By whom bred.
1879, Mar. 8, r. & w., B.C.	Florian	Kt. of Knowlmere, 38533	Mr. F. Greenwood

Florian, sold to Messrs. Downes and Co., for exportation to America.

FLORA, red, calved March 23, 1870, Vols. xxi., xxiv., and xxv. pp. 744, 475, 479. Bred by Mr. J. Greenwood, Swarcliffe; got by Grindelwald (26323), dam (Florence) by Gamecock (19807), &c.

Produce in	Names, &c.	By what Bull.	By whom bred.
1879, June 3, roan, C.C.	Fluellin	Kt. of Knowlmere, 38533	Mr. F. Greenwood

FOXGLOVE, red and white, calved May 15, 1876. Bred by Mr. F. B. Greenwood; got by High Constable (34158), dam (Flora) by Grindelwald (26323), &c. See Vol. xxiv. p. 475.

Produce in	Names, &c.	By what Bull.	By whom bred.
1879, Dec. 5, r. & w., C.C.	Field Flower	Kt. of Windermere, 38533	Mr. F. Greenwood

FRAXINELLA, roan, calved February 17, 1874, Vol. xxv. p. 479. Bred by Mr. J. Greenwood, Swarcliffe; got by High Constable (34158), dam (Flora) by Grindelwald (26323), &c.

Produce in	Names, &c.	By what Bull.	By whom bred.
1879, May 24, roan, C.C.	(dead)	Kt. of Windermere, 38533	Mr. F. Greenwood

LADY ROCHESTER, white, calved March 17, 1873, Vols. xxii., xxiv., and xxv. pp. 439, 476, 479. Bred by Mr. J. Greenwood, Swarcliffe; got by King James (28971), dam (Revival) by Reformer (18687), &c.

Produce in	Names, &c.	By what Bull.	By whom bred.
1879, July 14, white, B.C.	Obstructionist	Kt. of the Heather, 38524	Mr. F. Greenwood

SALSIFY, roan, calved February 24, 1875, Vols. xxiv. and xxv. pp. 476, 479. Bred by Mr. F. B. Greenwood; got by High Constable (34158), dam (Sylvia 4th) by Grindelwald (26323), &c.

Produce in	Names, &c.	By what Bull.	By whom bred.
1879, Oct. 4, roan, C.C.	Seakale	Kt. of Windermere, 38533	Mr. F. Greenwood

SCYLLA, roan, calved June 9, 1870. Bred by Mr. F. B. Greenwood; got by Peer of the Realm (27057), dam (Sorceress) by Grindelwald (26323), &c. See Vol. xxiii. p. 472.

Produce in	Names, &c.	By what Bull.	By whom bred.
1879, Jan. 16, white, C.C.	Sesame	Kt. of Windermere, 38533	Mr. F. Greenwood

STAUBACH, roan, calved September 6, 1874, Vol. xxiv. p. 476. Bred by Mr. F. B. Greenwood; got by Peer of the Realm (27057), dam (Simplon) by Grindelwald (26323), &c.

Produce in	Names, &c.	By what Bull.	By whom bred.
1879, May 2, white, B.C.	Bel Alp	Kt. of Windermere, 38533	Mr. F. Greenwood

Bel Alp, sold to Mr. G. Carter, Sandhill Farm, Bedale.

STYRIA, roan, calved February 5, 1873, Vols. xxiii., xxiv., and xxv. pp. 472, 476, 479.   Bred by Mr. J. Greenwood, Swarcliffe; got by Tenth Lord (29050), dam (Sylvia 4th) by Grindelwald (26323), &c.

Produce in	Names, &c.	By what Bull.	By whom bred.
1879, April 26, white, B.C.	Bulgarian	Kt. of Windermere, 38533	Mr. F. Greenwood
Bulgarian, sold to Mr. J. Brown, Fountaines, Ripon.			

SYRA, white, calved June 12, 1877.   Bred by Mr. F. B. Greenwood; got by Hautboy (36748), dam (Seringa) by High Constable (34158), &c.   See Vol. xxiv. p. 476.

1879, Oct. 19, white, C.C.	Sherpur	Kt. of Windermere, 38533	Mr. F. Greenwood

GRIFFIN, C. W.,
Werrington, Peterborough.

BLUSH 8TH, roan, calved December 9, 1876.   Bred by Mr. C. W. Griffin; got by Telemachus 10th (35728), dam (Blush 2nd) by Baron Torr (23380), &c. See "Bacchus 2nd," p. 9.

1879, May 23, roan, B.C.	Bacchus 2nd	Beaconsfield, 42767	Mr. Griffin

LADY BLANCHE 3RD, roan, calved April 24, 1876.   Bred by Mr. C. W. Griffin; got by Telemachus 9th (35727), dam (Lady Blanche 2nd) by Nestor (24648), or Sugar Cane (32625), &c.   See "Werrington 2nd," p. 265.

1878, Aug. 23, white, B.C.	Werrington	Barabbas, 37762	Mr. Griffin
1879, Sept. 14, roan, C.C.	Lady Blanche 5th	Telemachus 10th, 35728	do.

NETWORK, roan, calved September 12, 1877.   Bred by Mr. C. W. Griffin; got by Telemachus 10th (35728), dam (Nettie 2nd) by Duke of Kingscote (25981), g. d. (Netty) by Jasper (26456), &c.   See Vol. xix. p. 649.

1879, Dec. 15, roan, B.C.	(dead)	Beaconsfield, 42767	Mr. Griffin

GRIMES, J.,
Newton-on-Trent, Newark.

FAIRY 3RD, red, calved June 9, 1873, Vol. xxiv. p. 477.   Bred by Mr. J. Burgess, Edenham; got by Prince Boabdil (27120), dam (Fairy 2nd) by Prince Boabdil (27120), &c.

1878, April 3, red, C.C.	Fairy 4th	Bright Prince, 28083	Mr. Grimes

RANBY, roan, calved in 1875.   Bred by Mr. J. S. Walesby, Ranby; got by Prince Royal 2nd (35176), dam by Cock of the Walk (15782), g. d. by Alonzo (19219), gr. g. d. by Soho (32563), — by Splendid 3rd (32583), — by Belvedere 4th (3130).

1879, May 16, roan, B.C.	Reformer	Prince Royal 2nd, 35176	Mr. Grimes

GUMBLETON, R. J. M.,
Glanatore, Tallow, Co. Waterford.

CASKET, red and white, calved September 14, 1869, Vol. xxv. p. 480.   Bred by Mr. J. B. Chirnside, Newham; got by Royal Knight (25032), dam (Castanet) by Prince Arthur (13497), &c.

1879, Aug. 28, roan, C.C.	Constance	Captain Cook, 39548	Mr. Gumbleton

EMMA OPOPONAX, roan, calved January 3, 1877. Bred by Mr. R. J. M.
Gumbleton; got by Opoponax (34950), dam (Twinkle) by Duke of Flanders
(21590), &c. See Vol. xxi. p. 746.

Produce in	Names, &c.	By what Bull.	By whom bred.
1879, Aug. 13, white, B.C.	Lord Cook	Captain Cook, 39548	Mr. Gumbleton

LADY MATILDA, roan, calved January 11, 1876. Bred by Mr. R. J. M.
Gumbleton; got by Opoponax (34950), dam (Lady Martha) by Elfin King
(17796), &c. See Vol. xxiii. p. 475.

1879, Sept. 6, roan, C.C.	Lady Mabel	Captain Cook, 39548	Mr. Gumbleton

NEMESIS, red, calved June 6, 1871, Vol. xxiv. p. 478. Bred by Mr. J. G. Grove,
Castle Grove; got by Great Hope (24082), dam (Nonsense) by Sir Roger
(16991), &c.

1879, Mar. 20, r. & w., C.C.	Nemesis 3rd	Bismarck Baron, 30545	Mr. Gumbleton

PRIMROSE, roan, calved March 28, 1866, Vols. xix. and xx. pp. 674, 698. Bred
by Mr. J. W. Philips, Heybridge; got by Lord Liverpool (22168), dam
(Marian) by Prince Alfred (13494), &c.

1879, Mar. 28, roan, C.C.	Maria Louise	Royal Fitz-Rose, 37390	Mr. Gumbleton

SERAPHINA, roan, calved March 30, 1866, Vols. xix., xxi., xxiii., and xxv.
pp. 728, 746, 476, 480. Bred by Mr. J. Byrne, Wallstown Castle; got by
Ducrow (19571), dam (Serena) by Royal Oxford (18775), &c.

1879, Feb. 1, red, C.C.	Sophie	Bismarck Baron, 30545	Mr. Gumbleton

SYLPHIDE 5TH, red and white, calved April 27, 1876. Bred by Mr. R. J. M.
Gumbleton; got by Opoponax (34950), dam (Sylphide 2nd) by Lord Stanley
(24466), &c. See Vol. xxiii. p. 476.

1879, Sept. 26, roan, C.C.	Sylphide 10th	Captain Cook, 39548	Mr. Gumbleton

GUNNIS, George,
Bratoft, Burgh, Lincolnshire.

COSEY, red, calved November 2, 1876. Bred by Mr. G. Gunnis; got by Syrian
Prince (35715), dam (Coral) by Seventh Duke of York (17754), &c. See
"Candleshoe," p. 40.

1879, Aug. 27, red, B.C.	Candleshoe	Syrian Duke, 40782	Mr. Gunnis

GUNTER, Colonel,
The Grange, Wetherby, Yorkshire.

ACOMB 6TH, roan, calved February 23, 1877. Bred by Colonel Gunter; got
by Duke of Tregunter 5th (33743), dam (Ada) by Eighteenth Duke of Oxford
(25995), &c. See Vol. xxiv. p. 479.

1879, Aug. 1, r. & w., C.C.	Acomb 7th	D. of Clarence 5th, 38133	Colonel Gunter

BLANCHE ROSE 3RD, red and white, calved March 22, 1870, Vol. xxiii. p. 517.
Bred by Mr. E. H. Cheney, Gaddesby Hall; got by General Napier (24023),
dam (Olga) by Old Buck (15017), &c.

1879, Mar. 4, r. & w., C.C.	Blanche 12th	Thurston, 39222	Colonel Gunter

CLARENCE ROSE, white, calved October 18, 1873, Vol. xxiv. p. 479. Bred by Colonel Gunter; got by Duke of Clarence (19611), dam (Thorndale Rose) by Fourth Duke of Thorndale (17750), &c.

Produce in	Names, &c.	By what Bull.	By whom bred.
1878, Feb. 16, roan, B.C.	(dead)	18th D. of Oxford, 25995	Colonel Gunter
1879, March 2, roan, C.C.	Oxford Rose	do.	do.

DARLINGTINA 5TH, roan, calved July 16, 1875. Bred by Colonel Gunter; got by Duke of Siddington (28445), dam (Damsel) by Oxford Don (20451), &c. See Vol. xxii. p. 441.

Produce in	Names, &c.	By what Bull.	By whom bred.
1879, Aug. 16, roan, B.C.	Baron Darlington 2d	Duke Gwynne, 39704	Colonel Gunter

DARLINGTINA 6TH, roan, calved September 5, 1875. Bred by Colonel Gunter; got by Duke of Clarence (19611), dam (Darlingtina) by Archduke (25525), &c. See Vol. xxii. p. 441.

Produce in	Names, &c.	By what Bull.	By whom bred.
1879, Apr. 18, r. & w., C.C.	Darlingtina 10th	D. of Clarence 5th, 38133	Colonel Gunter

DARLINGTINA 9TH, white, calved October 1, 1876. Bred by Colonel Gunter; got by Fifth Duke of Tregunter (33743), dam (Darlingtina 2nd) by Third Duke of Wharfdale (21619), &c. See Vol. xxiii. p. 477.

Produce in	Names, &c.	By what Bull.	By whom bred.
1879, May 2, roan, B.C.	Baron Darlington	Duke Gwynne, 39704	Colonel Gunter

DUCHESS 111TH, roan, calved August 27, 1873, Vols. xxiii., xxiv., and xxv. pp. 478, 479, 481. Bred by Colonel Gunter; got by Second Duke of Barrington (30922), dam (Duchess 109th) by Second Duke of Claro (21576), &c.

Produce in	Names, &c.	By what Bull.	By whom bred.
1879, Dec. 27, r. & w., B.C.	D. of Wetherby 7th	D. of Underley 3rd, 38196	Colonel Gunter

EUTERPE, roan, calved March 21, 1876. Bred by the Earl of Zetland, Upleatham; got by Grand Monarch (28774), dam (Eleanor) by King of the Roses (22043), &c. See Vol. xxiv. p. 662.

Produce in	Names, &c.	By what Bull.	By whom bred.
1879, Mar. 16, roan, C.C.	Venilia	Duke Gwynne, 39704	Colonel Gunter

FLORA 2ND, roan, calved May 3, 1876. Bred by Colonel Gunter; got by Duke of Tregunter 5th (33743), dam (Floriline) by Third Duke of Wharfdale (21619), &c. See Vol. xxiii. p. 478.

Produce in	Names, &c.	By what Bull.	By whom bred.
1879, Mar. 3, roan, B.C.	Florist	Tregunter, 42516	Colonel Gunter

Florist, sold to Sir W. A. Lethbridge, Bart., Sandhill Park, Taunton.

LADY'S MAID, white, calved December 22, 1872, Vols. xxiv. and xxv. pp. 480, 482. Bred by Colonel Gunter; got by Duke Ferdinand 3rd (28354), dam (Milkmaid) by Marquis (20290), &c.

Produce in	Names, &c.	By what Bull.	By whom bred.
1879, July 23, roan, B.C.	Lanercost	18th D. of Oxford, 25995	Colonel Gunter

LILY BLANCHE, white, calved September 21, 1876. Bred by Mr. M. Kennedy, Stone Cross; got by Second Duke of Glo'ster (28392), dam (Oxford's Blanche) by Twenty-third Duke of Oxford (31001), &c. See Vol. xxiii. p. 518.

Produce in	Names, &c.	By what Bull.	By whom bred.
1879, Aug. 7, white, C.C.	Blanche 14th	D. of Clarence 5th, 38133	Colonel Gunter

MILD EYES 7TH, roan, calved December 28, 1871, Vols. xxii., xxiii., and xxiv. pp. 442, 478, 480. Bred by Colonel Gunter; got by Third Duke of Wharfdale (21619), dam (Bright Eyes 3rd) by Beau of Oxford (21254), &c.

Produce in	Names, &c.	By what Bull.	By whom bred.
1878, July 3, roan, B.C.	(dead)	18th D. of Oxford, 25995	Colonel Gunter
1879, July 24, roan, B.C.	Mowbray	do.	do.

MILD EYES 8TH, red and white, calved July 23, 1873, Vols. xxiii. and xxiv.
pp. 478, 480. Bred by Colonel Gunter; got by Fifth Duke of Wetherby
(31033), dam (Mild Eyes 5th) by Third Duke of Wharfdale (21619), &c.

Produce in	Names, &c.	By what Bull.	By whom bred.
1877, Oct. 2, red,	B.C.|(dead)	D. of Tregunter 5th, 33743	Colonel Gunter
1879, April 4, red,	B.C.|Duke of Gunterstone	Oxf'd Cherry D. 2nd, 34972	do.

MILD EYES 10TH, roan, calved August 22, 1876. Bred by Colonel Gunter;
got by Duke of Tregunter 5th (33743), dam (Mild Eyes 5th) by Third Duke
of Wharfdale (21619), &c. See "Duke of Gunterstone," p. 78.

1879, March 2, roan, B.C.|Lord Mild Eyes	Duke Gwynne, 39704	Colonel Gunter

MISS BEVERLEY 2ND, roan, calved March 31, 1869, Vols. xxii., xxiii., and
xxiv. pp. 400, 682, 688. Bred by Mr. G. Dixon, Mantle Hill; got by Lieutenant Oxford (24336), dam (Miss Beverley 18th) by Royal Butterfly 16th
(20724), &c.

1879, Oct. 17, roan, B.C.|Baron Beverley	Oxford's Prince, 34998	Colonel Gunter

Baron Beverley, sold to Mr. Tennant, Flint-Mill Grange, Wetherby.

MISS BEVERLEY 3RD, white, calved May 13, 1873, Vol. xxiv. p. 688. Bred
by Mr. G. Dixon, Mantle Hill; got by Duke of Tynedale (28462), dam (Miss
Beverley 2nd) by Lieutenant Oxford (24336), &c.

1879, Oct. 30, roan, B.C.|Lord Beverley	Oxford's Prince, 34998	Colonel Gunter

MISS BEVERLEY 4TH, roan, calved June 2, 1874, Vol. xxiv. p. 688. Bred by
Mr. G. Dixon, Mantle Hill; got by Cherubim 2nd (30727), dam (Miss Beverley
2nd) by Lieutenant Oxford (24336), &c.

1879, Oct. 21, roan, B.C.|Beverley	Oxford's Prince, 34998	Colonel Gunter

Beverley, sold to Mr. Tomlinson, Cowthorpe, Wetherby.

MISS BEVERLEY 27TH, red, calved November 5, 1874, Vols. xxiv. and xxv.
pp. 480, 482. Bred by Colonel Gunter; got by Eighteenth Duke of Oxford
(25995), dam (Miss Beverley 26th) by Royal Butterfly 16th (20724), &c.

1879, July 6, r. & w., C.C.|Miss Beverley 32nd	D. of Clarence 5th, 38133	Colonel Gunter

MISS BEVERLEY 28TH, red, calved March 23, 1876. Bred by Colonel Gunter;
got by Eighteenth Duke of Oxford (25995), dam (Miss Beverley 26th) by
Royal Butterfly 16th (20724), &c.

1879, March 4, red, C.C.|Miss Beverley 31st	Oxf'd Cherry D. 2nd, 34972	Colonel Gunter

MOONLIGHT, roan, calved February 26, 1877. Bred by the Earl of
Feversham, Duncombe Park; got by Twentieth Duke of Oxford (28432),
dam (Midnight) by Lord Nelson (26693), &c. See Vol. xxiv. p. 435.

1879, July 17, roan, C.C.|Blanche 13th	D. of Clarence 5th, 38133	Colonel Gunter

MOONSHINE, red, calved October 19, 1874. Bred by the Earl of Feversham,
Duncombe Park; got by Baron Napier (30471), dam (Midnight) by Lord
Nelson (26693), &c. See Vol. xxi. p. 705.

1879, Aug. 27, red, C.C.|Blanche 15th	D. of Clarence 5th, 38133	Colonel Gunter

ONDINE, roan, calved September 6, 1874, Vols. xxiv. and xxv. pp. 481, 482.
Bred by Colonel Gunter; got by Eighteenth Duke of Oxford (25995), dam
(Ondine's Last) by Nobility (22415), &c.

1879, Aug. 1, roan, C.C.|Ondinette	Village Squire, 37630	Colonel Gunter

TUSCARORA 2ND, roan, calved May 18, 1873, Vols. xxiii., xxiv., and xxv. pp. 626, 633, 646. Bred by Mr. J. A. Rolls, The Hendre; got by Towneley Wild Eyes (27674), dam (Tuscarora) by The Bully (23019), &c.

Produce in	Names, &c.	By what Bull.	By whom bred.
1879, July 9, white,	B.C.|Earl of Tregunter	D.of Siddington 2nd,33732	Colonel Gunter

GURNEY, F. J.,
Henlow, Biggleswade.

PATIENCE, roan, calved March 15, 1877. Bred by Mr. F. J. Gurney; got by Red Clover (35216), dam (Poverty) by Provost (27878), &c. See Vol. xx. p. 694.

Produce in	Names, &c.	By what Bull.	By whom bred.
1879, Oct. 2, roan,	C.C.|Perplexity	Prince Leopold, 37239	Mr. Gurney

HALE, Bernard,
Holly Hill, Hartfield, Tunbridge Wells.

ASTRÆA, roan, calved May 9, 1874, Vols. xxiv. and xxv. pp. 481, 482. Bred by Mr. B. Hale; got by Webster (27769), dam (Ariadne) by Red Butterfly (22690), &c.

Produce in	Names, &c.	By what Bull.	By whom bred.
1879, Aug. 12, roan,	B.C.|Asterion	Windermere, 42622	Mr. Hale

BIANCA, roan, calved December 21, 1867, Vols. xx., xxi., and xxiii., pp. 410, 749, 479. Bred by Sir J. Lubbock, Bart., High Elms; got by Red Butterfly (22690), dam (Bessy) by Louis (14861), &c.

Produce in	Names, &c.	By what Bull.	By whom bred.
1879, Jan. 24, roan,	C.C.|Biarritz	Baron Lampson, 39419	Mr. Hale

FAWSLEY 9TH, roan, calved December 11, 1868, Vols. xxi., xxiii., xxiv., and xxv. pp. 749, 479, 481, 483. Bred by Messrs. F. Leney and Sons, Wateringbury; got by Lord Oxford 2nd (20215), dam (Fawsley 5th) by Tippoo 2nd (23075), &c.

Produce in	Names, &c.	By what Bull.	By whom bred.
1879, April 28, roan,	C.C.|Fawsley 22nd	Baron Lampson, 39419	Mr. Hale

MAY DUCHESS 4TH, roan, calved February 15, 1871, Vols. xxi. and xxiii. pp. 750, 479. Bred by Messrs. F. Leney and Sons, Wateringbury; got by Grand Duke 15th (21852), dam (May Queen) by May Duke (13320), &c.

Produce in	Names, &c.	By what Bull.	By whom bred.
1879, April 7, roan,	B.C.|May Baron	Baron Lampson, 39419	Mr. Hale

HALES, Edward,
North Frith, Hadlow, Kent.

FERNANDE 2ND, red and white, calved April 15, 1873, Vols. xxiii. and xxiv. pp. 642, 647. Bred by Mr. H. J. Sheldon, Brailes House; got by Duke of Cerisia (30937), dam (Fernande) by Eighteenth Duke of Oxford (25995), &c.

Produce in	Names, &c.	By what Bull.	By whom bred.
1879, Oct. 25, roan,	C.C.|Ly.Rothesay Walnut	Duke of Rothesay, 36534	Mr. Hales

FRILL, red and white, calved December 30, 1876. Bred by Mr. E. Hales; got by Cherry Prince 4th (25765), dam (Florence) by MacDuff (24500), &c. See Vol. xxii. p. 443.

Produce in	Names, &c.	By what Bull.	By whom bred.
1879, Mar. 24, r. & w.,	C.C.|Fancy	Romulus, 37362	Mr. Hales

HOLKER GWYNNE, red, calved January 10, 1877. Bred by Mr. H. J. Sheldon, Brailes House; got by Baron Oxford 7th (36199), dam (Rival Gwynne) by Twenty-second Duke of Oxford (31000), &c. See Vol. xxiv. p. 647.

Produce in	Names, &c.	By what Bull.	By whom bred.
1879, Oct. 14, r. & w., B.C.	Ld. Rothesay Gwynne	Duke of Rothesay, 36534	Mr. Hales

MARY 13TH, red and white, calved March 6, 1876, Vol. xxv. p. 483. Bred by Mr. E. Hales; got by Redlands (35237), dam (Mary 11th) by Duke of Wrotham 2nd (31036), &c.

1879, July 25, r. & w., C.C.	Mary 16th	Romulus, 37362	Mr. Hales

PRINCESS OF BARRINGTON, red, calved July 9, 1870, Vols. xxii., xxiii., and xxv. pp. 560, 642, 660. Bred by Mr. H. J. Sheldon, Brailes House; got by Duke of Brailes (23724), dam (Countess of Barrington 2nd) by Ninth Duke of Oxford (17738), &c.

1879, Oct. 5, roan, C.C.	L'y R'say Barrington	Duke of Rothesay, 36534	Mr. Hales

RED STRAWBERRY, red, calved August 5, 1873, Vols. xxii., xxiii., and xxiv. pp. 597, 676, 682. Bred by Mr. G. Murton Tracy, Redlands; got by Red Belvedere (29724), dam (Roan Strawberry) by Rambler (27229), &c.

1878, Dec. 5, red, C.C.	Sweet Strawberry	Prince of the Blood, 37259	Mr. Hales

HAMER, C. M.,
Snitterfield, Stratford-on-Avon.

AIRDRIE'S KNIGHTLEY, white, calved December 16, 1876. Bred by Mr. G. Fox, Elmhurst Hall; got by Twenty-fourth Duke of Airdrie (36460), dam (Cambridge Knightley) by Christmas Duke (33378), &c. See Vol. xxiii. p. 442.

1879, April 4, white, C.C.	Rosy Knightley	Kalafat, 40038	Mr. Hamer

CHERRY BLANCHE 2ND, roan, calved September 18, 1867, Vols. xx., xxi., xxiii., and xxiv. pp. 441, 627, 481, 482. Bred by Mr. J. Harward, Winterfold; got by Charleston (21400), dam (Cherry Blanche) by Cherry Duke 4th (17552), &c.

1879, May 22, roan, B.C.	(Steer)	Kalafat, 40038	Mr. Hamer

CHERRY BLANCHE 4TH, roan, calved November 24, 1875, Vol. xxv. p. 483. Bred by Lord Chesham, Latimer; got by Duke of Oxford 28th (33710), dam (Cherry Blanche 3rd) by Baron Wastwater (30492), &c.

1879, July 3, roan, C.C.	Cherry Blanche 7th	Kalafat, 40038	Mr. Hamer

EDITHA, white, calved March 26, 1872. Bred by Mr. O. Viveash, Strensham; got by Monitor (24615), dam (Edith) by Mocassin (18406), &c. See "Clopton," Vol. xxv. p. 54.

1879, Dec. 2, white, C.C.	Ethel	Kalafat, 40038	Mr. Hamer

LADY BLANCHE 2ND, white, calved April 6, 1869, Vols. xx., xxi., xxiii., and xxiv. pp. 586, 753, 481, 483. Bred by Mr. E. H. Cheney, Gaddesby Hall; got by General Napier (24023), dam (Lady Blanche) by Charley (17541), &c.

1879, May 18, roan, B.C.	(Steer)	Kalafat, 40038	Mr. Hamer

HAMILTON, Hon. R. Baillie,
Langton, Dunse, N.B.

EMMA 10TH, red, calved March 7, 1876. Bred by Sir W.C. Trevelyan, Bart., Wallington; got by Oxford Beau 4th (34964), dam (Energy) by Lord Lally 3rd (24408), &c. See Vol. xxiii. p. 682.

Produce in	Names, &c.	By what Bull.	By whom bred.
1879, Aug. 6, r. & w., C.C.	Daria	Duke of Oxford 27th, 33709	Hn.R.B.Hamilton

OXFORD CHERRY 2ND, red, calved December 28, 1874, Vol. xxiv. p. 688. Bred by Sir W. C. Trevelyan, Bart., Wallington; got by Oxford Beau 4th (34964), dam (Grand Oxford Cherry) by Grand Duke 13th (21850), &c.

1879, Aug. 2, r. & w., C.C.	Langton Cherry	Duke of Oxford 27th, 33709	Hn.R.B.Hamilton

HAMOND, Anthony,
Westacre, Swaffham, Norfolk.

CINDERELLA, red and white, calved April 29, 1874, Vol. xxv. p. 484. Bred by Mr. A. Hamond; got by Prince Louis (29643), dam (Princess) by Prince Leopold (20557), &c.

1879, Aug. 21, roan, C.C.	Capsicum	Hotspur, 34190	Mr. Hamond

CLOUDY, roan, calved January 8, 1877. Bred by Mr. A. Hamond; got by Hotspur (34190), dam (Cinderella) by Prince Louis (29643), &c.

1879, Sept. 22, roan, C.C.	Cloudless	Heir Apparent, 36760	Mr. Hamond

CROCUS, roan, calved February 9, 1876. Bred by Mr. A. Hamond; got by Hotspur (34190), dam (Cinderella) by Prince Louis (29643), &c.

1878, April 27, roan, C.C.	Cowslip	Heir Apparent, 36760	Mr. Hamond
1879, May 14, roan, C.C.	Careless	do.	do.

KITTEN, white, calved June 4, 1875, Vol. xxiv. p. 483. Bred by Mr. A. Hamond; got by British Flag (30601), dam (Ketura 7th) by Second Duke of Collingham (23730), &c.

1879, April 2, white, C.C.	Kiss	Heir Apparent, 36760	Mr. Hamond

KITTY, white, calved June 8, 1877. Bred by Mr. A. Hamond; got by Hotspur (34190), dam (Kitten) by British Flag (30601), &c.

1879, Sept. 11, white, C.C.	Kate	Heir Apparent, 36760	Mr. Hamond

LA FRANCE, roan, calved September 18, 1876. Bred by Mr. A. Hamond; got by Hotspur (34190), dam (Fleur de Lis) by Duke of Edinburgh (23741), &c. See Vol. xxii. p. 444.

1879, Jan. 1, roan, C.C.	Eugenie	Heir Apparent, 36760	Mr. Hamond
1879, Dec. 8, roan, B.C.	St. Denis	do.	do.

LOLA MONTES, roan, calved January 13, 1876, Vol. xxv. p. 414. Bred by H.R.H. the Prince of Wales, Sandringham; got by General McNab (34012), dam (Friendship) by Theodorus (27639), &c.

1879, April 11, roan, C.C.	Mantilla	Heir Apparent, 36760	Mr. Hamond

MISS CONSTANCE, roan, calved April 6, 1875, Vol. xxv. p. 484. Bred by H.R.H. the Prince of Wales, Sandringham; got by General McNab (34012), dam (Satinet) by Duke of Edinburgh (23741), &c.

1879, Oct. 14, white, C.C.	Conny	Heir Apparent, 36760	Mr. Hamond

PANSY, roan, calved May 3, 1872, Vol. xxv. p. 484.   Bred by Mr. A. Hamond;
got by Prince Louis (29643), dam (Princess) by Prince Leopold (20557), &c.

Produce in	Names, &c.	By what Bull.	By whom bred.
1879, July 23, roan, B.C.	Pater	Hotspur, 34190	Mr. Hamond

PETTICOAT, roan, calved September 19, 1876.   Bred by Mr. A. Hamond; got
by Hotspur (34190), dam (Princess) by Prince Leopold (20557), &c.   See
Vol. xxv. p. 484.

| 1879, June 8, roan,   C.C. | Plaiting | Heir Apparent, 36760 | Mr. Hamond |

POSEY, white, calved October 18, 1876.   Bred by Mr. A. Hamond; got by
Hotspur (34190), dam (Pansy) by Prince Louis (29643), &c.   See " Pater,"
p. 183.

| 1879, May 10, white, C.C. | Pimpernel | Heir Apparent, 36760 | Mr. Hamond |

REBECCA, roan, calved March 12, 1875.   Bred by Mr. A. Hamond; got by
British Flag (30601), dam (Shepherdess) by Ant-Eater (19228), &c.   See
" Mr. Winkle," p. 172.

| 1879, Mar. 24, roan,   C.C. | Rita | Hotspur, 34190 | Mr. Hamond |

SARCENET, roan, calved January 29, 1873, Vol. xxii. p. 445.   Bred by Mr. H.
Aylmer, West Dereham Abbey; got by General Hopewell 2nd (24021), dam
(Satinet) by Duke of Edinburgh (23741), &c.

| 1879, Sept. 13, r. & w., B.C. | Sir Hal | Hotspur, 34190 | Mr. Hamond |

SILKY, roan, calved March 21, 1876, Vol. xxv. p. 485.   Bred by Mr. A. Hamond;
got by Sir Harry (35558), dam (Sarcenet) by General Hopewell 2nd (24021)
&c.

| 1879, June 1, roan.   B.C. | Spree | Heir Apparent, 36760 | Mr. Hamond |

WHINNY, white, calved February 1, 1876, Vol. xxv. p. 485.   Bred by Mr. A.
Hamond; got by Hotspur (34190), dam (Wretham) by Prince Louis (29643),
&c.

| 1879, April 4, white, C.C. | Watercress | Heir Apparent, 36760 | Mr. Hamond |

WHISPER, white, calved March 10, 1877.   Bred by Mr. A. Hamond; got by
Hotspur (34190), dam (Wretham) by Prince Louis (29643), &c.   See " Mr.
Winkle," p. 172.

| 1879, May 23, roan,   B.C. | Mr. Winkle | Heir Apparent, 36760 | Mr. Hamond |

HAMPSON, Mrs.,
Ullen Wood, Cheltenham.

SERAPHINA BELLA 5TH, roan, calved June 2, 1877.   Bred by Lord Sudeley,
Toddington; got by Baron Winsome 3rd (33108), dam (Booth's Seraphina)
by Baron Booth (21212), &c.   See Vol. xxi. p. 946.

| 1879, Oct. 5, roan, | C.C. | Royal Seraphina 2nd | G. Duke of Waterloo, 28766 | Mrs. Hampson |

HANDLEY, W.,
Green Head, Milnthorpe.

CHERRY BLOOM, roan, calved April 30, 1876, Vol. xxv. p. 485.   Bred by
Messrs. Angus, Broomley; got by Ben Brace (30524), dam (Cherry Queen) by
Merry Monarch (22349), &c.

| 1879, Dec. 24, r. & w., C.C. | Red Cherry | Alfred the Great, 36121 | Mr. Handley |

EARL'S FLORA, red and white, calved July 13, 1872, Vols. xxi. and xxiii.
pp. 754, 482. Bred by Mr. R. B. Brockbank, Burgh-by-Sands; got by Earl of
Eglinton (23832), dam (Flora Cobham) by Marquis of Cobham (22299), &c.

Produce in	Names, &c.	By what Bull.	By whom bred.
1879, Mar. 4, roan,	C.C.|Princess Flora	Alfred the Great, 36121	Mr. Handley

LAVENDER 3RD, white, calved November 10, 1875. Bred by Mr. W. Handley;
got by Sir Arthur Windsor (35541), dam (Lavender 2nd) by Prince Arthur
(29597), g. d. (Lavender) by Sir Walter Trevelyan (25179), gr. g. d. (Old
Lavender) by General Garibaldi (21813), &c. See "Farewell Lad," Vol. xxv.
p. 105.

Produce in	Names, &c.	By what Bull.	By whom bred.
1879, Sept. 20, { roan, C.C.|Lavender Leaf roan, C.C.|Lavender Leaf 2nd }	Alfred the Great, 36121	Mr. Handley

RED PRINCESS, red, calved March 23, 1873. Bred by Mr. W. Handley; got
by Prince Arthur (29597), dam (Venus) by Sir Walter Trevelyan (25179), &c.
See "Lord Mirth," Vol. xxiv. p. 154.

Produce in	Names, &c.	By what Bull.	By whom bred.
1878, Aug. 28, red,	C.C.|Red Princess 2nd	Alfred the Great, 36121	Mr. Handley

HANDS, Thomas,
Canley, Coventry.

LADY HILDA, white, calved January 12, 1868, Vols. xx., xxi., xxii., and xxiii.
pp. 598, 537, 293, 309. Bred by Mr. A. Dugdale, Rose Hill, the property of
Mr. T. Hands; got by Lord Liverpool (22168), dam (Lady Hudson) by Fourth
Duke of Oxford (11387), &c.

Produce in	Names, &c.	By what Bull.	By whom bred.
1879, May 9, white,	C.C.|Miss Hilda	2nd Duke of Rowley, 28441	Mr. Allen

Miss Hilda, sold to Mr. W. T. Wakefield, Fletchampstead Hall, Coventry.

MISS PEARL 19TH, roan, calved in May 1873, Vol. xxiii. p. 484. Bred by
Mr. T. Hands; got by Second Duke of Darlington (28378), dam (Miss Pearl
4th) by Viceroy (21019), &c.

Produce in	Names, &c.	By what Bull.	By whom bred.
1878, Mar. 3, roan,	C.C. Miss Pearl 35th	Duke of Elford 2nd, 36400	Mr. Hands
1879, Mar. 21, roan,	C.C. Miss Pearl 36th	Ld. Darlington 16th, 40151	do.

STATIRA 27TH, red and white, calved April 15, 1876. Bred by Mr. T.
Comber, Redcliffe; got by Fourth Duke of Grafton (28396), dam (Statira 7th)
by Fourth Duke of Thorndale (17750), &c. See Vol. xix. p. 738.

Produce in	Names, &c.	By what Bull.	By whom bred.
1879, June 25, r. & w.,	C.C.|Statira Wild Eyes	Royal Oxford, 39050	Mr. Hands

WOODBINE 7TH, red and white, calved July 12, 1877. Bred by Mr. T. Hands;
got by Cæsarewitch 4th (33263), dam (Woodbine 4th) by Second Duke of
Darlington (28378), g. d. (Woodbine 2nd) by Canley Duke (23506), gr. g. d.
(Woodbine) by Prince Pearl 5th (24855), — bred at Canley, and descended
from the stock of Mr. Bellamy, of Haseley.

Produce in	Names, &c.	By what Bull.	By whom bred.
1879, Oct. 19, roan,	B.C.|Baron Woodbine	Baron Shendish 4th, 41062	Mr. Hands

HANNAN, B.,
Riverstown, Killucan, Co. Westmeath.

JENNY LIND 12TH, roan, calved June 10, 1870, Vols. xxii., xxiv., and xxv.
pp. 446, 485, 486. Bred by Mr. N. M. Archdall, Crock-na-Crieve; got by
Abercorn (25484), dam (Jenny Lind 7th) by Prince of Lurg (22617), &c.

Produce in	Names, &c.	By what Bull.	By whom bred.
1879, Aug. 29, red,	C.C.|	Ruby King, 42345	Mr. Hannan

JENNY LIND 15TH, roan, calved September 9, 1873, Vols. xxiii. and xxiv. pp. 484, 485. Bred by Mr. N. M. Archdall, Crock-na-Crieve; got by Abercorn (25484), dam (Jenny Lind 7th) by Prince of Lurg (22617), &c.

Produce in		Names, &c.	By what Bull.	By whom bred.
1879, Feb. 2, roan,	B.C.	Royal Lind	St. Ronan, 35458	Mr. Hannan

JENNY LIND 18TH, red and white, calved March 10, 1874, Vol. xxiv. p. 485. Bred by Mr. N. M. Archdall, Crock-na-Crieve; got by Abercorn (25484), dam (Jenny Lind 5th) by Ancient Briton (19225), &c.

1879, Jan. 19, r. & w.,	C.C.	Jenny Lind 26th	St. Ronan, 35458	Mr. Hannan

JENNY LIND 19TH, red and white, calved March 21, 1875, Vols. xxiv. and xxv. pp. 485, 486. Bred by Mr. T. Conolly, Castletown; got by Abercorn (25484), dam (Jenny Lind 5th) by Ancient Briton (19225), &c.

1879, Oct. 17, r. & w.,	B.C.	Fusilier	Premier Lind, 40474	Mr. Hannan

JENNY LIND 22ND, roan, calved June 2, 1876. Bred by Mr. B. Hannan; got by Arronant (25530), dam (Jenny Lind 12th) by Abercorn (25484), g. d. (Jenny Lind 7th) by Prince of Lurg (22617), &c. See "Regal Lind," p. 209.

1879, Sept. 17, roan,	C.C.	·	Ruby King, 42345	Mr. Hannan

JENNY LIND 25TH, roan, calved March 24, 1878. Bred by Mr. B. Hannan; got by St. Ronan (35458), dam (Jenny Lind 17th) by Abercorn (25484), &c. See "Colonel Lind," p. 53.

1879, Nov. 9, roan,	B.C.		Ruby King, 42345	Mr. Hannan

HARDY, George,
Pickering Lodge, Timperley, Cheshire.

HAVERING LILY, red, calved April 6, 1875, Vols. xxiv. and xxv. pp. 556, 564. Bred by Mrs. McCulloch, Balloan of Cawdor, the property of Mr. G. Hardy; got by Star (37524), dam (Rosy 3rd) by Fawsley Prince (21733), &c.

1879, July 31, red,	C.C.	D'ss of Timperley	Duke of Havering, 33664	Mr. Davies

HARRETT, Robert,
Kirkwhelpington, Newcastle-on-Tyne.

GIPSY 2ND, roan, calved April 24, 1875. Bred by Mr. R. Harrett; got by Waterloo King (35946), dam (Gipsy) by Gipsy Prince (17965), &c. See Vol. xxi. p. 755.

1878, April 11, roan,	C.C.	Gipsy 3rd	D. of Holderness 2nd, 36502	Mr. Harrett
1879, May 12, roan,	B.C.	Lord Chelmsford	Pr. of Waterloo 2nd, 38944	do.

GIPSY-LE-GRAND, roan, calved August 26, 1873, Vol. xxiv. p. 487. Bred by Mr. R. Harrett; got by Oxford-le-Grand (29496), dam (Gipsy) by Gipsy Prince (17965), &c.

1878, April 6, red,	C.C.	Gipsy-le-Grand 3rd	D. of Holderness 2nd, 36502	Mr. Harrett
1879, April 6, roan,	C.C.	Gipsy-le-Grand 4th	Pr. of Waterloo 2nd, 38944	do.

LADY WHARFDALE 3RD, roan, calved July 23, 1875. Bred by Mr. R. Harrett; got by Grand Duke of Wallington (31300), dam (Lady Wharfdale) by Third Lord Wharfdale (26759), &c. See "Northern Duke 3rd," p. 177.

1878, July 25, white,	C.C.	Lady Wharfdale 6th	Pr. of Waterloo 2nd, 38944	Mr. Harrett
1879, Aug. 30, roan,	B.C.	Northern Duke 4th	do.	do.

HARRIS, Thomas,
Stonylane House, Bromsgrove.

ANNIE LAURIE, roan, calved February 10, 1876. Bred by Mr. T. Harris, the property of Mr. Z. Walker, Fox Hollies Hall; got by Waterloo Prince (35949), dam (Helena) by Lord Red Eyes 2nd (24460), &c. See Vol. xxiii. p. 487.

Produce in	Names, &c.	By what Bull.	By whom bred.
1879, Feb. 23, roan,	C.C. Lily	D. of Carolina 2nd, 38127	Mr. Walker

DEWDROP, roan, calved February 10, 1872, Vol. xxiii. p. 486. Bred by Mr. T. Harris; got by Lord Wharfdale (26761), dam (Majestic) by Archduke (21185), &c.

1879, Mar. 1, red,	C.C. Celeste	Waterloo Prince, 35949	Mr. Harris

DUCHESS OF YORK 3RD, roan, calved February 23, 1873, Vol. xxiii. p. 359. Bred by Mr. T. G. Curtler, Bevere House; got by Grand Duke of Clarence (28750), dam (Duchess of York) by Seventh Duke of York (17754), &c.

1879, Nov. 4, roan,	C.C. Duchess of York 5th	B'n Barrington 6th, 33008	Mr. Harris

FLOWER OF THE LINK, roan, calved March 29, 1871, Vol. xxi. p. 756. Bred by Mr. T. Harris; got by Festival (26147), dam (Belle of Braithwaite) by Prince George (13510), &c.

1879, Feb. 27, roan.	B.C. The Baron	Waterloo Prince, 35949	Mr. Harris

The Baron, sold to Mr. Dixon, Tardebig.

HELENA, red and white, calved March 5, 1871, Vols. xxii., xxiii., and xxiv. pp. 448, 487, 488. Bred by Mr. T. Harris, the property of Mr. M. Dixon, Tardebig; got by Lord Red Eyes 2nd (24460), dam (Majestic) by Archduke (21185), &c.

1879, Mar. 8, roan,	C.C. Annie	Waterloo Prince, 35949	Mr. Dixon

LADY ELIZABETH, roan, calved March 8, 1868, Vols. xx., xxi., xxii., xxiii., and xxv. pp. 592, 756, 449, 487, 489. Bred by Mr. T. Harris, the property of Mr. M. Dixon, Tardebig; got by Diplomatist (19571), dam (Edith) by Baron Berrington (17357), &c.

1879, Feb. 2, roan,	C.C. Bessie	Waterloo Prince, 35949	Mr. Dixon

LADY FRANCES, red, calved March 1, 1877. Bred by Mr. T. Harris; got by Waterloo Prince (35949), dam (Lady Georgiana) by Lord Red Eyes 2nd (24460), &c. See Vol. xxiv. p. 488.

1879, July 1, roan,	B.C. Prince Carolina	D. of Carolina 2nd, 38127	Mr. Harris

LADY GEORGIANA, roan, calved January 8, 1871, Vols. xxi., xxii., xxiii., xxiv., and xxv. pp. 757, 449, 487. 488. 489. Bred by Mr. H. F. Vernon, Hanbury Hall; got by Lord Red Eyes 2nd (24460), dam (Gertrude) by Diplomatist (19571), &c.

1879, Feb. 8, roan,	C.C. Lady Mabel	Waterloo Prince, 35949	Mr. Harris

NIOBE 18TH, roan, calved November 4, 1875. Bred by Mr. W. Strickland, Cokethorpe Park; got by North Star (31988), dam (Niobe 4th) by Oxartes (22471), &c. See "Paxton," p. 184.

1879, Jan. 30, red,	B.C. Paxton	Baron Winsome 3rd, 33108	Mr. Harris

NORAH, red and white, calved May 2, 1872, Vol. xxiii. p. 432. Bred by Mr. J. Evans, Uffington, the property of Mr. Z. Walker, Fox Hollies Hall; got by Manager (24521), dam (Erin) by Shamrock (20797), &c.

Produce in		Names, &c.		By what Bull.		By whom bred.
1878, May 14, red,	C.C.	Pandora		Waterloo Prince, 35949		Mr. Harris
1879, April 17, red,	C.C.	Violet		do.		Mr. Walker

HART, G. V.,
Kilderry, Muff, Co. Donegal.

AMETHYST, red, calved March 2, 1872, Vols. xxiii. and xxv. pp. 487, 490. Bred by Mr. G. V. Hart; got by Herdsman (26374), dam (Precious Jewel) by Northern Chief (22429), &c.

1879, Feb. 11, red,	B.C.	Admiral		Viscount Studley, 35903	Mr. Hart

Admiral, sold to Mr. J. Marshall, Newtown Cunningham, Co. Donegal.

DIADEM, roan, calved February 26, 1870, Vols. xxii., xxiv., and xxv. pp. 450, 490, 490. Bred by Mr. G. V. Hart; got by Herdsman (26374), dam (Diamond) by Northern Chief (22429), &c.

1879, Mar. 17, roan,	B.C.	Knight of St. Patrick	Viscount Studley, 35903	Mr. Hart

Knight of St. Patrick, sold to Mr. R. C. Edwards, Burt, Derry.

LADY BLUSH, roan, calved January 26, 1875, Vol. xxv. p. 490. Bred by Mr. G. V. Hart; got by Lord Westland (34700), dam (Bloom) by Knight Errant (31527), &c.

1879, Feb. 27, roan,	B.C.	Baron Studley	Viscount Studley, 35903	Mr. Hart

Baron Studley, sold to Mr. C. A. Smyth, Stradreagh, Co. Derry.

LADY CORAL, red, calved March 10, 1876. Bred by Mr. G. V. Hart; got by Lord Westland (34700), dam (Cornelian) by Herdsman (26374), &c. See Vol. xxiii. p. 487.

1879, Feb. 10, red,	C.C.	Coralline	Viscount Studley, 35903	Mr. Hart

LADY DI, red and white, calved February 10, 1874, Vols. xxiv. and xxv. pp. 490, 491. Bred by Mr. G. V. Hart; got by Lord Westland (34700), dam (Diadem) by Herdsman (26374), &c.

1879, Feb. 9, r. & w.,	B.C.	Defiance	Viscount Studley, 35903	Mr. Hart

Defiance, sold to Archdeacon Hamilton, Desertmartin, Magherafelt.

LADY DRYAD, red and white, calved January 25, 1876. Bred by Mr. G. V. Hart; got by Lord Westland (34700), dam (Diadem) by Herdsman (26374), &c. See Vol. xxiv. p. 490.

1879, Mar. 5, r. & w.,	C.C.	Daphne	Viscount Studley, 35903	Mr. Hart

Daphne, sold to Mr. A. Weir, Ballindrait, Strabane.

LADY ELEGANT, red, calved February 3, 1876. Bred by Mr. G. V. Hart; got by Lord Westland (34700), dam (Queen Esther) by Knight Errant (31527), &c. See Vol. xxiv. p. 491.

1879, Jan. 28. r. & w.,	B.C.	Excelsior	Viscount Studley, 35903	Mr. Hart

Excelsior, sold to Mr. S. Sproule, Ramelton, Co. Donegal.

LADY JASPER, red, calved May 3, 1874, Vol. xxv. p. 491. Bred by Mr. G. V. Hart; got by Lord Westland (34700), dam (Diamond) by Northern Chief (22429), &c.

1879, Mar. 2, red,	B.C.	Jeweller	Viscount Studley, 35903	Mr. Hart

Jeweller, sold to Captain E. S. Lecky, Eglinton, Co. Derry.

LADY PEDIGREE, roan, calved July 26, 1875, Vol. xxv. p. 491. Bred by
Mr. G. V. Hart; got by Lord Westland (34700), dam (Peggy Ban) by
Northern Chief (22429), &c.

Produce in		Names, &c.	By what Bull.	By whom bred.
1879, Mar. 31, roan,	B.C.	Paragon	Viscount Studley, 35903	Mr. Hart

Paragon, sold to Mr. H. M. Richardson, Rossfad, Ballycassidy, Co. Fermanagh.

PATTY, white, calved February 27, 1870, Vols. xxiii., xxiv., and xxv. pp. 488,
491, 491. Bred by Mr. G. V. Hart; got by Herdsman (26374), dam (Peggy
Ban) by Northern Chief (22429), &c.

Produce in		Names, &c.	By what Bull.	By whom bred.
1879, April 12, roan,	C.C.	Primrose	Viscount Studley, 35903	Mr. Hart

PIMPERNEL, roan, calved March 15, 1871, Vols. xxii., xxiv., and xxv. pp. 450,
491, 491. Bred by Mr. G. V. Hart; got by Herdsman (26374), dam (Peggy
Ban) by Northern Chief (22429), &c.

Produce in		Names, &c.	By what Bull.	By whom bred.
1879, Mar. 27, roan,	C.C.	Peony	Viscount Studley, 35903	Mr. Hart

ROSETTE, roan, calved March 8, 1867, Vols. xxiv. and xxv. pp. 491, 492. Bred
by Mr. S. Smyth, The Cross; got by Prince Patrick (40532), dam (White
Rose) by Ancient Briton (19225), &c.

Produce in		Names, &c.	By what Bull.	By whom bred.
1879, Mar. 22, red,	B.C.	Rubicund	Viscount Studley, 35903	Mr. Hart

Rubicund, sold to Mr. H. A. Johnstone, Altaghmore, Fintona, Co. Tyrone.

SYMMETRY, red and white, calved March 10, 1867, Vol. xxiv. p. 491. Bred
by Mr. S. Smyth, The Cross; got by Prince Patrick (40532), dam (Violet) by
Ancient Briton (19225), &c.

Produce in		Names, &c.	By what Bull.	By whom bred.
1879, Feb. 14, r. & w.,	C.C.	Sweetbrier	Viscount Studley, 35903	Mr. Hart

HARWARD, John, Exors. of,
Winterfold, Kidderminster.

DÉBRIS 3RD, roan, calved April 20, 1870, Vols. xxi. and xxii. pp. 759, 450.
Bred by Mr. J. Harward; got by Third Duke of Claro (23729), dam (Débris
2nd) by John O'Gaunt (21995), &c.

Produce in		Names, &c.	By what Bull.	By whom bred.
1876, Sept. 9, white,	C.C.	Débris 7th	Bn. Barrington 6th, 33008	Mr. Harward
1878, Aug. 12, roan,	B.C.	Lord Claro	do.	do.
1879, Aug. 7, white,	B.C.	(Steer)	do.	Exrs. Mr. Harward

DIDO 9TH, roan, calved May 1, 1868, Vols. xxi. and xxii. pp. 759, 451. Bred
by Mr. H. Caddy, Rougholm; got by Sir Windsor (22927), dam (Dido 7th)
by Knight of Distington (18158), &c.

Produce in		Names, &c.	By what Bull.	By whom bred.
1876, June 14, roan,	C.C.	Dido 12th	Bn. Barrington 6th, 33008	Mr. Harward
1877, June 8, white,	B.C.	(Steer)	do.	do.
1878, June 3, roan,	C.C.	Dido 14th	do.	do.
1879, June 3, roan,	C.C.	Dido 15th	do.	do.

DIDO 10TH, red and white, calved July 20, 1873, Vol. xxiii. p. 488. Bred by
Mr. J. Harward; got by Waterloo Cherry (27763), dam (Dido 9th) by Sir
Windsor (22927), &c.

Produce in		Names, &c.	By what Bull.	By whom bred.
1877, April 24, roan,	C.C.	Dido 13th	Bn. Barrington 6th, 33008	Mr. Harward
1878, April 19, roan,	B.C.	Lord Muncaster	do.	do.
1879, April 23, roan,	B.C.	Lord Muncaster 2nd	do.	do.

DIDO 12TH, roan, calved June 14, 1876. Bred by Mr. J. Harward; got by
Baron Barrington 6th (33008), dam (Dido 9th) by Sir Windsor (22927), &c.
See "Lord Muncaster 2nd," p. 151.

Produce in		Names, &c.	By what Bull.	By whom bred.
1879, May 25, red,	B.C.	(Steer)	Earl of Oxford, 39828	Mr. Harward

DUCHESS LALLY, roan, calved January 11, 1876. Bred by Mr. J. Harward;
got by Fifth Duke of Wetherby (31033), dam (Lally 10th) by Fifth Lord
Wild Eyes (26762), &c. See "Baron Oxford Barrington 2nd," p. 18.

Produce in		Names, &c.		By what Bull.		By whom bred.
1878, Sept. 8, red,	B.C.	Bn.Oxf.Barr'ton 2nd	B.T'croftOxford 4th, 37822	Mr. Harward

EMPRESS OF YORK 4TH, roan, calved September 13, 1871, Vols. xxi. and
xxv. pp. 661, 492. Bred by Mr. T. G. Curtler, Bevere House; got by Eighth
Duke of Geneva (28390), dam (Third Empress of York) by Third Duke of
Claro (23729), &c.

Produce in		Names, &c.	By what Bull.	By whom bred.
1879, July 8, roan,	B.C.	Emperor Barrington	Bn. Barrington 6th, 33009	Exrs. Mr. Harward

GONERIL 3RD, roan, calved January 30, 1869, Vols. xxi. and xxii. pp. 759,
451. Bred by Sir J. W. C. Hartopp, Bart., Burbage Fields; got by Lord
Burghersh (24358), dam (Goneril 2nd) by Lord Wild Eyes 3rd (22235), &c.

Produce in		Names, &c.	By what Bull.	By whom bred.
1876, Aug. 17, roan,	C.C.	Goneril 7th	Bn. Barrington 6th, 33008	Mr. Harward
1877, Aug. 9, roan,	C.C.	Goneril 8th	do.	do.
1878, Sept. 3, white,	C.C.	Goneril 10th	do.	do.

GONERIL 4TH, red and white, calved December 28, 1870, Vol. xxi. p. 759.
Bred by Sir J. W. C. Hartopp, Bart., Burbage Fields; got by Duke of Hazlecote 4th (25972), dam (Goneril 2nd) by Lord Wild Eyes 3rd (22235), &c.

Produce in		Names, &c.	By what Bull.	By whom bred.
1876, Feb. 17, roan,	C.C.	Goneril 6th	Bn. Barrington 6th, 33008	Mr. Harward
1878, Aug. 6, roan,	C.C.	Goneril 9th	do.	do.
1879, Aug. 1, roan,	C.C.	Goneril 13th	do.	Exrs. Mr. Harward

GONERIL 6TH, roan, calved February 17, 1876. Bred by Mr. J. Harward; got
by Baron Barrington 6th (33008), dam (Goneril 4th) by Duke of Hazlecote
4th (25972), &c.

Produce in		Names, &c.	By what Bull.	By whom bred.
1878, Sept. 17, roan,	C.C.	Goneril 11th	Earl Gwynne, 38213	Mr. Harward

GONERIL 7TH, roan, calved August 17, 1876. Bred by Mr. J. Harward; got
by Baron Barrington 6th (33008), dam (Goneril 3rd) by Lord Burghersh
(24358), &c. See Vol. xxii. p. 451.

Produce in		Names, &c.	By what Bull.	By whom bred.
1879, May 21, red,	C.C.	Goneril 12th	Earl of Oxford, 39828	Mr. Harward

JOAN OF ARC 3RD, white, calved March 26, 1873, Vol. xxiii. p. 489. Bred
by Mr. J. Harward; got by Third Duke of Claro (23729), dam (Joan of Arc)
by Cupid (14359), &c.

Produce in		Names, &c.	By what Bull.	By whom bred.
1877, Feb. 11, white,	B.C.	(Steer)	Bn. Barrington 6th, 33008	Mr. Harward
1878, Jan. 31, white,	C.C.	Joan of Arc 4th	do.	do.

LADY CHARLESTON 2ND, roan, calved February 22, 1876, Vol. xxv. p. 492.
Bred by Mr. J. Harward; got by Baron Barrington 6th (33008), dam (Lady
Charleston) by Eighth Duke of Geneva (28390), &c.

Produce in		Names, &c.	By what Bull.	By whom bred.
1879, Sept. 19, r. & w.,	C.C.	Lady Charleston 4th	Earl of Oxford, 39828	Exrs. Mr. Harward

LALLY 10TH, roan, calved June 22, 1868, Vols. xxi. and xxiv. pp. 759, 491.
Bred by Mr. C. W. Harvey, Walton-on-the-Hill; got by Fifth Lord Wild
Eyes (26762), dam (Lally 5th) by Duke of Wetherby (17753), &c.

Produce in		Names, &c.	By what Bull.	By whom bred.
1878, Aug. 20, white,	B.C.	Bn. Oxf. Barrington	Baron Oxford 4th, 25580	Mr. Harward

LIPPING LASSIE, white, calved December 25, 1871, Vol. xxii. p. 451. Bred
by Mr. I. Downing, Turner's Hill; got by Eighth Duke of Geneva (28390),
dam (Bovington Belle) by Didmarton Duke (21546), &c.

Produce in		Names, &c.	By what Bull.	By whom bred.
1877, Nov. 30, white,	C.C.	Barrington's Belle	Bn. Barrington 6th, 33008	Mr. Harward
1878, Nov. 15, white,	C.C.	Barrington's Belle 2d	do.	do.

MAID OF OXFORD 7TH, red and white, calved December 20, 1873. Bred by Mr. T. G. Curtler, Bevere House; got by Grand Duke of Clarence (28750), dam (Maid of Oxford 4th) by Lord Waterloo 2nd (26755), &c. See Vol. xxi. p. 662.

Produce in		Names, &c.	By what Bull.	By whom bred.
1878, Aug. 9, roan,	B.C.	Earl Barrington	Bn. Barrington 6th, 33008	Mr. Harward
1879, Aug. 4, roan,	C.C.	Maid of Oxford 16th	do.	Exrs.Mr. Harward

MAID OF OXFORD 13TH, roan, calved June 6, 1875. Bred by Mr. T. G. Curtler, Bevere House; got by Ragman (35198), dam (Maid of Oxford 3rd) by Third Duke of Claro (23729), &c. See Vol. xxi. p. 661.

Produce in		Names, &c.	By what Bull.	By whom bred.
1878, Feb. 10, white,	B.C.	(Steer)	Bn. Barrington 6th, 33008	Mr. Harward
1879, May 6, white,	B.C.	(Steer)	do.	do.

MARCHIONESS, red, calved May 6, 1875, Vol. xxiv. p. 491. Bred by Mr. J. Harward; got by Earl Mortimer (33780), dam (Kelstern's Marchioness 2nd) by Second Duke of Wetherby (21618), &c.

Produce in		Names, &c.	By what Bull.	By whom bred.
1879, Feb. 24, roan,	B.C.	Marquis Barrington	Bn. Barrington 6th, 33008	Mr. Harward

MARCHIONESS 2ND, white, calved May 1, 1876. Bred by Mr. J. Harward; got by Baron Barrington 6th (33008), dam (Kelstern's Marchioness 2nd) by Second Duke of Wetherby (21618), &c. See "Marquis Barrington," p. 164.

Produce in		Names, &c.	By what Bull.	By whom bred.
1879, Feb. 6, roan,	C.C.	Marchioness 5th	B. T'croftOxford4th, 37822	Mr. Harward

PRINCESS OF WALES, roan, calved March 10, 1875, Vol. xxv. p. 493. Bred by Mr. J. Harward; got by Earl Mortimer (33780), dam (Cambria 10th) by Eighth Duke of Geneva (28390), &c.

Produce in		Names, &c.	By what Bull.	By whom bred.
1879, July 2, white,	C.C.	Princess of Wales 2d	Bn. Barrington 6th, 33008	Exrs.Mr. Harward

SUNRISE, roan, calved September 29, 1871, Vol. xxii. p. 451. Bred by Mr. C. O. Eaton, Tixover Hall; got by Lord Nelson (26693), dam (Alexandra) by Cromwell (19528), &c.

Produce in		Names, &c.	By what Bull.	By whom bred.
1877, July 20, white,	C.C.	Countess Blanche 2nd	Bn. Barrington 6th, 33008	Mr. Harward
1878, July 8, white,	B.C.	(Steer)	do.	do.
1879, July 1, white,	C.C.	Countess Blanche 3rd	do.	Exrs.Mr. Harward

TWILIGHT, red and white, calved February 10, 1874, Vol. xxiii. p. 489. Bred by the Rev. J. Swarbrick, Thurnham; got by Grand Duke of Waterloo (28766), dam (Emily) by Second Duke of Collingham (23730), &c.

Produce in		Names, &c.	By what Bull.	By whom bred.
1877, Dec. 29, red,	B.C.	(dead)	Baron Oxford 7th, 36199	Mr. Harward
1879, Jan. 25, roan,	C.C.	Lady Mornington 2d	Bn. Barrington 6th, 33008	do.

WILD EYES, roan, calved September 2, 1873, Vol. xxiii. p. 489. Bred by Mr. T. Whiteside, Hesketh End; got by Oxford Cheerboy (32014), dam (Wild Rose) by Knight of the Harem (24278), &c.

Produce in		Names, &c.	By what Bull.	By whom bred.
1877, D 21, roan,	C.C.	Oxford's Wild Eyes	Baron Oxford 7th, 36199	Mr. Harward
1879, M 14, white,	C.C.	Bar'gton'sWildEyes	Bn. Barrington 6th, 33008	do.

HAWKES, W.,
Thenford, Banbury.

CHERRY, red, calved January 16, 1872, Vol. xxv. p. 493. Bred by Mr. W. Hawkes; got by Cherry Prince (28180), dam (Cestus) by Windsor Fitz-Windsor (25458), &c.

Produce in		Names, &c.	By what Bull.	By whom bred.
1879, April 4, roan,	C.C.	Cheerful	Harper, 38407	Mr. Hawkes

DELIGHT, red and white, calved April 20, 1871, Vol. xxi. p. 761. Bred by Mr. W. Hawkes ; got by Warden (27746), dam (Dutiful) by The General (25295), &c.

Produce in		Names, &c.	By what Bull.	By whom bred.
1877, Sept. 12, r. & w.,	B.C.	(Steer)	Fair Thane, 31127	Mr. Hawkes
1879, May 22, red,	C.C.	Dewdrop	Harper, 38407	do.

DIFFERENT, red, calved January 10, 1874. Bred by Mr. W. Hawkes; got by Fair Thane (31127), dam (Delight) by Warden (27746), &c. See Vol. xxi. p. 761.

1878, Dec. 20, red,	B.C.	(Steer)	Harper, 38407	Mr. Hawkes
1879, Dec. 16, roan,	C.C.	Drift	Harcourt 39977	do.

FAWSLEY GARLAND 8TH, red and white, calved March 13, 1871, Vol. xxv. p. 494. Bred by Mr. W. Hawkes; got by Warden (27746), dam (Fawsley Garland 6th) by Booth Royal (15673), &c.

1879, April 26, red,	C.C.	FawsleyGarland11th	Harper, 38407	Mr. Hawkes

FLORA, roan, calved June 22, 1877. Bred by Mr. W. Hawkes; got by Fair Thane (31127), dam (Frolicsome) by Lord Blithe (22126), &c. See " Fair Play," p. 97.

1879, Nov. 2, roan,	C.C.	Faith	Sir Swithin. 40722	Mr. Hawkes

HAPPINESS, red and white, calved December 6, 1875. Bred by Mr. W. Hawkes; got by Fair Thane (31127), dam (Heiress) by Security (22859), &c. See Vol. xxii. p. 452.

1878, Apr. 19, r. & w.,	B.C.	(Steer)	Foremark, 38308	Mr. Hawkes
1879, Aug. 10, roan,	C.C.	Heartless	Harcourt, 39977	do.

JEWEL, red, calved February 2, 1877. Bred by Mr. W. Hawkes; got by Tangible (35719), dam (Joyful) by Lord Lyon (40201), &c. See Vol. xxiv. p. 492.

1879, May 14, r. & w.,	C.C.	Jasmine	Sir Swithin, 40722	Mr. Hawkes

JOYFUL, roan, calved October 30, 1869, Vol. xxiv. p. 492. Bred by Mr. J. Laxton, Morborne; got by Lord Lyon (40201), dam (Julia) by Rifleman (18704), &c.

1879, March 17, red,	C.C.	Jollity	Harper, 38407	Mr. Hawkes

PATRONESS, roan, calved March 17, 1877. Bred by Mr. W. Hawkes; got by Heir of Lorne (36761), dam (Petala) by Fair Thane (31127), &c. See Vol. xxiv. p. 493.

1879, June 25, r. & w.,	C.C.	Piedness	Sir Swithin, 40722	Mr. Hawkes

PETALA, red and white, calved October 1, 1874, Vol. xxiv. p. 493. Bred by Mr. W. Hawkes; got by Fair Thane (31127), dam (Pattern) by Warden (27746), &c.

1879, March 29, red,	C.C.	Perfume	Harper, 38407	Mr. Hawkes

PHILLIS 21ST, roan, calved December 19, 1875, Vol. xxv. p. 494. Bred by Mr. H. Aylmer, West Dereham Abbey; got by Royal Magnate (32398), dam (Phillis 12th) by Royal Broughton (27352), &c.

1879, Aug. 28, white,	C.C.	Phillis Alba	Harcourt, 39977	Mr. Hawkes

PINK 20TH, white, calved July 24, 1872, Vols. xxii. and xxiv. pp. 425, 449. Bred by Mr. T. Garne, Broadmoor; got by Buccaneer (25693), dam (Pippin) by Cynric (19542), &c.

1879, April 9, white,	C.C.	Pine	Harper, 38407	Mr. Hawkes

PLAYFUL 5TH, white, calved December 1, 1876. Bred by Mr. W. Hawkes;
got by Cecrops (33311), dam (Playful 4th) by Fair Thane (31127), &c. See
Vol. xxiii. p. 489.

Produce in		Names, &c.		By what Bull.		By whom bred.
1879, May 24, roan,	C.C.	Playful 6th		Sir Swithin, 40722		Mr. Hawkes

PRUDENCE, roan, calved March 30, 1876. Bred by Mr. W. Hawkes; got by
Fair Thane (31127), dam (Picotee) by Warden (27746), &c. See Vol. xxiii.
p. 489.

| 1878, July 12, roan, | B.C. | (Steer) | | Harper, 38407 | | Mr. Hawkes |
| 1879, July 31, roan, | C.C. | Prioress | | Sir Swithin, 40722 | | do. |

SWEET ROSE, red and white, calved September 11, 1874. Bred by Mr. W.
Hawkes; got by Fair Thane (31127), dam (Sweetbrier 40th) by Second Earl
of Darlington (26056), &c. See Vol. xxi. p. 762.

| 1877, Aug. 6, r. & w., | B.C. | (Steer) | | Heir of Lorne, 36761 | | Mr. Hawkes |
| 1878, July 21, red, | B.C. | (Steer) | | Foremark, 38308 | | do. |

VIOLET, roan, calved December 27, 1871, Vols. xxiii. and xxiv. pp. 490, 493.
Bred by Mr. W. Hawkes; got by Warden (27746), dam (Princess of Wales)
by General Hopewell (17953), &c.

| 1879, July 7, roan, | C.C. | Viola | | Sir Swithin, 40722 | | Mr. Hawkes |

WREATH, red and white, calved September 20, 1872, Vols. xxi., xxii., and xxv.
pp. 851, 510, 494. Bred by Mr. H. A. Brassey, Preston Hall; got by Cherry
Grand Duke 3rd (28174), dam (Warrior's Crest) by Blinkhoolie (23428), &c.

| 1879, Nov. 2, roan, | C.C. | Wealthy | | Harcourt, 39977 | | Mr. Hawkes |

HAWKINS, E.,
Cruckfield, Ford, Salop.

DONNA MARIA 2ND, roan, calved March 26, 1876. Bred by the Rev. H.
O. Wilson, Church Stretton; got by King of Trumps (31512), dam (Donna
Maria) by Jupiter (24228), &c. See Vol. xxi. p. 996.

| 1879, Dec. 24, red, | C.C. | Donna Julia | | Duke of Stretton, 43128 | | Mr. Hawkins |

HEADFORT, Marquis of,
Headfort House, Kells, Ireland.

BRITANNIA 8TH, roan, calved April 12, 1870. Bred by Mr. W. S. Garnett,
Williamston; got by Prince Bertram (27119), dam (Britannia 7th) by Dr.
McHale (15887), g. d. (Britannia 3rd) by Rex (16833), &c. See "Duke Ber-
tram," Vol. xix. p. 77.

1877, Jan. 2, r. & w.,	C.C.	Britannia 9th		Lieut.-General, 31600		Marq's of Headfort
1878, April 13, roan,	C.C.	Britannia 10th		Bates-Booth, 33121		do.
1879, May 26, r. & w.,	C.C.	Britannia 11th		Finlarig, 33918		do.

FAIR MAID OF COLLINGHAM, roan, calved April 22, 1869, Vols. xxi.,
xxiii., and xxv. pp. 566, 490, 495. Bred by the Earl of Bective, Underley
Hall; got by Second Duke of Collingham (23730), dam (Fair Maid of York)
by Seventh Duke of York (17754), &c.

| 1879, June 3, r. & w., | C.C. | F. M. of Ballydurrow | | Finlarig, 33918 | | Marq's of Headfort |

JEWEL, roan, calved March 22, 1871. Bred by Major Hamilton, Brownhall; got by Third Earl of Cleveland (28497), dam (Red Riding Hood) by Earl of Cleveland (23828), g. d. (Heroine) by Comet (11298), &c. See Vol. xviii. p. 527.

Produce in		Names, &c.	By what Bull.	By whom bred.
1877, Nov. 17, r. & w.,	B.C.	Emerald	D. of Dentsdale 3rd, 33618	Marq's of Headfort
1879, Jan. 13, roan,	C.C.	Ruby	Finlarig, 33918	do.

LOUISA 23RD, red and white, calved March 29, 1875. Bred by Mr. W. S. Garnett, Williamston; got by Breakspear (33197), dam (Louisa 5th) by Dr. McHale (15887), &c. See Vol. xxiv. p. 394.

1878, Apr. 2, r. & w.,	B.C.	Duke of Dunancory	D. of Dentsdale 3rd, 33618	Marq's of Headfort

PENELOPE, red and white, calved April 3, 1875. Bred by Mr. W. S. Garnett, Williamston; got by England's Glory (23889), dam (Perseverance) by Prince Christian (22581), &c. See Vol. xxiii. p. 535.

1877, June 7, r. & w.,	C.C.	Rosy Morn	Rosy King, 35326	Marq's of Headfort
1878, Dec. 29, r. & w.,	C.C.	Jessie	Finlarig, 33918	do.

RED DAISY, red, calved February 11, 1872, Vol. xxiii. p. 589. Bred by Mr. J. Madden, Roslea Manor; got by Heir of Lothian (28841), dam (Daisy Chain 2nd) by Tally Ho (20927), &c.

1878, Sept. 18, r. & w.,	C.C.	Marguerite	Finlarig, 33918	Marq's of Headfort

HECTOR, A. E.,
Collyhill, Inverurie, N.B.

HARMONY, red, calved March 10, 1876. Bred by Mr. A. E. Hector; got by Chronometer (33385), dam (Helen 3rd) by Young Grand Guard (31305), g. d. (Helen 2nd) by Lord Granville (24395), &c. See "Butterfly," Vol. xxv. p. 41.

1878, Nov. 19, red,	B.C.	Wilfrid	British Champion, 36273	Mr. Hector

Wilfrid, sold to Mr. Myles, Bervie.

HELEN 7TH, red, calved February 23, 1876. Bred by Mr. A. E. Hector; got by Chronometer (33385), dam (Helen 2nd) by Lord Granville (24395), &c. See "Butterfly," Vol. xxv. p. 41.

1878, Nov. 15, red,	C.C.	Helena	British Champion, 36273	Mr. Hector
1879, Dec. 10, roan,	B.C.		Revenue, 40591	do.

HELEN 8TH, red, calved March 16, 1877. Bred by Mr. A. E. Hector; got by Chronometer (33385), dam (Helen 2nd) by Lord Granville (24395), &c.

1879, Apr. 11, red,	C.C.	Herb	British Champion, 36273	Mr. Hector

RACHEL 12TH, red and white, calved January 18, 1875. Bred by Mr. A. E. Hector; got by Chronometer (33385), dam (Rachel 11th) by Heir of Englishman (24122), g. d. (Rachel 6th) by Lord Lyons (22173), gr. g. d. (Rachel 5th) by Lord of Lorn (18258), — (Rachel 2nd) by Baron Renfrew (15624). — (Rachel) by Clarendon (14280), &c. See Vol. xiv. p. 661.

1877, Nov. 12, roan,	C.C.	Rachel 13th	British Champion, 36273	Mr. Hector
1878, Dec. 14, roan,	B.C.	Cicerone	do.	do.

Cicerone, sold to Mr. Bain, Oldmill Reformatory, Aberdeen.

VERBENA 3RD, roan, calved January 18, 1877. Bred by Mr. A. E. Hector; got by Chronometer (33385), dam (Verbena) by Golden Eagle (26267), g. d. (Violet) by Sir Thomas Stanley (25176), gr. g. d. (Duchess) by Marquis of Bute (18336), &c. See Vol. xvi. p. 426.

Produce in		Names, &c.	By what Bull.	By whom bred.
1879, Mar. 23, red,	B.C.	Pearl	British Champion, 36273	Mr. Hector

VIRGINIA, roan, calved April 22, 1877. Bred by Mr. A. E. Hector; got by British Champion (36273), dam (Verbena 2nd) by Rajah (32232), g. d. (Verbena) by Golden Eagle (26267), &c.

1879, Nov. 25, white,	C.C.	Virginia 2nd	Revenue, 40591	Mr. Hector

HEINEMANN, E.,
Ratton Park, Willingdon, Sussex.

GENEROUS, roan, calved in July 1871, Vols. xxii. and xxv. pp. 328, 495. Bred by Mr. J. B. Booth, Killerby Hall; got by Merry Monarch (22349), dam (Brevity) by Brigade Major (21312), &c.

1879, Mar. 21, roan,	C.C.	Georgia Regia	King of Trumps, 31512	Mr. Heinemann

MARSHAL'S GIFT, roan, calved October 31, 1873, Vols. xxiii. and xxv. pp. 603, 496. Bred by Mr. H. Pickersgill, Middleton Quernhow; got by Marshal Booth (31852), dam (New Year's Gift) by Thorntree (25312), &c.

1879, Feb. 26, r. & w.,	B.C.	Marshal Benedict	Prince Benedict, 38893	Mr. Heinemann

Marshal Benedict, sold to Mr. J. Ramsbotham, Crowborough Warren, Tunbridge Wells.

HEMMING, R.,
Bentley Manor, Bromsgrove.

ADA 4TH, roan, calved March 29, 1867, Vols. xx., xxi., and xxiv. pp. 384, 763, 494. Bred by Mr. R. Hemming; got by Grand Cazique (26282), dam (Ada 2nd) by Prince Carl (20538), &c.

1879, Mar. 22, r. & w.,	C.C.	Ada 40th	Gambados, 39911	Mr. Hemming

ADA 10TH, roan, calved January 10, 1871, Vols. xxiv. and xxv. pp. 494, 496. Bred by Mr. R. Hemming; got by Norgrove (29447), dam (Ada 4th) by Grand Cazique (26282), &c.

1879, Dec. 16, roan,	C.C.	Ada 42nd	Gambados, 39911	Mr. Hemming

ADA 14TH, red and white, calved April 26, 1872, Vol. xxv. p. 497. Bred by Mr. R. Hemming; got by Saturnus (29932), dam (Ada 8th) by Cavalier (23524), &c.

1879, April 14, roan,	C.C.	Ada 41st	Gambados, 39911	Mr. Hemming

FAIRY 5TH, roan, calved September 25, 1871, Vols. xxiii. and xxv. pp. 491, 497. Bred by Mr. R. Hemming; got by Norgrove (29447), dam (Fairy 3rd) by Grand Cazique (26282), &c.

1879, Dec. 5, red,	C.C.	Fairy 14th	Gambados, 39911	Mr. Hemming

FAIRY 9TH, red and white, calved February 12, 1876. Bred by Mr. R. Hemming; got by Normansels (34923), dam (Fairy 5th) by Norgrove (29447), &c. See Vol. xxiii. p. 491.

Produce in		Names, &c.	By what Bull.	By whom bred.
1879, Jan. 25, roan,	C.C.	Fairy 13th	Corrie, 38031	Mr. Hemming

HERRICK, Mrs. Perry,
Beau Manor Park, Loughborough.

AZALIA 2ND, roan, calved June 20, 1874. Bred by Mr. W. Perry Herrick, Beau Manor Park; got by Duke of Elvira (30950), dam (Cheney) by Duke of Magdala (28418), g. d. (Pretty Jane) by Thorndale Troutbeck (23071), gr. g. d. (Azalia) by Archduke (37738).

1878, May 8, white,	C.C.	Azalia 4th	L. Fitzclarence 20th, 38604	Mrs. Perry Herrick
1879, Sept. 21, white,	C.C.	Azalia 6th	do.	do.

BELLE, roan, calved January 8, 1872, Vol. xxiv. p. 495. Bred by Mr. W. Perry Herrick, Beau Manor Park; got by Duke of Magdala (28418), dam (Crocus) by Thorndale Troutbeck (23071), &c.

1878, Nov. 7, r. & w.,	C.C.	Bess 17th	L. Fitzclarence 20th, 38604	Mrs. Perry Herrick
1879, Nov. 19, red,	C.C.	Bess 23rd	D. of Worcester 6th, 43149	do.

BESS 2ND, roan, calved March 26, 1874. Bred by Mr. W. Perry Herrick, Beau Manor Park; got by Duke of Magdala (28418), dam (Crocus) by Thorndale Troutbeck (23071), &c.

1879, Jan. 10, roan,	C.C.	Bess 18th	L. Fitzclarence 20th, 38604	Mrs. Perry Herrick
1879, Dec. 15, r. & w.,	C.C.	Bess 25th	D. of Worcester 6th, 43149	do.

CASSANDRA, red and white, calved July 3, 1874, Vols. xxiii. and xxv. pp. 498, 506. Bred by Sir G. R. Philips, Bart., Weston Park; got by Cherry Grand Duke 5th (30712), dam (Clover) by Third Duke of Geneva (21592), &c.

1879, Feb. 4, r. & w.,	C.C.	Cassandra 2nd	Grand Duke 23rd, 34063	Mrs. Perry Herrick

DOLLY, roan, calved April 15, 1874, Vol. xxiii. p. 498. Bred by Sir G. R. Philips, Bart., Weston Park; got by Duke Polycherry (33763), dam (Dewlap) by Third Duke of Geneva (21592), &c.

1878, Sept. 15, roan,	C.C.	Dolly 2nd	Grand Duke 23rd, 34063	Mrs. Perry Herrick
1879, Oct. 7, red	B.C.		L. Fitzclarence 20th, 38604	do.

OLIVE 4TH, red, calved January 8, 1874. Bred by Mr. W. Perry Herrick, Beau Manor Park; got by Duke of Elvira (30950), dam (Olive 2nd) by Thorndale Troutbeck (23071), g. d. (Olive) by Archduke (37738), gr. g. d. (Bess) by Salopian (39076).

1879, Jan. 19, roan,	C.C.	Bess 19th	L. Fitzclarence 20th, 38604	Mrs. Perry Herrick
1879, Dec. 21, roan,	C.C.	Bess 26th	do.	do.

HETHERINGTON, J. J.,
Middle Farm, Brampton, Cumberland.

AGNES GWYNNE, roan, calved December 6, 1869, Vols. xxi., xxiv., and xxv. pp. 581, 497, 498. Bred by Mr. W. Bolton, The Island; got by Duke of Marlborough (23768), dam (Moll Gwynne) by Equinox (17810), &c.

1879, April 2, white	B.C.	Lanercost	Duke of Dentsdale, 33616	Mr. Hetherington

DUCHESS CAMILLA, white, calved April 9, 1874, Vols. xxiv. and xxv. pp. 497; 498. Bred by Mr. J. J. Hetherington; got by Eighth Duke of York (28480), dam (Camilla) by Fourteenth Duke of Oxford (21605), &c.

Produce in	Names, &c.	By what Bull.	By whom bred.
1879, April 6, roan, C.C.	Duchess Camilla 3d	Duke of Dentsdale, 33616	Mr. Hetherington

LADY LOUISA, roan, calved January 3, 1876, Vol. xxv. p. 498. Bred by Mr. F. Postlethwaite, The Hollins; got by Baron Tregunter (33085), dam (Oxford Lily) by Grand Duke of Kent 2nd (28759), &c.

1879, Oct. 25, roan, B.C.	Lord Howard	Lt'b'ne'sD.ofOxf.2d,38564	Mr. Hetherington

LADY MARY 7TH, white, calved May 25, 1875, Vol. xxv. p. 499. Bred by Mrs. and Mr. M. Fawcett, Scaleby Castle; got by Eighth Duke of York (28480), dam (Lady Mary 2nd) by Earl of Glo'ster (21644), &c.

1879, June 30, white, C.C.	Lady Mary 9th	Lt'b'ne'sD.ofOxf.2d,38564	Mr. Hetherington

NELL GWYNNE, roan, calved March 7, 1877. Bred by Mr. J. J. Hetherington; got by Duke of Dentsdale (33616), dam (Agnes Gwynne) by Duke of Marlborough (23768), &c. See Vol. xxiv. p. 497.

1879, Oct. 8, red, C.C.	Nell Gwynne 2nd	Lt'b'ne'sD.ofOxf.2d,38564	Mr. Hetherington

OXFORD BELLE, red and white, calved August 12, 1870, Vol. xxi. p. 956. Bred by Mr. R. B. Brockbank, Burgh-by-Sands; got by Fifth Duke of Oxford (23785), dam (Bessie Belle) by Mandarin (18317), &c.

1879, Mar. 9, roan, C.C.	Oxford Belle 2nd	Baron Barrington 4th, 33006	Mr. Hetherington

PEACH BLOSSOM 6TH, white, calved December 2, 1868, Vols. xx., xxi., and xxv. pp. 683, 701, 499. Bred by Mr. T. Bell, Brockton Hall; got by Earl of Glo'ster (21644), dam (Peach Blossom 2nd) by Baron Westbury (19287), &c.

1879, Mar. 11, white, B.C.	Baron West	Duke of Dentsdale, 33616	Mr. Hetherington

PEACH BLOSSOM 14TH, roan, calved July 23, 1876, Vol. xxv. p. 499. Bred by Mrs. and Mr. M. Fawcett, Scaleby Castle; got by Grand Duke of Kirk levington (34071), dam (Peach Blossom 12th) by Eighth Duke of York (28480), &c.

1879, Sept. 26, roan, B.C.	Duke of Irthing	Lt'b'ne'sD.ofOxf.2d,38564	Mr. Hetherington

POLLY GWYNNE 10TH, roan, calved August 18, 1873, Vol. xxv. p. 499. Bred by Mr. J. J. Hetherington; got by Grand Duke of Lightburne 2nd (26291), dam (Polly Gwynne 4th) by Emperor Maximillian (26100), &c.

1879, Oct. 7, roan, B.C.	Geltsdale	Lt'b'ne'sD.ofOxf.2d,38564	Mr. Hetherington

POLLY GWYNNE 11TH, red and white, calved April 12, 1875, Vols. xxiv. and xxv. pp. 497, 499. Bred by Mr. J. J. Hetherington; got by Eighth Duke of York (28480), dam (Polly Gwynne 6th) by Oxford 4th (29481), &c.

1879, Sept. 23, r. & w., C.C.	Polly Gwynne 15th	Lt'b'ne'sD.ofOxf.2d,38564	Mr. Hetherington

SWEETHEART 4TH, roan, calved September 12, 1871, Vols. xxi., xxii., xxiv., and xxv. pp. 956, 592, 498, 499. Bred by Mr. T. H. Parker, Warwick Hall; got by Seventeenth Duke of Oxford (25994), dam (Sweetheart 3rd) by Sir Walter Gwynne (22921), &c.

1879, Nov. 12, red, B.C.	Naworth	Lt'b'ne'sD.ofOxf.2d,38564	Mr. Hetherington

HEWER, George,
Ley Gore House, Northleach.

CRIMEA 5TH, red and white, calved October 10, 1872.   Bred by Mr. G. Hewer;
got by Paritor (27040), dam (Crimea 4th) by Prince George (24808), &c.   See
"Cyprus," p. 63.

Produce in		Names, &c.	By what Bull.	By whom bred.
1875, Feb. 20, roan,	B.C.	Prince Christian	Prince Alfred, 35083	Mr. Hewer
1876, Feb. 18, r. & w.,	C.C.	Crimea 7th	Prince of the Heath, 35161	do.
1876, Dec. 27, roan,	B.C.	Prince Charlie	do.	do.
1878, Jan. 9, roan,	C.C.	Cream of the Herd	do.	do.
1878, Dec. 21, red,	B.C.	Colonel Crispin	Colonel Geneva, 38012	do.

Prince Charlie, sold to Mrs. Crawshay, Caversham Park, Reading.

MYRTLE, roan, calved December 19, 1875.   Bred by Mr. G. Hewer; got by
Prince of the Heath (35161), dam (Amontillado) by Ranger (35203), &c.   See
Vol. xxiv. p. 498.

1878, Sept. 24, r. & w.,	C.C.	Mabel	Colonel Geneva, 38012	Mr. Hewer
1879, Sept. 5, roan,	C.C.	Muscatel	do.	do.

ROSA 7TH, red and white, calved February 10, 1875.   Bred by Mr. G. Hewer;
got by Christmas Box (33374), dam (Rosa 6th) by Prince George (24808), &c.
See Vol. xxiii. p. 492.

1877, Nov. 10, roan,	C.C.	Rosa 10th	Prince Patrick, 37268	Mr. Hewer
1878, Nov. 18, roan,	B.C.	Prince Rupert	Prince of Prussia, 40520	do.
1879, Nov. 30, r. & w.,	B.C.	Prince Richard	Prince of the Pinks, 42215	do.

ROSA 9TH, roan, calved January 17, 1876.   Bred by Mr. G. Hewer; got by
Prince of the Heath (35161), dam (Rosa 6th) by Prince George (24808), &c.
See Vol. xxiii. p. 492.

1878, Oct. 18, roan,	C.C.	Rosa 12th	Colonel Geneva, 38012	Mr. Hewer
1879, Aug. 27, { roan,	C.C.	Rosa 15th	} do.	do.
{ roan,	C.C.	Rosa 16th		

HICKS, Charles,
Felsteadbury, Chelmsford.

FAIR LASS, roan, calved January 22, 1873, Vol. xxiii. p. 493.   Bred by Mr.
D. A. Green, East Donyland; got by Cherry Prince (28182), dam (Rich Lass)
by Sir Charles (22892), &c.

1877, Sept. 4, white,	C.C.	Blanche	British Heir, 33218	Mr. Hicks

LADY JUGA, red and white, calved October 21, 1876.   Bred by Mr. C. Hicks;
got by Prior (35185), dam (Fair Lass) by Cherry Prince (28182), &c.   See
Vol. xxiii. p. 493.

1879, June 29, roan,	C.C.	Josephine	British Heir, 33218	Mr. Hicks

MAID MARIAN, roan, calved August 25, 1875.   Bred by Mr. C. Hicks; got
by Cherry Prince 8th (33358), dam (Maiden) by Rosicrucian (27337), &c.   See
Vol. xxii. p. 454.

1879, Jan. 12, roan,	C.C.	Maiden	British Heir, 33218	Mr. Hicks

ROAN STATIRA, roan, calved March 26, 1873, Vol. xxiv. p. 500.   Bred by Sir
A. de Rothschild, Bart., Aston Clinton; got by Fatherland (28574), dam
(Statira 6th) by Duke of York (23804), &c.

1878, July 28, roan,	C.C.	Jessica	British Heir, 33218	Mr. Hicks

TRUELOVE, red and white, calved February 1, 1876. Bred by Mr. W. How, Tottington ; got by Proctor (35186), dam (Titlark) by Lord of the Isles (24443), &c. See Vol. xxi. p. 782.

Produce in		Names, &c.	By what Bull.	By whom bred.
1879, May 1, roan,	C.C.	Tibbie	British Heir, 33218	Mr. Hicks

VIRTUE, red and white, calved March 22, 1873, Vol. xxiv. p. 500. Bred by Sir A. de Rothschild, Bart., Aston Clinton ; got by Prince Alexander (35082), dam (Virgin) by Duke of York (23804), &c.

1878, Nov. 16. r. & w.,	B.C.	Hamlet	British Heir, 33218	Mr. Hicks

HOBBS, Charles,
Maisey Hampton, Fairford.

BESSIE 6TH, white, calved March 26, 1875. Bred by Mr. C. Hobbs; got by Colonel Tregunter 7th (30769), dam (Bessie 3rd) by Lord Hastings (29115), &c. See "The Porte." Vol. xxv. p. 279.

1878, April 11, roan,	C.C.	Bessie 13th	D. of Hazlecote 36th, 33671	Mr. Hobbs
1879, June 3, roan,	B.C.	Benjamin	do.	do.

BESSIE 8TH, roan, calved April 4, 1876. Bred by Mr. C. Hobbs; got by Comet of 1874 (33416), dam (Bessie 3rd) by Lord Hastings (29115), &c.

1879, April 2, roan,	B.C.	Bruin	D. of Hazlecote 55th, 39746	Mr. Hobbs

BESSIE 9TH, roan, calved April 30, 1876. Bred by Mr. C. Hobbs ; got by Comet of 1874 (33416), dam (Bessie 4th) by Harry Clifton (31342), &c. See "Telephone," Vol. xxv. p. 276.

1879, Sept. 28, roan,	C.C.	Bessie 14th	G. D. of Waterloo, 34077	Mr. Hobbs

CHORUS 5TH, roan, calved December 29, 1876. Bred by Mr. C. Hobbs ; got by Duke of Hazlecote 36th (33671), dam (Chorus 3rd) by Bates Tertius (21249), &c. See "Minstrel Boy," p. 171.

1879, May 3, roan,	B.C.	Minstrel Boy	D. of Hazlecote 55th, 39746	Mr. Hobbs

COUNTESS, roan, calved August 4, 1875. Bred by Mr. L. Little, Harnhill ; got by Colonel Tregunter 7th (30769), dam (Countess 10th) by Lord Napier (29156), &c. See "Countess 11th," Vol. xxv. p. 558.

1879, Jan. 28, roan,	B.C.	Clarence	Pr. of Clarence 5th, 35139	Mr. Hobbs

COUNTESS 7TH, roan, calved December 11, 1876. Bred by Mr. W. Strickland, Cokethorpe Park ; got by Duke of Barrington 7th (33577), dam (Countess 3rd) by Earl of Verulam (26077), g. d. (Patience) by Waterloo Duke (21077), &c. See Vol. xx. p. 681.

1879, Oct. 31, roan,	C.C.	Poppy	G. D. of Waterloo, 34077	Mr. Hobbs

DARLINGTON 2ND, red, calved February 21, 1876. Bred by Mr. C. Hobbs; got by Comet of 1874 (33416), dam (Prettymaid) by Young Butterfly (23495), &c. See Vol. xxii. p. 455.

1879, April 16, roan,	B.C.	(slaughtered)	D. of Hazlecote 55th, 39746	Mr. Hobbs

FANNY 9TH, roan, calved May 31, 1876. Bred by Mr. C. Hobbs; got by Colonel Tregunter 7th (30769), dam (Fanny) by Duke of Lancaster (25983), &c. See Vol. xxii. p. 454.

1879, Mar. 25, roan,	C.C.	Fanny 12th	D. of Hazlecote 55th, 39746	Mr. Hobbs

JENNY LIND, red, calved February 4, 1877. Bred by Mr. J. B. Jenkins, Kingston House; got by North Star (31988), dam (Jenny Lind) by Lancaster (24302), &c. See Vol. xxi. p. 789.

Produce in		Names, &c.	By what Bull.	By whom bred.
1879, May 3, red,	B.C.	Kingston	D. of Hazlecote 55th, 39746	Mr. Hobbs

MATCHLESS 12TH, red and white, calved May 26, 1876. Bred by Mr. C. Hobbs; got by Comet of 1874 (33416), dam (Matchless 6th) by Lord Hastings (29115), &c. See Vol. xxii. p. 455.

1879, Mar. 18, roan,	B.C.	Merchant	D. of Hazlecote 55th, 39746	Mr. Hobbs

MELODY 5TH, red and white, calved September 25, 1873, Vols. xxiii. and xxiv. pp. 561, 567. Bred by Mr. J. Blyth, Woolhampton House; got by Prince Charlie (32116), dam (Melody) by Grand Duke 13th (21850), &c.

1879, Mar. 18, r. & w.,	B.C.	Wellesley 2nd	G. D. of Waterloo, 34077	Mr. Hobbs

Wellesley 2nd, sold to Mr. W. H. Poynder, Hartham Park, Chippenham.

MUSICAL 3RD, roan, calved March 16, 1876. Bred by Mr. C. Hobbs; got by Duke of Hazlecote 36th (33671), dam (Musical 2nd) by Lord Hastings (29115), &c. See "Minstrel Boy 3rd," p. 171.

1879, June 18, red,	B.C.	Minstrel Boy 3rd	D. of Hazlecote 55th, 39746	Mr. Hobbs

HODGSON, Edward,
Blands Wath, Little Musgrave, Brough, Westmoreland.

VERBENA 16TH, roan, calved January 24, 1874. Bred by Mr. A. Metcalfe, Park House; got by Peer of the Realm (27057), dam (Verbena 15th) by Fredericus 2nd (26196), &c. See "Lord Kellet," Vol. xxv. p. 169.

1878, Jan. 23, roan,	B.C.	Lord Kellet	True Briton, 35823	Mr. Metcalfe
1879, April 23, roan,	C.C.	Verbena 17th	Royal Gift, 39040	Mr. Hodgson

Lord Kellet, sold to Mr. Shepherd, Nether Kellet, Bolton-le-Sands.

HODGSON, Lumley,
Highthorn, Easingwold, Yorkshire.

CRANBERRY 9TH, roan, calved November 15, 1868, Vols. xxi., xxii., and xxiii. pp. 767, 456, 494. Bred by Mr. L. Hodgson; got by Napoleon 3rd (24638), dam (Cranberry) by Kirklevington (14774), &c.

1879, Mar. 25, roan,	B.C.	Cyprus	Lord Angram 3rd, 41818	Mr. Hodgson

CRANBERRY 11TH, red and white, calved December 16, 1870, Vol. xxiii. p. 494. Bred by Mr. L. Hodgson; got by The Bard (25276), dam (Cranberry 3rd) by The Duke (18982), &c.

1879, May 2, r. & w.,	C.C.	Cranberry 26th	Lord Angram 3rd, 41818	Mr. Hodgson

CRANBERRY 12TH, roan, calved October 13, 1871. Bred by Mr. L. Hodgson; got by David (28304), dam (Cranberry 8th) by Dalesman (21523), &c. See Vol. xx. p. 464.

1879, June 14, red,	C.C.	Cranberry 27th	Lord Angram 3rd, 41818	Mr. Hodgson

CRANBERRY 16TH, red, calved September 27, 1875, Vol. xxv. p. 503. Bred by Mr. L. Hodgson ; got by Statesman (35680), dam (Cranberry 13th) by Newburgh 2nd (29433), &c.

Produce in		Names, &c.		By what Bull.		By whom bred.
1879, Mar. 11, roan,	C.C.	Cranberry 25th		Lord Angram 3rd, 41818		Mr. Hodgson

CRANBERRY 18TH, white, calved April 20, 1876. Bred by Mr. L. Hodgson ; got by Statesman (35680), dam (Cranberry 8th) by Dalesman (21523), &c. See Vol. xxiii. p. 494.

1879, July 20, white,	B.C.	Crafty		Duke of Ryedale 2nd,41423		Mr. Hodgson

NELLY 21ST, roan, calved April 15, 1869, Vol. xxiii. p. 495. Bred by Mr. L. Hodgson ; got by Conrad (25826), dam (Nelly 10th) by Abbot 6th (19181), &c.

1879, Dec. 14, roan,	C.C.	Nelly 44th		Duke of Ryedale 2nd,41423		Mr. Hodgson

NELLY 27TH, roan, calved April 25, 1873. Bred by Mr. L. Hodgson ; got by David (28304), dam (Nelly 19th) by Napoleon 3rd (24638), &c. See Vol. xxi. p. 768.

1879, Feb. 16, white,	C.C.	Nelly 43rd		Lord Angram 3rd, 41818		Mr. Hodgson

NELLY 32ND, roan, calved February 18, 1875, Vol. xxv. p. 504. Bred by Mr. L. Hodgson ; got by Buckingham (28106), dam (Nelly 24th) by David (28304), &c.

1879, Nov. 30, roan,	B.C.	Nestor 2nd		Duke of Ryedale 2nd,41423		Mr. Hodgson

PEACH 3RD, roan, calved July 23, 1873, Vol. xxv. p. 504. Bred by Mr. W. Hutton, Gate Burton Hall ; got by Herr Joel (31367), dam (White Cherry) by White Lion (27795), &c.

1879, Apr. 20, roan,	B.C.	Wiseton Duke		Wild Hillhurst, 39321		Mr. Hodgson

STRAWBERRY 9TH, red and white, calved March 20, 1867, Vols. xix., xxii., xxiii., xxiv., and xxv. pp. 740, 456, 495, 503, 504. Bred by Mr. L. Hodgson ; got by Dalesman (21523), dam (Strawberry 3rd) by Chance (15756), &c.

1879, Apr. 16,r.&w.,	C.C.	Strawberry 15th		Lord Angram 3rd, 41818		Mr. Hodgson

STRAWBERRY 13TH, roan, calved March 11, 1875. Bred by Mr. L. Hodgson ; got by Newburgh 5th (34904), dam (Strawberry 9th) by Dalesman (21523), &c. See Vol. xxii. p. 456.

1879, Feb. 23, roan,	B.C.	Sir George		Lord Angram 3rd, 41818		Mr. Hodgson

HOLBOROW, D. B.,
Knockdown House, Tetbury.

BRACELET, red, calved March 26, 1876. Bred by Mr. D. B. Holborow ; got by Geneva Lad (34021), dam (Bertha) by Sunbeam (22994), &c. See "Blanche," Vol. xxiv. p. 503.

1879, Nov. 3, red,	B.C.			Lord Hillhurst, 38614		Mr. Holborow

GEM 4TH, roan, calved March 14, 1876. Bred by Mr. D. B. Holborow ; got by Geneva Lad (34021), dam (Mystery) by Second Earl of Walton (19672), &c. See "Master Cranmore," Vol. xxi. p. 316.

1879, Feb. 1, roan,	C.C.	Gentle		Lord Hillhurst, 38614		Mr. Holborow

HOLBOROW, H., ·
Willsley, Tetbury, Gloucestershire.

LADYLOVE, roan, calved December 27, 1875. Bred by Mr. H. Holborow; got by Northfleet (34926), dam (Lady) by Duke Carlo 5th (28349), &c. See Vol. xxii. p. 457.

Produce in		Names, &c.	By what Bull.	By whom bred.
1879, Feb. 6, r. & w.,	C.C.	Landlady	Emperor, 41508	Mr. Holborow
1879, Dec. 18, r. & w.,	C.C.	Lady-fair	do.	do.

LILA, roan, calved March 28, 1876. Bred by Mr. H. Holborow; got by Northfleet (34926), dam (Lovely) by Viscount Walton (23154), &c. See "Lover," Vol. xxiv. p. 166.

1879, Feb. 17, roan,	B.C.	Liberator	Emperor, 41508	Mr. Holborow

Liberator, sold to Mr. H. Richards, Chilbridge Farm, Wimborne, Dorset.

LYNX, roan, calved March 3, 1876. Bred by Mr. H. Holborow; got by Northfleet (34926), dam (Lizzy) by Viscount Walton (23154), &c. See "Labrador," Vol. xxiv. p. 134.

1879, Feb. 19, roan,	C.C.	Light	Emperor, 41508	Mr. Holborow

NECKTIE, roan, calved December 21, 1875. Bred by Mr. H. Holborow; got by Northfleet (34926), dam (Nightingale) by Viscount Walton (23154), &c. See Vol. xx. p. 672.

1879, Mar. 2, r. & w.,	C.C.	Necklet	Emperor, 41508	Mr. Holborow

NIC-NAC, red and white, calved March 9, 1876. Bred by Mr. H. Holborow; got by Northfleet (34926), dam (Neatness) by Duke Carlo 5th (28349), &c. See Vol. xxii. p. 457.

1879, Mar. 29, roan,	C.C.	Nonsuch	Emperor, 41508	Mr. Holborow

NIGHT-JAR, red, calved January 23, 1876. Bred by Mr. H. Holborow; got by Northfleet (34926), dam (Night-watch) by Duke Carlo 5th (28349), &c. See Vol. xxiii. p. 496.

1879, Feb. 16, r. & w.,	C.C.	Nightshade	Emperor, 41508	Mr. Holborow

NIX, white, calved December 6, 1876. Bred by Mr. H. Holborow; got by Northfleet (34926), dam (Numa) by Baron Walton (25590), &c. See Vol. xxi. p. 770.

1879, Dec. 5, roan,	C.C.	Ninguis	Emperor, 41508	Mr. Holborow

NIXEY, roan, calved March 2, 1876. Bred by Mr. H. Holborow; got by Northfleet (34926), dam (Novel) by Duke Carlo 5th (28349), &c. See Vol. xxii. p. 457.

1879, Feb. 20, r. & w.,	C.C.	Niggle	Emperor, 41508	Mr. Holborow

HOLDEN, Edward,
Laurel Mount, Shipley, Yorkshire.

LADY BARRINGTON 10TH, red, calved March 25, 1876, Vol. xxv. p. 506. Bred by Mr. E. Holden; got by Second Duke of Gloster (28392), dam (Lady Barrington 9th) by Wild Duke (27808), &c.

1879, Aug. 1, roan,	B.C.	Barrington Duke	D. of Tregunter 5th, 33743	Mr. Holden

LADY OF OXFORD 17TH, red, calved July 24, 1875, Vol. xxiv. p. 504. Bred
by Mr. E. H. Cheney, Gaddesby Hall; got by Ninth Duke of Geneva (28391),
dam (Lady of Oxford 13th) by Baron of Oxford (23371), &c.

Produce in	Names, &c.	By what Bull.	By whom bred.
1879, Nov. 2, roan,	B.C.|Oxford Duke	D. of Tregunter 5th, 33743|Mr. Holden

HOLFORD, T.,
Papillon Hall, Market Harborough.

AIRDRIE DUCHESS 6TH, red, calved August 29, 1876. Bred by Mr. A.
Crane, Durham Park, Marion Co.; Kansas, U.S.A ; got by Duke of Hillhurst
4th (41396), dam (Airdrie Duchess 3rd) by Duke of Geneva 11th (41385). &c.
See Vol. xxv. p. 314.

1879. Mar. 17, red,	C.C.|Duchess of Leicester |Viscount Oxford, 40876	Mr. Holford

*AIRDRIE DUCHESS 7TH, roan, calved December 5, 1876. Bred by Mr. A.
Crane, Durham Park, Marion Co., Kansas, U.S.A.; got by Duke of Hillhurst
2nd (39748), dam (Airdrie Duchess 2nd) by Fourteenth Duke of Thorndale
(28459), &c. See " Duke of Leicester," p. 80.

1879, Jan. 25, r. & w.,	B.C.|Duke of Leicester	Viscount Oxford, 40876	Mr. Holford

BARONESS OXFORD 3RD, red, calved November 12, 1873, Vols. xxiii., xxiv..
and xxv. pp. 497, 505, 506. Bred by the Duke of Devonshire, Holker Hall:
got by Duke of Hillhurst (28401), dam (Baroness Oxford) by Second Duke
of Claro (21576), &c.

1879, Jan. 5, red,	B.C.|Viscount Oxford 3rd|Grand Duke 23rd, 34063	Mr. Holford
1879, Dec. 1, red,	B.C.|Viscount Oxford 5th| do.	do.

BEAMING EYES, red, calved November 25, 1876. Bred by Mr. T. Holford;
got by Grand Duke 23rd (34063), dam (Lady Worcester 5th) by Fourth Duke
of Geneva (30958), &c. See Vol. xxiii. p. 498.

1879, Feb. 12, red,	B.C.|D. of Worcester 8th|Viscount Oxford, 40876	Mr. Holford

CHARMER DUCHESS ECHTER, roan, calved December 22, 1874, Vols.
xxiv. and xxv. pp. 505, 506. Bred by Mr. J. Fawcett, Scaleby Castle ; got
by Eighth Duke of York (28480), dam (Constancy) by Third Duke of Geneva
(21592), &c.

1879, June 9, r. & w.,	C.C.|Venusta 3rd	Grand Duke 23rd, 34063	Mr. Holford

LADY GENEVA WATERLOO, red and white, calved June 25, 1871, Vols.
, xxi., xxiv., and xxv. pp. 931, 505, 507. Bred by Lord Skelmersdale, Lathom
House ; got by Ninth Duke of Geneva (28391), dam (Lady Waterloo 16th)
by Third Viscount Waterloo (25387), &c.

1879, Dec. 23, red,	B.C.|Duke of Vittoria 4th|Grand Duke 23rd, 34063	Mr. Holford

LADY WORCESTER 5TH, red, calved April 20, 1871, Vols. xxi., xxiii., and
xxv. pp. 770, 498, 507. Bred by Messrs. Walcott and Campbell, New York
Mills, U.S.A.; got by Fourth Duke of Geneva (30958), dam (Lady Worcester
4th) by Second Duke of Wetherby (21618), &c.

1879, Aug. 20, red,	C.C.	Beaming Eyes 4th	Grand Duke 23rd, 34063	Mr. Holford

MISS PEARL 23RD, roan, calved in June, 1874. Bred by Mr. T. Hands, Canley; got by Second Duke of Darlington (28378), dam (Miss Pearl 4th) by Viceroy (21019), &c. See "Miss Pearl 27th," Vol xxiv. p. 679.

Produce in	Names, &c.	By what Bull.	By whom bred.
1878, April 10, red, B.C.	(Steer)	Cæsarewitch 4th, 33263	Mr. Holford
1879, June 14, r. & w., C.C.	Margarita	Grand Duke 23rd, 34063	do.

SISTER OF CHASTITY, roan, calved April 6, 1874, Vol. xxv. p. 507. Bred by Archdeacon Holbech, Farnborough Hall; got by Earl of Chatham (28495), dam (Sister of Mercy) by Earl of Walton (17787), &c.

1879, July 26, roan, C.C.	Oppia 2nd	Grand Duke 23rd, 34063	Mr. Holford

VISCOUNTESS OXFORD, red, calved January 31, 1876. Bred by Mr. T. Holford; got by Grand Duke 23rd (34063), dam (Baroness Oxford 3rd) by Duke of Hillhurst (28401), &c. See "Viscount Oxford 4th," p. 261.

1878, May 12, roan, C.C.	(dead)	D. of Underley 3rd, 38196	Mr. Holford
1879, July 1, roan, B.C.	Viscount Oxford 4th	D. of Rosedale 6th, 38176	do.

HOLLIDAY, Joseph,
The Tarns, Abbey Town, Carlisle.

DAME WARDEN, red and white, calved February 11, 1871. Bred by Mr. R. B. Brockbank, Burgh-by-Sands; got by Warden (30257), dam (Quickly) by Marquis of Cobham (22299), &c. See Vol. xx. p. 716.

1877, Mar. 25, r. & w., C.C.	Dame Warden 2nd	Baron Tregunter, 33085	Mr. Postlethwaite
1878, Mar. 12, r. & w., B.C.	(dead)	do.	Mr. Holliday
1879, April 8, r. & w., B.C.	Baron Warden	Baron Benson 4th, 36157	do.

Dame Warden 2nd, sold to Mr. Joseph Holliday, The Tarns.

DAME WARDEN 2ND, red and white, calved March 25, 1877. Bred by Mr. Postlethwaite, The Hollins; got by Baron Tregunter (33085), dam (Dame Warden) by Warden (30257), &c.

1879, Sep. 28, r. & w., C.C.	Lady Warden	Duke Butterfly, 39702	Mr. Holliday

PRINCESS OF LORN, roan, calved October 14, 1876. Bred by Mr. G. Ashburner, Low Hall; got by Duke of Oxford (31004), dam (Bride of Lorn) by Sockburn Lad (30024), &c. See Vol. xxiv. p. 304.

1879, Mar. 26, r. & w., C.C.	Princess of Lorn 2nd	D. of Oxford 41st, 38174	Mr. Holliday

ROSAMOND 7TH, white, calved October 24, 1871. Bred by Mr. A. Thompson, The Cross; got by Second Duke of Richmond (28438), dam (Rosamond 5th) by Cressidus Gwynne (25850), g. d. (Rosamond) by Cherry Farmer (15764), &c. See "Rufus," Vol. xxv. p. 253.

1879, April 24, white, C.C.	Fair Rosamond	Town Councillor, 44155	Mr. Holliday

ROSEBUD 2ND, roan, calved March 24, 1877. Bred by Mr. J. Holliday; got by Baron Benson 4th (36157), dam (Rosebud) by Earl of Stair (31083), g. d. (Christmas Rose) by Valentine (23107), gr. g. d. (Primrose) by General Havelock (16118), — Rosy, by a Bull of Sir Wilfrid Lawson's.

1879, Oct. 10, r. & w., B.C.	Duke of Holme	Duke Butterfly, 39702	Mr. Holliday

THORNDALE'S BUTTERFLY PRINCESS, red and white, calved November 17, 1876. Bred by Mr. G. J. Bell, The Nook; got by Duke of Thorndale (33739), dam (Butterfly Princess 18th) by Fourteenth Duke of Oxford (21605), &c. See Vol. xxii. p. 320.

1879, Aug. 6, roan, C.C.	Clifford's B'fly P'cess	Sir Clifford, 40698	Mr. Holliday

WHIMSICAL, roan, calved April 17, 1877. Bred by Mr. J. Holliday; got by
Baron Benson 4th (36157), dam (White Rose 5th) by Balmoral (39393), g. d.
(White Rose 3rd) by Valentine (23107), gr. g. d. (Primrose) by General
Havelock (16118), — Rosy, by a Bull of Sir Wilfrid Lawson's.

Produce in	Names, &c.	By what Bull.	By whom bred.
1879, Dec. 28, roan, C.C.	Lady Whimsical	British Interest, 41150	Mr. Holliday

HOLMES, John,
Strandabrosny, Donemana, Strabane, Ireland.

LADY MARY, white, calved August 8, 1875. Bred by Mr. C. Seaton, Black
Park; got by Bismarck (33155), dam (Moss Rose) by Lord Sefton (29203),
&c. See Vol. xxii. p. 555.

1878, June 1, white, B.C.	(Steer)	Mason, 37065	Mr. Holmes
1879, June 12, roan, C.C.	Lady Maud	do.	do.

HORNBY, Major E. G. S.,
Dalton Hall, Burton, Westmoreland.

LADY NANCY 10TH, roan, calved April 4, 1876. Bred by Major E. G. S.
Hornby; got by Oxford Cherry Duke 2nd (34971), dam (Lady Nancy 6th) by
General Napier (26238), &c. See "Nobleman," p. 176.

1879, Mar. 12. roan, B.C.	Nobleman	Royal Ensign, 37388	Major Hornby

SYLVAN 18TH, red and white, calved May 20, 1874. Bred by Major E. G.
S. Hornby; got by Brutus 2nd (30622), dam (Sylvan 12th) by General
Napier (26238), &c. See "Dalton Hero," p. 65.

1878, July 9, roan, C.C.	Sylvan 29th	Royal Ensign, 37388	Major Hornby
1879, July 6, roan, B.C.	Dalton Hero	do.	do.

SYLVAN 22ND, roan, calved March 26, 1876. Bred by Major E. G. S. Hornby;
got by Oxford Cherry Duke 2nd (34971), dam (Sylvan 13th) by General
Napier (26238), &c. See "Duke of Burton," p. 73.

1879, Nov. 9, roan, B.C.	Duke of Burton	Royal Ensign, 37388	Major Hornby

HORNE, W. Pybus,
Moulton, Richmond, Yorkshire.

DIANA, white, calved February 28, 1877. Bred by Mr. W. P. Horne; got by
Royal Windsor (35419), dam (Daphne) by Lothair (29231), &c. See Vol.
xxiv. p. 506.

1879, Dec. 22, roan, C.C.	Deinira	Sir John Ridd, 35577	Mr. Horne

HAWTHORN 7TH, red, calved June 8, 1877. Bred by Mr. W. P. Horne; got
by Ryedale Duke (35435), dam (Hawthorn 4th) by Sir Richard (32516), g. d.
(Hawthorn 3rd) by Frederick First Fruits (23989), &c. See "Hawthorn 5th,"
Vol. xxv. p. 509.

1879. Dec. 5, red, C.C.	Hawthorn 11th	Knight of Ebor, 43435	Mr. Horne

HAWTHORN 8TH, red, calved July 22, 1877. Bred by Mr. W. P. Horne; got
by Ryedale Duke (35435), dam (May Day) by Lord Beaconsfield (34490),
g. d. (White May) by Lothair (29231), gr. g. d. (Hawthorn 3rd) by Frederick
First Fruits (23989), &c.

1879, Aug. 14, roan, B.C.	Ryedale John	Abbot of Ryedale, 42688	Mr. Horne

HORSWELL and SONS,
Burns Hall, Lew Down, Devon.

BLOSSOM 12TH, roan, calved September 12, 1876. Bred by Messrs. Horswell and Sons; got by Duke of Gaddesby (30956), dam (Blossom 8th) by First General Barrington (26231), g. d. (Blossom 2nd) by Duke (15908), &c. See Vol. xx. p. 414.

Produce in	Names, &c.	By what Bull.	By whom bred.
1879, Dec. 24, red,	B.C.	Lally's Hillhurst D.,38538	Messrs. Horswell

DUCHESS OF GADDESBY, roan, calved April 1, 1876. Bred by Messrs. Horswell and Sons; got by Duke of Gaddesby (30956), dam (Oxford Duchess 4th) by Baron Oxford 2nd (23376), &c. See Vol. xxiii. p. 502.

1879, June 28, roan,	B.C. (Steer)	Oxford Duke 17th, 42089	Messrs. Horswell

GADDESBY BARRINGTON 2ND, red and white, calved June 2, 1876. Bred by Messrs. Horswell and Sons; got by Duke of Gaddesby (30956), dam (Oxford Barrington 2nd) by Baron Oxford 2nd (23376), &c. See Vol. xxi. p. 774.

1879, Mar. 24, r. & w.,	C.C.	Barrington 9th	D. of Florence 3rd, 33641	Messrs. Horswell

GADDESBY BARRINGTON 3RD, roan, calved October 10, 1876. Bred by Messrs. Horswell and Sons; got by Duke of Gaddesby (30956), dam (Oxford Barrington 1st) by Baron Oxford 2nd (23376), &c. See Vol. xxi. p. 774.

1879, Aug. 28, red,	B.C.	Active	Lally's Hillhurst D.,38538	Messrs. Horswell

GADDESBY BLOSSOM 2ND, roan, calved September 20, 1874. Bred by Messrs. Horswell and Sons; got by Duke of Gaddesby (30956), dam (Oxford Blossom 3rd) by Baron Oxford 2nd (23376), &c. See Vol. xxi. p. 774.

1879, Feb. 21, red,	B.C.	D. of Florence 3rd, 33641	Messrs. Horswell

GADDESBY BLOSSOM 3RD, white, calved August 19, 1875. Bred by Messrs. Horswell and Sons; got by Duke of Gaddesby (30956), dam (Oxford Blossom 3rd) by Baron Oxford 2nd (23376), &c.

1879, Mar. 25, roan,	C.C.	G'desby Blossom 7th	D. of Florence 3rd, 33641	Messrs. Horswell

GADDESBY BLOSSOM 5TH, roan, calved April 21, 1875. Bred by Messrs. Horswell and Sons; got by Duke of Gaddesby (30956), dam (Blossom 4th) by Hector (21911), &c. See Vol. xx. p. 414.

1879, Mar. 17, roan,	C.C.	G'desby Blossom 6th	D. of Florence 3rd, 33641	Messrs. Horswell

GADDESBY CERES 3RD, red, calved April 14, 1876. Bred by Messrs. Horswell and Sons; got by Duke of Gaddesby (30956), dam (Red Ceres 6th) by First General Barrington (26231), &c. See "Gaddesby's Duke of Ceres," Vol. xxiv. p. 103.

1879, Sept. 30, r. & w.,	B.C.	D. of Florence 3rd, 33641	Messrs. Horswell

GAINFUL 5TH, roan, calved January 21, 1876, Vol. xxv. p. 510. Bred by Messrs. Horswell and Sons; got by Duke of Gaddesby (30956), dam (Oxford Gainful 3rd) by Baron Oxford 2nd (23376), &c.

1879, Oct. 20, red,	B.C.	Lally's Hillhurst D.,38538	Messrs. Horswell

GAINFUL 6TH, roan, calved January 18, 1877. Bred by Messrs. Horswell and Sons; got by Duke of Gaddesby (30956), dam (Oxford Gainful 3rd) by Baron Oxford 2nd (23376), &c. See Vol. xxiii. p. 502.

1879, Dec. 30, red,	B.C.	Lally's Hillhurst D.,38538	Messrs. Horswell

OXFORD ADA 3RD, red and white, calved November 17, 1876. Bred by Messrs. Horswell and Sons; got by Duke of Gaddesby (30956), dam (Oxford Ada) by Baron Oxford 2nd (23376), &c. See Vol. xxii. p. 460.

Produce in		Names, &c.		By what Bull.		By whom bred.
1879, Oct. 21, roan,	B.C.			Lally's Hillhurst D.,38538	Messrs. Horswell

OXFORD CHERRY 3RD, red and white, calved March 3, 1877. Bred by Messrs. Horswell and Sons; got by Baron Oxford 2nd (23376), dam (Oxford Cherry 2nd) by Baron Oxford 2nd (23376), &c. See Vol. xxiv. p. 510.

1879, Aug. 17, r.&w., B.C.|(Steer)	Oxford Duke 17th, 42089	Messrs. Horswell

OXFORD DUCHESS 10TH, red and white, calved March 28, 1876. Bred by Messrs. Horswell and Sons; got by Baron Oxford 2nd (23376), dam (Duchess of Northumberland 2nd) by First Earl Ducie (23814), &c. See Vol. xxv. p. 330.

1879, Jan. 5, red, B.C.|Oxford Duke 28th	Oxford Duke 9th, 37162	Messrs. Horswell

OXFORD DUCHESS 11TH, red, calved August 24, 1876. Bred by Messrs. Horswell and Sons; got by Baron Oxford 2nd (23376), dam (Oxford Duchess 2nd) by Baron Oxford 2nd (23376), &c. See Vol. xxii. p. 461.

1879, May 6, r. & w., B.C.|(Steer)	Oxford Duke 17th, 42089	Messrs. Horswell

OXFORD GAINFUL 5TH, red and white, calved January 25, 1877. Bred by Messrs. Horswell and Sons; got by Duke of Gaddesby (30956), dam (Oxford Gainful 4th) by Baron Oxford 2nd (23376), &c. See Vol. xxiv. p. 510.

1879, Oct. 9, red, B.C.|	Lally's Hillhurst D.,38538	Messrs. Horswell

OXFORD VANQUISH 9TH, red, calved January 2nd, 1877. Bred by Messrs. Horswell and Sons; got by Oxford Duke 9th (37162), dam (Oxford Vanquish 3rd) by Baron Oxford 2nd (23376), &c. See Vol. xxi. p. 775.

1879, Nov. 11, red, C.C.	Oxf'd Vanquish 17th	Lally's Hillhurst D.,38538	Messrs. Horswell

HORTON, S. L.,
Park House, Shifnal.

COLUMBINE, red and white, calved July 7, 1875, Vol. xxv. p. 511. Bred by Mr. S. L. Horton; got by Abacot (32900), dam (Maria) by Second Duke of Wharfdale (19649), &c.

1879, May 29, roan, B.C.|Boscobel	M. of Blandford 6th,41983|Mr. Horton

LILY, white, calved February 28, 1873, Vols. xxiii., xxiv., and xxv. pp. 503, 511, 511. Bred by Mr. S. L. Horton; got by Prince Charming (27130), dam (Ellie) by Prince Albert (18579), &c.

1879, Dec. 20, roan, B.C.|Constantine 5th	M. of Blandford 6th,41983|Mr. Horton

LITTLE DUCHESS, red and white, calved November 25, 1872, Vols. xxiii. and xxv. pp. 503, 511. Bred by Mr. S. L. Horton; got by Prince Charming (27130), dam (Duchess of Wetherby) by Fifth Duke of Wharfdale (26033), &c.

1879, Dec. 11, roan, C.C.|Little Duchess 7th	M. of Blandford 6th,41983|Mr. Horton

LITTLE DUCHESS 3RD, red and white, calved September 15, 1876. Bred by Mr. S. L. Horton; got by Abacot (32900), dam (Little Duchess) by Prince Charming (27130), &c. See Vol. xxiii. p. 503.

1879, Feb. 10, roan, B.C.|(Steer)	M. of Blandford 6th, 41983|Mr. Horton

LITTLE DUCHESS 4TH, roan, calved September 15, 1876   Bred by Mr. S. L. Horton; got by Abacot (32900), dam (Little Duchess) oy Prince Charming (27130), &c. See Vol. xxiii. p. 503.

Produce in		Names, &c.	By what Bull.	By whom bred.
1879, Feb. 11, roan,	C.C.	Little Duchess 6th	M. of Blandford 6th, 41983	Mr. Horton

LORNA DOON, roan, calved March 12, 1877. Bred by Mr. S. L. Horton; got by Abacot (32900), dam (Ellie) by Prince Albert (18579), &c. See "Constantine 5th," p. 56.

Produce in		Names, &c.	By what Bull.	By whom bred.
1879, July 22, roan.	C.C.	Constance Doon	M. of Blandford 6th, 41983	Mr. Horton

MARIA FAWSLEY, roan, calved August 29, 1873, Vols. xxiii. and xxv. pp. 503, 511. Bred by Mr. S. L. Horton; got by Marquis (31830), dam (Maria) by Second Duke of Wharfdale (19649), &c.

Produce in		Names, &c.	By what Bull.	By whom bred.
1879, July 14, roan,	B.C.	Idsal 2nd	M. of Blandford 6th, 41983	Mr. Horton

WATERLOO CHERRY 4TH, red and white, calved July 6, 1873, Vols. xxiii. and xxv. pp. 504, 511. Bred by Mr. S. L. Horton; got by Idsal (31404), dam (Waterloo Cherry) by Kirbythore Waterloo (24263), &c.

Produce in		Names, &c.	By what Bull.	By whom bred.
1879, Jan. 28, r. & w.,	B.C.	Cherry Princess 6th	M. of Blandford 6th, 41983	Mr. Horton

WATERLOO CHERRY 7TH, red and white, calved October 14, 1876. Bred by Mr. S. L. Horton; got by Abacot (32900), dam (Waterloo Cherry 4th) by Idsal (31404), &c. See "Sir Bevis," p. 230.

Produce in		Names, &c.	By what Bull.	By whom bred.
1879, May 28. roan,	B.C.	Sir Bevis	M. of Blandford 6th, 41983	Mr. Horton

HOSKEN, W. and SON,
Loggans Mill, Hayle, Cornwall.

ALBERTA 3RD, roan, calved September 1, 1876. Bred by Messrs. W. Hosken and Son; got by Second Baron Wild Eyes (30497), dam (Alberta) by Second Earl of Oxford (23844), &c. See Vol. xxiii. p. 504.

Produce in		Names, &c.	By what Bull.	By whom bred.
1879, Mar. 31, roan,	C.C.	Alberta 5th	D. of Oxford 33rd, 36528	Messrs. Hosken

CARNATION 4TH, roan, calved March 29, 1875. Bred by Messrs. W. Hosken and Son; got by Second Baron Wild Eyes (30497), dam (Carnation) by Prince Frederick (16734), &c.. See Vol. xxii. p. 462.

Produce in		Names, &c.	By what Bull.	By whom bred.
1877, Nov. 3, r. & w.,	C.C.	(dead)	D. of Oxford 33rd, 36528	Messrs. Hosken
1878, Oct. 25, white,	B.C.	(Steer)	do.	do.
1879, Nov. 7, roan,	C.C.	Carnation 6th	do.	do.

GERTRUDE, white, calved July 31, 1870, Vol. xxi. p. 778. Bred by Messrs. W. Hosken and Son; got by Second Earl of Oxford (23844), dam (Grateful) by Thorndale Mason (23067), &c.

Produce in		Names, &c.	By what Bull.	By whom bred.
1875, April 16, roan,	B.C.	General Wild Eyes	2d Baron Wild Eyes, 30497	Messrs. Hosken
1876, Aug. 22, roan,	B.C.	Master Frederick	Sir Frederick, 35551	do.
1877, Oct. 2, white,	B.C.	Lord of the Soil	2d Baron Wild Eyes, 30497	do.
1879, Feb. 4, white,	C.C.	Gertrude 4th	D. of Oxford 33rd, 36528	do.

General Wild Eyes, sold to Mr. E. Symons, Winsdon, North Petherwin; Master Frederick, to Mr. O. Young, Parc Behan, Veryan; Lord of the Soil, to Lord Clinton, Heanton, Devon.

KATE 7TH, roan, calved July 31, 1876. Bred by Messrs. W. Hosken and Son; got by Duke of Oxford (31005), dam (Kate 3rd) by Prince Frederick 2nd (24806), &c. See Vol. xxiii. p. 504.

Produce in		Names, &c.	By what Bull.	By whom bred.
1879, Feb. 11, roan,	C.C.	Kate 8th	D. of Oxford 33rd, 36528	Messrs. Hosken

KATHLEEN 4TH, roan, calved March 21, 1875, Vol. xxv. p. 513. Bred by Messrs. W. Hosken and Son; got by Second Baron Wild Eyes (30497), dam (Kathleen) by Prince Frederick (16734), &c.

Produce in		Names, &c.	By what Bull.	By whom bred.
1879, July 3, roan,	C.C.	Kathleen 6th	D. of Oxford 33rd, 36528	Messrs. Hosken

KEEPSAKE 5TH, white, calved April 26, 1874, Vol. xxiv. p. 511. Bred by Messrs. W. Hosken and Son; got by Second Baron Wild Eyes (30497), dam (Keepsake 2nd) by Fifth Earl of Oxford (28515), &c.

1878, July 19, white,	B.C.	Sir Frederick 2nd	Sir Frederick, 35551	Messrs. Hosken
1879, July 20, roan,	C.C.	Keepsake 10th	D. of Oxford 33rd, 36528	do.

KEEPSAKE 6TH, white, calved August 4, 1876. Bred by Messrs. W. Hosken and Son; got by Duke of Oxford (31005), dam (Keepsake) by Thorndale Mason (23067), &c. See Vol. xxiii. p. 504.

1879, Mar. 18, roan,	C.C.	Keepsake 9th	D. of Oxford 33rd, 36528	Messrs. Hosken

MISS ADA 9TH, red and white, calved May 1, 1876. Bred by Messrs. W. Hosken and Son; got by Second Baron Wild Eyes (30497), dam (Miss Ada 2nd) by Second Earl of Oxford (23844), &c. See Vol. xxiii. p. 505.

1879, Feb. 7, roan,	C.C.	Miss Ada 13th	D. of Oxford 33rd, 36528	Messrs. Hosken

RUTH 4TH, red and white, calved May 14, 1876. Bred by Messrs. W. Hosken and Son; got by Second Baron Wild Eyes (30497), dam (Ruth) by Thorndale Mason (23067), &c. See Vol. xxiii. p. 505.

1879, Jan. 18, red,	C.C.	Ruth 5th	D. of Oxford 33rd, 36528	Messrs. Hosken

SYLVIA 3RD, roan, calved April 25, 1874; Vols. xxiv. and xxv. pp. 513, 514. Bred by Messrs. W. Hosken and Son; got by Second Baron Wild Eyes (30497), dam (Sylvia) by Thorndale Mason (23067), &c.

1879, Jan. 18, roan,	C.C.	Sylvia 7th	D. of Oxford 33rd, 36528	Messrs. Hosken

SYLVIA 4TH, roan, calved January 14, 1877. Bred by Messrs. W. Hosken and Son; got by Duke of Oxford 33rd (36528), dam (Sylvia 3rd) by Second Baron Wild Eyes (30497), &c. See Vol. xxiv. p. 513.

1879, Dec. 12, red,	C.C.	(dead)	D. of Oxford 33rd, 36528	Messrs. Hosken

HOW, James,
Broughton, Huntingdon.

CROCUS, roan, calved in September 1869, Vols. xxiv. and xxv. pp. 513, 515. Bred by Mr. W. Ladds, Ellington; got by Wellesley (25421), dam (Cress) by Paul Pry (20478), &c.

1879, June 1, white,	C.C.	Camilla	Proctor, 35186	Mr. How

FAIRY PRINCESS, red and white, calved April 29, 1877. Bred by Mr. J. How; got by Fairy Prince (36627), dam (Queen Victoria) by King Victor (28986), &c. See "Redwing," p. 208.

1879, Nov. 16, red,	B.C.	Redwing	Regal Booth, 35252	Mr. How

MERRY QUEEN, red and white, calved August 27, 1875, Vol. xxiv. p. 513. Bred by Mr. W. Sisman, Buckworth Lodge; got by Julius Cæsar (31452), dam (May Queen 2nd) by Claxton (21433), &c.

1878, Nov. 18, r. & w.,	B.C.	(Steer)	Proctor, 35186	Mr. How
1879, Oct. 17, red,	C.C.	Jolly Queen	Regal Booth, 35252	do.

PAULINE 17TH, roan, calved July 13, 1876. Bred by Mr. J. How; got by
King Victor (28986), dam (Pauline 10th) by Prince of the Realm (22627),
&c. See "Partisan," p. 183.

Produce in		Names, &c.	By what Bull.	By whom bred.
1879, Jan. 17, r. & w.,	C.C.	Pauline 19th	Proctor, 35186	Mr. How
1879, Nov. 18, r. & w.,	B.C.	Partisan	Regal Booth, 35252	do.

VILLAGE QUEEN, roan, calved November 6, 1877. Bred by Mr. J. How;
got by Proctor (35186), dam (Merry Queen) by Julius Cæsar (31452), &c.
See Vol. xxiv. p. 513.

1879, Nov. 29, roan,	C.C.		Regal Booth, 35252	Mr. How .

WATER NYMPH, red and white, calved May 24, 1872, Vols. xxii. and xxiv.
pp. 464, 513. Bred by Mr. T. Bracewell, Ribchester; got by Sir Windsor
Broughton (27507), dam (Water Wave) by Sir David (25135), &c.

1878, Mar. 7, r. & w.,	B.C.	(dead)	Proctor, 35186	Mr. How
1879, Feb. 22, roan,	C.C.	Water Nymph 3rd	do.	do.

WATER NYMPH 2ND, roan, calved March 31, 1877. Bred by Mr. J. How;
got by Fairy Prince (36627), dam (Water Nymph) by Sir Windsor Broughton
(27507), &c. See Vol. xxiv. p. 513.

1879, Oct. 6, r. & w.,	C.C.	Water Nymph 4th	Regal Booth, 35252	Mr. How

HOWARD, James,
Clapham Park, Bedfordshire.

BROUGHTON GWYNNE, white, calved April 29, 1871, Vol. xxiii. p. 318.
Bred by Mr. H. Aylmer, West Dereham Abbey; got by Royal Broughton
(27352), dam (Young Daffy Gwynne 5th) by Young Duke of Cambridge 2nd
(17709), &c.

1878, Sept. 23, roan,	C.C.	Clapham Gwynne	Earl of Leicester 3d, 33804	Mr. Howard
1879, Oct. 27, roan,	B.C.	Clapham Duke	Marquis of Oxford, 34786	do.

GIPSY GWYNNE, red and white, calved July 5, 1869, Vol. xxi. p. 692. Bred
by the Marquis of Exeter, Burghley Park; got by Royal Oxford (27380),
dam (Nancy Gwynne) by Fourth Duke of Thorndale (17750), &c.

1875, Feb. 4, r. & w.,	B.C.		Cambridge Dk. 5th, 30644	Marquis of Exeter
1876, Mar. 22, r. & w.,	C.C.	Gracie Gwynne	do.	do.
1877, Aug. 31, r. & w.,	B.C.	Baron Exeter	do.	Mr. Howard
1879, Feb. 8, r. & w.,	C.C.	Gipsy Countess	Earl of Leicester 5th, 36591	do.
1879, Dec. 22, r. & w.,	C.C.	Gipsy Princess	do.	do.

Gracie Gwynne, sold to Mr. James Howard.

GRACIE GWYNNE, red and white, calved March 22, 1876. Bred by the
Marquis of Exeter, Burghley Park; got by Cambridge Duke 5th (30644),
dam (Gipsy Gwynne) by Royal Oxford (27380), &c. See "Baron Exeter," p. 13.

1878, June 20, red,	C.C.	(dead)	Earl of Leicester 5th, 36591	Mr. Howard

HOWE, Earl,
Gopsall Hall, Atherstone.

BELLE 11TH, red and white, calved April 5, 1875. Bred by Earl Howe; got
by Duke of Hazlecote 24th (30971), dam (Belle 9th) by General Wetherby 2nd
(24027), &c. See Vol. xxiii. p. 505.

1878, June 30, r. & w.,	C.C.	Belle 14th	Ct. Bickerstaffe 6th, 38041	Earl Howe
1879, May 14, r. & w.,	C.C.	Belle 15th	do.	do.

JOSEPHINE 8TH, red and white, calved March 18, 1875. Bred by Earl Howe; got by Duke of Hazlecote 24th (30971), dam (Josephine 5th) by Royal Arch 8th (22771), &c. See "Josephine 6th," Vol. xxiii. p. 506.

Produce in		Names, &c.	By what Bull.	By whom bred.
1878, May 30, red,	B.C.	(Steer)	Ct. Bickerstaffe 6th, 38041	Earl Howe
1879, Mar. 4, red,	C.C.	Josephine 12th	do.	do.

MAY ROSE 24TH, red and white, calved June 15, 1872. Bred by Earl Howe; got by Burleigh (25697), dam (May Rose 17th) by Admiral (21146), &c. See "The Duke," p. 248.

Produce in		Names, &c.	By what Bull.	By whom bred.
1874, Nov. 4, r. & w.,	B.C.	(Steer)	D. of Hazlecote 24th, 30971	Earl Howe
1875, Oct. 4, roan,	B.C.	(Steer)	do.	do.
1877, Jan. 2, red,	B.C.	(Steer)	Oxford Boy, 34966	do.
1878, Feb. 1, roan,	B.C.	The Duke	D. of Hazlecote 38th, 36498	do.

The Duke, sold to Mr. R. Low, Ashby de la Zouch.

MAY ROSE 27TH, red and white, calved March 5, 1874. Bred by Earl Howe; got by Duke of Hazlecote 24th (30971), dam (May Rose 19th) by General Wetherby 2nd (24027), &c. See Vol. xxiii. p. 507.

Produce in		Names, &c.	By what Bull.	By whom bred.
1877, July 20, r. & w.,	C.C.	May Rose 38th	Oxford Boy, 34966	Earl Howe
1878, Aug. 30, r. & w.,	B.C.	(Steer)	Ct. Bickerstaffe 6th, 38041	do.
1879, July 11, red,	B.C.	(dead)	do.	do.

MAY ROSE 32ND, red and white, calved April 6, 1875. Bred by Earl Howe; got by Duke of Hazlecote 24th (30971), dam (May Rose 19th) by General Wetherby 2nd (24027), &c. See Vol. xxiii. p. 507.

Produce in		Names, &c.	By what Bull.	By whom bred.
1877, Dec. 16, roan,	C.C.	May Rose 39th	D. of Hazlecote 38th, 36498	Earl Howe
1879, Jan. 17, red,	C.C.	May Rose 40th	Ct. Bickerstaffe 6th, 38041	do.

MAY ROSE 33RD, red and white, calved May 18, 1875. Bred by Earl Howe; got by Oxford Boy (34966), dam (May Rose 23rd) by Burleigh (25697), &c. See Vol. xxiii. p. 507.

Produce in		Names, &c.	By what Bull.	By whom bred.
1877, Dec. 3, roan,	C.C.	(dead)	D. of Hazlecote 38th, 36498	Earl Howe
1879, Jan. 18, roan,	B.C.	(Steer)	Ct. Bickerstaffe 6th, 38041	do.

MONTHLY ROSE 10TH, roan, calved April 2, 1875. Bred by Earl Howe; got by Duke of Hazlecote 24th (30971), dam (Monthly Rose 6th) by Admiral (21146), &c. See Vol. xxiii. p. 507.

Produce in		Names, &c.	By what Bull.	By whom bred.
1877, Dec. 5, red,	C.C.	Monthly Rose 16th	D. of Hazlecote 38th, 36498	Earl Howe
1879, Jan. 31, red,	B.C.	(Steer)	Ct. Bickerstaffe 6th, 38041	do.

HOWIE, George,
Haughs, Turriff, N.B.

PARAGON, roan, calved March 10, 1866, Vols. xix., xx., xxiv., and xxv. pp. 660, 680, 544, 553. Bred by Mr. A. Longmore, Rettie; got by Sir Charles 2nd (20812), dam (Nonsuch) by Imperial Rome (16292), &c.

Produce in		Names, &c.	By what Bull.	By whom bred.
1879, Feb. 26, r. & w.,	C.C.	Rose of Haughs	Baron Havering, 33043	Mr. Howie

HUGHES, Stephen,
The Elms, Stoke Bishop, Bristol.

CHARMING DUCHESS 4TH, red, calved February 1, 1872, Vols. xxii., xxiii., xxiv., and xxv. pp. 558, 642, 646, 659. Bred by Mr. H. J. Sheldon, Brailes House; got by Seventeenth Duke of Oxford (25994), dam (Twin Duchess 3rd) by Knightley (22051), &c.

Produce in		Names, &c.	By what Bull.	By whom bred.
1879, Nov. 22, roan,	B.C.	Duke Charming	Duke of Rothesay, 36534	Mr. Hughes

SPECIMEN 3RD, roan, calved May 19, 1876. Bred by Mr. W. Angerstein, Weeting Hall ; got by Duke Lally (36457), dam (Specimen) by Grand Duke 15th (21852), &c. See Vol. xxiii. p. 339.

Produce in		Names, &c.		By what Bull.		By whom bred.
1879, Jan. 16, r.&w.,	C.C.	Oxford Charmer		D. of Oxford 32nd, 36527		Mr. Hughes

HUMPHREYS, John,
Hanley Hall, West Felton, Salop.

BONNY LASS, roan, calved March 10, 1872. Bred by Mr. T. Williams, Al-, brightlee ; got by Berwick (23411), dam (Lassy) by Emperor (23879), g. d. (Laura) by Hero of the West (8150), &c. See Vol. xvii. p. 586.

Produce in		Names, &c.	By what Bull.	By whom bred.
1876, Mar. 8, r.&w.,	C.C.	Buttermaid	Butterman, 33258	Mr. Williams
1877, Feb. 25, red,	B.C.	Bonny Lad	Broderick, 33234	Mr. Humphreys

Bonny Lad, sold to Mr. E. Wright, Halston, Oswestry.

DUCHESS 3RD, white, calved March 11, 1864. Bred by Messrs. G. and J. Perry, Acton Pigott ; got by Romeo (22754), dam (Duchess 1st) by Napier (14973), &c. See "Lady Grey," Vol. xxiii. p. 486.

Produce in		Names, &c.	By what Bull.	By whom bred.
1874, Jan. 4, white,	C.C.	Duplicate	Forest Knight, 28634	Mr. Humphreys
1874, Dec. 4, roan,	C.C.	Roan Duchess	Rockingham, 35298	do.

DUCHESS WETHERBY 5TH, white, calved November 28, 1870. Bred by the Hon. Colonel Duncombe, Waresley Park ; got by General Wetherby (24026), dam (Satin) by Volunteer (19087), &c. See "Wetherby Gwynne," Vol. xxiv. p. 275.

Produce in		Names, &c.	By what Bull.	By whom bred.
1878, Dec. 9, roan,	C.C.	D'ss Wetherby 6th	Broderick, 33234	Mr. Humphreys

FOLLY, red and white, calved April 26, 1872. Bred by Mr. W. Bradburn, Wednesfield ; got by Woodhouse (32887), dam (Miss Valentine) by Huntsman (21964), g. d. (Duchess of Argyll) by Hercules (14692), &c. See Vol. xvii. p. 461.

Produce in		Names, &c.	By what Bull.	By whom bred.
1875, Jan. 24, red,	C.C.	Frolick	Rockingham, 35298	Mr. Humphreys
1876, Jan. 20, roan,	C.C.	Frantic	Broderick, 33234	do.

HILDA, roan, calved January 10, 1873. Bred by Mr. W. G. Preece. Bicton ; got by Havelock (26350), dam (Hazle) by Hemlock (19942), &c. See "Lady Helen," Vol. xxv. p. 580.

Produce in		Names, &c.	By what Bull.	By whom bred.
1876, May 7, roan,	C.C.	Heliotrope	Duke of Richmond, 33719	Mr. Preece
1877, July 5, roan,	C.C.	Hollyhock	Broderick, 33234	Mr. Humphreys

LOTTY 18TH, roan, calved March 17, 1874. Bred by Mr. H. Killick, Walton Hall ; got by Prince Bismarck (35089), dam (Princess Louise) by Roan Prince 2nd (29791), g. d. (Lotty 6th) by Delhison (19557), &c. See "White Prince 11th," Vol. xxii. p. 279.

Produce in		Names, &c.	By what Bull.	By whom bred.
1877, Jan. 5, roan,	C.C.	Lotty Louise	Broderick, 33234	Mr. Humphreys

LUSTRE, roan, calved March 10, 1870. Bred by Mr. T. Williams, Albrightlee ; got by Worcester Duke (36032), dam (Lily Queen) by Leander (22085), g. d. (Lady Rockingham 2nd) by Prince Patrick (16760), &c. See Vol. xvii. p. 579.

Produce in		Names, &c.	By what Bull.	By whom bred.
1878, Mar. 30, roan,	C.C.	Lustre 2nd	Broderick, 33234	Mr. Humphreys
1879, Jan. 28, roan,	B.C.	Lustrous	do.	do.

PRINCESS 6TH, roan, calved May 21, 1871. Bred by Mr. W. Bradburn, Wednesfield; got by Wednesfield (30281), dam (Princess 3rd) by Romagna (29818), &c. See "White Satin 2nd," Vol. xx. p. 377.

Produce in		Names, &c.	By what Bull.	By whom bred.
1876, Jan. 29, roan,	C.C.	Princess 7th	Broderick, 33234	Mr. Humphreys
1877, Jan. 23, roan	C.C.	Princess 8th	do.	do.

HUNTER, James,
The Palms Farm, Slaley, Hexham.

RINGLET 3RD, roan, calved November 20, 1872. Bred by Messrs. Angus, Broomley; got by Prince Charlie (29606), dam (Ringlet) by Merry Monarch (22349), &c. See "Prince Leopold," p. 195.

1879, June 28, white,	B.C.	Prince Leopold	Prince Regent, 29676	Mr. Hunter

ROYAL DUCHESS, roan, calved April 5, 1875, Vol. xxv. p. 645. Bred by Mr. J. Roddam, Spittal Shield; got by Royal Duke (35357), dam (Maid of Derwent) by Earl of Derwent (28503), &c.

1879, April 29, roan,	C.C.	Royal Bride	Com. W'sor K.C.B., 33421	Mr. Roddam

Royal Bride, sold to Mr. J. Hunter, The Palms Farm.

HUTCHINSON, T. H.,
Manor House, Catterick, Yorkshire.

BLUSHING MAID, white, calved October 7, 1875, Vols. xxiv. and xxv. pp. 514, 516. Bred by Mr. T. H. Hutchinson; got by M. C. (31898), dam (Rustic Maid) by Drum Major (30898), &c.

1879, Dec. 18, roan,	C.C.	Modest Maid	Beau Benedict, 37841	Mr. Hutchinson

GRATEFUL, roan, calved December 26, 1874, Vol. xxiv. p. 514. Bred by Mr. T. H. Hutchinson; got by M. C. (31898), dam (Gerty 3rd) by Knight of the Shire (26552), &c.

1879, Mar. 4, r. & w., C.C.	(dead)	Pluto, 35050	Mr. Hutchinson

GRATIFICATION, white, calved November 14, 1875, Vol. xxv. p. 516. Bred by Mr. T. H. Hutchinson; got by M. C. (31898), dam (Gerty 3rd) by Knight of the Shire (26552), &c.

1879, Sept. 12, white, C.C.	Glory	British Knight, 33220	Mr. Hutchinson

GRATUITY, roan, calved February 23, 1876. Bred by Mr. T. H. Hutchinson; got by M. C. (31898), dam (Gertrude) by King James (28971), &c. See Vol. xxiii. p. 508.

1879, July 20, r. & w., C.C.	Gracilis	British Knight, 33220	Mr. Hutchinson

LADY AGATHA, white, calved September 30, 1875, Vol. xxv. p. 516. Bred by Mr. T. H. Hutchinson; got by M. C. (31898), dam (Lady Alicia) by King James (28971), &c.

1879, Mar. 21, roan,	C.C.	Lady Ada	Pluto, 35050	Mr. Hutchinson

LADY FROLICSOME, roan, calved May 1, 1875, Vol. xxv. p. 516. Bred by Mr. T. H. Hutchinson; got by M. C. (31898), dam (Lady Playful) by Merry Monarch (22349), &c.

1879, Sept. 9, white,	B.C.	Flashman	British Knight, 33220	Mr. Hutchinson

LADY GRACE, roan, calved April 15, 1872, Vols. xxii. and xxv. pp. 466, 516. Bred by Mr. T. H. Hutchinson; got by K. C. B. (26492), dam (Lady Graceful) by Knight Errant (18154), &c.

Produce in		Names, &c.	By what Bull.	By whom bred.
1879, July 3, roan,	C.C.	Lady Gray	British Knight, 33220	Mr. Hutchinson

LADY GRACELESS, red and white, calved May 25, 1874, Vols. xxiv. and xxv. pp. 514, 517. Bred by Mr. T. H. Hutchinson; got by M. C. (31898), dam (Lady Grace) by K. C. B. (26492), &c.

Produce in		Names, &c.	By what Bull.	By whom bred.
1879, Oct. 19, r. & w.,	B.C.	Knight of Kars	British Knight, 33220	Mr. Hutchinson

LADY GRACIOUS, red and white, calved April 23, 1875, Vol. xxv. p. 517. Bred by Mr. T. H. Hutchinson; got by British Lion (30609), dam (Lady Grace) by K. C. B. (26492), &c.

Produce in		Names, &c.	By what Bull.	By whom bred.
1879, July 10, roan,	B.C.	British Farmer	British Knight, 33220	Mr. Hutchinson

LADY LAURA, roan, calved February 20, 1873, Vols. xxii., xxiv., and xxv. pp. 466, 514, 517. Bred by Mr. T. H. Hutchinson; got by British Lion (30609), dam (Lady Louisa) by K. C. B. (26492), &c.

Produce in		Names, &c.	By what Bull.	By whom bred.
1879, Aug. 2, roan,	C.C.	Lady Lena	British Knight, 33220	Mr. Hutchinson

LADY PLAYFUL, roan, calved November 11, 1871, Vols. xxi. and xxii. pp. 784, 466. Bred by Mr. T. H. Hutchinson; got by Merry Monarch (22349), dam (Lady Sophia) by Brigade Major (21312), &c.

Produce in		Names, &c.	By what Bull.	By whom bred.
1879, April 6, r. & w.,	C.C.	Lady Mirthful	Bezique, 33148	Mr. Hutchinson

MERMAID, roan, calved February 23, 1875, Vol. xxv. p. 517. Bred by Mr. T. H. Hutchinson; got by M. C. (31898), dam (Pretty Maid) by Merry Monarch (22349), &c.

Produce in		Names, &c.	By what Bull.	By whom bred.
1879, May 6, roan,	C.C.	Charming Maid	Bezique, 33148	Mr. Hutchinson

PRETTY MAID, roan, calved November 6, 1871, Vols. xxi., xxii., and xxiii. pp. 784, 466, 509. Bred by Mr. T. H. Hutchinson; got by Merry Monarch (22349), dam (Dairy Girl) by Brigade Major (21312), &c.

Produce in		Names, &c.	By what Bull.	By whom bred.
1879, May 5, r. & w.,	B.C.	Pretty Boy	Bezique, 33148	Mr. Hutchinson

VANITY, roan, calved August 6, 1874, Vol. xxv. p. 517. Bred by Mr. T. H. Hutchinson; got by M. C. (31898), dam (Vexation) by Booth's Royal Signet (28061), &c.

Produce in		Names, &c.	By what Bull.	By whom bred.
1879, Dec. 19, r. & w.,	C.C.	Vain Beauty	Beau Benedict, 37841	Mr. Hutchinson

WHITE ROSE, white, calved January 18, 1875, Vol. xxiv. p. 514. Bred by Mr. J. Outhwaite, Bainesse; got by Royal Windsor (29890), dam (Moss Rose) by Baron Killerby (27949), &c.

Produce in		Names, &c.	By what Bull.	By whom bred.
1879, Mar. 18, roan,	C.C.	Queen of the Roses	King of Trumps, 31512	Mr. Hutchinson

HUTTON, Colonel G. M.,
Gate Burton Hall, Gainsborough.

HECUBA, white, calved March 20, 1873, Vol. xxv. p. 517. Bred by Mr. W. Hutton, Gate Burton Hall; got by White Lion (27795) dam (Lady Macbeth 3rd) by Windsor 2nd (25154), &c.

Produce in		Names, &c.	By what Bull.	By whom bred.
1879, Nov. 1, r. & w.,	B.C.	Hector	Oxford Gwynne, 42096	Colonel Hutton

LADY MACBETH 3RD, white, calved July 16, 1868, Vols. xxiii., xxiv., and
xxv. pp. 510, 515, 518. Bred by Mr. W. Hutton, Gate Burton Hall; got by
Windsor 2nd (25454), dam (Lady Macbeth) by Monarch (13347), &c.

Produce in	Names, &c.	By what Bull.	By whom bred.
1879, Nov. 13, roan, B.C.	Banquo	Oxford Gwynne, 42096	Colonel Hutton

OPHELIA, roan, calved July 6, 1868, Vols. xxii., xxiii., xxiv., and xxv. pp. 467,
510, 515, 518. Bred by Mr. W. Hutton, Gate Burton Hall; got by Windsor
2nd (25454), dam (Melpomene) by Waterloo Royal (19125), &c.

1879, Feb. 8, roan, C.C.	Gertrude	Bright Prince, 28083	Colonel Hutton

IRVINE, John,
Long Marton, Kirbythore, Penrith.

MAID OF MARTON, roan, calved July 31, 1876. Bred by Mr. J. Irvine; got
by Sir John (32507), dam (Christmas Rose) by Grand Duke of Westmoreland
(28767), &c. See "Crossfell 6th," Vol. xxiv. p. 577.

1879, Aug. 30, roan, C.C.	Fair Maid	Cherry King, 36352	Mr. Irvine

ISHERWOOD, Thomas,
Fryton, Slingsby, York.

DOROTHY, roan, calved April 29, 1874, Vols. xxiv. and xxv. pp. 515, 518.
Bred by Mr. T. Isherwood; got by General Monk (34013), dam (Dora) by
Sir Roger (35590), &c.

1879, April 7, roan, B.C.	Sergeant Garnett	L.Oxf'd Br't Eyes 2d, 36998	Mr. Isherwood

LADY MARY, roan, calved February 23, 1875, Vol. xxiv. p. 515. Bred by
Mr. T. Isherwood; got by Young Oxford (32010), dam (Lady Blanche) by
Major (34726), &c.

1879, Jan. 8, red, C.C.	Lady Abbess 5th	L.Oxf'd Br't Eyes 2d, 36998	Mr. Isherwood

MARY, roan, calved March 3, 1871, Vols. xxii., xxiii., and xxv. pp. 467, 500,
518. Bred by Mr. T. Isherwood; got by Major (34726), dam (Martha) by
Friar Tuck (33976), &c.

1879, Feb. 18, roan, C.C.	Delicacy 4th	L.Oxf'd Br't Eyes 2d, 36998	Mr. Isherwood

PRIORESS, red and white, calved December 10, 1874, Vol. xxiv. p. 516. Bred
by Mr. T. Isherwood; got by Young Oxford (32010), dam (The Nun) by Sir
Roger (35590), &c.

1879, Jan. 19, r. & w., C.C.	Lady Abbess 7th	L.Oxf'd Br't Eyes 2d, 36998	Mr. Isherwood

THE NUN, roan, calved October 27, 1870, Vols. xxi., xxiii., and xxiv. pp. 785,
510, 516. Bred by Mr. T. Isherwood; got by Sir Roger (35590), dam (Lady
Abbess) by Whiske Boy (23234), &c.

1879, Jan. 16, roan, C.C.	Lady Abbess 6th	L.Oxf'd Br't Eyes 2d, 36998	Mr. Isherwood

JACKSON, R. A.,
Clayfield, Pocklington.

KIRKLEVINGTON CLARA, roan, calved January 12, 1877. Bred by Mr. R.
A. Jackson; got by Grand Duke of Kirklevington 3rd (34073), dam (Clara's
Daughter) by Grand Duke of Cambridge 2nd (26285), &c. See "Hard
Times," p. 118.

1879, Oct. 9, r. & w., B.C.	Lancastrian	Silent Duke, 42382	Mr. Jackson

JARDINE, James,
Dryfeholme, Lockerbie, N.B.

LADY NYANZA, red and white, calved March 24, 1876. Bred by Mr. J. Jardine; got by Knight of Derwent (31536), dam (Miss McCalmont) by Hugh Cairns (26422), &c. See Vol. xxiv. p. 517.

Produce in		Names, &c.	By what Bull.	By whom bred.
1879, Jan. 20, red,	C.C.	Cleopatra	Frankland, 39895	Mr. Jardine

LAVINIA, roan, calved March 22, 1877. Bred by Mr. J. Jardine; got by Knight of Derwent (31536), dam (Miss Livingstone) by Magician (34720), &c. See Vol. xxiv. p. 516.

1870, May 4, white,	C.C.	Lent Lily	Inkermann, 43377	Mr. Jardine

MISS LIVINGSTONE, white, calved February 15, 1874, Vol. xxiv. p. 516. Bred by Mr. J. Jardine; got by Magician (34720), dam (Miss McCalmont) by Hugh Cairns (26422), &c.

1879, March 17, roan,	C.C.	Lucetta	Alfred, 42659	Mr. Jardine

JEFFERSON, J. J. Dunnington,
Thicket Priory, York.

BLANCHE 15TH, red and white, calved December 20, 1875, Vol. xxv. p. 520. Bred by the Duke of Devonshire, Holker Hall; got by Fifth Duke of Wetherby (31033), dam (Blanche 12th) by Baron Oxford 4th (25580), &c.

1879, Dec. 18, r. & w.,	C.C.	Blanche Flower	L'd of the Forth 2nd, 38639	Mr. Jefferson

CLEOPATRA 12TH, white, calved April 13, 1868, Vol. xxiv. p. 368. Bred by Mr. D. R. Davies, Mere Old Hall; got by Grand Duke of Essex 4th (24068), dam (Cleopatra 5th) by Ninth Duke of Oxford (17738), &c.

1878, Mar. 11, roan,	C.C.	Cleopatra 13th	3rd Duke of Glo'ster, 33653	Mr. Jefferson

COUNTESS OF JERSEY 2ND, white, calved January 26, 1872, Vols. xxiv. and xxv. pp. 517, 520. Bred by the Rev. J. D. Jefferson; got by Lord Waterloo (24475), dam (Countess of Jersey) by Duke of Waterloo (21616), &c

1879, Jan. 31, roan,	CC.	Jonquil	D. of Oxford 28th, 33710	Mr. Jefferson

JERSEY COUNTESS 3RD, roan, calved March 26, 1874, Vol. xxv. p. 520. Bred by the Rev. J. D. Jefferson; got by Waterloo Duke (32813), dam (Countess of Jersey) by Duke of Waterloo (21616), &c.

1879, May 29, roan,	B.C.	Jerry	D. of Oxford 28th, 33710	Mr. Jefferson

JESSAMINE, red, calved April 11, 1869, Vols. xxiv. and xxv. pp. 518, 521. Bred by Mr. J. J. D. Jefferson; got by Duke of Waterloo (21616), dam (Julia) by Ranter (18666), &c.

1879, Oct. 26, r. & w.,	C.C.	Jersey Lily	L'd of the Forth 2nd, 38639	Mr. Jefferson

LADY FEATHERS 4TH, red and white, calved December 9, 1875. Bred by Mr. T. Barber, Sproatley Rise; got by Baron Wild Eyes (33100), dam (Lady Feathers 2nd) by Oxford's Baronet (29499), &c. See Vol. xxii. p. 312.

1879, Feb. 25, red,	C.C.	Florence	D. of Oxford 28th, 33710	Mr. Jefferson

LADY TREGUNTER, red and white, calved December 12, 1872, Vols. xxii., xxiv., and xxv. pp. 359, 518, 521. Bred by Lord Chesham, Latimer; got by Second Duke of Tregunter (26022), dam (Gazelle 13th) by Grand Duke 13th (21850), &c.

1879, June 8, red,	C.C.	Lady Garnet	D. of Oxford 28th, 33710	Mr. Jefferson

LADY TREGUNTER 2ND, roan, calved August 28, 1875, Vol. xxv. p. 521.
Bred by Lord Chesham, Latimer; got by Baron Wastwater (30492), dam
(Lady Tregunter) by Second Duke of Tregunter (26022), &c.

Produce in		Names, &c.		By what Bull.		By whom bred.
1879, Oct. 1, roan,	C.C.	Lady Graceful		L'd of the Forth 2nd, 38639	Mr. Jefferson

URSULA OF COQUET 2ND, red and white, calved January 28, 1877. Bred
by Sir W. G. Armstrong, Cragside; got by Oxford Beau 4th (34964), dam
(Ursula of Coquet) by Duke of Edlingham (30948), &c. See Vol. xxiv.
p. 303.

1879, Nov. 23, r. & w., C.C.	Ulundi	W. D. of Geneva 2nd, 39318	Mr. Jefferson

WATER DUCHESS 4TH, roan, calved April 4, 1874. Bred by Mr. T. Barber,
Sproatley Rise; got by Oxford's Baronet (29499), dam (Water Duchess 2nd)
by Grand Duke 13th (21850), &c. See Vol. xxi. p. 560.

1877, Oct. 25, red,	B.C.	Water Duke	D. of Oxford 28th, 33710	Mr. Jefferson
1879, Mar. 23, white, C.C.		Water Ousel	do.	do.

WATERLOO COUNTESS, roan, calved in June, 1874, Vols. xxiv. and xxv.
pp. 518, 521. Bred by the Rev. J. D. Jefferson; got by Waterloo Duke
(32813), dam (Countess of Waterloo) by Veteran (13941), &c.

1879, April 5, red,	C.C.	Lady of Waterford	D. of Oxford 28th, 33710	Mr. Jefferson

WATERLOO COUNTESS 2ND, roan, calved March 3, 1877. Bred by Mr. J. J.
D. Jefferson; got by Duke of Oxford 28th (33710), dam (Waterloo Countess)
by Waterloo Duke (32813), &c.

1879, Sept. 15, r. & w., C.C.	Water Lily	L'd of the Forth 2nd, 38639	Mr. Jefferson

WATERLOO DUCHESS, roan, calved June 9, 1874, Vols. xxiv. and xxv.
pp. 518, 521. Bred by the Rev. J. D. Jefferson; got by Waterloo Earl
(32816), dam (Water Weed) by Duke of Waterloo (21616), &c.

1879, April 6, r. & w., C.C.	Water Baby	D. of Oxford 28th, 33710	Mr. Jefferson

WATERLOO DUCHESS 2ND, red and white, calved May 11, 1875. Bred by
the Rev. J. D. Jefferson; got by Waterloo Earl (32816), dam (Water Weed)
by Duke of Waterloo (21616), &c.

1878, Mar. 30, roan,	B.C.	Wellesley	D. of Oxford 28th, 33710	Mr. Jefferson
1879, July 23, r. & w., C.C.		Wellingtonia	do.	do.

WATER WEED, red and white, calved July 19, 1870, Vols. xxi., xxii., xxiii.,
xxiv., and xxv. pp. 787, 468, 511, 518, 521. Bred by the Rev. J. D. Jefferson;
got by Duke of Waterloo (21616), dam (Countess of Waterloo) by Veteran
(13941), &c.

1879, Sept. 26, red,	B.C.	Waterford	L'd of the Forth 2nd, 38639	Mr. Jefferson

JEFFERSON, R.,
Preston Hows, Whitehaven.

DOLLY VARDEN, roan, calved April 26, 1872, Vol. xxiii. p. 365. Bred by
Mr. W. Burnyeat, Grenahy; got by Windsor's Prince (32881), dam (Jane
Eyre) by Lord Frederick (22156), &c.

1876, Mar. 1, roan,	C.C.	Lonardi	Sir W'sor Broughton, 27507	Mr. Burnyeat.
1879, May 8, roan,	C.C.	Dolly Prinknash	Lord Prinknash, 34655	Mr. Jefferson

Lonardi, sold to Mr. R. Jefferson, Preston Hows.

HECUBA 2ND, red and white, calved April 29, 1875. Bred by Mr. M. Kennedy, Stone Cross; got by Knight of Windsor (31573), dam (Good Cressida) by Good Fitz (21844), &c. See "Bohemian," p. 30.

Produce in	Names, &c.	By what Bull.	By whom bred.
1877, Dec. 13, r. & w., C.C.	Hecuba 3rd	Prince Edward, 37229	Mr. Kennedy
1879, June 3, r. & w., C.C.	Hebe	do.	Mr. Jefferson

IVORY, roan, calved October 21, 1869. Bred by Messrs. Gaitskell, Hall Santon; got by Puff (27219), dam (Imogene) by Precious Metal (24769), g. d. (Illustrious) by Master Gwynne (16539), &c. See "Illustrious Lord," Vol. xxv. p. 137.

1877, Oct. 16, red, C.C.	Ivy Green	Flambard, 38305	Mr. Jefferson

JENNY'S SECRET, roan, calved March 23, 1869, Vol. xx. p. 568. Bred by Mr. R. Jefferson; got by King Charming (22033), dam (Jenny's Bawbee) by General Gwynne (19840), &c.

1875, Mar. 16, r. & w., C.C.	Jenny's Pride	Hubback Junior, 31395	Mr. Cousins
1879, June 3, roan, C.C.	Jenny's Gift	Bailiff, 42686	Mr. Jefferson

Jenny's Pride, sold to Mr. R. Jefferson, Preston Hows.

JENNINGS, Thomas, and Son,
Cockfield Hall, Sudbury, Suffolk.

BRACELET 11TH, red and white, calved January 23, 1870, Vols. xxi. and xxv. pp. 790, 522. Bred by Mr. R. E. Oliver, Sholebroke Lodge; got by Grand Duke 7th (19877), dam (Bracelet 3rd) by Touchstone (20986), &c.

1879, April 20, white, C.C.	Lady Bracelet 2nd	George Frederick, 34030	Messrs. Jennings

BRACELET 22ND, white, calved August 31, 1875, Vol. xxv. p. 522. Bred by Mr. H. J. Sheldon, Brailes House; got by Second Duke of Collingham (23730), dam (Cherry Bracelet) by Cherry Grand Duke 2nd (25758), &c.

1879, Oct. 21, white, C.C.	Lady Bracelet 3rd	George Frederick, 34030	Messrs. Jennings

BRIDAL ROSE, roan, calved November 15, 1876. Bred by Mr. T. Jennings, Cockfield Hall; got by George Frederick (34030), dam (Cathleen) by Sixth Duke of Oneida (30997), &c. See Vol. xxiii. p. 513.

1879, Nov. 30, roan, C.C.	Rosy Duchess	Grand Duke 23rd, 34063	Messrs. Jennings

CONCORD, red and white, calved September 20, 1862, Vol. xix. p. 451. Bred by Mr. E. Hales, North Frith, the property of Mr. F. Sartoris, Rushden Hall; got by Fourth Duke of Thorndale (17750), dam (Charmer 5th) by Garrick (11506), &c.

1876, Oct. 10, roan, C.C.	Lady Frith	Lord York Fawsley, 34709	Mr. Sartoris

Lady Frith, sold to Messrs. T. Jennings and Son, Cockfield Hall.

CORONELLA, roan, calved December 25, 1876. Bred by Mr. T. Jennings, Cockfield Hall; got by George Frederick (34030), dam (Claro's Coral) by Second Duke of Claro (21576), &c. See Vol. xxiii. p. 513.

1879, Oct. 26, r. & w., C.C.	Coral Countess 2nd	Grand Duke 23rd, 34063	Messrs. Jennings

COUNTESS OF SUFFOLK, roan, calved March 29, 1875, Vol. xxiv. p. 519. Bred by Mr. T. Jennings, Cockfield Hall; got by Monarque (37104), dam (Countess of Oxford) by Earl of Oxford (31079), &c.

1878, Dec. 25, r. & w., B.C.	(Steer)	George Frederick, 34030	Messrs. Jennings
1879, Dec. 23, roan, B.C.	Jolly Friar	do.	do.

FAREWELL, white, calved March 11, 1870. Bred by Lord Braybrooke, Audle
End; got by Claro's Rose (25784), dam (Memento) by Old Buck (15017), &
See "Snowflake," Vol. xxv. p. 266.

Produce in	Names, &c.	By what Bull.	By whom bred.
1878, June 8, white, B.C.	Snowflake	George Frederick, 34030	Messrs. Jennings
1879, Oct. 26, white, B.C.	(Steer)	do.	do.

FLORIDA, roan, calved January 31, 1869, Vols. xxi. and xxiii. pp. 791, 513.
Bred by Her Majesty the Queen; got by England's Glory (23889), dam
(Floret) by Douglas (12714), &c.

Produce in	Names, &c.	By what Bull.	By whom bred.
1878, Mar. 27, white, B.C.	Free Kirk	George Frederick, 34030	Messrs. Jennings
1879, Mar. 29, white, C.C.	(dead)	do.	do.

Free Kirk, sold to Mr. G. Gayford, jun., Barrow Green, Bury St. Edmonds.

GEORGIA, red, calved May 29, 1876. Bred by Mr. T. Jennings, Cockfield
Hall; got by Monarque (37104), dam (Florida) by England's Glory (23889),
&c.

Produce in	Names, &c.	By what Bull.	By whom bred.
1878, Nov. 10, r.& w., B.C.	(dead)	George Frederick, 34030	Messrs. Jennings
1879, Oct. 18, roan, B.C.	(Steer)	do.	do.

IDALIA, red and white, calved January 7, 1868, Vols. xx. and xxiv. pp. 562,
519. Bred by Mr. H. J. Sheldon, Brailes House; got by Duke of Darlington
(21586), dam (Countess) by Duke of Cambridge (12742), &c.

Produce in	Names, &c.	By what Bull.	By whom bred.
1878, Nov. 29, roan, B.C.	(dead)	George Frederick, 34030	Messrs. Jennings
1879, Nov. 16, roan, C.C.	Countess Idalia 2nd	do.	do.

LADY FANTAIL, red and white, calved August 20, 1869, Vols. xxii. and xxiv.
pp. 532, 306. Bred by Mr. J. K. Fowler, Prebendal Farm, the property of
Mr. W. Bliss, Chipping Norton; got by Hardicanute (26338), dam (Fantail
2nd) by Costa (21487), &c.

Produce in	Names, &c.	By what Bull.	By whom bred.
1877, Sept. 29, r. & w., C.C.	Lady Fletcher	Cherry Gr. Duke 5th, 30712	Sir G. R. Philips

Lady Fletcher, sold to Messrs. T. Jennings and Son, Cockfield Hall.

LADY FANTAIL 3RD, red, calved March 13, 1875. Bred by Sir G. R. Philips,
Bart., Weston Park; got by Cherry Fawsley (30711), dam (Lady Fantail) by
Hardicanute (26338), &c. See Vol. xxii. p. 532.

Produce in	Names, &c.	By what Bull.	By whom bred.
1878, Nov. 30, roan, C.C.	(dead)	George Frederick, 34030	Messrs. Jennings
1879, Dec. 20, roan, B.C.	Lord Fletcher	do.	do.

LADY FAWSLEY 6TH, red and white, calved February 25, 1875. Bred by
Mr. H. J. Sheldon, Brailes House; got by Duke of Barrington 6th (33576),
dam (Lady Fawsley 4th) by Eighteenth Duke of Oxford (25995), &c. See
Vol. xxii. p. 559.

Produce in	Names, &c.	By what Bull.	By whom bred.
1878, April 10, roan, B.C.	General Fawsley	George Frederick, 34030	Messrs. Jennings
1879, Mar. 18, roan, C.C.	Grace Fawsley	do.	do.

General Fawsley, sold to Mr. M. Witt, Thurston, Bury St. Edmonds.

RUBY'S CHARM, red, calved May 13, 1876. Bred by Mr. T. Jennings, Cockfield Hall; got by Monarque (37104), dam (Ruby 3rd) by Prince Gwynne
(20547), &c. See Vol. xxiii. p. 513.

Produce in	Names, &c.	By what Bull.	By whom bred.
1879, Jan. 12, roan, C.C.	Bridesmaid 2nd	George Frederick, 34030	Messrs. Jennings

VIOLET, roan, calved June 22, 1876. Bred by Mr. T. Jennings, Cockfield
Hall; got by Monarque (37104), dam (White Duchess) by Prolific 3rd
(27214), &c. See Vol. xxiv. p. 519.

Produce in	Names, &c.	By what Bull.	By whom bred.
1878, Oct. 16, roan, B.C.	Tom Tit	Lord Marcellus, 40202	Messrs. Jennings
1879, Dec. 13, { white, B.C. (Steer) / white, B.C. (dead) }		George Frederick, 34030	do.

JOHNSON, T. C.,
Tothby, Alford, Lincolnshire.

COUNTESS, red, calved December 26, 1870, Vols. xxi. and xxiv. pp. 620, 365. Bred by Mr. J. H. Casswell, Laughton·; got by Baron Panton (23377), dam (Cherry) by Grand Duke 6th (19876), &c.

Produce in	Names, &c.	By what Bull.	By whom bred.
1878, April 1, red, B.C.	(Steer)	Princeps 8th, 37272	Mr. Johnson

FANNY, red and white, calved February 9, 1874. Bred by Mr. T. C. Johnson; got by Fortunatus (33961), dam (Rosaline) by Nicholas (24658), &c. See Vol. xxi. p. 791.

Produce in	Names, &c.	By what Bull.	By whom bred.
1878, Feb. 9, r. & w., C.C.	Fan	Princeps 8th, 37272	Mr. Johnson
1879, Mar. 25, roan, C.C.	Becky	Prince Gwynne, 40499	do.

LADY FAWSLEY, red and white, calved June 7, 1867, Vols. xx., xxii., and xxiv. pp. 594, 354, 519. Bred by Mr. C. W. Packe, Prestwold Hall; got by Don Windsor 2nd (21550), dam (Gwendoline) by Old Buck (15017), &c.

Produce in	Names, &c.	By what Bull.	By whom bred.
1879, Mar. 13, r. & w., B.C.	(Steer)	Prince Gwynne, 40499	Mr. Johnson

LADY FAWSLEY BARRINGTON, red and white, calved May 16, 1875, Vol. xxiv. p. 520. Bred by Mr. J. H. Casswell, Laughton; got by Earl of Barrington 2nd (28494), dam (Lady Fawsley) by Don Fawsley 2nd (21550), &c.

Produce in	Names, &c.	By what Bull.	By whom bred.
1879, Feb. 26, r. & w., C.C.	Lady Gwynne	Prince Gwynne, 40499	Mr. Johnson

RHODA, roan, calved February 11, 1873, Vols. xxiii. and xxiv. pp. 513, 520. Bred by Mr. T. C. Johnson; got by Romulus (29823), dam (Rosa) by Nicholas (24658), &c.

Produce in	Names, &c.	By what Bull.	By whom bred.
1878, Mar. 16, roan, B.C.	(Steer)	Princeps 8th, 37272	Mr. Johnson
1879, Mar. 9, roan, B.C.	(Steer)	Prince Gwynne, 40499	do.

RHODA GWYNNE, red and white, calved February 8, 1876. Bred by Mr. T. C. Johnson; got by Oxford's Gwynne (32033), dam (Rhoda) by Romulus (29823), &c.

Produce in	Names, &c.	By what Bull.	By whom bred.
1879, Feb. 13, r. & w., C.C.	Miss Gwynne	Prince Gwynne, 40499	Mr. Johnson

ROAN GWYNNE, roan, calved March 9, 1875. Bred by Mr. T. C. Johnson; got by Oxford's Gwynne (32033), dam (Rosaline) by Nicholas (24658), &c. See Vol. xxiii. p. 514.

Produce in	Names, &c.	By what Bull.	By whom bred.
1878, May 25, red, C.C.	Red Gwynne	Princeps 8th, 37272	Mr. Johnson
1879, June 10, white, B.C.	(Steer)	Prince Gwynne, 40499	do.

ROSA, roan, calved January 22, 1870, Vols. xxi., xxiii., and xxiv. pp. 791, 514, 520. Bred by Mr. T. C. Johnson; got by Nicholas (24658), dam (Ringdove) by Comedian (15789), &c.

Produce in	Names, &c.	By what Bull.	By whom bred.
1878, April 6, red, B.C.	(Steer)	Princeps 8th, 37272	Mr. Johnson
1879, Mar. 11, roan, C.C.	Rosa Gwynne 3rd	Prince Gwynne, 40499	do.

ROSA GWYNNE 1st, red and white, calved January 18, 1875. Bred by Mr. T. C. Johnson; got by Oxford's Gwynne (32033), dam (Rosa) by Nicholas (24658), &c.

Produce in	Names, &c.	By what Bull.	By whom bred.
1878, Feb. 17, r. & w., C.C.	Princeps Gwynne	Princeps 8th, 37272	Mr. Johnson
1879, Mar. 31, roan, B.C.	(Steer)	Prince Gwynne, 40499	do.

ROSA GWYNNE 2ND, roan, calved March 3, 1876. Bred by Mr. T. C.
Johnson; got by Oxford's Gwynne (32033), dam (Rosa) by Nicholas (24658),
&c.

Produce in	Names, &c.	By what Bull.	By whom bred.
1879, Apr. 11, r. & w., B.C.	(Steer)	Prince Gwynne, 40499	Mr. Johnson

ROSALINE, roan, calved January 26, 1871, Vols. xxi., xxiii., and xxiv. pp. 791,
514, 520. Bred by Mr. T. C. Johnson; got by Nicholas (24658), dam (Ringdove) by Comedian (15789), &c.

1878, Apr. 15, red, B.C.	(Steer)	Princeps 8th, 37272	Mr. Johnson
1879, Apr. 27, r. & w., B.C.	(Steer)	Prince Gwynne, 40499	do.

RUBY, roan, calved November 19, 1867, Vols. xxi., xxiii., and xxiv. pp. 791, 514,
520. Bred by Mr. T. C. Johnson; got by Duke of Grafton (21594), dam
(Ringdove) by Comedian (15789), &c.

1878, Apr. 17, r. & w., C.C.	Gertrude	Princeps 8th, 37272	Mr. Johnson
1879, Mar. 4, { white, B.C.|(Steer) { roan, C.C.|Gatty	} Prince Gwynne, 40499	do.

RUBY GWYNNE 1ST, red and white, calved March 27, 1876. Bred by Mr. T.
C. Johnson; got by Oxford's Gwynne (32033), dam (Ruby) by Duke of
Grafton (21594), &c.

1879, Mar. 16, roan, B.C.	(Steer)	Prince Gwynne, 40499	Mr. Johnson

JOHNSON, William,
Prumplestown House, Carlow, Ireland.

INDIAN QUEEN, red, calved March 10, 1875. Bred by Mr. W. Davidson,
Esker, the property of Mr. W. Johnson; got by Indian Chief (34205), dam
(Lady St. Patrick) by The Red Duke (27643), &c. See "Uncas," p. 254.

1879, Mar. 10, r. & w., B.C.	Uncas	Frank, 43246	Mr. Davidson

Uncas, sold to Mr. W. Johnson, Prumplestown House.

JOHNSTONE, Sir Harcourt V. B., Bart.,
Hackness Hall, Scarborough.

SYMPHONY 2ND, roan, calved October 17, 1871, Vol. xxv. p. 523. Bred by
Mr. F. N. Sartoris, Rushden Hall; got by Duke of Kingscote (25981), dam
(Symphony) by Twelfth Duke of Oxford (19633), &c.

1879, Apr. 25, roan, C.C.	Symphony 5th	Duke of Nawton, 36524	Sir H. Johnstone

JONES, George,
Milford, Stafford.

CECILIA 9TH, roan, calved June 13, 1876. Bred by Major Rowley Conwy,
Bodrhyddan; got by Robin Hood (38998), dam (Cecilia) by New York
(22412), g. d. (Agnes) by Nonsuch (18460), &c. See "Nonsuch 2nd,"
Vol. xxiv. p. 187.

1879, June 9, roan, C.C.	Cecilia 10th	Edward Waterloo, 38246	Mr. Jones

CONSTANCE, red, calved January 19, 1876. Bred by Mr. C. Stubbs, Preston
Hill; got by Proud Preston (37286), dam (Cassia) by Charles Edward (25743),
&c. See Vol. xxiii. p. 394.

| 1879, July 2, roan, | C.C. | Sea Gull | Milford 2nd, 42020 | Mr. Jones |

DUCHESS OF NANTWICH, red and white, calved April 17, 1877. Bred by
Baron W. H. von Schroder, The Rookery; got by Count Bickerstaffe 5th
(36400), dam (Duchess 3rd) by Grand Duke of York (12966), &c. See "Lord
Tyrley," p. 159.

| 1879, Sept. 16, red, | C.C. | Duchess 4th | Milford 2nd, 42020 | Mr. Jones |

LILY 10TH, white, calved December 15, 1876. Bred by Mr. G. Jones; got by
Milford (34847), dam (Lily 9th) by White Prince (36762), &c. See Vol. xxv.
p. 524.

| 1879, May 16, white, | C.C. | Lily 13th | Milford 2nd, 42020 | Mr. Jones |

SPINSTER 4TH, red and white, calved June 17, 1870. Bred by Major Rowley
Conwy, Bodrhyddan; got by His Royal Highness (26399), dam (Spinster) by
Lord of the Hills (18267), &c. See Vol. xvii. p. 745.

| 1879, Mar. 7, roan, | C.C. | Spinster 5th | Edward Waterloo, 38246 | Mr. Jones |

JONES, James,
Ballyloughan, Richill, Co. Armagh.

HEROINE OF LOTHIAN, red and white, calved February 6, 1872. Bred by
Mr. J. Madden, Roslea Manor; got by Heir of Lothian (28841), dam (Leo)
by Earl of Cleveland (23828), &c. See "Lammas Boy," p. 136.

| 1878, May 17, r. & w., | B.C. | Gl'y of Lothian(dead) | All Glory, 39369 | Mr. Jones |
| 1879, Aug. 1, roan, | B.C. | Lammas Boy | White Boy, 40908 | do. |

Glory of Lothian, sold to Mr. J. Campbell, Woodhouse Street, Portadown.

LEONORA, roan, calved November 19, 1870. Bred by Mr. J. Madden, Roslea
Manor; got by Backwoodsman (21203), dam (Leo) by Earl of Cleveland
(23828), &c.

Produce in		Names, &c.	By what Bull.	By whom bred.
1877, May 4, roan,	C.C.	Princess Leonora	Prince Albert, 37211	Mr. Jones
1878, Mar. 27, r. & w.,	B.C.	Lionel	All Glory, 39369	do.
1879, Mar. 23, r. & w.,	C.C.	Lina	do.	do.

Princess Leonora, sold to Mr. Dunbar McMaster, Gilford, Co. Down; Lionel, to Mr. H.
Townsend, 12 Upper Ormond Quay, Dublin.

KEARNEY, M.,
The Ford, Lanchester, Durham.

VALENTINE 5TH, roan, calved March 4, 1874. Bred by Mr. J. Smith, Cote
Hill; got by Prince Imperial (29632), dam (Valentine 2nd) by New Year's
Gift (24657), &c. See "Duke of Abercorn 2nd," p. 71.

| 1879, Mar. 30, r. & w., | B.C. | D. of Abercorn 2nd | Duke of Abercorn, 39708 | Mr. Kearney |

VALENTINE 6TH, roan, calved May 5, 1875, Vol. xxiv. p. 521. Bred by Mr.
J. Smith, Cote Hill; got by Duke of Thorndale (33739), dam (Valentine 3rd)
by Flag of Truce (31172), &c.

| 1879, Jan. 9, roan, | C.C. | Rose Valentine | Duke of Abercorn, 39708 | Mr. Kearney |

KERFOOT, John,
Faenol Bach, St. Asaph.

AGNES 2ND, roan, calved December 9, 1873, Vols. xxiii. and xxiv. pp. 393, 384. Bred by Mr. T. Nash, Featherstone; got by Norman (31980), dam (Agnes) by Charles Edward (25743), &c.

Produce in	Names, &c.	By what Bull.	By whom bred.
1879, June 10, roan, B.C.	Wild Wattie	Edward Waterloo, 38246	Mr. Kerfoot

BONA 2ND, red and white, calved August 12, 1872, Vols. xxii., xxiii., and xxiv. pp. 371, 393, 384. Bred by Captain C. R. Conwy, Bodrhyddan; got by Oxford Wild Eyes 2nd (32038), dam (Charlotte 3rd) by New York (22412), &c.

1878, Oct. 23, r. & w., B.C.	Lord Dinorben	Favourite, 33895	Mr. Kerfoot
1879, Dec. 1, r. & w., C.C.	Eugenie	Chairman, 37960	do.

CHARLOTTE 4TH, roan, calved July 6, 1873, Vols. xxii. and xxiv. pp. 372, 385. Bred by Captain C. R. Conwy, Bodrhyddan; got by Oxford Wild Eyes 2nd (32038), dam (Charlotte 3rd) by New York (22412), &c.

1879, Jan. 11, roan, B.C.	Little Jimmy	Favourite, 33895	Mr. Kerfoot

COLUMBINE 5TH, red and white, calved September 15, 1872, Vols. xxii., xxiii., and xxiv. pp. 372, 394, 385. Bred by Captain C. R. Conwy, Bodrhyddan; got by Oxford Wild Eyes 2nd (32038), dam (Columbine) by New York (22412), &c.

1879, Mar. 7. r. & w., C.C.	Mary Jane	Edward Waterloo, 38246	Mr. Kerfoot

ROMPING GIRL, roan, calved March 7, 1873, Vols. xxiii. and xxiv. pp. 395, 386. Bred by Mr. C. Collard, Little Barton; got by Hotspur (28878), dam (Reckless) by Mandarin (26799), &c.

1879, Feb. 8, white, B.C.	Romping Jack	Edward Waterloo, 38246	Mr. Kerfoot

KEY, John,
Musley Bank, Malton.

JOY, roan, calved April 7, 1870, Vols. xxi. and xxii. pp. 792, 471. Bred by Mr. J. Key; got by Lord Jersey (31673), dam (Countess) by Prince Arthur (22574), &c.

1876, Dec. 22, roan, C.C.	Joy 2nd	Wyndham, 36036	Mr. Key
1878, Jan. 2, roan, C.C.	(dead)	D. of Marlborough, 36520	do.
1879, Jan. 11, roan, B.C.	Lord Lytton	do.	do.
1879, Dec. 25, red, B.C.	Misterton	do.	do.

LANCASTER ROSE, red and white, calved January 28, 1875. Bred by Mr. J. Key; got by Slingsby (35605), dam (Rose) by Lord Jersey (31673), &c. See Vol. xxiii. p. 519.

1877, June 5, r. & w., C.C.	Valerian	D. of Marlborough, 36520	Mr. Key
1878, June 12, r. & w., C.C.	Moss Rose	do.	do.

RED ROSE 3RD, red, calved May 19, 1876. Bred by Mr. J. Key; got by Wyndham (36036), dam (Red Rose 2nd) by Lord Jersey (31673), &c. See Vol. xxiii. p. 519.

1878, Oct. 29, red, B.C.	Hercules	D. of Marlborough, 36520	Mr. Key
1879, Sept. 13, red, B.C.	Neptune	do.	do.

ROSE, roan, calved April 30, 1872, Vol. xxiii. p. 519.   Bred by Mr. J. Key ; got
by Lord Jersey (31673), dam (Red Rose) by Prince (32089), &c.

Produce in	Names, &c.	By what Bull.	By whom bred.
1877, Feb. 22, white, C.C.	White Rose	Wyndham, 36036	Mr. Key
1878, Feb. 22, red, B.C.	Red Prince	D. of Marlborough, 36520	do.
1879, Jan. 19, red, B.C.	Plutus	do.	do.

Red Prince, sold to Mr. Bielby, Espersykes, Malton.

VICTORIA 4TH, red and white, calved September 5, 1874.   Bred by Mr. J.
Key ; got by Slingsby (35605), dam (Victoria 3rd) by Cherry B. (28167), &c.
See Vol. xxi. p. 793.

1878, Feb. 27, red, B.C.	(dead)	D. of Marlborough, 36520	Mr. Key
1879, Jan. 27, r. & w., C.C.	Salvia	do.	do.

KING, J. Pittman,
North Stoke, Wallingford.

BEILBY, roan, calved July 23, 1874.   Bred by Mr. J. P. King ; got by Grand
Prince of Claro (28781), dam (Boquet) by Royal Duke (25016), &c.   See
Vol. xxi. p. 793.

1879, Sept. 1, white, C.C.	Bluebell	Ld. Darlington 14th, 40149	Mr. King

BOQUET, white, calved February 27, 1868, Vols. xx., xxi., xxii., xxiii., and xxiv.
pp. 416, 793, 471, 519, 521.   Bred by Mr. W. W. Champion, Calcot ; got by
Royal Duke (25016), dam (Violet) by His Highness (14708), &c.

1879, July 16, white, C.C.	Bridesmaid	Ld. Darlington 14th, 40149	Mr. King

LOCKET, roan, calved October 20, 1873.   Bred by Mr. J. P. King ; got by
Forest King (28633), dam (Lucy) by Wolfsbane (15518), &c.   See Vol. xxi.
p. 793.

1879, July 15, red, C.C.	Laurel	Ld. Darlington 14th, 40149	Mr. King

ORANGE BLOSSOM, roan, calved October 10, 1870, Vols. xxii. and xxiii.
pp. 472, 519.   Bred by Mr. J. P. King ; got by His Lordship (26398), dam
(Orange) by Dukedom (12729), &c.

1879, June 21, white, C.C.	Orange Peel	Ld. Darlington 14th, 40149	Mr. King

ORANGE FLOWER, roan, calved January 27, 1876.   Bred by Mr. J. P. King ;
got by Grand Prince of Claro (28781), dam (Orange Blossom) by His Lordship
(26398), &c.

1879, July 8, red, C.C.	Orange Bud	Ld. Darlington 14th, 40149	Mr. King

KINGSCOTE, Colonel,
Kingscote, Wotton-under-Edge.

ARIEL LADY, roan, calved February 5, 1873, Vol. xxiv. p. 521.   Bred by
Colonel Kingscote ; got by Third Duke of Clarence (23727), dam (Ariel
Duchess) by Duke of Wharfdale (19648), &c.

1878, Mar. 22, roan, C.C.	Ariel Viscountess	Grand D. of Glo'ster, 36721	Colonel Kingscote
1879, Apr. 15, roan, B.C.	Areonaut 4th	Duke of Hillhurst, 28401	do.

Ariel Viscountess, sold to Colonel Luttrell, Badgworth Court.

CELESTE 7TH, roan, calved January 7, 1873.   Bred by Lord Sudeley, Toddington, the property of Colonel Kingscote ; got by Mandarin (29269), dam
(Celeste) by Imperial Oxford (18084), &c.   See Vol. xxi. p. 946.

1879, May 22, roan, C.C.	Countess Celeste 2nd	3rd Duke of Glo'ster, 33653	Mr. Cheney

CHUTNEE, roan, calved February 21, 1873, Vol. xxiii. p. 351. Bred by Colonel Kingscote; got by Third Duke of Clarence (23727), dam (Clove) by Duke of Wharfdale (19648), &c.

Produce in	Names, &c.	By what Bull.	By whom bred.
1879, Dec. 27, roan,	B.C.|Lord Calcot 6th	Grand Duke 24th, 34064	Colonel Kingscote

CLOVE, red, calved May 10, 1865, Vols. xviii., xxiii., and xxiv. pp. 432, 347, 339. Bred by Colonel Kingscote; got by Duke of Wharfdale (19648), dam (Cinnamon) by General Canrobert (12927), &c.

1878, June 3, roan,	C.C.|Caraway 2nd	3rd D. of Clarence, 23727	Colonel Kingscote
1879, July 9, red,	B.C.|Lord Calcot 5th	Duke of Hillhurst, 28401	do.

COWSLIP TEA, roan, calved April 21, 1876. Bred by Colonel Kingscote; got by Duke of Rosedale (33721), dam (Cowslip) by Grand Duke 7th (19877), &c. See Vol. xix. p. 460.

1879, Jan. 28, roan,	B.C.|Cowslip Lad 4th	Oxford Beau 6th, 40422	Colonel Kingscote

GEORGIE HILLHURST 6TH, red, calved March 29, 1875. Bred by Colonel Kingscote; got by Duke of Hillhurst (28401), dam (Georgina 6th) by Fourth Duke of Oxford (11387), &c. See "George," Vol. xxv. p. 121.

1878, May 12, red,	C.C.|Georgie Glo'ster	G. D. of Glo'ster, 36721	Colonel Kingscote
1879, May 12, red,	C.C.|Georgie's Belle	Oxford Beau 6th, 40422	do.

HONEY 39TH, white, calved February 25, 1872, Vols. xxi. and xxv. pp. 794, 527. Bred by Colonel Kingscote; got by Third Duke of Clarence (23727), dam (Honey 20th) by Duke of Wharfdale (19648), &c.

1879, Feb. 18, roan,	C.C.|Honey 75th	Duke of Hillhurst, 28401	Colonel Kingscote

HONEY 43RD, roan, calved April 8, 1873, Vols. xxiii. and xxv. pp. 520, 527. Bred by Colonel Kingscote; got by Duke of Hillhurst (28401), dam (Honeyless) by Caleb (15718), &c.

1879, Apr. 15, white,	B.C.|D. of Hazlecote 73rd	Oxford Beau 6th, 40422	Colonel Kingscote

HONEY 46TH, roan, calved February 7, 1874, Vol. xxv. p. 528. Bred by Colonel Kingscote; got by Duke of Hillhurst (28401), dam (Honey 20th) by Duke of Wharfdale (19648), &c.

1879, Mar. 25, { red,	C.C.|(dead) red, C.C.|(dead)	} Oxford Beau 6th, 40422	Colonel Kingscote

HONEY 53RD, roan, calved December 1, 1875. Bred by Colonel Kingscote, the property of Mrs. Booth, Peddington; got by Duke of Hillhurst (28401), dam (Honey 38th) by Grand Duke of Clarence (28750), &c. See Vol. xxii. p. 472.

1878, Dec. 4, white,	C.C.|Honey 72nd	Cowslip Boy, 38051	Colonel Kingscote

HONEY 54TH, roan, calved March 29, 1876. Bred by Colonel Kingscote; got by Duke of Rosedale 2nd (33722), dam (Honey 26th) by Second Duke of Wetherby (21618), &c. See "Duke of Hazlecote 40th," Vol. xxii. p. 73.

1879, Mar. 4, white,	B.C.|D. of Hazlecote 71st	Oxford Beau 6th, 40422	Colonel Kingscote

Duke of Hazlecote 71st, sold to Mr. James Garlick, Beverston, Tetbury.

HONEY 55TH, red, calved April 1, 1876. Bred by Colonel Kingscote, the property of Mr. T. W. Cadman, Ballifield Hall; got by Duke of Rosedale 2nd (33722), dam (Honey 32nd) by Third Duke of Clarence (23727), &c. See Vol. xx. p. 559.

1879, Feb. 16, roan,	C.C.|Honey 74th	Oxford Beau 6th, 40422	Colonel Kingscote

HONEY 56TH, red and white, calved May 3, 1876. Bred by Colonel Kingscote; got by Duke of Rosedale 2nd (33722), dam (Honey 36th) by Third Duke of Clarence (23727), &c. See Vol. xxiv. p. 522.

Produce in		Names, &c.	By what Bull.	By whom bred.
1879, Jan. 26, roan,	B.C.	D. of Hazlecote 70th	Oxford Beau 6th, 40422	Colonel Kingscote

HONEY 57TH, white, calved May 5, 1876. Bred by Colonel Kingscote; got by Baron Oxford 4th (25580), dam (Honey 23rd) by Second Duke of Wetherby (21618), &c. See Vol. xix. p. 547.

1879, Jan. 19, roan,	B.C.	D. of Hazlecote 68th	Oxford Beau 6th, 40422	Colonel Kingscote

HONEY 58TH, white, calved May 20, 1876. Bred by Colonel Kingscote; got by Third Duke of Claro (23729), dam (Honey 41st) by Second Duke of Glo'ster (28392), &c. See Vol. xxiii. p. 520.

1879, Jan. 20, white,	B.C.	D. of Hazlecote 69th	Oxford Beau 6th, 40422	Colonel Kingscote

Duke of Hazlecote 69th, sold to Mr. James Garlick, Beverston, Tetbury.

HONEY 61ST, red, calved September 28, 1876. Bred by Colonel Kingscote; got by Duke of Hillhurst (28401), dam (Honey 33rd) by Third Duke of Clarence (23727), &c. See Vol. xxv. p. 527.

1879, June 26, red,	C.C.	Honey 78th	Cowslip Boy, 38051	Colonel Kingscote

JESSICA, red and white, calved April 23, 1870, Vols. xxi. and xxv. pp. 568, 337. Bred by Mr. S. O. Priestley, Trefan; got by The Baron (25277), dam (Jessie) by Lord Raglan (13222), &c.

1879, July 22, red,	C.C.	Daisy Wreath	Duke of Glo'ster 7th, 39735	Colonel Kingscote

LADY SCAR HILL 4TH, red and white, calved February 27, 1872, Vol. xxv. p. 528. Bred by Mr. W. Playne, Minchinhampton; got by Duke of Fawsley (28387), dam (Wallflower 11th) by Second Earl of Walton (19672), &c.

1879, Feb. 21, red,	C.C.	W'flower's Sunshine	Duke of Hillhurst, 28401	Colonel Kingscote

LADY SECRET 2ND, red and white, calved April 8, 1876. Bred by Colonel Kingscote; got by Duke of Rosedale 2nd (33722), dam (Lady Surmise) by Duke of Brailes (23724), &c. See Vol. xxv. p. 528.

1879, March 21, roan,	C.C.	Lady Secret 3rd	Oxford Beau 6th, 40422	Colonel Kingscote

LADY SURMISE, red and white, calved January 27, 1869, Vols. xx. and xxv. pp. 615, 528. Bred by Mr. H. J. Sheldon, Brailes House; got by Duke of Brailes (23724), dam (Surmise 2nd) by May Duke (13320), &c.

1879, Feb. 20, red,	B.C.	Lord Suspicion 5th	Duke of Hillhurst, 28401	Colonel Kingscote

LADY WALTON 2ND, roan, calved July 20, 1868, Vols. xx. and xxv. pp. 616, 529. Bred by Mr. W. Wood, Hollyhurst; got by Earl of Glo'ster (21644), dam (Lady Walton) by Third Earl of Walton (19673), &c.

1879, March 7, red,	B.C.	Baron Dagpath 3rd	Duke of Hillhurst, 28401	Colonel Kingscote

OXFORD BELLE 2ND, roan, calved August 1, 1874, Vol. xxv. p. 529. Bred by Colonel Kingscote; got by Duke of Hillhurst (28401), dam (Countess of Oxford) by Seventh Duke of Airdrie (23718), &c.

1879, Mar. 14, white,	B.C.	Oxford Beau 8th	Ld.T'croftOxford2d,38668	Colonel Kingscote

SABRINA 2ND, roan, calved March 9, 1876. Bred by Colonel Kingscote, the property of Mrs. S. W. Hampson, Ullenwood; got by Prig (35073), dam (Seraphina 15th) by Sinbad (29981), &c. See Vol. xxiv. p. 522.

1879, Feb. 6, white,	C.C.	Sabrina 4th	Cowslip Boy, 38051	Colonel Kingscote

SERAPHINA 15TH, roan, calved June 20, 1871, Vols. xxi. and xxiv. pp. 795, 522. Bred by Mr. E. Clarke, Lillingstone Dayrell; got by Sinbad (29981), dam (Seraphine 2nd) by Dreadnought (12719), &c.

Produce in		Names, &c.	By what Bull.	By whom bred.
1878, Jan. 7, r. & w.,	B.C.	Lord Sandgrove 5th	Duke of Hillhurst, 28401	Colonel Kingscote
1878, Dec. 29, r. & w.,	B.C.	Lord Sandgrove 8th	do.	do.
1879, Dec. 13, red,	C.C.	Sabrina 5th	do.	do.

Lord Sandgrove 5th, sold to the Grand Duke Albrecht of Austria; Lord Sandgrove 8th, to Mr. James Peacey, Chedglow, Tetbury.

WALLFLOWER 9TH, white, calved February 28, 1863, Vols. xviii. and xxv. pp. 772, 529. Bred by Mr. W. Playne, Minchinhampton; got by Duke of Clarence (18611), dam (Wallflower 6th) by Grand Duke of Oxford (16184), &c.

1879, May 20, roan,	C.C.	W'flower's Sunbeam	Duke of Hillhurst, 28401	Colonel Kingscote

KIRBY, Aquila,
Market Weighton.

PRINCESS JANE, roan, calved September 23, 1868, Vol. xx. p. 703. Bred by Mr. R. Taylor, Sigglesthorne Manor, the property of Mr. A. Kirby; got by Cherry-Royal (23556), dam (Princess Rose) by Rustic Prince (18789), &c.

1879, March 5, roan,	C.C.	Princess Ida	Mantalini Chief, 34753	Mr. Wood

Princess Ida, sold to Mr. Aquila Kirby, Market Weighton.

KNAPTON, William,
Kelk, Lowthorpe, Hull.

ARCHDUCHESS 3RD, roan, calved March 14, 1873, Vol. xxv. p. 530. Bred by Mr. W. Knapton; got by Pretender (29581), dam (Archduchess 2nd) by Archduke (19239), &c.

1879, Feb. 14, white,	C.C.	Archduchess 8th	Mantalini Chief, 34753	Mr. Knapton

ARCHDUCHESS 4TH, red and white, calved May 10, 1876. Bred by Mr. W. Knapton; got by Kelk Manfred (31460), dam (Archduchess 2nd) by Archduke (19239), &c. See Vol. xxiii. p. 521.

1879, Aug. 4, r. & w.,	B.C.	Arcturus	Prince of the Tyne, 37262	Mr. Knapton

BRUNETTE, roan, calved March 1, 1877. Bred by Mr. J. J. Jefferson, Harpham; got by Oculist (34944), dam (Windsor Leaf 6th) by Booth's Kinsman (25658), &c. See Vol. xxiv. p. 523.

1879, Dec. 2, white,	B.C.	(dead)	Monarch, 40362	Mr. Knapton

EMILY, roan, calved June 4, 1876, Vol. xxv. p. 530. Bred by Mr. E. Robinson, Nafferton; got by Halloo (34105), dam (Eva 3rd) by Master Edmund (29322), &c.

1879, Aug. 4, roan,	B.C.	Everbright	Prince of the Tyne, 37262	Mr. Knapton

JANITA, roan, calved September 13, 1876. Bred by Mr. W. Knapton; got by Kelk Manfred (31460), dam (Lady Janet) by Knight of Osberton (29003), &c. See Vol. xxiii. p. 521.

1879, Jan. 11, roan,	C.C.	Jessie	Prince of the Tyne, 37262	Mr. Knapton

'LADY JANET, roan, calved February 28, 1874, Vols. xxiii. and xxv. pp. 521, 530. Bred by Mr. W. Knapton; got by Knight of Osberton (29003), dam (Augusta) by Lord Panton (22204), &c.

Produce in	Names, &c.	By what Bull.	By whom bred.
1879, Sept. 15, roan, C.C.	(dead)	Prince of the Tyne, 37262	Mr. Knapton

LADY PANTON, roan, calved April 30, 1872, Vols. xxiii. and xxv. pp. 521, 530. Bred by Mr. W. Knapton; got by Manfred (26801), dam (Augusta) by Lord Panton (22204), &c.

1879, Nov. 23, r. & w., B.C.	Pan	Prince of the Tyne, 37262	Mr. Knapton

LADY SARAH, red and white, calved March 28, 1873, Vol. xxiii. p. 521. Bred by Mr. W. Knapton; got by Knight of Osberton (29003), dam (Augusta) by Lord Panton (22204), &c.

1879, Mar. 4, r. & w., B.C.	(dead)	Prince of the Tyne, 37262	Mr. Knapton

NANNETTE, white, calved December 8, 1877. Bred by Mr. W. Knapton; got by Prince of the Tyne (37262), dam (Nursemaid) by Prince Alfred (29593), &c. See Vol. xxiv. p. 522.

1879, Dec. 25, roan, B.C.	Nabob	Monarch, 40362	Mr. Knapton

NURSELING, white, calved September 15, 1875, Vol. xxiv. p. 522. Bred by Mr. W. Knapton; got by Royal Killerby (32396), dam (Nursemaid) by Prince Alfred (29593), &c.

1879, March 15, roan, B.C.	(dead)	Monarch, 40362	Mr. Knapton

NURSEMAID, white, calved March 30, 1873, Vols. xxii., xxiv., and xxv. pp. 474, 522, 530. Bred by Mr. Robinson, Ulverston; got by Prince Alfred (29593), dam (Bridesmaid) by Prince Patrick (16760), &c.

1879, Nov. 10, white, C.C.	Neatness	Monarch, 40362	Mr. Knapton

RACHEL 3RD, white, calved October 10, 1877. Bred by Mr. W. Knapton; got by Prince of the Tyne (37262), dam (Rachel) by Kelk Manfred (31460), &c. See Vol. xxiv. p. 522.

1879, Dec. 23, white, B.C.	(dead)	Monarch, 40362	Mr. Knapton

REBECCA 4TH, roan, calved April 1, 1876. Bred by Mr. W. Knapton; got by Knight of Osberton (29003), dam (Rebecca 2nd) by Windsor 4th (23228), &c. See Vol. xxiii. p. 521.

1879, July 24, roan, C.C.	Rebecca 7th	Prince of the Tyne, 37262	Mr. Knapton

ROYAL BRIDE, red and white, calved May 18, 1875, Vol. xxv. p. 530. Bred by Mr. W. Knapton; got by Knight of Osberton (29003), dam (Royal Bridesmaid) by Robin (24968), &c.

1879, Sept. 24, roan, B.C.	Robin Adair	Prince of the Tyne, 37262	Mr. Knapton

LAMBART, G. W.,
Beau Parc, Co. Meath, Ireland.

ARTHURA, roan, calved March 9, 1875. Bred by Mr. G. W. Lambart; got by Fitz-Arthur (33926), dam (Blush) by British Sailor (23472), &c. See "Impudence," p. 123.

1879, Jan. 4, white, C.C.	White Arthura	Jupiter, 38477	Mr. Lambart

PRIDE OF THE SEA, roan, calved March 11, 1876. Bred by Mr. G. W. Lambart; got by Rupert (29902), dam (Ocean Pride) by British Sailor (23472), g. d. (Her Royal Highness) by Royal Standard (40644), &c. See "Spaniard," p. 239.

Produce in	Names, &c.	By what Bull.	By whom bred.
1879, Mar. 24, white, C.C.	Wild Wave	Jupiter, 38477	Mr. Lambart

LAMBE, Charles,
Aubourn, Lincoln.

ROSE OF THORNDALE, red, calved October 31, 1871, Vol. xxiii. p. 522. Bred by Mr. C. Lambe; got by Thorndale Lad (23066), dam (Rose of Summer) by Great Mogul (14651), &c.

1877, Jan. 3, red,	B.C.	(dead)	Thorndale Knight, 32714	Mr. Lambe
1878, Aug. 12, red,	C.C.	Rose of Thorndale 2d	D.Charm'g Land 2d, 36477	do.

LAMBE, William,
Aubourn, Lincoln.

HERMIA, white, calved March 18, 1874. Bred by Mr. W. Lambe; got by Thorndale Knight (32714), dam (Heroine) by Great Eastern (24081), &c. See Vol. xxi. p. 800.

1879, Sept. 30, roan, C.C.	Hippia	D.Charm'g Land 2d, 36477	Mr. Lambe

LALLA ROOKH, red, calved June 13, 1873, Vols. xxiv. and xxv. pp. 525, 532. Bred by Mr. W. Lambe; got by Lord of the Manor (29178), dam (Queen of Oude) by Great Mogul (14651), &c.

1879, July 13, red, C.C.	Ghuznee	D.Charm'g Land 2d, 36477	Mr. Lambe

PEACH 2ND, roan, calved December 19, 1873, Vol. xxiii. p. 523. Bred by Mr. W. Lambe; got by Tartar (30104), dam (Peach) by Sir Christopher (22895), &c.

1878, Jan. 4, roan,	B.C.	Parmesan	D.Charm'g Land 2d, 36477	Mr. Lambe
1879, Feb. 23, roan,	C.C.	Peach 4th	do.	do.

ROAN ROSE, roan, calved July 13, 1873. Bred by Mr. W. Lambe; got by Lord of the Manor (29178), dam (Red Rose) by Sir Christopher (22895), &c. See Vol. xxi. p. 800.

1876, Aug. 16, white,	B.C.	(Steer)	Roan Knight, 35287	Mr. Lambe
1879, Jan. 18, r. & w.,	C.C.	Rose of Whisby	Ruby King, 37407	do.

ROSE OF RICHMOND, roan, calved March 11, 1872, Vols. xxi. and xxiii. pp. 800, 523. Bred by Mr. W. Lambe; got by Thorndale Lad (23066), dam (Rose of Delhi) by Great Mogul (14651), &c.

1877, Oct. 30, roan,	C.C.	(dead)	Baron Thorndale, 37820	Mr. Lambe
1879, Feb. 18, red,	B.C.	Duke of Richmond	D.Charm'g Land 2d, 36477	do.

ROSE OF RICHMOND 2ND, roan, calved September 11, 1874. Bred by Mr. W. Lambe; got by Red Knight (32260), dam (Rose of Richmond) by Thorndale Lad (23066), &c. See "Duke of Richmond," p. 82.

1879, Jan. 15, r. & w., C.C.	Ruby Rose	Ruby King, 37407	Mr. Lambe
1879, Dec. 24, roan, C.C.	Rose of Richm'nd 4th	D.Charm'g Land 2d, 36477	do.

SUNSHINE, roan, calved June 15, 1873, Vol. xxiii. p. 524. Bred by Mr. W. Lambe; got by Mars (29306), dam (Sunbeam) by Great Eastern (24081), &c.

Produce in		Names, &c.	By what Bull.	By whom bred.
1878, Oct. 15, red,	C.C.	Sunshine 2nd	Baron Thorndale, 37820	Mr. Lambe

LAMBERT, Robert,
Weeton, Patrington, Hull.

EVALINE 10TH, roan, calved April 17, 1874. Bred by Mr. R. Lambert; got by Duke of Ferriby (33635), dam (Evaline 6th) by Royal Frederick (27365), g. d. (Evaline 2nd) by Baron (17352), gr. g. d. (Evaline) by Lord Ashley (14802), — (Eva 2nd) by Protection (11955), — (Eva) by Lord of Holderness (9324), — by a Bull from Mr. James Collins.

1877, May 21, red,	C.C.	Evaline 11th	Weeton Prince, 37661	Mr. Lambert
1878, Oct. 21, roan,	B.C.	Weeton Florence	Duke of Florence, 33640	do.
1879, Sept. 4, roan,	C.C.	Evaline 12th	Beverley Oxford 5th, 41105	do.

LAMBERT, T.,
Elrington Hall, Haydon Bridge.

BELLE OF DERWENT, roan, calved January 23, 1872, Vol. xxv. p. 535. Bred by Mr. R. Coulson, Coastley; got by Earl of Derwent (28503), dam (Belle) by Royal Butterfly 16th (20724), &c.

1879, Mar. 1, white,	C.C.	Princess Lily	Prince Regent, 29676	Mr. Lambert

LILY, white, calved January 10, 1876. Bred by Mr. W. Lambert, Elrington Hall; got by Heather Bred Rose (31346), dam (Trip Away) by Pizarro (20497), &c. See Vol. xxiii. p. 525.

1879, Jan. 22, roan,	C.C.	Rosa	Killerby Star, 40045	Mr. T. Lambert

LUCKY LASS, red and white, calved February 29, 1876. Bred by Mr. W. Lambert, Elrington Hall; got by Lucky Lad (34714), dam (Sonsie) by Wild Boy (25447), &c. See Vol. xxiii. p. 525.

1879, March 25, red,	C.C.	Princess Rosalind	Prince Regent, 29676	Mr. T. Lambert

PRINCESS MARIA, roan, calved April 7, 1877. Bred by Mr. T. Lambert; got by Prince Regent (29676), dam (White Socks) by Beauty's Butterfly (23399), &c. See Vol. xxiv. p. 526.

1879, Dec. 11, r. & w.,	C.C.	Regina	Rex, 37337	Mr. Lambert

WARLABY'S ROSEDALE, roan, calved April 22, 1875, Vol. xxv. p. 532. Bred by Mr. A. Raine, Bow Bank; got by Warlaby (32792), dam (Ringlet 2nd) by Whiff (30299), &c.

1879, Sept. 27, white,	C.C.	Regent's Rosedale	Prince Regent, 29676	Mr. Lambert

WHITE SOCKS, red and white, calved April 25, 1869, Vols. xxi., xxiii., xxiv., and xxv. pp. 801, 525, 526, 533. Bred by Mr. W. Lambert, Elrington Hall; got by Beauty's Butterfly (23399), dam (Trip the Daisy) by Ivanhoe (14735), &c.

1879, Oct. 20, r. & w.,	C.C.	Regent's White Socks	Prince Regent, 29676	Mr. T. Lambert

WILD EYES, red and white, calved February 6, 1872, Vols. xxi. and xxv. pp. 801, 533. Bred by Mr. W. Lambert, Elrington Hall; got by Wild Boy (25447), dam (Fashion) by Warlaby Knight (25403), &c.

1879, Nov. 21, r. & w.,	C.C.	Pr. Regent's Farewell	Prince Regent, 29676	Mr. T. Lambert

LAMPSON, Sir Curtis M., Bart.,
Rowfant, Crawley, Sussex.

DUCHESS OF GLO'STER, red, calved July 10, 1873, Vol. xxii. p. 475. Bred by Mr. E. H. Cheney, Gaddesby Hall; got by Ninth Duke of Geneva (28391), dam (Duchess of Airdrie 14th) by Tenth Duke of Thorndale (28458), &c.

Produce in		Names, &c.	By what Bull.	By whom bred.
1878, Mar. 18, red,	C.C.	Duchess of Rowfant	D. of Underley 2nd, 36551	Sir C. M. Lampson
1879, Oct. 13, r. & w.,	B.C.	Duke of Sussex 2nd	do.	do.

FANCY DUCHESS, roan, calved June 9, 1871, Vols. xxi. and xxiii. pp. 801, 526. Bred by Mr. J. Lynn, Stroxton; got by Ninth Duke of Geneva (28391), dam (Fancy 2nd) by Rowfant 1st (22767), &c.

Produce in		Names, &c.	By what Bull.	By whom bred.
1878, April 24, roan,	C.C.	SurmiseDuchess16th	D. of Underley 2nd, 36551	Sir C. M. Lampson
1879, Sept. 8, roan,	C.C.	SurmiseDuchess19th	do.	do.

FANTAIL'S DUCHESS 2ND, red and white, calved February 28, 1874, Vol. xxiv. p. 526. Bred by Mr. E. H. Cheney, Gaddesby Hall; got by Ninth Duke of Geneva (28391), dam (Fantail) by Barleycorn (17348), &c.

Produce in		Names, &c.	By what Bull.	By whom bred.
1878, Nov. 18, red,	C.C.	Rowfant Fantail	D. of Underley 2nd, 36551	Sir C. M. Lampson

GRAND DUCHESS OF OXFORD 11TH, red, calved July 6, 1867, Vols. xix., xx., xxi., and xxiii. pp. 538, 548, 675, 526. Bred by the Duke of Devonshire, Holker Hall; got by Grand Duke 10th (21848), dam (Grand Duchess of Oxford 5th) by Priam (18567), &c.

Produce in		Names, &c.	By what Bull.	By whom bred.
1879, Jan. 16, r. & w.,	B.C.	Rowfant D. of Oxford	D. of Underley 2nd, 36551	Sir C. M. Lampson

KIRKLEVINGTON DUCHESS 5TH, red, calved February 20, 1870, Vols. xx., xxi., and xxiii. pp. 580, 700, 526. Bred by Mr. R. P. Davies, Horton; got by Second Duke of Claro (21576), dam (Duchess of Kent) by Lord Liverpool (22168), &c.

Produce in		Names, &c.	By what Bull.	By whom bred.
1877, Sept. 11, red,	C.C.	Rowfant K'lev'ton 3d	3rd D. of Hillhurst, 30975	Sir C. M. Lampson
1879, Nov. 23, red,	C.C.	Rowfant K'lev'tn 5th	D. of Underley 2nd, 36551	do.

LADY SECRET, red and white, calved February 24, 1874. Bred by Colonel Kingscote, Kingscote; got by Duke of Hillhurst (28401), dam (Lady Surmise) by Duke of Brailes (23724), &c. See Vol. xxv. p. 528.

Produce in		Names, &c.	By what Bull.	By whom bred.
1878, Sept. 2, red,	C.C.	SurmiseDuchess17th	D. of Underley 2nd, 36551	Sir C. M. Lampson
1879, Aug. 7, r. & w.,	C.C.	SurmiseDuchess20th	do.	do.

LADY WILD EYES SURMISE, roan, calved December 11, 1875, Vol. xxv. p. 533. Bred by Mr. F. Sartoris, Rushden Hall; got by Wild Eyes Duke (36007), dam (Grand Duchess Surmise) by Grand Duke 10th (21848), &c.

Produce in		Names, &c.	By what Bull.	By whom bred.
1879, Dec. 29, r. & w.,	C.C.	Surmise Duchess 22d	D. of Underley 2nd, 36551	Sir C. M. Lampson

OXFORD FAWSLEY 4TH, white, calved March 14, 1872, Vol. xxii. p. 475. Bred by Messrs. F. Leney and Sons, Wateringbury; got by Grand Duke of Kent (26289), dam (Oxford Fawsley 2nd) by Lord Oxford 2nd (20215), &c.

Produce in		Names, &c.	By what Bull.	By whom bred.
1876, Oct. 1, white,	B.C.	Fawsley Duke 2nd	G. Duke of Geneva, 28756	Sir C. M. Lampson
1877, Nov. 4, roan,	C.C.	Fawsley Duchess	D. of Underley 2nd, 36551	do.
1878, Oct. 3, roan,	B.C.	Fawsley Duke 3rd	do.	do.
1879, Sept. 26, roan,	B.C.	Fawsley Duke 4th	do.	do.

Fawsley Duke 2nd, sold to Mr. W. Wicking, Temsbridge, Godstone.

PEACH BLOSSOM 12th, white, calved July 20, 1874, Vol. xxiv. p. 526. Bred by Mr. J. Fawcett, Scaleby Castle; got by Eighth Duke of York (28480), dam (Peach Blossom 6th) by Earl of Glo'ster (21644), &c.

Produce in	Names, &c.	By what Bull.	By whom bred.
1878, April 30, roan, C.C.	R'fant Peach 2d(dead)	D. of Underley 2nd, 36551	Sir C. M. Lampson
1879, June 15, roan, C.C.	Rowfant Peach 3rd	do.	do.

ROWFANT KIRKLEVINGTON 2nd, roan, calved March 8, 1877. Bred by Sir C. M. Lampson, Bart.; got by Twenty-second Duke of Oxford (31000), dam (Siddington 12th) by Second Duke of Tregunter (26022), &c. See Vol. xxiv. p. 526.

1879, Dec. 13, roan,	C.C.	R'fant K'lev'ton 6th	D. of Underley 2nd, 36551	Sir C. M. Lampson

ROWFANT OXFORD, red and white, calved December 7, 1876. Bred by Sir C. M. Lampson, Bart.; got by Third Duke of Hillhurst (30975), dam (Grand Duchess of Oxford 11th) by Grand Duke 10th (21848), &c. See "Rowfant Duke of Oxford 2nd," p. 215.

1879, May 25, red,	B.C.	R'fant D. of Oxf'd 2d	D. of Underley 2nd, 36551	Sir C. M. Lampson

SIDDINGTON 12th, roan, calved March 23, 1872, Vols. xxi. and xxiv. pp. 716, 526. Bred by Mr. E. Bowly, Siddington House; got by Second Duke of Tregunter (26022), dam (Siddington 2nd) by Fourth Duke of Oxford (11387), &c.

1878, April 18, r.&w., C.C.	R'fant K'lev'ton 4th	D. of Underley 2nd, 36551	Sir C. M. Lampson
1879, June 17, r.&w., B.C.	Duke of Huntsland	do.	do.

SURMISE DUCHESS 8th, roan, calved May 8, 1873, Vol. xxiii. p. 527. Bred by Sir C. M. Lampson, Bart.; got by Grand Duke of Geneva 2nd (31288), dam (Surmise Duchess 2nd) by Grand Duke 15th (21852), &c.

1877, Oct. 3, r.&w.,	C.C.	SurmiseDuchess 14th	D. of Underley 2nd, 36551	Sir C. M. Lampson
1879, Jan. 17, roan,	C.C.	SurmiseDuchess 18th	do.	do.

SURMISE DUCHESS 12th, roan, calved June 4, 1876. Bred by Sir C. M. Lampson, Bart.; got by Fourth Duke of Grafton (28396), dam (Alexandra) by Worth (23244), &c. See Vol. xxiii. p. 525.

1879, Dec. 17, red,	C.C.	SurmiseDuchess 21st	D. of Underley 2nd, 36551	Sir C. M. Lampson

LANCASTER, John,
Skygarth, Temple Sowerby, Penrith.

CHRISTMAS HERALD, roan, calved December 24, 1873, Vol. xxiv. p. 323. Bred by Mr. J. C. Bowstead, Hackthorpe Hall; got by Monarch (31930), dam (Family Herald) by Grand Herald (26301), &c.

1878, May 26, white, C.C.	Herald	King of Scotland, 34332	Mr. Lancaster

LANGHAM, Herbert H.,
Cottesbrooke Park, Northampton.

HAWTHORN, roan, calved December 1, 1872, Vol. xxv. p. 534. Bred by Mr. H. H. Langham; got by Earl of Darlington 4th (31058), dam (Honesty) by Chanter (19423), &c.

1879, Jan. 2, roan,	B.C.	Marquis 3rd	E. of Leicester 16th, 39826	Mr. Langham

JUNIPER, roan, calved February 23, 1875, Vol. xxv. p. 534. Bred by Mr. H. H. Langham; got by Earl of Darlington 4th (31058), dam (Janetta) by Duke of Waterloo 2nd (23800), &c.

Produce in		Names, &c.	By what Bull.	By whom bred.
1879, Oct. 25,	red, B.C.	The Knight 3rd	E.of Leicester 15th, 39826	Mr. Langham
	roan, B.C.	The Knight 4th		

SYCAMORE 1st, red and white, calved March 17, 1873. Bred by Mr. H. H. Langham; got by Earl of Darlington 4th (31058), dam (Sarah 4th) by Crown Prince (21510), &c. See Vol. xxi. p. 803.

1878, Feb. 18, r.&w., B.C.	Squire 4th	D. of Goscote 2nd, 33657	Mr. Langham
1879, Dec. 12, r.&w., B.C.	Squire 6th	E.of Leicester 15th, 39826	do.

Squire 4th, sold to Mr. H. Hales, Short Wood, Northampton.

SYCAMORE 2ND, red and white, calved May 15, 1874. Bred by Mr. H. H. Langham; got by Earl of Darlington 4th (31058), dam (Sarah 4th) by Crown Prince (21510), &c.

1877, Oct. 7, r.&w., B.C.	Squire 3rd	D. of Goscote 2nd, 33657	Mr. Langham
1878, Oct. 18, r.&w., B.C.	Squire 5th	D.of Barrington 2d, 30922	do.

Squire 3rd, sold to Mr. Underwood, Hollowell Grange, Northampton; Squire 5th, to Mr. J. Smeeton, Naseby, Northampton.

LANGHORN, W.,
East Mill Hills, Haydon Bridge.

DIADEM 1st, roan, calved February 26, 1871, Vols. xxiii. and xxiv. pp. 528, 525. Bred by Mr. W. Eshton, Chesterwood Grange, the property of Mrs. Eshton; got by Wild Boy (25447), dam (Dinah 4th) by Pizarro (20497), &c.

1878, Oct. 5, r.&w., C.C.	Lady Granville	Earl Granville, 33778	Mr. Eshton

Lady Granville, sold to Mr. W. Langhorn.

DIADEM 2ND, roan, calved January 15, 1876, Vol. xxv. p. 535. Bred by Mrs. Eshton, Chesterwood Grange; got by Lucky Lad (34714), dam (Diadem 1st) by Wild Boy (25447), &c.

1879, May 10, r.&w., B.C.	Red Gauntlet	Prince Regent, 29676	Mr. Langhorn

DINAH 5TH, roan, calved February 16, 1875, Vol. xxv. p. 535. Bred by Mrs. Eshton, Chesterwood Grange; got by Lucky Lad (34714), dam (Diadem 1st) by Wild Boy (25447), &c.

1879, Feb. 10, roan, C.C.	Young Dinah	Prince Regent, 29676	Mr. Langhorn

ORANGE BLOSSOM 2ND, red and white, calved October 19, 1870, Vol. xxv. p. 535. Bred by Mr. M. Stephenson, Fourstones; got by Manfred (26801), dam (Orange Blossom) by Voltigeur (13964), &c.

1879, April 9, r.&w., C.C.	Orange Girl	Prince Regent, 29676	Mr. Langhorn

PRIMROSE, roan, calved February 15, 1872. Bred by Mr. R. Coulson, Coastley; got by Earl of Derwent (28503), dam (Cowslip) by Royal Butterfly 16th (20724), &c. See "Royal Andrew," Vol. xxi. p. 405.

1879, May 18, r.&w., C.C.	Coastley Lass	Prince Regent, 29676	Mr. Langhorn

TRIP THE DAISY, roan, calved April 5, 1872. Bred by Mr. J. Davidson, Whinnetly; got by Wild Boy (25447), dam (Rosebud) by Pizarro (20497), g. d. (Rosemaid) by Master Annandale (14916), gr. g. d. (Rosemary) by Matchem (10517), — (Rose) by Snowy Down (8607).

1879, April 17, roan, C.C.	Cowslip	Prince Regent, 29676	Mr. Langhorn

LASCELLES, Hon. G. E.,
Sion Hill, Thirsk.

FAIR WIND, roan, calved February 22, 1876, Vols. xxv. p. 535. Bred by Mr. H. Chandos-Pole-Gell, Hopton Hall; got by Favourite (31145), dam (Zephyr 3rd) by Wolfran (25469), &c.

Produce in		Names, &c.	By what Bull.	By whom bred.
1879, Nov. 24, roan,	B.C.	Euroclydon	Sir Andrew, 42387	Hon.G.E.Lascelles

LOZENGE 2nd, roan, calved September 11, 1875, Vol. xxv. p. 535. Bred by the Hon. G. E. Lascelles; got by Merry King (40348), dam (Peppermint) by Orestes (22443), &c.

Produce in		Names, &c.	By what Bull.	By whom bred.
1879, May 1, white,	B.C.	(Steer)	Solomon's Seal, 42429	Hon.G.E.Lascelles

ROSE OF MAY, roan, calved May 10, 1872, Vol. xxv. p. 536. Bred by Mr. W. White, Burrill; got by Lord of the Hills (31723), dam (Rosetta) by Lord of the Hills (18267), &c.

Produce in		Names, &c.	By what Bull.	By whom bred.
1879, July 30, r.&w.,	C.C.	Noisette	Wallenstein, 39277	Hon.G.E.Lascelles

ROSY, roan, calved May 31, 1868, Vol. xx. p. 747.. Bred by Mr. W. White, Burrill; got by King Charles 2nd (24240), dam (Rosetta) by Lord of the Hills (18267), &c.

Produce in		Names, &c.	By what Bull.	By whom bred.
1879, April 4, r.&w.,	B.C.	(Steer)	Wallenstein, 39277	Hon.G.E.Lascelles

LAWSON, T.,
Stapleton Grange, Darlington.

ALICE, roan, calved November 26, 1872, Vols. xxii., xxiv., and xxv. pp. 479, 529, 536. Bred by Mr. T. Lawson; got by Squire Booth (30049), dam (Butterfly) by Mentana (26893), &c.

Produce in		Names, &c.	By what Bull.	By whom bred.
1879, May 11, roan,	C.C.	Cream	Prince Edward, 38901	Mr. Lawson

GRACIE, roan, calved June 2, 1874, Vol. xxiv. p. 529. Bred by Mr. T. Lawson; got by Sir Thomas (35993), dam (Fairy) by Sailor Boy (25077), &c.

Produce in		Names, &c.	By what Bull.	By whom bred.
1878, Jan. 13, r.&w.,	C.C.	Lady Slee	Baronet, 36175	Mr. Lawson
1879, Jan. 21, roan,	C.C.	Lady Frost	Hero, 38424	do.

PRINCESS MAY, roan, calved May 1, 1874, Vol. xxiii. p. 530. Bred by Mr. T. Lawson; got by Sir Thomas (35993), dam (Miss Tod) by Fitz Roy (21760), &c.

Produce in		Names, &c.	By what Bull.	By whom bred.
1878, Jan. 22, white,	B.C.	Prince Osman	Baronet, 36175	Mr. Lawson
1879, Feb. 17, white,	C.C.	Princess Lily	King Manfred, 38491	do.

TINY, roan, calved November 22, 1873, Vols. xxiv. and xxv. pp. 529, 536. Bred by Mr. T. Lawson; got by Waterloo Prince (32822), dam (Sweetbrier) by Peabody (29535), &c.

Produce in		Names, &c.	By what Bull.	By whom bred.
1879, July 6, roan,	C.C.	Orange Flower	Prince Edward, 38901	Mr. Lawson

LAWSON, Sir Wilfrid, Bart.,
Brayton, Carlisle.

BENSON 19th, roan, calved February 23, 1874, Vol. xxiv. p. 530. Bred by Sir W. Lawson, Bart.; got by Cambridge Duke (28120), dam (Old Fanny Benson) by Duke (14419), &c.

Produce in		Names, &c.	By what Bull.	By whom bred.
1878, Mar. 19, white,	B.C.	(dead)	Wild Eyes Duke, 36007	Sir W. Lawson
1879, July 31, r.&w.,	C.C.	Benson 25th	Baron Winsome, 39449	do.

BONNY LASS, red and white, calved December 30, 1873, Vol. xxv. p. 536. Bred by Sir W. Lawson, Bart. ; got by Baslow (30507), dam (Countess of Rose-berry) by Seventeenth Duke of Oxford (25994), &c.

Produce in		Names, &c.	By what Bull.	By whom bred.
1879, Aug. 14, roan,	C.C.	Blooming Girl	Wild Eyes Duke, 36007	Sir W. Lawson

DAIRY LASS, red, calved August 30, 1872, Vols. xxiii. and xxiv. pp. 531, 530. Bred by Sir W. Lawson, Bart. ; got by Cambridge Duke (28120), dam (Dairy Maid) by Grand Vizier (26313), &c.

Produce in		Names, &c.	By what Bull.	By whom bred.
1878, June 30, roan,	B.C.	Dairy Boy	Wild Eyes Duke, 36007	Sir W. Lawson
1879, June 27, roan,	B.C.	Farmer's Boy	do.	do.

DAISY, roan, calved June 3, 1876. Bred by Sir W. Lawson, Bart. ; got by Cambridge Duke (28120), dam (Daisy) by Perth (22513), &c. See "Duster," p. 88.

Produce in		Names, &c.	By what Bull.	By whom bred.
1879, Mar. 3, roan,	B.C.	Duster	Waterloo Lad, 40894	Sir W. Lawson

GRACEFUL DUCHESS, roan, calved January 23, 1872, Vols. xxi., xxii., xxiii., and xxv. pp. 855, 480, 531, 537. Bred by Mr. G. Moore, Whitehall ; got by Baron Oxford 4th (25580), dam (Duchess) by Grand Duke 15th (21852), &c.

Produce in		Names, &c.	By what Bull.	By whom bred.
1879, April 4, roan,	B.C.	(dead)	Duke of Ormskirk, 36526	Sir W. Lawson

LADY CLARE, roan, calved March 25, 1876. Bred by Sir W. Lawson, Bart.; got by Baron Oxford 6th (33075), dam (Lady Cambridge) by Royal Cam-bridge (25009), &c. See "Cambridge Prince," p. 40.

Produce in		Names, &c.	By what Bull.	By whom bred.
1879, Feb. 13, roan,	C.C.	Lady Carlisle	Waterloo Lad, 40894	Sir W. Lawson

PINK, red and little white, calved October 30, 1875. Bred by Sir W. Lawson, Bart. ; got by Baron Oxford 6th (33075), dam (Peony) by Parson (29526), &c. See Vol. xxii. p. 480.

Produce in		Names, &c.	By what Bull.	By whom bred.
1879, Nov. 1, r.&w.,	B.C.		D.of Barrington 6th,39714	Sir W. Lawson

RHODO 5TH, red, calved March 11, 1876. Bred by Sir W. Lawson, Bart. ; got by Cambridge Duke (28120), dam (Rhodo 2nd) by Perth (22513), &c. See Vol. xxiii. p. 531.

Produce in		Names, &c.	By what Bull.	By whom bred.
1879, Dec. 10, red,	B.C.	(dead)	Waterloo Lad, 40894	Sir W. Lawson

RHODO 6TH, red, calved August 26, 1876. Bred by Sir W. Lawson, Bart. ; got by Baron Oxford 6th (33075), dam (Rhodo 3rd) by Perth (22513), &c. See Vol. xx. p. 726.

Produce in		Names, &c.	By what Bull.	By whom bred.
1879, Sept. 26, red,	B.C.		D.of Barrington 7th,39714	Sir W. Lawson

ROYAL BENSON, roan, calved February 20, 1871, Vols. xxiv. and xxv. pp. 531, 537. Bred by Sir W. Lawson, Bart. ; got by Royal Cambridge (25009), dam (Benson 15th) by Kildonan (20051), &c.

Produce in		Names, &c.	By what Bull.	By whom bred.
1879, Mar. 4, roan,	C.C.	Benson 24th	Wild Eyes Duke, 36007	Sir W. Lawson

SONSIE 26TH, red, calved September 18, 1876. Bred by Sir W. Lawson, Bart. ; got by Baron Oxford 6th (33075), dam (Sonsie 22nd) by Royal Cam-bridge (25009), &c. See "Sonsie Lad 2nd," p. 239.

Produce in		Names, &c.	By what Bull.	By whom bred.
1879, Nov. 8, r. & w.,	B.C.		D.of Barrington 6th,39714	Sir W. Lawson

STRAWBERRY, roan, calved April 4, 1870, Vols. xxiii., xxiv., and xxv. pp. 531, 531, 538. Bred by Sir W. Lawson, Bart. ; got by Grand Vizier (26313), dam (Spotless) by Perth (22513), &c.

Produce in		Names, &c.	By what Bull.	By whom bred.
1879, June 19, red,	C.C.	Strawberry 6th	Waterloo Lad, 40894	Sir W. Lawson

STRAWBERRY 2ND, roan, calved February 28, 1874, Vol. xxv. p. 538. Bred by Sir W. Lawson, Bart. ; got by Wellington (32825), dam (Strawberry) by Grand Vizier (26313), &c.

Produce in		Names, &c.	By what Bull.	By whom bred.
1879, Mar. 12, roan,	B.C.	Strawberry Chief	Wild Eyes Duke, 36007	Sir W. Lawson

STRAWBERRY 3RD, red and white, calved March 7, 1876. Bred by Sir W. Lawson, Bart. ; got by Cambridge Duke (28120), dam (Strawberry) by Grand Vizier (26313), &c. See "Strawberry Chief," p. 243.

1879, Feb. 15, roan,	C.C.	Strawberry 5th	Doctor, 39687	Sir W. Lawson

SWEETBRIAR 2ND, red and white, calved October 15, 1875. Bred by Sir W. Lawson, Bart. ; got by Baron Oxford 6th (33075), dam (Sweetbriar) by Royal Cambridge (25009), &c. See Vol. xxi. p. 808.

1879, Jan. 13, roan,	C.C.	Sweetbriar 3rd	Wild Eyes Duke, 36007	Sir W. Lawson

WATERLOO 38TH, red, calved August 27, 1875. Bred by Sir W. Lawson, Bart. ; got by Baron Oxford 6th (33075), dam (Waterloo 35th) by Waterloo Chief (23184), &c. See "Woden," p. 273.

1879, Mar. 4, r. & w.,	B.C.	Woden	Doctor, 39687	Sir W. Lawson

WINSOME 14TH, roan, calved April 2, 1872, Vols. xxi., xxiv., and xxv. pp. 808, 531, 538. Bred by the Duke of Devonshire, Holker Hall ; got by Baron Oxford 4th (25580), dam (Winsome 4th) by Grand Duke 10th (21848), &c.

1879, Mar. 29, white,	B.C.	Brayton Winsome 2d	Duke of Ormskirk, 36526	Sir W. Lawson

LAYCOCK, Richard,
Winlaton, Blaydon on Tyne.

FLORA, roan, calved January 11, 1874. Bred by Mr. R. Laycock ; got by Marquis of Lorne (31842), dam (Tulip) by Combat (23598), &c. See Vol. xxi. p. 809.

1879, May 25, r. & w.,	C.C.	Flora 2nd	Prince Leopold, 37237	Mr. Laycock

LAZONBY, J. and D. D.,
Calthwaite House, Penrith.

LADY WINDSOR, red, calved January 16, 1876. Bred by Messrs. J. and D. D. Lazonby ; got by Royal Fantail (32383), dam (Alexandria Windsor) by Prince of Wales (24851), &c. See "Royal Windsor," p. 223.

1878, Oct. 29, roan,	B.C.	Royal Windsor	Bright Duke, 37893	Messrs. Lazonby
1879, Nov. 5, red,	B.C.	Minor Windsor	do.	do.

STRAWBERRY WINDSOR, roan, calved January 23, 1877. Bred by Messrs. J. and D. D. Lazonby ; got by Sir Lionel Deans (35583), dam (Alexandria Windsor) by Prince of Wales (24851), &c.

1879, Oct. 29, white,	B.C.	General Bright	Bright Duke, 37893	Messrs. Lazonby

LEES, Harold,
Pickhill Hall, Wrexham.

CACHUCHA, roan, calved March 25, 1876. Bred by Captain C. R. Conwy, Bodrhyddan ; got by Favourite (33895), dam (Columbine 2nd) by Churchwarden (23572), &c. See "Sir Garnet," p. 232.

1878, July 6, roan,	B.C.	Sir Garnet	Edward Waterloo, 38246	Mr. Lees

CAVATINA, white, calved March 4, 1876. Bred by Captain C. R. Conwy, Bodrhyddan; got by Favourite (33895), dam (Carrie) by New York (22412), &c. See Vol. xxiii. p. 394.

Produce in		Names, &c.	By what Bull.	By whom bred.
1879, June 13, roan,	C.C.	Bravura	Edward Waterloo, 38246	Mr. Lees

HOIDEN, white, calved October 1, 1876. Bred by Captain C. R. Conwy, Bodrhyddan; got by Earl of Killerby (33802), dam (Romping Girl) by Hotspur (28878), &c. See "Edward Senior," p. 93.

1878, Dec. 24, r. & w.,	B.C.	Edward Senior	Edward Waterloo, 38246	Mr. Lees

LENEY, F. and Sons,
Orpines, Wateringbury, Kent.

DUCHESS 5TH, red, calved March 29, 1873, Vol. xxiv. p. 532. Bred by Messrs. F. Leney and Sons; got by Grand Duke of Kent (26289), dam (Duchess) by Grand Duke 15th (21852), &c.

1879, Mar. 4, red,	C.C.	Duchess 11th	6th D. of Oneida, 30997	Messrs. Leney

HYPERIA, roan, calved December 14, 1875. Bred by Mr. G. Savill, Ingthorpe; got by Earl of Geneva (33794), dam (Hydrangea) by President (27088), &c. See Vol. xxii. p. 552.

1879, Nov. 25, red,	C.C.	Lady Furbelow	D. of Gloster 6th, 39734	Messrs. Leney

LADY HUDSON'S DUCHESS 5TH, red, calved December 24, 1874, Vol. xxv. p. 539. Bred by Messrs. F. Leney and Sons; got by Sixth Duke of Oneida (30997), dam (Lady Hudson's Duchess) by Grand Duke 15th (21852), &c.

1879, April 19, red,	C.C.	L'yHudson'sD'ss9th	6th D. of Oneida, 30997	Messrs. Leney

LADY KNIGHTLEY 4TH, red and white, calved January 15, 1870, Vols. xx., xxi., xxiii., and xxiv. pp. 602, 997, 619, 626. Bred by Mr. D. McIntosh, Havering Park; got by Third Duke of Geneva (23753), dam (Dewdrop) by Prince of Saxe Coburg (20576), &c.

1879, April 9, red,	C.C.	Lady Knightley 5th	D. of Oxford 26th, 33708	Messrs. Leney

OXFORD FAWSLEY 7TH, roan, calved April 26, 1876. Bred by Messrs. F. Leney and Sons; got by Sixth Duke of Oneida (30997), dam (Oxford Fawsley 2nd) by Second Lord Oxford (20215), &c. See Vol. xxiv. p. 533.

1879, Jan. 4, roan,	C.C.	Oxford Fawsley 8th	6th D. of Oneida, 30997	Messrs. Leney

WHITE ROSE, white, calved September 22, 1871, Vols. xxi. and xxiv. pp. 812, 534. Bred by Messrs. F. Leney and Sons; got by Grand Duke 15th (21852), dam (Maidenhair) by Mocassin (18406), &c.

1879, Feb. 24, roan,	C.C.	Florence 2nd	6th D. of Oneida, 30997	Messrs. Leney

LETT, G. H.,
Millpark, Enniscorthy, Co. Wexford.

COUNTESS 7TH, roan, calved March 15, 1873. Bred by Mr. J. Moffat, Ballyhyland; got by Rufus (29897), dam (Countess 4th) by Wiseacre (25465), &c. See "Coastguard," Vol. xxiv. p. 47.

1876, June 8, red,	C.C.	Countess 11th	Richard's Bolivar, 32300	Mr. Moffat
1878, Feb. 1, roan,	C.C.	Zoe	Royalist, 37396	Mr. Lett
1879, Oct. 28, roan,	C.C.	Constance	do.	do.

FRIGHT, red and little white, calved July 6, 1875. Bred by Mr. J. Moffat, Ballyhyland, the property of Mr. G. H. Lett; got by Richard's Bolivar (32300), dam (Frizzle) by Lord of the Manor (29181), &c. See Vol. xxv. p. 592.

Produce in	Names, &c.	By what Bull.	By whom bred.
1878, Feb. 21, roan,	B.C. (dead)	Royalist, 37396	Mr. Moffat
1879, Mar. 31, red,	C.C. Fear	do.	do.
Fear, sold to Mr. G. H. Lett.			

HONEYFLOWER, roan, calved June 4, 1876. Bred by Mr. S. Armstrong, Gally House, the property of Mr. G. H. Lett; got by Golden Cross (34045), dam (Honeysuckle) by Marine (24529), &c. See "Beeswax," p. 24.

1879, Feb. 9, r. & w.,	B.C. Hector	British Lad, 41151	Mr. Armstrong

LETT, William,
Rushock, Droitwich.

GAUNTLET, roan, calved February 26, 1871, Vol. xxiv. p. 534. Bred by Mr. E. Rich, Willesley; got by Priam (22569), dam (Gaunty) by Frankenstein (19778), &c.

1879, Mar. 5, white,	C.C. Snowdrop	Cherub 6th, 36355	Mr. Lett

MISS FRANCES, roan, calved April 10, 1876. Bred by Mr. T. Harris, Stonylane House; got by Gay Boy (31222), dam (Rose) by Lackey (24291), g. d. (Stately) by Castlereagh (19409), &c. See "Magnum," Vol. xxiii. p. 182.

1879, May 8, r. & w.,	C.C. Lady Mary	Prince John, 38911	Mr. Lett

LIDDON, Captain,
Acton Lodge, Iron Acton, Gloucestershire.

PRINCESS MAGENTA, red, calved October 19, 1876. Bred by the Rev. W. Holt Beever, Pencraig Court; got by Grand Duke of Clarence (28750), dam (Princess Maud) by Sir Knight (25167), &c. See "Earl of Glo'ster," p. 90.

1879, Aug. 19, roan,	B.C. Earl of Glo'ster	20th D. of Oxford, 28432	Capt. Liddon

LINDSAY, Colonel R. Loyd,
Lockinge Park, Wantage.

AZALIA, roan, calved July 24, 1875, Vol. xxiv. p. 535. Bred by Colonel R. Loyd Lindsay; got by Duke of Cerisia (30937), dam (Aurora) by Chanter (19423), &c.

1878, Nov. 4, roan,	C.C. Astra	Earl of Horton 11th, 36588	Col. Loyd Lindsay
1879, Nov. 11, r. & w.,	B.C. Acrobat	do.	do.

BURLESQUE, red, calved December 6, 1868, Vols. xx., xxi., and xxii. pp. 425, 814, 484. Bred by the Earl of Radnor, Coleshill; got by Fawsley Baronet (23920), dam (Britannia) by Master Coleshill (18344), &c.

1878, Oct. 23, red,	C.C. Barbara	Earl of Horton 11th, 36588	Col. Loyd Lindsay
1879, Oct. 30, red,	C.C. Bo Peep	do.	do.

CHERRY DUCHESS, red and white, calved February 16, 1876. Bred by Colonel R. Loyd Lindsay; got by Duke of Cerisia (30937), dam (Maid of Orleans) by Duke of Jamaica (23578), &c. See "Duke of Orleans," p. 81.

1879, Jan. 12, r. & w.,	C.C. White Duchess	Don Carlos, 39093	Col. Loyd Lindsay

CLOTILDA ROCK, red, calved October 23, 1875. Bred by Colonel R. Loyd
Lindsay ; got by Lord Rockville (34658), dam (Clotilde) by Grand Duke of
Kent 2nd (28759), &c. See "Cabul," p. 39.

Produce in		Names, &c.	By what Bull.	By whom bred.
1879, Nov. 14, roan,	B.C.	Cabul	Earl of Horton 11th, 36588	Col. Loyd Lindsay

GAY FLOWER, roan, calved January 5, 1876. Bred by Colonel R. Loyd
Lindsay ; got by Count Blanche (33455), dam (Gillyflower) by Lord Napier
(26691), &c. See Vol. xxiii. p. 533.

1879, Jan. 2, red,	C.C.	Red Garland	Don Carlos, 39693	Col. Loyd Lindsay
1879, Dec. 15, roan,	C.C.	Geranium	Earl of Horton 11th, 36588	do.

JESSAMINE, red, calved November 5, 1876. Bred by Mr. J. N. Beasley,
Chapel Brampton ; got by John of Oxford (34267), dam (Jessica) by J's Grand
Duke 2nd (28916). &c. See Vol. xxiii. p. 328.

1879, Dec. 23, red,	C.C.	The Jilt	Earl of Horton 11th, 36588	Col. Loyd Lindsay

JESSICA, red and white, calved January 13, 1874, Vol. xxiv. p. 814. Bred by
Colonel R. Loyd Lindsay ; got by Rob Roy (29806), dam (Judy) by Second
Duke of Airdrie (19600), &c.

1878, Sept. 30, roan,	C.C.	Jellahabad	Earl of Horton 11th, 36588	Col. Loyd Lindsay
1879, Dec. 22, roan,	B.C.	Janizary	do.	do.

ROCK ROSE, red and white, calved September 13, 1875. Bred by Colonel R.
Loyd Lindsay ; got by Lord Rockville (34658), dam (Romola) by Lord Napier
(26691), &c. See "General Roberts," p. 107.

1878, Aug. 22, roan,	B.C.	Rockingham	Earl of Horton 11th, 36588	Col. Loyd Lindsay
1879, Oct. 20, r. & w.,	C.C.	Lady Roseberry	do.	do.

RUBY, roan, calved January 1, 1874. Bred by Colonel R. Loyd Lindsay ; got
by Rob Roy (29806), dam (Rosetta) by Costa (21487), &c. See "Rubicon,"
p. 223.

1878, Mar. 10, red,	C.C.	Roseleaf	Earl of Horton 11th, 36588	Col. Loyd Lindsay
1879, Feb. 6, roan,	B.C.	Rubicon	do.	do.

VIRGINIAN CREEPER, roan, calved February 13, 1869, Vols. xx. and xxi.
pp. 810, 815. Bred by Colonel R. Loyd Lindsay ; got by Lord Claret (24376),
dam (Vesper) by His Majesty (19965), &c.

1879, Jan. 9, roan,	B.C.	Duke of Bordeaux	Earl of Horton 11th, 36588	Col. Loyd Lindsay
1879, Dec. 23, red,	C.C.	Victoria Vokes	do.	do.

LINTON, John,
Buckden Wood, Huntingdon.

ALMA, red and white, calved March 26, 1872. Bred by Mr. J. Linton ; got
by Prince Royal (40536), dam (Alpha) by Buckden (39520), g. d. (Roany) by
Cambridge Barrington 1st (14223), gr. g. d. (Roan Cow) by Lord Chancellor
(10435), — a Dairy Cow.

1876, Dec. 9, roan,	C.C.	Alice	King William, 40067	Mr. Linton
1877, Dec. 6, r. & w.,	C.C.	Amy	Blanche Duke, 39470	do.

RUBY CHARMER 2ND, red and white, calved January 23, 1876. Bred by Mr.
R. Loder, Whittlebury ; got by Grand Duke 22nd (34062), dam (Jacob's Ruby)
by Lord Gwynne (29113), &c. See Vol. xxiii. p. 536.

1879, Mar. 23, roan,	C.C.	Oxford Charmer	Oxford-de-Vere, 40424	Mr. Linton

SIDDINGTON GAZELLE, red and white, calved August 5, 1876. Bred by
Mr. W. Ladds, Ellington; got by Baron Havering 5th (33047), dam (British
Queen) by Earl of Walton (17787), &c. See Vol. xix. p. 422.

Produce in		Names, &c.	By what Bull.	By whom bred.
1879, July 16, roan,	C.C.	Oxford Gazelle	Oxford-de-Vere, 40424	Mr. Linton

WATERLOO 1ST, roan, calved June 27, 1874. Bred by Mr. W. Ladds, Ellington; got by Colonel Tregunter 7th (30771), dam (Water Hen) by Wellesley
(25421), &c. See "Buckden Waterloo," p. 37.

Produce in		Names, &c.	By what Bull.	By whom bred.
1879, Feb. 2, roan,	C.C.	Waterloo 6th	E. of Southampton, 38238	Mr. Linton

WATERLOO 2ND, roan, calved June 27, 1874. Bred by Mr. W. Ladds,
Ellington; got by Colonel Tregunter 7th (30771), dam (Water Hen) by Wellesley (25421), &c.

Produce in		Names, &c.	By what Bull.	By whom bred.
1877, May 11, roan,	C.C.	Waterloo 4th	Baron Havering 5th, 33047	Mr. Ladds
1879, Oct. 25, red,	B.C.	Waterloo Duke	G.D. of D'lington 2d, 41647	Mr. Linton

WATERLOO 3RD, white, calved May 1, 1876. Bred by Mr. W. Ladds,
Ellington; got by Colonel Tregunter 7th (30771), dam (Water Hen) by Wellesley (25421), &c.

Produce in		Names, &c.	By what Bull.	By whom bred.
1878, Dec. 23, roan,	B.C.	Buckden Waterloo	D. of Underley 3rd, 38196	Mr. Linton

WILDAIR, roan, calved January 1, 1877. Bred by Mr. J. Linton; got by
Julius Cæsar (31452), dam (Countess) by Bolden Gwynne (23437), &c. See
Vol. xxii. p. 563.

Produce in		Names, &c.	By what Bull.	By whom bred.
1879, Sept. 17, red,	B.C.	(Steer)	G.D. of D'lington 2d, 41647	Mr. Linton

LINTON, W.,
Sheriff Hutton, York.

FAME, red and white, calved March 19, 1873, Vol. xxii. p. 485. Bred by Mr.
W. Linton; got by Sergeant Major (29957), dam (Emily) by Earl Torquil
(23854), &c.

Produce in		Names, &c.	By what Bull.	By whom bred.
1879, Jan. 27, roan,	C.C.	Sheriff Hutton Rose	Sir Arthur Ingram, 32400	Mr. Linton

Sheriff Hutton Rose, sold to Mr. W. Linton, Aurora, York County, Ontario, Canada.

LADY PRIMROSE, red and white, calved November 18, 1872, Vols. xxii. and
xxiv. pp. 485, 537. Bred by Mr. W. Linton; got by Sergeant Major (29957),
dam (Primrose) by Baron Booth (23354), &c.

Produce in		Names, &c.	By what Bull.	By whom bred.
1879, Dec. 7, roan,	C.C.	Flora	Arthur Victor, 39380	Mr. Linton

LITTLE, William,
Burnfoot, Langholm, N.B.

MISS POLLY, roan, calved March 21, 1875. Bred by Mr. W. Little; got by
Wauchope (35953), dam (Mary 15th) by Grand Duke of Lightburne 2nd
(26291), &c. See Vol. xxii. p. 486.

Produce in		Names, &c.	By what Bull.	By whom bred.
1878, Feb. 22, r. & w.,	C.C.	Dainty	Eagle, 38207	Mr. Little

ROSEMARY, red, calved April 5, 1876. Bred by Mr. W. Little; got by
Romulus (35313), dam (Mary 15th) by Grand Duke of Lightburne 2nd
(26291), &c.

Produce in		Names, &c.	By what Bull.	By whom bred.
1879, April 8, roan,	B.C.	Nithsdale	Marq. of Annandale, 40306	Mr. Little

LITTLE, William,
Littleport, Ely, Cambridgeshire.

LADY GRACE, roan, calved November 20, 1876. Bred by Mr. R. Beldam, Witchford; got by Western Abbot (35967), dam (Lady Atherton) by Hamlet (31335), &c. See Vol. xxiii. p. 535.

Produce in	Names, &c.	By what Bull.	By whom bred.
1879, Sept. 7, r. & w., C.C.	Lady Mary	Lord Spencer, 41925	Mr. Little

LADY WESTERN SPENCER, roan, calved March 31, 1877. Bred by Mr. W. Little; got by Western Abbot (35967), dam (Lady Hamlet Spencer) by Hamlet (31335), &c. See Vol. xxiv. p. 539.

Produce in	Names, &c.	By what Bull.	By whom bred.
1879, Mar. 19, red, C.C.	Northern Spencer	Royal Plantagenet, 39053	Mr. Little

QUEEN OF BRIDPORT, red and white, calved July 31, 1872, Vols. xxi. and xxiii. pp. 816, 536. Bred by Mr. W. Little; got by Bridport (25668), dam (Queen of the Valley) by Bridport (25668), &c.

Produce in	Names, &c.	By what Bull.	By whom bred.
1878, Feb. 24, r. & w., C.C.	Queen Lydia	Western Abbot, 35967	Mr. Little
1879, Jan. 20, roan, C.C.	Queen Laura	do.	do.

QUEEN OF THE ABBEY, roan, calved December 1, 1874, Vols. xxiv. and xxv. pp. 539, 544. Bred by Mr. W. Little; got by Lord Abbot (29052), dam (Queen of Bridport) by Bridport (25668), &c.

Produce in	Names, &c.	By what Bull.	By whom bred.
1879, May 24, red, C.C.	Queen Bess	Baron Torr, 36207	Mr. Little

QUEEN OF THE MOORS, white, calved April 20, 1877. Bred by Mr. W. Little; got by Western Abbot (35967), dam (Queen of the Abbey) by Lord Abbot (29052), &c. See "King of Denmark," p. 131.

Produce in	Names, &c.	By what Bull.	By whom bred.
1879, Nov. 21, roan, B.C.	King of Denmark	King Hamlet, 34318	Mr. Little

WESTERN BRIDE, roan, calved April 25, 1877. Bred by Mr. W. Little; got by King Hamlet (34318), dam (Western Maid) by Lord Abbot (29052), &c. See Vol. xxiv. p. 539.

Produce in	Names, &c.	By what Bull.	By whom bred.
1879, Mar. 23, roan, B.C.	(dead)	Royal Plantagenet, 39053	Mr. Little

WESTERN MAID, white, calved April 5, 1875, Vol. xxiv. p. 539. Bred by Mr. W. Little; got by Lord Abbot (29052), dam (Maid of the West) by Sovereign (27538), &c.

Produce in	Names, &c.	By what Bull.	By whom bred.
1878, April 19, roan, B.C.	Western Prince	King Hamlet, 34318	Mr. Little
1879, May 30, roan, C.C.	Wheel of Fortune	Baron Torr, 36207	do.

LITTLEBOY, Charles,
Barberstown Castle, Straffan, Ireland.

PRINCESS MARIA, red and white, calved June 28, 1876. Bred by Mr. C. Littleboy; got by King Charles (24240), dam (A. out of Vesta) by Iron Duke (31420), &c. See Vol. xxiii. p. 536.

Produce in	Names, &c.	By what Bull.	By whom bred.
1879, Mar. 1, r. & w., B.C.	Ben Lomond	Ben Ledi, 37853	Mr. Littleboy

LODER, Robert,
Whittlebury, Northamptonshire.

DARLINGTON 17TH, red and white, calved January 29, 1868, Vol. xxii. p. 487. Bred by Lord Penrhyn, Penrhyn Castle; got by Grand Duke 11th (21849), dam (Darlington 13th) by Marmaduke 2nd (22287), &c.

Produce in	Names, &c.	By what Bull.	By whom bred.
1879, Feb. 26, red, C.C.	D'ss of Darlington 7th	Grand Duke 22nd, 34062	Mr. Loder

DUCHESS OF DARLINGTON, red and white, calved February 28, 1875, Vol. xxv. p. 546. Bred by Mr. R. Loder; got by Second Duke of Tregunter (26022), dam (Darlington 17th) by Grand Duke 11th (21849), &c.

Produce in	Names, &c.	By what Bull.	By whom bred.
1879. May 26, red,	C.C.|D's of Darlington 8th	Grand Duke 22nd, 34062	Mr. Loder

DUCHESS OF HILLHURST 3rd, red, calved December 25, 1875, Vol. xxv. p. 547. Bred by the Hon. M. H. Cochrane, Hillhurst, Compton, Canada; got by Second Duke of Hillhurst (39748), dam (Tenth Duchess of Airdrie) by Royal Oxford (18774), &c.

Produce in	Names, &c.	By what Bull.	By whom bred.
1879, Aug. 4, red,	C.C.|D's of Whittlebury 2d	D. of Connaught. 33604	Mr. Loder

GENEVA FAWSLEY, red and white, calved March 14, 1874, Vols. xxiii. and xxv. pp. 536, 547. Bred by Mr. Lancaster, Berriden; got by Eighth Duke of Geneva (28390), dam (Fawsley 3rd) by Grand Duke 4th (19874), &c.

Produce in	Names, &c.	By what Bull.	By whom bred.
1879, Oct. 1, r. & w.,	C.C.|G. D'ss Fawsley 9th	G. D. of Oxford 3rd, 39953	Mr. Loder

GRAND DUCHESS FAWSLEY, roan, calved December 12, 1875. Bred by Mr. R. Loder; got by Grand Duke 22nd (34062), dam (Treble Fawsley) by Grand Duke 15th (21852), &c. See Vol. xxii. p. 488.

Produce in	Names, &c.	By what Bull.	By whom bred.
1879, Nov. 14, roan,	C.C.|G. D'ss Fawsley 10th	G. D. of Oxford 3rd, 39953	Mr. Loder

GRAND DUCHESS GWYNNE, red, calved December 6, 1875. Bred by Mr. R. Loder; got by Grand Duke 22nd (34062), dam (Duchess Gwynne 4th) by Baron Oxford 5th (27958), &c. See "Baron Gwynne," p. 15.

Produce in	Names, &c.	By what Bull.	By whom bred.
1879, May 7, r. & w.,	C.C.|G. D'ss Gwynne 3rd	G. D. of Oxford 3rd, 39953	Mr. Loder

JUDITH, roan, calved January 12, 1873, Vols. xxii. and xxiv. pp. 487, 539. Bred by Colonel Kingscote, Kingscote; got by Grand Duke of Waterloo (28766), dam (Joan of Arc) by Duke of Brailes (23724), &c.

Produce in	Names, &c.	By what Bull.	By whom bred.
1879, Nov. 18, r. & w.,	C.C.|Duchess Craggs 4th	G. D. of Oxford 3rd, 39953	Mr. Loder

MINNIE 9th, red and white, calved March 2, 1874, Vol. xxv. p. 547. Bred by Mr. A. Renick, Clark Co., Kentucky, U.S.A.; got by Fourth Duke of Geneva (30958), dam (Minnie 3rd) by Airdrie (30365), &c.

Produce in	Names, &c.	By what Bull.	By whom bred.
1879, July 16, red,	C.C.|R. R. Whittlebury 2d	Grand Duke 22nd, 34062	Mr. Loder

TREBLE FAWSLEY, red, calved August 4, 1873, Vols. xxii., xxiv., and xxv. pp. 488, 539, 547. Bred by Mr. Lancaster, Berriden; got by Grand Duke 15th (21852), dam (Oxford Fawsley 2nd) by Lord Oxford 2nd (20215), &c.

Produce in	Names, &c.	By what Bull.	By whom bred.
1879, Dec. 19, r. & w.,	C.C.|G. D'ss Fawsley 11th	G. D. of Oxford 3rd, 39953	Mr. Loder

WINSOME WILD EYES 3rd, white, calved October 12, 1874. Bred by Mr. J. W. Philips, Heybridge; got by Lord Tregunter (31758), dam (Winsome 6th) by Grand Duke 10th (21848), &c. See Vol. xvi. p. 887.

Produce in	Names, &c.	By what Bull.	By whom bred.
1879, July 11, r. & w.,	C.C.|W'some Wild Eyes 5h	G. D. of Oxford 3rd, 39953	Mr. Loder

LODGE, Robert,
The Rookery, Bishopdale, Bedale.

IDA 7th, roan, calved February 22, 1876. Bred by Mr. J. Russell, Cark-in Cartmel; got by Baron Lightburne 3rd (36192), dam (Ida 6th) by Grand Duke of Lightburne 3rd (28761), &c. See "Lord of the North," p. 154.

Produce in	Names, &c.	By what Bull.	By whom bred.
1879, Jan. 30, roan,	B.C.|Lord of the North	D. of Oxford 36th, 38170	Mr. Lodge

KIRKLEVINGTON DUCHESS 13TH, roan, calved December 5, 1872, Vol. xxii.
p. 394. Bred by Mr. R. P. Davies, Horton ; got by Second Duke of Tregunter
(26022), dam (Kirklevington Duchess 4th) by Thirteenth Duke of Oxford
(21604), &c.

Produce in	Names, &c.	By what Bull.	By whom bred.
1879, Feb. 16, roan, B.C.	Kirklevington Prince	Earl of Horton 10th, 36587	Mr. Lodge

KIRKLEVINGTON PRINCESS 5TH, red and white, calved September 8, 1876.
Bred by Mr. J. W. Larking, Ashdown House ; got by Third Duke of Hillhurst (30975), dam (Kirklevington Duchess 9th) by Grand Duke of Clarence
(28750), &c. See Vol. xxiii. p. 529.

1879, June 19, r. & w., C.C.	K'lev'ton D'ss 27th	L'burne's D.of Oxf'd, 36896	Mr. Lodge

LADY WORCESTER WILD EYES, red and white, calved December 12, 1876.
Bred by Mr. G. Fox, Elmhurst Hall ; got by Grand Duke of Weston 3rd
(34079), dam (Lady Worcester 13th) by Third Duke of Hillhurst (30975), &c.
See Vol. xxiii. p. 444.

1879, April 30, red, C.C.	L'y Wor'ster W.E. 2d	B'n Siddington 2nd, 37816	Mr. Lodge

PRINCESS OF THE VALLEY, red and white, calved October 28, 1876.
Bred by Mr. R. Lodge ; got by Grand Duke of Thorndale 2nd (31298), dam
(Princess of Lightburne) by Twenty-third Duke of Oxford (31001), &c. See
Vol. xxiii. p. 537.

1879, May 7, r. & w., B.C.	Prince of the Dale	B'n Siddington 2nd, 37816	Mr. Lodge

LONGMAN, A. H.,
Shendish, Hemel Hempstead.

ACTRESS, red and white, calved December 25, 1870, Vols. xxi., xxiii., xxiv.,
and xxv. pp. 819, 538, 541, 549. Bred by Mr. S. Canning, Snitterfield ; got
by The Cardinal (27612), dam (Abbess) by Prince of Saxe Coburg (20576), &c.

1879, Sept. 15, red, C.C.	Adelina 3rd	D. of Hillhurst 2nd, 39748	Mr. Longman

ADELINA, roan, calved May 13, 1874, Vols. xxiii. and xxv. pp. 539, 549. Bred
by Mr. A. H. Longman ; got by Knightley Wellington (31533), dam (Actress)
by The Cardinal (27612). &c.

1879, Aug. 3, r. & w., B.C.	(Steer)	D. of Hillhurst 2nd, 39748	Mr. Longman

ALICE, roan, calved December 1, 1869, Vols. xxi., xxii., xxiii.. and xxv. pp. 819,
489, 539, 550. Bred by Mr. S. Canning, Snitterfield ; got by General Bragg
(26232), dam (Abbess) by Prince of Saxe Coburg (20576), &c.

1879, June 4, roan, C.C.	Apsley 6th	D. of Hillhurst 2nd, 39748	Mr. Longman

APSLEY, red and white, calved September 12, 1874, Vol. xxv. p. 550. Bred by
Mr. A. H. Longman ; got by Cremorne (33470), dam (Alice) by General
Bragg (26232), &c.

1879, Feb. 22, r. & w., B.C.	(Steer)	D. of Hillhurst 2nd, 39748	Mr. Longman
1879, Dec. 30, red, B.C.		do.	do.

APSLEY 3RD, roan, calved July 26, 1876. Bred by Mr. A. H. Longman ; got
by Oxford Duke (34977), dam (Alice) by General Bragg (26232), &c.

1879, Sept. 9, r. & w., C.C.	Apsley 7th	L.Shendish Fawsley, 43576	Mr. Longman

BLUE BELL 3RD, red, calved March 14, 1873, Vol. xxv. p. 550. Bred by Mr. S. Canning, Snitterfield; got by The Cardinal (27612), dam (Blue Bell 2nd) by Grand Duke of Wetherby (17997), &c.

Produce in		Names, &c.		By what Bull.		By whom bred.
1879, July 4, red,	C.C.	Blue Bell 5th		D. of Hillhurst 2nd, 39748		Mr. Longman

DAPHNE, roan, calved September 17, 1875, Vol. xxv. p. 550. Bred by Mr. A. H. Longman; got by Oxford Duke (34977), dam (Datura) by Fawsley Prince (31150), &c.

Produce in		Names, &c.		By what Bull.		By whom bred.
1879, April 9, red,	C.C.	Daphne 4th		D. of Hillhurst 2nd, 39748		Mr. Longman

DAPHNE 2ND, roan, calved October 6, 1876. Bred by Mr. A. H. Longman; got by Oxford Duke (34977), dam (Dainty) by Fawsley Prince (31150), &c. See Vol. xxiii. p. 539.

Produce in		Names, &c.		By what Bull.		By whom bred.
1879, Dec. 8, roan,	C.C.	Daphne 6th		Ld. of Strathtay 2d, 38636		Mr. Longman

DATURA, roan, calved in June, 1873, Vols. xxii., xxiii., xxiv., and xxv. pp. 489, 539, 541, 550. Bred by Mr. D. Fisher, Pitlochrie; got by Fawsley Prince (31150), dam (Dairymaid) by Scottish Chief (22850), &c.

Produce in		Names, &c.		By what Bull.		By whom bred.
1879, Dec. 7, red,	C.C.	Daphne 5th		D. of Hillhurst 2nd, 39748		Mr. Longman

FAWSLEY 5TH, red and white, calved March 19, 1873, Vols. xxii., xxiii., and xxiv. pp. 489, 539, 541. Bred by Sir W. S. Maxwell, Bart., Keir Mains; got by Second Duke of Collingham (23730), dam (Fawsley 4th) by Grand Duke 4th (19874), &c.

Produce in		Names, &c.		By what Bull.		By whom bred.
1879, April 13, red,	B.C.	L. Sh'dish Fawsley 2d	D. of Hillhurst 2nd, 39748		Mr. Longman

FENELLA, roan, calved March 8, 1871, Vols. xxiii., xxiv., and xxv. pp. 539, 541, 550. Bred by Lord Skelmersdale, Lathom House; got by Waterloo Boy (27762), dam (Farewell's White Rose) by Earl of Eglinton (23832), &c.

Produce in		Names, &c.		By what Bull.		By whom bred.
1879, Dec. 1, roan,	B.C.	(dead)		D. of Hillhurst 2nd, 39748		Mr. Longman

JACOB'S RUBY, roan, calved February 2, 1872, Vols. xxi., xxiii., and xxv. pp. 817, 536, 550. Bred by Mr. Lancaster, Berriden; got by Lord Gwynne (29113), dam (Ruby) by Lord of the Harem (16430), &c.

Produce in		Names, &c.		By what Bull.		By whom bred.
1879, Nov. 14, r. & w.,	B.C.			D. of Hillhurst 2nd, 39748		Mr. Longman

LADY CONSTANCY, roan, calved September 23, 1875, Vol. xxv. p. 550. Bred by Mr. A. H. Longman; got by Oxford Duke (34977). dam (Lady Celia 7th) by Lord Eglintoun (31652), &c.

Produce in		Names, &c.		By what Bull.		By whom bred.
1879, July 10, roan,	B.C.	(Steer)		D. of Hillhurst 2nd, 39748		Mr. Longman

LADY ROSY, red and white, calved October 30, 1875. Bred by Mr. A. H. Longman; got by Oxford Duke (34977), dam (Fawsley 5th) by Second Duke of Collingham (23730), &c. See "Lord Shendish Fawsley," p. 157.

Produce in		Names, &c.		By what Bull.		By whom bred.
1879, Aug. 4, red,	C.C.	Lady Rosy 3rd		D. of Hillhurst 2nd, 39748		Mr. Longman

OXFORD ROSE 3RD, roan, calved June 24, 1869, Vols. xx., xxiii., and xxiv. pp. 678, 539, 541. Bred by the Duke of Devonshire, Holker Hall; got by Baron Oxford 3rd (25579), dam (Rose of Raby) by Lumley (16478), &c.

Produce in		Names, &c.		By what Bull.		By whom bred.
1879, Feb. 2, roan,	C.C.	Shendish Rose 4th	D. of Hillhurst 2nd, 39748		Mr. Longman

RARITY 4TH, roan, calved October 17, 1876. Bred by Mr. A. H. Longman; got by Twenty-second Duke of Oxford (31000), dam (Charmer 14th) by Third Duke of Geneva (23753), &c. See Vol. xxiii. p. 539.

Produce in		Names, &c.		By what Bull.		By whom bred.
1879, Oct. 2, roan,	C.C.	Rarity 6th		Ld. of Strathtay 2d, 38636		Mr. Longman

SILENCE 4TH, red and white, calved May 2, 1865, Vols. xviii., xix., xxi., xxii., and xxiii. pp. 729, 730, 819, 489, 540. Bred by the Rev. W. Holt Beever, Pencraig Court; got by Frank (17874), dam (Silence 3rd) by Wild Duke (19148), &c.

Produce in	Names, &c.	By what Bull.	By whom bred.
1879, Jan. 23, r. & w., B.C.	(Steer)	D. of Hillhurst 2nd, 39748	Mr. Longman

SILENCE 6TH, red and white, calved July 20, 1870, Vols. xxi., xxiii., xxiv., and xxv. pp. 819, 540, 542, 551. Bred by the Rev. W. Holt Beever, Pencraig Court; got by Etoile du Nord (21710), dam (Silence 4th) by Frank (17874), &c.

Produce in	Names, &c.	By what Bull.	By whom bred.
1879, Oct. 7, r. & w., B.C.	(Steer)	D. of Hillhurst 2nd, 39748	Mr. Longman

SILENCE 7TH, red and white, calved June 5, 1873, Vols. xxii., xxiii., and xxv. pp. 490, 540, 551. Bred by Mr. G. Barton, Fundenhall Grange; got by The Speaker (32701), dam (Silence 6th) by Etoile du Nord (21710), &c.

Produce in	Names, &c.	By what Bull.	By whom bred.
1879, July 27, r. & w., B.C.	(Steer)·	D. of Hillhurst 2nd, 39748	Mr. Longman

SILENCE 8TH, red and white, calved June 25, 1873, Vols. xxii., xxiii., and xxiv. pp. 490, 540, 542. Bred by Mr. G. Barton, Fundenhall Grange; got by The Speaker (32701), dam (Silence 4th) by Frank (17874), &c.

Produce in	Names, &c.	By what Bull.	By whom bred.
1879, Jan. 21, r. & w., B.C.	(Steer)	D. of Hillhurst 2nd, 39748	Mr. Longman

SILENCE 10TH, red, calved October 7, 1874, Vols. xxiv. and xxv. pp. 542, 551. Bred by Mr. A. H. Longman; got by Seneca (35496), dam (Silence 4th) by Frank (17874), &c.

Produce in	Names, &c.	By what Bull.	By whom bred.
1879, Oct. 24, r. & w., C.C.	Silence 19th	D. of Hillhurst 2nd, 39748	Mr. Longman

SILENCE 11TH, red and little white, calved September 10, 1875, Vol. xxv. p. 551. Bred by Mr. A. H. Longman; got by Oxford Duke (34977), dam (Silence 4th) by Frank (17874), &c.

Produce in	Names, &c.	By what Bull.	By whom bred.
1879, June 24, red, B.C.	(Steer)	D. of Hillhurst 2nd, 39748	Mr. Longman

SILENCE 12TH, red and little white, calved November 11, 1875. Bred by Mr. A. H. Longman; got by Oxford Duke (34977), dam (Silence 8th) by The Speaker (32701), &c.

Produce in	Names, &c.	By what Bull.	By whom bred.
1879, July 10, red, B.C.	(Steer)	D. of Hillhurst 2nd, 39748	Mr. Longman

SILENCE 14TH, roan, calved September 28, 1876. Bred by Mr. A. H. Longman; got by Oxford Duke (34977), dam (Silence 8th) by The Speaker (32701), &c.

Produce in	Names, &c.	By what Bull.	By whom bred.
1879, May 29, red, B.C.	(Steer)	Ld. of Strathtay 2d, 38636	Mr. Longman

WETHERBY DUCHESS, roan, calved November 23, 1875, Vol. xxv. p. 551. Bred by Sir W. Lawson, Bart., Brayton; got by Fifth Duke of Wetherby (31033), dam (Graceful Duchess) by Baron Oxford 4th (25580), &c.

Produce in	Names, &c.	By what Bull.	By whom bred.
1879, Dec. 9, red. B.C.	L. Shendish Charmer	D. of Hillhurst 2nd, 39748	Mr. Longman

LONGMORE, Andrew,
Rettie, Banff, N.B.

BAROSSA, roan, calved March 13, 1876. Bred by Mr. A. Longmore; got by Victor Royal (35886), dam (Barmaid) by Osborne (22467), &c. See Vol. xxiii. p. 541.

Produce in	Names, &c.	By what Bull.	By whom bred.
1878, Nov. 16, roan, C.C.	Boadicea	Felgate, 41541	Mr. Longmore

CRAIGOWAN, roan, calved December 15, 1876. Bred by Mr. A. Longmore; got by Baron Havering (33043) dam (Caradena) by Lord Forth (26649), &c. See Vol. xxiv. p. 542.

Produce in	Names, &c.	By what Bull.	By whom bred.
1879, Nov. 10, white, C.C.	Clarinda	Duke of Cambridge, 41363	Mr. Longmore

EUPHROSYNE, red, calved January 9, 1868, Vols. xxi., xxiv., and xxv. pp. 820, 543, 552. Bred by Mr. A. Longmore; got by Halliday (24096), dam (Glycerine) by Sir Charles 2nd (20812), &c.

1879, May 1, roan, C.C.	Eurydice	Baron Havering, 33043	Mr. Longmore

FEODORA, red and little white, calved March 27, 1874, Vol. xxiv. p. 543. Bred by Mr. J. Longmore, Hilton; got by Victor Royal (35886), dam (Favourite) by Lord Forth (26649), &c.

1879, Feb. 16, { red, C.C. Flower { red, B.C. Brodie }		Baron Havering, 33043	Mr. A. Longmore

ISABEL, red, calved April 25, 1874, Vol. xxiv. p. 543. Bred by Mr. A. Longmore; got by Lord Forth (26649), dam (Indian Pink) by Beacon (21252), &c.

1879, March 17, red, C.C.	Iona	Duke of Cambridge, 41363	Mr. Longmore

LADY HAVERING, roan, calved February 20, 1875. Bred by Mr. A. Longmore; got by Baron Havering (33043), dam (Optima) by Royal Blossom (25002), &c. See Vol. xxiii. p. 542.

1879, Mar. 10, roan, C.C.	Lady Rettie	Duke of Cambridge, 41363	Mr. Longmore

LONDON'S BEAUTY, red and little white, calved April 18, 1877. Bred by Mr. A. Longmore; got by Baron Havering (33043), dam (London Lady) by Royal Blossom (25002), &c. See Vol. xxiv. p. 543.

1879, Nov. 8, r. & w., C.C.	London's Queen	Killiecrankie, 43411	Mr. Longmore

MADAM, roan, calved February 11, 1876. Bred by Mr. A. Longmore; got by Lord Forth (26649), dam (Madam Rachel) by Sir Charles 2nd (20812), &c. See Vol. xxiii. p. 542.

1878, Nov. 8, roan, C.C.	Mademoiselle	Caligula, 36303	Mr. Longmore

MADAM RACHEL, roan, calved May 30, 1868, Vols. xxi. and xxiii. pp. 821, 542. Bred by Mr. A. Longmore; got by Sir Charles 2nd (20812), dam (Xanthippe) by Inheritor (13065), &c.

1879, May 19, white, C.C.	Madam Tussaud	Baron Havering, 33043	Mr. Longmore

MAID OF SPARTA, red, calved January 3, 1873, Vol. xxiv. p. 544. Bred by Mr. A. Longmore; got by Lord Forth (26649), dam (Barmaid) by Osborne (22467), &c.

1879, April 10, roan, C.C.	Maid of Llangollen	Baron Havering, 33043	Mr. Longmore

MILLMAY, roan, calved November 26, 1870, Vol. xxiii. p. 542. Bred by Mr. A. Longmore; got by Royal Blossom (25002), dam (Barmaid) by Osborne (22467), &c.

1879, June 15, roan, C.C.	Minerva	Duke of Cambridge, 41363	Mr. Longmore

MISS LAURA, roan, calved February 23, 1871, Vols. xxi., xxiii., xxiv., and xxv. pp. 821, 542, 544, 553. Bred by Mr. A. Longmore; got by Donald Dinnie (28328), dam (Flower of the Boyn) by Sir Charles 2nd (20812), &c.

1879, March 31, roan, C.C.	Miss Mitford	Baron Havering, 33043	Mr. Longmore

MISS RACHEL, red and little white, calved June 3, 1877. Bred by Mr. A. Longmore; got by Baron Havering (33043), dam (Lady Rachel) by Lord Forth (26649), &c. See Vol. xxiv. p. 543.

Produce in		Names, &c.	By what Bull.	By whom bred.
1879, Oct. 29, red,	C.C.	Margaret	Duke of Cambridge, 41363	Mr. Longmore

OPTIMA, red, calved January 5, 1870, Vols. xxi., xxiii., and xxv. pp. 822, 542, 553. Bred by Mr. A. Longmore; got by Royal Blossom (25002), dam (Meg Merrilies) by Viceroy (19054), &c.

1879, March 22, red,	C.C.	Lady Havering 4th	Baron Havering, 33043	Mr. Longmore

PALM BRANCH, roan, calved December 23, 1876. Bred by Mr. A. Longmore; got by Baron Havering (33043), dam (Palm Leaf) by Donald Dinnie (28328), &c. See Vol. xxiv. p. 544.

1878, Dec. 26, red,	C.C.	Palm	Caligula, 36303	Mr. Longmore

ROSABELLE, red and little white, calved March 11, 1877. Bred by Mr. A. Longmore; got by Baron Havering (33043), dam (Claret) by Lord Forth (26649), &c. See Vol. xxiv. p. 542.

1879, Nov. 14, red,	C.C.	Roxy	Duke of Cambridge, 41363	Mr. Longmore

ROYAL RAKE, red, calved April 16, 1870, Vols. xxi., xxiv., and xxv. pp. 822, 544, 553. Bred by Mr. A. Longmore; got by Royal Blossom (25002), dam (Star o' the Gloamin) by Viceroy (19054), &c.

1879, May 5, roan,	C.C.	Royal Alice	Duke of Cambridge, 41363	Mr. Longmore

LOVAT, Lord,
Beaufort Castle, Inverness, N.B.

BEAUFORT ROSE 1st, roan, calved January 12, 1877. Bred by Lord Lovat; got by Highland Chief (38431), dam (Beaufort Rose) by Bachelor of Arts (32982), &c. See "Belladrum," p. 24.

1879, Dec. 26, roan,	B.C.	Earl of Groam	Duke of Beaufort, 38122	Lord Lovat

JULIA 2nd, roan, calved January 10, 1877. Bred by Lord Lovat; got by Highland Chief (38431), dam (Julia) by Allan (21172), &c. See "Marquis," p. 164.

1879, Dec. 19, r. & w.,	C.C.	Julia Duchess	Duke of Beaufort, 38122	Lord Lovat

LOVATT, Henry,
Low Hill, Bushbury, Wolverhampton.

COWSLIP 5th, roan, calved June 29, 1876. Bred by Colonel Lane, King's Bromley Manor; got by Grand Duke of Weston 2nd (34079), dam (Cowslip 4th) by Grand Duke 11th (21849), &c. See Vol. xxiii. p. 528.

1879, April 18, roan,	C.C.	Cowslip Cherry	E. of Horton 10th, 36587	Mr. Lovatt

KIRKLEVINGTON DUCHESS 16th, red, calved November 24, 1873, Vol. xxiii. p. 415. Bred by Mr. R. P. Davies, Horton; got by Second Duke of Glo'ster (28392), dam (Duchess of Kent) by Lord Liverpool (22168), &c.

1879, Apr. 30, { red, red,	B.C. C.C.	(dead) (dead) }	E. of Horton 10th, 36587	Mr. Lovatt

LADY MILD EYES, roan, calved July 8, 1875, Vol. xxiv. p. 506. Bred by
Mr. G. Moore, Whitehall; got by Eighth Duke of York (28480), dam (Lady
Worcester 10th) by Eighth Duke of Geneva (28390), &c.

Produce in	Names, &c.	By what Bull.	By whom bred.
1879, Sept. 4, red,	C.C.|Lady Worcester 23rd	B.T'croft Oxford 4th,37822	Mr. Lovatt

LADY WESTON WATERLOO, roan, calved July 14, 1875. Bred by Mr. G.
Fox, Elmhurst Hall; got by Grand Duke of Weston 3rd (34079), dam (Lady
Waterloo 15th) by Third Viscount Waterloo (25387), &c. See Vol. xxii. p. 419.

Produce in	Names, &c.	By what Bull.	By whom bred.
1879, Apr. 26, r. & w.,	C.C.|LadyOxf'd Waterloo	B.T'croft Oxford 4th,37822	Mr. Lovatt

LADY WORCESTER 2ND, red, calved June 16, 1866, Vols. xix., xx., and xxiii.
pp. 595, 618, 429. Bred by Mr. J. Harward, Winterfold; got by Charleston
(21400), dam (Clear Star) by Marton Duke (22307), &c.

Produce in	Names, &c.	By what Bull.	By whom bred.
1879, Dec. 19, red,	C.C.|(dead)	Marq's of Oxford 2nd,37055	Mr. Lovatt

LADY WORCESTER 9TH, red and white, calved August 19, 1871, Vols. xxi.
and xxii. pp. 685, 405. Bred by Mr. J. Harward, Winterfold; got by Third
Duke of Claro (23729), dam (Lady Worcester 2nd) by Charleston (21400), &c.

Produce in	Names, &c.	By what Bull.	By whom bred.
1879, April 7, red,	C.C.|Lady Worcester 22ud	B.T'croft Oxford 4th,37822	Mr. Lovatt

LADY WORCESTER 16TH, roan, calved May 23, 1875. Bred by the Earl of
Dunmore, Dunmore, the property of Mr. H. Lovatt; got by Sixth Duke of
Geneva (30959), dam (Lady Worcester) by Charleston (21400), &c. See Vol.
xxii. p. 405.

Produce in	Names, &c.	By what Bull.	By whom bred.
1879, June 16, r. & w.,	B.C.|	Marq's of Oxford 2nd,37055	Earl of Dunmore

LADY WORCESTER 17TH, roan, calved August 14, 1875. Bred by the Earl
of Dunmore, Dunmore, the property of Mr. H. Lovatt; got by Third Duke
of Hillhurst (30975), dam (Lady Worcester 3rd) by Third Duke of Wharfdale
(21619), &c. See Vol. xxii. p. 405.

Produce in	Names, &c.	By what Bull.	By whom bred.
1879, July 18, red,	B.C.|	Marq's of Oxford 2nd,37055	Earl of Dunmore

MAID OF OXFORD 11TH, red and white, calved February 9, 1876. Bred by
Mr. E. H. Cheney, Gaddesby Hall; got by Ninth Duke of Geneva (28391),
dam (Maid of Oxford 10th) by Fourth Duke of Geneva (30958), &c. See Vol.
xxiv. p. 545.

Produce in	Names, &c.	By what Bull.	By whom bred.
1879, Dec. 10, r. & w.,	C.C.|Countess of Oxford	B.T'croft Oxford 4th,37822	Mr. Lovatt

WILD EYES 37TH, roan, calved April 23, 1876. Bred by Mr. J. Robinson,
Fell Side; got by Lord Darlington 8th (34520), dam (Wild Eyes 30th) by
Don John (19583), &c. See "Wild Earl," p. 269.

Produce in	Names, &c.	By what Bull.	By whom bred.
1878, June 28, roan,	B.C.|Wild Prince	B.T'croft Oxford 4th,37822	Mr Lovatt
1879, July 13. roan.	B.C.|Wild Earl	do.	do.

LOW, Francis W.,
Kilshane, Tipperary, Ireland.

CHERRY RIPE 11TH, roan, calved March 29, 1876. Bred by Mr. F. W. Low;
got by Nobleman (31979), dam (Cherry Ripe 4th) by Woodcraft (25470), &c.
See Vol. xxiii. p. 543.

Produce in	Names, &c.	By what Bull.	By whom bred.
1879, Jan. 30, roan,	B.C.|John Atkinson	Wideawake 2nd, 44250	Mr. Low

DAGMAR 3RD, roan, calved January 12, 1871, Vols. xxi., xxiii., xxiv., and xxv. pp. 823, 543, 546, 555. Bred by Mr. F. W. Low; got by Woodcraft (25470), dam (Dagmar) by Cœur-de-Lion (15783), &c.

Produce in		Names, &c.	By what Bull.	By whom bred.
1879, Feb. 27, white,	B.C.	Jack Frost	Foreign Prince, 36656	Mr. Low

DAGMAR 6TH, white, calved January 18, 1876. Bred by Mr. F. W. Low; got by Nobleman (31979), dam (Dagmar 3rd) by Woodcraft (25470), &c.

1879, May 25, white,	B.C.	Prince Charley	Foreign Prince, 36656	Mr. Low

LIMERICK LASS 9TH, roan, calved April 3, 1873, Vols. xxiii., xxiv., and xxv. pp. 544, 547, 556. Bred by Mr. F. W. Low; got by Wideawake (27805), dam (Limerick Lass 4th) by Woodcraft (25470), &c.

1879, May 28, roan,	C.C.	Limerick Lass 19th	Foreign Prince, 36656	Mr. Low

LIMERICK LASS 13TH, roan, calved March 6, 1876. Bred by Mr. F. W. Low; got by Visigoth (35907), dam (Limerick Lass 9th) by Wideawake (27805), &c.

1879, Feb. 22, white,	C.C.	Limerick Lass 18th	Wideawake 2nd, 44250	Mr. Low

MATILDA, red, calved April 20, 1873, Vols. xxiii. and xxiv. pp. 385, 374. Bred by Messrs. Christy Brothers, Fort Union; got by The Earl (27623), dam (Effie Deans) by Fairy King (21716), &c.

1878, Aug. 29, red,	C.C.	Matilda 2nd		Lord Broughton, 31626	Messrs. Christy
1879, July 11,	r.&w.,B.C.	Huz	}	Verger, 35860	Mr. Low
	red, B.C.	Buz			

NARCISSA 2ND, red and white, calved January 11, 1873, Vols. xxiii., xxiv., and xxv. pp. 544, 547, 556. Bred by Mr. F. W. Low; got by Wideawake (27805), dam (Narcissa) by Narcissus (26948), &c.

1879, Feb. 24, roan,	C.C.	Narcissa 9th	Prince Gwynne, 40498	Mr. Low

NARCISSA 5TH, roan, calved January 30, 1876. Bred by Mr. F. W. Low; got by Nobleman (31979), dam (Narcissa) by Narcissus (26948), &c. See Vol. xxiii. p. 544.

1879, Mar. 3, roan,	B.C.	Billy	Wideawake 2nd, 44250	Mr. Low

LOW, George,
Birtown, Burghtown, Co. Kildare.

BRITANNIA, roan, calved June 9, 1873. Bred by Mr. G. Low; got by Escape (28556), dam (Clarina) by Napoleon 3rd (29417), &c. See "Britain's Fame," p. 34.

1876, Jan. 23, roan,	B.C.	Prophet	White King, 32853	Mr. Low
1878, Feb. 1, roan,	B.C.	(dead)	do.	do.
1879, Mar. 23, roan,	B.C.	Britain's Fame	Great Fame, 34087	do.

Prophet, sold to Lord De Vesci, Abbeyleix House, Abbeyleix.

LOY, S. H.,
Keld Head, Pickering, Yorkshire.

CERES, roan, calved July 20, 1874, Vol. xxiv. p. 547. Bred by Mr. S. H. Loy; got by Cambridge Gwynne (28122), dam (Psyche) by Tisamines (27670), &c.

1878, April 11, white,	C.C.	Proserpine	Cleveland 3rd, 39600	Mr. Loy
1879, Mar. 10, white,	C.C.	Rhea	do.	do.

HERMIONE, red and white, calved June 8, 1876. Bred by Mr. S. H. Loy; got by Duke of Rosedale 2nd (33727), dam (Psyche) by Tisamines (27670), &c. See Vol. xxiv. p. 548.

Produce in	Names, &c.	By what Bull.	By whom bred.
1879, Feb. 6, r. & w., B.C.	(Steer)	Ryedale Oxford, 40658	Mr. Loy

LAUREL OF CLEVELAND, red and white, calved February 17, 1875, Vol. xxiv. p. 548. Bred by Mr. S. H. Loy; got by Duke of Ryedale 2nd (33727), dam (Laurel Berry) by Cambridge Gwynne (28122), &c.

| 1879, Feb. 27, r. & w., C.C. | Laurel Wreath | Ryedale Oxford, 40658 | Mr. Loy |

ROSE OF GWYNNE, white, calved in June 1874, Vols. xxiv. and xxv. pp. 548, 557. Bred by Mr. S. H. Loy; got by Cambridge Gwynne (28122), dam (White Rose) by Shuttlecock (25126), &c.

| 1879, Sept. 2, white, C.C. | Rose of the Keld 2nd | Ryedale Oxford, 40658 | Mr. Loy |

VESTA, red and white, calved February 3, 1875, Vol. xxv. p. 557. Bred by Mr. S. H. Loy; got by Duke of Ryedale 2nd (33727), dam (Venus) by Tisamines (27670), &c.

| 1879, Nov. 5, red, B.C. | Victor 2nd | Beau of Acomb, 36228 | Mr. Loy |

VESTAL, red and white, calved March 30, 1877. Bred by Mr. S. H. Loy; got by Duke of Ryedale 2nd (33727), dam (Venetia) by Tisamines (27670), &c. See Vol. xxiv. p. 548.

| 1879, Dec. 13, roan, C.C. | Venus | Cyprus, 41313 | Mr. Loy |

WHITE ROSE, white, calved November 18, 1870, Vols. xxi., xxii., xxiv., and xxv. pp. 825, 491, 548, 557. Bred by Mr. S. H. Loy; got by Shuttlecock (25126), dam (White Hind) by Seventh Duke of Oxford (17741), &c.

| 1879, Feb. 2, red, B.C. | Lord Rose | Beau of Acomb, 36228 | Mr. Loy |

LYALL, Charles,
Old Montrose, Montrose, N.B.

DOLLY WEAR 7TH, roan, calved January 5, 1874. Bred by Mr. R. Arklay, Ethiebeaton; got by Lord Kinnaird (31679), dam (Dolly Wear 6th) by Jolly Prince (24222), &c. See "Nero," Vol. xxiii. p. 199.

| 1876, Dec. 4, r. & w., B.C. | Derby | Monarch of Lorne, 34869 | Mr. Lyall |
| 1877, Nov. 19, red, C.C. | Dolly Wear 8th | Quaker, 38960 | do. |

MYSIE 12TH, roan, calved March 4, 1872. Bred by Mr. R. Arklay, Ethiebeaton; got by Brother Booth (25689), dam (White Mysie 11th) by Jolly Prince (24222), &c. See "Minstrel," p. 171.

1877, Feb. 17, roan, B.C.	Mariner	Monarch of Lorne, 34869	Mr. Lyall
1878, Mar. 2, roan, B.C.	Minstrel	Leopold, 38561	do.
1879, April 25, red, C.C.	Mysie 13th	do.	do.

RED RUBY 6TH, red, calved March 10, 1872. Bred by Mr. R. Arklay, Ethiebeaton; got by Brother Booth (25689), dam (Red Friar Ruby) by Red Friar 2nd (27247), g. d. (Ruby 4th) by Representative (20666), &c. See "Jolly Tar," Vol. xxi. p. 525.

| 1876, April 10, roan, C.C. | Red Ruby 7th | Monarch of Lorne, 34869 | Mr. Lyall |

LYALL, Robert J.,
Powis, Montrose, N.B.

MARIGOLD, red, calved March 15, 1869, Vol. xxiii. p. 547. Bred by Mr. C. Lyall, Old Montrose; got by Crœsus (21506), dam (May Flower) by Little Go (18200), &c.

Produce in		Names, &c.	By what Bull.	By whom bred.
1877, Feb. 13, roan,	C.C.	Myrtle	D. of Westminster, 38200	Mr. R. J. Lyall
1878, Mar. 24, red,	B.C.	Marquis of Salisbury	Leopold, 38561	do.
1879, Mar. 20, red,	C.C.	Moss Rose	do.	do.

MOLLY 3RD, roan, calved February 20, 1872, Vol. xxiii. p. 547. Bred by Mr. R. Arklay, Ethiebeaton; got by Brother Booth (25689), dam (Molly 2nd) by Panmure 6th (29517), &c.

1877, June 4, red,	C.C.	Molly 5th	D. of Westminster, 38200	Mr. R. J. Lyall

LYLE, J. G. W. and W. R.,
Donaghmore House, Co. Tyrone.

FLORA'S PRIDE, red and white, calved January 23, 1873. Bred by Mr. A. M. Lyle, Donaghmore House, the property of Messrs. Lyle; got by Lord Wodehouse (29224), dam (Flora) by Alabama (30366), &c. See "Sir Louis," p. 234.

1875, May 8, red,	B.C.	Sir James	Sir Andrew, 37451	Mr. A. M. Lyle
1876, Nov. 9, r. & w.,	B.C.	Royal Knight	Prince Royal, 37277	do.
1878, Feb. 12, red,	C.C.	Flora's Hope	do.	do.
1879, Mar. 9, roan,	B.C.	Sir Louis	Lord of the Manor, 38641	do.

Royal Knight, sold to Mr. Wilson, Omagh, Co. Tyrone.

LADY OF THE MANOR, roan, calved September 30, 1872. Bred by Mr. A. M. Lyle, Donaghmore House, the property of Messrs. Lyle; got by Earl of Donaghmore (31064), dam (Perfume 2nd) by Best Hope (23414), &c. See "Umpire," p. 254.

1874, Nov. 30, red,	C.C.	Lady of the Vale	Sailor Prince, 35442	Mr. A. M. Lyle
1876, May 10, roan,	B.C.	Lord of the Manor	Sir Andrew, 37451	do.
1878, May 23, red,	B.C.	Lord of the Vale	Sir James, 37466	do.

Lord of the Vale, sold to Mr. R. Ferguson, Loughhill, Dundonald, Co. Down.

LYNN, John,
Church Farm, Stroxton, Grantham.

FAWSLEY ROSE, red and white, calved February 3, 1870, Vol. xxv. p. 557. Bred by Mr. F. Sartoris, Rushden Hall; got by Baron Oxford 2nd (23376), dam (Fawsley 6th) by Grand Duke 4th (19874), &c.

1879, Dec. 22, r. & w.,	B.C.		Cambridge Duke 8th, 33274	Mr. Lynn

HYDROMEL, roan, calved October 15, 1871, Vols. xxi., xxii., and xxiii. pp. 918, 552, 631. Bred by Mr. G. Savill, Ingthorpe, the property of Mr. J. Lynn; got by President (27088), dam (Hyacinth) by Potentate (22537), &c.

1879, Sept. 18, red,	C.C.		D. of Gloster 6th, 39734	Mr. Cheney

Cow Calf, sold to Mr. Allen, Thurmaston, Leicester.

LADY EMILY OXFORD, roan, calved October 23, 1871, Vols. xxii. and xxiii. pp. 552, 631. Bred by Mr. G. Savill, Ingthorpe, the property of Mr. J. Lynn; got by Eighteenth Duke of Oxford (25995), dam (Lady Emily 2nd) by Seventh Duke of York (17754), &c.

1879, June 13, r. & w.,	B.C.	Lord Emilius	D. of Gloster 6th, 39734	Mr. Cheney

Lord Emilius, sold to Colonel Burnaby, Baggrave, Leicester.

LADY ROSE SPENCER, red and white, calved November 27, 1873, Vols. xxiii., xxiv., and xxv. pp. 547, 550, 558. Bred by Mr. J. Lynn; got by Cambridge Duke 4th (25706), dam (Lady Sylvia Spencer) by Grand Duke of Lightburne (26290), &c.

Produce in	Names, &c.	By what Bull.	By whom bred.
1879, May 1, red,	B.C.|Lord Duke Spencer	CambridgeDuke8th,33274|Mr. Lynn
Lord Duke Spencer, sold to Mr. Hack, Buckminster, Grantham.			

QUEEN ROSY OXFORD, red and white, calved February 24, 1876. Bred by Mr. J. Lynn; got by Cambridge Duke 8th (33274), dam (Queen Oxford) by Fifth Lord Oxford (31738), &c. See Vol. xxiv. p. 556.

Produce in	Names, &c.	By what Bull.	By whom bred.
1879, Sept. 23, roan,	C.C.|Queen Rosy Duchess|Grand Duke 25th, 34065	Mr. Lynn

RED ROSE 8TH, red, calved July 21, 1875, Vol. xxv. p. 558. Bred by Mr. J. Lynn; got by Ninth Duke of Geneva (28391), dam (Red Rose 5th) by Grand Duke 4th (19874), &c.

Produce in	Names, &c.	By what Bull.	By whom bred.
1879, July 28, red,	C.C.|Red Rose 10th	3rd D. of Glo'ster, 33653	Mr. Lynn

RED ROSE 9TH, red, calved March 18, 1876. Bred by Mr. J. Lynn; got by Ninth Duke of Geneva (28391), dam (Red Rose 6th) by Cambridge Duke 4th (25706), &c. See "Cambridge Duke 10th," p. 39.

Produce in	Names, &c.	By what Bull.	By whom bred.
1879, Mar. 1, red,	B.C.|CambridgeDuke13th|3rd D. of Glo'ster, 33653	Mr. Lynn

SERAPHINA DUCHESS, roan, calved January 17, 1876. Bred by Sir C. M. Lampson, Bart., Rowfant, the property of Mr. J. Lynn; got by Eighth Duke of York (28480), dam (Seraphina 23rd) by Royal Cumberland (27358), &c. See Vol. xxiii. p. 527.

Produce in	Names, &c.	By what Bull.	By whom bred.
1879, Sept. 19	(dead calf)|	Lord Oxford 13th, 40230	Mr. Cheney

WILD EYES 30TH, roan, calved November 8, 1867, Vol. xxii. p. 297. Bred by Mr. C. W. Harvey, Walton on the Hill, the property of Mr. J. Lynn; got by Seventh Duke of York (17754), dam (Wild Eyes 24th) by Fourth Duke of Oxford (11387), &c.

Produce in	Names, &c.	By what Bull.	By whom bred.
1876, May 18, roan,	B.C. Lord Wild Eyes	9th Duke of Geneva, 28391	Mr. Cheney
Lord Wild Eyes, sold to Sir J. Whitworth, Bart., Stancliffe Hall, Matlock.			

WILD OXFORD, roan, calved January 10, 1864, Vols. xviii., xix., xx., and xxii. pp. 786, 787, 823, 299. Bred by Mr. T. Atherton, Speke, the property of Mr. J. Lynn; got by Lord Oxford 2nd (20215), dam (Wild Eyes 24th) by Lord Barrington 3rd (16382), &c.

Produce in	Names, &c.	By what Bull.	By whom bred.
1878, May 17, roan,	B.C.|Lord Wild Eyes 2nd |3rd D. of Glo'ster, 33653	Mr. Cheney
Lord Wild Eyes 2nd, sold to Mr. E. C. Tisdall, Holland Park Farm, Kensington.			

LYTHALL, Edmund,
Radford Hall, Leamington.

ALICIA, roan, calved December 10, 1873. Bred by Mr. E. Lythall; got by Duke of Lancaster (28411), dam (Alice Swinnerton) by Lord of the Isles (20206), &c. See "Virgil," p. 259.

Produce in	Names, &c.	By what Bull.	By whom bred.
1877, Jan. 31, roan,	C.C.|(dead)	Pan, 35008	Mr. Lythall
1878, Apr. 13, { roan, B.C.|Cherry Brandy / roan, C.C.|Princess Alice }		Cherry Prince, 37985	do.
1879, Mar. 18, roan,	B.C.|Virgil	Virginus, 40873	do.
Cherry Brandy, sold to Mr. Oakes, Riddings, Derbyshire.			

KILLERBY ROSE, roan, calved May 19, 1875. Bred by Mr. E. Lythall; got by Kingcraft (31478), dam (Killerby Lass) by Fitz Killerby (26166),&c. See Vol. xxi. p. 829.

Produce in		Names, &c.	By what Bull.	By whom bred.
1878, Mar. 10, r. & w.,	C.C.	Killerby Cherry	Cherry Prince, 37985	Mr. Lythall
1879, April 7, white,	C.C.	Rosette	Tablet, 37558	do.

LADY CRAFTY, roan, calved January 28, 1875. Bred by Mr. E. Lythall; got by Kingcraft (31478), dam (Lady Leeke) by Waverley 3rd (21083), &c. See " Remus," Vol. xxiv. p. 220.

Produce in		Names, &c.	By what Bull.	By whom bred.
1877, Dec. 7, r. & w.,	B.C.	Remus	Rob Roy, 29806	Mr. Lythall
1878, Oct. 13, red,	C.C.	Crafty Venus	Virginus, 40873	do.

Remus, sold to Mr. H. Davenport, Maer Hall.

LIVELY, roan, calved November 30, 1872. Bred by Mr. E. Lythall; got by Fitz Killerby (26166), dam (Lavender 2nd) by Waverley 2nd (21083), &c. See "Snowball," Vol. xxv. p. 266.

Produce in		Names, &c.	By what Bull.	By whom bred.
1877, Mar. 1, red,	C.C.	Lollypop	Pan, 35006	Mr. Lythall
1878, Mar. 22, white,	B.C.	Snowball	Cherry Prince, 37985	do.
1879, Feb. 8, red,	C.C.	Vanity	Virginus, 40873	do.

Snowball, sold to Mr. J. Gale, Ramsgate.

MAGNOLIA 3RD, white, calved November 9, 1869, Vol. xxii. p. 492. Bred by Mr. E. Lythall; got by Fitz Killerby (26166), dam (Magnolia 2nd) by Waverley 3rd (21083), &c.

Produce in		Names, &c.	By what Bull.	By whom bred.
1876, Feb. 17, roan,	C.C.	Magnolia 4th	Telemachus 5th, 35724	Mr. Lythall
1877, Oct. 29, roan,	C.C.	Magnolia 5th	Rob Roy, 29806	do.
1878, Sept. 10, { roan,	B.C.	(dead)	} Cherry Prince, 37985	do.
{ roan,	C.C.	(dead)		

QUEENLIKE, red, calved October 18, 1874, Vol. xxv. p. 558. Bred by Mr. E. Lythall; got by Kingcraft (31478), dam (Lavender 2nd) by Waverley 3rd (21083), &c.

Produce in		Names, &c.	By what Bull.	By whom bred.
1879, Oct. 24, red,	C.C.	Rowena	Tablet, 37558	Mr. Lythall

QUEEN MADGE, roan, calved September 1, 1873. Bred by Sir A. de Rothschild, Bart., Aston Clinton; got by Fatherland (28574), dam (Lady Madge) by Duke of York (23804), &c. See Vol. xxi. p. 670.

Produce in		Names, &c.	By what Bull.	By whom bred.
1877, Mar. 4, red,	C.C.	Meg Merrilies	Pan, 35006	Mr. Lythall
1878. Sept. 10, roan,	C.C.	Cherry Madge	Cherry Prince, 37985	do.

VIRGIN, red and white, calved November 21, 1868, Vol. xxi. p. 673. Bred by Sir A. de Rothschild, Bart., Aston Clinton; got by Duke of York (23804), dam (Vesta) by Highthorn (13028), &c.

Produce in		Names, &c.	By what Bull.	By whom bred.
1876, Dec. 6, r. & w.,	C.C.	Virginia	Ranger, 35203	Mr. Lythall

Virginia, sold to Herr von Skvbensky, Bresa.

WATERLOO CHERRY 5TH, roan, calved February 8, 1875. Bred by Mr. S. L. Horton, Park House; got by Abacot (32900), dam (Waterloo Cherry 3rd) by Prince Charming (27130), &c. See " Prince of Waterloo," Vol. xxv. p. 231.

Produce in		Names, &c.	By what Bull.	By whom bred.
1878, Mar. 1, r. & w.,	B.C.	Prince of Waterloo	Cherry Prince, 37985	Mr. Lythall
1879, Jan. 15, { roan,	C.C.	Cherry Ripe	} Virginus, 40873	do.
{ r. & w.,	B.C.	Brandy Cherry		

Prince of Waterloo, sold to Mr. T. Whitton, Caswell; Brandy Cherry, to M. Wilson, Stettin.

WHITE ROSE, white, calved October 23, 1873. Bred by Mr. E. Lythall; got
by Duke of Lancaster (28411), dam (Starling) by Lord of the Isles (20206),
&c. See "Pilot," Vol. xxiv. p. 199.

Produce in		Names, &c.	By what Bull.	By whom bred.
1877, Mar. 25, roan,	B.C.	Pilot	Pan, 35006	Mr. Lythall
1878, July 17, roan,	C.C.	White Heart	Cherry Prince, 37985	do.
1879, June 15, roan,	C.C.	Rose of the Vale	Virginus, 40873	do.

Pilot, sold to Mr. Grigg, Cornwall.

WOOD-NYMPH, roan, calved April 30, 1874, Vol. xxv. p. 558. Bred by Mr.
E. Lythall; got by Duke of Lancaster (28411), dam (Witchcraft) by Fitz
Killerby (26166), &c.

1878, Dec. 15, roan,	C.C.	Woodbine	Virginus, 40873	Mr. Lythall
1879, Oct. 24, red,	B.C.	Woodpecker	do.	do.

McCULLOCH, William,
Glenroy, Melbourne, Australia, late of Evington, Kent.

JASMINE, roan, calved March 7, 1875, Vol. xxiv. p. 711. Bred by Mr. T.
Wilson, Shotley Hall; got by Lord of Nunwick (26702), dam (Lady Villiers)
by Earl of the Valley (31086), &c.

1879, Jan. 4, red,	C.C.	Lady Jocelyn	D. of Oxford 31st, 33713	Mr. McCulloch

OXFORD'S DUCHESS 2ND, roan, calved December 1, 1876. Bred by Mr. R.
P. Davies, Horton; got by Third Duke of Clarence (23727), dam (Maid of
Oxford 6th) by Imperial Oxford (24185), &c. See Vol. xxiii. p. 415.

1879, Mar. 13, roan,	B.C.	Grand D. of Oxford	D. of Collingham 3rd. 38134	Mr. McCulloch

SOPHIA, roan, calved October 24, 1875. Bred by Lord Penrhyn, Wicken Park;
got by Oxford Waterloo (35002), dam (Sophy) by Cherry Duke (25752), &c.
See Vol. xxii. p. 530.

1879, Jan. 1, roan,	C.C.	Lady Penrhyn	G. D. of Oxford 2nd, 38381	Mr. McCulloch

WATER LILY, red, calved June 25, 1874, Vol. xxv. p. 507. Bred by the Earl
of Dunmore, Dunmore; got by Third Duke of Hillhurst (30975), dam
(Waterflower) by Sixth Duke of Geneva (30959), &c.

1879, Jan. 26, red,	C.C.	G. D'ss of Waterloo	Grand Duke 23rd, 34063	Mr. McCulloch

MACE, Mrs.,
Sherborne, Northleach, Gloucestershire.

MISS WETHERBY, roan, calved in December 1870, Vols. xxi., xxiii., and xxv.
pp. 833, 548, 560. Bred by Mr. T. Mace, Sherborne; got by Baron Wetherby
24th (27980), dam (Rosemary) by Butter Boy (21349), &c.

1879, Feb. 27, red,	C.C.	Kitty Wetherby	Geneva's Duke, 38349	Mrs. Mace

PARADOX 3RD, red, calved April 1, 1875, Vol. xxiv. p. 552. Bred by Mr. T.
Mace, Sherborne; got by Duke Geneva (30911), dam (Paradox 2nd) by Royal
Cambridge 2nd (25010), &c.

1879, Jan. 24, red,	C.C.	Paradox 5th	D. of Wellington 7th, 33753	Mrs. Mace

PRECEDENCE 2ND, roan, calved in March 1872, Vols. xxii. and xxiv. pp. 493,
552. Bred by Mr. T. Mace, Sherborne; got by Royal Cambridge 2nd (25010),
dam (Precedent) by Cynric (19542), &c.

1879, Feb. 15, { red,	C.C.	Precedent 6th	} D. of Wellington 7th, 33753	Mrs. Mace
{ red,	C.C.	Precedent 7th		

PRECEDENT 3RD, red, calved November 20, 1875, Vol. xxv.-p. 560. Bred by
Mr. T. Mace, Sherborne ; got by Oxford's Tony (35000), dam (Precedence 2nd)
by Royal Cambridge 2nd (25010), &c.

Produce in		Names, &c.		By what Bull.		By whom bred.
1879, Dec. 2, red,	C.C.	Precedent 8th		Geneva's Duke, 38349		Mrs. Mace

QUITE A PET, roan, calved in April, 1874. Bred by Mr. T. Mace, Sherborne;
got by Duke Geneva (30911), dam (White Pet) by Royal Cambridge 2nd
(25010), &c. See Vol. xxi. p. 834.

Produce in		Names, &c.	By what Bull.	By whom bred.
1879, Feb. 18, roan,	C.C.	Pettish Lass	D. of Wellington 7th, 33753	Mrs. Mace

RED CHERRY, red, calved March 3, 1873, Vols. xxii., xxiv., and xxv. pp. 493,
553, 560. Bred by Mr. T. Mace, Sherborne ; got by Duke Geneva (30911),
dam (Moss Rose) by Lord of the Manor (26714), &c.

Produce in		Names, &c.	By what Bull.	By whom bred.
1879, Mar. 21, red,	C.C.	Cherry Ripe	Thorndale Geneva, 39217	Mrs. Mace

RED ROSE, red, calved February 10, 1874, Vols. xxiv. and xxv. pp. 553, 561.
Bred by Mr. T. Mace, Sherborne ; got by Young Benedict (30525), dam
(Weston Rose) by Royal Cambridge 2nd (25010), &c.

Produce in		Names, &c.	By what Bull.	By whom bred.
1879, Mar. 3, red,	C.C.	Rosanna	Thorndale Geneva, 39217	Mrs. Mace

VIOLET 8TH, roan, calved in December, 1873. Bred by Mr. T. Mace, Sherborne;
got by Kenelm Butterfly (31462), dam (Violet 3rd) by Seventh Duke of York
(17754), &c. See "General Grant," p. 106.

Produce in		Names, &c.	By what Bull.	By whom bred.
1879, Nov. 5, roan,	C.C.	Venus	Remus, 37333	Mrs. Mace

WHITE PET, white, calved December 19, 1868, Vols. xxi. and xxiv. pp. 834,
553. Bred by Mr. T. Mace, Sherborne ; got by Royal Cambridge 2nd (25010),
dam (Roan Pet) by Royal Butterfly 14th (20722), &c.

Produce in		Names, &c.	By what Bull.	By whom bred.
1879, Mar. 13, roan,	C.C.	Priestess	Geneva's Duke, 38349	Mrs. Mace

McELDERRY, John,
Ballymoney, Co. Antrim.

FANNY 19TH, roan, calved April 14, 1874, Vol. xxv. p. 553. Bred by Mr. J.
Cooke, Ballyneal House; got by Lord of the Manor (29181), dam (Fanny
16th) by Prince Bertram (27119), &c.

Produce in		Names, &c.	By what Bull.	By whom bred.
1879, Jan. 10, roan,	C.C.	Fanny Fern	British Mantalini, 36281	Mr. McElderry

JOLLITY A., red, calved July 7, 1873, Vols. xxiv. and xxv. pp. 554, 561. Bred
by Messrs. T. and J. McElderry, Ballymoney; got by Abab (27850), dam
(Jollity 4th) by Sir Roderick (22914), &c.

Produce in		Names, &c.	By what Bull.	By whom bred.
1879, Dec. 3, r.&w.,	C.C.	Jollity E.	British Mantalini, 36281	Mr. McElderry

JOLLITY B., red and white, calved July 11, 1875, Vol. xxv. p. 562. Bred by
Messrs. T. and J. McElderry, Ballymoney; got by Rollo (39005), dam
(Jollity 4th) by Sir Roderick (22914), &c.

Produce in		Names, &c.	By what Bull.	By whom bred.
1879, Nov. 29, r.&w.,	C.C.	Jollity D.	British Mantalini, 36281	Mr. McElderry

PELERINE, red and white, calved June 19, 1876, Vol. xxv. p. 562. Bred by
Mr. G. Allen, Unicarville ; got by British Hero (30604), dam (Elfleda) by Blood
Royal (23429), &c.

Produce in		Names, &c.	By what Bull.	By whom bred.
1879, Dec. 26, roan,	B.C.	Bold Briton	Bright Baronet, 37891	Mr. McElderry

McGILL, Patrick,
Ballyduff, Kilmuckridge, Gorey, Ireland.

GLOSSY 7TH,⸱ red, calved May 15, 1870, Vols. xxi. and xxiv. pp. 583, 554.
Bred by Mr. W. Bolton, The Island; got by Manrico (26805), dam (Glossy
3rd) by Chieftain (21419), &c.

Produce in	Names, &c.	By what Bull.	By whom bred.
1879, Mar. 23, roan, C.C.	Gaiety	Albion, 36112	Mr. McGill

Gaiety, sold to Messrs. Lett, Mill Park, Enniscorthy.

GOLDEN BLOOM, roan, calved June 23, 1875. Bred by Mr. S. Armstrong,
Gally House; got by Prince of Munster (32174), dam (Golden Bud) by King
Richard (26523), &c. See Vol. xxiii. p. 312.

Produce in	Names, &c.	By what Bull.	By whom bred.
1879, Aug. 3, roan, C.C.	Golden Bracelet	Regal Buck, 38982	Mr. McGill

McINTOSH, David,
Havering Park, Romford, Essex.

AMATA 3RD, roan, calved December 20, 1874, Vols. xxiv. and xxv. pp. 554,
562. Bred by Mr. D. McIntosh; got by Third Duke of Geneva (23753), dam
(Amata) by Eighteenth Duke of Oxford (25995), &c.

Produce in	Names, &c.	By what Bull.	By whom bred.
1879, April 14, roan, C.C.	Amata 8th	Duke of Havering, 33664	Mr. McIntosh

BARONESS OXFORD 5TH, red and white, calved October 17, 1875, Vol. xxv.
p. 562. Bred by the Duke of Devonshire, Holker Hall; got by Fifth Duke
of Wetherby (31033), dam (Baroness Oxford) by Second Duke of Claro
(21576), &c.

Produce in	Names, &c.	By what Bull.	By whom bred.
1879, Nov. 17, r.&w., B.C.	Baron Oxford 3rd	Duke of Havering, 33664	Mr. McIntosh

CATHERINE GWYNNE, red and white, calved July 29, 1874, Vol. xxiii.
p. 549. Bred by Mr. G. Paine, Great Braxted Park; got by Second Duke
of Wellington (28465), dam (Frisky Gwynne) by Thorndale Knightley
(23065), &c.

Produce in	Names, &c.	By what Bull.	By whom bred.
1877, Dec. 29, roan, B.C.	(dead)	P. of Havering 3rd, 38930	Mr. McIntosh
1879, Jan. 5, r.&w., C.C.	Havering Gwy'ne 5th	Duke of Havering, 33664	do.

CHARMER 13TH, roan, calved October 6, 1869, Vols. xx., xxi., xxiii., and xxv.
pp. 438, 835, 550, 562. Bred by Mr. D. McIntosh; got by Third Duke of
Geneva (23753), dam (Science) by Chanter (19423), &c.

Produce in	Names, &c.	By what Bull.	By whom bred.
1879, May 11, roan, C.C.	Charmer 38th	Duke of Havering, 33664	Mr. McIntosh

CHARMER 19TH, roan, calved January 17, 1872, Vols. xxi., xxii., xxiii., and
xxv. pp. 836, 494, 550, 563. Bred by Mr. D. McIntosh; got by Third
Duke of Geneva (23753), dam (Charmer 11th) by Baron Killerby (23364), &c.

Produce in	Names, &c.	By what Bull.	By whom bred.
1879, May 17, red, B.C.	Baron Havering 21st	Duke of Havering, 33664	Mr. McIntosh

CHARMER 21ST, red and white, calved May 31, 1873, Vols. xxiii. and xxiv.
pp. 550, 554. Bred by Mr. D. McIntosh; got by Third Duke of Geneva
(23753), dam (Charmer 10th) by Baron Killerby (23364), &c.

Produce in	Names, &c.	By what Bull.	By whom bred.
1878, Mar. 7, red, B.C.	(dead)	Duke of Havering, 33664	Mr. McIntosh
1879, April 30, red, C.C.	Charmer 37th	do.	do.

CHARMER 25TH, roan, calved October 19, 1874, Vols. xxiv. and xxv. pp. 555,
563. Bred by Mr. D. McIntosh; got by Third Duke of Geneva (23753), dam
(Charmer 11th) by Baron Killerby (23364), &c.

Produce in	Names, &c.	By what Bull.	By whom bred.
1879, Jan. 3, red, C.C.	Charmer 36th	Duke of Havering, 33664	Mr. McIntosh

DOROTHY GWYNNE, red, calved May 29, 1874, Vols. xxiv. and xxv. pp. 555, 563. Bred by Mr. G. Paine, Great Braxted Park; got by Second Duke of Wellington (28465), dam (Flora Gwynne) by Oxford Gwynne (24711), &c.

Produce in	Names, &c.	By what Bull.	By whom bred.
1879, Sept. 13, red,	B.C.|Baron Gwynne	Duke of Havering, 33664	Mr. McIntosh

DUCHESS CAROLINA 2ND, red and white, calved September 25, 1872, Vols. xxi., xxiv., and xxv. pp. 836, 555, 563. Bred by Mr. D. McIntosh; got by Third Duke of Geneva (23753), dam (Carolina 5th) by Seventh Duke of York (17754), &c.

Produce in	Names, &c.	By what Bull.	By whom bred.
1879, Dec. 4, r. & w.,	B.C.|Duke of Carolina 5th|Duke of Havering, 33664	Mr. McIntosh

DUCHESS CAROLINA 4TH, red, calved February 21, 1877. Bred by Mr. D. McIntosh; got by Duke of Havering (33664), dam (Duchess Carolina 2nd) by Third Duke of Geneva (23753), &c.

Produce in	Names, &c.	By what Bull.	By whom bred.
1879, Sept. 12, red,	C.C.|Duchess Carolina 6th|Viscount Oxford, 40876	Mr. McIntosh

HAVERING FUCHSIA 2ND, red and white, calved January 6, 1877. Bred by Mr. D. McIntosh; got by Duke of Havering (33664), dam (Lady Fuchsia 3rd) by Lord Tregunter (31758), &c. See Vol. xxiv. p. 556.

Produce in	Names, &c.	By what Bull.	By whom bred.
1879, Sept. 18, red,	C.C.|Havering Fuchsia 3d|Viscount Oxford, 40876	Mr. McIntosh

HAVERING GWYNNE, red, calved November 21, 1876. Bred by Mr. D. McIntosh; got by Duke of Havering (33664), dam (Catherine Gwynne) by Second Duke of Wellington (28465), &c. See Vol. xxiii. p. 549.

Produce in	Names, &c.	By what Bull.	By whom bred.
1879, Jan. 3, roan,	C.C.	HaveringGwy'ne4th	P. of Havering 3rd, 38930	Mr. McIntosh

HAVERING LILY 3RD, red, calved July 28, 1875, Vol. xxv. p. 564. Bred by Mrs. McCulloch, Balloan of Cawdor; got by Star (37524), dam (Rosy 4th) by Fawsley Prince (21733), &c.

Produce in	Names, &c.	By what Bull.	By whom bred.
1879, Dec. 3, red,	C.C.|Havering Lily 13th |Duke of Havering, 33664	Mr. McIntosh

WATERLOO 38TH, red and white, calved August 27, 1872, Vol. xxv. p. 565. Bred by Lord Penrhyn, Penrhyn Castle; got by Grand Duke 11th (21849), dam (Waterloo 27th) by Duke of Geneva (19614), &c.

Produce in	Names, &c.	By what Bull.	By whom bred.
1879, Mar. 31, red,	C.C.|Havering W'loo 3rd |Duke of Havering, 33664	Mr. McIntosh

MACKAY, R. J.,
Burgie Lodge, Forres, N.B.

RED ROSE, red, calved March 10, 1876. Bred by Mr. R. J. Mackay; got by King Charming (34309), dam (Rosebud) by Baron Cecil (27921), &c. See " Chelmsford," p. 46.

Produce in	Names, &c.	By what Bull.	By whom bred.
1879, April 20, red,	B.C.|Chelmsford	Charm's Windsor, 36342	Mr. Mackay

Chelmsford, sold to Mr. J. Fletcher, Rosehaugh, Avoch, Ross-shire.

MACKIE, William,
Petty Fyvie, Aberdeen, N.B.

DAISY, red, calved January 1, 1875. Bred by Mr. W. Mackie; got by Towneley's Oxford (32729), dam (Venus) by Blair Athol (25638), &c. See "Longfellow," Vol. xxii. p. 144.

Produce in	Names, &c.	By what Bull.	By whom bred.
1877, Dec. 26, red,	B.C.|Merryman	Baron Lowther, 33082	Mr. Mackie
1879, Jan. 26, roan,	B.C.|I. X. L.	Fred.Fitz-Windsor, 31196| do.

Merryman, sold to Mr. T. McBey, Auchronie, Aberdeenshire.

FAIR MAID, roan, calved February 18, 1876.   Bred by Mr. W. Mackie ; got
by Wrangler (36035), dam (Dairymaid) by Lord Charles (31634), g. d. (Housemaid) by Blair Athol (25638), &c.   See Vol. xxi. p. 839.

Produce in		Names, &c.	By what Bull.	By whom bred.
1878, May 10, white,	B.C.	Bismarck	Baron Lowther, 33062	Mr. Mackie
1879, Dec. 21, roan,	B.C.	Belvidere	Fred. Fitz-Windsor,31196	do.

Bismarck, sold to Mr. J. Sharp, Bogdavie, Fyvie, Aberdeenshire.

MERMAID, roan, calved April 1, 1876.   Bred by Mr. W. Mackie; got by
Wrangler (36035), dam (Housemaid 2nd) by Lord Charles (31634), g. d.
(Housemaid) by Blair Athol (25638), &c.   See Vol. xxi. p. 839.

1878, Nov. 2, roan,	C.C.	Lowland Maid	Baron Lowther, 33062	Mr. Mackie

McLAREN, John,
Craggish, Comrie, Perthshire.

MERINO, roan, calved in April, 1870.   Bred by Mr. J. Dickson, Cambushinnie;
got by Strathallan (37539), dam (Gertrude) by Bridegroom (17441), g. d.
(Lady Mary) by Windsor (37683), &c.   See " Sir William," Vol. xxv. p. 265.

1878, Feb. 25, r. & w.,	C.C.	Flora	Prince of Schleswig, 35158	Mr. McLaren

McWILLIAM, J.,
Stoneytown, Keith, N.B.

DIDO 3RD, red, calved March 22, 1867, Vol. xxiii. p. 552.   Bred by the Duke
of Richmond and Gordon, Gordon Castle ; got by Duke of Bowland (21568),
dam (Dido 2nd) by Prince Arthur (16723), &c.

1877, Jan. 5, red,	B.C.	Beaconsfield	Spicer, 35655	Mr. McWilliam
1878, Mar. 10, roan,	B.C.	Duke of Richmond	Knickerbocker, 38510	do.
1879, May 30, roan,	B.C.	Duke of Gordon	do.	do.

Beaconsfield, sold to Lord Middleton, Applecross; Duke of Richmond, to Mr. Grant, Stank-
house, Elgin.

DIDO 4TH, red, calved March 6, 1875.   Bred by Mr. J. McWilliam; got by
Alfred (32935), dam (Dido 3rd) by Duke of Bowland (21568), &c.

1877, Nov. 24, red,	B.C.	Sir James	Spicer, 35655	Mr. McWilliam
1879, May 7, red,	C.C.	Dido 5th	Knickerbocker, 38510	do.

Sir James, sold to Mr. G. Richie, Mains of Cairnty, Keith.

GOLDIE 17TH, roan, calved March 8, 1874, Vol. xxiii. p. 552.   Bred by Mr. W.
S. Marr, Uppermill ; got by Heir of Englishman (24122), dam (Goldie 9th)
by Macduff (26733), &c.

1878, April 25, roan,	C.C.	Golden Rose	Knickerbocker, 38510	Mr. McWilliam
1879, May 1, white,	B.C.	(dead)	do.	do.

GOLD LEAF, red and white, calved October 11, 1876.   Bred by Mr. J.
McWilliam ; got by Spicer (35655), dam (Goldie 17th) by Heir of Englishman (24122), &c.

1879, Apr. 13, r. & w.,	C.C.	Gold Cherry	Knickerbocker, 38510	Mr. McWilliam

MAGNIAC, C.,
Colworth, Bedford.

AIRDRIE FLOWER, red and white, calved October 18, 1876.   Bred by Mr. C.
Magniac ; got by Twenty-fourth Duke of Airdrie (36460), dam (Cactus 2nd)
by Alaska (37718), &c.   See Vol. xxiii. p. 553.

1879, April 6, roan,	C.C.	Countess of Antrim	Oxford de Vere, 34424	Mr. Magniac

BLANCHE 12TH, roan, calved January 24, 1876, Vol. xxv. p. 567. Bred by
Sir W. H. Salt, Bart., Maplewell; got by Fifth Lord Oxford (31738), dam
(Blanche 3rd) by Marquis of York (34791), &c.

Produce in	Names, &c.	By what Bull.	By whom bred.
1879, Dec. 20, white, B.C.	Duke of Bedford 7th	Oxford de Vere, 40424	Mr. Magniac

ETHELFLEDA, red and white, calved January 15, 1877. Bred by Mr. W.
Lavender, Biddenham; got by Earl of Leicester 2nd (33803), dam (Fleda of
York) by Eighth Duke of York (28480), &c. See Vol. xxiii. p. 529.

Produce in	Names, &c.	By what Bull.	By whom bred.
1879, June 4, r. & w., C.C.	Red Lavender	Knightley York, 38511	Mr. Magniac

HONEYCOMB, red and white, calved May 20, 1873, Vols. xxiv. and xxv.
pp. 559, 567. Bred by Lord Skelmersdale, Lathom House; got by First
Duke of Oneida (30996), dam (Honeymoon) by Viscount Curly (25382), &c.

Produce in	Names, &c.	By what Bull.	By whom bred.
1879, May 19, r. & w., B.C.	Duke of Bedford 5th	Marquis of Oxford, 34786	Mr. Magniac

MINSTREL 4TH, white, calved November 11, 1872, Vols. xxii., xxiii., xxiv., and
xxv. pp. 497, 553, 559, 567. Bred by Mr. E. Bowly, Siddington House; got
by Second Duke of Tregunter (26022), dam (Minstrel) by Sixth Earl of
Walton (26078), &c.

Produce in	Names, &c.	By what Bull.	By whom bred.
1879, June 14, white, C.C.	(dead)	Marquis of Oxford, 34786	Mr. Magniac

PAGARES. red and white, calved October 21, 1875, Vol. xxv. p. 568. Bred by
Mr. C. Magniac; got by Duke of Hillhurst (28401), dam (Papaver) by Third
Duke of Clarence (23727), &c.

Produce in	Names, &c.	By what Bull.	By whom bred.
1879, Nov. 24, roan, C.C.	Pimpernel	Marquis of Oxford, 34786	Mr. Magniac

PAPAVER, roan, calved June 24, 1871, Vols. xxi., xxii., and xxv. pp. 795, 497,
568. Bred by Colonel Kingscote, Kingscote; got by Third Duke of Clarence
(23727), dam (Poppy Juice) by Second Earl of Walton (19672), &c.

Produce in	Names, &c.	By what Bull.	By whom bred.
1879, June 22, roan, C.C.	Flower of Yarm	Marquis of Oxford, 34786	Mr. Magniac

THE QUEEN, roan, calved November 1, 1869, Vols. xxi. and xxiv. pp. 628,
559. Bred by Messrs. F. Leney and Sons, Wateringbury; got by Lord Oxford
2nd (20215), dam (Princess Alice) by British Prince (14197), &c.

Produce in	Names, &c.	By what Bull.	By whom bred.
1879, July 26, roan, B.C.	Duke of Bedford 6th	Marquis of Oxford, 34786	Mr. Magniac

WATERLOO 22ND, roan, calved January 26, 1862, Vols. xviii., xix., xxii., and
xxiv. pp. 775, 776, 418, 559. Bred by Mr. G. Shepherd, Shethin; got by
Speculator (13775), dam (Waterloo 18th) by Bosquet (14183), &c.

Produce in	Names, &c.	By what Bull.	By whom bred.
1879, Aug. 13, roan, C.C.	Waterloo 49th	Oxford de Vere, 40424	Mr. Magniac

WATERLOO 46TH, roan, calved April 13, 1877. Bred by Mr. C. Magniac; got
by Twenty-second Duke of Oxford (31000), dam (Waterloo 22nd) by Speculator (13775), &c.

Produce in	Names, &c.	By what Bull.	By whom bred.
1879, Sept. 3, roan, C.C.	Waterloo 50th	Oxford de Vere, 40424	Mr. Magniac

MAHONY, Pierce,
Kilmorna, Listowel, Ireland.

DAWN OF LIGHT, roan, calved April 30, 1868, Vols. xx. and xxiv. pp. 476,
392. Bred by Mr. W. T. Crosbie, Ardfert Abbey; got by Northern Light
(24670). dam (Circassia) by Lamp of Lothian (16356), &c.

Produce in	Names, &c.	By what Bull.	By whom bred.
1879, Feb. 28, r. & w., C.C.	Rosy Dawn	Royal Fitz-Rose, 37390	Mr. Mahony

LADY ELIZABETH, roan, calved January 28, 1874. Bred by Mr. W. T. Crosbie, Ardfert Abbey ; got by Regal Booth (27262), dam (Lisa) by Northern Light (24670), &c. See Vol. xxi. p. 652.

Produce in	Names, &c.	By what Bull.	By whom bred.
1879, Mar. 2, r. & w., C.C.	Sister Bessie	Simpleton, 37447	Mr. Mahony

LUSTRE 4TH, red and white, calved January 29, 1877. Bred by Mr. J. Cooke, Ballyneal House; got by Ruby Star (35422), dam (Lustre) by Chief Baron (25769), g. d. (Lunette) by Master Harbinger (18352), &c. See Vol. xix. p. 609.

Produce in	Names, &c.	By what Bull.	By whom bred.
1879, June 9, r. & w., C.C.	Lady Fern	Royal Fern, 29865	Mr. Mahony

VICTORIA 50TH, roan, calved April 20, 1869, Vols. xx. and xxii. pp. 805, 373. Bred by Mr. J. Cooke, Ballyneal House; got by St. Ringan (27417), dam (Violet) by Ravenspur (20628), &c.

Produce in	Names, &c.	By what Bull.	By whom bred.
1878, July 20, roan, C.C.	Lady Victoria 2nd	Red Cross Knight, 35219	Mr. Mahony

VICTORIA 65TH, roan, calved April 9, 1873. Bred by Mr. J. Cooke, Ballyneal House; got by Lord of the Manor (29181), dam (Victoria 50th) by St. Ringan (27417), &c.

Produce in	Names, &c.	By what Bull.	By whom bred.
1878, Mar. 29, r. & w., C.C.	Lady Victoria 1st	Red Cross Knight, 35219	Mrs. Pery
1879, June 11, roan, B.C.	Victor's Crown	Royal Crown, 40628	Mr. Mahony

WINIFRED, red, calved March 15, 1874. Bred by the Hon. M. H. Cochrane, Hillhurst, Compton, Canada ; got by Cavalier (39560), dam (Welfare) by Royal Commander (28957), &c. See Vol. xxiv. p. 382.

Produce in	Names, &c.	By what Bull.	By whom bred.
1878, Aug. 13, r. & w., B.C.	Redwin	Red Cross Knight, 35219	Mr. Mahony

MANCHESTER, Duke of,
Kimbolton Castle, St. Neots, Hunts.

BEVERLEY DUCHESS 8TH, roan, calved May 16, 1876. Bred by Mr. R. Botterill, Wauldby ; got by Oxford le Grand (29496), dam (Miss Beverley 4th) by Van Thol (32755), g. d. (Miss Beverley 22nd) by Royal Butterfly 16th (20724), &c. See "Beverley Oxford 5th," Vol. xxv. p. 29.

Produce in	Names, &c.	By what Bull.	By whom bred.
1878, July 5, white, B.C.	(dead)	D. of Underley 3rd, 38196	D. of Manchester
1879, July 12, r. & w., C.C.	Beverley Duchess 9th	do.	do.

BRUNETTE 10TH, roan, calved April 9, 1874, Vols. xxiii. and xxv. pp. 526, 569. Bred by Sir C. M. Lampson, Bart., Rowfant ; got by Grand Duke of Geneva (28756), dam (Brunette) by Fourth Duke of Thorndale (17750), &c.

Produce in	Names, &c.	By what Bull.	By whom bred.
1879, July 14, white, C.C.	Blanche 2nd	D. of Underley 3rd, 38196	D. of Manchester

FLIRT, roan, calved February 13, 1873, Vols. xxiii., xxiv., and xxv. pp. 554, 560, 569. Bred by the Duke of Manchester ; got by Lord Stanley (29205), dam (Fancy) by Vizier (21042), &c.

Produce in	Names, &c.	By what Bull.	By whom bred.
1879, Dec. 20, r. & w., C.C.	Flirt 2nd	D. of Underley 3rd, 38196	D. of Manchester

GUINEVERE, red and white, calved March 16, 1869, Vols. xxiv. and xxv. pp. 560, 569. Bred by Mr. H. J. Sheldon, Brailes House; got by Duke of Brailes (23724), dam (Gionetta) by Sarawak (15238), &c.

Produce in	Names, &c.	By what Bull.	By whom bred.
1879, Sept. 20, roan, C.C.	Vivien	D. of Underley 3rd, 38196	D. of Manchester

JARDINIERE, roan, calved May 6, 1871. Bred by Mr. J. N. Beasley, Chapel Brampton; got by Jay (26457), dam (Jaqueline) by Dandilly Dan (23675), &c. See Vol. xx. p. 566.

Produce in	Names, &c.	By what Bull.	By whom bred.
1879, Nov. 9, roan, C.C.	Jaconette	D. of Underley 3rd, 38196	D. of Manchester

LADY MONTAGU, red, calved October 22, 1876. Bred by the Duke of Manchester; got by Duke of Hillhurst (28401), dam (Lady Worcester 10th) by Eighth Duke of Geneva (28390), &c. See " Marquis Monthermer 4th," p. 164.

Produce in	Names, &c.	By what Bull.	By whom bred.
1879, May 14, roan, B.C.	Mq. Monthermer 4th	D. of Underley 3rd, 38196	D. of Manchester

LADY WATERLOO SURMISE, white, calved May 18, 1877. Bred by Mr. F. N. Sartoris, Rushden Hall; got by Grand Duke of Waterloo (28766), dam (Lady Ferrers) by Duke of Kingscote (25981), &c. See Vol. xxv. p. 623.

Produce in	Names, &c.	By what Bull.	By whom bred.
1879, Aug. 1, roan, C.C.	Silentia 2nd	D. of Underley 3rd, 38196	D. of Manchester

LADY WORCESTER 11TH, white, calved October 2, 1872, Vols. xxiv. and xxv. pp. 560, 569. Bred by the Earl of Dunmore, Dunmore; got by Third Duke of Claro (23729), dam (Lady Worcester 3rd) by Third Duke of Wharfdale (21619), &c.

Produce in	Names, &c.	By what Bull.	By whom bred.
1879, June 15, roan, C.C.	Lady Montagu 2nd	D. of Underley 3rd, 38196	D. of Manchester

LALLY 16TH, white, calved September 9, 1872, Vols. xxiv. and xxv. pp. 560, 569. Bred by Mr. J. Harward, Winterfold; got by Third Duke of Claro (23729), dam (Lally 9th) by Seventh Duke of York (17754), &c.

Produce in	Names, &c.	By what Bull.	By whom bred.
1879, Jan. 10, white, B.C.	Tollendal	D. of Underley 3rd, 38196	D. of Manchester

LALLY OF ELLINGTON, roan, calved February 7, 1877. Bred by the Duke of Manchester; got by Duke of Hillhurst (28401), dam (Lally 16th) by Third Duke of Claro (23729), &c.

Produce in	Names, &c.	By what Bull.	By whom bred.
1879, Nov. 13, red, C.C.	Lally of Littlehurst	D. of Underley 3rd, 38196	D. of Manchester

LAURESTINA, roan, calved January 20, 1872, Vol. xxiv. p. 560. Bred by the Duke of Manchester; got by Second Earl of Barrington (28494), dam (Laurel) by Cambridge Barrington 1st (14223), &c.

Produce in	Names, &c.	By what Bull.	By whom bred.
1879, Aug. 14, roan, B.C.	Laurel	D. of Underley 3rd, 38196	D. of Manchester

MARCHIONESS OF OXFORD 3RD, white, calved March 3, 1873, Vols. xxii., xxiv., and xxv. pp. 498, 561, 570. Bred by the Earl of Dunmore, Dunmore; got by Second Duke of Collingham (23730), dam (Eighth Maid of Oxford) by Second Duke of Geneva (23752), &c.

Produce in	Names, &c.	By what Bull.	By whom bred.
1879, Feb. 25, white, C.C.	Oxford Augusta	D. of Underley 3rd, 38196	D. of Manchester

MERCIA 1ST, roan, calved March 17, 1877. Bred by the Duke of Manchester; got by Grand Duke of Waterloo (28766), dam (Poppy 4th) by Lord Stanley (29205), &c. See Vol. xxiv. p. 561.

Produce in	Names, &c.	By what Bull.	By whom bred.
1879, June 12, red, C.C.	Mercia 10th	D. of Underley 3rd, 38196	D. of Manchester

OXFORD LOUISE, white, calved September 26, 1875, Vol. xxv. p. 570. Bred by the Duke of Manchester; got by Duke of Connaught (33604), dam (Marchioness of Oxford 3rd) by Second Duke of Collingham (23730), &c.

Produce in	Names, &c.	By what Bull.	By whom bred.
1879, Feb. 23, roan, C.C.	Oxford Mary	D. of Underley 3rd, 38196	D. of Manchester

PEONY, red and white, calved December 24, 1874, Vol. xxv. p. 570. Bred by the Duke of Manchester; got by Wild Eyes Duke (36007), dam (Poppy 5th) by Lord Stanley (29205), &c.

Produce in	Names, &c.	By what Bull.	By whom bred.
1879, Aug. 21, r. & w., C.C.	Mercia 11th	D. of Underley 3rd, 38196	D. of Manchester

POLLY 4TH, red and white, calved April 2, 1872, Vols. xxiv. and xxv. pp. 561, 570. Bred by the Duke of Manchester; got by Lord Stanley (29205), dam (Polly 2nd) by Major (13283), &c.

Produce in	Names, &c.	By what Bull.	By whom bred.
1879, June 9, r. & w., C.C.	Mercia 9th	D. of Underley 3rd, 38196	D. of Manchester

POPULACE, red and white, calved April 24, 1874, Vol. xxv. p. 570. Bred by the Duke of Manchester; got by Warrior (32799), dam (Poppy 3rd) by Lucifer (16468), &c.

Produce in		Names, &c.	By what Bull.	By whom bred.
1879, Jan. 4, red,	B.C.	Earl of Mercia 3rd	G. D. of Morecambe, 36722	D. of Manchester
1879, Dec. 21, red,	C.C.	Mercia 12th	D. of Underley 3rd, 38196	do.

Earl of Mercia 3rd, sold to Mr. Walker, Cranford, Kettering.

PRINCESS ALEXANDRA, roan, calved October 29, 1876. Bred by the Duke of Manchester; got by Second Duke of Tregunter (26022), dam (Princess 4th) by Baron Oxford 3rd (25579), &c. See Vol. xxiii. p. 554.

Produce in	Names, &c.	By what Bull.	By whom bred.
1879, May 10, white, C.C.	Princess of Wales	D. of Underley 3rd, 38196	D. of Manchester

PRINCESS SILENCE, roan, calved July 17, 1876, Vol. xxv. p. 570. Bred by the Duke of Manchester; got by Wild Eyes Duke (36007), dam (Princess) by May Duke (13320), &c.

Produce in	Names, &c.	By what Bull.	By whom bred.
1879, Oct. 12, roan, B.C.	Trappist 4th	D. of Underley 3rd, 38196	D. of Manchester

PRINCESS SURMISE, roan, calved September 3, 1875, Vol. xxv. p. 570. Bred by the Duke of Manchester; got by Wild Eyes Duke (36007), dam (Princess) by May Duke (13320), &c.

Produce in	Names, &c.	By what Bull.	By whom bred.
1879, Feb. 14, roan, B.C.	Trappist 3rd	D. of Underley 3rd, 38196	D. of Manchester

SONSIE 24TH, red and white, calved March 1, 1875, Vol. xxv. p. 571. Bred by Sir W. Lawson, Bart., Brayton; got by Baron Oxford 6th (33075), dam (Sonsie 22nd) by Royal Cambridge (25009), &c.

Produce in	Names, &c.	By what Bull.	By whom bred.
1879, Jan. 31, roan, C.C.	Sonsie 25th	D. of Underley 3rd, 38196	D. of Manchester

WATERLOO 33RD, roan, calved September 5, 1873, Vol. xxiii. p. 554. Bred by Lord Fitzhardinge, Berkeley Castle; got by Duke of Hillhurst (28401), dam (Waterloo 32nd) by Seventh Duke of York (17754), &c.

Produce in	Names, &c.	By what Bull.	By whom bred.
1879, Sept. 25, r. & w., B.C.	Wellesley 2nd	D. of Underley 3rd, 38196	D. of Manchester

WATERLOO 34TH, red and white, calved November 30, 1876. Bred by the Duke of Manchester; got by Duke of Underley (33745), dam (Waterloo 33rd) by Duke of Hillhurst (28401), &c.

Produce in	Names, &c.	By what Bull.	By whom bred.
1879, July 10, r. & w., C.C.	Waterloo 35th	D. of Underley 3rd, 38196	D. of Manchester

WELLINGTONIA 3RD, white, calved July 26, 1871, Vol. xxiii. p. 554. Bred by Messrs. F. Leney and Sons, Wateringbury; got by Grand Duke of Kent (26289), dam (Wellingtonia 2nd) by Twelfth Duke of Thorndale (20020), &c.

Produce in	Names, &c.	By what Bull.	By whom bred.
1879, Jan. 1, white, C.C.	Sequoiah 3rd	D. of Underley 3rd, 38196	D. of Manchester

MANN, Thomas,
Thelveton Hall, Scole, Norfolk.

PERDITA, red, calved August 20, 1875. Bred by Mr. J. Stratton, Alton Priors; got by Royal (35331), dam (Persephone) by Eighth Duke of York (23808), &c. See Vol. xxii. p. 579.

Produce in		Names, &c.	By what Bull.	By whom bred.
1878, Jan. 29, roan,	B.C.	Patrician	Royal James, 35387	Mr. Stratton
1879, Feb. 2, red,	C.C.	Poppy	Abelard, 39348	Mr. Mann

MARR, W. S.,
Uppermill, Tarves, N.B.

ALEXANDRINA 7TH, roan, calved April 2, 1874. Bred by Mr. W. S. Marr;
got by Heir of Englisman (24122), dam (Alexandrina 5th) by Macduff
(26773), &c. See Vol. xxiv. p. 562.

Produce in		Names, &c.	By what Bull.	By whom bred.
1878, Feb. 26, white	C.C.	Alexandrina 10th	Cherub 4th, 33359	Mr. Marr
1879, Mar. 21, roan,	B.C.	Anchor	D. of Glo'ster, 33652	do.
Anchor, sold to Mr. Henderson, Rattray.				

PRINCESS ROYAL 17TH, roan, calved March 7, 1874. Bred by Mr. W. S.
Marr; got by Heir of Englishman (24122). dam (Princess Royal 11th) by
King of the Isles (31506), &c. See Vol. xxiii. p. 556.

1878, April 10, roan,	C.C.	Princess Royal 21st	Nonsuch, 40389	Mr. Marr
1879, Mar. 24, roan,	B.C.	Promise	Cherub, 33359	do.
Promise, sold to Mr. W. Duthie, Collynie, Tarves, N.B.				

MARSHALL, Rev. Charles,
Ripley Court, Woking Station, Surrey.

CHERRY RIPE 1ST, roan, calved May 11, 1871, Vols. xxi., xxiv., and xxv.
pp. 841, 562, 575. Bred by the Rev. C. Marshall; got by Sweet William 2nd
(27590), dam (Cherry Royal) by Knightley Grand Duke (24268), &c.

1879, July 30, roan,	C.C.	Cherry Ripe 7th·	The Friar, 35766	Rev. C. Marshall

CHERRY RIPE 2ND, roan, calved April 16, 1872, Vols. xxiv. and xxv. pp. 563,
575. Bred by the Rev. C. Marshall; got by Sweet William 2nd (27590), dam
(Cherry Royal) by Knightley Grand Duke (24268), &c.

1879, Nov. 29, r. & w., B.C.	(Steer)	The Friar, 35766	Rev. C. Marshall

CHERRY RIPE 3RD, roan, calved April 21, 1874, Vols. xxiv. and xxv. pp. 563,
575. Bred by the Rev. C. Marshall; got by Mameluke (34741), dam (Cherry
Ripe 1st) by Sweet William 2nd (27590), &c.

1879, Jan. 2, roan,	B.C.	(Steer)	The Friar, 35766	Rev. C. Marshall
1879, Nov. 30, roan,	C.C.	Cherry Ripe 9th	do.	do.

CHERRY RIPE 4TH, red, calved February 6, 1877. Bred by the Rev. C.
Marshall; got by The Friar (35766), dam (Cherry Ripe 3rd) by Mameluke
(34741), &c. See Vol. xxiv. p. 563.

1879, Sept. 7, red,	C.C.	Cherry Ripe 8th	Duke, 41344	Rev. C. Marshall

CHERRY ROYAL 2ND, roan, calved December 21, 1871, Vols. xxiv. and xxv.
pp. 563, 575. Bred by the Rev. C. Marshall; got by Sweet William 2nd
(27590), dam (Cherry Blossom) by Duke of Cambridge (21574), &c.

1879, Apr. 14, r. & w., B.C.	(Steer)	The Friar, 35766	Rev. C. Marshall

CHRISTIE, red, calved May 23, 1874, Vols. xxiv. and xxv. pp. 563, 575. Bred
by the Rev. C. Marshall; got by Dublin (28340), dam (Credit) by Lord
Lennox (20189), &c.

1879, Oct. 22, red,	C.C.	Christie 4th	The Friar, 35766	Rev. C. Marshall

CREDIT 2ND, roan, calved November 25, 1876. Bred by the Rev. C. Marshall;
got by Albion (32931), dam (Credit) by Lord Lennox (20189), &c. See
Vol. xxiii. p. 557.

1879, Dec. 22, white,	B.C.	(Steer)	The Friar, 35766	Rev. C. Marshall

ELEGY, roan, calved March 8, 1875, Vol. xxv. p. 575. Bred by the Rev. C. Marshall; got by Dublin (28340), dam (Epitaph 8th) by Borrowby Lad (12486), &c.

Produce in	Names, &c.	By what Bull.	By whom bred.
1879, Feb. 19, roan, B.C.	(Steer)	The Friar, 35766	Rev. C. Marshall

ROSABELLA, roan, calved April 16, 1874, Vols. xxiv. and xxv. pp. 563, 576. Bred by the Rev. C. Marshall; got by Dublin (28340), dam (Rosaline) by Knightley Grand Duke (24268), &c.

1879, April 22, roan, C.C.	Rosabella 3rd	The Friar, 35766	Rev. C. Marshall

ROSABUND, roan, calved December 4, 1872, Vols. xxiii. and xxv. pp. 557, 576. Bred by the Rev. C. Marshall; got by The Baronet (25283), dam (Rosaline) by Knightley Grand Duke (24268), &c.

1879, Dec. 16, r. & w., B.C.	(Steer)	The Friar, 35766	Rev. C. Marshall

ROSABUND 2ND, red and white, calved November 21, 1876. Bred by the Rev. C. Marshall; got by The Friar (35766), dam (Rosabund) by The Baronet (25283), &c.

1879, July 29, red, C.C.	Rosabund 5th	Duke, 41344	Rev. C. Marshall

ROSALIA 2ND, roan, calved November 12, 1876. Bred by the Rev. C. Marshall; got by The Friar (35766), dam (Rosalia) by Sweet William 2nd (27590), &c. See Vol. xxiii. p. 557.

1879, Nov. 26, white, C.C.	Rosalia 3rd	Duke, 41344	Rev. C. Marshall

ROSEDALE, red and white, calved June 11, 1876. Bred by the Rev. C. Marshall; got by The Friar (35766), dam (Rosetta) by Dublin (28340), &c. See Vol. xxiii. p. 558.

1879, June 16, roan, B.C.	(Steer)	Duke, 41344	Rev. C. Marshall

ROSETTA, roan, calved November 7, 1873, Vols. xxiii., xxiv., and xxv. pp. 558, 563, 576. Bred by the Rev. C. Marshall; got by Dublin (28340), dam (Rosalia) by Sweet William 2nd (27590), &c.

1879, July 26, roan, C.C.	Rosetta 3rd	The Friar, 35766	Rev. C. Marshall

SYLPH 3RD, roan, calved February 6, 1871, Vols. xxi., xxiii., xxiv., and xxv. pp. 842, 558, 563, 576. Bred by the Rev. C. Marshall; got by Sweet William 2nd (27590), dam (Sylph) by Knightley Grand Duke (24268), &c.

1879, Dec. 10, red, B.C.	(Steer)	The Friar, 35766	Rev. C. Marshall

SYLPH 4TH, white, calved May 29, 1872, Vols. xxi., xxii., xxiv., and xxv. pp. 842, 500, 564, 576. Bred by the Rev. C. Marshall; got by Sweet William 2nd (27590), dam (Sylph 2nd) by Knightley Grand Duke (24268), &c.

1879, Aug. 22, roan, B.C.	(Steer)	The Friar, 35766	Rev. C. Marshall

SYLPH 6TH, roan, calved February 18, 1874, Vol. xxiv. p. 564. Bred by the Rev. C. Marshall; got by Dublin (28340), dam (Sylph 3rd) by Sweet William 2nd (27590), &c. .

1879, Jan. 15, roan, B.C.	(Steer)	The Friar, 35766	Rev. C. Marshall

SYLPH 9TH, roan, calved November 7, 1875, Vol. xxv. p. 576. Bred by the Rev. C. Marshall; got by Albion (32931), dam (Sylph 4th) by Sweet William 2nd (27590), &c.

1879, June 20, r. & w., B.C.	(Steer)	The Friar, 35766	Rev. C. Marshall

SYLPH 11th, red and white, calved April 26, 1876. Bred by the Rev. C.
Marshall ; got by Dublin (28340), dam (Sylph 2nd) by Knightley Grand
Duke (24268), &c.

Produce in		Names, &c.		By what Bull.		By whom bred.
1879, Mar. 24, red,	C.C	Sylph 17th		The Friar, 35766		Rev. C. Marshall

MARSHALL, H. J.,
Poulton Priory, Hereford.

MIDGE, red and white, calved January 26, 1875. Bred by Her Majesty the
Queen; got by King Tom (31521), dam (Milkmaid) by Prince of Saxe Coburg
(20576), &c. See Vol. xxii. p. 286.

1879, Mar. 11, roan, · C.C.	Midge 2nd	Sir Windsor 2nd, 39135	Mr. Marshall

RED LASS, red, calved in April, 1872. Bred by Mr. W. Hewer, Sevenhampton;
got by Marksman (26814), dam (Red Lady) by Experience (23900), &c. See
"Robespierre," p. 212.

1879, Apr. 5, r. & w.,	B.C.	Robespierre	The Barrister, 35740	Mr. Marshall

MARTIN, E.,
East Down Lodge, Woodlands, Sevenoaks, Kent.

CASSANDRA, red and white, calved July 1, 1876. Bred by Mr. E. Randall,
East Down Lodge, the property of Mr. E. Martin; got by Catesby (37955),
dam (Beenham Lass) by Snowball (30020), &c. See Vol. xxii. p. 540.

1879, June 30, roan,	C.C.	Cassandra 2nd	Fred Vokes, 43248	Mr. Randall

Cassandra 2nd, sold to Mr. E. Martin.

SUNBEAM, roan, calved December 16, 1876. Bred by Mr. E. Randall, East
Down Lodge, the property of Mr. E. Martin ; got by Spectator (37508), dam
(Sunflower) by Brazenose (15686), &c. See "Sunrise," p. 244.

1879, July 22, roan,	B.C.	Sunrise	Fred Vokes, 43248	Mr. Randall

Sunrise, sold to Mr. E. Martin.

SUNSHINE, roan, calved October 7, 1875. Bred by Mr. E. Randall, East
Down Lodge, the property of Mr. E. Martin; got by Horsa (34188), dam
(Sunflower) by Brazenose (15686), &c.

1877, Nov. 9, roan,	C.C.	Sunshade	Spectator, 37508	Mr. Randall
1879, July 4, roan,	C.C.	Sunset	Fred Vokes, 43248	do.

Sunshade and Sunset, sold to Mr. E. Martin.

MARTIN, John,
Hawkshead Hall, *viâ* Ambleside.

FLOSSY GWYNNE 3rd, white, calved February 10, 1875, Vols. xxiv. and xxv.
pp. 564, 577. Bred by Sir W. H. Salt, Bart., Maplewell; got by Marquis of
York (34791), dam (Flossy Gwynne) by Twelfth Duke of York (19633), &c.

1879, Oct. 17, white,	B.C.	Gwynne Prince	Prince of L'tb'ne 2nd, 37258	Mr. Martin

LADY WATERLOO 26th, red and white, calved May 11, 1873, Vols. xxiii.
and xxiv. pp. 559, 564. Bred by Mr. E. H. Cheney, Gaddesby Hall; got by
Ninth Duke of Geneva (28391), dam (Lady Waterloo 21st) by General Napier
(24023), &c.

1879, Nov. 1, r. & w.,	C.C.	Lady Waterloo 28th	D. of Oxford 34th, 36529	Mr. Martin

MINSTREL 6TH, red and white, calved November 21, 1874, Vols. xxiv. and xxv. pp. 565, 578. Bred by Mr. W. Slater, Park Farm ; got by Sixth Duke of Kirklevington (30982), dam (Minstrel 5th) by Hematite 2nd (24129), &c.

Produce in	Names, &c.	By what Bull.	By whom bred.
1879, July 24, r. & w., B.C.	Minstrel Prince	D. of Oxford 34th, 36529	Mr. Martin

WILD EYES 29TH, red and white, calved April 2, 1868, Vols. xxii., xxiv., and xxv. pp. 547, 565, 578. Bred by Mr. J. Robinson, Fell Side ; got by Knight of the Harem (24278), dam (Wild Eyes 27th) by Gainford 5th (12913), &c.

1879, Jan. 27, r. & w., C.C.	Wild Eyes Belle 2nd	2nd D. of Glo'ster, 28392	Mr. Martin

MASON, John,
Dishforth, Thirsk.

MAY DEW, red and white, calved May 10, 1870, Vol. xxi. p. 846. Bred by Mr. J. Mason, the property of Mr. R. Earle, Catterick ; got by Master Frank (31865), dam (Mayflower 3rd) by Mussulman (34887), &c.

1876, May 27, r. & w., B.C.	Mayman	Clarion, 33393	Mr. Mason
1877, June 19, r. & w., C.C.	(dead)	Lord Mirth, 34602	do.
1878, Aug. 7, r. & w., C.C.	May Dew 3rd	do.	do.

Mayman, sold to Mr. Spence, Aldfield, Ripon.

MATTHEWS, A. T.,
Church Hanborough, Eynsham, Oxon.

EPITAPH 29TH, roan, calved June 23, 1877. Bred by the Exors. of Mr. W. Hewer, Sevenhampton ; got by Numa (34931), dam (Epitaph 26th) by Westrop (35969), g. d. (Epitaph 22nd) by Lord Lyon (24424), gr. g. d. (Epitaph 16th) by General Johnson (21815), &c. See " Evasion," Vol. xx. p. 120.

1879, Nov. 2, { roan, C.C.	Epitaph 30th	Duke Gwynne, 39706	Mr. Matthews
{ roan, B.C.	Epicure		

REGALIA, roan, calved June 5, 1875. Bred by Mr. W. Arkell, Dudgrove ; got by Pompey (35059), dam (Red Empress) by Duke of Wateringbury 2nd (26024), &c. See " Royal Clarence," p. 217.

1879, Mar. 15, roan, B.C.	Royal Clarence	Ld. Fitzclarence16th,36943	Mr. Matthews

RUBY, red and white, calved October 17, 1871, Vols. xxi., xxiii., and xxv. pp. 688, 452, 465. Bred by Mr. W. H. Dunn, Elcot Park ; got by Hercules (28843), dam (Romance) by Lord Lyon (24424), &c.

1879, Nov. 18, roan, C.C.	Ruth	Sir Robert Frogmore, 40719	Mr. Matthews

SILENCE 9TH, red, calved April 16, 1874, Vols. xxiv. and xxv. pp. 450, 463. Bred by Mr. G. Barton, Fundenhall Grange; got by Senator (32470), dam (Silence 5th) by Comte de Vermandois (19507), &c.

1879, Jan. 9, r. & w., B.C.	Silent Duke	G. D. of Geneva 2nd, 31288	Mr. Matthews
1879, Dec. 25, roan, C.C.	Princess Silence	Duke Gwynne, 39706	do.

MATTHEWS, F. C.,
Easterfield House, Driffield.

HAWTHORN, red, calved July 18, 1876. Bred by Mr. E. Robinson, Nafferton ; got by Halloo (34105), dam (Leonora) by Master Edmund (29322), &c. See Vol. xxiii. p. 623.

1879, Feb. 17, r. & w., C.C.	Ozone	Grand Duke, 48305	Mr. Matthews

LADY AGATHA, red and white, calved August 25, 1868. Bred by Sir Tatton
Sykes, Bart., Sledmere; got by Duke of Towneley (21615), dam (Selina)
by Photograph (20492), &c. See "Monument," p. 173.

Produce in		Names, &c.	By what Bull.	By whom bred.
1879, Jan. 28, roan,	B.C.	Monument	Ignoramus, 28887	Mr. Matthews

SELINA, roan, calved October 30, 1864. Bred by Lord Feversham, Duncombe
Park; got by Photograph (20492), dam (Soprano) by Vice Chancellor (17180),
&c.

1868, Aug. 25, r. & w.,	C.C.	Lady Agatha	Duke of Towneley, 21615	Sir Tatton Sykes

Lady Agatha, sold to Mr. F. C. Matthews.

MAXWELL, Sir W. S., Bart., Trustees of,
Keir Mains, Dunblane, N.B.

COLD CREAM 5TH, roan, calved in July, 1871, Vol. xxv. p. 581. Bred by
Mr. D. Fisher, Pitlochrie; got by Good Hope (31274), dam (Cold Cream 3rd)
by England's Glory (23889), &c.

1879, Jan. 20,	{ white, C.C. Cream Girl { roan, B.C. Cream Man	Fandango, 33879	{ Trustees of Sir W. { S. Maxwell

Cream Girl, sold to Colonel Murray, Polmaise, Stirling; Cream Man, to Mr. J. Stirling,
Kippendavie, Kippenross, Dunblane.

DAIRYMAID, roan, calved December 1, 1868, Vols. xxi., xxii., xxiii., and xxv.
pp. 847, 506, 562, 581. Bred by Sir W. S. Maxwell, Bart.; got by Keir
Butterfly 1st (24235), dam (Nurserymaid) by Young John O'Groat (16324),
&c.

1879, Feb. 28, roan,	C.C.	Dairy Keeper	Fandango, 33879	Ts. Sir W. Maxwell

YOUNG FEROOZA, roan, calved April 12, 1875. Bred by Sir W. S. Maxwell,
Bart.; got by Banner Bearer (27907), dam (Ferooza) by Knight Errant
(18154), &c. See Vol. xxiii. p. 562.

1878, June 24, white, B.C.	(dead)	Fandango, 33879	Ts. Sir W. Maxwell
1879, May 16, white, B.C.	Bearer	do.	do.

Bearer, sold to Mr. Thomson, Nyadd, Stirling.

FLOWER LASS, roan, calved May 6, 1870, Vols. xxi., xxii., and xxv. pp. 848,
507, 582. Bred by Sir W. S. Maxwell, Bart.; got by The Chieftain (20942),
dam (Flower Girl) by Baron Killerby (19280), &c.

1879, Feb. 6, white,	B.C.	White Florist	Fandango, 33879	Ts. Sir W. Maxwell

White Florist, sold to Mr. Allan, Kinnon Park, Perth.

FLOWER OF THE RHINE, roan, calved June 12, 1874, Vol. xxv. p. 582.
Bred by Mr. W. Torr, Aylesby Manor; got by Knight of the Shire (26552),
dam (Flower of Germany by Breast Plate (19337), &c.

1879, Jan. 6, white,	C.C.	Flower of the Allan	Baron Errant, 37788	Ts. Sir W. Maxwell

GLOSSY 4TH, roan, calved June 2, 1868, Vols. xx., xxi., xxii., and xxv. pp. 541,
848, 507, 582. Bred by Mr. W. Bolton, The Island; got by Duke of
Marlboro' (23768), dam (Miss Glossy) by Equinox (17810), &c.

1879, May 31, red,	B.C.	Varnisher	Fandango, 33879	Ts. Sir W. Maxwell

Varnisher, sold to Mr. J. Dunn, Newton Mains, Doune.

GLOSSY 5TH, red, calved May 5, 1874. Bred by Sir W. S. Maxwell, Bart.; got by Lieutenant-General (31600), dam (Glossy 4th) by Duke of Marlboro' (23768), &c.

Produce in	Names, &c.	By what Bull.	By whom bred.
1877, Jan. 21, r. & w., B.C.	Burnisher	Fandango, 33879	Sir W. S. Maxwell
1879, Jan. 13, r. & w., B.C.	Painter	do.	Ts. Sir W. Maxwell

Burnisher, sold to Mr. G. Duff-Dunbar, Ackergill Tower, Wick; Painter, to Mr. A. Mackie, Bandeath, Stirling.

HEBE 29TH, roan, calved November 10, 1870, Vols. xxiii. and xxv. pp. 562, 582. Bred by Messrs. Dudding, Panton House; got by Robin (24968), dam (Hebe 21st) by King John (14763), &c.

1879, Feb. 2, roan, B.C.	Roseberry	Fandango, 33879	Ts. Sir W. Maxwell

Roseberry, sold to Mr. W. Reid, Craigarnhill, Bridge of Allan.

HOPE OF THE GIPSIES, white, calved March 12, 1874, Vol. xxv. p. 582. Bred by Lord Kinnaird, Rossie Priory; got by Prince of the Gipsies (27179), dam (Hope of Baroness) by Great Hope (24082), &c.

1879, July 3, white, B.C.	Gipsy Lad	Baron Errant, 37788	Ts. Sir W. Maxwell

Gipsy Lad, sold to Mr. Fleming, Lower Ballaird, Balfron.

JESSIE GROAT 2ND, red, calved May 6, 1876. Bred by the Trustees of Mr. J. Dickson, Cambushinnie; got by Bywell (33261), dam (Amelia Groat) by Magenta (20253), g. d. (Jeannie Groat) by John O'Groat (13090), gr. g. d. (Alma) by Baron of Ravensworth (7811), &c. See Vol. xi. p. 313.

1879, Mar. 15, roan, B.C.	Young Bywell	Baron Errant, 37788	Ts. Sir W. Maxwell

Young Bywell, sold to Mr. C. H. Dundas, Gerrichue, Dunira, Crieff.

LADY LOUISA'S DUCHESS 3RD, roan, calved May 5, 1870, Vols. xx. and xxv. pp. 605, 582. Bred by Messrs. F. Leney and Sons, Wateringbury; got by Cambridge Duke 3rd (23503), dam (Lady Louisa's Duchess 1st) by Grand Duke 4th (19874), &c.

1879, June 18, roan, B.C.	Humber	Fandango, 33879	Ts. Sir W. Maxwell

Humber, sold to Mr. L. Drew, Merryton, Hamilton.

MOUNTAIN VIOLET, roan, calved April 28, 1874. Bred by Mr. A. Cruickshank, Sittyton; got by Ben Wyvis (30528), dam (Violet's Pride) by Scotland's Pride (25100), &c. See Vol. xx. p. 809.

1876, Nov. 19, white, C.C.	White Violet	Lord Lancaster, 26666	Sir W. S. Maxwell
1879, Feb. 16, roan, B.C.	Forthside	Fandango, 33879	Ts. Sir W. Maxwell

White Violet, sold to Messrs. J. T. S. Paterson, Plean Farm, Bannockburn; Forthside, to Mr. J. Balfour, Balbirnie, Markinch, Fife.

MEADE, W. R.,
Ballymartle, Ballinhassig, Ireland.

FAN 2ND, red, calved June 11, 1874. Bred by Mr. W. R. Meade; got by Royal Red (32410), dam (Fan) by Soubadar 2nd (25201), &c. See "Red Knight," Vol. xxi. p. 388.

1877, April 10, red, C.C.	Fan 3rd	Lord Percy, 34653	Mr. Meade
1878, April 2, red, B.C.	Chaka	Great Prince, 39964	do.
1879, April 14, red, C.C.	Fan 4th	do.	do.

POPPY, red, calved April 16, 1873.   Bred by Mr. W. R. Meade; got by Royal
Red (32410), dam (Peony) by Danny Man (21527), &c.   See "Osman,"
p. 179.

Produce in		Names, &c.	By what Bull.	By whom bred.
1876, June 8, red,	B.C.	Fitz Percy	Lord Percy, 34653	Mr. Meade
1877, April 30, red,	C.C.	Poppina	do.	do.
1878, Mar. 25, red,	B.C.	Cetewayo	Great Prince, 39964	do.
1879, Apr. 21, { red,	B.C.	Sultan	} do.	do.
{ red,	C.C.	(dead)		

PRINCESS, roan, calved January 13, 1869.   Bred by Mr. W. R. Meade; got
by Soubadar 2nd (25201), dam (Pink) by Danny Man (21527), g. d. (Polyanthus) by Australian (12414), &c.   See "Osman," p. 179.

1874, April 29, red,	C.C.	Princissessa	Royal Red, 32410	Mr. Meade
1875, April 17, roan,	C.C.	Princissessa 2nd	Archduke, 36131	do.
1876, Jan. 27, roan,	B.C.	(dead)	Lord Percy, 34653	do.
1877, Jan. 30, roan,	B.C.	Ignatieff	do.	do.
1878, Feb. 15, roan,	B.C.	Afghan	Great Prince, 39964	do.
1879, Feb. 22, roan,	C.C.	Princissessa 3rd	do.	do.

MEADE-WALDO, E. W.,
Stonewall Park, Edenbridge, Kent.

FLOWER BUD, roan, calved November 2, 1872, Vols. xxii. and xxiv. pp. 507,
569.   Bred by Mr. E. W. Meade-Waldo; got by Beadsman (27998), dam
(Flower Bloom) by Blinkhoolie (23428), &c.

1879, May 22, white,	C.C.	Fleur-de-lys	King Harry, 36841	Mr. Meade-Waldo

KATHERINE 5TH, roan, calved March 11, 1870, Vols. xxi., xxii., and xxv.
pp. 976, 508, 583.   Bred by Mr. G. Pell, Rothersthorpe; got by Prince Regent
(35169), dam (Katherine) by Royal Hamlet (18769), &c.

1879, June 17, white,	C.C.	Kitty	King Harry, 36841	Mr. Meade-Waldo

LADY OF THE MERE, roan, calved February 5, 1872, Vol. xxv. p. 583.
Bred by the Rev. T. Staniforth, Storrs; got by England's Glory (23889), dam
(Lady of the Manor) by British Crown (21322), &c.

1879, Oct. 7, r. & w.,	C.C.	Lady of the Burn	Baron Aylesby, 39397	Mr. Meade-Waldo

ROSALIND, roan, calved October 16, 1876.   Bred by Sir E. S. Hardinge, Bart.,
Fowler's Park; got by Sirius (39121), dam (Rosamond) by Royal Commander
(29857), &c.   See Vol. xxiii. p. 485.

1879, Oct. 8, roan,	C.C.	Rosy Light	Baron Aylesby, 39397	Mr. Meade-Waldo

MELROSE, P.,
West Loch, Eddlestone, N.B.

BEAUTY, roan, calved March 4, 1874.   Bred by Mr. J. Turnbull, Burnfoot;
got by Fitz Errant (28605), dam (Ruby) by Red Friar (24913), g. d. (Brilliant) by Majestic (22263), gr. g. d. (Pearl) by King Tom (18152), — (Diamond) by King Orrie (18147), — by Belted Will (6780), by Clansman, —
by Shedlaw (2615), — by Commodore (3447), — by Latham (364), — by
Simon (5133).

1876, June 6, roan,	C.C.	Teena	Dalzell, 33502	Mr. Melrose
1877, May 15, roan,	C.C.	Jessie	Viscount Thorndale, 37637	do.
1878, Mar. 10, roan,	C.C.	(slaughtered)	do.	do.
1879, April 15, red,	B.C.	King John	Knight of Harden, 40077	do.

JESSIE, roan, calved May 15, 1877.   Bred by Mr. P. Melrose; got by Viscount Thorndale (37637), dam (Beauty) by Fitz Errant (28605), &c.

Produce in		Names, &c.	By what Bull.	By whom bred.
1879, July 25, roan,	C.C.	Jane	Knight of Harden, 40077	Mr. Melrose

TEENA, roan, calved June 6, 1876.   Bred by Mr. P. Melrose; got by Dalzell (33502), dam (Beauty) by Fitz Errant (28605), &c.

1879, May 12, red,	B.C.	Prince Charles	Knight of Harden, 40077	Mr. Melrose

METCALFE, Anthony,
Park House, Ravenstonedale, Kirkby Stephen.

GAIETY, red, calved December 31, 1868, Vol. xx. p. 533.   Bred by Messrs. Angus, Bromley; got by Merry Monarch (22349), dam (Rachel) by Monarch (18412), &c.

1879, Oct. 14, r. & w.,	C.C.	Angus Gaiety	Royal Gift, 39040	Mr. Metcalfe

GOLDEN GIFT, red and white, calved October 8, 1873, Vols. xxiii., xxiv., and xxv. pp. 563, 570, 584.   Bred by Mr. A. Metcalfe; got by Peer of the Realm (27057), dam (Gift 8th) by King Richard (26523), &c.

1879, Oct. 15, r. & w.,	B.C.	Noble Gift	Earl of Sheffield, 33812	Mr. Metcalfe

Noble Gift, sold to Mr. J. Barker, Gunnerthwaite, Arkholme, Carnforth.

GOLDEN QUEEN, red and white, calved February 2, 1869, Vols. xx., xxi., xxiii., xxiv., and xxv. pp. 543, 973, 564, 570, 584.   Bred by Mr. W. Dobbyn, Abbey House; got by Asteroid (21193), dam (Golden Arch) by The Druid (18981), &c.

1879, April 17, roan,	B.C.	Golden Time	Royal Gift, 39040	Mr. Metcalfe

ROYAL ROSE, white, calved November 17, 1870, Vols. xxiv. and xxv. pp. 381, 584.   Bred by Messrs. Walcott and Campbell, New York Mills, U.S.A.; got by Royal Briton (27351), dam (White Rose) by Mountain Chief (20383), &c.

1879, July 4, roan,	C.C.	Miss Rose	Royal Gift, 39040	Mr. Metcalfe

SISTER CHERRY, roan, calved December 21, 1869, Vol. xxi. p. 672.   Bred by Sir A. de Rothschild, Bart., Aston Clinton; got by Sir Grace (27474), dam (Cherry Lass) by Officer (20432), &c.

1879, Jan. 10, roan,	C.C.	Harttorth Cherry	Prince Warlaby, 38954	Mr. Metcalfe
1879, Nov. 29, white,	C.C.	HartforthCherry 2nd	Royal Gift, 39040	do.

MICKLE, John,
Stoneshiel, Ayton, N.B.

ROSEBUD 10TH, roan, calved March 12, 1870, Vols. xxi., xxiv., and xxv. pp. 756, 487, 584.   Bred by Mr. J. Wilson, Cumledge; got by Red Knight (24916), dam (Rosebud 7th) by Snowflake (18888), &c.

1879, May 4, roan,	B.C.	Monomeath	Lord Thorndale, 37010	Mr. Mickle

MILES, Sir Philip, Bart.,
Leigh Court, Bristol.

BLUEBELL, red and white, calved January 19, 1876.   Bred by Sir W. Miles, Bart., Leigh Court; got by Proud Youth (32224), dam (Tulip) by Stockwood (27574), &c.   See Vol. xxiv. p. 574.

1879, May 10, red,	C.C.	(dead)	The Prior, 42493	Sir P. Miles, Bart.

CHERRY, roan, calved October 3, 1874. Bred by Sir W. Miles, Bart., Leigh
Court; got by Proud Youth (32224), dam (Crocus) by Lygon (21494), &c.
See " Shere Ali," p. 227.

Produce in	Names, &c.	By what Bull.	By whom bred.
1879, Jan. 17, roan, C.C.	(dead)	The Prior, 42493	Sir P. Miles, Bart.

COQUETTE 2ND, red and white, calved March 29, 1876. Bred by Sir W.
Miles, Bart., Leigh Court; got by Proud Youth (32224), dam (Pride) by
Hopeful (26410), &c. See " The Abbot," p. 247.

Produce in	Names, &c.	By what Bull.	By whom bred.
1879, Mar. 21, r. & w., B.C.	The Abbot	The Prior, 42493	Sir P. Miles, Bart.

DARING 10TH, roan, calved March 26, 1876. Bred by Sir W. Miles, Bart.,
Leigh Court; got by Proud Youth (32224), dam (Snowdrop) by Meteor
(29364), g. d. (Daring 8th) by Stockwood (27574), &c. See Vol. xx. p. 475.

Produce in	Names, &c.	By what Bull.	By whom bred.
1879, Mar. 20, white, C.C.	(dead)	The Prior, 42493	Sir P. Miles, Bart.

MERMAID, roan, calved January 14, 1876. Bred by Sir W. Miles, Bart.,
Leigh Court; got by Neptune (40380), dam (Careless) by Meteor (29364),
&c. See " Shere Ali," p. 227.

Produce in	Names, &c.	By what Bull.	By whom bred.
1879, Mar. 21, roan, C.C.	Constance 2nd	The Prior, 42493	Sir P. Miles, Bart.

PLEVNA, red and white, calved March 26, 1876. Bred by Sir W. Miles, Bart.,
Leigh Court; got by Proud Youth (32224), dam (Bravery) by Hopeful (26410),
&c. See Vol. xxiv. p. 571.

Produce in	Names, &c.	By what Bull.	By whom bred.
1879, Apr. 2, r. & w., C.C.	Isandula	The Prior, 42493	Sir P. Miles, Bart.

MILLER, T. Horrocks,
Singleton Park, Poulton le Fylde, Lancashire.

BRIDAL GIFT, white, calved September 3, 1872, Vol. xxiv. p. 589. Bred by
the Mytton Farming Company, Whalley; got by King Charles (24240), dam
(Bride of Windsor) by Roan Windsor (24967), &c.

1878, Jan. 20, roan, C.C.	Bride Elect	Shooting Star, 35519	Mytton Farm'g Co.
1879, May 28, roan, B.C.	Bridegroom Elect	Star Chieftain, 35672	Mr. Miller

BRIGHT BELLE, red, calved February 5, 1875, Vol. xxv. p. 585. Bred by
Mr. T. H. Miller; got by Royal Benedict (27348), dam (Bright Ringlet) by
Ringleader (15164), &c.

1879, Aug. 8, roan, B.C.	Bright Monarch	Fredk. the Great, 36665	Mr. Miller

FANNY GWYNNE, red, calved March 18, 1875, Vol. xxv. p. 587. Bred by
Mr. W. Bolton, The Island; got by King Charles (24240), dam (Flora
Gwynne) by Grey Gauntlet (19908) or Achilles (23257), &c.

1879, July 11, white, C.C.	Frederica Gwynne	Fredk. the Great, 36665	Mr. Miller

RINGLET 4TH, roan, calved August 4, 1872, Vols. xxii., xxiii., xxiv., and xxv.
pp. 511, 566, 574, 586. Bred by Mr. T. H. Miller; got by White Duke
(32849), dam (Ringlet 2nd) by Bywell Victor (21353), &c.

1879, July 22, r. & w., B.C.	Royal Farmer	Fredk. the Great, 36665	Mr. Miller

RINGLET 5TH, roan, calved October 27, 1873, Vols. xxiii. and xxiv. pp. 566,
574. Bred by Mr. T. H. Miller; got by Flag of Ireland (28613), dam
(Ringlet 2nd) by Bywell Victor (21353), &c.

1879, Jan. 23, white, C.C.	Ringlet 9th	Fredk. the Great, 36665	Mr. Miller

RINGLET 7TH, red and white, calved December 14, 1876. Bred by Mr. T. H. Miller; got by Braithwaite Booth (33192), dam (Ringlet 2nd) by Bywell Victor (21353), &c. See "Royal Farmer," p. 218.

Produce in		Names, &c.	By what Bull.	By whom bred.
1879, Nov. 20, roan,	C.C.	Ringlet 10th	King James, 28971	Mr. Miller

ROSE OF THE WYRE, roan, calved July 8, 1875. Bred by Mr. T. H. Miller; got by Flag of Ireland (28613), dam (Diamond Rose) by Sir David (25135), &c. See Vol. xxii. p. 511.

1879, Jan. 28, roan,	C.C.	Rose of the Ribble	Fredk. the Great, 36665	Mr. Miller

MITCHELL, A. and A.,
Alloa, Clackmannan, N.B.

BELLADONNA, red, calved February 24, 1875, Vols. xxiv. and xxv. pp. 576, 587. Bred by Messrs. A. and A. Mitchell; got by Foggathorpe 1st (31179), dam (Rover's Belle) by Red Rover (29754), &c.

1879, Aug. 16, roan,	C.C.	Bella	Brocklesby, 36288	Messrs. Mitchell

CHRISTIANA, roan, calved February 18, 1874, Vol. xxiv. p. 577. Bred by Messrs. A. and A. Mitchell; got by Foggathorpe 1st (31179), dam (Clotilda) by Knight Errant (18154), &c.

1879, April 11, roan,	C.C.	Constance	Brocklesby, 36288	Messrs. Mitchell

CORISANDE, red, calved July 27, 1869. Vols. xx., xxi., xxii., xxiii., xxiv., and xxv. pp. 459, 852, 512, 569, 577, 587. Bred by Mr. J. Wood, Stanwick Park; got by Lord Plymouth (24455), dam (Coral) by Valasco (15443), &c.

1879, June 24, roan,	B.C.	Lothair	Brocklesby, 36288	Messrs. Mitchell

DAINTY, roan, calved February 23, 1876, Vol. xxv. p. 587. Bred by Messrs. A. and A. Mitchell; got by Foggathorpe 1st (31179), dam (Dorothy) by Brother Windsor (25690), &c.

1879, July 8, roan,	B.C.	Dandy	Brocklesby, 36288	Messrs. Mitchell

DOROTHY, roan, calved in June 1869, Vols. xxii. and xxiii. pp. 512, 569. Bred by Mr. D. Fisher, Pitlochrie; got by Brother Windsor (25690), dam (Diamond) by Chieftain (20942), &c.

1879, Feb. 19, roan,	B.C.	Bencleuch	Brocklesby, 36288	Messrs. Mitchell

EMERALD, roan, calved May 5, 1874, Vol. xxiii. p. 569. Bred by Mr. J. Whyte, Clinterty; got by K. C. B. (26492), dam (Ruby) by Prince George (13510), &c.

1878, April 11, roan,	B.C.	Lincoln	Brocklesby, 36288	Messrs. Mitchell
1879, Aug. 30, white,	C.C.	Opal	do.	do.

Lincoln, sold to Sir R. Anstruther, Bart., Balcaskie, Fifeshire.

PRIMROSE, roan, calved January 20, 1873, Vol. xxiii. p. 569. Bred by Mr. J. C. Toppin, Musgrave Hall; got by Knight of Killerby (29000), dam (Princess) by Count of the Realm (23640), &c.

1878, Feb. 4, roan,	B.C.	Bright Baron	Brocklesby, 36288	Messrs. Mitchell
1879, Mar. 15, roan,	B.C.	Benledi	do.	do.

ROAN MOSS ROSE, roan, calved March 19, 1874. Bred by Messrs. A. and A. Mitchell; got by Foggathorpe 1st (31179), dam (Moss Rose) by Colin (25795), &c. See Vol. xxi. p. 852.

1877, Nov. 17, white,	B.C.	(dead)	Brocklesby, 36288	Messrs. Mitchell
1879, July 10, white,	C.C.	Rose	do.	do.

MITCHELL, James,
Howgill Castle, Penrith.

DAINTY 6TH, red, calved March 14, 1877. Bred by Mr. J. Mitchell; got by Kirkbythore Wild Eyes 3rd (36864), dam (Little Dainty) by Heart of Gold (26355), &c. See Vol. xxii. p. 513.

Produce in	Names, &c.	By what Bull.	By whom bred.
1879, Sept. 22, roan, C.C.	Dainty 11th	Crossfell 16th, 39654	Mr. Mitchell

DUCHESS URSULA 2ND, roan, calved August 23, 1875. Bred by Mrs. Fawcett, Scaleby Castle; got by Eighth Duke of York (28480), dam (Ursula 27th) by Grand Duke 13th (21850), &c. See Vol. xxi. p. 702.

Produce in	Names, &c.	By what Bull.	By whom bred.
1879, April 6, white, C.C.	Duchess Ursula 3rd	Crossfell 7th, 39653	Mr. Mitchell

GRANGE 2ND, roan, calved May 20, 1877. Bred by Mr. J. Mitchell; got by Kirkbythore Wild Eyes 3rd (36864), dam (Grange) by Trevelyan d'Eden (35813), &c. See Vol. xxii. p. 578.

Produce in	Names, &c.	By what Bull.	By whom bred.
1879, Sept. 15, white, C.C.	Grange 3rd	Crossfell 16th, 39654	Mr. Mitchell

MARGERY 4TH, red, calved June 2, 1877. Bred by Mr. J. Mitchell; got by Kirkbythore Wild Eyes 3rd (36864), dam (Margery) by Heart of Gold (26355), &c. See "Crossfell 48th," p. 61.

Produce in	Names, &c.	By what Bull.	By whom bred.
1879, Sept. 15, roan, B.C.	Crossfell 48th	Crossfell 16th, 39654	Mr. Mitchell

ROACHY 2ND, roan, calved May 15, 1877. Bred by Mr. J. Mitchell; got by Kirkbythore Wild Eyes 3rd (36864), dam (Young Roachy) by Trevelyan d'Eden (35813), &c. See "Crossfell 50th," p. 61.

Produce in	Names, &c.	By what Bull.	By whom bred.
1879, Oct. 22, white, B.C.	Crossfell 50th	Crossfell 16th, 39654	Mr. Mitchell

ROSE OF HACKTHORPE, red, calved February 21, 1877. Bred by Mr. J. Mitchell; got by Kirkbythore Wild Eyes 3rd (36864), dam (Princess) by Prince of Wales (29665), &c. See Vol. xxiv. p. 578.

Produce in	Names, &c.	By what Bull.	By whom bred.
1879, Dec. 5, roan, C.C.	(dead)	Crossfell 16th, 39654	Mr. Mitchell

ROYAL ROSE, white, calved June 16, 1877. Bred by Messrs. T. and J. H. Bird, Halefield; got by King of Scotland (34332), dam (Christmas Herald) by Monarch (31930), &c. See "Crossfell 51st," p. 61.

Produce in	Names, &c.	By what Bull.	By whom bred.
1879, Dec. 8, white, B.C.	Crossfell 51st	Crossfell 17th, 39655	Mr. Mitchell

TEATY 2ND, roan, calved May 29, 1877. Bred by Mr. J. Mitchell; got by Kirkbythore Wild Eyes 3rd (36864), dam (Teaty) by Trevelyan d'Eden (35813), &c. See "Crossfell 49th," p. 61.

Produce in	Names, &c.	By what Bull.	By whom bred.
1879, Sept. 28. white, B.C.	Crossfell 49th	Crossfell 16th, 39654	Mr. Mitchell

MITCHELL, W.,
Cleasby, Darlington.

LADY BLOOM, white, calved December 28, 1874, Vol. xxiv. p. 579. Bred by Mr. W. Mitchell; got by Squire Booth (30049), dam (Ladyship) by Major (26790), &c.

Produce in	Names, &c.	By what Bull.	By whom bred.
1879, June 10, roan, B.C.	Royal Manfred	King Manfred, 38491	Mr. Mitchell

LADY BOOTH 4TH, roan, calved July 25, 1876. Bred by Mr. W. Mitchell;
got by Manfred Booth (34749), dam (Lady Booth 2nd) by Major (26790), &c.
See Vol. xxiii. p. 570. .

Produce in	Names, &c.	By what Bull.	By whom bred.
1879, June 19, roan, C.C.	Lady Booth 6th	King Harold, 40053	Mr. Mitchell

LADY JAMES, roan, calved August 12, 1873, Vol. xxv. p. 589. Bred by Mr.
F. Heugh, Broomfield House; got by King James (28971), dam (Strawberry) .
by The Sutler (23061), &c.

1879, April 6, r. & w., C.C.	Lady Harold	King Harold, 40053	Mr. Mitchell

LADYSHIP, roan, calved May 24, 1870, Vols. xx., xxi., xxiii., and xxv. pp. 614,
853, 570, 589. Bred by Mr. W. Mitchell; got by Major (26790), dam (Lady
Ann) by Sir Christopher (22895), &c.

1879, Mar. 23, roan, B.C.	Lord Manfred	King Manfred, 38491	Mr. Mitchell

QUEEN OF THE PALACE, roan, calved January 19, 1873, Vols. xxii., xxiii.,
and xxiv. pp. 513, 570, 579. Bred by Mr. W. Mitchell; got by Squire Booth
(30049), dam (Queen of the Day) by Sir Christopher (22895), &c.

1879, Mar. 23, roan, B.C.	Lord Harold	King Harold, 40053	Mr. Mitchell

QUEEN OF THE ROSES, white, calved May 9, 1874, Vols. xxiii., xxiv., and
xxv. pp. 571, 579, 590. Bred by Mr. W. Mitchell; got by Squire Booth
(30049), dam (Queen of the Day) by Sir Christopher (22895), &c.

1879, April 23, white, B.C.	Knight Harold	King Harold, 40053	Mr. Mitchell

MITCHELL, W. A.,
Auchnagathle, Whitehouse, N.B.

ALMA, red and white, calved March 21, 1872, Vols. xxiii. and xxiv. pp. 571,
579. Bred by Mr. A. Cruickshank, Sittyton; got by Prince Alfred (27107),
dam (Adeline) by Cæsar Augustus (25704), &c.

1879, Feb. 9, red, C.C.	Amethyst	D. of Chamburgh, 36052	Mr. Mitchell

CARCULAS, roan, calved May 1, 1869, Vols. xxiii., xxiv., and xxv. pp. 571,
580, 590. Bred by Mr. W. Benton, Harthill; got by Lord Lincoln (34538),
dam (Williamina) by Old England (24681), &c.

1879, Dec. 15, roan, B.C.	Cardinal	D. of Chamburgh, 36052	Mr. Mitchell

HAWTHORN, roan, calved February 1, 1871, Vols. xxi. and xxv. pp. 853, 591.
Bred by Mr. W. A. Mitchell; got by Forth 4th (28636), dam (Lady Forbes,
by Loyalty (26768), &c.

1879, Mar. 29, r. & w., C.C.	Onyx	D. of Chamburgh, 36052	Mr. Mitchell

JUNO, red, calved in April, 1868, Vol. xxv. p. 591. Bred by Mr. W. A.
Mitchell; got by Loyalty (26768), dam (Rosy) by Bertram (17408), &c.

1879, June 21, roan, C.C.	Jasper	D. of Chamburgh, 36052	Mr. Mitchell

LADY JANE, roan, calved May 29, 1869, Vols. xxi., xxiv., and xxv. pp. 570,
581, 591. Bred by Mr. W. Benton, Harthill; got by Lord Lincoln (34583),
dam (Bessie) by Old England (24681), &c.

1879, Mar. 25, r. & w., B.C.	Lord James	D. of Chamburgh, 36052	Mr. Mitchell

Lord James, sold to Colonel A. S. L. Hay, Leith Hall, Kennethmore, N.B.

LADY LINCOLN, white, calved April 14, 1874, Vols. xxiv. and xxv. pp. 581, 591. Bred by Mr. W. Benton, Harthill; got by Shuttlecock (35520), dam (Lady Jane) by Lord Lincoln (34583), &c.

Produce in	Names, &c.	By what Bull.	By whom bred.
1879, Dec. 19, roan,	C.C.|Lady Lyle	D. of Chamburgh, 36052	Mr. Mitchell

MOON, J. Stocks,
Starborough Castle, Edenbridge, Kent.

LADY CRESSIDA, roan, calved November 2, 1873. Bred by Messrs. Dudding, Panton House; got by Buckingham (33246), dam (Carnation) by Wellesley (25421), &c. See Vol. xxi. p. 684.

1879, Oct. 9, red,	C.C.|Annie 1st	Bromley, 36289	Mr. Moon

MORETON, Lord,.
Tortworth Court, Falfield, Gloucestershire.

GLADYS 2ND, red and white, calved August 9, 1873, Vol. xxiii. p. 638. Bred by Mr. H. J. Sheldon, Brailes House; got by Duke of Cerisia (30937), dam (Gladys) by Duke of Brailes (23724), &c.

1879, May 7, red,	C.C.|Griselda	Lally's Hillhurst D.,38538	Lord Moreton

JAVA, red and white, calved March 18, 1876. Bred by Lord Moreton; got by Cherry Grand Duke 5th (30712), dam (Japonica) by Lodowick (20136), &c. See "Jupiter," Vol. xxv. p. 144.

1879, Mar. 13, r. & w.,	B.C.|(Steer)	Jupiter, 41737	Lord Moreton

JURA, red, calved February 24, 1877. Bred by Lord Moreton; got by Coronation (30796), dam (Jessica) by Lodowick (20136), &c. See "Coronet 2nd," Vol. xxiv. p. 52.

1879, Oct. 6, roan,	B.C.|Jeweller	L.T'croft Oxford 2d.38668	Lord Moreton

KIRKLEVINGTON DUCHESS 9TH, red, calved December 21, 1871, Vols. xxi., xxii., xxiii., and xxv. pp. 804, 477, 529, 593. Bred by Mr. R. P. Davies, Horton; got by Grand Duke of Clarence (28750), dam (Duchess of Kent) by Lord Liverpool (22168), &c.

1879, Nov. 24, roan,	C.C.'M'ss. of K'lev'ton 2d|20th Duke of Oxford,28432	Lord Moreton

LADY LOUISA'S DUCHESS 6TH, roan, calved February 11, 1875, Vol. xxv. p. 594. Bred by Messrs. F. Leney and Sons, Wateringbury; got by Sixth Duke of Oneida (30997), dam (Lady Louisa's Duchess 4th) by Grand Duke 15th (21852), &c.

1879, Mar. 15 red,	C.C.|Lavender	Oxford's Prince, 34998	Lord Moreton

PRINCESS ALEXANDRA 2ND, red, calved January 26, 1875, Vol. xxv. p. 594. Bred by Mr. E. H. Cheney, Gaddesby Hall; got by Ninth Duke of Geneva (28391), dam (Princess Alexandra) by Eighth Duke of Oxford (15939), &c.

1879, July 29, red,	C.C.|Furbelow Duchess 2d|Duke of Connaught, 33604	Lord Moreton

SIDDINGTON 10TH, roan, calved March 22, 1871, Vol. xxi. p. 674. Bred by Mr. E. Bowly, Siddington House; got by Second Duke of Tregunter (26022), dam (Siddington 2nd) by Fourth Duke of Oxford (11387), &c.

1879, June 6, roan,	B.C.|E. of Siddington 2nd|Oxford's Prince, 34998	Lord Moreton

SIDDINGTON 15TH, white, calved August 9, 1874, Vol. xxv. p. 594. Bred by Mr. E. Bowly, Siddington House; got by Third Duke of Clarence (23727), dam (Siddington 11th) by Second Duke of Tregunter (26022), &c.

Produce in	Names, &c.	By what Bull.	By whom bred.
1879, Oct. 7, red, B.C.	E. of Siddington 3rd	Duke of Hillhurst, 28401	Lord Moreton

TUBE ROSE OF BRATTLEBORO 4TH, red, calved September 3, 1873. Bred by Mr. W. Angerstein, Weeting Hall; got by Earl of Grasshill (36584), dam (Tube Rose of Brattleboro 3rd) by Sheridan (37442), g. d. (Tube Rose 7th) by Tornado (42510), gr. g. d. (Tube Rose 3rd) by Third Duke of Cambridge (5941), &c. See " Baron Morley 2nd," Vol. xxv. p. 19.

1879, April 16, r. & w., B.C.	Pr. of Tortworth 2d	Oxford's Prince, 34998	Lord Moreton

MORTIMER, W. B.,
Hay Carr House, Lancaster.

RED ROSE, red, calved March 16, 1876. Bred by Mr. W. B. Mortimer; got by Sir Richard (42407), dam (Sweet Rose) by Grand Count 2nd (24059), &c. See Vol. xxv. p. 594.

1878, Mar. 13, red, B.C.	Red Duke	Earl Blanche, 36566	Mr. Mortimer
1879, April 12, red, B.C.	Prince Regent	L.A.Wild Eyes 2nd,36912	do.

Red Duke, sold to Mr. Whinneray, Nateby Hall, Garstang; Prince Regent, to Mr. J. Shorrock, Warleys, Kirkham.

SUNFLOWER 4TH, roan, calved March 10, 1876. Bred by Mr. W. B. Mortimer; got by Cambridge Barrington (30642), dam (Sunflower 3rd) by Lord Chancellor (26622), g. d. (Sunflower) by General Fairfax (14594), &c. See Vol. xv. p. 743.

1879, Jan. 30, roan, B.C.	Lord of Crimbles	L. A. Wild Eyes 2nd,36912	Mr. Mortimer

Lord of Crimbles, sold to Mr. R. Mason, Crimbles, Cockerham.

SWEET ROSE 2ND, red and white, calved March 11, 1873, Vol. xxv. p. 594. Bred by Mr. T. Lamb, Hay Carr House; got by Golden Duke (26266), dam (Sweet Rose) by Grand Count 2nd (24059), &c.

1879, Nov. 22, r. & w., C.C.	Rose of Ashton 2nd	Lord of the Lakes, 40219	Mr. Mortimer

WILD ROSE 2ND, roan, calved January 10, 1873, Vol. xxv. p. 595. Bred by Mr. I. Smith, Halton; got by Golden Duke (26266), dam (Wild Rose) by Majestic (20264), &c.

1879, April 14, roan, C.C.	(dead)	L. A. Wild Eyes 2nd,36912	Mr. Mortimer

MORTON, John,
Fences Farm, Stow, Downham Market.

ANNIE LISLE, roan, calved November 20, 1876. Bred by Mr. J. Morton; got by Broughton Royal (33236), dam (Annie Wyndham) by General Wyndham (31240), &c. See Vol. xxiv. p. 583.

1879, Jan. 21, roan, C.C.	Annie Lizzie	Hogarth 3rd, 40003	Mr. Morton
1879, Dec. 13, r. & w., C.C.	Annie	Hesperus, 39994	do.

BROUGHTON SPOT, red, calved November 21, 1876. Bred by Mr. J. Morton; got by Broughton Royal (33236), dam (Prim Spot) by General Prim (34017), g. d. (Spot) by Grand Prince (16187), gr. g. d. (Markham Favourite) by Lord of Lindsay (13210), — by Sharp's Bull.

1879, April 19, roan, B.C.	Bardolph	Hesperus, 39994	Mr. Morton

HAPPY NEW YEAR, red, calved January 1, 1875. Bred by Mr. J. Morton;
got by Broughton Royal (33236), dam (Honeysuckle) by Marquis of Corn-
wallis (18337), &c. See Vol. xxii. p. 515.

Produce in		Names, &c.	By what Bull.	By whom bred.
1877, Aug. 21, roan,	B.C.	Happy Prince	Prince Arthur, 37218	Mr. Morton
1879, Nov. 11, roan,	C.C.	Hesperier	Hesperus, 39994	do.

Happy Prince, sold to Mr. W. W. Hopkin, Home Farm, Stow.

LANCASTER 59TH, roan, calved September 3, 1872. Bred by Mr. E. Hall,
Shallcross Hall; got by Lancaster Hero (29019), dam (Lancaster 27th) by
Lancaster Comet (22072), &c. See Vol. xx. p. 620.

1876, Nov. 4, red,	C.C.	Lancaster Louise	Duke of Lorn, 25985	Mr. Morton
1879, Dec. 23, r.&w.,	C.C.	Lancaster Lady	Hogarth 3rd, 40003	do.

MORTON, J., Exors. of,
Skelsmergh Hall, Kendal.

FAIR FANNY 3RD, roan, calved March 28, 1874. Bred by Mr. J. Morton,
Skelsmergh Hall; got by Oxford's Victor (35001), dam (Fair Fanny) by
Wallace (23166), &c. See Vol. xxi. p. 859.

1878, Nov. 3, roan,	C.C.	(dead)	D.of Barrington 2nd,36463	Exrs.of Mr.Morton
1879, Dec. 9, roan,	B.C.	Duke of Skelsmergh	do.	do.

HEBE, roan, calved April 25, 1876. Bred by Mr. J. Morton, Skelsmergh Hall;
got by Oxford's Victor (35001), dam (Henrietta) by Freedom (17884), &c.
See Vol. xviii. p. 526.

1879, Sept. 10, r. & w.,C.C.	Holly	D.of Barrington 2nd,36463	Exrs.of Mr.Morton

LADY CLARE, white, calved March 20, 1876. Bred by Mr. J. Morton,
Skelsmergh Hall; got by Oxford's Victor (35001), dam (Lady Oxford) by
Atherton's Oxford (21195), &c. See Vol. xix. p. 590.

1879, Sept. 15, roan, C.C.	Lady Clara	D.of Barrington 2nd,36463	Exrs.of Mr.Morton

LADY VICTOR, roan, calved November 19, 1873. Bred by Mr. J. Morton,
Skelsmergh Hall; got by Oxford's Victor (35001), dam (Lady Oxford) by
Atherton's Oxford (21195), &c.

1877, Nov. 20, white,	B.C.	Vandyke	D.of Barrington 2nd,36463	Mr. Morton
1879, May 3, roan,	C.C.	Lady Violet	do.	Exrs.of Mr.Morton

Vandyke, sold to Mr. J. Rigg, Kit Cragg, Selside, Kendal.

OXFORD'S BEAUTY, red and white, calved November 19, 1875. Bred by
Mr. J. Morton, Skelsmergh Hall; got by Oxford's Victor (35001), dam
(Oxford's Belle) by Atherton's Oxford (21195), &c. See Vol. xxii. p. 515.

1879, Nov. 6, r.&w., C.C.	Olivia	D.of Barrington 2nd,36463	Exrs.of Mr.Morton

OXFORD'S FAIRY, roan, calved July 29, 1874. Bred by Mr. J. Morton,
Skelsmergh Hall; got by Oxford's Victor (35001), dam (Oxford's Faith) by
Atherton's Oxford (21195), &c. See Vol. xxi. p. 860.

1879, May 9, roan, C.C.	Fragrance	D.of Barrington 2nd,36463	Exrs.of Mr.Morton

PRINCESS 4TH, roan, calved March 20, 1875. Bred by Mr. J. Morton, Skelsmergh Hall; got by Oxford's Victor (35001), dam (Oxford's Princess) by
Atherton's Oxford (21195), &c. See "Prince Imperial," p. 194.

1878, Oct. 2, r.&w.,	B.C.	Prince Imperial	D.of Barrington 2nd,36463	Exrs.of Mr.Morton
1879, Dec. 23, roan,	B.C.	(Steer)	do.	do.

Prince Imperial, sold to Mr. Leach, Northsceugh.

RINGLET 4th, roan, calved September 6, 1875. Bred by Mr. J. Morton, Skelsmergh Hall; got by Oxford's Victor (35001), dam (Ringlet) by Atherton's Oxford (21195), g. d. (Robina) by Coxcomb (19527), &c. See "Robin Adair," p. 212.

Produce in	Names, &c.	By what Bull.	By whom bred.
1879, Aug. 5, r.&w., B.C.	(Steer)	D.of Barrington 2nd,36463	Exrs.of Mr.Morton

ROAN OXFORD, roan, calved July 24, 1875. Bred by Mr. J. Morton, Skelsmergh Hall; got by Oxford's Heir (35001), dam (Red Oxford) by Atherton's Oxford (21195), &c. See Vol. xxi. p. 860.

1879, Nov. 6, roan, B.C.	Lord Ronald	D.of Barrington 2nd,36463	Exrs.of Mr.Morton

ROSE OF SKELSMERGH, roan, calved May 10, 1875. Bred by Mr. J. Morton, Skelsmergh Hall; got by Oxford's Victor (35001), dam (Rose of Oxford) by Atherton's Oxford (21195), &c. See Vol. xxi. p. 861.

1879, Apr. 18, { r.&w.,B.C.(Steer) / r.&w.,B.C.(Steer) }		D.of Barrington 2nd,36463	Exrs.of Mr.Morton

STRADELLA, roan, calved January 9, 1876. Bred by Mr. J. Morton, Skelsmergh Hall; got by Oxford's Victor (35001), dam (Stella) by Atherton's Oxford (21195), &c. See Vol. xxi. p. 861.

1879, Sept. 24, roan, C.C.	Sunbeam	D.of Barrington 2nd,36463	Exrs.of Mr.Morton

VICTOR'S ROSE, white, calved November 8, 1873. Bred by Mr. J. Morton, Skelsmergh Hall; got by Oxford's Victor (35001), dam (Rose of Oxford) by Atherton's Oxford (21195), &c. See Vol. xxi. p. 861.

1877, Nov. 24, roan, B.C.	Royal Barrington	D.of Barrington 2nd,36463	Mr. Morton
1879, April 28, white, C.C.	Rosebud	do.	Exrs.of Mr.Morton

Royal Barrington, sold to Mr. Richardson, Thrimby, Penrith.

MORTON, R. C.,
Lane House, Burton, Westmoreland.

DOWAGER DUCHESS 4th, red and white, calved May 5, 1873. Vol. xxiv. p. 555. Bred by Lord Penrhyn, Wicken Park; got by Grand Duke 11th (21849), dam (Dowager) by Duke of Geneva (19614), &c.

1879, Sept. 2, r.&w., B.C.	Defender	Duke of Havering, 33664	Mr. Morton

DOWAGER DUCHESS 5th, red and white, calved March 4, 1877. Bred by Mr. D. McIntosh, Havering Park; got by Third Duke of Geneva (23753), dam (Dowager Duchess 4th) by Grand Duke 11th (21849), &c.

1879, July 14, r.&w., B.C.	Duke of Farleton	Duke of Havering, 33664	Mr. Morton

MONICA, roan, calved April 12, 1877. Bred by Mr. R. C. Morton; got by Baron Barrington 4th (33006), dam (Royal Rose) by Royal George (32387), &c. See "St. Augustin," p. 224.

1879, Dec. 16, white, B.C.	St. Augustin	Duke of Holker, 38153	Mr. Morton

QUICKLY 4th, roan, calved April 12, 1876, Vol. xxv. p. 596. Bred by the Rev. J. Storer, Hellidon; got by Young Knightley (31529), dam (Wine) by Wolfsbane (15518), &c.

1879, Oct. 30, white, B.C.	Master Quickly 2nd	B. Barrington 4th, 33006	Mr. Morton

WALNUT 6TH, roan, calved January 25, 1877. Bred by Mr. D. McIntosh, Havering Park ; got by Duke of Havering (33664), dam (Walnut) by Third Duke of Geneva (23753), &c. See " Warrior," p. 262.

Produce in		Names, &c.	By what Bull.	By whom bred.
1879, Aug. 30, roan,	B.C.	Warrior	Viscount Oxford, 40876	Mr. Morton

MOSS, F. J.,
Stainfield Hall, Wragby.

CAMPFOLLOWER'S DAUGHTER, roan, calved January 11, 1872, Vol. xxii. p. 403. Bred by Messrs. Dudding, Panton House ; got by Manfred (26801), dam (Campfollower) by Friar Tuck (17892), &c.

Produce in		Names, &c.	By what Bull.	By whom bred.
1877, July 7, roan,	C.C.	Campfollower'sBride	M. C., 31898	Mr. Moss
1879, Feb. 9, roan,	C.C.	Campfollower's Gem	General Wharfdale, 31237	do.

GIPSY QUEEN, red, calved February 22, 1874. Bred by Mr. R. Moss, Whisby ; got by Dunbar (31045), dam (Daisy Queen) by Chieftain (21421), &c. See " Fairy King," p. 97.

Produce in		Names, &c.	By what Bull.	By whom bred.
1877, June 1, roan	C.C.	May Queen	Ruby King, 37407	Mr. F. J. Moss
1879, Feb. 26, roan,	C.C.	Gipsy Wharfdale	Gen. Wharfdale, 31237	do.

MOUTRAY, Rev. J. J.,
Favour Royal, Aughnacloy, Co. Tyrone.

BLOSSOM, roan, calved April 7, 1869, Vol. xxi. p. 862. Bred by the Rev. W. Moutray, Favour Royal ; got by Knight of the Grand Cross (31558), dam (Violet) by Boreas (30566), &c.

Produce in		Names, &c.	By what Bull.	By whom bred.
1876, Aug. 12, white,	B.C.	Baker Pasha	Prince of the Woods, 32185	Rev. J. J. Moutray
1877, Nov. 2, white,	C.C.	Snowdrop	Gladiator, 36698	do.

Baker Pasha, sold to Miss Rose, Mullaghmore, Monaghan, Co. Monaghan.

CHEMISETTE, roan, calved June 15, 1872, Vol. xxii. p. 516. Bred by the Rev. J. J. Moutray, the property of Mr. J. Madden, Roslea Manor ; got by Lord Wodehouse (29224), dam (Chaumontel) by Agamemnon (23278), &c.

Produce in		Names, &c.	By what Bull.	By whom bred.
1876, Mar. 14, roan,	C.C.	Chenille	Prince of the Woods, 32185	Rev. J. J. Moutray
1877, June 5, r. & w.,	C.C.	Cleopatra	Gladiator, 36698	do.
1878, Aug. 19, white,	C.C.	Carnation	do.	do.

Chenille, sold to Mr. A. M. Millar, Findermore, Clogher, Co. Tyrone.

GOVERNESS, roan, calved January 27, 1870, Vol. xxi. p. 863. Bred by the Rev. W. Moutray, Favour Royal, the property of Lord Clermont, Clermont Park ; got by The Governor (32681), dam (Norma) by Knight of the Grand Cross (31558), &c.

Produce in		Names, &c.	By what Bull.	By whom bred.
1876, Sept. 13, roan,	B.C.	Osman	Prince of the Woods, 32185	Rev. J. J. Moutray
1877, Aug. 4, red,	B.C.	The Lawyer	Gladiator, 36698	do.
1878, July 25. roan,	C.C.	Godolphine	do.	do.

Osman, sold to the Duke of Abercorn, Barouscourt, Omagh, Co. Tyrone ; The Lawyer, to Mr. A. M. Millar, Findermore, Clogher, Co. Tyrone.

GOVERNESS 2ND, white, calved September 9, 1874, Vol. xxiv. p. 585. Bred by the Rev. J. J. Moutray ; got by Prince of the Woods (32185), dam (Governess) by The Governor (32681), &c.

Produce in		Names, &c.	By what Bull.	By whom bred.
1878, Aug. 10, white,	B.C.	Knight Templar	Gladiator, 36698	Rev. J. J. Moutray

NORMA 2ND, red, calved July 16, 1874. Bred by the Rev. J. J. Moutray; got by Count Robert (30812), dam (Norma) by Knight of the Grand Cross (31558), &c. See "Prince Humbert," p. 194.

Produce in	Names, &c.	By what Bull.	By whom bred.
1878, June 10, r. & w.,B.C.	Brilliant	Gladiator, 36698	Rev. J.J.Moutray
1879, July 17, roan, C.C.	Evelyn	do.	do.

Brilliant, sold to Mr. A. Moutray, Killybrick, Favour Royal, Aughnacloy, Co. Tyrone.

NORMA 3RD, red and white, calved July 25, 1875. Bred by the Rev. J. J. Moutray; got by Prince of the Woods (32185), dam (Norma) by Knight of the Grand Cross (31558), &c.

Produce in	Names, &c.	By what Bull.	By whom bred.
1878, May 21, r. & w.,C.C.	Norah	Gladiator, 36698	Rev. J.J.Moutray
1879, May 12, roan, C.C.	Belladonna	do.	do.

Norah, sold to Mr. B. Dickson, Gilford House, Gilford, Co. Down.

MOWBRAY and STOURTON, Lord,
Stourton, Knaresborough, Yorkshire.

BESSIE 11TH, roan, calved February 20, 1869, Vols. xxii. and xxiv. pp. 578, 586. Bred by Lord Mowbray and Stourton; got by Golden Horn (31267), dam (Bessie 10th) by Rose Duke (22760), &c.

Produce in	Names, &c.	By what Bull.	By whom bred.
1878, Dec. 24, roan, C.C.	Bessie 16th	L. Oxf. Bright Eyes,34648	Lord Mowbray

JUNO 15TH, red, calved January 20, 1869, Vols. xxi., xxii., and xxiv. pp. 943, 579, 586. Bred by Lord Mowbray and Stourton; got by Golden Horn (31267), dam (Juno 12th) by Rose Duke (22760), &c.

Produce in	Names, &c.	By what Bull.	By whom bred.
1878, Mar. 20, roan, B.C.	Rose Duke 8th	L. Oxf. Bright Eyes,34648	Lord Mowbray
1879, April 5, roan, C.C.	Juno 30th	do.	do.

Rose Duke 8th, sold to Mr. Jacques, Staddlethorpe, Malton.

JUNO 18TH, roan, calved March 4, 1873, Vol. xxiv. p. 586. Bred by Lord Mowbray and Stourton; got by Sprightly Lord (30044), dam (Juno 12th) by Rose Duke (22760), &c.

Produce in	Names, &c.	By what Bull.	By whom bred.
1878, Aug. 12, r. & w.,C.C.	Juno 28th	L. Oxf. Bright Eyes,34648	Lord Mowbray
1879, Sept. 1, roan, C.C.	Juno 31st	do.	do.

JUNO 19TH, roan, calved March 23, 1873, Vol. xxiv. p. 587. Bred by Lord Mowbray and Stourton; got by Sprightly Lord (30044), dam (Juno 15th) by Golden Horn (31267), &c.

Produce in	Names, &c.	By what Bull.	By whom bred.
1878, Mar. 23, red, B.C.	(Steer)	L. Oxf. Bright Eyes,34648	Lord Mowbray
1879, July 1, white, B.C.		do.	do.

JUNO 21ST, red and white, calved April 8, 1875. Bred by Lord Mowbray and Stourton; got by Sandown (35470), dam (Juno 16th) by Golden Horn (31267), &c. See Vol. xxii. p. 579.

Produce in	Names, &c.	By what Bull.	By whom bred.
1878, July 6, roan, B.C.	Rose Duke 9th	L. Oxf. Bright Eyes,34648	Lord Mowbray
1879, May 4, r. & w., B.C.		do.	do.

Rose Duke 9th, sold to Mr. J. Key, Musley Bank, Malton.

JUNO 22ND, roan, calved March 26, 1875. Bred by Lord Mowbray and Stourton; got by Sandown (35470), dam (Juno 12th) by Rose Duke (22760), &c. See Vol. xxiv. p. 586.

Produce in	Names, &c.	By what Bull.	By whom bred.
1878, April 12, red, B.C.		L. Oxf. Bright Eyes,34648	Lord Mowbray
1879, Mar. 21, red, B.C.	(Steer)	do.	do.

JUNO 23RD, roan, calved February 12, 1876. Bred by Lord Mowbray and Stourton; got by Sandown (35470), dam (Juno 16th) by Golden Horn (31267), &c. See Vol. xxiv. p. 586.

Produce in		Names, &c.	By what Bull.	By whom bred.
1879, Jan. 22, roan,	C.C.	Juno 29th	L. Oxf. Bright Eyes, 34648	Lord Mowbray

JUNO 24TH, red and white, calved March 29, 1876. Bred by Lord Mowbray and Stourton; got by Sandown (35470), dam (Juno 12th) by Rose Duke (22760), &c. See Vol. xxiv. p. 586.

Produce in		Names, &c.	By what Bull.	By whom bred.
1879, July 12, red,	B.C.	(Steer)	L. Oxf. Bright Eyes, 34648	Lord Mowbray

LAVINIA 9TH, roan, calved June 15, 1872, Vol. xxiv. p. 587. Bred by Lord Mowbray and Stourton; got by Sprightly Lord (30044), dam (Lavinia 6th) by Rose Duke (22760), &c.

Produce in		Names, &c.	By what Bull.	By whom bred.
1878, Mar. 29, white,	B.C.	(Steer)	L. Oxf. Bright Eyes, 34648	Lord Mowbray
1879, April 1, roan,	C.C.	Lavinia 13th	do.	do.

LAVINIA 10TH, red and white, calved March 29, 1876. Bred by Lord Mowbray and Stourton; got by Sandown (35470), dam (Lavinia 9th) by Sprightly Lord (30044), &c.

Produce in		Names, &c.	By what Bull.	By whom bred.
1879, July 29, roan,	B.C.	Rose Duke 10th	L. Oxf. Bright Eyes, 34648	Lord Mowbray

ROSAMOND 7TH, roan, calved May 15, 1871, Vols. xxii. and xxiv. pp. 579, 587. Bred by Lord Mowbray and Stourton; got by Golden Horn (31237), dam (Rosamond 6th) by Rose Duke (22760), &c.

Produce in		Names, &c.	By what Bull.	By whom bred.
1878, May 29, white,	C.C.	Rosamond 13th	L. Oxf. Bright Eyes, 34648	Lord Mowbray
1879, April 28, roan,	B.C.	Mauleverer 7th	do.	do.

Mauleverer 7th, sold to Mr. J. Kerr, Lythe Hall, Whitby, Yorkshire.

ROSAMOND 8TH, roan, calved May 22, 1872, Vols. xxii. and xxiv. pp. 579, 587. Bred by Lord Mowbray and Stourton; got by Sprightly Lord (30044), dam (Rosamond 6th) by Rose Duke (22760), &c.

Produce in		Names, &c.	By what Bull.	By whom bred.
1878, April 10, roan,	B.C.	Baron Segrave	L. Oxf. Bright Eyes, 34648	Lord Mowbray
1879, Mar. 7, red,	B.C.	(Steer)	do.	do.

Baron Segrave, sold to Mr. J. Wildon, Knapton Grange, Rillington.

ROSAMOND 10TH, white, calved April 4, 1875. Bred by Lord Mowbray and Stourton; got by Sandown (35470), dam (Rosamond 7th) by Golden Horn (31237), &c.

Produce in		Names, &c.	By what Bull.	By whom bred.
1878, Mar. 21, roan,	B.C.	Mauleverer 6th	L. Oxf. Bright Eyes, 34648	Lord Mowbray
1879, April 19, white,	B.C.	(Steer)	do.	do.

Mauleverer 6th, sold to Mr. W. Day, Eversley Garth, South Milford.

MUIR, John,
Dryhope, Selkirk, N.B.

COLLIER JEAN, red and white, calved January 14, 1874. Bred by Mr. N. Milne, Dryhope, the property of Mr. J. Muir; got by Emerald Knight (28547), dam (Collier Cherry) by Oxford Cherry (24709), g. d. (Collier Lassie) by Collier Laddie (21443), &c. See "Brave Douglas," Vol. xix. p. 35.

Produce in		Names, &c.	By what Bull.	By whom bred.
1879, June 7, r. & w.,	C.C.	Collier Betty	Earl of Kelso, 38228	Mr. Milne

PRINCESS THORNDALE, roan, calved February 13, 1873, Vol. xxiv. p. 576. Bred by Mr. N. Milne, Dryhope, the property of Mr. J. Muir; got by Thorndale Duke 2nd (32712), dam (Lady Cherry) by Oxford Cherry (24709), &c.

Produce in		Names, &c.	By what Bull.	By whom bred.
1879, June 24, roan,	C.C.	Thorndale Marion	Earl of Kelso, 38228	Mr. Milne

MUNTON, William,
Banbury, Oxon.

COUNSELLOR'S MEMENTO, roan, calved March 6, 1877. Bred by Mr. W.
Munton ; got by Prince Puck (35168), dam (Counsellor's Heiress) by Roan
Silk (27307), &c. "See Geneva Prince 9th," p. 108.

Produce in	Names, &c.	By what Bull.	By whom bred.
1879, Dec. 30, r. & w., B.C.	Geneva Prince 9th	Pr. of Geneva 4th, 37250	Mr. Munton

FLEUR D'AMOUR 3RD, red, calved June 25, 1874, Vol. xxiv. p. 587. Bred
by Mr. W. Munton; got by Pretender (29577), dam (Fleur d'Amour) by
Waverley 3rd (21083), &c.

1879, Jan. 12, r. & w., B.C.	Geneva Prince 5th	Pr. of Geneva 4th, 37250	Mr. Munton

Geneva Prince 5th, sold to Mr. J. H. Gardner, Upper Boddington, Northamptonshire.

LOUISE 2ND, roan, calved November 23, 1873, Vol. xxv. p. 598. Bred by
Mr. W. Munton; got by Prince of Wales (32189), dam (Louise) by Fashion
(23913), &c.

1879, May 20, roan, B.C.	Geneva Prince 7th	Pr. of Geneva 4th, 37250	Mr. Munton

MOSS ROSE 4TH, roan, calved March 4, 1876, Vol. xxv. p. 599. Bred by
Mr. W. Munton; got by Prince Puck (35168), dam (Moss Rose) by Guy
Fawkes (21886), &c.

1879, Dec. 26, roan, C.C.	Moss Rose 7th	Pr. of Geneva 4th, 37250	Mr. Munton

MURRAY, G. W.,
Colleonard, Banff, N.B.

INDIANA, red, calved May 13, 1877. Bred by Mr. A. Longmore, Rettie ; got
by Baron Havering (33043), dam (Isabel) by Lord Forth (26649), &c. See
Vol. xxiv. p. 543.

1879, Dec. 10, red, C.C.	Colleonard	Killiecrankie, 43411	Mr. Murray

MUSGRAVE, Sir R. C., Bart.,
Eden Hall, Penrith.

WILD EYES 32ND, red, calved September 30, 1876. Bred by Sir R. C. Musgrave, Bart.; got by Duke of Underley (33745), dam (Wild Eyes 30th) by
Fourteenth Duke of Oxford (21605), &c. See "Valentine Duke," p. 255.

1879, June 10, roan, C.C.	Wild Eyes 33rd	Rl. Cambridge 4th, 40624	Sir R. C. Musgrave

WINSOME WILD EYES 2ND, roan, calved December 5, 1871, Vol. xxiii.
p. 578. Bred by Mr. J. W. Philips, Heybridge ; got by Bolton (25650), dam
(Winsome 3rd) by Thirteenth Duke of Oxford (21604), &c.

1877, July 3, roan, B.C.	(dead)	Duke of Underley, 33745	Sir R. C. Musgrave
1878, June 11, roan, C.C.	Win. Wild Eyes 4th	do.	do.
1879, May 26, red, C.C.	Win. Wild Eyes 5th	Rl. Cambridge 4th, 40624	do.

MUSTERS, J. C.,
Annesley Park, Nottingham.

BLANCHE ROSE, red, calved July 1, 1873, Vol. xxv., p. 599. Bred by Mr. J.
Snodin, Stonesby ; got by Cambridge Duke 4th (25706), dam (Blancheflower)
by May Duke 2nd (18372), &c.

1879, Mar. 8, red, C.C.	Wild Blanche Rose	Sir Wild Eyes, 35597	Mr. Musters

FANTINE 4TH, red, calved December 24, 1873, Vol. xxv. p. 599. Bred by Mr. J. H. Casswell, Laughton; got by Lord Nelson (31707), dam (Fantine 3rd) by Baron Panton (23377), &c.

Produce in		Names, &c.	By what Bull.	By whom bred.
1879, Jan. 19, red,	C.C.	Lady Fan	Lord Chaworth, 36924	Mr. Musters

PRINCESS KELLY, red and white, calved September 30, 1876. Bred by Mr. J. C. Musters; got by Oxonian (38840), dam (Princess of Windsor 4th) by Lord Lorne (34587), &c. See "Prince George," p. 193.

Produce in		Names, &c.	By what Bull.	By whom bred.
1879, Aug. 5, r. & w.,	B.C.	Prince George	D. of Glo'ster 6th, 39734	Mr. Musters

PRINCESS OF WINDSOR 4TH, roan, calved December 2, 1873, Vol. xxiii. p. 578. Bred by Mr. J. C. Musters; got by Lord Lorne (34587), dam (Princess of Windsor 2nd) by Duke of Burghley (36472), &c.

Produce in		Names, &c.	By what Bull.	By whom bred.
1877, Sept. 11, roan,	B.C.	Prince Charlie (Steer)	Lord Chaworth, 36924	Mr. Musters
1878, Aug. 8, roan,	B.C.	Oxonian, 38840	do.
1879, July 3, { r.&w.,C.C. roan, B.C.		Lady Mary (Steer)	} Lord Chaworth, 36924	do.

Prince Charlie, sold to Captain Jarvis, Doddington Hall, Lincoln.

PRIZE LEAF 2ND, red, calved June 12, 1876. Bred by Mr. J. C. Musters; got by Oxonian (38840), dam (Prize Leaf) by Lord Chancellor (20160), &c. See "The Squire," p. 250.

Produce in		Names, &c.	By what Bull.	By whom bred.
1879, May 27, roan,	B.C.	The Squire	Lord Chaworth, 36924	Mr. Musters

MYTTON, Captain D. H.,
Garth, Welshpool.

AGNES, roan, calved April 17, 1874. Bred by Captain Mytton; got by Vespasian (32759), dam (Gem) by Berwick (23411), &c. See "Admiral Rodney," p. 3.

Produce in		Names, &c.	By what Bull.	By whom bred.
1876, Dec. 29, roan,	B.C.	Cicero	Constantine 2nd, 33439	Captain Mytton
1877, Dec. 1, white,	B.C.	Catiline	do.	do.
1878, Oct. 10, red,	B.C.	Tacitus	do.	do.
1879, Dec. 30, r. & w.,	B.C.	Admiral Rodney	Admiral Hornby, 42647	do.

Cicero, sold to Mr. H. Wilson, Wrottesley Lodge, Wolverhampton; Catiline, to Lord Harrington, Elvaston Castle, Derbyshire.

EARLY ROSE, roan, calved April 29, 1876. Bred by Mr. J. Evans, Uffington; got by Richmond (35269), dam (Eglantine) by Manager (24521), &c. See "Diogenes," p. 68.

Produce in		Names, &c.	By what Bull.	By whom bred.
1878, April 23, white,	B.C.	Diogenes	Geneva Prince, 36691	Captain Mytton
1879, Mar. 23, r. & w.,	B.C.	Lord Cobham	Constantine 2nd, 33439	do.

Diogenes, sold to Mr. Whittingham, Llandrinio, Llandysillio, Oswestry.

ETIQUETTE, roan, calved May 10, 1874. Bred by Mr. J. Evans, Uffington; got by Duke of York (31038), dam (Eglantine) by Manager (24521), &c. See "Sir Charlton," p. 231.

Produce in		Names, &c.	By what Bull.	By whom bred.
1878, April 26, white,	C.C.	Joan	Geneva Prince, 36691	Captain Mytton
1879, Mar. 7, roan,	B.C.	Sir Charlton	Constantine 2nd, 33439	do.

GEM, red and white, calved February 10, 1870, Vols. xxii. and xxiii. pp. 517, 579. Bred by Mr. W. Nevett, Yorton Villa; got by Berwick (23411), dam (Glossary 2nd) by Hardware (19919), &c.

Produce in		Names, &c.	By what Bull.	By whom bred.
1877, Feb. 16, roan,	C.C.	Virginia	Constantine 2nd, 33439	Captain Mytton
1878, Mar. 4, roan,	C.C.	Jenny	do.	do.

HEBE, roan, calved March 7, 1869, Vol. xxii. p. 517.   Bred by Mr. W. Nevett, Yorton Villa ; got by Berwick (23411), dam (Hecuba) by Hector (14685), &c.

Produce in		Names, &c.	By what Bull.	By whom bred.
1876, May 13, roan,	B.C.	Nero	Vespasian, 32759	Captain Mytton
1877, Mar. 17, roan,	C.C.	May	Constantine 2nd, 33439	do.
1878, Mar. 18, roan,	B.C.	Pompey	do.	do.
1879, July 19, roan,	B.C.	Sir Griffith	Blanche's Oxford, 42796	do.

Nero, sold to Mr. Whittingham, Llandrinio, Llandysillio, Oswestry ; Pompey, to Mr. W. Gilbert, Shawell, Rugby.

HOPBINE, red and white, calved May 24, 1870, Vol. xxii. p. 408.   Bred by Mr. J. Evans, Uffington ; got by Manager (24521), dam (Mistletoe Berry) by St. Helena (35450), &c.

1879, Mar. 22, roan,	B.C.	Lord Tankerville	Constantine 2nd, 33439	Captain Mytton

JANET, red and white, calved March 7, 1874.   Bred by Captain Mytton ; got by Vespasian (32759), dam (Hebe) by Berwick (23411), &c.   See " Prince of Powys," p. 200.

1877, Jan. 27, red,	B.C.	Scipio	Constantine 2nd, 33439	Captain Mytton
1878, Jan. 13, red,	B.C.	Leo	do.	do.
1879, Jan. 26, roan,	B.C.	Prince of Powys	do.	do.

Scipio, sold to Mr. J. Denny, Muscott, Weedon ; Leo, to Mr. Dunnamore, Edgbaston.

JOYCE, red and white, calved January 29, 1876.   Bred by Captain Mytton ; got by Vespasian (32759), dam (Gem) by Berwick (23411), &c.   See " Caractacus," p. 41.

1878, Oct. 13, roan,	B.C.	Caractacus	Constantine 2nd, 33439	Captain Mytton

MAUD, roan, calved March 17, 1875.   Bred by Captain Mytton ; got by Vespasian (32759), dam (Gem) by Berwick (23411), &c.   See " Breidden," p. 32.

1878, Jan. 12, roan,	C.C.	Alice	Constantine 2nd, 33439	Captain Mytton
1879, Jan. 10, white,	B.C.	Breidden	do.	do.

MAY, roan, calved March 17, 1877.   Bred by Captain Mytton ; got by Constantine 2nd (33439), dam (Hebe) by Berwick (23411), &c.

1879, Dec. 21, roan,	C.C.	Cicely	Admiral Hornby, 42647	Captain Mytton

MISS DOROTHY, roan, calved April 25, 1873.   Bred by Captain Mytton ; got by Beacon (36225), dam (Gem) by Berwick (23411), &c.   See " Cæsar," p. 39.

1877, Nov. 6, roan,	B.C.	Cæsar	Constantine 2nd, 33439	Captain Mytton
1878, Nov. 13, roan,	B.C.	Offa	do.	do.

Cæsar, sold to Mr. Hall, Ireland ; Offa, to the Earl of Powis, Powis Castle, Welshpool.

NASH, Thomas,
Featherstone, Wolverhampton.

AGATHA, red and white, calved December 7, 1875.   Bred by Mr. T. Nash ; got by Fourth Duke of Claro (30940), dam (Agnes) by Charles Edward (25743), &c.   See Vol. xxii. p. 519.

1879, Jan. 1, r. & w.,	C.C.	Amy	Duke of Goldsmith, 41386	Mr. Nash

EMPRESS, roan, calved April 17, 1876.   Bred by Mr. W. Bradburn, Wednesfield ; got by Phosphate (32064), dam (Helen) by Paris (29523), &c.   See " Emperor," p. 94.

1879, Jan. 12, white,	B.C.	Emperor	Duke of Goldsmith, 41386	Mr. Nash

NEWTON, H. V.,
Polstrong, Camborne, Cornwall.

BEAUTY OF BERKS, white, calved June 28, 1866, Vols. xx. and xxi. pp. 404,
868. Bred by Mr. J. Blyth, Woolhampton House; got by Tam O'Shanter
(20930), dam (Fair Lady) by Earl of Walton (17787), &c.

Produce in	Names, &c.	By what Bull.	By whom bred.
1879, Mar. 13, r. & w., C.C.	Louise Margaret	Lord Chief Justice, 34507	Mr. Newton

NELLIE, red and white, calved July 16, 1874, Vol. xxv. p. 600. Bred by Mr. H.
V. Newton; got by Duke of Kennet (30977), dam (Beauty of Berks) by Tam
O'Shanter (20930), &c.

1879, Nov. 2, white, C.C.	Nina	Colonel Wild Eyes, 36383	Mr. Newton

NICHOLS, John,
Iron Acton, Bristol.

COUNTESS, roan, calved April 11, 1871. Bred by Mr. J. Dove, Hambrook;
got by Reflector (29760), dam (Clio) by Coronet (23624), &c. See "Count de
Acton," p. 57.

1874, Mar. 1, r. & w., B.C.	(dead)	Lord Aberdeen, 31610	Mr. Nichols
1875, Mar. 6, roan, C.C.	Acton Cream	do.	do.
1876, Feb. 5, roan, B.C.	Count de Acton	Weldred, 35959	do.
1877, Mar. 9, roan, C.C.	Capsicum	Lord Aberdeen, 31610	do.
1878, Mar. 14, white, B.C.	Acton Comet	do.	do.
1879, Feb. 26, roan, B.C.	Acton Chief	Viceroy, 44184	do.

NICHOLS, Thomas,
The Grange, Raunds, Thrapston.

FORMOSA, roan, calved April 22, 1868, Vols. xix. and xxi. pp. 522, 868. Bred
by Mr. H. J. Sheldon, Brailes House; got by Duke of Brailes (23724), dam
(Gionetta) by Sarawak (15238), &c.

1873, Feb. 2, white, B.C.	Baron Brailes	D. of Kingscote, 25981	Mr. Nichols
1874, April 26, roan, C.C.	(dead)	Baron Oxford 5th, 27958	do.
1875, Mar. 11, roan, C.C.	(dead)	Wild Eyes Duke, 36007	do.
1876, Feb. 19, roan, C.C.	(dead)	do.	do.
1877, Feb. 13, red, C.C.	Formosa 5th	D. of Wellington 9th, 33754	do.
1879, Feb. 21, red, B.C.	Baron Brailes	L. Clarence Waterloo, 36926	do.

LADY BRAILES, red, calved January 18, 1872, Vol. xxi. p. 869. Bred by
Mr. T. Nichols; got by Duke of Brailes (23724), dam (Formosa) by Duke of
Brailes (23724), &c.

1875, May 18, roan, C.C.	Lady Brailes 3rd	Wild Eyes Duke, 36007	M. Nichols
1876, July 2, r. & w., C.C.	Lady Brailes 4th	Grand Duke 21st, 34061	do.
1877, July 11, roan, C.C.	Lady Brailes 5th	D. of Wellington 9th, 33754	do.
1878, Nov. 1, roan, B.C.	Lord of the Grange	G. D. of Morecambe, 36722	do.
1879, Nov. 8, red, C.C.	Lady Brailes 8th	L. Clarence Waterloo, 36926	do.

LADY BRAILES 2nd, red, calved May 12, 1874, Vol. xxii. p. 582. Bred by
Mr. T. Nichols; got by Baron Oxford 5th (27958), dam (Lady Brailes) by
Duke of Brailes (23724), &c.

1877, Mar. 28, roan, C.C.	Lady Brailes 6th	D. of Wellington 9th, 33754	Mr. Nichols
1878, July 5, roan, C.C.	Lady Brailes 7th	do.	do.
1879, July 26, red, B.C.	Rushden	L. Clarence Waterloo, 36926	do.

LADY BRAILES 3RD, roan, calved May 18, 1875. Bred by Mr. T. Nichols; got by Wild Eyes Duke (36007), dam (Lady Brailes) by Duke of Brailes (23724), &c. See " Lord Shipston," p. 157.

Produce in		Names, &c.	By what Bull.	By whom bred.
1878, Aug. 12, red,	B.C.	Lord Shipston	D. of Wellington 9th, 33754	Mr. Nichols

LADY BRAILES 4TH, red and white, calved July 2, 1876. Bred by Mr. T. Nichols ; got by Grand Duke 21st (34061), dam (Lady Brailes) by Duke of Brailes (23724), &c.

1879, April 6, roan,	B.C.	Northampton	L. Clarence Waterloo, 36926	Mr. Nichols

LADY EDITH 2ND, red, calved March 3, 1874, Vol. xxiii. p. 582. Bred by Lord Penrhyn, Wicken Park ; got by Fourth Duke of Barrington (30924), dam (Lady Edith) by Duke of Brailes (23724), &c.

1877, Sept. 29, red,	C.C.	Lady Edith 3rd	D. of Wellington 9th, 33754	Mr. Nichols
1878, Nov. 27, r. & w.,	C.C.	Lady Edith 4th	G. D. of Morecambe, 36722	do.
1879, Nov. 4, red,	B.C.	Lord Raunds	L. Clarence Waterloo, 36926	do.

NICHOLSON, James,
Murton, Berwick-on-Tweed.

LADY FAIR, white, calved September 19, 1876. Bred by Mr. R. Jefferson, Preston Hows ; got by Jacob's Ladder (34237), dam (Lady Faith) by England's Fame (31111), g. d. (Lady Hope) by Vain Hope (23102), &c. See " British Hope," Vol. xxiv. p. 30.

1879, Mar. 29, roan,	C.C.	Lady Charity	Ben Lomond, 37854	Mr. Nicholson

QUEEN BOADICEA, roan, calved January 15, 1875, Vol. xxiv. p. 591. Bred by Messrs. Atkinson, Peepy ; got by Royal Killerby (32396), dam (Jess) by Bumper 2nd (28107), &c.

1879, Mar. 6, roan,	C.C.	Joanna	Ben Lomond, 37854	Mr. Nicholson

REBECCA, red, calved April 19, 1876, Vol. xxv. p. 601. Bred by Mr. R. Jefferson, Preston Hows ; got by Prince Thomas (35183), dam (Rosebud) by Royal Heir (35377), &c.

1879, Mar. 26, r. & w.,	C.C.	Reality	Ben Lomond, 37854	Mr. Nicholson

NICHOLSON, W.,
Basing Park, Alton, Hants.

PRINCESS 4TH, roan, calved August 1, 1871. Bred by Mr. W. Nicholson ; got by Mentor (24576), dam (Princess 1st) by Prince Albert (22560), &c. See Vol. xix. p. 677.

1875, Aug. 10, red,	B.C.	(dead)	Lord Geneva 2nd, 34556	Mr. Nicholson

PRINCESS 5TH, red and white, calved May 12, 1873. Bred by Mr. W. Nicholson ; got by Gilsland (31249), dam (Princess 2nd) by Mentor (24576), &c. See Vol. xxi. p. 869.

1876, Sept. 6, r. & w.,	B.C.		Lord Geneva 2nd, 34556	Mr. Nicholson
1877, July 10, red,	C.C.	(dead)	do.	do.

PRINCESS 9TH, red, calved September 24, 1875. Bred by Mr. W. Nicholson ; got by Lord Geneva 2nd (34556), dam (Princess 3rd) by Mentor (24576), &c. See Vol. xxi. p. 869.

1878, Sept. 6, roan,	C.C.	Princess 17th	Denison. 43039	Mr. Nicholson
1879, July 28, roan,	C.C.	Princess 19th	Morris Dancer, 43673	do.

NICHOLSON, W. A.,
Pembroke House, Penrith.

GENERAL'S DAUGHTER, red, calved in 1867. Bred by Mr. J. Atkinson, Whinfell Park; got by General Haynau (11520), dam (Red Rose 6th) by Gainford 2nd (10255), &c. See "Lord De Clifford," p. 144.

Produce in		Names, &c.	By what Bull.	By whom bred.
1870, April, r. & w.,	C.C.	Ada	Edwin, 21671	Mrs. Atkinson
1871, Oct. 20, white,	C.C.	Lady De Clifford	do.	do.
1872, Oct. 6, roan,	C.C.	Farewell	Roy.Westmoreland, 35416	do.

Ada, sold to Mr. J. Wright, Greengill Head, Penrith; Lady De Clifford, to Mr. W. A. Nicholson, Pembroke House, Penrith; Farewell, to Mr. R. Thompson, Inglewood Bank, Penrith.

NOEL, C. Perrot,
Bell Hall, Stourbridge.

LAURUSTINUS, roan, calved February 17, 1876. Bred by Mr. C. P. Noel; got by Mercury (34836), dam (Lydia) by Viscount Middleham (25385), g. d. (Favourite) by Castlereagh (19409), &c. See "Young Hero," p. 120.

1879, Dec. 25, roan,	C.C.	Mistletoe	Bumble Bee, 39523	Mr. Noel

LUPINE, red, calved May 12, 1871, Vols. xxii. and xxv. pp. 521, 603. Bred by Mr. C. P. Noel; got by Lord Red Eyes 2nd (24460), dam (Lydia) by Viscount Middleham (25385), &c.

1879, July 16, r. & w.,	C.C.	Lily	King Hal, 40052	Mr. Noel

NORTHUMBERLAND, Duke of,
Alnwick Castle.

CHERRY BUD, roan, calved March 31, 1877. Bred by the Duke of Northumberland; got by Fitz Roland (33936), dam (Red Blossom) by Rifleman (27283), &c. See "Sir Arthur," p. 229.

1879, Oct. 9, roan,	B.C.	Sir Arthur	Sir Raymond, 40716	D. of North'land

CHERRY STONE, roan, calved August 30, 1876. Bred by the Duke of Northumberland; got by Fitz Roland (33936), dam (Clara) by Mayor of Windsor (31897), &c. See Vol. xxiii. p. 584.

1879, June 25, roan,	C.C.	Cherry Ripe	Sir Raymond, 40716	D. of North'land

DAISY'S GEM, roan, calved December 13, 1875. Bred by the Exors. of Mr. G. Angus, Broomley; got by Ben Brace (30524), dam (Daisy 4th) by Prince Charlie (29606), &c. See Vol. xxi. p. 541.

1879, Sept. 14, red,	C.C.	Daisy's Pride	Sir Raymond, 40716	D. of North'land

JULIA, roan, calved October 23, 1875, Vol. xxv. p. 604. Bred by the Duke of Northumberland; got by White Prince (35988), dam (Constance) by Mayor of Windsor (31897), &c.

1879, Mar. 15, white,	B.C.	Glacier	Sir Raymond, 40716	D. of North'land

Glacier, sold to the owners of Backworth Colliery, Newcastle.

LADY SELINA, roan, calved December 15, 1872, Vols. xxiii. and xxv. pp. 585, 604. Bred by Mr. Rothwell, Foxholes; got by Golden Duke (26266), dam (Selina 7th) by Royal Gwynne (22784), &c.

1879, April 21, roan,	C.C.	Lady Alice	Sir Raymond, 40716	D. of North'land

LENA, red, calved June 20, 1874, Vols. xxiii. and xxv. pp. 585, 604. Bred by
the Duke of Northumberland; got by Mayor of Windsor (31897), dam
(Selina 7th) by Royal Gwynne (22784), &c.

Produce in		Names, &c.	By what Bull.	By whom bred.
1879, Jan. 9, red,	C.C.	Princess Selina	Sir Raymond, 40716	D. of North'land

LUCRETIA 4TH, roan, calved September 25, 1874, Vols. xxiv. and xxv. pp. 593,
605. Bred by the Duke of Northumberland; got by Mayor of Windsor
(31897), dam (Lucretia 2nd) by Royal Butterfly 23rd (27355), &c.

Produce in		Names, &c.	By what Bull.	By whom bred.
1879, Mar. 3, red,	C.C.	Lucretia's Ruby	Sir Raymond, 40716	D. of North'land

LUCRETIA 5TH, white, calved February 8, 1876. Bred by the Duke of
Northumberland; got by Duke of Tyne (33744), dam (Lucretia 2nd) by Royal
Butterfly 23rd (27355), &c. See "Lucifer," p. 161.

Produce in		Names, &c.	By what Bull.	By whom bred.
1879, June 5, roan,	B.C.	Lucifer	Sir Raymond, 40716	D. of North'land

LUCRETIA 6TH, white, calved February 16, 1877. Bred by the Duke of
Northumberland; got by Fitz Roland (33936), dam (Lucretia 4th) by Mayor
of Windsor (31897), &c. See "Sir William," p. 237.

Produce in		Names, &c.	By what Bull.	By whom bred.
1879, Oct. 13, roan,	B.C.	Sir William	Sir Raymond, 40716	D. of North'land

MAID OF WINDSOR, roan, calved August 27, 1874, Vols. xxiv. and xxv.
pp. 593, 605. Bred by the Duke of Northumberland; got by Mayor of
Windsor (31897), dam (Village Maid) by Jeweller (26460), &c.

Produce in		Names, &c.	By what Bull.	By whom bred.
1879, Dec. 4, roan,	B.C.	Knight of Windsor	Sir Raymond, 40716	D. of North'land

MODESTY, roan, calved August 10, 1872, Vols. xxii. and xxiv. pp. 523, 593.
Bred by the Duke of Northumberland; got by Brigadier (28078), dam
(Rosette) by Briton (25686), &c.

Produce in		Names, &c.	By what Bull.	By whom bred.
1879, Feb. 11,	roan, B.C.	Huz	Sir Raymond, 40716	D. of North'land
	r. &w., B.C.	Buz		

PRIMROSE 2ND, roan, calved March 30, 1875, Vol. xxiv. p. 372. Bred by
Mr. L. C. Chrisp, Hawkhill, the property of the Duke of Northumberland;
got by Duke of Aosta (28356), dam (Rose 2nd) by Peak (24733), &c.

Produce in		Names, &c.	By what Bull.	By whom bred.
1879, June 24, roan,	B.C.	Sir Rose	Sir Raymond, 40716	Mr. Chrisp

Sir Rose, sold to Dr. Marshall, Chatton Park, Belford.

PRUNA, roan, calved January 20, 1876. Bred by Mr. A. H. Browne, Callaly
Castle, the property of the Duke of Northumberland; got by Rosario (35315),
dam (Primula) by Duke of Aosta (28356), &c. See Vol. xxiv. p. 352.

Produce in		Names, &c.	By what Bull.	By whom bred.
1879, Mar. 23, roan,	C.C.	Prudery	Muscovite, 38773	Mr. Browne

RED BLOSSOM, red, calved December 23, 1868, Vols. xx., xxi., xxiii., xxiv.,
and xxv. pp. 718, 871, 585, 594, 605. Bred by the Duke of Northumberland;
got by Rifleman (27283), dam (Maid of Aln) by Melsonby (18380), &c.

Produce in		Names, &c.	By what Bull.	By whom bred.
1879, Sept. 3, roan,	B.C.	Alexander the Great	Sir Raymond, 40716	D. of North'land

ROSEBUD 12TH, red, calved December 20, 1873, Vol. xxiv. p. 594. Bred by
Mr. J. Wilson, Cumledge; got by Hotspur (28876), dam (Rosebud 11th) by
Brigand (28080), &c.

Produce in		Names, &c.	By what Bull.	By whom bred.
1879, April 7, red,	B.C.	Rufus	Sir Raymond, 40716	D. of North'land

ROSE OF TYNE, roan, calved August 22, 1875, Vol. xxv. p. 605. Bred by
the Duke of Northumberland; got by Duke of Tyne (33744), dam (Marion)
by Vanguard (23115), &c.

Produce in		Names, &c.	By what Bull.	By whom bred.
1879, Jan. 21, roan,	B.C.	Royal Commissioner	Sir Raymond, 40716	D. of North'land
1879, Dec. 2, red,	C.C.	Rose of Coquetdale	do.	do.

SELINA 8TH, roan, calved August 22, 1875, Vol. xxv. p. 606. Bred by the Duke of Northumberland; got by Duke of Tyne (33744), dam (Selina 7th) by Royal Gwynne (22784), &c.

Produce in		Names, &c.	By what Bull.	By whom bred.
1879, May 20, white,	B.C.	Iceberg	Sir Raymond, 40716	D. of North'land

SNOWFLAKE 2ND, white, calved February 5, 1878. Bred by the Exors. of Mr. G. Angus, Broomley; got by Hawthorn (36751), dam (Snowflake) by Ben Brace (30524), &c. See "Wellington," p. 264.

Produce in		Names, &c.	By what Bull.	By whom bred.
1879, July 10, white,	B.C.	Wellington	Waterloo, 40892	D. of North'land

SUNBEAM, red and white, calved October 20, 1873, Vols. xxiii. and xxv. pp. 586, 606. Bred by the Duke of Northumberland; got by Mayor of Windsor (31897), dam (Dewdrop) by President (20510), &c.

Produce in		Names, &c.	By what Bull.	By whom bred.
1879, April 15, roan,	C.C.	Sunshade	Sir Raymond, 40716	D. of North'land

SWEET DAISY, roan, calved September 29, 1875, Vol. xxv. p. 606. Bred by the Duke of Northumberland; got by Duke of Tyne (33744), dam (Red Daisy) by Jeweller (26460), &c.

Produce in		Names, &c.	By what Bull.	By whom bred.
1879, Jan. 4, roan,	B.C.	William Tell	Sir Raymond, 40716	D. of North'land

VILLAGE MAID, roan, calved March 2, 1871, Vols. xxi., xxiii., and xxv. pp. 871, 586, 606. Bred by the Duke of Northumberland; got by Jeweller (26460), dam (Harriet) by Vice President (23125), &c.

Produce in		Names, &c.	By what Bull.	By whom bred.
1879, Jan. 12, roan,	C.C.	Maid of Bassington	Sir Raymond, 40716	D. of North'land

WINDSOR'S PEARL, roan, calved July 24, 1876. Bred by the Duke of Northumberland; got by Mayor of Windsor (31897), dam (Dewdrop) by President (20510), &c. See "Pearl Diver," p. 185.

Produce in		Names, &c.	By what Bull.	By whom bred.
1879, Jan. 30, roan,	B.C.	Pearl Diver	Sir Raymond, 40716	D. of North'land

NORTON, P. R.,
Turvey House, Donabate, Ireland.

CYBELE 3RD, red and white, calved February 25, 1878. Bred by Mr. W. F. Budds, Courtstown; got by Duke of Cranagh (39726), dam (Cybele) by Prince Arthur (29598), &c. See "Turvey," p. 254.

Produce in		Names, &c.	By what Bull.	By whom bred.
1879, May 17, roan,	B.C.	Turvey	Grand Lama, 41656	Mr. Norton

OASTLER, J.,
Loxwood House, Billingshurst, Sussex.

JUDY, red, calved February 1, 1871. Bred by Mr. W. Burton, Northill; got by Pyramus (27223), dam (Julia) by Hero (28848), &c. See "Prince Alexander William," Vol. xxv. p. 220.

Produce in		Names, &c.	By what Bull.	By whom bred.
1875, Feb. 13, red,	B.C.	Beldorny	Keithmore, 34297	Mr. Sewell
1876, Jan. 17, red,	C.C.	Lady Juliet	do.	Mr. Oastler
1877, May 1, r. & w.,	C.C.	Sheriff's Daughter	High Sheriff, 34163	do.
1878, July 18, r. & w.,	B.C.	Prince Alex. William	Prince Alfred, 37215	do.

Beldorny, sold to Mr. J. Oastler.

LADY BELL 7TH, roan, calved December 13, 1874. Bred by Mr. H. W. Martin, Upton House; got by Lord Abbot (29052), dam (Lady Bell 4th) by Mainstay (24507), &c. See Vol. xxi. p. 843.

Produce in		Names, &c.	By what Bull.	By whom bred.
1877, Dec. 18, roan,	B.C.	Baron Bell	Beldorny, 36234	Mr. Oastler
1879, Mar. 7, roan,	C.C.	(dead)	do.	do.

SPLENDOUR 26th, red, calved February 21, 1873. Bred by Mr. H. Sewell, Stow Maries; got by Keithmore (34297), dam (Splendour 16th) by Remus (20665), &c. See "Royal Splendour," p. 222.

Produce in	Names, &c.	By what Bull.	By whom bred.
1876, Dec. 26, r. & w., B.C.	Baron Splendour	High Sheriff, 34163	Mr. Oastler
1878, Feb. 27, r. & w., C.C.	Lady Splendour	Beldorny, 36234	do.
1879, Feb. 10, red, B.C.	Royal Splendour	Royal Bell, 39030	do.

Baron Splendour, sold to Mr. Wills, Moreton Hampstead, Devon.

OGLE, George F.,
Top House, Rawcliffe, Selby.

DUCHESS OF WALLINGTON, roan, calved February 8, 1876. Bred by Mr. G. F. Ogle; got by Grand Duke of Wallington (31300), dam (Red Rose) by Hawkseye (34119), &c. See Vol. xxiii. p. 587.

1879, May 30, red, B.C.	(Steer)	L. Ox. Br't Eyes 4th, 40231	Mr. Ogle

STRAWBERRY 4th, red, calved April 2, 1876. Bred by Mr. G. F. Ogle; got by Grand Duke of Wallington (31300), dam (Strawberry 2nd) by Hawkseye (34119), &c. See Vol. xxi. p. 872.

1879, July 10, roan, C.C.	Strawberry 5th	L. Ox. Br't Eyes 4th, 40231	Mr. Ogle

OLIVER, Robert E.,
Sholebroke Lodge, Towcester.

CHERRY GRAND DUCHESS 10th, white, calved June 4, 1876. Bred by Mr. R. E. Oliver; got by Grand Duke 25th (34065), dam (Cherry Grand Duchess 6th) by Grand Duke 19th (28746), &c. See "Cherry Grand Duke 9th," p. 48.

1879, April 10, roan, B.C.	Cherry G. Duke 9th	Grand Duke 30th, 38373	Mr. Oliver

DAPHNE, red and white, calved January 10, 1876. Bred by Mr. A. J. Robarts, Lillingstone Dayrell; got by Caractacus (28141), dam (Diamond) by Second Duke of Claro (21576), &c. See "Lord Dacre," p. 144.

1879, Jan. 31, r. & w., B.C.	Lord Dacre	Grand Duke 30th, 38373	Mr. Oliver

ELYSIUM, roan, calved May 16, 1874. Bred by Mr. R. E. Oliver; got by Grand Duke 19th (28746), dam (Elia) by Sam of Oxford (25084), &c. See Vol. xxi. p. 873.

1879, April 1, roan, C.C.	Elysian	Duke of Cerisia 3rd, 36476	Mr. Oliver

FRISKY, white, calved September 29, 1871. Bred by Mr. R. E. Oliver; got by Lord of the Forest (26704), dam (Frivolity) by Cherry Butterfly (23550), &c. See Vol. xx. p. 532.

1879, May 20, white, B.C.	(Steer)	Duke of Cerisia 3rd, 36476	Mr. Oliver

GRAND DUCHESS 17th, red and white, calved September 16, 1864, Vols. xviii., xix., xxi., and xxii. pp. 518, 537, 873, 524. Bred by Mr. J. Hegan, Dawpool; got by Imperial Oxford (18084), dam (Grand Duchess 10th) by Grand Duke 3rd (16182), &c.

1879, Jan. 22, red, C.C.	Grand Duchess 38th	D. of Underley 3rd, 38196	Mr. Oliver

GRAND DUCHESS 25TH, white, calved March 2, 1872, Vols. xxi. and xxv.
pp. 873, 607. Bred by Mr. R. E. Oliver; got by Second Duke of Tregunter
(26022), dam (Grand Duchess 23rd) by Grand Duke 7th (19877), &c.

Produce in	Names, &c.	By what Bull.	By whom bred.
1879, Mar. 13, roan,	B.C. Grand Duke 36th	Grand Duke 30th, 38373	Mr. Oliver

GRAND DUCHESS 27TH, roan, calved April 18, 1873, Vols. xxii. and xxiii.
pp. 525, 588. Bred by Mr. R. E. Oliver; got by Duke of Hillhurst (28401),
dam (Grand Duchess 22nd) by Grand Duke 7th (19877), &c.

1879, April 11, red,	C.C. Grand Duchess 39th	Grand Duke 30th, 38373	Mr. Oliver

GRAND DUCHESS 28TH, roan, calved August 5, 1874, Vols. xxiii. and xxv.
pp. 588, 607. Bred by Mr. R. E. Oliver; got by Third Duke of Clarence
(23727), dam (Grand Duchess 17th) by Imperial Oxford (18084), &c.

1879, June 25, r. & w.,	B.C. Grand Duke 39th	D. of Underley 3rd, 38196	Mr. Oliver

GRAND DUCHESS 31ST, roan, calved November 17, 1875. Bred by Mr. R.
E. Oliver; got by Third Duke of Clarence (23727), dam (Grand Duchess
17th) by Imperial Oxford (18084), &c. See "Grand Duke 40th," p. 112.

1879, Sept. 15, roan,	B.C. Grand Duke 40th	Grand Duke 30th, 38373	Mr. Oliver

GRAND DUCHESS 32ND, roan, calved November 15, 1876. Bred by Mr. R.
E. Oliver; got by Grand Duke 25th (34065), dam (Grand Duchess 27th) by
Duke of Hillhurst (28401), &c. See Vol. xxiii. p. 588.

1879, June 24, r. & w.,	B.C. Gd. Duke 38th (dead)	Grand Duke 30th, 38373	Mr. Oliver

GRAND DUCHESS OF BARRINGTONIA 2ND, roan, calved September 23,
1873, Vol. xxiii. p. 588. Bred by Mr. R. E. Oliver; got by Third Duke
of Clarence (23727), dam (Grand Duchess of Barrington) by Grand Duke
7th (19877), &c.

1879, Aug. 31, r. & w.,	B.C. G.D.of B'rngtonia 5th	Grand Duke 30th, 38373	Mr. Oliver

GRAND DUCHESS OF BARRINGTONIA 3RD, roan, calved August 26, 1874,
Vol. xxv. p. 607. Bred by Mr. R. E. Oliver; got by Grand Duke 21st (34061),
dam (Grand Duchess of Barrington) by Grand Duke 7th (19877), &c.

1879, Nov. 8, red,	B.C. G.D.of B'rngtonia 6th	Grand Duke 30th, 38373	Mr. Oliver

LADY ADA, roan, calved January 22, 1873. Bred by Mr. R. E. Oliver; got
by Cherry Grand Duke 4th (28175), dam (Lady Adelaide) by Northern Duke
(22431), &c. See Vol. xxi. p. 873.

1879, Jan. 23, roan,	C.C. Lady Amelia	Duke of Cerisia 3rd, 36476	Mr. Oliver

SILVERDALE, roan, calved April 21, 1872, Vol. xxv. p. 607. Bred by Mr. R.
E. Oliver; got by Lord of the Forest (26704), dam (Sweetheart 12th) by
Duke of Albemarle (21560), &c.

1879, June 12, roan,	C.C. Silver Cloud 2nd	Grand Duke 30th, 38373	Mr. Oliver

WATERLOO 43RD, roan, calved May 20, 1876. Bred by Lord Penrhyn, Penrhyn Castle; got by Grand Duke of Oxford (31293), dam (Waterloo 35th) by
Grand Duke 11th (21849), &c. See "Grand Duke of Waterloo 3rd," p. 115.

1879, April 4, roan,	B.C. G.D. of Waterloo 3rd	Grand Duke 30th, 38373	Mr. Oliver

WATERLOO BIENVENUE, roan, calved August 10, 1873, Vol. xxiii. p. 588.
Bred by Lord Skelmersdale, Lathom House; got by Oxford Beau (29485),
dam (Waterloo 33rd) by Grand Duke 11th (21849), &c.

1879, June 10, r. & w.,	B.C. G.D. of Waterloo 4th	Grand Duke 30th, 38373	Mr. Oliver

O'REILLY, Major,
Knock Abbey, Dundalk, Ireland.

BANGLET, white, calved July 28, 1875. Bred by Major M. O'Reilly; got by Remus (35258), dam (Bracelet) by Breast Plate (19337), &c. See Vol. xxii. p. 525.

Produce in		Names, &c.	By what Bull.	By whom bred.
1877, Oct. 9, white,	C.C.	(dead)	Lord Remus, 37002	Major O'Reilly
1879, April 21, roan,	B.C.	Rhine-guard	Red Rhine, 38976	do.

DOUBLE DAISY, red, calved February 9, 1876. Bred by Mr. J. Madden, Roslea Manor; got by Westland's Glory (35968), dam (Red Daisy) by Tally Ho (28841), &c. See Vol. xxiii. p. 589.

Produce in		Names, &c.	By what Bull.	By whom bred.
1877, Dec. 23, roan,	C.C.	(dead)	Lord Remus, 37002	Major O'Reilly
1879, May 6, red,	C.C.	Rhine Daisy	Red Rhine, 38976	do.

WATER KELPIE, roan, calved May 6, 1874. Bred by Major M. O'Reilly; got by King Richard 2nd (31514), dam (Water Spirit) by Regal Dane (22710), &c. See Vol. xx. p. 816.

Produce in		Names, &c.	By what Bull.	By whom bred.
1878, May 6, white,	B.C.	White Prince	Grand Prince, 34085	Major O'Reilly
1879, Mar. 24, red,	C.C.	Rhine Fairy	Red Rhine, 38976	do.

WATER QUEEN, roan, calved July 5, 1876. Bred by Major M. O'Reilly; got by Prince Royal (32202), dam (Water Spirit) by Regal Dane (22710), &c.

Produce in		Names, &c.	By what Bull.	By whom bred.
1878, June 28, roan,	B.C.	Water Prince	Grand Prince, 34085	Major O'Reilly
1879, July 12, roan,	B.C.	Rhine King	Red Rhine, 38976	do.

OSMASTON, John,
Osmaston Manor, Derby.

YOUNG CHERRY, roan, calved January 3, 1875. Bred by the Earl of Lichfield, Shugborough; got by Sir Henry (35559), dam (Cherry) by Knight Templar 3rd (34412), g. d. (Clematis 2nd) by Knight Gwynne (34368), &c. See Vol. xxi. p. 813.

Produce in		Names, &c.	By what Bull.	By whom bred.
1878, May 25, white,	C.C.	Young Cherry 2nd	Sir Arthur, 44013	Earl of Lichfield

Young Cherry 2nd, sold to Mr. J. Osmaston, Osmaston Manor.

COUNTESS OF YORK, white, calved June 1, 1869. Bred by Mr. R. Betts, Holbeck Lodge; got by Grand Duke of Essex 2nd (21860), dam (Princess of York) by Baron Crossley (19269), g. d. (Princess of York) by Mac Turk (14872), &c. See Vol. xxi. p. 573.

Produce in		Names, &c.	By what Bull.	By whom bred.
1879, Aug. 14, roan,	C.C.	Suez	Taciturn, 42473	Mr. Betts

Suez, sold to Mr. J. Osmaston, Osmaston Manor.

LADY HELEN 4TH, white, calved June 2, 1877. Bred by Sir R. Peel, Bart., Drayton Manor, the property of Mr. J. Osmaston; got by The Druid (35753), dam (Pride) by Count Bickerstaffe (23630), &c. See Vol. xxiv. p. 604.

Produce in		Names, &c.	By what Bull.	By whom bred.
1879, Sept. 9, white,	C.C.	Lady Helen 5th	Saturn, 40673	Mr. Lyon

SNOWBALL, white, calved April 19, 1874. Bred by the Earl of Lichfield, Shugborough; got by Sir Henry (35559), dam (Clematis 2nd) by Knight Gwynne (34368), &c. See Vol. xxi. p. 813.

Produce in		Names, &c.	By what Bull.	By whom bred.
1878, Feb. 3, white,	C.C.	Young Snowball	Sir Arthur, 44013	Earl of Lichfield

Young Snowball, sold to Mr. J. Osmaston, Osmaston Manor.

SYREN, roan, calved June 4, 1873. Bred by Mr. R. Betts, Holbeck Lodge; got by Major (26793), dam (Violet 2nd) by Grand Duke of Essex 2nd (21860), g. d. (Violet) by Baron Crossley (19269), &c. See "Iron Clad," Vol. xx. p. 165.

Produce in	Names, &c.	By what Bull.	By whom bred.
1877, April 20, white, C.C.	Cynthia	Duke of Hainault, 33661	Mr. Betts

Cynthia, sold to Mr. J. Osmaston, Osmaston Manor.

OSSINGTON, Viscountess,
Ossington, Newark.

ADMIRATION 3RD, roan, calved February 7, 1877. Bred by Viscountess Ossington; got by Glo'ster (36701), dam (Admiration 2nd) by Twenty-first Duke of Oxford (30999), &c. See Vol. xxiv. p. 597.

1879, Dec. 9, red,	B.C.	Earl of Wragby 4th	Sherwood Chieftain, 39098	Visct'ss Ossington

CLOTILDE 5TH, red and white, calved May 26, 1877. Bred by Viscountess Ossington; got by Twenty-first Duke of Oxford (30999), dam (Clotilde 3rd) by Duke of Hazlecote (25969), &c. See Vol. xxiii. p. 589.

1879, Aug. 30, r.&w., C.C.	Clotilde 8th	Sherwood Chieftain, 39098	Visct'ss Ossington

COQUETTE 9TH, roan, calved January 29, 1877. Bred by Viscountess Ossington; got by Glo'ster (36701), dam (Coquette 5th) by Lord Harley (31665), &c. See Vol. xxiv. p. 597.

1879, Mar. 17, red,	C.C.	Coquette 13th	Sherwood Chieftain, 39098	Visct'ss Ossington

COQUETTE 10TH, red, calved May 2, 1877. Bred by Viscountess Ossington; got by Twenty-first Duke of Oxford (30999), dam (Coquette 3rd) by Duke of Hazlecote (25969), &c. See Vol. xxi. p. 875.

1879, April 27, red,	C.C.	Coquette 14th	Sherwood Chieftain, 39098	Visct'ss Ossington

COQUETTE 11TH, roan, calved May 29, 1877. Bred by Viscountess Ossington; got by Twenty-first Duke of Oxford (30999), dam (Coquette) by Kelstern Grand Duke (22024), &c. See "Badminton 18th," p. 9.

1879, Aug. 20, roan, C.C.	Coquette 15th	Sherwood Chieftain, 39098	Visct'ss Ossington

PRETTY MAID 6TH, red, calved February 23, 1877. Bred by Viscountess Ossington; got by Twenty-first Duke of Oxford (30999), dam (Pretty Maid) by Robin (24968), &c. See "Sherwood Chieftain 2nd." p. 228.

1879, April 12, red,	B.C.	(dead)	Sherwood Chieftain, 39098	Visct'ss Ossington

PRETTY MAID 7TH, red and white, calved August 27, 1877. Bred by Viscountess Ossington; got by Glo'ster (36701), dam (Pretty Maid 4th) by Twenty-first Duke of Oxford (30999), &c. See Vol. xxiv. p. 598.

1879, Nov. 17, r.&w., C.C.	Pretty Maid 12th	Sherwood Chieftain, 39098	Visct'ss Ossington

RUBY 3RD, red, calved January 17, 1877. Bred by Viscountess Ossington; got by Twenty-first Duke of Oxford (30999), dam (Ruby 2nd) by The Stuart (27650), &c. See Vol. xxii. p. 525.

1879, Mar. 25, red,	C.C.	Ruby 4th	Sherwood Chieftain, 39098	Visct'ss Ossington

OUTHWAITE, John,
Bainesse, Catterick.

MARCIA, red and white, calved July 22, 1867, Vols. xx., xxi., and xxv. pp. 641,
1014, 609.   Bred by the Earl of Zetland, Upleatham; got by King of the
Roses (22043), dam (Music) by Sir James the Rose (15290), &c.

Produce in		Names, &c.	By what Bull.	By whom bred.
1879, April 8, roan,	C.C.	Marcia 2nd	Royal Windsor, 29890	Mr. Outhwaite

OWEN, F. B.,
Deefield, Ellesmere, Salop.

DOT, red and little white, calved July 26, 1874, Vol. xxv. p. 671.   Bred by the
Rev. W. Sneyd, Keele Hall; got by Baron Wellington (30495), dam (Cathe-
rina 3rd) by Duke of Lorn (25985), &c.

1879, Apr. 17, r. & w., B.C.	(Steer)	Welcome Guest, 42584	Mr. Owen

HAWTHORN, red, calved March 15, 1876.   Bred by Mr. J. Evans, Uffington;
got by Richmond (35269), dam (Hopbine) by Manager (24521), &c.   See Vol.
xxii. p. 408.

1879, Feb. 21, r. & w., B.C.	(Steer)	Welcome Guest, 42584	Mr. Owen

SWEET PEA 5TH, roan, calved in June 1875.   Bred by Mr. W. Sheraton,
Broom House, the property of Mr. F. B. Owen; got by Darius (30860), dam
(Sweet Pea 2nd) by May Duke (26885), g. d. (Sweet Pea) by Don Windsor
2nd (21550), &c.   See Vol. xxi. p. 926.

1879, Mar. 28, white, C.C.	Lady Louisa Heald	The Sultan, 40811	Mr. Sheraton

WALWORTH LADY 16TH, red and white, calved June 1, 1875.   Bred by Mr.
W. Sheraton, Broom House, the property of Mr. F. B. Owen; got by Lilles-
hall (34456), dam (Walworth Lady 10th) by Lord of the Nile (34635), g. d.
(Walworth Lady 6th) by Prince Arthur (27117), &c.   See Vol. xxi. p. 927.

1879, Mar. 16, r. & w., C.C.	Walworth Lady 18th	The Sultan, 40811	Mr. Sheraton

OWEN, J. Dorsett,
Plasyn Grove, Ellesmere, Salop.

ANTONIA'S RED DUCHESS, red, calved March 28, 1877.   Bred by Mr. W.
Sheraton, Broom House; got by Lilleshall (34456), dam (Antonia's Red and
White Duchess) by May Duke (26885), g. d. (Antonia 4th) by Richard
(16834), &c.   See Vol. xxi. p. 925. .

1879, Dec. 15, roan,	C.C.	Antonia Princess	The Sultan, 40811	Mr. Owen

BONA 3RD, roan, calved November 26, 1875.   Bred by Captain C. R. Conwy,
Bodrhyddan; got by Favourite (33895), dam (Bona 2nd) by Oxford Wild
Eyes 2nd (32038), &c.   See Vol. xxii. p. 371.

1879, April 20, r. & w., C.C.	Bona Waterloo	Edward Waterloo, 38246	Mr. Owen

HAWARDEN LADY 6TH, white, calved June 20, 1875.   Bred by Mr. W.
Sheraton, Broom House, the property of Mr. J. D. Owen; got by Darius
(30860), dam (Hawarden Lady 3rd) by May Duke (26885), g. d. (Hawarden
Lady 2nd) by Union Jack (27691), &c.   See Vol. xxv. p. 661.

1878, April 17, roan,	C.C.	Hawarden Lady16th	Lilleshall, 34456	Mr. Sheraton
1879, Mar. 28, red,	C.C.	Hawarden Rose	The Sultan, 40811	do.

Hawarden Rose, sold to Mr. J. D. Owen, Plasyn Grove, Ellesmere.

2 P

LADY FITZWOLERAN 3RD, roan, calved March 31, 1877. Bred by Mr. W.
Sheraton, Broom House; got by Lilleshall (34456), dam (Winston Lady) by
Knight of the Garter (34400), &c. See Vol. xxii. p. 561.

Produce in		Names, &c.	By what Bull.	By whom bred.
1879, Dec. 31, roan,	B.C.	Grand Vizier	The Sultan, 40811	Mr. Owen

LADY JANE SEATON, roan, calved July 8, 1876. Bred by Mr. W. Sheraton,
Broom House; got by Lilleshall (34456), dam (Lady Mary Seaton) by
Alliance 3rd (32938), g. d. (Lady Hamilton 2nd) by May Duke (26885), &c.
See Vol. xxii. p. 560.

1878, Dec. 6, white,	B.C.		The Sultan, 40811	Mr. Sheraton
1879, Dec. 30, white,	C.C.	Lady Anne Seaton	do.	Mr. Owen

Bull Calf, sold to Mrs. Walley, Frankton, Oswestry.

PAIN, J. K.,
Eridge, Tunbridge Wells.

FLORENCE 2ND, roan, calved June 20, 1876. Bred by Mr. J. W. Temple,
Leyswood, the property of Mr. H. B. Collins, Tunbridge Wells; got by Baron
Lightburne (30468), dam (Florence) by Sultan (27583), &c. See Vol. xxiii.
p. 669.

1879, Jan. 10, roan,	C.C.	Florence 3rd	Fugleman, 36670	Mr. Temple

Florence 3rd, sold to Mr. J. K. Pain, Eridge, Tunbridge Wells.

PALMER, W. I.,
Kendrick Villa, Reading.

ALICE SWINNERTON, roan, calved November 10, 1868, Vols. xxi. and xxv.
pp. 828, 727. Bred by Mr. E. Lythall, Radford Hall, the property of Mr.
W. I. Palmer; got by Lord of the Isles (20206), dam (Alice) by Alma
(14085), &c.

1879, May 6, roan,	C.C.	Lady Grazeley	Remus, 40584	Mr. Wren

Lady Grazeley, sold to Mr. W. I. Palmer, Kendrick Villa.

LADY ALICE, roan, calved October 8, 1875. Bred by Mr. W. Wren,
Grazeley Court, the property of Mr. W. I. Palmer; got by Telemachus 5th
(35724), dam (Alice Swinnerton) by Lord of the Isles (20206), &c. See
Vol. xxv. p. 727.

1879, Mar. 10, white,	C.C.	Lady Wren	Remus, 40584	Mr. Wren

Lady Wren, sold to Mr. W. I. Palmer, Kendrick Villa.

PARKER, John,
Ingleby, Lincoln.

MERINO, roan, calved April 28, 1875. Bred by the Earl of Zetland, Upleatham; got by Grand Monarch (28774), dam (Marigold) by Maximilian
(20322), &c. See Vol. xxii. p. 629.

1879, Aug. 3, roan,	C.C.	Merino 2nd	Duke of Kent, 33680	Mr. Parker

SIMPLE, red, calved April 16, 1873. Bred by Mr. J. Parker; got by Wellington (30288), dam (Sylph) by Lord Macdonald (22175), &c. See " Frugal," p. 103.

Produce in		Names, &c.	By what Bull.	By whom bred.
1876, April 9, red,	B.C.		D. of Barrington 3rd,30923	Mr. Parker
1877, Mar. 6, red,	B.C.	(Steer)	do.	do.
1878, Feb. 23, r. & w.,	C.C.	Simple 2nd	Pretender, 38884	do.
1879, Jan. 7, { r. & w.,	C.C.	Fruitful	} Duke of Kent, 33680	do.
{ red,	B.C.	Frugal		

1879 Bull Calf, sold to Mr. F. G. Marshall, Riseholme, Lincoln.

PARKER, Rowland,
Moss End, Burton, Westmoreland.

JESSIE 31st, roan, calved March 6, 1871, Vols. xxiii. and xxiv. pp. 590, 599. Bred by Mr. R. Parker; got by Royalist (27371), dam (Jessie 28th) by Oxford 3rd (24705), &c.

1879, May 6, roan,	B.C.	Honest Tom	Bates Duke, 36223	Mr. Parker

LUCY, red and white, calved March 7, 1875. Bred by Mr. R. Parker; got by Grand Duke of Beaumont (31283), dam (Countess of Dalton) by Prince of Dalton (24827), &c. See Vol. xxiii. p. 589.

1879, Mar. 15, roan,	C.C.	Isabella	Bates Duke, 36223	Mr. Parker

MISS SURPRISE, roan, calved March 21, 1872, Vol. xxiii. p. 590. Bred by Mr. R. Parker; got by Fawsley Baronet (26142), dam (Surprise) by Royal Duke (16865), &c.

1879, May 18, r. & w.,	B.C.	Farleton Hero	Bates Duke, 36223	Mr. Parker

PARKIN, William,
Blaithwaite, Aspatria, Cumberland.

BRIDESMAID, red, calved October 16, 1871. Bred by Mr. T. Watson, High House; got by England's Hero (26104), dam (Gertrude) by Sir John (16983), &c. See Vol. xx. p. 539.

1878, July 22, red,	C.C.	Red Bud	Marquis of Lorne, 29298	Mr. Parkin

COWSLIP 22nd, red and white, calved February 24, 1873, Vols. xxii., xxiii., and xxiv. pp. 619, 713, 710. Bred by Mr. T. Wilson, Shotley Hall; got by Lord of Nunwick (26702), dam (Cowslip 12th) by Beverley Butterfly (28021), &c.

1879, Sept. 14, red,	C.C.	Lowther Cowslip	Prince Saladin, 40542	Mr. Parkin

ELLIE CARDIGAN, roan, calved February 21, 1875. Bred by Mr. G. Dixon, Mantle Hill; got by Cherubim 2nd (30727), dam (Gowan Cardigan 3rd) by Majestic (22263), &c. See " Wideawake," Vol. xxiv. p. 277.

1878, Jan. 6, roan,	B.C.	Baron Cardigan	Beverley Baron, 39467	Mr. Parkin
1879, Sept. 14, r. & w.,	B.C.	(Steer)	Wigton, 35995	do.

Baron Cardigan, sold to Mr. J. Sharpe, The Beck, Abbey Town.

EVE, roan, calved April 20, 1877. Bred by Mr. J. Johnstone, Halleaths; got by Magician (34720), dam (Miss Eliza) by Rifleman (32304), &c. See Vol. xxiv. p. 520.

1879, Dec. 1, white,	B.C.	Abel	Captain Radstock, 42878	Mr. Parkin

FANLIGHT, red, calved August 3, 1876, Vol. xxv. p. 548.   Bred by Sir G. R. Philips, Bart., Weston Park ; got by Cherry Grand Duke 5th (30712), dam (Lady Fantail 2nd) by Lord Darlington 2nd (29096), &c.

Produce in		Names, &c.	By what Bull.	By whom bred.
1879, Oct. 30, roan,	B.C.	Factotum	BaronSiddington 2d, 37816	Mr. Parkin

GOWAN CARDIGAN 3RD, roan, calved May 5, 1867.   Bred by Mr. A. Haddon, Honeyburn ; got by Majestic (22263), dam (Gowan Gwynne) by Silk and Scarlet (18825), &c.   See " Wideawake," Vol. xxiv. p. 277.

1878, May 18, r. & w.,	B.C.	Lord of Crookdake	Star of the West, 39165	Mr. Parkin
1879, June 4, red,	C.C.	(dead)	Wigton, 35995	do.

GOWAN CARDIGAN 5TH, white, calved February 27, 1871.   Bred by Mr. G. Dixon, Mantle Hill ; got by Lieutenant Oxford (24336), dam (Gowan Cardigan 2nd) by Majestic (22263), &c.   See " Lord Cardigan," Vol. xxiii. p. 163.

1877, Oct. 4, roan,	C.C.	Lady Cardigan	Star of the West, 39165	Mr. Parkin
1878, Sept. 25, roan,	C.C.	Jinny Cardigan	Wigton, 35995	do.

LADY BLANCHE 2ND, roan, calved February 2, 1871, Vol. xxii. p. 399.   Bred by Mr. G. Dixon, Mantle Hill ; got by Lieutenant Oxford (24336), dam (Lady Blanche Hollyhock) by Fourteenth Duke of Oxford (21605), &c.

1878, Feb. 11, roan,	B.C.	Captain Radstock	Lord Radstock, 34656	Mr. Parkin
1879, April 18, roan,	C.C.	Strawberry Blanche	Wigton, 35995	do.

Captain Radstock, sold to Messrs. R. and W. Hornsby, Abbey House, Silloth.

LADY BLANCHE 3RD, roan, calved June 28, 1873, Vol. xxiii. p. 682.   Bred by Mr. G. Dixon, Mantle Hill ; got by Duke of Tynedale (28462), dam (Lady Blanche 2nd) by Lieutenant Oxford (24336), &c.

1877, Nov. 8, red,	C.C.	Alice Blanche	Lord Radstock, 34656	Mr. Parkin
1879, Mar. 8, roan,	B.C.	Commissioner	Wigton, 35995	do.

LOUISA CARDIGAN, red, calved June 11, 1875.   Bred by Mr. T. Horn, Baggrow ; got by Marquis of Lorne (31847), dam (Lily Gwynne Cardigan 2nd) by Lord Bates (29065), g. d. (Lily Gwynne Cardigan) by Majestic (22263), gr. g. d. (Lily Gwynne) by Silk and Scarlet (18825), &c.   See Vol. xvii. p. 593.

1878, Nov. 16, red,	C.C.	Rose Cardigan	Wigton, 35995	Mr. Parkin

MAGGIE CARDIGAN. roan, calved January 24, 1876.   Bred by Mr. G. Dixon, Mantle Hill ; got by Lord Radstock (34656), dam (Gowan Cardigan 6th) by Duke of Tynedale (28462), &c.   See " Agility," Vol. xxiv. p. 3.

1878, Dec. 28, r. & w.,	C.C.	Nanny Cardigan	Wigton, 35995	Mr. Parkin

MATCHLESS 13TH, roan, calved April 12, 1875.   Bred by Sir W. Lawson, Bart., Brayton ; got by Cambridge Duke (28120), dam (Oxford Matchless 14th) by Fourteenth Duke of Oxford (21605), g. d. (Matchless 10th) by Richard (16834), &c.   See Vol. xviii. p. 610.

1878, Mar. 10, roan,	B.C.	Spring	Star of the West, 39165	Mr. Parkin

Spring, sold to Mr. T. Clark, Akehead, Wigton.

PEARL NECKLACE, roan, calved November 19, 1872.   Bred by Mr. G. Dobson, Williamsgill ; got by Bountiful (30571), dam (Necklace) by Augustus (19247), &c.   See " Royal Glo'ster," Vol. xxi. p. 411.

1877, July 5, r. & w.,	C.C.	Bracelet	Solway, 37500	Mr. Parkin
1878, June 23, roan,	B.C.	Midsummer	Star of the West, 39165	do.
1879, May 9, red,	C.C.	Diamond Necklace	Wigton, 35995	do.

Midsummer, sold to Mr. H. Drape, Kingside Hill, Abbey Town.

PRINCESS ALICE 8TH, red and white, calved September 15, 1873. Bred by
Mr. Nelson, Catgill Hall; got by Duke of Ellendale (30949), dam (Princess
Alice 7th) by Gay Lad (21800), &c. See Vol. xix. p. 678.

Produce in	Names, &c.	By what Bull.	By whom bred.
1879, Oct. 10, r. & w., B.C.	Delegate	Baron Winsome, 39449	Mr. Parkin

RUBY 22ND, red, calved November 2, 1876. Bred by Mr. J. Blackstock,
Hayton Castle; got by Baslow (30507), dam (Ruby 10th) by Bates Duet
(21248), g. d. (Ruby 6th) by Skyrocket (18874), &c. See "Red Duke," Vol. xx.
p. 287.

1879, Dec. 8, red, C.C.	Redbreast	Captain Radstock, 42878	Mr. Parkin

RUTH, red and white, calved December 31, 1874, Vol. xxiv. p. 300. Bred by
the Exors. of Mr. G. Angus, Broomley; got by Royal Frederick (35366),
dam (Rachel 2nd) by Prince Charlie (29606), &c.

1879, May 30, roan, C.C.	Rebecca	Ben Brace, 30624	Mr. Parkin

PATERSON, J. T. S.,
Plean Farm, Bannockburn, N.B.

JESSAMINE, roan, calved September 18, 1873. Bred by Mr. W. J. Legh,
Lyme Park; got by Lord Nelson (31709), dam (Jessica) by Valorous (27701),
g. d. (Jeannie) by Asteroid (21193), &c. See "Julienne," Vol. xxiii. p. 604.

1877, Dec. 19, roan, B.C.		Lord Spencer 2nd, 37006	Mr. Paterson
1879, Jan. 26, red, C.C.	Jessamine 2nd	Marquis of Oxford 2d, 37055	do.

ROSEMARY 13TH, red and white, calved August 12, 1875. Bred by Mr. J. T.
S. Paterson; got by Third Duke of Hillhurst (30975), dam (Rosemary 12th)
by Brigand (28080), &c. See Vol. xxiv. p. 600.

1877, May 26, r. & w., C.C.	Rosemary 15th	Wild Chieftain 2nd, 37678	Mr. Paterson
1878, May 31, roan, C.C.	Rosemary 17th	Earl of Connaught, 38219	do.

PATERSON, J. W. J.,
Terrona, Langholm, N.B.

LILY HEARTBREAKER, red and white, calved November 22, 1876. Bred
by Mr. J. W. J. Paterson; got by Duke of Thorndale (33739), dam (Lucy
Bertram) by Heartbreaker (28826), &c. See Vol. xxiii. p. 591.

1879, Nov. 7, r. & w., B.C.	Stonebreaker	Marq. of Annandale, 40306	Mr. Paterson

PAUL, Sir R. J., Bart.,
Ballyglan, Waterford, Ireland.

ANNASTASIA, red, calved in May, 1870, Vols. xxi., xxiii., and xxv. pp. 878,
591, 612. Bred by the Lord Bishop of Cashel, Bishopscourt; got by Knight
of the Empire (22061), dam (Annie Laurie) by Berrington Boy (11173), &c.

1879, April 17, red, C.C.	Ariadne	Red-wood, 42265	Sir R. J. Paul

REINE, roan, calved March 11, 1873, Vols. xxiii. and xxv. pp. 592, 613. Bred
by Sir R. J. Paul, Bart.; got by Troubadour (32736), dam (Royalty) by Glory
(24043), &c.

1879, April 15, red, C.C.	Regency	Red-wood, 42265	Sir R. J. Paul

REPRISAL, red and white, calved March 4, 1876. Bred by Sir R. J. Paul, Bart.; got by Pirate (37191), dam (Reine) by Troubadour (32736), &c. See "Rosewood," p. 214.

Produce in		Names, &c.	By what Bull.	By whom bred.
1879, Feb. 28, red,	B.C.	Rosewood	Red-wood, 42265	Sir R. J. Paul

SATCHEL, roan, calved April 4, 1875, Vol. xxv. p. 613. Bred by Sir R. J. Paul, Bart.; got by Troubadour (32736), dam (Sable) by Glory (24043), &c.

1879, Mar. 10, roan,	B.C.	Ringwood	Red-wood, 42265	Sir R. J. Paul

STRESA, red, calved March 6, 1876. Bred by Sir R. J. Paul, Bart.; got by Troubadour (32736), dam (Silk Jacket 3rd) by Honiton (41691), g. d. (Silk Jacket 1st) by Laurustinus (41788), gr. g. d. (Silkworm) by Sir Charles (7501), — Silkworm.

1879, Mar. 5, r. & w.,	C.C.	Spindle	Red-wood, 42265	Sir R. J. Paul

PAULL, J. W.,
Knott Oak House, Ilminster.

FROGMORE 4TH, roan, calved January 4, 1876. Bred by Mr. J. W. Paull; got by Duke of Kennet (30977), dam (Frogmore 3rd) by Snowball (30020), g. d. (Ringlet) by Prince Albert (22560), &c. See Vol. xx. p. 727.

1879, Jan. 17, r. & w.,	B.C.	Somerset	Lord Townsend, 38667	Mr. Paull

PEACEY, William,
Chedglow, Tetbury, Gloucestershire.

CARRIE WALTON 2ND, roan, calved March 4, 1876. Bred by Mr. W. Peacey; got by Crown Prince (30829), dam (Polly Walton) by General Clarence (28689), &c. See Vol. xxiv. p. 600.

1879, Mar. 8, white,	C.C.	Lily	Mandarin 14th, 40289	Mr. Peacey

MISS WALTON, roan, calved March 12, 1875. Bred by Mr. W. Peacey; got by Crown Prince (30829), dam (Polly Walton) by General Clarence (28689), &c. See Vol. xxii. p. 527.

1878, Mar. 29, white,	B.C.	(Steer)	L. Fitzclarence 11th, 34550	Mr. Peacey
1879, Mar. 11, roan,	C.C.	Miss Walton 2nd	Mandarin 14th, 40289	do.

PATIENCE, roan, calved December 4, 1874. Bred by Mr. W. Peacey; got by Crown Prince (30829), dam (Passive) by Cynric (19542), &c. See "Principal," p. 203.

1878, Mar. 12, roan,	B.C.	Principal	L. Fitzclarence 11th, 34550	Mr. Peacey
1879, Feb. 26, roan,	C.C.	Patience 2nd	Mandarin 14th, 40289	do.

PENSIVE, red and little white, calved December 23, 1873. Vol. xxiv. p. 600. Bred by Mr. W. Peacey; got by Ranger (35203), dam (Passive) by Cynric (19542), &c.

1878, Mar. 6, red,	B.C.	(dead)	L. Fitzclarence 11th, 34550	Mr. Peacey
1879, Feb. 3, roan,	C.C.	Primula	Sonsie Boy, 40740	do.

PYE, red and little white, calved December 29, 1875. Bred by Mr. W. Peacey; got by Crown Prince (30829), dam (Passive) by Cynric (19542), &c. See "Principal," p. 203.

1878, Oct. 7, red,	B.C.	(Steer)	Mandarin 14th, 40289	Mr. Peacey
1879, Sept. 24, roan,	C.C.	Pye 2nd	do.	do.

SONSIE, red and white, calved February 24, 1876. Bred by Mr. W. Peacey;
got by Crown Prince (30829), dam (Rose of Dumbleton 2nd) by Grand Duke
of Glo'ster (19882), &c. See Vol. xxiv. p. 600.

Produce in	Names, &c.	By what Bull.	By whom bred.
1878, Sept. 30, roan, B.C.	(Steer)	Mandarin 14th, 40289	Mr. Peacey
1879, Oct. 24, r.&w., C.C.	Sonsie 2nd	do.	do.

PEARS, T.,
Hackthorne, Lincoln.

AGNES, roan, calved July 6, 1873, Vol. xxiv. p. 600. Bred by Mr. T. Pears;
got by Knight of Killerby (28999), dam (Attraction) by Robin (24968), &c.

1879, Oct. 22, red, C.C.	Attractive Maid	King of Trumps, 31512	Mr. Pears

ALPINE MANTALINI, roan, calved February 17, 1876. Bred by Mr. A.
Raine, Bow Bank; got by Warlaby (32792), dam (Lady Mowbray) by Whiff
(30299), &c. See " Merope," p. 169.

1879, Jan. 11, r.&w., B.C.	Merope	Lichtenstein, 41801	Mr. Pears

ATTRACTION, red and white, calved November 14, 1869, Vols. xxii., xxiv., and
xxv. pp. 527, 601, 613. Bred by Messrs. W. and H. Dudding, Panton House;
got by Robin (24968), dam (Alice Buckingham) by Royal Buckingham
(20718), &c.

1879, Dec. 12, r.&w., B.C.	Attractive King	King of Trumps, 31512	Mr. Pears

BUTTERCUP 4TH, white, calved in March, 1875. Bred by Mr. J. Pears, Mere;
got by Booth Satellite (30564), dam (Young Buttercup) by Fitz-Realm (33935),
g.d. (Buttercup 2nd) by Orthodox 24th (18491), &c. See Vol. xvii. p. 396.

1879, Mar. 24, white, C.C.	Bloomer	Fanatic, 38269	Mr. Pears

CHATHAM ROSE, red, calved in May, 1871. Bred by Mr. R. Searson, Cranmore Lodge; got by Lord Chatham (26625), dam (Bonny Rose) by Falstaff
(21720), &c. See Vol. xx. p. 416.

1877, Jan. 26, r.&w., B.C.	Revolter	Bezique, 33148	Mr. Pears
1878, Feb. 22, r.&w., B.C.	Randolph	Earl of Ashfield, 36571	do.
1879, July 6, r.&w., B.C.	Rousseau	King of Trumps, 31512	do.

Revolter, sold to Mr. J. Byron, Kirkby Green, Sleaford; Randolph, to Mr. J. Davy,
Worlaby.

DOUBLE ROSE 1ST, roan, calved June 23, 1876. Bred by the Hon. M. H.
Cochrane, Hillhurst, Compton, Canada; got by Sirius (39121), dam (Royal
Rose) by Royal Briton (27351), &c. See Vol. xxiv. p. 381.

1879, Dec. 29, roan, C.C.	Roberta	King of Trumps, 31512	Mr. Pears

DOUBLE ROSE 2ND, roan, calved June 23, 1876. Bred by the Hon. M. H.
Cochrane, Hillhurst, Compton, Canada; got by Sirius (39121), dam (Royal
Rose) by Royal Briton (27351), &c. See Vol. xxiv. p. 381.

1879, July 13, roan, C.C.	Roxana	King of Trumps, 31512	Mr. Pears

LADY DUCHESS 2ND, red and white, calved October 18, 1871, Vol. xxv.
p. 613. Bred by the Rev. J. Storer, Hellidon; got by Earl of Clare (28496),
dam (Lady Duchess) by Prince of Rosedale (24837), &c.

1879, May 16, r.&w., C.C.	Danae	Earl of Ashfield, 36571	Mr. Pears

LIGHT, red and white, calved October 23, 1865, Vol. xxiii. p. 592. Bred by
Mr. S. Wiley, Brandsby; got by Earl of Derby (21638), dam (Luna) by
Prince George (13510), &c.

Produce in		Names, &c.	By what Bull.	By whom bred.
1878, Aug. 3, r.&w.,	B.C.	Lightfoot	Earl of Ashfield, 36571	Mr. Pears
1879, Oct. 1, r.&w.,	B.C.	Lancelot	King of Trumps, 31512	do.

LUSTRE, red and white, calved July 17, 1874. Bred by Mr. T. Pears; got by
Knight of Killerby (28999), dam (Light) by Earl of Derby (21638), &c. See
Vol. xxiii. p. 592.

1879, Sept. 22, r.&w.,	B.C.	Lepidus	King of Trumps, 31512	Mr. Pears

MANTALINI BREEZE, red, calved July 5, 1876. Bred by Mr. A. Raine,
Bow Bank; got by Whiff (30299), dam (Maid of Connington) by Quis
(20618), &c. See "Machaon," p. 161.

1879, Jan. 4, r.&w.,	B.C.	Machaon	Earl of Ashfield, 36571	Mr. Pears

MARCHIONESS MARGARET, red and white, calved July 13, 1874. Bred by
Mr. T. Pears; got by Knight of Killerby (28999), dam (Mace 4th) by
Paradox 1st (22487), &c. See Vol. xxii. p. 527.

1876, Aug. 19, roan,	C.C.	Margaret Black	Pearl Diver, 35025	Mr. Pears
1878, Dec. 19, r.&w.,	C.C.	Myra	Earl of Ashfield, 36571	do.
1879, Dec. 8, r.&w.,	C.C.	Milk Maid	King of Trumps, 31512	do.

MATTIE, white, calved April 4, 1873, Vols. xxiv. and xxv. pp. 610, 614. Bred
by Mr. T. Pears; got by Knight of Knowlmere (22055), dam (Matchless
12th) by Duke of Windsor (19651), &c.

1879, Mar. 21, roan,	C.C.	Minnie Warren	Earl of Ashfield, 36571	Mr. Pears

ROSAMOND, roan, calved June 23, 1873. Bred by Mr. T. Pears; got by
Knight of Killerby (28999), dam (Winter Rose) by Duke of Devonshire
(21588), &c. See Vol. xxii. p. 527.

1879, Mar. 16, roan,	B.C.	Ricimer	Earl of Ashfield, 36571	Mr. Pears

SUMMER MORN, red and white, calved in July, 1873. Bred by Mr. D.
Fisher, Pitlochrie; got by Valentine Vox (32752), dam (Rosy Morn) by
Brother Windsor (25690), &c. See "Mino Taurus," p. 171.

1879, Mar. 3, r.&w.,	B.C.	Mino Taurus	Earl of Ashfield, 36571	Mr. Pears

WOOD NYMPH, red and white, calved April 25, 1875, Vol. xxv. p. 614. Bred
by Mr. J. How, Broughton; got by Great Hope (24082), dam (Water
Nymph) by Sir Windsor Broughton (27507), &c.

1879, Dec. 23, r.&w.,	C.C.	Woodland Queen	King of Trumps, 31512	Mr. Pears

PEEL, Jonathan,
Knowlmere Manor, Clitheroe.

ELFIN QUEEN, white, calved October 19, 1873, Vols. xxii. and xxv. pp. 518,
616. Bred by Mr. J. P. Haslam, Gilnow House; got by Sidus (29969), dam
(Gipsy Queen) by Roan Windsor (24967), &c.

1879, March, roan,	C.C.	Elaine	Star Chieftain, 35672	Mr. Peel

MARINARA, roan, calved June 3, 1875, Vol. xxv. p. 616. Bred by Mr. J.
Peel; got by Germanicus (28706), dam (Marina) by Sacristan (29911), &c.

1879, Aug. 17, r.&w.,	C.C.	Marissa	Baron Stackhouse, 30488	Mr. Peel

TOTTIE, red, calved March 6, 1874. Bred by Mr. G. J. Day, Horsford Hall; got by Lord Blithesome (29067), dam (Telltale) by Lord of the Isles (24443), g. d. (Truth) by Rajah Brooke (20621), &c. See Vol. xx. p. 792.

Produce in		Names, &c.	By what Bull.	By whom bred.
1879, Dec. 23, roan,	B.C.	(dead)	Dunstan, 41461	Mr. Peel

PEEL, Sir Robert, Bart.,
Drayton Manor, Fazeley, Tamworth.

COUNTESS BICKERSTAFFE 2ND, red and white, calved July 17, 1873, Vols. xxiii. and xxiv. pp. 595, 603. Bred by Sir R. Peel, Bart.; got by Red Knight (29741), dam (Cherry) by Count Bickerstaffe (23630), &c.

1879, Mar. 31, r. & w.,	C.C.	C'ss Bickerstaffe 13th	The Druid, 35753	Sir R. Peel, Bart.

COUNTESS BICKERSTAFFE 3RD, roan, calved June 16, 1874. Bred by Sir R. Peel, Bart.; got by Red Knight (29741), dam (Cherry) by Count Bickerstaffe (23630), &c. See Vol. xxi. p. 880.

1879, Mar. 8, roan,	C.C.	C'ss Bickerstaffe 12th	The Druid, 35753	Sir R. Peel, Bart.

GARLAND 9TH, red and white, calved May 14, 1874, Vols. xxiii. and xxiv. pp. 595, 603. Bred by Sir R. Peel, Bart.; got by Red Knight (29741), dam (Garland 7th) by Count Bickerstaffe (23630), &c.

1878, Dec. 19, roan,	C.C.	Garland 13th	The Druid, 35753	Sir R. Peel, Bart.
1879, Dec. 23, r. & w.,	B.C.	The Rover	do.	do.

ISABELLA 10TH, red, calved June 30, 1874, Vol. xxiv. p. 603. Bred by Sir R. Peel, Bart.; got by Red Knight (29741), dam (Perry) by Tulip (27683), &c.

1879, Jan. 19, r. & w.,	C.C.	Isabella 18th	The Druid, 35753	Sir R. Peel, Bart.

LADY ELEANOR 2ND, red, calved August 4, 1873, Vols. xxiii. and xxiv. pp. 595, 603. Bred by Sir R. Peel, Bart.; got by Red Knight (29741), dam (Lightback) by Count Bickerstaffe (23630), &c.

1879, Mar. 17, roan,	C.C.	Lady Eleanor 10th	The Druid, 35753	Sir R. Peel, Bart.

LADY TRENT, roan, calved December 17, 1871, Vols. xxii. and xxiv. pp. 528, 604. Bred by Sir R. Peel, Bart.; got by Red Knight (29741), dam (Young Tulip) by Baron Trent (19285), &c.

1879, Aug. 20, r. & w.,	B.C.	Roger	The Druid, 35753	Sir R. Peel, Bart.

PRIDE, white, calved May 4, 1870, Vols. xxi., xxiii., and xxiv. pp. 881, 596, 604. Bred by Sir R. Peel, Bart.; got by Count Bickerstaffe (23630) dam (Nancy 2nd) by Tulip (27683), &c.

1879, Mar. 3, white,	C.C.	Lady Helen 6th	The Druid, 35753	Sir R. Peel, Bart.

PENRHYN, Lord,
Penrhyn Castle, Bangor, North Wales.

ARCHDUCHESS OF OXFORD, roan, calved May 11, 1875, Vol. xxiv. p. 604. Bred by Lord Penrhyn, Penrhyn Castle; got by Grand Duke 20th (31281), dam (Grand Duchess of Oxford 7th) by Lord Oxford (20214), &c.

1879, Jan. 26, roan,	B.C.	Archduke of Oxford	Duke of Glo'ster 7th, 39735	Lord Penrhyn

BARONESS, roan, calved June 26, 1871, Vols. xxii. and xxiv. pp. 545, 627.
Bred by Mr. A. J. Robarts, Lillingstone Dayrell; got by Caractacus (28141),
dam (Lady Barrington 7th) by Baron Tarves (17387), &c.

Produce in		Names, &c.	By what Bull.	By whom bred.
1879, Jan. 2, white,	C.C.	Baroness 2nd	G. Duke of Oxford, 31293	Lord Penrhyn

FOURTH BELLE OF OXFORD, red and white, calved February 20, 1872,
Vols. xxi., xxiv., and xxv. pp. 881, 604, 616. Bred by Lord Penrhyn, Penrhyn
Castle; got by Grand Duke 11th (21849), dam (Third Belle of Oxford) by
Marmaduke (14897), &c.

1879, Jan. 16, roan,	C.C.	Belle of Oxford 10th	G. Duke of Oxford, 31293	Lord Penrhyn
1879, Dec. 20, red,	B.C.	Beau of Oxford 6th	do.	do.

CAMBRIA 10TH, roan, calved June 12, 1874, Vols. xxiv. and xxv. pp. 604,
617. Bred by Lord Penrhyn, Penrhyn Castle; got by Grand Duke 20th
(31281), dam (Cambria 8th) by Third Duke of Claro (23729), &c.

1879, Dec. 31, roan,	B.C.	(Steer)	G. Duke of Oxford, 31293	Lord Penrhyn

CAMBRIA 11TH, red and white, calved June 28, 1875, Vol. xxv. p. 617. Bred
by Lord Penrhyn, Penrhyn Castle; got by Grand Duke 20th (31281), dam
(Cambria 8th) by Third Duke of Claro (23729), &c.

1879, May 4, red,	C.C.	Cambria 15th	G. Duke of Oxford, 31293	Lord Penrhyn

CAMBRIA 12TH, roan, calved August 24, 1876. Bred by Lord Penrhyn,
Penrhyn Castle; got by Grand Duke 20th (31281), dam (Cambria 8th) by
Third Duke of Claro (23729), &c. See Vol. xxiii. p. 597.

1879, June 1, roan,	B.C.	(Steer)	Beau of Oxford 3rd, 33130	Lord Penrhyn

CHARMING GENEVA, roan, calved July 3, 1873, Vols. xxiv. and xxv.
pp. 605, 617. Bred by Mr. J. K. Fowler, Prebendal Farm; got by King
Charming (28962), dam (Lady Geneva 2nd) by Duke of Cumberland
(21584), &c.

1879, Dec. 4, red,	B.C.	Geneva's Duke 6th	G. Duke of Oxford, 31293	Lord Penrhyn

CHERRY DUCHESS 19TH, roan, calved July 11, 1872, Vols. xxiv. and xxv.
pp. 605, 617. Bred by Lord Penrhyn, Penrhyn Castle; got by Oxford Beau
(29485), dam (Cherry Duchess 14th) by Third Duke of Wharfdale (21619), &c.

1879, Sept. 30, r.&w.,	C.C.	(dead)	G. Duke of Oxford, 31293	Lord Penrhyn

CHERRY DUCHESS 23RD, roan, calved May 6, 1873, Vols. xxiii. and xxv.
pp. 597, 617. Bred by Lord Penrhyn, Penrhyn Castle; got by Oxford Beau
(29485), dam (Cherry Duchess 15th) by Second Duke of Grafton (25968), &c.

1879, May 15, r.&w.,	B.C.	Cherry Duke 11th	G. Duke of Oxford, 31293	Lord Penrhyn

CHERRY DUCHESS 26TH, red and white, calved March 12, 1876. Bred by
Lord Penrhyn, Penrhyn Castle; got by Grand Duke of Oxford (31293), dam
(Cherry Duchess 23rd) by Oxford Beau (29485), &c. See "Cherry Duke
11th," p. 47.

1879, Jan. 30, red,	C.C.	Cherry Duchess 29th	Duke of Rosedale 6th, 38176	Lord Penrhyn
1879, Dec. 27 { roan,	B.C.	(dead)	Beau of Oxford 3rd, 33130	do.
roan,	C.C.	(dead)		

DOWAGER DUCHESS 6TH, red and white, calved July 18, 1874, Vols. xxiv.
and xxv. pp. 605, 617. Bred by Lord Penrhyn, Penrhyn Castle; got by Grand
Duke 20th (31281), dam (Dowager) by Duke of Geneva (19614), &c.

1879, April 8, roan,	C.C.	Dowager Duchess 10th	G. Duke of Oxford, 31293	Lord Penrhyn

DOWAGER DUCHESS 8TH, roan, calved November 15, 1875, Vol. xxv.
p. 617. Bred by Lord Penrhyn, Penrhyn Castle; got by Grand Duke of
Oxford (31293), dam (Dowager) by Duke of Geneva (19614), &c.

Produce in	Names, &c.	By what Bull.	By whom bred.
1879, July 15, r.&w., C.C.	DowagerDuchess11th	Beau of Oxford 3rd, 33130	Lord Penrhyn

DUCHESS OF ARVON, red, calved December 11, 1873, Vols. xxiv. and xxv.
pp. 605, 617. Bred by Lord Penrhyn, Penrhyn Castle; got by Grand Duke
20th (31281), dam (Duet) by Cherry Duke (25752), &c.

1879, Oct. 31, r. & w., B.C.	Duke of Arvon 2nd	G. Duke of Oxford, 31293	Lord Penrhyn

ELLA, roan, calved January 20, 1869, Vols. xx. and xxv. pp. 500, 618.` Bred
by Lord Southampton, Whittlebury Lodge; got by Oliver 1st (24683), dam
(Laura) by Marmaduke (14897), &c.

1879, June 1, r.&w., C.C.	Ella 3rd	G. D. of Oxford 2nd, 38381	Lord Penrhyn

GRAND DUCHESS OF OXFORD 3RD, white, calved June 7, 1871, Vols. xxii.
and xxv. pp. 529, 618. Bred by Lord Braybrooke, Audley End; got by
Thorndale Duke (27661), dam (Grand Duchess of Oxford) by Grand Duke
7th (19877), &c.

1879, Feb. 3, roan, B.C.	Duke of Grafton 13th	G. Duke of Oxford, 31293	Lord Penrhyn

GRAND DUCHESS OF OXFORD 7TH, roan, calved May 23, 1864, Vols.
xvii., xviii., xix., and xxii. pp. 516, 519, 538, 529. Bred by the Duke of
Devonshire, Holker Hall; got by Lord Oxford (20214), dam (Grand Duchess
of Oxford) by Grand Duke 3rd (16182), &c.

1879, April 28, roan, B.C.	G. D. of Oxford 5th	Grand Duke 20th, 31281	Lord Penrhyn

GUINEVERE 2ND, red and white, calved September 7, 1873, Vols. xxiv. and
xxv. pp. 606, 618. Bred by Lord Penrhyn, Penrhyn Castle; got by
Eighteenth Duke of Oxford (25995), dam (Duchess of Brailes) by Duke of
Brailes (23724), &c.

1879, Dec. 23, r.&w., B.C.	(Steer)	G. Duke of Oxford. 31293	Lord Penrhyn

HYLDA, red, calved December 11, 1869, Vols. xxii., xxiii., xxiv., and xxv. pp.
529, 597, 606, 618. Bred by Lord Southampton, Whittlebury Lodge; got
by Sam of Oxford (25084), dam (Laura) by Marmaduke (14897), &c.

1879, Oct. 13, r.&w., C.C.	Hylda 8th	G. D. of Oxford 2nd, 38381	Lord Penrhyn

HYLDA 2ND, red and white, calved April 17, 1875. Bred by Lord Penrhyn,
Wicken Park; got by Grand Duke 21st (34061), dam (Hylda) by Sam of
Oxford (25084), &c. See Vol. xxii. p. 529.

1879, April 2, r.&w., C.C.	Hylda 7th	G. D. of Oxford 2nd, 38381	Lord Penrhyn

JAVA, red, calved January 2, 1876, Vol. xxv. p. 618. Bred by Lord Penrhyn,
Wicken Park; got by Oxford Waterloo (35002), dam (Jubilee) by Cherry
Duke (25752), &c.

1879, Dec. 15, red, C.C.	Java 2nd	G. D. of Oxford 2nd, 38381	Lord Penrhyn

JOVIAL, roan, calved December 10, 1872, Vols. xxiii. and xxv. pp. 597, 618.
Bred by Lord Penrhyn, Wicken Park; got by Oxford Beau (29485), dam
(Jemima) by Cherry Duke (25752), &c.

1879, Jan. 1, roan, B.C.	(Steer)	G. D. of Oxford 2nd, 38381	Lord Penrhyn
1879, Dec. 9, roan, C.C.	Jovial 3rd	do.	do.

LADY EDITH, red and white, calved May 1, 1871, Vols. xxi., xxiv., and xxv. pp. 883, 606, 619. Bred by Mr. H. J. Sheldon, Brailes House; got by Duke of Brailes (23724), dam (Edith of Fawsley) by Prince Christian (22582), &c.

Produce in	Names, &c.	By what Bull.	By whom bred.
1879, Dec. 19, red,	C.C.|Lady Edith 4th	G. D. of Oxford 2nd, 38381	Lord Penrhyn

LADY GENEVA 2ND, red and white, calved March 23, 1869, Vols. xx., xxi., xxiv., and xxv. pp. 596, 718, 606, 619. Bred by Mr. H. J. Meakin, Shobnal Grange; got by Duke of Cumberland (21584), dam (Lady Geneva) by Duke of Geneva (19614), &c.

Produce in	Names, &c.	By what Bull.	By whom bred.
1879, Aug. 30, r. & w.,	B.C.|Geneva's Duke 5th	G. Duke of Oxford, 31293	Lord Penrhyn

PRINCESS JANETTE, red and white, calved February 11, 1870, Vols. xx., xxii., and xxv. pp. 703, 529, 619. Bred by Mr. J. N. Beasley, Chapel Brampton; got by Juvenile (22021), dam (Queen Janette) by Royal Butterfly 5th (18756), &c.

Produce in	Names, &c.	By what Bull.	By whom bred.
1879, Dec. 7, r. & w.,	B.C.|Prince John 4th	G. D. of Oxford 2nd, 38381	Lord Penrhyn

PRINCESS JANETTE 3RD, red and white, calved February 22, 1875, Vol. xxv. p. 619. Bred by Lord Penrhyn, Wicken Park; got by Cherry Duke (25752), dam (Princess Janette) by Juvenile (22021), &c.

Produce in	Names, &c.	By what Bull.	By whom bred.
1879, Nov. 7, red,	C.C.|Princess Janette 5th	G. D. of Oxford 2nd, 38381	Lord Penrhyn

PRINCESS JOSEPHINE, roan, calved February 10, 1873, Vols. xxiii., xxiv., and xxv. pp. 598, 606, 619. Bred by Mr. T. Allen, Thurmaston Lodge; got by Duke of Waterloo (28464), dam (Jacinth 2nd) by Damascus (25871), &c.

Produce in	Names, &c.	By what Bull.	By whom bred.
1879, April 14, red,	C.C.|P'cess Josephine 5th	G. D. of Oxford 2nd, 38381	Lord Penrhyn

PRINCESS JOSEPHINE 2ND, roan, calved January 9, 1876, Vol. xxv. p. 619. Bred by Lord Penrhyn, Wicken Park; got by Oxford Waterloo (35002), dam (Princess Josephine) by Duke of Waterloo (28464), &c.

Produce in	Names, &c.	By what Bull.	By whom bred.
1879, Dec. 20, red,	B.C.|Prince Joseph	G. D. of Oxford 2nd, 38381	Lord Penrhyn

ROSE OF SUMMER, roan, calved July 30, 1871, Vols. xxi., xxiii., and xxiv. pp. 805, 529, 528. Bred by Mr. J. W. Larking, Ashdown House; got by Third Duke of Waterloo (23801), dam (Lady Knightley) by Sixth Duke of Airdrie (19602), &c.

Produce in	Names, &c.	By what Bull.	By whom bred.
1879, Oct. 15, red,	B.C.|Summer Sun	G. Duke of Oxford, 31293	Lord Penrhyn

RUBY, red, calved February 20, 1876. Bred by Mr. C. M. Hamer, Snitterfield; got by Cherry Grand Duke 2nd (25758), dam (Ruth) by Friponnier (26208), &c. See Vol. xxiii. p. 598.

Produce in	Names, &c.	By what Bull.	By whom bred.
1879, Mar. 2, red,	C.C.|Ruby 2nd	G. D. of Oxford 2nd, 38381	Lord Penrhyn

RUBY LIPS 2ND, roan, calved October 11, 1873, Vols. xxiii. and xxv. pp. 537, 619. Bred by Mr. J. Lancaster, Berriden; got by Grand Duke of Kent (26289), dam (Ruby Lips) by Sir Charles Knightley (27466), &c.

Produce in	Names, &c.	By what Bull.	By whom bred.
1879, Oct. 12, red,	B.C.|Rupert	G. D. of Oxford 2nd, 38381	Lord Penrhyn

SERAPHINA'S DUCHESS 2ND, roan, calved December 8, 1873, Vols. xxiv. and xxv. pp. 606, 620. Bred by Mr. E. H. Cheney, Gaddesby Hall; got by Ninth Duke of Geneva (28391), dam (Seraphina 22nd) by Royal Cumberland (27358), &c.

Produce in	Names, &c.	By what Bull.	By whom bred.
1879, Dec. 22, roan,	B.C.|(Steer)	G. Duke of Oxford, 31293	Lord Penrhyn

WATERLOO 39TH, red and white, calved April 19, 1874, Vols. xxiv. and xxv.
pp. 607, 620. Bred by Lord Penrhyn, Penrhyn Castle; got by Grand Duke
20th (31281), dam (Waterloo 34th) by Grand Duke 11th (21849), &c.

Produce in	Names, &c.	By what Bull.	By whom bred.
1879, April 23, roan, B.C.	D.of Wellington14th	G. Duke of Oxford, 31293	Lord Penrhyn

WATERLOO 41ST, red, calved April 15, 1875, Vol. xxv. p. 620. Bred by Lord
Penrhyn, Penrhyn Castle; got by Grand Duke 20th (31281), dam (Waterloo
34th) by Grand Duke 11th (21849), &c.

| 1879, April 6, red, | C.C. | Waterloo 48th | G. Duke of Oxford, 31293 | Lord Penrhyn |

WILD DUCHESS 4TH, roan, calved August 14, 1874. Bred by Messrs. F.
Leney and Sons, Wateringbury; got by Sixth Duke of Oneida (30997), dam
(Wild Duchess 3rd) by Lord Oxford 2nd (20215), &c. See "Oxford Wild Eyes
3rd," p. 182.

| 1879, June 29, white, B.C. | OxfordWildEyes3rd | G. Duke of Oxford, 31293 | Lord Penrhyn |

WILD EYES 32ND, roan, calved December 25, 1874, Vol. xxv. p. 621. Bred
by Mr. W. Angerstein, Weeting Hall; got by Third Duke of Claro (23729),
dam (Wild Eyes 30th) by Seventh Duke of York (17754), &c.

| 1879, July 25, white, C.C. | Wild Eyes 34th | G. Duke of Oxford, 31293 | Lord Penrhyn |

PERRY, G. and J.,
Acton Pigott, Condover, Salop.

AGNES CLARO, roan, calved June 1, 1871, Vols. xxi., xxiii., and xxiv. pp.
614, 368, 360. Bred by Mr. W. Caless, Bodicote House; got by Lord Claro
(29085), dam (Artful) by Francis 1st (21772), &c.

| 1879, May 20, white, C.C. | Annie Claro | Moonstone, 37107 | Messrs. Perry |

ANEMONE 2ND, red and white, calved March 5, 1875. Bred by Messrs. G.
and J. Perry; got by Sir Jonathan (32508), dam (Appletreewick) by Emperor
of the Isles (23887), &c. See Vol. xxiii. p. 598.

| 1879, June 9, roan, C.C. | Ada | Fawsley Knight, 38284 | Messrs. Perry |

ANEMONE 3RD, roan, calved May 20, 1876. Bred by Messrs. G. and J.
Perry; got by Sir Jonathan (32508). dam (Appletreewick) by Emperor of the
Isles (23887), &c. See Vol. xxiii. p. 598.

| 1879, May 27, roan, C.C. | Annora | Fawsley Knight, 38284 | Messrs. Perry |

BLUSHING BRIDE, red, calved January 27, 1871, Vol. xxiii. p. 598. Bred
by Messrs. G. and J. Perry; got by Manager (24521), dam (Bridesmaid) by
Romeo (22754), &c.

| 1877, May 17, red, | B.C. | (Steer) | Ancient Britain, 37733 | Messrs. Perry |
| 1879, July 10, roan, | C.C. | Blooming Bride | Fawsley Knight, 38284 | do. |

CLARO 1ST, white, calved May 9, 1875. Bred by Mr. W. Caless, Bodicote House;
got by Lord Hatherley (34564). dam (Agnes Claro) by Lord Claro (29085),
&c. See Vol. xxiii. p. 368.

| 1879, Aug. 1, white, C.C. | Claro 2nd | Moonstone, 37107 | Messrs. Perry |

LADY COLERIDGE, white, calved August 5, 1875. Bred by Mr. W. Caless,
Bodicote House; got by Lord Hatherley (34564), dam (Lady Love) by Huntsman (21963), &c. See Vol. xxiii. p. 369.

| 1879, Mar. 26, roan, C.C. | Lady Coleridge 2nd | Fawsley Knight, 38284 | Messrs. Perry |

LADY HATHERLEY, roan, calved June 18, 1873, Vol. xxiv. p. 360. Bred by Mr. W. Caless, Bodicote House; got by Best Man (30532), dam (Lady Love) by Huntsman (21963), &c.

Produce in		Names, &c.	By what Bull.	By whom bred.
1879, May 5, roan,	C.C.	Lady Harriet	Fawsley Knight, 38284	Messrs. Perry

MARCHIONESS 4TH, red and white, calved June 20, 1876. Bred by Messrs. G. and J. Perry; got by Lord Lind (31692), dam (Marchioness 3rd) by Dusty (23811) or Marquis of Cobham (22299), &c. See Vol. xxiii. p. 599.

Produce in		Names, &c.	By what Bull.	By whom bred.
1879, July 26, roan,	C.C.	Marchioness 5th	Fawsley Knight, 38284	Messrs. Perry

PETUNIA 6TH, red, calved April 10, 1875, Vol. xxiv. p. 607. Bred by Messrs. G. and J. Perry; got by Sir Jonathan (32508), dam (Petunia 4th) by Manager (24521), &c.

Produce in		Names, &c.	By what Bull.	By whom bred.
1878, June 4, red,	C.C.	Petunia 9th	Fawsley Knight, 38284	Messrs. Perry
1879, May 7, roan,	C.C.	Petunia 10th	do.	do.

QUEEN OF WARLABY, roan, calved February 5, 1871, Vol. xxiii. p. 599. Bred by Captain Winnall, Burton House; got by Baron Fawsley (27936), dam (Queen of the Empire) by Prince of the Empire (20578), &c.

Produce in		Names, &c.	By what Bull.	By whom bred.
1877, Aug. 16, roan,	B.C.	Prince Imperial	Ancient Britain, 37732	Messrs. Perry
1878, June 10, roan,	C.C.	Queen of Warlaby 3d	Fawsley Knight, 38284	do.

Prince Imperial, sold to Mrs. Wells, Booth Ferry House, Howden, Yorks.

TOPSY 8TH, roan, calved April 21, 1876. Bred by Messrs. G. and J. Perry; got by Lord Lind (31692), dam (Topsy 7th) by Prince Gwynne (20547), &c. See Vol. xxiii. p. 599.

Produce in		Names, &c.	By what Bull.	By whom bred.
1879, May 20, red,	C.C.	Topsy 10th	Fawsley Knight, 38284	Messrs. Perry

VANITY 3RD, roan, calved February 27, 1871. Bred by Messrs. G. and J. Perry; got by Duke of Lancaster (28411), dam (Vanity 2nd) by Napier (14973), &c. See "Knight of the Garter," Vol. xxi. p. 250.

Produce in		Names, &c.	By what Bull.	By whom bred.
1878, April 14, roan,	C.C.	Vanity 4th	Fawsley Knight, 38284	Messrs. Perry
1879, May 19, roan,	C.C.	Vanity 5th	do.	do.

PERY, Mrs.,
Coolcronan House, Ballina, Co. Mayo.

ANNA 6TH, roan, calved January 10, 1871. Bred by the Rev. J. Storer, Hellidon; got by Earl of Rosedale (26072), dam (Anna 3rd) by Mantalini Prince (22276), &c. See Vol. xx. p. 393.

Produce in		Names, &c.	By what Bull.	By whom bred.
1879, Mar. 2, roan,	C.C.	Anna Bella	Don Diego, 33539	Mrs. Pery

BLITHE BUTTERFLY, red, calved April 27, 1871, Vols. xxi., xxii., and xxv. pp. 780, 464, 621. Bred by Mr. J. How, Broughton; got by Lord Blithe (22126), dam (Alice Butterfly) by Master Butterfly (13311), &c.

Produce in		Names, &c.	By what Bull.	By whom bred.
1879, Oct. 13, red,	C.C.	Royal Madeline	Royal Crown, 40628	Mrs. Pery

LADY VIOLET, roan, calved December 19, 1876. Bred by Mrs. Pery; got by Don Diego (33539), dam (Lady Love) by The Earl (27623), &c. See "Royal Arch," p. 216.

Produce in		Names, &c.	By what Bull.	By whom bred.
1879, Mar. 24, red,	C.C.	Lady Edith	Royal Crown, 40628	Mrs. Pery

ROYAL MARY, roan, calved September 2, 1872, Vols. xxiv. and xxv. pp. 608, 621. Bred by Mr. W. Bolton, The Island; got by King Richard (26523), dam (Royal Mistress) by Royal Buckingham (20718), &c.

Produce in	Names, &c.	By what Bull.	By whom bred.
1879, May 8, red, C.C.	Royal May	Royal Crown, 40628	Mrs. Pery

TITANIA 4TH, roan, calved April 28, 1866, Vols. xxi., xxiv., and xxv. pp. 680, 608, 621. Bred by Mr. H. M. Richardson, Rossfad; got by Prince of Lurg (22617), dam (Titania 2nd) by St. Patrick (15230), &c.

Produce in	Names, &c.	By what Bull.	By whom bred.
1879, Jan. 5, red, C.C.	Floral Crown	Royal Crown, 40628	Mrs. Pery

PHILIPS, Sir G. R., Bart.,
Weston Park, Shipston-on-Stour.

CHARMING COUNTESS, white, calved December 16, 1873, Vols. xxiv. and xxv. pp. 609, 621. Bred by Sir G. R. Philips, Bart.; got by Chorister (30737), dam (Countess 3rd) by Grand Duke 15th (21852), &c.

Produce in	Names, &c.	By what Bull.	By whom bred.
1879, Sept. 19, white, B.C.	Charming Count 2nd	Grand Duke 29th, 38372	Sir G. R. Philips

COUNTESS 3RD, white, calved February 5, 1871, Vols. xxii. and xxiv. pp. 531, 609. Bred by Messrs. F. Leney and Sons, Wateringbury; got by Grand Duke 15th (21852), dam (Chorus) by Fourth Duke of Thorndale (17750), &c.

Produce in	Names, &c.	By what Bull.	By whom bred.
1879, Jan. 26, white, B.C.	Commander 2nd	Grand Duke 29th, 38372	Sir G. R. Philips

Commander 2nd, sold to Mr. F. C. Davis, Whichford, Shipston-on-Stour.

DUCHESS OF BRAILES 3RD, roan, calved May 4, 1871, Vols. xxii. and xxiv. pp. 531, 609. Bred by Messrs. F. Leney and Sons, Wateringbury; got by Grand Duke 15th (21852), dam (Miss Knightley) by Bull's Run (19368), &c.

Produce in	Names, &c.	By what Bull.	By whom bred.
1879, Jan. 12, r. & w., C.C.	Walnut Duchess 3rd	Grand Duke 29th, 38372	Sir G. R. Philips

FAIREST, roan, calved August 25, 1876. Bred by Sir G. R. Philips, Bart.; got by Cherry Grand Duke 5th (30712), dam (Farewell) by Eighteenth Duke of Oxford (25995), &c. See Vol. xxiv. p. 609.

Produce in	Names, &c.	By what Bull.	By whom bred.
1879, Sept. 17, red, C.C.	Fairy	Grand Duke 29th, 38372	Sir G. R. Philips

FAREWELL, roan, calved March 29, 1872, Vols. xxii., xxiv., and xxv. pp. 532, 609, 622. Bred by Sir G. R. Philips, Bart.; got by Eighteenth Duke of Oxford (25995), dam (Edith of Fawsley) by Prince Christian (22582), &c.

Produce in	Names, &c.	By what Bull.	By whom bred.
1879, Oct. 31, r. & w., B.C.	Fearless 2nd	Grand Duke 29th, 38372	Sir G. R. Philips

FAWSLEY CHERRY 2ND, red and white, calved March 13, 1872, Vol. xxiv. p. 609. Bred by Sir G. R. Philips, Bart.; got by Third Cherry Duke (28171), dam (Fawsley Cherry) by Third Duke of Geneva (21592), &c.

Produce in	Names, &c.	By what Bull.	By whom bred.
1879, Feb. 12, roan, B.C.	Ruby's G. Duke 2nd	Grand Duke 29th, 38372	Sir G. R. Philips

FAWSLEY CHERRY 4TH, red, calved December 24, 1874, Vol. xxv. p. 622. Bred by Sir G. R. Philips, Bart.; got by Cherry Grand Duke 5th (30712), dam (Fawsley Cherry) by Third Duke of Geneva (21592), &c.

Produce in	Names, &c.	By what Bull.	By whom bred.
1879, May 13, r. & w., B.C.	Fa'sleyCherry's G.D.	Grand Duke 29th, 38372	Sir G. R. Philips

GRACE COSTA 2ND, roan, calved November 6, 1873, Vols. xxiv. and xxv. pp. 610, 622. Bred by Sir G. R. Philips, Bart.; got by Cherry Fawsley (30711), dam (Grace Costa) by Costa (21487), &c.

Produce in	Names, &c.	By what Bull.	By whom bred.
1879, Dec. 21, roan, C.C.	Weston Walnut 3rd	Grand Duke 29th, 38372	Sir G. R. Philips

JARDINIERE, roan, calved March 2, 1876. Bred by Mr. J. N. Beasley,
Chapel Brampton; got by John of Oxford (34267), dam (Junket) by Jacket
(24199), &c. See "Jackadandy," p. 126.

Produce in		Names, &c.		By what Bull.		By whom bred.
1879, April 7, white, B.C.		Jackadandy		Grand Duke 29th, 38372		Sir G. R. Philips

JESDEMONA 5TH, red, calved November 18, 1875, Vol. xxv. p. 622. Bred by
Sir G. R. Philips, Bart.; got by Cherry Grand Duke 5th (30712), dam
(Jesdemona 2nd) by Third Cherry Duke (28171), &c.

Produce in		Names, &c.		By what Bull.		By whom bred.
1879, Sept. 28, red, B.C.		Janus 2nd		Grand Duke 29th, 38372		Sir G. R. Philips

JOYFUL 2ND, roan, calved December 27, 1873, Vols. xxiii. and xxv. pp. 600,
622. Bred by Sir G. R. Philips, Bart.; got by Oxford Beau (29485), dam
(Joyful) by Cherry Duke (25752), &c.

Produce in		Names, &c.		By what Bull.		By whom bred.
1879, Feb. 16, roan, C.C.		Joybells 2nd		Grand Duke 29th, 38372		Sir G. R. Philips

JOYFUL 3RD, red, calved February 25, 1875. Bred by Sir G. R. Philips, Bart.;
got by Cherry Fawsley (30711), dam (Joyful) by Cherry Duke (25752), &c.
See Vol. xxii. p. 532.

Produce in		Names, &c.		By what Bull.		By whom bred.
1879, April 16, roan, B.C.		(dead)		Grand Duke 29th, 38372		Sir G. R. Philips

JOYFUL 4TH, red, calved June 6, 1876, Vol. xxv. p. 623. Bred by Sir G. R.
Philips, Bart.; got by Cherry Grand Duke 5th (30712), dam (Joyful 2nd)
by Oxford Beau (29485), &c.

Produce in		Names, &c.		By what Bull.		By whom bred.
1879, Nov. 11, r. & w., B.C.		Jove 2nd		Grand Duke 29th, 38372		Sir G. R. Philips

JOYOUS, red, calved April 19, 1876. Bred by Sir G. R. Philips, Bart.; got
by Cherry Grand Duke 5th (30712), dam (Joyful) by Cherry Duke (25752),
&c. See "Jingo," p. 127.

Produce in		Names, &c.		By what Bull.		By whom bred.
1879, Jan. 22, roan, B.C.		Jingo		Grand Duke 29th, 38372		Sir G. R. Philips

JUANITA, red and white, calved September 28, 1870, Vols. xxii. and xxiii.
pp. 316, 328. Bred by Mr. J. N. Beasley, Chapel Brampton; got by Jay
(26457), dam (Duchess Janette) by Second Duke of Airdrie (19600), &c.

Produce in		Names, &c.		By what Bull.		By whom bred.
1879, Mar. 20, r. & w., B.C.		Jorrocks		Grand Duke 29th, 38372		Sir G. R. Philips

JUNKET, roan, calved July 28, 1873. Bred by Mr. J. N. Beasley, Chapel
Brampton; got by Jacket (24199), dam (Judith) by Juvenile (22021), &c. See
Vol. xxi. p. 562.

Produce in		Names, &c.		By what Bull.		By whom bred.
1876, Mar. 2, roan, C.C.		Jardiniere		John of Oxford, 34267		Mr. Beasley
1878, Feb. 28, roan, C.C.		Joyous		Jester, 36807		do.
1879, Oct. 24, roan, C.C.		Jardiniere 2nd		Grand Duke 29th, 38372		Sir G. R. Philips

LADY FERRERS, white, calved August 25, 1870, Vol. xxv. p. 623. Bred by
Mr. F. N. Sartoris, Rushden Hall; got by Duke of Kingscote (25981), dam
(Fancy 3rd) by Worth (23244), &c.

Produce in		Names, &c.		By what Bull.		By whom bred.
1879, April 28, white, B.C.		G. Duke of Surmise		Grand Duke 29th, 38372		Sir G. R. Philips

POLYCHARM, roan, calved January 29, 1877. Bred by Sir G. R. Philips,
Bart.; got by Duke Polycherry (33763), dam (Polyanthus) by Third Duke of
Geneva (21592), &c. See "Polypus," p. 188.

Produce in		Names, &c.		By what Bull.		By whom bred.
1879, May 23, r. & w., B.C.		Polypus		Grand Duke 29th, 38372		Sir G. R. Philips

POLYGLOT, roan, calved December 22, 1873, Vols. xxiii. and xxv. pp. 601, 623.
Bred by Sir G. R. Philips, Bart.; got by Chorister (30737), dam (Polyanthus)
by Third Duke of Geneva (21592), &c.

Produce in		Names, &c.		By what Bull.		By whom bred.
1879, Feb. 13, roan, B.C.		(dead)		Grand Duke 29th, 38372		Sir G. R. Philips

ROSE OF JUNE, red, calved June 28, 1874. Bred by Mr. J. W. Larking,
Ashdown House; got by Grand Duke of Geneva (28756), dam (Rose of
Autumn) by Lord Oxford 2nd (20215), &c. See Vol. xxi. p. 805.

Produce in	Names, &c.	By what Bull.	By whom bred.
1879, April 12, r. & w., C.C.	Rose of Sussex 2nd	Grand Duke 29th, 38372	Sir G. R. Philips

ROSE OF THE FOREST, roan, calved May 29, 1876, Vol. xxv. p. 623. Bred
by Mr. J. W. Larking, Ashdown House; got by Grand Duke of Geneva
(28756), dam (Rose of Summer) by Third Duke of Waterloo (23801), &c.

1879, Aug. 5, white, B.C.	King of the Forest	Grand Duke 29th, 38372	Sir G. R. Philips

WALNUT, roan, calved February 4, 1877. Bred by Sir G. R. Philips, Bart.
got by Duke Polycherry (33763), dam (Duchess of Brailes 3rd) by Grand
Duke 15th (21852), &c. See "Cherry Walnut," p. 50.

1879, Nov. 2, white, B.C.	(dead)	Grand Duke 29th, 38372	Sir G. R. Philips

WESTON WALNUT, roan, calved June 16, 1877. Bred by Sir G. R. Philips,
Bart.; got by Duke Polycherry (33763), dam (Grace Costa 2nd) by Cherry
Fawsley (30711), &c. See Vol. xxiv. p. 610.

1879, Dec. 4, roan,	B.C.	Walnut's Grand D.	Grand Duke 29th, 38372	Sir G. R. Philips

PHILIPS, J. W.,
Heybridge, Tean, Stoke-on-Trent.

FAIR JUTE, roan, calved February 20, 1868, Vol. xix. p. 502. Bred by Mr.
W. Torr, Aylesby Manor; got by Breast Plate (19337), dam (Fair Dane) by
Fitz-Clarence (14552), &c.

1879, Feb. 16, roan,	C.C.	Flower of the Scheldt	His Lordship 38433	Mr. Philips

FLOWER OF BELGIUM, roan, calved June 17, 1872, Vols. xxiii. and xxiv.
pp. 601, 611. Bred by Mr. W. Torr, Aylesby Manor; got by Royal Prince
(27384), dam (Flower of Denmark) by Fitz-Clarence (14552), &c.

1878, April 1, white, B.C.	Egmont	His Lordship, 38433	Mr. Philips
1879, Mar. 16, white, B.C.	Esmond	Royal Saxon, 39057	do.

LADY OF THE MOOR, roan, calved January 9, 1873, Vol. xxv. p. 624.
Bred by the Rev. T. Staniforth, Storrs; got by High Sheriff (26392), dam
(Lady of the Manor) by British Crown (21322), &c.

1879, May 14, roan,	C.C.	Lady of the Bourne	His Lordship, 38433	Mr. Philips

MAID OF THE TAYNE, roan, calved March 1, 1874. Bred by Mr. J. W.
Philips; got by Prince Christian (22581), dam (Zephyr 2nd) by Manfred
(24522), &c. See Vol. xxi. p. 888.

1879, May 14, roan,	C.C.	Maid of the Floss	His Lordship, 38433	Mr. Philips

MEDORA 3RD, roan, calved April 9, 1872, Vols. xxii., xxiv., and xxv. pp. 533,
611, 624. Bred by Mr. R. Blackwell, Tansley; got by Knight of Knowlmere (22055), dam (Bagatelle) by Sultan (17053), &c.

1879, Mar. 15, roan,	C.C.	Mercia	Royal Saxon, 39057	Mr. Philips

ROULETTE, roan, calved October 25, 1873, Vol. xxiii. p. 380. Bred by Mr.
H. Chandos-Pole-Gell, Hopton Hall; got by Favourite (31145), dam (Bagatelle) by Sultan (17053), &c.

1879, Feb. 27, roan,	C.C.	Minna	Royal Saxon, 39057	Mr. Philips

PHILLIPS, G. T.,
Sheriff Hales Manor, Newport, Salop.

YOUNG BUTTERCUP, white, calved December 29, 1870.   Bred by Mr. G. T. Phillips; got by Prince Charming (27130), dam (Buttercup) by Royal Butterfly 15th (20723), &c.   See "Ruby Duke," Vol. xxiv. p. 232.

Produce in	Names, &c.	By what Bull.	By whom bred.
1877, July 20, r. &. w., B.C.	Ruby Duke	Marsh'l MacMahon, 34792	Mr. Phillips
1879, May 14, roan, C.C.	Buttercup 2nd	Viscount Cherry, 37635	do.

CHARMING PRINCESS, roan, calved August 26, 1872, Vol. xxi. p. 888.   Bred by Mr. G. T. Phillips; got by Prince Charming (27130), dam (Honeysuckle) by Duke of Manchester (15934), &c.

Produce in	Names, &c.	By what Bull.	By whom bred.
1875, Dec. 26, white, B.C.	Lord Horton	Earl of Horton, 36585	Mr. Phillips
1876, Oct. 25, roan, C.C.	Honeysuckle 2nd	Marsh'l MacMahon, 34792	do.
1877, Nov. 21, r. & w., B.C.	Rowley Duke	2nd D. of Rowley, 28441	do.
1879, May 24, roan, B.C.	Sutherland	Viscount Cherry. 37635	do.

Lord Horton, sold to Mr. J. Lea, Chester; Rowley Duke, to the Duke of Leinster.

HONEYSUCKLE 2ND, roan, calved October 25, 1876.   Bred by Mr. G. T. Phillips; got by Marshal MacMahon (34792), dam (Charming Princess) by Prince Charming (27130), &c.   See "Lilleshall," p. 139.

Produce in	Names, &c.	By what Bull.	By whom bred.
1879 June 11, roan, B.C.	Lilleshall	Ruby Duke, 40654	Mr. Phillips

MISS IDSAL, red and white, calved July 20, 1873.   Bred by Mr. G. T. Phillips; got by Idsal (31404), dam (Honeysuckle) by Duke of Manchester (15934), &c.   See Vol. xxi. p. 888.

Produce in	Names, &c.	By what Bull.	By whom bred.
1879, July 24, roan, C.C.	Miss Idsal 2nd	Rowley Duke, 40615	Mr. Phillips

PRINCESS CHARMING, red and white, calved June 2, 1874, Vol. xxiv. p. 612.   Bred by Mr. G. T. Phillips; got by Young Idsal (34197), dam (Charming Princess) by Prince Charming (27130), &c.

Produce in	Names, &c.	By what Bull.	By whom bred.
1879, May 28, roan, C.C.	Miss Charming	M. of Blandford 6th, 41983	Mr. Phillips

SNOWDROP, white, calved August 4, 1876.   Bred by Mr. G. T. Phillips; got by Marshal MacMahon (34792), dam (Buttercup) by Royal Butterfly 15th (20723), g. d. (Honeysuckle) by Duke of Manchester (15934), &c.   See "Sutherland," p. 244.

Produce in	Names, &c.	By what Bull.	By whom bred.
1879, April 8, white, C.C.	Snowdrop 2nd	M. of Blandford 6th, 41983	Mr. Phillips

PHIPPS, P.,
Collingtree Grange, Northampton.

DAWN, red and white, calved February 20, 1876.   Bred by Mr. P. Phipps; got by Rosehill (29833), dam (Dahlia 7th) by Saracen (25091), &c.   See Vol. xxiii. p. 601.

Produce in	Names, &c.	By what Bull.	By whom bred.
1879, May 29, r. & w., C.C.	Dinah	Rose of Collingtree, 39021	Mr. Phipps

DIDO, roan, calved April 27, 1875.   Bred by Mr. P. Phipps; got by Rosehill (29833), dam (Duchess of Blaston) by Duke of Wateringbury (23799), &c.   See "George Frederick," Vol. xxi. p. 525.

Produce in	Names, &c.	By what Bull.	By whom bred.
1879, May 27, r. & w., C.C.	Deliah	Rose of Collingtree, 39021	Mr. Phipps

FAREWELL, white, calved January 23, 1874. Bred by Mr. W. Faulkner, Rothersthorpe; got by Royalty (35412), dam (Fancy) by Knight of Branches (20076), &c. See Vol. xxi. p. 694.

Produce in	Names, &c.	By what Bull.	By whom bred.
1879, Mar. 25, roan, C.C.	Fairy	Rose of Collingtree, 39021	Mr. Phipps

FLOUNCE, roan, calved February 25, 1875, Vol. xxv. p. 624. Bred by Mr. P. Phipps; got by Fawsley Oxford 2nd (33904), dam (Fashion 25th) by Second Earl of Darlington (26056), &c.

| 1879, Aug. 28, r. & w., C.C. | Flora | Rose of Collingtree, 39021 | Mr. Phipps |

FOREST ROSE, red and white, calved February 15, 1874, Vol. xxiv. p. 612. Bred by Mr. W. Faulkner, Rothersthorpe; got by Royalty (35412), dam (Faith) by Romeo (24985), &c.

| 1879, Mar. 29, r. & w., B.C. | Forest Lad | Rose of Collingtree, 39021 | Mr. Phipps |

ROSE 11TH, roan, calved April 10, 1868, Vols. xx., xxi. and xxiii. pp. 735, 695, 601. Bred by Mr. W. Faulkner, Rothersthorpe; got by Saracen (25091), dam (Rose) by August (12412), &c.

| 1879, Feb. 27, r. & w., B.C. | Romulus | Rose of Collingtree, 39021 | Mr. Phipps |

SWEETBRIER 58TH, roan, calved January 2, 1874, Vol. xxiv. p. 612. Bred by Mr. W. Faulkner, Rothersthorpe; got by Royalty (35412), dam (Sweetbrier 46th) by Athelstane (23331), &c.

| 1879, Mar. 22, r. & w., B.C. | Sullivan | Rose of Collingtree, 39021 | Mr. Phipps |

PIGOT, Lady,
West Hall, Weybridge, Surrey.

ELEZE GWYNNE, red and white, calved April 6, 1874, Vols. xxiv. and xxv. pp. 613, 625. Bred by Mr. W. Bolton, The Island; got by Lieutenant General (31600), dam (Eveline Gwynne) by Grey Gauntlet (19908), &c.

| 1879, Dec. 26, roan, C.C. | Effie Gwynne | Knight of Malta, 38518 | Lady Pigot |

MAY QUEEN, roan, calved November 5, 1876. Bred by Lady Pigot; got by Red Cross Knight (35219), dam (May Belle) by Hardicanute (26338), &c. See Vol. xxiii. p. 604.

| 1879, Sept. 18, roan, C.C. | | Nobilis, 38795 | Lady Pigot |

VICTORIA FULGIDA, red, calved February 20, 1873, Vols. xxiii. and xxiv. pp. 605, 614. Bred by Lady Pigot; got by Constellation (28243), dam (Victoria Victrix) by Sidus (29969), &c.

| 1879, Nov. 10, r. & w., B.C. | | Opoponax, 34950 | Lady Pigot |

PINDER, R.,
Whitwell, Oakham.

DAME LIVELY, red and white, calved January 11, 1877. Bred by Mr. R. Pinder; got by Fiery Star (33914), dam (Dame Prim) by Constellation (28243), &c. See "Cedric," p. 44.

| 1879, Aug. 19, red, C.C. | Dame Frolic | Burghley, 36296 | Mr. Pinder |

DAME MARY, red and white, calved March 3, 1875, Vol. xxiv. p. 614. Bred by Mr. R. Pinder; got by Royal Warrior (35415), dam (Dame Prudence) by Prince of Rosedale (24837), &c.

Produce in		Names, &c.	By what Bull.	By whom bred.
1878, Mar. 5, r.&w.,	C.C.	Dame Durden	Burghley, 36296	Mr. Pinder
1879, Mar. 1, r.&w.,	C.C.	(dead)	do.	do.

DAME PRIM, red and white, calved November 30, 1872, Vols. xxi. and xxiv. pp. 891, 614. Bred by Lady Pigot, Branches Park; got by Contsellation (28243), dam (Dame Prudence) by Prince of Rosedale (24837), &c.

Produce in		Names, &c.	By what Bull.	By whom bred.
1878, Oct. 24, r.&w.,	C.C.	Dame Pride	Burghley, 36296	Mr. Pinder
1879, Nov. 25, red,	B.C.	Cedric	do.	do.

PRINCESS KITTY, red, calved October 1, 1870, Vol. xxiv. p. 615. Bred by Mr. R. Pinder; got by Paragon (27037), dam (Queen of the Glen) by Lord of the Valley (14837), &c.

Produce in		Names, &c.	By what Bull.	By whom bred.
1879, Feb. 23, red,	C.C.	Queen of the West	Prince William, 37283	Mr. Pinder

QUEEN OF THE MEADOW, red, calved August 20, 1874, Vol. xxiv. p. 615. Bred by Mr. R. Pinder; got by Red Cross Knight (35219), dam (Queen of the Glen) by Lord of the Valley (14837), &c.

Produce in		Names, &c.	By what Bull.	By whom bred.
1878, Mar. 18, red,	B.C.	Baron Drummond	Prince William, 37283	Mr. Pinder
1879, Apr. 16, { r.&w.B.C.		(Steer)	} do.	do.
{ red,	C.C.	Queen of the Castle	}	

Baron Drummond, sold to Mr. Wortley, Ridlington, Uppingham.

REBECCA, red and white, calved March 15, 1876. Bred by Mr. R. Pinder; got by Fiery Star (33914), dam (Rowena) by The Stuart (27650), &c. See "Wamba," p. 262.

Produce in		Names, &c.	By what Bull.	By whom bred.
1879, Nov. 20, red,	B.C.	Wamba	Prince William, 37283	Mr. Pinder

ROWENA, red, calved August 28, 1873, Vol. xxiv. p. 615. Bred by Mr. R. Pinder; got by The Stuart (27650), dam (Cleodora) by Falstaff (21720), &c.

Produce in		Names, &c.	By what Bull.	By whom bred.
1878, Aug. 8, red,	C.C.	Jewess	Prince William, 37283	Mr. Pinder

SANDPIPER 2ND, red, calved January 21, 1871, Vols. xxi. and xxiv. pp. 891, 615. Bred by Mr. R. Pinder; got by President (27088), dam (Sandpiper) by The Briar (15376), &c.

Produce in		Names, &c.	By what Bull.	By whom bred.
1878, Aug. 10, red,	B.C.	Sir Oswald	Prince William, 37283	Mr. Pinder

Sir Oswald, sold to Mr. W. Harris, Cottesmore, Oakham.

SANDPIPER 3RD, red, calved November 10, 1873, Vol. xxiv. p. 615. Bred by Mr. R. Pinder; got by The Stuart (27650), dam (Sandpiper 2nd) by President (27088), &c.

Produce in		Names, &c.	By what Bull.	By whom bred.
1878, Aug. 27, red,	C.C.	Sandpiper 7th	Prince William, 37283	Mr. Pinder
1879, July 23, red,	B.C.	(dead)	do.	do.

SANDPIPER 6TH, red and white, calved August 20, 1876. Bred by Mr. R. Pinder; got by Fiery Star (33914), dam (Sandpiper 3rd) by The Stuart (27650), &c. See "Baveno," p. 23.

Produce in		Names, &c.	By what Bull.	By whom bred.
1879, Nov. 2, red,	B.C.	Baveno	Prince William, 37283	Mr. Pinder

VICTORIA LA BELLE, red and white, calved May 7, 1877. Bred by Mr. R. Pinder; got by Opoponax (34950), dam (Victoria Pulcherima) by Bythis (25700), &c. See Vol. xxiv. p. 615.

Produce in		Names, &c.	By what Bull.	By whom bred.
1879, Nov. 17, r.&w.,	C.C.	Victoria La Bonne	Burghley, 36296	Mr. Pinder

VICTORIA PULCHERIMA, roan, calved August 14, 1872, Vol. xxiv. p. 615.
Bred by Lady Pigot, Branches Park; got by Bythis (25700), dam (Victoria
Pulchra) by Charles-le-Beau (23542), &c.

Produce in		Names, &c.	By what Bull.	By whom bred.
1878, Mar. 31, roan,	B.C.	Lord Eric	Burghley, 36296	Mr. Pinder
1879, April 20, roan,	C.C.	Victoria La Reine	do.	do.

PLATT, Major,
Gorddinog, Bangor, North Wales.

COLUMBINE 7TH, red and white, calved January 9, 1877. Bred by Captain
C. R. Conwy, Bodrhyddan; got by Favourite (33895), dam (Columbine 5th)
by Oxford Wild Eyes 2nd (32038), &c. See Vol. xxiv. p. 385.

1879, Apr. 12, r.&w.,	C.C.	Columbine 8th	Edward Waterloo, 38246	Major Platt

POCHIN, H. D.,
Bodnant Hall, Conway, North Wales.

MISCHIEF, roan, calved July 14, 1874, Vol. xxiv. p. 616. Bred by Captain
C. R. Conwy, Bodrhyddan; got by Charles 1st (30685), dam (Mishap) by
Harold (26341), &c.

1879, Aug. 29, roan,	C.C.	Lily	Carlos, 36322	Mr. Pochin

ROMPING LASS, white, calved August 27, 1877. Bred by Captain C. R.
Conwy, Bodrhyddan; got by Favourite (33895), dam (Romping Girl) by
Hotspur (28878), &c. See Vol. xxiv. p. 386.

1879, Dec. 30, white,	C.C.	Rose	Nonsuch 2nd, 40390	Mr. Pochin

POE, H. R.,
Solsborough, Nenagh, Co. Tipperary.

DEODORA, roan, calved July 6, 1870, Vols. xxi. and xxiii. pp. 643, 395. Bred
by Sir G. R. Philips, Bart., Weston Park, the property of Mrs. Power Lalor,
Long Orchard; got by Third Duke of Wetherby (26030), dam (Diadem) by
Touchstone (20986), &c.

1879, Jan. 25, roan,	C.C.	Dido 10th	Knight of Raby, 34393	Mr. Welsh

Dido 10th, sold to Mr. H. R. Poe, Solsborough, Nenagh.

HONEY BELL 3RD, roan, calved February 25, 1874, Vol. xxiii. p. 700. Bred
by Major Hamilton, Brownhall, the property of Mr. H. R. Poe; got by
Baron Underley (30490), dam (Honey Bell 2nd) by Third Duke of Clarence
(23727), &c.

1879, Mar. 7, r.&w.,	B.C.	Duke of Nenagh	Duke of Underley, 33745	Mr. Bolton

PRINCESS ISABEL, roan, calved April 8, 1876. Bred by Mr. W. T. Crosbie,
Ardfert Abbey, the property of Mr. H. R. Poe; got by Lord Blithesome
(29067), dam (Sovereign's Queen) by Irish Baron (31417), &c. See Vol. xxiii.
p. 404.

1879, Mar. 16, roan,	B.C.	Lord Blithesome 2nd	Red Cross Knight, 35219	Mr. Welsh

Lord Blithesome 2nd, sold to Mr. J. Gairdner, Lisheg, Eyrecourt, Ballinasloe.

POLWARTH, Lord,
Mertoun House, St. Boswells, N.B.

BROUGHTON BUTTERFLY, red, calved May 30, 1869, Vols. xx., xxi., and xxiv. pp. 424, 780, 616. Bred by Mr. J. How, Broughton; got by Victorious (25378), dam (Alice Butterfly) by Master Butterfly (13311), &c.

Produce in		Names, &c.	By what Bull.	By whom bred.
1878, Aug. 9, roan,	B.C.	Splendid Butterfly	Rapid Rhone, 35205	Lord Polwarth
1879, Sept. 14, r.&w.,	C.C.	Golden Butterfly	do.	do.

DOUBLE BUTTERFLY 3RD, roan, calved November 9, 1869, Vol. xx. p. 483. Bred by Mr. Eastwood, Thorneyholme; got by Victorious (25378), dam (Double Butterfly) by Royal Butterfly (16862), &c.

Produce in		Names, &c.	By what Bull.	By whom bred.
1878, July 22, roan,	C.C.	Bright Butterfly	Rapid Rhone, 35205	Lord Polwarth
1879, Sept. 16, roan,	B.C.	Peacock Butterfly	do.	do.

EASTTHORPE STRAWBERRY 6TH, red and white, calved February 13, 1872, Vols. xxii. and xxiv. pp. 304, 312. Bred by Mr. H. Aylmer, West Dereham Abbey; got by Royal Broughton (27352), dam (Eastthorpe Strawberry 4th) by Ravenspur (20628), &c.

Produce in		Names, &c.	By what Bull.	By whom bred.
1878, July 31, roan,	B.C.	Superb	Rapid Rhone, 35205	Lord Polwarth
1879, July 21, r.&w.,	B.C.	The Sheriff	do.	do.

ECHO, red and white, calved March 5, 1870, Vol. xxi. p. 893. Bred by the Earl of Zetland, Upleatham; got by Chatsworth (23546), dam (Emmy) by Emilius (21682), &c.

Produce in		Names, &c.	By what Bull.	By whom bred.
1878, April 5, roan,	B.C.	British Empire	Rapid Rhone, 35205	Lord Polwarth
1879, June 14, roan,	C.C.	Emerald Gwynne	do.	do.

FAIRY QUEEN, red and white, calved March 22, 1872. Bred by Mr. T. E. Pawlett, Beeston; got by Majestic (29255), dam (Flora) by Booth Royal (15673), &c. See Vol. xxi. p. 841.

Produce in		Names, &c.	By what Bull.	By whom bred.
1874, April 7, r.&w.,	C.C.	Frances	Brighton, 33200	Mr. Bowman
1879, Feb. 19, roan,	B.C.	Good Hope	Peter the Great, 38863	Lord Polwarth

FLOWER OF THE FIELD, roan, calved June 25, 1877. Bred by Mr. H. Wardle, Highfield; got by Royal Victor (35414), dam (Flower of Holland) by Breast Plate (19337), &c. See Vol. xxiv. p. 695.

Produce in		Names, &c.	By what Bull.	By whom bred.
1879, Sept. 11, r.&w.,	B.C.	Feudal Chief	Lord Ringlet, 38655	Lord Polwarth

MAGGIE GWYNNE, white, calved July 15, 1871, Vols. xxi. and xxiii. pp. 890, 604. Bred by Mr. W. Bolton, The Island; got by King Richard (26523), dam (Margery Gwynne) by Grey Gauntlet (19908), &c.

Produce in		Names, &c.	By what Bull.	By whom bred.
1877, June 3, white,	C.C.	Snowy Gwynne	Royal Commander, 29857	Lord Polwarth
1879, May 3, white,	C.C.	Mayflower Gwynne	Rapid Rhone, 35205	do.

RIBY FOGGATHORPE, red and white, calved July 8, 1872, Vol. xxiii. p. 607. Bred by Mr. W. Torr, Aylesby Manor; got by Breast Plate (19337), dam (Georgie) by Elfin King (17796), &c.

Produce in		Names, &c.	By what Bull.	By whom bred.
1877, Nov. 3, roan,	B.C.	Rapid Foggathorpe	Rapid Rhone, 35205	Lord Polwarth
1879, Nov. 21. r.&w.,	C.C.	Tweedside Fog'horpe	do.	do.

WAVE FOAM, roan, calved January 16, 1873, Vols. xxii. and xxiv. pp. 317, 617. Bred by Mr. J. Beattie, Newbie House; got by Manfred (26801), dam (Wave Breast) by Breast Plate (19337), &c.

Produce in		Names, &c.	By what Bull.	By whom bred.
79, Nov. 11, roan,	B.C.	The Warder	Rapid Rhone, 35205	Lord Polwarth

WAVE SURF, roan, calved July 1, 1875, Vol. xxiv. p. 617. Bred by Mr. J.
Beattie, Newbie House; got by Knight of Knowlmere 2nd (31542), dam
(Wave Foam) by Manfred (26801), &c.

Produce in	Names, &c.	By what Bull.	By whom bred.
1878, June 5, r.&w., C.C.	Wave of the Ocean	Rapid Rhone, 35208	Lord Polwarth
1879, April 11, roan, C.C.	Wave of Atlantic	do.	do.

POPE, F. E.,
Great Toller, Dorchester.

GENTIAN, red and white, calved February 12, 1868, Vols. xix. and xxii. pp. 529,
536. Bred by Mr. R. Stratton, Burderop; got by Eighth Duke of York
(23808), dam (Gertrude) by Windsor Castle (21118), &c.

Produce in	Names, &c.	By what Bull.	By whom bred.
1877, Jan. 6, r.&w., C.C.	Lady Gentian 2nd	Lord Harry, 31666	Mr. Pope
1878, Jan. 23, roan, B.C.	General Harry 4th	do.	do.
1879, Jan. 25, r.&w., B.C.	General Cossington	Baron Cossington, 37784	do.
1879, Dec. 30, roan, C.C.	Lady Gentian 3rd	Baron Chilfrome, 42706	do.

General Harry 4th, sold to Mr. Meikle, Godlingstone, Swanage, Dorset.

LADY GENTIAN, roan, calved January 29, 1875. Bred by Mr. F. E. Pope;
got by Lord Harry (31666), dam (Gentian) by Eighth Duke of York (23808),
&c. See " Baron Chilfrome," p. 12.

Produce in	Names, &c.	By what Bull.	By whom bred.
1878, Jan. 20, white, B.C.	Baron Chilfrome	Baron Cossington, 37784	Mr. Pope
1879, Mar. 6, roan, C.C.	Gentiana	do.	do.

Baron Chilfrome, sold to Mrs. Symes, Kingston Russell, Dorchester.

PORTER, John,
Blackhow, Seascale, Carnforth.

COUNTESS, roan, calved January 19, 1872, Vol. xxii. p. 537. Bred by Mr. J.
Bowman, High House, Sandwith; got by Bismarck (28039), dam (Lady
Sarah) by Frenchman (28657), &c.

Produce in	Names, &c.	By what Bull.	By whom bred.
1878, Feb. 19, roan, B.C.	British Squire	Monarch, 37101	Mr. Porter
1879, Mar. 7, roan, C.C.	Countess 2nd	Earl of Fawsley 7th, 38222	do.

British Squire, sold to Mr. J. McIntyre, Belmont, Northwich.

POTTERTON, W. H.,
Boughton Grange, Northampton.

JUNE DUCHESS, roan, calved June 8, 1873, Vols. xxiv. and xxv. pp. 617, 627.
Bred by the Rev. J. Storer, Hellidon; got by Prince of Fawsley (32171), dam
(Booth Duchess 4th) by Earl of Rosedale (26072), &c.

Produce in	Names, &c.	By what Bull.	By whom bred.
1879, Nov. 9, roan, B.C.	Fair Duke	Flame, 38308	Mr. Potterton

MARCH DUCHESS, roan, calved March 18, 1877. Bred by Mr. W. H. Pot-
terton; got by Abbot of Windsor (32903), dam (June Duchess) by Prince of
Fawsley (32171), &c. See Vol. xxiv. p. 617.

Produce in	Names, &c.	By what Bull.	By whom bred.
1879, Sept. 15, roan, B.C.	Duke	Flame, 38308	Mr. Potterton

ROMER, roan, calved January 8, 1873, Vols. xxiii. and xxv. pp. 608, 627. Bred
by Mr. W. Faulkner, Rothersthorpe; got by Young Hector (28830), dam
(Rosedale) by Watchman (27757), &c.

Produce in	Names, &c.	By what Bull.	By whom bred.
1879, Mar. 19, white, C.C.	Rhoda	Fair Thane, 31127	Mr. Potterton

PRICE, W.,
Goldcliff, Newport, Mon.

LADY GOLDCLIFF, red and white, calved April 24, 1875. Bred by Mr. W. Price; got by Master Broomstick (34803), dam (Promised Land 4th) by James 2nd (24203), g. d. (Promised Land) by Star of Gwent (25227), &c. See "Star of Goldcliff," p. 242.

Produce in	Names, &c.	By what Bull.	By whom bred.
1878, May 20, red,	C.C. Lady Goldcliff 2nd	Prince of Goldcliff, 43827	Mr. Price
1879, May 28, red,	B.C. (Steer)	do.	do.

PROMISED LAND 5TH, roan, calved June 12, 1873. Bred by Mr. W. Price; got by James 2nd (24203), dam (Promised Land) by Star of Gwent (25227), &c. See "Star of Goldcliff," p. 242.

1876, May 20, roan,	C.C. Promised Land 7th	Cromwell, 33480	Mr. Price
1877, June 12, white,	C.C. Promised Land 8th	do.	do.
1878, June 3, roan,	C.C. Promised Land 9th	Prince of Goldcliff, 43827	do.
1879, June 26, r. & w.,	B.C. Star of Goldcliff	do.	do.

PROMISED LAND 7TH, roan, calved May 20, 1876. Bred by Mr. W. Price; got by Cromwell (33480), dam (Promised Land 5th) by James 2nd (24203). &c.

1878, June 4, roan,	C.C. Ladyday	Prince of Goldcliff, 43827	Mr. Price
1879, May 10, red,	B.C. (dead)	do.	do.

ROSEBUD 2ND, red and white, calved in July 1874. Bred Mr. W. Price; got by Master Broomstick (34803), dam (Rosebud) by Lord Darlington 2nd (29096), &c. See "Duke of Goldcliff," p. 78.

1877, April 23, r. & w.,	C.C. Rosebud 3rd	Cromwell, 33480	Mr. Price
1878, April 8, r. & w.,	C.C. Rosebud 4th	Prince of Goldcliff, 43827	do.
1879, May 22, r. & w.,	B.C. Duke of Goldcliff	do.	do.

PRYSE, J. P. V,
Bwlchbychan, Llanybyther, South Wales.

LADY STIRLING, red and white, calved February 8, 1874. Bred by Mr. D. Thomas, Llanfair; got by Duke of Kirklevington (39757), dam (Lady Ann) by Prince of the Empire (20578), &c. See Vol. xxv. p. 420.

1879, Feb. 1, r. & w.,	C.C. Lady Red Eyes	Baron Red Eyes, 39434	Mr. Pryse

PRINCESS IMPERIAL 2ND, red and white, calved March 23, 1876. Bred by Mr. J. P. V. Pryse; got by Wild Welshman (32869), dam (Peggy) by Royal Wharfdale 2nd (27393), &c. See Vol. xxiv. p. 619.

1879, May 29, roan,	C.C. Lady Red Eyes 2nd	Baron Red Eyes, 39434	Mr. Pryse

PRYSE, Sir Pryse, Bart.,
Gogerddan, Bowstreet, *via* Shrewsbury.

CHERRY, white, calved April 28, 1877. Bred by Sir Pryse Pryse. Bart.; got by Prince Alfred (32096), dam (Crocus) by April Fool (32955), g. d. (Cactus) by Lysidas (20249), &c. See Vol. xxiii. p. 608.

1879, Dec. 12, roan,	C.C. Cherry Queen	Prince Rufus, 40540	Sir Pryse Pryse

CHRISTMAS ROSE, red and white, calved December 26, 1873. Bred by Mr. T. Swingler, Langham; got by Wisdom (30331), dam (Picotee) by Garibaldi (24006), &c. See Vol. xxi. p. 950.

1879, Mar. 24, r. & w.,	C.C. Christmas Rose 1st	Clodhopper, 42945	Sir Pryse Pryse

CORAL, roan, calved February 1, 1876. Bred by Sir Pryse Pryse, Bart.; got
by Prince Alfred (32096), dam (Crocus) by April Fool (32955), g. d. (Cactus)
by Lysidas (20249), &c. See Vol. xxiii. p. 608.

Produce in	Names, &c.	By what Bull.	By whom bred.
1879, May 14. r. & w., B.C.	(Steer)	Prince Rufus, 40540	Sir Pryse Pryse

STOCK No. 1, roan, calved March 6, 1877. Bred by Sir Pryse Pryse, Bart.;
got by Marquis of Geneva (31838), dam (Stock) by Master Red Eyes (29339),
&c. See Vol. xxiii. p. 608.

Produce in	Names, &c.	By what Bull.	By whom bred.
1879, Oct. 26, r. & w., C.C.	Stock No. 2	Prince Rufus, 40540	Sir Pryse Pryse

SUNFLOWER, roan, calved March 27, 1876. Bred by Sir Pryse Pryse, Bart.;
got by Prince Alfred (32096), dam (Stock) by Master Red Eyes (29339), &c.
See Vol. xxiii. p. 608.

Produce in	Names, &c.	By what Bull.	By whom bred.
1879, May 25, r. & w., C.C.	Bright Eyes	Prince Rufus, 40540	Sir Pryse Pryse

PUGH, David,
Manoravon, Llandilo, Carmarthenshire.

BERGAMOT 7TH, red and white, calved January 11, 1875. Bred by Mr. D.
Pugh; got by Sir Hildebrand (29993), dam (Bergamot 6th) by Second Earl
of Walton (19672), &c. See Vol. xxv. p. 628.

Produce in	Names, &c.	By what Bull.	By whom bred.
1878, Jan. 8, white, B.C.	(Steer)	Marquis 2nd, 34775	Mr. Pugh
1879, Dec. 18, roan, B.C.	Viscount Cennen	Falmouth, 38268	do.

CHARMING BEAD 2ND, roan, calved August 17, 1876. Bred by Mr. D.
Pugh; got by Marquis 1st (34774), dam (Charming Bead) by Beadsman
(27998), &c. See Vol. xxv. p. 628.

Produce in	Names, &c.	By what Bull.	By whom bred.
1878, July 18, r. & w., C.C.	Charming Ruby	Falmouth, 38268	Mr. Pugh
1879, May 16, white, C.C.	Charming Beauty	do.	do.

CZARINA 11TH, roan, calved July 31, 1875. Bred by Mr. D. Pugh; got by
Sir Hildebrand (29993), dam (Czarina 9th) by Falconer (23907), &c. See
Vol. xxv. p. 629.

Produce in	Names, &c.	By what Bull.	By whom bred.
1878, July 26, roan, C.C.	Crocus	Falmouth, 38268	Mr. Pugh
1879, July 25, roan, C.C.	Czarina Manoravon	do.	do.

CZARINA 12TH, roan, calved August 3, 1876. Bred by Mr. D. Pugh; got by
Duke of Albemarle (28355), dam (Czarina 9th) by Falconer (23907), &c. See
Vol. xxv. p. 629.

Produce in	Names, &c.	By what Bull.	By whom bred.
1879, Sept. 26, white, B.C.	(Steer)	Falmouth, 38268	Mr. Pugh

FAREWELL 1ST, roan, calved December 14, 1876. Bred by Mr. D. Pugh;
got by Duke of Albemarle (28355), dam (Farewell) by Mantalini Prince
(29273), &c. See Vol. xxv. p. 629.

Produce in	Names, &c.	By what Bull.	By whom bred.
1879, Aug. 4, roan, B.C.	Duke of Manoravon	Falmouth, 38268	Mr. Pugh

MARCHIONESS, red and white, calved September 10, 1869, Vol. xx. p. 641.
Bred by Mr. J. Peel, Knowlmere Manor; got by Lord Lyons (26677), dam
(Mistress Mary) by Baron Warlaby (7813), &c.

Produce in	Names, &c.	By what Bull.	By whom bred.
1873, July 18, roan, B.C.	Marquis 1st	Sir Hildebrand, 29993	Mr. Pugh
1874, July 30, roan, B.C.	Marquis 2nd	do.	do.
1875, Aug. 9, r. & w., C.C.	Marchioness 2nd	do.	do.
1876, Aug. 1, roan, B.C.	Marquis 3rd	do.	do.
1877, Nov. 11, roan, C.C.	Marchioness 3rd	Falmouth, 38268	do.

MARCHIONESS 2ND, red and white, calved August 9, 1875. Bred by Mr. D. Pugh; got by Sir Hildebrand (29993), dam (Marchioness) by Lord Lyons (26677), &c.

Produce in	Names, &c.	By what Bull.	By whom bred.
1878, July 13, roan,	C.C. March'ness Towyside	Falmouth, 38268	Mr. Pugh
1879, Nov. 16, roan,	B.C. Marquis Towyside	do.	do.

MARCHIONESS MANORAVON 2ND, roan, calved April 27, 1874. Bred by Mr. D. Pugh; got by Sir Hildebrand (29993), dam (Marchioness Manoravon 1st) by Lord Abbot (29052), &c. See Vol. xxi. p. 896.

1877, May 19, white, C.C.	Marchioness Towy	Falmouth, 38268	Mr. Pugh
1878, April 17, roan, B.C.	(Steer)	do.	do.
1879, Mar. 4, white, B.C.	Marquis Towy	do.	do.

MARCHIONESS MANORAVON 3RD, roan, calved July 16, 1876. Bred by Mr. D. Pugh; got by Sir Hildebrand (29993), dam (Marchioness Manoravon 1st) by Lord Abbot (29052), &c. See Vol. xxv. p. 629.

1879, Dec. 11, red,	B.C. Marquis Towyvale	Falmouth, 38268	Mr. Pugh

ZOE, roan, calved July 14, 1876. Bred by Mr. D. Pugh; got by Marquis 1st (34774), dam (Czarina 10th) by Duke of Albemarle (28355), &c. See Vol. xxiii. p. 608.

1879, Feb. 1, white,	C.C. White Zoe	Falmouth, 38268	Mr. Pugh

PURKIS, T.,
The Grange, West Wratting, Cambridgeshire.

LADY FURBELOW 3RD, red, calved April 8, 1876, Vol. xxv. p. 630. Bred by Mr. C. M. Hamer, Snitterfield; got by Cherry Grand Duke 2nd (25758), dam (Harmonia) by Grand Duke of Essex 4th (24068), &c.

1879, Aug. 23, roan,	C.C. Lady Furbelow 6th	Wharfdale Oxford, 27786	Mr. Purkis

SWEETHEART 35TH, roan, calved March 13, 1870, Vols. xx., xxi., xxiii., xxiv., and xxv. pp. 783, 898, 609, 621, 630. Bred by Mr. G. Murton Tracy, Redlands; got by Patrician (24728), dam (Sweetheart 11th) by The Baron (13833), &c.

1879, Oct. 6, red,	C.C. Sweetheart 40th	Cambridge Duke 5th, 30644	Mr. Purkis

SWEETHEART 37TH, roan, calved January 5, 1876, Vol. xxv. p. 630. Bred by Mr. T. Purkis; got by Wharfdale Oxford (27786), dam (Sweetheart 35th) by Patrician (24728), &c.

1879, Nov. 25, roan,	C.C. Sweetheart 41st	Cambridge Duke 5th, 30644	Mr. Purkis

THORNDALE BLANCHE 5TH, white, calved March 7, 1874, Vols. xxii. and xxiv. pp. 538, 621. Bred by Mr. T. Purkis; got by Baron Acomb (30418), dam (Thorndale Blanche 2nd) by Sir Rainald (27485), &c.

1879, Sept. 25, roan,	C.C. Th'dale Blanche 10th	Cambridge Duke 5th, 30644	Mr. Purkis

THORNDALE BLANCHE 6TH, roan, calved December 18, 1875, Vol. xxv. p. 630. Bred by Mr. T. Purkis; got by Wharfdale Oxford (27786), dam (Thorndale Blanche 5th) by Baron Acomb (30418), &c.

1879, April 30, roan,	C.C. Th'ndale Blanche 9th	Saccharine, 40660	Mr. Purkis

WHARFDALE OXFORD'S BEAUTY 1st, white, calved October 4, 1874, Vol. xxiv. p. 621. Bred by Mr. T. Purkis; got by Wharfdale Oxford (27786), dam (Maggie Lind) by Bohemia (23434), &c.

Produce in	Names, &c.	By what Bull.	By whom bred.
1879, Nov. 18, roan, C.C.	Marguerite	Cambridge Duke 5th, 30644	Mr. Purkis

ZENOBIA 13th, roan, calved January 8, 1876. Bred by Mr. T. Purkis; got by Wharfdale Oxford (27786), dam (Zenobia 6th) by Baron Killian (21226), &c. See Vol. xxiii. p. 609.

Produce	Names	By what Bull	By whom bred
1879, Dec. 23, r. & w., C.C.	Zenobia 15th	Cambridge Duke 5th, 30644	Mr. Purkis

PYM, G. E.,
Reigate, Surrey.

FAIRY, roan, calved February 16, 1864, Vols. xx., xxi., and xxii. pp. 510, 899, 538. Bred by Mr. E. W. Smyth, Wadhurst Castle; got by Touchstone (20987), dam (Freedom) by Don Juan (15896), &c.

Produce	Names	By what Bull	By whom bred
1876, Mar. 25, roan, B.C.		Puritan, 32227	Mr. Pym
1877, April 7, r. & w., C.C.	Fairy Rose	Thorndale Rose, 37595	do.
Bull Calf, sold to Mr. Ellis, Brockhamhurst.			

FANCY, roan, calved June 1, 1872, Vol. xxiii. p. 609. Bred by Mr. G. E. Pym; got by Boston (23443), dam (Fairy) by Touchstone (20987), &c.

Produce	Names	By what Bull	By whom bred
1877, Aug. 29, r. & w., C.C.	Fancy Rose	Thorndale Rose, 37595	Mr. Pym
1878, Oct. 28, red, C.C.	Fancy Rose 2nd	do.	do.

SYLVIA 2nd, red and white, calved January 8, 1873, Vol. xxiii. p. 609. Bred by Mr. G. E. Pym; got by Napier (26942), dam (Sylvia) by Vampire (19043), &c.

Produce	Names	By what Bull	By whom bred
1878, Mar. 9, r. & w., B.C.	Sylvanus	Thorndale Rose, 37595	Mr. Pym
1879, Mar. 11, r. & w., C.C.	Sylvia 6th	Fugleman, 36670	do.
Sylvanus, sold to Mr. J. Selmes, Bletchingley.			

RADCLIFFE, John,
Stearsby, Easingwold, Yorkshire.

BRACELET 2nd, roan, calved December 20, 1875, Vol. xxv. p. 630. Bred by Mr. J. Radcliffe; got by Bismarck (28040), dam (Necklace) by British Crown (21322), &c.

Produce	Names	By what Bull	By whom bred
1879, Mar. 1, r. & w., B.C.	Breastplate	Sir Hugo Irwin, 35563	Mr. Radcliffe

MAY ROSE, red and white, calved May 21, 1875. Bred by Mr. J. Radcliffe; got by Chancellor (30679), dam (Midsummer Rose) by The General (27632), &c. See Vol. xxii. p. 599.

Produce	Names	By what Bull	By whom bred
1878, April 1, r. & w., B.C.	(dead)	Sergeant, 37434	Mr. Radcliffe
1879, Mar. 11, r. & w., C.C.	March Rose	Hopeful, 41692	do.

RAINE, Albert,
The Grove, Bow Bank, Middleton-in-Teesdale.

LADY ANN, red, calved March 11, 1869, Vols. xxiii. and xxv. pp. 415, 420. Bred by Mr. W. E. B. Gwyn, Plas-Cwrt-Hyr; got by Prince of the Empire (20578), dam (Lady Pigot) by Duke of Knowlmere (19623), &c.

Produce	Names	By what Bull	By whom bred
1879, Aug. 14, roan, C.C.	Lady Elinor	Whiff, 30299	Mr. Raine

LADY KILLERBY, roan, calved December 14, 1874. Bred by Mr. A. Raine; got by Royal Killerby (32396), dam (Violet 3rd) by Whiff (30299) or Baron Mantalini (21951), &c. See Vol. xxii. p. 540.

Produce in		Names, &c.	By what Bull.	By whom bred.
1878, Feb. 7, roan,	B.C.	Regent's Prince	Prince Regent, 29676	Mr. Raine
1879, Feb. 16, r. & w.,	C.C.	Lady Wenlock	Lord Wenlock, 38671	do.

Regent's Prince, sold to Mr. G. Thompson, Coastley, Hexham.

LADY MOWBRAY, roan, calved May 1, 1873, Vols. xxiii. and xxiv. pp. 610, 622. Bred by Mr. A. Raine; got by Whiff (30299), dam (Maid of Connington) by Quis (20618), &c.

1878, June 15, roan,	B.C.	Mantalini King	Warlaby, 32792	Mr. Raine
1879, June 20, roan,	C.C.	Rose of Autumn	do.	do.

LADY ROSEDALE, roan, calved July 16, 1876. Bred by Mr. A. Raine; got by Sir James (35566), dam (Ringlet 2nd) by Whiff (30299), &c. See Vol. xxiii. p. 611.

1878, July 13, red,	B.C.	(dead)	Crown of the Realm, 38056	Mr. Raine
1879, Nov. 13, roan,	C.C.	Rose Duchess	Warlaby, 32792	do.

LADY SUTLER, roan, calved July 4, 1875. Bred by Mr. A. Raine; got by Sir James (35566), dam (Lady Grateful) by The Sutler (23061), &c. See Vol. xxiii. p. 610.

1879, Jan. 31, white,	C.C.	Warlaby's Lady	Warlaby, 32792	Mr. Raine
1879, Dec. 13, roan,	C.C.	Lady Gaiety	Whiff, 30299	do.

Lady Gaiety, sold to Mr. J. Peel, Knowlmere Manor, Clitheroe.

NECKLACE, roan, calved in 1871, Vols. xxi., xxii., xxiii., and xxv. pp. 900, 539, 610, 630. Bred by Mr. J. Haxby, Easingwold; got by British Crown (21322), dam (Polly) by Sir Samuel (15302), &c.

1879, Sept. 4,	red, B.C. Prince Arthur roan, C.C. Bracelet	Whiff, 30299	Mr. Raine

RANKIN, James,
Bryngwyn, Hereford.

DOWKE 10TH, red and white, calved November 10, 1869. Bred by Captain Winnall, Eccleswall Court; got by Lord Craggs (26632), dam (Dowke 9th) by Duke of Liverpool (23766), g. d. (Dowke 8th) by Sir John (18846), gr. g. d. (Dowke 7th) by Masterman (18356), &c. See Vol. xiii. p. 423.

1875, Apr. 20, r. & w.,	B.C.	Saxon 2nd	Saxon, 35477	Mr. Rankin
1876, Apr. 22, r. & w.,	B.C.	Saxon 3rd	do.	do.
1877, Nov. 7, r. & w.,	C.C.	Dowke 12th	Red Duke, 35224	do.
1878, Oct. 9, r. & w.,	C.C.	Dowke 13th	Claro 6th, 39597	do.
1879, Sept. 6, r. & w.,	B.C.	Blushing Duke 3rd	Blushing Duke 2nd, 42802	do.

DOWKE 11TH, red and white, calved May 7, 1874. Bred by Mr. J. Rankin; got by Saxon (35477), dam (Dowke 10th) by Lord Craggs (26632), &c.

1876, Aug. 8, r. & w.,	C.C.	Duchess 1st	Red Duke, 35224	Mr. Rankin
1877, Oct. 17, r. & w.,	C.C.	Duchess 2nd	do.	do.
1878, Feb. 9, roan,	C.C.	Duchess 3rd	Claro 6th, 39597	do.

DUCHESS 1ST, red and white, calved August 8, 1876. Bred by Mr. J. Rankin; got by Red Duke (35224), dam (Dowke 11th) by Saxon (35477), &c.

1878, Aug. 7, r. & w.,	B.C.	Blushing Earl	Blushing Duke 2nd, 42802	Mr. Rankin

Blushing Earl, sold to Mr. R. Chitson, Lawns Farm, Llanwarne, Ross.

LAUNDRESS 4th, red and white, calved May 20, 1876. Bred by Mr. J. Rankin; got by Regent (35254), dam (Laundress 2nd) by Duke of Liverpool (23766), g. d. (Laundress) by Duke of Cambridge (15921), &c. See " Laundress 3rd," Vol. xxi. p. 643.

Produce in		Names, &c.	By what Bull.	By whom bred.
1878, July 6, red,	B.C.	Otago	Blushing Duke 2nd, 42802	Mr. Rankin
1879, June 19, roan,	B.C.	Canterbury	Claro 6th, 39597	do.

Otago, sold to Mr. R. Chitson, Lawns Farm, Llanwarne, Ross.

RATHDONNELL, Lord,
Lisnavagh, Tullow, Co. Carlow.

CLOCHETTE, roan, calved April 20, 1877. Bred by Messrs. Mitchell, Alloa; got by Foggathorpe 1st (31179), dam (Clarissa) by Beau (28000), &c. See Vol. xxiv. p. 577.

1879, Dec. 8, white, B.C. (dead)	Brocklesby, 36288	Lord Rathdonnell

JAMES'S NANCY, roan, calved April 2, 1874. Bred by Mr. R. Chaloner, King's Fort; got by King James (28971), dam (Red Nancy) by Sovereign (27538), &c. See " Master Nobody," p. 167.

1879, Aug. 24, {	roan, B.C. Colonel Slick	} Anchor, 32947, or Lieu- }	Lord Rathdonnell
	roan, B.C. Master Nobody	tenant, 43466 }	

RAWLINSON, Robert,
Docker Hall, Kendal.

BUTTERCUP 15th, red and white, calved January 29, 1874. Bred by Mr. R. Rawlinson; got by Earl of Clare (28496), dam (Buttercup 13th) by Perseverance (27062), &c. See Vol. xxiv. p. 622.

1878, Mar. 25, r. & w., B.C. (Steer)	Francisco, 38315	Mr. Rawlinson
1879, Mar. 14, r. & w., C.C. British Girl	British Duke, 37901	do.

JESSAMINE 9th, roan, calved May 26, 1875. Bred by Mr. J. Bowman, High House; got by Knight of the Ribble (31563), dam (Jessamine 8th) by Second Duke of Richmond (28438), g. d. (Jessamine 5th) by Cressidus Gwynne (25850), &c. See " Regent," Vol. xxiv. p. 220.

1878, Oct. 4, roan, C.C. Jemima	British Duke, 37901	Mr. Rawlinson
1879, Sept. 26, roan, C.C. Jenny	do.	do.

MISS JEFFERSON, roan, calved March 2, 1876. Bred by Mr. R. Jefferson, Preston Hows; got by Silver Cloud (35528), dam (Mystic Tie) by Earl of Eglinton (23832), g. d. (Marquis's Flower) by Marquis of Cobham (22299), &c. See " Last Hope," Vol. xxi. p. 255.

1878, Dec. 22, roan, C.C. Miss Mary	British Duke, 37901	Mr. Rawlinson
1879, Dec. 27, red, B.C. (dead)	do.	do.

ROSEBERRY BELLE, roan, calved July 26, 1874. Bred by Messrs. J. and P. Robertson, Moor Gate; got by British Flag (33216), dam (Roseberry 29th) by Waterloo Cherry (27763), g. d. (Roseberry 20th) by Sir Windsor (22027), gr. g. d. (Roseberry 15th) by Knight of Distington (18158), &c. See Vol. xviii. p. 700.

1877, May 17, white, C.C. Lily of the Valley	Cassius, 33300	Mr. Rawlinson
1878, Apr. 28, white, C.C. Lily of the Forest	Francisco, 38315	do.
1879, Apr. 14, r. & w., C.C. British Lass	British Duke, 37901	do.

RAWSTORNE, Lawrence,
Hutton Hall, Preston.

CHRISTMAS GWYNNE, white, calved April 22, 1875, Vol. xxv. p. 633.
Bred by Mr. G. Fox, Elmhurst Hall; got by The Christmas Duke (32663),
dam (Fauna Gwynne) by Grand Duke 5th (19875), &c.

Produce in	Names, &c.	By what Bull.	By whom bred.
1879, May 23, roan, B.C.	Knight of Gwynne	Duke of Underley, 33745	Mr. Rawstorne

LADY WELLESLEY 2ND, roan, calved May 11, 1875, Vol. xxv. p. 633.
Bred by Mr. E. H. Cheney, Gaddesby Hall; got by Ninth Duke of Geneva
(28391), dam (Lady Wellington) by Duke of Putney (33717), &c.

1879, Mar. 25, roan, B.C.	(dead)	Duke of Underley, 33745	Mr. Rawstorne

SONSIE BELLE, red, calved March 14, 1877. Bred by Sir W. Lawson, Bart.,
Brayton; got by Baron Oxford 6th (33075), dam (Sonsie 18th) by Royal
Cambridge (25009), &c. See "Sonsie Boy," p. 239.

1879, Oct. 14, red, B.C.	Sonsie Boy	D. Wellington 12th, 38199	Mr. Rawstorne

REID, Alexander,
Cruivie, Cupar, Fife, N.B.

LAURA GWYNNE, red and white, calved December 31, 1874. Bred by Mr.
J. Unthank, Netherscales; got by British Boy (30597), dam (Vesper Gwynne
2nd) by Warwick (27755), &c. See Vol. xxi. p. 973.

1879, Jan. 20, red, C.C.	Ruth Gwynne	Earl, 38208	Mr. Reid
1879, Dec. 13, red, B.C.	Prior	do.	do.

NELL, roan, calved April 17, 1876. Bred by Mr. A. Reid; got by Stackhouse
(32591), dam (Cherry Brandy) by Prelate (29570), g. d. (Pale Brandy) by
Dunallan (31044), &c. See Vol. xxi. p. 973.

1878, Dec. 24, roan, B.C.	Samson	Earl, 38208	Mr. Reid

SHORT TAIL, white, calved July 16, 1877. Bred by Mr. A. Reid; got by
Stackhouse (32591), dam (Annie Laurie) by Beverley Lad (25630), &c. See
Vol. xxi. p. 973.

1879, Feb. 5, white, C.C.	Snowdrop	Bright Prince, 39496	Mr. Reid

REID, Nathaniel,
Danestown, Aberdeen, N.B.

ANNE OF LANCASTER 2ND, red, calved March 15, 1876. Bred by Mr. N.
Reid; got by Hope of Britain (34179), dam (Anne of Lancaster) by Lord
Raglan (13244), &c. See Vol. xvi. p. 337.

1879, Jan. 9, red, C.C.	Anne of L'ncaster 3d	Revenue, 40591	Mr. Reid

DAHLIA, red and white, calved in May, 1873. Bred by Mr. D. Fisher,
Pitlochrie; got by Fawsley Prince (31150), dam (Dewberry) by Brother
Windsor (25690), &c. See Vol. xxii. p. 343.

1879, Mar. 28, r. & w., B.C.	Marquis	Kt. of Killerby 2nd, 36875	Mr. Reid

Marquis, sold to Mr. Brown, Knockollochy, Aberdeenshire.

SPICY 4th, red, calved March 12, 1875. Bred by Mr. N. Reid ; got by Duke
of Lancaster (30986), dam (Spicy 3rd) by Duke (28342), &c. See Vol. xxii.
p. 541.

Produce in	Names, &c.	By what Bull.	By whom bred.
1877, Jan. 21, r. & w., B.C.	Prince	Kt. of Killerby 2nd, 36875	Mr. Reid
1878, April 12, red, C.C.	Spicy 6th	do.	do.

REYNELL, Richard,
Killynon, Killucan, Ireland.

HAWTHORN, roan, calved April 15, 1871, Vol. xxi. p. 902. Bred by Mr.
R. Reynell; got by Royal Prince (27384), dam (Sweetbrier) by Nimrod
(13388), &c.

1877, July 12, roan, B.C.	Pilot	Lord Westland, 34701	Mr. Reynell
1878, Aug. 27, white, B.C.	Actor	Albion, 36112	do.

Pilot, sold to Mr. T. Battersley, Newcastle, Oldcastle.

ISABELLA BROUGHTON, roan, calved March 20, 1874, Vol. xxiv. p. 623.
Bred by Mr. E. J. Smith, Islanmore; got by Lord Broughton (31626), dam
(Richard's Blossom) by King Richard (26523), &c.

1878, Oct. 12, roan, C.C.	Isabella Westland	Lord Westland, 34701	Mr. Reynell
1879, Nov. 14, roan, B.C.	Artizan	Agamemnon, 39357	do.

SWEETBRIER, roan, calved June 8, 1874, Vol. xxiv. p. 623. Bred by Mr. R.
Reynell; got by King James (28971), dam (Hawthorn) by Royal Prince
(27384), &c.

1878, Apr. 19, white, B.C.	Favourite	Albion, 36112	Mr. Reynell
1879, May 30, r. & w., B.C.	Royalist	Royal Baron, 40617	do.

VIOLET 3rd, roan, calved June 13, 1866, Vols. xix. and xxiv. pp. 770, 289.
Bred by Mr. J. P. Kearney, Milltown House; got by Dr. McHale (15887),
dam (Portia) by Paddy Hopewell (16677), &c.

1878, Mar. 4, roan, B.C.	Prince Mason	Prince James, 37236	Mr. Reynell
1879, Mar. 4, r. & w., C.C.	Violet 8th	Agamemnon, 39357	do.

Prince Mason, sold to Mr. J. Wadsworth, Dumralla, Newton Butler.

VIOLET 4th, roan, calved January 21, 1872, Vol. xxii. p. 542. Bred by Mr.
R. Reynell; got by Lord Spencer (26738), dam (Violet 3rd) by Dr. McHale
(15887), &c.

1876, Dec. 6, white, B.C.	Cato	Lieutenant-General, 31600	Mr. Reynell

Cato, sold to Mr. T. Rothwell, Rockfield, Kells.

VIOLET 5th, roan, calved January 8, 1875, Vol. xxiv. p. 623. Bred by Mr. R.
Reynell; got by King James (28971), dam (Violet 4th) by Lord Spencer
(26738), &c.

1879, June 18, roan, B.C.	Adjutant	Agamemnon, 39357	Mr. Reynell

WOODBINE, roan, calved August 3, 1876. Bred by Mr. R. Reynell; got by
Lieutenant-General (31600), dam (Hawthorn) by Royal Prince (27384), &c.
See " Royal," p. 215.

1879, Feb. 12, roan, B.C.	Royal	Royal Baron, 40617	Mr. Reynell

RHODES, A.,
Lundholme Cottage, *viâ* Ingleton.

BEAUTY'S OXFORD, roan, calved November 28, 1872, Vol. xxii. p. 542. Bred by Mr. A. Rhodes; got by Skelsmergh Oxford (30015), dam (Royal Beauty) by Royal Scotforth (25042), &c.

Produce in	Names, &c.	By what Bull.	By whom bred.
1877, Mar. 16, r. & w., B.C.	Garland Oxford	Garland Duke, 36676	Mr. Rhodes
1878, Mar. 19, r. & w., B.C.	Royal Garland	do.	do.
1879, Feb. 14, r. & w., C.C.	Beauty 3rd	do.	do.

Garland Oxford, sold to Mr. Lamb, Sand Villa; Royal Garland, to Mr. Laycock, Eldroth Hall.

NONPAREIL, roan, calved April 17, 1876. Bred by Mr. A. Rhodes; got by The Squire (35791), dam (Sunflower) by Third Duke of Lancaster (28409), &c. See Vol. xxv. p. 635.

1878, Nov. 13, roan, C.C.	Nonpareil 3rd	Anthony Maynard, 37736	Mr. Rhodes
1879, Nov. 29, white, B.C.	Baron Nonpareil 2nd D. of Ulster 2nd, 43131		do.

ROSALBINA, roan, calved March 2, 1876. Bred by Mr. A. Rhodes; got by Royal Geneva (39037), dam (Rose Alice 2nd) by Prince of Wales (27186), &c. See Vol. xxii. p. 542.

1878, Nov. 7, r. & w., B.C.	Rob Roy	Anthony Maynard, 37736	Mr. Rhodes
1879, Nov. 30, r. & w., B.C.	Romulus	do.	do.

Rob Roy, sold to Mr. T. Rhodes, Conder Green.

ROSE ALICE 2ND, roan, calved March 15, 1872, Vol. xxii. p. 542. Bred by Mr. Wilkinson, Whinney Clough; got by Prince of Wales (27186), dam (Alice Hawthorn) by Prince James (20555), &c.

1876, Mar. 2, roan,	C.C. Rosalbina	Royal Geneva, 39037	Mr. Rhodes
1877, Mar. 4, roan,	C.C. Rosina	Garland Duke, 36676	do.
1879, Mar. 8, roan,	B.C. Royalist	do.	do.

ROSE OF RICHMOND, red and white, calved February 13, 1873, Vol. xxiii. p. 615. Bred by Mr. C. L. Whalley, Richmond House; got by Royal Hatley (37394), dam (Wild Rose) by Napoleon (20395), &c.

1877, Dec. 24, roan,	C.C. Wild Rose 2nd	Garland Duke, 36676	Mr. Rhodes
1879, Jan. 11, red,	C.C. Wild Rose 3rd	do.	do.

ROYAL BEAUTY, white, calved March 13, 1869, Vol. xx. p. 748. Bred by Mr. J. Thompson, Scotforth; got by Royal Scotforth (25042), dam (Beauty) by White Hamlet (15508), &c.

1875, May 20, roan,	B.C. Golden Edgar	Golden Constantine, 36703	Mr. Rhodes
1877, April 30, white, B.C.	Royal Duke	Garland Duke, 36676	do.
1878, April 6, roan,	C.C. Beauty 2nd	do.	do.
1879, Mar. 4, roan,	C.C. Beauty 4th	Anthony Maynard, 37736	do.

Golden Edgar, sold to Mr. R. Mason, Great Crimbles; Royal Duke, to Mr. Gorst, Westmore.

RICH, Edmund,
Willesley, Tetbury, Gloucestershire.

BEAUTY, red, calved April 28, 1876. Bred by Mr. E. Rich; got by Count Bickerstaffe 4th (38040), dam (Bridesmaid) by Umpire (30189), &c. See Vol. xxii. p. 542.

1879, Mar. 4, red,	B.C. (Steer)	Prince Fulgens, 37231	Mr. Rich

HOLLY, red and white, calved March 2, 1876. Bred by Mr. E. Rich ; got by
Count Bickerstaffe 4th (38040), dam (Harriett 2nd) by Viscount Walton
(23154), &c. See Vol. xxiii. p. 615.

Produce in	Names, &c.	By what Bull.	By whom bred.
1879, Mar. 6, r. & w., B.C.	(Steer)	Prince Fulgens, 37231	Mr. Rich

NORA, red and white, calved March 4, 1876. Bred by Mr. E. Rich ; got by
Count Bickerstaffe 4th (38040), dam (Nellie) by Umpire (30189), &c. See
Vol. xxiii. p. 615.

1879, Feb. 16, r. & w., C.C.	Nightingale	Prince Fulgens, 37231	Mr. Rich

RICHARDSON, Joseph,
Ober Green, Hutton Rudby, Yarm.

ARCHDUCHESS 2ND, red and white, calved March 20, 1877. Bred by Mr.
J. Richardson ; got by Prince Charlie (37224), dam (Archduchess) by Chilton
(25774), &c. See Vol. xxiii. p. 616.

1879, Aug. 6, roan, C.C.	Citrona	Zetland, 40946	Mr. Richardson

IVY QUEEN, white, calved October 17, 1876. Bred by Mr. J. Richardson ;
got by Archduke (36132), dam (Ivy) by Cherry Bud (33339), &c. See Vol.
xxiii. p. 616.

1879, Oct. 30, white, B.C.	Ivy King	Zetland, 40946	Mr. Richardson

RICHMOND and GORDON, Duke of,
Gordon Castle, Fochabers, N.B.

DUCHESS 16TH, roan, calved June 2, 1875. Bred by the Duke of Richmond
and Gordon ; got by Royal Hope (32392), dam (Duchess 5th) by Duke of
Bowland (21568), &c. See " Baron Hope," Vol. xxv. p. 17.

1878, Feb. 8, r. & w., B.C.	Vulcan	Chief Officer, 36359	Duke of Richmond
1879, April 30, roan, C.C.	Duchess 17th	do.	do.

FLIRT 14TH, roan, calved August 16, 1876. Bred by the Duke of Richmond
and Gordon ; got by Royal Hope (32392), dam (Flirt 6th) by Baron Colling
(25560), &c. See Vol. xxiii. p. 616.

1879, Mar. 8, roan, B.C.	Heir of Promise	R. A. D., 38963	Duke of Richmond

LADY EVELYN HOPE, roan, calved May 1, 1877. Bred by the Duke of
Richmond and Gordon ; got by Royal Hope (32392), dam (Wimple 8th) by
Baron Colling (25560), g. d. (Wimple 5th) by Fifteenth Duke of Oxford
(23776), gr. g. d. (Wimple 3rd) by Whipper In (19139), &c. See " Wimple 4th,"
Vol. xxiii. p. 617.

1879, April 11, red, B.C.	Alliance	R. A. D., 38963	Duke of Richmond

LUSTRE 19TH, roan, calved December 13, 1875. Bred by the Duke of Richmond
and Gordon ; got by Montrose (34873), dam (Oxford Lustre) by Fifteenth
Duke of Oxford (23776), &c. See Vol. xxi. p. 905.

1878, July 27, roan, B.C.	Officer	Chief Officer, 36359	Duke of Richmond
1879, July 5, roan, C.C.	Lady Violet	Royal Hope, 32392	do.

PEACH BLOSSOM 8TH, roan, calved December 5, 1876. Bred by the Duke
of Richmond and Gordon ; got by White Duke (32849), dam (Peach Blossom
5th) by Royal Hope (32392), &c. See Vol. xxiv. p. 624.

1879, April 7, r. & w., B.C.	Radix	R. A. D., 38963	Duke of Richmond

Radix, sold to Mr. Adam, Unthank, Duffus, Elgin.

QUEEN ESTHER, roan, calved January 6, 1873, Vol. xxiii. p. 617. Bred by
Mr. J. Bowman, High House; got by Squire Booth (30049), dam (Queen of
the Valley) by Sir Christopher (22895), &c.

Produce in		Names, &c.	By what Bull.	By whom bred.
1878, May 29, roan,	C.C.	Queen Esther 1st	Royal Hope, 32392	Duke of Richmond
1879, May 7, r. & w.,	B.C.	Royal General	do.	do.

RIDLEY, Sir M. W., Bart.,
Blagdon, Cramlington.

LADY FLORA, roan, calved January 2, 1872, Vols. xxiii., xxiv., and xxv.
pp. 618, 625, 636. Bred by Messrs. Atkinson, Peepy; got by King Charles
(24240), dam (Sylvia) by Killerby Lad (20052), &c.

1879, May 26, roan,	C.C.	Lady Sylvia	Garterley Bell, 33991	Sir M. W. Ridley

LADY PRIMROSE, roan, calved July 11, 1876. Bred by Sir M. W. Ridley,
Bart.; got by Garterley Bell (33991), dam (Lady Ruby 1st) by Lord Charles
(29080), &c. See Vol. xxiii. p. 618.

1879, July 25, roan,	B.C.	Lord of the Manor	Hymen, 34195	Sir M. W. Ridley

LADY RUBY, roan, calved May 12, 1871, Vols. xxiii., xxiv., and xxv. pp. 618,
625, 636. Bred by Sir M. W. Ridley, Bart.; got by Lord Bloomfield (29068),
dam (Ruby) by Knight of Killerby (22054), &c.

1879, Dec. 22, roan,	B.C.	Baron Ruby	Garterley Bell, 33991	Sir M. W. Ridley

LADY RUBY 2ND, red, calved March 23, 1874, Vols. xxiii., xxiv., and xxv.
pp. 618, 625, 636. Bred by Sir M. W. Ridley, Bart.; got by Lord Charles
(29080), dam (Ruby) by Knight of Killerby (22054), &c.

1879, Dec. 26, roan,	B.C.	General Roberts	Garterley Bell, 33991	Sir M. W. Ridley

RIGG, Jonathan,
Wrotham Hill Park, Sevenoaks, Kent.

BEAUTY 2ND, roan, calved March 18, 1873, Vols. xxii., xxiii., xxiv., and xxv.
pp. 544, 619, 625, 637. Bred by Messrs. F. Leney and Sons, Wateringbury;
got by Grand Duke of Oxford (28764), dam (Sultana 2nd) by Man in the
Moon (18320), &c.

1879, July 22, white,	C.C.	Lady Albino	Sir L. Barrington, 35582	Mr. Rigg

CAMBRIDGE FANTAIL, red and white, calved May 16, 1877. Bred by
Mr. J. Rigg; got by Duke of Rosedale 2nd (33722), dam (Fantail 8th) by
Seventeenth Duke of Oxford (25994), &c. See Vol. xxiv. p. 626.

1879, Oct. 27, roan,	C.C.	CambridgeFantail 2d	Sid'ngton Kirkl'ton, 42379	Mr. Rigg

DUCHESS, red, calved November 1, 1868, Vols. xx., xxi., xxiv., and xxv.
pp. 486, 855, 625, 637. Bred by Messrs. F. Leney and Sons, Wateringbury;
got by Grand Duke 15th (21852), dam (Countess) by Knightley Grand
Duke (24268), &c.

1879, July 30, roan,	C.C.	Duchess of Cyprus	Duke of Oxford 26th, 33708	Mr. Rigg

DUCHESS ANNIE 5TH, red and white, calved February 20, 1871, Vols. xxii.,
xxiii., xxiv., and xxv. pp. 544, 619, 625, 637. Bred by Mr. C. Lyall, Old
Montrose; got by Falconer (23907), dam (Duchess Annie) by Little Go
(18200), &c.

1879, Apr. 7, r. & w.,	B.C.	Lord Aberdeen 3rd	Duke of Oxford 26th, 33708	Mr. Rigg

DUCHESS ANNIE 6TH, red and white, calved September 20, 1872, Vols. xxii., xxiii., and xxiv. pp. 544, 619, 625. Bred by Mr. C. Lyall, Old Montrose ; got by Crown Prince (28281), dam (Duchess Annie 3rd) by Midas (22355), &c.

Produce in		Names, &c.	By what Bull.	By whom bred.
1879, Jan. 24, roan,	C.C.	Duchess Annie 8th	Duke of Oxford 26th, 33708	Mr. Rigg

GRAND DUCHESS CAROLINA 2ND, roan, calved June 12, 1873, Vols. xxiv. and xxv. pp. 626, 637. Bred by Mr. W. Angerstein, Weeting Hall ; got by Eighth Duke of Geneva (28390), dam (Grand Duchess Carolina) by Grand Duke 10th (21848), &c.

Produce in		Names, &c.	By what Bull.	By whom bred.
1879, Aug. 10, roan,	C.C.	G. D'ss Carolina 3rd	Duke of Oxford 26th, 33708	Mr. Rigg

KIRKLEVINGTON 25TH, roan, calved June 9, 1872, Vols. xxii. and xxiv. pp. 338, 443. Bred by Mr. I. Downing, Turner's Hill ; got by Third Duke of Claro (23729), dam (Kirklevington 17th) by Lord Lally (22161), &c.

Produce in		Names, &c.	By what Bull.	By whom bred.
1879, Jan. 28, roan,	C.C.	Kirklevington Gem	Duke of Oxford 39th, 38173	Mr. Rigg

LADY LAWRENCE, red and white, calved February 11, 1877. Bred by Mr. J. Rigg ; got by Sir Lawrence Barrington (35582), dam (Oxford Donna) by Didmarton Duke (21546), &c. See Vol. xxiv. p. 626.

Produce in		Names, &c.	By what Bull.	By whom bred.
1879, Nov. 3, roan,	B.C.	Lord Siddington	Sid'ngton Kirkl'ton, 42379	Mr. Rigg

LIGHTBURNE WINSOME, roan, calved June 25, 1874, Vols. xxiv. and xxv. pp. 626, 638. Bred by Mr. A. Brogden, Lightburne Park ; got by Twenty-third Duke of Oxford (31001), dam (Winsome 7th) by Grand Duke 10th (21848), &c.

Produce in		Names, &c.	By what Bull.	By whom bred.
1879, Sept. 8, r. & w.,	B.C.	D. of Wrotham 3rd	Duke of Oxford 26th, 33708	Mr. Rigg

NANNY BARRINGTON, red and white, calved May 6, 1875, Vol. xxv. p. 638. Bred by Mr. J. Robinson, Northchurch Farm ; got by Barrington Duke (27985), dam (Duchess of Oxford 3rd) by Second Duke of Claro (21576), &c.

Produce in		Names, &c.	By what Bull.	By whom bred.
1879, Feb. 24, roan,	B.C.	(dead)	Duke of Oxford 26th, 33708	Mr. Rigg

OXFORD DONNA, red, calved August 14, 1866, Vols. xix., xxi., xxii., xxiv., and xxv. pp. 656, 855, 544, 626, 638. Bred by Mr. I. Downing, Turner's Hill ; got by Didmarton Duke (21546), dam (Lady Oxford) by Imperial Oxford (18084), &c.

Produce in		Names, &c.	By what Bull.	By whom bred.
1879, July 3, roan,	C.C.	Oxford Donna 2nd	Duke of Oxford 26th, 33708	Mr. Rigg

ROSE OF WATERLOO, red and white, calved November 30, 1875. Bred by Mr. J. Rigg ; got by Second Duke of Glo'ster (28392), dam (Lady Waterloo 18th) by Second Lord of Waterloo (22198), &c. See Vol. xxii. p. 544.

Produce in		Names, &c.	By what Bull.	By whom bred.
1879, Jan. 13, r. & w.,	C.C.	Rose of Waterloo 2nd	Duke of Oxford 26th, 33708	Mr. Rigg

SWEETHEART, roan, calved September 2, 1873, Vols. xxiii. and xxv. pp. 620, 638. Bred by Messrs. F. Leney and Sons, Wateringbury ; got by Eighth Duke of Geneva (28390), dam (Sylph) by Knightley Grand Duke (24268), &c.

Produce in		Names, &c.	By what Bull.	By whom bred.
1879, Sept. 29, roan,	B.C.	Charmer Duke 2nd	Duke of Oxford 26th, 33708	Mr. Rigg

VERBENA 7TH, red and white, calved November 5, 1871, Vols. xxi. and xxiii. pp. 887, 620. Bred by Mr. J. W. Philips, Heybridge ; got by Bolton (25650), dam (Verbena 2nd) by Lord Liverpool (22168), &c.

Produce in		Names, &c.	By what Bull.	By whom bred.
1879, July 17, roan,	C.C.	Lady Verbena 2nd	Sir L. Barrington, 35582	Mr. Rigg

WATERLOO OF ONEIDA, red and white, calved May 6, 1875, Vol. xxiv.
p. 636. Bred by Mr. J. D.'A. Samuda, Chillies; got by Sixth Duke of Oneida
(30997), dam (Waterloo 37th) by Oxford Beau (29485), &c.

Produce in	Names, &c.	By what Bull.	By whom bred.
1879, Jan. 19, red, B.C.	Duke Oneida	Duke of Ormskirk, 36526	Mr. Rigg

Duke Oneida, sold to Mr. H. Webb, Linton, Cambs.

WILD EYES WINSOME, red and white, calved February 1, 1877. Bred by
Mr. J. Rigg; got by Duke of Underley 2nd (36551), dam (Lightburne Winsome) by Twenty-third Duke of Oxford (31001), &c. See Vol. xxiv. p. 626.

| 1879, Nov. 22, r. & w., C.C. | W. E. Winsome 2nd | Sid'ngton Kirkl'ton, 42379 | Mr. Rigg |

RILEY, Thomas,
Ewood Hall, Mytholmroyd, Manchester.

PEARL OF BEAUTY, roan, calved July 27, 1874, Vol. xxv. p. 638. Bred by
Mr. T. Riley; got by Junior Lord (31453), dam (Countess Granville) by Lord
Belmore (31617), &c.

| 1879, June 21, roan, B.C. | Lord of Ewood 4th | Grandee, 36726 | Mr. Riley |

ROBARTS, A. J.,
Lillingstone Dayrell, Buckingham.

SALVIA, red, calved September 27, 1873, Vols. xxiii., xxiv., and xxv. pp. 621,
627, 638. Bred by Mr. A. J. Robarts; got by Caractacus (28141), dam
(Stephanotis) by Second Duke of Claro (21576), &c.

| 1879, April 7, roan, C.C. | Salsify | Lord Siddington, 38664 | Mr. Robarts |

SERAPHINA 10TH, roan, calved July 6, 1869, Vols. xxi. and xxv. pp. 906, 639.
Bred by Mr. E. Clarke, Lillingstone Dayrell; got by Red Prince (27254),
dam (Seraphina 6th) by Prince of Cambridge (22610), &c.

| 1879, July 20, roan, C.C. | Salamis | Lord Siddington, 38664 | Mr. Robarts |

SERVIA, roan, calved October 9, 1875, Vol. xxv. p. 639. Bred by Mr. A. J.
Robarts; got by Caractacus (28141), dam (Stephanotis) by Second Duke of
Claro (21576), &c.

| 1879, Aug. 16, roan, B.C. | Seawell | Lord Siddington, 38664 | Mr. Robarts |

Seawell, sold to Mr. Bird, Foxley, Towcester.

SNOWBALL, white, calved November 16, 1871, Vols. xxi., xxiii., and xxiv.
pp. 906, 621, 627. Bred by Mr. A. J. Robarts; got by Sinbad (29981), dam
(Seraphina 4th) by Royal Essex (18767), &c.

| 1879, June 9, roan, C.C. | Syria | Lord Siddington, 38664 | Mr. Robarts |

SNOWBERRY, white, calved August 13, 1876. Bred by Mr. A. J. Robarts;
got by Caractacus (28141), dam (Snowball) by Sinbad (29981), &c. See Vol.
xxiii. p. 621.

| 1879, April 2, white, C.C. | Snowy | Lord Siddington, 38664 | Mr. Robarts |

SNOWFLAKE, roan, calved July 31, 1874, Vol. xxv. p. 639. Bred by Mr. A.
J. Robarts; got by Caractacus (28141), dam (Snowball) by Sinbad (29981),
&c.

| 1879, May 14, roan, C.C. | Shannon | Lord Siddington, 38664 | Mr. Robarts |

STELLA, white, calved August 10, 1874, Vols. xxiv. and xxv. pp. 628, 639.
Bred by Mr. A. J. Robarts; got by Caractacus (28141), dam (Stephanotis) by
Second Duke of Claro (21576), &c.

Produce in		Names, &c.	By what Bull.	By whom bred.
1879, July 8, roan,	C.C.	Seaforth	Lord Siddington, 38664	Mr. Robarts

STEPHANOTIS, roan, calved March 28, 1869, Vols. xxi., xxii., xxiii., xxiv., and
xxv. pp. 906, 545, 621, 628, 639. Bred by Mr. A. J. Robarts; got by Second
Duke of Claro (21576), dam (Seraphina 3rd) by Royal Essex (18767), &c.

1879, Sept. 22, roan,	B.C.	Steamer 2nd	Lord Siddington, 38664	Mr. Robarts

Steamer 2nd, sold to Mr. Whiting, Castle Thorpe, Stony Stratford.

SUNSHINE, red and white, calved February 28, 1874, Vol. xxv. p. 640. Bred
by Mr. A. J. Robarts; got by Caractacus (28141), dam (Seraphine 9th) by
Red Prince (27254), &c.

1879, Jan. 23, roan,	C.C.	Sunflower	Lord Siddington, 38664	Mr. Robarts

SWEETBRIAR, red and white, calved March 16, 1875, Vol. xxiv. p. 628. Bred
by Mr. A. J. Robarts; got by Caractacus (28141), dam (Strawberry) by Sinbad
(29981), &c.

1879, Jan. 17, red,	C.C.	Sweetmeat	Lord Siddington, 38664	Mr. Robarts

ROBERTSON, Thomas,
Narraghmore, Athy, Co. Kildare.

DREAM OF GOLD, roan, calved April 11, 1876. Bred by Mr. J. Byrne,
Wallstown Castle; got by Valorous (27701), dam (Cloth of Gold) by Premier
(32075), g. d. (Duchess of Thorndale 4th) by Hero of Thorndale (18061),
gr. g. d. (Gold Lace) by Soubadar (18901), &c. Vol. xxii. p. 361.

1879, Jan. 27, roan,	B.C.	King's Prince	King of England, 43422	Mr. Robertson

FANNY VALOROUS, red, calved April 14, 1876. Bred by Mr. J. Byrne,
Wallstown Castle; got by Valorous (27701), dam (Fanny 2nd) by Hero of
Thorndale (18061), g. d. (Fanny) by Jasper 2nd (14738), gr. g. d. (Effie) by
Duke of Marchmont (14446), &c. See "Ruby," Vol. xxiv. p. 643.

1879, Jan. 10, red,	B.C.	King's Duke	King of England, 43422	Mr. Robertson
1879, Dec. 28, red,	C.C.	King's Duchess	do.	do.

LADY MARCHMONT, roan, calved June 27, 1876. Bred by Mr. J. Byrne,
Wallstown Castle; got by Valorous (27701), dam (Duchess of Marchmont)
by Hero of Thorndale (18061), g. d. (Effie) by Duke of Marchmont (14446),
&c. See "Ruby," Vol. xxiv. p. 643.

1879, Jan. 23, r. & w.,	C.C.	King's Princess	King of England, 43422	Mr. Robertson

ROBINSON, E.,
West End, Nafferton, Hull.

BRENDA 11TH, roan, calved May 9, 1876. Bred by Mr. E. Robinson; got by
Dalesman (33499), dam (Brenda 3rd) by Lord Valentine (26751), &c. See
Vol. xxiii. p. 623.

1879, Aug. 13, roan,	C.C.	Bella	Halloo, 34105	Mr. Robinson

EVA 3RD, red and white, calved January 12, 1874, Vols. xxiii. and xxv. pp. 623, 641. Bred by Mr. E. Robinson; got by Master Edmund (29322), dam (Eva) by Lord Valentine (26751), &c.

Produce in	Names, &c.	By what Bull.	By whom bred.
1879, June 11, r. & w., C.C.	Birthday 2nd	Halloo, 34105	Mr. Robinson

LADY MAW 7TH, red, calved May 14, 1874, Vols. xxiii., xxiv., and xxv. pp. 623, 629, 641. Bred by Mr. E. Robinson; got by Rose Duke 2nd (32341), dam (Lady Maw 5th) by Lord Valentine (26751), &c.

| 1879, March 12, red, C.C. | Lady Maw 11th | Halloo, 34105 | Mr. Robinson |

WHITE ROSE, white, calved March 21, 1877. Bred by Mr. E. Robinson; got by Halloo (34105), dam (Brenda 4th) by Lord Valentine (26751), &c. See Vol. xxii. p. 547.

| 1879, July 21, roan, C.C. | Brenda 14th | Imperial Turchil, 40019 | Mr. Robinson |

ROBINSON, Joseph,
Broughton House, Aylesbury.

BARONESS, roan, calved April 7, 1872, Vol. xxv. p. 642. Bred by Lord Chesham, Latimer; got by Baron Oxford 4th (25580), dam (Bland) by Sir James (22902), &c.

| 1879, Aug. 10, r. & w., C.C. | Baroness Lochaber | L'd of Lochaber 2nd, 36983 | Mr. Robinson |

BARONESS BLANCHE, red and white, calved March 24, 1877. Bred by Mr. J. Robinson; got by Oxford Duke (34977), dam (Baroness) by Baron Oxford 4th (25580), &c. See Vol. xxv. p. 642.

| 1879, Aug. 12, roan, C.C. | Baroness Hillhurst | M'quis of Hillhurst, 40309 | Mr. Robinson |

BLAND 1ST, roan, calved April 20, 1874, Vol. xxv. p. 642. Bred by Lord Chesham, Latimer; got by Baron Wastwater (30492), dam (Bland) by Sir James (22902), &c.

| 1879, Nov. 10, roan, C.C. | (dead) | M'quis of Hillhurst, 40309 | Mr. Robinson |

CAMBRIDGE MOSS ROSE, roan, calved September 14, 1873, Vol. xxv. p. 642. Bred by Lord Chesham, Latimer; got by Baron Wastwater (30492), dam (Moss Rose 6th) by Grand Duke of Lancaster (19883), &c.

| 1879, July 22, r. & w., B.C. | Cambridge | Ld. of Lochaber 2nd, 36983 | Mr. Robinson |

CARNATION, roan, calved January 11, 1873, Vol. xxv. p. 642. Bred by Mr. T. G. Curtler, Bevere House; got by Grand Duke of Clarence (28750), dam (Camilla) by Touchstone (20986), &c.

| 1879, Dec. 31, roan, B.C. | Cicero | T'ndale Grand Duke, 37593 | Mr. Robinson |

Cicero, sold to Mr. C. A. Barnes, Solesbridge, Rickmansworth.

CHARMER 2ND, red, calved December 22, 1868, Vols. xx., xxi., and xxv. pp. 437, 661, 642. Bred by Mr. T. G. Curtler, Bevere House; got by Wild Boy (23219), dam (Pride of Churchill) by Lord of the Harem (16430), &c.

| 1879, June 9, r. & w., C.C. | Charming Thorndale | T'ndale Grand Duke, 37593 | Mr. Robinson |

CHRISTINE 6TH, roan, calved April 13, 1876, Vol. xxv. p. 643. Bred by Mr. G. Graham, The Oaklands; got by Coronation (30796), dam (Christine 4th) by Prince Gwynne (20547), &c.

| 1879, Oct. 19, roan, C.C. | (dead) | T'ndale Grand Duke, 37593 | Mr. Robinson |

COUNTESS FAWSLEY, red and white, calved February 21, 1871, Vol. xxv. p. 643. Bred by Mr. T. G. Curtler, Bevere House; got by Lord Waterloo 2nd (26755), dam (Camilla) by Touchstone (20986), &c.

Produce in	Names, &c.	By what Bull.	By whom bred.
1879, March 21, roan, B.C.	Baron Fawsley	King Charming, 28962	Mr. Robinson

CRINOLINE 2ND, red, calved January 6, 1874, Vol. xxv. p. 643. Bred by Mr. T. G. Curtler, Bevere House; got by Grand Duke of Clarence (28750), dam (Crinoline) by Wild Boy (23219), &c.

Produce in	Names, &c.	By what Bull.	By whom bred.
1879, Feb. 2, roan, B.C.	(Steer)	King Charming, 28962	Mr. Robinson

CROCUS 2ND, red and white, calved April 1, 1869, Vols. xxi. and xxv. pp. 631, 643. Bred by Mr. S. Canning, Snitterfield; got by University (27693), dam (Crocus) by Comet (19484), &c.

Produce in	Names, &c.	By what Bull.	By whom bred.
1879, April 3, roan, C.C.	Queen Dido	Royal Duke 6th, 39035	Mr. Robinson

DIDONA 4TH, red and white, calved September 24, 1874. Bred by Sir G. R. Philips, Bart., Weston Park; got by Cherry Grand Duke 5th (30712), dam (Empress 7th) by Florist (23962), &c. See Vol. xxii. p. 531.

Produce in	Names, &c.	By what Bull.	By whom bred.
1878, July 17, red, B.C.	Duke of Beaufort	Elmhurst Prince, 41503	Mr. Bliss
1879, July 1, r.&w., C.C.	Doubtful	do.	Mr. Robinson
1879, July 1, r.&w., B.C.	(slaughtered)		

DUCHESS, roan, calved March 9, 1873, Vols. xxii. and xxiv. pp. 293, 292. Bred by Mr. G. Allen, Knightley Hall, the property of Mr. J. Robinson; got by Second Duke of Wetherby (21618), dam (Annette 3rd) by Lord Oxford 2nd (20215), &c.

Produce in	Names, &c.	By what Bull.	By whom bred.
1879, March 24, roan, C.C.	Rowley's Duchess 2d	2nd Duke of Rowley, 28441	Mr. Allen

Rowley's Duchess 2nd, sold to Mr. J. Robinson, Broughton Pastures.

JESSAMINE, red and white, calved December 28, 1876. Bred by Sir G. R. Philips, Bart., Weston Park; got by Cherry Grand Duke 5th (30712), dam (Jesdemona 2nd) by Third Cherry Duke (28171), &c. See Vol. xxv. p. 453.

Produce in	Names, &c.	By what Bull.	By whom bred.
1879, July 21, red, C.C.	Jessie Elmhurst	Elmhurst Prince, 41503	Mr. Robinson

JESSICA 5TH, white, calved October 10, 1872, Vol. xxv. p. 643. Bred by Mr. G. Graham, The Oaklands; got by Lord Thorndale (29210), dam (Jessica 3rd) by Janitor (24204), &c.

Produce in	Names, &c.	By what Bull.	By whom bred.
1879, Feb. 8, roan, C.C.	Jessie Thorndale	T'ndale Grand Duke, 37593	Mr. Robinson

LADY OXFORD CRAGGS, red, calved December 14, 1876. Bred by Mr. J. Robinson; got by Duke of Oxford 26th (33708), dam (Mary Stuart) by University (27693), &c. See Vol. xxiii. p. 623.

Produce in	Names, &c.	By what Bull.	By whom bred.
1879, June 30, red, C.C.	Red Rose Craggs	Ld. of Lochaber 2nd, 36983	Mr. Robinson

LADY STUART 2ND, red, calved April 6, 1875. Bred by Mr. G. Garne, Churchill Heath, the property of Mr. J. Robinson; got by Lord of the Valley (31733), dam (Lady Stuart) by John O'Groat (18115), &c. See Vol. xxiii. p. 448.

Produce in	Names, &c.	By what Bull.	By whom bred.
1879, April 8, red, C.C.	Lady Stuart 3rd	Elmhurst Prince, 41503	Mr. Bliss

Lady Stuart 3rd, sold to Mr. J. Robinson, Broughton Pastures.

MARY STUART, red, calved April 29, 1871, Vols. xxi., xxii., xxiii., and xxv. pp. 910, 548, 623, 643. Bred by Mr. S. Canning, Snitterfield; got by University (27693), dam (Miss Stuart) by Standard (22963), &c.

Produce in	Names, &c.	By what Bull.	By whom bred.
1879, Apr. 26, r. & w., C.C.	Ly. T'ndale Craggs 2d	T'ndale Grand Duke, 37593	Mr. Robinson

NORA, red and white, calved November 20, 1871, Vol. xxiii. p. 492. Bred by
Mr. G. Hewer, Ley Gore House; got by Paritor (27040), dam (Nannette) by
Sir John (22904), &c.

Produce in	Names, &c.	By what Bull.	By whom bred.
1878, Oct. 18, r. & w., B.C.	Colonel	Elmhurst Prince, 41503	Mr. Bliss
1879, Nov. 15, r. & w., B.C.	Nicodemus	do.	Mr. Robinson

ROSE OF RABY 3RD, roan, calved May 22, 1875, Vol. xxv. p. 644. Bred by
Mr. W. Ashburner, Conishead Grange; got by Second Duke of Glo'ster
(28392), dam (Maid of Lorn) by Sixth Duke of Airdrie (19602), &c.

1879, Dec. 24, r. & w., B.C.	(dead)	Tindale Grand Duke, 37593	Mr. Robinson

WESTON'S FLOWER, roan, calved May 23, 1876, Vol. xxv. p. 342. Bred by
Mr. G. Fox, Elmhurst Hall; got by Grand Duke of Weston 3rd (34079),
dam (Earl's Flower) by Earl of Eglinton (23832), &c.

1879, Dec. 17, roan, C.C.	Elmhurst's Flower	Elmhurst Prince, 41503	Mr. Robinson

ROBINSON, W.,
Edge Hill, Ulverston.

BARMPTON ROSE 2ND, white, calved October 24, 1869, Vols. xx. and xxiv.
pp. 400, 629. Bred by Mr. W. Ashburner, Netherhouse; got by Perseverance
(27062), dam (Puffin) by Hematite (21917), &c.

1878, May 4, roan, C.C.	Barmpton Rose 6th	British King, 36279	Mr. Robinson

BARMPTON ROSE 3RD, roan, calved January 23, 1875. Bred by Mr. W.
Robinson; got by Julius Cæsar (31452), dam (Barmpton Rose 2nd) by Perseverance (27062), &c. See Vol. xxiv. p. 629.

1879, Aug. 27, roan, B.C.	(dead)	Sir Roland, 40721	Mr. Robinson

BRIGHT COUNTESS, red and white, calved September 24, 1872, Vol. xxiii.
p. 624. Bred by the Hon. J. Massy, Milford House; got by Earl of Rosedale (26072), dam (Bright Duchess) by Royal Duke (25014), &c.

1877, Jan. 20, roan, B.C.	Prince of Salthouse	Prince Alfred, 29593	Mr. Robinson
1878, Mar. 30, roan, C.C.	Bright Countess 2nd	do.	do.
1879, Mar. 10, roan, B.C.	(dead)	do.	do.

Prince of Salthouse, sold to Mr. W. Kendal, Lane House, Ulverstone.

EXPECTED GUEST, red and white, calved January 5, 1875. Bred by Mr.
H. N. Fraser, Hay Close; got by Marquis of Lorne (31847), dam (Comely
Wench) by Vain Hope (23102), &c. See Vol. xxv. p. 566.

1877, Apr. 16, r. & w., B.C.	(dead)	Prince Edward, 37229	Mr. Robinson
1879, May 2, roan, C.C.	Cheerful Guest	Faunus, 43216	do.

FLOWER BUD, roan, calved April 28, 1871, Vols. xxi. and xxiv. pp. 910, 630.
Bred by Mr. W. Torr, Aylesby Manor; got by Lord Napier (26688), dam
(Flower Child) by Cherry Prince (23555), &c.

1878, April 29, roan, C.C.	Flower Bud 4th	British King, 36279	Mr. Robinson

FLOWER BUD 2ND, red and white, calved January 12, 1875, Vol. xxiv. p. 630.
Bred by Mr. W. Robinson; got by Julius Cæsar (31452), dam (Flower Bud)
by Lord Napier (26688), &c.

1879, Sept. 29, roan, C.C.	Flower Bud 5th	Prince Alfred, 29593	Mr. Robinson

FLOWER BUD 3RD, red and white, calved April 1, 1877. Bred by Mr. W. Robinson; got by Prince Edward (37229), dam (Flower Bud 2nd) by Julius Cæsar (31452), g.d. (Flower Bud) by Lord Napier (26688), &c. See "Faunus," p. 98.

Produce in	Names, &c.	By what Bull.	By whom bred.
1879, Dec. 8, r. & w., C.C.	Flower Bud 6th	Sir Roland, 40721	Mr. Robinson

FLOWER CHILD, roan, calved May 12, 1868, Vol. xxi. p. 910. Bred by Mr. W. Torr, Aylesby Manor; got by Cherry Prince (23555), dam (Flower Damsel) by Breast Plate (19337), &c.

Produce in	Names, &c.	By what Bull.	By whom bred.
1875, April 16, roan, B.C.	Prince of Flowers	Prince Alfred, 29593	Mr. Robinson
1876, April 11, roan, B.C.	Prince Regent	do.	do.
1877, April 3, white, B.C.	Prince John	do.	do.
1878, Oct. 4, roan, B.C.	Prince Harry	do.	do.

Prince of Flowers, sold to Mr. R. Postlethwaite, Pennington; Prince Regent, to Mr. J. Atkinson, Millom; Prince John, to Mr. J. Holmes, Mansrigg; Prince Harry, to Mrs. Lawrence, Flookburgh.

KATHLEEN 3RD, roan, calved July 5, 1871. Bred by Messrs. Atkinson, Peepy; got by King Charles (24240), dam (Kathleen 2nd) by Yellow Jack (23248), &c. See Vol. xxii. p. 440.

Produce in	Names, &c.	By what Bull.	By whom bred.
1878, Feb. 17, roan, C.C.	Kathleen 4th	Prince Alfred, 29593	Mr. Robinson
1879, Mar. 20, roan, C.C.	Kathleen 5th	do.	do.

QUEEN NECKLACE, roan, calved July 31, 1872. Bred by Mr. E. A. Fawcett, Childwick Hall; got by Prince Frederick (35110), dam (Katherine 5th) by Prince Regent (35169), &c. See Vol. xxi. p. 976.

Produce in	Names, &c.	By what Bull.	By whom bred.
1878, Oct. 24, r. & w., B.C.	(dead)	Prince Edward, 37229	Mr. Robinson
1879, Dec. 30, r. & w., B.C.	Sir Fred	Sir Roland, 40721	do.

ROSABELLA 2ND, white, calved in April, 1870. Bred by Mrs. Grice, Kiskin, Bootle; got by Hyperion (21966), dam (Princess) by Warwick (15486), &c. See "Prince of Witley," p. 201.

Produce in	Names, &c.	By what Bull.	By whom bred.
1877, Mar. 13, roan, C.C.	Rosabella 3rd	King Henry, 36842	Mr. Ashburner
1878, Mar. roan, C.C.	Rosabella 4th	British King, 36279	do.
1879, April 13, roan, B.C.	Prince of Witley	Prince Waterloo, 40546	Mr. Robinson

Prince of Witley, sold to Mr. J. Mason, Hillhampton, Little Witley, Worcester.

WEAL BUD, red and white, calved January 18, 1869, Vol. xxiii. p. 624. Bred by Mr. W. Torr, Aylesby Manor; got by Breast Plate (19337), dam (Weal Blossom) by Fitz-Clarence (14552), &c.

Produce in	Names, &c.	By what Bull.	By whom bred.
1878, April 6, roan, C.C.	Weal Bud 3rd	Prince Alfred, 29593	Mr. Robinson

WEAL BUD 2ND, roan, calved December 17, 1876. Bred by Mr. W. Robinson; got by Prince Alfred (29593), dam (Weal Bud) by Breast Plate (19337), &c. See Vol. xxiii. p. 624.

Produce in	Names, &c.	By what Bull.	By whom bred.
1879, Oct. 1, r. & w., C.C.	Weal Bud 4th	Sir Roland, 40721	Mr. Robinson

WILD SPRAY, red and white, calved January 18, 1872, Vol. xxiii. p. 624. Bred by Mr. H. A. Brassey, Preston Hall; got by Cherry Grand Duke 3rd (28174), dam (Wave Rise) by Blinkhoolie (23428), &c.

Produce in	Names, &c.	By what Bull.	By whom bred.
1877, Oct. 24, roan, C.C.	Wild Spray 2nd	Prince Alfred, 29593	Mr. Robinson
1878, Oct. 23, roan, B.C.	Prince of Martley	do.	do.

Prince of Martley, sold to Mr. J. Robinson, Rodge Hill, Martley, Worcester.

ROBSON, John,
Newton, Bellingham, Northumberland.

JESSY, red, calved February 1, 1876.   Bred by Mr. T. Wilson, Shotley Hall;
got by Oxford Beau 3rd (32013), dam (Lady Villiers) by Earl of the Valley
(31086), &c.   See Vol. xxiii. p. 713.

Produce in	Names, &c.	By what Bull.	By whom bred.
1879, March 12, roan, C.C.	Janet	D. of Oxford 31st, 33713	Mr. Robson

ROCKETT, J. H.,
Snaith Hall, Selby, Yorkshire.

CHERRY LUCK, roan, calved July 16, 1870.   Bred by Mr. W. Gomersall,
Otterburn; got by Golden Cherry (31263), dam (Lady Miney) by Baron
Wetherell (19289), &c.   See "Jester King," p. 127.

Produce	Names	Bull	Bred
1877, Oct. 20. r. & w., C.C.	Cherry Flower	Jester, 41718	Mr. Rockett

KENT BUTTERFLY 3RD, roan, calved January 18, 1868, Vols. xix. and xxi.
pp. 565, 911.   Bred by Messrs. F. Leney and Sons, Wateringbury; got by
Knightley (22051), dam (Kent Butterfly) by Second Duke of Kent (19620), &c.

Produce	Names	Bull	Bred
1877, May 5. r. & w., C.C.	Kent Buttercup	Jester, 41718	Mr. Rockett
1878, Feb. 27, r. & w., C.C.	Kent Beauty	do.	do.

NOT OUT, red and white, calved July 18, 1876.   Bred by Mr. J. H. Rockett;
got by Speedwell (35652), dam (Kent Butterfly 3rd) by Knightley (22051),
&c.   See Vol. xxi. p. 911.

Produce	Names	Bull	Bred
1878, July 9, r. & w., C.C.	Nosegay	Jester, 41718	Mr. Rockett
1879, Sept. 17, r. & w., B.C.	Statesman	Linden, 41808	do.

Nosegay, sold to Mr. Cattley, York; Statesman, to Mr. W. Latham, Cowick, Selby.

OXFORD LOZENGE, roan, calved November 10, 1873.   Bred by Mr. J. H.
Rockett; got by Twentieth Duke of Oxford (28432), dam (Lozenge) by Lord
Towneley (26749), &c.   See Vol. xxi. p. 911.

Produce	Names	Bull	Bred
1878, June 30, r. & w., C.C.	Peppermint Lozenge	Jester, 41718	Mr. Rockett

SULTANA, red and little white, calved July 20, 1873.   Bred by Major Stapyl-
ton, Myton Hall, the property of Mr. W. Coupland, Bolton House; got by
Colonist (28227), dam (Kent Butterfly 3rd) by Knightley (22051), &c.   See
Vol. xxi. p. 911.

Produce	Names	Bull	Bred
1877, Oct. 10, r. & w., C.C.	Starlight	Jester, 41718	Mr. Rockett
1878, Dec. 31, r. & w., C.C.	Sweetheart	do.	do.

ROLLS, John Allan,
The Hendre, Monmouth.

CLARENCE DAISY, red and white, calved February 4, 1877.   Bred by Mr.
J. A. Rolls; got by Grand Duke of Clarence (28750), dam (Criccieth) by The
Baron (25277), &c.   See Vol. xxiv. p. 631.

Produce	Names	Bull	Bred
1879, Nov. 1, roan, C.C.	Clarence Daisy 3rd	D. of Siddington 2nd, 33732	Mr. Rolls

DAFFODIL, red and white, calved January 3, 1872, Vols. xxiii., xxiv., and xxv.
pp. 625, 631, 645.   Bred by Lord Chesham, Latimer; got by Prince of the
Realm (22627), dam (Dauntless) by Great Mogul (14651), &c.

Produce	Names	Bull	Bred
1879, April 21, roan, C.C.	Daffodil Daisy 2nd	D. of Siddington 2nd, 33732	Mr. Rolls

DAHLIA, red and white, calved March 25, 1870, Vols. xxi., xxiv., and xxv. pp. 768, 631, 645. Bred by Archdeacon Holbech, Farnborough Hall; got by Second Earl of Cleveland (26054), dam (Daffodil) by Eleventh Duke of Oxford (19632), &c.

Produce in	Names, &c.	By what Bull.	By whom bred.
1879, Sept. 26, roan,	C.C.|Dahlia Daisy	|20th D. of Oxford, 28432	|Mr. Rolls

DAISY, red and white, calved April 19, 1876, Vol. xxv. p. 645. Bred by Lord Chesham, Latimer; ·got by Duke of Oxford 28th (33710), dam (Daffodil) by Prince of the Realm (22627), &c.

Produce in	Names, &c.	By what Bull.	By whom bred.
1879, Aug. 14, roan,	B.C.,Pr. of the Daisies	|20th D. of Oxford, 28432	|Mr. Rolls

HILDA 2ND, red, calved April 19, 1872, Vols. xxii., xxiii., xxiv., and xxv. pp. 549, 625, 631, 645. Bred by Mr. J. A. Rolls; got by Towneley Wild Eyes (27674), dam (Hilda) by The Baron (25277), &c.

Produce in	Names, &c.	By what Bull.	By whom bred.
1879, Mar. 21, roan,	C.C.|Hilda Daisy 3rd	|D. of Siddington 2nd,33732	Mr. Rolls

KIRKLEVINGTON EMPRESS 2ND, red and white, calved March 1, 1877. Bred by Lord Fitzhardinge, Berkeley Castle; got by Duke of Connaught (33604), dam (Siddington 7th) by Seventh Duke of York (17754), &c. See Vol. xx. p. 764.

Produce in	Names, &c.	By what Bull.	By whom bred.
1879, July 23, r.&w.,	B.C.|Duke of Monmouth	|D. of Oxford 45th, 39775	|Mr. Rolls

LADY BICKERSTAFFE 3RD, red, calved September 4, 1874. Bred by Colonel Kingscote, Kingscote Park; got by Oxford Beau 3rd (32013), dam (Baroness Bickerstaffe) by Third Duke of Clarence (23727), &c. See Vol. xxi. p. 793.

Produce in	Names, &c.	By what Bull.	By whom bred.
1879, June 28, roan,	C.C.|Miss Bickerstaffe	|Oxford Beau 6th, 40422	|Mr. Rolls

LINDA, roan, calved January 27, 1868, Vols. xx., xxi., xxii., xxiii., xxiv., and xxv. pp. 627, 912, 549, 625, 632, 645. Bred by Mr. Priestley, Trefan; got by The Bully (23019), dam (Geraldine) by Sir Henry Havelock (16975), &c.

Produce in	Names, &c.	By what Bull.	By whom bred.
1879, May 17, white,	B.C.|Lucknow	|D. of Siddington 2nd,33732	Mr. Rolls

LINDA 3RD, roan, calved April 23, 1872, Vols. xxii., xxiii., and xxiv. pp. 549, 626, 632. Bred by Mr. J. A. Rolls; got by Towneley Wild Eyes (27674), dam (Linda) by The Bully (23019), &c.

Produce in	Names, &c.	By what Bull.	By whom bred.
1879, Feb. 9, roan,	C.C.|Marguerite	|D. of Siddington 2nd,33732| Mr. Rolls

LINDA 7TH, red, calved May 27, 1876, Vol. xxv. p. 646. Bred by Mr. J. A. Rolls; got by Grand Duke of Clarence (28750), dam (Linda) by The Bully (23019), &c.

Produce in	Names, &c.	By what Bull.	By whom bred.
1870, Oct. 14, red,	C.C.|Linda Daisy 3rd	|D. of Siddington 2nd,33732 Mr. Rolls

LINDA 8TH, roan, calved April 4, 1876. Bred by Mr. J. A. Rolls; got by Grand Duke of Clarence (28750), dam (Linda 3rd) by Towneley Wild Eyes (27674), &c. See Vol xxiii. p. 626.

Produce in	Names, &c.	By what Bull.	By whom bred.
1879, July 29, roan,	C.C.|(dead)	|20th D. of Oxford, 28432	|Mr. Rolls

MIGNON, red and white, calved February 9, 1869, Vols. xxi., xxiv., and xxv. pp. 912, 632, 646. Bred by Mr. Priestley, Trefan; got by The Baron (25277), dam (Welcome) by The Bully (23019), &c.

Produce in	Names, &c.	By what Bull.	By whom bred.
1879, Dec. 13, roan,	C.C.|Mignon Daisy 4th	|D. of Siddington 2nd,33732|Mr. Rolls

MIGNON DAISY, red, calved February 11, 1877. Bred by Mr. J. A. Rolls ;
got by Grand Duke of Clarence (28750), dam (Mignon 2nd) by Butterfly
Baron (30632), &c. See Vol. xxiv. p. 632.

Produce in		Names, &c.	By what Bull.	By whom bred.
1879, Oct. 26, red,	C.C.	Mignon Daisy 3rd	D.of Siddington 2nd,33732	Mr. Rolls

MINE OWN, roan, calved March 24, 1872, Vols. xxii., xxiii., and xxiv. pp. 549,
626, 633. Bred by Mr. J. A. Rolls ; got by Towneley Wild Eyes (27674),
dam (Edna) by The Bully (23019), &c.

1879, Nov. 11, white, B.C.			D.of Siddington 2nd,33732	Mr. Rolls

Bull Calf, sold to Mr. J. Price, The Tump, Rockfield, Monmouth.

TREFAN 3RD, red and white, calved May 6, 1876, Vol. xxv. p. 646. Bred by
Mr. J. A. Rolls ; got by Grand Duke of Clarence (28750), dam (Trefan) by
The Baron (25277), &c.

1879, Dec. 13, roan,	C.C.	Trefan 4th	D.of Siddington 2nd,33732	Mr. Rolls

ZOE 3RD, roan, calved May 23, 1875, Vol. xxiv. p. 633. Bred by Mr. J. A.
Rolls ; got by Butterfly Baron (30632), dam (Zoe) by The Baron (25277),
&c.

1879, Nov. 10, roan,	C.C.	Zoe's Daisy	D.of Siddington 2nd,33732	Mr. Rolls

ROSE, Rev. H. F.,
Holme Rose, Fort George Station, N.B.

BLINKBONNY 5TH, red, calved February 5th, 1872, Vols. xxiii., xxiv., and
xxv. pp. 627, 633, 647. Bred by Mr. J. Geddes, Orbliston ; got by Earl of
Buchan (31054), dam (Blinkbonny 4th) by Wizard (25467), &c.

1879, May 23, roan,	B.C.	Prince Charlie	Buccleuch, 36292	Rev. H. F. Rose

BLINKBONNY 6TH, red, calved April 27, 1876. Bred by the Rev. H. F. Rose ;
got by Baron Lowther (33062), dam (Blinkbonny 5th) by Earl of Buchan
(31054), &c. See Vol. xxiii. p. 627.

1879, Mar. 17, red,	B.C.	Buchan Lad	Alexander 2nd, 39364	Rev. H. F. Rose

FLORENCE 2ND, roan, calved February 23, 1869, Vols. xxiii., xxiv., and xxv.
pp. 627, 634, 647. Bred by Mr. D. McBean, Nairnside ; got by Duke of
Cornwall (23732), dam (Florence) by Magnum Bonum (13277), &c.

1879, April 29, white,	B.C.	Holme Rose	Buccleuch, 36292	Rev. H. F. Rose

KATE 2ND, roan, calved April 7, 1867, Vols. xxii., xxiii., xxiv., and xxv. pp. 549,
627, 634, 647. Bred by Mr. D. McBean, Nairnside ; got by Duke of Cornwall (23732), dam (Kate) by Garibaldi (17916), &c.

1879, Mar. 6, white,	C.C.	Kate 9th	Buccleuch, 36292	Rev. H. F. Rose

KATE 5TH, white, calved April 29, 1876. Bred by the Rev. H. F. Rose ; got
by Buccleuch (36292), dam (Kate 2nd) by Duke of Cornwall (23732), &c.
See Vol. xxiii. p. 627.

1879, Mar. 22, roan,	C.C.	Kate 10th	Alexander 2nd, 39364	Rev. H. F. Rose

KATE 6TH, roan, calved March 26, 1876. Bred by the Rev. H. F. Rose ; got
by Buccleuch (36292), dam (Kate 3rd) by Fawsley Prince 2nd (36635), &c.
See Vol. xxiii. p. 627.

1879, Mar. 8, r. & w.,	B.C.	Fawsley Lad	Alexander 2nd, 39364	Rev. H. F. Rose

ROSE, Thomas,
Melton Magna, Wymondham, Norfolk.

BRIDESMAID, red, calved January 18, 1873, Vols. xxiii. and xxv. pp. 627, 647.
Bred by Mr. H. A. Brassey, Preston Hall; got by Cherry Grand Duke 3rd
(28174), dam (Bright Ringlet) by Ringleader (15164), &c.

Produce in	Names, &c.	By what Bull.	By whom bred.
1879, Dec. 13, r. & w., C.C.	Bright Urania	Bright Damon, 36268	Mr. Rose

CHERRY BLOSSOM, roan, calved February 14, 1874. Bred by Mr. T. Rose;
got by Lord Blithesome (29067), dam (Cherry) by Ravenspur (20628), &c.
See "Cherry Bounce," p. 46.

1879, Jan. 24, roan, C.C.	Cherry Princess	Bright Damon, 36268	Mr. Rose

ROWLANDSON, P. H.,
Eden Bank, Kirkby Stephen, Westmoreland.

LADY BOOTH 3RD, roan, calved July 10, 1873, Vol. xxiii. p. 628. Bred by
Mr. W. Mitchell, Cleasby; got by Squire Booth (30049), dam (Lady Booth)
by Lord of the Hills (18267), &c.

1877, Mar. 30, roan, C.C.	Lady Booth 4th	Warlaby, 32792	Mr. Rowlandson
1878, Mar. 6, r. & w., C.C.	Lady Booth 5th	Earl of Sheffield, 33812	do.
1879, Feb. 1, white, C.C.	Lady Booth 6th	do.	do.

Lady Booth 4th, sold to Mr. J. C. Toppin, Musgrave Hall, Skelton, Penrith.

RUBY 2ND, red and white, calved September 28, 1870, Vols. xxi. and xxiii.
pp. 913, 628. Bred by Mr. M. Thompson, Stobars Hall; got by Big Ben
(23417), dam (Ruby) by Watchman (17216), &c.

1875, July 17, roan, C.C.	Ruby 4th	General Booth, 33999	Mr. Rowlandson
1876, June 16, r. & w., B.C.	Ruby King	Warlaby, 32792	do.
1877, Apr. 24, r. & w., C.C.	Ruby 5th	do.	do.

Ruby 4th, sold to Mr. J. C. Toppin, Musgrave Hall, Skelton, Penrith.

ROWLEY, John,
Went Bank House, Stubbs Walden, Pontefract.

CHARMER, red and little white, calved February 8, 1877. Bred by Mr. J.
Rowley; got by Pluto (35050), dam (Charity) by Waltron (30255), &c.
See Vol. xxv. p. 647.

1879, Sept. 30, r. & w., B.C.	Master Arthur	Lord Arthur, 40135	Mr. Rowley

DUCHESS OF YETHOLME 2ND, roan, calved January 11, 1873, Vols. xxii.,
xxiii., and xxiv. pp. 600, 681, 686. Bred by Sir W. C. Trevelyan, Bart.,
Wallington; got by Duke of Fussbox (28389), dam (Queen of Yetholme) by
Second Earl of Oxford (23843), &c.

1879, April 20, r. & w., B.C.	Royal Yetholme	D. of Oxford 27th, 33700	Mr. Rowley

LADY LAVENDER 20TH, roan, calved January 23, 1872. Bred by Mr. J.
Raickstraw, Nateby; got by Lord Lyons (26676), dam (Lavender 10th) by
Red Duke (20640), &c. See Vol. xx. p. 623.

1879, Mar. 25, white, C.C.	Lady Windsor	Windsor's Gift, 39331	Mr. Rowley

SELINA 3RD, white, calved January 4, 1875. Bred by Sir G. O. Wombwell,
Bart., Newburgh Park; got by Newburgh 5th (34904), dam (Selina) by
Colonist (28227), &c. See Vol. xxiii. p. 715.

1877, Nov. 25, roan, C.C.	Selina Winsome	Winsome Duke, 36019	Sir G.O. Wombwell
1879, July 7, roan, C.C.	Lady Selina	Waterloo Grand Dk., 39291	Mr. Rowley

RUSSELL, Dr. G. B.,
St. James's Place, Fermoy, Co. Cork.

ROAN LILAC, roan, calved January 3, 1863, Vols. xix., xx., and xxii. pp. 699, 728, 550. Bred by Mr. J. H. Jones, Mullinabro'; got by Master Harbinger (18352), dam (Red Lilac) by Claret (10057), &c.

Produce in		Names, &c.		By what Bull.	By whom bred.
1879, May 8,	red, B.C. roan, B.C.	Flamingo Roan Prince	}	Prince of Killerby, 37256	Dr. Russell

RUTHERFORD, G.,
Printonan, Coldstream, N.B.

CHERRY RIPE, red, calved April 1, 1877. Bred by Mr. G. Torrance, Sisterpath, the property of Mr. G. Rutherford; got by Royal Duke 2nd (35361), dam (Cherry) by Red Friar (24913), &c. See "Cherry Stone," Vol. xxiv. p. 43.

1879, Jan. 15, red,	B.C.	Major 2nd	Major, 43607	Mr. Torrance

Major 2nd, sold to Mr. G. Rutherford, Printonan, Coldstream.

YOUNG PRESTON, red and white, calved February 20, 1877. Bred by Mr. G. Torrance, Sisterpath, the property of Mr. G. Rutherford; got by Royal Duke 2nd (35361), dam (Rosebloom 7th) by Brigand (28080), g. d. (Rosebloom 3rd) by Silver Star (22885), &c. See "Rosebloom 8th," Vol. xxiv. p. 599.

1879, Jan. 9, roan,	C.C.	Young Preston 2nd	Major, 43607	Mr. Torrance

Young Preston 2nd, sold to Mr. G. Rutherford, Printonan, Coldstream.

SALT, Sir W. H., Bart.,
Maplewell, Loughborough.

CARELESS 6TH, red and white, calved April 20, 1871, Vols. xxiii. and xxiv. pp. 628, 634. Bred by Mr. A. B. Winnall, The Hawthorns; got by Cherry Duke (25750), dam (Careless 5th) by Wellington (21090), &c.

1879, Feb. 4, red,	C.C.	Duchess Craggs 3rd	5th D. of Glo'ster, 36494	Sir W. H. Salt

GRAND DUCHESS CAROLINA 3RD, red, calved July 20, 1876, Bred by Mr. W. Angerstein, Weeting Hall; got by Duke Lally (36457), dam (Grand Duchess Carolina 2nd) by Eighth Duke of Geneva (28390), &c. See "Duke Craggs," p. 70.

1879, May 16, red,	B.C.	Duke Craggs	5th D. of Glo'ster, 36494	Sir W. H. Salt

Duke Craggs, sold to Mr. R. Taylor, Sigglesthorne Manor, Hull.

LADY DARLINGTON, roan, calved April 14, 1871, Vols. xxiii. and xxiv. pp. 629, 634. Bred by Mr. A. B. Winnall, The Hawthorns; got by Cherry Duke (25750), dam (Darling) by Fourth Duke of Oxford (11387), &c.

1879, May 15, roan,	C.C.	Lady Darlington 4th	5th D. of Glo'ster, 36494	Sir W. H. Salt

LADY SURMISE, roan, calved June 26, 1876. Bred by the Hon. M. H. Cochrane, Hillhurst, Compton, Canada; got by Duke of Hillhurst 2nd (39748), dam (Surmise Duchess 5th) by Grand Duke of Geneva (28756), &c. See Vol. xxiv. p. 381.

1879, Mar. 27, red,	C.C.	Lady Surmise 2nd	5th D. of Glo'ster, 36494	Sir W. H. Salt

LADY WORCESTER 19th, red, calved May 6, 1876. Bred by the Earl of Dunmore, Dunmore ; got by Sixth Duke of Geneva (30959), dam (Lady Worcester 2nd) by Charleston (21400), &c. See "Duke Wild Eyes 2nd," p. 88.

Produce in	Names, &c.	By what Bull.	By whom bred.
1879, July 8, red,	B.C.|Duke Wild Eyes 2nd	5th D. of Glo'ster, 36494	Sir W. H. Salt

Duke Wild Eyes 2nd, sold to the Earl of Carysfort, Warmington, Oundle.

MAPLEWELL, red and white, calved October 22, 1874, Vol. xxv. p. 649. Bred by Sir W. H. Salt, Bart. ; got by Second Duke of Glo'ster (28392), dam (Kirklevington 18th) by Third Lord Oxford (22200), &c.

Produce in	Names, &c.	By what Bull.	By whom bred.
1879, July 17, red,	B.C.|Baron Maplewell 2nd	5th D. of Glo'ster, 36494	Sir W. H. Salt

Baron Maplewell 2nd, sold to Mr. R. Arnold, Shackerstone, Atherstone.

MAPLEWELL 2nd, red, calved April 18, 1876. Bred by Sir W. H. Salt, Bart. ; got by Fifth Lord Oxford (31738), dam (Kirklevington 18th) by Third Lord Oxford (22200), &c. See "Baron Maplewell," p. 16.

Produce in	Names, &c.	By what Bull.	By whom bred.
1879, May 18, red,	B.C.|Baron Maplewell	5th D. of Glo'ster, 36494	Sir W. H. Salt

Baron Maplewell, sold to Mr. J. Barrs, Nailstone, Hinckley.

MARCHIONESS OF BARRINGTON, roan, calved November 11, 1875, Vol. xxv. p. 649. Bred by the Hon. M. H. Cochrane, Hillhurst, Compton, Canada ; got by Grand Duke 22nd (34062), dam (Grand Duchess of Barringtonia) by Eighteenth Duke of Oxford (25995), &c.

Produce in	Names, &c.	By what Bull.	By whom bred.
1879, Nov. 3, red,	B.C.|Duke of Barrington	5th D. of Glo'ster, 36494	Sir W. H. Salt

Duke of Barrington, sold to Mr. G. Ward, Nuneaton Fields, Nuneaton.

PRINCESS 7th, roan, calved July 10, 1874, Vol. xxiv. p. 634. Bred by the Earl of Bective, Underley Hall ; got by Second Duke of Tregunter (26022), dam (Princess 4th) by Baron Oxford 3rd (25579), &c.

Produce in	Names, &c.	By what Bull.	By whom bred.
1879, May 9, red,	B.C.|Pr. of Maplewell 2nd	5th D. of Glo'ster, 36494	Sir W. H. Salt

Prince of Maplewell 2nd, sold to Mrs. Barrs, Odstone Hall, Atherstone.

RED ROSE OF BRAEMAR, red and white, calved July 3, 1871, Vols. xxi., xxiii., and xxv. pp. 686, 429, 650. Bred by the Earl of Dunmore, Dunmore ; got by Eleventh Duke of Thorndale (31024), dam (Red Rose of Breadalbane) by Duke Frederick (30910), &c.

Produce in	Names, &c.	By what Bull.	By whom bred.
1879, April 14, red,	C.C.|R. R. of Maplewell 2d	5th D. of Glo'ster, 36494	Sir W. H. Salt

WILD DUCHESS OF GLO'STER, roan, calved April 20, 1877. Bred by Mr. E. H. Cheney, Gaddesby Hall ; got by Third Duke of Glo'ster (33653), dam (Wild Oxford) by Lord Oxford 2nd (20215), &c. See "Duke Wild Eyes," p. 87.

Produce in	Names, &c.	By what Bull.	By whom bred.
1870, May 0, roan,	D.C.|Duke Wild Eyes	5th D. of Glo'ster, 36494	Sir W. H. Salt

Duke Wild Eyes, sold to Mr. W. Murray, Chesterfield, Ontario, Canada.

SANDERSON, H. M.,
Moulton Hall, Richmond, Yorkshire.

ESSIE, red and white, calved January 21, 1877. Bred by Mr. D. Sanderson, Moulton Hall ; got by Star Regent (35679), dam (Ella) by Booth Satellite (30564), &c. See Vol. xxiv. p. 636.

Produce in	Names, &c.	By what Bull.	By whom bred.
1879, Nov. 6, r.&w.,	B.C.|Brennus	British Knight, 33220	Mr. H. M. Sanderson

RED ROSEBUD, red and white, calved October 13, 1873, Vols. xxiii. and xxiv.
pp. 629, 636. Bred by Mr. R. Burdon, Castle Eden, the property of Mr. H.
M. Sanderson; got by Emperor Maximillian (26100), dam (Moss Rosebud)
by Premier (27084), &c.

Produce in		Names, &c.	By what Bull.	By whom bred.
1879, Jan. 12, r. & w.,	C.C.	Ribston Pippin	Reserve Force, 40588	Mr. D. Sanderson

VEXATION, red and white, calved May 21, 1871, Vol. xxiv. p. 636. Bred by
Mr. T. H. Hutchinson, Manor House, the property of Mr. H. M. Sanderson;
got by Booth's Royal Signet (28061), dam (Veronica) by Vain Hope (23102),
&c.

| 1878, May 8, red, | C.C. | Vera | Reserve Force, 40588 | Mr. D. Sanderson |
| 1879, Sept. 18, roan, | C.C. | Villette | British Knight, 33220 | do. |

SAVAGE, S. P.,
Lays Farm, Wotton-under-Edge.

AMELIA 2ND, red and white, calved April 26, 1875. Bred by Mr. S. P.
Savage; got by Fulgens (33980), dam (Amelia) by Surprise (20922), g. d.
(Acacia) by Count de Gourcy (17632), &c. See Vol. xvii. p. 357.

| 1878, Nov. 7, r. & w., | C.C. | (dead) | Oxford's Prince, 34998 | Mr. Savage |
| 1879, Nov. 20, r. & w., | C.C. | Amelia 5th | Lord Winsome, 40266 | do. |

CAROLINA 7TH, roan, calved September 14, 1876. Bred by Mr. S. P. Savage;
got by Lord Red Eyes 4th (26730), dam (Carolina 6th) by Lord Lally (22161),
&c. See "Cariophylus," Vol. xxii. p. 42.

| 1879, June 27, r. & w., | B.C. | (dead) | Lord Winsome, 40266 | Mr. Savage |

SAVIDGE, M.,
Sarsden Lodge Farm, Chipping Norton, Oxon.

FRESH START 3RD, red, calved in June 1869, Vols. xx., xxi., xxiii., and xxv.
pp. 531, 917, 630, 651. Bred by Mr. M. Savidge; got by Royal Butterfly
20th (25007), dam (Fresh Start) by Duke of Towneley (21615), &c.

| 1879, Feb. 20, roan, | C.C. | Miss Clarence | L. Fitzclarence 19th, 38603 | Mr. Savidge |

GOOD LUCK, red, calved June 22, 1872, Vols. xxiii. and xxiv. pp. 630, 637.
Bred by Mr. M. Savidge; got by Earl of Warwickshire 3rd (28524), dam
(Fresh Start 3rd) by Royal Butterfly 20th (25007), &c.

| 1878, Mar. 18, roan, | B.C. | Grand Eclipse | L. Fitzclarence 19th, 38603 | Mr. Savidge |
| 1879, April 12, roan, | C.C. | Lady Silvina | do. | do. |

LADY ROSY, red, calved November 12, 1874. Bred by Mr. M. Savidge; got
by Lord Rosy (34661), dam (Fresh Start) by Duke of Towneley (21615), &c.
See "Master Rosy," p. 168.

| 1878, Feb. 12, roan, | B.C. | Grand Umpire | L. Fitzclarence 19th, 38603 | Mr. Savidge |
| 1879, Apr. 4, roan, | C.C. | Lady Rosa | do. | do. |

LADY ROSY 2ND, red, calved November 8, 1875. Bred by Mr. M. Savidge;
got by Lord Rosy (34661), dam (Fresh Start) by Duke of Towneley (21615),
&c. See "Master Rosy," p. 168.

| 1879, May 15, roan, | B.C. | Master Rosy | Rag Merchant, 43865 | Mr. Savidge |

Master Rosy, sold to Mr. J. B. White, Street End House, Canterbury.

ROSA, red and white, calved December 26, 1875. Bred by Mr. M. Savidge ;
got by Lord Rosy (34661), dam (Fresh Start 3rd) by Royal Butterfly 20th
(25007), g. d. (Fresh Start) by Duke of Towneley (21615), &c. See " Master
Rosy," p. 168.

Produce in	Names, &c.	By what Bull.	By whom bred.
1879, May 25, roan,	C.C. Rosette	Rag Merchant, 43865	Mr. Savidge

ROSEBUD, red and white, calved September 18, 1874. Bred by Mr. M.
Savidge ; got by Lord Rosy (34661), dam (Fresh Start 3rd) by Royal Butterfly 20th (25007), &c. See Vol. xxi. p. 917.

1878, Jan. 2, roan,	B.C. (dead)	L. Fitzclarence19th, 38603	Mr. Savidge
1879, Jan. 24, roan,	B.C. Sarsden Clipper	do.	do.
1879, Dec. 29, r. & w.,	B.C. Sarsden Hero (Steer)	Grand Junction, 43327	do.

SAYER, Henry,
Close House, Kirkbythore, Penrith.

CHRISTMAS ROSE, roan, calved January 1, 1872. Bred by Mr. J. Irvine,
Long Marton ; got by Grand Duke of Westmoreland (28767), dam (Autumn
Rose) by Earl of Waterloo (23849), &c. See " Crossfell 34th," Vol. xxv. p. 65.

1877, July 20, white,	C.C. Cherry Blossom	Cherry King, 36352	Mr. Sayer
1878, Aug. 18, roan,	C.C. Cherry Princess	do.	do.
1879, Aug. 12, roan,	C.C. Cherry Duchess	do.	do.

GOLDEN DAISY, white, calved April 7, 1874. Bred by Mr. H. Sayer ; got
by Pure Gold (32226), dam (Queen of the Daisies) by Panton Squire (29520),
g. d. (Daisy) by British Prince (19354), &c. See " British Plate," Vol. xx. p. 41.

1877, May 20, roan,	B.C. (dead)	Kirk'thore W. E. 2d, 34382	Mr. Sayer
1878, June 10, roan,	C.C. Golden Daisy 2nd	Cherry King, 36352	do.
1879, July 3, white,	C.C. White Daisy	do.	do.

OXFORD ROSE, red, calved January 28, 1877. Bred by Mr. H. Sayer ; got
by Oxford (34961), dam (Autumn Rose) by Earl of Waterloo (23849), &c.
See " Crossfell 34th," Vol. xxv. p. 65.

1879, May 26, red,	C.C. Moss Rose	Cherry King, 36352	Mr. Sayer

QUEEN OF THE DAISIES 2ND, red and white, calved May 10, 1876. Bred
by Mr. H. Sayer ; got by Sir John (32507), dam (Queen of the Daisies) by
Panton Squire (29520), g. d. (Daisy) by British Prince (19354), &c. See
" British Plate," Vol. xx. p. 41.

1879, April 11, roan,	B.C. King of the Daisies	Cherry King, 36352	Mr. Sayer

SCHRÖDER, Baron W. H. von,
The Rookery, Nantwich, Cheshire,

FLORA 13TH, red, calved December 20, 1875. Bred by Mr. J. Gordon, Cluny
Castle ; got by Friar of Knowlmere (33975), dam (Flora 8th) by Windsor
Booth (32875), &c. See Vol. xxii. p. 432.

1879, April 21, red,	C.C. D'ss of Nantwich 2d	S. W'dsor Broughton, 27507	Baron Schröder

MAID OF OXFORD, roan, calved February 27, 1876. Bred by Mr. E. Musgrove, Aughton Old Hall ; got by Baron Oxford 4th (25580), dam (Grateful)
by Third Duke of Wharfdale (21619), &c. See Vol. xxiii. p. 578.

1879, April 16, r. & w.,	C.C. Lady Cross	Sultan's Fame, 35899	Baron Schröder

MISS GOLDIE 2ND, red and white, calved February 14, 1876. Bred by Mr. J. Gordon, Cluny Castle; got by Baron of Knowlmere (30474), dam (Miss Goldie) by Golden Era (31266), &c. See "Earl of Nantwich," p. 91.

Produce in	Names, &c.	By what Bull.	By whom bred.
1879, Dec. 31, r. & w., B.C.	Earl of Nantwich	Sultan's Fame, 35699	Baron Schröder

VAIL 14TH, white, calved December 22, 1875. Bred by Mr. J. Gordon, Cluny Castle; got by Friar of Knowlmere (33975), dam (Vail 9th) by Baron of Knowlmere (30474), &c. See Vol. xxii. p. 433.

1879, June 12, white, C.C.	C'tess of Nantwich	Lollius Booth, 34470	Baron Schröder

SCHROETER, C. W.,
Tedfold, Billingshurst, Sussex.

CYGNET, red, calved November 15, 1871. Bred by the Earl of Dunmore, Dunmore; got by Second Duke of Collingham (23730), dam (Baroness Oxford) by Baron Oxford 2nd (23376), &c. See Vol. xx. p. 401.

1878, Aug. 6, roan, B.C.	Cyprus	Prince Alfred, 37215	Mr. Schroeter
1879, Aug. 29, roan, C.C.	Cynthia	Knight of Geneva 3rd, 38515	do.

Cyprus, sold to Mr. King, Friam, Storrington, Sussex.

SCOBY, George,
Beadlam Grange, Nawton, York.

DUCHESS OF YETHOLME, red and white, calved January 13, 1871, Vols. xxiii., xxiv., and xxv. pp. 681, 686, 697. Bred by Sir W. C. Trevelyan, Bart., Wallington; got by Third Duke of Wharfdale (21619), dam (Queen of Yetholme) by Second Earl of Oxford (23843), &c.

1879, April 8, roan, C.C.	Empress of Yetholme	Duke of Oxford 27th, 33709	Mr. Scoby

ELLEN, roan, calved June 7, 1872, Vol. xxii. p. 628. Bred by the Earl of Zetland, Upleatham; got by Grand Monarch (28774), dam (Eveline) by King of the Roses (22043), &c.

1878, Jan. 16, r. & w., B.C.	Cyprus	20th Duke of Oxford, 28432	Mr. Scoby
1879, Jan. 16, white, B.C.	Ellen's Duke	D. of Tregunter 5th, 33743	do.

Cyprus, sold to Mr. T. Parkes, Swansey House, Pickering.

SCOBY, William,
Hob Ground House, Kirby Moorside, Yorkshire.

LADY CLEVELAND, red, calved May 22, 1874, Vol. xxiv. p. 638. Bred by Mr. T. Feetenby, Stonegrave; got by Duke of Cleveland (33599), dam (Red Rose) by General Scarlet (31235), &c.

1878, Aug. 12, roan, B.C.	(Steer)	Cleveland 3rd, 39600	Mr. Scoby
1879, Aug. 15, red, C.C.	Lady Cleveland 3rd	Victor Manfred, 40866	do.

LADY DANBY, red, calved March 13, 1873, Vol. xxiii. p. 632. Bred by Mr. W. White, Burrill; got by Lord Danby (31644), dam (Alice's Farewell) by Manfred (26801), &c.

1878, Aug. 31, red, C.C.	Lady Dartmouth	Victor 2nd, 37625	Mr. Scoby
1879, Sept. 22, roan, B.C.	Dannewerke	Victor Manfred, 40866	do.

RED ROSE, red, calved September 29, 1874, Vol. xxiv. p. 638. Bred by Mr. T. Feetenby, Stonegrave; got by Duke of Cleveland (33599), dam (Citron) by Cistercian (28202), &c.

Produce in		Names, &c.	By what Bull.	By whom bred.
1878, Feb. 18, roan,	B.C.	(Steer)	Cleveland 3rd, 39600	Mr. Scoby
1879, Jan. 27, roan,	C.C.	Rosamond	Victor Manfred, 40866	do.

VICTORINE 4TH, roan, calved August 18, 1870, Vol. xxi. p. 919. Bred by Messrs. Tindall, Knapton Hall; got by Cecil (25725), dam (Integrity) by Fawsley (14540), &c.

1876, Jan. 31, r.&w.,	C.C.	Victorine 6th	Sir Wiley, 35598	Mr. Scoby
1877, Mar. 13, roan,	B.C.	Victor Manfred	Manfred, 26801	do.
1878, Mar. 22, r.&w.,	C.C.	Violet	Gaberlunzie, 38323	do.
1879, Feb. 26, r.&w.,	C.C.	Veronica	do.	do.

VICTORINE 5TH, roan, calved December 19, 1872, Vol. xxii. p. 553. Bred by Messrs. Tindall, Knapton Hall; got by Pantaloon (29518), dam (Victorine 4th) by Cecil (25725), &c.

1876, Mar. 10, roan,	B.C.	(dead)	Sir Wiley, 35598	Mr. Scoby
1877, Sept. 21, roan,	B.C.	(Steer)	Victor Gwynne, 35882	do.
1879, Nov. 5, roan,	B.C.	Viking	Victor Manfred, 40866	do.

VICTORINE 6TH, red and white, calved January 31, 1876. Bred by Mr. W. Scoby; got by Sir Wiley (35598), dam (Victorine 4th) by Cecil (25725), &c. See Vol. xxi. p. 919.

1878, Oct. 17, r.&w.,	C.C.	Venus	Gaberlunzie, 38323	Mr. Scoby
1879, Sept. 5, r.&w.,	C.C.	Vesta	do.	do.

SCOTT, Alexander,
Towie Barclay, Turriff, N.B.

AVERNE, red, calved April 9, 1875. Bred by Mr. R. Bruce, Newton of Struthers, the property of Mr. A. Scott; got by Red Prince (32267), dam (Countess) by Hydra (19995), &c. See Vol. xxii. p. 343.

1879, April 11, roan,	C.C.	Amanda	Bromley, 36289	Mr. G. Marr

BELLA 2ND, roan, calved April 21, 1875, Vol. xxiv. p. 638. Bred by Mr. A. Scott; got by Lord Warden (31766), dam (Bella) by Lord Forth (26649), &c.

1878, April 9, roan,	C.C.	Bella 5th	Viscount, 37632	Mr. Scott
1879, Oct. 28, r.&w.,	C.C.	Bella 6th	General Gourko, 39922	do.

CAMILLA 6TH, red, calved April 30, 1870. Bred by Mr. J. Wood, Midtown; got by Vanguard (30204), dam (Camilla 4th) by Prince Louis (20560), &c. See Vol. xix. p. 427.

1878, June 21, roan,	B.C.	Hymen	The Admiral, 37566	Mr. Scott
1879, Nov. 29, r.&w.,	B.C.	Jacobi	Gallantry, 39908	do.

CECILIA 3RD, roan, calved June 23, 1874, Vol. xxiii. p. 633. Bred by Mr. A. Scott; got by Oxford (34960), dam (Cecilia) by Golden Eagle (26267), &c.

1878, Mar. 27, white,	B.C.	Hotspur	Fawsley, 33899	Mr. Scott
1879, Mar. 15, red,	C.C.	Cecilia 7th	Roseberry, 39016	do.

Hotspur, sold to Mr. Argo, Gateside, Belhelvie.

CECILIA 5TH, red, calved January 4, 1876. Bred by Mr. A. Scott; got by Fawsley (33899), dam (Cecilia 2nd) by Jupiter (31456), &c. See Vol. xxiii. p. 633.

1878, Mar. 23, red,	C.C.	Cecilia 6th	Roseberry, 39016	Mr. Scott
1879, April 18, red,	C.C.	Cecilia 8th	Gallantry, 39908	do.

CINDERELLA, roan, calved January 13, 1875. Bred by Mr. A. Cruickshank, Sittyton ; got by Royal Duke of Glo'ster (29864), dam (Charmer) by Champion of England (17526), &c. See Vol. xix. p. 436.

Produce in		Names, &c.	By what Bull.	By whom bred.
1878, Mar. 21, white,	B.C.	Homer	Viscount, 37632	Mr. Scott
1879, Dec. 14, roan,	C.C.	Cinderella 2nd	Roseberry, 39016	do.

Homer, sold to Mr. Robertson, Ellon.

CLARA 25TH, red, calved April 17, 1874, Vol. xxiii. p. 633. Bred by Mr. A. Scott ; got by Oxford (34960), dam (Clara 4th) by Prince Hopewell (29631), &c.

Produce in		Names, &c.	By what Bull.	By whom bred.
1878, Mar. 6, roan,	C.C.	Clara 28th	Viscount, 37632	Mr. Scott
1879, Mar. 14, roan,	B.C.	Invincible	General Gourko, 39922	do.

CLARA 26TH, roan, calved March 28, 1876. Bred by Mr. A. Scott ; got by Albert (32926), dam (Clara 23rd) by Brigand (21314), &c. See " Victor," Vol. xix. p. 364.

Produce in		Names, &c.	By what Bull.	By whom bred.
1878, Mar. 19, red,	C.C.	Clara 29th	Roseberry, 39016	Mr. Scott
1879, May 14, white,	C.C.	Clara 30th	Gallantry, 39908	do.

EDITH R., roan, calved April 11, 1876. Bred by Mr. A. Scott ; got by Commodore (33420), dam (Fair Maid of Perth) by Duke (30902), &c. See Vol. xxiii. p. 633.

Produce in		Names, &c.	By what Bull.	By whom bred.
1878, April 8, red,	C.C.	Edith R. 2nd	Roseberry, 39016	Mr. Scott
1879, Mar. 10, r. & w.,	B.C.	Ingersoll	Gallantry, 39908	do.

ETHEL, white, calved April 10, 1875. Bred by Mr. A. Scott ; got by Oxford (34960), dam (Fair Maid of Perth) by Duke (30902), &c. See Vol. xxiii. p. 633.

Produce in		Names, &c.	By what Bull.	By whom bred.
1878, Mar. 8, white,	C.C.	Ethel 2nd	Caractacus, 37945	Mr. Scott
1879, Mar. 11, roan,	B.C.	Innocent	Roseberry, 39016	do.

Innocent, sold to Sir R. Abercromby, Bart., Birkenbog and Forglen.

HANNAH, red, calved March 27, 1875. Bred by Mr. J. Durno, Sunnyside ; got by Cherry Prince (33355), dam (Mignonette) by Lothbury (24483), g. d. (Verbena) by Viceroy (19054), gr. g. d. (Lass O'Doune) by Inheritor (13065), &c. See Vol. xvii. p. 585.

Produce in		Names, &c.	By what Bull.	By whom bred.
1878, Mar. 29, roan,	B.C.	Horatio	Pr.Fred.of Ca'bridge,29621	Mr. Scott
1879, Mar. 23, red,	B.C.	Iscanus	General Gourko, 39922	do.

Horatio, sold to Mr. J. King, Burnside, Maryculter, Aberdeenshire.

JEALOUSY 14TH, roan, calved March 18, 1875. Bred by Mr. J. Cochrane, Little Haddo ; got by Prince Frederick of Cambridge (29621), dam (Jealousy 3rd) by Lord Buckingham (20151), &c. See " Inverthernie," p. 124.

Produce in		Names, &c.	By what Bull.	By whom bred.
1877, Feb. 22, roan,	B.C.	Gonzalo	Royal Heir, 35378	Mr. Scott
1879, Feb. 11, red,	B.C.	Inverthernie	Roseberry, 39016	do.

Gonzalo, sold to Mr. Thom, Quithelhead, Kincardineshire.

KATIE, roan, calved March 10, 1871, Vols. xxiii. and xxv. pp. 634, 651. Bred by Mr. A. Scott ; got by Merlin (31904), dam (Fair Maid of Perth) by Duke (30902), &c.

Produce in		Names, &c.	By what Bull.	By whom bred.
1879, Mar. 28, roan,	C.C.	Katie 7th	Roseberry, 39016	Mr. Scott

LEONORA, roan, calved May 8, 1874, Vol. xxiv. p. 639. Bred by Mr. A. Scott ; got by Lord Warden (31766), dam (Lady Kintore) by Prince Louis (20560), &c.

Produce in		Names, &c.	By what Bull.	By whom bred.
1878, Mar. 4, red,	B.C.	Hermogenes	Roseberry, 39016	Mr. Scott
1879, Mar. 25, roan,	B.C.	Iroquois	do.	do.

Hermogenes, sold to Mr. Low, Easter. Kinmundy, Skene.

LOUISA, red and white, calved March 21, 1873, Vols. xxiii. and xxiv. pp. 634, 639. Bred by Mr. A. Scott ; got by Star of Peace (30063), dam (Lady Ann Hope) by Lord Cobham (22141), &c.

Produce in		Names, &c.	By what Bull.	By whom bred.
1878, Mar. 29, red,	B.C.	Howard	Roseberry, 39016	Mr. Scott
1879, Feb. 27, roan,	B.C.	Ingot	Marseillaise, 41989	do.

Howard, sold to Mr. J. Fetch, Fen, Slaines.

LOVELACE, roan, calved March 15, 1871, Vol. xxiii. p. 635. Bred by Mr. A. Longmore, Rettie ; got by Lord Forth (26649), dam (Early Morn) by Earl of Aberdeen (12800), &c.

Produce in		Names, &c.	By what Bull.	By whom bred.
1877, Feb. 27, roan,	B.C.	Gladstone	Chancellor, 36335	Mr. Scott
1879, Nov. 5, red,	C.C.	Lady Lucy	Roseberry, 39016	do.

Gladstone, sold to Mr. John Milne, Mains of Laithers, Turriff.

LOVELY 17TH, red, calved December 21, 1873. Bred by Mr. A. Cruickshank, Sittyton ; got by Viceroy (32764), dam (Lovely 15th) by Scotland's Pride (25100), &c. See Vol. xxi. p. 657.

Produce in		Names, &c.	By what Bull.	By whom bred.
1878, Mar. 21, r.&w.,	C.C.	Lovely 18th	Viscount, 37632	Mr. Scott

LOVELY LASS, roan, calved April 1, 1873, Vol. xxiii. p. 635. Bred by Mr. A. Scott ; got by Victor Royal (35886), dam (Lovelace) by Lord Forth (26649), &c.

Produce in		Names, &c.	By what Bull.	By whom bred.
1878, Feb. 7, roan,	C.C.	Lady Caroline	Viscount, 37632	Mr. Scott
1879, Mar. 5, { roan, B.C. / r.&w.,B.C.		Inverurie / Inverary	} Roseberry, 39016	do. .

MARION 2ND, roan, calved March 15, 1873, Vol. xxiii. p. 635. Bred by Mr. W. Mackie, Petty Fyvie ; got by Lord Charles (31634), dam (Marion) by Blair Athol (25638), &c.

Produce in		Names, &c.	By what Bull.	By whom bred.
1878, Jan. 10, roan,	C.C.	Marion 4th	Fawsley, 33899	Mr. Scott
1879, Feb. 28, roan,	B.C.	Inheritor	Roseberry, 39016	do.

MOLLY 2ND, white, calved December 19, 1874. Bred by Mr. A. Scott ; got by Oxford (34960), dam (Molly) by Imperator (28888), &c. See Vol. xxiii. p. 635.

Produce in		Names, &c.	By what Bull.	By whom bred.
1878, Mar. 16, roan,	C.C.	Molly 3rd	Viscount, 37632	Mr. Scott

MONOGRAM 5TH, roan, calved February 20, 1874. Bred by Mr. A. Scott ; got by Lord Warden (31766), dam (Monogram 2nd) by General Grant (36683), &c. See Vol. xxiii. p. 635.

Produce in		Names, &c.	By what Bull.	By whom bred.
1878, Jan. 15, white,	B.C.	Handy Andy	Fawsley, 33899	Mr. Scott
1878, Dec. 31, roan,	B.C.	Ibrox	Roseberry, 39016	do.

Handy Andy, sold to Mr. J. Craighead, Ballenscoth, Aberdeenshire.

MYSIE 27TH, roan, calved January 20, 1871. Bred by Mr. J. Wood, Midtown ; got by Imperator (28888), dam (Mysie 24th) by Vanguard (30204), g. d. (Mysie 23rd) by Matchless (29345), &c. See Vol. xix. p. 645.

Produce in		Names, &c.	By what Bull.	By whom bred.
1878, May 4, roan,	B.C.	Humbert	The Admiral, 37568	Mr. Wood
1879, April 4, roan,	C.C.	Mysie 30th	Viscount, 37632	Mr. Scott

Humbert, sold to Mr. Ross, Kinnabud, Ross-shire.

MYSIE 28TH, roan, calved March 3, 1873. Bred by Mr. J. Wood, Midtown ; got by Royal Heir (35376), dam (Mysie 24th) by Vanguard (30204), g. d. (Mysie 23rd) by Matchless (29345), &c. See Vol. xix. p. 645.

Produce in		Names, &c.	By what Bull.	By whom bred.
1878, Mar. 1, roan,	B.C.	Harry Lorrequer	The Admiral, 37568	Mr. Wood
1879, Mar. 30, red,	C.C.	Mysie 29th	Roseberry, 39016	Mr. Scott

PRINCESS ROYAL 20TH, roan, calved December 21, 1875. Bred by Mr. W. S. Marr, Uppermill; got by The Baron (35738), dam (Princess Royal 10th) by Humbolt (31398), &c. See "Princess Royal 19th," Vol. xxv. p. 573.

Produce in		Names, &c.	By what Bull.	By whom bred.
1879, Jan. 27, roan,	C.C.	Princess Alice	Viscount, 37632	Mr. Scott

VELVET ROSE, red, calved May 11, 1874, Vol. xxiii. p. 636. Bred by Mr. A. Cruickshank, Sittyton; got by Millionaire (31917), dam (Village Flower) by Scotland's Pride (25100), &c.

1878, May 1, red,	B.C.	Huntly	Roseberry, 39016	Mr. Scott
1879, April 9, red,	B.C.	Ivo	Viscount, 37632	do.

Huntly, sold to Mr. Morrison, Petterden, Cornhill-up-Park.

VENUS 15TH, roan, calved April 20, 1869. Bred by Mr. J. Wood, Midtown; got by Vanguard (30204), dam (Venus 12th) by Speculator (13775), &c. See Vol. xix. p. 762.

1878, Mar. 20, r. & w.,	B.C.	Herbert	The Admiral, 37566	Mr. Wood
1879, Mar. 23, roan,	B.C.	Ingo	Roseberry, 39016	Mr. Scott

Herbert, sold to Mr. J. Johnstone, Darley, Auchterless.

VESTA 2ND, roan, calved June 30, 1870. Bred by Mr. J. Wood, Midtown; got by Lord Forth (26649), dam (Vesta) by Sir Charles 2nd (20812), &c. See Vol. xix. p. 764.

1878, Jan. 10, white,	B.C.	Hannibal	The Admiral, 37566	Mr. Wood
1879, Jan. 3, roan,	B.C.	Igor	do.	Mr. Scott

Hannibal, sold to Mr. Dingwall, Aboyne.

VINILLA, roan, calved October 11, 1874, Vol. xxiv. p. 640. Bred by Mr. A. Cruickshank, Sittyton; got by Lord Lancaster (26666), dam (Vivacity) by Cæsar Augustus (25704), &c.

1878, Dec. 4, red,	C.C.	Vinilla 3rd	Roseberry, 39016	Mr. Scott

SCOTT, Walter,
Glendronach, Huntly, N.B.

HELENA, red and white, calved April 12, 1875. Bred by Mr. W. Scott; got by Jeweller (34254), dam (Lady Jane) by Valiant 2nd (25353), &c. See "General Roberts," p. 107.

1878, Mar. 13, { roan,	C.C.	Milkmaid	} Norseman, 34924	Mr. Scott
{ roan,	C.C.	(dead)		
1879, April 7, red,	B.C.	General Roberts	do.	do.

JEMIMA, roan, calved February 18, 1875. Bred by Mr. W. Scott; got by Waverley (35955), dam (Annie) by Valiant 2nd (25353), &c. See Vol. xxii. p. 553.

1878, Feb. 18, red,	C.C.	The Peeress	Norseman, 34924	Mr. Scott
1879, Mar. 6, red,	B.C.	Aberdeen	do.	do.

SCRATTON, D. R.,
Ogwell, Newton Abbot, Devonshire.

ALICE GWYNNE, red, calved April 29, 1873, Vol. xxiii. p. 636. Bred by Mr. G. Paine, Great Braxted Park; got by Second Duke of Wellington (28465), dam (Flora Gwynne) by Oxford Gwynne (24711), &c.

1879, Feb. 5, red,	C.C.	Bar'gtonGwynne 4th	Lally's Hillhurst D., 38538	Mr. Scratton

BARRINGTON GWYNNE, red and white, calved August 26, 1876. Bred by
Mr. D. R. Scratton; got by Grand Duke of Barringtonia (34069), dam
(Double Gwynne) by Rufus (27397), &c. See " Cherry Gwynne 3rd," p. 48.

Produce in	Names, &c.	By what Bull.	By whom bred.
1879, Jan. 18, roan,	C.C. Barr'gtonGwynne 3d	Lally'sH'hurstD.2d,38539	Mr. Scratton

CHERRY KIRKLEVINGTON, red and little white, calved April 15, 1877.
Bred by Mr. D. R. Scratton; got by Sixth Cherry Duke (30705), dam (Kirklevington 20th) by Fifth Lord Wild Eyes (26762), &c. See Vol. xxiv. p. 641.

1879, Aug. 27, r. & w., C.C.	B'nessKirklevington	Baron Oxford 2nd, 23376	Mr. Scratton

CHERRY NANCY, roan, calved February 18, 1877. Bred by Mr. D. R.
Scratton; got by Sixth Cherry Duke (30705), dam (Duchess of Clarénce 3rd)
by Grand Duke 6th (19876), &c. See Vol. xxiv. p. 641.

1879, July 8, roan,	C.C. Barrington Nancy	Lally'sH'hurstD.2d,38539	Mr. Scratton

DUCHESS OF CLARENCE 3rd, roan, calved April 5, 1869, Vols. xx., xxi.,
xxii., xxiv., and xxv. pp. 488, 559, 311, 641, 653. Bred by Mr. T. Barber,
Sproatley Rise; got by Grand Duke 6th (19876), dam (Duchess of Clarence)
by Duke of Clarence (19611), &c.

1879, June 10, roan,	B.C. Clarence 2nd	Lally'sH'hurstD.2d,38539	Mr. Scratton

KIRKLEVINGTON 20th, roan, calved April 26, 1868, Vols. xxiv. and xxv.
pp. 641, 653. Bred by Mr. C. W. Harvey, Walton-on-the-Hill; got by Fifth
Lord Wild Eyes (26762), dam (Kirklevington 17th) by Lord Lally (22161), &c.

1879, June 6, r. & w.,	C.C. Lally Kirklevington	Lally'sH'hurstD.2d,38539	Mr. Scratton

LADY WILD EYES 9th, red, calved January 1, 1876, Vol. xxv. p. 447. Bred
by Lord Fitzhardinge, Berkeley Castle; got by Grand Duke of Waterloo
(28766), dam (Lady Wild Eyes 4th) by Cherry Grand Duke 2nd (25758), &c.

1879, Nov. 17, r. & w., C.C.	Lady Wild Eyes A.	Duke of Connaught, 33604	Mr. Scratton

LADY WILD EYES 10th, roan, calved February 20, 1876. Bred by
Lord Fitzhardinge, Berkeley Castle; got by Duke of Siddington 2nd (33732),
dam (Lady Wild Eyes 8th) by Grand Duke of Waterloo (28766), &c. See
" Berkeley," p. 27.

1879, May 10, r. & w., B.C.	Berkeley	Ld.T'croftOxford 2d,38668	Mr. Scratton

LALLY 7th, red and white, calved September 30, 1866, Vols. xix. and xx.
pp. 596, 618. Bred by Mr. C. W. Harvey, Walton-on-the-Hill; got by Third
Lord Oxford (22200), dam (Lally 2nd) by Malachite (18313), &c.

1879, Nov. 30, red,	C.C. Oxford Lally 2nd	Baron Oxford 2nd, 23376	Mr. Scratton

LALLY GENEVA, roan, calved July 29, 1875, Vol. xxiv. p. 641. Bred by Mr.
J. W. Larking, Ashdown House; got by Grand Duke of Geneva (28756), dam
(Lally 7th) by Third Lord Oxford (22200), &c.

1879, Jan. 20, roan,	B.C. Hayle	Duke of Oxford 33rd.36528	Mr. Scratton

SERGISON, Captain W.,
Cuckfield Park, Sussex.

PICOTEE, red and white, calved January 4, 1869, Vols. xxii. and xxv. pp. 306,
654. Bred by Mr. L. B. Bagshaw, Newton; got by Prince Royal (24859),
dam (Gaudy) by Hero of Kars (19956), &c.

1879, May 5, r. & w., C.C.	Pink	Iron Duke, 40021	Captain Sergison

RUBY, roan, calved March 26, 1871, Vols. xxii. and xxv. pp. 306, 654. Bred
by Mr. L. B. Bagshaw, Newton; got by Satan (27430), dam (Negus) by
Merlin (24581), &c.

Produce in	Names, &c.	By what Bull.	By whom bred.
1879, Jan. 11, r. & w., C.C.	Emerald	Iron Duke, 40021	Captain Sergison

TOPAZ, roan, calved February 28, 1873, Vols. xxiii. and xxv. pp. 322, 654.
Bred by Mr. L. B. Bagshaw, Newton; got by Satan (27430), dam (Locket)
by Ernest Gray (21699), &c.

1879, Jan. 13, r. & w., C.C.	Coral	Iron Duke, 40021	Captain Sergison
1879, Nov. 27, roan, C.C.	Jasper	do.	do.

SERJEANTSON, G. J.,
Camp Hill, Bedale, Yorkshire.

ECSTACY, red and white, calved December 19, 1876. Bred by Mr. G. J.
Serjeantson; got by Masterman (31873), dam (Example) by Masterman
(31873), &c. See Vol. xxiii. p. 639.

1879, May 5, r. & w., C.C.	Echo	Briony Berry, 36271	Mr. Serjeantson

EXAMPLE, roan, calved in March, 1873, Vols. xxii., xxiii., and xxv. pp. 555,
639, 654. Bred by Mr. G. J. Serjeantson; got by Masterman (31873), dam
(Friendship) by Whitelock (32856), &c.

1879, Aug. 16, roan, C.C.	Express	Briony Berry, 36271	Mr. Serjeantson

KISS, roan, calved in December, 1872, Vol. xxii. p. 556. Bred by Mr. G. J.
Serjeantson; got by Masterman (31873), dam (Kindness) by Gladstone
(31252), &c.

1879, Feb. 18, r. & w., C.C.	Keepsake	Briony Berry, 36271	Mr. Serjeantson
1879, Dec. 29, r. & w., C.C.	Kinsfolk	do.	do.

PRAISE, roan, calved in September, 1874, Vol. xxv. p. 655. Bred by Mr. G.
J. Serjeantson; got by Little Danby (34461), dam (Precious) by Whitelock
(32856), &c.

1879, May 18, r. & w., C.C.	Proxy	Briony Berry, 36271	Mr. Serjeantson

PRICE, red, calved in August, 1875, Vol. xxv. p. 655. Bred by Mr. G. J.
Serjeantson; got by Little Danby (34461), dam (Precious) by Whitelock
(32856), &c.

1879, Dec. 13, r. & w., C.C.	Prodigy	Briony Berry, 36271	Mr. Serjeantson

SEVERNE, Mrs.,
Thenford House, Banbury, Oxon.

LAUNDRESS 2ND, roan, calved November 10, 1871, Vols. xxi. and xxv.
pp. 923, 655. Bred by Mr. F. Lythall, Spittal Farm; got by Hailstorm
(28807), dam (Lady Killerby) by Fitz-Killerby (26166), &c.

1879, Mar. 4, roan, C.C.	Lady Rouse	Petrarch, 42136	Mrs. Severne

SCHOOL MISTRESS, red, calved in January, 1872, Vol. xxv. p. 656. Bred
by Mr. W. M. Severne, Thenford House; got by Warden (27746), dam (School
Mistress) by The General (25295), &c.

1879, Aug. 6, red, C.C.	Sweetbriar	Petrarch, 42136	Mrs. Severne

SHARPLEY, Coates,
Kelstern Hall, Louth, Lincolnshire.

AMY GWYNNE, roan, calved March 19, 1876. Bred by Mr. C. Sharpley; got by Cherry Grand Duke 4th (28175), dam (Lady Ann Gwynne) by Friponnier (26208), &c. See "Duke Gwynne 2nd," p. 71.

Produce in	Names, &c.	By what Bull.	By whom bred.
1879, Mar. 28, r. & w., C.C.	Ada Gwynne	Grand Duke 27th, 34067	Mr. Sharpley

DUCHESS GWYNNE 2ND, roan, calved January 14, 1873, Vol. xxiv. p. 643. Bred by Mr. H. Sharpley, Acthorpe; got by Grand Duke 19th (28746), dam (Fancy Gwynne) by Grand Duke 5th (19875), &c.

1879, Apr. 29, r. & w., C.C.	Duchess Gwynne 7th	Grand Duke 27th, 34067	Mr. C. Sharpley

DUCHESS NANCY, roan, calved February 19, 1874, Vol. xxiv. p. 643. Bred by Mr. H. Sharpley, Acthorpe; got by Heydon Duke 2nd (31370), dam (Grand Duchess of Oxford 2nd) by Fourth Duke of Geneva (25964), &c.

1879, Mar. 27, roan, B.C.	Nanson Grand D. 2nd	Grand Duke 27th, 34067	Mr. C. Sharpley

DUCHESS NANCY 2ND, roan, calved August 16, 1876. Bred by Mr. C. Sharpley; got by Grand Duke 21st (34061), dam (Duchess Nancy) by Heydon Duke 2nd (31370), &c. See Vol. xxiv. p. 643.

1879, Feb. 27, roan, B.C.	Nanson Grand Duke	Grand Duke 27th, 34067	Mr. Sharpley

LADY ANN GWYNNE, red and white, calved May 1, 1871, Vol. xxiv. p. 643. Bred by the Earl of Aylesford, Packington Hall; got by Friponnier (26208), dam (Polly Gwynne 3rd) by Duke of Cumberland (21584), &c.

1878, April 24, red, C.C.	Rosie Gwynne	Grand Duke 27th, 34067	Mr. Sharpley
1879, May 13, roan, B.C.	Duke Gwynne 2nd	do.	do.

Duke Gwynne 2nd, sold to Mr. L. W. Arkwright, Parndon Hall, Essex.

SOUVENIR, red, calved November 28, 1872. Bred by Mr. R. Betts, Holbeck Lodge; got by Major (26793), dam (Sweetheart 15th) by Baron of Rathcool (21233), &c. See "Duke of Cerisia 3rd," Vol. xxiii. p. 78.

1878, June 2, red, C.C.	Charmer	Grand Duke 27th, 34067	Mr. Sharpley

SHARPLEY, Henry,
Acthorpe, Louth, Lincolnshire.

DUCHESS GWYNNE, white, calved January 8, 1872, Vols. xxii., xxiii., and xxv. pp. 556, 640, 656. Bred by Mr. H. Sharpley; got by Grand Duke 7th (19877), dam (Fairy Gwynne) by Grand Duke 5th (19875), &c.

1879, Oct. 18, roan, C.C.	Prin'ss Gwynne 12th	Grand Duke 27th, 34007	Mr. Sharpley

DUCHESS GWYNNE 2ND, roan, calved September 5, 1875, Vol. xxv. p. 656. Bred by Mr. H. Sharpley; got by Grand Duke 22nd (34062), dam (Duchess Gwynne) by Grand Duke 7th (19877), &c.

1879, Sept. 24, white, B.C.	(Steer)	Grand Duke 27th, 34067	Mr. Sharpley

DUCHESS OF GENOA, roan, calved November 28, 1876. Bred by Mr. H. Sharpley; got by Grand Duke 19th (28746), dam (Grand Duchess of Oxford 2nd) by Fourth Duke of Geneva (25964), &c. See Vol. xxiii. p. 640.

1879, Oct., roan, B.C.	(dead)	Grand Duke 27th, 34067	Mr. Sharpley

FAIR GWYNNE, roan, calved October 31, 1873, Vols. xxiii. and xxv. pp. 640, 656. Bred by Mr. H. Sharpley; got by Cherry Grand Duke 4th (28175), dam (Favourite Gwynne) by Grand Duke of Lightburne (26290), &c.

Produce in		Names, &c.	By what Bull.	By whom bred.
1879, Jan. 30, roan,	B.C.	Prince Gwynne 4th	Grand Duke 27th, 34067	Mr. Sharpley

FAITHFUL GWYNNE, roan, calved June 3, 1871, Vols. xxii., xxiii., xxiv., and xxv. pp. 557, 640, 644, 657. Bred by Mr. H. Sharpley; got by Gamos (28672), dam (Fancy Gwynne) by Grand Duke 5th (19875), &c.

Produce in		Names, &c.	By what Bull.	By whom bred.
1879, Nov. 3,	{ roan, C.C. / white, B.C.	Pr'ss Gwynne 13th } (Steer)	Grand Duke 27th, 34067	Mr. Sharpley

FAITHFUL GWYNNE 2ND, red and white, calved October 31, 1874, Vol. xxiv. p. 644. Bred by Mr. H. Sharpley; got by Cherry Grand Duke 4th (28175), dam (Faithful Gwynne) by Gamos (28672), &c.

Produce in		Names, &c.	By what Bull.	By whom bred.
1879, April 18, roan,	C.C.	Prin'ss Gwynne 11th	Grand Duke 27th, 34067	Mr. Sharpley

FAVOURITE GWYNNE, red and white, calved December 30, 1870, Vols. xxii., xxiii., and xxv. pp. 557, 640, 657. Bred by Mr. H. Sharpley; got by Grand Duke of Lightburne (26290), dam (Fairy Gwynne) by Grand Duke 5th (19875), &c.

Produce in		Names, &c.	By what Bull.	By whom bred.
1879, Feb. 22, roan,	C.C.	Prin'ss Gwynne 10th	Grand Duke 27th, 34067	Mr. Sharpley

HAINI, roan, calved April 25, 1875. Bred by Mr. H. Sharpley; got by Cherry Grand Duke 4th (28175), dam (Duchess of Brailes 2nd) by Duke of Brailes (23724), &c. See Vol. xxii. p. 556.

Produce in		Names, &c.	By what Bull.	By whom bred.
1879, May 6, white,	C.C.	D'ss of Brailes 6th	Grand Duke 27th, 34067	Mr. Sharpley

LADY BLANCHE 2ND, red and white, calved February 18, 1874. Bred by Mr. H. Sharpley; got by Cherry Grand Duke 5th (30712), dam (Eugenia) by The Yeoman (25305), &c. See Vol. xx. p. 506.

Produce in		Names, &c.	By what Bull.	By whom bred.
1877, Feb. 21, r. & w.,	C.C.	Lady Blanche 5th	Grand Duke 27th, 34067	Mr. Sharpley
1878, Sept. 10, roan,	B.C.	Alexander	do.	do.
1879, Sept. 11, roan,	B.C.	Alexander 2nd	do.	do.

Alexander, sold to Mr. J. Brown, Park Cottage, Burton Constable, Hull.

LADY BLANCHE 4TH, red and white, calved January 9, 1875. Bred by Mr. H. Sharpley; got by Cherry Grand Duke 4th (28175), dam (Eugenia) by The Yeoman (25305), &c. See Vol. xx. p. 506.

Produce in		Names, &c.	By what Bull.	By whom bred.
1878, Nov. 8, roan,	B.C.	Lord Louth	Grand Duke 27th, 34067	Mr. Sharpley

Lord Louth, sold to Mr. J. Brown, Park Cottage, Burton Constable, Hull.

OXFORD'S WATERLOO 4TH, roan, calved February 28, 1874, Vol. xxiv. p. 644. Bred by Sir T. C. Constable, Bart., Burton Constable; got by Oxford's Baronet (29499), dam (Oxford's Waterloo 3rd) by Imperial Oxford (18084), &c.

Produce in		Names, &c.	By what Bull.	By whom bred.
1879, Nov. 3, roan,	B.C.	G. D. of Waterloo	Grand Duke 27th, 34067	Mr. Sharpley

PATRICIA, roan, calved June 14, 1874. Bred by Mr. R. Betts, Holbeck Lodge; got by Hardicanute (26338), dam (Sybil) by Major (26793), &c. See Vol. xxv. p. 340.

Produce in		Names, &c.	By what Bull.	By whom bred.
1879, Oct. 20, roan,	C.C.	Sylvan Charmer	Grand Duke 27th, 34067	Mr. Sharpley

PRINCESS GWYNNE 4TH, white, calved October 5, 1876. Bred by Mr. H. Sharpley; got by Grand Duke 25th (34065), dam (Fair Gwynne) by Cherry Grand Duke 4th (28175), &c. See Vol. xxiii. p. 640.

Produce in		Names, &c.	By what Bull.	By whom bred.
1879, Sept. 11, white,	C.C.	(dead)	Grand Duke 27th, 34067	Mr. Sharpley

SILENT, roan, calved January 3, 1870, Vols. xx., xxii., xxiii., xxiv., and xxv.
pp. 765, 557, 640, 644, 657. Bred by Mr. J. Clayden, Littlebury; got by
Earl of Warwickshire (26079), dam (Surmise 2nd) by May Duke (13320), &c.

Produce in	Names, &c.	By what Bull.	By whom bred.
1879, April 7, red,	B.C.|Earl of Louth 2nd	Grand Duke 27th, 34067	Mr. Sharpley

SHAW, Samuel,
Brooklands, Halifax, Yorkshire.

BLANCHE 5TH, roan, calved April 17, 1874. Bred by Sir W. H. Salt, Bart.,
Maplewell, the property of Mr. R. Man, Thornhill; got by Marquis of York
(34791), dam (Bantling) by Britannicus (17452), &c. See " Baron Sockburn,"
p. 20.

1879, Apr. 12, r. & w., C.C.|Blanche 6th	Lord Angram 1st, 38573	Mr. Man
Blanche 6th, sold to Mr. Shaw, Brooklands, Halifax.		

BLANCHE 7TH, roan, calved June 10, 1874, Vol. xxiii. p. 641. Bred by Sir
W. H. Salt, Bart., Maplewell; got by Marquis of York (34791), dam (Lady
Blanche 2nd) by Third Duke of Claro (23729), &c.

1879, May 1, roan,	C.C.|Blanche 8th	L.Oxf'd Sockburn2d,38648|Mr. Shaw

COUNTESS OF WINDSOR 2ND, roan, calved January 20, 1875, Vol. xxiv.
p. 644. Bred by Mr. R. C. Richards, Clifton Lodge; got by Baron Clifton
(30433), dam (Countess of Windsor) by Marquis of Windsor (22305), &c.

1879, Mar. 26, white, C.C.|White Windsor	L.Oxf'd Sockburn2d,38648|Mr. Shaw

DOLLY GWYNNE 4TH, roan, calved October 15, 1876. Bred by Mr. S. Shaw;
got by Fifth Lord Oxford (31738), dam (Dolly Gwynne 2nd) by Sultan
(30088), &c. See Vol. xxiii. p. 641.

1879, June 10, r. & w., C.C.|Dolly Gwynne 5th	L.Oxf'd Sockburn2d,38648|Mr. Shaw

DUCHESS 10TH, roan, calved March 4, 1876. Bred by Mr. R. C. Richards,
Clifton Lodge; got by Baron Winsome (30498), dam (Duchess 9th) by Grand
Duke of Oxford (24070), &c. See " Duke Sockburn," p. 87.

1879, Dec. 23, r. & w., B.C.	Duke Sockburn	L.Oxf'd Sockburn2d,38648 Mr. Shaw
Duke Sockburn, sold to Mr. C. Howgate, Lightcliff, Halifax.		

LADY BLANCHE 2ND, red, calved November 21, 1876. Bred by Mr. S.
Shaw; got by Fifth Duke of Glo'ster (36494), dam (Blanche 7th) by Marquis
of York (34791), &c. See Vol. xxiii. p. 641.

1879, Sept. 19, roan, C.C.|Blanche Duchess	L.Oxf'd Sockburn2d,38648|Mr. Shaw

MYTH 4TH, red and white, calved February 10, 1876. Bred by Mr. R. C.
Richards, Clifton Lodge; got by Baron Wesham 4th (33097), dam (Myth) by
Good Fitz (21844), &c. See Vol. xxiv. p. 645.

1879, April 7, r. & w., C.C.	Myth 5th	L.Oxf'd Sockburn2d,38648 Mr. Shaw

ROAN BETSY, roan, calved April 6, 1876. Bred by Mr. R. C. Richards, Clifton
Lodge; got by Baron Wesham 4th (33097), dam (Betsy) by Merlin (24582),
&c. See " Richard the Third," Vol. xxiii. p. 236.

1879, Feb. 13, roan, C.C.|Lady Betsy	L.Oxf'd Sockburn2d,38648|Mr. Shaw

SILKY GWYNNE 3RD, red and white, calved October 25, 1876. Bred by
Mr. S. Shaw; got by Fifth Duke of Glo'ster (36494), dam (Silky Gwynne
2nd) by Duke of Waterloo (28464), &c. See Vol. xxiii. p. 641.

1879, Apr. 26, r. & w., C.C.|Silky Gwynne 4th	L.Oxf'd Sockburn2d,38648|Mr. Shaw

SHEEPSHANKS, Rev. T.,
Arthington Hall, Otley.

CORAL, roan, calved May 11, 1872, Vols xxi. and xxv. pp. 698, 658. Bred by Mr. H. Fawcett, Old Bramhope ; got by Duke of Florence (28388), dam (Cordelia) by Mars (24543), &c.

Produce in	Names, &c.	By what Bull.	By whom bred.
1879, June 9, r. & w., C.C.	Coralline	Baron Hawksworth, 41036	Rev. T. Sheepsh'ks

ROYAL LADY 2ND, red, calved October 2, 1874, Vol. xxv. p. 658. Bred by Mr. A. Fawkes, Farnley Hall ; got by Royal Crown (32372), dam (Lady Victoria) by Lord Darlington (26633), &c.

1879, May 17, red, C.C.	Royal Lady 5th	Lord Eglintoun, 38602	Rev. T. Sheepsh'ks

VENETIA, red and white, calved April 3, 1876. Bred by Mr. A. Fawkes, Farnley Hall ; got by Baron Winsome 6th (33111), dam (First Duchess) by Reformer (24930), &c. See Vol. xxii. p. 413.

1879, June 8, roan, C.C.	Lady Primrose	Earl Roseberry, 33825	Rev. T. Sheepsh'ks

VILLETTE, red and white, calved May 15, 1876. Bred by Mr. A. Fawkes, Farnley Hall ; got by Baron Winsome 6th (33111), dam (Lady Vane) by Lord Cobham (20164), &c. See " Studley," p. 243.

1879, June 12, roan, B.C.	Studley	Earl Roseberry, 33825	Rev. T. Sheepsh'ks

SHELDON, H. J.,
Brailes House, Shipston-on-Stour.

AFRICA, red and white, calved April 2, 1869, Vol. xxii. p. 557. Bred by Mr. H. J. Sheldon ; got by Duke of Brailes (23724), dam (America) by Marmaduke (14897), &c.

1876, Dec. 29, roan, B.C.	(dead)	Duke of Connaught, 33604	Mr. Sheldon
1879, Mar. 25, red, B.C.	Duke of Brailes 10th	Duke of Rothesay, 36534	do.

CHARMER 20TH, roan, calved July 8, 1872, Vols. xxiii. and xxiv. pp. 642, 645. Bred by Mr. D. McIntosh, Havering Park ; got by Third Duke of Geneva (23753), dam (Charmer 10th) by Baron Killerby (23364), &c.

1878, Mar. 4, white, B.C.	D.Ch'mingLand10th	Duke of Rothesay, 36534	Mr. Sheldon
1879, Feb. 27, roan, C.C.	Charming Geneva 3rd	do.	do.

Duke of Charming Land 10th, sold to Mr. Cottam ; Charming Geneva 3rd, to Sir G. R. Philips, Bart., Weston Park, Shipston-on-Stour.

CHARMING DUCHESS 8TH, red and white, calved July 8, 1876. Bred by Mr. H. J. Sheldon ; got by Duke of Rosedale (33721), dam (Duchess 2nd) by Grand Duke 15th (21852), &c. See Vol. xxiii. p. 642.

1879, Jan. 16, r. & w., B.C.	D.Ch'mingLand16th	Duke of Rothesay, 36534	Mr. Sheldon

CHERRY COUNTESS 2ND, roan, calved June 26, 1874, Vols. xxiv. and xxv. pp. 646, 659. Bred by Mr. H. J. Sheldon ; got by Second Duke of Collingham (23730), dam (Cherry Countess) by Grand Duke 6th (19876), &c.

1879, July 11, r. & w., C.C.	Cher'yD'ssBrailes4th	Duke of Rothesay, 36534	Mr. Sheldon

CHERRY COUNTESS 3RD, red and white, calved April 28, 1877. Bred by Mr. H. J. Sheldon ; got by Duke of Charming Land 2nd (36477), dam (Cherry Countess 2nd) by Second Duke of Collingham (23730), &c. See Vol. xxiv. p. 646.

1879, Dec. 27, roan, B.C.	Duke of Cerisia 5th	Prince of Brailes, 42193	Mr. Sheldon

CORAL DUCHESS, red, calved March 26, 1875, Vols. xxiv. and xxv. pp. 646, 659. Bred by Mr. H. J. Sheldon, the property of Mr. W. Bliss, Chipping Norton ; got by Second Duke of Collingham (23730), dam (Lucinda) by Priam (18567), &c.

Produce in	Names, &c.	By what Bull.	By whom bred.
1879, July 2, r. & w., C.C.	Coral Duchess 8th	Duke of Rothesay, 36534	Mr. Sheldon

CORAL DUCHESS 2ND, red and white, calved May 22, 1876. Bred by Mr. H. J. Sheldon, the property of Mr. T. Holford, Papillon Hall ; got by Duke of Rosedale (33721), dam (Lucinda) by Priam (18567), &c. See Vol. xxiii. p. 642.

Produce in	Names, &c.	By what Bull.	By whom bred.
1878, Deo. 25, roan, B.C.	D.Ch'mingLand15th	Duke of Rothesay, 36534	Mr. Sheldon

COUNTESS OF DARLINGTON, red and white, calved September 9, 1870, Vols. xxii. and xxiv. pp. 558, 646. Bred by Mr. H. J. Sheldon, the property of Mr. Evan Baillie, Filleigh ; got by Duke of Brailes (23724), dam (Darlington 12th) by Duke of Geneva (19614), &c.

Produce in	Names, &c.	By what Bull.	By whom bred.
1878, July 12, roan, B.C.	D. of Darlington 6th	Duke of Rothesay, 36534	Mr. Sheldon
1879, July 10, r. & w., B.C.	D. of Darlington 7th	do.	do.

Duke of Darlington 6th, sold to Mr. C. Hobbs, Maisey Hampton ; Duke of Darlington 7th, to Messrs. Sturgeon and Sons, Grays Hall, Essex.

DUCHESS OF BARRINGTON, roan. calved January 7, 1877. Bred by Mr. H. J. Sheldon ; got by Duke of Connaught (33604), dam (Countess of Barrington 4th) by Lord Oxford (20214), &c. See " Duke of Barrington 10th," p. 72.

Produce in	Names, &c.	By what Bull.	By whom bred.
1879, Aug. 22, r. & w., B.C.	D.of Barrington 10th	Duke of Rothesay, 36534	Mr. Sheldon

ERIN, roan, calved August 27, 1870, Vol. xxi. p. 704. Bred by Lord Braybrooke, Audley End ; got by Thorndale Duke (27661), dam (Emerald) by Second Duke of Claro (21576), &c.

Produce in	Names, &c.	By what Bull.	By whom bred.
1879, Mar. 7, roan, B.C.	Lord Tregunter Rosy	D. of Tregunter 5th, 33743	Mr. Sheldon

FANNY, roan, calved August 11, 1876. Bred by Mr. H. J. Sheldon, the property of Mr. Rees ; got by George Frederick (34030), dam (Fernande 2nd) by Duke of Cerisia (30937), &c. See Vol. xxiii. p. 642.

Produce in	Names, &c.	By what Bull.	By whom bred.
1879, Jan. 21, roan, C.C.	Fernande 5th	Duke of Rothesay, 36534	Mr. Sheldon

Fernande 5th, sold to Colonel North, Wrotton Abbey.

FERNANDE 2ND, red and white, calved April 15, 1873, Vols. xxiii. and xxiv. pp. 642, 647. Bred by Mr. H. J. Sheldon, the property of Mr. E. Hales, North Frith ; got by Duke of Cerisia (30937), dam (Fernande) by Eighteenth Duke of Oxford (25995), &c.

Produce in	Names, &c.	By what Bull.	By whom bred.
1878, July 31, roan, B.C.	L. Rothesay Walnut	Duke of Rothesay, 36534	Mr. Sheldon

GLADYS, red and white, calved October 23, 1870, Vols. xxi. and xxv. pp. 924, 659. Bred by Mr. H. J. Sheldon, the property of Mr. W. Bliss, Chipping Norton ; got by Duke of Brailes (23724), dam (Harebell) by Fourth Duke of Thorndale (17750), &c.

Produce in	Names, &c.	By what Bull.	By whom bred.
1879, Mar. 5, r. & w., B.C.	Duke Furbelow 2nd	Duke of Rothesay, 36534	Mr. Sheldon

LADY FAWSLEY 3RD, roan, calved September 13, 1871, Vols. xxii. and xxiv. pp. 559, 647. Bred by Mr. H. J. Sheldon ; got by Eighteenth Duke of Oxford (25995), dam (Hyampia) by Æsop (19197), &c.

Produce in	Names, &c.	By what Bull.	By whom bred.
1878, Sept. 13, roan, B.C.	Earl of Fawsley 8th	Duke of Rothesay, 36534	Mr. Sheldon
1879, Nov. 7, white, B.C.	Earl of Fawsley 9th	do.	do.

LADY FLORENCE 2ND, red and white, calved July 11, 1871, Vols. xxii. and xxiv. pp. 559, 647. Bred by Mr. H. J. Sheldon ; got by Eighteenth Duke of Oxford (25995), dam (Lady Florence) by Duke of Brailes (23724), &c.

Produce in	Names, &c.	By what Bull.	By whom bred.
1878, Mar. 29, r. & w., B.C.	Lord Garland 6th	Duke of Rothesay, 36534	Mr. Sheldon
1879, Mar. 17, r. & w., C.C.	Lady Florence 10th	do.	do.

Lord Garland 6th, sold to Mr. J. G. Attwater, Britford, Salisbury ; Lady Florence 10th, to Mr. J. H. Blundell, Woodside, Luton.

LADY FLORENCE 7TH, roan, calved April 25, 1876. Bred by Mr. H. J. Sheldon ; got by Duke of Rosedale (33721), dam (Lady Florence) by Duke of Brailes (23724), &c. See Vol. xxiii. p. 642.

1878, Sept. 9, roan,	B.C.	Lord Garland 7th	Duke of Rothesay, 36534	Mr. Sheldon

Lord Garland 7th, sold to Mr. J. C. Adkins, Milcote, Stratford-on-Avon.

LADY MILCOTE 4TH, roan, calved May 1, 1876. Bred by Mr. F. Sartoris, Rushden Hall ; got by Baron Oxford Fawsley (33076), dam (Lady Milcote) by Twelfth Duke of Oxford (19633), g. d. (Concord) by Fourth Duke of Thorndale (17750), &c. See Vol. xix. p. 451.

1879, Mar. 27, white, B.C.	L. Rothesay Milcote	Duke of Rothesay, 36534	Mr. Sheldon

LADY SUSAN 3RD, roan, calved October 14, 1867, Vol. xx. p. 615. Bred by Mr. H. G. White, Framingham, Mass., U.S.A., the property of Mr. H. J. Sheldon ; got by Ninth Duke of Thorndale (31023), dam (Lady Susan 2nd) by Thorndale (32709), &c.

1877, April 4, roan,	B.C.	Earl of Leicester 16th	3rd Duke of Glo'ster, 33653	Mr. Cheney

Earl of Leicester 16th, sold to Mr. J. Banks Stanhope, Revesby Abbey, Boston.

LADY WELLINGTON, roan, calved December 28, 1868, Vol. xxi. p. 626. Bred by Mr. A. M. Winslow, Putney, Vermont, U.S.A., the property of Mr. H. J. Sheldon ; got by Duke of Putney (33717), dam (Lady Sale 10th) by Rising Star (35277), &c.

1875, May 11, roan,	C.C.	Lady Wellesley 2nd	9th Duke of Geneva, 28391	Mr. Cheney
1876, Nov. 22, r. & w., C.C.	Lady Wellesley 3rd	do.	do.
1878, April 17, red,	C.C.	Lady Wellesley 4th	3rd Duke of Glo'ster, 33653	do.

Lady Wellesley 2nd and Lady Wellesley 3rd, sold to Mr. Rawstorne, Hutton Hall, Preston ; Lady Wellesley 4th, to the Earl of Bective, Underley Hall, Carnforth.

PRINCESS OF GENEVA 2ND, roan, calved October 15, 1872, Vols. xxiv. and xxv. pp. 647, 660. Bred by Mr. E. H. Cheney, Gaddesby Hall ; got by Ninth Duke of Geneva (28391), dam (Princess) by General Havelock (17952), &c.

1879, April 6, white, B.C.	Prince of Brailes 3rd	Duke of Rothesay, 36534	Mr. Sheldon

PRINCESS OF THE VALLEY 2ND, roan, calved October 25, 1873, Vol. xxv. p. 660. Bred by Mr. J. W. Wadsworth, Geneseo, New York, U.S.A. ; got by Baron Morley 2nd (41047), dam (Damask 2nd) by Millbrook (34851), &c.

1879, June 9, white, C.C.	Princess of Brailes	Duke of Rothesay, 36534	Mr. Sheldon

RIVAL GWYNNE, roan, calved May 3, 1874, Vol. xxiv. p. 647. Bred by Mr. J. P. Foster, Killhow ; got by Twenty-second Duke of Oxford (31000), dam (Ross Gwynne) by Royal Cambridge (25009), &c.

1878, May 12, roan,	B.C. (dead)		Duke of Rothesay, 36534	Mr. Sheldon
1879, Dec. 2, roan,	C.C.	Rothesay Gwynne	do.	do.

TUBE ROSE 43RD, roan, calved September 5, 1871, Vol. xxii. p. 358.   Bred by
Mr. T. L. Harison, Albany, New York, U.S.A., the property of Mr. H. J.
Sheldon; got by Saladin (35461), dam (Tube Rose 40th) by Zanoni (37700), &c.

Produce in	Names, &c.	By what Bull.	By whom bred.
1877, Mar. 12, r. & w., C.C.	Lady Angelina 2nd	3rd Duke of Glo'ster, 33653	Mr. Cheney
1879, Feb. 26, roan, B.C.	Earl of Leicester 19th	do.	do.

Lady Angelina 2nd, sold to Mr. T. Holford, Papillon Hall, Market Harborough; Earl of Leicester 19th, to Mr. J. Linton, Buckden Wood, Hunts.

TUBE ROSE 44TH, red, calved October 21, 1872, Vol. xxii. p. 359.   Bred by
Mr. T. L. Harison, Albany, New York, U.S.A., the property of Mr. E. H.
Cheney, Gaddesby Hall; got by Saladin (35461), dam (Tube Rose 40th) by
Zanoni (37700), &c.

Produce in	Names, &c.	By what Bull.	By whom bred.
1876, Aug. 2, red, C.C.	Lady Angelina	9th Duke of Geneva, 28391	Mr. Cheney
1878, Apr. 16, r. & w., C.C.	Lady Angelina 3rd	3rd Duke of Glo'ster, 33653	do.

Lady Angelina, sold to the Earl of Bective, Underley Hall, Carnforth; Lady Angelina 3rd, to Mr. H. J. Sheldon, Brailes House.

SHERATON, W.,
Broom House, Ellesmere, Shropshire.

LADY HARPER 5TH, red and white, calved February 13, 1871.   Bred by
Mr. W. Sheraton, the property of Major H. F. Cockayne Cust, Ellesmere
House; got by May Duke (26885), dam (Lady Harper) by Prince Arthur
(27117), &c.   See Vol. xxi. p. 926.

Produce in	Names, &c.	By what Bull.	By whom bred.
1878, April 6, red, C.C.	Lady Harper 16th	Lilleshall, 34456	Mr. Sheraton
1879, Oct. 18, red, C.C.		The Sultan, 40811	do.

Lady Harper 16th, sold to Mr. K. G. Salter, Brynallt, Ellesmere; 1879 Cow Calf, to Major H. F. Cockayne Cust, Ellesmere House, Ellesmere, Shropshire.

LADY ISABELLA HAMILTON 2ND, roan, calved June 1, 1876.   Bred by
Mr. W. Sheraton; got by Lilleshall (34456), dam (Lady Isabella Hamilton)
by Darius (30860), g. d. (Lady Hamilton 3rd) by Knight of the Garter
(34400), &c.   See "Baron Hamilton," p. 15.

Produce in	Names, &c.	By what Bull.	By whom bred.
1878, July 26, white, C.C.	L. Isabella H'lton 3rd	Baron Winston, 42757	Mr. Sheraton
1879, July 18, white, C.C.	L. Isabella H'lton 4th	The Sultan, 40811	do.

Lady Isabella Hamilton 3rd, sold to Mr. T. J. Provis, The Grange, Ellesmere; Lady Isabella Hamilton 4th, to Mr. J. Bateman, Croxton Hanmer, Whitchurch, Salop.

LADY MARGARET HAMILTON, roan, calved May 15, 1876.   Bred by Mr.
W. Sheraton, the property of Mr. T. J. Provis, The Grange; got by Lilleshall
(34456), dam (Lady Hamilton 3rd) by Knight of the Garter (34400), &c.
See "Baron Hamilton," p. 15.

Produce in	Names, &c.	By what Bull.	By whom bred.
1878, Dec. 28, roan, C.C.	Lady Lucy Hamilton	The Sultan, 40811	Mr. Sheraton

SHINGLER, W. and J. H.,
Birch Hall, Ellesmere, Shropshire.

LAUREL LEAF, red and white, calved March 14, 1877.   Bred by Messrs. W.
and J. H. Shingler; got by King of the Homestead (31504), dam (Snowdrop)
by Highborn (26388), g. d. (Snowflake) by Dexterity (21538), gr. g. d.
(Primrose) by Harold (14670).

Produce in	Names, &c.	By what Bull.	By whom bred.
1879, Dec. 24, r. & w., C.C.	Lilac	D. of Ulverstone 2nd, 43132	Messrs. Shingler

SINDALL, T. Y.,
Fulney Hall, Spalding.

FLASH, roan, calved June 5, 1875. Bred by Mr. T. Y. Sindall; got by Lord
Windsor (31771), dam (Flame) by Henry 5th (19944), &c. See " Duke of
Fulney," Vol. xxv. p. 80.

Produce in	Names, &c.	By what Bull.	By whom bred.
1877, Nov. 22, r. & w., C.C.	Flower of Fulney	Forester, 36659	Mr. Sindall
1878, Nov. 21, roan, B.C.	D. of Fulney (steer)	do.	do.

TELLURIA PRINCESS, roan, calved March 4, 1874. Bred by Mr. J.
Burgess, Edenham; got by Job (31438), dam (Chilton Princess) by Prince
Boabdil (27120), &c. See " Don Alfonso," Vol. xxiv. p. 62.

1877, May 13, roan, B.C.	Don Alfonso (steer)	Lord Windsor, 31771	Mr. Sindall
1878, April 30, roan, C.C.	Telluria Rose	Forester, 36659	do.

SINGLETON, J. R.,
Great Givendale, Pocklington, Yorkshire.

ATE, roan, calved February 8, 1869, Vols. xx., xxi., xxiii., and xxv. pp. 396,
1012, 644, 662. Bred by the Earl of Zetland, Upleatham; got by Chatsworth
(23546), dam (Alert) by Surrey (17067), &c.

1879, Feb. 3, roan, B.C.	Lord Surrey	Duke of Oxford 37th, 38171	Mr. Singleton

BELIEF, red and white, calved July 13, 1874, Vols. xxiv. and xxv. pp. 650,
662. Bred by the Earl of Zetland, Upleatham; got by Grand Monarch
(28774), dam (Barnet) by King of the Roses (22043), &c.

1879, Feb. 4, r. & w., C.C.	Confidence 2nd	Duke of Oxford 37th, 38171	Mr. Singleton

BLOOMING ROSE, red and white, calved July 16, 1873, Vols. xxiii., xxiv.,
and xxv. pp. 644, 650, 662. Bred by Mr. J. R. Singleton; got by Bobby
(25648), dam (Bloomer) by Marquis (26825), &c.

1879, June 17, r. & w., B.C.	Blooming Boy	Duke of Oxford 37th, 38171	Mr. Singleton

CONFIDENCE, red, calved March 2, 1877. Bred by Mr. J. R. Singleton; got
by Grand Duke 26th (34066), dam (Belief) by Grand Monarch (28774), &c.
See Vol. xxiv. p. 650.

1879, Sept. 21, roan, C.C.	Confidence 3rd	Silent Duke, 42382	Mr. Singleton

DUCHESS OF TOWNELEY 4TH, roan, calved April 16, 1874, Vols. xxiii.,
xxiv., and xxv. pp. 644, 650, 663. Bred by Mr. J. R. Singleton; got by Red
Duke (32253), dam (Duchess of Towneley 3rd) by Nestor (26957), &c.

1879, Dec. 7, roan, B.C.	Duke Anthony	Duke of Oxford 37th, 38171	Mr. Singleton

DUCHESS OF TOWNELEY 5TH, roan, calved June 19, 1876, Vol. xxv.
p. 663. Bred by Mr. J. R. Singleton; got by Beau Siddington (30515), dam
(Duchess of Towneley 4th) by Red Duke (32253), &c.

1879, July (dead), C.C.		Silent Duke, 42382	Mr. Singleton

FLORA SIDDINGTON, roan, calved February 27, 1875, Vol. xxv. p. 663.
Bred by Mr. J. R. Singleton; got by Beau Siddington (30515), dam (Oxford
Cherry) by Second Earl of Oxford (23843), &c.

1879, Feb. 14 (dead), C.C.		Duke of Oxford 37th, 38171	Mr. Singleton

MABEL RUTH 2ND, roan, calved November 30, 1875, Vol. xxv. p. 663. Bred by Mr. J. R. Singleton; got by Bob Cherry (33172), dam (Mabel) by Chatsworth (25346), &c.

Produce in	Names, &c.	By what Bull.	By whom bred.
1879, May 28, roan, C.C.	Mabel Ruth 3rd	Th'ndale CherryLad,40818	Mr. Singleton

MARY RUTH, red and white, calved March 20, 1872, Vols. xxi., xxiii., and xxv. pp. 929, 644, 663. Bred by Mr. J. R. Singleton; got by Bobby (25648), dam (Maude) by Patriot (20475), &c.

Produce in	Names, &c.	By what Bull.	By whom bred.
1879, Mar. 10, r. & w., B.C.	Bob Ruth	Duke of Oxford 37th, 38171	Mr. Singleton

Bob Ruth, sold to Mr. A. Kirby, Market Weighton.

MARY RUTH 2ND, red and white, calved October 22, 1874, Vols. xxiv. and xxv. pp. 650, 663. Bred by Mr. J. R. Singleton; got by Beau Siddington (30515), dam (Mary Ruth) by Bobby (25648), &c.

Produce in	Names, &c.	By what Bull.	By whom bred.
1879, June 14, r. & w., B.C.	Thorndale Ruth	Th'ndale CherryLad,40818	Mr. Singleton

MARY RUTH 3RD, red and white, calved November 14, 1876. Bred by Mr. J. R. Singleton; got by Bob Cherry (33172), dam (Mary Ruth) by Bobby (25648), &c. See "Bob Ruth," p. 30.

Produce in	Names, &c.	By what Bull.	By whom bred.
1879, Feb. 25, r. & w., B.C.	(slaughtered)	Duke of Oxford 37th, 38171	Mr. Singleton

OXFORD CHERRY'S RUBY, red, calved January 27, 1876, Vol. xxv. p. 664. Bred by Mr. J. R. Singleton; got by Duke of Rubies (33724), dam (Oxford Cherry) by Second Earl of Oxford (23843), &c.

Produce in	Names, &c.	By what Bull.	By whom bred.
1879, May 9, red, C.C.	O.Cherry's Ruby 2nd	Duke of Oxford 37th, 38171	Mr. Singleton

QUEEN CHERRY ROSE, red, calved April 22, 1875, Vol. xxv. p. 664. Bred by Mr. J. Lynn, Stroxton; got by Cambridge Duke 4th (25706), dam (Queen Cherry) by Cherry Duke (25752), &c.

Produce in	Names, &c.	By what Bull.	By whom bred.
1879, Feb. 2, (dead) C.C.		Duke of Oxford 37th, 38171	Mr. Singleton

QUEEN DOUBLE ROSE, red and white, calved June 23, 1874, Vol. xxv. p. 664. Bred by Mr. J. Lynn, Stroxton; got by Cambridge Duke 4th (25706), dam (Queen of the Roses) by Cambridge Duke 4th (25706), &c.

Produce in	Names, &c.	By what Bull.	By whom bred.
1879, Jan. 12, r. & w., B.C.	(dead)	Duke of Oxford 37th, 38171	Mr. Singleton

ROSE OF RYEDALE 3RD, red and white, calved April 9, 1871, Vols. xxi. and xxv. pp. 705, 664. Bred by the Earl of Feversham, Duncombe Park; got by Coriolanus (30795), dam (Carnation) by Vesuvius (21017), &c.

Produce in	Names, &c.	By what Bull.	By whom bred.
1879, Dec. 5, red, B.C.	Earl of Ryedale	Silent Duke, 42382	Mr. Singleton

SINGLETON, John,
Teresa Cottage, Pocklington, Yorkshire.

BELLA 2ND, roan, calved July 4, 1876. Bred by Mr. J. P. Foster, Killhow; got by Duke of Hillhurst (28401), dam (Ariadne) by Earl of Eglinton (23832), &c. See Vol. xxiii. p. 435.

Produce in	Names, &c.	By what Bull.	By whom bred.
1879, May 9, roan, C.C.	Bella 3rd	D. of Tregunter 5th, 33743	Mr. Singleton

COMELY, red, calved March 30, 1874. Bred by the Earl of Zetland, Upleatham; got by Alexis (30378), dam (Custard) by Maximilian (20322), &c. See Vol. xxi. p. 1014.

Produce in	Names, &c.	By what Bull.	By whom bred.
1879, July 11, r. & w., C.C.	Comely 2nd	Lord Cockburn, 38594	Mr. Singleton

COWSLIP 30TH, white, calved April 1, 1876. Bred by Mr. T. Wilson, Shotley
Hall; got by Waterloo Duke (32815), dam (Cowslip 20th) by Lord of Nunwick (26702), &c. See Vol. xxiii. p. 713

Produce in	Names, &c.	By what Bull.	By whom bred.
1879, Mar. 30, roan, C.C.	Cowslip 44th	Duke of Oxford 31st, 33713	Mr. Singleton

COWSLIP 33RD, red, calved November 19, 1876. Bred by Mr. T. Wilson,
Shotley Hall; got by Duke of Oxford 31st (33713), dam (Cowslip 22nd) by
Lord of Nunwick (26702), &c. See Vol. xxiii. p. 713.

1879, July 11, roan, C.C.	Cowslip 45th	Lord Cockburn, 38594	Mr. Singleton

GRAND BLANCHE, roan, calved November 24, 1873. Bred by Mr. M.
Kennedy, Stone Cross; got by Grand Duke of Oxford (28764), dam (Blanche
3rd) by Tenth Duke of Oxford (17739), &c. See Vol. xxiii. p. 517.

1879, July 29, white, C.C.	Grand Blanche 3rd	Lord Cockburn, 38594	Mr. Singleton

JESSY 41ST, roan, calved April 4, 1876. Bred by Mr. J. Martin, Bardsea; got
by Baron Barrington 7th (33009), dam (Jessy 37th) by Grand Duke of Cambridge 2nd (26285), &c. See Vol. xxv. p. 578.

1879, Sept. 15, white, C.C.	Jessy Cockburn	Lord Cockburn, 38594	Mr. Singleton

LADY BEAUMONT 6TH, roan, calved May 19, 1874. Bred by Mr. J.
Singleton; got by Beau Siddington (30515), dam (Lady Beaumont 5th) by
Grand Duke of Lightburne 2nd (26291), &c. See Vol. xxi. p. 928.

1879, Sept. 20, roan, C.C.	Lady Beaumont 9th	Lord Cockburn, 38594	Mr. Singleton

LADY FLORA 3RD, white, calved August 5, 1876. Bred by Mr. J. Singleton;
got by Baron Jocelyn (33055), dam (Lady April) by Sir Roger Gwynne
(27496), &c. See Vol. xxiii. p. 643.

1879, June 27, white, C.C.	Lady Flora 5th	Lord Cockburn, 38594	Mr. Singleton

SISK, James,
Acres, Fermoy, Ireland.

VALERIA, red and white, calved February 20, 1873, Vol. xxii. p. 402. Bred
by Mr. J. Downing, Ashfield; got by Lord Stanley (24466), dam (Vidonia
4th) by Western Wonder (17225), &c.

1879, Aug. 14, r. & w., C.C.	Violet	Robert Stephenson, 32313	Mr. Sisk

SISSON, John,
Newfield, Ledsham, South Milford.

COUNTESS CLARENCE, white, calved April 18, 1872, Vols. xxi. and xxiv.
pp. 588, 336. Bred by Mr. R. Botterill, Wauldby; got by Duke of Clarence
(19611), dam (Miss Pyrgo) by Clement Cleveland (15775), &c.

1878, Mar. 6, white, C.C.	C'ntess Clarence 2nd	Oxford-le-Grand, 29496	Mr. Sisson

MARIE, red, calved August 6, 1871. Bred by Sir Tatton Sykes, Bart., Sledmere; got by Heydon Duke (28850), dam (Lady Di) by Duke of Towneley
(21615), &c. See Vol. xx. p. 590.

1877, Oct. 12, r. & w., C.C.	Newfield Marie	Ignoramus, 28887	Sir T. Sykes

Newfield Marie, sold to Mr. J. Sisson, Newfield.

SKELMERSDALE, Lord,
Lathom House, Ormskirk.

DUCHESS OF ORMSKIRK, roan, calved August 17, 1876. Bred by Lord Skelmersdale; got by Duke of Tregunter 5th (33743), dam (Eleventh Duchess of Oneida) by Duke of Oneida 2nd (33702), &c. See Vol. xxiii. p. 645.

Produce in	Names, &c.	By what Bull.	By whom bred.
1879, Mar. 9, white. C.C.	D'ss of Ormskirk 3rd	D. of Underley 3rd, 38196	Lord Skelmersdale

FELICITÉ GWYNNE, red and white, calved February 25, 1872, Vols. xxiii., xxiv., and xxv. pp. 645, 652, 664. Bred by Lord Skelmersdale; got by Cherry Grand Duke 2nd (25758), dam (Florence Gwynne) by Grand Duke 5th (19875), &c.

Produce in	Names, &c.	By what Bull.	By whom bred.
1879, Mar. 8, roan, C.C.	Freda Gwynne	Baron Oxford 4th, 25580	Lord Skelmersdale

LADY WILD EYES 2ND, red and white, calved January 24, 1876. Bred by Lord Skelmersdale; got by Second Duke of Glo'ster (28392), dam (Lady Wild Eyes) by Eighth Duke of Geneva (28390), &c. See "Wild Prince 7th," p. 270.

Produce in	Names, &c.	By what Bull.	By whom bred.
1879, Mar. 8, r. & w., C.C.	Lady Wild Eyes 5th	D. of Rosedale 6th, 38176	Lord Skelmersdale

LALLY 14TH, roan, calved April 16, 1871, Vols. xxi., xxiv., and xxv. pp. 931, 652, 665. Bred by Mr. J. Harward, Winterfold; got by Eighth Duke of Geneva (28390), dam (Lally 10th) by Fifth Lord Wild Eyes (26762), &c.

Produce in	Names, &c.	By what Bull.	By whom bred.
1879, Dec. 21, roan, C.C.	Lally Barrington 5th	D. of Rosedale 6th, 38176	Lord Skelmersdale

MINSTREL MAID, red and white, calved October 26, 1874. Bred by Mr. J. P. Foster, Killhow; got by Baron Oxford 6th (33075), dam (Geneva's Minstrel) by Ninth Duke of Geneva (28391), &c. See Vol. xxi. p. 706.

Produce in	Names, &c.	By what Bull.	By whom bred.
1879, Apr. 24, r. & w., C.C.	Minstrel Gwynne 2d	D. of Rosedale 6th, 38176	Lord Skelmersdale

PRINCESS OF LIGHTBURNE 2ND, roan, calved January 9, 1875, Vol. xxiv. p. 652. Bred by Mr. A. Brogden, Ulverston; got by Grand Duke of Kent 2nd (28759), dam (Princess 4th) by Royal Cambridge (25009), &c.

Produce in	Names, &c.	By what Bull.	By whom bred.
1879, Aug. 7, roan, C.C.	Pr'cess of Blythe 2d	D. of Rosedale 6th, 38176	Lord Skelmersdale

SMITH, Charles,
Norris House, Aughton, Lancashire.

GEORGIANA 21ST, roan, calved March 13, 1876. Bred by Mr. G. Allen, Knightley Hall; got by Duke of Wetherby 6th (33756), dam (Georgiana 13th) by Eighth Duke of York (28480), &c. See Vol. xxiii. p. 309.

Produce in	Names, &c.	By what Bull.	By whom bred.
1879, Mar. 28, roan, C.C.	Georgiana Rowley	2nd Duke of Rowley, 24441	Mr. Smith

LADY WHARFDALE 4TH, roan, calved October 8, 1876. Bred by Mr. R. Harrett, Kirkwhelpington; got by Chief Baron (33365), dam (Lady Wharfdale) by Third Lord Wharfdale (26759), &c. See "Northern Duke 3rd," p. 177.

Produce in	Names, &c.	By what Bull.	By whom bred.
1879, Aug. 2, roan, C.C.	Lady Wharfdale 5th	Pr. of Waterloo 2d, 38944	Mr. Smith

SMITH, Humphry,
Castlebrack, Mountmellick, Ireland.

BRITISH GIRL, white, calved April 14, 1870, Vol. xxv. p. 666. Bred by Mr. H. Smith; got by British Boy (25675), dam (Dowager) by British Duke (19350), &c.

Produce in	Names, &c.	By what Bull.	By whom bred.
1879, April 17, roan, C.C.	British Princess	Earl of Aylesby, 38215	Mr. Smith

DAYBREAK 2ND, red and white, calved March 16, 1873, Vols. xxiii. and xxv. pp. 647, 667. Bred by Mr. H. Smith; got by Lord Claud (29086), dam (Daybreak) by British Duke (19350), &c.

Produce in	Names, &c.	By what Bull.	By whom bred.
1879, Mar. 12, r. & w., C.C.	Daybreak 11th	Earl of Aylesby, 38215	Mr. Smith

DAYBREAK 3RD, roan, calved March 1, 1874, Vol. xxv. p. 667. Bred by Mr. H. Smith; got by Lord Claud (29086), dam (Daybreak) by British Duke (19350), &c.

1879, Dec. 23, r. & w., C.C.	Daybreak 12th	Earl of Aylesby, 38215	Mr. Smith

DAYBREAK 6TH, red, calved July 2, 1876. Bred by Mr. H. Smith; got by Conqueror (30786), dam (Daybreak 2nd) by Lord Claud (29086), &c.

1878, July 18, r. & w., C.C.	Daybreak 10th	Cupid, 38065	Mr. Smith
1879, Nov. 21, r. & w., C.C.	Daybreak 13th	Earl of Aylesby, 38215	do.

DUNLAVIN 5TH, red and white, calved May 27, 1877. Bred by Mr. H. Smith; got by Conqueror (30786), dam (Dunlavin 2nd) by Sheet Anchor 2nd (25121), &c. See "Druid," p. 69.

1879, Nov. 15, roan, C.C.	Dunlavin 7th	England's Fame, 39845	Mr. Smith

LOVESOME 2ND, red and white, calved December 30, 1866, Vols. xix., xx., and xxiii. pp. 606, 631, 647. Bred by Mr. J. Barcroft, Kilbogget; got by Favourite (21728), dam (Lovesome) by Great Mogul (14651), &c.

1877, May 12, r. & w., B.C.	Lord Lovel	Conqueror, 30786	Mr. Smith
1878, Apr. 22, r. & w., B.C.	Lord Castlebrack	do.	do.
1879, May 13, r. & w., C.C.	Lovesome 12th	Earl of Aylesby, 38215	do.

Lord Lovel, sold to Mr. W. Phillips, Woodbrook, Portarlington; Lord Castlebrack, to Mrs. O'Brien, Rathgronan, Ardagh, Co. Limerick.

LOVESOME 7TH, red and white, calved March 22, 1876. Bred by Mr. H. Smith; got by Conqueror (30786), dam (Lovesome 5th) by Royal Duke (25014), &c. See "Lover," p. 161.

1879, Sept. 10, r. & w., C.C.	Lovesome 14th	Earl of Aylesby, 38215	Mr. Smith

MUSIDORA, roan, calved March 25, 1872, Vols. xxii., xxiii., and xxiv. pp. 401, 423, 418. Bred by Mr. J. Meadows, Thornville; got by Prince Mason (29645), dam (Graceful) by Duke of Moscow (14447), &c.

1879, June 15, roan, C.C.	Muriel	Robert Stephenson, 32313	Mr. Smith

PRIMROSE 2ND, red, calved January 7, 1874. Bred by Mr. H. Smith; got by Bright Spur (25671), dam (Primrose) by Red Gauntlet (35227), &c. See Vol. xxii. p. 566.

1876, June 30, red, C.C.	Primrose 3rd	Red Knight, 37322	Mr. Smith
1879, June 2, r. & w., C.C.	Primrose 6th	Earl of Aylesby, 38215	do.

Primrose 3rd, sold to Mr. W. Kemmis, Las Rosas, Argentine Republic.

ROSE OF THE HEATH, red and white, calved May 6, 1870, Vol. xxiii. p. 647. Bred by Mr. J. Downing, Ashfield; got by Prince Bertram (27119), dam (Young Moss Rose) by Dr. McHale (15887), &c.

1877, Jan. 9, r. & w., B.C.	Roseberry	Conqueror, 30786	Mr. Smith
1878, Feb. 15, r. & w., B.C.	Rose King	Earl of Aylesby, 38215	do.
1879, Jan. 31, red, C.C.	Rose of Castlebrack	Conqueror, 30786	do.

Roseberry and Rose King, sold to Mr. W. Kemmis, Las Rosas, Argentine Republic.

SAUCE 4TH, roan, calved August 28, 1876. Bred by Mr. H. Smith; got by
Rosy King (35326), dam (Sauce 2nd) by Royal Prince (27384), &c. See
Vol. xxv. p. 668.

Produce in	Names, &c.	By what Bull.	By whom bred.
1879, Aug. 14, r. & w., C.C.	Sauce 6th	Earl of Aylesby, 38215	Mr. Smith

TITANIA 5TH, white, calved April 10, 1870, Vols. xxi. and xxii. pp. 932, 567.
Bred by Mr. N. M. Archdale, Crock-na-Crieve; got by Napoleon 3rd (29417),
dam (Titania 4th) by Prince of Lurg (22617), &c.

1877, Nov. 2, roan, B.C.	Double Fame	Conqueror, 30786	Mr. Smith
1879, Dec. 18, roan, C.C.	Truthful	Earl of Aylesby, 38215	do.

Double Fame, sold to Mr. J. Watt, Claragh, Ramelton, Co. Donegal.

TRUE LOVE 12TH, roan, calved September 22, 1874. Bred by Mr. E. J.
Smith, Islanmore; got by Lord Broughton (31626), dam (True Love 5th) by
Prince Bertram (27119), g. d. (True Love 2nd) by Dr. McHale (15887), &c.
See "True Earl," p. 253.

1877, Mar. 7, r. & w., C.C.	(dead)	Rosy King, 35326	Mr. H. Smith
1878, Mar. 12, r. & w., B.C.	True Lover	Earl of Aylesby, 38215	do.
1879, Nov. 12, red, C.C.	True Love 13th	do.	do.

True Lover, sold to Mr. W. Kemmis, Las Rosas, Argentine Republic.

TULIP, red and white, calved April 13, 1877. Bred by Mr. H. Smith; got by
Conqueror (30786), dam (Thrifty) by Abercorn (25484), &c. See Vol. xxiv.
p. 653.

1879, June 10, roan, C.C.	Aylesby Tulip	Earl of Aylesby, 38215	Mr. Smith

VICTORIA 43RD, white, calved February 16, 1872, Vols. xxi., xxiv., and xxv.
pp. 933, 653, 668. Bred by Mr. E. J. Smith, Islanmore; got by The Earl
(27623), dam (Victoria 41st) by King Richard (26523), &c.

1879, July 5, roan, C.C.	Victoria 49th	Earl of Aylesby, 38215	Mr. H. Smith

VICTORIA 45TH, red and white, calved July 14, 1872. Bred by Mr. E. J.
Smith, Islanmore; got by Prince Christian (22581), dam (Victoria 39th) by
Ravenspur (20628), &c. See Vol. xx. p. 805.

1878, Jan. 2, r. & w., B.C.		Lord Prinknash, 34655	Mr. Burnyeat
1879, Sept. 13, r. & w., C.C.	Victoria 50th	Earl of Aylesby, 38215	Mr. H. Smith

VICTORIA 47TH, roan, calved December 1, 1873, Vol. xxiv. p. 701. Bred by
Mr. E. J. Smith, Islanmore; got by Lord Broughton (31626), dam (Victoria
Alberta) by Symmetry (20926), &c.

1879, Mar. 26, white, C.C.	Victoria 48th	Robert Stephenson, 32313	Mr. H. Smith

YOUNGER SISTER, roan, calved July 11, 1871, Vols. xxiii. and xxv. pp. 648,
669. Bred by Mr. J. P. Tynte, Tynte Park; got by Bright Spur (25671),
dam (Hebe) by Hohenlohe (18074), &c.

1879, Dec. 15, r. & w., C.C.	Young Aylesby	Earl of Aylesby, 38215	Mr. Smith

SMITH, John,
Balmain, Laurencekirk, N.B.

AVERNE 2ND, red, calved May 15, 1877. Bred by Mr. G. Marr, Cairnbrogie,
the property of Mr. J. Smith; got by Ben Nevis (39462), dam (Averne) by
Red Prince (32267), &c. See "Norman," p. 176.

1879, April 21, roan, B.C.	Norman	Cossack, 39633	Mr. Marr

Norman, sold to Mr. J. Smith.

BELLONA, red, calved June 16, 1868, Vols. xx. and xxii. pp. 408, 567. Bred by Mr. A. Cruickshank, Sittyton; got by Champion of England (17525), dam (Beauty) by Lord Raglan (13244), &c.

Produce in		Names, &c.	By what Bull.	By whom bred.
1876, March 27, roan,	C.C.	Bella	Sir James, 35567	Mr. Smith
1877, March 7, red,	C.C.	Rose	do.	do.
1878, Feb. 9, roan,	C.C.	Bessy Lee	do.	do.
1879, April 4, roan,	C.C.	Verona	The Scotsman, 39211	do.

CHARM, roan, calved April 19, 1874. Bred by Mr. J. Smith; got by Master of the Mint (31878), dam (Chance) by Premier (32074), &c. See "Champion," p. 44.

1877, Jan. 16, roan,	B.C.	Champion	Sir James, 35567	Mr. Smith
1879, Mar. 28, white,	C.C.	Charm 3rd	The Scotsman, 39211	do.

Champion, sold to Mr. Wallace, Balbegne.

SMITH, William,
Melkington, Cornhill, Northumberland.

PRIMROSE, red, calved March 6, 1876. Bred by Mr. N. Henderson, Lowick; got by Duke of Florence (33638), dam by Northern Chief (26982), g. d. by Knight of Towneley (24281), gr. g. d. by Count Persigny (17633), — by Brizlee (37913).

1878, Jan. 29, roan,	C.C.	Primrose 2nd	Prince Rosario, 40535	Mr. Smith
1879, March 9, roan,	C.C.	Primrose 3rd	do.	do.

PRINCESS 6TH, white, calved March 16, 1877. Bred by Mr. W. Smith; got by Duke of Florence (33638), dam (Princess 3rd) by Welsh Prince (30291), &c. See Vol. xxiv. p. 654.

1879, Feb. 26, white,	C.C.	Princess 7th	Fitz-Rose, 41555	Mr. Smith

PRINCESS SALLY 2ND, roan, calved January 6, 1874, Vol. xxiv. p. 654. Bred by Mr. T. Simson, Blainslie; got by Windsor's Bridegroom (30325), dam (Princess Sally) by Allan (25507), &c.

1878, Feb. 25,	red, B.C.	Antony	Bondsman, 36252	Mr. Smith
	white, C.C.	Princess Sally 7th		
1879, March 7, roan,	B.C.	Buccleuch	Prince Rosario, 40535	do.

Antony, sold to Mr. Catton, Aberdeen.

PRINCESS SALLY 3RD, roan, calved March 7, 1875, Vol. xxiv. p. 654. Bred by Mr. W. Smith; got by Borrower (33187), dam (Princess Sally) by Allan (25507), &c.

1878, March 22, roan,	C.C.	Princess Sally 8th	Bondsman, 36252	Mr. Smith
1879, Feb. 26, white,	B.C.	Bambrough	Prince Rosario, 40353	do.

PRINCESS SALLY 4TH, roan, calved February 15, 1876. Bred by Mr. W. Smith; got by Borrower (33187), dam (Princess Sally) by Allan (25507), &c. See Vol. xxiv. p. 654.

1878, Jan. 22, roan,	B.C.	Ajax	Prince Rosario, 40535	Mr. Smith
1879, Mar. 7, r. & w.,	B.C.	Baron Crewe	do.	do.

Ajax, sold to Mr. Brown.

PRINCESS SALLY 5TH, roan, calved March 7, 1876. Bred by Mr. W. Smith; got by Earl of Studley (31085), dam (Princess Sally 2nd) by Windsor's Bridegroom (30325), &c. See Vol. xxiv. p. 654.

1878, Jan. 18, white,	C.C.	Princess Sally 6th	Prince Rosario, 40535	Mr. Smith
1879, Mar. 26, white,	B.C.	Burton	do.	do.

ROSEBUD, roan, calved November 10, 1873, Vol. xxiv. p. 654. Bred by Mr. N. Henderson, Lowick; got by Maximilian (37081), dam by Northern Chief (26982), &c.

Produce in		Names, &c.	By what Bull.	By whom bred.
1878, March 21, red,	C.C.	Rosebud 4th	Bondsman, 36252	Mr. Smith
1879, March 3, roan,	B.C.	Bossue	Prince Rosario, 40535	do.

ROSEBUD 2ND, roan, calved April 15, 1876. Bred by Mr. N. Henderson, Lowick; got by Duke of Florence (33638), dam (Rosebud) by Maximilian (37081), &c. See Vol. xxiv. p. 654.

1878, Jan. 29, roan,	B.C.	Apollo	Prince Rosario, 40535	Mr. Smith
1879, Feb. 27, roan,	C.C.	Rosebud 6th	do.	do.

ROSEBUD 3RD, roan, calved April 4, 1877. Bred by Mr. W. Smith; got by Bondsman (36252) dam (Rosebud) by Maximilian (37081), &c. See Vol. xxiv. p. 654.

1879, Feb. 13, white,	C.C.	Rosebud 5th	Prince Rosario, 40535	Mr. Smith

SALLY 3RD, roan, calved March 16, 1874, Vol. xxiv. p. 655. Bred by Mr. W. Smith; got by Windsor's Bridegroom (30325), dam (Sally 2nd) by Duke of Edinburgh (25956), &c.

1878, Mar. 8, white,	B.C.	Achilles	Bondsman, 36252	Mr. Smith
1879, Mar. 5, r. & w.,	B.C.	Bridegroom	Prince Rosario, 40535	do.

Achilles, sold to Mr. Hogg, Kyloe, Beal.

SALLY 4TH, roan, calved March 9, 1876. Bred by Mr. W. Smith; got by Earl of Studley (31085), dam (Sally 3rd) by Windsor's Bridegroom (30325), &c. See Vol. xxiv. p. 655.

1878, Jan. 8, white,	C.C.	Sally 6th	Prince Rosario, 40535	Mr. Smith
1879, Feb. 18, roan,	C.C.	Sally 7th	do.	do.

SALLY 5TH, roan, calved March 14, 1877. Bred by Mr. W. Smith; got by Duke of Florence (33638), dam (Sally 3rd) by Windsor's Bridegroom (30325), &c. See Vol. xxiv. p. 655.

1879, Mar. 11, white,	B.C.	Bradford	Fitz-Rose, 41555	Mr. Smith

SMITH, William,
The Grange, Goole, Yorkshire.

MARTHA GWYNNE, roan, calved November 11, 1876. Bred by Mr. W. Smith; got by Royal Charlie (35348), dam (Maude Gwynne) by Lieutenant-General (31600), &c. See Vol. xxiii. p. 649.

1879, Nov. 12, white,	C.C.	White Gwynne	Wave Prince, 40897	Mr. Smith

NANCY, white, calved March 19, 1877. Bred by Mr. W. Smith; got by Plutarch (35049), dam (Nickle) by Royal Benedict (27348), &c. See "Benedictus," Vol. xxiii. p. 28.

1879, Nov. 18, white,	C.C.	White Rose	Wave Prince, 40897	Mr. Smith

ROCK ROSE 67TH, red and white, calved August 2, 1875. Bred by Messrs. R. Fisher and Son, Leconfield; got by Rosary Monk (35316), dam (Rock Rose 56th) by Roan Prince (32310), g. d. (Rock Rose 49th) by Lord Greta (20174), &c. See Vol. xxi. p. 709.

1878, Jan. 14, roan,	B.C.	Ben	Benedictus, 37851	Mr. Smith
1879, May 2, roan,	C.C.	Wild Rose	King Brian, 34308	do.

SMITH, W.,
The Grange, Tadcaster.

SILVER SPRING 12TH, roan, calved July 28, 1871, Vols. xxi., xxii., and xxv.
pp. 933, 568, 669. Bred by Mr. J. Smith, Tadcaster, the property of Mr. W.
Smith; got by Duke of Clarence (19611), dam (Silver Spring 2nd) by Archduke 2nd (15588), &c.

Produce in		Names, &c.	By what Bull.	By whom bred.
1879, July 25, red,	C.C.	Silver Spring 19th	D. of Flanders 6th, 38144	Mr. J. Smith

SMITHSON, George,
Allen's Cross, Northfield, Birmingham.

QUEEN OF DIAMONDS, roan, calved in March, 1873.   Bred by Major
Hatherell, Radford House; got by Oxford Lad (29495), dam (British Queen)
by Knightley (22051), &c.   See Vol. xx. p. 424.

1878, July, white,	B.C.	(Steer)	Non-pedigree Bull	Mr. Lees
1879, June 16, roan,	B.C.	Charming Lad	Virginus, 40873	Mr. Smithson

SMYTH, Sir J. H. Greville, Bart.,
Ashton Court, Bristol.

BRACELET 22ND, roan, calved January 1, 1877.   Bred by Mr. H. A. Brassey,
Preston Hall; got by Grand Duke 24th (34064), dam (Bracelet 14th) by
Grand Duke 15th (21852), &c.   See Vol. xxiv. p. 343.

1879, Aug. 5, roan,	C.C.	Ashton Bracelet	D. of Oxford 44th, 39774	Sir J. H. G. Smyth

LADY ASHTON, white, calved December 15, 1874, Vol. xxiv. p. 655.   Bred
by Sir J. H. G. Smyth, Bart.; got by Bates Duke 4th (30509), dam (Lady
Penrhyn) by Oxford Duke (27019), &c.

1878, Oct. 26, roan,	C.C.	Lady Ashton 5th	Pearl Diver, 37182	Sir J. H. G. Smyth

LADY ASHTON 2ND, roan, calved January 12, 1876.   Bred by Sir J. H. G.
Smyth, Bart.; got by Duke of Leicester 3rd (33804), dam (Lady Penrhyn)
by Oxford Duke (27019), &c.   See Vol. xxiii. p. 649.

1878, Sept. 13, red,	C.C.	Lady Ashton 4th	D. of Connaught, 33604	Sir J. H. G. Smyth
1879, Sept. 28, white,	B.C.	Lord Somerset 4th	Protector, 32221	do.

LADY PENRHYN, white, calved December 31, 1871, Vols. xxi., xxiii., and
xxiv. pp. 934, 649, 655.   Bred by Mr. R. Fookes, Milton Abbas; got by Oxford Duke (27019), dam (Nosegay) by Earl of Darlington (21636), &c.

1879, Mar. 25, white,	B.C.	Lord Somerset 3rd	D. of Connaught, 33604	Sir J. H. G. Smyth

Lord Somerset 3rd, sold to Mr. W. Gibbs, Redcliffe Hill. Bristol.

MUSICAL 2ND, white, calved February 28, 1872, Vols. xxi., xxii., and xxv.
pp. 721, 419, 424.   Bred by Mr. R. B. Hetherington, Park Head; got by
Baron Oxford 4th (25580), dam (Musical) by Grenadier (21876), &c.

1879, Mar. 16, roan,	B.C.	Musical Boy	5th D. of Wetherby, 31033	Sir J. H. G. Smyth

WINSOME 20TH, roan, calved July 20, 1876.   Bred by the Duke of Devonshire, Holker Hall; got by Fifth Duke of Wetherby (31033), dam (Lady
Bright Eyes 3rd) by Seventh Duke of York (17754), &c.   See Vol. xxiii.
p. 420.

1878, Dec. 26, roan,	C.C.	Ashton Winsome	5th D. of Wetherby, 31033	Sir J. H. G. Smyth

SMYTHE, Sir C. F., Bart.,
A'cton Burnell Park, Shrewsbury.

CHILTON QUEEN, roan, calved April 6, 1873, Vols. xxiii., xxiv., an 1 xxv.
pp. 649, 655, 670. Bred by Mr. J. How, Broughton ; got by Prince Boabdil
(27120), dam (May Dew) by Claxton (21433), &c.

Produce in	Names, &c.	By what Bull.	By whom bred.
1879, Nov. 29, { white,B.C. red, C.C.	Cassius }	Paul Potter, 38854	Sir C. F. Smythe

DAME MARGARET, roan, calved February 2, 1872, Vols. xxi., xxiv., and xxv.
pp. 754, 486, 670. Bred by Mr. T. Rose, Melton Magna ; got by England's
Glory (23889), dam (Ruth 15th) by Sir Roger (16991), &c.

1879, May 18, white, B.C. Mars		Moonstone, 37107	Sir C. F. Smythe

DAMSEL, roan, calved May 12, 1862, Vols. xvii., xviii., xx., xxiii., xxiv., and
xxv. pp. 447, 450, 473, 649, 656, 670. Bred by Her Majesty the Queen ; got
by Lord Hopewell (18239), dam (Datura) by Fitz-Clarence (14552), &c.

1879, Mar. 1, r. & w., B.C. Nestor		Moonstone, 37107	Sir C. F. Smythe

DOUBTFUL, red, calved November 25, 1870, Vols. xxi. and xxv. pp. 669, 670.
Bred by Sir A. de Rothschild, Bart., Aston Clinton ; got by Duke of Leinster
(28413), dam (Dewdrop) by Captain Cherry (21363), &c.

1879, Feb. 24, roan, C.C. Dido		Moonstone, 37107	Sir C. F. Smythe

HEBE, red, calved February 13, 1871. Bred by Sir C. F. Smythe, Bart. ; got
by King of Trumps (31512), dam (Heroine) by British Hero (30604), &c.
See Vol. xxiv. p. 656.

1879, May 17, roan, B.C. Hector		Moonstone, 37107	Sir C. F. Smythe

LADY COPLEY, white, calved December 6, 1872, Vols. xxiii. and xxv. pp. 650,
671. Bred by Mr. J. J. Wilkinson, Brookfield ; got by Knight of Knowlmere
(22055), dam (Booth Duchess) by Viscount Windsor (19082), &c.

Produce in	Names, &c.	By what Bull.	By whom bred.
1879, Apr. 8, { white, C.C. white, B.C.	(dead) Leander }	Moonstone, 37107	Sir C. F. Smythe

LADY GOLIGHTLY, roan, calved September 7, 1876. Bred by Sir C. F.
Smythe, Bart. ; got by King of Trumps (31512), dam (Lady Copley) by
Knight of Knowlmere (22055), &c. See Vol. xxiii. p. 650.

1879, Jan. 21, roan, B.C. Laocoon		Moonstone, 37107	Sir C. F. Smythe

NELL, roan, calved January 16, 1877. Bred by Sir C. F. Smythe, Bart. ; got
by King of Trumps (31512), dam (Damsel) by Lord Hopewell (18239), &c.
See Vol. xxiv. p. 656. .

1879, June 2, white, C.C. Niobe		Moonstone, 37107	Sir C. F. Smythe

OXFORD PAGEANT, red, calved June 30, 1867, Vols. xxii., xxiii., and xxiv.
pp. 569, 650, 656. Bred by Mr. E. H. Cheney, Gaddesby Hall ; got by Beau
of Oxford (21254), dam (Pageant) by Duke of Clarence (19611), &c.

1879, May 19, roan, C.C. Lady Pageant		Moonstone, 37107	Sir C. F. Smythe

PEARL BEAUTY, roan, calved March 27, 1876. Bred by Sir C. F. Smythe,
Bart. ; got by King of Trumps (31512), dam (Pearlfeather 2nd) by Zealot
(25480), &c. See Vol. xxiii. p. 650.

1879, Mar. 17, white, C.C. Psyche		Moonstone, 37107	Sir C. F. Smythe

PRINCESS 9TH, roan, calved May 20, 1875, Vol. xxv. p. 671. Bred by Mr.
R. Hewer, Sandhill; got by Plebeian (29556), dam (Princess 3rd) by Phœnix
(24743), &c.

Produce in	Names, &c.	By what Bull.	By whom bred.
1879, March 8, roan, C.C.	Prudence	Moonstone, 37107	Sir C. F. Smythe

PRINCESS SOCKBURN, roan, calved September 28, 1874, Vol. xxv. p. 671.
Bred by Messrs. Atkinson, Peepy; got by Royal Killerby (32396), dam
(O. B.'s Justicia) by Baron Oxford (23375), &c.

Produce in	Names, &c.	By what Bull.	By whom bred.
1879, March 11, red, B.C.	Solon	D. of Oxford 31st, 33713	Sir C. F. Smythe

SNARRY, James,
Marramatt Farm, Sledmere, York.

DAISY, roan, calved May 15, 1873. Bred by Mr. J. Snarry; got by Rival
(29781), dam (Princess 3rd) by Duke of Towneley (21615), &c. See Vol. xxii.
p. 569.

Produce in	Names, &c.	By what Bull.	By whom bred.
1879, Aug. 22, white, C.C.	Primrose	Oxford Ryedale 2nd, 34991	Mr. Snarry

EASTERN PRINCESS, red and white, calved August 2, 1876. Bred by Mr.
J. Snarry; got by Ignoramus (28887), dam (Princess 3rd) by Duke of
Towneley (21615), &c. See "Prince of Ryedale," p. 200.

Produce in	Names, &c.	By what Bull.	By whom bred.
1879, Aug. 20, roan, B.C.	Prince of Ryedale	Lord of Ryedale, 40214	Mr. Snarry

PRINCESS 3RD, red and white, calved May 14, 1869, Vols. xxii. and xxiv.
pp. 569, 656. Bred by Mr. J. Snarry; got by Duke of Towneley (21615),
dam (Red Princess) by Gipsy Prince (17965), &c.

Produce in	Names, &c.	By what Bull.	By whom bred.
1878, March 8, roan, C.C.	Princess Mathilda	Oxford Ryedale 2nd, 34991	Mr. Snarry
1879, April 13, roan, B.C.	Count Towneley 2nd	do.	do.

WILD ROSE, white, calved November 21, 1873, Vol. xxiv. p. 656. Bred by
Sir T. Sykes, Bart., Sledmere; got by Ignoramus (28887), dam (Early Rose)
by Duke of Towneley (21615), &c.

Produce in	Names, &c.	By what Bull.	By whom bred.
1879, April 1, roan, C.C.	Moss Rose	Oxford Ryedale 2nd, 34991	Mr. Snarry

SNEYD, Rev. Walter,
Keele Hall, Newcastle, Staffs.

MAID OF THE MIST, roan, calved November 10, 1869, Vol. xx. p. 639. Bred
by Mr. R. Sneyd, Keele Hall; got by Heir of York (21915), dam (Josephine)
by Grand Duke of Speke 2nd (19886), &c.

Produce in	Names, &c.	By what Bull.	By whom bred.
1879, June 2, white, C.C.	Duchess of Keele 28th	Prince of Hillhurst, 35145	Rev. W. Sneyd

STAMPER, Thomas,
Highfield House, Oswaldkirk, York.

GIPSY GIRL, roan, calved April 4, 1877. Bred by Mr. T. Stamper; got by
Duke of Rubies (33724), dam (Countess) by Wathstone's Hero (25417), &c.
See Vol. xix. p. 458.

Produce in	Names, &c.	By what Bull.	By whom bred.
1879, Sept. 20, roan, B.C.	Baron Windsor	D. of Nawton 3rd, 39764	Mr. Stamper

MOSS ROSE 2ND, red, calved March 22, 1876. Bred by Mr. T. Feetenby,
Stonegrave; got by Duke of Cleveland (33599), dam (Rosa) by Grand Duke
Vladimir (28769), &c. See Vol. xxiv. p. 659.

Produce in	Names, &c.	By what Bull.	By whom bred.
1879, Feb. 6, r. & w., C.C.	Moss Rose 3rd	D. of Nawton 3rd, 39764	Mr. Stamper

STAR 2ND, roan, calved March 13, 1877. Bred by Mr. T. Stamper; got by
Snowball (37492), dam (Sunbeam) by Secret Guide (35492), &c. See Vol. xxiii.
p. 652.

Produce in	Names, &c.	By what Bull.	By whom bred.
1879, Sept. 9, r. & w., B.C.	Grimaldi	D. of Nawton 3rd, 39764	Mr. Stamper

SURPRISE, roan, calved December 14, 1875. Bred by Mr. T. Stamper; got
by Saxton (35478), dam (Special Stranger) by Duke of Waterloo (21616), &c.
See Vol. xxii. p. 571.

1879, Jan. 7, roan,	C.C.	Lady of Hazlecote	D. of Hazlecote 23rd, 30970 Mr. Stamper

WINTER ROSE, roan, calved April 8, 1876. Bred by Sir J. Rolt, Ozleworth
Park ; got by Duke of Hazlecote 23rd (30970), dam (White Frost) by Royal
Butterfly 17th (22774), &c. See Vol. xxiv. p. 350.

1879, May 30, white, B.C.	Photograph	D. of Nawton 3rd, 39764	Mr. Stamper

STANIFORTH, Rev. T.,
Storrs, Windermere.

BRITISH QUEEN, roan, calved December 1, 1869, Vols. xxiv. and xxv.
pp. 659, 674. Bred by the Hon. M. H. Cochrane, Hillhurst, Canada ; got
by Sovereign (27538), dam (British Maid) by British Prince (14197), &c.

1879, Dec. 11, white, B.C.	British Viceroy	Royal Stuart, 40646	Rev. T. Staniforth

FOREIGN EMPRESS, roan, calved January 14, 1870, Vols. xxi. and xxiv.
pp. 962, 351. Bred by Mr. W. Torr, Aylesby Manor ; got by Fitz-Royal
(26167), dam (Foreign Princess) by Prince of Warlaby (15107), &c.

1879, Nov. 28, roan, B.C.	Crowned Head	Lord Prinknash, 34855	Rev. T. Staniforth

PROBATION, roan, calved November 8, 1870, Vols. xxi., xxii., xxiii., and xxv.
pp. 937, 573, 653, 675. Bred by the Rev. T. Staniforth ; got by High Sheriff
(26392), dam (Noviciate) by Juryman (20043), &c.

1879, Nov. 15, roan, B.C.	Postulant	Royal Stuart, 40646	Rev. T. Staniforth

STRAWBERRY LEAF, roan, calved February 17, 1874, Vol. xxiv. p. 370.
Bred by Mr. H. Chandos-Pole-Gell, Hopton Hall ; got by Prince Christian
(22581), dam (Young Strawberry) by Knight of Windsor (16349), &c.

1879, April 23, roan, C.C.	Strawberry Syrup	Royal Saxon, 39057	Rev. T. Staniforth

STAPYLTON, Major,
Myton Hall, Helperby, Yorkshire.

SNOWDROP, white, calved November 30, 1875, Vol. xxv. p. 676. Bred by
Major Stapylton ; got by Newburgh 5th (34904), dam (Dewdrop) by Earl of
Walton (17787), &c.

1879, June 2. white, C.C.	Dewdrop 2nd	Sir Robert, 37477	Major Stapylton

SWEETBRIAR, white, calved January 6, 1877. Bred by Major Stapylton ;
got by Newburgh 5th (34904), dam (Dewdrop) by Earl of Walton (17787),
&c. See Vol. xxiv. p. 660.

1879, May 30, white, B.C.	Snowball	Sir Robert, 37477	Major Stapylton

STARKIE, Mrs. C.,
Ashton Hall, Lancaster.

KNIGHTLEY GRAND DUCHESS, roan, calved July 7, 1875, Vol. xxv.
p. 676. Bred by Mr. W. W. Slye, Beaumont Grange; got by Grand Duke
of Thorndale 2nd (31298), dam (Lady Orphan Knightley) by Grand Duke of
Kent 2nd (28759), &c.

Produce in	Names, &c.	By what Bull.	By whom bred.
1879, March 12, red, C.C.	(dead)	L.AshtonW.Eyes2d,36912	Mrs. C. Starkie

LADY ASHTON WILD EYES, roan, calved October 5, 1872, Vols. xxii., xxiv.,
and xxv. pp. 573, 661, 676. Bred by Mrs. C. Starkie; got by Baron Oxford
4th (25580), dam (Winsome 10th) by Eighteenth Duke of Oxford (25995),
&c.

Produce in	Names, &c.	By what Bull.	By whom bred.
1879, July 26, roan, B.C.	Earl Ashton 2nd	Duke of Ormskirk, 36526	Mrs. C. Starkie

STEPHENSON, J. H.,
Sancton Grange, Brough, Yorkshire.

LADY LE GRAND, roan, calved February 28, 1877. Bred by Mr. J. H.
Stephenson; got by Oxford le Grand (29496), dam (Vanity) by Thorndale
Lad (23066), &c. See Vol. xxv. p. 676.

Produce in	Names, &c.	By what Bull.	By whom bred.
1879, June 24, r. & w., C.C.	Lady le Grand 2nd	Ferdinando, 39872	Mr. Stephenson

STEVENSON, Peter,
Rainton, Thirsk, Yorkshire.

CHRISTMAS GWYNNE, roan, calved December 25, 1872, Vols. xxi., xxiv., and
xxv. pp. 888, 662, 677. Bred by Mr. H. Caddy, Rougholm; got by Waterloo
Cherry (27763), dam (Dainty Gwynne) by Sir Windsor (22927), &c.

Produce in	Names, &c.	By what Bull.	By whom bred.
1879, March 25, red, C.C.	Lady-Day Gwynne	D. of Edlingham 3rd,36489	Mr. Stevenson

CHRISTMAS WAIT, red and white, calved December 25, 1876. Bred by Mr.
P. Stevenson; got by Diamond Duke (33522), dam (Copperplate) by Lord
Lally 3rd (24408), &c. See Vol. xxiii. p. 655.

Produce in	Names, &c.	By what Bull.	By whom bred.
1879, July 18, red, C.C.	Choice Fairing	D. of Edlingham 3rd,36489	Mr. Stevenson

COCOA NUT, red and white, calved April 25, 1874, Vol. xxiv. p. 662. Bred
by Mr. P. Stevenson; got by Lackland (34413), dam (Coral) by Lord Lally
3rd (24408), &c.

Produce in	Names, &c.	By what Bull.	By whom bred.
1878, May 9, r. & w., B.C.	Conyngham	D. of Edlingham 3rd,36489	Mr. Stevenson
1879, June 15, r. & w., B.C.	Cunningham	do.	do.

COPPER DUCHESS, red, calved August 4, 1875. Bred by Mr. P. Stevenson;
got by Oxford's Duke 2nd (34994), dam (Copperplate) by Lord Lally 3rd
(24408), &c. See Vol. xxii. p. 574.

Produce in	Names, &c.	By what Bull.	By whom bred.
1878, April 17, red, C.C.	(dead)	D. of Edlingham 3rd,36489	Mr. Stevenson
1879, March 10, red, B.C.	Chatham	do.	do.

COPPERPLATE, red, calved September 30, 1869, Vols. xxi., xxii., and xxiii.
pp. 939, 574, 655. Bred by Mr. P. Stevenson; got by Lord Lally 3rd (24408),
dam (Cordelia) by Red Duke (18676), &c.

Produce in	Names, &c.	By what Bull.	By whom bred.
1878, March 23, red, C.C.	Copper Countess	D. of Edlingham 3rd,36489	Mr. Stevenson
1879, Feb. 28, red, B.C.	Coatham	do.	do.

COPPER QUEEN, roan, calved April 26, 1874, Vol. xxv. p. 677. Bred by
Mr. P. Stevenson; got by Prince of Thorndale (32187), dam (Copperplate)
by Lord Lally 3rd (24408), &c.

Produce in		Names, &c.	By what Bull.	By whom bred.
1879, June 11, red,	C.C.	Copper Maiden	D. of Edlingham 3rd, 36489	Mr. Stevenson

FROLIC 9TH, red and white, calved June 22, 1874. Bred by Sir W. H. Salt,
Bart., Maplewell; got by Cambridge Duke 4th (25706), dam (Frolic 4th) by
Duke of Belmont (38123), &c. See Vol. xxiv. p. 442.

Produce in		Names, &c.	By what Bull.	By whom bred.
1878, Nov. 20, r. & w.,	C.C.	(dead)	oth D. of Glo'ster, 36494	Sir W. H. Salt
1879, Nov. 8, r. & w.,	C.C.	Frolic's Duchess	do.	Mr. Stevenson

HASKA, roan, calved October 5, 1876. Bred by Mr. H. Pickersgill, Middleton, Quernhow; got by Hades (34101), dam (New Year's Gift) by Thorntree
(25312), &c. See Vol. xxiii. p. 603.

Produce in		Names, &c.	By what Bull.	By whom bred.
1879, Aug. 3, red,	C.C.	Hava	D. of Edlingham 3rd, 36489	Mr. Stevenson

LADY BELL, roan, calved May 31, 1872, Vols. xxii. and xxiv. pp. 575, 662.
Bred by Mr. P. Stevenson; got by Diamond Eyes (28318), dam (Lackawanna)
by Merryman (20341), &c.

Produce in		Names, &c.	By what Bull.	By whom bred.
1878, July 15, roan,	B.C.	Langham	D. of Edlingham 3rd, 36489	Mr. Stevenson
1879, July 10, red,	C.C.	Lady Edlingham	do.	do.

LAURA BELL, roan, calved May 31, 1870, Vols. xxi. and xxiv. pp. 940, 662.
Bred by Mr. P. Stevenson; got by Lord Lally 3rd (24408), dam (Lackawanna)
by Merryman (20341), &c.

Produce in		Names, &c.	By what Bull.	By whom bred.
1878, April 9, roan,	B.C.	(dead)	D. of Edlingham 3d, 36489	Mr. Stevenson
1879, April 28, red,	C.C.	Lucky Bell	do.	do.

LAZY BELL, roan, calved April 25, 1874, Vol. xxiv. p. 662. Bred by Mr. P.
Stevenson; got by Prince of Thorndale (32187), dam (Lackawanna) by
Merryman (20341), &c.

Produce in		Names, &c.	By what Bull.	By whom bred.
1879, Mar. 27, roan,	C.C.	Larky Bell	D. of Edlingham 3rd, 36489	Mr. Stevenson

LOBISA, red, calved May 26, 1875. Bred by Mr. P. Stevenson; got by
Oxford's Duke 2nd (34994), dam (Loo Duchess) by Duke Royal (21623), &c.
See "Effingham," p. 93.

Produce in		Names, &c.	By what Bull.	By whom bred.
1878, Feb. 19, red,	B.C.	Effingham	D. of Edlingham 3rd, 36489	Mr. Stevenson
1879, June 5, red,	C.C.	Lobisa Bell	do.	do.

Effingham, sold to Mr. G. Brown, Greenfield House, Raskelfe.

LOFTY DUCHESS, red and white, calved March 13, 1875. Bred by Mr. P.
Stevenson; got by Oxford's Duke 2nd (34994), dam (Lottie Bell) by Lustre
(29236), &c. See Vol. xxii. p. 575.

Produce in		Names, &c.	By what Bull.	By whom bred.
1878, June 26, red,	C.C.	(dead)	D. of Edlingham 3rd, 36489	Mr. Stevenson
1879, Nov. 24, r. & w.,	C.C.	Loyal Duchess	do.	do.

LOLA BELL, roan, calved June 6, 1874. Bred by Mr. P. Stevenson; got by
Prince of Thorndale (32187), dam (Loo Duchess) by Duke Royal (21623), &c.
See "Effingham," p. 93.

Produce in		Names, &c.	By what Bull.	By whom bred.
1878, Jan. 9, red,	C.C.	(dead)	D. of Edlingham 3rd, 36489	Mr. Stevenson
1879, Feb. 26, r. & w.,	B.C.	Lingham	do.	do.

LOO LALLY, red and white, calved March 10, 1868, Vols. xx., xxiii., and xxiv.
pp. 628, 656, 663. Bred by Mr. P. Stevenson; got by Lord Lally 3rd (24408),
dam (Loo) by Cromwell (17640), &c.

Produce in		Names, &c.		By what Bull.		By whom bred.
1879, Feb. 8, r. & w.,	C.C.	Loo Edlingham		D.of Edlingham 3rd, 36489		Mr. Stevenson

LOO LASSIE, red, calved October 8, 1874. Bred by Mr. P. Stevenson; got by
Lackland (34413), dam (Loo) by Cromwell (17640), &c. See "Effingham,"
p. 93.

1877, Dec. 14, red,	B.C. (dead)		D.of Edlingham 3rd, 36489	Mr. Stevenson
1879, April 6, red,	C.C.	Loo Swinburne	do.	.do.

LOTTIE BELL, red and white, calved August 11, 1871, Vols. xxi., xxii., and
xxiv. pp. 940, 575, 663. Bred by Mr. P. Stevenson; got by Lustre (29236),
dam (Loo Duchess) by Duke Royal (21622), &c.

1879, Feb. 8, r. & w.,	C.C.	Lofty Bell		D.of Edlingham 3rd, 36489	Mr. Stevenson

LUCY BELL, roan, calved June 30, 1874, Vol. xxiv. p. 663. Bred by Mr. P.
Stevenson; got by Prince of Thorndale (32187), dam (Laura Bell) by Lord
Lally 3rd (24408), &c.

1879, May 8, red,	C.C.	Lucinda		D.of Edlingham 3rd, 36489	Mr. Stevenson

STEWART, Samuel,
Sandhole, Fraserburgh, N.B.

ALICE, red and white, calved March 27, 1876. Bred by Mr. S. Stewart; got
by Disraeli (33529), dam (Ruth) by Alliance (32937), &c. See Vol. xxii.
p. 576.

1878, Jan 3, r. & w.,	B.C.	Baronet	Constellation, 38027	Mr. Stewart
1879, June 29, roan,	B.C.	Sir Garnet	Alphonso Fawsley, 36126	do.

Baronet, sold to Mr. Park, Mains of Kindrought, Strichen.

STOBO, Andrew,
Porterstown, Thornhill, N.B.

LADY ANN, roan, calved June 6, 1875. Bred by Mr. A. Stobo; got by Prince
Arthur Patrick (29600), dam (Lady Redkirk) by Royal Charlie (29851), g. d.
(Lively) by Sir William (22925), &c. See "Closeburn," p. 52.

1878, Jan. 21, white,	C.C.	Snowdrop	Duke of Tyne, 38195	Mr. Stobo
1879, Jan. 6, r. & w.,	C.C.	Annie	Captain Hardy, 39549	do.

ROYAL DUCHESS 5th, red and white, calved April 12, 1874. Bred by Mr.
A. Stobo; got by Prince Arthur Patrick (29600), dam (Royal Duchess 2nd)
by Statesman (27569), &c. See Vol. xxi. p. 941.

1877, Jan. 30, r. & w.,	C.C.	Royal Duchess 9th	Sir Francis, 37459	Mr. Stobo
1878, Jan. 13, r. & w.,	C.C.	Royal Duchess 10th	do.	do.
1879. Jan. 12, r. & w.,	B.C.	Symmetry	Captain Hardy, 39549	do.

ROYAL PET, red and white, calved May 2, 1875. Bred by Mr. A. Stobo; got
by Forest Prince (33950), dam (Royal Lady) by Royal Charlie (29851), &c.
See Vol. xxv. p. 679.

1878, Feb. 9, r. & w.,	B.C.	Kirkbean	Duke of Tyne, 38195	Mr. Stobo
1879, Feb. 25, r. & w.,	B.C.	Tosh	Captain Hardy, 39549	do.

Kirkbean, sold to Mr. J. McGill, Torrora, Dumfries.

STONE, J. J., Executors of,
Stoneleigh Park, Ewell, Surrey.

ASTARTE 3RD, red, calved May 12, 1873, Vol. xxiii. p. 657. Bred by Mr. H. J. Sheldon, Brailes House; got by Duke of Cerisia (30937), dam (Astarte) by Duke of Brailes (23724), &c.

Produce in		Names, &c.	By what Bull.	By whom bred.
1878, Jan. 12, roan,	C.C.	Astarte 4th	R. Cambridge 2nd, 25010	Mr. Stone
1879, Jan. 21, r. & w.,	C.C.	Astarte 5th	Chy. Gd. Duke 5th, 30712	do.

BLANCHE DUCHESS 3RD, roan, calved May 26, 1875. Bred by Mr. J. J. Stone; got by Oxford Duke (32022), dam (Blanche Duchess 2nd) by Thirteenth Duke of Oxford (21604), &c. See Vol. xxii. p. 577.

1878, July 25, roan,	C.C.	Blanche Duchess 4th	R. Cambridge 2nd, 25010	Mr. Stone
1879, July 1, roan,	B.C.	(dead)	Chy. Gd. Duke 5th, 30712	do.

DARLINGTON DUCHESS, red and white, calved May 4, 1874, Vol. xxiii. p. 657. Bred by Mr. J. J. Stone; got by Oxford Duke (32022), dam (Prioress) by Grand Duke 7th (19877), &c.

1877, Sept. 2, r. & w.,	C.C.	Darlington D'ss 4th	R. Cambridge 2nd, 25010	Mr. Stone
1878, July 21, r. & w.,	C.C.	Darlington D'ss 5th	do.	do.

DARLINGTON DUCHESS 2ND, red, calved March 20, 1875. Bred by Mr. J. J. Stone; got by Oxford Duke (32022), dam (Prioress) by Grand Duke 7th (19877), &c. See Vol. xxiii. p. 658.

1878, July 10, r. & w.,	B.C.	Darlington Duke 3rd	R. Cambridge 2nd, 25010	Mr. Stone
1879, July 10, red,	C.C.	Darlington D'ss 7th	Chy. Gd. Duke 5th, 30712	do.

GANZA 3RD, roan, calved April 13, 1872, Vol. xxiv. p. 665. Bred by Mr. J. J. Stone; got by Duke of Florence (28388), dam (Gertrude) by Littlebury (24341), &c.

1878, June 15, roan,	C.C.	Ganza 6th	R. Cambridge 2nd, 25010	Mr. Stone
1879, May 26, r. & w.,	B.C.	Gamester	do.	do.

GANZA 4TH, red, calved January 31, 1874. Bred by Mr. J. J. Stone; got by Oxford Duke (32022), dam (Ganymede) by Sir Rainald (30001), &c. See Vol. xxi. p. 941.

1877, Sept. 20, red,	B.C.	(Steer)	R. Cambridge 2nd, 25010	Mr. Stone
1879, Feb. 13, r. & w.,	C.C.	Ganza 7th	Chy. Gd. Duke 5th, 30712	do.

LAPWING 3RD, red and white, calved June 6, 1872, Vols. xxiii. and xxv. pp. 658, 680. Bred by Mr. J. J. Stone; got by Third Duke of Geneva (21592), dam (Lapwing 2nd) by Grand Duke 13th (21850), &c.

1879, May 29, r. & w.,	C.C.	Lapwing 8th	Chy. Gd. Duke 5th, 30712	Mr. Stone

LAPWING 4TH, red and white, calved December 24, 1874. Bred by Mr. J. J. Stone; got by Oxford Duke (32022), dam (Lapwing 2nd) by Grand Duke 13th (21850), &c. See Vol. xxi. p. 942.

1879, May 14, red,	C.C.	Lapwing 7th	R. Cambridge 2nd, 25010	Mr. Stone

LAPWING 5TH, red and white, calved February 6, 1876. Bred by Mr. J. J. Stone; got by Oxford Duke (32022), dam (Lapwing 3rd) by Third Duke of Geneva (21592), &c.

1879, Feb. 15, r. & w.,	B.C.	Lapidist	Stoneleigh Duke 4th, 39172	Mr. Stone

RUBY 3RD, red, calved in December, 1870, Vols. xxii. and xxv. pp. 577, 680. Bred by Sir C. F. Smythe, Bart., Acton Burnell Park; got by Lord Warwick (26753), dam (Ruby) by Lucknow (16471), &c.

1879, July 24, red,	C.C.	Ruby 6th	R. Cambridge 2nd, 25010	Mr. Stone

STONELEIGH DUCHESS 10TH, roan, calved April 23, 1874, Vol. xxiv.
p. 665. Bred by Mr. J. J. Stone; got by Stoneleigh Duke (32613), dam
(Stoneleigh Duchess 6th) by Second Duke of Wetherby (21618), &c.

Produce in		Names, &c.		By what Bull.		By whom bred.
1878, Aug. 1, roan,	C.C.	Stoneleigh D'ss 14th	R. Cambridge 2nd, 25010			Mr. Stone
1879, June 29, r. & w.,	B.C.	Stoneleigh Duke 7th	Chy. Gd. Duke 5th, 30712			do.

STONELEIGH DUCHESS 11TH, red and white, calved July 25, 1875. Bred
by Mr. J. J. Stone; got by Oxford Duke (32022), dam (Stoneleigh Duchess
7th) by Duke of Monmouth (28425), &c. See Vol. xxiii. p. 658.

1878, Sept. 7, r. & w.,	B.C.	Stoneleigh Duke 6th	R. Cambridge 2nd, 25010			Mr. Stone
1879, Sept. 16, r. & w.,	C.C.	Stoneleigh D'ss 16th	Chy. Gd. Duke 5th, 30712			Exors. Mr. Stone

VIOLET 2ND, roan, calved November 25, 1873, Vol. xxv. p. 680. Bred by Mr.
J. J. Stone; got by Stoneleigh Duke (32613), dam (Duchess of York 5th) by
Royal Butterfly 17th (22774), &c.

1879, April 6, roan,	C.C.	Violet 4th		Chy. Gd. Duke 5th, 30712		Mr. Stone

WELCOME 5TH, red, calved July 20, 1873, Vol. xxiv. p. 666. Bred by Mr. J.
J. Stone; got by Third Duke of Geneva (21592), dam (Welcome 2nd) by
Chanter (19423), &c.

1878, Aug. 3, roan,	C.C.	Welcome 7th	R. Cambridge 2nd, 25010			Mr. Stone
1879, July 23, red,	B.C.	Lord Welcome 5th	do.			do.

WILD CHERRY DUCHESS, roan, calved April 25, 1874, Vol. xxiv. p. 666.
Bred by Mr. J. J. Stone; got by Oxford Duke (32022), dam (Wild Cherry)
by Cherry Grand Duke 2nd (25758), &c.

1878, June 3, white,	C.C.	Wild Cherry D'ss 6th	R. Cambridge 2nd, 25010			Mr. Stone
1879, May 19, red,	C.C.	Wild Cherry D'ss 7th	Chy. Gd. Duke 5th, 30712			do.

STONHAM, Henry,
Thurnham, Maidstone.

MISS NIEL 2ND, red, calved January 29, 1876, Vol. xxv. p. 681. Bred by
Mr. H. Stonham; got by Marshal Niel (41991), dam (Miss Wrotham 2nd)
by Duke of Wrotham (23803), &c.

1879, Sept. 16, r. & w.,	C.C.	Miss Tenterden 4th	Tenterden, 42479			Mr. Stonham

STOTT, Miss S. G.,
Pool House, Otley, Yorkshire.

LADY LILY, red and white, calved September 15, 1876. Bred by Mr. A.
Fawkes, Farnley Hall; got by Baron Winsome 6th (33111), dam (Lioness
2nd) by Bismarck Baron (30545), &c. See Vol. xxiii. p. 435.

1879, Oct. 26, r. & w.,	B.C.	Lionel Rothschild	Rothschild, 40612			Miss Stott

STRATTON, F.,
Merdon, Hursley, Winchester.

PREMIUM 2ND, red, calved February 22, 1877. Bred by Mr. F. Stratton; got
by Merry Monarch (37091), dam (Candia) by Knight of the Forest (31556),
&c. See Vol. xxiv. p. 666.

1879, Sept. 17, red,	C.C.	Premium 3rd	Abelard, 39346			Mr. Stratton

WESTLAND ROSE, roan, calved July 8, 1877. Bred by Mr. F. Stratton; got by Royal James (35387), dam (Last Rose of Westland) by King James (28971), &c. See "Duke of Westland," p. 86.

Produce in		Names, &c.		By what Bull.		By whom bred.
1879, Dec. 23, roan,	C.C.	Merdon Rose		Abelard, 39346		Mr. Stratton

STRATTON, George,
Husbands Bosworth, Rugby.

OLIVIA, red, calved March 25, 1876. Bred by Mr. G. Stratton; got by Earl of Waterloo 3rd (39833), dam (Oxford's Butterfly) by Seventeenth Duke of Oxford (25994), g. d. (Fair Butterfly Princess) by Fourteenth Duke of Oxford (21605), &c. See Vol. xx. p. 508.

Produce in		Names, &c.		By what Bull.		By whom bred.
1879, Nov. 15, red,	B.C.	Oliver Cromwell		Industry, 43372		Mr. Stratton

OXFORD'S BUTTERFLY, roan, calved May 23, 1872. Bred by Mr. E. H. Cheney, Gaddesby Hall; got by Seventeenth Duke of Oxford (25994), dam (Fair Butterfly Princess) by Fourteenth Duke of Oxford (21605), &c.

Produce in		Names, &c.		By what Bull.		By whom bred.
1876, Mar. 25, red,	C.C.	Olivia		Earl of Waterloo 3rd, 39833		Mr. Stratton
1877, Nov. 25, red,	C.C.	Orchid		do.		do.
1878, Dec. 9, roan,	C.C.	Orange Blossom		do.		do.
1879, Nov. 13, red,	B.C.	Prince of Orange		Industry, 43372		do.

STRATTON, Joseph,
Alton Priors, Marlborough.

GENUINE, roan, calved August 13, 1877. Bred by Mr. J. Stratton; got by Cardinal (36318), dam (Glitter) by Endymion (31109), &c. See Vol. xxiii. p. 659.

Produce in		Names, &c.		By what Bull.		By whom bred.
1879, Dec. 11, roan,	C.C.	Grace		Proteus, 40552		Mr. Stratton

JUNO, red, calved April 12, 1877. Bred by Mr. J. Stratton; got by Royal James (25387), dam (Jenny Deans) by Moonraker (22383), &c. See Vol. xxiii. p. 659.

Produce in		Names, &c.		By what Bull.		By whom bred.
1879, July 28, roan,	C.C.	Jewel		Hudibras, 38441		Mr. Stratton

STRATTON, R.,
The Duffryn, Newport, Mon.

BADINAGE, red, calved August 13, 1876. Bred by Mr. R. Stratton; got by Charles 1st (33322), dam (Brownie) by Majesty (26789), &c. See "Banter," p. 10.

Produce in		Names, &c.		By what Bull.		By whom bred.
1879, July 30, red,	B.C.	Banter		Pearl Diver, 37182		Mr. Stratton

HEATHER BELL, white, calved July 11, 1876. Bred by Mr. R. Stratton; got by Charles 1st (33322), dam (Heather) by James 3rd (26450), &c. See Vol. xxii. p. 580.

Produce in		Names, &c.		By what Bull.		By whom bred.
1879, Jan. 11, white,	C.C.	Hare Bell		Expectation, 38264		Mr. Stratton

HECUBA 3RD, roan, calved March 30, 1875. Bred by Mr. R. Stratton; got by Charles 1st (33322), dam (Hecuba 2nd) by Miracle (24602), &c. See Vol. xxii. p. 580.

Produce in		Names, &c.		By what Bull.		By whom bred.
1878, Mar. 15, white,	B.C.	(Steer)		Ashfield, 42676		Mr. Stratton
1879, Mar. 23, red,	C.C.	Hecuba 6th		Pearl Diver, 37182		do.

LADY MAUD, red, calved April 25, 1876. Bred by Mr. R. Stratton; got by
Charles 1st (33322), dam (Lady Mary) by Festival (26147), &c. See Vol.
xxii. p. 580.

Produce in	Names, &c.	By what Bull.	By whom bred.
1879, Aug. 11, roan, C.C.	Lady Love	Hampden, 36739	Mr. Stratton

MEDEA, roan, calved August 24, 1870. Bred by Mr. R. Stratton, Burderop;
got by James 1st (24202), dam (Mary) by Eighth Duke of York (23808),
g. d. (Martha) by Roderick (18730), &c. See Vol. xix. p. 619.

Produce in	Names, &c.	By what Bull.	By whom bred.
1879, Jan. 23, red, C.C.	Medusa	Pearl Diver, 37182	Mr. Stratton

MERRY MAIDEN, white, calved May 8, 1876. Bred by Mr. R. Stratton;
got by Charles 1st (33322), dam (Merry Maid) by James 1st (24202), &c.
See Vol. xxii. p. 580.

Produce in	Names, &c.	By what Bull.	By whom bred.
1879, June 6, roan, C.C.	Merry Thought	Hampden, 36739	Mr. Stratton

MERRY MAY, roan, calved May 28, 1875. Bred by Mr. R. Stratton; got by
Charles 1st (33322), dam (Merry Maid) by James 1st (24202), &c.

Produce in	Names, &c.	By what Bull.	By whom bred.
1879, May 3, red, C.C.	Mayfair	Expectation, 38264	Mr. Stratton

PRIMITIVE, roan, calved August 12, 1875. Bred by Mr. R. Stratton; got by
Rob Roy (29806), dam (Prima) by James 1st (24202), &c. See Vol. xxii. p. 581.

Produce in	Names, &c.	By what Bull.	By whom bred.
1879, Aug. 18, white, B.C.	Grandee	Hampden, 36739	Mr. Stratton

ROSALIE, roan, calved July 20, 1874. Bred by Mr. R. Stratton; got by
Protector (32221), dam (Rosa) by Lamp of Lothian (16356), &c. See "Lifeboat," p. 139.

Produce in	Names, &c.	By what Bull.	By whom bred.
1879, Aug. 12, red, B.C.	Grouse	Pearl Diver, 37182	Mr. Stratton

ROSETTE, red, calved July 28, 1875. Bred by Mr. R. Stratton; got by Rob
Roy (29806), dam (Rosa) by Lamp of Lothian (16356), &c.

Produce in	Names, &c.	By what Bull.	By whom bred.
1879, Mar. 1, red, C.C.	Red Rosette	Expectation, 38264	Mr. Stratton

RUBINA, red and white, calved January 7, 1877. Bred by Mr. R. Stratton;
got by Rob Roy (29806), dam (Ruby) by James 1st (24202), &c. See Vol.
xxi. p. 945.

Produce in	Names, &c.	By what Bull.	By whom bred.
1879, Oct. 28, red, C.C.	Rubicund	Expectation, 38264	Mr. Stratton

VICTORESS, white, calved in September, 1872, Vol. xxiii. p. 696. Bred by Mr.
R. Attenborough, Whitley Grove; got by Rose Butterfly (24993), dam
(Victorine) by His Eminence (19961), &c.

Produce in	Names, &c.	By what Bull.	By whom bred.
1879, Dec. 30, roan, C.C.	Victoria	Lord Ringlet, 38655	Mr. Stratton

STURGE, Charles,
Summer House, Bewdley.

HILL TULIP 4TH, roan, calved December 8, 1876. Bred by Mr. C. Sturge;
got by Marshal Windsor (34795), dam (Tulip 3rd) by Saracen (25091), &c.
See " Hill Tulip Lad 5th," p. 120.

Produce in	Names, &c.	By what Bull.	By whom bred.
1878, June 20, white, B.C.	(Steer)	Foster Brother, 36661	Mr. Sturge

HILL TULIP 5TH, roan, calved September 1, 1876. Bred by Mr. C. Sturge;
got by Marshal Windsor (34795), dam (Hill Tulip 2nd) by Coralys (33445),
g. d. (Tulip 3rd) by Saracen (25091), &c.

Produce in	Names, &c.	By what Bull.	By whom bred.
1878, Aug. 20, red, B.C.	(dead)	Foster Brother, 36661	Mr. Sturge

STURGEON and SONS,
South Ockendon Hall, Romford, Essex.

THORNDALE LILY, roan, calved in March, 1873, Vol. xxv. p. 683. Bred by Mr. E. Burnell, Shudy Camps; got by Sunshine (32632), dam (Thorndale Buttercup) by Thorndale Butterfly (25308), &c.

Produce in		Names, &c.	By what Bull.	By whom bred.
1879, June 22, roan,	C.C.	Ly'T'dale But'rfly2d	Lord Metheglin, 38627	Messrs. Sturgeon

SUFFOLK AND BERKSHIRE, Earl of,
Charlton Park, Malmesbury.

AMY 3RD, red and white, calved January 29, 1876. Bred by the Earl of Suffolk and Berkshire; got by Lord Lind 2nd (36969), dam (Amy 2nd) by Duke of Dursley (25953), &c. See Vol. xxi. p. 947.

1879, Feb. 26, r. & w.,	B.C.	(Steer)	Lord Lovat, 40196	Earl of Suffolk

GRISELDA, roan, calved February 27, 1876. Bred by the Earl of Suffolk and Berkshire; got by Lord Lind 2nd (36969), dam (Patience 11th) by Duke of Hazlecote 17th (30965), &c. See Vol. xxiv. p. 668.

1879, Feb. 9, roan,	C.C.	Grizzel	Lord Lovat, 40196	Earl of Suffolk
1879, Dec. 24, roan,	C.C.	Elaine	Protector, 42241	do.

HELIOTROPE, roan, calved July 27, 1876. Bred by the Earl of Suffolk and Berkshire; got by Lord Lind 2nd (36969), dam (Hyacinth Marchioness) by Third Duke of Clarence (23727), &c. See Vol. xxi. p. 947.

1879, Feb. 26, roan,	B.C.	(Steer)	Lord Lovat, 40196	Earl of Suffolk

LADY 8TH, roan, calved May 12, 1876. Bred by the Earl of Suffolk and Berkshire; got by Lord Lind 2nd (36969), dam (Lady 6th) by Duke of Hazlecote 17th (30965), &c. See Vol. xxiii. p. 661.

1879, Mar. 13, roan,	C.C.	Lady Love	Lord Lovat, 40196	Earl of Suffolk

LADY JOHNSTONE, roan, calved January 21, 1876. Bred by the Earl of Suffolk and Berkshire; got by Lord Lind 2nd (36969), dam (Patience 5th) by Duke of Dursley (25953), &c. See Vol. xxi. p. 948.

1879, Feb. 9, white,	B.C.	Snowball	Lord Lovat, 40196	Earl of Suffolk

LADY WESTON 2ND, roan, calved May 25, 1877. Bred by Mr. E. Bowly, Siddington House; got by Beau of Oxford 2nd (33129), dam (Queen of Weston 3rd) by Cherry Fawsley (30711), &c. See Vol. xxiv. p. 341.

1879, June 24, roan,	B.C.	Westender	Duke of Holker 2nd, 39749	Earl of Suffolk

PATIENCE 19TH, red, calved January 18, 1877. Bred by the Earl of Suffolk and Berkshire; got by Lord Lind 2nd (36969), dam (Patience 5th) by Duke of Dursley (25953), &c. See Vol. xxi. p. 948.

1879, Dec. 29, red,	C.C.	Lady Patience	Lord Lovat, 40196	Earl of Suffolk

PATIENCE 20TH, red and white, calved January 23, 1877. Bred by the Earl of Suffolk and Berkshire; got by Lord Lind 2nd (36969), dam (Patience 9th) by Honeysuckle Marquis (21953), &c. See Vol. xxiii. p. 662.

1879, Dec. 1, red,	C.C.	Peony	Red Rufus, 42263	Earl of Suffolk

SUTTON, Sir R., Bart.,
Benham Park, Newbury.

BIRTHDAY, red and white, calved April 10, 1875, Vol. xxv. p. 684. Bred by Captain Young, Ashey Farm; got by Prince of Clarence 3rd (32166), dam (Queen) by Robin Hood (32318), &c.

Produce in		Names, &c.	By what Bull.	By whom bred.
1879, April 1, roan,	C.C.	Daybreak	Mendelssohn, 37088	Sir R. Sutton

FLUTTER, roan, calved April 17, 1874, Vols. xxiv. and xxv. pp. 668, 684. Bred by Mr. R. Bruce, Newton of Struthers; got by Gang Forward (33988), dam (Flavia) by Fashion (21724), &c.

Produce in		Names, &c.	By what Bull.	By whom bred.
1879, Feb. 19, roan,	B.C.	Royal Balmoral	Lord Balmoral, 38576	Sir R. Sutton

MISS HAWTHORN, roan, calved March 7, 1870, Vols. xxiii. and xxiv. pp. 583, 668. Bred by Mr. E. Lythall, Radford Hall; got by Fitz Killerby (26166), dam (Lady Hawthorn) by Lord of the Isles (20206), &c.

Produce in		Names, &c.	By what Bull.	By whom bred.
1879, Jan. 6, roan,	C.C.	Hawthorn Lassie	Alphonso, 32943	Sir R. Sutton
1879, Dec. 11, roan,	C.C.	Hawthorn Maid	Duke Irwin, 39707	do.

RUBY 2ND, roan, calved April 5, 1874, Vol. xxiv. p 668. Bred by Mr. Y. R. Graham, Yardley Stud Farm; got by Prince Gwynne (20547), dam (Ruby) by Second Duke of Cumberland (23735), &c.

Produce in		Names, &c.	By what Bull.	By whom bred.
1879, Jan. 31, roan,	C.C.	Oxford Ruby	Oxford's Butterfly, 34993	Sir R. Sutton
1879, Dec. 6. white,	C.C.	Guelder Rose	Duke Irwin, 39707	do.

SWANN, John,
Bedlington, R. S. O., Northumberland.

SUNFLOWER, roan, calved October 21, 1872, Vols. xxii. and xxv. pp. 584, 685. Bred by Mr. J. Stirling, Bridekirk; got by England's Champion (28554), dam (Sunny Smile) by King Charming (22033), &c.

Produce in		Names, &c.	By what Bull.	By whom bred.
1879, May 21, roan,	C.C.	Sunflower 3rd	C'bridge'sP.ofMona,41179	Mr. Swann

SWINBURNE, Sir John, Bart.,
Capheaton, Newcastle-on-Tyne.

FANTAIL 5TH, red and white, calved June 19, 1868, Vols. xxii. and xxv. pp. 296, 685. Bred by Mr. J. P. Foster, Killhow; got by Royal Cambridge (25009), dam (Fantail 3rd) by Touchstone (20986), &c.

Produce in		Names, &c.	By what Bull.	By whom bred.
1879, Apr. 11, r. & w.,	C.C.	Lady Fantail	D. of Oxford 27th, 33709	Sir J. Swinburne

FANTAIL'S DUCHESS 3RD, red and white, calved November 4, 1875. Bred by Sir J. Swinburne, Bart.; got by Oxford Beau 4th (34964), dam (Fantail's Duchess) by Ninth Duke of Geneva (28391), &c. See "Fantail Duke," p. 97.

Produce in		Names, &c.	By what Bull.	By whom bred.
1879, Jan. 2, r. & w.,	C.C.	Fantail's D'ss 5th	D. of Oxford 27th, 33709	Sir J. Swinburne

GRAND WATERLOO, red, calved February 21, 1872, Vols. xxii. and xxiv. pp. 600, 687. Bred by Sir W. C. Trevelyan, Bart., Wallington, the property of Mr. J. J. Hetherington, Brampton; got by Grand Duke 13th (21850), dam (Waterloo 22nd) by Kildonan (20051), &c.

Produce in		Names, &c.	By what Bull.	By whom bred.
1879, Jan. 11, roan,	C.C.	Grand Waterloo 4th	D. of Oxford 27th, 33709	Sir W.C. Trevelyan

Grand Waterloo 4th, sold to Sir J. Swinburne, Bart.

LADY FLORENCE 4TH, roan, calved May 20, 1875, Vol. xxv. p. 685. Bred by Sir J. Swinburne, Bart.; got by Oxford Beau 4th (34964), dam (Lady Florence 3rd) by Duke of Edlingham (30948), &c.

Produce in	Names, &c.	By what Bull.	By whom bred.
1879, April 1, white, C.C.	Lady Florence 7th	D. of Oxford 27th, 33709	Sir J. Swinburne

LADY MARY, roan, calved April 6, 1872, Vols. xxii., xxiii., and xxiv. pp. 600, 682, 687. Bred by the Earl of Dunmore, Dunmore, the property of Mr. W. Cruddace, Benson's Fell; got by Baron Oxford 5th (27958), dam (Lady Bird 6th) by Third Duke of Geneva (23753), &c.

Produce in	Names, &c.	By what Bull.	By whom bred.
1879, April 22, roan, C.C.	Lady Bird 16th	D. of Oxford 27th, 33709	Sir W.C. Trevelyan

Lady Bird 16th, sold to Sir J. Swinburne, Bart.

OXFORD CHERRY 3RD, red, calved March 23, 1876. Bred by Sir W. C. Trevelyan, Bart., Wallington, the property of the Earl of Durham, Lambton Castle; got by Oxford Beau 4th (34964), dam (Grand Oxford Cherry) by Grand Duke 13th (21850), &c. See Vol. xxiii. p. 682.

Produce in	Names, &c.	By what Bull.	By whom bred.
1879, Jan. 17, r. & w., C.C.	Lady Oxford Cherry	D. of Oxford 27th, 33709	Sir W.C. Trevelyan

Lady Oxford Cherry, sold to Sir J. Swinburne, Bart.

PRINCESS SALE, red, calved May 23, 1874, Vol. xxiv. p. 669. Bred by the Earl of Bective, Underley Hall; got by Second Duke of Tregunter (26022), dam (Lady Sale of Putney) by Ninth Duke of Thorndale (31023), &c.

Produce in	Names, &c.	By what Bull.	By whom bred.
1879, Jan. 30, white, C.C.	Princess Sale 4th	D. of Oxford 27th, 33709	Sir J. Swinburne

SWINGLER, Thomas,
Langham, Oakham.

BLOOM 2ND, red, calved October 28, 1873. Bred by Mr. J. Woods, Langham; got by Union Jack (32746), dam (Bloom) by Margrave (20276), &c. See "Lord Hardwicke," Vol. xxv. p. 167.

Produce in	Names, &c.	By what Bull.	By whom bred.
1877, Jan. 12, red, B.C.	(dead)	Baron Lincoln, 33059	Mr. Woods
1878, Apr. 3, r. & w., C.C.	Baroness	Lord Hardwick, 41866	do.
1879, June 30, red, B.C.	Cherry Boy	do.	Mr. Swingler

CLARA NOVELLO 5TH, roan, calved July 10, 1877. Bred by Lord Chesham, Latimer; got by Oxford Duke (34977), dam (Clara Novello 4th) by Royal Cambridge (25009), g. d. (Clara Novello 2nd) by Charleston (21400), &c. See Vol. xxi. p. 627.

Produce in	Names, &c.	By what Bull.	By whom bred.
1879, Dec. 9, roan, C.C.	Clara Priscilla	Ld. of Lochaber 2nd, 36983	Mr. Swingler

TABER, James,
Rose Cottage, Rivenhall, Witham, Essex.

EMPRESS, roan, calved January 2, 1877. Bred by Mr. T. Pool, Black Notley Hall; got by Starlight (40754), dam (Dagmar) by Grand Vizier (21871), &c. See Vol. xx. p. 469.

Produce in	Names, &c.	By what Bull.	By whom bred.
1879, June 11, white, C.C.	Alba	Osman Pasha, 42078	Mr. Taber

LADY AUGUSTA, red and white, calved August 1, 1876. Bred by Mr. J. Upson, Rivenhall; got by Oxford's Baron (32030), dam (Princess Augusta) by Baron Roxwell (21240), &c. See Vol. xx. p. 701.

Produce in	Names, &c.	By what Bull.	By whom bred.
1879, July 10, r. & w., B.C.	Levi	Osman Pasha, 42078	Mr. Taber

MARGERY, roan, calved December 29, 1876.   Bred by Mr. N. Catchpole, Bramford ; got by Oxford Prize 2nd (40430), dam (Minnie) by Young Hector (28830), &c.   See Vol. xxiv. p. 367.

Produce in	Names, &c.	By what Bull.	By whom bred.
1879, April 19, white, C.C.	Ada	Osman Pasha, 42078	Mr. Taber

TANKERVILLE, Earl of,
Chillingham Castle, Alnwick, Northumberland.

GAIETY 3RD, roan, calved June 28, 1875, Vol. xxv. p. 686.   Bred by the Exors. of Mr. G. Angus, Broomley ; got by Ben Brace (30524), dam (Gaiety) by Merry Monarch (22349), &c.

1879, Nov. 18, white, B.C.	Gay Roland	Fitz-Roland, 33936	Earl of Tankerville

PEERESS 10TH, roan, calved March 3, 1873, Vols. xxii. and xxv. pp. 587, 686. Bred by the Earl of Tankerville ; got by Colonel (25798), dam (Peeress 9th) by Knight of Hawkhill (26539), &c.

1879, Aug. 20, white, C.C.	Peeress 16th	Fitz-Roland, 33936	Earl of Tankerville

PEERESS 11TH, red, calved October 1, 1874, Vols. xxiii. and xxiv. pp. 666, 670.   Bred by the Earl of Tankerville ; got by Red Cross Knight (32248), dam (Peeress 9th) by Knight of Hawkhill (26539), &c.

1879, Feb. 23, red, C.C.	Peeress 15th	Fitz-Roland, 33936	Earl of Tankerville

ROYAL GUEST, roan, calved March 9, 1876.   Bred by the Earl of Tankerville ; got by Monarch (31930), dam (Honoured Guest) by Knight of the Shire (26552), &c.   See Vol. xxiii. p. 666.

1879, April 27, white, B.C		Fitz-Roland, 33936	Earl of Tankerville

WELL BORN, white, calved February 5, 1874, Vol. xxv. p. 686.   Bred by Mr. W. Torr, Aylesby Manor; got by Knight of the Shire (26552), dam (Weal Royal) by Booth Royal (15673), &c.

1879, April 22, white, B.C.	Lord Bennet	Sir Raymond, 40716	Earl of Tankerville

Lord Bennet, sold to the Duke of Northumberland, Alnwick Castle.

TAUNTON, Miss,
Home Farm, Ashley, Stockbridge, Hampshire.

HEBE 4TH, white, calved March 10, 1875, Vol. xxv. p. 687.   Bred by Miss Taunton ; got by Pizarro (32069), dam (Hebe) by Guardian of Oxford (26328), &c.

1879, Sept. 24, roan, C.C.	Hebe of Ashley	Holcus, 40004	Miss Taunton

HONEYSUCKLE, white, calved May 9, 1873.   Bred by Miss Taunton ; got by Duke of Hazlecote 15th (33665), dam (Hecate) by Caleb (15718), &c.   See "Honey Dew." Vol. xxv. p. 135.

1879, Mar. 18. r. & w., B.C.	Hatchwarren	Holcus, 40004	Miss Taunton

Hatchwarren, sold to Mr. H. Cleese. Hatchwarren, Basingstoke.

LAURESTINA, roan, calved October 12, 1869, Vols. xxi., xxiii., and xxv. pp. 951, 666, 687.   Bred by Miss Taunton ; got by Hector Wetherby (26358), dam (Laura) by Viceroy (13945), &c.

1879, July 28. r. & w.. C.C.	Laurestina Ashley	Holcus, 40004	Miss Taunton

LUCY, roan, calved October 3, 1873, Vols. xxiii. and xxv. pp. 666, 687. Bred
by Miss Taunton; got by Duke of Hazlecote 15th (33665), dam (Lauretta)
by Third Marquis of Wetherby (26842), &c.

Produce in	Names, &c.	By what Bull.	By whom bred.
1879, July 24, roan, B.C.	Lord Lake	Holcus, 40004	Miss Taunton

TAYLOR, George,
Sewerby Cottage, Sewerby, Hull.

FOREST FLOWER, red, calved October 7, 1876. Bred by Mr. G. Taylor; got
by Azor (32979), dam (Forest Maid) by Baron Percy (30477), &c. See
Vol. xxiii. p. 667.

1879, May 29, r. & w., C.C.	Queen of the Day	Prince Alfonso, 37214	Mr. Taylor

TAYLOR, George,
Stanton Prior. Bristol.

BELINDA, roan, calved April 4, 1874, Vols. xxiv. and xxv. pp. 487, 488. Bred
by Mr. T. Harris, Stonylane House; got by Winterfold (36020), dam
(Lavender 12th) by Lord Red Eyes 2nd (24460), &c.

1878, Dec. 19, { white, C.C.	Princess Alice	} Waterloo Prince, 35949	Mr. Taylor
{ roan, B.C.	Prince Louis		

LAVENDER 13TH, roan, calved January 10, 1875. Bred by Mr. T. Harris,
Stonylane House; got by Winterfold (36020), dam (Lavender 12th) by Lord
Red Eyes 2nd (24460), &c. See " Prince Louis," p. 196.

1879, April 12, roan, C.C.	Lavender 20th	Waterloo Prince, 35949	Mr. Taylor

TAYLOR, R.,
New House, Burnside, Kendal.

BLITHER YET, roan, calved June 19, 1875. Bred by Mr. R. Jefferson,
Preston Hows; got by King Richard 2nd (28984), dam (Blithe Beauty) by
Lord Blithe (22126), g.d. (Britannia 15th) by Wild Duke 4th (21107), &c.
See Vol. xvii. p. 394.

1878, Jan. 12, roan, C.C.	(dead)	Major Irwin, 34735	Mr. Taylor
1879, Feb. 10, white, C.C.	Blithe Queen	do.	do.

TAYLOR, R.,
Sigglesthorne Manor, Hull.

CHERRY COUNTESS, white, calved November 16, 1872, Vols. xxiii., xxiv.,
and xxv. pp. 667, 671, 687. Bred by Mr. R. Taylor; got by Netta (31970),
dam (Cherry Countess) by Cherry Royal (23556), &c.

1879, July 30, white, B.C.	Wild Count	Wild Oxford, 40926	Mr. Taylor

CHERRY COUNTESS 4TH, roan, calved March 23, 1876. Bred by Mr. R.
Taylor; got by Oneida Prince (34948), dam (Cherry Countess 2nd) by Netta
(31970), &c. See Vol. xxiii. p. 667.

1879, May 28, white, B.C.	Oneida Cherry	Wild Oxford, 40926	Mr. Taylor

ETHEL, roan, calved October 21, 1875, Vol. xxv. p. 687. Bred by Mr. R. Taylor; got by Roan King (35285), dam (Eva 2nd) by Royal Prince (29874), &c.

Produce in	Names, &c.	By what Bull.	By whom bred.
1879, Aug. 9, white, B.C.	Oxford King	Wild Oxford, 40926	Mr. Taylor

PENELOPE 2ND, roan, calved May 1, 1872, Vols. xxii., xxiii., and xxv. pp. 587, 667, 688. Bred by Mr. R. Taylor; got by Waltron (30255), dam (Penelope) by Jerry (28910), &c.

Produce in	Names, &c.	By what Bull.	By whom bred.
1879, July 4, roan, B.C.	Lord Hawthorn	Wild Oxford, 40926	Mr. Taylor

Lord Hawthorn, sold to the Marquis of Bute, Cardiff Castle.

PENELOPE 7TH, red, calved April 24, 1876, Vol. xxv. p. 688. Bred by Mr. R. Taylor; got by Oneida Prince (34948), dam (Penelope) by Jerry (28910), &c.

Produce in	Names, &c.	By what Bull.	By whom bred.
1879, Nov. 16, roan, C.C.	Penelope 10th	Wild Oxford, 40926	Mr. Taylor

PENELOPE 8TH, roan, calved June 30, 1876. Bred by Mr. R. Taylor; got by Oneida Prince (34948), dam (Penelope 2nd) by Waltron (30255), &c. See "Lord Hawthorn," p. 148.

Produce in	Names, &c.	By what Bull.	By whom bred.
1879, Aug. 22, white, C.C.	Penelope 9th	Wild Oxford, 40926	Mr. Taylor

PRINCESS ROSE 3RD, red, calved December 17, 1876. Bred by Mr. R. Taylor; got by Oneida Prince (34948), dam (Princess Beatrice) by Netta (31970), &c. See Vol. xxiii. p. 667.

Produce in	Names, &c.	By what Bull.	By whom bred.
1879, Sept. 1, roan, C.C.	Princess Rose 6th	Wild Oxford, 40926	Mr. Taylor

ROAN QUEEN, roan, calved March 17, 1875, Vol. xxv. p. 688. Bred by Mr. R. Taylor; got by Roan King (35285), dam (Ruby) by Grenadier (24085), &c.

Produce in	Names, &c.	By what Bull.	By whom bred.
1879, Sept. 2, white, C.C.	Oxford Queen	Wild Oxford, 40926	Mr. Taylor

ROSE 1ST, roan, calved August 9, 1876, Vol. xxv. p. 662. Bred by Mr. E. Holden, Laurel Mount; got by Oxford Cherry Duke 2nd (34972), dam (Rose) by Prince of Thorndale (32187), &c.

Produce in	Names, &c.	By what Bull.	By whom bred.
1879, Oct. 25, r. & w., B.C.	Baron Cockburn	Lord Cockburn, 38594	Mr. Taylor

SYLPH, white, calved October 4, 1872, Vols. xxii., xxiii., and xxv. pp. 588, 668, 688. Bred by Mr. W. Clapham, Hatfield Magna; got by Royal Prince (29874), dam (Symmetry) by Royal Butterfly 3rd (18754), &c.

Produce in	Names, &c.	By what Bull.	By whom bred.
1879, July 10, { white, C.C. / white, B.C.	Sylph 4th / Oneida Royal	} Oneida Prince, 34948	Mr. Taylor

SYLPH 2ND, roan, calved April 1, 1875, Vol. xxv. p. 689. Bred by Mr. R. Taylor; got by Barrington Duke (33115), dam (Sylph) by Royal Prince (29874), &c.

Produce in	Names, &c.	By what Bull.	By whom bred.
1879, Aug. 4, roan, C.C.	Sylph 5th	Wild Oxford, 40926	Mr. Taylor

WHITE CLARENCE, white, calved May 10, 1873, Vols. xxiii., xxiv., and xxv. pp. 668, 672, 689. Bred by Mr. R. Taylor; got by Netta (31970), dam (White Lily) by Lord of Clarence (24439), &c.

Produce in	Names, &c.	By what Bull.	By whom bred.
1879, May 22, white, B.C.	(dead)	Oneida Prince, 34948	Mr. Taylor

THEAKSTON, Thomas,
Masham, Bedale, Yorkshire.

ROSY LASS, red and white, calved July 2, 1876. Bred by Mr. H. Pickersgill,
Middleton Quernhow; got by Hades (34101), dam (Braithwaite Lass) by
Booth's Royal Signet (28061), &c. See Vol. xxiii. p. 602.

Produce in	Names, &c.	By what Bull.	By whom bred.
1879, Sept. 19, r. & w.,C.C.	Rosy May	Rollicking Laddie, 39004	Mr. Theakston

THEAKSTON, T. and R.,
Masham, Bedale, Yorkshire.

SAUCY JADE, red, calved October 5, 1873, Vol. xxiv. p. 673. Bred by Mr. W.
White, Burrill; got by Lord Danby (31644), dam (Saucy Jane) by Manfred
(26801), &c.

1878, Nov. 23, red,	C.C.	Saucy Jenny	Star Regent, 35679	Messrs. Theakston

THOMPSON, John,
Badminton, Chippenham.

DARLINGTON 22ND, roan, calved April 5, 1874, Vol. xxiv. p. 674. Bred by
Mr. J. Thompson; got by Second Duke of Glo'ster (28392), dam (Darlington
15th) by Grand Duke of York (24071), &c.

1879, Sept. 2, r. & w., C.C.	Darlington 30th	Earl of Horton 10th, 36587	Mr. Thompson

DARLINGTON 24TH, red and white, calved June 23, 1875, Vol. xxv. p. 689.
Bred by Mr. J. Thompson; got by Grand Duke of Clarence (28750), dam
(Darlington 17th) by Gr'nd Duke of York (24071), &c.

1879, May 22. roan,	C.C.	Darlington 29th	Oxford's King, 34997	Mr. Thompson

DARLINGTON 25TH, roan, calved April 10, 1876. Bred by Mr. J. Thompson;
got by Lord Darlington 7th (34519), dam (Darlington 15th) by Grand Duke
of York (24071), &c. See Vol. xxiii. p. 669.

1879, Mar. 19, roan,	B.C.	(dead)	Earl of Horton 10th, 36587	Mr. Thompson

OXFORD DUCHESS 5TH, red and white, calved September 8, 1874, Vol. xxv.
p. 689. Bred by Messrs. Horswell and Sons, Burns Hall; got by Baron
Oxford 2nd (23376), dam (Duchess of Northumberland 3rd) by Duke of
Oxford and Glo'ster (28436), &c.

1879, Jan. 1, r. & w., C.C.	Oxford Duchess 17th	Baron Gaddesby, 41028	Mr. Thompson

THOMPSON, John,
Sandymount, Tipperary, Ireland.

LIMERICK LASS 12TH, white, calved March 10, 1875. Bred by Mr. F. W.
Low, Kilshane; got by Nobleman (31979), dam (Limerick Lass 5th) by Wide-
awake (27805), g. d. (Limerick Lass) by Little Wonder (18204), &c. See
Vol. xx. p. 626.

1878, April 8, white,	C.C.	Narra Mattah	Garryowen, 43264	Mr. Thompson
1879, Mar. 25, roan,	B.C.	Hiawatha	Prince, 43778	do.

THOMPSON, Joseph,
Anlaby, Hull, Yorkshire.

CECILIA, red, calved October 11, 1875. Bred by Mr. J. Thompson; got by
Samson (39079), dam (Chance) by Lord Arthur (38574), g. d. (Carry) by
Second Squire of Waterloo (25218), gr. g. d. (Cassia 2nd) by Grand Guard
(17998), &c. See Vol. xxiii. p. 669.

Produce in	Names, &c.	By what Bull.	By whom bred.
1879, Jan. 10, roan, C.C.	Cora	Costermonger, 39635	Mr. Thompson

CHARLOTTE, red, calved May 8, 1872. Bred by Mr. J. Thompson; got by
Sir Tatton (27497), dam (Cassia 2nd) by Grand Guard (17998), &c.

1875, May 18, roan, C.C.	Countess	Samson, 39079	Mr. Thompson
1878, Mar. 25, roan, C.C.	Carry 2nd	Beverley Oxf'd 3rd, 39468	do.

EMMA, roan, calved July 1, 1874. Bred by Mr. J. Thompson; got by Baron
Ruth (30482), dam (Empress) by Sir Tatton (27497), &c. See Vol. xxv.
p. 690.

1877, Apr. 25, r. & w., C.C.	Economy	Marshal Wellington, 34794	Mr. Thompson
1879, Feb. 14, white, B.C.	Enoch	Beverley Oxf'd 3rd, 39468	do.

THOMPSON, M., and Sons,
Kirkhouse, Milton, Carlisle.

FANNY DEANS, roan, calved February 28, 1875. Bred by Mr. J. Lamb,
Burrell Green; got by Hubback Junior (31395), dam (Effy Deans) by Edgar
(19680), &c. See Vol. xxi. p. 799.

1878, April 18, white, B.C.	(dead)	Jacobite, 40027	Messrs. Thompson
1879, April 22, roan, B.C.	Midlothian	Kirk'thore W.E.5th, 38505	do.

JUANITA, roan, calved February 15, 1874, Vols. xxiii. and xxiv. pp. 670, 675.
Bred by Mr. J. Fawcett, Scaleby Castle; got by Eighth Duke of York
(28480), dam (Jenny Lind 2nd) by Rover (22765), &c.

1878, Dec. 18, white, C.C.	Julia	Kirk'thore W.E.5th, 38505	Messrs. Thompson

LADY BEVERLEY 17TH, red and white, calved April 7, 1874, Vol. xxiv.
p. 675. Bred by Messrs. M. Thompson and Sons; got by Belted Will (36236),
dam (Lady Beverley 15th) by Constitution (28244), &c.

1878, Feb. 9, r. & w., C.C.	(dead)	Duke of Devonshire, 36484	Messrs. Thompson
1879, Jan. 16, roan, B.C.	(dead)	Kirk'thore W.E.5th, 38505	do.
1879, Dec. 31, roan, C.C.	Lady Beverley 21st	do.	do.

LILY 5TH, red, calved March 12, 1872, Vols. xxiii. and xxiv. pp. 670, 675.
Bred by Messrs. M. Thompson and Sons; got by Crown Prince (28290), dam
(Lily 4th) by Inglewood (20006), &c.

1878, May 28, r. & w., B.C.	(dead)	Jacobite, 40027	Messrs. Thompson
1879, April 29, red, B.C.	Cetewayo	Kirk'thore W.E.5th, 38505	do.

OXFORD LASS, red, calved April 24, 1872. Bred by Mr. G. Moore, Whitehall; got by Seventeenth Duke of Oxford (25994), dam (Oxford Witch) by
Imperial Oxford (18084), &c. See Vol. xx. p. 679.

1876, July 30, r. & w., B.C.	(Steer)	Lord Spencer, 37005	Messrs. Thompson
1878, May 14, roan, B.C.	Oxford Lad	Jacobite, 40027	do.
1879, April 19, red, C.C.	(dead)	Duke of Devonshire, 36484	do.

PRINCESS OF CAMBRIDGE 2ND, roan, calved April 6, 1872, Vol. xxiv. p. 676. Bred by Mr. J. Fawcett, Scaleby Castle ; got by Royal Cumberland (27358), dam (Princess Royal) by Rob Roy (22740), &c.

Produce in	Names, &c.	By what Bull.	By whom bred.
1879, Jan. 26, white, B.C.	Royal Duke	Duke of Devonshire, 36484	Messrs. Thompson
1879, Dec. 23, white, B.C.	Prince Teck	Kirk'thoreW.E.5th, 38505	do.

QUEEN OF THE FOREST, white, calved October 23, 1873, Vols. xxiii. and xxiv. pp. 671, 676. Bred by Mr. J. Davidson, Greengill ; got by Gamester (31216), dam (Newton) by Baronet (19274), &c.

1879, Jan. 22, white, B.C.	(Steer)	Duke of Devonshire, 36484	Messrs. Thompson

SONSIE 22ND, red, calved January 15, 1872, Vols. xxii., xxiii., and xxiv. pp. 591, 671, 676. Bred by Messrs. M. Thompson and Sons ; got by Crown · Prince (28290), dam (Sonsie 21st) by Border Union (19329), &c.

1878, April 18, roan, B.C.	(dead)	Jacobite, 40027	Messrs. Thompson
1879, April 3, red, B.C.	(Steer)	Kirk'thoreW.E.5th, 38505	do.

SONSIE 23RD, red, calved February 2, 1874, Vol. xxiv. p. 676. Bred by Messrs. M. Thompson and Sons ; got by Belted Will (36236), dam (Sonsie 21st) by Border Union (19329), &c.

1878, Dec. 12, r. & w., B.C.	(Steer)	Kirk'thoreW.E.5th, 38505	Messrs. Thompson

THOMPSON, R.,
Inglewood Bank, Penrith.

AGNES GWYNNE, roan, calved November 9, 1872, Vols. xxii. and xxiv. pp. 591, 676. Bred by Mr. J. M. Richardson, Hutton House ; got by Lord Bates (29065), dam (Princess Gwynne) by Knight of Santon (24275), &c.

1878, Oct. 9, r. & w., C.C.	Glady's Gwynne	Brilliant Butterfly, 36270	Mr. Thompson

BUTTERFLY'S MEMENTO 5TH, red and white, calved January 9, 1873, Vol. xxiii. p. 671. Bred by Colonel Towneley, Towneley Park ; got by Earl of Thorndale (28521), dam (Butterfly's Memento 2nd) by Baron Oxford (23375), &c.

1879, Feb. 8, r. & w., C.C.	Rose Butterfly	Brilliant Butterfly, 36270	Mr. Thompson

CICELY, roan, calved January 19, 1875. Bred by Mr. R. Thompson ; got by Grand Duke of Fawsley 3rd (31286), dam (Jess) by Fifth Duke of Oxford (23785), &c. See Vol. xxi. p. 955.

1878, Oct. 23, r. & w., C.C.	Sweet Cicely	Brilliant Butterfly, 36270	Mr. Thompson
1879, Oct. 12, roan, C.C.	Braw Cicely	do.	do.

FAIR MILLICENT, white, calved January 5, 1876. Bred by Mr. R. Thompson ; got by Grand Duke of Fawsley 3rd (31286), dam (Moss Rose 4th) by Royal Gwynne (22784), &c. See "Master Farnley," Vol. xxiv. p. 175.

1878, Oct. 4, roan, C.C.	(dead)	Brilliant Butterfly, 36270	Mr. Thompson
1879, Nov. 1, roan, C.C.	Fair Millicent 2nd	do.	do.

FARNLEY PRINCESS, roan, calved April 8, 1874. Bred by Mr. R. Thompson ; got by Prince of Perth (32176), dam (Cambridge Moss Rose 2nd) by Barrington Oxford (25607), &c. See " Farnley Prince 2nd," p. 98.

1878, May 24, r. & w., C.C.	Farnley Princess 2nd	Brilliant Butterfly, 36270	Mr. Thompson
1879, April 15, r. & w., B.C.	Farnley Prince 2nd	do.	do.

LADY EGLINTON, roan, calved March 3, 1872, Vols. xxii. and xxiv. pp. 592, 677. Bred by Mr. R. B. Brockbank, Burgh-by-Sands; got by Earl of Eglinton (23832), dam (Lively) by Master Annandale (14916), &c.

Produce in		Names, &c.	By what Bull.	By whom bred.
1878, March 3, roan,	C.C.	Lady Staveley	Brilliant Butterfly, 36270	Mr. Thompson
1879, Feb. 7, roan,	C.C.	Lady Comely	do.	do.

LOVE TOKEN, white, calved April 20, 1875, Vol. xxiv. p. 677. Bred by Mr. R. Thompson; got by Grand Duke of Fawsley 3rd (31286), dam (Farewell) by Royal Westmoreland (35416), &c.

1878, Oct. 7, roan,	C.C.	Inglewood Pet	Brilliant Butterfly, 36270	Mr. Thompson
1879, Oct. 14, roan,	B.C.	Belted Knight	do.	do.

PEARL 9TH, roan, calved February 16, 1868, Vols. xx. and xxii. pp. 684, 592. Bred by Mr. J. M. Richardson, Hutton House; got by British Cherry (23461), dam (Pearl 6th) by Welcome Guest (15497), &c.

1878, Jan. 20, r. & w.,	C.C.	(dead)	Brilliant Butterfly, 36270	Mr. Thompson
1879, Jan. 30, r. & w.,	C.C.	Pearl Drop	do.	do.

RED ROSE 4TH, roan, calved October 12, 1873, Vol. xxiii. p. 671. Bred by Mr. J. J. Hetherington, Middle Farm; got by Baron Deepdale (30438), dam (Red Rose 1st) by Grand Duke of Lightburne 2nd (26291), &c.

1878, April 8, roan,	C.C.	Mildred Millicent	Brilliant Butterfly, 36270	Mr. Thompson
1879, May 16, roan,	C.C.	May Millicent	do.	do.

ROSAMOND GWYNNE, white, calved April 15, 1870. Bred by Mr. J. Caddy, Rougholm; got by Sir Windsor (22927), dam (Rebecca Gwynne) by Knight of Distington (18158), &c. See "Cressida Gwynne," p. 59.

1878, May 19, roan,	B.C.	Cressida Gwynne	British Duke, 37901	Mr. Thompson
1879, March 30, roan,	B.C.	(Steer)	Brilliant Butterfly, 36270	do.

THOMPSON, William,
Moresdale Hall, Kendal, Westmoreland.

ANDROMACHE 2ND, red and white, calved November 15, 1875. Bred by Mr. M. Kennedy, Stone Cross; got by Knight of Windsor (31573), dam (Macco) by Paris (20469), &c. See Vol. xxiii. p. 518.

1879, Sept. 27, r. & w.,	C.C.	Cousin Cressida	British Duke, 37901	Mr. Thompson

THOMSON, James,
Newseat of Dumbreck, Udny, N.B.

BLYTHESOME, red, calved March 9, 1874. Bred by Mr. J. Thomson; got by Albert (30370), dam (Butterfly 29th) by Senator (27441), g. d. (Butterfly 3rd) by The Baron (13833), &c. See Vol. xvii. p. 396.

1876, Mar. 18, red,	C.C.	Blythesome 2nd	Blucher, 33170	Mr. Thomson
1877, May 29, red,	C.C.	Blythesome 3rd	do.	do.
1878, Apr. 6, r. & w.,	C.C.	Britannia	British Prince, 36282	do.

CECIL'S MATCHLESS, red, calved March 29, 1876. Bred by Mr. J. Thomson; got by Lord Cecil (26621), dam (Matchless 12th) by Lord Raglan (13244), &c. See "Matchless 17th," Vol. xxiii. p. 708.

1878, Nov. 2, red,	C.C.	Mature	British Prince, 36282	Mr. Thomson

CIRCE, roan, calved July 11, 1873. Bred by Mr. R. Burdon, Castle Eden ; got by Emperor Maximilian (26100), dam (Syren) by Knight of Eden (26537), &c. See "Mountain Chief." Vol. xxiii. p. 195.

Produce in		Names, &c.	By what Bull.	By whom bred.
1876, Apr. 27, roan,	B.C.	Mountain Chief	Mountain Hero, 31944	Mr. Thomson
1879, Apr. 9, red,	C.C.	Titiens	Titus, 40822	do.

FLORETTE, red, calved February 25, 1874. Bred by Mr. J. Thomson ; got by Albert (30370), dam (Flora) by Allan (21172), &c. See "Lord Roseberry," p. 156.

1877, Mar. 15, roan,	C.C.	Florette 2nd	Blucher, 33170	Mr. Thomson
1878, Mar. 7, r. & w.,	B.C.	Lord Roseberry	do.	do.
1879, Feb. 22, roan,	C.C.	Florette 3rd	Mountain Chief, 38767	do.

Lord Roseberry, sold to Mr. A. Milne, Corse of Kinmoir, Huntly.

MATILDA 14TH, red, calved February 2, 1874. Bred by Mr. J. Thomson ; got by Albert (30370), dam (Matilda 11th) by Nelson (22401), &c. See "Malcolm," Vol. xxiii. p. 183.

1877, Jan. 20, r. & w.,	C.C.	Matilda 16th	Blucher, 33170	Mr. Thomson
1879, Apr. 15, roan,	C.C.	Marvel	Mountain Chief, 38767	do.

MATILDA 15TH, red, calved February 24, 1875. Bred by Mr. J. Thomson ; got by Albert (30370), dam (Matilda 11th) by Nelson (22401), &c.

1878, Mar. 27, red,	C.C.	Marjoram	British Prince, 36282	Mr. Thomson
1879, Mar. 21, red,	C.C.	Maxim	Mountain Chief, 38767	do.

ORANGE BLOSSOM 13TH, red, calved April 29, 1869, Vol. xx. p. 676. Bred by Mr. A. Cruickshank, Sittyton ; got by Allan (21172), dam (Orange Blossom 2nd) by The Baron (13833), &c.

1874, May 16, red,	C.C.	Orange Leaf	Ld. Lansdowne, 29128	Mr. Thomson
1877. Mar. 10, roan,	C.C.	Orchard	Blucher, 33170	do.

ORANGE LEAF 2ND, roan, calved April 30, 1876. Bred by Mr. J. Thomson ; got by Blucher (33170), dam (Orange Leaf) by Lord Lansdowne (29128), g. d. (Orange Blossom 13th) by Allan (21172), &c.

1879, Jan. 17, red,	C.C.	Orange Ripe	Mountain Chief, 38767	Mr. Thomson

ROAN DUCHESS 3RD, roan, calved April 12, 1874. Bred by Mr. J. Thomson ; got by Golden Cross (31264), dam (Roan Duchess 2nd) by Nelson (22401), g. d. (Roan Duchess) by Prince of Coburg (15100), &c. See "Richmond," Vol. xxii. p. 222.

1879, Mar. 18, white,	B.C.	White Chief	Mountain Chief, 38767	Mr. Thomson

WANTON 15TH, roan, calved March 5, 1872. Bred by Mr. J. Thomson ; got by Golden Cross (31264), dam (Wanton 9th) by Lieutenant (24335), g. d. (Wanton 7th) by Harvester (21004), &c. See "Wellington," Vol. xxii. p. 277.

1878, Mar. 31, roan,	B.C.	King Lear	British Prince, 36282	Mr. Thomson
1879, Mar. 14, roan,	B.C.	Warrior Chief	Mountain Chief, 38767	do.

King Lear, sold to Mr. Ledingham, Hayhillock, Ellon, N.B.

WINIFRED, red, calved March 11, 1876. Bred by Mr. J. Thomson ; got by Royal Hope (32392), dam (Wimple 7th) by Baron Laurie 2nd (25570), g. d. (Wimple) by Prince Arthur (16723), &c. See "Wimple 4th," Vol. xxiii. p. 617.

1878, Oct. 29. roan.	C.C.	Winifred 2nd	Mountain Chief, 38767	Mr. Thomson

WINSOME, roan, calved January 31, 1873. Bred by Mr. J. Thomson ; got by
Albert (30370), dam (Wanton 12th) by Crusade (30831), g. d. (Wanton 9th)
by Lieutenant (24335), gr. g. d. (Wanton 7th) by Harvester (21904), &c. See
" Wellington," Vol. xxii. p. 277.

Produce in		Names, &c.	By what Bull.	By whom bred.
1878, March 8, red,	B.C.	Waddington	British Prince, 36282	Mr. Thompson
1879, Apr. 7, r. & w.,	B.C.	Weather Guage	Mountain Chief, 38767	do.

Waddington, sold to Mr. Wilson, Huntly.

THURSFIELD, T. H.,
Barrow, Broseley, Shropshire.

MARIAN, red and white, calved April 27, 1876. Bred by the Rev. H. O. Wilson,
Church Stretton ; got by King of Trumps (31512), dam (Maid Marian) by
Lord Warwick (26753), &c. See Vol. xxi. p. 996.

1878, Dec. 17, roan,	B.C.	Moonshine	Moonstone, 37107	Rev. H. O. Wilson
1879, Dec. 27, r. & w.,	C.C.	Marian 2nd	Matchfield, 42004	Mr. Thursfield

TINDALL, J. and E.,
Knapton Hall, Rillington, Yorkshire.

CELANDINE 4TH, roan, calved October 22, 1874. Bred by Messrs. J. and E.
Tindall ; got by Melrose (29357), dam (St. Hilda) by Lord Granville (26653),
&c. See Vol. xxi. p. 958.

1877, Feb. 5, roan,	B.C.	(Steer)	Sampiero, 35466	Messrs. Tindall
1878, Jan. 28, r. & w.,	C.C.	Golden Reed	do.	do.

CELANDINE 5TH, roan, calved December 10, 1874, Vol. xxiv. p. 677. Bred
by Messrs. J. and E. Tindall ; got by Melrose (29357), dam (Celandine) by
Cecil (25725), &c.

1878, April 26, roan,	C.C.	Derwent Queen	X. L., 37697	Messrs. Tindall
1879, July 1, white,	C.C.	Ivy Leaf	Astral Prince, 39386	do.

FOREST PALM 3RD, roan, calved June 21, 1877. Bred by Messrs. J. and E.
Tindall ; got by Sampiero (35466), dam (Golden Palm) by Melrose (29357),
&c. See Vol. xxiv. p. 678.

1879, Nov. 16, roan,	B.C.	Glenivor	Astral Prince, 39386	Messrs. Tindall

Glenivor, sold to Mr. J. Wildon, Knapton Grange, Rillington.

GOLDEN PALM, roan, calved July 12, 1872, Vols. xxii., xxiii., and xxiv.
pp. 593, 672, 678. Bred by Messrs. J. and E. Tindall ; got by Melrose (29357),
dam (Palm Flower) by Cecil (25725), &c.

1878, May 30, r. & w.,	C.C.	Forest Palm 4th	X. L., 37697	Messrs. Tindall

PALM FLOWER 2ND, roan, calved February 6, 1871, Vols. xxi., xxiii., and
xxiv. pp. 957, 672, 678. Bred by Messrs. J. and E. Tindall ; got by Cecil
(25725), dam (Palm Leaf) by Royal Charlie (25012), &c.

1878, Feb. 16, r. & w.,	C.C.	Palm Flower 5th	Sampiero, 35466	Messrs. Tindall
1879, Aug. 26, roan,	B.C.	(Steer)	Astral Prince, 39386	do.

ROSETTE 3RD, roan, calved June 10, 1870, Vols. xx., xxi., and xxiv. pp. 746,
957, 678. Bred by Messrs. J. and E. Tindall ; got by Cecil (25725), dam
(Rosette) by Prince Louis (20563), &c.

1878, Feb. 18, roan,	B.C.	Coronet	Sampiero, 35466	Messrs. Tindall
1879, Apr. 8, red,	C.C.	Alimondi	Astral Prince, 39386	do.

ST. HILDA 2ND, roan, calved October 18, 1868, Vols. xx., xxi., and xxiv.
pp. 755, 958, 678. Bred by Messrs. J. and E. Tindall; got by Lord Carlisle
(24367), dam (Bloom) by My Lord (14972), &c.

Produce in	Names, &c.	By what Bull.	By whom bred.
1878, June 10, r. & w., B.C.	(Steer)	Sampiero, 35466	Messrs. Tindall

ST. HILDA 3RD, white, calved December 28, 1872, Vols. xxiii. and xxiv.
pp. 673, 678. Bred by Messrs. J. and E. Tindall; got by Melrose (29357),
dam (St. Hilda 2nd) by Lord Carlisle (24367), &c.

1878, Aug. 25, white, C.C.	St. Hilda 7th	Sampiero, 35466	Messrs. Tindall
1879, Aug. 12, white, C.C.	St. Hilda 8th	Astral Prince, 39386	do.

SILVER QUEEN, roan, calved June 9, 1876. Bred by Messrs. J. and E. Tindall; got by Comedian (33412), dam (Victorine 3rd) by Lord Carlisle (24367),
&c. See Vol. xxiii. p. 673.

1879, Feb. 12, roan, C.C.	Golden Star	Astral Prince, 39386	Messrs. Tindall

VICTORINE 2ND, red and white, calved June 16, 1867, Vols. xx., xxi., and xxiv.
pp. 807, 958, 678. Bred by Messrs. J. and E. Tindall; got by Lord Granville (26653), dam (Integrity) by Fawsley (14540), &c.

1878, May 27, r. & w., C.C.	Derwent Lass	Sampiero, 35466	Messrs. Tindall

VICTORINE 3RD, white, calved June 14, 1868, Vols. xx., xxi., xxiii., and xxiv.
pp. 807, 958, 673, 678. Bred by Messrs. J. and E. Tindall; got by Lord
Carlisle (24367), dam (Integrity) by Fawsley (14540), &c.

1878, Dec. 5, white, C.C.	White Star 2nd	Sampiero, 35466	Messrs. Tindall

WHITE STAR, white, calved April 7, 1872, Vols. xxiii. and xxiv. pp. 673, 678.
Bred by Messrs. J. and E. Tindall; got by Earl of Derby 2nd (31061), dam
(Victorine 3rd) by Lord Carlisle (24367), &c.

1878, Mar. 30, roan, B.C.	Derwent King	X. L., 37697	Messrs. Tindall
1879, July 16, white, B.C.	St. Swithin	Astral Prince, 39386	do.
St. Swithin, sold to Mr. J. Wildon, Knapton Grange, Rillington.			

WHITE THORN, white, calved January 15, 1870, Vols. xx., xxi., xxiii., and
xxiv. pp. 821, 958, 673, 679. Bred by Messrs. J. and E. Tindall; got by Cecil
(25725), dam (Miss Wiley) by Cavendish (15745), &c.

1878, May 7, roan, C.C.	Derwent Queen 2nd	Sampiero, 35466	Messrs. Tindall

WILD VIOLET, roan, calved June 27, 1870. Vols. xxi. and xxiii. pp. 958, 673.
Bred by Messrs. J. and E. Tindall; got by Cecil (25725), dam (Miss Spearman) by Sir Charles (16949), &c.

1877, Nov. 9, white, B.C.	Cavalier	Sampiero, 35466	Messrs. Tindall
1878, Oct. 15, r. & w., C.C.	Maple Leaf	do.	do.
1879, Dec. 27, roan, C.C.	Xmas Rose	Astral Prince, 39386	do.

TISDALL, E. C.,

Holland Park Farm, Kensington, Middlesex; and Horton, Epsom, Surrey.

ALIX 3RD, roan, calved October 11, 1868, Vols. xx., xxi., and xxv. pp. 389,
529, 691. Bred by Her Majesty the Queen; got by Waterloo Duke (21077),
dam (Alix) by Earl of Dublin (10178), &c.

1879, July 4, r. & w., C.C.	Alix 8th	E. of Leicester 6th, 38230	Mr. Tisdall

ARIEL 3RD, roan, calved September 2, 1874. Bred by Mr. W. J. Edmonds, Southrop House; got by Lord Walton (31765), dam (Ariel 2nd) by Sylvan King (27593), &c. See "Arian," Vol. xxv. p. 8.

Produce in	Names, &c.	By what Bull.	By whom bred.
1877, Dec. 11, roan, B.C.	Arian	Earl of Leicester 6th, 38230	Mr. Tisdall
1879, May 20, r.&w., B.C.	Aries	May Duke, 40336	do.

Aries, sold to Mr. F. Hale, Horton, Epsom.

DUCHESS OF LANCASTER 11TH, roan, calved February 26, 1873, Vols. xxiii., xxiv., and xxv. pp. 673, 679, 691. Bred by Mr. W. Curtis, Fernham; got by Duke of Flamborough (25960), dam (Duchess of Lancaster 10th) by Lord Stanley (22217), &c.

1879, Dec. 31, roan, C.C.	D's of Lancaster 14th	Briton, 39516	Mr. Tisdall

JAPONICA 2ND, red and white, calved April 13, 1877. Bred by Mr. W. Arkell, Dudgrove; got by Wandering Minstrel (35924), dam (Japonica) by Purple Emperor (27222), &c. See Vol. xxv. p. 334.

1879, Oct. 18, roan, B.C.	Jasper	Sultan 2nd, 44106	Mr. Tisdall

Jasper, sold to Mrs. Prescot, The Birches, Tenbury.

JESSAMINE, white, calved March 21, 1875. Bred by Mr. W. Arkell, Dudgrove; got by Pompey (35059), dam (Japonica) by Purple Emperor (27222), &c. See Vol. xxv. p. 334.

1879, Oct. 15, roan, C.C.	Jessica	L. Fitzclarence 16th, 36943	Mr. Tisdall

LADY KENDAL, roan, calved March 7, 1874, Vol. xxiv. p. 679. Bred by the Earl of Bective, Underley Hall; got by Grand Duke of Kent 2nd (28759), dam (Lady Oxford) by Lord Chancellor (20160), &c.

1879, Feb. 18, red, B.C.	Kenric	E. of Leicester 6th, 38230	Mr. Tisdall

Kenric, sold to Mr. T. Whitbourn, Epsom, Surrey.

MAY FAIR, roan, calved February 1, 1876. Bred by Mr. E. C. Tisdall; got by The Bursar (35742), dam (Mayflower) by Duke John (30913), &c. See Vol. xxiii. p. 674.

1879, Feb. 5, roan, C.C.	May Blossom	E. of Leicester 6th, 38230	Mr. Tisdall

NANCY DUCHESS, red, calved March 2, 1875. Bred by Mr. J. Rigg, Wrotham Hill Park; got by Red Duke (37316), dam (Duchess Annie 6th) by Crown Prince (28281), &c. See Vol. xxii. p. 544.

1878, Mar. 6, roan, C.C.	(dead)	D. of Oxford 26th, 33708	Mr. Tisdall
1879, Jan. 29, red, B.C.	(Steer)	E. of Leicester 6th, 38230	do.
1879, Dec. 18, { r.&w. B.C. roan, C.C.	Duke Nancy } Duchess Nancy }	May Duke, 40336	do.

PEARL 12TH, red, calved August 5, 1876. Bred by Mr. E. C. Tisdall; got by Sockburn Duke (35621), dam (Pearl 10th) by British Cherry (23461), &c. See Vol. xxiii. p. 674.

1879, Oct. 13, roan, B.C.	Pearl-spar	May Duke, 40336	Mr. Tisdall

ROSABEL, roan, calved April 27, 1873. Bred by Mr. W. Arkell, Dudgrove, the property of Mr. E. C. Tisdall; got by Viceroy (30216), dam (Roseleaf) by Oxford Don (20451), g. d. (Rosebud) by Proctor (27210), &c. See Vol. xviii. p. 701.

1876, April 9, white, C.C.	Rosabel 2nd	Rustic, 35433	Mr. Arkell
1877, April 18, white, C.C.	Rosabel 3rd	Pompey, 35059	do.
1878, April 12, r.&w., C.C.	Rosabel 4th	Roderick, 35300	do.
1879, Sept. 8, roan, C.C.	Rosalie	Baron Lee 3rd, 36189	do.

Rosalie, sold to Mr. E. C. Tisdall, Holland Park Farm.

SNOWDROP, white, calved January 15, 1875, Vol. xxv. p. 693. Bred by Mr. T. Simonds, Carters Hill; got by Berks Butterfly (33141), dam (Ducky) by His Majesty (19965), &c.

Produce in		Names, &c.	By what Bull.	By whom bred.
1879, Dec. 23, roan,	B.C.	Snowflake	Earl of Leicester 6th, 38230	Mr. Tisdall

VENUS 3RD, roan, calved July 9, 1875, Vol. xxiv. p. 680. Bred by Mr. E. C. Tisdall; got by Lord Hastings (29115), dam (Venus) by Bates Tertius (21249), &c.

Produce in		Names, &c.	By what Bull.	By whom bred.
1879, Oct. 1, roan	C.C.	Venus 5th	Bridegroom, 39484	Mr. Tisdall

TOD, T. M.,
West Brackley, Kinross, N.B.

LADY JEAN, red and white, calved June 12, 1875. Bred by Mr. T. M. Tod; got by Merry Duke (34840), dam (Lady Havelock) by General Havelock (16130), &c. See Vol. xxii. p. 594.

Produce in		Names, &c.	By what Bull.	By whom bred.
1879, March 13, r. & w.,	C.C.	Fisher Lass	Glosser, 41624	Mr. Tod

TODD, John,
Scalthwaiterigg Stocks, Kendal.

BRENDA, roan, calved April 23, 1876. Bred by Mr. J. Todd; got by Prince of Perth (32176), dam (Minna) by White Oxford (27797), &c. See Vol. xxiii. p. 675.

Produce in		Names, &c.	By what Bull.	By whom bred.
1878, Dec. 3, roan,	B.C.	Conrad	British Duke, 37901	Mr. Todd

Conrad, sold to Mr. J. Clark, Sizergh Castle Farm, Kendal.

MINNA, roan, calved September 23, 1873, Vol. xxiii. p. 675. Bred by Mr. J. Todd; got by White Oxford (27797), dam (Spot) by Grenadier 2nd (28794), &c.

Produce in		Names, &c.	By what Bull.	By whom bred.
1877, May 10, r. & w.,	C.C.	Rose	Tenant Farmer, 35731	Mr. Todd
1878, April 27, white,	B.C.	Harold	Count Oxford, 33460	do.
1879, April 3, red,	C.C.	Flora	British Duke, 37901	do.

Harold, sold to Messrs. Dixon, Brund Rigg, Kendal.

TOPHAM, John,
Middleham House, Bedale, Yorkshire.

BONA, white, calved February 21, 1874. Bred by Mr. J. Topham; got by Booth's Royal Signet (28061), dam (Daisy) by Booth's Kinsman (25658), &c. See Vol. xxi. p. 960.

Produce in		Names, &c.	By what Bull.	By whom bred.
1879, March 6, white,	C.C.	Katie	Heir-at-Law, 34124	Mr. Topham

COWSLIP, roan, calved November 15, 1871, Vol. xxiii. p. 675. Bred by Mr. J. Topham; got by Booth's Royal Signet (28061), dam (Middleham Lassie) by Yorkshireman (17263), &c.

Produce in		Names, &c.	By what Bull.	By whom bred.
1879, Jan. 14, white,	B.C.	Sir Ulshaw	Heir-at-Law, 34124	Mr. Topham
1879, Dec. 19, red,	C.C.	Rosie	Star Regent, 35679	do.

Sir Ulshaw, sold to Mr. F. Robinson, Ulshaw Farm, Middleham, Bedale.

TOPPIN, J. C.,
Musgrave Hall, Skelton, Penrith.

GOLDEN DUCHESS, roan, calved June 18, 1876. Bred by Mr. J. Unthank, Netherscales; got by Iron Duke (31420), dam (Golden Queen) by Asteroid (21193), &c. See Vol. xxiii. p. 564.

Produce in	Names, &c.	By what Bull.	By whom bred.
1879, Feb. 19, r.&w., C.C.	Golden Necklace	Norman Fame, 34922	Mr. Toppin

LADY SHEFFIELD, red and white, calved May 7, 1877. Bred by Mr. J. Close, Holmescales; got by Earl of Sheffield (33812), dam (Lady Manfred) by Manfred (26801), &c. See Vol. xxiv. p. 379.

1879, June 6, r.&w., C.C.	Fair Lady	Warrior's Fame, 40889	Mr. Toppin

RUBY 4TH, roan, calved July 17, 1875. Bred by Mr. P. H. Rowlandson, Eden Bank; got by General Booth (33999), dam (Ruby 2nd) by Big Ben (23417), &c. See Vol. xxiii. p. 628.

1879, June 9, red	C.C.	Ruby 5th	Earl of Sheffield, 33812	Mr. Toppin

WARRIOR'S HOPE, red, calved April 17, 1875, Vol. xxiv. p. 681. Bred by Mr. J. Beattie, Newbie House; got by Knight of Knowlmere 2nd (31542), dam (Warrior's Prize) by Bentinck (28016), &c.

1879, Nov. 13, red,	C.C.	Warrior's Gift	Norman Fame, 34922	Mr. Toppin

TRACY, G. Murton,
Redlands, Edenbridge, Kent.

CHERRY BRANCH, red, calved December 27, 1871, Vols. xxii. and xxiii. pp. 596, 676. Bred by Mr. G. Murton Tracy; got by Cherry Prince 4th (25765), dam (Eba) by May Duke (13321), &c.

1879, Jan. 13, red,	B.C.	Wealden Cherry	King of the Weald, 38499	Mr. Tracy
1879, Dec. 16, red,	B.C.	Grand Cherry	HillhurstCh'yDuke,39999	do.

CHERRY EMPRESS, roan, calved October 22, 1872, Vols. xxii., xxiii., and xxv. pp. 596, 676, 696. Bred by Mr. G. Murton Tracy; got by Cherry Prince 4th (25765), dam (Cherry Blanche) by Dairy Prince (17655), &c.

1879, May 18, white, B.C.	Royal Cherry	King of the Weald, 38499	Mr. Tracy

CHERRY EMPRESS 2ND, red and white, calved January 13, 1875, Vols. xxiv. and xxv. pp. 682, 696. Bred by Mr. G. Murton Tracy; got by Horsa (34188), dam (Cherry Empress) by Cherry Prince 4th (25765), &c.

1879, April 17, r.&w., B.C.	Hillhurst Cherry	HillhurstCh'yDuke,39999	Mr. Tracy

CHERRY EMPRESS 3RD, red, calved March 26, 1877. Bred by Mr. G. Murton Tracy; got by Cherry Duke of Glo'ster (36348), dam (Cherry Empress 2nd) by Horsa (34188), &c. See " Imperial Cherry," p. 122.

1879, July 8, r. & w., B.C.	Imperial Cherry	HillhurstCh'yDuke,39999	Mr. Tracy

CHERRY FLOWER, roan, calved November 9, 1875, Vol. xxv. p. 696. Bred by Mr. G. Murton Tracy; got by Cherry Emperor (33347), dam (Cherry Leaf) by Red Belvedere (29724), &c.

1879, Feb. 6, r. & w., C.C.	Cherry Blossom	HillhurstCh'yDuke,39999	Mr. Tracy

CHERRY LEAF, red, calved July 24, 1873, Vols. xxii., xxiii., and xxiv. pp. 596
676, 682. Bred by Mr. G. Murton Tracy; got by Red Belvedere (29724),
dam (Eba) by May Duke (13321), &c.

Produce in	Names, &c.	By what Bull.	By whom bred.
1879, Jan. 28, r. & w., B.C.	Pioneer	HillhurstCh'yDuke,39999	Mr. Tracy

CHERRY QUEEN 2ND, red, calved June 22, 1877. Bred by Mr. G. Murton
Tracy; got by Cherry Duke of Glo'ster (36348), dam (Cherry Queen) by
Horsa (34188), &c. See " Regal Cherry," p. 209.

Produce in	Names, &c.	By what Bull.	By whom bred.
1879, Dec. 27, r. & w., B.C.	Regal Cherry	HillhurstCh'yDuke,39999	Mr. Tracy

CHERRY STALK, red and white, calved November 12, 1876. Bred by Mr. G.
Murton Tracy; got by Cherry Emperor (33347), dam (Cherry Leaf) by Red
Belvedere (29724), &c. See Vol. xxiii. p. 676.

Produce in	Names, &c.	By what Bull.	By whom bred.
1879, March 3, r. & w., C.C.	Cherry Stone	King of the Weald, 38499	Mr. Tracy

CHERRY TWIG, red and white, calved October 20, 1876. Bred by Mr. G.
Murton Tracy; got by Cherry Emperor (33347), dam (Cherry Branch) by
Cherry Prince 4th (25765), &c. See " Grand Cherry," p. 111.

Produce in	Names, &c.	By what Bull.	By whom bred.
1879, March 25, r. & w., C.C.	Cherry Tree	HillhurstCh'yDuke,39999	Mr. Tracy

PRINCESS MAYNARD, roan, calved September 8, 1875, Vol. xxv. p. 696.
Bred by Mr. G. Murton Tracy; got by Horsa (34188), dam (Cherry Princess)
by Cherry Prince 4th (25765), &c.

Produce in	Names, &c.	By what Bull.	By whom bred.
1879, Dec. 19, r. & w., C.C.	Lady Maynard	HillhurstCh'yDuke,39999	Mr. Tracy

TREDEGAR, Lord,
Tredegar Park, Newport, Mon.

BEAUTIFUL ROSE, red, calved December 20, 1875. Bred by Lord Tredegar;
got by Prince of the Blood (35160), dam (Beauty 2nd) by James 2nd (24203),
&c. See Vol. xxii. p. 597.

Produce in	Names, &c.	By what Bull.	By whom bred.
1879, Jan. 28, r. & w., B.C.	(Steer)	Royal John, 43953	Lord Tredegar

COMELY, roan, calved March 17, 1876. Bred by Lord Tredegar; got by
Prince of the Blood (35160), dam (Countess) by Lord Adolphus (20142), &c.
See " Prize Flower," Vol. xxiii. p. 677.

Produce in	Names, &c.	By what Bull.	By whom bred.
1879, Feb. 5, white, C.C.	C'tess of Monmouth	Royal John, 43953	Lord Tredegar

GILLYFLOWER 1ST, roan, calved February 16, 1876. Bred by Lord Tredegar;
got by Prince of the Blood (35160), dam (Gillyflower) by James 2nd (24203),
&c. See Vol. xxiv. p. 683.

Produce in	Names, &c.	By what Bull.	By whom bred.
1879, May 30, red, B.C.	Bassalleg	Mustapha, 34888	Lord Tredegar

GRASSHOPPER, white, calved March 10, 1875. Bred by Lord Tredegar;
got by James 2nd (24203), dam (Cricket) by Prince of Gwent (24830), &c.
See Vol. xxi. p. 966.

Produce in	Names, &c.	By what Bull.	By whom bred.
1878, Mar. 21, white, C.C.	Grass Leaf	Prince James, 37235	Lord Tredegar

LAVENDER BLOSSOM, red and white, calved April 15, 1874. Bred by
Lord Tredegar; got by James 2nd (24203), dam (Lavender) by Prince of
Gwent (24830), &c. See Vol. xxi. p. 967.

Produce in	Names, &c.	By what Bull.	By whom bred.
1878, Oct. 21, r. & w., B.C.	(Steer)	Mustapha, 34888	Lord Tredegar
1879, Dec. 21, roan, B.C.	Ivor Hael	Royal John, 43953	do.

LAVENDER GIRL, roan, calved September 18, 1875. Bred by Lord Tredegar; got by James 2nd (24203), dam (Lavender) by Prince of Gwent (24830), &c. See Vol. xxi. p. 967.

Produce in	Names, &c.	By what Bull.	By whom bred.
1879, April 21, roan, B.C.	(Steer)	Mustapha, 34888	Lord Tredegar

LITTLE BEAUTY, red and white, calved February 6, 1875. Bred by Lord Tredegar; got by James 2nd (24203), dam (Beauty) by Lord Adolphus (20142), &c. See " Sir Hugh," Vol. xxi. p. 443.

| 1878, March 17, red, C.C. | Beauty 3rd | Prince Arthur, 37219 | Lord Tredegar |
| 1879, March 18, red, C.C. | Beauty 4th | Mustapha, 34888 | do. |

MARIGOLD 3RD, roan, calved January 6, 1875. Bred by Lord Tredegar; got by James 2nd (24203), dam (Marigold) by Prince of Gwent (24830), &c. See Vol. xxi. p. 967.

| 1878, March 26, white, C.C. | Marigold 6th | Prince James, 37235 | Lord Tredegar |
| 1879, March 9, r.&w., B.C. | Ragman | Mustapha, 34888 | do. |

MARIGOLD 4TH, roan, calved November 17, 1876. Bred by Lord Tredegar; got by Duke of Craven (33609), dam (Marigold) by Prince of Gwent (24830), &c. See Vol. xxi. p. 967.

| 1879, March 16, white, C.C. | Marigold 9th | Royal John, 43953 | Lord Tredegar |

PRINCESS OF WALES, red, calved December 4, 1873. Bred by Lord Tredegar; got by Vampire (30200), dam (Princess) by Lord Adolphus (20142), &c. See "Gillyflower," Vol. xxiv. p. 683.

| 1878, Mar. 17, r.&w., C.C. | Pr'cess of Wales 2nd | Mustapha, 34888 | Lord Tredegar |
| 1879, Nov. 2, r.&w., C.C. | Pr'cess of Wales 3rd | Royal John, 43953 | do. |

PRUDENCE, red and white, calved March 12, 1877. Bred by Lord Tredegar; got by Prince James (37235), dam (Lucy Grey) by James 2nd (24203), &c. See Vol. xxiv. p. 683.

| 1879, Nov. 5, r.&w., C.C. | Pride | Royal John, 43953 | Lord Tredegar |

ROSAMOND, red, calved January 18, 1875. Bred by Lord Tredegar; got by James 2nd (24203), dam (Rosa) by Prince of Gwent (24830), &c. See Vol. xxii. p. 598.

| 1878, March 15, roan, B.C. | (Steer) | Prince Arthur, 37219 | Lord Tredegar |
| 1879, Sep. 24, { roan, B.C. / roan, C.C. | | } Royal John, 43953 | do. |

ROSAMOND 2ND, red and white, calved October 24, 1876. Bred by Lord Tredegar; got by Duke of Craven (33609), dam (Rosa) by Prince of Gwent (24830), &c. See Vol. xxii. p. 598.

| 1879, April 4, r.&w., C.C. | Rosamond 3rd | Royal John, 43953 | Lord Tredegar |

ROSE OF MAY, roan, calved March 8, 1877. Bred by Lord Tredegar; got by Duke of Craven (33609), dam (Rosy) by James 2nd (24203), &c. See Vol. xxiii. p. 678.

| 1879, Dec. 9, white, B.C. | (Steer) | Royal John, 43953 | Lord Tredegar |

SANDY GIRL, white, calved April 3, 1876. Bred by Lord Tredegar; got by Prince of the Blood (35160), dam (Snowball) by James 2nd (24203), &c. See " Prince Charlie," Vol. xxiv. p. 203.

| 879, Feb. 5, white, B.C. | (Steer) | Mustapha, 34888 | Lord Tredegar |

WATER LILY, white, calved March 17, 1876. Bred by Lord Tredegar ; got by Prince of the Blood (35160), dam (Marigold 2nd) by James 2nd (24203), &c. See Vol. xxiii. p. 677.

Produce in	Names, &c.	By what Bull.	By whom bred.
1879, Dec. 28, roan, C.C.	Marigold 10th	Royal John, 43053	Lord Tredegar

TRETHEWY, W.,
Tregoose, Grampound Road, Cornwall.

RUTH 90TH, red and white, calved June 3, 1875. Bred by Mr. W. Trethewy ; got by Crœsus (30820), dam (Ruth 48th) by Lord Montgomery (26686), g. d. (Ruth 18th) by Sir Roger (18863), &c. See Vol. xxii. p. 599.

1878, May 25, roan, C.C.	Ruth 122nd	Wallenstein, 39277	Mr. Trethewy
1879, June 13, r. & w., C.C.	Ruth 146th	M. C., 31898	do.

RUTH 95TH, red and white, calved December 26, 1875. Bred by Mr. W. Trethewy ; got by British Lion (30609), dam (Ruth 45th) by Lord Montgomery (26686), &c. See " Lord Tehidy," p. 158.

1878, June 28, r. & w., B.C.	Lord Truro	M. C., 31898	Mr. Trethewy
1879, May 19, r. & w., C.C.	Ruth 143rd	do.	do.

Lord Truro, sold to Mr. J. Nicholls, Treleage, Ruan Minor, Helstone, Cornwall.

RUTH 99TH, red and white, calved January 3, 1876. Bred by Mr. W. Trethewy ; got by British Lion (30609), dam (Ruth 65th) by Lord Montgomery (26686), g. d. (Ruth 23rd) by Rufus (35423), &c. See " Carn Brea," p. 42.

1878, Aug. 23, roan, C.C.	Ruth 132nd	M. C., 31898	Mr. Trethewy
1879, Nov. 7, roan, C.C.	Ruth 152nd	do.	do.

RUTH 102ND, roan, calved January 18, 1876. Bred by Mr. W. Trethewy ; got by British Lion (30609), dam (Ruth 71st) by Lord Montgomery (26686), g. d. (Ruth 31st) by Duke of Manchester (33690), &c. See Vol. xxii. p. 599.

1878, July 11, white, C.C.	Ruth 129th	M. C., 31898	Mr. Trethewy
1879, June 21, roan, C.C.	Ruth 147th	do.	do.

TURNER, John,
The Grange, Ulceby, Lincolnshire.

BONNY BELLE, red, calved May 18, 1873, Vol. xxiv. p. 689. Bred by Mr. J. Turner ; got by Marquis of Thorndale (29304), dam (Young Beauty 2nd) by Paradox 1st (22487), &c.

1879, March 3, r. & w., C.C.	Mordem Belle	G. Duke of Glo'ster, 36721	Mr. Turner

CAROLINA, roan, calved June 7, 1877. Bred by Mr. J. Turner ; got by Geneva's Oxford (34025), dam (Creole) by General (19836), &c. See Vol. xxiv. p. 690.

1879, Oct. 14, roan, B.C.	Uncle Jonathan	G. Duke of Glo'ster, 36721	Mr. Turner

CHLOE, red, calved February 28, 1874, Vol. xxiv. p. 690. Bred by Mr. J. Turner ; got by Marquis of Thorndale (29304), dam (Creole) by General (19836), &c.

1879, March 6, red, B.C.	Uncle Tom	G. Duke of Glo'ster, 36721	Mr. Turner

FAWSLEY CHERRY 5TH, red, calved November 9, 1874, Vols. xxiv. and
xxv. pp. 690, 700. Bred by Sir G. R. Philips, Bart., Weston Park; got by
Cherry Grand Duke 5th (30712), dam (Fawsley Cherry 2nd) by Third Cherry
Duke (28171), &c.

Produce in	Names, &c.	By what Bull.	By whom bred.
1879, July 16, roan, C.C.	Fawsley Garland 3rd	G. Duke of Glo'ster, 36721	Mr. Turner

MAY'S ROSETTE, roan, calved December 11, 1876. Bred by Mr. J. Turner;
got by Geneva's Oxford (34025), dam (Rosette 4th) by Marquis of Thorndale (29304), g. d. (Rosette 3rd) by Grand Duke 6th (19876), &c. See
" Roland," p. 213.

Produce in	Names, &c.	By what Bull.	By whom bred.
1879, July 17, roan, B.C.	(Steer)	G. Duke of Glo'ster, 36721	Mr. Turner

MORNING STAR, red, calved January 2, 1875. Bred by Mr. J. Turner;
got by King of the Greeks (31503), dam (Beautiful Star) by Duke Ulric
(28482), &c. See Vol. xxii. p. 602.

Produce in	Names, &c.	By what Bull.	By whom bred.
1879, March 28, red, B.C.	Aldebaran	G. Duke of Glo'ster, 36721	Mr. Turner

PRINCESS BEATRICE, roan, calved August 11, 1874, Vol. xxiv. p. 350.
Bred by Lord Chesham, Latimer; got by Eighth Duke of Geneva (28390),
dam (The Queen) by Lord Oxford 2nd (20215), &c.

Produce in	Names, &c.	By what Bull.	By whom bred.
1879, April 15, white, C.C.	Lady Mary	G. Duke of Glo'ster, 36721	Mr. Turner

PRINCESS ELEANOR, roan, calved November 26, 1876. Bred by Mr. J.
Turner; got by Twenty-fourth Duke of Airdrie (36460), dam (Princess May)
by Second Duke of Glo'ster (28392), &c. See Vol. xxiii. p. 685.

Produce in	Names, &c.	By what Bull.	By whom bred.
1879, Aug. 1, roan, C.C.	Princess Eleanor 2nd	G. Duke of Glo'ster, 36721	Mr. Turner

PRINCESS MAY, white, calved January 23, 1874, Vols. xxiii. and xxv.
pp. 685, 700. Bred by Mr. J. Turner; got by Second Duke of Glo'ster
(28392), dam (Princess Claro) by Second Duke of Claro (21576), &c.

Produce in	Names, &c.	By what Bull.	By whom bred.
1879, April 10, white, B.C.	Glo'ster's Prince	G. Duke of Glo'ster, 36721	Mr. Turner

PRINCESS SIBYL, white, calved August 17, 1876. Bred by Mr. J. Turner;
got by Geneva's Oxford (34025), dam (Roan Princess) by Grand Duke of
Clarence (28750), &c. See Vol. xxiii. p. 685.

Produce in	Names, &c.	By what Bull.	By whom bred.
1879, July 29, roan, C.C.	Princess Sibyl 2nd	G. Duke of Glo'ster, 36721	Mr. Turner

QUEEN MAB, red, calved May 29, 1872, Vols. xxii., xxiii., and xxv. pp. 602,
685, 700. Bred by Mr. J. Turner; got by Earl of Thorndale (28520), dam
(Fairy Queen) by Duke of Ulster (12774), &c.

Produce in	Names, &c.	By what Bull.	By whom bred.
1879, Nov. 2, r. & w., B.C.	Erl King	G. Duke of Glo'ster, 36721	Mr. Turner

ROAN PRINCESS, roan, calved January 9, 1872, Vols. xxi. and xxiii. pp. 971,
685. Bred by Mr. R. P. Davies, Horton; got by Grand Duke of Clarence
(28750), dam (White Princess) by Grand Duke 3rd (16182), &c.

Produce in	Names, &c.	By what Bull.	By whom bred.
1879, Mar. 31, red, C.C.	Princess Dol	G. Duke of Glo'ster, 36721	Mr. Turner

SIBYL'S ROSETTE, roan, calved December 5, 1875. Bred by Mr. J. Turner;
got by Geneva's Oxford (34025), dam (Rosette 3rd) by Grand Duke 6th
(19876), &c. See " Roland," p. 213.

Produce in	Names, &c.	By what Bull.	By whom bred.
1879, May 7, r. & w., C.C.	Sibyl's Rosette 2nd	G. Duke of Glo'ster, 36721	Mr. Turner

SUNSET, roan, calved July 13, 1876. Bred by Mr. J. Brown, Burton Constable; got by Duke of Edinburgh (33629), dam (Sunrise) by Mameluke
(13299), &c. See Vol. xxiv. p. 350.

Produce in	Names, &c.	By what Bull.	By whom bred.
1879, Aug. 22, white, B.C.	(Steer	G. Duke of Glo'ster, 36721	Mr. Turner

SYRIAN BELLE, red and white, calved September 19, 1866, Vols. xix., xx., xxii., xxiii., xxiv., and xxv. pp. 750, 787, 602, 685, 691, 700. Bred by Mr. S. Rich, Didmarton ; got by Lord Lally (22161), dam (Queen of Tyre) by Archduke (17316), &c.

Produce in		Names, &c.	By what Bull.	By whom bred.
1879, Oct. 29, roan,	B.C.	Grand Vizier	G. Duke of Glo'ster, 36721	Mr. Turner

UNA, red and white, calved October 8, 1871, Vols. xxi., xxii., xxiii., xxiv., and xxv. pp. 971, 602, 685, 691, 700. Bred by Mr. J. Turner ; got by Marquis of Thorndale (29304), dam (Ursula 32nd) by Second Duke of Waterloo (23800), &c.

1879, June 6, r. & w.,	B.C.	Sir Tristram	G. Duke of Glo'ster, 36721	Mr. Turner

UNA 2ND, red and white, calved March 3, 1873, Vols. xxiii., xxiv., and xxv. pp. 686, 691, 700. Bred by Mr. J. Turner ; got by Marquis of Thorndale (29304), dam (Ursula 32nd) by Second Duke of Waterloo (23800), &c.

1879, Jan. 10, roan,	C.C.	Una 8th	G. Duke of Glo'ster, 36721	Mr. Turner

UNA 3RD, red and white, calved April 24, 1874, Vols. xxiii. and xxv. pp. 686, 700. Bred by Mr. J. Turner ; got by King of the Greeks (31503), dam (Una) by Marquis of Thorndale (29304), &c.

1879, Feb. 3, red,	B.C.	(dead)	G. Duke of Glo'ster, 36721	Mr. Turner

UNA 4TH, roan, calved May 21, 1875. Bred by Mr. J. Turner ; got by Geneva's Oxford (34025), dam (Una) by Marquis of Thorndale (29304), &c. See " Sir Tristram," p. 236.

1879, Aug. 2, roan,	C.C.	Una 9th	G. Duke of Glo'ster, 36721	Mr. Turner

TWEDDELL, John,
Dunstan West Farm, Whickham, Gateshead-on-Tyne.

VICTORIA 5TH, red, calved May 21, 1872, Vol. xxv. p. 701. Bred by Mr. R. Coulson, Coastley ; got by Earl of Derwent (28503), dam (Victoria 4th) by Cherry Prince (17554), &c.

1879, July 9, {	roan, C.C.	Victoria 7th	} Lion of Flanders, 34460	Mr. Tweddell
	roan, C.C.	Victoria 8th		

TWEEDIE, Richard,
The Forest, Catterick, Yorkshire.

GIPSY COUNTESS, red, calved February 11, 1875, Vol. xxv. p. 701. Bred by Mr. R. Tweedie ; got by Job (31438), dam (Countess) by Bumper (19371), &c.

1879, Mar. 20, red,	C.C.	Gipsy Maid	British Herald, 39504	Mr. Tweedie

RARE TULIP, red and white, calved June 3, 1876. Bred by Mr. J. Tweedie, Deuchrie ; got by Job (31438), dam (Red Tulip) by Raymond Booth (29720), &c. See Vol. xxiii. p. 686.

1878, Aug. 7, r. & w.,	C.C.	Regal Tulip	British Grenadier, 33217	Mr. J. Tweedie
1879, Oct. 10, r. & w.,	C.C.	Royal Tulip	do.	Mr. R. Tweedie

RED BELL, red, calved February 23, 1872, Vol. xxii. p. 604. Bred by Mr. R. Tweedie ; got by Royal Clan Booth (29853), dam (Red Blossom) by Professor Miller (22657), &c.

1878, Feb. 26, r. & w.,	B.C.	(Steer)	Job, 31438	Mr. Tweedie
1879, Mar. 13, red,	C.C.	Red Flower	British Herald, 39504	do.

ROSE CLAN ALPINE, red and white, calved January 30, 1875. Bred by Mr.
R. Tweedie; got by Job (31438), dam (Rose Clan) by Royal Clan Booth
(29853), &c. See Vol. xxii. p. 604.

Produce in		Names, &c.		By what Bull.		By whom bred.
1879, Feb. 12, red,	C.C.	Rose of the Glen		British Herald, 39504		Mr. Tweedie

ROSE McPHEDRAN, roan, calved March 9, 1877. Bred by Mr. R. Tweedie;
got by Job (31438), dam (Rose McIvor) by Booth Satellite (30564), &c. See
Vol. xxiv. p. 692.

1879, April 3, r. & w., C.C.	Rose McPhie	British Herald, 39504	Mr. Tweedie

TYACKE, John,
Merthen, Helston, Cornwall.

LINDA GWYNNE, roan, calved June 14, 1876. Bred by Mr. J. Tyacke; got
by Monarch Gwynne (37103), dam (Lady Bird) by Don Pedro (25910), &c.
See Vol. xxiii. p. 687.

1878, Dec. 20, roan, C.C.	Leona Dare	Vain Glory, 39248	Mr. Tyacke

SYRINGA, white, calved June 22, 1876. Bred by Mr. J. Tyacke; got by
Monarch Gwynne (37103), dam (Syren 2nd) by Don Pedro (25910), &c. See
"Snowdon," p. 238.

1879, May 11, white, B.C.	Snowdon	Plantaganet, 38871	Mr. Tyacke

TYNTE, J. P.,
Tynte Park, Dunlavin, Co. Wicklow.

BLOOMER, red and white, calved April 15, 1872, Vol. xxiii. p. 688. Bred by
Mr. J. P. Tynte; got by British Boy (25675), dam (Sweetheart) by Bright
Spur (25671), &c.

1877, May 7, red, B.C.	Frill	Royal Fern, 29865	Mr. Tynte
1879, Mar. 10, r. & w., C.C.		do.	do.

Bull Calf, sold to Mr. Buchanan, Harristown, Brannockstown.

BONNY BELL, roan, calved October 10, 1871, Vol. xxii. p. 605. Bred by Mr.
J. P. Tynte; got by British Boy (25675), dam (Dew Drop) by British Duke
(19350), &c.

1878, Feb. 7, r. & w., C.C.	Golden Bell	Leogaire, 38559	Mr. Tynte
1879, June 15, roan, C.C.	Marriage Bell	Martello, 40319	do.

BRACKEN, roan, calved April 24, 1872, Vol. xxiii. p. 688. Bred by Mr. J. P.
Tynte; got by British Boy (25675), dam (Maidenhair) by Ravenspur
(20628), &c.

1877, Dec. 31, roan, C.C.	(dead)	British Peer, 33224	Mr. Tynte
1879, May 21, white, C.C.	Lamaria	do.	do.

BRIGANTINE, red and white, calved March 31, 1872, Vol. xxii. p. 605. Bred
by Mr. J. P. Tynte; got by British Boy (25675), dam (Starlight) by Sheet
Anchor (18820), &c.

1878, Jan. 28, red, B.C.	Felucca	Royal Fern, 29865	Mr. Tynte
1879, Feb. 19, red, C.C.		do.	do.

Bull Calf, sold to Mr. T. Davidson, Ballyneal, Queen's Co.

FAITHFUL, roan, calved January 18, 1876. Bred by Mr. J. P. Tynte; got by Royal Fern (29865), dam (Sybil) by Bright Spur (25671), g. d. (Hebe) by Hohenlohe (18074), &c. See "Landlord," p. 137.

Produce in	Names, &c.	By what Bull.	By whom bred.
1879, Aug. 24, roan, C.C.	Loyalty	British Peer, 33224	Mr. Tynte

FLOUNCE, red, calved April 10, 1875. Bred by Mr. J. P. Tynte; got by Royal Fern (29865), dam (Bloomer) by British Boy (25675), &c. See Vol. xxiii. p. 688.

Produce in	Names, &c.	By what Bull.	By whom bred.
1878, Feb. 19, r. & w., B.C.		Leoguire, 38559	Mr. Tynte
1879, May 5, r. & w., C.C.	Lappet	British Peer, 33224	do.

Bull Calf, sold to Mr. T. Davidson, Ballyneal, Queen's County.

FRANGIPANNI, roan, calved February 6, 1875, Vol. xxv. p. 703. Bred by Mr. J. P. Tynte; got by Royal Fern (29865), dam (Blanc Mange) by British Boy (25675), &c.

Produce in	Names, &c.	By what Bull.	By whom bred.
1879, May 14, roan, C.C.	Lozenge	British Peer, 33224	Mr. Tynte

FROLIC, red, calved August 2, 1873, Vol. xxv. p. 704. Bred by Mr. J. P. Tynte; got by Royal Fern (29865), dam (Playful) by The Peer (23045), &c.

Produce in	Names, &c.	By what Bull.	By whom bred.
1879, April 5, roan, C.C.	Lively	British Peer, 33224	Mr. Tynte

LADY FLORA, red, calved September 5, 1873, Vol. xxv. p. 704. Bred by Mr. J. P. Tynte; got by Royal Fern (29865), dam (Florence) by British Flag (19351), &c.

Produce in	Names, &c.	By what Bull.	By whom bred.
1879, May 8, roan, C.C.	Lady Loo	British Peer, 33224	Mr. Tynte

MIRTH, red and white, calved December 5, 1873, Vol. xxv. p. 704. Bred by Mr. J. P. Tynte; got by Mac Cullum Mohr (31789), dam (Blithe) by British Boy (25675), &c.

Produce in	Names, &c.	By what Bull.	By whom bred.
1879, Aug. 4, r. & w., C.C.	Merry Lass	Martello, 40319	Mr. Tynte

SOLO, roan, calved June 11, 1871, Vol. xxii. p. 606. Bred by Mr. J. P. Tynte; got by Bright Spur (25671), dam (Pasta) by The Peer (23045), &c.

Produce in	Names, &c.	By what Bull.	By whom bred.
1878, July 13, roan, B.C.		Royal Fern, 29865	Mr. Tynte
1879, Aug. 14, red. C.C.	Maritana	Martello, 40319	do.

Bull Calf, sold to Mr. Kelly, Craigue, Co. Carlow.

SUNRISE, red, calved June 27, 1868, Vols. xxiii. and xxv. pp. 689, 704. Bred by Mr. J. P. Tynte; got by Bright Spur (25671), dam (Dawn) by British Duke (19350), &c.

Produce in	Names, &c.	By what Bull.	By whom bred.
1879, Mar. 3, red, C.C.	Fair Morning	Royal Fern, 29865	Mr. Tynte

USHAW, William,
Lissett, Lowthorpe, Hull.

LADY BEAUMONT 5TH, roan, calved September 23, 1871, Vols. xxi., xxii., and xxv. pp. 928, 561, 662. Bred by Mr. R. B. Hetherington, Park Head; got by Grand Duke of Lightburne 2nd (26291), dam (Lady Beaumont 2nd) by Sir Walter Gwynne (22921), &c.

Produce in	Names, &c.	By what Bull.	By whom bred.
1879, July 4, roan, C.C.	Lady Beaumont 9th	Lord Cockburn, 38594	Mr. Ushaw

UTTING, H. A.,
Hockering, East Dereham, Norfolk.

FRAGRANCE, red and white, calved March 26, 1877. Bred by Mr. H. A. Utting; got by Sultan (35697), dam (Farewell) by Robert Peel (29796), &c. See Vol. xxiv. p. 693.

Produce in		Names, &c.	By what Bull.	By whom bred.
1879, Oct. 3, red,	B.C.	(Steer)	Bright Damon, 36268	Mr. Utting

SYLPH, red and white, calved October 5, 1876. Bred by Mr. H. A. Utting; got by Sultan (35697), dam (Scarlet) by Robert Peel (29796), &c. See Vol. xxii. p. 607.

1878, Dec. 14, red,	B.C.	(dead)	Bright Damon, 36268	Mr. Utting
1879, Dec. 1, r. & w.,	C.C.	Sparkle	do.	do.

VIVIAN, H. Hussey,
Park Wern, Swansea, Glamorganshire.

FAIRY BELLE, roan, calved April 9, 1873. Bred by Mr. W. Arkell, Dudgrove, the property of Mr. H. H. Vivian; got by Viceroy (30216), dam (Formosa) by Research (27270), g. d. (Folly) by Royal Arch 3rd (27346), gr. g. d. (Flourish) by The Slave (27647), &c. See Vol. xviii. p. 500.

1879, April 7, white,	C.C.	Fairy Belle 2nd	L. Fitzclarence 16th, 36943	Mr. Arkell

FORMOSA, red and white, calved April 21, 1869. Bred by Mr. W. Arkell, Dudgrove, the property of Mr. H. H. Vivian; got by Research (27270), dam (Folly) by Royal Arch 3rd (27346), &c.

1873, April 9, roan,	C.C.	Fairy Belle	Viceroy, 30216	Mr. Arkell
1879, Feb. 22, r. & w.,	C.C.	Forget-me-not	Baron Lee 4th, 37799	do.

SWEETBRIAR 2ND, roan, calved April 10, 1877. Bred by Mr. W. Arkell, Dudgrove; got by Theodore (37586), dam (Sunshine) by Fitz Walton (28609), g. d. (Satellite) by Sixteenth Baron Wetherby (25597), gr. g. d. (Snowdrop) by Brazenose (15686), &c. See Vol. xviii. p. 732.

1879, Oct. 21, white,	B.C.	(dead)	Baron Lee 5th, 39420	Mr. Vivian

TRIP AWAY, roan, calved March 26, 1875. Bred by Mr. W. Arkell, Dudgrove; got by Duke of Fawsley (28387), dam (Thrifty) by Dalesman (25869), &c. See "Baron Fitzclarence," p. 14.

1879, Sept. 21, roan,	B.C.	Baron Fitzclarence	L. Fitzclarence 16th, 36943	Mr. Vivian

WAIND, John,
Ankness, Kirby Moorside, Yorkshire.

GIPSY GIRL, roan, calved May 23, 1877. Bred by Mr. J. Waind; got by Duke of Rubies (33724), dam (Gipsy Maid) by Snowstorm (35616), &c. See "Vladimir's Ling Cropper," p. 261.

879, Dec. 25, r. & w.,	B.C.	Vladimir's L.Crop'er	Count Vladimer, 33461	Mr. Waind

WAKEFIELD, W. H.,
Sedgwick, Kendal.

GARNETT 9TH, roan, calved April 19, 1876. Bred by Mr. W. H. Wakefield;
got by Baron Barrington 4th (33006), dam (Garnett 4th) by Dunrobin (28486),
&c. See "Garnett Lad," p. 105.

Produce in		Names, &c.	By what Bull.	By whom bred.
1878, Aug. 28, roan,	B.C.	Garnett Lad	Duke of Holker, 38153	Mr. Wakefield

GARNETT 10TH, roan, calved May 8, 1877. Bred by Mr. W. H. Wakefield; got
by Baron Barrington 4th (33006), dam (Garnett 3rd) by Frederick Warlaby
(23990), &c. See "Squire Garnett," p. 240.

| 1879, Aug. 30, roan, | C.C. | Garnett 16th | Duke of Holker, 38153 | Mr. Wakefield |

GARNETT 11TH, roan, calved July 31, 1877. Bred by Mr. W. H. Wakefield; got
by Baron Barrington 4th (33006), dam (Garnett 2nd) by Frederick Warlaby
(23990), &c. See "Sandy," Vol. xxiv. p. 235.

| 1879, Nov. 24, white, | B.C. | (Steer) | Duke of Holker, 38153 | Mr. Wakefield |

GARNETT 12TH, roan, calved July 31, 1877. Bred by Mr. W. H. Wakefield; got
by Baron Barrington 4th (33006), dam (Garnett 2nd) by Frederick Warlaby
(23990), &c.

| 1879, Oct. 25, white, | C.C. | Garnett 14th | Duke of Holker, 38153 | Mr. Wakefield |

ROANY 3RD, roan, calved April 5, 1876. Bred by Mr. W. H. Wakefield; got
by Lord of Garsdale (34617), dam (Roany 2nd) by Dunrobin (28486), &c.
See "Count Moltke," p. 57.

| 1879, June 12, roan, | B.C. | Count Moltke | Duke of Holker, 38153 | Mr. Wakefield |

SEDGWICK 2ND, roan, calved September 20, 1875. Bred by Mr. W. H.
Wakefield; got by Baron Barrington 4th (33006), dam (Sedgwick Lass) by
Dunrobin (28486), &c. See "Sedgwick Lad 3rd," p. 226.

| 1879, Aug. 29, red, | C.C. | Sedgwick Lass 3rd | Duke of Holker, 38153 | Mr. Wakefield |

WELCOME 2ND, roan, calved March 24, 1877. Bred by Mr. W. H. Wakefield;
got by Baron Barrington 4th (33006), dam (Welcome) by Dunrobin (28486),
g. d. (Banks 3rd) by Frederick Warlaby (23990), &c. See "Sedgwick Lad
3rd," p. 226.

| 1879, July 18, roan, | C.C. | Welcome 4th | Duke of Holker, 38153 | Mr. Wakefield |

WAKEFIELD, W. T.,
Fletchampstead Hall, Coventry.

BLOSSOM, roan, calved March 11, 1871, Vols. xxi., xxiii., and xxv. pp. 537,
308, 310. Bred by Mr. G. Allen, Knightley Hall, the property of Mr. C. A.
Barnes, Charleywood; got by Eighth Duke of York (28480), dam (Peach
Blossom 2nd) by Baron Westbury (19287), &c.

| 1879, Sept. 12, roan, | C.C. | Rowley's Blossom | 2nd Duke of Rowley, 28441 | Mr. Allen |

Rowley's Blossom, sold to Mr. W. T. Wakefield.

BRITISH QUEEN, roan, calved February 1, 1873, Vols. xxii. and xxiii. pp. 607,
691. Bred by Mr. W. T. Wakefield; got by Duke of Darlington 2nd (28378),
dam (Gipsy Queen) by Grand Duke of Essex 5th (26286), &c.

| 1878, May 26, roan, | C.C. | (dead) | Lord Darlingt'n 13th, 38597 | Mr. Wakefield |
| 1879, June 6, white, | B.C. | White Prince | Lord Darlingt'n 16th, 40151 | do. |

VINE LEAF, red and white, calved December 16, 1876. Bred by Mr. W. T. Wakefield; got by Duke of Brailes 2nd (30930), dam (Victoria) by Grand Duke 19th (19879), &c. See "Prince Victor," p. 203.

Produce in	Names, &c.	By what Bull.	By whom bred.
1879, Jan 20, r. & w., B.C.	Prince Victor	Thorndale G. D. 2nd, 39218	Mr. Wakefield

VINTAGE, red, calved November 14, 1876. Bred by Mr. W. T. Wakefield; got by Duke of Brailes 2nd (30930), dam (Vanity) by Grand Duke 9th (19879), &c. See Vol. xxiii. p. 691.

1879, June 24, r. & w., C.C.	Vine 7th	Lord Darlingt'n 16th, 40151	Mr. Wakefield

VIRTUE, red, calved January 10, 1876, Vol. xxv. p. 705. Bred by Mr. W. T. Wakefield; got by Duke of Brailes 2nd (30930), dam (Victoria) by Grand Duke 9th (19879), &c.

1879, April 14, { roan, C.C.	(dead)		Baron Shendish 4th, 41062	Mr. Wakefield
roan, B.C.	Marquis			

WALKER, T. Eades,
Studley Castle, Warwickshire.

COUNTESS, red and white, calved February 25, 1875, Vol. xxiv. p. 694. Bred by Mr. T. E. Walker; got by Lord Royal (34662), dam (Ursula 29th) by Grand Duke 13th (21850), &c.

1878, Oct. 11, roan, C.C.	Countess 6th	G.D. of Studley 2nd, 39954	Mr. Walker

COUNTESS 2ND, white, calved December 31, 1875. Bred by Mr. T. E. Walker; got by Lord Royal (34662), dam (Ursula 29th) by Grand Duke 13th (21850), &c. See Vol. xxiii. p. 692.

1878, April 18, roan, B.C.	Bannerman	G.D. of Studley 2nd, 39954	Mr. Walker

Bannerman, sold to Mr. R. D. Howarth, Bradwall, Sandbach.

DEODORA, roan, calved June 22, 1868, Vols. xxiii. and xxiv. pp. 692, 694. Bred by Mr. Mann, The Asps; got by Bull's Bay (23490), dam (Dinorah) by Baron Fleda (21221), &c.

1878, Dec. 29, white, C.C.	Deodora 5th	G.D. of Studley 2nd, 39954	Mr. Walker

DEODORA 2ND, roan, calved May 2, 1874, Vol. xxv. p. 706. Bred by Mr. T. E. Walker; got by Lord Royal (34662), dam (Deodora) by Bull's Bay (23490), &c.

1879, Aug. 4, white, B.C.	Cedrus	G.D. of Studley 2nd, 39954	Mr. Walker

DEODORA 3RD, red and white, calved March 3, 1876. Bred by Mr. T. E. Walker; got by Lord Royal (34662), dam (Deodora) by Bull's Bay (23490), &c. See Vol. xxiii. p. 692.

1879, Feb. 5, roan, B.C.	Atlantica	G.D. of Studley 2nd, 39954	Mr. Walker

EMILY, roan, calved May 23, 1872, Vols. xxiii. and xxv. pp. 692, 706. Bred by Mr. T. Walker, Berkswell Hall; got by Friponnier (26208), dam (Olga) by Second Duke of Cumberland (23735), &c.

1879, Jan. 25, white, C.C.	Emily 4th	G.D. of Studley 2nd, 39954	Mr. T. E. Walker

MINNIE, roan, calved January 5, 1870, Vols. xxi., xxiii., and xxiv. pp. 757, 487, 488. Bred by Mr. T. Harris, Stonylane House; got by Festival (26147), dam (Gertrude) by Diplomatist (19571), &c.

1879, Jan. 7, white, B.C.	Bromsgrove	Waterloo Prince, 35949	Mr. Walker

URSULA 29TH, roan, calved April 15, 1868, Vols. xx., xxiii., and xxv. pp. 796, 692, 706. Bred by Mr. S. Rich, Didmarton ; got by Grand Duke 13th (21850), dam (Ursula 12th) by General Canrobert (12927), &c.

Produce in	Names, &c.	By what Bull.	By whom bred.
1879, April 3, white, C.C.	Countess 6th	G.D. of Studley 2nd, 39954	Mr. Walker

WALKER, W. T.,
Melbourne Grange, Pocklington, Yorkshire.

DUCHESS 2ND, roan, calved December 20, 1872. Died by Mr. R. Tennant, Scarcroft Lodge, the property of Mr. J. Wilkinson ; got by Duke of Clarence (19611), dam (Miss Farewell) by Second Duke of Wharfdale (19649), &c. See Vol. xxiv. p. 673.

| 1878, Apr. 13, roan, C.C. | Bell Duchess 2nd | D. of Tregunter 5th, 33743 | Mr. Bell |

Bell Duchess 2nd, sold to Mr. W. T. Walker.

OXFORD ROSE 3RD, roan, calved March 23, 1874, Vols. xxiv. and xxv. pp. 303, 319. Bred by Sir W. G. Armstrong, Cragside ; got by Oxford le Grand (29496), dam (Wharfdale Rose) by Third Lord Wharfdale (26759), &c.

| 1879, Nov. 29, r. & w., C.C. | D'ss of Melbourne | W. D. of Geneva 2nd, 39318 | Mr. Walker |

WALSINGHAM, Lord,
Merton Hall, Thetford, Norfolk.

MANCHESTER LASS 4TH, roan, calved August 7, 1873, Vols. xxiii. and xxiv. pp. 693, 695. Bred by Lord Walsingham ; got by Fancy Boy (38270), dam (Manchester Lass 2nd) by Oxford Marshal (27022), &c.

| 1879, Aug. 8, red, C.C. | Manchester Lady | D. of Britford 3rd, 38125 | Lord Walsingham |

WALTER, John,
Bearwood, Wokingham.

JULIA, roan, calved April 8, 1876. Bred by Mr. J. Stratton, Alton Priors ; got by Royal James (35387), dam (Jenny Deans) by Moonraker (22383), &c. See Vol. xxiii. p. 659.

| 1878, March 19, roan, B.C. | Julius | Caractacus, 36315 | Mr. Walter |
| 1879, Oct. 19, roan, B.C. | Bearwood Butterfly | Charlton Butterfly, 37968 | do. |

LADY IN WAITING, roan, calved October 26, 1875. Bred by Archdeacon Holbech, Farnborough Hall ; got by Earl of Chatham (28495), dam (Lady in White) by Count Bickerstaffe (23630), &c. See Vol. xx. p. 599.

| 1878, Sept. 2, white, B.C. | Lord in Waiting 2nd | Caractacus, 36315 | Mr. Walter |
| 1879, June 27, white, B.C. | (Steer) | do. | do. |

Lord in Waiting 2nd, sold to Mr. J. Sharp, White Waltham, Twyford.

MAIDEN, white, calved March 7, 1877. Bred by Archdeacon Holbech, Farnborough Hall ; got by The Bursar (35742), dam (Memory) by Eleventh Duke of Oxford (19632), &c. See Vol. xxv. p. 515.

| 1879, Dec. 3, white, C.C. | Melon | Charlton Butterfly, 37968 | Mr. Walter |

MERRY MAID, red and white, calved October 6, 1875.   Bred by Mr. J. Walter;
   got by Evanus (31120), dam (Maid of Bearwood) by Favourite (26140), &c.
   See Vol. xxiii. p. 694.

Produce in	Names, &c.	By what Bull.	By whom bred.
1878, Dec. 21, r. & w., C.C.	Maid of Honour	Caractacus, 36315	Mr. Walter

PRAIRIE BUD, red and white, calved January 7, 1877.   Bred by Archdeacon
   Holbech, Farnborough Hall; got by The Bursar (35742), dam (Prairie Clove)
   by Second Earl of Cleveland (26054), &c.   See Vol. xxi. p. 769.

1879, Oct. 6, roan, B.C.	Peter	Charlton Butterfly, 37968	Mr. Walter

WARD, J. Gardner,
The Villa, Oxhill, Kineton, Warwick.

BOUNCE, red, calved March 8, 1876.   Bred by Mr. J. G. Ward; got by Joe
   (31439), dam (Bella Boyd) by Goliah (24053), &c.   See Vol. xxi. p. 979.

1879, April 11, red, B.C.	Frank	Jago, 34239	Mr. Ward

Frank, sold to Mr. Gardner, Tysoe, Warwickshire.

FRY, red and white, calved March 12, 1874.   Bred by Mr. J. G. Ward; got by
   Joe (31439), dam (Fritter) by Goliah (24053), &c.   See Vol. xxi. p. 979.

1877, April 30, roan, C.C.	Friendship	Jago, 34239	Mr. Ward
1878, April 27, roan, C.C.	Frontlet	do.	do.
1879, Mar. 19, r. & w., B.C.	Foster	do.	do.

Friendship, sold to Mr. Buttermer, for exportation to The Cape: Frontlet, to Mr. H. T.
   Chamberlayne, Stoney, Thorpe; Foster, to Mr. W. Bull, Woolford, Shipston-on-Stour.

LARKSPUR, red, calved May 4, 1876.   Bred by Mr. J. G. Ward; got by Joe
   (31439), dam (Lonely) by Frightenem (28658), &c.   See Vol. xxi. p. 980.

1879, April 3, roan. B.C.	Landseer	Jago, 34239	Mr. Ward

Landseer, sold to Mr. G. Canning, Shottery, Stratford, Warwickshire.

VICTORIA 3RD, red and white, calved March 24, 1875.   Bred by Mr. J. G.
   Ward; got by Joe (31439), dam (Victoria) by Goliah (24053), &c.   See
   Vol. xxi. p. 980.

1878, May 4, r. & w., B.C.	Viceroy	Jago, 34239	Mr. Ward
1879, Apr. 10, r. & w., B.C.	(dead)	do.	do.

Viceroy, sold to Dr. Fyfe, Kineton, Warwick.

WARD, Robert,
Sodstone House, Narberth.

ALICE CLARENCE 2ND, red, calved in June, 1875.   Bred by Mr. R. P. Davies,
   Horton, the property of Mr. R. Ward; got by Royal Claro (29854), dam
   (Alice Clarence) by Grand Duke of Clarence (28750), &c.   See Vol. xxiii. p. 414.

1879, May 15, r. & w., C.C.	Princess Alice	Squire Barrington, 39160	Mr. Davies

ALICE CLARO, white, calved October 16, 1870, Vol. xxi. p. 664.   Bred by R.
   P. Davies, Horton, the property of Mr. R. Ward; got by Second Duke of
   Claro (21576), dam (Alice Wetherby) by General Wetherby (24026), &c.

1879, Mar. 26, white, B.C.	Don Claro	Squire Barrington, 39160	Mr. Davies

GWENDOLINE, roan, calved July 9, 1874. Bred by Mr. A. Baird, Robeston Hall, the property of Mr. R. Ward ; got by Lord Darlington 3rd (31646), dam (Garland) by Ernest (31116), &c. See Vol. xxi. p. 556.

Produce in	Names, &c.	By what Bull.	By whom bred.
1879, Dec. 9, roan,	B.C. Ganymede	Earl of Horton 5th, 33797	Mr. Baird

HONEYSUCKLE 4TH, red and white, calved March 10, 1876. Bred by Mr. R. P. Davies, Horton, the property of Mr. R. Ward ; got by Earl of Horton 5th (33797), dam (Honeysuckle 3rd) by Royal Claro (29854), g. d. (Honeysuckle) by Archbishop (17312), &c. See Vol. xxi. p. 665.

1879, Feb. 11, red,	B.C. Romeo	Squire Barrington, 30160	Mr. Davies

WARD, S. R. Chapman,
Neasham Hill, Darlington.

LADY WORCESTER, red and white, calved March 13, 1874, Vol. xxiv. p. 483. Bred by Mr. J. W. Wilson, Broadway, the property of Mr. S. R. C. Ward ; got by Severn Lad (29959), dam (Woman in Red) by Duke of Darlington (21586), &c.

1879, Jan. 1, red,	C.C. Lady Sockburn	Duke of Sockburn, 33734	Mr. Lyon

LEODINE 3RD, red and white, calved March 11, 1876. Bred by Mr. J. Gordon, Cluny Castle ; got by Friar of Knowlmere (33975), dam (Leodine) by King Charles (24240), &c. See Vol. xxiii. p. 461.

1879, Mar. 20, r. & w.,	C.C. The Nun	Sir W'sor Broughton, 27507	Mr. Ward

PEARL, roan, calved May 21, 1874, Vol. xxv. p. 707. Bred by Mr. D. Nesham, Hall Farm, Gainford ; got by Boaz (30552), dam (Young Pauline) by Vesuvian (35862), &c.

1879, Aug. 8, roan,	C.C. Neasham Pearl	Lord Ripon, 38656	Mr. Ward

ROAN DUCHESS, roan, calved November 23, 1874, Vol. xxv. p. 707. Bred by Mr. D. Nesham, Hall Farm, Gainford ; got by Boaz (30552), dam (Gentle Duchess) by Marmaduke 2nd (22287), &c.

1879, March 24, roan,	C.C. Neasham Star	Lord Ripon, 38656	Mr. Ward

SCOTHERN DUCHESS 2ND, roan, calved December 10, 1876. Bred by Mr. A. Garfit, Scothern ; got by Lord of Scothern (34626), dam (Red and White Duchess) by Prince Imperial (27150), &c. See "Neasham Duke," p. 174.

1879, Jan. 27, roan,	C.C. Scothern Duchess 4th	Grand Duke 25th, 34065	Mr. Garfit

Scothern Duchess 4th, sold to Mr. S. R. C. Ward.

WARREN, E. L.,
Lodge Park, Freshford, Co. Kilkenny.

MIDGE, roan, calved March 30, 1871, Vols. xxi., xxii., xxiii., xxiv., and xxv. pp. 981, 608, 697, 697, 709. Bred by Mr. E. L. Warren ; got by Rimmel (29779), dam (Frederica) by Master Harbinger (18352), &c.

1879, Dec. 7, white,	C.C. Unity	Mehemet Ali, 42008	Mr. Warren

QUARREL, red, calved April 19, 1875, Vol. xxv. p. 709. Bred by Mr. E. L. Warren ; got by Oberon (34942), dam (Martha) by Rimmel (29779), &c.

1879, April 25, roan,	C.C. Uproar	Mehemet Ali, 42008	Mr. Warren

ROCHELLE, roan, calved April 9, 1876. Bred by Mr. E. L. Warren; got by
Oberon (34942), dam (Midge) by Rimmel (29779), &c. See Vol. xxiii. p. 697.

Produce in	Names, &c.	By what Bull.	By whom bred.
1879, April 4, roan, C.C.	Upsala	Mehemet Ali, 42008	Mr. Warren

ROMANCE, roan, calved April 16, 1876. Bred by Mr. E. L. Warren; got by
Paddy Carey (35003), dam (Safety) by Lord Nelson (22178), &c. See Vol.
xxiii. p. 697.

1879, May 17, white, C.C.	Utopia	Mehemet Ali, 42008	Mr. Warren

WATT, James,
Claragh, Ramelton, Co. Donegal.

YOUNG CIRCLET, roan, calved in April, 1873. Bred by Mr. P. Donovan,
Frogmore; got by Prince George 2nd (43798), dam (Circlet) by Northern
Light (24670), g. d. (Daisy Wreath) by Nobleman (18457), gr. g. d. (Daisy
Crown) by Dargan (17666), &c. See Vol. xvi. p. 412.

1878, March 19, roan, C.C.	Circlet 2nd	Simpleton, 37447	Mr. Watt
1879, Feb. 2, r. & w., B.C.	Coronet	Baron Gwynne, 33041	do.

COUNTESS, red, calved March 11, 1871, Vols. xxii. and xxv. pp. 565, 667.
Bred by Mr. H. Smith, Mountmellick; got by Lord Claud (29086), dam
(Daybreak) by British Duke (19350), &c.

1879, May 17, red, B.C.	Viscount	Conqueror, 30786	Mr. Watt

FANCY GWYNNE, roan, calved May 13, 1876. Bred by Mr. W. Bolton, The
Island; got by King James (28971), dam (Florence Gwynne) by Lieutenant
General (31600), &c. See Vol. xxiii. p. 341.

1879, Jan. 8, roan, B.C.	Lord Albion Gwynne	Albion, 36112	Mr. Watt

FRISKY, red and little white, calved June 29, 1876. Bred by Mr. H. Smith,
Mountmellick; got by Conqueror (30786), dam (Francisca) by Prince Francis
(32131), &c. See Vol. xxv. p. 667.

1879, Feb. 11, r. & w., B.C.	Lord Aylesby	Earl of Aylesby, 38215	Mr. Watt

GWYNNORA, roan, calved May 18, 1874, Vol. xxiv. p. 393. Bred by Mr. W.
T. Crosbie, Ardfert Abbey; got by Irish Baron (31417), dam (Lady Gwynne
Booth) by Regal Booth (27262), &c.

1879, Dec. 1, red, B.C.	Sir Gwy'ne de Gwy'ne	Baron Gwynne, 33041	Mr. Watt

ISABELLA BROUGHTON, roan, calved April 9, 1876. Bred by Messrs.
Christy Brothers, Fort Union; got by Lord Broughton (31626), dam (Isabella)
by Fairy King (21716), &c. See Vol. xxiii. p. 384.

1879, May 25, roan, C.C.	Princess Isabella	Verger, 35860	Mr. Watt

RED BREAST 4TH, red and white, calved January 5, 1876. Bred by Mr. H.
Smith, Mountmellick; got by Conqueror (30786), dam (Red Breast) by Sheet
Anchor 2nd (25121), &c. See Vol. xxv. p. 668.

1879. Mar. 2, r. & w., C.C.	(dead)	Earl of Aylesby, 38215	Mr. Watt

YOUNGSTER, red, calved March 22, 1876. Bred by Mr. H. Smith, Mountmellick; got by Conqueror (30786), dam (Younger Sister) by Bright Spur
(25671), &c. See Vol. xxiii. p. 648.

1879, April 6, red, C.C.	(dead)	Earl of Aylesby, 38215	Mr. Watt

WEBB, Lieut.-Colonel,
Elford House, Tamworth.

ISABELLA 8TH, red and white, calved July 20, 1873, Vols. xxiii. and xxiv. pp. 595, 603. Bred by Sir R. Peel, Bart., Drayton Manor; got by Red Knight (29741), dam (Young Isabella) by Count Bickerstaffe (23630), &c.

Produce in	Names, &c.	By what Bull.	By whom bred.
1879, Feb. 18, roan, C.C.	Druidess	The Druid, 35753	Lieut.-Col. Webb

LADY 4TH, red, calved January 20, 1875. Bred by Lieut.-Colonel Webb; got by Grand Duke of Lightburne (26299), dam (Lady 3rd) by Senator (29950), &c. See "Lord Edgcott 3rd," p. 145.

1879, Dec. 19, red, C.C.	Lady 10th	D.of Barrington 4th, 39712	Lieut.-Col. Webb

LADY 5TH, red and white, calved January 14, 1877. Bred by Lieut.-Colonel Webb; got by Marquis 6th (34777), dam (Lady 3rd) by Senator (29950), &c.

1879, July 21, { roan, C.C. red, B.C.	Lady 9th (Steer) }	D.of Barrington 4th, 39712	Lieut.-Col. Webb

PRETTY GWYNNE, red and white, calved June 13, 1874, Vol. xxiv. p. 698. Bred by Lieut.-Colonel Webb; got by Baron Deepdale (30438), dam (Polly Gwynne 5th) by Barrington Oxford 2nd (27987), &c.

1879, Dec. 15, r. & w., C.C.	Pretty Gwynne 2nd	D.of Barrington 4th, 39712	Lieut.-Col. Webb

WEBB, Henry,
Streetly Hall, Linton, Cambs.

BABRAHAM BEAUTY, red, calved January 2, 1870, Vols. xx., xxiv., and xxv. pp. 398, 698, 710. Bred by Mr. H. Webb; got by Prolific 2nd (27213), dam (Beauty) by Cheltenham (12588), &c.

1879, Jan. 2, roan, B.C.	Babraham Baron	Havering Baron, 38413	Mr. Webb

Babraham Baron, sold to Mr. M. Slater, jun., Trumpington, Cambs.

BABRAHAM BEAUTY 2ND, red and white, calved February 16, 1877. Bred by Mr. H. Webb; got by Hairy Boy (34102), dam (Babraham Beauty) by Prolific 2nd (27213), &c.

1879, March 13, red, B.C.	(dead)	Viscount Worcester, 40881	Mr. Webb

BARRINGTON STAR, red, calved June 26, 1876. Bred by Mr. H. Webb; got by Duke of Barrington 5th (33575), dam (Star of the Evening) by Lord Chancellor (20160), &c. See "Star of Worcester," p. 242.

1879, March 7, red, B.C.	Star of Worcester	Viscount Worcester, 40881	Mr. Webb

DUCHESS OF GRAFTON 3RD, red, calved April 6, 1877. Bred by Mr. H. Webb; got by Heydon Duke 4th (36769), dam (Duchess of Grafton 2nd) by His Lordship (34166), &c. See Vol. xxiv. p. 699.

1879, March 15, red, C.C.	Duch'ss of Worcester	Viscount Worcester, 40881	Mr. Webb

HOURI, roan, calved December 22, 1870, Vols. xxi., xxii., xxiii., and xxv. pp. 983, 609, 698, 711. Bred by Lord Braybrooke, Audley End; got by Heydon Duke (28850), dam (Fair Maid) by Second Duke of Claro (21576), &c.

1879, Nov. 19, roan, B.C.	Aston Beau	Baron Aston, 37770	Mr. Webb

YOUNG MAY FLOWER 2ND, roan, calved October 10, 1873, Vols. xxiii. and xxv. pp. 698, 711. Bred by Mr. H. Webb; got by Archduke Charles (30395), dam (Young May Flower) by Costa King (28254), &c.

Produce in	Names, &c.	By what Bull.	By whom bred.
1879, Aug. 30, red, B.C.	(Steer)	Baron Aston, 37770	Mr. Webb

MAY FLOWER 4TH, roan, calved December 31, 1876. Bred by Mr. H. Webb; got by Hairy Boy (34102), dam (Young May Flower 2nd) by Archduke Charles (30395), &c. See Vol. xxiii. p. 698.

Produce in	Names, &c.	By what Bull.	By whom bred.
1879, Feb. 14, red, C.C.	Valentine	Havering Baron, 38413	Mr. Webb
1879, Dec. 31, roan, C.C.	(dead)	Baron Aston, 37770	do.

MOSS ROSE, roan, calved May 9, 1876, Vol. xxv. p. 711. Bred by Mr. D. A. Green, East Donyland; got by Baron Richard (37812), dam (Babraham Rose 2nd) by Heydon Duke 2nd (31370), &c.

Produce in	Names, &c.	By what Bull.	By whom bred.
1879, July 21, roan, C.C.	Moss Rosebud 2nd	Baron Aston, 37770	Mr. Webb

PHILLIS FILLPAIL, white, calved October 27, 1877. Bred by Mr. J. Webb, Horseheath, the property of Mr. H. Webb; got by Hairy Boy (34102), dam (Phillis 7th) by Costa King (28254), &c. See Vol. xxiv. p. 699.

Produce in	Names, &c.	By what Bull.	By whom bred.
1879, Dec. 7, white, C.C.	Phillis Fillpail 2nd	Baron Aston, 37770	Mr. J. Webb

Phillis Fillpail 2nd, sold to Mr. H. Webb.

STAR OF THE EVENING, red, calved December 12, 1870, Vol. xxiii. p. 698. Bred by Mr. R. Wood, Clapton; got by Lord Chancellor (20160), dam (Saturn) by Sambo (20781), &c.

Produce in	Names, &c.	By what Bull.	By whom bred.
1879, Nov. 20, { (dead)B.C. / (dead)C.C. }		Baron Aston, 37770	Mr. Webb

WELCOME LADY, roan, calved July 2, 1877. Bred by Mr. H. Webb; got by Hairy Boy (34102), dam (Welcome Lassie 3rd) by Costa King (28254), &c. See "Welcome Worcester," p. 264.

Produce in	Names, &c.	By what Bull.	By whom bred.
1879, Apr. 7, r. & w., B.C.	Welcome Worcester	Viscount Worcester, 40881	Mr. Webb

WELCOME LASSIE 5TH, white, calved December 6, 1874, Vols. xxiv. and xxv. pp. 699, 711. Bred by Mr. H. Webb; got by Prolific 3rd (27214), dam (Welcome Lassie) by Costa (21487), &c.

Produce in	Names, &c.	By what Bull.	By whom bred.
1879, Jan. 9, roan, C.C.	Wel. Wickham La'sie	Havering Baron, 38413	Mr. Webb

WEBB, John,
Horseheath, Linton, Cambs.

CHERRY 8TH, roan, calved March 30, 1873, Vol. xxv. p. 711. Bred by Mr. J. Webb; got by Prolific 3rd (27214), dam (Cherry) by Lord Royston 2nd (16449), &c.

Produce in	Names, &c.	By what Bull.	By whom bred.
1879, Dec. 31, white, C.C.	Cherry White	Baron Aston, 37770	Mr. Webb

CHERRY PIE, red, calved November 23, 1874. Bred by Mr. J. Webb; got by Prolific 3rd (27214), dam (Cherry 3rd) by Clumber (17577), &c. See Vol. xxi. p. 984.

Produce in	Names, &c.	By what Bull.	By whom bred.
1879, Nov. 25, red, C.C.	Cherry Tart	Baron Aston, 37770	Mr. Webb

LUCKY 2ND, roan, calved March 8, 1876. Bred by Mr. J. Webb; got by Pretty Pink (35071), dam (Lucy) by Prolific 2nd (27213), &c. See Vol. xxiii. p. 698.

Produce in	Names, &c.	By what Bull.	By whom bred.
1879, Oct. 30, r. & w., C.C.	Lucky Aston	Baron Aston, 37770	Mr. Webb

LUCY 2ND, roan, calved March 22, 1873, Vols. xxii. and xxiv. pp. 609, 699.
Bred by Mr. J. Webb; got by Prolific 3rd (27214), dam (Lucy) by Prolific
2nd (27213), &c.

Produce in	Names, &c.	By what Bull.	By whom bred.
1879. Oct. 24, r. & w., C.C.	Lucy Aston	Baron Aston, 37770	Mr. Webb

PHILLIS 7TH, roan, calved October 22, 1871, Vols. xxii., xxiv., and xxv.
pp. 610, 699, 711. Bred by Mr. J. Webb; got by Costa King (28254), dam
(Phillis 2nd) by Sir Charles (22892), &c.

Produce in	Names, &c.	By what Bull.	By whom bred.
1879, Oct. 24, white, C.C.	Streetly Phillis	Baron Aston, 37770	Mr. Webb

Streetly Phillis, sold to Mr. H. C. Webb, Streetly Hall, Linton.

RUBY RING, roan, calved January 6, 1877. Bred by Mr. J. Webb; got by
Hairy Boy (34102), dam (Ruby 3rd) by Prolific 2nd (27213), &c. See
Vol. xxiv. p. 700.

Produce in	Names, &c.	By what Bull.	By whom bred.
1879, Mar. 31, r. & w., C.C.	Ruby Worcester	Viscount Worcester, 40881	Mr. Webb

WEBB, Jonas,
Melton Ross, Ulceby, Lincolnshire.

BABRAHAM FOGGATHORPE, red and white, calved January 3, 1876,
Vol. xxv. p. 712. Bred by Mr. J. Webb; got by Young Heydon Duke
(34156), dam (Fraulein Foggathorpe) by Wilson (27811), &c.

Produce in	Names, &c.	By what Bull.	By whom bred.
1879, Dec. 15, r. & w., C.C.	Bar'gton Fog'thorpe	D. of Barrington 5th. 33575	Mr. Webb

SYCAMORE 10TH, red and white, calved August 2, 1876. Bred by Mr. R.
Botterill, Wauldby; got by Oxford le Grand (29496), dam (Sycamore 5th) by
Nineteenth Duke of Oxford (23431), &c. See Vol. xxiii. p. 346.

Produce in	Names, &c.	By what Bull.	By whom bred.
1878, Nov. 11, roan, C.C.	Sycamore Princess	Foggathorpe Beau, 39889	Mr. Webb
1879, Dec. 9, red, C.C.	Sycamore Belle	D. of Barrington 5th, 33575	do.

WELSBY, R. C.,
Criccin, Rhuddlan, North Wales.

AGNES 4TH, red, calved April 28, 1875. Bred by Captain C. R. Conwy, Bodrhyddan; got by Favourite (33895), dam (Agnes 2nd) by New York (22412), &c.
See "Criccin Waterloo," p. 60.

Produce in	Names, &c.	By what Bull.	By whom bred.
1878, Apr. 22, r. & w., C.C.	Agnes 5th	Sir Fk. Gwynne 3rd, 39117	Captain Conwy
1879, Apr. 2, red, B.C.	Criccin Waterloo	Edward Waterloo, 38246	Mr. Welsby

CADER, red and white, calved June 20, 1872, Vols. xxi., xxii., xxiii., and
xxiv. pp. 640, 372, 393, 384. Bred by Captain C. R. Conwy, Bodrhyddan; got
by Oxford Wild Eyes 2nd (32038), dam (Lady Sarah) by Harold (26341), &c.

Produce in	Names, &c.	By what Bull.	By whom bred.
1879, Feb. 14, r. & w., C.C.	Sarah Waterloo	Edward Waterloo, 38246	Mr. Welsby

CAPER, red and white, calved January 21, 1876. Bred by Captain C. R. Conwy,
Bodrhyddan, the property of Mr. R. C. Welsby; got by Favourite (33895),
dam (Columbine 5th) by Oxford Wild Eyes 2nd (32038), &c. See "Lyons,"
p. 161.

Produce in	Names, &c.	By what Bull.	By whom bred.
1878, July 6, r. & w., B.C.	Lyons	Sir Fk. Gwynne 3rd, 39117	Captain Conwy

Lyons, sold to Mr. R. C. Welsby, Criccin.

WELSTED, Richard,
Ballywalter, Castle Town Roche, Ireland.

COWSLIP 14TH, white, calved February 21, 1868, Vols. xx., xxi., xxii., and xxiii.
pp. 463, 986, 610, 701. Bred by Mr. R. Welsted ; got by Sir James (16980),
dam (Cowslip 10th) by Elfin King (17796), &c.

Produce in	Names, &c.	By what Bull.	By whom bred.
1879, Nov. 7, roan, C.C.	Cowslip 33rd	Master of Arts, 34816	Mr. Welsted

COWSLIP 18TH, roan, calved April 8, 1872, Vol. xxiii. p. 701. Bred by Mr.
R. Welsted ; got by Prince Christian (22581), dam (Cowslip 14th) by Sir
James (16980), &c.

1878, Nov. 12, r. & w., B.C.	Lord Chelmsford	Master of Arts, 34816	Mr. Welsted
1879, Nov. 2, roan, C.C.	Cowslip 32nd	do.	do.

Lord Chelmsford, sold to Mr. J. Shackleton, West House, Leeds.

COWSLIP 20TH, red, calved November 16, 1872, Vols. xxii. and xxv. pp. 610,
714. Bred by Mr. R. Welsted ; got by England's Glory (23889), dam (Cowslip 13th) by Sir James (16980), &c.

1879, May 27, red, C.C.	Cowslip 31st	Master of Arts, 34816	Mr. Welsted

ENGLAND'S GEM, red, calved May 29, 1874, Vol. xxiv. p. 702. Bred by Mr.
R. Welsted ; got by England's Glory (23889), dam (Red Gem) by Uncle Ned
(19026), &c.

1879, Jan. 12, r. & w., C.C.	Scotland's Gem	Master of Arts, 34816	Mr. Welsted

ENGLAND'S KATE, red, calved February 2, 1874, Vol. xxv. p. 714. Bred
by Mr. R. Welsted ; got by England's Glory (23889), dam (Lady Kate) by
Sir James (16980), &c.

1879, May 2, r. & w., C.C.	Irish Kate	Master of Arts, 34816	Mr. Welsted

ENGLAND'S LYRA, white, calved November 17, 1874. Bred by Mr. R.
Welsted ; got by England's Glory (23889), dam (Lyra) by Prince of Warlaby
(15107), &c. See " Sir Lancelot," Vol. xxv. p. 263.

1878, Nov. 9, roan, B.C.	Sir Lancelot	Master of Arts, 34816	Mr. Welsted
1879, Dec. 27, roan, C.C.	Roan Lyra	do.	do.

Sir Lancelot, sold to Mr. P. Cowtry, Anna, Churchtown.

GOLDEN DAWN, roan, calved March 3, 1873, Vols. xxii. and xxiii. pp. 611,
702. Bred by Mr. R. Welsted ; got by England's Glory (23889), dam (Golden
Light) by Roseberry (22757), &c.

1879, April 3, red, C.C.	Golden Shade	Master of Arts, 34816	Mr. Welsted

GOLDEN LIGHT, roan, calved March 17, 1869, Vol. xxi. p. 987. Bred by
Colonel Fisher, Castle Crogan ; got by Roseberry (22757), dam (Golden Halo)
by Hopewell (10332), &c.

1879, Aug. 5, r. & w., C.C.	Golden Cloud	King William, 34358	Mr. Welsted

GOLDEN RED, red, calved January 26, 1874. Bred by Mr. R. Welsted ; got
by England's Glory (23889), dam (Golden Ray) by Booth Royal (15673), &c.
See " Golden Rufus," Vol. xxv. p. 124.

1878, June 18, red, B.C.	Golden Rufus	Master of Arts, 34816	Mr. Welsted
1879, July 29, red, C.C.	Golden Ring	do.	do.

Golden Rufus, sold to Mr. R. A. Chearnley, Salterbridge, Cappoquin, Co. Waterford.

LADY KATE, roan, calved February 10, 1868, Vols. xx., xxi., and xxii. pp. 601, 987, 611. Bred by Mr. R. Welsted ; got by Sir James (16980), dam (Keepsake) by Elfin King (17796), &c.

Produce in		Names, &c.	By what Bull.	By whom bred.
1877, Nov. 11, roan,	B.C.	Kingcraft	Master of Arts, 34816	Mr. Welsted
1879, June 14, roan,	C.C.	Katy	do.	do.

Kingcraft, sold to Captain Croker, The Grange, Croom, Co. Limerick.

LADYLIKE 7TH, white, calved February 26, 1875, Vol. xxiv. p. 702. Bred by Mr. R. Welsted ; got by England's Glory (23889), dam (Ladylike 5th) by Prince Christian (22581), &c.

Produce in		Names, &c.	By what Bull.	By whom bred.
1879, Feb. 20, roan,	B.C.	Lord Chesterfield	Master of Arts, 34816	Mr. Welsted

Lord Chesterfield, sold to Mr. R. McDougall, Arundel, Keillor, Melbourne, Australia.

LADY LUCIA, roan, calved February 14, 1875, Vol. xxv. p. 714. Bred by Mr. R. Welsted ; got by England's Glory (23889), dam (Princess Lyra) by Prince Christian (22581), &c.

Produce in		Names, &c.	By what Bull.	By whom bred.
1879, May 25, roan,	C.C.	Queen of Love	King William, 34358	Mr. Welsted

LADY LYRA, red, calved April 6, 1876. Bred by Mr. R. Welsted ; got by England's Glory (23889), dam (Princess Lyra) by Prince Christian (22581), &c. See Vol. xxiii. p. 703.

Produce in		Names, &c.	By what Bull.	By whom bred.
1879, Mar. 13, red,	C.C.	Lady Laura	Master of Arts, 34816	Mr. Welsted

LADY PRIMROSE, red and white, calved November 24, 1874, Vol. xxiv. p. 702. Bred by Mr. R. Welsted ; got by England's Glory (23889), dam (Princess Primrose) by Uncle Ned (19026), &c.

Produce in		Names, &c.	By what Bull.	By whom bred.
1878, Dec. 6, r. & w.,	B.C.	Lord Primrose	Master of Arts, 34816	Mr. Welsted
1879, Dec. 18, r. & w.,	C.C.	Ada Primrose	do.	do.

Lord Primrose, sold to Mr. A. Greer, Dripsey Castle, Co. Cork.

MISS PRIMROSE, roan, calved January 8, 1872, Vols. xxii. and xxiv. pp. 611, 702. Bred by Mr. R. Welsted ; got by Prince Christian (22581), dam (Princess Primrose) by Uncle Ned (19026), &c.

Produce in		Names, &c.	By what Bull.	By whom bred.
1879, Mar. 25, r. & w.,	C.C.	Pretty Primrose	Master of Arts, 34816	Mr. Welsted

PRICELESS, red, calved October 8, 1873, Vol. xxiv. p. 702. Bred by Mr. R. Welsted ; got by England's Glory (23889), dam (Royal Peeress) by Prince Christian (22581), &c.

Produce in		Names, &c.	By what Bull.	By whom bred.
1878, Mar. 22, red,	B.C.	Prince Precious	Master of Arts, 34816	Mr. Welsted
1879, Oct. 10, red,	C.C.	Perfection	do.	do.

Prince Precious, sold to Mr. R. A. Chearnley, Salterbridge, Cappoquin, Co. Waterford.

PRINCESS ELIZABETH, red and white, calved March 11, 1871, Vol. xxiii. p. 702. Bred by Mr. R. Welsted ; got by Prince Christian (22581), dam (Lady Elizabeth) by Sir James (16980), &c.

Produce in		Names, &c.	By what Bull.	By whom bred.
1879, Nov. 17, r. & w.,	C.C.	Princess Eleanor	Master of Arts, 34816	Mr. Welsted

PRINCESS LYRA, red and white, calved February 11, 1870, Vols. xxii., xxiii., xxiv., and xxv. pp. 611, 703, 703, 715. Bred by Mr. R. Welsted ; got by Prince Christian (22581), dam (Lyra) by Prince of Warlaby (15107), &c.

Produce in		Names, &c.	By what Bull.	By whom bred.
1879, Dec. 31, red,	C.C.	Red Lyra	Master of Arts, 34816	Mr. Welsted

PRINCESS PRIMROSE, red and white, calved May 3, 1866, Vols. xx., xxi., xxii., and xxiv. pp. 705, 988, 611, 703. Bred by Mr. R. Welsted ; got by Uncle Ned (19026), dam (Primrose) by Orson (13432), &c.

Produce in		Names, &c.	By what Bull.	By whom bred.
1879, July 3, red,	C.C.	Patty Primrose	Master of Arts, 34816	Mr. Welsted

PRINCESS ROYAL 2ND, roan, calved March 3, 1871, Vols. xxii. and xxv. pp. 612, 715. Bred by Mr. R. Welsted; got by Prince Christian (22581), dam (Primrose) by Orson (13432), &c.

Produce in	Names, &c.	By what Bull.	By whom bred.
1879, Mar. 16, r. & w., C.C.	Royal Princess	Master of Arts, 34816	Mr. Welsted

ROSE OF ERIN, roan, calved January 5, 1873, Vols. xxii. and xxv. pp. 612, 715. Bred by Mr. R. Welsted; got by England's Glory (23889), dam (Rosabel) by Sir James (16980), &c.

Produce in	Names, &c.	By what Bull.	By whom bred.
1879, June 20, red, C.C.	Queen Rose	King William, 34358	Mr. Welsted

WHARTON, J. T.,
Skelton Castle, Marske-by-the-Sea.

SOPHIA JANE, roan, calved October 8, 1876. Bred by Mr. J. T. Wharton; got by Scots Fusilier (35483), dam (Agatha) by Dunkeld (26042), &c. See Vol. xxiii. p. 704.

Produce in	Names, &c.	By what Bull.	By whom bred.
1879, Feb. 28, roan, C.C.	Countess	Upleatham, 44173	Mr. Wharton

WHITE, William,
Grimston Park, Tadcaster.

RED ROSE, red, calved June 11, 1868, Vols. xx. and xxv. pp. 723, 716. Bred by Mr. W. White; got by King Charles (24240), dam (Rose) by Baron Barton (14123), &c.

Produce in	Names, &c.	By what Bull.	By whom bred.
1879, June 16, red, B.C.	Knave of Trumps	King of Trumps, 31512	Mr. White

WHYTE, James,
Aldborough, Darlington.

BAINESSE ROSE, roan, calved January 16, 1876. Bred by Mr. J. Outhwaite, Bainesse; got by Lord Godolphin (36065), dam (Moss Rose) by Baron Killerby (27949), &c. See "King of the Roses," Vol. xxiv. p. 129.

Produce in	Names, &c.	By what Bull.	By whom bred.
1879, Aug. 23, roan, C.C.	Rosemary	Sir W'sor Broughton, 27507	Mr. Whyte

BARONESS KNOWLMERE, roan, calved February 6, 1875. Bred by Mr. Martin, Littleport; got by Lord Abbot (29052), dam (Baroness Hopewell 2nd) by King Hopewell (26506), &c. See Vol. xxii. p. 501.

Produce in	Names, &c.	By what Bull.	By whom bred.
1878, Feb. 26, r. & w., C.C.	Wellingtonia	Star, 32601	Mr. Whyte
1879, Apr. 26, r. & w., C.C.	Wisteria	Hyperion, 34196	do.

DUCHESS OF KNIGHTLEY 4TH, red and white, calved May 13, 1874, Vol. xxiii. p. 705. Bred by Mr. D. Fisher, Pitlochrie; got by Valentine Vox (32752), dam (Damsel) by Lord Hopewell (18239), &c.

Produce in	Names, &c.	By what Bull.	By whom bred.
1878, May 21, r. & w., B.C.	Hanibal	Handel, 38396	Mr. Whyte
1879, Apr. 6, r. & w., B.C.	Mars	Gunpowder, 28801	do.

PHILLIS 13TH, roan, calved December 17, 1871, Vols. xxi. and xxii. pp. 553, 305. Bred by Mr. H. Aylmer, West Dereham Abbey; got by Royal Broughton (27352), dam (Phillis 6th) by General Hopewell (17953), &c.

Produce in	Names, &c.	By what Bull.	By whom bred.
1878, Apr. 6, roan, C.C.	Ophelia	Handel, 38396	Mr. Whyte
1879, July 13, r. & w., C.C.	Polly Windsor	Sir W'sor Broughton, 27507	do.

WILKES, J. P.,
Oakley House, Droitwich.

AMELIA, roan, calved October 2, 1876. Bred by Mr. J. P. Wilkes; got by Second Duke of Rowley (28441), dam (Aurelia) by Highland Chief (24142), g.d. (Annie Lisle) by Golden Crown (21835), gr. g. d. (Amelia) by Count Cavour (19522), &c. See Vol. xvii. p. 364.

Produce in		Names, &c.	By what Bull.	By whom bred.
1879, Feb. 25, red,	C.C.	Annie	L. Fitzclarence 10th, 34549	Mr. Wilkes

MINNIE, roan, calved June 20, 1874. Bred by Mr. J. P. Wilkes; got by Master Oxford (31881), dam (Marion) by Conqueror (25823), &c. See Vol. xxi. p. 991.

1878, Apr. 25, roan,	C.C.	Myrtle	L. Fitzclarence 10th, 34549	Mr. Wilkes
1879, Apr. 30, roan,	B.C.	Monarch	do.	do.

YOUNG PRIDE, red, calved October 3, 1874. Bred by Mr. J. P. Wilkes; got by Master Oxford (31881), dam (Pride) by Conqueror (25823), &c. See Vol. xxi. p. 991.

1877, Dec. 15, r. & w.,	C.C.	Prude	L. Fitzclarence 10th, 34549	Mr. Wilkes
1879, Mar. 25, r. & w.,	C.C.	Prude 2nd	do.	do.

WILKINSON, John,
Carlton Grange, Skipton.

COUSIN B., red and white, calved October 23, 1875, Vol. xxv. p. 718. Bred by Colonel Towneley, Towneley Park; got by Second Hubback (28880), dam (Lady Florence) by Baron Stanley (27969), &c.

1879, Oct. 12, roan,	C.C.	Miss Mary	Monitor, 42025	Mr. Wilkinson

LADY FLORENCE, roan, calved February 28, 1872, Vols. xxi., xxii., and xxv. pp. 964, 595, 718. Bred by Mr. R. Parker, Burnley; got by Baron Stanley (27969), dam (Wild Brier) by Royal Scotforth (25042), &c.

1879, Oct. 20, r. & w.,	C.C.	Miss Bessie	Monitor, 42025	Mr. Wilkinson

WILLIAMS, Charles,
Pilton House, Barnstaple, Devon.

DAISY, red, calved February 24, 1875, Vol. xxv. p. 718. Bred by Mr. J. Garsed, The Moorlands; got by Barmpton (32994), dam (Dido) by Sir John (35572). &c.

1879, Sept. 9, red,	B.C.	The Dandler	L. Ashton Wild Eyes, 34485	Mr. Williams

DELIGHT, red and white, calved March 23, 1876, Vol. xxv. p. 718. Bred by Mr. J. Garsed, The Moorlands; got by Barmpton (32994), dam (Dido) by Sir John (35572), &c.

1879, May 4, r. & w.,	C.C.	Dainty Daisy	L. Ashton Wild Eyes, 34485	Mr. Williams

DUCHESS, red and white, calved September 6, 1873, Vol. xxv. p. 719. Bred by Mr. J. Garsed, The Moorlands; got by Imperial Prince (24186), dam (Dido) by Sir John (35572), &c.

1879, Feb. 6, roan,	C.C.	Devon Daisy	Kentish Charmer, 38482	Mr. Williams

LADY GWYNNE, roan, calved March 22, 1870, Vols. xx., xxii., and xxv. pp. 597, 614, 719. Bred by the Earl of Aylesford, Packington Hall; got by Duke of Cambridge (25940), dam (Jenny Gwynne) by Duke of York (14461), &c.

Produce in	Names, &c.	By what Bull.	By whom bred.
1879, May 31, r. & w., B.C.	Devon Gwynne	L. Ashton Wild Eyes, 34485	Mr. Williams

Devon Gwynne, sold to Mr. H. Humfrey, Kingstone Farm, Shrivenham.

NORNA, red and white, calved February 25, 1870, Vols. xxi. and xxiii. pp. 568, 331. Bred by Mr. S. O. Priestley, Trefan; got by Sir Knight (25157), dam (Welcome) by The Bully (23019), &c.

1879, Oct. 25, roan, C.C.	Autumn Daisy	20th D. of Oxford, 28432	Mr. Williams

PILTON GWYNNE, red and white, calved September 3, 1875. Bred by Mr. C. Williams; got by Pencraig (35027), dam (Lady Gwynne) by Duke of Cambridge (25940), &c. See "Devon Gwynne," p. 67.

1879, Mar. 4, red, C.C.	Heanton Gwynne	L. Ashton Wild Eyes, 34485	Mr. Williams

ROSS GWYNNE 3RD, roan, calved April 4, 1877. Bred by Mr. C. Williams; got by Twenty-second Duke of Oxford (31000), dam (Ross Gwynne) by Royal Cambridge (25009), &c. See Vol. xxiv. p. 705.

1879, Nov. 1, roan, C.C.	Rose Gwynne 5th	L. Ashton Wild Eyes, 34485	Mr. Williams

SUGARPLUM 2ND, roan, calved March 20, 1866, Vols. xxi. and xxii. pp. 992, 615. Bred by Captain J. E. Winnall, Burton House; got by Duke of Liverpool (23766), dam (Sugarplum) by Second Duke of Bolton (12739), &c.

1876, Oct. 27, roan, B.C.	Sweet Tooth	Clarence of that Ilk, 33391	Mr. Williams
1879, Feb. 12, roan, C.C.	Spun Sugar	do.	do.

Sweet Tooth, sold to Mr. Brailey, Tutshill, Barnstaple.

WILLIAMS, Rev. E. T.,
Caldicot Parsonage, Chepstow, Monmouthshire.

IRAS, roan, calved November 19, 1876. Bred by the Rev. E. T. Williams; got by Earl of Horton 8th (33800), dam (Irony) by Baron Oxford 4th (25582), &c. See Vol. xxi. p. 993.

1879, Oct. 20, white, C.C.	Ixia	Orlando, 40411	Rev. E. T. Williams

OPHELIA, red, calved May 16, 1876. Bred by the Rev. E. T. Williams; got by Earl of Horton 8th (33800), dam (Oxford Iris) by Baron Oxford 2nd (23376), &c. See "Orlando," Vol. xxiv. p. 190.

1879, May 27, red, B.C.	Oswald	Orlando, 40411	Rev. E. T. Williams

WILLIAMS, Walter,
Sugnall Hall, Eccleshall, Staffordshire.

DIANTHUS, red and white, calved November 26, 1873, Vols. xxiii. and xxv. pp. 707, 720. Bred by Mr. J. Robinson, Berkhampstead; got by Charming King (30698), dam (Dahlia) by Londonderry (13169), &c.

1879, May 16, r. & w., C.C.	Dianthus Siddington	B. T'croft Siddington, 37823	Mr. Williams

HONEYDEW 2ND, red, calved November 7, 1874. Bred by Mr. G. Fox, Elmhurst Hall; got by Baron Holker (30459), dam (Honeydew) by Orestes (22443), &c. See Vol. xxi. p. 720.

Produce in	Names, &c.	By what Bull.	By whom bred.
1879, Sept. 18, roan, B.C.	Honeydew's Baron	Baron Sugnall, 41066	Mr. Williams

SPECIMEN 2ND, red, calved June 10, 1875. Bred by Mr. W. Angerstein, Weeting Hall; got by Cavendish (42893), dam (Specimen) by Grand Duke 15th (21852), g. d. (Spangle) by Chanter (19423), &c. See Vol. xix. p. 735.

Produce in	Names, &c.	By what Bull.	By whom bred.
1879, March 10, red, C.C.	Sugnall Specimen	Baron Sugnall, 41066	Mr. Williams

WESTON'S HONEYDEW, white, calved January 13, 1876, Vol. xxv. p. 720. Bred by Mr. G. Fox, Elmhurst Hall; got by Grand Duke of Weston 3rd (34079), dam (Honeydew) by Orestes (22443), &c.

Produce in	Names, &c.	By what Bull.	By whom bred.
1879, May 17, red, C.C.	Sugnall Honeydew	Baron Sugnall, 41066	Mr. Williams

WILLIAMSON, J.,
East Pitdoulsie, Turriff, N.B.

BELLA 3RD, roan, calved February 20, 1871, Vols. xxi., xxiii., and xxv. pp. 994, 707, 720. Bred by Mr. J. Williamson; got by Nobility (34912), dam (Missie) by Banker (19255), &c.

Produce in	Names, &c.	By what Bull.	By whom bred.
1879, Feb. 6, roan, C.C.	Miss Priestly	Comet, 41250	Mr. Williamson

COUNTESS CAVOUR, roan, calved June 6, 1874, Vols. xxiii. and xxv. pp. 707, 720. Bred by Mr. J. Williamson; got by Oregon (34953), dam (Eglantine) by Nobility (34912), &c.

Produce in	Names, &c.	By what Bull.	By whom bred.
1879, Mar. 19, white, B.C.	Rutherford	Comet, 41250	Mr. Williamson

EGLANTINE, red, calved March 19, 1871, Vols. xxi., xxiii., and xxv. pp. 994, 707, 720. Bred by Mr. J. Williamson; got by Nobility (34912), dam (Bella 2nd) by Earl Russell (33826), &c.

Produce in	Names, &c.	By what Bull.	By whom bred.
1879, Feb. 13, roan, B.C.	Dr. A. Duff	Comet, 41250	Mr. Williamson

FRIDAY, red, calved March 30, 1877. Bred by Mr. J. Williamson; got by Cassius (33301), dam (Missie) by Banker (19255), &c. See Vol. xxiii. p. 707.

Produce in	Names, &c.	By what Bull.	By whom bred.
1879, Dec. 22, red, B.C.	Lord Neaves	Comet, 41250	Mr. Williamson

LADY HOWE, red, calved May 17, 1876. Bred by Mr. J. Williamson; got by Cassius (33301), dam (Eglantine) by Nobility (34912), &c.

Produce in	Names, &c.	By what Bull.	By whom bred.
1879, Jan. 28, roan, B.C.	Sir Francis	Comet, 41250	Mr. Williamson

MISSIE 2ND, white, calved February 22, 1873, Vols. xxiii. and xxv. pp. 708, 721. Bred by Mr. J. Williamson; got by Oregon (34953), dam (Missie) by Banker (19255), &c.

Produce in	Names, &c.	By what Bull.	By whom bred.
1879, May 7, roan, C.C.	Schiller	Comet, 41250	Mr. Williamson

QUEEN OF BELGIUM, roan, calved December 16, 1876. Bred by Mr. J. Williamson; got by Cassius (33301), dam (Countess Cavour) by Oregon (34953), &c.

Produce in	Names, &c.	By what Bull.	By whom bred.
1879, Mar. 14, white, B.C.	Humbert	Comet, 41250	Mr. Williamson

WILLIS, Thomas,
Manor House, Carperby, Bedale.

BASHFUL BRIDE, roan, calved August 27, 1875, Vols. xxiv. and xxv. pp. 706, 721. Bred by Mr. T. Willis ; got by K. C. B. (26492), dam (Blushing Bride) by Fitz-Clarence (14552), &c.

Produce in	Names, &c.	By what Bull.	By whom bred.
1879, Sept. 2, roan,	B.C.|Fitz-Clar'ce|Benedict	Windsor Benedict, 40933	Mr. Willis

CAROLINE WINDSOR, roan, calved January 26, 1876. Bred by Mr. T. Willis ; got by Admiral Windsor (32912), dam (Charlotte Windsor) by Windsor Fitz-Windsor (25458), &c. See "Benedict," p. 25.

| 1879, June 26, red, | B.C.|Benedict | Windsor Benedict, 40933 | Mr. Willis |

DIAMOND NECKLACE, roan, calved August 8, 1874, Vol. xxv. p. 721. Bred by Mr. T. Willis ; got by Booth's Royal Seal (28060) dam (Windsor's Necklace) by Windsor Fitz-Windsor (25458), &c.

| 1879, May 24, red, | B.C.|Benedict Booth | Windsor Benedict, 40933 | Mr. Willis |

LILY EMPRESS, roan, calved April 16, 1876. Bred by Mr. T. Willis ; got by Admiral Windsor (32912), dam (Imperial Lily) by Windsor Fitz-Windsor (25458), &c. See Vol. xxiii. p. 709.

| 1879, June 29, red, | C.C.|Lily Countess | Windsor Benedict, 40933 | Mr. Willis |

QUEEN OF THE LILIES, roan, calved May 22, 1875. Bred by Mr. T. Willis ; got by K. C. B. (26492), dam (Regal Lily) by Lord Frederick (22156), &c. See Vol. xxii. p. 617.

| 1879, Jan. 21, white, | C.C.|Lily Queen | Vice Admiral, 39257 | Mr. Willis |

ROSE CLARENCE, red and white, calved January 28, 1862, Vols. xvi., xvii., xviii., xix., xx., xxi., xxiii., xxiv., and xxv. pp. 668, 716, 704, 707, 737, 995, 709, 707, 722. Bred by Mr. T. Willis ; got by Fitz-Clarence (14552), dam (Moss Rose) by Gavazzi (11508), &c.

| 1879, June 20, roan, | C.C.|Rose of Spring | Major Windsor, 34739 | Mr. Willis |

ROSE FITZ-CLARENCE, roan, calved August 9, 1876. Bred by Mr. T. Willis ; got by Major Windsor (34739), dam (Rose Clarence) by Fitz-Clarence (14552), &c. See "Benezet," p. 26.

| 1879, Jan. 16, roan, | B.C.|Vice Adm. Windsor | Vice Admiral, 39257 | Mr. Willis |
| 1879, Dec. 12, red, | B.C.|Benezet | Windsor Benedict, 40933 | do. |

ROSE OF THE VALE, roan, calved April 22, 1870, Vols. xxi., xxii., xxiv., and xxv. pp. 995, 617, 707, 722. Bred by Mr. T. Willis ; got by Windsor Fitz-Windsor (25458), dam (Rose of the Valley) by Lord of the Valley (14837), &c.

| 1879, July 8, red, | B.C.|Earl Benedict | Windsor Benedict, 40933 | Mr. Willis |

RUBY ROSE, red and white, calved March 27, 1876. Bred by Mr. T. Willis ; got by Abbot of Windsor (32903), dam (Rose of Cashmere) by Lord Frederick (22156), &c. See "Benedict Booth," p. 25.

| 1879, May 7, roan, | C.C.|Rose of the Day | Major Windsor, 34739 | Mr. Willis |

WINDSOR'S BRACELET, roan, calved February 9, 1868, Vols. xxi. and xxiii. pp. 995, 709. Bred by Mr. T. Willis ; got by Windsor Fitz-Windsor (25458), dam (Rose of Cashmere) by Lord Frederick (22156), &c.

| 1879, Feb. 4, roan, | C.C.|Bracelet | Vice Admiral, 39257 | Mr. Willis |

WINDSOR'S BRITANNIA, roan, calved February 29, 1872, Vols. xxii., xxiii.,
and xxv. pp. 617, 710, 722. Bred by Mr. T. Willis ; got by Windsor's Prince
(32881), dam (Britannia Windsor) by Windsor Fitz-Windsor (25458), &c.

Produce in	Names, &c.	By what Bull.	By whom bred.
1879, April 29, white, C.C.	Princess Louise	Major Windsor, 34739	Mr. Willis

WINDSOR'S FAREWELL, roan, calved October 9, 1873, Vol. xxiv. p. 708.
Bred by Mr. T. Willis ; got by Windsor Fitz-Windsor (25458), dam (Rose of
the Vale) by Windsor Fitz-Windsor (25458), &c.

1878, May 1, white, B.C.	(dead)	Heir-at-Law, 34124	Mr. Willis
1879, June 1, roan, C.C.	Welcome Rose	Windsor Benedict, 40933	do.

WINDSOR'S HYACINTH, roan, calved April 6, 1872, Vols. xxii., xxiii., xxiv.,
and xxv. pp. 617, 710, 708, 722. Bred by Mr. T. Willis ; got by Windsor's
Prince (32881), dam (Camelia Windsor) by Windsor Fitz-Windsor (25458), &c.

1879, Oct. 20, r. & w., B.C.	Admiral of the Fleet	Admiral Windsor, 32912	Mr. Willis

ZENOBIA, roan, calved June 26, 1875. Bred by Mr. T. Willis ; got by K. C. B.
(26492), dam (Zone) by Windsor Fitz-Windsor (25458), &c. See " Zenobius,"
p. 274.

1879, June 2, roan, B.C.	Zenobius	Major Windsor, 34739	Mr. Willis

WILSON, Jacob,
Woodhorn Manor, Morpeth.

BLOOMING QUEEN, roan, calved August 23, 1871, Vols. xxi., xxii., and xxiii.
pp. 996, 618, 711. Bred by Mr. Jacob Wilson ; got by Merry Monarch
(22349), dam (Bloom) by Knight of Killerby (22054), &c.

1879, Jan. 27, white, C.C.		Hymen, 34195	Mr. Wilson

DOTTIE, roan, calved August 27, 1875. Bred by Mr. Jacob Wilson ; got by
Jubilant (34276), dam (Dora) by Prowler (22662), &c. See Vol. xxii. p. 618.

1879, Feb. 16, red, C.C.		Hymen, 34195	Mr. Wilson

YOUNG JANET, roan, calved in April, 1869, Vols. xxi., xxiii., and xxv.
pp. 997, 712, 723. Bred by Miss Calvert, Sandysike ; got by Snowdrift
(32558), dam (Janet) by Rufus (22811), &c.

1879, Sept. 2, white, C.C.	Hymen's Janet	Hymen, 34195	Mr. Wilson

WEAL WEAL, roan, calved April 8, 1876, Vol. xxv. p. 724. Bred by Mr. Jacob
Wilson ; got by Knight of the Shire (26552), dam (Weal Royal) by Booth
Royal (15673), &c.

1879, Nov. 2, white, C.C.		Hymen, 34195	Mr. Wilson

WILSON, James,
Mains of Scotstown, Aberdeen.

FRAGRANT 5TH, roan, calved in June, 1873. Bred by Mr. N. Reid, Danestown, the property of Mr. J. Wilson ; got by Duke of Lancaster (30986),
dam (Fragrant 4th) by Baronet (30448), &c. See " Baron," p. 11.

1876, April , B.C.	Heather Jock	Hope of Britain, 34179	Mr. Reid
1877, May , red, C.C.	Lady Isabella	Merry Baron, 43649	Mr. Innes
1879, Feb. 3, roan, B.C.	Baron	do.	do.

Heather Jock, sold to Mr. Gray, Oldtold, Aberdeen.

WINN, J. R.,
Lower Coundon, Coventry.

BARONESS FAWSLEY 5TH, red, calved April 5, 1874. Bred by Messrs. F.
Leney and Sons, Wateringbury; got by Sixth Duke of Oneida (30997), dam
(Baroness Fawsley 3rd) by Grand Duke 15th (21852), &c. See Vol. xxi.
p. 908.

Produce in		Names, &c.	By what Bull.	By whom bred.
1878, Apr. 14, r. & w., C.C.		Bertha Fawsley	5th D. of Glo'ster, 36404	Mr. Canning
1879, June 1, r. & w., B.C.		Baron of Coundon	D. of Barrington6th, 33576	Mr. Winn

CIRCE, red and white, calved May 8, 1870, Vols. xx. and xxi. pp. 447, 692.
Bred by the Marquis of Exeter, Burghley Park; got by Nestor (24648), dam
(Louisa 9th) by Prince Albert (18579), &c.

Produce in		Names, &c.	By what Bull.	By whom bred.
1878, Jan. 31, roan,	C.C.	Circe 2nd	Telemachus 6th, 35725	Mr. Winn
1879, May 18, red,	B.C.	(dead)	Duke Dentsdale, 33561	do.

DUCHESS OF FLORENCE, roan, calved April 5, 1869, Vols. xx. and xxi.
pp. 489, 618. Bred by Mr. R. H. Carter, Fern Hill; got by Second Duke of
Collingham (23730), dam (Florentia 5th) by Third Duke of Lancaster
(19624), &c.

Produce in		Names, &c.	By what Bull.	By whom bred.
1878, April 25, red,	C.C.	D'ss of Florence 2nd	Oxford's Fane, 34995	Mr. Winn
1879, May 18, red,	B.C.	Duke of Florence	Duke Dentsdale, 33561	do.

MISS PEARL 28TH, red and white, calved January 12, 1876. Bred by Mr. T.
Hands, Canley; got by Cæsarewitch 4th (33263), dam (Miss Pearl 12th) by
Grand Duke 9th (19879), &c. See Vol. xxiii. p. 483.

Produce in		Names, &c.	By what Bull.	By whom bred.
1879, Nov. 18, roan,	C.C.	Miss Pearl 29th	Duke Dentsdale, 33561	Mr. Winn

WODEHOUSE, W. H.,
Woolmers Park, Hertford.

CHLOE, red and white, calved March 9, 1876. Bred by Mr. W. H. Wodehouse; got by Woolmers Duke (32890), dam (Corisande) by Vulcan (27739),
&c. See Vol. xxii. p. 620.

Produce in		Names, &c.	By what Bull.	By whom bred.
1879, May 8, red,	B.C.	Cadet	Lord Janicula, 41873	Mr. Wodehouse

COUNTESS MARY, roan, calved December 10, 1876. Bred by Mr. W. H.
Wodehouse; got by Royal Havering 2nd (35375), dam (Countess) by Archdale (21183), &c. See Vol. xxiii. p. 715.

Produce in		Names, &c.	By what Bull.	By whom bred.
1879, Sept. 17, {	red, B.C. / roan, C.C.	Romeo / Juliet	} Lord Janicula, 41873	Mr. Wodehouse

VERONA, roan, calved September 6, 1876. Bred by Mr. W. H. Wodehouse;
got by Royal Havering 2nd (35375), dam (Vesper) by Vulcan (27739), &c.
See Vol. xxii. p. 621.

Produce in		Names, &c.	By what Bull.	By whom bred.
1879, May 11, white, C.C.		Venus	Lord Janicula, 41873	Mr. Wodehouse

WOOD, James,
Humbleton Hall, Hull.

AMELIA 16TH, roan, calved September 28, 1873, Vols. xxiii. and xxv. pp. 716,
725. Bred by Mr. T. Barber, Sproatley Rise; got by Oxford's Baronet
(29499), dam (Amelia 14th) by Cherry Duke (25753), &c.

Produce in		Names, &c.	By what Bull.	By whom bred.
1879, July 16, r. & w., C.C.		Amelia 20th	Prince of Glo'ster, 40517	Mr. Wood

AMELIA 18TH, red and white, calved July 18, 1876. Bred by Mr. J. Wood ; got by Baron Wild Eyes (33100), dam (Amelia 16th) by Oxford's Baronet (29499), &c. See Vol. xxiii. p. 716.

Produce in	Names, &c.	By what Bull.	By whom bred.
1879, Dec. 5, roan,	B.C. Yorkshireman	D. of Holderness 3rd, 43106	Mr. Wood

MAID OF ATHENS, roan, calved March 2, 1875, Vol. xxiv. p. 713. Bred by Mr. T. Barber, Sproatley Rise ; got by Baron Wild Eyes (33100), dam (Lady of Athens) by Oxford's Baronet (29499), &c.

1879, May 31, r. & w.,	B.C. Lord of Athens	Baron W. Eyes 2nd, 36214	Mr. Wood

ORANGE BLOSSOM, white, calved March 9, 1868, Vols. xxi., xxii., xxiii., xxiv., and xxv. pp. 589, 332, 345, 338, 725. Bred by Mr. F. Sartoris, Rushden Hall ; got by Baron Geneva (25568), dam (Blossom) by Sir James (22902), &c.

1879, Mar. 29, white,	B.C. Baron Blanche 2nd	Baron Wild Eyes, 33100	Mr. Wood

RED ROSE, roan, calved May 1, 1872, Vols. xxii., xxiii., xxiv., and xxv. pp. 621, 716, 713, 726. Bred by Lord Londesborough, Grimston Park ; got by Snowball (32553), dam (Rose) by Third Duke of Flanders (23750), &c.

1879, Aug. 29, roan,	C.C. Red Rose 5th	Baron W. Eyes 2nd, 36214	Mr. Wood

WOOD, Rowland,
Clapton, Thrapstone, Northamptonshire.

BRUNETTE 12TH, red, calved April 25, 1876. Bred by Sir C. M. Lampson, Bart., Rowfant ; got by Duke of Rowfant (31015), dam (Brunette 7th) by Grand Duke of Geneva (28756), &c. See Vol. xxiii. p. 526.

1879, Mar. 10, red,	C.C. Underley Brunette	D. of Underley 2nd, 36551	Mr. Wood

CHARMING WOMAN, red, calved March 19, 1877. Bred by Mr. W. Sisman, Buckworth Lodge ; got by Julius Cæsar (31452), dam (Charity) by Prince Royal (29682), dam (Charm) by Pilot (22520), &c. See Vol. xviii. p. 419.

1879, Mar. 20, red,	C.C. Charming Duchess	D. of Vittoria 2nd, 41440	Mr. Wood

LADY BARRINGTON BUTTERFLY, red, calved January 20, 1876. Bred by Mr. R. Wood ; got by Duke of Barrington 5th (33575), dam (Butterfly) by Claret Duke (33392), g. d. (Queen Butterfly) by Rose Butterfly (24993), &c. See Vol. xxiv. p. 527.

1879, Feb. 3, red,	C.C. L.B'gton Butterfly 2d	D. of Vittoria 2nd, 41440	Mr. Wood

LAURA, red, calved March 19, 1873. Bred by Mr. W. B. Stopford-Sackville, Drayton House ; got by Hardicanute (26338), dam (Lovely) by George 1st (19848), &c. See "Lively," p. 306.

1877, Jan. 2, r. & w.,	C.C. Laura 2nd	Kettledrum, 34302	Mr. Wood
1878, Feb. 16, roan,	C.C. Laura 3rd	do.	do.

QUEEN OF THORNDALE 7TH, red, calved August 1, 1876. Bred by Mr. R. Wood ; got by Lord Aberdeen 3rd (34477), dam (Queen of Thorndale 3rd) by Lord Chancellor (20160), &c. See Vol. xxi. p. 1003.

1878, Nov. 12, r. & w.,	B.C. (Steer)	Kettledrum, 34302	Mr. Wood
1879, Dec. 26, red.	C.C. Q. of Thorndale 8th	D. of Vittoria 2nd, 41440	do.

SWEETMEAT, red, calved January 4, 1870, Vol. xxii. p. 621. Bred by Mr. R. Searson, Cranmore Lodge; got by Duke of Devonshire (21588), dam (Sweet Rose) by Chieftain (21421), &c.

Produce in		Names, &c.		By what Bull.		By whom bred.
1876, June 20, red,	B.C.	Baron Barrington		D. of Barrington 5th, 33575		Mr. Wood
1877, July 10, red,	C.C.	Sweetbrier		Kettledrum, 34302		do.
1878, Oct. 16, roan,	B.C.	Duke of Kent		do.		do.

WEE PET, red and white, calved June 30, 1875. Bred by Mr. W. Sisman, Buckworth Lodge; got by Prince Paul (43848), dam (Wild Rose) by Prince Royal (29682), g. d. (White Tail) by Pilot (22520), gr. g. d. (Welcome Lass) by Sir Edward (20819), — (Water Lily) by Abelard (15543), — (Florence) by Paul Pry (20478), — (Fancy) by Colonel (14296), &c. See Vol. xv. p. 486.

1878, June 22, r. & w.,	C.C.	Wedlock		Kettledrum, 34302		Mr. Wood
1879, May 1. red.	C.C.	Welcome		D. of Vittoria 2nd, 41440		do.

WOODHOUSE, John,
Scale Hall, Skerton, Lancaster.

MARIE STUART, red, calved December 3, 1874. Bred by Mr. Whittam. Mount Pleasant; got by Second Hubback (28880), dam (Fancy 16th) by Festus 3rd (26148), &c. See "Royal Hubback," p. 219.

1879, Feb. 20, roan,	B.C.	Royal Hubback	Fennel Duke, 36639		Mr. Woodhouse

PRINCESS AMELIA, red and white, calved April 12, 1869, Vol. xxii. p. 622. Bred by Mr. T. Lamb, Hay Carr House; got by Grand Count 2nd (24059). dam (Empress) by Majestic (20264), &c.

1879, Dec. 28, roan,	B.C.	Young Fennel	Fennel Duke, 36639		Mr. Woodhouse

WOODROFFE, W. S.,
Beaumont Grange, Halton, Lancaster.

DERWENT LADY 2ND, red and white, calved September 11, 1874, Vol. xxiv. p. 714. Bred by Mr. W. S. Woodroffe; got by Vice Roi (30214), dam (Derwent Lady) by The Premier (27640), &c.

1878, July 12, roan,	C.C.	Derwent Lady 3rd	Kt.of Knowlmere 2d,31542		Mr. Woodroffe
1879, May 28, r. & w.,	C.C.	Derwent Queen	Baron Stackhouse, 30488		do.

DUFFIELD QUEEN, red and white, calved April 10, 1873. Bred by Mr. W. S. Woodroffe; got by The Premier (27640), dam (Damask) by Baron Killerby (23364), &c. See "Vice Roi," Vol. xix. p. 362.

1877, Dec. 27, roan,	C.C.	Duffield Queen 2nd	Kt.of Knowlmere 2d,31542		Mr. Woodroffe
1879, Feb. 12, r. & w.,	B.C.		Baron Stackhouse, 30488		do.

LADY BOLIVAR, white, calved May 10, 1873. Bred by Mr. W. S. Woodroffe; got by Derbyshire Hero (33517), dam (Soldier's Daughter) by Brigade Major (21312), &c. See Vol. xxiv. p. 715.

1878, Jan. 16, white,	C.C.	Lady Beaumont	Kt.of Knowlmere 2d,31542		Mr. Woodroffe
1879, Jan. 10, roan,	B.C.	Lord Lovel	Baron Stackhouse, 30488		do.

MISS CHARLOTTE, roan, calved February 27. 1868, Vols. xxi. and xxiv. pp. 724, 714. Bred by Mr. A. Stables, Kirkbank; got by Croupier (23656), dam (Currer Bell) by Royal Alfred (18748). &c.

1878, Oct. 16. white.	C.C.	Miss Charlotte 2nd	Kt.of Knowlmere 2d,31542		Mr. Woodroffe

SOLDIER'S GIRL, red and white, calved September 1, 1874. Bred by Mr. W. S. Woodroffe; got by Vice Roi (30214), dam (Soldier's Daughter) by Brigade Major (21312), &c. See Vol. xxiv. p. 715.

Produce in	Names, &c.	By what Bull.	By whom bred.
1878, Feb. 10, roan, B.C.	Sergeant Major	Kt.of Knowlmere 2d,31542	Mr. Woodroffe
1879, Jan. 28, r. & w., B.C.	Sam Slick	Baron Stackhouse, 30488	do.

WORSLEY, Sir W. C., Bart.,
Hovingham, York.

LADY OF CONNAUGHT, white, calved April 30, 1876. Bred by the Earl of Feversham, Duncombe Park; got by Duke of Connaught (33604), dam (Gwendoline) by Second Duke of Collingham (23730), &c. See Vol. xxiii. p. 435.

1879, April 16, white, C.C.	Lady Leman	D. of Tregunter 5th, 33743	Sir W. C. Worsley

WRAITH, L. H.,
Newfield, Lower Darwen, Lancashire.

CHERRY QUEEN, roan, calved October 2, 1872, Vols. xxiv. and xxv. pp. 527, 727. Bred by the Earl of Bective, Underley Hall; got by Baron Oxford 5th (27958), dam (Cherry Princess) by General Napier (24023), &c.

1879, Mar. 6, red, B.C.	Ch.D.of Hillhurst 2d	3rd D. of Hillhurst, 30975	Mr. Wraith

FAVOURITE, roan, calved January 20, 1869, Vols. xx., xxii., xxiii., xxiv., and xxv. pp. 518, 624, 717, 716, 727. Bred by the Rev. R. E. Taylor, Moreton Hall; got by Duke of Clarence (19611), dam (Jennet) by Orion 2nd (22445), &c.

1879, July 4, red, B.C.	Newfield Baron 3rd	B.T'croft Oxford 2d, 33087	Mr. Wraith

Newfield Baron 3rd, sold to Mr. A. Holden, Lower Darwen.

RED DUCHESS, red, calved July 9, 1873, Vols. xxiv. and xxv. pp. 716, 727. Bred by Mr. G. Moore, Whitehall; got by Seventeenth Duke of Oxford (25994), dam (Duchess) by Grand Duke 15th (21852), &c.

1879, April 12, red, B.C.	T'gunt'r Char'ng Dk.	D. of Tregunter 7th, 38194	Mr. Wraith

WISTFUL EYES, red, calved October 29, 1875, Vol. xxv. p. 727. Bred by the Earl of Bective, Underley Hall; got by Second Duke of Tregunter (26022), dam (Winsome 8th) by Grand Duke 17th (24064), &c.

1879, May 12, roan, C.C.	Winsome Lassie	Baron Oxford 4th, 25580	Mr. Wraith

WRIGHT, Charles,
Oglethorpe Hall, Tadcaster.

JENNY LIND 28TH, roan, calved May 4, 1873, Vol. xxiii. p. 717. Bred by Mr. C. Wright; got by Duke of Oglethorpe (28428), dam (Jenny Lind 19th) by Duke of Clarence (19611), &c.

1878, April 16, roan, B.C.	Duke of York 2nd	Duke of Oxford, 38168	Mr. Wright
1879, Mar. 18, roan, B.C.	Duke of York 3rd	do.	do.

Duke of York 3rd, sold to Mr. Thomlinson, Cowthorpe, Wetherby.

LADY OXFORD, roan, calved November 1, 1874, Vol. xxv. p. 728. Bred by
Mr. C. Wright ; got by Eighteenth Duke of Oxford (25995), dam (Theodora)
by Duke of Clarence (19611), &c.

Produce in		Names, &c.	By what Bull.	By whom bred
1879, May 6, roan,	C.C.	Flora 2nd	Duke of Waterloo, 39788	Mr. Wright

MOSS ROSE 16TH, roan, calved October 28, 1869, Vols. xxi. and xxii. pp. 1009,
625. Bred by Mr. C. Wright ; got by Grand Duke of Wharfdale (26295),
dam (Moss Rose 14th) by Prince Imperial (22596), &c.

Produce in		Names, &c.	By what Bull.	By whom bred
1876, Nov. 27, red,	B.C.	D. of Oglethorpe 3rd	D. of Tregunter 5th, 33743	Mr. Wright
1877, Nov. 22, roan,	B.C.	D. of Oglethorpe 4th	do.	do.
1879, Mar. 15, roan,	C.C.	Moss Rose 25th	Oxford Cherry D. 2d, 34972	do.

MOSS ROSE 20TH, roan, calved December 5, 1874. Bred by Mr. C. Wright ;
got by Mac Mahon (31790), dam (Moss Rose 16th) by Grand Duke of Wharfdale (26295), &c. See Vol. xxi. p. 1009.

Produce in		Names, &c.	By what Bull.	By whom bred
1878, June 11, roan,	B.C.	(Steer)	Duke of Oxford, 38168	Mr. Wright
1879, May 13, r. & w.,	C.C.	Moss Rose 28th	Duke of Waterloo, 39788	do.

WRIGHT, John,
Green Gill Head, Penrith.

COUNTESS, white, calved January 5, 1877. Bred by Mr. J. Wright ; got by
Luck Estate (34713), dam (Ada) by Edwin (21671), &c. See "Royal
Carlisle," p. 216.

Produce in		Names, &c.	By what Bull.	By whom bred
1879, July 26, white,	B.C.	Royal Carlisle	Game Bird, 39912	Mr. Wright

GENEVA, roan, calved October 10, 1876. Bred by Mr. J. Wright ; got by
Man's Estate (26806), dam (Genevieve) by Royal Cambridge Gwynne (29849),
&c. See Vol. xxi. p. 1009.

Produce in		Names, &c.	By what Bull.	By whom bred
1879, Dec. 1, white,	C.C.	Geneva 2nd	British Squire, 37908	Mr. Wright

LIZZIE LORTON, roan, calved November 7, 1876. Bred by Mr. R. Thompson,
Inglewood Bank ; got by Master Dragonfly (34807), dam (Lady Lacy) by
Earl of Eglinton (23832), &c. See Vol. xxi. p. 955.

Produce in		Names, &c.	By what Bull.	By whom bred
1879, July 3, r. & w.,	C.C.	Lady Lorton	Prince Royal, 40537	Mr. Wright

WYNNE, Thomas,
Lislea, Armagh, Ireland.

ISLAND QUEEN, red, calved January 13, 1874. Bred by Earl Fitzwilliam,
Coollattin Park, the property of Mr. T. Wynne, got by Robert Burns
(29795), dam (Irish Girl 8th) by Chief Justice (28188), &c. See "Island
Prince," p. 125.

Produce in		Names, &c.	By what Bull.	By whom bred
1878, June 17, red,	B.C.	Island Prince	British Buck, 42840	Mr. Houston

Island Prince, sold to Mr. T. Wynne.

YEATS, George,
Studley, Ripon, Yorkshire.

CHERRY QUEEN, roan, calved May 26, 1869, Vol. xx. p. 443. Bred by Mr. G. Yeats; got by Emperor (23875), dam (Windsor Cherry) by Royal Windsor (18784), &c.

Produce in		Names, &c.	By what Bull.	By whom bred.
1873, May 20, roan,	B.C.	Statesman	Bismarck, 28040	Mr. Yeats
1874, April 1, roan,	C.C.	(dead)	do.	do.
1875, April 4, roan,	B.C.	(dead)	British Hope, 30607	do.
1876, Nov. 28, roan,	B.C.	Royal Manfred	Manfred, 26801	do.
1878, Feb. 1, r. & w.,	B.C.	Chief Constable	High Constable, 34158	do.
1879, May 12, white,	C.C.	Merry Queen	Lord Mirth, 34602	do.

Statesman, sold to Mr. L. Hodgson, Highthorne, Easingwold; Royal Manfred, sold for exportation to Australia.

CHRISTMAS ROSE, roan, calved January 27, 1872, Vols. xxi. and xxiv. pp. 1010, 717. Bred by Mr. G. Yeats; got by Christmas Present (28201), dam (Snowdrop) by May Snowflake (26888), &c.

1878, July 16, red,	C.C.	Rose Wreath	Royal Rufus, 37402	Mr. Yeats
1879, Aug. 30, r. & w.,	C.C.	Star of Braithwaite	St.SwithinStarDrop, 40667	do.

CLARICE, roan, calved March 30, 1876. Bred by Mr. G. Yeats; got by Clarion (33393), dam (Isabella) by Christmas Present (28201), &c. See "Fitz-Royal," p. 100.

1878, May 19, r. & w.,	C.C.	(dead)	Royal Halnaby, 39041	Mr. Yeats
1879, April 20, roan,	B.C.	Fitz-Royal	Royal Manfred, 40640	do.

ETHELINDA, roan, calved March 24, 1876. Bred by Mr. G. Yeats; got by Ethelred (33862), dam (Verbena) by Christmas Present (28201), &c. See Vol. xxiii. p. 718.

1878, May 15, white,	B.C.	Royal Gift	Royal Saxon, 40643	Mr. Yeats
1879, Aug. 20, white,	C.C.	White Verbena	Royal Manfred, 40640	do.

ISABELLA, white, calved September 25, 1872, Vols. xxii., xxiii., and xxv. pp. 625, 718, 729. Bred by Mr. G. Yeats; got by Christmas Present (28201), dam (Bella) by Emperor (23875), &c.

1879, Mar. 18, roan,	C.C.	Queen Hatty	King Harold, 40053	Mr. Yeats

ROSE ETHEL, roan, calved January 28, 1876. Bred by Mr. G. Yeats; got by Ethelred (33862), dam (Venus) by Bismarck (28040), &c. See Vol. xxiii. p. 718.

1878, April 10, roan,	B.C.	Royal Rose	Royal Saxon, 40643	Mr. Yeats
1879, April 6, r. & w.,	B.C.	Royal Hope	Royal Manfred, 40640	do.

YOOL, Thomas,
Coulard Bank, Lossiemouth, N.B.

CELANDINE, red and white, calved June 3, 1876. Bred by Mr. T. Yool; got by Gunpowder (28801), dam (Charlotte 10th) by Baron Coutts (27926), &c. See Vol. xxiii. p. 719.

1878, Aug. 2, roan,	B.C.	Pythius	Damon, 33503	Mr. Yool
1879, June 18, red,	C.C.	Constance	Riby Prince, 42278	do.

CHARLOTTE 18TH, red and white, calved April 25, 1874. Bred by Mr. T.
Yool; got by Gunpowder (28801), dam (Charlotte 9th) by Haberdasher
(28806), &c. See Vol. xxiii. p. 719.

Produce in		Names, &c.		By what Bull.		By whom bred.
1878, June 17, roan,	C.C.	Rose of June		Damon, 33503		Mr. Yool
1879, June 13, red,	B.C.	Prince Alfred		Riby Prince, 42278		do.

CHARLOTTE 19TH, red and white, calved May 27, 1874, Vol. xxiv. p. 718.
Bred by Mr. T. Yool; got by Gunpowder (28801), dam (Charlotte 10th) by
Baron Coutts (27926), &c.

1878, April 19, roan,	C.C.	Lady Claribel		Damon, 33503		Mr. Yool
1879, Mar. 9, roan,	B.C.	Sir Christopher		do.		do.

PRIMROSE, roan, calved March 4, 1874, Vol. xxiv. p. 718. Bred by Mr. T.
Yool; got by Gunpowder (28801), dam (Agnes) by Water King (13980), &c.

1878, Aug. 10, white,	C.C.	White Rose		Damon, 33503		Mr. Yool
1879, Aug. 2, white,	C.C.	White Rose 2nd		do.		do.

QUEEN OF THE MAY, red and white, calved May 20, 1873, Vol. xxiv. p. 718.
Bred by Mr. T. Yool; got by Gunpowder (28801), dam (Lady May) by
Cherry Prince 4th (25765), &c.

1878, June 3, roan,	C.C.	Queen Mab		Damon, 33503		Mr. Yool
1879, April 26, roan,	C.C.	Queen Rose		do.		do.

YORKE, John,
Bewerley Hall, Pateley Bridge, Yorkshire.

ROSTRA 2ND, roan, calved February 24, 1876, Vol. xxv. p. 730. Bred by Mr.
J. Yorke; got by Forester (33951), dam (Rosabel) by General (26230), &c.

1879, Aug. 13, r.&w.,	C.C.	Rosina		Wildfire, 40924		Mr. Yorke

WHITE ROSE, white, calved October 8, 1875. Bred by Mr. J. Yorke; got
by Forester (33951), dam (Wild Cherry) by Fourth Lord (29044), &c. See
Vol. xxii. p. 626.

1879, Mar. 9, roan,	C.C.	Wild Winsome		Sergeant Windsor, 37435		Mr. Yorke

WILD BRIER, roan, calved November 7, 1875, Vol. xxv. p. 730. Bred by
Mr. J. Yorke; got by Forester (33951), dam (Wild Rose) by Reformer
(18687), &c.

1879, Sept. 12, white,	C.C.	Wild Lass		Wildfire, 40924		Mr. Yorke

WILD CHERRY, roan, calved February 17, 1873, Vols. xxii., xxiii., xxiv., and
xxv. pp. 626, 720, 719, 731. Bred by Mr. J. Yorke; got by Fourth Lord
(29044), dam (Wild Rose) by Reformer (18687), &c.

1879, Aug. 23, r.&w.,	C.C.	Wild Strawberry		Sergeant Windsor, 37435		Mr. Yorke

WILD FLOWER, roan, calved December 27, 1874, Vol. xxv. p. 734. Bred by
Mr. J. Yorke; got by Vehement (35853), dam (Wild Rose) by Reformer
(18687), &c.

1879, April 13, roan,	C.C.	(dead)		Sergeant Windsor, 37435		Mr. Yorke

ZETLAND, Earl of,
Aske Hall, Richmond, Yorkshire.

ACONITE 4TH, roan, calved October 6, 1875. Bred by the Earl of Zetland,
Aske Hall; got by Telemachus 4th (35723), dam (Aconite 2nd) by Dunkeld
(26042), &c. See Vol. xxii. p. 627.

Produce in		Names, &c.		By what Bull.		By whom bred.
1879, Mar. 19, red,	B.C.	(Steer)		Grand D. Gwynne, 41644		Earl of Zetland

ACONITE 5TH, red, calved October 10, 1876. Bred by the Earl of Zetland,
Aske Hall; got by Cynthius 2nd (33495), dam (Aconite 2nd) by Dunkeld
(26042), &c.

1879, May 16, roan,	B.C.	Benedict		Benedictine, 42784		Earl of Zetland

ADELE, red, calved June 3, 1877. Bred by the Earl of Zetland, Upleatham;
got by Scots Fusilier (35483), dam (Alonette) by Sir James the Rose (15290),
&c. See "Bandinell," p. 9.

1879, Aug. 16, roan,	B.C.	Bandinell		Blucher, 39475		Earl of Zetland

ALLSPICE 2ND, roan, calved November 28, 1876. Bred by the Earl of Zetland,
Aske Hall; got by Cynthius 2nd (33495), dam (Allspice) by George Peabody
(28710), &c. See Vol. xxiii. p. 723.

1879, July 12, white,	C.C.	Allspice 3rd		Benedictine, 42784		Earl of Zetland

BEDALE 17TH, red and white, calved November 12, 1875. Bred by the Earl
of Zetland, Aske Hall; got by Bates (30508), dam (Bedale 14th) by Dunkeld
(26042), &c. See Vol. xxii. p. 628.

1879, Jan. 25, roan,	B.C.	(Steer)		Grand D. Gwynne, 41644		Earl of Zetland

BLOOM 5TH, red and white, calved November 8, 1874, Vols. xxiv. and xxv.
pp. 720, 734. Bred by the Earl of Zetland, Upleatham; got by Egremont
(39836), dam (Bloom 2nd) by Chatsworth (23546), &c.

1879, Sept. 16, roan,	C.C.	Bloom 8th		Blucher, 39475		Earl of Zetland

CYNTHIA 7TH, roan, calved December 27, 1875. Bred by the Earl of Zetland,
Aske Hall; got by Lord Godolphin (36065), dam (Cynthia 2nd) by Fitz-
James (19755), &c. See "Gwynne Prince," p. 117.

1870, Oct. 19, roan,	B.C.	Lottery		Benedictine, 42784		Earl of Zetland

SYBIL 2ND, white, calved November 21, 1875. Bred by the Earl of Zetland,
Aske Hall; got by Telemachus 4th (35723), dam (Sybil) by Dunkeld (26042),
&c. See Vol. xxii. p. 629.

1879, Jan. 20, white,	C.C.	Sybil 4th		Grand D. Gwynne, 41644		Earl of Zetland
1879, Dec. 31, white,	C.C.	Sybil 5th		Benedictine, 42784		do.

INDEX

TO

COWS AND HEIFERS.

Shorthorn Society of The United Kingdom of Great Britain and Ireland.

OFFICES:—12 HANOVER SQUARE, LONDON, W.

Patroness.

HER MOST GRACIOUS MAJESTY THE QUEEN.

President.

THE DUKE OF MANCHESTER,
KIMBOLTON CASTLE, ST. NEOTS.

Vice-President.

THE EARL OF BECTIVE, M.P.,
UNDERLEY HALL, CARNFORTH.

Council.

ACKERS, B. ST. JOHN, Prinknash Park, Painswick, Gloucestershire.

*AYLMER, HUGH, West Dereham Abbey, Stoke Ferry, Norfolk.

BEAUFORD, H. W., Sudborough House, Thrapston, Northamptonshire.

*BECTIVE, Earl of, M.P., Underley Hall, Carnforth.

BOOTH, JOHN B., Killerby Hall, Catterick.

BOWLY, E., Siddington House, Cirencester, Gloucestershire.

CHANDOS-POLE-GELL, H., Hopton Hall, Wirksworth, Derbyshire.

*CROSSBIE, W. TALBOT, Ardfert Abbey, Ardfert, Ireland.

CRUICKSHANK, J. W., Lethenty, Inverurie, N.B.

DUNMORE, Earl of, Dunmore, Stirling, N.B.

FEVERSHAM, Earl of, Duncombe Park, Helmsley, Yorkshire.

*FOLJAMBE, J. F. S., M.P., Osberton Hall, Worksop, Nottinghamshire.

*FOSTER, S. PORTER, Killhow, Mealsgate, Carlisle.

GUNTER, Colonel, Wetherby Grange, Wetherby, Yorkshire.

HOWARD, CHARLES, Biddenham, Bedford.

*KINGSCOTE, Colonel, C.B., M.P., Kingscote, Wotton-under-Edge, Gloucestershire.

LATHOM, Earl of, Lathom House, Ormskirk, Lancashire.

*LAWSON, Sir WILFRID, Bart., M.P., Brayton, Carlisle, Cumberland.

*LINDSAY, Colonel LOYD, M.P., Lockinge Park, Wantage, Berkshire.

---

* Those Members of Council whose names are prefixed by an asterisk retire at the next Annual General Meeting, but are eligible for re-election.

3 D

*McINTOSH, DAVID, Havering Park, Romford, Essex.

MANCHESTER, Duke of, Kimbolton Castle, St. Neots, Huntingdonshire.

MEADE-WALDO, E. W., Stonewall Park, Edenbridge, Kent.

*MITCHELL, A., The Walk House, Alloa, N.B.

MORETON, Lord, M.P., Tortworth Court, Wotton-under-Edge, Gloucestershire.

PENRHYN, Lord, Penrhyn Castle, Bangor, North Wales.

RATHDONNELL, Lord, Lisnavagh, Tullow, Co. Carlow.

RICHMOND AND GORDON, K.G., Duke of, Gordon Castle, Fochabers, Banff, N.B.

SHELDON, H. J., Brailes House, Shipston-on-Stour.

STANIFORTH, Rev. T., Storrs, Windermere, Westmoreland.

STRATTON, R., The Duffryn, Newport, Monmouthshire.

TRACY, G. MURTON, Redlands, Edenbridge, Kent.

WELSTED, RICHARD, Ballywalter, Castletown Roche, Ireland.

*WILSON, JACOB, Woodhorn Manor, Morpeth, Northumberland.

Secretary.
H. J. HINE.

Editing Committee.

KINGSCOTE, Colonel.	CHANDOS POLE-GELL, H.
BEAUFORD, H. W.	CRUICKSHANK, J. W.

General Purposes Committee.

LATHOM, Earl of.	HOWARD, C.
KINGSCOTE, Colonel.	McINTOSH, D.
AYLMER, HUGH.	TRACY, G. MURTON.
BEAUFORD, H. W.	WILSON, JACOB.

Solicitor.
SHEPHERD, J. B., Stourbridge.

Counsel.
HILL, A. STAVELEY, Q.C., M.P., 3 Garden Court, Temple, E.C.

Auditor.
EDMONDS, W. J., Southrop House, Lechlade.

Accountants.
Messrs. QUILTER, BALL, & Co., 3 Moorgate Street, E.C.

Bankers.
THE LONDON AND WESTMINSTER BANK, Marylebone Branch, Stratford Place, W.

---

* Those Members of Council whose names are prefixed by an asterisk retire at the next Annual General Meeting, but are eligible for re-election.

LIST OF MEMBERS.

[Life Members are distinguished thus †.]

†HER MAJESTY THE QUEEN, WINDSOR CASTLE.
†H.R.H. THE PRINCE OF WALES, K.G., SANDRINGHAM, KING'S LYNN.

A.

Abell, John, The Valley Farm, Higham, Hinckley

†Ackers, B. St. John, Prinknash Park, Painswick

Adam, Thomas, Lynegar, Wick, Caithness

Adams, Thomas, Mains of Eden and Bowiebank, Banff

Addison, S. T., Ellenhall, Eccleshall, Staffs

Adkins, George C., The Lightwoods, Birmingham

†Adkins, John Caleb, Milcote, Stratford-on-Avon

Aitcheson, Peter, West Garleton, Haddington

†Alexander, A. J., Woodburn, Spring Station, Kentucky

Alexander, S. M., Roe Park, Limavady, Ireland

Allan, Alexander, Kinnon Park, Perth, N.B.

†Allan, John, Billie Mains, Ayton, N.B.

Allan, John, Crieff Vechter, Crieff, Perthshire

†Allen, B. Haigh, Clifford Priory, Hereford

Allen, George (no address)

†Allen, George, Unicarville, Comber, Co. Down

Allen, Stephen H., Eastover, Andover

Allsopp, Sir Henry, Bart., Hindlip Hall, Worcester

†Angas, John Howard, Collingrove, Adelaide, South Australia

†Angerstein, W., Weeting Hall, Brandon, Norfolk

Angus, John, Bearl, Stocksfield-on-Tyne

Annandale, W. M., Lintz Ford, Lintz Green, Newcastle-on-Tyne

Annett, J. W., Houndalee, Acklington, Northumberland

†Arabin, Julien, Preswylfa, Neath

†Arkell, William, Hollies, Lechlade

†Arkell, W., jun., Hatherop, Fairford

Arklay, Robert, Ethiebeaton, Dundee

†Armistead, J. Fisher, Cob Wall House, Blackburn

†Armstrong, Samuel, Shingay, Royston

†Armstrong, S., Vinegar Hill, Enniscorthy, Ireland

†Armstrong, Sir William G., C.B., Cragside, Newcastle-on-Tyne

Armytage, Sir George, Bart., Kirklees Park, Brighouse, Yorkshire

Arnold, Thomas, The Brewery, Wickwar, Gloucestershire

†Ashburner, George, Low Hall, Kirkby Ireleth, Carnforth

†Ashburner, W., Conishead Grange, Ulverston

Ashby, Capt. G. Ashby, Naseby Woolleys, Rugby

†Ashton, Charles, Delrow House, Watford

Ashwin, M. C., Cross o' the Hill Farm, Stratford-on-Avon

†Ashworth, Alfred, Egerton Hall, Bolton-le-Moors

Askew, J. W., Ash Meadow, Arnside, near Milnthorpe

Atkinson, George, Stanley House, Crook, Darlington

Atkinson, William, Burneside Hall, Kendal

†Attenborough, Richard, Whitley Grove, Reading

Attwater, John Gay, Britford, Salisbury

Australian Agricultural Company, 196 Gresham House, Old Broad Street, E.C.; W. Robinson, Secretary

†Aylmer, Hugh, West Dereham Abbey, Stoke Ferry

†Aylmer, J. B., Fincham Hall, Downham, Norfolk

B.

†Bailey, Crawshay, Maindiff Court, Abergavenny

†Bailey, James, C.B., Bleak House, Howick, Auckland, New Zealand

†Bailey, Stephen, Hornshay Farm, Nynehead, Wellington, Somerset

Baillie, Evan, Dochfour, Inverness, N.B.

†Baillie, Evan, Filleigh, Chudleigh, Devon

Bainton, Thomas, Arram Hall, Seaton, Hull

Baird, Alexander, Robeston Hall, Milford Haven

Baker, Thomas, Harpole, Weedon

Baldwin, Thomas, Glasnevin, Dublin

Baldwyn, William, Ashton-under-Hill, Tewkesbury

·†Balfour, Arthur James, M.P., Whittinghame, Prestonkirk, N.B.

†Balfour, Colonel, Balfour Castle, Kirkwall, N.B.

Bangor, Viscount, Castleward, Downpatrick, Ireland

†Banks, I. J., Lanesfoot, Kendal

Banyard, Thomas, Poplar Hall, Horningsea, Cambs.

Barber, Thomas, Sproatley Rise, Hull

†Barchard, Francis, Horsted, Uckfield, Sussex

Barford, William, Gayhurst House, Peterborough

†Barnes, C. A., Solesbridge, Rickmansworth, Herts.

Barnett, Miss F. S. A., Costislost, Bodmin, Cornwall

†Barroby, Francis, Dishforth, Thirsk

†Barrow, Bridgman Langdale, Sydnope Hall, Matlock

†Barton, Major H. L., Straffan House, Straffan Station, Co. Kildare

Barton, Richard, Caldy Manor, Birkenhead

†Bates, Thomas, Heddon, Wylam, Northumberland

Baxter, H. P., Southall Green · Farm Southall

Bayes, Charles, Kettering

†Beak, Captain W. E., Manor House, Somerford, Chippenham

Beards, Joseph, Worksop Mill, Worksop

†Bensley, J. Noble, Chapel Brampton, Northampton

†Beattie, James, Newbie House, Annan, Dumfries

†Beattie, Simon, Preston Hall, Annan, Dumfries

†Beauchamp, Earl, Madresfield Court, Malvern

Beauchamp, Edward Beauchamp, Trevince, Scorrier, Cornwall

†Beauford, H. W., Sudborough House, Thrapston

Beckwith, Rev. H., Eaton Constantine, Ironbridge, Salop.

†Bective. Earl of, M.P., Underley Hall, Carnforth

Beever, Rev. W. Holt, Pencraig Court, Ross

Bell, Alexander, Barrowford Brewery, Nelson, Burnley

†Bell, G. J., The Nook, Irthington, Brampton

Bell, Robert, Standen Hall, Clitheroe

†Bell, William, Bodrhyddan, Rhyl

Benton, William, Cattie, Keig, Aberdeenshire

†Benyon, Richard, Englefield House, Reading

†Berwick, H. W. B., Ardgowan, Greenock

†Bethune, A., Blebo, Cupar, Fifeshire

Bettridge, Henry, East Hanney, Wantage, Berks.

Betts, R., Holbeck Lodge, Horncastle

Betts, W. Hammond, Frenze Hall, Diss, Norfolk

Bird, George, Volis, Kingston, Taunton

Bird, John, jun., Conquest House, Farcet, Peterborough

Blackler, William, Wottons, Broadhempston, Totnes

†Blackwell, George, Hazlecote, Kingscote, Wotton-under-Edge

Blackwell, Richard, Nottingham Road Corn-Mills, Derby

Bland, Charles, Gaddesby, Leicester

Bland, George, Coleby Hall, Lincoln

†Bland, T. H., Dingley Grange, Market Harborough

Blantern, George G., Haston Hadnall, Salop

†Blathwayt, Capt., Dyrham Park, Chipping Sodbury

Blenkarn, Thomas, sen., Keisley, Appleby

†Blezard, Robert, Pool Park, Ruthin

Bliss, Stephen V., Weston Underwood, Olney

Bliss, William, Chipping Norton, Oxon

†Bluett, Rev. W. J., Holcombe, Leeston Canterbury, New Zealand

†Blundell, J. H., Woodside, Luton

†Blyth, R. Burn, Woolhampton, Reading

†Bolton, Lord, Bolton Hall, Bedale

†Bolton, William, The Island, Oulart, Co. Wexford

Bond, George Morton, Alrewas House, Ashbourne, Derbyshire

†Boog, Thomas Elliott, Timpendean, Jedburgh, N.B.

Booth, Mrs., Warlaby, Northallerton

†Booth, J. B., Killerby Hall, Catterick

†Borthwick, William, Monkwray, Whitehaven

Bott, Joseph F., Cammars Hall, Morrel Roding, Dunmow

Botterill, Richard Wauldby, Brough, Yorkshire

Bourke, W. Campbell, Tleclash, Fermoy, Co. Cork

Bousfield, John, Soulby, Brough, *via* Penrith, Westmoreland

Bowers, William, Harewood Park, Cheadle, Staffordshire

†Bowly, E., Siddington House, Cirencester

†Bowman, John, Victoria, Kansas, U.S.A.

Bowness, Moses, Vale View, Ambleside, Westmoreland

Bowstead, J. C., Binbury Manor, Maidstone

†Boyd, John, Simprim Mains, Coldstream

†Boyes, Matthew, Wandale House, Slingsby, York

†Brackenbury, W. T., Thorpe Hall, Downham, Norfolk

†Braikenridge, John H., The Rookery, Chew Magna, Somerset

†Bramley, J. Curtois, Langrick, Boston

†Brassey, Albert, Heythrop Park, Chipping Norton

†Brassey, H. A., M.P., Preston Hall, Aylesford

†Braybrooke, Lord, Audley End, Saffron Walden

†Briggs, David Grant, Kelstern Grange, Louth

†Brigham, James, Slingsby, York

Briscoe, John, Hill Croome, Severn Stoke, Worcestershire

Brise, Colonel, M.P., Spains Hall, Braintree

†Britten, George, Overstone Farm, Northampton

†Brockbank, R. Bowman, Crosby, near Maryport

†Brocklebank, T., Springwood, Liverpool

Brogden, A., M.P., Lightburne Park, Ulverston

Bromet, W. R., Oocksford, Tadcaster

Bromley, James, Forton, Garstang

Bromley, John, Lancaster

Brooke, Sir Victor, Bart., Colebrooke, Brookeboro', Co. Fermanagh

†Brooke-Hunt, A. E., Peers Court, Dursley, Gloucestershire

Brooksbank, B. H., Tickhill, Rotherham

Brown, E., Prince of Wales Lake Hotel, Grasmere, Windermere

Brown, George, Willow Hill, Morhanger, Sandy, Beds.

Brown, John, Park Cottage, Burton Constable, Hull

†Brown, Joseph John, Shangton, Leicester

Brown, Richard, Ruyton-of-the-Eleven-Towns, Shrewsbury

Brown, William, The Villa, Ackworth, Pontefract

†Brown, W. H., Drayton House, Belbroughton, Stourbridge

†Browne, A. H., Callaly Castle, Alnwick

Browne, Colvile, 69 Millbank Street, Westminster, S.W.

Browne, Samuel, Haughton Hill, Shifnal

Browne, T. Beale, Salperton Park, Harleton, Cheltenham

†Brownlow, Earl, Belton House, Grantham

Bruce, David C., Broadland, Huntly, N.B.

Bruce, James, Burnside, Fochabers, N.B.

Bruce, John, Barmoor Castle, Beal, Northumberland

†Bruce, Robert, Manor House Farm, Great Smeaton, Northallerton

†Bruen, Henry, Oak Park, Carlow, Ireland

Bruere, Raymond S., Braithwaite Hall, Middleham

Brundell, R. S., Leicester House, Doncaster

Brunyee, Samuel C., Sand Hall, Crowle, Lincolnshire

†Buccleuch, Duke of, K.G., Dalkeith Park, N.B.

Buchanan, Philips, Hales Hall, Market Drayton, Salop.

Budds, William Frederick, Courtstown, Freshford, Co. Kilkenny

Bult, James S., Dodhill House, Kingston, Taunton

†Burbidge, Charles, Chittern St. Mary, Heytesbury, Wilts.

Burbury, John, Wootton Grange, Kenilworth

Burnaby, J. D. A., Ashfordby Hall, Melton Mowbray

†Burnyeat, W., jun., 14 Tangier Street, Whitehaven

Burra, Abraham, Raisbeck, Orton, Westmoreland

†Burrow, Robert, Wrayton Hall, Kirkby Lonsdale

Burton, William, Eastoft Hall, Goole

Burtt, Henry, Fulbeck Grange, Grantham

Burtt, J. B., Fawsley House, Gloucester

Butcher, William, Gosmere, Selling, Faversham

†Byron, E., Coulsdon Court, Surrey

C.

†Caddy, Henry, Rougholm, Bootle

†Cadman, Peter, Ballahutchin, Union Mills, Isle of Man

Cadman, Thomas Watson, Ballifield Hall, Sheffield

†Calthorpe, Lord, Elvetham, Winchfield

Camm, Thomas, Ossington, Newark

Campbell, Archibald, Dunmore, Stirling

†Campbell, Sylvester, Kinellar, Blackburn, N.B.

†Campbell, T. F. (no address)

†Cannon, Henry Le Grand, Over Lake, Burlington, Vermont, U.S.A.

Cantlie, C. A., Keithmore, Dufftown, N.B.

†Carbery, Lord, Laxton Hall, Wansford

†Carrington, W. T., Croxden Abbey, Uttoxeter

†Carter, R. H., Fernhill, Cradley, Malvern

Carter, W. H., Brockholme, Hornsea, Hull

†Cartwright, F., Temple Grange, Navenby, Lincolnshire

Cartwright, Thomas Dallow, St. Stephens, St. Albans

Carysfort, Earl of, Warmington, Oundle

†Casswell, J. H., Laughton, Folkingham

Catchpole, Nathaniel, Bramford, Ipswich

Cather, George, Carrichue, Londonderry

Cattley, John, Stearsby, Easingwold

†Cawdor, Earl of, Stackpole Court, Pembroke

Cazalet, Edward, Fairlawn, Shipbourne, Tonbridge, Kent

Chaffey, Major John, Prince Hill, Worton, Devizes

Chalk, Thomas, Linton, Cambs.

Chaloner, Admiral Thomas, Longhull, Guisborough

Chamley, Captain Thomas, Warcop House, Penrith

†Chandos-Pole-Gell, H., Hopton Hall, Wirksworth

Chapman, George, Brook Farm, Exton, Oakham

Charley, William, Seymour Hill, Dunmurry, Co. Antrim

†Chearnley, R. A., Salterbridge, Cappoquin, Co. Waterford

Cheney, E. H., Gaddesby Hall, Leicester

†Chesham, Lord, Latimer, Chesham

Chester, George, Waltham, Melton Mowbray

Chirnside, John B., The Lodge, Clifton-on-Dunsmore, Rugby

†Chrisp, L. C., Hawkhill, Alnwick

Christie, James, Bankend Farm, Stirling, N.B.

†Christy, George, Buckhurst Lodge, Westerham

†Christy, James, Boynton Hall, Chelmsford

Christy, Luke, Carrigeen, Croom, Co. Limerick

Clark, Isaac, The Manor, Heddington, Calne, Wilts.

†Clark, John Percy, North Ferriby, Brough, E. Yorkshire

†Clarke, Robert, Bolinda Vale, Lancefield Road, Bourke, Victoria, Australia

†Clay, John, jun., Winfield, Berwick-on-Tweed

†Clear, Albert P., Maldon, Essex

Cleasby, William, Eden Place, Kirkby Stephen

Clerk, Edmund Hugh, Burford, Shepton Mallet, Somerset

Clifton, J. Talbot, Lytham House, Lytham

Close, James, Holmescales, Milnthorpe

Close, Jarvis, Smardale Hall, Kirkby Stephen

Coates, Christopher, Cleasby, Darlington

Cobb, Frederick, Walton, Warwick

Cochran, Richard, Spring Hill, Quigley's Point, Co. Donegal

Cochrane, Major A. H., Aldwark Manor, Easingwold, Yorks.

Cochrane, James, Little Haddo, Newburgh, N.B.

†Cochrane, James A., Hillhurst, Compton, Canada

†Cock, C. H., Bridgefoot, Barnet

Cock, Henry, Coat Green Farm, Burton, Westmoreland

†Colcombet, A., 15 Quai Tilsitt, Lyons, France

†Coleman, Edward J., Stoke Park, Slough

†Coleman, J., Park Nook, Derby
Coleman, John, Carnaby, Hull
†Collard, Charles, Little Barton, Canterbury
†Collingwood, Mrs., Clayworth, Bawtry
Colthurst, J., Chew Court, Chew Magna, Somerset
†Connon, A. W., Cremore, Kilmuckridge, Gorey, Co. Wexford
Constable, Sir Talbot Clifford, Bart., North Ferriby, Yorkshire
†Conwy, Major Conwy Rowley, Bodrhyddan, Rhyl
Cook, Francis, Thixendale, York
†Cooper, John, East Hadden, Northampton
†Cooper, Percy H., Bulwell Hall, Nottingham
Cope, J. A. M., Drummilly, Loughgall, Co. Armagh
†Cosby, Capt. R. G., Stradbally Hall, Queen's County, Ireland
Coupland, William, Bolton House, Yeadon
†Courtown, Earl of, Courtown House, Gorey, Ireland
Coventry, Earl of, Croome Court, Severn Stoke, Worcestershire
Cowton, Robert, Potter Brompton, Ganton, York
†Crabb, R. H., Great Baddow Place, Chelmsford
Craddock, T. M., Petwick Farm, Faringdon, Berks.
Cradock, Christopher, Hartforth, Richmond, Yorkshire
Cragg, W. Smith, Arkholme, Carnforth
†Craig, John R., Secretary, Agriculture and Arts Association, Toronto, Canada
†Cramer, M. C., Rathmore, Kinsale, Ireland
†Cran, John, Kirkton, Bunchrew Station, Highland Railway, N.B.
Crawford, Miss E. A., Hill House, Farnsfield, Southwell, Notts.
Crickmore, William, Seething, Brooke, Norfolk
Crisp, Arthur W., Orford, Wickham Market, Suffolk
Croome, James Capel, Dagendon, Cirencester
†Crosbie, William Talbot, Ardfert Abbey, Ardfert, Ireland
Cross, J., Chiswell Hall Farm, Edenbridge
†Croudson, John, Urswick, Ulverston
Crowe, Robert, Speeton, Bempton, Hull
Cruickshank, A., Sittyton, Aberdeen, N.B.
†Cruickshank, Edward, Lethenty, Inverurie, N.B.

Cruickshank, George, Comisty, Huntly, N.B.
†Cruickshank, J. W., Lethenty, Inverurie, N.B.
Cruse, Jabez, Cleave Farm, Brandiscorner, North Devon
Crust, John, Garton Field, Driffield
Crust, Joseph, Middle Street, Driffield
Culshaw, Joseph, Towneley, Burnley
†Currie, James, Halkerston, Gorebridge, Edinburgh
†Curtis, Major-General, Ogdensburg, New York, U.S.A.
†Curtis, William, Fernham, Faringdon, Berks.

D.

D'Aeth, N. H., Knowlton Court, Wingham, Kent
†Dalgety, F. G., Lockerly Hall, Romsey, Hants.
Dalton, John, Cummersdale Mills, Carlisle
Danby, Robert, Stamford Bridge, York
Daniell, G. W., St. Leonard's, Blandford, Dorset
†Darby, E. W., Little Ness, Shrewsbury
†Dargue, Thomas, Whale Farm, Askham, Penrith
Darke, C. P., Caversham, Reading
†Darley, Henry, Aldby Park, York
†Darling, John, Beau Desert, Rugeley, Staffs.
Darling, Robert, Plawsworth, Chester-le-Street, Durham
Daubuz, J. C., Killiow, Truro, Cornwall
†Davey, J. Sydney, Bochym, Helstone, Cornwall
†Davidson, Alexander, Mains of Cairnbrogie, Tarves, N.B.
†Davidson, J., Shepherd's Hill, Penrith
†Davidson, James, Bank House, Acklington, Northumberland
†Davies, D. Reynolds, Agden Hall, Lymm, Warrington
†Davies, R. P. (no address)
Dawson, George, Thorncliffe Iron Works, Sheffield
Dawson, Thos., jun., Poundsworth, Driffield
†Day, Gerard J., Horsford Hall, Norwich
Denchfield, Edward, Burston House, Aylesbury
Denchfield, John, Burston House, Aylesbury
Dent, A. C., Wharton Hall, Kirkby Stephen
†Dent, John Dent, Ribston Hall, Wetherby

Dent, William, Kaber Fold, Brough, Westmoreland

†de Vitre, H. Denis, Charlton House, Wantage

†Devonshire, Duke of, K.G., Holker Hall, Carke-in-Cartmel, Carnforth

Dickinson, James, Balcony Farm, Upholland, Wigan

Dickinson, John A., Brough Sowerby, Penrith

†Dickson, Ben., Gilford House, Gilford, Co. Down

†Dickson, J. H., Saughton Mains, Edinburgh

Dixon, George, Mantle Hill, Bellingham, Northumberland

Dixon, George Moore, Bradley Hall, Ashbourne

Dixon, T. C., Barff House, Brandsburton, Beverley, Yorks.

Dobson, Anthony, Williamsgill, Temple-Sowerby, Penrith

Dodd, Francis, Rush Court, Wallingford

Dodds, T., Mount Pleasant, Wakefield

Donaldson, Alexander (no address)

†Dormer, C. U. Cottrell, Rousham, Oxford

Douglas, G. S., Berryhill, Kelso, N.B.

†Downing, John, Ashfield, Fermoy, Ireland

†Doyne, C. M., Wells, Gorey, Co. Wexford

†Drake, Thomas T., Shardeloes, Amersham

Drewry, George, Holker House, Carke-in-Cartmel

†Druce, Joseph, Eynsham, Oxford

†Drummond, John, Blackruthven, Perth

†Duchateau, Florimond, Quevaucamps, Hainault, Belgium

†Dudding, Henry, Panton House, Wragby, Lincolnshire

Dudgeon, Robert Francis, jun., Cargen, Dumfries.

†Dun, Finlay, 2 Portland Place, London, W.

†Duncombe, Hon. Cecil, Nawton Grange, York

Dundas, Charles Henry, Gerrichrew, Dunria, Crieff, N.B.

†Dunmore, Earl of, Dunmore, Stirling, N.B.

Dunn, Frederick, Ebor House, Keyingham, Hull

†Dunn, J. H., Springfield, Gillingham, Dorset

†Dunn, William Hew, Elcot Park, Hungerford, Berks.

†Dunning, J. H., Court Barton, Creech St. Michael, Taunton

Durham, Charles, Aldenham Abbey, Watford

†Duthie, William, Collynie, Tarves, N.B.

E.

†Easton, Thomas, Storrs Farm, Windermere

†Edmonds, W. J., Southrop House, Lechlade

Edmondson, Bernard, Lower Hood House, Burnley

†Edmondson, William, Highfield, Denton, Otley, Yorks.

†Edwards, John, jun., Holly Lodge, Buckworth, Huntingdon

Edwards, Joseph W., Park Farm, Maesbury, Oswestry

†Eland, Thomas John, The Poplars, Fotherby, Louth

†Ellesmere, Earl of, Worsley Hall, Manchester

Elliot, James, Park View, Staffield, Penrith

Elliot, John, Polnicol, Parkhill, Ross-shire

Ellis, Charles, Oak Villa, Meldreth, Royston

Ellis, Colonel J. J., Ellistown, Leicester

Elwell, John, Timberley, Castle Bromwich

†Errington, Robson, Scotby Farm. Carlisle

†Evans, Humphrey, Woodburn, Kentucky, U.S.A.

Evans, John, Uffington, Shrewsbury

Ewbanke, Mrs., Stainmore, Borrenthwaite Brough, Westmoreland

†Exeter, Marquis of, K.G., Burghley House, Stamford

F.

Fair, Jacob Wilson, Lytham, Lancashire

†Fanning, Major, Parkwood, Farningham, Kent

Farmer, James, 6 Porchester Gate, Hyde Park, W.

Farrer, John, Thorneyholme, Burnley

†Faulkner, William, Rothersthorpe, Northampton

Fawcett, E. A., Childwick Hall, St. Albans

Fawcett, H., Kirkstall Road, Leeds

†Fawcett, Mrs., Scaleby Castle, Carlisle

†Fawkes, Ayscough, Farnley Hall, Otley

Fenton, John, Redkirk, Annan, N.B.

Fergusson, John, Brettenham Manor, Thetford

†Feversham, Earl of, Duncombe Park, Helmsley

Ffolliott, Colonel John, Hollybrook, Boyle, Ireland
†Fielden, John, Dobroyd Castle, Todmorden
†Fieldsend, Charles R., Kirmond, Market Rasen, Lincolnshire
Findlater, James, Milltack, King Edward, Aberdeen
†Fisher, Donald, Pitlochrie, Perthshire, N.B.
†Fisher, Edward Knapp, Hill Crest, Market Harborough
†Fisher, Robert, jun., Leconfield, Beverley
†Fitzhardinge, Lord, Berkeley Castle, Glos'
†Fitzwilliam, Earl, Wentworth, Woodhouse, Rotherham
Fitzwilliam, Hon. C. W. W., M.P., Alwalton, Peterborough
†Foljambe, F. J. S., M.P., Osberton Hall, Worksop
Fortescue, A. Irvine, Kingcausie, Aberdeen
Fosbery, George R., Kilgobbin, Patrick's Well, Co. Limerick
†Foster, Samuel P., Killhow, Mealsgate, Carlisle
†Foster, William, Westward Park, Wigton, Cumberland
†Foster, W. O., Apley Park, Bridgnorth
Fowler, F., Henlow, Biggleswade
Fowler, John K., Willowbank, Aylesbury
Fowler, Willingham, Cottesmore, Oakham
†Fox, George, Elmhurst Hall, Lichfield
†Fox, William, Abbey, St. Bees, Carnforth
Frank, R. Hayston, Ashbourne Hall, Ashbourne, Derbyshire
Frank, Thomas, Fylingdales, Whitby
Frank, T. Peirson, High Hall, Kirby Moorside, Yorkshire
†Fraser, W. A., Brackla, Nairn, N.B.
Freeman, Edwin, Chilton, Thame, Oxon
Frier, Matthew, Nether Kidston, Peebles
Frost, Frederick, Bowforth, Kirby Moorside, Yorkshire
†Frost, Robert, Lime Grove, Chester
Frudd, J. M., Bloxholm Moor, Sleaford
Furness, Captain Matthew W., Clifton-on-Dunsmore, Rugby
†Fytche, Major-General, Pyrgo Park, Havering-Atte-Bower, Essex

G.

†Gaitskell, J., Hall Santon, Holmrook, Carnforth
†Gamble, John, Shouldham Thorpe, Downham Market

†Gandy, Captain Henry, Castle Bank, Appleby
Ganly, James, Usher's Quay, Dublin
Garbutt, Isaac, Manor House, Sinnington, Pickering
†Garfit, Arthur, Scothern, Lincoln
†Garne, George, Churchill Heath, Chipping Norton
†Garne, R., Aldsworth, Northleach
Garne, Thomas, Great Barrington, Burford, Oxon.
Garne, W., South Corney, Cirencester
†Garne, W. G., Broadmoor, Northleach
†Garner, Charles Carter, The Wolds, Snitterfield, Stratford-on-Avon
Garnett, Robert, Wyreside, Lancaster
Garsed, John, The Moorlands, Cowbridge
Garth, Francis, Crackpot, Reeth, Yorkshire
Gatley, C. Ralph, Polsue Barton, St. Erme, Truro
Geekie, Alexander, Baldowrie, Cupar Angus, N.B.
†Geekie, Robert, jun., Rosemount, Blairgowrie, N.B.
Gell, Evan, White House, Kirk Michael, Isle of Man
†German, William, Measham Lodge, Atherstone
Gibbs, W. Slocombe, The Manor House, Cothelestone, Taunton
†Gibson, Arthur S., Bulwell, Nottingham
†Gibson, Joseph, Whelprigg, Kirkby Lonsdale
†Gibson, Richard, London, Ontario, Canada
†Gibson, Richard, 66 Queen Street, Melbourne, Australia
Gilbert, Rev. G., Claxton Grange, Norwich
Gillow, Rev. Charles, Ushaw College, Durham
Godson, J. S., Saundby, Gainsborough
Godson, John W., Edge Hill House, Banbury
†Godwin, J. S. S., Hazlewood, Hadlow, Tonbridge
Goodson, Thomas, Eastwell, Melton Mowbray
†Gordon, James A., Udale House, Invergordon, N.B.
Gordon, Mrs., Cluny Castle, Aberdeen
†Gorringe, Hugh, Kingston - by - Sea, Brighton
Gould, R. H., Didmarton, Chippenham
Goulder, H. W., Wimbotsham, Downham Market

Goulding, John, Aik Bank, Calthwaite, Penrith

Goulter, James, Littleton Drew, Chippenham

†Gow, Thomas, Cambo, Newcastle-on-Tyne

†Graham, Arthur, Kirkbythore Hall, Kirkbythore, Westmoreland

†Graham, George, The Oaklands, Birmingham

Graham, Miss I., 3 Westbourne Road, Edgbaston, Birmingham

Graham, James, Calthwaite Hall, Penrith

Graham, J. Maxtone, Cultoquhey, Perth, N.B.

†Graham, Rev. P., Turncroft, Over Darwen, Lancashire

†Graham, William, Eden Grove, Bolton, Westmoreland

†Graham, Col. W., Mossknow, Ecclefechan, N.B.

Graham, William, Waterloo Farm, East Witton, Bedale

†Graham, Y. R., 3 Westbourne Road, Edgbaston, Birmingham

Grahame, Thomas, 20 Chiswick Street, Carlisle

Granger, John, Pitcur, Coupar Angus, Forfar

Grant, George, Pollo Farm, Invergordon, N.B.

Grant, W. J., Hope End, Ledbury, Herefordshire

Graves, Septimus Perry, Southam, Warwickshire

†Green, D. A., East Donyland, Colchester

Greenway, G. C., Ashorne Hill, Leamington

†Greenwood, Charles, Clayworth, Bawtry

†Greenwood, F. B., Swarcliffe Hall, Ripley, Yorkshire

Grenfell, Arthur R., 4 Queen's Terrace, Windsor

†Grey, Edward, Eastham, Cheshire

Griffin, C. W., Werrington, Peterboro'

Grime, G. A., East Keal, Spilsby

†Grimes, J., Newton-on-Trent, Newark

†Grissell, Thomas D., Norbury Park, Dorking

†Groom, B. B., Vine Wood, Winchester, Kentucky, U.S.A.

Grundy, J. M., Normanton, Hinckley

†Gulliver, W. H., Manor Farm, Millbrook, Southampton

Gumbleton, R. J. M., Glanatore, Tallow, Co. Waterford

Gunnis, G. Bratoft, Burgh, Lincolnshire

†Gunter, Col., Wetherby Grange, Wetherby

H.

'Haffenden, J. Wilson, Homewood, Tenterden

Hale, Bernard, Holly Hill, Hartfield, Tunbridge Wells

†Hales, Edward, North Frith, Tonbric'ge

Hall, John, Hegdale, Bampton, Penrith

†Hamer, C. M., Snitterfield, Stratford-on-Avon

Hamilton, Hon. R. Baillie, Largton, Dunse, N.B.

†Hamond, Anthony, West Acre, Brandon

Hampson, Mrs. S. W., Ullenwood, Cheltenham

†Hanbury, Edgar, Eastrop Grange, Highworth

Handley, William, Green Head, Milnthorpe, Westmoreland

Hands, Thomas, Canley, Coventry

†Hannan. Benjamin, Riverstown, Killucan, Co. Westmeath

†Harcourt, Colonel E. W., M.P., Nuneham Park, Oxford

Harding, Henry, The Grove, East Melbury, Shaftesbury

†Hardinge, Sir E. S., Bart., Fowlers Park, Hawkhurst, Kent

†Hardyside, G., Cambo, Newcastle-on-Tyne

†Harewood, Earl of, Harewood, Leeds

†Hargreaves, Samuel, Hazelhurst, Knutsford, Cheshire

†Harison, Thomas L., Morley, St. Lawrence County, N.Y.

Harker, Robert, Kirkby Stephen, Westmoreland

Harland, William, Blois Hall, Ripon

Harrett, Robert, Kirkwhelpington, Newcastle-on-Tyne

Harris, J. H., Greengill, Penrith

Harris, Thomas, Stonylane House, Bromsgrove

†Harris, William, Tirinie, Aberfeldie, N.B.

Harrison, John, Much Hoole, Preston

Hart, G. V., Kilderry, Londonderry, Ireland

Harter, Percival L. T., Lytham, Lancaster

Hartley, David, Westerdale, Yarm, Yorkshire

Harvey, Frederick, Churcham House, Gloucester

†Haslam, J. P., Gilnow House, Bolton

†Hatchett, John, Preswylfa, Neath, Glamorganshire

†Hawkes, W., Thenford, Banbury

Hawkins, Edmund, Dinthill Ford, Shrewsbury

Headfort, Marquis of, Headfort House, Kells, Co. Meath

Heasman, W., Coltsford Mills, Oxted, Godstone, Surrey

Heaton, Capt. W., Worsley, Manchester

†Heaver, John, Broadbridge, Bosham, Chichester

Hector, Andrew E., Collyhill, Inverurie, N.B.

Hedges, David, Yardley, Birmingham

†Heinemann, E., Ratton Park, Willingdon, Sussex

Hemming, Richard, Bentley Manor, Bromsgrove

Henderson, R., East Elrington, Haydon Bridge

Hepburn, Sir Thomas B., Bart., Smeaton, Hepburn, Prestonkirk, N.B.

Herbert, W. H., Havenfield Lodge, Great Missenden

†Herrick, Mrs. S. Perry, Beau Manor Park, Loughborough

Heskett, William, 24 King Street, Crown Square, Penrith

Hetherington, J. J., Middle Farm, Brampton, Cumberland

Hewer, George, Leygore, Northleach

Hewer, Thomas, Inglesham, Lechlade, Wilts.

Hewer, William, Sevenhampton, Highworth

Hewett, W. H., Norton Court, Taunton

Heys, John, Facit, Rochdale, Lancashire

Heywood-Lonsdale, A. P., 23 Grosvenor Square, W.

Hicken, John, Dunchurch, Rugby

†Hicks, Charles, Felsteadbury, Chelmsford

Hicks, Walter, St. Austell, Cornwall

Highfield, John, Frandley House, Seven Oaks, Northwich, Cheshire

†Hill, A. Staveley, Q.C., M.P., Oxley Manor, Wolverhampton

Hoare, Charles A. R., Kelsey, Beckenham, Kent

†Hobbs, Charles, Maisey Hampton, Fairford

†Hoddinott, Benjamin, Moor Court, Romsey

†Hodgkinson, Charles W., Witton Lodge, Blyth, Worksop

Hodgkinson, E. A., Ogle Flats, Ponteland, Newcastle-on-Tyne

Hodgson, Edward, Blands Wath, Brough, Westmoreland

Hodgson, John, jun., Eamont Bridge, Penrith

Hodgson, Lumley, Highthorne, Easingwold

†Holbech, Archdeacon, Farnborough Hall, Banbury

Holborow, D. B., Knockdown House, Tetbury

†Holborow, Henry, Willesley, Tetbury

Holden, Edward, Laurel Mount, Shipley, Yorkshire

|Holford, R. S., Westonbirt, Tetbury

†Holford, T., Papillon Hall, Market Harborough

Holland, Robert, Norton Hill, Runcorn, Chester

Holland, Trevor, Oak Fields, Rugeley, Staffs.

Holliday, Joseph, The Tarns, Abbey Town, Cumberland

Holmes, John, Strandabrosny, Donemana, Strabane

Hoole, Captain W. W., Ravenfield Park, Rotherham

†Hope, A. Peterkin, Oxwell Mains, Dunbar, N.B.

†Hope, John, Markham, Ontario, Canada

Hopkins, T. M., Lower Wick, Worcester

Hornby, Major E. G. Stanley, Dalton Hall, Burton, Westmoreland

Horne, W. Pybus, Moulton, Richmond, Yorkshire

Horsfall, A., Whitworth, Rochdale

Horsfall, Thomas, Hollin Bank, Brierfield, Lancashire

Horsley, E. H., Colton Manor House, Rugeley

†Horswell, James, Exwick House, Exeter

Horton, S. L., Park House, Shifnal

†Hosken, Samuel, Loggans Mill, Hayle, Cornwall

Housman, William, 19 Gayton Road, Hampstead, N.W.

†How, James, Broughton, Huntingdon

†Howard, Charles, Biddenham, Bedford

Howard, James, M.P., Clapham Park, Bedford

Howe, Earl, Gopsall, Atherstone

Hulbert, T. R., North Cerney, Cirencester

Humphreys, John, Hanley Hall, West Felton, Salop.

Hunt, Henry, Lillington, Leamington

Hunter, E. H. (no address)

Hunter, James, The Palms Farm, Slaley, Hexham

†Hurt, Albert Frederic, Alderwasley, Derby

Hutchinson, John, Brougham Castle Farm, Penrith

†Hutchinson, Teasdale H., Manor House, Catterick

Hutton, Colonel G. Morland, Gate Burton, Gainsborough

I.

Isherwood, Thomas, Fryton, Slingsby, York

†Ivimy, E. T., 5 Pavilion Buildings, Brighton

J.

James, Mrs. Margaret, Bridgetown, Stratford-on-Avon

Jamieson, Thomas, Mains of Waterton, Ellon, N.B.

Jardine, James, Dryfeholm, Lockerbie, Dumfries

Jefferson, J. J., Harpham, Hull

†Jefferson, J. J. Dunnington, Thicket Priory, York

†Jefferson, Robert, Preston Hows, Whitehaven

†Jenkinson, Sir George S., Bart., Eastwood Park, Falfield

Jenkyns, Arthur, Charlestown, St. Austell, Cornwall

†Jennings, Frederick H., Cockfield Hall, Sudbury, Suffolk

†Jervis, William Hudson, Woodford, Thrapston

Johnson, Thomas C., Tothby, Alford, Lincoln

Johnson, William, Prumplestown House, Carlow, Ireland

Johnstone, Sir Harcourt, Bart., Hackness Hall, Scarborough

Johnstone, John, Halleaths, Lockerbie, Dumfries

Jones, George, Milford, Stafford

†Joseph, T., Broomhill, Taibach

Juckes, George, Beslow Hall, Wroxeter, Salop.

K.

†Kello, James, Horton, Chipping Sodbury

Kelsey, Henry, Crowhurst, East Grinstead

Kennard, Rev. R. B., Marnhull Rectory, Blandford

Kent, Frederick George, Island View, Dunmore East, Waterford

†Kerfoot, John, Faenol Bach, St. Asaph

Key, John, Musley Bank, Malton

Key, Thomas William, Casterton Old Hall, Kirkby Lonsdale

King, J. Pittman, North Stoke, Wallingford

†King, Michael, Strangemore, Londonderry

†Kingscote, Colonel, C.B., M.P., Kingscote, Wotton under-Edge

Kirkbride, John, Plumpton, Penrith

Kirkham, Thomas, Biscathorpe, Louth

Knapton, W., Great Kelk, Lowthorpe, Hull

L.

Lace, Thomas, Grenaby, Ramsey, Isle of Man

Lamb, Henry, Norwood House, March, Cambs.

Lamb, John, Soulby, Pooley Bridge, Penrith

Lamb, M., Caley Hall Farm, Otley

Lambe, William, Aubourn, Lincoln

Lambert, Robert, Weeton, Patrington, Hull

Lambert, Thomas, Elrington Hall, Haydon Bridge

Lamplugh, George, Dringhoe, Lowthorpe, Hull

†Lampson, Sir Curtis M., Bart., Rowfant, Crawley, Sussex

Lancaster, George, Morton Grange, Northallerton

Lancaster, John, Skygarth, Penrith

†Lane, Colonel H. B., Kings Bromley Manor, Lichfield

†Langham, H. H., Cottesbrooke Park, Northampton

Langhorn, William, East Mill Hills, Haydon Bridge

Larking, J. W., The Firs, Lee, Kent

†Larkworthy, Falconer, 1 Queen Victoria Street, London, E.C.

Lascelles, Hon. G. E., Sion Hill, Thirsk

Latham, Thomas, Wittenham, Abingdon

†Lathom, Earl of, Lathom House, Ormskirk

†Lavender, William, Biddenham, Bedford

Lawrence, W., Pirton Court, Churchdown, Gloucestershire

Lawson, Alexander, Braelossie, Elgin, N.B.

†Lawson, Sir John, Bart., Brough Hall, Catterick

Lawson, Thomas, Stapleton, Darlington

†Lawson, Sir Wilfrid, Bart., M.P., Brayton, Carlisle

†Laycock, Joseph, Low Gosforth, Newcastle-on-Tyne

Laycock, Richard, Hallgarth House, Winlaton, Blaydon-on-Tyne

Laycock, Robert, Wiseton House, Bawtry

Lazonby, Joseph, Calthwaite House, Penrith

Leadbeater, John B., Thorpe Satchville, Melton Mowbray

†Leatham, Gerald A. W., Thorganby Hall, York

Lees, Harold, Pickhill Hall, Wrexham

†Leney, Frederick, Orpines, Wateringbury, Maidstone

†Leslie, Hon. G. Waldegrave, Leslie House, Leslie, Fife, N.B.

Lett, George Henry, Millpark, Enniscorthy

Lett, William, Rushock, Droitwich

Levett, Lieut.-Col. T. G., M.P., Wichnor Park, Burton-on-Trent

†Lindsay, Col. Loyd, V.C., M.P., Lockinge Park, Wantage

†Linton, John, Buckden Wood, Buckden, Hunts.

Linton, William, Sheriff Hutton, York

†Lismore, Viscount, Shanbally Castle, Clogheen, Ireland

Little, Henry, Boroughbury House, Peterborough

†Little, William, Littleport, Ely

†Lloyd, A. H., Harewoods, Bletchingley, Surrey ·

†Lloyd, John, Kingsbury, St. Albans, Herts

Locke, T. B., Hessle, Hull

†Loder, Robert, M.P., Whittlebury, Towcester

†Lodge, Robert, The Rookery, Bishopdale, Bedale, Yorkshire

†Lofthouse, T. G., Boroughbridge, Yorks.

Logan, Robert, Calgarth, Windermere

Long, Alexander W., Mint Cottage, Kendal

Long, D. F., Oldbury-on-the-Hill, Chippenham

†Longman, A. H., Shendish, Hemel Hempstead

Longmore, Andrew, Rettie, Banff, N.B.

Longstaffe, William, Smardale, Kirkby Stephen

†Lovat, Lord, Beaufort Castle, Beauly, N.B.

†Lovatt, Henry, Low Hill, Bushbury Wolverhampton

Low, Francis Wise, Kilshane, Tipperary, Ireland

†Low, Gavin, 50 Prussia Street, Dublin

Loy, Samuel H., Keldhead, Pickering

†Loyd, William Jones, Langleybury, Watford

†Lucas, B., Hasland Hall, Chesterfield

†Lucas, Thomas, Eastwick Park, Leatherhead

Ludlow, Thomas S., Roseland Farm, Mansfield

Lumsden, John, Castle Heaton, by Coldstream, N.B.

†Luttrell, Col., Badgworth Court, Axbridge

Lyall, Charles, Old Montrose, Montrose, N.B.

Lyall, J. G., Spittal House, Market Rasen, Lincolnshire

Lyle, J. G. Winder, Donaghmore House, Donaghmore, Co. Tyrone, Ireland

†Lynn, John, Church Farm, Stroxton, Grantham

†Lyon, C. E., Johnson Hall, Eccleshall, Staffs.

Lythall, Edmund, Radford Hall, Leamington

Lythall, John B., Bingley Hall, Birmingham

M.

†McConnel, David C., Cressbrook, Ipswich, Queensland, Australia

†McCulloch, William, Glenroy, Melbourne Victoria, Australia

†Macdonald, Ranald, Cluny Castle, Aberdeen, N.B.

Mace, Mrs., Sherborne, Northleach, Gloucestershire

McElderry, John, Ballymoney, Co. Antrim

†Machin, John Vessey, Gateford Hill, Worksop

MacInnes, Miles, Rickerby, Carlisle

†McIntosh, David, Havering Park, Romford

Mackay, Richard J., Burgie, Forres, N.B.

McKenna, Bernard, Lea Grange, Blackley, Manchester

Mackie, William, Petty, Fyvie, N.B.

†Mackinder, Draper, Sempringham House, Folkingham

Mackinder, Robert, Langton, Spilsby

McKinnon, Lauchlan, jun., 239 Union Street, Aberdeen

Macvicar, Neil, Kirmond-le-Mire, Market Rasen

McWilliam, James, Stoneytown, Keith, Banff

Madden, John, Roslea Manor, Clones, Co. Fermanagh

Magniac, Charles, M.P., Colworth House, Bedford

†Mahony, Pierce, Kilmorna, Listowel, Co. Kerry

†Mallock, Richard, Cockington Court, Torquay

†Manchester, Duke of, Kimbolton Castle, St. Neots

†Marjoribanks, Sir Dudley Coutts, Bart., M.P., Guisachan, Beauly, N.B.

Marr, William S., Uppermill, Tarves, N.B.

Marsh, H., Holmbush House, Ashington, Pulboro', Sussex

Marsh, John, jun., Devizes

†Marsh, Richard, Little Offley, Hitchin

Marsh, W. J., Loridge, Berkeley, Glos.

Marshall, Rev. Charles, Ripley Court, Woking Station, Surrey

Marshall, H. J., Poulton Priory, Fairford, Glos.

Marshall, John, Chalton Park, Belford, Northumberland

Marshall, Thomas, Howes, Annan, N.B.

Martin, Edmund, East Down Lodge, Woodlands, Sevenoaks, Kent ·

†Martin, J., Hawkshead Hall, *viâ* Ambleside

Mason, Charles, Dishforth, Thirsk

†Mason, John, Dishforth, Thirsk

Matthews, Alfred Thomas, Church Hanborough, Eynsham, Oxon.

†Maxwell, R. Perceval, Finnebrogue, Downpatrick, Ireland

†Maxwell, W. Perceval, Moor Hill, Tallow, Co. Waterford

Meade, W. R., Ballymartle, Ballinhassig, Co. Cork

†Meade-Waldo, E. W., Stonewall Park, Edenbridge, Kent

Metcalfe, Antony, Park House, Ravenstonedale, Kirkby Stephen

Mickle, John, Stoneshiel, Ayton, N.B.

†Miles, Sir Philip, Bart., M.P., Leigh Court, Bristol

†Miller, T. Horrocks, Singleton Park, Poulton-le-Fylde

†Miller, William M., Brougham, Pickering, Ontario

†Miller, W. Pitt, Thiselton, Kirkham

Milligan, Colonel Charles, Caldwell Hall, Burton-on-Trent

Milne, Miss M., Otterburn, Kelso, N.B.

†Mitchell, A., The Walk House, Alloa, N.B.

Mitchell, H. B., Leiniog Castle, Beaumaris

Mitchell, James, Howgill Castle, Penrith

Mitchell, Robinson, Agricultural Hall, Fairfield, Cockermouth

Mitchell, William, Cleasby, Darlington

Mitchell, W. A., Auchnagathle, Keig, N.B.

Moffat, James, Ballyhyland, Enniscorthy, Ireland

Moir, Captain J. G., The Manor House, Colley, Reigate

Monkhouse, J. C., Eggleston, Barnard Castle

†Montgomery, A. Shirley, Kilmer, Ballivor, Co. Meath

†Moreton, Lord, M.P., Tortworth Court, Wotton-under-Edge

Morice, Francis, Springfield, Bunratty, Ireland

Morley, William, East Gate Farm, East Gate in Weardale, *viâ* Darlington

†Morrin, Thomas, Auckland, New Zealand

Morris, Christopher, Upton Lawn, Chester

†Morris, J. G., Allerton Priory, Woolton, Liverpool

Morris, Thomas, Maisemore Court, Gloucester

†Morrison, John, Bushmead Priory, St. Neots

Mortimer, William Brook, Hay Carr, Ellel, Lancaster

Morton, Mrs. F., Skelsmergh Hall, Kendal

†Morton, J., Stow, Downham Market

†Morton, Richard C., Lane House, Burton, Westmoreland

Moser, F. R., Carbery, Christchurch

Moss, Richard, Whisby, Lincoln

Moutray, Ankitill, Killywick, Favour Royal, Aughnacloy, Ireland

Mowbray and Stourton, Lord, Stourton, Knaresborough

Mumford, J. A., Brill House, Thame

Munton, William, Melbourne Villa, Banbury

†Murchison, K. R., Brokehurst, East Grinstead

Murdoch, William, Oldwhat, New Deer, Aberdeen

Murray, James, Fauchfolds, Turriff, N.B.

†Musgrave, Sir R. C., Bart., M.P., Eden Hall, Penrith

Musgrove, Edgar, Aughton Old Hall, Ormskirk

†Musters, J. Chaworth, Annesley Park, Nottingham

†Mytton, Captain D. H., Garth, Welshpool

N.

Nash, Thomas, Featherstone, Wolverhampton

Naylor, R. C., Kelmarsh Hall, Northampton

Nelson, George, Great Salkeld, Penrith

Nelson, Jacob, Higher Brockholes, Preston, Lancashire

†Nelson, Joseph, Maiden Hill, Penrith

Nevett, William, Yorton Villa, Harmer Hill, Shropshire

Newton, H. V., Polstrong, Camborne, Cornwall

Nichols, John, Iron Acton, Bristol

Nichols, Thomas, Raunds, Thrapston

†Nicholson, James, Murton, Berwick-on-Tweed

Nicholson, John, Kirkbythore Hall, Kirkbythore, Penrith

Nicholson, Robert G., Crosby, Ravensworth, Shap, Westmoreland

Nicholson, W. J., Willoughton Grange, Kirton in Lindsey, Lincoln

Nidd, Clement William, Creeton Manor Farm, Stamford

Nimmo, G. A., Beechwood, Castle Eden, Durham

†Noel, C. P., Bell Hall, Belbroughton, Stourbridge

Norman, Christopher H., Mains Farm, Kirkoswald, Penrith

†Northumberland, Duke of, Alnwick Castle

Norton, P. R., Turvey House, Donabate, Dublin

O.

Oastler, Jonah, Loxwood House, Billingshurst, Sussex

Ogle, Edwin, Heck Hall, Selby, Yorkshire

Ogle, G. F., Top House, Rawcliffe, Selby

†Oliver, R. E., Sholebroke Lodge, Towcester

Osmaston, John, Osmaston Manor, Derby

Outhwaite, John, Bainesse, Catterick

†Owen, F. B., Deefield, Ellesmere, Salop

†Owen, John Dorsett, Plasyn Grove, Ellesmere, Salop

†Owen, William Thomas, Featherston, Wanganui, New Zealand

P.

Paddison, E., Ingleby, Lincoln

Page, Mark, Newbold Grounds, Daventry

Palmer, W. I., Kendrick House, Reading

Park, F. W., Grove, Retford

Parker, John, Ingleby, Lincoln

Parker, Richard, The Tarn, Bootle, Cumberland

Parker, T. T. Townley, Cuerden Hall, Preston, Lancashire

Parker, William, Carleton Hill, Penrith

Parkin, William, Blaithwaite, Aspatria, Cumberland

Parnell, John, Rugby

Parry, T. R., Barras Hall, Wrexham

Part, Thomas, Aldenham Lodge, Watford

Paterson, J. T. S., Plean Farm, Bannockburn, Stirling

†Paterson, J. W. J., Terrona, Langholm

Patrick, Captain Charles, Clough Fold, Manchester

†Paul, Sir Robert J., Bart., Ballyglan, Waterford

Paull, James W., Knott Oak House, Ilminster

Peacey, William, Chedglow, Tetbury

†Pears, Thomas, Hackthorne, Lincoln

Pearson, Thomas, Harbour Flatt, Appleby

Peel, Jonathan, Knowlmere, Clitheroe

Peel, Right Hon. Sir Robert, Bart., Drayton Manor, Tamworth

Pender, W. Rous Tresilian, Budockvean, Falmouth

†Penrhyn, Lord, Penrhyn Castle, Bangor

†Perry, Graddon, Acton Pigott, Condover, Salop

Pery, E. H. C., Coolcronan House, Foxford, Co. Mayo

Pery, Mrs., Coolcronan House, Foxford, Co. Mayo

Peter, James, Ham Villa, Berkeley, Glos.

†Peter, John, Kingscote, Wotton-under-Edge, Glos.

†Philipps, Charles E. G., Picton Castle, Haverfordwest, Pembrokeshire

†Philips, Sir George R., Bart., Weston Park, Shipston-on-Stour

†Philips, J. W., Heybridge, Cheadle, Staffordshire

Phillips, Guy Taylor, Sheriff Hales Manor, Newport, Salop

Phipps, P., Collingtree, Northampton

†Phipps, R., Spencer Parade, Northampton

†Pickrell, James H., Harristown, Macon County, Illinois, U.S.A.

Pigot, Lady, West Hall, Weybridge

Pilgrim, S. C., The Outwoods, Hinckley

Pilkington, Leonard, Widnes, Lancashire

†Pinder, Robert, Whitwell, Oakham

†Platt, Major Henry, Gorddinog, Bangor

Pollard, R. W., Blagdon, Paignton

†Polwarth, Lord, Mertoun House, St. Boswells, N.B.

†Poole, John, Ford House, Ulverston

Pope, F. E., Great Toller, Dorchester

Porter, John, Blackhow, Seascale, Carnforth

†Potterton, William Higgins, Boughton Grange, Northampton

†Pressland, John, Harleston, Northampton

Price, Mrs., Tibberton Court, Gloucester

Price, William, Goldcliff, Newport, Monmouth

†Pryse, Sir Pryse, Bart., Gogerddan, Bowstreet, *viâ* Shrewsbury and Tam T. P. O.

†Puckridge, A. F., Grange Hill, Chigwell, Essex

†Pugh, David, Manoravon, Llandilo, Carmarthen

Pulley, Joseph, M.P., Lower Eaton, Hereford

†Punchard, F., Underley Estates Office, Carnforth

Purdon, Edward, Bachelors Walk, Dublin

Purkis, Thomas, The Grange, West Wratting, Linton, Cambs.

†Pym, G. E., Reigate, Surrey

R.

Radcliffe, John, Stearsby, Easingwold

Raine, A., The Grove, Bow Bank, Middleton-in-Teesdale

Ramsay, Alexander, Banff, N.B.

†Rankin, James, M.P., Bryngwyn, Hereford

†Rathdonnell, Lord, Lisnavagh, Tullow, Ireland

Rawlinson, Robert, Docker Hall. Kendal

†Rawstorne, Lawrence, Hutton Hall, Preston, Lancashire

Ray, William B., Guy Hill, Bentham, Lancashire

Raynbird, Hugh E., Basingstoke

Reed, Matthew, Ludwell, East Gate, Co. Durham

Reid, Alexander, Cruivie, Cupar, Fife

Reid, Nathaniel, Danestown, Bridge of Don, N.B.

†Reynell, Richard, Killynon, Killucan, Co. Westmeath

Rhodes, Andrew, Lundholme, Thornton-in-Lonsdale, *viâ* Ingleton

†Richardson, John, Rowfant, Crawley

†Richardson, John, The Oaks, Dalston, Carlisle

Richardson, Joseph, Potto Hall, Swainby, Northallerton

†Richmond and Gordon, Duke of, K.G., Gordon Castle, Fochabers, Banff

Riddell, James, Hindlip Court Farm, Worcester

Riddell, J. T., Grange House, Kilkenny

Ridley, James, Merry Shields, Stocksfield-on-Tyne

†Ridley, Sir M. W., Bart., M.P., Blagdon, Cramlington

†Rigg. Jonathan, Wrotham Hill Park, Sevenoaks

Riley, John, Putley Court, Ledbury

Riley, Thomas, Ewood Hall, Mytholmroyd

†Robarts, A. J., Lillingstone Dayrell, Buckingham

†Robbins, Richard, The Hollies, Kenilworth

Roberts, Edward, Cockley House, Farthinghoe, Brackley, Northampton

†Roberts, Joseph, Lower Clopton, Chipping Campden

Robertson, James, Whitehaven

Robinson, Edward, Nafferton, Hull

Robinson, H. T., The Cliff, Wensleydale, Bedale

Robinson, John, Bye Beck, Tebay, Westmoreland

Robinson, Joseph, Broughton House, Aylesbury

†Robinson, J. Salkeld, Mount Falinge, Rochdale

Robinson, W., Ulverston

Robson, John, jun., Newton, Bellingham, Northumberland

Robson, J. W., Ascot House, Strandtown, Co. Down

Rockett, J. H., Snaith Hall, Selby

Roddam, John S., Spital Shield, Hexham

†Rolls, John Allan, M.P., The Hendre, Monmouth

Roper, John, Greenway Court, Hollingbourne

†Rose, Rev. H. F., Holme Rose, Fort George Station, Inverness

Rose, Thomas, Melton Magna, Wymondham, Norfolk

Rowlandson, Peter H., Kirkby Stephen, Westmoreland

Rowley, John, Stubbs Walden, Pontefract

†Roynon. John, Havering Park Farm, Romford, Essex

Russell, Nathaniel, Northallerton, Yorkshire

Russell, Thomas, 22 Kensington Palace Gardens, W.

†Rutherford, George. Printonan, Coldstream, N.B.

S.

†Salt, Sir W. H., Bart., Maplewell, Loughborough

†Samuda, Joseph D'A., Chillies, Buxted, Sussex

†Sanday, G. H., Wensley House, Bedale

Sanderson, H. M., Moulton Hall, Richmond, Yorkshire

Sandes, T. W., Sallowglan, Tarbert, Co. Kerry

Sandford, Mark, Hever Lodge, Edenbridge

†Sartoris, F., Rushden Hall, Higham Ferrers

†Saunders, C. R., Nunwick Hall, Penrith

Savage, Saul Powell, Leys Farm, Wotton-under-Edge

Savage, William, Hanging Bank, Bolton, Penrith

†Savidge, Matthew, Sarsden Lodge Farm, Chipping Norton

†Savill, George, Ingthorpe, Stamford

Sayer, Henry, Close House, Kirkbythore, Westmoreland

Scarth, W. T., Staindrop House, Darlington

Schröder, Baron W. H. von, The Rookery, Nantwich, Cheshire

Schroeter, C. W., Tedfold, Billingshurst

Scoby, George, Beadlam Grange, Nawton, York

Scoby, William, Hob Ground House, Kirby Moorside, York

†Scott, Alexander, Towie Barclay, Turriff, N.B.

†Scott, Arthur J., Rotherfield Park, Alton, Hants

Scott, Walter, Glendronach, Huntly, N.B.

†Scott, Sir William, Bart., Ancrum, Jedburgh, N.B.

Scott, W. C., Thorpe, Chertsey

†Scratton, D. R., Ogwell, Newton Abbot

†Senhouse, Humphrey P., Nether Hall, Maryport, Cumberland

Sergison, Captain Warden, Cuckfield Park, Hayward's Heath, Sussex

Serjeantson, George J., Camp Hill, Bedale

Sharp, John Jervis, Broughton, Kettering

†Sharpley, Coates, Kelstern Hall, Louth

†Sharpley, Henry, Acthorpe, Louth

Shaw, Henry, Whittingham Hall, Preston

Shaw, James, Tillyching, Banchory, N.B.

Shaw, Samuel, Brooklands, Halifax

†Sheldon, H. J., Brailes House, Shipston-on-Stour

Shelton, W. G., The Grange, Wergs, Wolverhampton

Shepherd, George, Shethin, Tarves, N.B.

Shepherd, W. F., Douthwaite Lodge, Kirby Moorside

†Sheraton, William, Broom House, Ellesmere, Salop

†Shiels, George, Balgove, St. Andrews, N.B.

Simpson, John, Vincent Place, Hunslet Road, Leeds

Sindall, Thomas Yarrad, Fulney Hall, Spalding

Singleton, J. R., Givendale, Pocklington

Singleton, John, Teresa Cottage, Pocklington

Sisman, William, The Lodge, Buckworth, Huntingdon

Slattery, Denis F., Coolnagour, Dungarvan, Ireland

†Smith, E. J., Clonard, Dundrum, Co. Dublin

†Smith, F. N., Wingfield Park, Derby

Smith, Henry, The Grove, Cropwell Butler, Nottingham

Smith, Henry, The Chesnuts, Leamington

†Smith, H. F., Lamwath House, Sutton, Hull

Smith, Humphrey, Mountmellick, Portarlington, Ireland

Smith, James, Harthill, Kiveton Park, Sheffield

Smith, John, Balmain, Lawrencekirk, N.B.

†Smith, John, Coathill, Brampton

Smith, J. G., Minmore, Glenlivat, Ballindalloch, Banffshire

Smith, William, Goole Grange, Goole, Yorkshire

Smith, William, jun., Melkington, Cornhill, Northumberland

†Smyth, Sir J. H. G., Bart., Ashton Court, Bristol

†Smythe, Sir C. F., Bart., Acton Burnell, Shrewsbury

Sneyd, Rev. W., Keele Hall, Newcastle, Staffordshire

Spence, John S., Springfield, Skirlaugh, Hull

†Spencer, Earl, K.G., Althorp, Northampton

†Spencer, Robert, Livesey, Blackburn

Spencer, Sanders, Holywell, St. Ives, Hunts.

Stamper, Thomas, Highfield House, Oswaldkirk, York

Stanford, Joseph, Haxted Mills, Lingfield, Edenbridge

†Stanhope, J. Banks, Revesby Abbey, Boston

†Staniforth, Rev. Thomas, Storrs, Windermere

†Stapylton, Major, Myton Hall, Borough-bridge

Starkie, C., Ashton Hall, Lancaster

†Stephenson, Christopher, Naworth Estate Office, Brampton, Cumberland

†Stephenson, Clement, Sandyford Villa, Newcastle-on-Tyne

Stern, S. T., Littlegrove, East Barnet, Herts

†Stevenson, Peter, Rainton, Thirsk

Stewart, Samuel, Sandhole, Fraserburgh, N.B.

Stilgoe, Henry, Clopton, Stratford-on-Avon

†Stilgoe, N. P., Manor Farm, Adderbury, Oxon

Stobo, Andrew, Porterstown, Thornhill, N.B.

Stone, F. W., 7 Stone Buildings, Lincoln's Inn, W.C.

Stonham, Henry, Thurnham, Maidstone, Kent

Stopford-Sackville, Mrs., Drayton House, Thrapstone

Stott, John, Lower Powburn, Lawrence-kirk, N.B.

†Strafford, Henry, 13 Euston Square, London, N.W.

Stratton, Frederick, Merdon, Hursley, Winchester

Stratton, George, Wheler Lodge, Husbands Bosworth, Rugby

†Stratton, Joseph, Alton Priors, Marlborough

†Stratton, R., The Duffryn, Newport, Monmouthshire

†Streator, S. R., East Cleveland, Ohio, U.S.A.

†Strickland, Miss, Apperley Court, Tewkesbury

Strickland, Rev. N. C., Reighton, Bempton, Hull

Strickland, T., Depôt Hotel, Thirsk Junction

Strong, John, Culgaith, Penrith, Cumberland

†Sturgeon, Charles, Grays Hall, Essex

Suffolk and Berkshire, Earl of, Charlton Park, Malmesbury

Sullivan, Francis, Castle Bamford, Kilkenny, Ireland

Sutherland, Charles Leslie, Coombe, Croydon, Surrey

Sutton, Sir Richard, Bart., Benham Park, Newbury, Berkshire

Swann, John, Bedlington, Northumberland

†Swinburne, Sir John, Bart., Capheaton, Newcastle-on-Tyne

. Swingler, Thos., Langham, Oakham

T.

†Taber, James, Rivenhall, Witham, Essex

Tait, Henry, Prince Consort's Shaw Farm, Windsor

†Tallant, Francis, Easebourne Priory, Midhurst

Tankerville, Earl of, Chillingham Castle, Alnwick

†Tanner, Thomas, Hawkes Bay, New Zealand

Taunton, Miss Frances, Ashley, Stockbridge, Hants

†Taylor, E., Clogheen, Cahir, Ireland

†Taylor, George, Stanton Prior, Bristol

Taylor, Henry, Grange Farm, Wetherby

Taylor, Richard, Newhouse, Burnside, Kendal

Taylor, Richard, Sigglesthorne Manor, Hull

Taylor, Thomas, Hall Garth, Brough, Westmoreland

†Taylor, William, Clanrolla House, Lurgan, Ireland

Temple, Captain E., Saltergill, Yarm, Yorkshire

†Theakston, Thomas, Masham, Bedale, Yorkshire

Thom, James, Kirkbythore, Penrith

Thompson, A., The Cross, Whitehaven

†Thompson, John, Badminton, Chippenham

Thompson, John, Sandymount, Tipperary

†Thompson, Joseph, Anlaby, Hull

†Thompson, Robert, Inglewood Bank, Penrith

Thompson, Thomas Charles, Milton Hall, Carlisle

†Thompson, William, Moresdale Hall, Kendal

†Thomson, J., Newseat of Dumbreck, Tarves, N.B.

Thomson, R., Burnbank Farm, Blairdrummond, N.B.

Thorburn, James, Stonedge, Hawick, N.B.

†Thornton, John, 7 Princes Street, London, W.

Thurgood, James, Abbott's Roothing-Ongar, Essex

Thurgood, John H., Harlow, Essex

†Tindall, C. W., Aylesby Manor, Grimsby

Tindall, W., 31 Hallgate, Doncaster

Tisdall, E. C., Holland Park Farm, Kensington, W.

Todd, John, Mereside, Aspatria, Cumberland

Tombs, John, Langford, Lechlade

Topham, John, Middleham House, Middleham, Bedale

Toppin, J. C., Musgrave Hall, Skelton, Penrith

†Torrance, George, Sisterpath, Dunse, Berwick

Torromé, Francisco, 4 Jeffrey's Square, St. Mary Axe, E.C.

Townshend, Capt. Henry, Caldecote Hall, Nuneaton

†Tracy, G. Murton, Redlands, Edenbridge

†Treadwell, John, Upper Winchendon, Aylesbury

Treadwell, Richard, Shalstone, Buckingham

†Tredegar, Lord, Tredegar Park, Newport, Monmouthshire

Tregaskis, S. T., St. Issey, Cornwall

Tremaine, William, Polsue, Grampound, Cornwall

Trethewy, William, Tregoose, Probus

Trotter, William, South Acomb, Stocksfield-on-Tyne

Tulloch, W., Slaugham Park, Crawley

Tunnicliffe, E. T., Bromley Hall, Eccleshall, Staffordshire

†Turbervill, Lieut.-Colonel, Ewenny Priory, Bridgend

†Turner, John, The Grange, Ulceby

Tweddell, William, Edington, Whalton, Northumberland

Tweedie, James, Deuchrie, Prestonkirk, N.B.

†Tweedie, R., The Forest, Catterick

Tyacke, John, Merthen, Penryn

†Tynte, J. Pratt, Tynte Park, Dunlavin, Co. Wicklow

U.

Underwood, George, Little Gaddesden, Great Berkhampstead

Upson, James, Rivenhall, Witham

Utting, H. A., Hockering, Norwich

V.

†Vavasour, H. D., Tataraimaka, New Plymouth, Taranaki, New Zealand

Vickers, William, Howl John, Stanhope, Darlington

Vivian, Henry Hussey, M.P., Park Wern, Swansea

W.

†Wadsworth, Charles F., Geneseo, New York, U.S.A.

Waind, John, Ankness, Kirby Moorside, Yorkshire

Waite, Richard, Duffield, Derby

Wakefield, W. H., Sedgwick, Kendal

Wakefield, W. T., Fletchampstead Hall, Coventry

Walker, Thomas, Stowell Park, Northleach, Glos.

†Walker, T. E., Studley Castle, Warwickshire

Walker, William Thomlinson, Clifton Grove, York

Wallis, Robert, Old Ridley, Stocksfield-on-Tyne

†Walsingham, Lord, Merton Hall, Thetford, Norfolk

†Walter, John, M.P., Bearwood, Wokingham, Berks

Ward, James G., The Villa, Oxhill, Kineton

Ward, Robert, Sodston, Narberth, Pembrokeshire

†Ward, S. R. Chapman, Neasham Hill, Darlington

Wardle, Henry, Highfield, Burton-on-Trent

†Ware, Joseph, Minjah, Caramat, Villiers, Victoria, Australia

Waring, Henry, Beenham House, Reading

Warren, Edward L., Lodge Park, Freshford, Co. Kilkenny

Watson, John F., Crowle Wharf, Doncaster

Watt, James, Claragh, Ramelton, Co. Donegal

Waugh, William, Allerby Mill, Maryport

†Webb, Lieut.-Colonel C. J., Elford House, Tamworth

Webb, Henry, Streetly Hall, Linton

Webb, John, Horseheath, Linton

†Webb, Jonas, Melton Ross, Ulceby

Webster, C., Uxbridge Common, Middlesex

Wells, Mrs., Booth Ferry House, Howden, East Yorkshire

†Wells, William, Holmewood, Peterborough

†Welsted, Richard, Ballywalter, Castletown Roche, Ireland

West, Samuel, Upwell, Wisbech

†Wharton, J. T., Skelton Castle, Marske-by-the-Sea

Wheeler, A. O., Upton Hill, Gloucester

†Whichcote, Sir Thomas, Bart., Aswarby Park, Folkingham

White, William, Home Farm, Grimston, Tadcaster

†Whiteside, Thomas (no address)

†Whitmore, F. A. W., Kingslee, Farndon, Chester

Whitteron, S., Braham Hall, Wetherby

Whitwell, John, Springfield House, Peterborough

Whitworth, Sir Joseph, Bart., Stancliffe Hall, Matlock Bath

†Whyte, James, Aldborough, Darlington

Wilkes, J. P., Oakley House, Droitwich

Wilkes, Richard Hammond, Showhill Farm, Wolverhampton

Wilkinson, John, Carleton Grange, Carleton, Skipton, Yorkshire

†Williams, Charles, Pilton House, Barnstaple

Williams, Rev. E. T., Caldicot Parsonage, Chepstow

Williams, George, Scorrier House, Scorrier, Cornwall

†Williams, Walter, Sugnall Hall, Eccleshall, Staffs.

Williamson, Colonel D. R., Lawers House, Crieff, N.B.

†Williamson, Edward, Ramsdell Hall, Lawton, Chester

Williamson, James, East Pitdoulsie, Turriff, N.B.

Willis, Joseph Deans, Bapton, Codford, Wilts

†Willis, Thomas, jun., Manor House, Carperby, Bedale

Willson, Mrs. M., Rauceby, Sleaford, Lincolnshire

Wilson, Charles, Shotley Park, Shotley Bridge

†Wilson, Christopher W., High Park, Kendal

†Wilson, Jacob, Woodhorn Manor, Morpeth

Wilson, James, Mains of Scotstown, Aberdeen

†Wilson, John Wilson, Farmers' Club, Inns of Court Hotel, Holborn, W.O.

Wilson, Joseph, Corn Market, Penrith, Cumberland

Wilson, Richard, Ellonby, Penrith, Cumberland

†Winn, John Russell, Lower Coundon, Coventry

Withington, Thomas Ellames, Holton Cottage, Wheatley, Oxon

Wodehouse, W. Herbert, Woolmers Park, Hertford

Wood, James, Humbleton Hall, Hull

Wood, James, Midtown, Banff

†Wood, J. B., The Hall, Wirksworth, Derbyshire

Wood, Rowland, Clapton, Thrapston

Woodburne, Thomas, Thurstonville, Ulverston

†Woodhouse, E. B., Mount Gilead, Campbelltown, Sydney, New South Wales

Woodhouse, John, Scale Hall, Lancaster

Woodroffe, W. S., Beaumont Grange, Halton, Lancaster

†Worsley, Sir W. C., Bart., Hovingham, York

†Wraith, Lawrence H., Newfield House, Lower Darwen, Blackburn

†Wren, Walter, 7 Powis Square, London, W.

Wright, Charles, Oglethorpe Hall, Tadcaster

Wright, George, The Hall, Park Lane, Doncaster

†Wright, John, jun., Green Gill Head, Penrith, Cumberland

†Wright, William, Wollaton, Nottingham

†Wynn, Sir Watkin W., Bart., M.P., Wynstay, Ruabon

Wynne, Thomas, Lislea, Armagh

Y.

Yeats, George, Studley, Ripon

Yool, Thomas, Calcots, Elgin, N.B.

Yorke, John, Bewerley Hall, Pateley Bridge, Leeds

Yorkshire Agricultural Society, Secretary of, York

†Young, Captain, Ashey Farm, Brading, Isle of Wight

Z.

†Zetland, Earl of, Aske Hall, Richmond, Yorkshire

SHORTHORN SOCIETY OF THE UNITED KINGDOM OF GREAT BRITAIN AND IRELAND.

ABSTRACT OF THE MEMORANDUM AND ARTICLES OF ASSOCIATION, BYE-LAWS, &c.

OBJECTS OF THE SOCIETY.

1. To maintain unimpaired the purity of the breed of cattle known as SHORT-HORNS, and to promote impartially the breeding of all the various tribes, families, and strains of such cattle.

2. To further the above objects :—
   (*a*) By continuing the issue of the publication called " Coates's Herd Book," and acquiring the unsold stock thereof.
   (*b*) By acquiring and either continuing (with or without any kind of modification) or suspending the issue of any other publication dealing with or bearing upon such objects.
   (*c*) By collecting, verifying, and publishing in the United Kingdom and abroad, information relating to the pedigrees of Shorthorn Cattle.
   (*d*) By investigating and reporting upon cases of doubtful or suspected pedigrees of such Cattle.
   (*e*) By undertaking the arbitrament upon and settlement of disputes and questions relating to or connected with Shorthorn Cattle and the breeding thereof.
   (*f*) By corresponding with and affiliating Corporations, Societies, and persons in the United Kingdom and abroad interested in or professing objects similar to or connected with the objects of this Society.

3. To receive subscriptions, &c., and issue copies of the publications of the Society.

4. To acquire for the purposes of the Society any lands, tenements, &c., and to sell, let, and dispose of the same.

5. To make and frame Bye-laws, and to do all other things incidental or conducive to the attainment of the above objects or any of them.

MEMORANDUM AND ARTICLES OF ASSOCIATION, AND BYE-LAWS.

A copy of the Memorandum and Articles (price 1*s.*), and of the Bye-laws (gratis), giving full information as to the constitution of the Society, powers and duties of the Council, &c., may be obtained on application to the Secretary.

MEMORANDA.

NEW MEMBERS.—Application for Membership must be made through a Member of the Society. The form required to be filled up can be obtained at the Society's office, or will be sent by post on application.

SUBSCRIPTIONS.—1. *Life Members.* In addition to an Entrance Fee of One Guinea, Life Members pay on entrance a subscription of Ten Guineas.

2. *Annual Members.* In addition to an Entrance Fee of One Guinea, Annual Members shall pay on entrance and thenceforth annually, IN ADVANCE, ON THE 1ST DAY OF JANUARY IN EACH YEAR, a Subscription of One Guinea. Annual

Members whose Subscriptions are not in arrear may compound for their Annual Subscriptions by a single payment of Ten Guineas.

Annual Subscriptions shall be in arrear if unpaid after the 1st day of June.

PAYMENTS.—Subscriptions &c. may be paid to the Secretary either at the office of the Society, No. 12 Hanover Square, London, W., between the hours of 10 and 4 (Saturdays 10 and 2), or by means of a cheque, or a Money Order on the Vere-Street Office, made payable to the Secretary, H. J. HINE.

MEMBERS' PRIVILEGES.—Volumes of the Herd Book which are purchased by them for purposes connected with their herds, at a reduced price. The Annual Vols. of the Herd Book, which belong to the year for which their Subscription has been paid, transmitted by post, free of charge, to their address. The entry in the Herd Book of the pedigrees of their Shorthorn Cattle at a reduced rate. The privilege of referring disputes or questions relating to or connected with Shorthorn Cattle and the breeding thereof, for arbitrament and settlement by the Council.

No Member in arrear of his subscription is entitled to any of the privileges of the Society.

RESIGNATION.—Members who shall have paid the Subscriptions and other Moneys which may have become due from them, may retire from Membership on giving *three* months' notice in writing to that effect; but unless such notice is given on or before the 1st day of June in any year, the retiring Member shall, if an Annual Member, pay, notwithstanding his retirement, the annual subscription for the then ensuing year.

HERD BOOK.—No SHORTHORN CATTLE shall be entered in the Herd Book unless application for entry shall be made whilst they are in the United Kingdom, or unless satisfactory reason be given for such application not having been so made.

No BULL is eligible for insertion in the Herd Book unless it has FIVE CROSSES, and no Cow unless it has FOUR CROSSES of Shorthorn Blood, which are, or are eligible to be, inserted in the Herd Book. *If any Cross is not registered* its pedigree—if eligible for insertion—together with the fee, must accompany the entry of the animal it is required to complete.

ALL ENTRIES MUST BE MADE ON THE SOCIETY'S PRINTED FORMS, and must be certified by the breeder or owner, his accredited agent or representative. *Entries must be accompanied with the* NECESSARY FEES *as follows* :—

	s.	d.
To a MEMBER OF THE SOCIETY (whose subscription for the current year is not unpaid)—		
For inserting the pedigree of a Bull	5	0
Ditto ditto Cow with not more than two Calves as produce	2	6
For *each* additional Calf above two	2	6
To a NON-MEMBER—		
For inserting the pedigree of a Bull	10	0
Ditto ditto Cow with not more than two Calves as produce	5	0
For *each* additional Calf above two	2	6

*** All communications to the Society shall be in writing, addressed to the Secretary, and shall be sent to the offices of the Society, at No. 12 Hanover Square, London, W.